D1549519

WITHDRAWN

LIVERPOOL JMU LIBRARY
3 1111 01053 0010

Tools in Fluvial Geomorphology

Tools in Fluvial Geomorphology

Editors

G. Mathias Kondolf
University of California, Berkeley

and

Hervé Piégay
CNRS

WILEY

Telephone (+44) 1243 779777

Email (for orders and customer service enquiries): cs-books@wiley.co.uk
Visit our Home Page on www.wileyeurope.com or www.wiley.com

Other Wiley Editorial Offices

John Wiley & Sons Inc., 111 River Street, Hoboken, NJ 07030, USA

Jossey-Bass, 989 Market Street, San Francisco, CA 94103-1741, USA

Wiley-VCH Verlag GmbH, Boschstr. 12, D-69469 Weinheim, Germany

John Wiley & Sons Australia Ltd, 33 Park Road, Milton, Queensland 4064, Australia

John Wiley & Sons (Asia) Pte Ltd, 2 Clementi Loop #02-01, Jin Xing Distripark, Singapore 129809

John Wiley & Sons Canada Ltd, 22 Worcester Road, Etobicoke, Ontario, Canada M9W 1L1

Wiley also publishes its books in a variety of electronic formats. Some content that appears in print may not be available in electronic books.

British Library Cataloguing in Publication Data

A catalogue record for this book is available from the British Library

ISBN 0-471-49142-X

Typeset in 9 on 11pt Times by Kolam Information Services Pvt. Ltd, Pondicherry, India
Printed and bound in Great Britain by Antony Rowe Ltd, Chippenham, Wiltshire
This book is printed on acid-free paper responsibly manufactured from sustainable forestry in which at least two trees are planted for each one used for paper production.

Contents

List of Contributors

Dr James P. Bennett	US Geological Survey, Denver Federal Center, PO Box 25046, MS 413, Lakewood, CO 80225, USA jbennett@usgs.gov
Dr Gudrun Bornette	UMR 5023 C.N.R.S., "Ecology of Fluvial Hydrosystems", Université Claude Bernard Lyon I, 69622 Villeurbanne cedex, France Tel.: +33-472-431-294; fax: +33-472-431-141. bornette@avosnes.univ-lyon1.fr
Dr Anthony G. Brown	Department of Geography, School of Geography and Archaeology, Amory Building, Rennes Drive, University of Exeter, Exeter EX4 4RJ, UK a.g.brown@exeter.ac.uk
Dr Robert Bryant	Department of Geography, University of Sheffield, SN102TN UK r.g.bryant@sheffield.ac.uk
Dr Janine Castro	US Fish and Wildlife Service, 2600 SE 98th Avenue Suite 100, Portland, OR 97266, USA
Dr Pierre Clément	Laboratoire de Géographie Physique de l'Environnement, Université Lyon 2, 5 avenue Pierre Mendès-France, 69 676 Bron cedex, France pmjclement@club-internet.fr
Dr Stephen E. Darby	Department of Geography, University of Southampton, Highfield, Southampton SO17 1BJ, UK s.e.darby@soton.ac.uk
Dr Peter W. Downs	Stillwater Sciences, 2532 Durant Avenue, Berkeley, CA 94704, USA Department of Geography, University of Nottingham, University Park, Nottingham NG7 2RD, UK downs@stillwatersci.com
Dr Thomas Dunne	Donald Bren School of Environmental Sciences and Management and Department of Geological Sciences, 4670 Physical Science North, University of California, Santa Barbara, CA 93106-5131, USA tdunne@bren.ucsb.edu
Dr Peter Ergenzinger	Geographisches Institut, FU Berlin, B.E.R.G., Malteserstr. 74-100, 12249 Berlin Tel.: +49-30-7792-253/252; fax: +49-30-767-06439. B.E.R.G., Geographisches Institut, FU Berlin, Grunewaldstr.35, 12165 Berlin, Germany perg@geog.fu-berlin.de
Dr David Gilvear	Department of Environmental Science, University of Stirling, Stirling FK9 4LA, UK d.j.gilvear@stir.ac.uk

Dr Basil Gomez	Department of Geography, Indiana University, 4885 Fieldstone Trail, Indianapolis, IN 46254, USA gegomez@isugw.indstate.edu
Professor Angela M. Gurnell	Department of Geography, King's College London, Strand, London WC2R 2LS, UK School of Geography and Environmental Sciences, University of Birmingham, Edgbaston, Birmingham B15 2TT, UK a.m.gurnell@bham.ac.uk
Dr Marwan A. Hassan	Department of Geography, University of British Columbia, Vancouver, BC, Canada V6T 1Z2. Tel.: +604-822-5894; fax: +604-822-6150. Department of Geography, Hebrew University of Jerusalem, Mount Scopus, 91905 Jerusalem, Israel mhassan@geog.ubc.ca
Dr D. Murray Hicks	NIWA, PO Box 8602, Christchurch, New Zealand m.hicks@niwa.cri.nz
Dr Cliff R. Hupp	United States Geological Survey, 430 National Center, Reston, VA 20192, USA Tel.: +001-703-648-5207; fax: +001-703-648-5484. crhupp@usgs.gov
Dr Robert B. Jacobson	US Geological Survey ECRC, 4200 New Haven Road, Columbia, MO 65201, USA robb_jacobson@usgs.gov
Dr Allan James	Department of Geography, University of South Carolina, Callcott Building, Columbia, SC 29208, USA ajames@garnet.cla.sc.edu
Dr G. Mathias Kondolf	Department of Landscape Architecture and Environmental Planning and Department of Geography, University of California, Hearst Field Annex B, Berkeley, CA 94720-2000, USA kondolf@uclink.berkeley.edu
Dr Thomas E. Lisle	U.S. Forest Service Redwood Science Laboratory, 1700 Bayview Drive, Arcata, CA 95521, USA tel7001@axe.humboldt.edu
Dr David R. Montgomery	Department of Geological Sciences AJ-20, University of Washington, Seattle, WA 98195, USA dave@geology.washington.edu
Dr Jonathan M. Nelson	US Geological Survey, Denver Federal Center, PO Box 25046, MS 413, Lakewood, CO 80225, USA jmn@usgs.gov
Dr James E. O'Connor	US Geological Survey, 10615 S.e. Cherry Blossom Drive, Portland, OR 97216, USA oconnor@usgs.gov
Dr Takashi Oguchi	Department of Geography, Center for Spatial Information Science, University of Tokyo, 7-3-1 Hongo, Bunkyo-Ku, Tokyo 113-0033, Japan oguchi@geogr.s.u-tokyo.ac.jp
Professor Jean-Luc Peiry	Laboratoire de Geographie Physique, UMR 6042 CNRS – "Geodynamique des Milieux Naturels et Anthropises", Maison de la Recherche – Universite Blaise Pascal, 4, rue Ledru, 63057 – Clermont-Ferrand cedex 1, France j-luc.peiry@univ-bpclermont.fr

Dr François Petit	Institute de Geographie, Université de Leige, Sart Tilman, Batiment 11, 4000 Leige, Belgium francois.petit@ulg.ac.be
Professor Geoffrey E. Petts	School of Geography, Earth and Environmental Sciences, University of Birmingham, Birmingham B15 2TT, UK pettsge@mis3.bham.ac.uk
Dr Hervé Piégay	UMR 5600, CNRS, 18, rue Chevreul, 69 362 Lyon cedex 07, France piegay@sunlyon3.univ-lyon3.fr;piegay@univ-lyon3.fr
Dr James E. Pizzuto	Department of Geology, University of Delaware, Newark, DE 17916, USA pizzuto@udel.edu
Dr Gary Priestnall	Department of Geography, University of Nottingham, University Park, Nottingham NG7 2RD, UK gary.priestnall@nottingham.ac.uk
Dr Leslie M. Reid	USDA Forest Service Pacific Southwest Research Station, Redwood Sciences Laboratory, Arcata, CA, USA lmr7001@axe.humboldt.edu
Dr Laurent Schmitt	Faculté de Géographie, Histoire, Histoire de l'Art, Tourisme, Université Lyon 2, 5, avenue Pierre Mendès-France, 69676 Bron cedex, UMR 5600, France laurent.schmitt@univ.lyon2.fr
Professor Stanley A. Schumm	Department of Earth Resources, Colorado State University, Fort Collins, CO 80523-1482, USA stans@mussei.com
Dr David A. Sear	Department of Geography, University of Southampton, Highfield, Southampton SO17 1BJ, UK d.sear@soton.ac.uk
Dr Andrew Simon	USDA Agricultural Research Service, NSL, 430 Highway 7, N. Oxford, MS 38655, USA asimon@ars.usda.gov
Dr Stephen Stokes	School of Geography and the Environment, University of Oxford, Mansfield Road, Oxford OX1 3TB, UK stephen.stokes@geog.ox.ac.uk
Dr Marco J. Van de Wiel	Institute of Geography and Earth Sciences, University of Wales, Aberystwyth, Wales, UK
Professor Des E. Walling	Department of Geography, University of Exeter, Rennes Drive, Exeter EX4 4RJ, UK d.e.walling@exeter.ac.uk
Dr Peter J. Whiting	Department of Geological Science, Case Western Reserve University, Cleveland, OH 44106, USA pjws@po.cwru.edu
Dr Stephen M. Wiele	US Geological Survey, Tucson, AZ, USA smwiele@usgs.gov
Professor Gordon M. Wolman	Department of Geography and Environmental Engineering, John Hopkins University, Ames Hall, Baltimore, MD 21218, USA wolman@jhunix.hcf.jhu.edu

Part I

Background

1

Tools in Fluvial Geomorphology: Problem Statement and Recent Practice

G. MATHIAS KONDOLF[1] AND HERVÉ PIÉGAY[2]
[1]*University of California, Berkeley, CA, USA*
[2]*UMR 5600, CNRS, Lyon, France*

As explained by Wolman (1995) in a manuscript titled "Play: the handmaiden of work", much geomorphological research is applied, and geomorphologists compete with other disciplines for funding from public agencies. Moreover, geomorphologists are increasingly in demand to participate in ecological restoration projects because of the spatial and temporal scales at which they analyze channel change and sensitivity, which can provide insights for solving problems in river engineering (Giardino and Marston 1999) and ecological river restoration (Brookes and Shields 1996). As do all scientists, fluvial geomorphologists employ tools in their research, but the range of tools is probably broader in this field than others because of its position on the intersection of geology, geography, and river engineering. Increasingly, the tools of fluvial geomorphology have been adopted, used, and sometimes modified by non-geomorphologists, such as scientists in allied fields seeking to incorporate geomorphic approaches in their work, managers who prescribe a specific tool be used in a given study, and consultants seeking to package geomorphology in an easy-to-swallow capsule for their clients. Frequently, the lack of geomorphic training shows in the questions posed, which may often be at spatial and temporal scales smaller than the underlying cause of the problem. For example, to address a bank erosion problem, we have frequently seen costly structures built to alter flow patterns within the channel. While the designers may have employed hydraulic formulae to design the structures, they may have neglected to look at processes at the basin scale, such as increased runoff from land-use change, which may be driving channel widening. In such a case, controlling bank erosion through mechanical means will at best provide only temporary and local relief from a system-wide trend. In such cases, geomorphic tools may be used to address the problem, but the results are ultimately ineffective (or at least not sustainable) because the question was poorly posed at the outset and a limited range of tools was employed.

The purpose of this book is to review the range of tools employed by geomorphologists, and to clearly link the choice of tool to the question posed, thereby providing guidance to scientists in allied fields and to decision makers about various methods available to address questions in the field, and the relative advantages and disadvantages of each. This book is the result of a collective effort, involving contributors from diverse ages, disciplinary expertise, professional experience, and geographic origins to illustrate the range of tools in the field and their application to problems in other fields or in management issues.

1.1 TOOLS AND FLUVIAL GEOMORPHOLOGY: THE TERMS

Webster's dictionary defines a tool as anything used for accomplishing a task or purpose (Random House 1996). By a tool, we refer comprehensively to concepts, theories, methods, and techniques. The distinction among these four terms is not always clear, depending on the level of thinking and abstraction. Moreover, definitions vary somewhat with dictionaries (e.g., Merriam 1959 versus Random House 1996), and definitions of one term may include the other terms. In our usage, a concept is defined as a mental representation of a reality, and a theory is an explicit formulation of relationships among concepts. Both are tools because they provide the framework within which problems are approached and techniques and methods deployed. As scientific theories can be important elements in the methods of science (Brown 1996), they can be considered as tools. A method involves an approach, a set of steps taken to solve a problem, and would often include more than one technique. As suggested by Webster's Dictionary (Random House 1996), it is an orderly procedure or process, a regular way or manner of doing something. Techniques are the most concrete and specific tools, referring to discrete actions that yield measurements, observations, or analyses.

As an illustration, a researcher can base her approach on the fluvial system theory and, within this general framework, one of the field's seminal concepts, the notion of dominant and bankfull discharge. To test the relation between dominant discharge and channel dimensions, she will act step by step, identifying a general methodological protocol. She may survey channel slope and cross-sectional geometry, and measure water flow and velocity, or if field measurements of flow were not possible, she might estimate flow characteristics from the surveyed geometry and hydraulic equations. If she measures flow in the field, she can choose among several methods, such as using a portable weir, salt dilution, or current meter method. The last method would involve various techniques, such as techniques to measure flow depth and velocity (e.g., using Pryce AA or other current meters, wading with top-setting wading rods or suspending the meter from a cableway or bridge), clearing aquatic vegetation and otherwise improving the cross-section for measurement, details of placing the current meter in the water, accounting for flow angles, and estimating the precision of the measurement. At least, as suggested by Wharton *et al.* (1989), channel capacity discharge has to be related to the long-term flow frequency. One can use gauging station data wherever possible, and only if that is not available would one try to collect data for the purpose.

While some tools are specific to fluvial geomorphology, others are borrowed from sister disciplines, and some (such as mathematical modeling, statistical analysis, and inductive or hypothetico-deductive reasoning) are used by virtually all sciences (Bauer 1996, Osterkamp and Hupp 1996). Our aim is not to describe generic tools used by all scientists, but to focus on tools currently used by fluvial geomorphologists.

We define fluvial geomorphology in its broadest sense, considering channel forms and processes, and interactions among channel, floodplain, network, and catchment. A catchment-scale perspective, at least at a network level, is needed to understand channel form and adjustments over time. Of particular relevance are links among various components of the fluvial system, controlling the transfer of water and sediment, states of equilibrium or disequilibrium, reflecting changes in climate, tectonic activity, and human effects, over timescales from Pleistocene (or earlier) to the present. Thus, to understand rivers can involve multiple questions and require application of multiple methods and data sources. As a consequence, we consider fluvial geomorphology at different spatial and temporal scales within a nested systems perspective (Schumm 1977). Analysis of fluvial geomorphology can involve application of various approaches from reductionism to a holistic perspective, two extremes of a continuum of underlying scientific approach along which the scientist can choose tools according to the question posed.

1.2 WHAT IS A TOOL IN FLUVIAL GEOMORPHOLOGY?

Roots and Tools

Fluvial geomorphology is a discipline of synthesis, with roots in geology, geography, and river engineering, and which draws upon fields such as hydrology, chemistry, physics, ecology, human, and natural history. The choice of tools by geomorphologists has been influenced largely by the disciplinary training of the investigators. The geologically trained fluvial geomorphologist may be more likely to apply tools such as stratigraphic analysis of floodplains, or to incorporate analysis of large-scale tectonic influences. For example, recent papers in the *Geological Society of America Bulletin* have concerned topics such as bedrock benches and boulder bars, stream morphology in response to uplift, and Holocene entrenchment in British uplands. In contrast, the investigator trained in river hydraulics and physics is more likely to apply tools such as numerical modeling and mechanics. Recent papers in *Water Resources Research* have concerned topics such as numerical simulations of river widening and braiding, micro-mechanics of bedload transport, and predictions of velocity distribution in channels. Geographers tend to focus on comparisons of fluvial forms and processes according to the characteristics of the basin or bioclimatic regions within which they are observed, the influence of human activities, vegetation cover, or geological settings. Some recent papers in *Geomorphology* and *Earth Surface Processes and Landforms* have reflected combination of geographical approaches in fluvial geomorphology with remote sensing (Bryant and Gilvear 1999), historical perspective (Brooks and Brierley 1997, Leys and Werrity 1999, Liébault and Piégay 2002), and analysis of longitudinal or inter-system complexity (Petit and Pauquet 1997, Walling and He 1998, Madej 1999). Increasing interactions with biology have led to use of a new term to characterize this branch of discipline: biogeomorphology (Viles 1988, Gregory 1992).

A more holistic investigation of rivers requires a multidisciplinary approach and therefore fluvial geomorphology increasingly interacts with other disciplines such as engineering (e.g., Thorne *et al.* 1997, Gilvear 1999), ecology (Hupp *et al.* 1995), and environmental science and management (e.g., Brookes 1995, Thorne and Thompson 1995), and is increasingly recognized as a key element in successful river restoration (Kondolf and Larson 1995, Bravard *et al.* 1999). These interactions are two-way, in that geomorphology is not only applied to these allied fields, but that tools from the allied fields are applied to fluvial geomorphic problems. Geomorphological techniques, such as sediment sampling and channel facies/habitat assessment, are applied to ecological problems such as assessments of fish habitat, and biological techniques (e.g., dendrochronology, biochemistry analysis or biometrics) are applied to geomorphological problems, such as dating deposits and surfaces or highlighting variability in forms and processes. More sophisticated statistical analyses developed for understanding complex social or biological objects are now applied to geomorphic data sets. Likewise, geomorphology's interactions with archeology have yielded benefits to both fields. As a result of its multiple roots and extensive interactions with other disciplines, the set of tools used in fluvial geomorphology is unusually rich and diverse, and many tools are now no longer confined to a single discipline. For example, a tool developed in one discipline or sub-discipline (e.g., lichenometry for dating recent glacial moraines) can be applied in fluvial geomorphology to date the sides of certain river channels and to date flow events (Gregory 1977).

From Conceptual to Working Tools

As any other discipline, geomorphology is characterized by internal debates about theories and methods used, and about its history and development (Smith 1993, Rhoads and Thorn 1996, Yatsu 2002). Amongst the most influential theories have been the cycle of erosion (Davis 1899), and the magnitude–frequency and effective discharge (Wolman and Miller 1960). But as underlined by Knighton (1984), the field of fluvial geomorphology has developed relatively few original theories, tending rather to import theories from allied fields such as hierarchical theory, system theory, chaos theory, probabilistic theory and their associated concepts.

Among methods used in this interdisciplinary field, we can distinguish methods of thought that structure the way we do research, and working methods used during the research process, each with its specific techniques (Table 1.1). The inductive method involves generalizations developed from a set of observations. For example, in historical geomorphology, we do not know in advance what we will find, so the field data (e.g., date of deposits provided by archeological artifacts) drive the research. As another example, the empirical relationships established between the fluvial forms and flow regime have led to formulation of many new scientific questions. As empirical data have accumulated, the conceptual models of flow-channel form relations have been modified based on the new findings. In contrast, in the deductive method, the research process is driven by a preliminary hypothesis, which may be invalidated, using traditional statistical tests. The deductive approach can be purely experimental, with the researcher reducing artificially, in laboratory as well as in field, the number of acting variables, to establish and validate the basic links among some of them. It can also be based on comparisons between spatial objects whose existing conditions are used for testing and validating an a priori hypothesis (in natura experience) for which specific areas as well as specific data are selected.

A restrictive definition of science, as one proposed by Bernard (1890), which excludes humanities and requires a strict trinome of hypothesis, experience, and conclusion applied to a simple or simplified object does not apply well to geomorphology. Laboratory experiments are often used in fluvial geomorphology to complement field studies, but controlled experimentation in the manner of pure physics is not possible for most geomorphological concerns (Baker 1996). More fundamentally, some geomorphic questions cannot be solved by testing of hypotheses posed a priori, and complex new questions have emerged that cannot be simplified without losing relevance. Similarly, problems are brought to geomorphologists from other

Table 1.1 A few examples of thought and working methods

Thought methods

Inductive method. Generalization from data collected in the field, laboratory, literature, etc. Often an exploratory method from which some hypotheses can be developed

Hypothetico-deductive method. A preliminary hypothesis or conceptual model is modified, confirmed or rejected based on results of the (usually field or laboratory) studies. It can be applied by using comparative methods or experimental ones:

> *Systemic/comparative methods.* Simultaneous observations of multiple rivers or reaches, sometimes at multiple scales and involving different levels of a drainage network, from which the scientist attempts to identify distinct forms or types of functioning, sensitivity to changes, and potential thresholds. A pair of spatial objects (one control and one observed) is the basic step of doing natural experiments. Working methods and associated techniques developed in inductive approaches can be used, but a preliminary hypothesis has been posed and is tested
>
> *Experimental methods.* Controlled conditions are created in the laboratory (e.g., flume) or in the field when possible (e.g., erosion plots with artificial rain). This approach is based on a specific framework of working methods and associated techniques

Working methods and associated techniques

Pre-field methods. Any approach developed in a preliminary step to select the thought methods and design a data collection framework, in some cases a sampling protocol. This is on the question posed, when, where, and what one does in the field

Field methods. Any approach to measure processes, forms, and deposits in the field or to collect archival data or any spatial information

Laboratory methods. Any approach performed in a laboratory on field samples to measure physical, chemical and biotic characteristics

Post-field methods. Any approach used to treat the data and interpret the results

fields: problems that are frequently posed at spatial and temporal scales smaller and shorter than those needed to understand the fluvial processes involved.

By virtue of their complexity, fluvial systems are difficult to study and we cannot modify them to create controlled situations without modifying their nature. With development of new technologies and larger databases, it is now possible to pose new questions at different spatial levels. It becomes possible to consider complexity, and to work with convergence of evidence instead of conclusive proofs, comparisons among multiple sites instead of between treated and control sites, and enlargement of the idea of experimentation to include directed, organized observations over large number of sites, partial models (accepting that it is impossible to fully model complex systems), and clearly articulated conceptual models. A comparative analysis becomes increasingly important, especially to consider geomorphological questions holistically.

In this context, there is a clear challenge to mix holistic and reductionist approaches, the first to integrate the studied object in its temporal and spatial context; the second to highlight the physical laws controlling the forms and processes. The inductive and deductive methods can be complementary, and by using both, one can avoid problems of overgeneralizing on one hand, or reaching conclusions that are only narrowly applicable on the other. Experimentation, conducted in tandem with field observation, can significantly advance our understanding of process (Schumm *et al.* 1987).

Multidisciplinary approaches, such as coupling hydraulics and geomorphology, have facilitated application of physics and mechanics to the field. This has resulted in better understanding of the acting processes, limits of validity of given laws, and limitations of numerical models. Using bank erosion as an example, geomorphological research has identified complexity of geographical contexts and of physical processes controlling the phenomena, underlined potential consequences of bank protections and their often limited life span, and has proposed other solutions, such as the streamway concept that requires an interdisciplinary framework, e.g., with legal scholars to address property rights, sociologists to understand reactions of landowners, and economists to evaluate the long-term economics of various alternatives. This evolution of the research perspective has been accompanied by increasing participation in decision-making by citizens, landowners, governmental and non-governmental agencies, and other stakeholders.

Working methods are diverse because there are many ways of approaching fluvial geomorphological questions: in the field or experimentally, from archives and historical images, at various spatial and temporal scales, in various man-made and natural contexts. We propose a rough classification based on the stage at which the methods are used: pre-field, field and post-field methods, with "field" being considered here in a larger sense not only for data collection in the landscape but also in archives, etc. (Table 1.1). Sampling methods, sites, frequency, etc., must be determined before collecting data. Once these preliminary questions have been answered, methods are used in the field to collect information, potentially reinforced by laboratory techniques to measure quantities, concentrations, or dates. At the post-field stage, other methods (e.g., statistical, graphics, mapping, imagery analysis) are used to treat the data and interpret the results. Whatever the stage, the methods and techniques used depend strongly on the question posed and the thought method chosen, whether to describe, to explain, to simulate, or to predict.

The organization in Table 1.1 is obviously only one of many ways to classify these, but it provides an overview of current approaches in the discipline. Under each working method (as defined in Table 1.1), a number of techniques may be used, depending on the characteristics of the field site and the nature of the question posed. For example, there are multiple methods for measuring discharge, one of which is the current meter method, involving measurements of depth and velocity across the channel, another being the salt dilution method (Chapter 12). The method of bedload sampling can involve techniques such as bedload traps, Helley–Smith sampling, or tracer gravels (Chapter 13). However, the line between method and technique is not always clear, as the more one knows about a tool, its components, and variants, the more one is inclined to call it a method rather than a technique. For example, to the non-specialist, dating or assessing overbank sedimentation rates from C^{137} concentration measurement in the soil profile of a floodplain appears at first to be a technique, but to the specialist it is a method that can involve several techniques, such as sampling (from coring and slice cutting to getting sediment samples or digging, bulk sample or profile analysis), as well as measuring radioactivity (high versus low resolution spectrometer, alpha versus gamma spectrometry) (Chapter 9).

In this book, we focus not only on the field/laboratory methods and techniques as they refer to the specific tools of the discipline but also to key concepts and methods that are fundamental for the geomorphological thinking, the way of approaching the applied problems. Because pre-field and post-field methods and techniques are more generic tools in science, we focus less on these. Moreover, we have organized the book according to key geomorphological topics rather than to key tools because one of our main messages is that the geomorphological question is the key. The tools themselves are secondary, and follow directly from the question. Accordingly we introduce the tools based on the question posed, considering five main types of geomorphological questions and the associated tools:

- *the historical framework* and the methods and associated techniques to date and assess historical geomorphological trends;
- *the spatial framework* and the concepts, methods and associated techniques that reveal spatial structure and nested character of fluvial forms;
- *the chemical, physical, and biological methods* for dating and the study of spatial structure and fluvial processes;
- *the analysis of processes and forms*, the traditional heart of the discipline based on field surveys and measurements of sediment and water flow;
- *the future framework* for which methods and techniques exist for discriminating, simulating and modeling processes and trends.

The aim is not to describe any specific technique in detail, but rather to focus on the geomorphological methods within which techniques are applied. The techniques have been well described in specific papers, as well as in more comprehensive works (e.g., Dackcombe and Gardiner 1983, Goudie 1990, Thorne 1998). The greatest contribution of this book, then, is probably to better develop the context within which techniques are chosen in a better way, and to enrich the description of methods and techniques by contrasting examples. Two chapters are also specifically devoted to conceptual approaches, such as the fluvial system theory and the sediment budget concept. Through these treatments we seek to show the manner and spirit in which the geomorphologist works.

Tools and Questions

Concepts, methods, and techniques are tools used to answer to questions (Figure 1.1). The key element, then, is the question. This is true even for an inductive approach, in which there is an implicit question posed

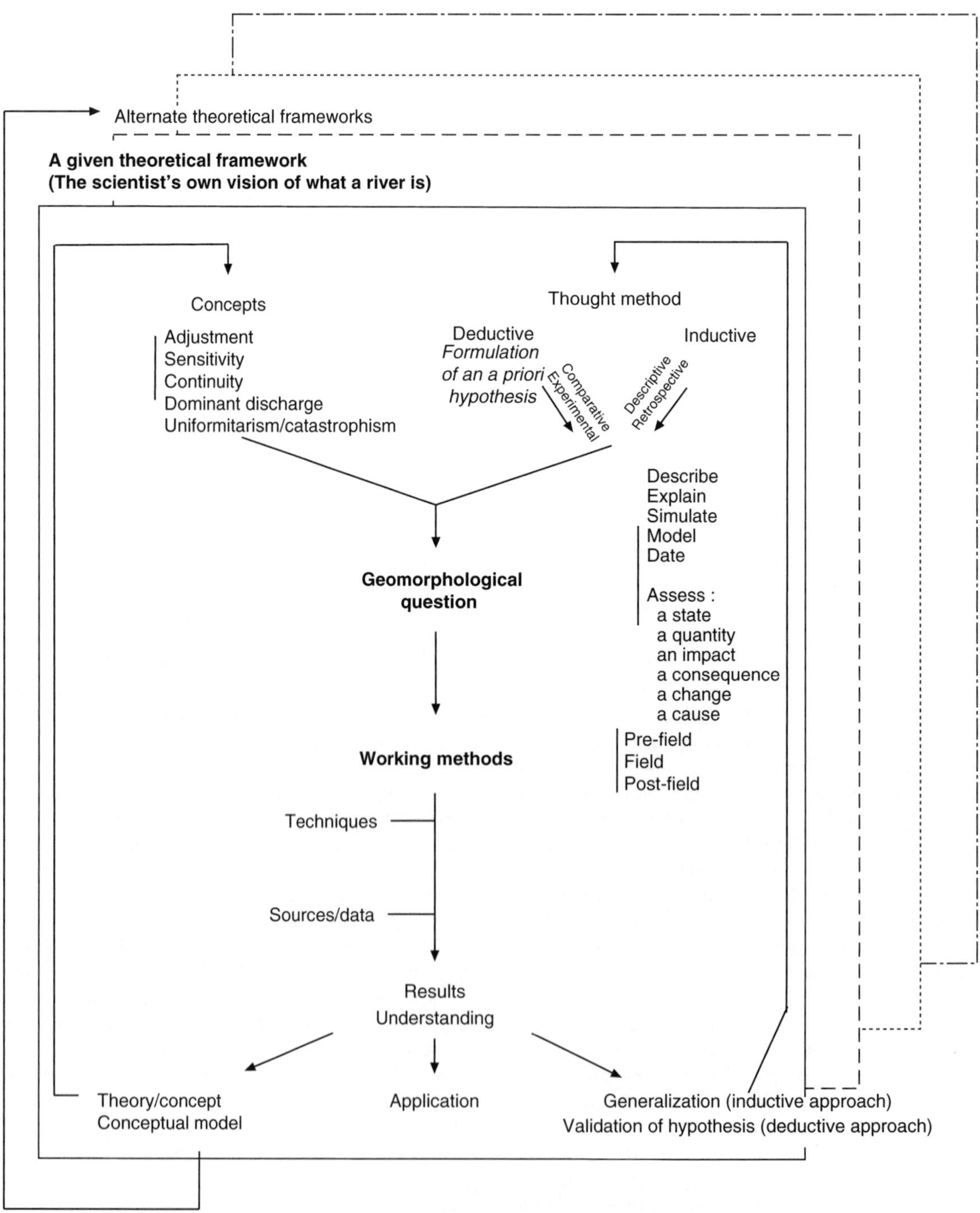

Figure 1.1 General framework of the way a geomorphological question is posed in the research process. Different tools are then used: theoretical framework, concepts, thought methods and working methods with their associated techniques variously dependant on the sources used

of what kind of geomorphological forms and processes trends occur. The efficacy of any geomorphic research depends much less on the choice of method than the quality of the research question posed (Leopold and Langbein 1963). Once the question is posed, based on deductive or inductive approaches and supported by a given set of concepts, supposing that it is valid, the second step is to define the working methods and potential data sources. Next (or simultaneously), methods and associated techniques are identified within a given conceptual framework, and with spatial and temporal resolution appropriate to the scale at which the question is posed. When one considers the river as a system, one's questions are usually less time and scale restrictive, and one tends to pose specific questions about links among catchment sub-divisions.

Concepts can be both the result of a research program (question → result → generalization → conceptualization) and also tools with which to carry out the research; as once established, concepts allow us to organize data and guide our subsequent research. The graded river concept (Mackin 1948), the concept of dynamic equilibrium (Hack 1960, Chorley and Kennedy 1971), and the concept of reaction and relaxation times (Graf 1977, Brunsden 1980) are all a result of generalization provided by previous research. These concepts also led to development of other research questions, which in turn were tested in various environments in order to better understand the sensitivity of regional contexts and the variability of thresholds. But as with other tools, concepts may be applicable in some situations but not others. For example, the concept of dominant discharge as a frequently occurring flood (such as the $Q_{1.5}$) is a useful concept in humid climate or snowmelt streams, but generally is of little use in semi-arid climate channels (Wolman and Gerson 1978). We can also step back to larger-scale conceptual frameworks or conceptual models that guide our research: the continuum concept (Leopold and Wolman 1957, Vannote *et al.* 1980), the fluvial system concept (Schumm 1977), the hydrosystem concept (Roux 1982), and the sediment budget (Dietrich and Dunne 1978).

In most cases, there is no perfect tool to answer the question posed. Instead, we usually must employ a set of (often diverse) methods to approach a question. Ironically, it seems that non-geomorphologists sometimes assume that a perfect tool must exist to easily answer their questions, and thus they may readily adopt tools billed by proponents as ideally suited to address management concerns. For example, Bevenger and King (1997: 1393) argued that there was a need for "well-designed monitoring protocols" using "tools that are relatively simple to implement, that can be used directly and consistently by field personnel, and are sensitive enough to provide a measure of impact". While probably no one would disagree with the desirability of such methods, there is no a priori reason to assume that they exist, and certainly the zigzag method of sampling bed material promoted by Bevenger and King (1995, 1997) fell far short of such an ideal (Kondolf 1997a,b) (see discussion in Chapter 13 of this book).

In trying to make sense of the tools used in the field of fluvial geomorphology, we have reviewed the recent geomorphic literature to ascertain what tools have been used to address what questions, and at what temporal and spatial scales. In this chapter, we report on the results of this survey and briefly describe the organization of this book.

1.3 LITERATURE REVIEW: METHODS

We identified tools in fluvial geomorphology used in papers published in the 11-year period 1987–1997 in the journals *Catena*, *Earth Surface Processes and Landforms* (ESPL), *Géographie Physique et Quaternaire* (GPQ), *Geomorphology*, *Zeitschrift für Geomorphologie* (ZfG). We also studied *Water Resources Research* and *Geological Society of America Bulletin,* which publish some papers in the field and *Géomorphologie*, published since 1995 by the *Groupe Français de Géomorphologie* (French Group of Geomorphology) for the period 1996–1999.

We conducted a quantitative analysis of the papers published in the first five journals only, as they were the journals primarily devoted to geomorphology publishing over the last decade, and other journals had far fewer papers in fluvial geomorphology. *Earth Surface Processes* had the most papers in fluvial geomorphology (20 papers per year, ppy, or 33% of the total papers), followed by *Geomorphology* (11.7 ppy, or 31%), *Water Resources Research* (9.6 papers out of more than 100 published annually in the whole range of topics considered by this journal), *Catena* (5.5 ppy, or 13%), and *Zeitschrift für Geomorphologie* (5.4 ppy, or 18%), followed by *Géographie Physique et Quaternaire* (2.4 ppy, or 9%), and *Geological Society of America Bulletin* (two papers out of 130 published annually). Of the papers published by *Géomorphologie* between 1996 and 1999, 30% were on fluvial geomorphology but this may reflect the influence of special issues from a fluvial geomorphology

Table 1.2 Characteristics recorded for each paper reviewed

Journal/issue
Year of publication
Country of authors
Location of study
Timescale: 1. Present (within the year to a few year survey) 2. Decade 3. Century 4. Holocene 5. Pleistocene 6. >Quaternary 7. Not applicable
Spatial scale: 1. Channel 2. Floodplain 3. Terrace 4. Network 5. Drainage basin 6. Not applicable
Type of tool applied (see Table 1.3 for the categories and their meaning)

conference entitled River Basins, Channels, and Floods (*Crues, Versants et Lits Fluviaux*) held in Paris in March 1995.

For each paper in fluvial geomorphology reviewed in the first five journals, we noted year of publication, authors' country, timescale, spatial scale, and tool(s) used (Table 1.2). We classified the papers according to the temporal and spatial scales studied, using seven temporal and six spatial categories, which reflected our attempt to boil down a large number of potential categories to a number that covered the spread of the data but in a manageable number, and in a way that minimized the amount of interpretation required as we reviewed the papers. Under spatial scale, e.g., some flume studies would be classified as "not applicable", even though it might be argued that the questions addressed by a given flume study would relate primarily to problems at certain spatial scales. Only a few papers were directly concerned with more than one spatial scale or temporal scale, not enough to warrant complicating the data set by identifying these separately.

We recorded all tools used, as most papers used more than one tool (see the list in Table 1.3). Some of the listed tools can be considered techniques, others methods, others "sources" because some techniques and methods are very specific to the sources used. When concepts and theories were clearly identified as tools in papers, they were censed, recorded as "Conceptual Modeling". Only one tool is then identified for this complex setting. Some authors based their approach on previous data, notably when using hydraulic formulae. We also defined a tool "Review of bibliographic data" to acknowledge this common approach. Obviously, this is only one approach and others might develop different categories, but it provides one way to organize the range of tools and see in what context they have been used.

In many papers, the methods and materials were clearly presented and described. From these papers it was easy to extract the necessary information. However, some papers were vague on methods, and we sometimes had to inspect the entire paper before figuring out what tools were used. For example, from the results reported we might infer that aerial photographs were analyzed, or that hydrologic analyses were undertaken, although they were not specifically reported in the methods section. Many papers mixed previously published data with original data for comparison without clearly and explicitly distinguishing the two, nor fully addressing possible differences in data collection. The origin of archival data and their potential limitations were not always clearly explained, compromising the validity of results and limiting the usefulness of the data for comparison. For discharge, water level, and water depth, it was not always clear whether the authors themselves measured flow or extracted data from existing databases. For maps and air photos, it was often not clear whether the authors themselves undertook measurements on them. Papers on hydraulic modeling were often not clear regarding input data used, focusing mostly on the modeling methods.

We built a database from these observations with 496 rows (corresponding to each reviewed paper) and 46 columns (corresponding to the tools recorded, each being considered as present, 1, or absent, 0); we removed nine tools amongst the 53 previously censed whose frequency was too low (less than five times censed), and conducted statistical analyses using the software ADE 4 (Chessel and Dolédec 1996). To summarize the table, we first conducted a correspondence analysis (CA) (Lebart *et al.* 1995), and retained the six first factorial factors loading (i.e., explaining) 25% of the total variance (or inertia). We performed a cluster analysis (Hierarchical Ascendant Classification) on the six axes and produced a dendrogram to show the tools that tended to be used in combination with others.

Table 1.3 Alphabetical list of codes used for the type of tools in Figures 1.2, 1.3, 1.5, and 1.6

Code	Meaning	Code	Meaning
1D	1D hydraulic modeling	Lwd	LWD surveys
2D	2D–3D hydraulic modeling	Mag	Magnetic measurement
Aan	Archival analysis (topographic data, reports)	Map	Geomorphological mapping
Aph	Vertical aerial photographies	Mca	Movie camera survey
Arc	Archeological evidence, paleontology	Mea	Field/topographic measurements
Bib	Review/bibliographic data	Mes	Suspended sediment sampling and turbulence meas
Bpr	Measurements of bank profile and geotech. prop.	Met	Metal concentration
C14	C^{14}	Min	Mineralogic analysis
Che	Chemistry and biochemistry analysis*	Mob	Particule mobility measurements
Chs	Climate and hydrological series	Mon	Piezometry, rainfall and discharge monitoring
Com	Conceptual models	Mul	Multivariate and cluster analysis
Cor	Sediment core, freeze core, excavation	Num	Numerical modeling
Cpb	Cs^{137} and Pb^{210}	Obs	Geomorphic/flow/field observations
Dem	Digital elevation modeling	Opd	Optical dating, long period radionucl. (U^{234}, K–Ar)
Dep	Water depth measurements	Oph	Ground and air oblique photographies
Ero	Erosion pins, floodplain accretion measurements	Pho	Photogrammetry
Flu	Physical modeling (flume experiment)	Pol	Pollen analysis
For	Hydraulic formulae*	Sam	Bedload sampling (traps, Helley–Smith sampler)
Fra	Fractal analysis and geostatistics	Soi	Soil profile analysis, soil density, soil moisture
Gis	GIS (geographic information systems)	Sta	Bivariate statistical modeling
Gma	Geological map	Sto	Stochastical modeling
GrS	Grain size measurements	Tep	Tephrochronology
Ima	Satellite image, radar, image analysis	Tma	Topographical map (obs., morphometric meas.)
Int	Interview of resident	Tre	Tree-ring analysis
Iso	Cosmogenic isotopes	Veg	Vegetal survey
Lic	Lichenometry	Vel	Water velocity measurements
Lit	Lithology, petrography, stratigraphy observations		

*Also conductivity, solute concentration, temperature.

We also analyzed the data set to determine if certain tools were used for answering questions at particular spatial scale and timescale, whether certain tools could be associated with particular journals or geographic origins of the authors, and whether any changes in tool use could be detected over the studied decade.

In order to estimate if tools were used according to specific context, we performed between-class CA (Lebart *et al.* 1995), which can be defined as discriminant analysis on categorical data. The aim of the between-class CA is to identify, from previous factorial axes summarizing the co-occurrence of tools, those that maximize the variance between some distinct groups (e.g., journal, spatial object under study, timescale, year of publication, first author's country). For spatial scale, we initially made separate groupings for papers dealing with drainage basin and drainage network, and for floodplain and terrace, but we found that the groups overlapped and thus could be lumped in the CA. Concerning the temporal scales, only four groups were also retained, "Holocene", "Pleistocene", and "older than Quaternary" being grouped and papers without timescale being removed (only a few). Moreover, we created a new variable with 12 categories based on both scales.

In this procedure, the first discriminant axis has the highest between-class inertia while the second axis, which is not correlated to the first one, has the second highest between-class inertia. The number of discriminant axis is equal to the number of groups and the sum of inertia for each axis is equal to the between-class inertia. To statistically validate the difference between the groups, we conducted a randomization test (Manly 1991). For each random distribution of individuals (e.g., the studied papers) within the classes ($n = 10\,000$ in our case), we calculated the ratio "between-class inertia/total inertia" and tested whether the result (the computed between-class inertia) was higher than that would be expected from random runs. A positive test indicates that the difference between groups was greater than that would be expected if individuals had been assigned randomly to groups.

Lastly, a co-inertia analysis was performed to study the co-structure of the two tables, highlighting the correspondences between the tools used and the characteristics of the papers (author's country, timescale, spatial scale, journal). This procedure searches for factorial axis (e.g., co-inertia axis) that maximize the covariance of projection coordinates of the data sets "papers × tools" and "papers – characters of papers" for which each structure was previously studied with factorial analysis (Tucker 1958, Chevenet *et al.* 1994). A randomization test (a so-called Monte Carlo test) was used to test the statistical validity of the computed co-structure by comparing it to other co-structures resulting from simultaneous random permutations of the papers within the two tables.

1.4 LITERATURE REVIEW: RESULTS

Overview of Tools Used

We identified 496 papers on fluvial geomorphology out of a total of 2228 published in the five journals from 1987 to 1997. Of the papers dealing with fluvial geomorphology, an average of 45 being published annually, most utilized more than one tool (the average was 2.5 tools used per paper), with 25% using only one tool, and 60% using two. The maximum number of tools censed in a paper was 10.

The dendrogram (Figure 1.2) shows that there is no clear classification of tools, just a rough ordination. One of the reasons is that many papers used only a few tools. Some tools, usually new techniques and those in development (e.g., fractal analysis, numerical and physical modeling, isotopes, short-life radionuclides) tended not to be associated closely with other tools, not because they were specific in terms of geomorphological questions posed, but because the studies employing these new tools tended to focus on methodological aspects, sometimes explicitly to test the usefulness of the tools. Moreover, the authors focused so much on the tool itself that they may have neglected to describe the material they used to apply their technique and how it was produced. Moreover, we can expect that several combinations of tools could be possible depending on geomorphological question posed (timescale and spatial scale, process-orientated or form-orientated). If some tools such as measures of flow are clearly associated with a specific approach in terms of temporal and spatial scales, some others such as geomorphological mapping or multivariate statistics can be expected to be more widely used whatever the geomorphological question posed.

Certain tools tended to be grouped and then to be used together, more frequently. We identified five distinct associations of tools; each of them could correspond to sub-fields of fluvial geomorphology. Archeological evidence, C^{14} dating, pollen analysis, stratigraphic analysis, and sediment coring or trenching tend to combine more frequently and would illustrate the sub-field of palaeo-fluvial geomorphology. Tools for spatial analysis tended to be used in combination, such as digital elevation modeling (DEM), geographic information systems (GIS), satellite imagery, with analysis of existing documents such as geological or topographic maps and oblique photography, with multivariate analysis. Optical dating and some radionuclides (Ar–K) were associated with this set, because they were often integrated in large-scale approaches to date surfaces previously identified from satellite or airborne image.

Tools used in studies at the river reach and mainly based on field measurements tended to fall into two groups. Channel process studies tended to use velocity, depth, discharge measurements, suspended sediment sampling, bedload sampling, and particle mobility analysis. Another group that deals with channel adjustment, associate two sub-sets of tools. The first one tended to use hydraulic formulae, bivariate statistical models, air photo analysis (e.g., for measures of sinuosity and other pattern descriptors), and geomorphological mapping. Some of these tools are not only traditionally used in hydraulic geometry analysis but also used in channel form description and channel change assessment. Hydraulic modeling was not integrated in this group, mainly because a lot of the papers did not explicitly explain how their data collection was undertaken. Moreover, numerical modeling can be

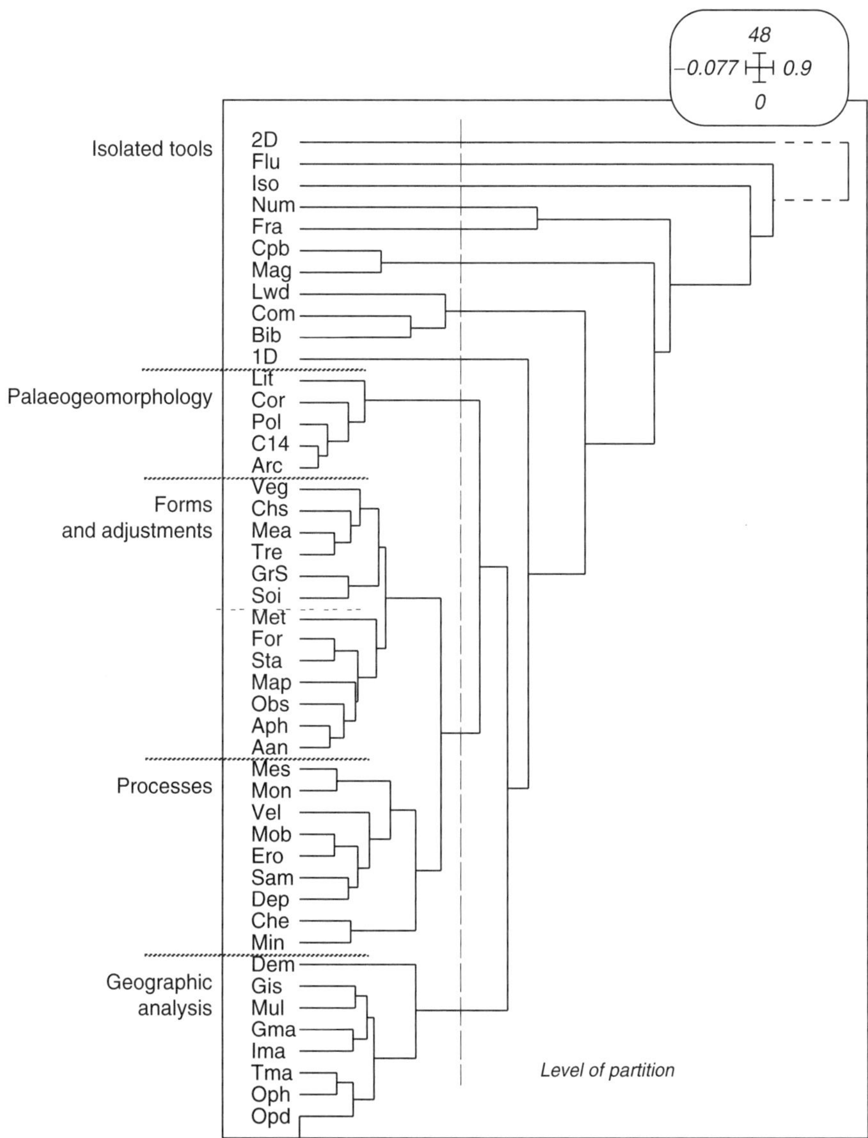

Figure 1.2 Dendrogram showing associations of tools used in the reviewed papers. Five distinct groupings emerged from the analysis, each of them being interpreted as a sub-field of geomorphology. For meaning of abbreviations, see Table 1.3

based on "working data" or even arbitrary values (e.g., arbitrary channel geometry) without clear relationship with field examples. The second sub-set is characterized by particular tools such as vegetation surveys, analyses of soil profiles and tree-ring measurements, which would suggest a sub-field titled biogeomorphology. Climatologic and hydrologic series analyses as well as grain size and topographic measurements were also associated with this set of tools. This underlines that biological tools are not used independently of more conventional tools (e.g., topographic measurements and grain size), but usually in tandem to link biological and geomorphological parameters in the sense of the term "biogeomorphology".

Figure 1.3 shows the frequency of occurrence for the 46 tools, grouped by the categories developed by

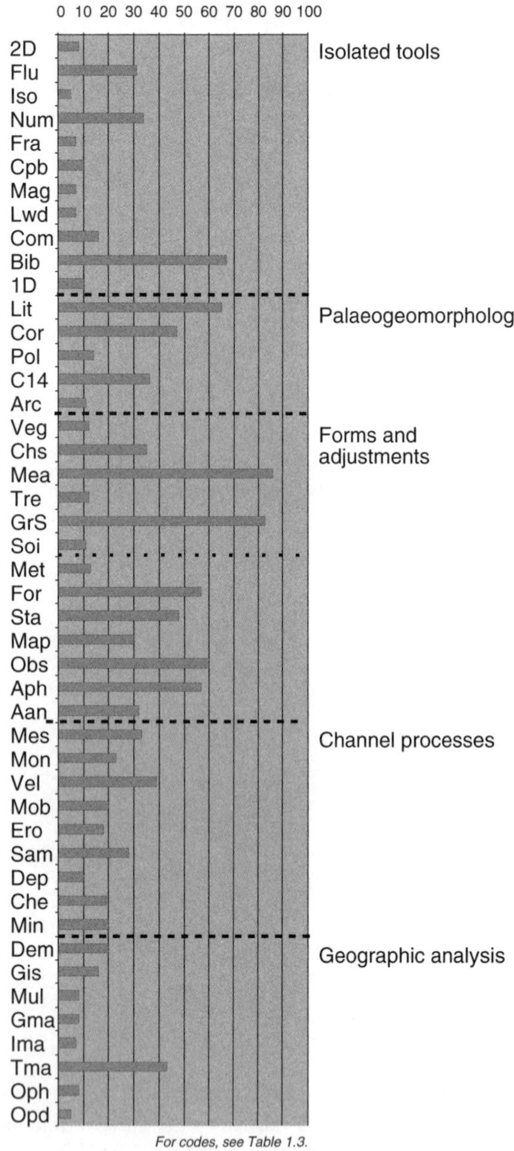

Figure 1.3 Frequency of tools distinguished from our literature review of 496 papers published in fluvial geomorphology between 1987 and 1997. The tools have been classified according to dendrogram ordination of Figure 1.2. For meaning of abbreviations, see Table 1.3

the dendrogram (Figure 1.2). The most frequently employed were traditional tools of the geomorphologist, e.g., the field tools (geol./geom.), observations and measurements (Obs and Mea), sediment (GrS), and stratigraphical analysis (Lit). Topographic surveys were the most widely used of all (86 times, concerning 17% of the papers); grain size measurements were employed 83 times, and geomorphological observations 65 times (Figure 1.3). Bibliographic review is also one of the most frequent approaches (67 times). Among the others that occurred in at least 10% of the manuscripts were hydraulic formulae (For), bivariate statistics (Sta), and vertical air photos (Aph).

Geographic tools were much less used than those dealing with form characterization. Measurements of flow, transport processes, and form dynamics (e.g., bedload sampling, velocity measurement, erosion pins) were less frequently used than might be expected, perhaps reflecting a relative decline as these have been replaced by more recently developed tools such as modeling and geographic analyses. Another interpretation would suggest that the most frequently used tools are basic tools used in every sub-field (e.g., observations, topographic measurements) whereas the tools used for characterizing channel processes are already specialized in terms of time and spatial scales. Hydraulic modeling (e.g., 1D, 2D and part of the Num category) accounts for almost 50 papers out of 496, making this one of the more heavily used tools in the field. Papers based on flume experiment are also relatively frequent (32 times) compared to other tools such as GIS or DEM that are less common (less than 20 times cited) than we would have expected. It is probable that the popularity of such tools will increase with the recent progress in computer and data collection capability.

Correspondence Between Tools and the General Characters of the Reviewed Papers

We tested, using the between-group analysis and the co-inertia analysis, the correspondences existing between the tools and a few variables describing the papers: the year of publication, country origin of authors, and journal concerned, and the spatial and temporal scales. The whole censing of the papers underlined that most of the works dealt with channel scale and present time (e.g., within the year to 1–5 year surveys) (Figure 1.4). Fifty eight percent of the papers concerned the channel, 23% concerned the drainage basin/network, and 14% concerned the floodplain. For temporal scale, 61% of the papers concerned the present, 15% concerned the decadal scale, 9% concerned the Pleistocene, 8% the Holocene, and 6% the century. After grouping the spatial scale in three groups and the temporal scale in four groups, we

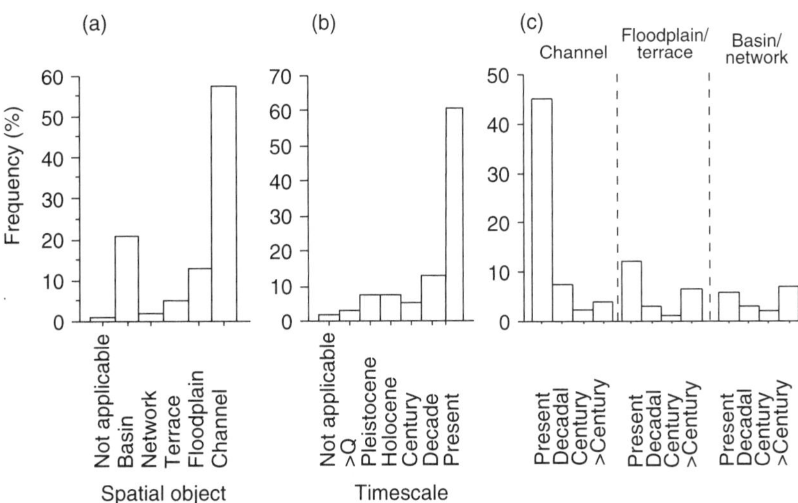

Figure 1.4 Frequency of spatio-temporal scales distinguished from our literature review of 496 papers published in fluvial geomorphology between 1987 and 1997: (a) spatial scales, (b) temporal scales, and (c) combination of both scales

observed that almost one paper in two concerned the present and the channel. The next most frequent spatio-temporal pair was floodplain present processes (13%), followed by channels considered at decadal scale. The first two categories (Figure 1.4c) contributed 60% of the papers with the others having less than 10% frequency of occurrence.

The randomization test did not indicate any change in types of tools used over the 11-year period studied, indicating that the evolution in methodologies and techniques that might have occurred during that period is too subtle to be detected by the test. Also, a decade may not be long enough to detect any clear evolution of practices in geomorphology and geology, where the Science Citation Index (2000) indicates the half-life of published papers is relatively long compared to other disciplines.

Concerning both spatial and temporal scales, the randomization tests validated correspondence between some tools and the scale levels. The ratio between the frequency of simulations for which the simulated inertia is lower than the observed one and those for which the simulated inertia is higher than the observed one is, respectively, 4% and 1%. Correspondence analyses between the tools used in the published papers and either the journals in which they were published or the author's nationality (i.e., addresses given) were not statistically validated by a randomization test for a ratio of 5%, but the results suggested some trends.

The co-inertia coupling the table of tools (496 papers × 46 tools) with the table of paper characteristics (496 × 26 modalities of four basic variables—country of the first author, journal, timescale, and spatial objects) summarized then the main links among them (Figures 1.5 and 1.6). The first four factors of the co-inertia explained 61% of the total variance. The permutation test confirmed that the two tables have a co-structure.

The first map of the co-inertia highlights the spatio-temporal framework and the corresponding tools (Figure 1.5), whereas the second map considers the correspondence between some tools and some nationalities and journals (Figure 1.6). For each, two maps can be compared, the map of the tools on the right side and the map of the characters of the papers on the left side, each variable and its modalities being distinguished from others to facilitate the reading and interpretation. It is then possible to see what tool combination can be expected in what journal, for what nationality, and how it is related to any spatial and temporal scales.

Authors from some countries tended to use particular sets of tools. Authors from Australia, China, France, and India, tended to use tools suitable for analyzing a long timescale, such as archival data, optical dating, image analysis, pollen analysis, lithologic analysis, whereas Canadian and Japanese authors evinced more interest in biogeomorphologic tools, such as vegetation surveys, tree-ring analysis,

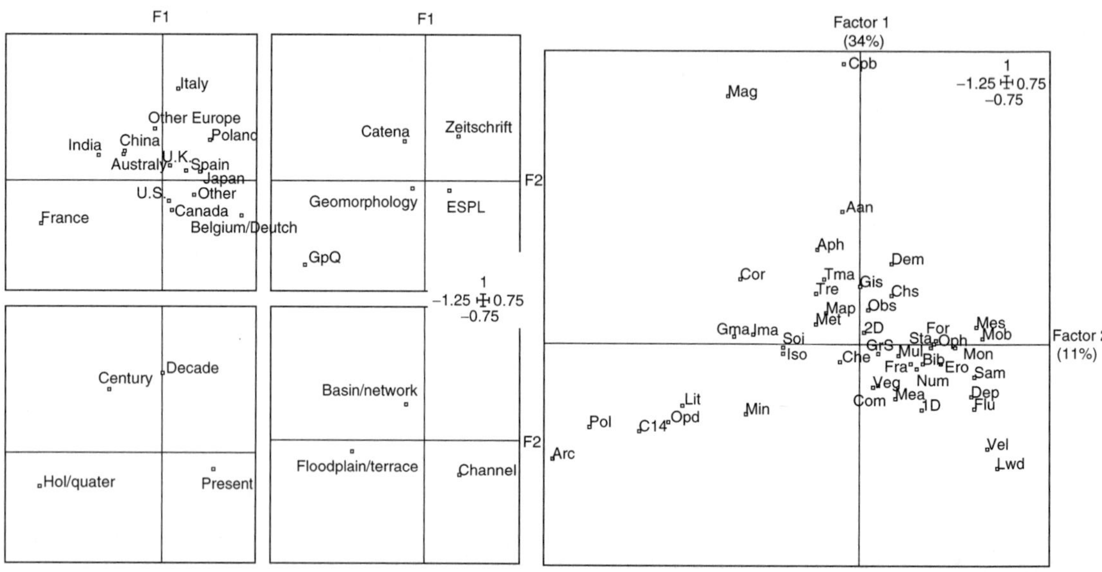

Figure 1.5 Projections on the first co-inertia map (factors 1 and 2 of the analysis) of the tools (on the right side; for meaning of abbreviations, see Table 1.3) and the four main characters of the 496 papers reviewed (e.g., the four small maps located on the left side of the figure). The countries appear on the upper left, the journals on the upper right, the temporal scales on the lower left, and the spatial scales on the lower right

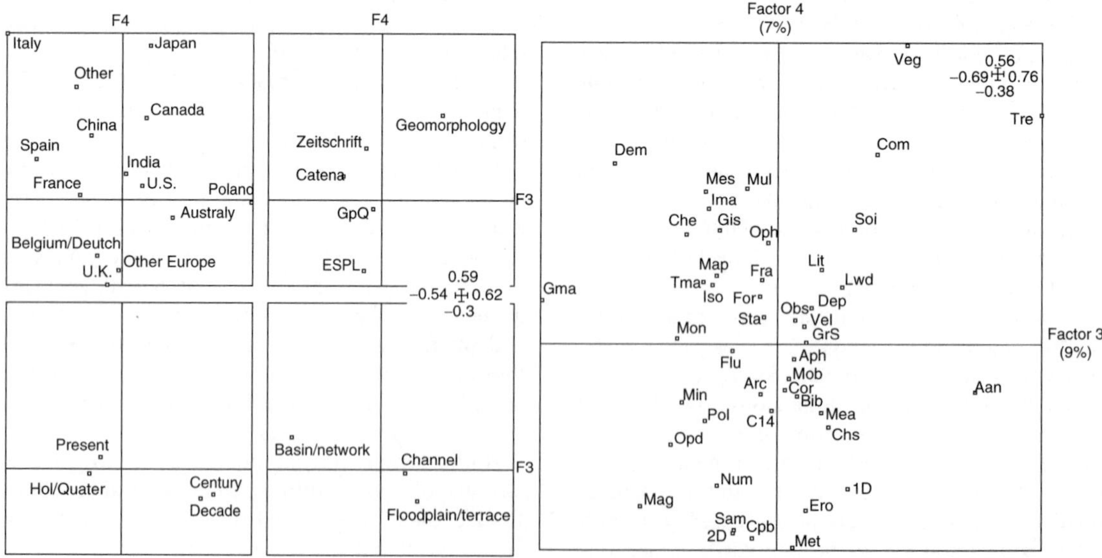

Figure 1.6 Projections on the second co-inertia map (factors 3 and 4 of the analysis) of the tools (on the right side) and the four main characters of the 496 papers reviewed (e.g., the four small maps located on the left side of the figure). The countries appear on the upper left, the journals on the upper right, the temporal scales on the lower left, and the spatial scales on the lower right

large woody debris surveys, and soil analyses. Authors from Belgium, the Netherlands, and the UK tended to use hydraulic modeling, bibliographic data, heavy metal concentrations, Cs^{137}, and Pb^{210}. Authors from US do not show any preferences in terms of tool used. The geomorphic community there is so large that it encompasses all sub-disciplines and can publish more widely. In Europe, the national geomorphological communities tended to work at certain spatial and temporal scales, to publish in certain journals, and to use certain tools.

The CA between tools and journals showed some tools were more closely associated with some journals. For example, *Géographie Physique et Quaternaire* was dominated by tools used in historical geomorphology, such as pollen analysis, archaeological evidence, C^{14}, optical dating, and isotopes. Papers in *Zeitschrift für Geomorphologie* and *Catena* are characterized by similar tools. On both maps, it is also not easy to distinguish the two journals in terms of authorship, timescale, or spatial scale. They are both characterized by a European authorship, a particular timescale (decadal/century scale) and spatial scale (mainly basin and hydrographic network), but also particular tools such as hydraulic formulae, imagery, multivariate analysis, as well as archival sources and DEM. The "geoecology" signature of *Catena* is not so clear and the journal shared this approach with *Zeitschrift* and *Geomorphology* as underlined by the distribution of "biogeomorphological tools" (soil analyses, tree-ring analyses, vegetation surveys and multivariate analysis, this latter being a tool more frequently used in ecology than in fluvial geomorphology) on the second factorial map (Figure 1.6). The *Earth Surface Processes and Landforms* barycenter was located on the right side of the first co-inertia map, and many of its papers reported on processes acting within the channel during the present time, using tools such as flume experiments, velocity measurements, and numerical modeling. *Geomorphology* plotted more centrally on the first factorial map, reflecting larger spatial and temporal scales in the tools used and the largest international authorship. On the second co-inertia map (Figure 1.6), *Geomorphology* is located in the upper part underlining particular authorship (Italy, Japan, Canada) when using particular tools such as conceptual modeling, tree-ring analysis, vegetation censing.

French authors tended to work on long timescales and on terrace/floodplain features, published in *Géographie Physique et Quaternaire* (in French), and used tools such as archival data, pollen analysis, C^{14} dating, optical dating, and lithologic analysis. Other European authors (e.g., from Poland, Italy, Spain, as well as "other Europe") worked mainly on the basin and century/decadal scales, published in *Catena* and *Zeitschrift für Geomorphologie*, and used archives, aerial photography, DEMs, and GIS. Another well-identified group were Belgian and Dutch authors and also those from the UK who tended to work more frequently on present channel features, published in *Earth Surface Processes and Landforms*, and used flume studies and measurements of water depth and velocity.

When focusing on timescales, the strongest correspondence appears between dating tools and certain scales, as some dating tools cover particular timescales (e.g., C^{14}, optimal stimulated luminescence—thermo-luminescence, Cs^{137}/Pb^{210}). Some other tools have a particular temporal signature; tree-ring, archives, air photos, climate, and hydrological series are clearly key tools to understand fluvial processes at decadal and also century scales. Most of the tools were used to describe present features and processes, such as measurement of suspended sediment concentration, velocity measurements, monitoring of rainfall and flow, numerical modeling, and fractal analysis. At longer timescales, Holocene and Early Quaternary studies did not differ in terms of tools used, drawing upon dating methods such as C^{14}, optical dating, archeological evidence, and field geological analyses such as lithology, sedimentary structure, pollen, and rock magnetism. Older features have been studied using spatial approach at the landscape level and then using small-scale imagery (e.g., satellite imagery).

There is a clear statistical discrimination among the three spatial groups in tools used, as well as associated timescales. Certain tools were applied in combination only at certain landform and specific timescales. C^{14} dating, mineral analysis, archeological evidence, and stratigraphic observations were applied in combination at the floodplain/terrace scale at the longest timescale. The basin/network scale was approached at the century/decadal scale using particular set of tools such as Cs^{137} and Pb^{210}, magnetic measures, vertical air photos, archives analysis, DEM, and GIS. Large woody debris surveys, particle mobility analyses, water depth and velocity measurements, and physical models were applied at the channel scale. Some tools do not characterize any clear timescale or spatial scale such as field observations, geomorphological mapping, or grain size analysis. While this survey found that many tools have been applied primarily to one scale only, this is probably largely a matter of custom, as large woody debris surveys could be extended to the floodplain scale to better understand interactions of wood with overbank flow processes and sediment distribution. DEM has been applied mostly at the basin scale, but could be usefully combined with field topographical measurements to study channel and floodplain forms. Image analysis thus far has not been associated with channel scale studies, but new developments in low-elevation imagery (from sensors mounted on airplanes, drones or stag-beetles) should create opportunities to use image analysis at the channel scale.

1.5 DISCUSSION

The results of the literature review indicate distinct sub-disciplines within fluvial geomorphology, reflecting the diverse roots of the field. The geological approach tends to use sedimentological tools and longer timescales, the engineering-hydraulic approach tends to use physical and numerical models, the larger-scale geographical approach employs GIS, and the biologically oriented approach tends to relate landform and vegetation. Channel studies have reflected two schools: process-oriented studies based on field measurements of process, and an emphasis on hydraulic geometry and channel change assessment utilizing measurements of channel form over time or space, but these two approaches have tended to converge in addressing controlling factors.

We conducted our quantitative analysis only on the five journals with the most papers published in fluvial geomorphology. Although there are many papers that could equally well appear in any of the five journals, we observed that each journal had a specialty: *Catena* and *Zeitschrift für Geomorphologie* for basin-scale studies, *Earth Surface Processes and Landforms* for hydraulic development, and *Géographie Physique et Quaternaire* for longer timescales. Of the 112 fluvial geomorphic papers in *Water Resources Research* during the period studied, nearly half used either hydraulic formulae or flume experiments as their principal tools, and over a third used numerical modeling, simulation, geostatistics, stochastic modeling, or fractal analysis as the principal tool (Table 1.4). *Geomorphology* seemed to be the least restricted in terms of tools used, authorship and temporal and spatial scales, and it was a preferred outlet for biogeomorphological studies.

Table 1.4 Classification of papers published on fluvial geomorphic topics in *Water Resources Research* from 1987 to 1997 by principal tool used

	Frequency	Percentage
Hydraulic formulae	38	34
Bedload sampling	3	3
Flume experiment	9	8
Numerical modeling and simulation	10	9
Geostatistics and stochastic modeling	6	5
Fractal analysis	16	14
Inferential statistics	11	10
Review, field obs. and meas.	19	17
Total	112	100

If the base of analyzed journals were expanded, this might change the results, but would also require more careful consideration (and development of criteria) to decide which papers were to be included in the analysis (e.g., hydraulic engineering studies). Also, we analyzed only an 11-year period near the end of the 20th century, and thus our study provides only a snapshot of practices at one point in time. It might be interesting to conduct this analysis over a longer period of time, reaching back to earlier in the century to capture the evolution of the literature since 1950. This would, perforce, also require expanding the analysis to include a broader range of journals, as the two journals yielding the majority of papers analyzed are specialized geomorphic journals founded in the last quarter of the 20th century. If this analysis was conducted over a longer period, we would expect to see an evolution in tools used, reflecting both, availability of new technologies as well as emergence of new concepts and research questions based on them. Finally, in our literature review, we did not attempt to identify how exactly the hypothesis/conceptual problem was addressed, although this would be a key piece of information to help us in the field.

Our finding that researchers in sub-disciplines tend to use a relatively narrow range of tools at certain temporal and spatial scales is not surprising, but does suggest potential opportunities to apply different tools at different temporal or spatial scales than is customarily done at present. Among tools primarily used to study present processes, some should be suited to integrating data from other timescales, such as integrating historical layers in GIS analyses, analyzing successional stages in vegetation analyses, using bibliographic sources to compare a historical state with the current, and adding various historical layers from topographical maps into multivariate analyses. Inspecting Figure 1.6, it is surprising that remote sensing (Ima) was so little used to analyze present features and processes (see discussion of this by Gilvear and Bryant in Chapter 6). Moreover, analysis of lithology and geological structures could be more widely applied to the study of channel pattern, because of the strong control of pattern by underlying geology, often underestimated because of preconceived notions of hydraulic geometry on the river continuum. Also, there is a strong challenge to better integrate some tools, still in development, with more traditional tools. We might gain in efficiency and in application if numerical modeling (1D, 2D) was confronted with some reality, such as observed historical evolutions. This approach could be applied on channel pattern changes which can be accurately evaluated from air photo analysis and topographical records. Some simulations could be done on older patterns for testing whether it is possible to reproduce existing patterns.

1.6 SCOPE AND ORGANIZATION OF THIS BOOK

As suggested by the literature review presented in this first chapter, the multiple disciplinary roots of fluvial geomorphology, and the field's increasing interaction with other disciplines and applications to management problems, have resulted in an array of tools that is diverse and becoming more so. This book presents summaries of the tools used in various areas of fluvial geomorphology, written at a level that falls between broad generalization and highly specific instruction on technique. The aim of the chapters is to help managers or scientists in other fields to understand the capabilities and limitations of various geomorphic tools, to aid in choosing methods appropriate to the questions posed. Of course, this requires the understanding of how various tools fit within the conceptual framework of the field (big picture context), and it requires some explanation of how the methods are actually carried out, the equipment and resources required, accuracy and precision, etc. However, for detailed instructions and descriptions of equipment and supplies needed, we refer the reader to more specialized works. Most chapters include case studies to illustrate applications of the tools described. While the scope of this book is quite broad, it neither covers geophysical methods nor flume experiments, not because these methods are considered less important, but simply to limit the book to a more manageable size and scope.

This book is organized into an introduction, five main sections, and a conclusion. Following this (introductory) chapter, the second section concerns the temporal framework, moving from mainly physical evidence and longer timescales to more recent and anthropic evidence. The section begins with Chapter 2, in which Robert B. Jacobson, James E. O'Connor, and Takashi Oguchi review surficial geological tools, such as floodplain stratigraphy and slackwater deposits, from which past hydrologic and geomorphic events (such as floods) can be interpreted and dated, changes in land use inferred, etc. In Chapter 3, Anthony G. Brown, François Petit, and Allen James discuss the use of archeology and human artifacts (such as mining waste) to measure and date fluvial geomorphic processes and events. In Chapter 4, Angela M. Gurnell, Jean-Luc Peiry, and Geoffrey E. Petts review the use of historical records to document and date geomorphic changes, mostly in recent centuries and decades.

The next (third) section addresses the spatial framework, emphasizing spatial structure and the nested character of fluvial systems. In Chapter 5, Hervé Piégay and Stanley A. Schumm review the systems approach in fluvial geomorphology from its roots in strictly physical processes through more recent systems approaches that integrate ecological processes. In Chapter 6, David Gilvear and Robert Bryant review the applications of aerial photography and other remotely sensed data to fluvial geomorphology, from traditional stereoscopic air photo interpretation to more recently developed remote-sensing techniques. In Chapter 7, G. Mathias Kondolf, David R. Montgomery, Hervé Piégay, and Laurent Schmitt review the uses and limitations of geomorphic channel classification systems, tools that have become extremely popular recently among non-geomorphologists, especially as applied to management questions. Concluding the spatial framework section, Peter W. Downs and Gary Priestnall review approaches to modeling catchment processes in Chapter 8.

The fourth section covers chemical, physical, and biological evidences, i.e., the applications of methods in these allied fields to fluvial geomorphic problems. In Chapter 9, Stephen Stokes and Des E. Walling review chemical and physical methods, with a substantial section on isotopic methods for dating, with their revolutionary effect on the field. Cliff R. Hupp and Gudrun Bornette detail biological methods, such as dendrochronology and vegetative evidence of past floods, in Chapter 10.

The fifth section includes analyses of processes and forms. In Chapter 11, Andrew Simon and Janine Castro describe methods to analyze channel form, emphasizing field survey and measurement techniques. Peter J. Whiting details methods of flow and velocity measurement in Chapter 12. In Chapter 13, Mathias Kondolf, Thomas E. Lisle, and Gordon M. Wolman review methods of bed sediment measurement (surface and subsurface) in light of various possible research objectives. Tracers, such as painted gravels, magnetic rocks, and clasts fitted with radio transmitters, are reviewed by Marwan A. Hassan and Peter Ergenzinger in Chapter 14. Methods of measuring and calculating sediment transport, suspended, bedload, and dissolved, are reviewed by Murray D. Hicks and Basil Gomez in Chapter 15. Sediment budgets are increasingly used as an organizing framework in fluvial geomorphology, especially in studies of impacts of human actions such as dams. In Chapter 16, Leslie M. Reid and Thomas Dunne draw upon their pioneering work in this area to provide guidance on how to approach sediment budget construction under various objectives and field situations.

The next (sixth) section concerns tools for discriminating, simulating, and modeling processes and trends. In Chapter 17, Stephen E. Darby and Marco J. Van de Wiel lay out general considerations for models in fluvial geomorphology. Jonathan M. Nelson, James P. Bennett, and Stephen M. Wiele provide a thorough review of the broad topic of hydraulic and sediment transport modeling methods in Chapter 18. Methods for modeling channel changes are described by James E. Pizzuto in Chapter 19. In Chapter 20, Pierre Clément and Hervé Piégay review statistical tools in fluvial geomorphology, not only commonly used tools such as regression but also statistical analyses often applied in allied fields such as ecology but rarely in geomorphology.

Most of the tools described in this book can be used to answer applied questions, and given the increasing demand for geomorphic input to river management, a wider range of tools deserve to be employed in support of management decisions. The concluding chapter (Chapter 21) considers the bridge between geomorphology and management, and presents illustrations from the US, UK, and France of fluvial geomorphology used to help river ecologists, planners, and managers to answer to their own questions, and in some cases, to redefine their questions on a larger spatial and temporal scales.

Obviously, a survey of tools in this field could be organized in different ways, and even within the chosen structure there were a number of tools that could logically have gone in different chapters, and chapters that could have gone in different sections. For example, Cs^{137} and Pb^{210} analyses are usually used in a temporal sense (e.g., to assess the variability of sedimentation rate over time), but they can also be used at a catchment scale to distinguish erosional from depositional areas. Aerial photography can be used to support a range of studies, from historical channel evolution to mapping of spatial patterns over large areas at one point in time. Like ecology or medicine, fluvial geomorphology is a synthesis science, analogous to the composite sciences as visualized by Osterkamp and Hupp (1996), meaning that it is based on a range of methods. Fluvial geomorphology is a thematic area where some scientific disciplines can interact and produce real interdisciplinary insights.

As a consequence, we cannot adopt one way of approaching geomorphological problems and neglect all others. In combination, multiple methods can be helpful in appreciating problems and addressing societal needs. Fluvial geomorphology can be useful in river management, especially as managers begin to think at different timescales and spatial scales (as implied when one adopts sustainability as a goal). Probably, all geomorphologists would agree that it is necessary to specify the problem as clearly as possible and to use the most appropriate tools from the great range now available. This book is intended to help in the realization of these aims.

ACKNOWLEDGMENTS

We are indebted to Pierre Clément, Ken Gregory, Fred Liébault, and Didier Pont for review comments on this chapter.

For the book as a whole, each chapter was peer reviewed by two reviewers, usually one external and one contributor, and by the two editors. We are indebted to the following reviewers who contributed to the improvement of the book: Vic Baker, James Bathurst, Tony Brown, Pierre Clément, John Buffington, Mike Church, Nic Clifford, Peter Downs, Jonathan Friedman, David Gilvear, Ken Gregory, Basil Gomez, Gordon Grant, Angela Gurnell, Jud Harvey, Marwan Hassan, Nicolas Lamouroux, John Laronne,

Eric Larsen, Stuart Lane, Fred Liébault, Mike Macklin, Andrew Miller, David Montgomery, Gary Parker, François Petit, Geoffrey Petts, Didier Pont, Michel Pourchet, Ian Reid, Steve Rice, David Sear, Stanley Trimble, Peter Wilcock, and Ellen Wohl.

REFERENCES

Baker, V.R. 1996. Hypothesis and geomorphological reasoning. In: Rhoads, B.L. and Thorn, C.E., eds., *The Scientific Nature of Geomorphology*, Chichester, UK: John Wiley and Sons, pp. 57–85.

Bauer, B.O. 1996. Geomorphology, geography and science. In: Rhoads, B.L. and Thorn, C.E., eds., *The Scientific Nature of Geomorphology*, Chichester, UK: John Wiley and Sons, pp. 381–413.

Bernard, C. 1890. *La science expérimentale*, 3rd edition, Paris: J.B. Baillière et Fils, 448 p.

Bevenger, G.S. and King, R.M. 1995. *A Pebble Count Procedure for Assessing Watershed Cumulative Effects*, USDA Forest Service Research Paper RM-RP-319, Fort Collins, Colorado: Rocky Mountain Forest and Range Experiment Station.

Bevenger, G.S. and King, R.M. 1997. Discussion of "Application of the pebble count: notes on purpose, method, and variants", by G.M. Kondolf. *Journal of the American Water Resources Association* 33(6): 1393–1394.

Bravard, J.P., Kondolf, G.M. and Piégay, H. 1999. Environmental effects of incision and mitigation strategies. In: Simon, A. and Darby, S., eds., *Incised River Channels*, Chichester, UK: John Wiley and Sons, pp. 303–341.

Brown, H.I. 1996. The methodological roles of theory in science. In: Rhoads, B.L. and Thorn, C.E., eds., *The Scientific Nature of Geomorphology*, Chichester, UK: John Wiley and Sons, pp. 3–20.

Brookes, A. 1995. Challenges and objectives for geomorphology in UK river management. *Earth Surface Processes and Landforms* 20: 593–610.

Brookes, A. and Shields, F.D. 1996. *River Channel Restoration: Guiding Principles for Sustainable Projects*, Chichester, UK: John Wiley and Sons.

Brooks, A.P. and Brierley, G.J. 1997. Geomorphic responses of lower Bega River to catchment disturbance, 1851–1926. *Geomorphology* 18: 291–304.

Brunsden, D. 1980. Applicable models of long term landform evolution. *Zeitschrift für Geomorphologie* 36: 16–26.

Bryant, R. and Gilvear, D.J. 1999. Quantifying geomorphic and riparian land cover changes either side of a large flood event using airborne remote sensing: river Tay, Scotland. *Geomorphology* 23: 1–15.

Chessel, D. and Dolédec, S. 1996. ADE version 4.0: Hypercard© Stacks and Quickbasic Microsoft© Programme Library for the Analysis of Environmental Data. Ecologie des eaux douces et des grands fleuves – URA CNRS 1451, Université Lyon I, 69622 Villeurbanne, France.

Chevenet, F., Dolédec, S. and Chessel, D. 1994. A fuzzy coding approach for the analysis of long term ecological data. *Freshwater Biology* 31: 295–309.

Chorley, R.J. and Kennedy, B.A. 1971. *Physical Geography: A System Approach*, London: Prentice-Hall, 370 p.

Dackcombe, R.V. and Gardiner, V. 1983. *Geomorphological Field Manual*, London: George Allen and Unwin.

Davis, W.M. 1899. The geographical cycle. *Geographical Journal* 14: 481–504.

Dietrich, W.E. and Dunne, T. 1978. Sediment budget for a small catchment in mountainous terrain. *Zeitschrift für Geomorphologie Supplement Band* 29: 191–206.

Giardino, J.R. and Marston, R.A., eds. 1999. Changing the Face of the Earth: Engineering Geomorphology. *Geomorphology* 31(1–4): 1–439 (special issue).

Gilvear, D.J. 1999. Fluvial geomorphology and river engineering: future roles utilizing a fluvial hydrosystems framework. *Geomorphology* 31: 229–245.

Goudie, A. 1990. *Geomorphological Techniques*, 2nd edition, London: Unwin Hyman.

Graf, W.L. 1977. The rate law in fluvial geomorphology. *American Journal of Science* 277: 178–191.

Gregory, K.J., ed. 1977. River channel changes, New York: John Wiley and Sons, 448 p.

Gregory, K.J. 1992. Vegetation and river channel processes. In: Boon, P.J., Calow, P. and Petts, G.E., eds., *River Conservation and Management*. Chichester, UK: John Wiley and Sons, pp. 255–269.

Hack, J.T. 1960. Interpretation of erosional topography in humid temperate regions. *American Journal of Science* 258A: 80–97.

Hupp, C.R., Osterkamp, W.R. and Howard, A.D., eds. 1995. *Biogeomorphology – Terrestrial and Freshwater Systems*, Amsterdam: Elsevier Science, 347 p.

Knighton, D. 1984. *Fluvial Forms and Processes*, London: Edward Arnold, 218 p.

Kondolf, G.M. 1997a. Application of the pebble count: reflections on purpose, method, and variants. *Journal of the American Water Resources Association* (formerly, *Water Resources Bulletin*) 33(1): 79–87.

Kondolf, G.M. 1997b. Reply to discussion by Gregory S. Bevenger and Rudy M. King on "Application of the pebble count: reflections on purpose, method, and variants". *Journal of the American Water Resources Association* (formerly, *Water Resources Bulletin*) 33(6): 1395–1396.

Kondolf, G.M. and Larson, M. 1995. Historical channel analysis and its application to riparian and aquatic habitat restoration. *Aquatic Conservation* 5: 109–126.

Lebart, L., Morineau, A. and Piron, M. 1995. *Statistique Exploratoire Multidimensionnelle*, Paris: Dunod, 439 p.

Leys, K.F. and Werrity, W.A. 1999. River channel planform change: software for historical analysis. *Geomorphology* 29: 107–120.

Leopold, L.B. and Langbein, W.B. 1963. Association and indeterminacy in geomorphology. In: Albritton, C.C., ed., *The Fabric of Geology*, Palo Alto, California: Cooper and Co., pp. 184–192.

Leopold, L.B. and Wolman, M.G. 1957. River channel patterns; braided, meandering and straight. *US Geological Survey Professional Paper* 282-b: 39–85.

Liébault, F. and Piégay, H. 2002. Causes of 20th century channel narrowing in mountain and piedmont rivers and streams of Southeastern France. *Earth Surface Processes and Landforms* 27: 425–444.

Mackin, J.H. 1948. Concept of the graded river. *Bulletin of the Geological Society of America* 59: 463–512.

Madej, M.A. 1999. Temporal and spatial variability in thalweg profiles of a gravel-bed river. *Earth Surface Processes and Landforms* 24: 1153–1169.

Manly, B.F.J. 1991. *Randomization and Monte Carlo Methods in Biology*, London: Chapman and Hall, 281 p.

Merriam, 1959. *Webster's New Collegiate Dictionary*, Springfield, Massachusetts: G. & C. Merriam Co.

Osterkamp, W.R. and Hupp, C.R. 1996. The evolution of geomorphology, ecology and other composite sciences. In: Rhoads, B.L. and Thorn, C.E., eds., *The Scientific Nature of Geomorphology*, Chichester, UK: John Wiley and Sons, pp. 415–441.

Petit, F. and Pauquet, A. 1997. Bankfull discharge recurrence interval in gravel-bed rivers. *Earth Surface Processes and Landforms* 22(7): 685–694.

Random House, 1996. *Webster's Dictionary*, 2nd edition, New York: Ballantine Books.

Roux, A.L., coord. 1982. *Cartographie polythématique appliquée à la gestion écologique des eaux: étude d'un hydrosystème fluvial: le Haut-Rhône français*, Lyon: CNRS-Piren, 113 pp.

Rhoads, B.L. and Thorn, C.E. 1996. *The Scientific Nature of Geomorphology*, Chichester, UK: John Wiley and Sons, 481 p.

Schumm, S.A. 1977. *The Fluvial System*, New York: John Wiley and Sons.

Schumm, S.A., Mosley, M.P. and Weaver, W.E. 1987. *Experimental Fluvial Geomorphology*, New York: John Wiley and Sons, 413 p.

Smith, D.G. 1993. Fluvial Geomorphology: Where do we go from here? *Geomorphology* 7: 251–262.

Thorne, C.R. and Thompson, A., eds. 1995. Geomorphology at Work, *Earth Surface Processes and Landforms* 20(7): 583–705 (special issue).

Thorne, C.R., Hey, R.D. and Newson, M.D. 1997. *Applied Fluvial Geomorphology for River Engineering and Management*, Chichester, UK: John Wiley and Sons, 376 p.

Thorne, C.R. 1998. *Stream Reconnaissance Handbook. Geomorphological Investigation and Analysis of River Channels*, Chichester, UK: John Wiley and Sons, 133 p.

Tucker, L.R. 1958. An inter-battery method of factor analysis, *Psychometrika* 23: 111–136.

Vannote, R.L., Minshall, G.W., Cummins, K.W., Sedell, J.R. and Cushing, C.E. 1980. The river continuum concept. *Canadian Journal of Fisheries and Aquatic Science* 37: 130–137.

Viles, H.A., ed. 1988. *Biogeomorphology*, Oxford, UK: Basil Blackwell.

Walling D.E. and He, Q. 1998. The spatial variability of overbank sedimentation on river floodplains. *Geomorphology* 24: 209–223.

Wharton, G., Arnell, N.W., Gregory, K.J. and Gurnell, A.M. 1989. River discharge estimated from channel dimensions. *Journal of Hydrology* 106: 365–376.

Wolman, M.G. and Gerson, R. 1978. Relative scales of time and effectiveness of climate in watershed geomorphology. *Earth Surface Processes and Landforms* 3: 189–208.

Wolman, M.G. and Miller, J.P. 1960. Magnitude and frequency of forces in geomorphic processes. *Journal of Geology* 68: 54–74.

Wolman, M.G. 1995. Play: the handmaiden of work. *Earth Surface Processes and Landforms* 20: 585–591.

Yatsu, E. 2002. *Fantasia in Geomorphology*, Tokyo: Sozosha, 215 pp. (reprint of "To Make Geomorphology More Scientific" and its supplemental discussion).

Part II

The Temporal Framework: Dating and Assessing Geomorphological Trends

2

Surficial Geologic Tools in Fluvial Geomorphology

ROBERT B. JACOBSON[1], JAMES E. O'CONNOR[2] AND TAKASHI OGUCHI[3]
[1]*US Geological Survey, Columbia, MO, USA*
[2]*US Geological Survey, Portland, OR, USA*
[3]*Center for Spatial Information Science, University of Tokyo, Tokyo, Japan*

2.1 INTRODUCTION

Increasingly, environmental scientists are being asked to develop an understanding of how rivers and streams have been altered by environmental stresses, whether rivers are subject to physical or chemical hazards, how they can be restored, and how they will respond to future environmental change. These questions present substantive challenges to the discipline of fluvial geomorphology, especially since decades of geomorphologic research have demonstrated the general complexity of fluvial systems. It follows from the concept of complex response that synoptic and short-term historical views of rivers will often give misleading understanding of future behavior. Nevertheless, broadly trained geomorphologists can address questions involving complex natural systems by drawing from a tool box that commonly includes the principles and methods of geology, hydrology, hydraulics, engineering, and ecology.

A central concept in earth sciences holds that "the present is the key to the past" (Hutton 1788, cited in Chorley *et al.* 1964); that is, understanding of current processes permits interpretation of past deposits. Similarly, an understanding of the past can be key to understanding the future. A river's history may indicate stability or instability. It may indicate trends, or episodic behavior that can be attributed to particular environmental causes. It may indicate the role of low-frequency events like floods in structuring a river and its flood plain. A river's history may provide an understanding of the natural characteristics of a river to serve as reference condition for assessments and restoration. The geologic information in river deposits results from real processes that, if properly interpreted, provide a reality-based approach to scientific reasoning (Baker 1996).

Questions about how rivers behave can involve a broad range of spatial and temporal scales. Questions may involve stability of individual habitat patches to continental scale evolution of drainage basins. Time frames may range from seconds to geologic eras. Surficial geologic tools can scale to contribute to understanding over all these time frames, and surficial geologic tools are uniquely capable of addressing long-term questions.

The surficial geologic record contained in river deposits is incomplete and biased, and it presents numerous challenges of interpretation. The stratigraphic record in general has been characterized as "... a lot of holes tied together with sediment" (Ager 1993). Yet this record is a critical component in the development of integrated understanding of fluvial geomorphology because it provides information that is not available from other sources. The surficial geologic record may present information that predates historical observation, and in many cases contains information that is highly complementary to historical records. Although river deposits are rarely complete enough to form precise predictive models, they provide contextual information that can constrain predictions and help guide choices of appropriate processes to study more closely. In paleohydrological investigations, Baker (1996) described the primary importance of flood-plain and flood deposits in providing hypotheses and inferences about how earth systems operate.

Flood-plain chronicles of earth history can also provide datasets for calibration and verification of predictive geomorphic models.

The purpose of this chapter is to introduce and discuss surficial geologic tools that can be used to improve understanding of fluvial geomorphology. We present general descriptions of geologic tools, provide selected field-based examples, and discuss the expectations and limitations of geologic approaches. Geologic investigations of fluvial deposits have been pursued for many decades and typically involve many disciplines. We do not attempt in this chapter to discuss techniques in detail or to review the entire literature on techniques or examples. Instead, our emphasis is on the conceptual basis of how geologic tools are used in geomorphological reasoning. This chapter begins with some general descriptions of stratigraphic, sedimentologic, and pedologic tools, followed by examples of how these tools are applied to geomorphologic analysis of fluvial systems. The last section presents discussion of the expectations and limitations of surficial geologic approaches to fluvial geomorphology.

2.2 SURFICIAL GEOLOGIC APPROACHES

Analysis of deposits left behind by rivers involves approaches from many disciplines, each of which has its own technical lexicon. For clarity, we will begin with some common definitions. *Surficial geology* refers to the study of the rocks and mainly unconsolidated materials that lie at or near the land surface (Ruhe 1975). In our usage, surficial geology includes as a minimum, the application of sedimentology, geochronology, pedology, and stratigraphy to study of surficial deposits and geomorphology. *Alluvium* is the detrital sediment deposited by rivers, ranging from clay size (<0.002 mm) to boulder size (>256 mm) materials, including detrital organic material. Alluvium is used interchangeably with *fluvial deposits*. Alluvium typically occurs on the landscape in modern channel and bar deposits, and in deposits that underlie adjacent flood plains, terraces, and alluvial fans.

The term *flood-plain deposits* will be defined here restrictively to denote deposits adjacent to a river channel that are being deposited under the current hydrologic regime, typically by flow events with frequencies of 0.5–1 year^{-1} and higher (Leopold *et al.* 1964, p. 319). It should be noted, however, that in some environments flood plains may be primarily constructed by flow of much lower frequency (for example, Baker 1977). *Flood plain* is used by different professions in different ways. Definitions range from the entire valley bottom outside of the channel to particular statistical definitions such as the 100- or 500-year flood plain. In this chapter, *flood plain* refers to the geomorphic surface underlain by flood-plain deposits, that is, those sediments being deposited by relatively frequent floods under the current hydrologic regime. Other fluvial deposits adjacent to a river will be referred to as *terrace deposits*, or in combination with flood-plain deposits, as undifferentiated *valley-bottom alluvium*. The term *terrace* will apply to surfaces of abandoned flood plains (Leopold *et al.* 1964). The term *soil* is used in this chapter in the pedogenic sense: mineral and organic material at the earth's surface that has been altered by weathering processes and living organisms (Holliday 1990). Hence, a soil forms from post-depositional alteration of parent material. In the case of alluvium, the characteristics of post-depositional alteration are useful sources of information for inferring age and soil-forming environment.

Sedimentology

Sedimentology can be considered as the encompassing study of all the aspects of sedimentary deposits (Pettijohn 1975), but in this chapter we will emphasize the aspects of grain size distributions, sedimentary structures, and facies assemblages that can be applied to geomorphologic interpretations. The literature on sedimentologic studies of ancient and modern fluvial systems is extensive. Readers interested in greater detail than provided here are referred to works by Carling and Dawson (1996), Jones *et al.* (1999), Marzo and Puigdefabregas (1993), Reading (1978), Allen (1982a,b, 1985), and Walker and James (1992).

Particle Size, Sedimentary Structures, Facies, and Provenance

Particle size of sediment in deposits is an indicator of the hydrologic, hydraulic, and sediment supply regime of the river. The size of sediment entrained or suspended can be related to calculable shear stresses and sediment fabric (imbrication, amount of matrix) can be related to hydraulic conditions or stream power (Allen 1985, Chapters 9 and 17). The size and sorting of sediments relate in part to the size and sorting of sediment delivered to the system. The sorting and bed scale variability of particle size may relate to hydrologic variability.

Mechanical sedimentary structures—including bedding, internal bedding, bedding-plane markings, and deformed bedding—are also amenable to hydraulic interpretations (Gregory and Maizels 1991, Allen 1985). Bedding dimensions and grain size can be used to estimate water depth and constrain possible values of Froude number and velocity. In addition, flow-direction indicators in bedforms and internal structures can be useful in reconstructing flow patterns.

Frequently, the sedimentologic information in alluvium is simplified by grouping sediments into *lithofacies* (or facies), units defined to have relatively uniform grain-size and types of sedimentary structures (Walker and James 1992). Facies for fluvial systems have been defined based primarily on particle size and secondarily on sedimentary structures and organic content (Miall 1992). Facies can have particular genetic interpretations associated with them, indicating the type of hydraulic environment in which they form. For example, massive mud with freshwater molluscs might be interpreted as a channel-fill facies. Facies can also be aggregated into related facies assemblages in order to simplify analysis of alluvial deposits. For example, a channel facies might be defined as a combination of massive gravel, plane-bedded gravel, and trough-cross-bedded gravel facies associated with channel and point-bar environments, or overbank and channel-fill facies could be combined to define a topstratum unit (Figure 2.1). Definitions of facies, and the degree of splitting or lumping of facies assemblages, ultimately depend on the utility to answer specific questions.

There are substantial practical difficulties in sampling alluvial deposits statistically or representatively, for quantitative sedimentologic analyses (see, for example, Wolcott and Church 1991, Rice and Church 1998). The particle size distribution of the sediment deposited by a river is a complex function of the transport capacity—as determined by available discharge and channel hydraulic conditions—and sediment availability—including quantity and sorting. Transport capacity and sediment availability typically vary spatially through the channel and flood plain, creating a three-dimensional mosaic of facies through which Figure 2.1 provides a two-dimensional vertical slice. To address a question about trends in sediment supply over time, for example, one would first have to determine age-equivalent units (sections following), and then follow one or more of the following strategies:

(A) The age-equivalent unit could be randomly sampled in three dimensions to provide an

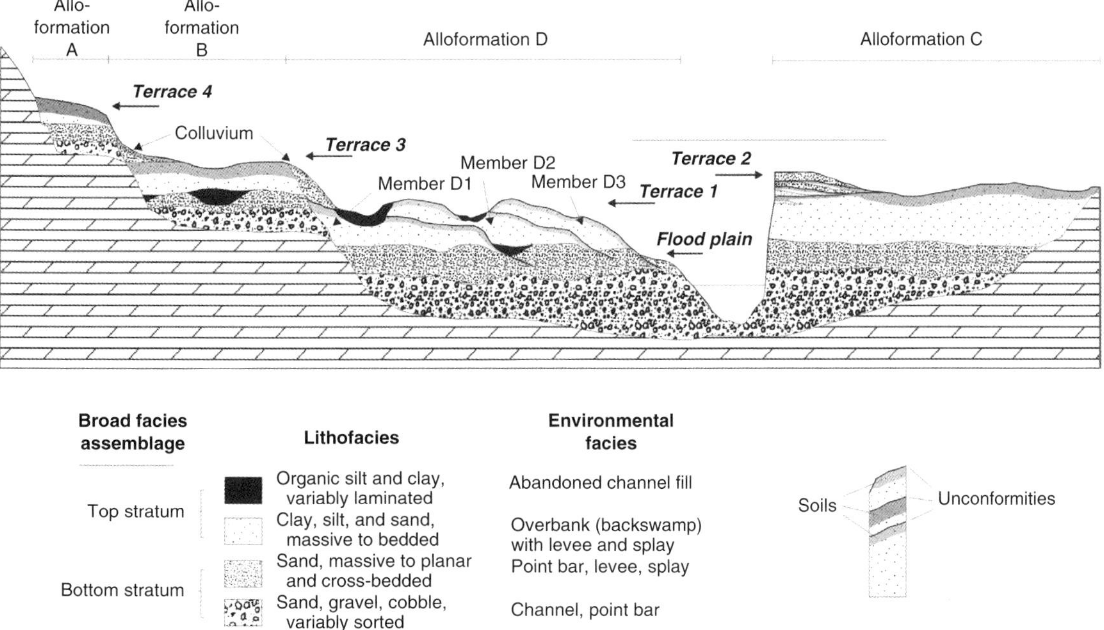

Figure 2.1 Diagrammatic cross-section of alluvial strata showing allostratigraphic units, weathering profiles, and terrace levels

unbiased estimate of total particle-size distribution or sedimentary features. This approach is extremely time-intensive and may be practically impossible.

(B) The age-equivalent units can be separated into facies units before random sampling (that is, stratified, random sampling), thereby reducing total variance and increasing efficiency.

(C) After mapping ages and facies from the best available field data, representative samples can be selected from facies common to all units. Such samples form a basis for comparison, but do not provide a basis for establishing the statistical significance of variation among units. Nevertheless, given the costs and logistic constraints on field studies, the representative sample approach is the most common.

In some sedimentological studies the provenance of the sediment is a central question, perhaps indicating shifting sources for the sediment. Lithologic, mineralogic, chemical, and particle-size characteristics can be compared to distinct sediment sources to infer proportional contributions (for example, Nanson *et al.* 1995). Chemical and mineralogical characteristics of sediment are especially useful in assessing the nature of downstream sediment routing by delineating recent, contaminated sediments (for example, Owens *et al.* 1999, Magilligan 1985, Marron 1992).

Paleohydraulic and Hydrologic Interpretations

During the past few decades, stratigraphic, sedimentologic, and geomorphic approaches have been developed to decipher quantitatively past river flow conditions. The approaches range from empirically relating evidence of past channel morphology and facies architecture to formative flow conditions, to reconstruction of site-specific shear stresses associated with movement of an individual clast. These paleohydraulic tools have been used to achieve understanding of a variety of geomorphic, ecologic, paleoclimatic, and hazard issues.

As reviewed by Baker (1989, 1991), there have been three fundamental approaches to retrodict past channel and flow conditions from geologic and geomorphic considerations [Table 2.1—summary of methods, after Baker (1989)]. The first, termed *regime-based* (Baker 1989), relies on empirically derived relations between channel morphologic or sedimentologic characteristics and past-flow conditions. Another classical approach to estimate past flow is through use of *flow-competence* criteria, which takes advantage of empirical and theoretical relations between some measure of flow strength and the size of clasts transported by the flow. The third approach, which received significant attention in the 1980s and 1990s, has been the analysis of the surficial geologic record of individual flood events preserved in *slack-water deposits* or other evidence of flood stage. Each of these approaches to deciphering past fluvial condi-

Table 2.1 Comparison of paleohydrologic and paleohydraulic approaches

Attribute	Approach		
	Regime	Flow competence	Slack-water deposits
River type	Alluvial (deformable boundaries)	Alluvial and stable boundary channels	Stable boundary channels
Scale of analysis	A reach of river or channel cross-section	Individual deposit	Individual or multiple deposits within a reach of river
Commonly retrodicted properties	Mean annual discharge, "bankfull" discharge for a channel reach or cross-section	Shear stress, velocity, unit stream power associated with individual deposit	Rare and high magnitude floods for a channel reach
Estimated accuracy under ideal conditions	±100%	±100%	± 25%
Reviews of method	(Williams 1988, Dury 1985)	(Komar 1996, Maizels 1983)	(Baker 1989, Kochel and Baker 1988)
Example applications	(Williams 1984b, Dury 1976)	(Williams 1983, Costa 1983)	(O'Connor *et al.* 1994, Ely *et al.* 1993)

tions has unique powers as well as shortcomings, summarized in the following.

Regime-based methods of retrodicting past-flow conditions have been used on a variety of alluvial systems where there is a surficial geologic record of channel deposits, or stratigraphic or geomorphic evidence of plan-view or cross-sectional channel geometry. These methods usually result in an estimate of high-frequency discharges (such as the mean annual flood or bankfull discharge) that are thought to control channel morphology in alluvial systems. One regime-based approach is determine to the type of channel (for example, meandering, braided, or straight) from sedimentology of the deposits, and then relate the inferred channel type to empirically defined fields of hydraulic and sediment load factors that distinguish various channel morphologies. A simple application would be to retrodict limiting channel slope and bankfull discharge conditions for deposits of either a braided or a meandering river by invoking the threshold hydraulic conditions between these two channel patterns defined by Leopold and Wolman (1957). More complex hydraulic criteria separating braided, meandering, and straight channels have also been proposed (Schumm and Khan 1972, Parker 1976), which also can be used to constrain paleohydraulic conditions such as width/depth ratio, channel slope, and flow velocity. For steep alluvial streams where there is independent evidence of channel cross-section morphology, Grant's (1997) hypothesis that such streams tend to adopt cross-section geometries that convey flow at or near critical flow conditions can serve as a basis to retrodict velocity and discharge.

A more commonly used regime-based approach is to use empirical relations between channel cross-section or meander dimensions and formative discharges to determine past-flow conditions. Classic examples of relating meander wavelengths to paleodischarge are provided by Dury (1954, 1965, 1976) and Schumm (1967, 1968). Good examples and discussions of relating high-frequency discharge (such as bankfull or mean annual flow) to cross-section dimensions (such as depth and width) are provided by Williams (1978, 1984b) and Rotnicki (1983).

In general, regime-based retrodiction is subject to large uncertainties arising from:

1. errors in assignment of the predictor variables, such as misinterpretation of channel type, cross-section shape, and meander wavelength (Dury 1976, Rotnicki 1983); and
2. the large standard errors of the empirical relations between predictor variable and discharge, which, in the most favorable cases, result in a 50% chance of error of greater than 24% (Dury 1985).

Nevertheless, regime-based flow estimates can be useful for addressing questions of broad environmental change resulting in regional changes in channel behavior. Useful and complete discussions of the various regime-based methods and their uncertainties are provided by Ethridge and Schumm (1978), Rotnicki (1983), Dury (1985), and (Williams 1988). Williams (1984a, 1988) provides a list of many equations used in regime-based analyses and comments on their sources and applicability.

Flow competence refers to the largest grain transported by a given discharge (Gilbert 1914). Flow strength is usually described by some measure of velocity, shear stress (force exerted by the flow parallel to the bed), or stream power (time rate of energy dissipation by the flow). Gilbert developed this concept to predict the effects of future flows, but the concept has been used since to retrodict paleohydraulic conditions from coarse-clast deposits in the surficial geologic record. Over the last 30 years, theoretical and empirical relations between clast size and flow conditions have been used to reconstruct the hydraulic conditions associated with fluvial deposits of individual flows in a wide variety of environments (Figure 2.2), including Pleistocene outburst floods (Malde 1968, Birkeland 1968, Baker 1973, Kehew and Teller 1994, Lord and Kehew 1987), Holocene flood and outwash deposits (Church 1978, Bradley and Mears 1980, Williams 1983, Costa 1983, O'Connor *et al.* 1986, Waythomas *et al.* 1996), and Miocene turbidite deposits (Komar 1970, 1989). Paleohydraulic studies have used relations between particle size and flow conditions proposed by Baker and Ritter (1975), Church (1978), Costa (1983), Williams (1983), Komar (1989), and O'Connor (1993). Reviews of flow-competence methods, including their application and uncertainties, are provided by Maizels (1983), Williams (1983), Komar (1989, 1996), Komar and Carling (1991), Wilcock (1992), and O'Connor (1993).

Flow-competence methods are most suitable for reconstruction of local hydraulic conditions at the site of deposits from an individual flow. This method can be applied over a broad range of fluvial environments wherever there are coarse-clastic deposits. But as most of the empirical relations between particle size and flow strength yield predictions of local shear

Figure 2.2 Photographs of sites of paleohydrologic analysis. (A) Slack-water flood deposits preserved in an alcove along the Escalante River, Utah. Such deposits were used to reconstruct a history of large floods for the last 2000 years (Webb *et al.* 1988) (US Geological Survey photograph courtesy of Robert Webb). (B) Site of boulders deposited by the Late Pleistocene Bonneville Flood near Swan Falls, Idaho. Measurements of Bonneville Flood boulder diameters were compared to reconstructed flow conditions to develop flow-competence relations for large floods (O'Connor 1993)

stress, stream power, or velocity, the discharge of the flow can only be determined if there is independent information of the channel geometry. Furthermore, key assumptions must be met for a valid analysis:

1. the analyzed particles must have indeed been transported and must closely represent the maximum size that could have been transported by the flow, and
2. the analyzed clasts must have been transported by a water flow rather than a debris flow or other type of mass movement.

Uncertainties in flow retrodiction from competence criteria are generally large and hard to quantify, primarily resulting from:

1. difficulty in adequately sampling the largest particles in a deposit (Church 1978, Wilcock 1992),
2. large standard errors associated with the empirical relations between clast size and flow conditions (Church 1978, Costa 1983), and
3. uncertainty as to the timing of the deposit (and its retrodicted hydraulic conditions) in relation to the general inference that the deposits represent peak flow conditions (O'Connor 1993).

Since the late 1970s, it has been recognized that fine-grained slack-water deposits preserved in stratigraphic sequences along the margins of river channels can provide detailed and complete records of flood events that extend back to several thousand years (Baker *et al.* 1979, Patton *et al.* 1979) (Figure 2.2). Slack-water deposits form from clay, silt, and sand carried in suspension by large floods and deposited in zones of local velocity reduction. Common depositional environments include recirculation zones associated with valley constrictions or bends, tributary mouths (Baker and Kochel 1988), alcoves and caves in bedrock walls (O'Connor *et al.* 1986, 1994), and on top of high alluvial or bedrock surfaces that flank the channel (Ely and Baker 1985). The sedimentary records contained in these slack-water deposits can, in certain cases, be supplemented with botanical evidence (Hupp 1988) and erosional evidence of large floods (Ely and Baker 1985).

Most of the earliest studies of slack-water deposits were from the arid southwestern US (Costa 1978, Baker *et al.* 1979, Patton and Dibble 1982, Kochel *et al.* 1982, Kochel and Baker 1982, Ely and Baker 1985, Webb 1985), but in recent years the scope of application has expanded to most continents (for example, Wohl 1988, Ely *et al.* 1996) and into more humid environments (Knox 1988, 1993). Studies of flood stratigraphy have been motivated by questions of dam safety (for example, Ely and Baker 1985, Partridge and Baker 1987, Levish and Ostenna 1996), climate change (Ely *et al.* 1993), and geomorphic effects of floods (Webb 1985, O'Connor *et al.* 1986). Baker (1989) reviewed the components of a slack-water deposit study, including stratigraphic analysis and correlation, geochronology, flood discharge determination, and flood frequency analysis. Detailed discussions of methods are also provided in chapters within Baker and Kochel (1988). O'Connor *et al.* (1994) provide a recent example of how these components were woven together to provide an analysis of long-term flood frequency of the Colorado River in the Grand Canyon, United States.

Complications of using slack-water deposits as flood-stage indicators have been emphasized in a study of extreme flooding in 1985 on the Cheat River in West Virginia (Kite and Linton 1991). These authors compared measured discharges with calculated discharges based on slack-water stage indicators and demonstrated that in a bedrock canyon in the humid eastern US, slack-water deposits consistently under estimated peak stages. The differences were attributed to possible interactions between tributary and main-stem flow, lack of a stable cross-section during the flood event, and/or lack of a temporal correspondence between peak flow and peak sediment load.

Studies of slack-water deposits yield information on the timing and magnitude of individual flood events for a reach of river. These analyses are most effective at extending flood frequency and magnitude relations beyond the limits of existing gage records (Stedinger and Baker 1987)—information that can be useful for addressing a variety of environmental questions, ranging from determining the specific sequence of large flows that may have affected a fluvial system to more general questions of how flood frequency has changed with time. Slack-water studies are most successfully applied in river systems with stable boundaries, such as bedrock rivers, where there are large stage changes with fluctuating discharge, and the chances for formation and preservation of long-lived slack-water deposits are greater. Each component of a slack-water study can present uncertainties that need to be carefully addressed if there are to be robust determinations of flood frequency. Commonly, the largest sources of uncertainty involve questions regarding the resolution and completeness of the stratigraphic record, the age of deposits, and assumptions and methods required to calculate the discharge of the flows responsible for a sequence of deposits. Nevertheless, an analysis of slack-water deposits can yield the most detailed and quantitative information of past flows than any of the currently available techniques for paleohydrologic and paleohydraulic interpretation of surficial geologic and geomorphic evidence.

Geochronology of Alluvium

Geochronological methods for determining numerical ages for alluvial strata are numerous. Recent reviews of Quaternary dating methods provide comprehensive discussion (Mahaney 1984, Easterbrook 1988a, Gruen 1994). The following summary emphasizes methods applicable to the late Quaternary and typical alluvial sediments.

Dating methods can be divided into three general categories: relative, numerical, and hybrid (Table 2.2).

Table 2.2 Geochronological methods, notes, and resources

Type	Method	Notes
Relative		
	Stratigraphic superposition	Highly reliable method for determining sequence of deposition. Requires good exposures or drilling, trenching observations
	Weathering characteristics	In regions with established age trends of pedogenesis, weathering rinds on clasts, desert varnish, etc., weathering characteristics can be used to determine relative age, constrain age limits, and correlate units spatially (Pinter *et al.* 1994, Knuepfer 1994, Dorn 1994)
	Morphologic criteria	Relative elevations of alluvial terraces can be used to determine local sequence of deposition, if there is a one-to-one relation between terrace and allostratigraphic units; if not, caution is advised. Good for regional correlations of large events, if complemented with weathering chronology and numerical dates. Degree of erosion of terraces can be indicative of relative age, as well (Coates 1984, Pinter *et al.* 1994)
Numerical		
	Radiocarbon	Most highly used radiometric method for dating alluvial sediments. Careful sampling and processing are required to reduce contamination errors. Interpretation should account for type of organic matter (that is, soil organic fractions, leaf litter, charcoal, or wood), and probable effects of inherited carbon. Dendrochronologically based calibration of radiocarbon years to calendar years (Stuiver 1982) is recommended for correcting dates for secular variations in radiocarbon production (Taylor *et al.* 1992, Bowman 1990)
	Photoluminescence	Useful for dating sediments or artifacts that have been zeroed by heat or sunlight. Techniques using thermal or optically stimulated luminescence (OSL) vary in precision and reliability, and the techniques are evolving fast. For sediments, most reliable dating has been using loess rather than alluvium (Gruen 1994, Duller 1996)
	^{210}Pb	^{210}Pb generated in the atmosphere, scavenged by precipitation and adsorbed to particulates decays with a half-life of 22.3 years, providing a geochronometer for recent sediments. Calculation requires assumptions of uniform deposition rate of sediment and ^{210}Pb; the former constraint is rarely met in alluvial deposits but may be met with flood-plain lakes or abandoned channels (Durham and Joshi 1984)

(*Continues*)

Table 2.2 (*Continued*)

Type	Method	Notes
	Cosmogenic isotopes	^{10}Be, ^{26}Al, ^{36}Cl, ^{3}He, ^{21}Ne, ^{14}C, ^{41}Ca cosmogenic nuclides for exposure age dating of some materials (1 ka to 10 Ma). Requires sophisticated chemical extraction and accelerator or conventional mass spectrometry (Kurz and Brook 1994)
	Dendrochronology	Tree rings provide detailed chronometers for dating surfaces, sedimentation rates, and individual floods over short time frames. See Chapter 7 (this volume)
Hybrid-correlative		
	Palynology	Pollen is extremely useful for developing environmental and climatic conditions, and can be used by correlation to date events, such as the settlement/post-settlement boundary. Pollen is poorly preserved in many alluvial settings, however. Best results are from adjacent flood-plain lakes or abandoned channels, and when supplemented with radiocarbon numerical dates. For example, Royall *et al.* (1991)
	Paleomagnetism	Magnetic properties of sediments can be used to correlate based on measures of susceptibility, or remanent magnetism of sediments or heated sediments or artifacts can be compared to secular variation of earth's magnetic field. Correlations by secular variation have been demonstrated from 0.1 to 20 ka, but requires an independently dated sequence. Works best in lacustrine depositional environments (Stupavsky and Gravenor 1984, Lund 1996)
	Archaeology	Independently dated archaeological artifacts can provide tools for relative and absolute dating, and for correlation. See Chapter 3 (this volume)
	Tephrochronology	Tephra found in alluvium can be correlated by chemical or petrographic techniques, combined with stratigraphic sequence, with independently dated volcanic deposits (for example, Sarna-Wojcicki *et al.* 1984, 1991).

Relative methods are useful for establishing whether stratigraphic units are older or younger than others, and in some cases can be usefully calibrated to interpolate or extrapolate ages. Relative methods based on a combination of weathering characteristics, topographic position, and morphology are useful for developing field criteria for regional correlations of numerically dated strata. These concepts will be discussed in more detail in the following discussion of pedology and morphostratigraphy.

Numerical dating methods provide estimates of time elapsed since deposition of alluvial strata. Probably, the most useful method for dating alluvium is radiocarbon dating based on the progressive decay of ^{14}C in plant or animal material once the plant has died. With the use of accelerator mass spectrometry (AMS), very small amounts of sample ($\sim$1 mg) can be used to calculate dates in the time frame 200–55 000+ years BP. Conventional radiocarbon dates (in radiocarbon years before present) should be calibrated to calendar years to account for secular variations in radiocarbon production in the atmosphere (Stuiver 1982). Secular variations in ^{14}C pose special problems for samples less than 500 years old, making it difficult to define precise age estimates for this time period. Additional errors in radiocarbon dates relate to laboratory statistical counting, sample preparation, and estimates of laboratory reproducibility; these are usually reported as $\pm$ values in laboratory results. However, much greater errors can be introduced in sampling, especially in the inherited age of the carbon in the sample. A radiocarbon date from a piece of charcoal from a tree that was 800 years old when it died, will overestimate the age of associated sediments by at least 800 years. Resistant materials like charcoal and bone, in fact, may be eroded and redeposited in even younger sediments. These inherited age errors (old wood errors) are very difficult to control. One strategy is to avoid sampling resistant materials in favor of materials that would likely contribute small inherited ages, like twigs and leaves. Radiocarbon dates should be interpreted as a maximum limiting age for the enclosing deposit. Complete discussion of assumptions and cautions of using various radiocarbon dateable materials can be found in Taylor *et al.* (1992).

In addition to the methods listed in Table 2.2, several other numerical methods deserve comment. Photoluminescence dating in alluvial sediments is based on the accumulation of a thermoluminescence (TL) or optically stimulated luminescence (OSL) signal with time after burial. Luminescence signals of materials are assumed zeroed by strong heating, or from precipitation by biogenic or chemical processes (Berger 1988). The TL signal then accumulates over time due to migration of electrons into crystal defects (traps) because of ambient radiation in the burial environment. Laboratory analyses of the movement (and luminescence) of electrons from traps, as the sample is heated, and estimates of burial radiation dosage are used to calculate elapsed time of burial. The TL signal in sediments may be completely or only partially zeroed by exposure to sunlight, leaving a residual that must be determined in the laboratory. Lately, emphasis has focused on using OSL for non-burned sediments that may have not been completely zeroed (Aitken 1997). TL and OSL extend to greater age than radiocarbon dating (as much as 800 ka) but with typical precision of ±5–10% (Aitken 1997). This makes luminescence dating useful for strata that are too old or lack material for radiocarbon dating. See Chapter 6 for details and examples.

The accumulation of ^{210}Pb dating can provide a useful dating method for short time intervals if slack-water sedimentation sites are available. ^{210}Pb is produced as a decay product of ^{222}Rn in the atmosphere, and accumulates in sediments with atmospheric fallout. The half-life is only 22.36 years, so the method provides high precision over short (10–150-year) time frames. Calculation of age of the sediment requires an assumption of a constant sedimentation rate and negligible bioturbation (Durham and Joshi 1984), hence the method is restricted to slow-water facies. ^{137}Cs is another short-lived isotope, with a half-life of 30 years. ^{137}Cs was produced in abundance in the atmosphere from nuclear testing 1954–1971, with a peak about 1963 in North America. Detection of the ^{137}Cs spike in sediments can provide pre- and post-1963 relative ages. See Chapter 6 for more information on this method.

The use of cosmogenic isotopes (^{3}He, ^{10}Be, ^{14}C, ^{21}Ne, ^{26}Al, and ^{36}Cl) for exposure-age dating has been increasing dramatically in recent years (Bierman 1994). Theoretically, cosmogenic isotopes can be used to date surfaces from as little as 1 ka (^{3}He) to as much as 10 Ma (^{21}Ne) (Kurz and Brook 1994). Since exposure calculations should start with a zeroed surface or known starting inventory, most applications have been on eroded or volcanic bedrock (Weissel and Seidel 1998) or sedimentary surfaces for which inherited ages can be assumed to be small (Bierman 1994). Inherited cosmogenic isotopes can be problematic in dating alluvial sediments that move slowly through drainage basins. Hallet and Putkonen (1994) discuss some of the complications of applying cosmogenic surface dating to actively eroding surfaces.

Hybrid methods are those that can be used to estimate numerical ages through application of calibrated models. For example, calibrated models of weathering rind thickness (Durham and Joshi 1984), lichen growth (Matthews 1994), or desert varnish geochemistry (Dorn 1994, Schneider and Bierman 1997) can be used to estimate ages of undated surfaces through various measurements of these properties. Presence or absence of diagnostic pollen, diagnostic tephra, macrofossils, or archaeological artifacts can also provide constraints on age estimates.

Included in this hybrid group is paleomagnetism. In the late Quaternary time frame, independently dated secular variation in magnetic field strength and orientation can provide a master curve for comparison with magnetic inclination and declination of magnetic minerals in alluvial deposits. Paleomagnetism in this time frame is mostly performed on heated sediments, typically found in hearths buried with alluvial sediments; magnetism of heated sediments is referred to as thermal remanent magnetism (Sternberg and McGuire 1990). Remanent magnetism can also be measured from fine-grained sediments deposited in still water (detrital remanent magnetism) (Easterbrook 1988b, Verosub 1988). Such deposits, free of bioturbation, are much more likely to be found in lakes and coastal zones, but might exist for some flood-plain lakes.

Pedology

Pedology is the study of soil-formation processes. Physical, chemical, and biological processes transform freshly deposited alluvial sediments into soil profiles with characteristics that reflect the five classic soil-forming factors: climate, topography, parent material, biologic influences, and time (Jenny 1941). Although complex in interaction, each of the factors is governed individually by more-or-less systematic processes. In cases where the scale and scope of study allow one or more of these factors to be considered constant (for example, climate or biologic influences), the soil-forming processes can provide sufficiently systematic variations in soil profiles that the properties of the profiles can impart valuable information about geomorphic processes.

In fluvial geomorphologic studies, the time factor is often of greatest interest. If variation in pedogenesis with time can be separated from the effects of other factors, then pedogenic characteristics can be used for relative dating, correlation, and for estimating deposition dates when supported by independent numerical age control (Birkeland 1984). In a dissenting opinion, Daniels and Hammer (1992) argue that it is effectively impossible to hold other factors constant—that the complexities of surface processes, different parent materials, and drainage influences contribute too much variation in pedogenic characteristics to usefully extract age information. Daniels and Hammer's (1992) discussion underscores the need for a careful field study so pedological sampling sites are chosen to minimize variation in erosional history, parent material, and drainage.

In studies where the geomorphologic questions being addressed are sufficiently broad, pedogenic characteristics of alluvium can be useful age indicators. For example, in arid and semi-arid areas, the accumulation of soil carbonate over time has been used very successfully in correlation and age estimation (for example, Vincent *et al.* 1994) over time frames of 10^4 years. In humid environments, accumulations of clay and iron and aluminum oxides have been found systematic over 10^4–10^7-year time frames (Markewich and Pavich 1991).

Pedogenic characteristics also are useful for interpretation of environment and environmental change. Although not as useful as pollen for general climate-change assessment, soil mineralogy can provide an integrated understanding of local moisture conditions, which can be interpreted in terms of changing drainage or water table configuration. For example, micromorphological examination of concentrically zoned, secondary accumulations of Fe and Al oxides in soils have been used to document changes in soil drainage (Birkeland 1984, Kemp 1985).

One of the more useful pedologic perspectives on processes of alluvial sedimentation comes from the recognition of cumulative soil profiles. Cumulative soil profiles are undergoing simultaneous soil formation and sediment deposition (Nikiforoff 1949). Cumulative soils can be considered part of a continuum relating degree of horizon differentiation and sedimentation rate (Figure 2.3). Where alluvial sedimentation rates are rapid, periods of stable sub-aerial exposure are short or non-existent. In these cases, pedogenic alteration and bioturbation are minimal and, consequently, sedimentologic information is best preserved. At the other extreme, where sedimentation rates are very slow—for example, on an alluvial terrace—pedogenic processes dominate over depositional processes and weathering information is best preserved. On low terraces and flood plains, it is common to have alternating periods of deposition and sub-aerial exposure resulting in cumulative soil profiles. These profiles are characterized by overthickened A horizons with high organic content and massive to weak pedogenic structure. Identification of the spatial and stratigraphic distribution of cumulative soil profiles may indicate substantive changes in river behavior over time.

Measures of pedogenesis can be combined with lithofacies to define pedofacies: laterally contiguous bodies of alluvium that differ in pedogenic attributes as a result of differing sedimentation rates (Kraus and Brown 1988). Pedofacies units (Figures 2.1 and 2.3) can be a useful concept in describing relative sedimentation rates of a channel and flood plain. The concept has greatest applicability to interpretation of the ancient sedimentary record in aggrading environments where relative proportions of pedofacies can be compared over long geologic time intervals.

Given the inherent complexity of soil forming processes, the utility of soil chronosequences or environmental indicators may be subject to overstatement. Utility of the methods, of course, ultimately depends on the questions being addressed. Pedogenic characteristics of alluvial strata probably never will be sufficiently precise to date high-frequency geomorphic events like individual floods. On the other hand, pedogenic characteristics can provide useful filters for constraining geomorphic understanding. For example, Bettis (1992) used a simple set of soil properties as regional indicators of broad age classes of Holocene alluvium, and applied this filter to map archaeological potential of alluvial deposits. Pedogenesis in this case was sufficiently systematic to sort out Early-Middle Holocene, Late Holocene, and Historic strata.

Stratigraphy

In this chapter, *stratigraphy* is used restrictively to denote the sequence and spatial framework of construction of the geologic column (Pettijohn 1975).

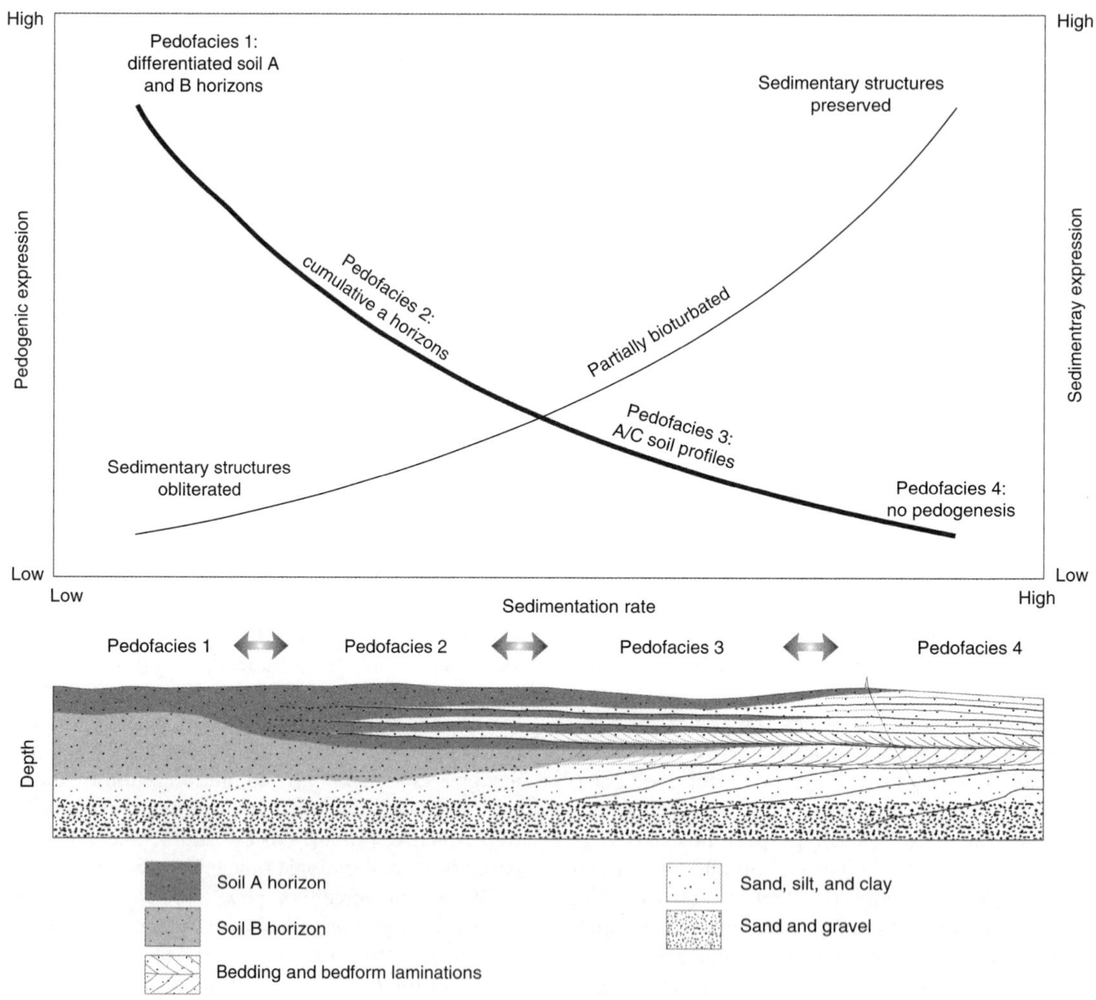

Figure 2.3 Diagram of tradeoffs between sedimentation rate and expression of pedogenic and sedimentologic features

Stratigraphy serves to integrate sedimentology, pedology, and other disciplinary approaches into a systematic understanding of how the alluvial record was constructed. Since textbooks on stratigraphy tend to emphasize long intervals of geologic time and regional to continental spatial scales, they can be of limited use in geomorphologic applications. Some of the best general resources in alluvial stratigraphy are in textbooks and volumes devoted to geoarcheology (for example, Lasca and Donahue 1990, Brown 1997).

Allostratigraphic Units

A useful, basic unit for describing and mapping alluvial deposits is the allostratigraphic unit, a "... mappable stratiform body of sedimentary rock that is defined and identified on the basis of its bounding discontinuities" (North American Commission on Stratigraphic Nomenclature 1983). Allostratigraphic units are similar in concept to synthems, as defined by International Subcommission on Stratigraphic Classification, although synthems have been used to describe unconformity-bounded stratigraphic units of regional to continental scale (International Subcommission on Stratigraphic Classification 1994).

Allostratigraphic units are well suited for describing alluvial deposits because their definition allows the upper boundary to be a sub-aerial geomorphic surface (Figure 2.1), and no constraints are put on internal characteristics, age, or genesis. Hence, allostratigraphic units can be subdivided by facies (with a specified range of lithologic, mineralogic, particle-size, or sedimentologic features), age, or weathering. Allostratigraphic units can be diachronous (that is, different parts of the unconformity bounded unit are different ages) or isochronous (that is, given the available resolution of dating method, all parts of the unit were deposited over the same time interval). Allostratigraphic units are not defined by inferred time spans, but age relations may influence choice of unit boundaries (North American Commission on Stratigraphic Nomenclature 1983). Allostratigraphic units may have pedogenic soils formed in them and soils may conform to upper and lower bounding unconformities; hence, one or more pedostratigraphic units may be defined within an allostratigraphic unit, and pedostratigraphic units or surface soils may be defined across several allostratigraphic units. Allostratigraphic units are usually defined as alloformations, and may be aggregated into allogroups or disaggregated into allomembers (North American Commission on Stratigraphic Nomenclature 1983).

Scale of allostratigraphic units is unlimited by definition, but subject to the resolution of techniques used for measuring, tracing, and mapping the units. According to the North American Code of Stratigraphic Nomenclature (North American Commission on Stratigraphic Nomenclature 1983), the only scale limitation on allostratigraphic units is that they must be mappable. Therefore, the fine-scale definition and use of alloformations may be limited by availability of base maps and the scale used by precedent stratigraphic studies. Geographic extent of allostratigraphic units is limited by the ability to trace them continuously, or to correlate from place to place based on fossil content, tephras, pedogenesis, numerical ages, or topographic position.

The concept of allostratigraphic unit, therefore, provides a useful framework for describing and analyzing alluvial deposits. Definition of the units is based on the bounding unconformities, therefore emphasis is on the sequence of erosional and depositional events; usually, these are of critical interest in geomorphologic analysis. In many applications and for many alluvial deposits, it is possible and advantageous to choose alloformation boundaries based on determined ages, thereby imparting chronostratigraphic attributes to the alloformation. Differentiation of depositional lithofacies within an alloformation can be used to infer variations in depositional processes among alloformations. Differential pedogenesis of alloformations can be used to aid in assignment of relative ages and in tracing and correlation of allostratigraphic units. Autin (1992) provided a particularly complete example of the use of alloformations in analyzing the Holocene geomorphology of a large, low-gradient river in Louisiana.

Morphostratigraphy

The concept of alloformations has added useful rigor to the conventional geomorphic tool of mapping and correlating fluvial geomorphic events by the landforms they leave behind. Stratigraphic correlations can also be achieved by reference to characteristic morphology, that is, the shape and relative position of fluvial landforms. Characteristic depositional morphologies can be used to infer process origins or to correlate units. For example, levee splays from a particular flood may be manifested as mappable, lobate landforms on flood plains. Morphological correlations are much stronger, however, if supported with stratigraphic, sedimentologic, and pedologic information.

The practice of mapping and correlating terrace surfaces has underlain a great deal of geomorphologic analysis, particularly at the scale of tectonic and eustatic controls on base level (Miller 1970, Bull 1991, Pazzaglia *et al.* 1998). The typical—but not universal—observation that alluvial terrace deposits at lower elevation are younger than those at higher elevations is a morphostratigraphic basis for assigning sequence and relative age (Ruhe 1975). Surface morphology also changes with age, allowing correlation based on morphostratigraphic parameters such as degree of erosional dissection and progressive erosion of depositional landforms (Pinter *et al.* 1994).

Obtaining Surficial Geologic Data

Surficial geologic data can only be compiled by looking into and sampling beneath the ground surface. In a typical project, the data requirements are balanced with logistical constraints of time and money, and it is rare that the scientist is satisfied that all the pertinent observations have been made. Several types of subsurface data can be considered. Most river reaches or segments will have natural exposures of flood plains and terraces in cutbanks. These present a low-cost but highly biased subsurface view of alluvium. Natural exposures should be observed, measured, and sampled first, and the knowledge gained from them should be applied to subsequent subsurface exploration. In many landscapes, man-made features such as gravel pits, pipeline trenches, and road embankments also provide opportunities for observing the subsurface.

Subsurface exploration techniques present tradeoffs that need to be considered in terms of the questions being addressed and the evolving understanding of the complexity of the alluvial deposits. Hand and power augers provide for extensive probing and sampling of alluvial deposits. Fine-scale sedimentary and pedogenic features can be sampled with hydraulic split-barrel or tube samplers (Figure 2.4). Greater depth and coarser materials require large equipment to power hollow-stem augers, and even then it is rare to recover intact samples of non-cohesive sediments, and it may not be possible to drill to bedrock. Shallow

Figure 2.4 Shallow borehole drilling in valley-bottom alluvium. (A) Exploratory drilling of alluvium with a 4 in. (10 cm) auger. (B) Split spoon sample of alluvium obtained by hydraulic probing

seismic refraction, ground penetrating radar, and electrical resistivity can be efficient means to correlate units and map the alluvium/bedrock contact, especially when geophysical data can be compared with adjacent boreholes (US Army Corps of Engineers 1998). A complete review of applicable geophysical techniques is beyond the scope of this chapter. The interested reader is referred to the texts by Sharma (1997) and Reynolds (1997) for complete discussions.

Borehole logs and seismic data arranged along surveyed topographic cross-sections may provide sufficient information to correlate units and understand the stratigraphic architecture, but boreholes generally lack the breadth of exposure needed for interpretation of many sedimentary and pedogenic features. Backhoe trenches (Figures 2.5 and 2.6) can provide long, complete exposures of near-surface strata for complete descriptions and sampling. Exposures in trenches can show meter-plus-scale bedforms, details of stratigraphic contacts, continuity of units, and they provide much more efficient prospecting for datable materials or artifacts. Evaluation of soil structure, continuity of soil horizons, and interpretation of environmental indicators is much easier in a trench than in a 10–25 mm diameter core. Placement of trenches in key places on borehole transects can improve stratigraphic understanding without requiring extensive trenching of a valley bottom. For example, trenches may be placed preferentially to sample representative features of a formation, or to provide detail where contacts or facies changes occur.

Figure 2.5 Exploratory trench in flood-plain alluvium, Big Piney Creek, Missouri

Use of power equipment can involve considerable risk. Boring and augering equipment presents hazards from heavy and powerful equipment. Proper personal safety equipment and kill switches are recommended, and in many localities, required. Trenching also can present considerable hazard from cave-in. In the US, the Occupational Safety and Health Administration, for example, requires shoring of the walls of any open trench greater than 5 ft (1.5 m) deep, or stepping of the trench wall to a slope of no steeper than 34° from horizontal.

In addition, trenching and boring can create environmental hazards by delivering sediment to streams or opening up preferential pathways for contamination of shallow groundwater. These environmental hazards should be mitigated by using approved methods for filling and sealing excavations and boreholes. Many localities require permits for shallow exploratory drilling. Meeting safety and environmental requirements can add considerable cost and complexity to subsurface investigations.

Geologic Reasoning—Putting it Together

Interpretation of sedimentologic, geochronologic, pedologic, and stratigraphic data can lead to enhanced understanding of fluvial geomorphic processes if the data collection effort is carefully designed to address the question at hand and if the data are organized in a useful fashion. The task can seem daunting, but models of fluvial processes and facies architecture can help provide context. In a typical field situation, some "laws" of stratigraphic reasoning can help. Steno's law of superposition states that successively younger units overlie older units (cited in, Dott and Batten 1976). Trowbridge's law of ascendancy states that terraces at higher elevations are older than those at lower elevations (cited in, Ruhe 1975). Walther's law of facies states that facies that were formed in laterally adjacent environments can be found in conformable vertical sequence (cited in, Reading 1978). With these concepts and good field

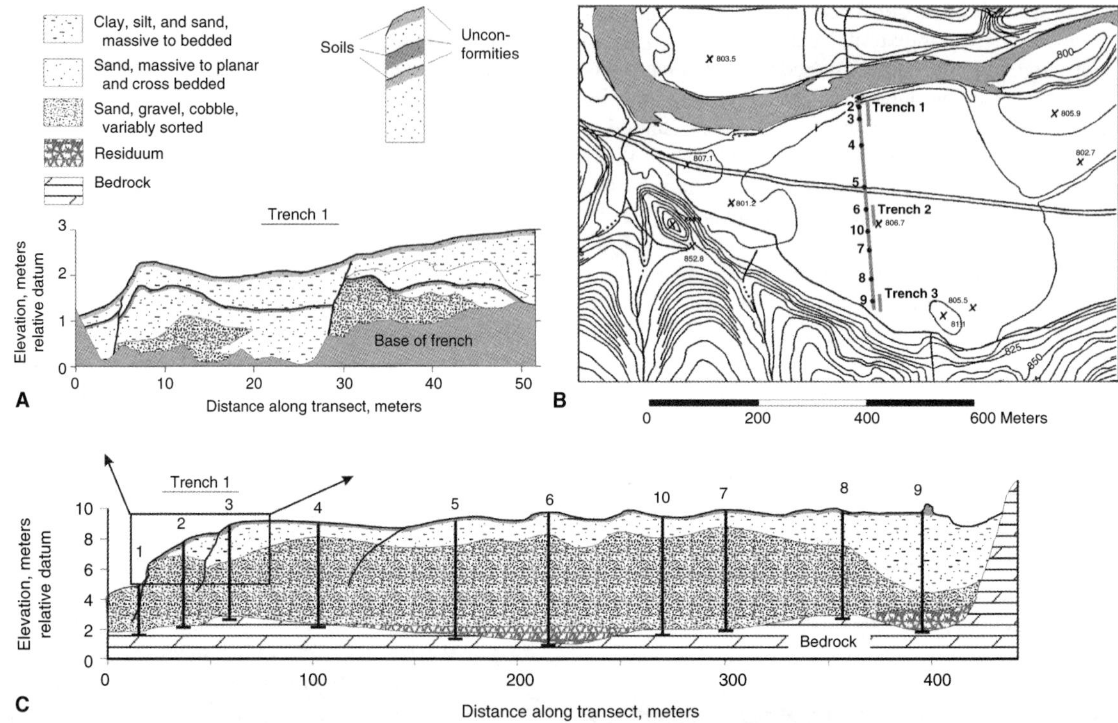

Figure 2.6 (A) Map of subsurface exploration strategy, showing locations of boreholes and trenches. (B) Close-up section of Trench 1, showing lithofacies, unconformities, and buried soil profiles. This level of detail is lost in larger cross-sections compiled from boreholes. (C) Modified from Albertson *et al.* (1995)

data, lithofacies and allostratigraphic units can be assembled and put in stratigraphic sequence.

Delineation of the stratigraphic units that chronicle geomorphic adjustments of a river can be accomplished best by mapping based on unconformity-bounded units. Sedimentologic and pedologic characteristics provide the keys to mapping allostratigraphic units. Pedologic characteristics are doubly important because soils help to define unconformities as well as yielding information on environmental conditions.

Historically, the sequence and magnitude of fluvial geomorphic events has been inferred mainly from sequences of terrace surfaces. Such analysis is based on the assumption that depositional (cut and fill) terraces have one-to-one relations with the stratigraphic units that underlie them. Detailed stratigraphic studies of flood plains and terraces have shown that continuous terrace surfaces can overlie multiple allostratigraphic units because of onlapping or planation (Taylor and Lewin 1996). In detailed stratigraphic studies on Duck River in Tennessee, for example, Brakenridge (1984) documented that single surfaces could overlie multiple unconformity-bounded units of vertically accreted silt and clay.

The importance of delineating allostratigraphic units within terrace deposits depends on the time frame of the questions being addressed. Many recent studies have demonstrated that alluvial stratigraphic histories exhibit two dominant orders of response behavior. Over the long term, a first-order response results in cut and fill terraces as a result of external forcing events such as climate change or tectonism (Bull 1991). Over a shorter time frame, internal threshold responses of alluvial systems can result in second-order cut and fill sequences that may or may not form distinct topographic surfaces depending on magnitudes of the events (Bull 1991, Schumm and Parker 1973). Hence, some allostratigraphic units may have no external forcing event, and some terrace surfaces may be underlain by multiple second-order allostratigraphic units. In addition, cut and fill stratigraphic units can form by lateral migration of a system that has surpassed intrinsic thresholds or has otherwise been unaffected by external forcing events (Ferring 1992). These units—herein called third-order cut and fill units—may be bounded by significant unconformities where the channel has migrated back into previously deposited sediment, but the unconformities are not necessarily evidence of episodic behavior.

Added to the "noise" of second- and third-order cut and fill responses is variation in timing of cut and fill within a drainage basin. Time lags in sediment transport in a drainage basin can result in non-synchronous deposition, or so-called diachronous terrace distributions (Brown 1990, Bull 1990, 1991). For example, alluvial stratigraphic studies in smaller drainage basins in the Great Plains of the US have shown strong correlations between moist, humid climatic conditions, and aggradation and stability of Holocene deposits (Fredlund 1996). With increasing drainage area, however, the terraces become diachronous because of the lagged transport of sediment through the drainage basins and correlations with paleoclimate diminish (Martin 1992).

Interpretations of the alluvial stratigraphic record are confounded in general by erosion of older units. The primary determinant of preservation potential of alluvial strata is the regional or tectonic context. Rivers in eroding parts of drainage basins will have low preservation potential and the alluvial record will be short and fragmented. Preservation occurs as downcutting and migration leave alluvial deposits behind in protected positions. Large rivers in large valleys or deltaic areas or subsiding basins will have longer and more complete sedimentary sequences that will tend to be preserved in the geologic record, although access to these records may be more difficult because of depth of burial. Sedimentology and stratigraphy of low-gradient, aggradational rivers have been studied extensively because of their importance in the geologic record (for example, Marzo and Puigdefabregas 1993). Facies and stratigraphic models developed for such rivers emphasize vertical aggradation of channel and back-swamp facies over time (for example, Bridge and Mackey 1993).

In eroding river systems, the probability of preservation of a stratigraphic unit associated with an external forcing event is inversely related to the time since deposition; details of the relation are dependent on factors like mode of channel migration, bedrock characteristics that might shelter deposits from erosion, and tectonism. The probability of preservation also should be directly proportional to the size of the forcing event. So, all other things being equal, the alluvial stratigraphic record will be biased toward recent and large events. The bias is illustrated in Figure 2.7, which shows a frequency distribution of radiocarbon dates from the Ozarks (Albertson *et al.* 1995, Haynes 1985, Jacobson unpublished data). Although the actual form of the steady-state age distribution is not known, an inverse relation as hypothesized provides a reasonable example of a background distribution against which potential forcing events must be compared. The size of the most recent events, as reflected in the frequency of radiocarbon dates, is biased compared to earlier events. Another potential bias in the stratigraphic record relates to preservation of evidence of extreme floods. If rivers do not migrate actively and create new flood plains, the record of floods is progressively filtered because only sediments from larger floods are preserved (Wells 1990).

2.3 APPLICATIONS OF SURFICIAL GEOLOGIC APPROACHES TO GEOMORPHOLOGIC INTERPRETATION

In this section, we present examples of how surficial geologic tools have been applied to some geomorphologic problems. Our emphasis is on illustrating the use of surficial geologic tools rather than completely reviewing the field.

Paleohydrologic Interpretations from Surficial Geologic Data

Surficial geologic investigations of alluvium in Japan demonstrate how the stratigraphic record can be used to evaluate sensitivity of the landscape to climate change and to gain insight into long-term flood frequency. Systematic changes in gravel facies in Japanese alluvial fans have been related to climatic change. At present, the Japanese Islands are characterized by frequent heavy storms. The daily maximum rainfall record exceeds 300 mm for most of Japan, among the world's highest (Matsumoto 1993). About every 10 years hourly rainfall exceeds 50 mm (Iwai and Ishiguro 1970), triggering widespread slope failure in mountainous areas (Oguchi 1996). Heavy rains occur during the typhoon season (mostly August–October) and during the Japanese rainy season (June–July) when the Polar front stays over Japan. The frequent storms lead to abundant supply of clastic materials from mountain slopes, rapidly transported to piedmont areas. Consequently, alluvial

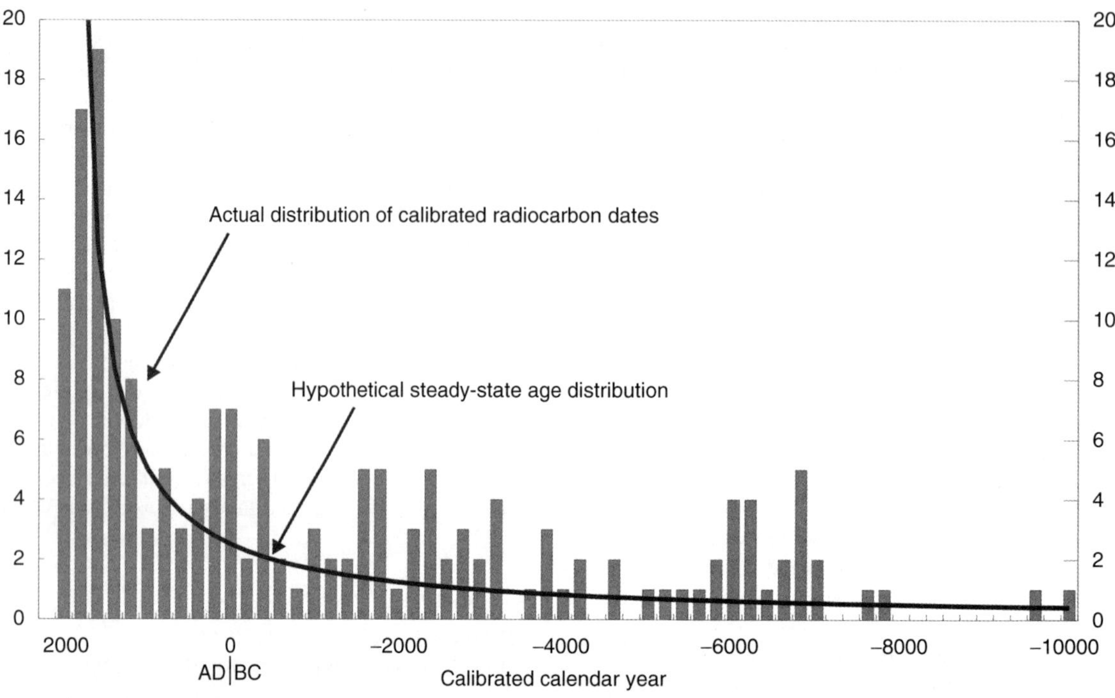

Figure 2.7 Histogram of numbers of calibrated radiocarbon dates from the Ozarks of Missouri, showing preservation bias against older units and possible peaks from paleoclimatic events. Data from Albertson *et al.* (1995), Haynes (1985), and Jacobson (unpublished data)

fans are abundant in Japan: 490 alluvial fans each with an area of more than $2\,km^2$ occur within the Japanese Islands (Saito 1988).

Surficial geologic studies of Japanese alluvial fans indicate that Holocene climatic conditions are substantially different from Pleistocene conditions. An extensive investigation on the 490 large alluvial fans in Japan revealed that particle-size distributions of alluvial fan deposits dating from the Last Glacial Maximum are generally smaller than those of Holocene deposits (Saito 1988). Borehole data (Figure 2.8) at the Karasu Alluvial Fan in an intermontane basin of central Japan show how gravel sizes differ between the Last Glacial and post-glacial time. The fan deposits have been supplied from the Northern Japan Alps, which consist mostly of steep hillslopes with a modal angle of about 35° (Katsube and Oguchi 1999). The Holocene fan deposits include abundant coarse gravel with sandy matrix, reflecting the fact that about 80% of contemporary alluvial fan sediments in mountain areas of Japan are transported as bed load (Oguchi 1997). By contrast, the Last Glacial deposits are characterized by finer matrix including silt and clay as well as smaller gravel sizes.

The contrasting lithofacies of these units is used to deduce the effect of the Pleistocene—Holocene climatic transition in Japan. Around the Last Glacial Maximum, the southward shift of frontal zones led to significantly reduced storm intensity in Japan, because both typhoons and the Polar front did not attack the Islands (Suzuki 1971). The decrease in heavy rainfall resulted in smaller sediment supply from hillslopes, lower tractive force of stream flow, and the reduced sizes of transported gravel, compared to the Holocene (Sugai 1993). The marked change in gravel sizes also is useful to estimate the rate of post-glacial sedimentation at alluvial fans. Subsurface contours representing the boundary between the Last Glacial and post-glacial fan deposits have been drawn for the eastern foot of the Japan Alps using data from approximately 120 boreholes (Tokyo Bureau of International Trade and Industry 1984,

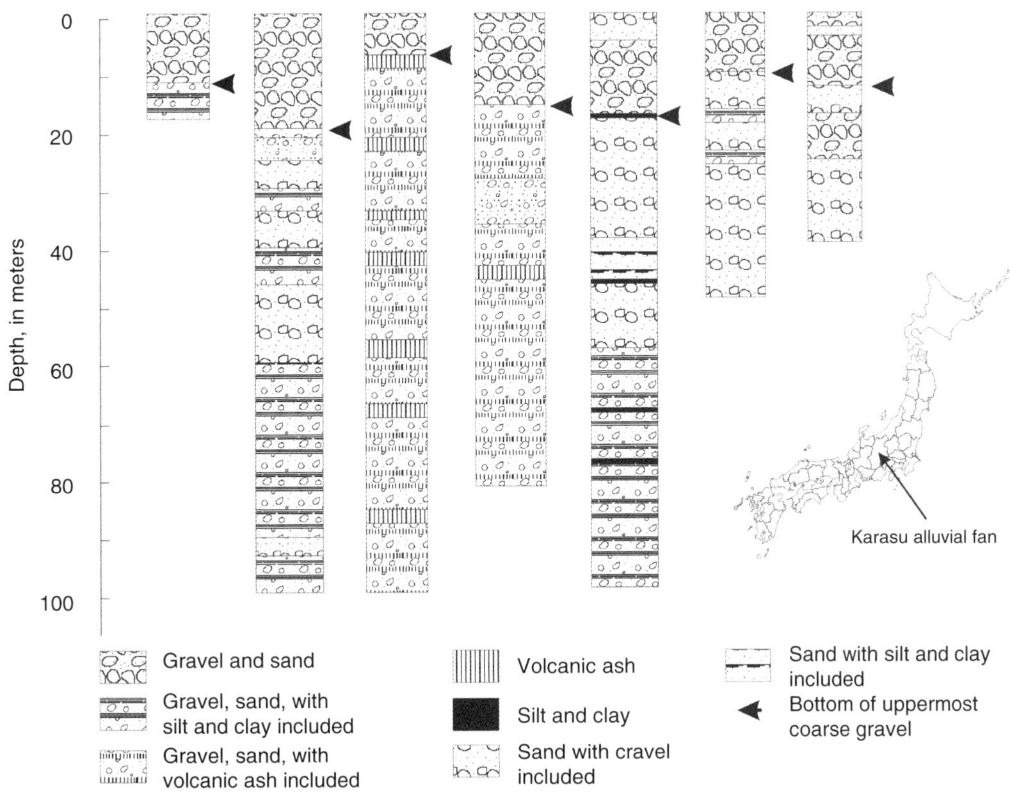

Figure 2.8 Columnar sections of alluvial fan deposits along the Karasu River, Japan

Oguchi 1997). The volume between the boundary and the present earth surface for each alluvial fan can be computed to estimate post-glacial sediment storage. The volumetric comparison between the storage and inferred sediment supply from upstream areas (Oguchi 1997) suggests that a significant portion of the sediment supplied during the Holocene has been stored in the alluvial fans. This is due to a large percentage of coarse bed load in post-glacial fan sediments, which are not easily transported downstream from alluvial fans.

Although clear stratigraphic evidence of slack-water deposits is thought to be rare in humid regions because of disturbance by bioturbation and pedogenesis (Kochel *et al.* 1982, Baker 1987), a recent study on the Nakagawa River in central Japan revealed that well-preserved Holocene slack-water deposits can occur in a humid region with abundant rainfall (Jones *et al.* 2001). The field section of the deposits is exposed on the outer bend of a meander in a gorge that cuts into Late Pleistocene river terraces. The section is about 25 m in length and 8 m in height (Figure 2.9). The sediments consist mainly of sand with numerous fine laminations and thin beds, although gravel units occur intermittently throughout the section. Radiocarbon dating and sedimentologic analyses indicate that the deposits were accumulated by more than 30–40 flood events during the last 500 years. The inferred recurrence interval of paleofloods is much shorter than that in arid to semi-arid regions, and the sedimentation rate of the deposits is much higher, which can be explained by the frequency of large storms and their associated sediment loads. Indeed, three big floods in 1986, 1992, and 1998 caused repeated riverside sedimentation in the watershed of the Nakagawa River. Despite the possibility for rapid bioturbation and pedogenesis under a humid climate, their effects are limited at the Nakagawa section because of very fast and frequent sedimentation.

Figure 2.9 Photograph of the field section on Nakagawa River, showing bedsets and laminasets used for reconstructing flood history

Catastrophic Events: Exceptional Floods and Channel and Valley-bottom Morphology on the Deschutes River, Oregon

The Deschutes River of central Oregon drains 28 000 km^2 of north-central Oregon, joining the Columbia River 160 km east of Portland (Figure 2.10). Three hydroelectric dams impound the river 160–180 km upstream from the Columbia confluence, and the effects of these dams on channel geomorphology and aquatic habitat have been studied by McClure (1998), Fassnacht (1998), McClure *et al.* (1997), Fassnacht *et al.* (1998), and O'Connor *et al.* (1998). There are few clear effects on the channel and valley bottom that can linked to the nearly 50 years of impoundment because:

1. there has been little alteration of the hydrologic regime;
2. sediment yield from the catchment is low, so the effect of trapping sediment behind the dams is less here than elsewhere; and
3. much of the present channel and valley bottom has been shaped by exceptional floods much larger than the largest historic meteorological floods of 1964 and 1996.

This section summarizes preliminary studies of the "outhouse" flood; a large, Late Holocene flood on whose deposits numerous campsite outhouses (pit toilets) have been built. Through this discussion, we illustrate a variety of surficial geologic and geomorphic methods for investigation the timing, magnitude, and effects of a large flood.

At several locations along the 160-km length of the Deschutes River canyon between the dam complex and the Columbia River, high cobble and boulder bars provide compelling evidence for a Holocene flood (or floods) much larger than the largest gaged floods of December 1964 and February 1996. The bar forms left by the outhouse flood and their relation to maximum stages of the February 1996 flood are particularly clear at Harris Island at River Mile (RM) 11 (Figure 2.11) where the tops of the cobbly bar crests are 5–6 m above summer water levels, and 1–2 m above the highest February 1996 inundation. Outhouse flood bars and trimlines are 1–7 m higher than February 1996 flood stages at many other locations as

Figure 2.10 Location map for Deschutes River Basin showing the three hydroelectric structures of the Pelton-Round Butte project and the Harris Island study site. The most downstream facility, the reregulation dam, is at River Mile 100.1

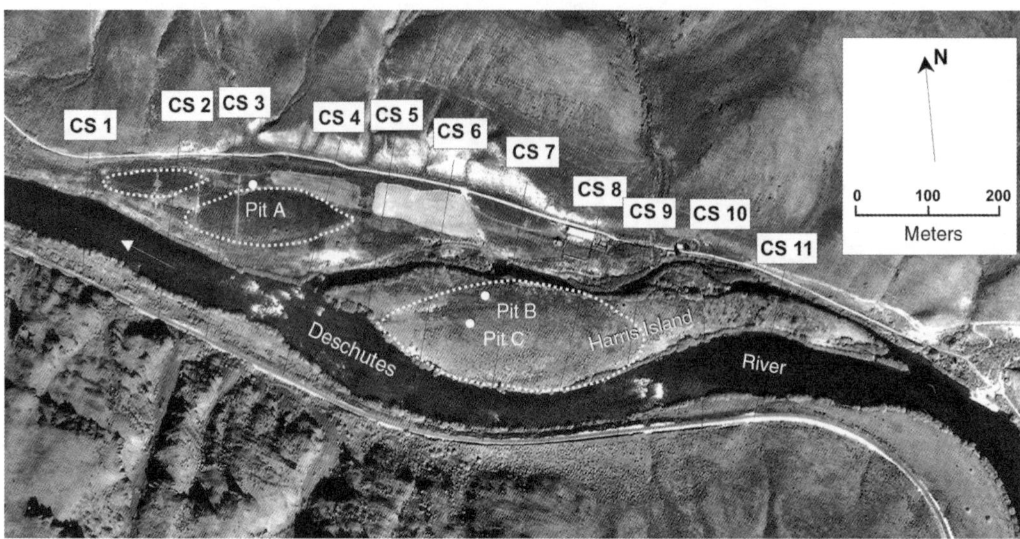

Figure 2.11 Portion of a vertical aerial photograph (WAC-95OR; 10–85; March 28 1995) of the area around Harris Island (RM 12) showing surveyed cross-sections, locations of trenches, and outlines of three flood bars that formed in this valley expansion

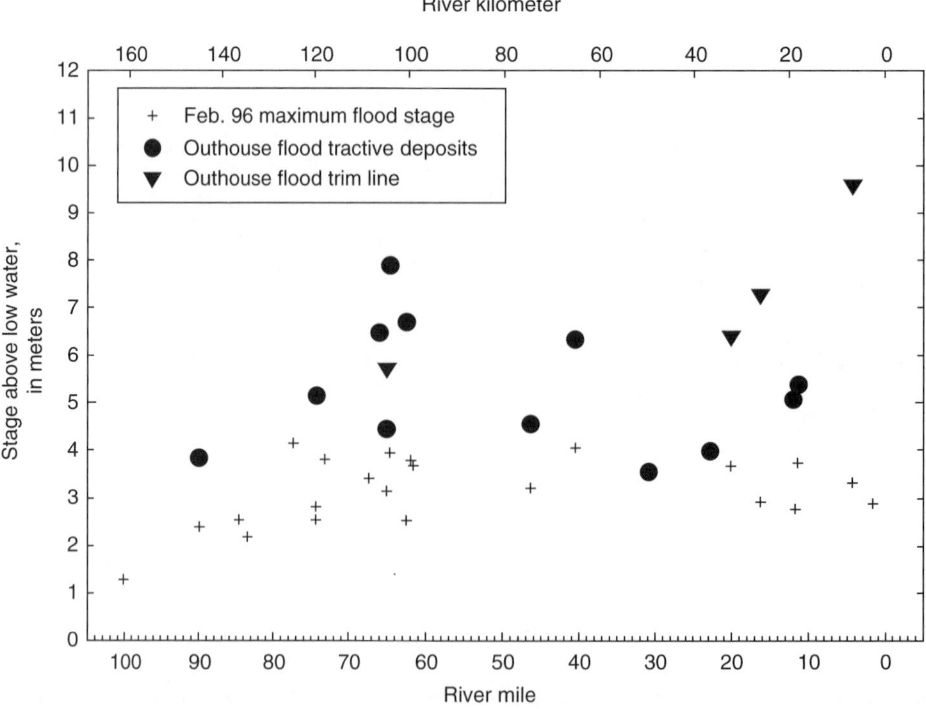

Figure 2.12 Maximum elevation of the February 1996 flood, outhouse flood deposits, and prominent trim lines formed in Pleistocene alluvial deposits above outhouse flood bars. Elevations are referenced to river level, which varied less than 0.3 m during the times of surveys. Surveys were conducted by tape and inclinometer. Also included are stages of February 1996 flood recorded at US Geological Survey gages "Deschutes River near Madras" (Station 14092500, River Mile 100.1) and "Deschutes River at Moody, near Biggs, Oregon" (Station 14103000, River Mile 1.4)

well (Figure 2.12). The positions of coarse outhouse flood deposits along the inside of canyon bends, at canyon expansions, and upstream of tight canyon constrictions are as would be expected considering the hydraulics of a large flow occupying the entire valley bottom (for example, Bretz 1928, Malde 1968). The rounded morphologies with boulder-covered surfaces that rise in the downstream direction rule out the possibility that they are terraces.

The age of the outhouse flood is only loosely constrained. Outhouse flood deposits sampled from a backhoe trench at Harris Island (Figure 2.13) contain pumice grains from the 7700 BP (calendar years) eruption from Mt. Mazama. Likely outhouse flood deposits at RM 62 are stratigraphically below a hearth which yielded a radiocarbon age of 2910 ± 50 ^{14}C years BP (Beta-131837, equivalent to 1220–1030 BC in calibrated calendar years) (Stuiver and Kra 1986).

Constraints on the peak discharge for this flood were estimated using the Mannings equation at a surveyed cross-section at Harris Island (Figure 2.14). An n value was selected to match gaged discharge of the February 8 1996 flood. The top of the flood bar at Harris Island requires that the outhouse flood had a maximum stage of at least 5.5 m above the summer low water surface. Assuming the present valley and channel bottom geometry and the slope and roughness parameters noted above, the discharge of the outhouse flood exceeded 5000 m^3/s. A more realistic discharge estimate of 12 500 m^3/s is obtained by assuming that water surface was about 2.5 m higher than the top of the bar and achieved a maximum stage of about 8 m above the summer water surface—a value consistent with the bouldery composition of the tops of the bars and the elevations of trim lines upstream and downstream of Harris Island (Figure 2.11). These calculations can be considered conservative because they assume no valley or channel scour and use a relatively large Mannings n value (0.045). These estimates indicate that the outhouse flood was 2.5–5 times as large as the largest historic flow recorded in nearly 100 years of record.

Figure 2.13 Photograph of backhoe trench excavated into outhouse flood bar at Harris Island (Deschutes River Mile 11; pit C of Figure 2.11). The deposit is composed of rounded to sub-rounded basalt clasts stratified into sub-horizontal, clast-supported layers distinguishable by variations in maximum clast size. Pumice grains collected from the sandy deposit matrix match tephra produced by the 7700 calendar year BP eruption of Mt. Mazama (Adrei Sarna-Wojcicki, US Geological Survey, written communication 1999), indicating that the deposits are younger than 7700 years BP. Gradations on the stadia rod are 0.3 m (1 ft)

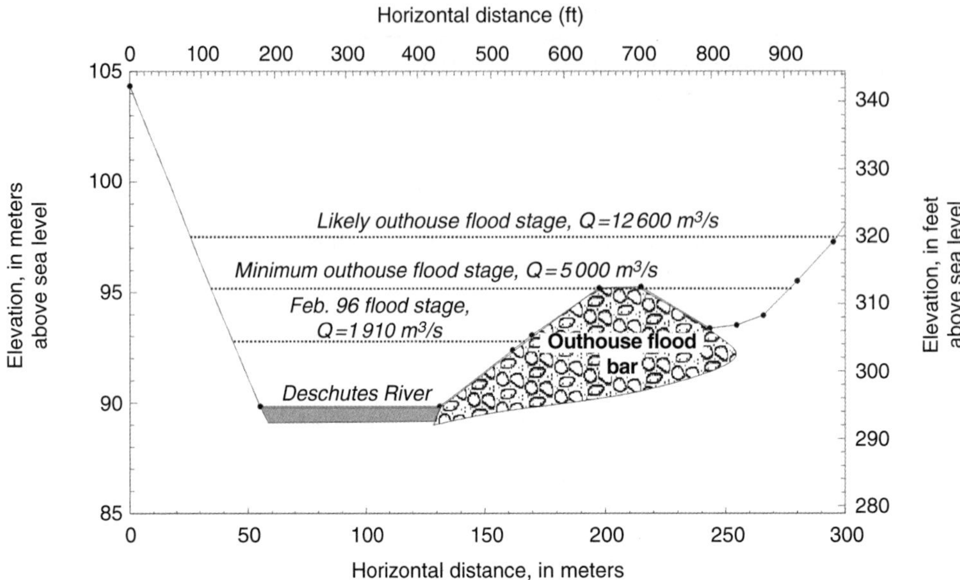

Figure 2.14 Cross-section and stages used for the Mannings equation estimates of the discharge of the outhouse flood at Harris Island. The "likely outhouse flood stage" was estimated from local elevation of prominent trim lines above nearby outhouse flood bars (Figure 2.13). Also shown is the maximum stage and discharge for the February 1996 flood, which was gaged at 1910 m^3/s at the Moddy Gage 19 km downstream. Outhouse flood discharges were calculated using a measured reach-scale slope of 0.02 and a Mannings n value of 0.045, which was derived based on the known stage and discharge of the February 96 flood. The cross-section corresponds to cross-section 3 of Figure 2.12

We have no evidence for the source of the outhouse flood, but the distribution of similar high boulder deposits along the entire Deschutes River canyon below the Pelton Round Butte Dam complex leads us to conclude that the flood came from upstream of the complex rather than from a landslide breach or some other impoundment within the canyon. The exceptionally large discharge seems greater than could plausibly result from a meteorologic event like the 1964 and 1996 floods, although we cannot rule out that possibility.

The effects of the outhouse flood on the present valley bottom are clear and substantial. Thirty-five percent of the valley bottom between the dam complex and the Columbia River confluence is composed of cobbly and bouldery alluvium interpreted to have been deposited by the outhouse flood. Additionally, the five largest islands in the Deschutes River downstream of the dam complex are large mid-channel flood bars left by this one ancient flood.

The bars deposited by the outhouse flood have left a lasting legacy that is relevant to assessing effects of the dam on the channel. The clasts composing these large bars are larger than can be carried by modern floods, and large portions of these bars stand above maximum modern flood stages. Only locally these large bars are eroded where main flow threads attack bar edges, but nowhere does it appear that cumulative erosion has exceeded more than a few percent of their original extent. Consequently, for many locations along the Deschutes River Valley, the present channel is essentially locked into its present position by the coarse bars, and modern processes—associated with pre-dam or post-dam conditions—are unable to substantially modify the valley-bottom morphology. This case study emphasizes the importance of understanding the surficial geologic and paleohydrologic context of individual river systems before one can fully assess the potential of environmental stresses to cause changes in channel or valley-bottom conditions.

Land-use Effects and Restoration

Land-use changes can affect runoff, sediment supply, or both, resulting in extensive changes to rivers and the ecosystems they support (for example, Wolman 1967, Nolan *et al.* 1995, Trimble and Lund 1982, Trimble 1974, Collier *et al.* 1996, Arnold *et al.* 1982).

Restoration of a river requires two critical concepts: a reference condition to define restoration goals—often taken as a natural, pre-anthropogenic disturbance state, and a process-based understanding of how to attain the goals. Information in the surficial geologic record can be used to construct both of these concepts. The surficial geologic record is sometimes the only source of historical information to define reference conditions in highly disturbed river basins. In particular, the surficial geologic record is a critical source of information for determining whether restoration is necessary by providing a long-term record of natural variation. The surficial geologic record can also be used to diagnose what has happened to degrade a river system, and thereby to develop an understanding of how to restore it.

In the Ozarks of Missouri (Figure 2.15), for example, there has been a pervasive belief that streams have too much gravel in them, indicated by the large number of extensive, unstable gravel bars. The abundance of gravel has been attributed to massive erosion associated with timber harvest 1880–1920 (Love 1990, Kohler 1984). Surficial geologic investigations of valley bottoms have provided a better understanding of gravel in Ozarks streams and how streams have responded to land-use changes (Table 2.3) (Jacobson and Primm 1997, Albertson *et al.* 1995, Jacobson and Pugh 1992). These investigations have documented:

1. There have been large quantities of gravel deposited in Ozarks streams throughout the Holocene (Figure 2.16A). This context indicates that present-day gravel distributions are not extreme aberrations.
2. Nonetheless, stratigraphic sections indicate that in 4th–5th-order streams, greater quantities of gravel have been deposited over the last 60–130 years than previously, an observation that corroborates popular perceptions that the streams have been quantitatively altered by land-use changes.
3. A more dramatic and unexpected effect, however, has been decreased deposition of fine sediment (silt and clay) over the same time interval. This observation has focused attention on the role of riparian land-use in providing hydraulic roughness and trapping fine sediments (McKenney *et al.* 1995).
4. The dominant mode of aggradation of land-use derived gravel has been lateral accumulation of extensive inset point bars with greater thicknesses than before settlement. Lateral aggradation of coarse sediment is favored in these watersheds because of the great quantities of chert produced by weathering of Paleozoic carbonate rocks.

The stratigraphic history of sedimentation in Ozarks streams has been an integral part of studies linking land-use changes to stream habitats and

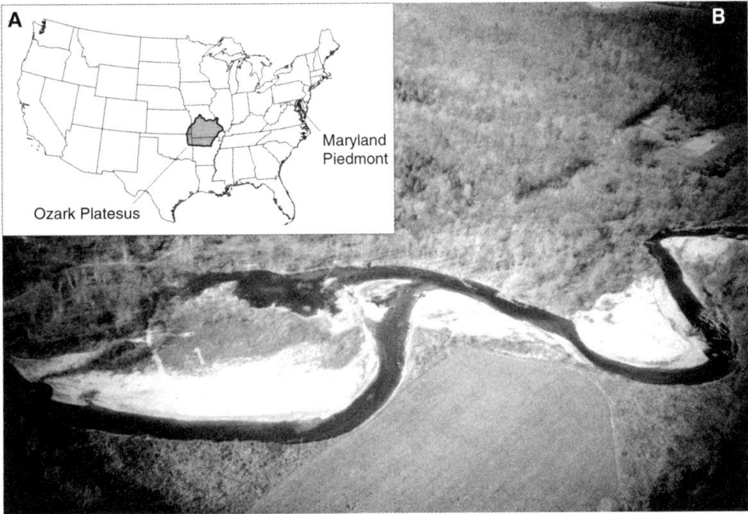

Figure 2.15 (A) Location map of land-use effects examples. (B) Photograph of a typical Ozarks stream, with extensive, unstable gravel bars

ecological processes (Jacobson and Gran 1999, Jacobson and Pugh 1997, Jacobson and Primm 1997, McKenney 1997, Peterson 1996). Surficial geologic tools were especially important for providing a qualitative understanding of what Ozarks streams looked like prior to European settlement, and for identifying changes in channel processes.

The Ozarks example of gravel aggradation provides a useful comparison to other studies of land-use-induced aggradation. Most documented stream responses to agricultural land-use changes in the humid, eastern half of the US have been dominated by aggradation of fine sediment (for example, Costa 1975, Trimble 1974). Jacobson and Coleman (1986) documented vertical aggradation of flood plains in several stream valleys in the Maryland Piedmont (Figure 2.16B). Vertical aggradation of overbank sediments was dominant because of the abundance of fine sediment produced by weathering of metasedimentary rocks in these watersheds, and because of proportionately greater increases in sediment supply compared to increases in runoff for a given increase in agricultural land use. In the past 50 years or so, vertical aggradation has been replaced by lateral aggradation of sand and gravel as soil-erosion controls have decreased fine sediment loads. In addition to focusing attention on the role of sediment supply in basin instability, the alluvial stratigraphic record in the Piedmont indicated which valley-bottom surfaces were appropriate to use for measuring bankfull channel dimensions (Coleman 1982). Moreover, the surficial geologic history documented the large quantity of flood-plain sediment that could be remobilized and delivered rapidly to streams as a result of lateral migration.

2.4 SUMMARY AND CONCLUSIONS

Surficial geologic tools fill an important role in geomorphologic studies. Sedimentology, geochronology, pedology, and stratigraphy in combination can extend the record of river behavior and provide essential context for predictive understanding. The chronicle of geomorphic changes preserved in alluvium, although subject to gaps and requiring interpretation, is primary evidence for how a river system has behaved in the past because of environmental stresses. Understanding of past behavior should constitute a solid foundation for assessing the present state of the river and for constraining predictions of future behavior.

A holistic understanding of alluvial deposits requires application of many disciplines, each of which has substantial complexity of its own. One of the most critical decisions in application of surficial geologic tools is how to limit a study, to collect data that are most efficient in addressing the geomorphologic question at hand. Some geomorphologic questions can involve relatively simple approaches. For example, creation of a flood hazard map may require simply mapping out Recent and Late Holocene deposits, and lumping the remainder of valley-bottom deposits into a low-hazard category. Definition of Recent and Late Holocene allostratigraphic units could be accomplished, perhaps, with description and sampling of available cutbank exposures, a few radiocarbon dates, dendrochronologic dating, and reference to aerial photographs. Construction of the maps could be accomplished by hand augering of soils and correlations to defined units based on simple indices of

Table 2.3 Alloformations defined for south-central Ozarks alluvial deposits. From Albertson *et al.* (1995)

Alloformation	Lithologic and pedologic features	Age range, calibrated calendar years
Cooksville	Actively aggrading point bars, alternate bars, and channel; sand, gravel, and cobbles; negligible pedogenesis	Present, 1850 AD
Happy Hollow	Stratified sand and gravel on low valley-bottom surfaces; very weak soil structure; multiple A–C horizons	Present, 1650 AD
Ramsey	Sandy silt overbank and gravel bottom stratum; cambic B horizons. A-Bw-C1 soil profiles	1650–550 AD
Dundas	Loamy overbank and gravel bottom stratum; weak argillic B horizons, partly oxidized with 7.5YR4/4–5/4 colors. Ap-Bw-Bt-C soil profiles	50 AD to 1050 BC
Miller	Silty, thick overbank deposits and gravel bottom stratum; moderate to well-developed soil structure with oxidized B horizons with 7.5YR5/6 colors. Ap-Bt-C soil profiles	2350–8050 BC

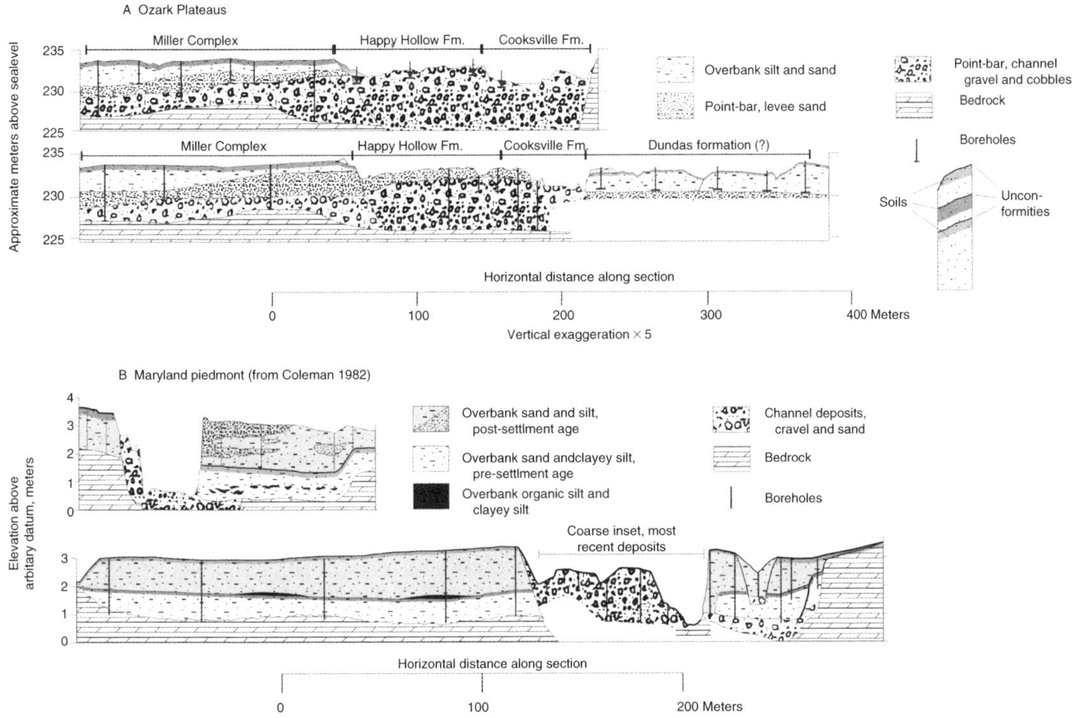

Figure 2.16 Stratigraphic cross-sections from: (A) Missouri Ozarks showing post-settlement aggradation of coarse-grain, point-bar and channel facies, and loss of fine, overbank facies. Land-use induced aggradation is shown to be dominantly lateral accumulation of coarse-grained Cooksville and Happy Hollow alloformations, Table 2.2. (B) Maryland Piedmont showing post-settlement, vertical flood-plain aggradation by fine, overbank facies. Vertical flood-plain aggradation was succeeded by deposition of coarse channel and flood-plain deposits when sediment supply was decreased (Jacobson and Coleman 1986) (reproduced by permission of Derrick Coleman)

pedogenic alteration. In this situation, a little investment in surficial geology could produce substantial increases in understanding.

In contrast, paleohydrologic and paleohydraulic studies to reconstruct the effects of climate change over the last 15 000 years would require a great deal more information from a similar flood plain. Such investigation could include detailed lithofacies maps, detailed definitions of allostratigraphic units, paleochannel and paleohydraulic reconstructions, numerical dates from multiple sources, and environmental indicators of climatic conditions. In this type of situation, a substantial investment of time and effort could yield large quantities of detailed information on timing and magnitude of geomorphologic changes.

Inevitably, however, there must be diminishing marginal returns of information for the investment in data collection. At some point, the geologist runs into the holes between the sediment (Ager 1993), or finds that past conditions are inadequate analogs for future conditions. At this point, the geomorphologist must turn to more complete sedimentary records (such as lakes or oceans) or other analytic tools to develop predictive understanding. In our experience, we have found that predictive tools have much greater utility when they are chosen and designed based on a solid understanding of the past.

REFERENCES

Ager, D.V. 1993. *The Nature of the Stratigraphical Record*, Chichester, UK: John Wiley and Sons, 151 p.

Aitken, M.J. 1997. Luminescence dating. In: Taylor, R.E. and Aitken, M.J., eds., *Chronometric Dating in Archaeology*, New York: Plenum Press, pp. 183–216.

Albertson, P.E., Meinert, D. and Butler, G. 1995. Geomorphic evaluation of Fort Leonard Wood: Vicksburg, Mississippi, US Army Corps of Engineers, Waterways Experiment Station, Technical Report GL-95-19, 241.

Allen, J.R.L. 1982a. Sedimentary structures – their character and physical basis. In: *Developments in Sedimentology* 30A, Volume 1, New York: Elsevier, 593 p.

Allen, J.R.L. 1982b. Sedimentary structures – their character and physical basis. In: *Developments in Sedimentology* 30B, Volume 2, New York: Elsevier, 663 p.

Allen, J.R.L. 1985. *Principles of Physical Sedimentology*, London: George Allen and Unwin, 272 p.

Arnold, C.A., Boison, P.J. and Patton, P.C. 1982. Sawmill brook: an example of rapid geomorphic change related to urbanization. *Journal of Geology* 90: 115–166.

Autin, W.J. 1992. Use of alloformations for defintion of Holocene meander belts in the middle Amite River, southeastern Louisiana. *Geological Society of America Bulletin* 104: 233–241.

Baker, V.R. 1973. *Paleohydrology and Sedimentology of Lake Missoula Flooding in Eastern Washington*, Special Paper 144, Boulder, Colorado: Geological Society of America, 79 p.

Baker, V.R. 1977. Stream-channel responses to floods, with examples from central Texas. *Geological Society of America Bulletin* 88: 1057–1071.

Baker, V.R. 1987. Paleoflood hydrology and extraordinary flood events. *Journal of Hydrology* 96: 79–99.

Baker, V.R. 1989. Magnitude and frequency of palaeofloods. In: Beven, K. and Carling, P., eds., *Floods – Hydrological, Sedimentological, and Geomorphological Implications*, Chichester, UK: John Wiley and Sons, pp. 171–183.

Baker, V.R. 1991. A bright future for old flows. In: Starkel, L., Gregory, K.J. and Thornes, J.B., eds., *Temperate Palaeohydrology*, New York: John Wiley and Sons, pp. 497–520.

Baker, V.R. 1996. Discovering Earth's future in its past: palaeohydrology and global environmental change. In: Branson, J. Brown A.G. and Gregory K.J., eds., *Global Continental Changes: The Context of Palaeohydrology*, London: The Geological Society, pp. 73–83.

Baker, V.R. and Kochel, R.C. 1988. Paleoflood analysis using slackwater deposits. In: Baker, V.R., Kochel, R.C. and Patton, P.C., eds., *Flood Geomorphology*, New York: John Wiley and Sons, pp. 357–376.

Baker, V.R. and Ritter, D.F. 1975. Competence of rivers to transport coarse bed load material. *Geological Society of America Bulletin* 86: 975–978.

Baker, V.R., Kochel, R.C. and Patton, P.C. 1979. Long-term flood frequency analysis using geological data. In: *The Hydrology of Areas of Low Precipitation*, Publication 128, Louvain: International Association of Hydrological Sciences, pp. 3–9.

Berger, G.W. 1988. Dating Quaternary events by luminescence. In: Easterbrook, D.J., ed., *Dating Quaternary Sediments*, Special Paper 227, Boulder, Colorado: Geological Society of America, pp. 13–59.

Bettis, A.E., III. 1992. The soil morphologic properties and weathering zone characteristics as age indicators in Holocene alluvium in the upper Midwest. In: Holliday, V.T., ed., *Soils in Archeology, Landscape Evolution, and Human Occupation*, Washington, DC: Smithsonian Institution, pp. 119–144.

Bierman, P.R. 1994. Using in situ produced cosmogenic isotopes to estimate rates of landscape evolution: a review from the geomorphic perspective. *Journal of Geophysical Research* 99: 13 885–13 896.

Birkeland, P.W. 1968. Mean velocities and boulder transport during Tahoe-age floods of the Truckee River, California-Nevada. *Geological Society of America Bulletin* 79: 137–142.

Birkeland, P.W. 1984. *Soils and Geomorphology*, New York: Oxford University Press, 372 p.

Bowman, S. 1990. *Radiocarbon Dating*, London: Trustees of the British Museum, 64 p.

Bradley, W.C. and Mears, A.I. 1980. Calculations of flows needed to transport coarse fraction of Boulder Creek alluvium at Boulder, Colorado. *Geological Society of America Bulletin, Part II* 91: 1057–1090.

Brakenridge, G.R. 1984. Alluvial stratigraphy and radiocarbon dating along the Duck River, Tennessee: implications regarding flood-plain origin. *Geological Society of America Bulletin 95*: 9–25.

Bretz, J H. 1928. Bars of the channeled scabland. *Geological Society of America Bulletin* 39: 643–702.

Bridge, J.S. and Mackey, S.D. 1993. A revised alluvial stratigraphy model. In: Marzo, M. and Puigdefabregas, C., eds., *Alluvial Sedimentation*, International Association of Sedimentologists, Special Publication 17, Oxford: Blackwell Scientific Publications, pp. 319–336.

Brown, A.G. 1990. Holocene floodplain diachronism and inherited downstream variations in fluvial processes: a study of the river Perry, Shropshire, England. *Journal of Quaternary Science* 5: 39–51.

Brown, A.G. 1997. *Alluvial Geoarchaeology*, Cambridge: Cambridge University Press, 377 p.

Bull, W.B. 1990. Stream-terrace genesis: implications for soil development. *Geomorphology* 3: 351–367.

Bull, W.B. 1991. *Geomorphic Responses to Climatic Change*, New York: Oxford University Press, 326 p.

Carling, P.A. and Dawson, M.R. 1996. *Advances in Fluvial Dynamics and Stratigraphy*, Chichester, UK: John Wiley and Sons, 521 p.

Chorley, R.J., Dunn, A.J. and Beckinsale, R.P. 1964. *The History of the Study of Landforms or the Development of Geomorphology*, Volume 1, London: Methuen and Co., 678 p.

Church, M. 1978. Paleohydrological reconstructions from a Holocene valley fill. In: Miall, A.D., ed., *Fluvial Sedimentology*, Memoir 5, Canadian Society of Petroleum Geologists, pp. 743–772.

Coates, D.R. 1984. Landforms and landscapes as measures of relative time. In: Mahaney, W.C., ed., *Quaternary Dating Methods*, New York: Elsevier, pp. 247–267.

Coleman, D.J. 1982. An Examination of Bankfull Discharge Frequency in Relation to Floodplain Formation, Baltimore, Maryland: Johns Hopkins University Press, 192 p.

Collier, M., Webb, R.H. and Schmidt, J.C. 1996. *Dams and Rivers, Tucson, Arizona*, US Geological Survey Circular 1126, US Geological Survey, 94 p.

Costa, J.E. 1975. Effects of agriculture on erosion and sedimentation in the Piedmont Province, Maryland. *Geological Society of America Bulletin* 86: 1281–1286.

Costa, J.E. 1978. Holocene stratigraphy in flood frequency analysis. *Water Resources Research* 14: 626–632.

Costa, J.E. 1983. Paleohydraulic reconstruction of flash-flood peaks from boulder deposits in the Colorado Front Range. *Geological Society of America Bulletin* 94: 986–1004.

Daniels, R.B. and Hammer, R.D. 1992. *Soil Geomorphology*, New York: John Wiley and Sons, 236 p.

Dorn, R.I. 1994. Surface exposure dating with rock varnish. Beck, C., ed., *Dating in Exposed and Surface Contexts*, Albuquerque, New Mexico: University of New Mexico Press, pp. 77–113.

Dott, R.H. and Batten, R.L. 1976. *Evolution of the Earth*, New York: McGraw-Hill, 504 p.

Duller, G.A.T. 1996. Recent developments in luminescence dating of Quaternary sediments. *Progress in Physical Geography* 20: 127–145.

Durham, R.W. and Joshi, S.R. 1984. Lead-210 dating of sediments from some northern Ontario lakes. In: Mahaney, W.C., ed., *Quaternary Dating Methods*, New York: Elsevier, pp. 75–85.

Dury, G.H. 1954. Contribution to a general theory of meandering valleys. *American Journal of Science* 252: 193–224.

Dury, G.H. 1965. *Theoretical Implications of Underlift Streams*, Professional Paper 452-C, US Geological Survey, 43 p.

Dury, G.H. 1976. Discharge prediction, present and former, from channel dimensions. *Journal of Hydrology* 30: 219–245.

Dury, G.H. 1985. Attainable standards of accuracy in the retrodiction of palaeodischarge from channel dimensions. *Earth Surface Processes and Landforms* 10: 205–213.

Easterbrook, D.J. 1988a. *Dating Quaternary Sediments*, Special Paper 227, Boulder, Colorado, Geological Society of America, 165 p.

Easterbrook, D.J. 1988b. Paleomagnetism of Quaternary deposits. In: Easterbrook, D.J., ed., *Dating Quaternary Sediments*, Special Paper 227, Boulder, Colorado, Geological Society of America, pp. 111–122.

Ely, L.L. and Baker, V.R. 1985. Reconstructing palaeoflood hydrology with slackwater deposits – Verde River Arizona. *Physical Geography* 6: 103–126.

Ely, L.L., Enzel, Y., Baker, V.R. and Cayan, D.R. 1993. A 5000-year record of extreme floods and climate change in the southwestern United States. *Science* 262: 410–412.

Ely, L.L., Enzel, Y., Baker, V.R., Kale, V.S. and Mishra, S. 1996. Changes in the magnitude and frequency of Late Holocene monsoon floods on the Narmada River, central India. *Geological Society of America Bulletin* 108: 1134–1148.

Ethridge, F.G. and Schumm, S.A. 1978. Reconstructing paleochannel morphologic and flow characteristics—methodology, limitations, and assessment. In: Miall, A.D., ed., *Fluvial Sedimentology*, Memoir 5, Canadian Society of Petroleum Geologists, pp. 703–721.

Fassnacht, H. 1998. Frequency and magnitude of bedload transport downstream of the Pelton-Round Butte Dam Complex, Lower Deschutes River, Oregon, M.S. Thesis, Oregon State University, Corvallis, Oregon, 311 p.

Fassnacht, H., Klingeman, P.C. and Grant, G.E. 1998. Effects of a hydroelectric dam complex on bedload transport, lower Deschutes River, Oregon. In: *EOS* (American Geophysical Union) 79: F374 (abstract).

Ferring, C.R. 1992. Alluvial pedology and geoarcheological research. In: Holliday, V.T., ed., Soils in *Archeology, Landscape Evolution, and Human Occupation*, Washington, DC: Smithsonian Institution, pp. 1–39.

Fredlund, G.G. 1996. Late Quaternary geomorphic history of lower Highland Creek, Wind Cave National Park, South Dakota. *Physical Geography* 17: 446–464.

Gilbert, G.K. 1914. *The transportation of debris by running water*, US Geological Survey Professional Paper 86, 263 p.

Grant, G.E. 1997. Critical flow constrains flow hydraulics in mobile-bed streams—a new hypothesis. *Water Resources Research* 33: 349–358.

Gregory, J.J. and Maizels, J.K. 1991. Morphology and sediments: typological characteristics of fluvial forms and deposits. In: Starkel, L., Gregory, K.J. and Thornes, J.B., eds., *Temperate Palaeohydrology*, Chichester, UK: John Wiley and Sons, pp. 31–59.

Gruen, R. 1994, Quaternary geochronology: *Quaternary Science Reviews* 13(2): 95–181.

Hallet, B. and Putkonen, J. 1994. Surface dating of dynamics landforms; young boulders on aging moraines. *Science* 265: 937–940.

Haynes, C.V. 1985. *Mastodon-bearing Springs and Late Quaternary Geochronology of the Lower Pomme de Terre Valley Missouri*, Special Paper 204, Boulder, Colorado: Geological Society of America, 35 p.

Holliday, V.T. 1990. *Pedology in Archaeology*. In: Lasca, N.P. and Donahue, J., eds., *Archaeological Geology of North America*, Centennial Special Volume 4, Boulder, Colorado: Geological Society of America, pp. 525–540.

Hupp, C.R. 1988. Plant ecological aspects of flood geomorphology and paleoflood history. In: Baker, V.R., Kochel, R.C. and Patton, P.C., eds., *Flood Geomorphology*, New York: John Wiley and Sons, pp. 335–356.

Hutton, J. 1788. Theory of the Earth. *Royal Society of Edinburgh, Transactions* 1: 209–304.

International Subcommission on Stratigraphic Classification. 1994. *A Guide to Stratigraphic Classification, Terminology, and Procedure*, Trondheim, Norway; Boulder, Colorado: The International Union of Geological Sciences and The Geological Society of America, 214 p.

Iwai, S. and Ishiguro, M. 1970. *Applied Statistics for Hydrology*. Tokyo: Morikita Shuppan, 369 p. (in Japanese).

Jacobson, R.B. and Coleman, D.J. 1986. Stratigraphy and recent evolution of Maryland Piedmont flood plains. *American Journal of Science* 286: 617–637.

Jacobson, R.B. and Gran, K.B. 1999. Gravel routing from widespread, low-intensity landscape disturbance, Current River Basin, Missouri. *Earth Surface Processes and Landforms* 24: 897–917.

Jacobson, R.B. and Primm, A.T. 1997. *Historical Land-use Changes and Potential Effects on Stream Disturbance in the Ozark Plateaus*, Water-Supply Paper 2484, Missouri: US Geological Survey, 85 p.

Jacobson, R.B. and Pugh, A.L. 1992. Effects of land use and climate shift on channel disturbance, Ozark Plateaus, USA, In: *Proceedings of the Workshop on the Effects of Global Climate Change on Hydrology and Water Resources at the Catchment Scale*, Japan–US committee on Hydrology, Water Resources and Global Climate Change: 432–444.

Jacobson, R.B. and Pugh, A.L. 1997. *Riparian Vegetation and the Spatial Pattern of Stream-channel Instability, Little Piney Creek*, Water Supply Paper 2494, Missouri: US Geological Survey, 33 p.

Jenny, H. 1941. *Factors of Soil Formation*, New York: McGraw-Hill, 281 p.

Jones, A.P., Tucker, M.E. and Hart, J.K. 1999. *The Description and Analysis of Quaternary Stratigraphic Field Sections*, Technical Guide No. 7, London: Quaternary Research Association, 286 p.

Jones, A.P., Shimazu, H., Oguchi, T., Okuno, M. and Tokutake, M. 2001. Late Holocene slackwater deposits on the Nakagawa River, Tochigi Prefecture, Japan. *Geomorphology* 39: 39–51.

Katsube, K. and Oguchi, T. 1999. Altitudinal changes in slope angle and profile curvature in the Japan Alps: a hypothesis regarding a characteristic slope angle. *Geographical Review of Japan* 72B: 63–72.

Kehew, A.E. and Teller, J.T. 1994. Glacial-lake spillway incision and deposition of a coarse-grained fan near Watrous, Saskatchewan. *Canadian Journal of Earth Sciences* 31: 544–553.

Kemp, P.A. 1985. *Soil Micromorphology and the Quaternary*, Amsterdam: Elsevier, 79 p.

Kite, J.S. and Linton, R.C. 1991. Depositional aspects of the November 1985 flood on Cheat River and Black Fork, West Virginia. In: Jacobson, R.B., ed., *Geomorphic Studies of the Storm and Flood of November 3–5 1985, in the Upper Potomac and Cheat River Basins in West Virginia and Virginia*, US Geological Survey Bulletin 1981, pp. D21–D24.

Knox, J.C. 1988. Climatic influence on upper Mississippi Valley floods. In: Baker, V.R., Kochel, R.C. and Patton, P.C., eds., *Flood Geomorphology*, New York: John Wiley and Sons, pp. 279–300.

Knox, J.C. 1993. Large increases in flood magnitude in response to modest changes in climate. *Nature* 361: 420–432.

Knuepfer, P.L. 1994. Use of rock weathering rinds in dating geomorphic surfaces. In: Beck, C., ed., *Dating in Exposed and Surface Contexts*, Albuquerque, New Mexico: University of New Mexico Press, pp. 15–28.

Kochel, R.C. and Baker, V.R. 1982. Paleoflood hydrology. *Science* 215: 353–361.

Kochel, R.C. and Baker, V.R. 1988. Paleoflood analysis using slackwater deposits. In: Baker, V.R., Kochel, R.C. and Patton, P.C., eds., *Flood Geomorphology*, New York: John Wiley and Sons, pp. 357–376.

Kochel, R.C., Baker, V.R. and Patton, P.C. 1982. Paleohydrology of southwestern Texas. *Water Resources Research* 18: 1165–1183.

Kohler, S. 1984. *Two Ozark Rivers: The Current and the Jacks Fork*, Columbia, Missouri: University of Missouri Press, 130 p.

Komar, P.D. 1970. The competence of turbidity current flow. *Geological Society of America Bulletin* 81: 1555–1562.

Komar, P.D. 1989. Flow-competence evaluations of the hydraulic parameters of floods – an assessment of the technique. In: Beven, K. and Carling, P., eds., *Floods – Hydrological, Sedimentological, and Geomorphological Implications*, Chichester, UK: John Wiley and Sons, pp. 107–134.

Komar, P.D. 1996. Entrainment of sediments from deposits of mixed grain sizes and densities. In: Carling, P.A. and Dawson, M.R., eds., *Advance in Fluvial Dynamics and Stratigraphy*, Chichester, UK: John Wiley and Sons, pp. 107–134.

Komar, P.D. and Carling, P.A. 1991. Grain sorting in gravel-bed streams and the choice of particle sizes for flow-competence evaluations. *Sedimentology* 38: 489–502.

Kraus, M.J. and Brown, T.M. 1988. Pedofacies analysis: a new approach to reconstructing ancient fluvial sequences. In: Reinhardt, J. and Sigleo, W.R., eds., *Paleosols and Weathering through Geologic Time: Principles and Applications*, Special Paper 216, Boulder, Colorado: Geological Society of America, pp. 143–152.

Kurz, M.D. and Brook, E.J. 1994. Surface exposure dating with cosmogenic nuclides. In: Beck, C., ed., *Dating in Exposed and Surface Contexts*, Albuquerque, New Mexico: University of New Mexico Press, pp. 139–159.

Lasca, N.P. and Donahue, J. 1990. *Archaeological Geology of North America*, Decade of North American Geology, Centennial Special Volume 4, Boulder, Colorado: Geological Society of America, 633 p.

Leopold, L.B. and Wolman, M.G. 1957. *River Channel Patterns: Braided, Meandering, and Straight*, US Geological Survey Professional Paper No. 282-B, 39 p.

Leopold, L.B., Wolman, M.G. and Miller, J.P. 1964. *Fluvial Processes in Geomorphology*, San Francisco, CA: W.H. Freeman and Co., 522 p.

Levish, D.R. and Ostenna, D.A. 1996. Applied paleoflood hydrology in north-central Oregon (Bureau of Reclamation Seismotectonic Report 96-7), Portland, Oregon, Guidebook for Field Trip 2. In: *Cordilleran Section Meeting of the Geological Society of America*, unpaginated.

Lord, M.L. and Kehew, A.E. 1987. Sedimentology and paleohydrology of glacial-lake outburst deposits in southeastern Saskatchewan and northwestern North Dakota. *Geological Society of America Bulletin* 99: 663–673.

Love, K. 1990. Paradise lost. *Missouri Conservationist* 51: 31–35.

Lund, S.P. 1996. A comparison of Holocene paleomagnetic secular variation records from North America. *Journal of Geophysical Research, B, Solid Earth and Planets* 101: 8007–8024.

Magilligan, F.J. 1985. Historical floodplain sedimentation in the Galena River Basin, Wisconsin and Illinois. *Annals of the Association of American Geographers* 75: 583–594.

Mahaney, W.C. 1984. Quaternary Dating Methods, New York, Elsevier, 431 p.

Maizels, J.K. 1983. Palaeovelocity and paleodischarge determination for coarse gravel deposits. In: Gregory, K.J., ed., *Background to Palaeohydrology*, Chichester, UK: John Wiley and Sons, pp. 101–139.

Malde, H.E. 1968. *The Catastrophic Late Pleistocene Bonneville Flood in the Snake River Plain, Idaho*, US Geological Survey Professional Paper 596, 52 p.

Markewich, H.W. and Pavich, M.J. 1991. Soil chronosequence studies in temperate to subtropical, low-latitude, low-relief terrain with data from the Eastern United States. In: Pavich, M.J., ed., *Weathering and Soils, Geoderma*, 51, 1–4, pp. 213–239.

Marron, D.C. 1992. Floodplain storage of mine tailings in the Belle Fourche River system a sediment budget approach. *Earth Surface Processes and Landforms* 17: 675–685.

Martin, C.W. 1992. The response of fluvial systems to climate change; an example from the central Great Plains. *Physical Geography* 13: 101–114.

Marzo, M. and Puigdefabregas, C. 1993. *Alluvial Sedimentation*, International Association of Sedimentologists, Special Publication 17, Oxford: Blackwell Scientific Publications, 586 p.

Matsumoto, J. 1993. Global distribution of daily maximum precipitation. *Bulletin of the Department of Geography, University of Tokyo* 25: 43–48.

Matthews, J.A. 1994. Lichenometric dating: a review with particular reference to 'Little Ice Age' moraines in southern Norway. In: Beck, C., ed., *Dating in Exposed and Surface Contexts*, Albuquerque, New Mexico: University of New Mexico Press, pp. 185–212.

McClure, E.M. 1998. Spatial and temporal trends in bed material and channel morphology below a hydroelectric dam complex, Deschutes River, Oregon, M.S. Thesis, Oregon State University, Corvallis, Oregon, 85 p.

McClure, E.M., Grant, G.E. and Jones, J.A. 1997. Longitudinal patterns of bed material size following impoundment of the lower Deschutes River, Oregon. *Geological Society of America Abstracts With Programs* 29: 314 (abstract).

McKenney, R. 1997. Formation and maintenance of hydraulic habitat units in the streams of the Ozark Plateaus, Missouri and Arkansas, Ph.D. Dissertation, State College, The Pennsylvania State University, 347 p.

McKenney, R., Jacobson, R.B. and Wertheimer, R.C. 1995. Woody vegetation and channel morphogenesis in low-gradient, gravel-bed streams in the Ozarks Region, Missouri and Arkansas. *Geomorphology* 13: 175–198.

Miall, A.D. 1992. Alluvial deposits. In: Walker, R.G., James N.P., eds., *Facies Models – Response to Sea Level Change*, St. John's, Newfoundland: Geological Association of Canada, pp. 119–142.

Miller, H. 1970. Methods and results of river terracing. *N:* Dury, G.H., ed., *Rivers and River Terraces*, New York: Praeger, pp. 19–35.

Nanson, G.C., Chen, X.Y. and Price, D.M. 1995. Aeolian and fluvial evidence of changing climate and wind patterns during the past 100 ka in the western Simpson Desert, Australia. *Palaeogeography, Palaeoclimatology, Palaeoecology* 113: 87–102.

Nikiforoff, C.C. 1949. Weathering and soil evolution. *Soil Science* 67: 219–223.

Nolan, K.M., Kelsey, H.M. and Marron, D.C. 1995. *Geomorphic Processes and Aquatic Habitat in the Redwood Creek Basin, northwestern California*, Professional Paper 1454, US Geological Survey, pp. A1–A6.

North American Commission on Stratigraphic Nomenclature. 1983. North American Stratigraphic Code. *American Association of Petroleum Geologists Bulletin* 67: 841–875.

O'Connor, J.E. 1993. *Hydrology, Hydraulics, and Geomorphology of the Bonneville Flood*, Special Paper 274, Boulder, Colorado: Geological Society of America, 83 p.

O'Connor, J.E., Ely, L.L., Wohl, E.E., Stenvens, L.E., Melis, T.S., Kale, V.S. and Baker, V.R. 1994. A 4500-year record of large floods on the Colorado River in the Grand Canyon, Arizona. *Journal of Geology* 102: 1–9.

O'Connor, J.E., Grant, G.E., Curran, J.H., Fassnacht. H. and Brink, M. 1998. Geomorphic processes within the lower Deschutes River, Oregon. In: *EOS* (American Geophysical Union) 79: F374 (abstract).

O'Connor, J.E., Webb, R.H. and Baker, V.R. 1986. Paleohydrology of pool-and-riffle pattern development: Boulder Creek, Utah. *Geological Society of America Bulletin* 97: 410–420.

Oguchi, T. 1996. Hillslope failure and sediment yield in Japanese regions with different storm intensity. *Bulletin of the Department of Geography, University of Tokyo* 28: 45–54.

Oguchi, T. 1997. Late Quaternary sediment budget in alluvial-fan-source-basin systems in Japan. *Journal of Quaternary Science* 12: 381–390.

Owens, P.N., Walling, D.E. and Leeks, G.J.L. 1999. Use of floodplain sediment cores to investigate recent historical changes in overbank sedimentation rates and sediment sources in the catchment of the River Ouse, Yorkshire. *Catena* (Giessen) 36: 21–47.

Parker, G. 1976. On the cause and characteristic scales of meandering and braiding in rivers. *Journal of Fluid Mechanics* 76: 457–480.

Partridge, J. and Baker, V.R. 1987. Paleoflood hydrology of the Salt River, Arizona. *Earth Surface Processes and Landforms* 12: 109–125.

Patton, P.C. and Dibble, D.S. 1982. Archeologic and geomorphic evidence for paleohydrologic record of the

Pecos River in west Texas. *American Journal of Science* 282: 97–121.

Patton, P.C., Baker, V.C. and Kochel, R.C. 1979. Slack-water deposits; a geomorphic technique for the interpretation of fluvial paleohydrology. In: Rhodes, D.D. and Williams, G.P., eds., *Adjustments of the Fluvial Fystem, A Proceedings Volume of the Annual Geomorphology Symposia Series, Binghampton, New York, Tenth Annual Geomorphology Symposium*, pp. 225–253.

Pazzaglia, F.J., Gardner, T.W. and Merritts, D.J. 1998. Bedrock fluvial incision and longitudinal profile development over geologic time scales determined by fluvial terraces. In: Tinkler, K.J. and Wohl, E.E., eds., *Rivers Over Rock: Fluvial Process in Bedrock Channels*, Geophysical Monograph 107, Washington, DC: American Geophysical Union, pp. 207–235.

Peterson, J.T. 1996. The evaluation of a hydraulic unit-based habitat system, Ph.D. Dissertation, University of Missouri, Columbia, 397 p.

Pettijohn, F.J. 1975. *Sedimentary Rocks*, New York: Harper and Row, 628 p.

Pinter, N., Keller, E.A. and West, R.B. 1994. Relative dating of terraces of the Owens River, northern Owens Valley, California, and correlation with moraines of the Sierra Nevada. *Quaternary Research* 42: 266–276.

Reading, H.G. 1978. *Sedimentary Environments and Facies*, New York: Elsevier, 557 p.

Reynolds, J.M. 1997. *An Introduction to Applied and Environmental Geophysics*, New York: John Wiley and Sons, 796 p.

Rice, S. and Church, M. 1998. Grain size along two gravel-bed rivers; statistical variation, spatial pattern and sedimentary links. *Earth Surface Processes and Landforms* 23: 345–363.

Rotnicki, K. 1983. Modelling past discharges of meandering rivers. In: Gregory, K.J., ed., *Background to Palaeohydrology*, Chichester, UK: John Wiley and Sons, pp. 321–354.

Royall, P.D., Delcourt, P.A. and Delcourt, H.R. 1991. Late Quaternary paleoecology and paleoenvironments of the central Mississippi alluvial valley. *Geological Society of America Bulletin* 103: 157–170.

Ruhe, R.V. 1975. *Geomorphology – Geomorphic Processes and Surficial Geology*, Boston: Houghton Mifflin Company, 246 p.

Saito, K. 1988. *Alluvial Fans in Japan*, Tokyo: Kokon-Shoin, 280 p. (in Japanese).

Sarna-Wojcicki, A.M., Bowman, H.R., Meyer, C.E., Russell, P.C., Woodward, M.J., McCoy, G., Rowe, J.J., Jr., Baedecker, P.A., Asaro, F. and Michael, H. 1984. *Chemical Analyses, Correlations, and Ages of Upper Pliocene and Pleistocene Ash Layers of East-central and Southern California*. US Geological Survey Professional Paper 1293, US Geological Survey, 40 p.

Sarna-Wojcicki, A.M., Lajoie, K.R., Meyer, C.E., Adam, D.P. and Rieck, H.J. 1991. Tephrochronologic correlation of upper Neogene sediments along the Pacific margin, conterminous United Sates. In: Morrison, R.B., ed., *Quaternary Nonglacial Geology; Conterminous U.S.*, Boulder, Colorado: Geological Society of America, The Geology of North America, K-2, pp. 117–140.

Schneider, J.S. and Bierman, P.R. 1997. Surface dating using rock varnish. In: Taylor, R.E. and Aitken, M.J., eds., *Chronometric Dating in Archaeology*, New York: Plenum Press, pp. 357–388.

Schumm, S.A. 1967. Meander wavelength of alluvial rivers. *Science* 157: 1549–1550.

Schumm, S.A. 1968. River Adjustment to Altered Hydrologic Regimen—Murrumbidgee River and Paleochannels, Australia, Professional Paper 598, US Geological Survey 65 p.

Schumm, S.A. and Khan, H.R. 1972. Experimental study of channel patterns. *Geological Society of America Bulletin* 83: 1755–1770.

Schumm, S.A. and Parker, R.S. 1973. Implications of complex response of drainage systems for Quaternary alluvial stratigraphy. *Nature* 243: 99–100.

Sharma, P.V. 1997. *Environmental and Engineering Geophysics*, Cambridge University Press, 475 p.

Stedinger, J.R. and Baker, V.C. 1987. Surface water hydrology—historical and paleoflood information. *Reviews of Geophysics* 25: 119–124.

Sternberg, R.S. and McGuire, R.H. 1990. Secular variation in the American southwest. In: Eighmy, J.L. and Sternberg, R.S., eds., *Archaeomagnetic Dating*, Tucson, Arizona: University of Arizona Press, pp. 199–225.

Stuiver, M. 1982. A high-precision calibration of the AD radiocarbon time scale. *Radiocarbon* 24: 1–26.

Stuiver, M. and Kra, R.S. 1986. Calibration issue. *Radiocarbon* 28: 805–1030.

Stupavsky, M. and Gravenor, C.P. 1984. Paleomagnetic dating of Quaternary sediments: a review. In: Mahaney, W.C., ed., *Quaternary Dating Methods*, New York: Elsevier, pp. 123–140.

Sugai, T. 1993. River terrace development by concurrent fluvial processes and climatic changes. *Geomorphology* 6: 243–252.

Suzuki, H. 1971. Climatic zones of the Wurm Glacial Age. *Bulletin of the Department of Geography, University of Tokyo* 3: 35–46.

Taylor, M.P. and Lewin, J. 1996. River behavior and Holocene alluviation: the River Severn at Welshpool, mid-Wales, UK *Earth Surfaces Processes and Landforms* 21: 77–91.

Taylor, R.E., Long, A. and Kra, R.S. 1992. *Radiocarbon After Four Decades – An Interdisciplinary Perspective*, New York: Springer-Verlag, 596 p.

Tokyo Bureau of International Trade and Industry. 1984. *Investigation report toward appropriate ground water usage in the Minimiazumi District, Nagano Prefecture*, Tokyo: Bureau of International Trade and Industry, 120 p. (in Japanese).

Trimble, S.W. 1974. *Man-induced Soil Erosion on the Southern Piedmont 1700–1970*, Ankeny, Iowa: Soil Conservation Society of America, 180 p.

Trimble, S.W. and Lund, S.W. 1982. *Soil Conservation and Reduction of Erosion and Sedimentation in the Coon Creek Basin, Wisconsin*, Professional Paper 1234, US Geological Survey, 35 p.

US Army Corps of Engineers. 1998. *Geophysical Exploration for Engineering and Environmental Investigations*, Reston, Va.: American Society of Civil Engineers Press, 204 p.

Verosub, K.L. 1988. Geomagnetic secular variation and the dating of Quaternary sediments. In: Easterbrook, D.J., ed., *Dating Quaternary Sediments*, Special Paper 227, Boulder, Colorado: Geological Society of America, pp. 123–138.

Vincent, K.R., Bull, W.B. and Chadwick, O.A. 1994. Construction of a soil chronosequence using the thickness of pedogenic carbonate coatings. *Journal of Geological Education* 42: 316–324.

Walker, R.G. and James, N.P. 1992. *Facies Models – Response to Sea Level Change*, St. John's, Newfoundland: Geological Society of Canada, 454 p.

Waythomas, C.F., Walder, J.S., McGimsey, R.G. and Neal, C.A. 1996. A catastrophic flood caused by drainage of a caldera lake at Aniakchak Volcano, Alaska, and implications for volcanic hazards assessment. *Geological Society of America Bulletin* 108: 861–871.

Webb, R.H. 1985. Late Holocene flooding on the Escalante River, southcentral Utah, Ph.D. Dissertation, University of Arizona, Tucson, Arizona, 204 p.

Webb, R.H., O'Connor, J.E. and Baker, V.R. 1988. Paleohydrologic reconstruction of flood frequency on the Escalante River, south-central Utah. In: Baker, V.R., Kochel, R.C. and Patton, P.C., eds., *Flood Geomorphology*, New York: John Wiley and Sons, pp. 403–418.

Weissel, J.K. and Seidel M.A. 1998. Inland propagation of erosional escarpments and river profile evolution across the southeast Australia passive continental margin. In: Tinkler, K.J. and Wohl, E.E., eds., *Rivers Over Rock: Fluvial Process in Bedrock Channels*, Washington, DC: American Geophysical Union, pp. 189–296.

Wells, L.W. 1990. Holocene history of the El Nino phenomenon as recorded in flood sediments of northern coastal Peru. *Geology* 18: 1134–1137.

Wilcock, P.R. 1992. Flow competence: a criticism of a classic concept. *Earth Surface Processes and Landforms* 7: 289–298.

Williams, G.P. 1978. Bankfull discharge of rivers. *Water Resources Research* 14: 1141–1154.

Williams, G.P. 1983. Paleohydrological methods and some examples from Swedish fluvial environments. I. Cobble and boulder deposits. *Geografiska Annaler* 65A: 227–243.

Williams, G.P. 1984a. Paleohydrologic equations for rivers. In: Costa, J.E. and Fleisher, P.J., eds., *Developments and Applications of Geomorphology*, New York: Springer-Verlag, pp. 343–367.

Williams, G.P. 1984b. Paleohydrological methods and some examples from Swedish fluvial environments. II. River meanders. *Geografiska Annaler* 66A: 89–102.

Williams, G.P. 1988. Paleofluvial estimates from dimensions of former channels and meanders. In: Baker, V.R., Kochel, R.C. and Patton, P.C., eds., *Flood Geomorphology*, New York: John Wiley and Sons, pp. 321–334.

Wohl, E.E. 1988. Northern Australian paleofloods as paleoclimatic indicators, Ph.D. Dissertation, University of Arizona, Tucson, Arizona, 285 p.

Wolcott, J. and Church, M. 1991. Strategies for sampling spatially heterogeneous phenomena; the example of river gravels. *Journal of Sedimentary Petrology* 61: 534–543.

Wolman, M.G. 1967. A cycle of sedimentation and erosion in urban river channels. *Geografiska Annaler* 49A: 385–395.

3

Archaeology and Human Artefacts

ANTHONY G. BROWN[1], FRANÇOIS PETIT[2] AND ALLAN JAMES[3]
[1]*Department of Geography, School of Geography and Archaeology, University of Exeter, UK*
[2]*Institut de Géographie, Université de Liège, Sart Tilman, Belgium*
[3]*Department of Geography, University of South Carolina, USA*

3.1 INTRODUCTION

Geomorphology and archaeology have strong historical and methodological links. Indeed, the origins of both geomorphology and archaeology lie in 18th and 19th century geology. The sub-discipline of geoarchaeology, defined as the use of geological and geomorphological methods in archaeology, has a long history (Zeuner 1945), even if the term is relatively new and of North American origin. Geoarchaeology and its variants, archaeogeology (Renfrew 1976) and archaeological geology (sensu, Herz and Garrison 1998) all seek to answer archaeological problems using techniques from the Earth Sciences (Water 1992). The subject of this chapter is subtly different, as it is the use of archaeological evidence to answer questions concerning earth surface processes and history, i.e., geomorphology. Whilst the most obvious way in which archaeology can be a tool in geomorphology is by dating sedimentation or erosion and thereby establishing rates of flux, there are many more applications including the identification and reconstruction of forcing factors on the earth system (e.g., climate) and the history of human influences on earth surface processes.

Archaeology can date erosion or deposition at the 10^3–10^4-year timescale, and occasionally the temporal phasing of sites can be converted into a spatial phasing of erosional or depositional segments of the landscape providing rates of erosion or deposition through site formation and destruction processes (Schiffer 1987). Examples include the use of tells or house mounds for estimating erosion rates (Kirby and Kirkby 1976), the use of site distribution for erosional surveys (Thornes and Gilman 1983) or the use of artefacts and sites in the studies of river channel changes (Brown *et al.* 2001). A second value of archaeological data is its potential to provide information concerning processes and environmental change. This approach has a venerable history in North America and particularly the American South West (Antevs 1935, 1955, Water 1992). Environmental change, and particularly denudation history, in the Mediterranean has also been approached using archaeological data (Vita-Finzi 1969) whilst only more recently have archaeological tools been extensively used in northwest Europe and the rest of the world. Environmental histories often reveal the role of humans in modifying the physical environment, although this is not inevitable, as climate can be the overwhelming factor.

3.2 GENERAL CONSIDERATIONS IN USING ARCHAEOLOGICAL EVIDENCE IN GEOMORPHOLOGY

Artefacts can give geomorphologists a datum point and sometimes a date either imprecise (e.g., Mesolithic) or remarkably precise—sometimes even a calendar date (e.g., from inscriptions). However, the value of the datum and date is dependent upon the origin of the artefact—if in-situ it may record a landsurface, whereas if transported a depositional event. Clearly, the transport history of an artefact is a function of its mass and origin; small pottery shards are easily transported by rivers whereas blockwork from stone bridge piers rarely travels far. It must also be remembered that the discovery and use of such data is often a function of system behaviour such as the retreat of a river bank section or gully incision. As archaeological evidence is commonly found in floodplain sediments, tools used for analysing floodplain sedimentation (Chapter 2) are applicable. Organic artefacts may also be found, either preserved through desiccation (e.g., from human bodies to seeds) or waterlogging preventing anaerobic decay. Examples include textiles and basketwork such as those associated with fishing at Noyen-sur-Seine, France (Mordant and Mordant 1992) and Grimsby in the Humber wetlands, UK (Van der Noort, pers. comm.), and wooden artefacts such as bowls, tools and figurines (Coles and Coles 1989). Environmental materials such as timbers may be dated by dendrochronology. Animal and human bones may also be preserved as in the Bronze Age log jam in the river Trent, UK (Howard *et al.* 2000). There is clearly a preservation bias here and this greatly reduces the occurrence of such data in seasonally variable and tropical climates.

The most common and most valuable source of archaeological data is the exposure or section face. Visual searching can in the case of coins or other metal objects be augmented by the use of a metal detector. Finds should be located precisely on field drawings and depth logged. Artefacts on the ground surface, residual in archaeological terminology, have little dating value. Spot dating of pottery can be difficult or impossible and it should be remembered that much archaeological dating relies on associations between artefacts and is therefore not completely reliable.

3.3 ARCHAEOLOGICAL TOOLS

A summary of the most common archaeological tools is given in Table 3.1 along with a summary of their potential advantages and disadvantages. A more detailed discussion including examples is given below.

Hearths and Lithics

In many sedimentary sequences in both the Old and New Worlds hearths, charcoal from hearths and stone

Table 3.1 A summary of archaeological tools in geomorphology

Archaeological data	Geomorphic use	Advantages	Disadvantages
Pottery	Dating	Can be precise and the only dating evidence available	Can greatly overestimate the age or be unreliable
	Tracing	Can indicate potential sediment sources	More easily transported than the equivalent sediment size; may have been reworked
Coins	Dating	Can be precise and the only dating evidence available	May have been reworked and or in circulation for many years
Hearths	Dating	Can provide relatively pure carbon for ^{14}C dating; can provide information on the prevailing ecology	Must be in-situ otherwise the date will be overestimated
	Alluviation rate	Provides a datum for calculation of the accumulation rate	Must be in-situ otherwise the date will be overestimated
Bone	Dating	Extraction of collagen provides C for ^{14}C dating	May be transported and can be contaminated
Earthworks	Geomorphic stability	Can indicate the age and conditions of a slope and so limit estimates of landscape change	The slope evolution will be affected by the earthwork and location cannot be regarded as random
	Erosion rate estimation	Degradation or gullying of the feature can provide an estimation of the erosion rate	The age and initial height must be known and the assumption made that the artificial slope is in some way representative of natural slopes
Middens	Shoreline location and sea level estimation	Middens composed of discarded Mollusca can indicate the location of past shorelines of lakes or the sea and can be used to constrain height estimations	Must be accurately dated and this can be difficult; have not been subjected to neotectonic effects

(*Continues*)

Table 3.1 (*Continued*)

Archaeological data	Geomorphic use	Advantages	Disadvantages
Land divisions: banks and walls	Erosion and colluviation rate estimation	Banks and walls can act as sediment traps, the depth of which can provide a minimum estimate of within-field erosion and translocation	The wall or bank pre-dated the accumulation of sediment; the date of a wall or bank is often unknown
Stonework	Weathering rate estimation	The depth and character of stone weathering can be estimated in relation to areas of surface protection or metallic components	Cut stone weathering is related to natural weathering rates; the age of the stonework must be known; there has been no past re-cutting of faces or cleaning
Structures: buildings	Ground surface	Most buildings can be related to a past ground surface level from steps, doorways, etc., if buried this can indicate the accumulation rate if now elevated an erosion rate	The date of the building must be known and particularly the period that relates to the ground surface. Some structures can be deliberately constructed below ground level and others above ground level
Structures: bridges	Channel and ground level information	Will indicate a past channel location, can constrain estimates of channel width, depth and even discharge and can give the contemporary floodplain height	The bridge has not determined the local geomorphic rates; there was no other channel on the floodplain; bridge locations are representative of river reaches
Structures: quays, wharves, jetties	Shoreline location and sea level estimation	These structures can indicate past shorelines and provide obvious evidence for past relative sea levels	They can be dated; there relationship to sea level (or the tidal) is known or can be estimated; neotectonic effects are recognized
Structures: wooden	Dating	Then they may be able to provide a calendrical date by dendrochronological dating and this can allow the phasing of sedimentation with unparalleled temporal precision	It requires suitable species (e.g., oak) and wood of sufficient diameter (15 cm min^{-1}), wood can be reworked and re-used
	Palaeoclimatic reconstruction	Using the variation in ring width it is possible to reconstruct past growing season conditions particularly precipitation	Only possible for a few species (e.g., oak) and difficult to calibrate unless part of the chronology can be related to a documentary series (e.g., precipitation or streamflow record)
Structures: wells, cisterns, aqueducts, drains, etc.	Palaeohydrological estimation	Can provide indications of past groundwater levels and possibly surface water discharge	Many assumptions: both chronological and hydrological, for cisterns, aqueducts, drains, etc., it has been assumed that the design Q was an accurate estimation of the prevailing hydrological regime (see text)

tools can be used as stratigraphic markers. Although charcoal is chemically ideal for conventional radiocarbon dating, the most common problem is reworking so the charcoal must be in-situ. Flints can be dated using electron spin resonance but it is rarely of sufficient precision for geomorphological studies. An example of a study, which uses charcoal dating of hearths, is Baker *et al.*'s (1985) work on slack-water deposits in Northern Territories, Australia. Indeed in most studies of slack-water deposits charcoal is the major material dated. It has also been extensively used in the determination of alluvial chronologies of arroyos in the SW of the USA (Water and Nordt 1995).

Pottery and Small Artefacts

Pottery is often the most common artefact in sediments after ceramic civilisations appear. If large enough and of a diagnostic type it can be dated, although, there are problems with 'spot dating' as it can be unreliable. When interpreting occasional pottery shards in a sediment body some thought should be given as to their provenance and therefore their possible antiquity prior to incorporation into the sediment body. Abraded shards that have been transported downstream may well have been eroded from older sediments and can only ever yield a maximum possible date of the sediment. However, in floodplain and colluvial sequences unabraded pottery is likely to have been incorporated into the sediment directly and it less likely to significantly predate incorporation. There also exists a serious sampling problem with pottery. One or two shards can easily give a false impression of antiquity, when a systematic search can provide a wide range of pottery ages (Brown, in press a). In this case, the youngest pottery can be used to provide a maximum age of deposition. If pottery cannot be dated stylistically then it can be dated using thermo-luminescence (see Chapter 6 for a description of this method). A pottery date derived in this way can then be compared to direct sediment dating using optically stimulated luminescence (Brown 1997).

Structures

Structures record a stable landsurface, or in the case of quays and dock a relative sea level. House floors or foundations can provide a datum in an aggrading sedimentary sequence. Bridges are particularly useful since they record both ground level, the location of a channel and in some cases some idea of the size of the channel. This is exploited in one of the case studies described later in this chapter. In the Old World, Roman bridges are particularly common and can be used to estimate the post-Roman overbank deposition (Figure 3.1). One cautionary point is that some bridges, those with stone piers in or at the edge of the channel, do influence channel processes and so may lead to a biased view of river behaviour. Quays and docks can be related to a sea level, and (whereas commonly the case in the Mediterranean) they are now submerged or elevated, then relative sea level must have changed. A good example of this is the 600 m long jetty of the Roman port of Leptiminus in Tunisia, which now lies 0.6 m below sea level due to neotectonic activity (Brown, in press b). The inland preservation of old quays, docks and ports is some of the strongest evidence of high sedimentation rates in the Classical and post-Classical Mediterranean. Structures may also provide evidence of geological events with geomorphological implications. The most obvious is earthquake damage to ancient buildings. In some cases the earthquake history can be linked to changes in drainage patterns (Jackson *et al.* 1996) and other geomorphic events such as landslides. Indeed both neotectonic studies in the Old and New Worlds have frequently used archaeological evidence (Vita-Finzi 1988, Keller and Pinter 1996).

Palaeohydrological Data from Archaeology

Since many archaeological structures were originally designed to carry or store water under certain circumstances, their dimensions can be used to estimate some parameter of the past hydrological regime. Examples of such structures include dams, aqueducts, water supply structures (leets), wells, cisterns, connecting tunnels, hydraulic-mining channels and drainage ditches. There are, however, significant methodological problems and convincing studies are rare. Assuming the structure can be accurately dated, there remain several assumptions:

1. The function of the structure is in no doubt.
2. The design capacity is a reasonable indicator of prevailing hydrological conditions.
3. The full instantaneously functioning system is known (e.g., the number of functioning channels or pipes).

Dams have been used to estimate rainfall in arid regions associated with floodwater farming (Gilbertson 1986) and Gale and Hunt (1986) attempted to use floodwater farming structures in Libya to reconstruct water supply during the Roman period. They used the Darcy–Weisbach equation and an expression for roughness in turbulent flow in rough channels. Aqueducts, leets and hydraulic-mining channels can be used to calculate discharge and thereby estimate

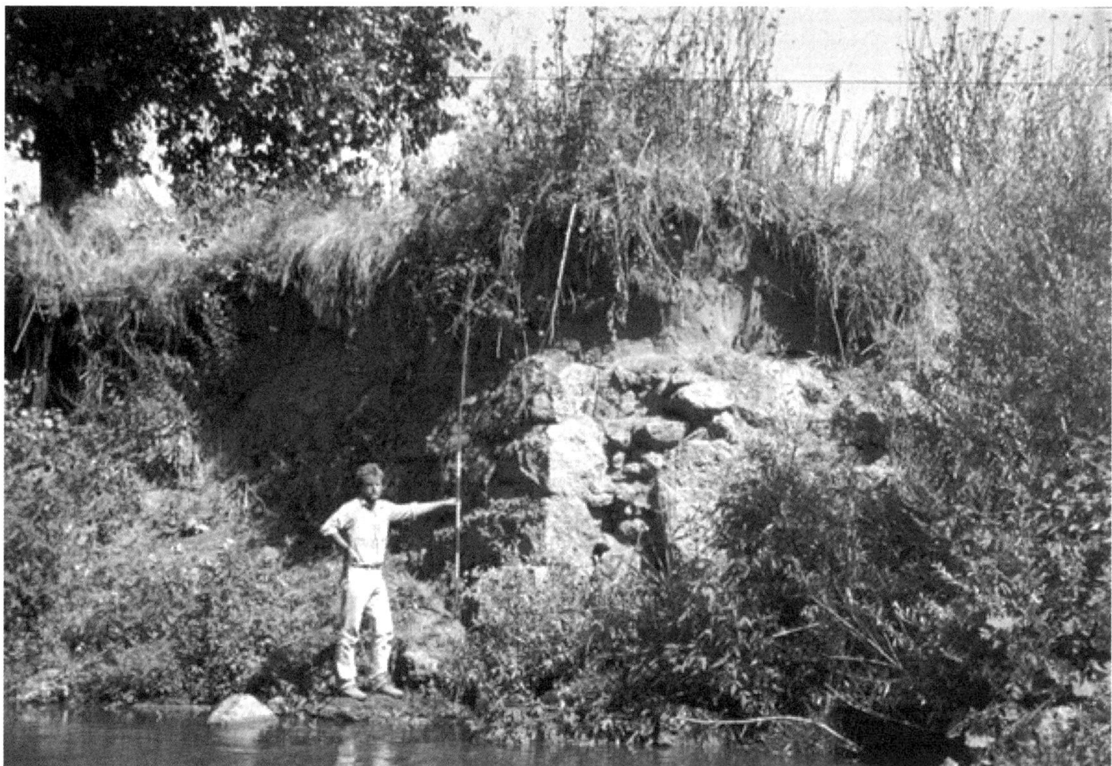

Figure 3.1 The remains of a Roman bridge across the river Treia in Central Italy

minimum precipitation. Drainage and artificial channels used in mineral processing should also reflect the prevailing hydrological conditions and water supply capacities of the system. Bradley (1990) attempted to use the dimensions and slope of tin streaming (alluvial mining) channels to estimate Late Medieval stream powers in SW England and he used this estimate to support the observation that downstream floodplain and channel sediments were relatively enriched in cassiterite content of the >63 μ particle size fraction. Masonry drainage structures can also provide palaeohydrological data, an example being Ortloff and Crouch's (1998) hydraulic analysis of a complex outlet structure in the Hellenistic city of Priene in modern Turkey that provided evidence for a lower-bound estimate of steady state water supply to the city. Shallow wells have considerable potential to indicate past watertable height (or a minimum altitude), but a regionally based realisation of this potential has yet to be attempted. One of the values of using archaeological data is that it may provide estimates of the most variable and least accurately modelled component of global climate models, namely, precipitation.

Artefacts and Fluvial Processes

The characteristics of archaeological tracers and the deposits in which they occur may indicate important aspects of their source, mode of transport and age. Careful observations should be made to determine the condition of artefacts, whether they occur in primary positions of human deposition or in secondary deposits and the geomorphic setting. For example, the amount of abrasion on individual artefacts may indicate distance from their source. Downstream abrasion increases downstream as has been shown with modern facsimiles of flint hand axes (Harding *et al.* 1987, Macklin 1995). Concentrations of tracer materials generally decrease with distance downstream due to dilution by barren sediment from local storage sites and from tributaries.

Mining Debris as Tracers

The link between cultural activities and sedimentation is particularly well expressed by mining. Mining sediment not only provides evidence of fluvial processes, but also provides prime examples of fluvial responses to human alterations of the environment. All extractive mining and mineral processing produces some waste, which is either separated using rivers, deliberately added to rivers, or eventually enters rivers via natural geomorphic processes. This line of geomorphological research can be traced back to Gilbert's (1917) classic study of mining in the Sierra Nevada. Several workers have distinguished between two types of sediment transport: *active transformation* where the fluvial system is transformed by the introduced waste (e.g., Gilbert's study) and *passive dispersal* where sediment markers are passed downstream mixed with the natural sediment without causing a substantial change in channel morphology (Lewin and Macklin 1986). While useful, this distinction does pose certain problems as it is fundamentally a function of the degree of fluvial change and the sensitivity of our detection of that change. Secondly, mining is often accompanied by other land-use changes, often, agricultural intensification in order to feed a growing local population, and so sediment loads may increase indirectly and from other sources. Some of these questions are discussed in the case studies presented below.

Mining sediment has often been studied because it forms distinctive stratigraphic units that can be recognised throughout a river course, dated and related to specific cultures or activities. Mining often amplifies background sediment loads by more than an order of magnitude as was shown in a basin-wide analysis by Gilbert (1917) and was demonstrated in a paired-watershed study of strip mining in Kentucky (Collier and Musser 1964, Meade *et al.* 1990). Several studies have documented severe alluvial sedimentation and channel morphologic changes below mines in Great Britain (Lewin *et al.* 1977, Lewin and Macklin 1987) and North America (Gilbert 1917, Graf 1979, James 1989, Hilmes and Wohl 1995).

Mining sediment is often rich in metals (Reece *et al.* 1978, Leenaers *et al.* 1988). The distinct signature of heavy metals associated with many mines often allows a local metal stratigraphy to be developed downstream of mines. For example, Knox (1987) was able to correlate floodplain strata with elevated concentrations of lead and zinc with periods of mining in southwest Wisconsin. Wolfenden and Lewin (1977) and Graf *et al.* (1991) developed similar chronologies for rivers in Wales and Arizona, respectively. Sediment sampling for evaluation of metals requires an understanding of fluvial transport processes and depositional environments. Heavy metals are often concentrated in the fine fraction of sediment due to sorting processes of the denser metalliferous particles; i.e., the principle of *hydrodynamic equivalency* (Rubey 1938). In mining sediment, however, the presence of multiple populations, including coarse metal particles, fine metal particles, and coatings on or inclusions in particles of various sizes and densities, may complicate this relationship. The importance of particle coatings varies with the metals being sampled and ephemeral environmental factors such as pH, which encourage speciation into oxide, hydrous oxide and other phases. Most studies perform chemical analysis on a sand fraction isolated by sieving. Sampling and sieving should be performed with a minimum use of metal tools to avoid contamination. In a comparison of laboratory methods, Matei *et al.* (1993) found that metal concentrations were homogeneous in the very fine sand grade, that splitting samples into quarters was not necessary, and that crushing followed by sieving should not be done prior to chemical analysis.

Changes in metal concentrations below a source are often modelled as a simple downstream logarithmic decay function (Wertz 1949, Lewin *et al.* 1977, Wolfenden and Lewin 1978). Marcus (1987) showed that the downstream decay in copper was largely due to dilution by sediment from non-mining tributaries. Graf (1994) described the complexities involved in mapping downstream changes in plutonium and demonstrated a general decrease in concentrations downstream in tributary canyons to the Rio Grande (Figure 3.2). At the channel-reach scale, metal

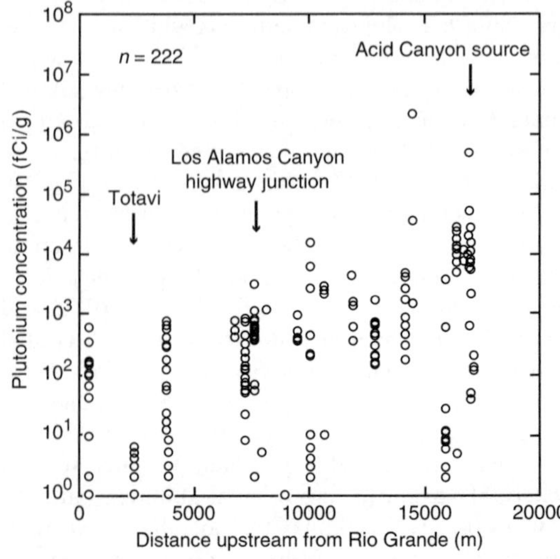

Figure 3.2 The along-stream distribution of plutonium in Acid, Pueblo and Los Alamos canyons. Reproduced from Graf (1994)

concentrations may vary greatly with geomorphic position. For example, Ladd *et al.* (1998) sampled 12 metals in seven morphologic units of a cobble-bed stream in Montana, USA. They found that concentrations varied between units; e.g., eddy drop zones and attached bars had high concentrations while low- and high-gradient riffles and glides had low concentrations.

3.4 USING ARCHAEOLOGICAL DATA: CASE STUDIES

In this part of the chapter, a series of case studies are used to illustrate how archaeological data can be used to answer geomorphological questions. The examples are taken from different climatic regions and cover different timescales.

Case Study 1: Fluvial Reconstruction From Bridge Structures on the River Trent UK

Archaeological finds can provide both the opportunity and raison d'etre for the reconstruction of past geomorphic and hydraulic conditions. The exploitation of aggregate from large areas of the Middle Trent floodplain in Central England has allowed the excavation and recording of hundreds of archaeological finds including, human and animal skeletal remains, log-boats, fish weirs, anchor weights, revetments, bridges and a mill. Together with 'natural' finds such as tree-trunks, organic palaeochannel sediments and flood debris, this has allowed a geomorphological reconstruction of the Holocene evolution of the Middle Trent floodplain based on both an archaeological and radiocarbon chronology. From Hemington and surrounding investigations, a partial fluvial history is be postulated in Table 3.2.

The Middle Trent has been characterised by channel change throughout the Holocene. In the absence of a high slope (in fact a low slope, Hemington–Sawley average 0.0006 mm^{-1}), this is most likely due to a rapid downstream increase in discharge from the four major tributaries that enter the main channel in under 40 km, the flood characteristics of two of these tributaries and an abundant supply of unconsolidated or cemented sandy gravels provided by the low and wide Devensian terraces. The unusual width of the Late Devensian (IO stages 3 and 2) gravel terraces here is proximity to the Devensian ice margin, which was just less than 30 km upstream. The Early to Middle Holocene data is largely derived from palaeochannels whilst the Late Holocene period is known in most detail due to the occurrence of buried bridges, a mill and weir and abundant other evidence of channel change. Several geomorphologists have attempted to use archaeological structures to quantify geomorphological parameters. This is based on the rationalist assumption that structures such as bridges, weirs, etc., were built to contain a certain flow, functioned by containing a run of flows. In some cases, destruction of the structure by a flood can also be used to estimate the magnitude of the event. Geomorphological studies of three Medieval bridges buried under gravels in the floodplain of the Middle Trent have employed several of these techniques including simple slope area calculations of discharge from channel dimensions, HEC-II flow modelling and palaeohydraulic calculations based upon transported and non-transported clasts (Brown and Salisbury, in press). The sedimentology, the archaeology and the pattern of palaeochannel fragments suggest that the reach was highly unstable during the Holocene and especially during the last 1000 years. The sedimentology suggests relatively shallow unstable channels eroding and depositing sand and gravel. The predominant sedimentological features, horizontal and low angle bedding with shallow channels suggests a locally braided river and this is in agreement with the low sinuosity of the channels during the Early Medieval period. However, the preservation of old palaeochannels and archaeological features (such as the mill and bridges) and the avulsion of the channel sometime between the 15th and 16th/17th centuries suggest river underwent a braided and anastomosing phase before returning to a single-channel meandering form. The typical form of both braided and anastomosing reaches has been used in a generalised geomorphological model of the reach from 8th to 19th centuries (Figure 3.3). It is impossible to accommodate both the archaeology and the palaeochannels unless the reach has at least two (preferably three) functioning channels, and the evidence suggests that there was a migration of channels eastwards leaving a Prehistoric meander core, but that there remained a functioning westerly channel (even if small) until an avulsion sometime between the 15th and 16th/17th centuries led to the abandonment of the easterly channel (old Trent) and conversion of the westerly channel into the only permanent channel in the reach. Associated with this change in channel numbers is a drop in main channel sinuosity as would be expected during a period of braiding and high bedload movement through the reach. This model therefore suggests that this reach of the Trent went from being a meandering single-channel

Table 3.2 A tentative chronology of the channel change Hemington–Sawley reach of the Middle Trent derived largely from geoarchaeological studies

Period	Channel type	Sites	Notes
Windermere interstadial	Meandering	*Hemington*, basal channel peat (Brown *et al.* 2001)	Down-cutting into terraces and bedrock
Loch Lomond Readvance	Braided	*Hemington* basal gravels (Brown *et al.* 2001), *Church Wilne* (Coope and Jones 1977, Jones *et al.* 1977a,b), *Attenborough* (BGS, Brown unpub.)	Deposition of basal 'Devensian' gravels and intense frost action creating polygons
Mesolithic	Low sinuosity, possibly multiple-channel (anastomosing)	*Shardlow*-stocking palaeochannels (Challis 1992, Knight and Howard 1994), *Repton* (Greenwood and Large 1992), *A6 Derby By-pass* (Brown unpub.), *Attenborough* (BGS, Brown unpub.)	Some avulsion leaving linear palaeochannels, which are often over a kilometre from the present channel
Neolithic	Multiple channel-braided, low sinuosity	*Hemington* (Clay and Salisbury 1990), *Colwick* (Salisbury *et al.* 1984), *Langford* and *Besthorpe* (Knight and Howard 1994)	Fish weirs and black oaks in small-shallow channels
Bronze Age	Meandering?	*Colwick* (Salisbury *et al.* 1984), *Collingham* (Greenwood, pers. comm.)	Little evidence except at Colwick and downstream
Iron Age and Roman	Meandering, sinuous, point-bar sediments	*Holme Pierrepoint* (Cummins and Rundell 1969)	Palaeochannel associated with settlement at Sawley and evidence of settlement on the terraces, excavated site RB site at Breaston (Todd 1973)
6th–9th centuries AD	Meandering, highly sinuous	*Hemington* (Ellis and Brown 1999)	Large palaeochannel dated by radiocarbon and palaeo-magnetics
11th–13th centuries AD	Braided, unstable	*Hemington, Colwick* (Salisbury *et al.* 1984), *Sawley* palaeochannel	Channels associated with the bridges
17th–19th centuries AD	Anastomosing to single-channel, moderate to low sinuosity	*Hemington*	Avulsion sometime between 15th–17th centuries from the old Trent to the modern Trent
19th–21st centuries AD	Meandering, stabilised	Map and documentary evidence	Embanked, partially regulated and engineered, construction of the Trent and Mersey canal, Sawley cut and Beeston canal

river in the Early Medieval period (6th–9th centuries) to a braided river in the 10th–11th centuries back to a single-thread meandering river by the end of the 17th century probably passing through an early wandering-gravel bed phase and a later transitional anastomosing phase. This is a classic example of medium to long-term metamorphosis of a river channel and floodplain. The processes responsible for this

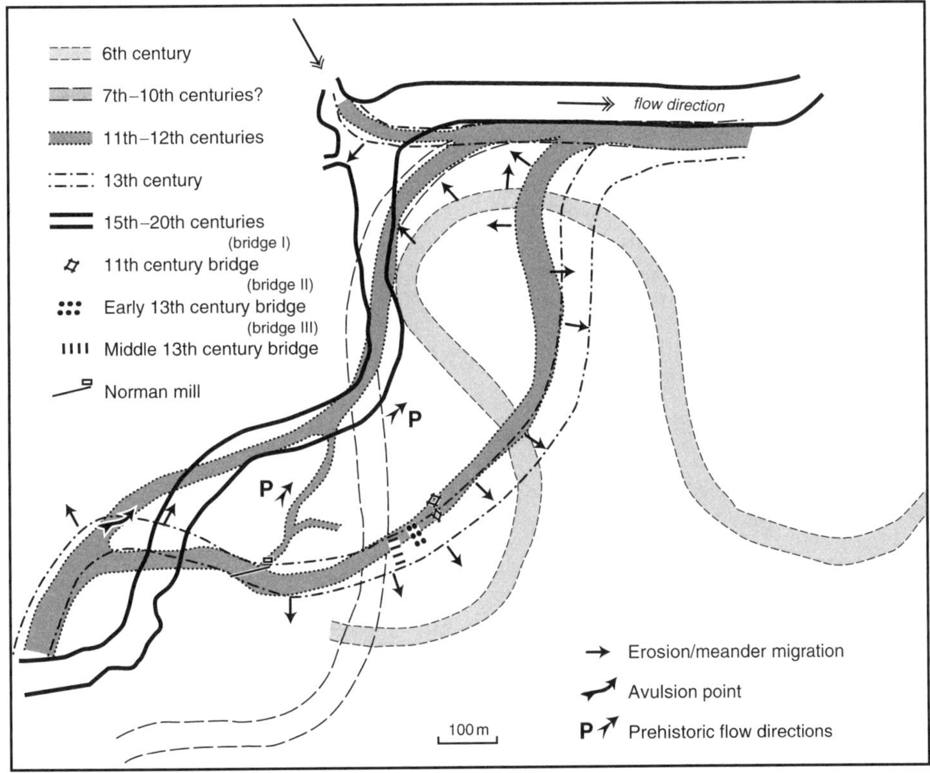

Figure 3.3 A model of channel change in the river Trent in the Hemington reach over the last 1000 years. The location and dating of each channel position is derived from archaeological data includes the Medieval bridges, fish weirs, a mill and anchor stones as well as sedimentological data (from Brown *et al.* 2001)

change are large floods, particularly those generated in the Pennine uplands and an increase in the transport of bedload into and through the reach. The trigger for this remains unclear but may have been floods, probably rain on snow that occurred during the 11th–13th centuries, a period which has been labelled the 'crusader cold period' and is part of the 'Late Medieval climatic deterioration'. The cycle of channel change is clearly related to abundant bedload supply and high sediment transport rates, and can be viewed as channels adjustment to a pulse of sediment which was through channel metamorphosis deposited into floodplain storage. The nature of the reach (shallow channels) was taken advantage of for the construction of bridges, the builders being presumable unaware of the transitory nature of the channel conditions or constrained by the geography of the route. The climate changes of the Late Medieval period and early modern period are now considered to have probably been the most dramatic in the Holocene (Rumsby and Macklin 1996) and the Middle Trent is particularly sensitive to changes in hydrometeorology. This is not to say that there were no human impacts on these events, as deforested uplands are far more likely to produce large rain on snow events due to the increased depth of the snowpack that can accumulate over grass as opposed to tree cover. Likewise there is little doubt that runoff generation times have been decreased, and therefore peak flows increased, by drainage and land-use change (Higgs 1987).

Bridges provide the most obvious evidence of channel change in either the case of bridges over palaeochannels or over contemporary channels. There is of course a conceptual problem with channel evidence from bridges, first because they may not be randomly located along channels, most replaced fords and in many cases were clearly located where the channel and floodplain was constricted and a terrace or high bank could be used in construction. This will also depend upon the state of bridge technology, early bridges with restricted single spans restricted to narrow or divided channels and later bridges requiring solid foundations on terraces or bedrock. However, the geographical location of a bridge depends upon the population pattern with routes linking towns or villages by the shortest or most practical route. So bridge *site* and geomorphic history is fundamentally geomorphologically controlled (or unavoidable) whilst bridge *location* is generally dictated by routes linking centres of population, at least in lowlands. It has also been argued that bridges prevent channel change, whilst this is certainly true in the case of lateral migration where it depends upon continued capital investment in the structure, it is not true where avulsion is a major cause of channel change.

Case Study 2: Slags, Bedload and Hydraulic Sorting in Belgium

Bedload progression has been evaluated in rivers using slags coming from old ironworks settled in the south Ardennes valleys at the early 17th century (Sluse and Petit 1998). During these periods, the slags were disposed of into the rivers. They are still being transported even if the factories have been closed for a considerable period. The slags are easily recognisable, thanks to their visual characteristics. Their average density is 2.1. The slags have been sampled in 19 riffles situated along the river Rulles, in its tributaries where ironworks have been installed and downstream, in the river Semois into which the river Rulles flows. Figure 3.4A shows the trend of the size of the 10 biggest particles measured by the *b*-axis (corresponding to the D_{90}), along the river Rulles course, using a cumulative distance from the most downstream of the iron factories (explaining the decrease of the size in sites 1–3). The slags brought down by tributaries explain the increase of the slag size in the Rulles (examples: site 4, and sites 7–10). Slags have also been found in the Semois (site 15) but none 4 km downstream of this last site.

A relationship is drawn between the slag size and the distance from the ironworks where these slags have been discarded into the river (Figure 3.4B). This curve shows a rough decrease in particle size, which fall from 80 to 20–30 mm in diameter in less than 5 km; afterwards the slag size only decreases slowly. The slags refining in the first kilometres downstream the ironworks does not result from modifications in hydraulic characteristics of the river or a diminution of its competence. Indeed, the unit stream powers remain identical along its course. This slag size reduction does not result from abrasion, neither from granular disintegration nor from gelifraction effects (Sluse and Petit 1998). It results from a hydraulic sorting occurring in the few kilometres downstream of the input sites.

The slag size, which, after 5 km, remains almost constant regardless of the distance, represents the actual competence of the river (the particle size transported along substantial distances and evacuated out the catchment). The particle size (12 mm maximum with regards to equivalent diameters using a density of 2.65) is relatively small, but is justified by the low values of the unit stream powers (25–30 W m^{-2} at the bankfull discharge). Higher competence causes the hydraulic sorting, but this is exerted only locally and during intense events.

Several slags (10–14 mm in diameter or 9–12 mm using equivalent diameters) have been found 12.5 km downstream the closer iron factory, which produces a bedload wave progression of 3.3 km century^{-1} (Figure 3.4A). The most upstream site in the river Semois where no slag has been found shows that the bedload wave progression is less than 17 km since the middle of the 17th century (less than 3.9 km century^{-1}). Such progression is low in comparison with others studies (between 10 and 20 km century^{-1}) although most of these have been mountain rivers with strong energy (Tricart and Vogt 1966, Salvador 1991).

Case Study 3: Artefactual Evidence of Floodplain Deposition and Erosion in Belgium

The rate of floodplains formation has been estimated in Ardennes rivers using stratigraphical markers identified by Henrotay (1973). These consist of scoria (smaller than 105 μ) produced by the previous metal industry set up in Ardennes valleys from middle of the 13th century. The debris from these factories was dumped into the rivers so that the presence of microscopic scoria in alluvial deposits affirms that the floodplain was built after the 13th century. As shown in Figure 3.5A, dealing with the river Amblève, the whole floodplain contains microscopic scoria deposited after the 13th century. The thickness of recent flood silt exceeds generally 1 m and reaches frequently 2 m which gives a rate of accumulation of 28 cm century^{-1}. Henrotay has prospected different rivers of the Ourthe Basin and the river Meuse downstream Liege (Table 3.3). The rate of sedimentation exceeds generally 20 cm century^{-1}. Everywhere the thickness of silt deposited since 13th century is greater than the layer of old silt deposited prior to scoria depos-

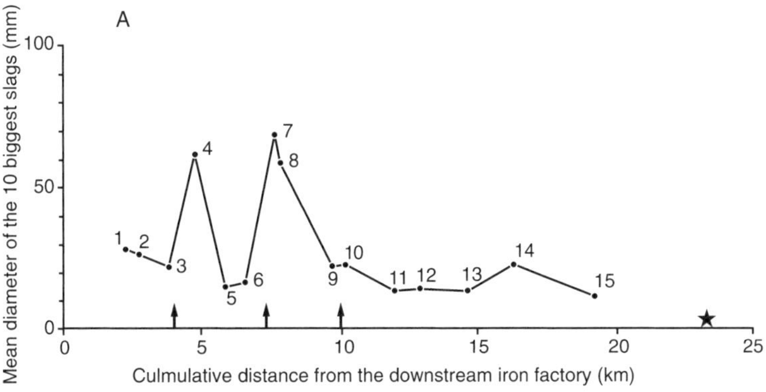

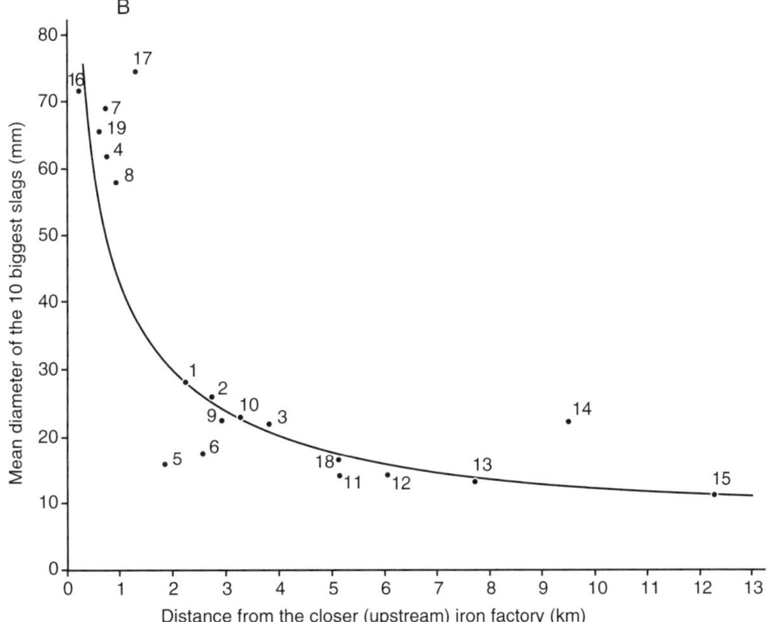

Figure 3.4 (A) Trend of the mean diameter of the 10 biggest slags measured by the *b*-axis, along the Rulles river and the Semois River, using a cumulative distance from the most downstream iron foundry located on the Rulles. The arrows on the *x*-axis indicate the junctions of the tributaries where ironworks were located. The star symbol on the *x*-axis indicated the upstream limit of slag deposition in the Semois. (B) The diameter of the 10 biggest slags measured by their *b*-axes in relation to distance from the closest iron foundry. Modified from Sluse and Petit (1998) (reproduced with permission from Les Presses de l'universite de Montreal)

ition. Human activities (deforestation and expansion of the area under tillage) have probably played a dominant role in the silt accumulations in the valleys. The same technique has been used in the south of the Ardenne by Sluse (1996). The rates of sedimentation are slightly less than in the north of the Ardenne (Table 3.3). Two reasons explain this difference. Deforestation in the south-Ardenne catchments has been less important and in the present time the land use of these watersheds is dominated by forests and pastures, so that the soil erosion is less than in the north part of the Ardenne. Furthermore, the less deposits are less thicker in the south-Ardenne and there is therefore less material to erode.

The microscoriae allow equally to evaluate the importance of lateral erosion of these rivers (Henrotay

Table 3.3 Sedimentation and erosion rates determined using microscopic scoria deposited in floodplain sediments

River	Catchment area (km^2)	Date of ironworks	Sedimentation rate (cm century^{-1})	Lateral erosion rate (m century^{-1})
North Ardenne (from Henrotay 1973)				
Amblève	1044	1250	23.5	14.6
Ourthe	1597	1250	28–33	6.3
Ourthe	2691	1250	28	–
Somme	38	1400	8–18	3.9
Meuse	20 802	1250	21	42
South Ardenne (from Sluse 1996)				
Rulles	96	1540	14.4	5.5
Rulles	134	1540	9.1 (6)	4.4
Mellier	63	1620	19.6 (5)	5.4
Rulles	220	1540	24.9 (5)	18.0
Semois	378	1540	19.8 (5)	33.0

1973). As shown in Figure 3.5A, the silt contains microscopic scoria and was thus deposited after the middle of the 13th century, along all the width of the floodplain. Silt without scoria (before the 13th century) has been eroded which implies that from that time, the river has swept away, at least once, all its floodplain across a width of 100 m. This gives valuable indications of lateral erosion. In this case, it achieves an average rate close to 15 m century^{-1}. In contrast to the river Amblève, the Ourthe has not systematically swept the totality of its floodplain since one can find old silt on which rests the recent silt (Figure 3.5B). This, nevertheless, allows us to observe a lateral erosion of at least 45 m. The rates of lateral erosion are similar in south-Ardenne rivers (Table 3.3). Using this method, it is clear that the lateral erosion can be underestimated because the river may have passed the zone where the old silt was eroded several times and this may explain low lateral erosion values. However, the rates agree with measurements taken from old maps (Petit 1995).

Case Study 4: Metal Mining and Fluvial Response: in the Old and New Worlds

Alluvial tin mining produces large amounts of sediment, which is directly input to rivers along with an increase in competent flows. Tin mining also has a long history, since it is one of the constituents of bronze and has been mined in Europe since the beginning of the so-called Bronze Age (third millennium BC). Both archaeologists and geomorphologists have a shared interest in the period before written records; the archaeologist in using sediments to search for pre-Medieval tin mining and geomorphologists in both dating alluvial deposits and understanding river behaviour at the 10^3 years timescale. A geochemical survey of rivers draining Dartmoor, SW England, was undertaken by Thorndycraft *et al.* (1999) in order to address both these questions. In this case, archaeological evidence of pre-Medieval tin mining is unlikely due to the almost complete reworking of any earlier deposits by Late and post-Medieval tin mining and streaming. Floodplain sedimentary successions, that had not themselves been mined, but are downstream of known areas of tin streaming, were found to retain a geochemical record of the mining activities because the early tin streaming released large quantities of mine waste tailings. Radiocarbon dating of these sequences has shown an excellent match with the documentary record confirming a first phase of streaming commencing in the 12–13th centuries AD, reaching a maximum in the 16th century, and a later phase in the 19th and early 20th centuries AD (Figure 3.6). A combination of XRF on particle size fractionated sediment and SEM/EDS studies of density separated samples allowed the geochemical characterisation of, and distinction between, streaming waste and naturally tin-enhancement sediments. In the Avon, Teign and Erme Valleys there is considerable overbank sediment aggradation coupled with the tin enhancement and this was probably associated with changes in channel pattern and morphology.

A more recent example of the use of archaeological/historical data in fluvial geomorphology is James' (1989) study of hydraulic gold mining sediments in the Bear River, California, USA. In the lower Bear

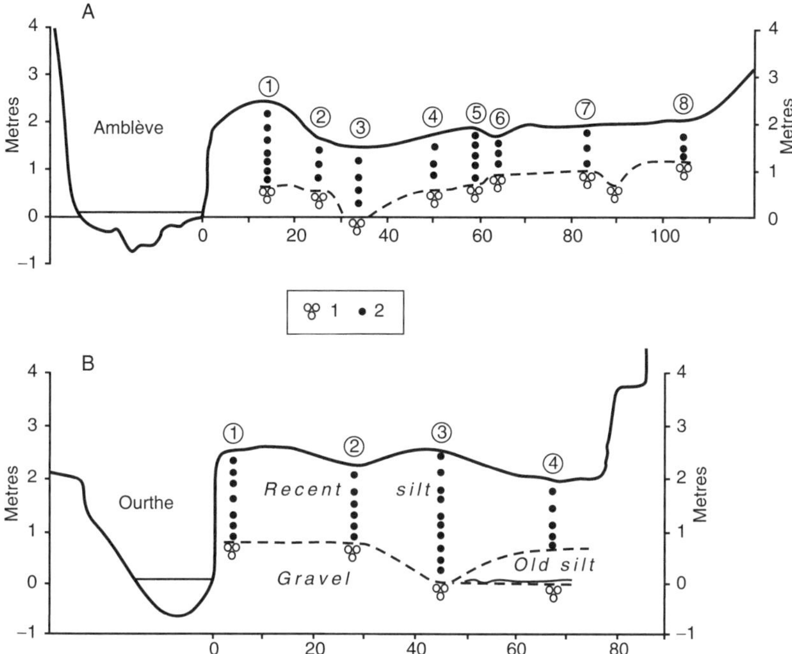

Figure 3.5 (A) Transverse profile of the Amblève River and the Ourthe River (B) with the presence of microscopic scoria in the floodplain fill, the depths of which provide an estimation of the sedimentation rates since the 13th century. Key: (1) gravel, (2) silt with scoria. ①–⑧: cores

Basin subsurface coring indicated that about 106 million m^3 of mining sediment remained stored 100 years after the cessation of gold mining. This estimate was more than double previous estimates and indicated that over 90% of the lower basin deposits remain in storage. Both topographic and historical evidence was used to illustrate the continued reworking of mining sediment as relatively frequent flows are competent to move channel-bed material derived from mining sediment. As sediment loads are still greater than pre-mining value in contrast to Gilbert's (1917) symmetrical wave model of geomorphic response under which there is a rapid return to pre-mining values; this suggests that the empirical foundation of the symmetrical wave model is biased. Channel incision and hence sustained erosion and deposition in the Bear River has been promoted by several factors in addition to decreased sediment supply, probably a function of catchment and valley geomorphic topography and geological conditions.

Another clear example of active transformation and the persistence of anthropogenic sediment in this case associated with tin mining is Knighton's (1989, 1991) work on the Ringarooma Basin in Tasmania. Mining in the basin lasted for over 100 years from 1875 to 1982 during that time 40 million m^3 of sediment were added to the river. The result was channel metamorphosis with bed aggradation, an increase in width where the channel was not confined and the development of a multiple channel pattern. Only now is degradation in the upstream reaches returning the river to something like its pre-mining condition. Similar results have come from studies of the fluvial response to lead mining in upland Britain and in particular the combined effects of increased sediment supply and climate change in the form of perturbations in the flood frequency/magnitude (Macklin, *et. al* 1992, Hudson-Edwards *et al.* 1999).

Both the archaeological and historical studies of mining sediments in the Old and New Worlds lead to two geomorphic conclusions. First, the Gilbert symmetrical wave model and river response itself is a function of basin conditions including basin topography, channel pattern, and long-term geomorphic trends such as neotectonically induced incision/

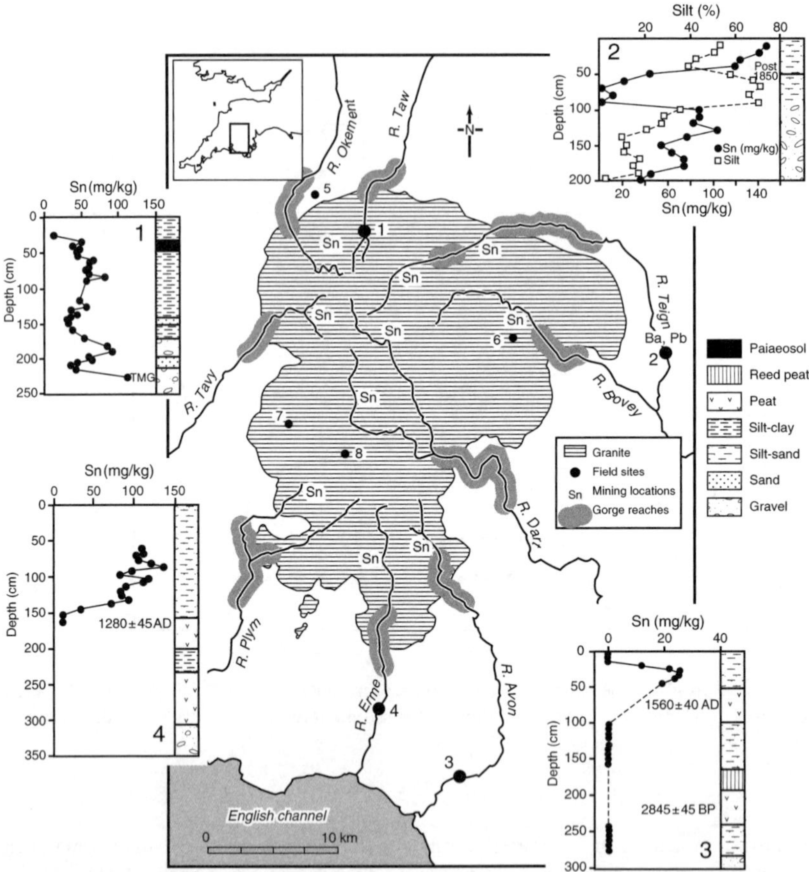

Figure 3.6 The distribution of Medieval tin mining sites (tin streaming sites) and geochemical profiles of alluvial sections in the floodplains of the rivers draining Dartmoor in SW England

aggradation in addition to post-mining flood history. Second, that in some, but not all basins, the residence time of mining derived sediments is truly geological being on the millennial rather than decade–century timescale. This finding has important implications for the release of stored contaminants from floodplains in response to changing forcing conditions such as global warming.

3.5 CONCLUSIONS

Archaeology can provide far more valuable information than just dating, indeed dating has now become the prerogative of the geomorphologists with artefact typological chronologies being re-evaluated as a result of the development of sediment-based dating techniques. Archaeology can provide rapid evidence of landsurfaces, sediment reworking and palaeoenvironmental conditions. It can also under favourable conditions set parameters, which can be used, in the modelling of past processes. So, archaeological data—including artefacts and the methods developed for their study—have led to the development of a set of tools that can be used by geomorphologists to study past fluvial processes and hydrological change. Conversely, fluvial geomorphology provides a series of tools if properly understood and applied can be used to study both the timing and environmental context of cultural impact on the landscape (Howard and Macklin 1999). The explanation for such strong linkages between archaeological and geomorphic methods arises from interactions between human societies and fluvial landforms; i.e., river channels and floodplains. These interactions include anthropogenic alterations of fluvial processes and magnitude–frequency relationships as well as the incorporation of human relics in alluvium.

In Europe, Asia and Africa substantial anthropogenic environmental disruptions began in the Middle Holocene and the clear cultural record allows the application of these tools over a relatively long period of time. In the Americas and Australia, the early cultural record is more subtle and extensive agriculture and deforestation came much later, leaving an abrupt boundary late in the stratigraphic record. There are advantages to both situations. In the Old World we can learn about long-term effects of multiple intermittent anthropogenic perturbations, while in the New World we can study the effects of the sudden introduction of environmental exploitation (e.g., the geomorphic response to mining). Both of these lessons are essential to an understanding of the future potential for human impacts on the environment and global environmental changes. It is unfortunate that one of the driving forces of increasing links between archaeologists and geomorphologists has been the relentless drift of funding towards applied and short-timescale studies in geomorphology. While such process-oriented studies are important, they cannot replace the need for an empirically based understanding of Earth-surface processes over millennial timescales.

REFERENCES

Antevs, E.V. 1935. Age of the Clovis Lake beds. *Proceedings of the Academy of Natural Science, Philadelphia* 87: 304–312.

Antevs, E.V. 1955. Geologic-climatic dating in the West. *American Antiquity* 20, 317–335.

Baker, V.R., Kochel, R.C. and Polach, H.A. 1985. Radiocarbon dating of flood events, Katherine Gorge, Northern Territory, Australia. *Geology* 13: 344–347.

Bradley, S.B. 1990. Characteristics of tin-streaming channels on Dartmoor, UK. *Geoarchaeology* 5: 29–41.

Brown, A.G. 1997. *Alluvial Geoarchaeology: Floodplain Archaeology and Environmental Change*, Cambridge University Press.

Brown, A.G. in press a. The environment. In: Barker, G., ed., *Tuscania*.

Brown, A.G. in press b. The environment of Lepti Minus. In: Humphreys, J., ed., *Lepti Minus a Roman port in North Africa*, Ann Arbor: University of Michigan.

Brown, A.G. and Salisbury, C.R. in press. The geomorphology of the Hemington reach. In: Cooper, L. and Ripper, S., eds., *The Hemington Bridges: The Excavation of Three Medieval Bridges at Hemington, Castle Donington, Leicestershire*.

Brown, A.G., Cooper, L., Salisbury, C.R. and Smith, D.N. 2001. Late Holocene channel changes of the Middle Trent: channel response to a thousand year flood record. *Geomorphology* 39: 69–82.

Challis, K. 1992. *Archaeological Investigations of a Proposed Stocking Area for Aggregate at Shardlow, Derbyshire*. Nottingham: Trent and Peak Archaeological Trust.

Clay, P. and Salisbury, C.R. 1990. A Norman mill dam and other sites at Hemington fields, Castle Donington, Leicestershire. *The Archaeological Journal* 147: 276–307.

Coles, B. and Coles, J. 1989. *People of the Wetlands: Bogs, Bodies and Lake-Dwellers*, London: Thames and Hudson.

Collier, C.R. and Musser, J.J. 1964. Sedimentation. In: *Influences of Strip Mining on the Hydrologic Environment of Parts of Beaver Creek Basin, Kentucky, 1955–59,* US Geological Survey Professional paper 427-B, pp. 48–64.

Coope, G.R. and Jones, P.J. 1977. Church Wilne. In: *The English Midlands*. International Quaternary Union Guidebook for Excursion A2, 25.

Cummins, W.A. and Rundle, A.J. 1969. The geological environment of the dug-out canoes from Holme Pierrepoint, Nottinghamshire. *Mercian Geologist* 3: 177–188.

Ellis, C.E. and Brown, A.G. 1999. Alluvial micro-fabrics, anisotropy of magnetic susceptibility and overbank processes. In: Brown, A.G. and Quine, T.S., eds., *Fluvial Processes and Environmental Change*, Chichester: Wiley, pp. 181–206.

Gale, S.J. and Hunt, C.O. 1986. The hydrological characteristics of a floodwater farming system. *Applied Geography* 6: 33–42.

Gilbert, G.K. 1917. *Hydraulic-mining Debris in the Sierra Nevada*, US Geological Survey Professional Paper 105.

Gilbertson, D.D. 1986. Runoff (floodwater) farming and the rural water supply in arid lands. *Applied Geography* 6: 5–12.

Graf, W.L. 1994. *Plutonium and the Rio Grande: Environmental Change and Contamination in the Nuclear Age*, Oxford: Oxford University Press.

Graf, W.L. 1979. Mining and channel response. *Annals of the Association of American Geographers* 69(2): 262–275.

Graf, W.L., Clark, S.L., Kammerer, M.T., Lehman, T., Randall, K. and Schröeder, R. 1991. Geomorphology of heavy metals in the sediments of Queen Creek, Arizona, USA. *Catena* 18: 567–582.

Greenwood, M.T. and Large, A.R.G. 1992. Insect sub-fossils from the Trent floodplain near Repton. *Journal of the Entomological Society* 107: 12–17.

Harding, P., Gibbard, P.L., Lewin, J., Macklin, M.G. and Moss, E.H. 1987. The transport and abrasion of flint handaxes in a gravel-bed river. In: de Sieveking, G.G. and Newcomer, M.H., eds., *The Human Uses of Flint and Chert*. Cambridge: Cambridge Univ. Press, pp. 115–126.

Henrotay, J. 1973. La sédimentation de quelques rivières belges au cours des sept derniers siècles. *Bulletin de la Sociète Géographique de Liège* 9: 101–115.

Herz, N. and Garrison, E.G. 1998. *Geological Methods for Archaeology*, Oxford: Oxford University Press.

Higgs, G. 1987. Environmental change and hydrological response: flooding in the Upper Severn catchment. In: Gregory, K.J., Lewin, J. and Thornes, J.B., eds., *Palaeohydrology in Practice*, Chichester: Wiley, pp. 131–159.

Hilmes, M.M. and Wohl, E.E. 1995. Changes in channel morphology associated with placer mining. *Physical Geography* 16: 223–242.

Howard, A.J. and Macklin, M.G. 1999. A generic geomorphological approach to archaeological interpretation and prospection in British river valleys: a guide for archaeologists investigating Holocene landscapes. *Antiquity* 73: 527–541.

Howard, A.J., Smith, D.N., Garton, D., Hillam, J. and Pearce, M. 2000. Middle to Late Holocene environments in the Middle to Lower Trent Valley. In: Brown, A.G. and Quine, T.A., eds., *Fluvial Processes and Environmental Change*, Chichester: Wiley, pp. 165–177.

Hudson-Edwards, K.A., Macklin, M.G., Finlayson, R. and Passmore, D.P. 1999. 2000 years of sediment-borne heavy metal storage in the Yorshire Ouse Basin, northeast England. *Hydrological Processes* 13: 1087–1102.

Jackson, J., Norris, R. and Youngson, J. 1996. The structural evolution of active fault and fold systems in central Otago, New Zealand: evidence revealed by drainage patterns. *Journal of Structural Geology* 18: 217–234.

James, L.A. 1989. Sustained storage and transport of hydraulic gold mining sediment in the Bear River, California. *Annals of the Association of American Geographers* 79: 570–592.

Jones, G.T., Bailey, D.G. and Beck, C. 1997a. Source provenance of andesite artefacts using non-destructive XRF analysis. *Journal Archaeological Science* 24(10): 929–943.

Jones, P.F., Salisbury, C.R., Fox, J.F. and Cummins, W.A. 1977b. Quaternary terrace sediments of the middle Trent Basin. *Mercian Geologist* 7: 223–229.

Keller, E.A. and Pinter, N. 1996. Active *Tectonics: Earthquakes, Uplift and Landscape*, New Jersey: Prentice-Hall.

Kirby, A. and Kirkby, M.J. 1976. Geomorphic processes and the surface survey of archaeological sites in semi-arid areas. In: Davidson, D.A. and Shackley, M.L., eds., *Geoarchaeology: Earth Science and the Past*, London: Duckworth, pp. 229–254.

Knight, D. and Howard, A.J. 1994. *Archaeology and Alluvium in the Trent Valley*. Nottingham: Trent and Peak Archaeological Trust.

Knighton, A.D. 1989. River adjustment to changes in sediment load: the effects of tin mining on the Ringarooma River, Tasmania, 1875–1984. *Earth Surface Processes and Landforms* 14: 333–359.

Knighton, A.D. 1991. Channel bed adjustment along mine-affected rivers of northeast Tasmania. *Geomorphology* 4: 205–219.

Knox, J. 1987. Historical valley floor sedimentation in the upper mississippi valley. *Annals of the Association of American Geographers* 77 224–244.

Ladd, S.C., Marcus, W.A. and Cherry, S. 1998. Differences in trace metal concentrations among fluvial morphologic units and implications for sampling. *Environmental Geology* 36: 259–270.

Leenaers, H., Schouten, C.J. and Rang, M.C. 1988. Variability of the metal content of flood deposits. *Environmental Geology and Water Science* 11: 95–106.

Lewin, J. Macklin, M.G. 1986. Metal mining and floodplain sedimentation in Britain. In: Gardiner, V., ed., *International Geomorphology, Part I*, New York: John Wiley and Sons, pp. 1009–1027.

Lewin, J. and Macklin, M.M. 1987. Metal mining and floodplain sedimentation in Britain. In: Gardiner, V., ed., *International Geomorphology, Part I*, New York: John Wiley and Sons, pp. 1009–1027.

Lewin, J. Davies, B.E. and Wolfenden, P.J. 1977. Interactions between channel change and historic mining sediments. In: Gregory, K.J., ed., *River Channel Changes*, New York: John Wiley and Sons, pp. 353–367.

Macklin, M.G. 1995. Archaeology and the river environment in Britain: a prospective review. In: Barham, A.J. and Macphail, R.I., eds., *Archaeological Sediments and Soils: Analysis, Interpretation and Management, 10th Anniv. Conf. Assn. Environmental Archaeology, 1989*, London: Institute of Archaeology, University College, pp. 205–220.

Macklin, M.G., Rumsby, B.T. and Newson, M.D. 1992. Historical floods and vertical accretion of fine-grained alluvium in the lower Tyne Valley, Northeast England. In: Billi, P., Hey, R.D., Thorne, C.R. and Tacconi, P., eds., *Dynamics of Gravel-bed Rivers*, Chichester: Wiley, pp. 564–580.

Marcus, W.A. 1987. Copper dispersion in ephemeral stream sediments. *Earth Surface Processes and Landforms* 12: 217–228.

Matei, E.J., Ernst, R. and Zhou, Y. 1993. Comparison of metal homogeneity in grab, quatered, and crushed-sieved portions of stream sediments and metal content change resulting from crushing-sieving activity. *Environmental Geology* 22: 186–190.

Meade, R.H., Yuzyk, T.R. and Day, T.J. 1990. Movement and storage of sediment in rivers of the United States and Canada. In: Wolman, M.G. and Riggs, H.C., eds., *Surface Water Hydrology, The Geology of North America*, Volume O-1, Boulder, CO: Geological Society of America, pp. 255–280.

Mordant, D. and Mordant, C. 1992. Noyen-Sur-Seine: a mesolithic waterside settlement. In: Coles, B., ed., *The Wetland Revolution in Prehistory*, 55–64, Wetland Archaeological Research Project Occasional Paper 6, Exeter.

Ortloff, C.R. and Crouch, D.P. 1998. Hydraulic analysis of a self-cleaning drainage outlet at the Hellenistic city of Priene. *Journal of Archaeological Science* 25: 1211–1220.

Petit, F. 1995. Régime hydrologiques et dynamique fluviale des rivières ardennaises. In: Demoulin, A., ed., *L'Ardenne, Essai de Géopgraphie Physiques*, Livre en Hommage au Prof. A. Pissart, Univ. Liège, pp. 194–223.

Reece, D.E., Felkey, J.R. and Wai, C.M. 1978. Heavy metal pollution in the sediments of the Coeur d'Alene River, Idaho. *Environmental Geology* 2: 289–293.

Renfrew, A.C. 1976. Archaeology and the earth sciences. In: Davidson, D.A. and Shackley, M.L., eds., *Geoarachaeology: Earth Science and the Past*, London: Duckworth, pp. 1–5.

Rubey, W.W. 1938. *The Force Required to Move Particles on a Stream Bed*, US Geological Survey Professional Paper 189-E, pp. 121–140.

Rumsby, B.T. and Macklin, M.G. 1996. River response to the last neoglacial (the 'Little Ice Age') in norther, western and central Europe. In: Branson, J., Brown, A.G. and Gregory, K.J., eds., *Global Continental Changes: The Context of Palaeohydrology*, Special Publication 115, London: The Geological Society, pp. 217–233.

Salisbury, C.R., Whitley, P.J., Litton, C.D. and Fox, J.L. 1984. Flandrian courses of the river Trent at Colwick, Nottingham. *Mercian Geologist* 9: 189–207.

Salvador, P.-G., 1991. Le thème de la métamorphose fluviale dans les plaines alluviales du Rhône et de IIsère: Bassin de Malville et ombilic de Moirans (Isère, France). Thèse Géographie et Aménagement, Univ. Lyon III, 498 p.

Schiffer, M.B. 1987. *Formational Processes of the Archaeological Record*, Albuquerque: University of New Mexico Press.

Sluse, P. 1996. *Évolution de la Rulles, de la Semois et de la Mellier au cours des cinq derniers siécles grâce aux résidus métallurgiques de l'industrie du fer et par l'étude des cartes anciennes*, Mémoire de licence en Sciences Géographiques, Univ. Liège, 206 p.

Sluse, P. and Petit, F. 1998. Evaluation de la vitesse de déplacement de la charge de fond caillouteuse dans le lit de rivières ardennaises au cours des trois derniers siècles, à partir de l'études des scories métallurgiques, *Géographie Physique et Quaternaire* 52: 373–380.

Thorndycraft, V.R., Pirrie, D. and Brown, A.G. 1999. Tracing the record of early alluvial tin mining on Dartmoor, UK. In: Pollard, A.M., ed., *Geoarchaeology: Exploitation, Environments, Resources*, Special Publications 165, London: Geological Society, pp. 91–102.

Thornes, J.B. and Gilman, A. 1983. Potential and actual erosion around archaeological sites in South-east Spain. *Catena Supplement* 4: 91–113.

Todd, M. 1973. *The Coritani*, London: Duckworth.

Tricart, J. and Vogt, H. 1967. Quelques aspects du transport des alluvions grossières et du paçonnement des lits fluviaux. *Geografiska Annaler* 49: 350–366.

Vita-Finzi, C. 1969. *The Mediterranean Valleys: Geological Changes in Historical Times*, Cambridge: Cambridge University Press.

Vita-Finzi, C. 1988. *Recent Earth Movements: An Introduction to Neotectonics*, London: Academic Press.

Water, M.R. 1992. *Principles of Geoarchaeology*, Tucson: University of Arizona Press.

Water, M.R. and Nordt, L.C. 1995. Late Quaternary floodplain history of the Brazos River in East-Central Texas. *Quaternary Research* 43: 311–319.

Wertz, J.B. 1949. Logarithmic pattern in river placer deposits. *Economic Geology* 44: 193–209.

Wolfenden, P.J. and Lewin, J. 1977. Distribution of metal pollutants in floodplain sediments. *Catena* 4: 309–317.

Wolfenden, P.J. and Lewin, J. 1978. Distribution of metal pollutants in active stream sediments. *Catena* 5: 67–78.

Zeuner, F.E. 1945. *The Pleistocene Period: Its Climate, Chronology and Faunal Succession*, Royal Society Publications no. 130, London.

4

Using Historical Data in Fluvial Geomorphology

ANGELA M. GURNELL[1], JEAN-LUC PEIRY[2] AND GEOFFREY E. PETTS[3]
[1]*Department of Geography, King's College London, UK*
[2]*Laboratoire de Géographie Physique, Maison de la Recherche, Université Blaise Pascal, Clermont-Ferrand, France*
[3]*School of Geography, Earth and Environmental Sciences, University of Birmingham, UK*

4.1 INTRODUCTION

Rarely, long-term monitoring sites have been established to document changes in the landscape with time. The most notable example is the Vigil Network (Emmett and Hadley 1968), an international network of areas including stream channels, hillslopes, reservoirs, precipitation and vegetation, on which periodic measurements are made and preserved. However, most studies of fluvial systems extend for periods of less than 5 years and, at best, provide detailed snapshots of the system or a small part of it. Therefore, for most studies, the only way to gain insights into the temporal variability of river channels in the longer term is by using historical analysis. Knowledge of previous conditions in the catchment, along the river corridor, and in the channel can provide valuable insights into channel behaviour. Historical analyses are required to establish channel and catchment conditions at one or more times in the past and to define times of major catchment, riparian and channel impacts, such as land-use change, channelisation and other engineering works. Consequently, many contemporary problems in fluvial geomorphology require a historical perspective whether the concern is to understand natural patterns of channel form variation, to establish the nature of human impacts, or to define benchmark conditions for channel restoration and management. For example, historical information can be useful in dating channel and catchment changes, documenting the nature and in some cases the rate of channel change, documenting changes in catchment conditions and human pressure on fluvial processes and forms, documenting channel response to and recovery from large floods and other disturbances, etc. It is only in the context of an understanding of the channel's evolution that we can confidently interpret current conditions.

Useful information on a range of geomorphological questions can be provided by analysis of historical sources (Table 4.1). Cooke and Reeves (1976) provide a particularly useful demonstration of the potential of historical analysis, for example: establishing channel widths from early maps; using old buildings and other structures to determine erosion rates; using repeat stage-discharge rating records to demonstrate cross-sectional changes; and reconstructing livestock densities from census reports, travel accounts and other sources to illuminate land-use changes. More recently, geomorphologists have benefited from advances in both historical geography (e.g. Hooke and Kain 1982), waterfront archaeology (e.g. Milne and Hobley 1981), and palaeohydrology (Gregory 1983, Gregory *et al.* 1987, Starkel *et al.* 1991) and through the latter's links with geoarchaeology (Davidson and Shackley 1976) and palaeoecology (Berglund 1986). Petts *et al.* (1989) bring together information from a wide range of sources to examine the history of European rivers and Trimble and Cooke (1991) have critically reviewed historical data sources for studies of geomorphological change in the United States. Other useful papers include: Patrick *et al.*'s (1982) review of methods for studying accelerated fluvial change; Trimble's (1998) review of dating fluvial processes from historical data and artefacts; Hooke's (1997) review of styles of channel change; Large and Petts' (1996) reconstruction of a channel-floodplain system; Sear *et al.* (1994)

Table 4.1 Some examples of the use of different information sources for historical analysis

Information	Source	Example
Stream widths	Land surveys	Knox (1977)
Channel migration	Land surveys	Kondolf and Curry 1986), Galatowitsch (1990)
Channel cutoff	Botanical evidence	Everett (1968)
	Surveys and travel accounts	Erskine *et al.* (1992)
Cross-section changes	Bridge surveys	Kondolf and Swanson (1993)
	Stage-discharge ratings and discharge measurement notes	Williams and Wolman (1984), Collins and Dunne (1989), Smelser and Schmidt (1998)
	Repeat channel surveys	Petts and Pratts (1983)
Channel change	Level and location of buildings	Womack and Schumm (1977)
	Historical ground photos	Laymon (1984)
Channel incision	Topographic records	Piégay and Peiry (1997)
Long profile/bedform	Navigation surveys	Large and Petts (1996)
Nature and distribution of large wood and fallen trees	Travel accounts	Maser and Sedell (1994)
Sedimentation	Datable artefacts	Costa (1975)
Sediment yield	Lake sediment chronologies	Davis (1975), Foster *et al.* (1985)
	Reservoir storage changes	Trimble and Carey (1992)
Palaeofloods	Slackwater deposits	Kochel and Baker (1988)
Flood dates	Botanical evidence	Hupp (1988)
Hydrological conditions	Diaries, journals and newspapers	Snell (1938)
	Water level and flood records on buildings	Pfister (1992)
Climate change	Diaries, log books and newspapers	Bradley and Jones (1992)

and Kondolf and Larson's (1995) demonstrations of the role of historical analysis in river restoration.

Historical analyses depend upon the integration of information from a variety of sources. The most important of these sources for fluvial geomorphological studies are topographic surveys (specifically river cross-sections, long profiles and topographic maps) and, more recently, air photographs; but many other historical documents can provide invaluable information. These include reports of exploration and land potential, census data, ecclesiastical records, fiscal documents and local and national government documents. Furthermore, recent technological advances in remote-sensing, have ensured an explosion of geomorphologically relevant information over the last 30 years, including the potential to identify relict features of the past.

This chapter, focuses on early documentary evidence, topographic survey, and cartographic sources, and complements the separate evaluations of archaeological data (Chapter 3), remotely sensed information (Chapter 6) and sediment budgets (Chapter 16). For historical sources prior to 1850, the nature, quantity, reliability, resolution and accuracy of information drawn is infinitely variable, and we do not attempt to provide specific and focused guidelines on their detailed potential for geomorphological application. Instead, we illustrate the range of opportunities presented by such sources for geomorphological research and provide some general cautions on their use. The sections on "using the topographic record" (Section 4.4) and "using the cartographic record" (Section 4.5) focus upon the last 100–150 years when increasingly precise information sources, allow us to illustrate their geomophological applications, and to provide more specific guidance on source accuracy and on the degree to which these sources can be used to derive quantitative as well as qualitative information. The chapter then addresses the problem of error propagation when information derived from historical sources is analysed using geographic information systems (GIS).

4.2 THE EARLY RECORD (1650–1850)

There are two primary categories of historical information. First, there are observations of fluvial phenomena per se. These include measurements of channel width; surveys of channel planform and long profile and water levels associated with land, drainage and navigation surveys; observations on the occurrence and impact of floods by early diarists; and records of the riverine environment including riparian vegetation, and catchment characteristics, by travellers and explorers. Secondly, there are records of fluvial-dependent phenomena. These include legal reports on land disputes caused by channel migration; disputes over water rights, navigation passage, and fisheries; and estate surveys and census reports. Often, non-fluvial information is required to interpret fluvial changes: where, when and over what period of time did channel change occur?

Bradley and Jones (1992) suggest the beginning of the 16th century was a watershed for the collection of spatial information: following the age of discovery, the stage was set for the advance of colonialism. In practice, records for historical analysis begin from about the 17th century when population growth, early capitalism, large-scale administration, and new surveying and measuring techniques required and enabled the production of detailed records, profiles and plans. Earlier documentary records may be found, for example, from the two core areas of merchant capitalism in Europe, upper Italy and Flanders. Here, during the 15th century, early innovation in the control of water flows and river regulation led to surveys with the aim of draining lowland plains and controlling flooding (Cosgrove 1990). Cosgrove refers to Cristoforo Sorte's text of 1593, which proposed to manage the upper Adige by both large-scale regulation and the control of deforestation in the higher Alpine catchment areas. However, many of these early works were local, small-scale or unrecorded.

From the 17th century, emerging nation states were able to legislatively and financially underwrite large-scale, integrated and enduring schemes. For states such as Britain, France and Spain to be successful, they required maps, plans and statistical knowledge of their territories to tax, administer and defend them. Smaller territories were obliged to emulate their practices.

In France, Napoleon introduced a system of cadastral surveys to improve efficiency of tax collection. The law was passed in 1807, some surveys were done 1809–1811, but most surveys were done around 1830. These provide excellent documentation of former channel and floodplain conditions (Piégay 1995, Piégay and Salvador 1997). The power of the large states was reinforced by improvements in technology. The 16th century had seen the rise of "mechanical practitioners" (Taylor 1967): navigators, instrument makers, estate managers, and naval architects with skills for accurate surveying, levelling and map-making; calculation of water volumes; and the design of pumps, sluices, land cuts and aqueducts. This technological renaissance enabled and sustained European imperial expansion into Asia and Africa during the 19th century and sustained the commercial expansion into the colonial lands (Heffernan 1990). Narrative accounts—travel logs, newspaper accounts, government records, and paintings—provide information on channel, riparian and catchment conditions at the time of European colonisation.

This period of exploration, colonisation and empowerment is characterised by a large amount of documentary material on such matters as landownership, land use, and agricultural potential; data on crop returns, areas of land under different crops, and the types and numbers of livestock; and plans and surveys for land drainage and navigation. Thus, for USA, Butlin (1993) refers to Gates (1960, p. 51): "Uncle Sam's acres were numerous and far flung by 1815. The business of surveying, sectioning, advertising, selling and collecting the proceeds constituted the largest single area of economic activity in the country and the major obligation of the federal government". Such data can provide valuable insights into catchment characteristics as well as details of channel form.

From the late 17th century considerable attention was directed to problems of floods and channel change in three situations: within piedmont rivers; in estuaries; and in wetlands. By the middle of the 19th century many advances had been made in developing solutions to river regulation, including, for example: Brooks (1841) "Treatise on the Improvement of the Navigation of Rivers", Cagliardi (1849) "Programme du Nouveau Régime des grandes rivières pour empécher les ruptures les changements de lit, et les inondations", Calver (1853) "The Conservation and Improvement of Tidal Rivers" and Ellet (1853) "The Mississippi and Ohio Rivers". These provide important qualitative information on the type, location and scale of river engineering works being promoted at a specific period

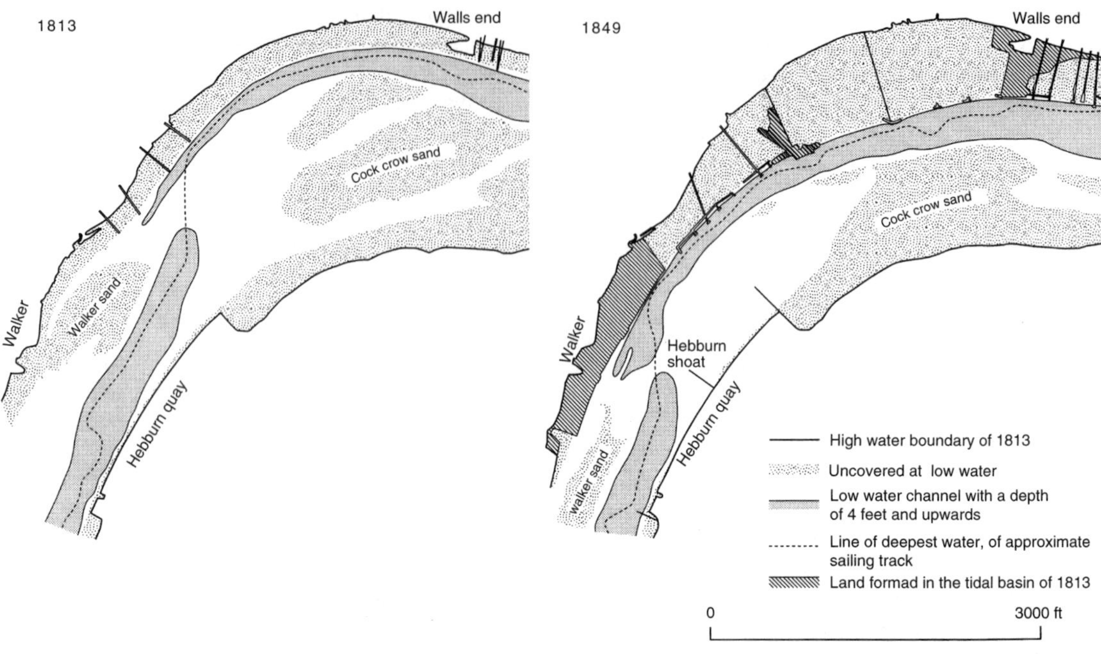

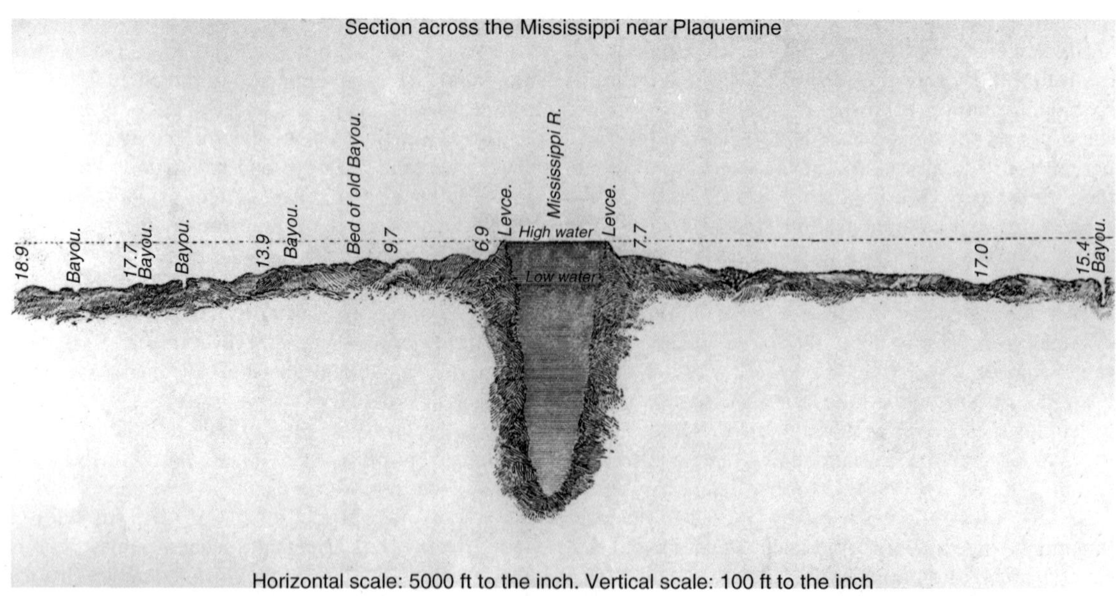

Figure 4.1 Examples of 19th century field-survey data. Above: surveys of the lower river Tyne in north-east England from Rennie survey of 1813 and Calvers survey of 1849 (from Calver 1853). Below: cross-profile of the lower Mississippi from Ellett (1853)

in time. Some include detailed plans and surveys of river reaches: examples of Calver's plans of a reach of the river Tyne, UK, including Rennie's survey of 1813, and of cross-sections of the Mississippi presented by Ellett are shown in Figure 4.1. Useful information may also be found in the notebooks and diaries of the engineers charged with the development of specific engineering works.

Problems of Data Reliability and Accuracy

A feature of this period is the unavailability and inconsistency of information in both space and time. For example, maps documenting topographic and other information at an acceptable accuracy for many geomorphological applications have been available for over 100 years for many parts of the World. However, early maps (for example the analysis of channel changes on the river Po, Italy, by Castaldini and Piacente 1995) may have restricted use because of their limited spatial coverage, low spatial resolution or poor spatial accuracy. There are three key aspects to consider. First, detailed information is often available only for parts of catchments or some reaches of a river, or for some rivers, and there are questions about the representativeness of the information. Such information often focuses upon river reaches experiencing particular problems for flood control, land-drainage or navigation. Secondly, there are problems caused by changes over time in survey or recording conventions, such as the areal units for which data were collected and changes in measurement technique or recording procedures between surveys. Experience in reconstructing climate records (Bradley and Jones 1992) shows that problems can arise because of changing calendar conventions. Thirdly, discrepancies in descriptions and records can reflect the changing perspectives and differing cultural attitudes towards natural resources (Hooke and Kain 1982).

Clearly, all sources require careful scrutiny and verification. Harley (1982) considered the scholarly evaluation of historical evidence must involve reference to the context of that evidence—why and for what purpose was it collected? Topographic and map survey data are discussed later and so discussion here is restricted to the use of qualitative data sources. Of prime importance is data verification. Often documents contain a mixture of both valuable and worthless information. The latter includes inaccurate or uncertain dating of events, or distortions or amplifications of original observations. Only if the observations are faithful in both time and space are they likely to be reliable and valuable. However, even if information can be verified, Bradley and Jones (1992, p. 6) emphasise the difficulty of ascertaining exactly what the information means. Terms such as "flood", "frozen river", "drought" and "summer" or "winter" used in the past may not be equivalent to terms of modern-day observations. In some cases, for example, the term "winter" has been used for the period of snow cover rather than for specific months. Qualifying words such as "unprecedented", "extreme", "in living memory", "extensive", or "deep" can be ambiguous. Bradley and Jones (1992) suggest a solution to this problem, using content analysis to help isolate the most pertinent and unequivocal aspects of the historical source. Content analysis provides an objective approach to assess the frequency of use of descriptive terms and the use made of qualifying terms, leading to statistical analysis (see Pfister 1992 for an application to climate data).

Early Documentary Records in Britain

The nature of documentary evidence prior to 1850 varies enormously, but the following British examples give a flavour of the variety and potential as well as the quality of such records. Documentary evidence includes much early information related to property transactions, rents, tithes, etc.; contracts and disputes; and legal cases regarding the breach of these. The larger and more wealthy a landowner, the more likely it was that documents would have been generated and the more likely it is that they will have survived. Thus, rivers within the largest, wealthiest estates can have a large deposit of relevant documents including letters (both personal and business), property deeds, surveys with some maps, and rentals, etc. However, estate surveys were expensive and were, therefore, only made at extraordinary times, such as on change of ownership or in response to major agricultural improvements. Ecclesiastical documents can be valuable sources and often include information on fisheries, ferries, mill operations and floods. The abbeys had major interests in economic farming practices, and weather events affecting crops were of vital interest. Consequently, many abbeys produced important Chronicles, which record storms, floods and droughts from the 13th century.

Thus, in Dugdale (1772, p. 150) there is evidence of the channelisation of the Ancholme River and of the draining of its associated marshes (Figure 4.2). In response to an enquiry from the king:

> a jury being impanelled accordingly, and sworn, did say upon their oaths, that it would not be to the damage of the said king, nor any other, but rather for the common benefit of the whole county of Lincoln, if the course of that river, obstructed, in part, in diverse places from Bishop's Brigge to the river of Humber, were open. And they farther said, that by this means, not only the meadows and pastures would be drained, but that ships and boats laden with corn, and other things, might then more commodiously pass with corn and other things from the said river Humbre, into parts of Lindsey, than they at that time could do, and as they had done formerly.

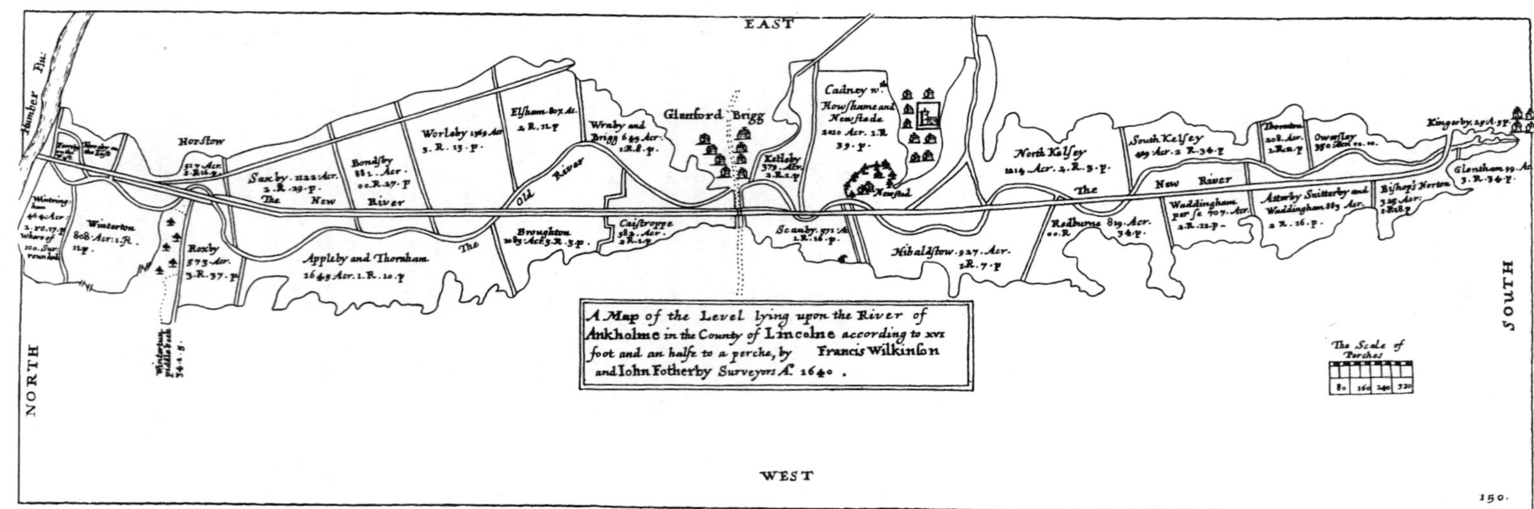

Figure 4.2 A survey of 1640 of the river Ankholme (now Anchholme) in eastern England showing an early channelisation scheme and the natural course of the river (reproduced from Dugdale 1772)

The records include specific instructions: "scouring the said channel from Glaunford brigge to the river of Humbre, to the breadth of XI feet, as it ought and want to be" (Dugdale 1772, p. 150).

A second example is provided by the reclamation of the wetland fens in eastern England, which once extended for 3400 km^2 (Butlin 1990). Camb (1586) records the general character of the area prior to drainage:

> All this country in the winter-time, and sometimes for the greatest part of the year, laid underwater by the rivers Ouse, Grant, Nene, Welland, Glene, and Witham . . . it affords great quantities of turf and Sedge for firing; Reeds for thatching; Elders also and other water shrubs, especially willows, either growing wild, or else set on the banks of rivers to prevent their overflowing . . .

Of particular importance in England is the later tithe survey (ca. 1830–1850), relating to the commutation of tithes previously paid originally in kind, as dues to support the local church. These comprise a large-scale map and a survey including the names of landholders, tenants and cottage holders; acreages, land use, fieldnames, parcel numbers and rental value. Other important documents of local administration were the parliamentary enclosure awards dating from the period 1750–1830, which combined an accurately surveyed map with a document of apportionment giving each landowner and tenant the parcels of land allocated to them. The Acts themselves are rather long, dense legal documents but the associated correspondence can be informative, providing information on land use and value. Other useful Acts, together with associated correspondence, include Acts to improve the navigability of rivers; to build bridges; and to build canals. The navigable rivers Acts of 1699 were particularly important for generating information on English rivers.

Often documents generated during the formulation of such bills included detailed plans and surveys of the river supported by explanatory text. The historical maps that are available for Britain are listed by Hooke (1997, pp. 240–241). In addition, maps or charts for navigation date from 1795 when the office of Hydrographer was established at the Admiralty and the second Hydrographer, Captain Hurd, originated the "Charts of the Coasts and Harbours in all Parts of the World". Surveys were also undertaken of the lower reaches of navigable rivers, and in England between 1810 and 1835, John Rennie published detailed channel plans as the basis for training rivers including the Tyne, Ouse, Nene, Welland and Witham (see Figure 4.1 and Petts 1995, p. 7). All these historical sources provide information on the date and extent of channel modification, floodplain and wetland drainage, and land-use change along the river corridor. Some early surveys, as well as later ones, also present opportunities for quantitative analysis of channel planform and location.

The River Trent

Historical information available for the corridor of the river Trent includes local government, national government and ecclesiastical sources, as well as family archives (Large and Petts 1996) (Table 4.2). The Trent was one of four "royal" rivers. Rights of navigation were founded in a royal decree of Edward the Confessor of 1065 and disputes between navigators and mill and fishery interests ensured a long history of documentation. Legal cases in the medieval and early modern period that were used to establish legal precedent proved valuable, such as documents concerning disputes over land in the 15th century. There are a number of large estates. As a result, a large amount of associated historical information survives, including the Harper-Crewe papers relating to the Calke Abbey estate, for which rental details exist back to the 16th century, and the Every Papers, which contain some good inscribed deeds and indentures relating to the period 1250–1600. Detailed surveys were also carried out at the time of extensive economic change and agricultural improvement in the 18th century.

The earliest known surveys of the river relate to the 1699 "Act for making and keeping the river Trent in

Table 4.2 Documentary sources available for the river Trent corridor

1200	*Deeds and private legal papers; ecclesiastical sources*
	Monastic Chronicles (1200–1713; incomplete record)
	Every Papers (1620–1890)
	Personal correspondence (1630–1890)
	Tithe surveys (1830–1850)
1699	*National Government Acts*
	To improve navigation (1699, 1740, 1781, 1783)
	To build bridges (1758, 1835)
	To build canals (1766, 1777, 1793)
1750	*Local Government surveys*
	Enclosure surveys (1750–1830)
	Topographical reports (1800–1820)

the County of Leicester, Derby and Stafford navigable"—later known as the Paget Act. The Paget Act of 1699 provided for the making of a tow-path by which barges could be hauled, effectively changing the character of the riparian zone and requiring the maintenance of a morphologically smooth bank profile. The first detailed surveys of the river were carried out between 1761 and 1792 in order to develop the river for inland navigation. The surveys located and provided detailed low-flow depth soundings of 67 shoals along the river over a 90-km reach in the low-flow months of August and September (Figure 4.3). The surveyor, William Jessop, made important observations on the fluvial geomorphology of the Trent (Petts 1995) and recommended works not only to self-scour the river but also to encourage overbank siltation; encouraging a natural process of channelisation (Large and Petts 1996).

4.3 SURVEYS 1850–1990

Particularly important for geomorphological research are surveys and reports commissioned during the second half of the 19th century. These benefit not only from advances in surveying techniques with the high level of detail necessary to underpin engineering schemes but also from direct measurements of river discharge, especially following the introduction of continuous flow recorders from 1889. Also, during this period the coverage of information expanded to larger areas of the globe. The technical and scientific renaissance, which both enabled and sustained the dramatic expansion in European commercial and industrial power, was led by groups of military surveyors and engineers as well as entrepreneurs. European engineers pushed the arteries of European trade across the vastness of Asia, Africa, Australia and the Americas.

Documentary evidence of river corridors includes material to support large-scale national surveys. Much information was produced in response to specific economic problems including land ownership, navigation, bridge construction, dam construction and land drainage or irrigation development for agriculture. Within Europe, for example, survey and statistical data were collected as the basis for developing navigation and water resources, and for increasing the area of agricultural land. Thus, since 1846 on the Rhône and the Saône Rivers, long profiles of the water line have been regularly surveyed for shipping by the "Service de la Navigation" in the French Alps. Knowledge of the hydroelectric power potential led the "Service des Grandes Forces Hydrauliques" to survey long profiles of all main rivers and their tributaries during the first quarter of the 20th Century. A characteristic of such records is their spatial and temporal discontinuity but when they exist, this kind of data can provide an irreplaceable source for studying channel changes, as is illustrated in the following two examples.

The Nile Surveys

Driven by the twin challenges of taming the river Nile and greening the desert, Britain invested in the survey of the Nile. A series of reports were published by William Wilcocks including: "The Nile Reservoir Dam at Assuân" (1901) and "The Nile in 1904" (1904). These

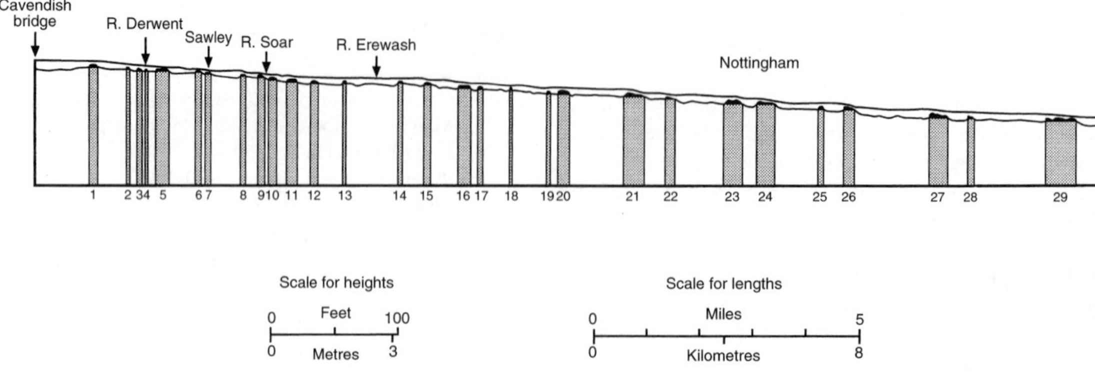

Figure 4.3 A section of William Jessop's survey of the river Trent showing the location and spacing of 29 shoals in 1792 (Jessop 1782)

works include detailed surveys of channel profiles and cross-sections (Figure 4.4) as well as information on discharges and flow velocities.

In 1915, a Physical Department was created within the Ministry of Public Works in Egypt and this had a major role in the development of the river Nile, along with the Irrigation Department. A series of publications from 1920 reported on hydrological and meteorological measurements. After the Second World War, attention was again directed to projects for increasing the cultivated area of Egypt and for flood protection, and a "catchment management" approach was advanced by Hurst *et al.* (1946). The latter includes annual hydrographs for 1870–1945 and assessments of the impacts of abstractions.

The Mississippi River Surveys

An Act of Congress, approved June 28, 1879, provided for the appointment of the Mississippi River Commission, which reported to the Senate through the Secretary of War. The third section of this law required the Commission to:

> Direct and complete such surveys of said river between the Head of the Passes, near its mouth, to its headwaters as may now be in progress, and to make such additional surveys, examinations, and investigations, topographical, hydrographical, and hydrometrical, of said river and its tributaries as may be deemed necessary by said Commission to carry out the objects of this act.

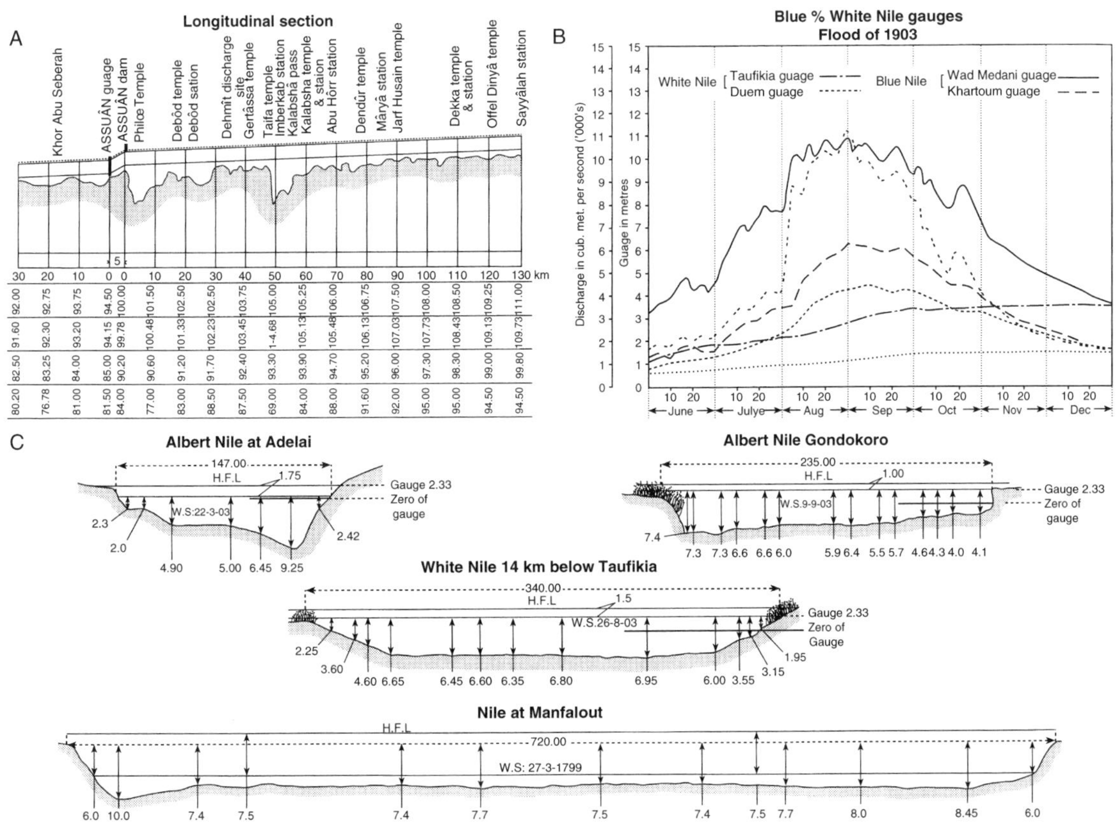

Figure 4.4 A selection of data from the river Nile surveys published by Wilcocks (1904) showing: (A) a section of the long profile with distances from Assuân with (in sequence from top to bottom) ground level, maximum flood level (1878), mean low flow level (from 1872 to 1902), and bed level in 1891; (B) flow and gauge levels of the 1904 flood on the Blue and White Nile; (C) selected cross-sections showing channel width and depths. All values are in metres

The Reports of the Commission include details of the triangulation of the river, of topographic surveys of the river corridor, of changes in the river bed, sediment loads and discharges, and of sediment sequences from boreholes. This information was used to plan a river regulation scheme to improve inland navigation and to prevent destructive floods. The value of these surveys is not only that they provide the basis for comparisons with more recent studies, but also that they often include comparative observations with even earlier studies: thus, page 11 of a report dated November 25, 1881, states:

> there is some evidence in a comparison of the results of Young, Poussin, and Tuttle's examinations of 1821 with later surveys, that the width of the river has increased since that date.

The Reports also provide detailed information on the methods employed on all aspects of the surveys: cross-section surveys, longitudinal surveys, slope, discharge observations, vertical velocity observations, transverse velocity observations, suspended sediment samples, dredgings for bed material samples, and the area and volume of caving banks. For example, the survey of one reach, Plum Point, between November 1879 and November 1880 (Figure 4.5) yielded: 43 694 located soundings; 201 discharge observations; 178 suspended sediment samples; 63 bed material samples; and 595 measurements of caving banks (Report of the Mississippi River Commission November 25th 1881, Appendix D). Part 6 demonstrates scour and fill on rising and falling flood stages, respectively. The average variation in the level of the river bed between high and low water was 6.5 ft. The data on caving for the 10 months ending October 1, 1880, record bank retreat by about 40 ft, on average, with the loss of 72.6 acres and erosion of 119 361 000 ft^3 of sediment. Supporting maps and observations are also useful (e.g. Figure 4.6), including, for example, comments on the ages of the riparian trees (e.g. boring 27: "...judging from the growth of timber, the formation would not be considered a very old one... N.B. the formation is a comparatively old one, the young growth of timber noted resulting from the desertion of an old cultivated field." and at boring 25: "at a depth of 35 feet the pipe struck a cottonwood log.").

The above brief introduction to the wealth of historical information available for geomorphological research shows that it is possible to establish baseline conditions for river landscapes prior to the modern period of industrialisation and environmental degradation for comparison with the present day. For many rivers, it is also possible to piece together the sequence of major changes in channel form over the past 100–300 years. However, even the best reconstructed sequences are likely to be incomplete both in time and space, and caution will be required in interpolating between reaches and extrapolating between points in time.

4.4 USING THE TOPOGRAPHIC RECORD

Topographic surveys over the last 100–150 years have produced two kinds of data that are of particular use in the geomorphological study of river channels: cross-sectional and long profile surveys. Cross-sections perpendicular to the flow direction provide information on channel shape and provide the basis from which morphometric indices (width; depth; thalweg, water line and bottom altitudes; channel asymmetry...) or hydraulic indices (bankfull cross-section area, hydraulic radius...) can be calculated (e.g. Gurnell 1997a). A comparison of cross-sections surveyed at different dates also allows the investigation of both vertical and lateral migration of channels (e.g. Petts and Pratts 1983). Downward (1995) explored the accuracy of cross-sectional surveys that are subject to both inherent and operational errors. Inherent errors are generated by the operators who survey cross-sections, both through their interpretation of the cross-sectional form and thus their selection of survey points, and through the precision with which they are able to relocate sections in the long term, possibly decades or more after a previous survey. Topographical landmarks to relocate cross-sections precisely are very useful, but do not always survive, particularly if the river channel is mobile.

Long profiles are intended for the study of slopes (channel bed slope, slope of the energy line). They may be directly constructed from a longitudinal topographic survey or they may be derived from cross-sections surveys that have been regularly distributed along the river channel. In both cases, the horizontal distance (X axis) by which every point of altitude is referenced is the distance along the channel centre line derived from direct measurements in the field or from estimates from large-scale maps. In contrast, the value of altitude presented in the long profiles (Y axis) may vary:

(i) altitudes almost always represent the water surface when the long profile has been surveyed along the river from upstream to downstream;

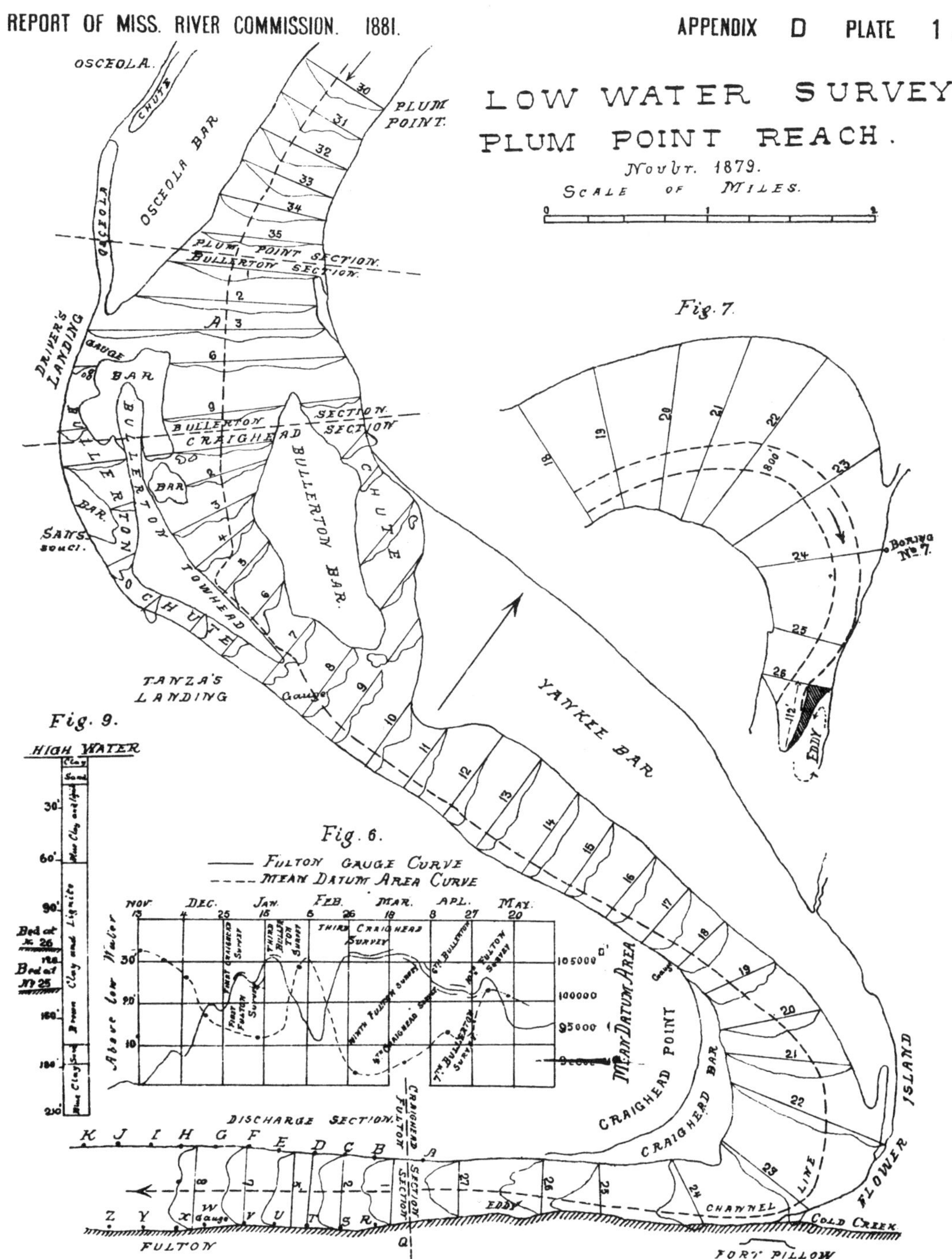

Figure 4.5 A reproduction of a page (Appendix D Plate 1) from the report of the River Mississippi Commission of 1881 showing a detailed plan of the Plum Point Reach with inserts showing: *Fig. 6.*, the Fulton Gauge Curve and Mean Datum Area Curve indicating periods of scour and fill from November 1879 to May 1880; *Fig. 7* shows the Fort Pillow eddy (shaded) on January 3, 1880; and *Fig. 9* shows results from a borehole located in *Fig. 7.* of this page (Mississippi River Commission 1881)

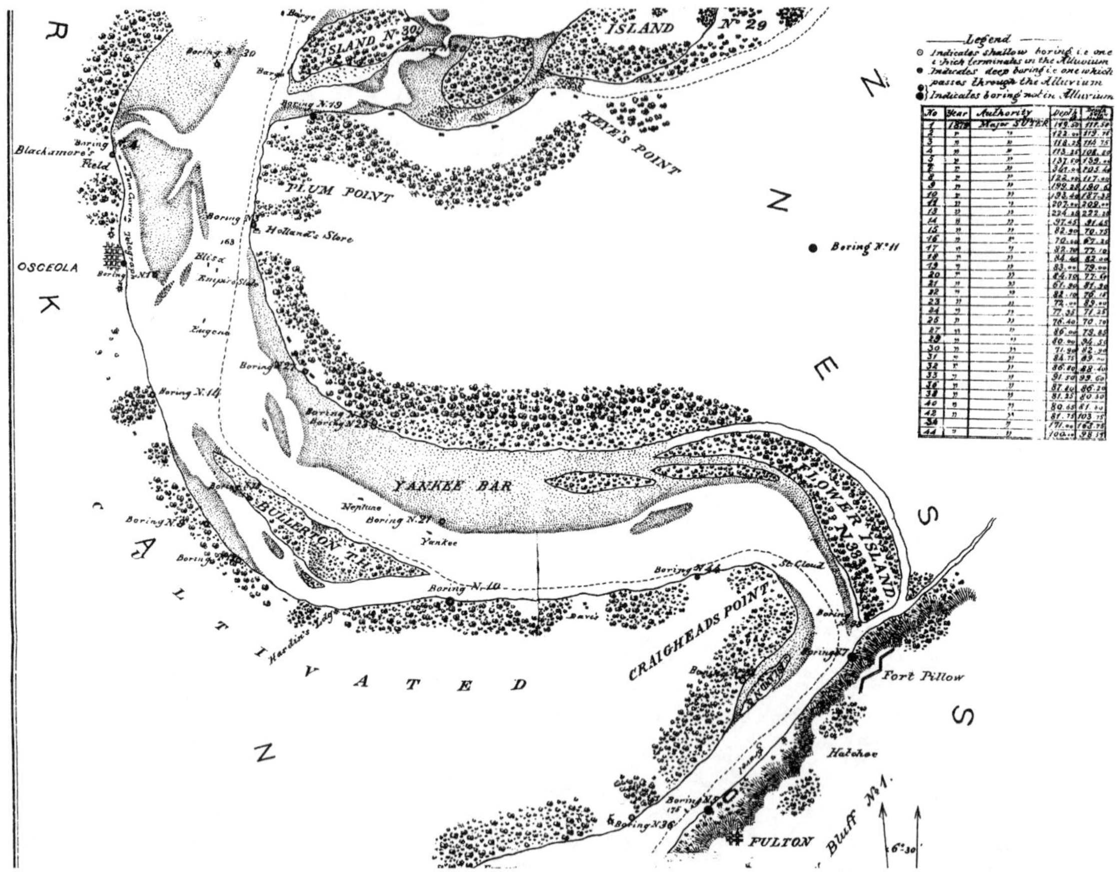

Figure 4.6 A reproduction of Appendix J Plate 2 from the Mississippi River Commission Report 1881 showing a sketch of the Mississippi River in the vicinity of Plum Point with locations of borings made in 1878 under the direction of Major C.H.R. Suter (scale of 1 in. to 1.33 miles). Insert lists boreholes with year of survey (1879), depth of borehole and depth of borehole below river

(ii) when data are derived from cross-sections, the altitude may represent the water level, the average level of the bed, or more rarely the altitude of the thalweg (Figure 4.7).

The water level is strongly dependent on the hydrological regime and hydrometeorological events. For reasons of convenience, historical topographic surveys were generally made at low flows, unless flood levels were the focus of the study, as for example when the survey was to be used to calibrate a hydraulic model. The average bed level is an altitude which smooths out the shape variability or asymmetry of the channel. It is frequently used when cross-sections are available to underpin estimation of the average bed level. The thalweg altitude, or altitude of the lowest point of a cross-section, is rarely used because on alluvial rivers it is subject to strong spatial variations associated with riffle-pool sequences.

Comparing Topographic Records

Diachronic comparison of topographic records requires sets of comparable data. Unfortunately, several difficulties are frequently met when researchers or engineers have to compare historical data. First, the reference system for altitude may have changed between survey dates. For example, in France, three successive systems of levelling were set up from the middle of the 19th century (Table 4.3). Between 1857 and 1864, the building of the first railway lines and the extension of navigable canals led to the establishment

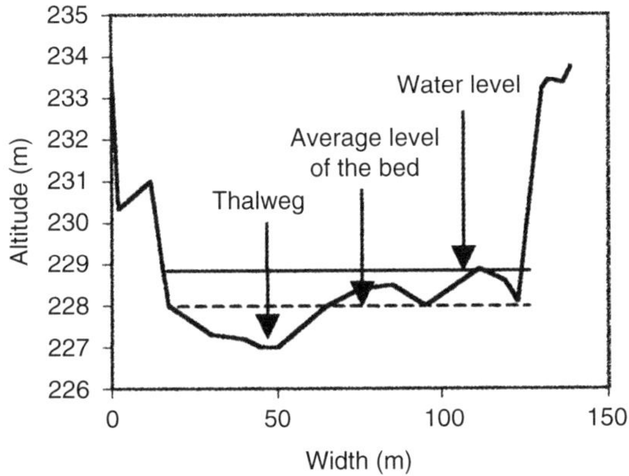

Figure 4.7 Data surveyed or calculated on a cross-section

of a first levelling network, which covered the whole country. The territory was covered by 38 polygons and zero altitude was the average level of the Mediterranean Sea at Marseille. On two later occasions, the network was changed through the replacement of geodetic landmarks and to increase the network accuracy (Landon 1999). Therefore, prior to any comparison of topographic records, it is essential to be sure that the altitude reference is identical for every set of data. In France, for example, maps are available for altitude conversion so that former values can be transposed to be compatible with the system used today. The conversion values are not constant in space, but increase from the South to the North, reaching a maximum + 60 cm in northern France.

Secondly, a lack of data homogeneity is a serious obstacle to long profile comparison. To avoid errors in geomorphological interpretation, it is preferable to compare topographical data of the same type, such as water surface levels with water surface levels, average bed levels with average bed levels, etc. Long profiles constructed from average bed levels allow the most accurate comparisons. Long profiles of the water surface at low flows are strongly influenced by river discharge at the time of survey. The lack of discharge data for the time of survey is a frequent limitation to the use of this type of historical data, although in some cases, water level–discharge relationships are available and can be used to correct the profile for this hydrological effect. Comparison of thalweg profiles are rather rare but their interpretation should be made carefully, because the migration of bedforms over time can lead to strong local variations in the thalweg line, which are independent of the general evolution of the river.

Thirdly, between two georeferenced points whose spatial location does not change over time (e.g. two bridges), the channel length may change with changes in river sinuosity. This is frequently the case on actively meandering rivers or on channels experiencing fluvial metamorphosis (e.g. from braiding to meandering). Under such circumstances, it becomes impossible to superimpose long profiles without first correcting the channel length. The best way to solve this problem is to calculate the ratio of channel sinuosity between the two dates and to then to adjust the horizontal distance scale along the long profile using this ratio. The ratio can be calculated reach by reach along a river valley, in order to ensure that the length correction is closely adapted to the fluvial pattern.

Some Examples of Using Topographic Records to Study Channel Change

A common use of historical topographic records is to describe channel aggradation or incision linked to hydrological or sediment load changes. Park (1995) reviewed channel cross-sectional changes in detail; thus, the following discussion will focus more specifically on longitudinal changes in river bed profiles.

Initially, long profile comparisons were used for studying complex readjustment of channel morphology below reservoirs (Petts 1979, Williams and Wolman 1984). More recently, fluvial geomorphologists have explored historical topographic data from archives to derive indices of natural or anthropogenic river metamorphosis. For example, Bravard (1987, 1994)

Table 4.3 Levelling networks in France from 1857 to today

Network name	Year of set-up	Number of polygons	Network length (km)	Altitude accuracy (cm km^{-1})	Altitude (centre of Paris) (m)	Difference in altitude (m)
Bourdapouë	1857–1864	38	15 000	±1.00	131.00	
Lallemand	1884–1931	32	12 715	±0.17	130.36	−0.64
IGN69	1963–1969	39	?	±0.13	130.70	+0.34

demonstrated aggradation of the upper Rhône River over the 19th and 20th centuries in association with a progradation of the braiding pattern of former glaciated basins. In the French Alps between 1840 and 1950, the longitudinal embankment of most Rivers at a time of abundant bedload supply, associated with the climatic degradation of the Little Ice Age, frequently led to channel aggradation (Gemaehling and Chabert 1962). On the Isère River close to the city of Grenoble, this phenomenon has been particularly well documented by civil engineers. Topographic records allow the channel aggradation to be quantified at 1–2 cm year^{-1} between 1880 and 1950 (Blanic and Verdet 1975). In the last 10 years, topographic records have been mainly used to document channel incision and its spatial distribution. Such records have been particularly effective in documenting rapid, deep incision, which, for example, has reached up to 2 m, mainly as a result of the impact of gravel mining (Peiry 1987, Peiry *et al.* 1994, Kondolf 1995, Piégay 1995).

From a technical point of view, a classical way to undertake a diachronic analysis of geomorphological changes is to superimpose long profiles on the same graph (Figure 4.8). However, when differences in altitude between two profiles are moderate, it is often more effective to graph positive and negative deviations in altitude. These differences in altitude are extrapolated from long profiles systematically, at a constant horizontal interval (e.g. from every 250 m to 1 km according to the river length). Positive and negative differences in altitude are shown by mapping the river line in plan and superimposing the deviations above and beneath the line (Landon and Piégay 1994, Piégay and Peiry 1997, Landon 1999) (Figure 4.9). Care needs to be taken to ensure that changes are not artefacts of differences in the spacing of survey points.

A combination of changes in altitude derived from long profiles and changes in channel width measured from maps or air photographs is also used to quantify bedload budgets and to study their spatial distribution. Approximate volumetric changes between two dates are calculated, reach by reach, by using the following formula:

$$V = \sum_{i=1}^{n} I_i \times L_i \times l_i$$

where V is the volume estimation (m^3), I_i the channel incision (m), L_i the length of the reach (m), l_i the channel width (m) and n is the number of reaches.

The superposition of sets of regularly spaced cross-sections allows more accurate assessment of alluvial sediment storage. On the embanked Isère River (French Alps), Vautier (1999) used five sets of cross-sections surveyed, respectively, in 1949, 1965, 1972, 1984 and 1990, to establish progressive channel degrad-

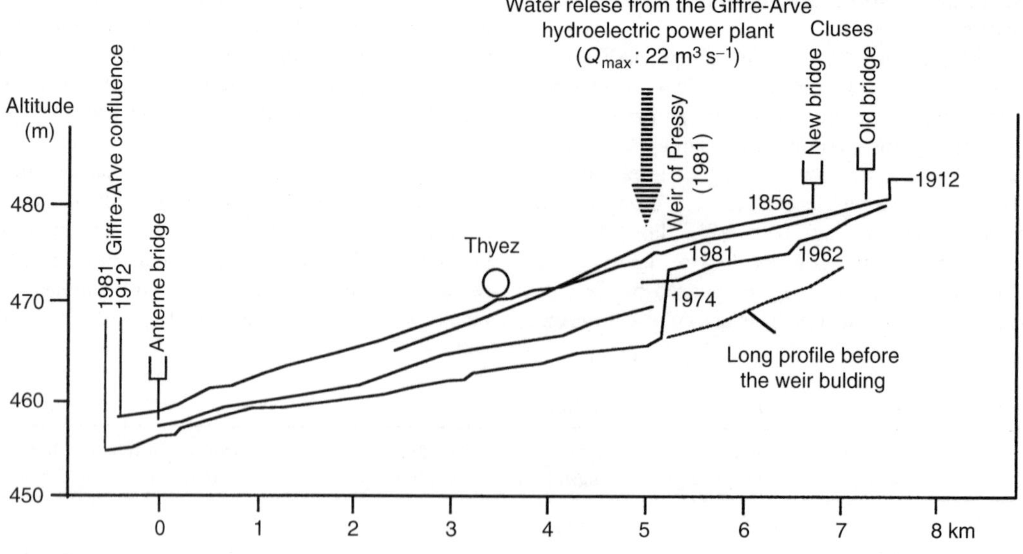

Figure 4.8 Long profile change on the Arve River, French Alps (modified from Peiry 1987)

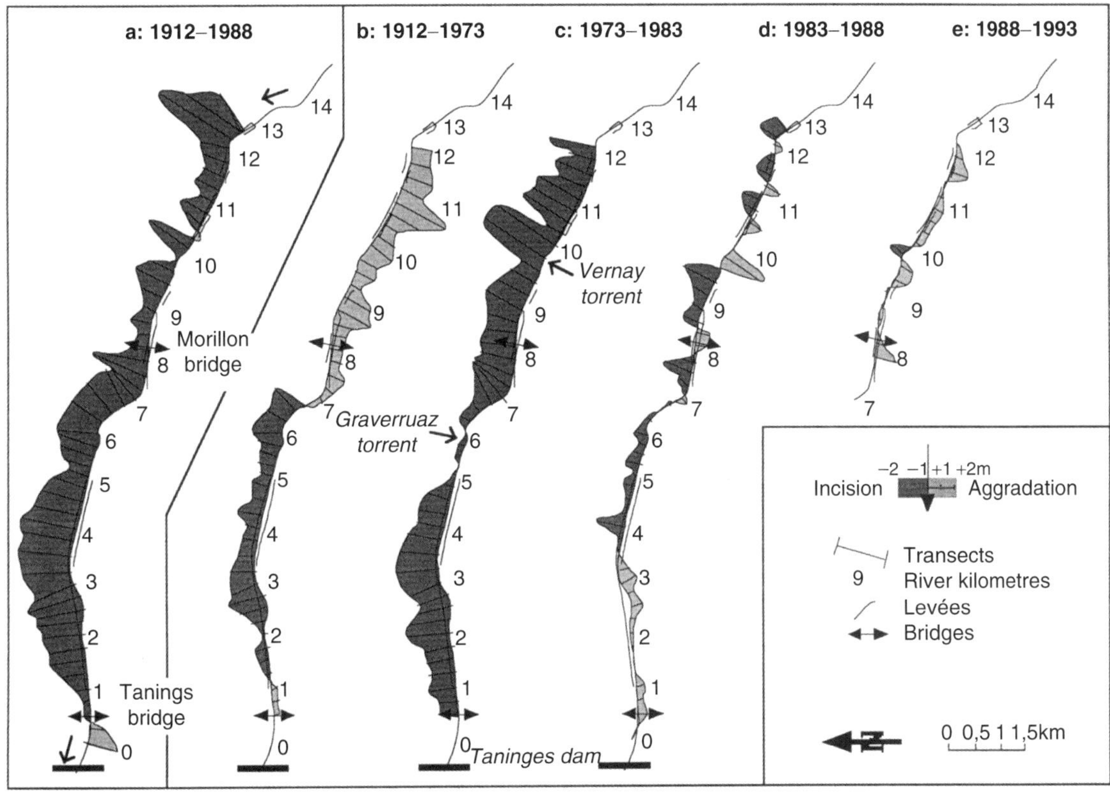

Figure 4.9 Long profile change on the Giffre River, French Alps (modified from Piégay and Peiry 1997)

ation that was mainly due to gravel mining (Figure 4.10).

Further information on constructing sediment budgets is given in Chapter 16.

4.5 USING THE CARTOGRAPHIC RECORD

Historical information derived from plan sources including maps, air photographs and satellite imagery, can provide a wealth of information for geomorphological research. Each type of source is capable of providing a different spectrum of information, is available with varying spatial and temporal resolution, and is representative of widely different historical time spans. Since chapter 6 is on air photography and other remotely sensed data, much of this section will be concerned with printed maps. The aim is to indicate generic problems associated with the use of map sources and to review the range of fluvial geomorphological applications for which they have been used.

General Issues of Accuracy

Maps are simply abstractions or generalisations of reality that have been produced with a specific purpose in mind. Therefore, it is important to avoid attempts to extract more information from a map than was there in the first place. National and regional map-making agencies usually provide detailed manuals on the survey and mapping conventions used in map production and frequently give estimates of the accuracy of their products. These sources should be used to assess whether the purpose of a particular analysis can be met by the information provided in particular maps. For example, only water courses that are 5 m or more wide are shown to scale with two lines marking their banks on UK Ordnance Survey 1:10 000 scale maps, whereas

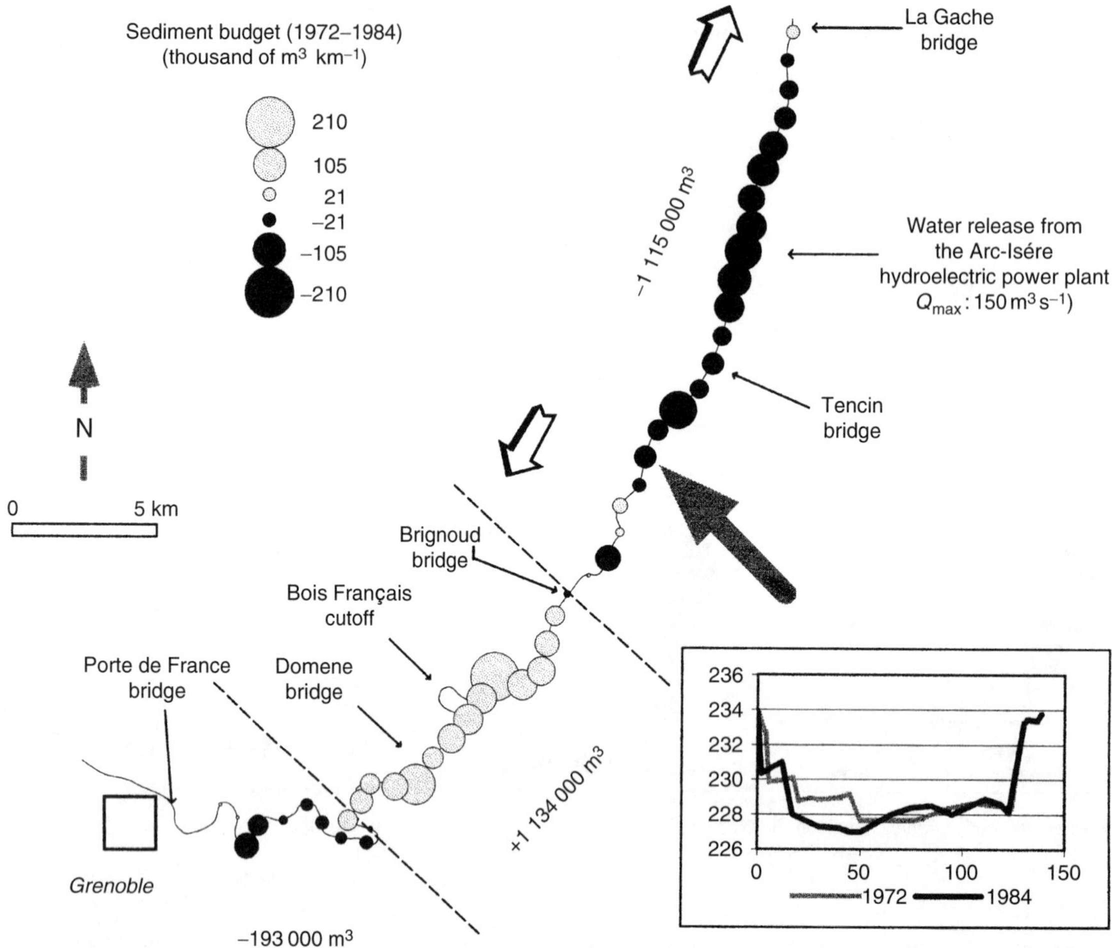

Figure 4.10 Bedload budget from cross-sections surveyed on the Isère River, French Alps (modified from Vautier 1999)

the threshold is 1 m on 1:1250 scale maps (Harley 1975). A careful consideration of the accuracy and conventions built into map production can provide the basis for the extraction of quantitative information from the most unlikely sources. For example, Gurnell *et al.* (1996) describe the use of a cover-abundance scale to extract quantitative spatial information from spatially distorted River Corridor Survey maps, which are essentially sketch maps produced for the UK Environment Agency to describe the biogeomorphological characteristics of 500 m stretches of river course. However, even if the intentional limitations of maps are taken into account, a variety of other errors can be introduced inadvertently at various stages in map production, which may have importance for geomorphological interpretation.

There are three fundamental dimensions of spatial data: space, attribute and time (Chrisman 1991) or "where something was observed", "what was observed" and "when was it observed" (Flowerdew 1991). The following account indicates some of the intentional and inadvertent errors associated with all of these three dimensions, which accumulate into a total map error.

Positional Accuracy

The first constraint on positional accuracy is the technical limitations of the surveying equipment and the methodology employed at the time of the original field survey. However, perhaps more important are the conceptual errors that may have been introduced during field survey, air photograph interpretation or the interpretation of information from other sources. The surveyor frequently has to make decisions about the location of features or boundaries. Such decisions are particularly difficult in relation to natural features, which rarely have crisp boundaries. Some features, such as agricultural fields, may have clearly defined boundaries. However, other features, such as soils and vegetation often grade gradually from one type to the next across transition zones, but the surveyor is still required to map a boundary. Even natural features with apparently crisp boundaries are usually "fuzzy" in practice. For example, it may be straightforward to identify the position of a river bank where the bank is vertical, but difficulties can arise where the river bank consists of a gently sloping aquatic–terrestrial transition, or a sequence of benches, slumps and terraces. As a result, conventions are usually devised to define boundaries. For example, river channel boundaries are defined by the UK Ordnance Survey in relation to the "normal winter level" (Harley 1975), but there is still great potential for error in applying such conventions. The timing of the field survey is likely to be an important influence on the accuracy with which the "normal winter level" is determined. "If, therefore, the stream is surveyed in summer, it is the permanent channel, eroded of vegetation, rather than the water width, which is measured" (Harley 1975: p. 44). Similarly, in semi-arid regions where channels are strongly influenced by infrequent, larger floods, definition of the unvegetated, active channel may vary in width depending upon the time of the survey in relation to the time of the last large flood.

Once the information for the map has been assembled, there is a range of error sources associated with translating the information into a map. All maps have a spatial reference system, based on a map projection, which translates latitude and longitude on the curved surface of the earth onto a flat map sheet. Thus, maps for the same area and at the same scale but based on different map projections are not directly comparable and, indeed, may vary in their spatial scale from one part of the map to another. Similar, but more severe problems arise when using information derived from photographs. If the photographs are oblique, projection problems arise as a result of varying spatial scale over the photographic image. Even with vertical photographs, significant distortions occur with increasing distance from the centre of the image and with differences in altitude of the terrain.

Another source of positional error relates to the map scale. Scale determines the smallest area that can be drawn and recognized on a map. It is not possible to locate any object more accurately than the width of one line on the map. This determines the resolution of the map which, assuming a minimum line width or point size of 0.5 mm, is 5, 25 and 50 m, respectively, for map scales of 1:10 000, 1:50 000 and 1:100 000 (Fisher 1991). Clearly, this places a limit on the accuracy with which locations can be measured from a map, but there are many other factors, which further degrade the locational accuracy of the map. For example, the map scale also influences whether or not features are shown on a map. Thus, maps of soil, vegetation or rock types, which may be extremely variable over small areas, have to be based on a minimum mapping unit—the smallest area that can be represented on the map. Features smaller than the minimum mapping unit must either be merged with adjacent areas so that the map reflects dominant classes, or if they are particularly important to the map theme, they can be represented by symbols or can be spatially exaggerated so that they can be mapped. This leads to the issue of information generalisation, which is used to ensure the visual clarity of a particular map. For example, information may be omitted or spatially smoothed, even when it relates to areas significantly greater than the minimum mapping unit if inclusion of the information is detrimental to map clarity. As a result, not only will different types of thematic map at the same spatial scale represent the same information to different levels of detail, but also different editions of the same thematic map may present very different quantities of information on the same features. For example, Gardiner (1975) showed that the length of streams depicted on 1:25 000 scale UK Ordnance Survey topographic maps varied greatly with the map edition. The stream length ratio between the Second Series and the Provisional Edition varied between 1.10 and 1.80 for a sample of map sheets from different areas of Great Britain. Furthermore, a series of papers (Ovenden and Gregory 1980, Burt and Gardiner 1982, Burt and Oldman 1986) has explored the accuracy with which headwater stream networks are depicted on Ordnance Survey 1:10 560 and 1:10 000 scale maps. These papers illustrate that extreme care must be taken in interpreting such information from different map editions and for different geographical locations.

A final point relates to the boundaries of map sheets. Traditional map series were often designed as series of individual map sheets with no guarantee of conformity across the margins of the maps. This can lead to many anomalies on map sheet margins, which simply reflect decisions relating to the generalisation and presentation of features on the individual sheets. All of these factors illustrate that although the resolution of a map is fundamentally dictated by map scale, there are a range of other factors, which vary within and between maps, and which influence the positional accuracy of the features that are depicted.

Attribute Accuracy

The accuracy of mapped attributes varies according to the measurement scale employed. If the attribute is measured on a continuous scale (e.g. precipitation), it can only be recorded on the map to a given level of accuracy. Particular problems arise for features, such as elevation, which occur everywhere and which are often represented on maps by isolines. The first problem relates to the precision of the attribute estimates on the mapped isolines. For example, the technical specification of the contours of the UK Ordnance Survey Land-Form PROFILE digital product, which is based upon data at a nominal 1:10 000 scale created from an archive of graphic contours, is given as ±1.5 m (contours are provided at a 5-m vertical interval). Even if the attribute values along the isolines are completely reliable, values of the attribute and error margins for points on the map that are located between isolines are difficult to assess. Fryer *et al.* (1994) provide a discussion of the potential accuracy of heighting derived from air photographs and maps, and Moore *et al.* (1991) discuss the quality of digital elevation models. If the attribute is categorical, exact recording is possible. However, as in the case of soil maps, mapped categories are frequently based on a classification, which may not represent the level of discrimination required by the user, and which is also open to inaccurate interpretation by the surveyor.

Temporal Accuracy

Every map relates to a particular survey date and so is always out-of-date by the time it is published. Because surveys are undertaken in different places at different times, the degree to which any particular map is out-of-date varies between different map sheets, even within the same thematic map series. Whereas these sources of temporal inaccuracy can be determined from information provided with the map, other sources of temporal (in)accuracy are more difficult to detect. Many maps are declared to be partial revisions of their predecessors or, more seriously, Carr (1962) provides examples of the use of information from previous maps, without acknowledgement. In both these cases, even assuming that the partial resurvey is accurate, there is no guarantee that the information depicted on the map is from the indicated date of survey.

A further time-related source of error in paper maps results from shrinkage and distortion of the paper over time, and distortion resulting from the use of copies of the original map.

Assessing Accuracy

The above discussion illustrates that it is important to devote some consideration to map accuracy if spurious geomorphological conclusions are to be avoided. Although most of the comments made above relate to printed maps, it is important to remember that digital map products are subject to the same surveyor errors. Furthermore, these products are frequently derived from paper maps and so incorporate all of the potential errors discussed above, with the addition of digitising error. In addition, many digital products are provided in a grid format, which has frequently been interpolated from non-gridded data derived from printed maps, and so interpolation error is yet another addition to the list of possible error sources.

Chrisman (1991) describes how the accuracy of maps may be assessed. Positional accuracy can be tested by comparing a sample of mapped positions against some measure of true position. This generates a series of displacements, which can be analysed for both bias and random error components. The former can often be removed by geometrical transformation, whereas the latter can then be quantified to provide "error margins" for positional information extracted from the map. This type of approach is used by mapping agencies to check the accuracy of their products (Harley 1975). Hooke and Perry (1976) illustrate the application of the approach to one of the earliest large-scale map sources for much of England and Wales, the Tithe maps. Many of these surveys were undertaken between 1837 and 1845 and, although the mapping scale varies, they were typically produced at a scale of three chains to one inch (approximately 1:2375). Hooke and Perry (1976) evaluated the accuracy of a sample of tithe maps and found a mean absolute linear error of the order of

2.7%. The problem with this type of approach to error assessment is that it is usually based on measurements at a series of well-defined points, and so the error margins are also associated with the identification of well-defined points. As discussed above, additional uncertainty arises when the features of interest are ill-defined and so do not have sharp boundaries. In this case, "ground-truth" information is required to estimate appropriate additional error margins relating to positional uncertainty.

Attribute accuracy, where the attribute is continuous (e.g. elevation), can be tested in a similar manner to positional accuracy. Where the attribute is categorical, the construction of a mis-classification matrix based upon map and "ground truth" information for the same sites can help to assign percentage errors to different attribute classes.

Some Examples of Using Maps to Study Channel Change

Figure 4.11 illustrates how a sample of research applications has employed historical maps in conjunction with other plan sources to investigate properties of fluvial systems. It provides information on the other plan sources that have been used in conjunction with maps; the channel dimension and planform properties that have been extracted; and the time period and channel properties used to explore channel dynamics. Although this figure does not attempt to be comprehensive, some

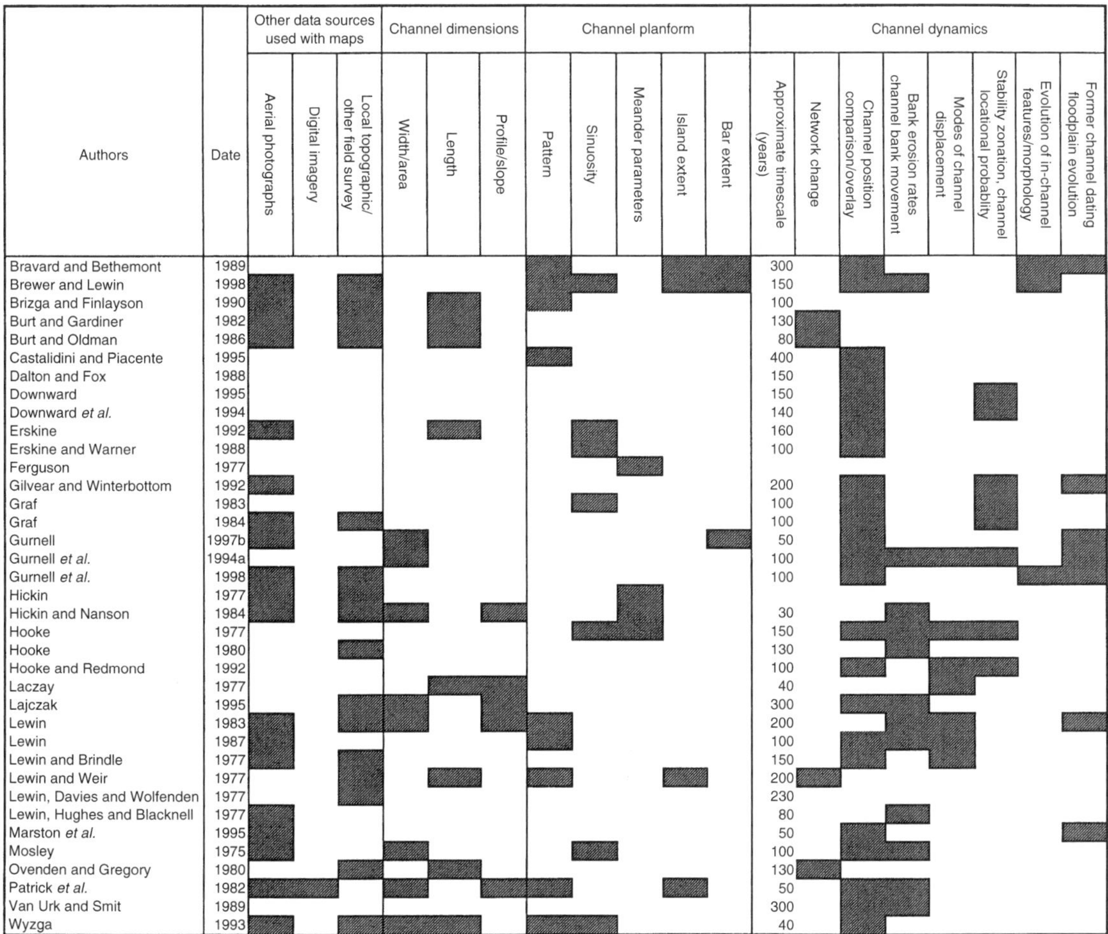

Figure 4.11 A sample of published fluvial geomorphological studies, which have been based upon information extracted from maps

generalisations can be drawn from the information presented. In relation to the sources employed, one-third of the studies were based entirely on map sources, over a half used air photographs in conjunction with maps, 40% used local topographic and other surveys, and approximately 30% used both air photographs and local surveys.

Of particular interest is that only one of the cited papers referred to digital imagery. This type of data is discussed fully in Chapter 6, but a brief mention is appropriate here. Although remotely sensed imagery from satellite platforms, which provide routinely repeated land cover information across the entire globe, has been available since the early 1970s, its spatial resolution is fairly coarse. For example, Figure 4.12 provides a comparison of a river network as it is depicted on a 1:10 560 scale Ordnance Survey map (Figure 4.12A) in comparison with the same network interpreted from simulated SPOT panchromatic (Figure 4.12B: 10 m resolution) and SPOT XS (Figure 4.12C: 20 m resolution) images of the same river stretch. There are major contrasts in both the widths and lengths of river channel presented in the three diagrams. To some extent the differences between Figure 4.12A and Figures 4.12B and C may reflect differences in the convention used to identify river channels, as well as the river stage and maintenance of the small channels (which form part of an old water meadow system), at the time of the map and remotely sensed surveys. However, the differences between Figures 4.12B and C, which represent the river at the same date, largely reflect differences in the spatial resolution of the imagery.

Although analysis of the spectral information from satellite platforms can provide a useful thematic complement to the information derived from historical maps, its spatial resolution restricts its application in fluvial geomorphology. Whereas the greater spatial and spectral resolution in the information available from airborne platforms has enormous potential for geomorphological research, this type of data is expensive to acquire and is not universally available. As a result, the present authors have found no published fluvial geomorphological studies that have combined historical map analysis with the analysis of airborne imagery, although Milton *et al.* (1995) refer to an unpublished example.

In relation to the geomorphological information drawn from historical map sources, Figure 4.11 illustrates that the most commonly derived indices are channel width, length, pattern and sinuosity. However, some researchers have attempted to extract more detailed quantitative information relating to the structure of river systems, including parameters describing the structure of meanders and the frequency and extent of bars and islands. Where repeat map and other

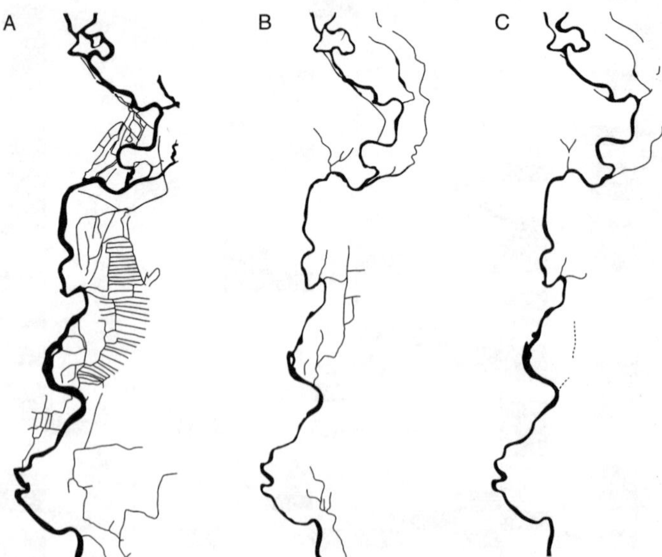

Figure 4.12 A comparison of the same river reach: (A) as it is depicted on a 1:10 560 scale Ordnance Survey map; (B) interpreted from simulated SPOT panchromatic (10 m resolution); (C) simulated SPOT XS (20 m resolution) images (image interpretation by A.M. Gurnell and M.J. Clark) (reproduced by permission of John Wiley and Sons, Ltd.)

surveys have been used to describe the dynamics of river channels, these have most frequently resulted in a simple overlay of maps from different dates. Some authors have then developed models or descriptions of channel stability, channel displacement and floodplain evolution from the patterns depicted in the overlays. When information from historical maps has been combined with information drawn from historical air photographs, it has been possible not only to quantify channel change, but also to identify some of the processes, related to sediment movement and vegetation colonisation, which facilitate change.

4.6 DATA INTEGRATION AND GIS

The focus of this chapter has been on the character, availability and potential use of historical information sources in fluvial geomorphological investigations. Over the last two decades, the rapid increase in the computer power available to fluvial geomorphologists has led to an enormous increase in the potential to integrate historical and contemporary information sources through GIS. Indeed, many of the analyses listed in Figure 4.11 were achieved using GIS software and so it is appropriate to provide a few comments and words of caution concerning the application of GIS to the historical study of fluvial geomorphology. The use of GIS has not only revolutionised the speed and ease with which historical analyses can be undertaken, it has also permitted more detailed, quantitative descriptions to be drawn from historical map overlays than was previously possible, including the estimation of spatially distributed bank erosion rates, modes of channel change and channel locational probabilities.

GIS undoubtedly provides a framework for integrating and analysing information from disparate sources in exciting, novel ways. However, the power of GIS as a data integrating, manipulating and visualising tool should not be allowed to disguise the dependence of the output on the accuracy of the input data. The previous two sections on the topographic and cartographic record have both explored various aspects of the accuracy of historical sources. Integration of information derived from such sources within a GIS must not ignore these errors in the input data or, more seriously, the propagation and magnification of the errors as the data sets are integrated and analysed. Whilst GIS software is capable of supporting exciting analyses and presenting the results in a visually appealing way, the potential for those results to be spurious is high. The sophistication of the technology should not be allowed to disguise the very important issue of error propagation. The problems of error propagation and error handling are explored further by Downward (1995), Grayson *et al.* (1993), Gurnell *et al.* (1994a,b, 1998) and Kemp (1993).

4.7 CONCLUSION

Historical studies (Table 4.4) involve the analysis of data sets of varying scales and degrees of completeness and accuracy that were rarely produced with the needs of a geomorphologist in mind—although the geomorphologist of the future will benefit from data collected following the "measurement revolution" of the 1950s. Nevertheless, *all* studies will need to consider the three key aspects of historical data: availability, accuracy and interpretation (Butlin 1993). Even with regard to modern maps, which provide an excellent historical information source, scientifically rigorous conclusions can only be drawn if the limitations of the map sources are fully understood. The same is true for all information sources. The main challenge of historical studies is to determine the accuracy of the derived information in order that genuine spatial or temporal patterns can be differentiated from those that are an artefact of the observer, recorder or cartographer.

Table 4.4 Steps in compiling historical data (developed from Hooke 1997)

1	*Establish research sources and dates available*, and if possible use all available material including complementary archaeological and remote-sensing sources
2	*Check background and general reliability* of sources
3	*Investigate document quality/accuracy/applicability*: source original documents or verify compilations based upon secondary sources; note purpose of records; undertake content analysis if appropriate
4	*Investigate topographic survey quality/accuracy/applicability*: check accuracy of individual surveys, including planimetric accuracy and content accuracy, and identify points of common detail to enable comparison of different surveys
5	*Create time sequence of catchment, river or reach conditions* using both qualitative and quantitative methods as appropriate
6	*Field check changes* indicated

ACKNOWLEDGEMENTS

Harold Potter and George Revill helped with the research on the river Trent during the 1980s. The manuscript was improved by comments from anonymous reviewers.

REFERENCES

Berglund, B.E., ed. 1986. *Handbook of Holocene Palaeoecology and Palaeohydrology*, Chichester: Wiley.

Blanic, R. and Verdet, G. 1975. Quelques travaux de correction sur le cours de l'Isère. *La Houille Blanche* 2/3: 191–197.

Bradley, R.S. and Jones, P.D., eds. 1992. *Climate Since A.D.1500*, London: Routledge.

Bravard, J.-P. 1987. *Le Rhône du Léman à Lyon*. Lyon: La Manufacture Ed.

Bravard, J.-P. 1994. La charge de fond du Haut-Rhône français, mise en perspective historique. *Dossier de la Revue de Géographie Alpine, Grenoble* 12: 163–169.

Bravard, J.-P. and Bethemont, J. 1989. Cartography of rivers in France. In: Petts, G.E., Muller, H. and Roux, A.L., eds., *Historical Change of Large Alluvial Rivers: Western Europe*, Chichester: Wiley, 95–111.

Brewer, P.A. and Lewin, J. 1998. Planform cyclicity in an unstable reach: complex fluvial response to environmental change. *Earth Surface Processes and Landforms* 23, 989–1008.

Brizga, S.O. and Finlayson, B.L. 1990. Channel avulsion and river metamorphosis: the case of the Thompson river, Victoria, Australia. *Earth Surface Processes and Landforms* 15: 391–404.

Brooks, W.A. 1841. *On the Improvement of Rivers*, London: John Weale.

Burt, T.P. and Gardiner, A.T. 1982. The permanence of stream networks in Britain: some further comments. *Earth Surface Processes and Landforms* 7: 327–332.

Burt, T.P. and Oldman, J.C. 1986. The permanence of stream networks in Britain: further comments. *Earth Surface Processes and Landforms* 11: 111–113.

Butlin, R.A. 1990. Drainage and landuse in the Fenlands and Fen-edge of northeast Cambridgeshire in the seventeenth and eighteenth centuries. In: Cosgrove, D. and Petts, G.E., eds., *Water, Engineering and Landscape*, London: Belhaven, pp. 54–76.

Butlin, R.A. 1993. *Historical Geography*, London: Edward Arnold.

Cagliardi, J. 1849. *Programme de Nouveau Regime des Grandes Rivières*, Lisbonne, France: Borroni & Scotti.

Calver, E.K. 1853. *Conservation and Improvement of Tidal Rivers*, London: John Weale.

Camben, J. 1586. Brittania: citation is from the Edmund Gibson 1695 ed., from the facsimile edited by S. Piggott, Newton Abbott (1971) cited in Butlin, R.A. 1990. Drainage and land use in the Fenlands and Fen-edge of northeast Cambridgeshire in the seventeenth and eighteenth centuries. In: Cosgrove, D. and Petts, G.E., eds., *Water, Engineering and Landscape*, London: Belhaven, pp. 54–76.

Carr, A.P. 1962. Cartographic record and historical accuracy. *Geography* 47: 135–145.

Castaldini, D. and Piacente, S. 1995. Channel changes on the Po River, Mantova Province, Northern Italy. In: Hickin, E.J., eds., *River Geomorphology*, Chichester: Wiley, pp. 193–207.

Chrisman, N.R. 1991. The error component in spatial data. In: Goodchild, M.F. and Gopal, S., eds., *Geographical Information Systems. Principles and Applications*, London: Longman, pp. 165–174.

Collins, B. and Dunne, T. 1989. Gravel transport, gravel harvesting, and channel-bed degradation in rivers draining the Southern Olympic Mountains, Washington, USA. *Enviornmental Geology and Water Science* 13: 213–224.

Cooke, R.U. and Reeves, R.W. 1976. *Arroyos and Environmental Change in the American South-west*, Oxford: Clarendon Press.

Cosgrove, D. 1990. Platonism and practicality: hydrology, engineering and landscape in sixteenth-century Venice. In: Cosgrove, D. and Petts, G.E., eds., *Water, Engineering and Landscape*, London: Belhaven Press, pp. 35–53.

Costa, J.E. 1975. Effects of agriculture on erosion and sedimentation in the Piedmont province, Maryland. *Bulletin of Geological Society of America* 86: 1281–1286.

Dalton, R.T. and Fox, H.R. 1988. Channel change on the river Dove. *East Midlands Geographer* 11: 40–47.

Davidson, D.A. and Shackley, M.L., eds. 1976. *Geoarchaeology*, London: Duckworth.

Davis, M.B. 1975. Erosion rates and land use history in southern Michigan. *Environmental Conservation* 3: 139–148.

Downward, S.R. 1995. Information from topographic survey. In: Gurnell, A.M. and Petts, G.E., *Changing River Channels*, Chichester: Wiley, pp. 303–323.

Downward, S.R., Gurnell, A.M. and Brookes, A. 1994. A methodology for quantifying river planform change using GIS. In: Olive, L.J., Loughran, R.J. and Kesby, J.A., eds., *Variability in Stream Erosion and Sediment Transport*, International Association of Hydrological Sciences Publication 224, pp. 449–456.

Dugdale, W. 1772. *History of Imbanking and Draining of Divers Fens and Marshes Both in Foreign Parts and in this Kingdom*, London: Geast.

Ellett, C. 1853. *The Mississippi and Ohio Rivers*, Philadelphia, USA: Lippincott, Grambo & Co.

Emmett, W.W. and Hadley, R.F. 1968. *The Vigil Network – Preservation and Access of Data*, US Geological Survey Circular 460–C, pp. 1–21.

Erskine, W.D. 1992. Channel response to large scale river training works: Hunter river, Australia. *Regulated Rivers: Research and Management* 7: 261–278.

Erskine, W.D. and Warner, R.F. 1988. Geomorphic eects of alternating flood- and drought-dominated regimes on NSW coastal rivers. In: Warner, R.F., ed., *Fluvial Geomorphology of Australia*, Sydney: Academic Press, pp. 223–244.

Erskine, W.C., McFadden C. and Bishop, P. 1992. Alluvial cutoffs as indicators of former channel conditions. *Earth Surface Processes and Landforms* 17: 23–27.

Everett, B.L. 1968. Use of cottonwood in the investigation of the recent history of a flood plain. *American Journal of Science* 206: 417–439.

Ferguson, R.I. 1977. Meander migration: equilibrium and change. In: Gregory, K.J., ed., *River Channel Changes*, Chichester: Wiley, pp. 235–248.

Fisher, P. 1991. Spatial data sources and data problems. In: *Geographical Information Systems. Principles and Applications*, London: Longman, pp. 175–189.

Flowerdew, R. 1991. Spatial data integration. In: Maguire, D.J., Goodchild, M.F. and Rhind, D.W., eds., *Geographical Information Systems. Principles and Applications*, London: Longman, pp. 375–387.

Foster, I.D.L., Dearing, J.A., Simpson, A. and Carter, A.D. 1985. Lake-catchment based studies of natural waters and soil denudation. *Earth Surface Processes and Landforms* 10: 45–68.

Fryer, J.G., Chandler, J.H. and Cooper, M.A.R. 1994. On the accuracy of heighting from aerial photographs and maps: implications to process modellers. *Earth Surface Processes and Landforms* 19: 577–583.

Galatowitsch, S.M. 1990. Using the original Laud Survey Notes to reconstruct presettlement landscapes in American west great Basin, *Naturalist* 50: 181–192.

Gardiner, V. 1975. Drainage basin morphometry. *British Geomorphological Research Group Technical Bulletin* 14: 48 p.

Gates, P.W. 1960. The farner's age: agriculture 1815–1860. In: *Economic History of the United States*, volume III, New York: Harper and Row (cited in Butlin 1993).

Gemaehling, C. and Chabert, J. 1962. *Transport solide et modification du lit d'une rivière à forte pente: la Drôme*, Bari: IAHS Publication, 59, pp. 259–272.

Gilvear, D.J. and Winterbottom, S.J. 1992. Channel change and flood events since 1783 on the regulated river Tay, Scotland: implications for flood hazard management. *Regulated Rivers: Research and Management* 7: 247–260.

Graf, W.L. 1983. Flood-related channel change in an arid-region river. *Earth Surface Processes and Landforms* 8: 125–139.

Graf, W.L. 1984. A probabilistic approach to the spatial assessment of river channel instability. *Water Resources Research* 20: 953–962.

Grayson, R.B., Blöschl, G., Barling, R.D. and Moore, I.D. 1993. Process, scale and constraints to hydrological modelling. In: Kovar, K. and Nachtnebel, H.P., eds., *HydroGIS 93: Applications of Geographic Information Systems in Hydrology and Water Resources*, International Association of Hydrological Sciences Publication 211, pp. 83–92.

Gregory, K.J., ed. 1983. *Background io Palaeohydrology*, Chichester: Wiley.

Gregory, K.J., Lewin, J. and Thornes, J.B., eds. 1987. *Palaeohydrology in Practice*, Chichester: Wiley.

Gurnell, A.M. 1997a. Adjustments in river channel geometry associated with hydraulic discontinuities across the fluvial-tidal transition of a regulated river. *Earth Surface Processes and Landforms 22*: 967–985.

Gurnell, A.M. 1997b. Channel change on the river Dee meanders, 1946–1992, from the analysis of air photographs. *Regulated Rivers: Research and Management* 13: 13–26.

Gurnell, A.M., Downward, S.R. and Jones, R. 1994a. Channel planform change on the River Dee meanders, 1876–1992. *Regulated Rivers: Research and Management* 9 187–204.

Gurnell, A.M., Angold, P. and Gregory, K.J. 1994b. Classification of River Corridors: issues to be addressed in developing an operational methodology. *Aquatic Conservation* 4: 219–231.

Gurnell, A.M., Angold, P.G. and Edwards, P.J. 1996. Extracting information from River Corridor Surveys. *Applied Geography* 16: 1–19.

Gurnell, A.M., Bickerton, M., Angold, P., Bell, D., Morrissey, I., Petts, G.E. and Sadler, J. 1998. Morphological and ecological change on a meander bend: the role of hydrological processes and the application of GIS. *Hydrological Processes* 12: 981–993.

Harley, J.B. 1975. *Ordnance Survey Maps*, A Descriptive Manual, Southampton: Ordnance Survey.

Harley, J.B. 1982. Historical geography and its evidence: reflections on modelling sources. In: Baker, A.R.H. and Billinge, M., eds., *Period and Place. Research Methods in Historical Geography*, Cambridge: Cambridge University Press, pp. 261–273.

Heffernan, M.J. 1990. Bringing the desert to bloom: French ambitions in the Sahara desert during the late nineteenth century – the strange case of 'la mer interieure'. In: Cosgrove, D. and Petts, G., eds., *Water, Engineering and Landscape*, London: Belhaven, pp. 94–114.

Hickin, E.J. 1977. The analysis of river planform responses to changes in discharge. In: Gregory, K.J., ed., *River Channel Changes*, Chichester: Wiley, pp. 249–263.

Hickin, E.J. and Nanson, G.C. 1984. Lateral migration rates of river bends. *Journal of Hydraulic Engineering* 110: 1557–1567.

Hooke, J.M. 1977. The distribution and nature of changes in river channel patterns: the example of Devon. In: Gregory, K.J., ed., *River Channel Changes*, Chichester: Wiley, pp. 265–280.

Hooke, J.M. 1980. Magnitude and distribution of rates of river bank erosion. *Earth Surface Processes* 5: 143–157.

Hooke, J.M. 1997. Styles of channel change. In: Thorne, C.R., Hey, R.D. and Newson, M.D., eds., *Applied Fluvial Geomorphology for River Engineering and Management*, Chichester: Wiley, pp. 237–268.

Hooke, J.M. and Kain, J.P. 1982. *Historical Changes in the Physical Environment: A guide to sources and techniques*, London: Butterworth.

Hooke, J.M. and Perry, R.A. 1976. The planimetric accuracy of tithe maps. *Cartographic Journal* 13: 177–183.

Hooke, J.M. and Redmond, C.E. 1992. Causes and nature of river planform change. In: Billi, P., Hey, R.D., Thorne, C.R. and Tacconi, P., eds., *Dynamics of Gravel-bed Rivers*, Chichester: Wiley, pp. 557–571.

Hupp, C.R. 1988. Plant ecological aspects of flood geomorphology and palaeoflood history. In: Baker, V.R., Kochel, R.C. and Patton, P.C., eds., *Flood Geomorphology*, New York: Wiley, pp. 335–356.

Hurst, H.E., Black, R.P. and Simaika, Y.M. 1946. *The Nile Basin. Volume VII. The Future Conservation of the Nile.* Cairo: S.O.P. Press.

Knox, J.C. 1977. Human impacts on Wisconsin stream channels. *Annals of the Association of American Geographers* 67: 323–342.

Jessop, W. 1782. *Report of William Jessop, Engineer, on a Survey of the River Trent, in the Months of August and September 1782, Relative to a Scheme for Improving Its Navigation*, Nottingham: Burbage and Son.

Kemp, K.K. 1993. Environmental modelling and GIS: dealing with spatial continuity. In: Kovar, K. and Nachtnebel, H.P., eds., *HydroGIS 93: Applications of Geographic Information Systems in Hydrology and Water Resources*, International Association of Hydrological Sciences Publication 211, pp. 107–115.

Kochel, R.C. and Baker, V.R. 1988. Palaeoflood analysis using slackwater deposits. In: Baker, V.R., Kochel, R.C. and Patton, P.C., eds., *Flood Geomorphology*, New York: Wiley, pp. 357–376.

Kondolf, G.M. 1995. Managing bedload sediment in regulated rivers: examples from California, USA. In: Costa, J.E., Miller, A.J., Potter, K.W. and Wilcock, P.R., eds., *Natural and Anthropogenic Influences in Fluvial Geomorphology*, The American Geophysical Union, pp. 165–176.

Kondolf, G.M. and Curry, R.R. 1986. Channel erosion along the Carmel River, Monterey County, California. *Earth Surface Processes and Landforms* 11: pp. 307–319.

Kondolf, G.M. and Larson, M. 1995. Historical channel analysis and its application to riparian and aquatic habitat restoration. *Aquatic Conservation* 5: 109–126.

Kondolf, G.M. and Swanson, M.L. 1993. Channel adjustments to reservoir construction and instream gravel mining, Stony Creek, California. *Environmental Geology and Water Science* 21: 256–269.

Laczay, I.A. 1977. Channel pattern changes of Hungarian rivers: the example of the Hernád River. In: Gregory, K.J., ed., *River Channel Changes*, Chichester: Wiley, pp. 185–192.

Lajczak, A. 1995. The impact of river regulation, 1850–1990, on the channel and floodplain of the Upper Vistula river, Southern Poland. In: Hickin, E.J., ed., *River Geomorphology*, Chichester: Wiley, pp. 209–233.

Landon, N. 1999. L'évolution contemporaine du profil en long des afffluents du Rhône moyen. Constat régional et analyse d'un hydrosystème complexe, la Drôme. Unpublished thesis in Geography, Univ. Paris 4 – Sorbonne.

Landon, N. and Piégay, H. 1994. L'incision de deux affluents méditerranéens du Rhône: la Drôme et l'Ardèche. *Revue Géogr. Lyon* 69: 63–72.

Large, A.R.G. and Petts, G.E. 1996. Historical channel-floodplain dynamics along the river Trent. *Applied Geography* 16: 191–209.

Laymon, S.A. 1984. Photo documentation of vegetation and landform change on a riparian site 1880–1980, Dog Island, Red Bluff, California. In: Wainer, R.E. and Hendrix, K.M., eds., *California Riparian Systems*, Berkeley: University of California press, pp. 150–158.

Lewin, J. 1983. Changes of channel patterns and floodplains. In: Gregory, K.J., ed., *Background to Palaeohydrology*, Chcihester: Wiley, pp. 303–319.

Lewin, J. 1987. Historical river channel changes. In: Gregory, K.J., Lewin, J. and Thornes, J.B., eds., *Palaeohydrology in Practice*, Chichester: Wiley, pp. 161–175.

Lewin, J. and Brindle, B.J. 1977. Confined meanders. In: Gregory, K.J., ed., *River Channel Changes*, Chichester: Wiley, pp. 221–238.

Lewin, J. and Weir, M.J.C. 1977. Morphology and recent history of the Lower Spey. *Scottish Geographical Magazine* 93: 45–51.

Lewin, J., Davies, B.E. and Wolfenden, P.J. 1977. Confined meanders. In: Gregory, K.J., ed., *River Channel Changes*, Chichester: Wiley, pp. 353–367.

Lewin, J., Hughes, D. and Blacknell, C. 1977. Incidence of river erosion. *Area* 9: 177–180.

Marston, R.A., Girel, J., Pautou, G., Piégay, H., Bravard, J.-P. and Arneson, C. 1995. Channel metamorphosis, floodplain disturbance, and vegetation developpment: Ain river, France. *Geomorphology* 13: 121–131.

Maser, C. and Sedell, J.R. 1994. *From the Forest to the Sea*, Florida: St Lucie Press.

Milne, G. and Hobley, B. 1981. *Waterfront Archaeology in Britain and Northern Europe*, CBA Research Report 41, London: Council for British Archaeology.

Milton, E.J., Gilvear, D.J., Hooper, I.D. 1995. Investigating change in fluvial systems using remotely-sensed data. In: Gurnell, A.M. and Petts, G.E., eds., *Changing River Channels*, Chichester: Wiley, pp. 277–302.

Mississippi River Commission. 1881. *Progress Report*. Washington: US War Department.

Moore, I.D., Grayson, R.B., Ladson, A.R. 1991. Digital terrain modelling: a review of hydrological, geomorphological and biological applications. *Hydrological Processes* 5: 3–30.

Mosley, M.P. 1975. Channel changes on the River Bollin, Cheshire, 1872–1973. *East Midland Geographer* 6: 185–199.

Ovenden, J.C. and Gregory, K.J. 1980. The permanence of stream networks in Britain. *Earth Surface Processes* 5: 47–60.

Park, C.C. 1995. Channel cross-sectional change. In: Gurnell, A.M. and Petts, G.E., eds., Changing River Channels, Chichester: Wiley, pp. 117–145.

Patrick, D.M., Smith, L.M. and Whitten, C.B. 1982. Methods for studying accelerated fluvial change. In: Hey, R.D., Bathurst, J.C. and Thorne, C.E., eds., *Gravel-bed Rivers*, Chichester: Wiley, pp. 783–815.

Peiry, J.-L. 1987. Channel degradation in the middle Arve River (France). *Regulated Rivers; Research and Management* 1/2: 183–188.

Peiry, J.-L., Salvador, P.-G. and Nouguier, F. 1994. L'incision des rivières dans les Alpes du Nord : état de la question. *Rev. Géogr. Lyon* 69(1): 47–56.

Petts, G.E. 1979. Complex response of river channel morphology subsequent to reservoir construction. *Progress in Physical Geography* 3: 329–362.

Petts, G.E. 1995. Changing river channels: the geographical tradition. In: Gurnell, A.M. and Petts, G.E., eds., *Changing River Channels*, Chichester: Wiley, pp. 1–23.

Petts, G.E. and Pratts, J.D. 1983. Channel changes following reservoir construction on a lowland English river. *Catena* 10: 77–85.

Petts, G.E., Muller, H. and Roux, A.L. 1989. *Historical Change of Large Alluvial Rivers: Western Europe*, Chichester: Wiley.

Pfister, C. 1992. Monthly temperature and precipitation in central Europe from 1525–1979: quantifying documentary evidence on weather and its effects. In: Bradley, R.S. and Jones, P.D., eds., *Cliamet Since A.D. 1500*, London: Routledge, pp. 118–142.

Piégay, H. 1995. Dynamiques et gestion de la ripisylve de cinq cours d'eau à charge grossière du bassin du Rhône (l'Ain, l'Ardèche, le Giffre, l'Ouvèze et l'Ubaye), 19ème-20ème siècles. Unpublished thesis in geography, Uinv. Paris 4 – Sorbonne.

Piégay, H. and Peiry, J.-L. 1997. Long profile evolution of an intra-mountain stream in relation to gravel load management: example of the middle Giffre River (French Alps). *Environmental Management* 21: 909–919.

Piégay, H. and Salvador, P.G. 1997. Contemporary floodplain forest evolution along the middle Ubaye River, *Global Ecology and Biogeography Letters* 6/5: 397–406.

Sear, D.A., Darby, S.E., Thorne, C.R. and Brookes, A.B. 1994. Geomorphological approach to stream stabilization and restoration: case study of the Mimmshall Brook, Hertfordshire. *Regulated Rivers* 9: 205–224.

Smelser, M.G. and Schmidt, V.C. 1998. An assessment methodology for determining historical changes in mountain streams, USDA Forest Service General Technical Report RMRS-GTR-6, Rocky Mountain Research Station, Fort Collins, Colorado.

Snell, F.C. 1938. *The Intermittent (or Nailbourne) Streams of Kent*, Canterbury: Hunt, Snell and Co.

Starkel, L., Gregory, K.J. and Thornes, J.B., eds. 1991. *Temperate Palaeohydrology*, Chichester: Wiley.

Taylor, E.G.R. 1967. *The Mathematical Practitioners of Tudor and Stuart England*, Cambridge: Cambridge University Press.

Trimble, S.W. 1998. The use of historical data in fluvial geomorphology. *Catena* 31: 283–304.

Trimble, S.W. and Carey, W.P. 1992. *A Comparison of the Brune and Churchill Methods for Computing Sediment Yields Applied to a Reservoir System*, US Geological Survey Water Supply Paper 2340, pp. 195–202.

Trimble, S.W. and Cooke, R.U. 1991. Historical sources for geomorphological research in the US. *Professional Geographer* 43: 212–228.

Van Urk, G. and Smit, H. 1989. The lower Rhone geomorphological changes. In: Petts, G.E., Muller, H. and Roux, A.L., eds., *Historical Change of Large Alluvial Rivers: Western Europe*, Chichester: Wiley, pp. 167–182.

Vautier, F. 1999. Dynamique fluviale et végétalisation des cours d'eau alpins endigués: l'exemple de l'Isère dans la vallée du Grésivaudan. Thesis in Geography, Université Joseph Fourier, Grenoble.

Wilcocks, W. 1901. *The Nile Reservoir Dam at Assuân*, London: E. & F.N. Spon.

Wilcocks, W. 1904. *The Nile in 1904*, London: E. & F.N. Spon.

Williams, G.P. and Wolman, M.G. 1984. *Downstream Effect of Dams on Alluvial Rivers*. United States Geological Survey Prof. Paper, 1286, 86 p.

Womack, W.R. and Schumm, S.A. 1977. Terraces of Douglas Creek, north-western Colorado: an example of episodic erosion. *Geology* 5: 72–76.

Wyzga, B. 1993. River response to channel regulation: case study of the Raba River, Carpathians, Poland. *Earth Surface Processes and Landforms* 18: 541–556.

Part III

The Spatial Framework: Emphasizing Spatial Structure and Nested Character of Fluvial Forms

5

System Approaches in Fluvial Geomorphology

HERVÉ PIÉGAY[1] AND STANLEY A. SCHUMM[2]
[1]*UMR 5600 CNRS, Lyon, France*
[2]*Department of Earth Resources, Colorado State University, CO, USA*

5.1 SYSTEM, FLUVIAL SYSTEM, HYDROSYSTEM

The System, a Widespread Concept

The system concept, increasingly used in environmental sciences during the last three decades to link physical, chemical, and biotic processes, has had an important influence on fluvial geomorphology (Chorley and Kennedy 1971, Bennett and Chorley 1978). Many textbooks dealing with the environment in general and with fluvial processes in particular are system-oriented (e.g., White *et al.* 1992, Bravard and Petit 1997).

As applied to fluvial geomorphology, the system concept has illuminated interactions with river systems and the links among geomorphology, its sister disciplines (ecology, hydrology, and human geography), and river management (Hack 1960, Chorley and Kennedy 1971, Schumm 1977). As a consequence, research in fluvial geomorphology is now strongly influenced by the system concept and has a strong interdisciplinary focus, notably in France, after the work by Roux (1982) on the "hydrosystem" and elaborated by Amoros and Petts (1993).

The concept of a system has become a tool in the sense that it is used to organize research. While providing important insights into processes, reductionist's approaches typically cannot bring a general understanding of landscapes and their evolution as holistic approaches can do. In this context, the system concept appears as a framework to develop an integrated picture of geomorphic processes and forms on larger time and spatial scales, which have appeal for river managers who seek to implement the concept of "sustainable development" and to better integrate scientific insights into management (Piégay *et al.* 1996).

The Fluvial System

A system can be defined as a meaningful combination of things that form a complex whole, with connections, interrelations, and transfers of energy and matter among them. The term fluvial is from the Latin word *fluvius*, a river, but when carried to its broadest interpretation, a fluvial system not only involves stream channels but also entire drainage networks and depositional zones of deltas and alluvial fans, and also to the hillslope sources of run-off and sediments.

The fluvial system is a complex adaptive process-response system with two main physical components, the morphologic system of channels, floodplains, hillslopes, deltas, etc., and the cascading system of the flow of water and sediment (Chorley and Kennedy 1971). The fluvial system changes progressively through geologic time, as a result of normal erosional and depositional processes, and it responds to changes of climate, base level, tectonics, and human impacts (Figure 5.1). Since at least the beginning of the Neolithic, human activities have played a major role in fluvial system evolution, affecting vegetation cover, base level, as well as water, sediment, and organic matter inputs at timescales which may be very short compared to those on which climate and tectonic changes are usually acting (Park 1981, Gregory 1987a). Therefore, there can be considerable variability of fluvial-system morphology and dynamics through time. In addition, there is great variability in space, as a result of different geologic, climatic, topographic, and societal environment. The prediction and postdiction of fluvial system behavior is greatly complicated by this variability (Figure 5.1).

At the channel scale, we conventionally summarize the fluvial system as a set of variables, some being the control/external/independent variables (e.g., Q_s, the

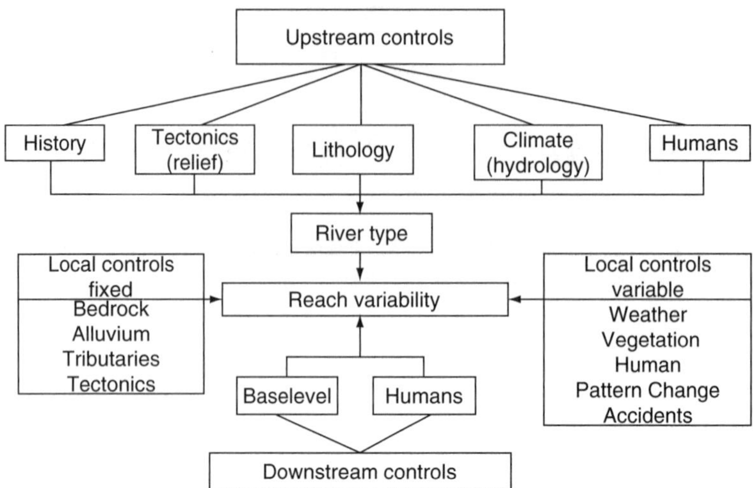

Figure 5.1 The fluvial system, a conceptual model of the geomorphological functioning of a river basin focusing on the channel reach variability associated with upstream, local, and downstream controls

sediment load and Q_P, the peak discharge) and the others are the adjustable/internal/dependant variables (e.g., channel pattern, meander wavelength, channel slope, width, and depth). The river is then seen to be in a dynamic equilibrium when the adjustable variables vary slightly around an average through time. When the control factors change, the fluvial system undergoes a correlative change, the dependant variables adjusting to a new equilibrium. This stage is called a fluvial metamorphosis and is often illustrated in scientific literature by a pattern change.

Some systems adjust rapidly to changes within the catchment, while others are more resistant. The thresholds for change (i.e., an erosional or depositional adjustment) vary from one system to another. Within the French Alps, gravel mining has been very active during the last three centuries, propagating channel degradation for kilometers upstream and downstream of the mining sites. Degradation has exceeded 10 m in reaches where shallow (ca 4 m) gravel layers are underlain by fine lacustrine deposits of the post-Würmian glaciation (Peiry 1987, Peiry *et al.* 1994). In Würmian periglacial areas, channel degradation was rapidly stopped upstream and downstream of the mining sites because it exhumed blocks transported by paleofloods, which now armor the bed.

Changes in conditions along the channel margin can also induce channel changes independently of upstream controls (Figure 5.1). In the Drôme watershed (1640 km^2), also in France, a major rockslide in the year 1442 at km 81 (measured upstream from its confluence with the Rhône) has strongly influenced the upstream channel characteristics. Here, the channel has a more gentle slope (0.003 versus 0.005 at the confluence with the Rhône) and a stable single-bed meandering channel. Without the damming effect of the landslide, a steep slope and a braided channel would be expected in this reach. Change in vegetation cover within the floodplain can also induce a channel metamorphosis. Riparian vegetation loss from fire, grazing, or mechanical removal can increase bank erosion and favor channel widening and shifting (Orme and Bailey 1970). Conversely, channel narrowing was observed in many rivers in France during the 20th century due to increase in bank resistance by the establishment of riparian vegetation after abandonment by agriculture (Liébault and Piégay 2002).

Fluvial systems range in scale from that of the vast Amazon River system (draining nearly 7 million km^2) to small badland watersheds of a few square meters. Fluvial systems can also be viewed over time periods ranging from a few minutes of present-day activity to channel changes of the past century, and even to the geologic time required for the development of the billion-year-old gold-bearing paleo-channels of the Witwatersrand conglomerate of South Africa.

To simplify discussion of the complex assemblage of landforms that comprise a fluvial system, its longitudinal dimension is traditionally subdivided into three zones (Schumm 1977). Zone 1 is the drainage basin, watershed, or sediment-source area. This is the area from which water and sediment are derived. It is primarily a zone of sediment production, although sediment storage does occur there in important ways. Zone 2 is the transfer zone, where, for a stable or graded channel, input of sediment can equal output. Zone 3 is the sediment sink or area of deposition (delta, alluvial fan). These three subdivisions of the fluvial system may appear artificial because obviously sediments are stored, eroded, and transported in all the zones; nevertheless, within each zone one process is normally dominant through time. However, the sequence of sediment source zone to transport zone, and possibly deposition zone, can be repeated many times along a river with active sediment sources.

Each zone of the fluvial system, as defined above, is an open system. Each has its own set of morphological attributes, which can be related to water discharge and sediment movement. The components of the Zone 1 morphological system are related to each other. For example, valley-side slopes and stream gradients are directly related to drainage density (Strahler 1950) and components of the morphological system (channel width, depth, drainage density) can be related statistically to the cascading system (water and sediment movement, shear forces, etc.).

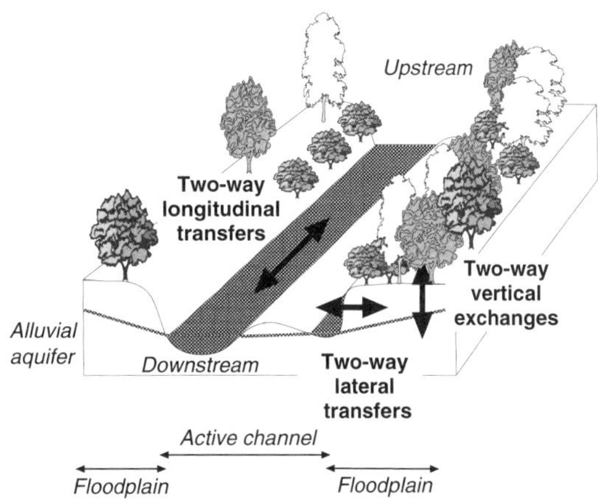

Figure 5.2 The hydrosystem, a complex system with three bidirectional axes: upstream/downstream, channel/margins, surficial/underground environments

The Hydrosystem Concept

As geomorphologists increasingly interact with other environmental scientists, geomorphic processes are considered in relation to biological processes and human actions. Predicting human effects on river systems at a timescale of multiple decades allows us to better evaluate the societal costs of human actions, to understand trade-offs between societal uses (e.g., leisure activities, navigation) and natural resources (water, gravel, forest, fish, hydroelectricity) supplied by the river.

The concept of the "hydrosystem" provides a framework within which to evaluate such interactions (Roux 1982, Amoros and Petts 1993). The hydrosystem can be defined as a 3D system dependent on longitudinal (upstream/downstream), lateral (channel/margins) and vertical (surficial/underground) transfers of energy, material, and biota (Figure 5.2). Its integrity depends on the dynamic interactions of hydrological, geomorphological, and biological processes acting in these three dimensions over a range of timescales. The system components are interrelated in the sense that many fluxes may be bidirectionnal.

The longitudinal dimension is defined by upstream–downstream relationships. For example, alluvial channel form is controlled by the sediment input from upstream, and changes in sediment supply can lead to aggradation, no change in bed elevations, or incision. In turn, channel changes such as degradation and armoring can influence fish habitat. Downstream factors may also affect upstream ones. A drop in base level (e.g., from sea level lowering or in-channel gravel mining) can induce regressive erosion upstream, which in turn can expose rock outcrops and undermine check dams, which can become barriers to anadromous fish migration.

In the lateral dimension, the bidirectional links between the main channel and its margins are particularly complex within alluvial corridors. In the valley bottom, paleoforms (terraces, alluvial fans, screes) commonly influence channel characteristics. The geological setting influences slope and width of the valley floor, and consequently channel slope and pattern. The lower valley of the Ubaye River, a tributary of the Durance in the southern French Alps, is characterized by an unusual successional pattern (Figure 5.3). It is braiding across a large valley cut in marl, but becomes progressively more meandering and then straight downstream as it traverses more resistant rocks, becoming narrower, with a higher gradient and coarser bed load (Piégay *et al.* 2000a). Channel behavior controls the floodplain architecture, and consequently its biological diversity. A freely meandering river has the capacity to create cut-off channels, within which perifluvial aquatic zones support exceptionally diverse ecosystems. Their life span at the aquatic stage depends on their efficiency in terms of sediment trapping

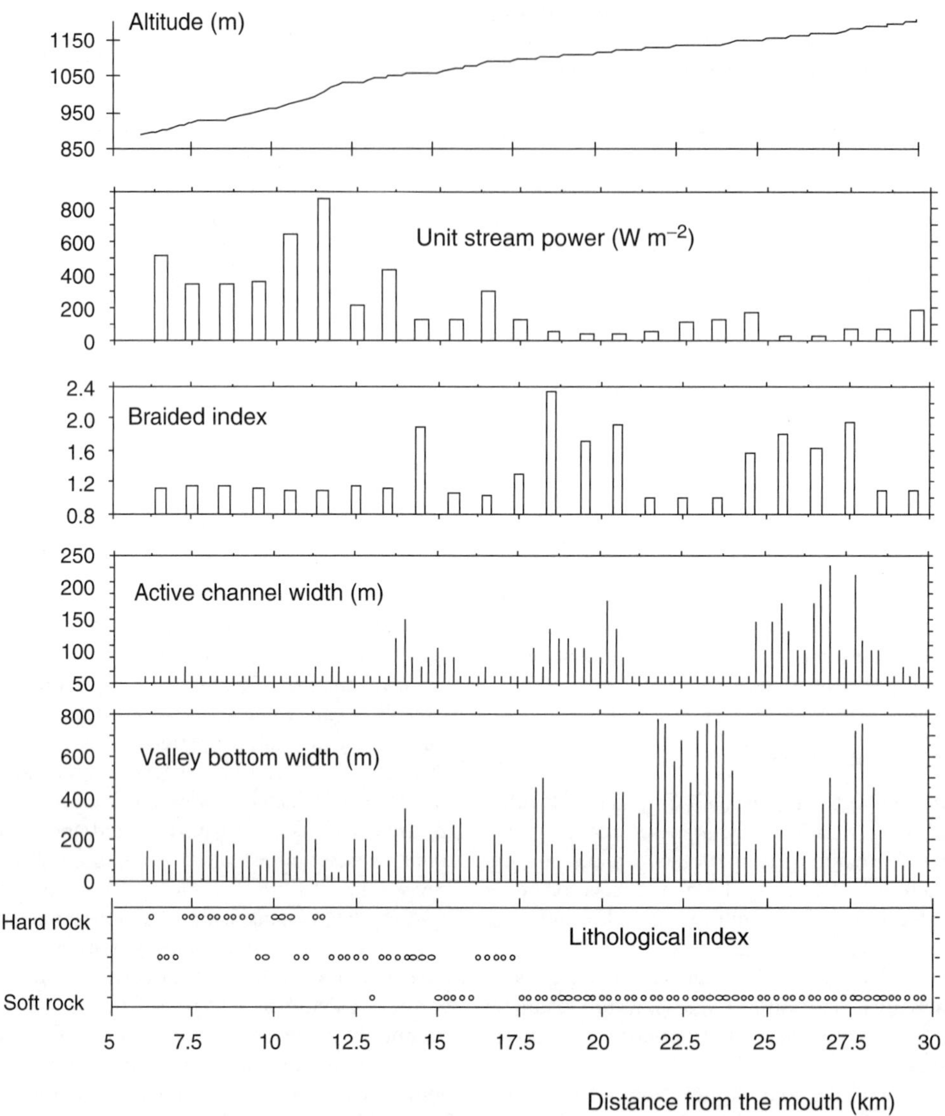

Figure 5.3 The longitudinal continuum observed along the downstream reaches of the Ubaye River (France): a channel slope increase at about 11 km and width decrease associated with modifications of lateral controls (valley narrowing and increase in coarse sediment input downstream) (modified from Piégay *et al.* 2000a)

(frequency of flooding and critical shear stress) and by the upstream characteristics of the basin (sediment supply, flood magnitude, and frequency).

The third dimension corresponds to vertical interrelationships. For example, channel degradation or aggradation may induce changes in the biological and chemical functioning of the floodplain. Channel degradation, e.g., may induce water table decline, terrestrialization of perifluvial aquatic zones, which consequently increases vegetation establishment and then sediment trapping. Several examples are given in Chapter 10 showing how ecological changes can provide information about the geomorphic adjustment of the river channel.

The hydrosystem concept can be considered as an extension of the fluvial system concept (Schumm

1977), as applied to large rivers with well-developed floodplains. It not only involves geomorphological parameters but also chemical and biological ones. While the fluvial system emphasizes temporal and longitudinal dimensions, the hydrosystem concept emphasizes the lateral and vertical dimensions, which are most important on large floodplains, and which strongly influence alluvial groundwater storage, ecological richness, and regeneration of riparian vegetation. Moreover, large floodplains are often heavily developed, so various uses and engineered structures affect natural processes, channel forms, and floodplains.

5.2 COMPONENTS OF THE FLUVIAL SYSTEM

Scales of Analysis and the Range of Influencing Factors

The fluvial system and its components can be considered at different spatial scales and in greater or lesser detail depending upon the objective of the observer. For example, a large segment, the dendritic drainage pattern, is a component of obvious interest to the geologist and geomorphologist. At a finer scale, the river reach is of interest to those who are concerned with what the channel pattern reveals about river history and behavior and to engineers who are charged with maintaining navigation and preventing channel erosion. A single meander can be the dominant feature of interest, for geomorphologists and hydraulic engineers for the information that it provides on flow hydraulics, sediment transport, and rate of meander migration. Within the channel itself is a sand bar, the composition of which is of concern to the sedimentologist, as are the bedforms (ripples and dunes) on the surface of the bar and the details of their sedimentary structure. These, of course, are composed of the individual grains of sediment which can provide information on sediment sources, sediment loads, and the feasibility of mining the sediment for construction purposes or as a placer deposit.

The fluvial system is characterized by an asymmetry of controls in the sense that the broader scale levels influence the smaller scale levels (e.g., influence of basin-scale on reach scales), while the inverse is rarely true. In the same way, changes affecting a given reach influence the structure and functioning of the sedimentological facies and the vegetation units. For example, a dam may provoke downstream incision and bed coarsening (boulder exhumation), and simplification of the channel pattern. If the broader levels are not considered, the ecologist may be unable to explain fish abundance and diversity at a reach scale relative to those elsewhere, or to design sustainable restoration or mitigation actions.

Various components of the fluvial system can be investigated at different scales, but no component should be totally ignored, because hydrology, hydraulics, geology, and geomorphology interact at all scales. This emphasizes that the entire fluvial system should not be ignored, even when only a small part of it is under investigation.

Interactions Among Zones: Generalized Examples

To further emphasize the interdependence of the three zones, the variability of the fluvial system under the influence of only three controls; stage of development, relief, and climate are summarized in Figure 5.4. Two idealized examples are shown. Example 1 is either a young, high-relief, or dry-climate drainage basin. Example 2 is either an old, low-relief or humid-climate drainage basin. Geologic conditions (lithology, structure) are the same for both examples.

For the youthful, high-relief, dry and sparsely vegetated drainage basin, the drainage density (D, ratio of total channel length to drainage area) will be high, and both hillslope inclination and stream gradient (S) will be steep. A fully developed drainage network will produce high discharge per unit area (Q), high peak discharge (Q_p), relatively low base flow (Q_b), and high sediment load and sediment yield (Q_s). The fine-textured drainage network will permit the rapid movement of water and sediment from Zone 1 to Zone 2. In Zone 2 the high sediment load, high bed load, and the highly variable (flashy) nature of the discharge will produce a bed-load channel of steep gradient, large width-to-depth ratio, low sinuosity (P, ratio of channel length to valley length) and a braided pattern. Channel shifting and change will be common. Downstream in Zone 3 the large quantity of coarse sediment will form an alluvial fan or fan delta. Deposition will be rapid, and the sedimentary deposit will contain many discontinuities and numerous sand bodies.

At the other extreme is Example 2, an old, low-relief, humid, well-vegetated drainage basin that has a low drainage density (D), gentle slopes, and low discharge per unit area (Q). A high percentage of the precipitation infiltrates or is lost to evapotranspiration. Peak discharge (Q_p) will be relatively low, and groundwater will be abundant, leading to a high base flow (Q_b). Sediment load and sediment yield will be low. This produces a suspended-load channel in Zone 2, which

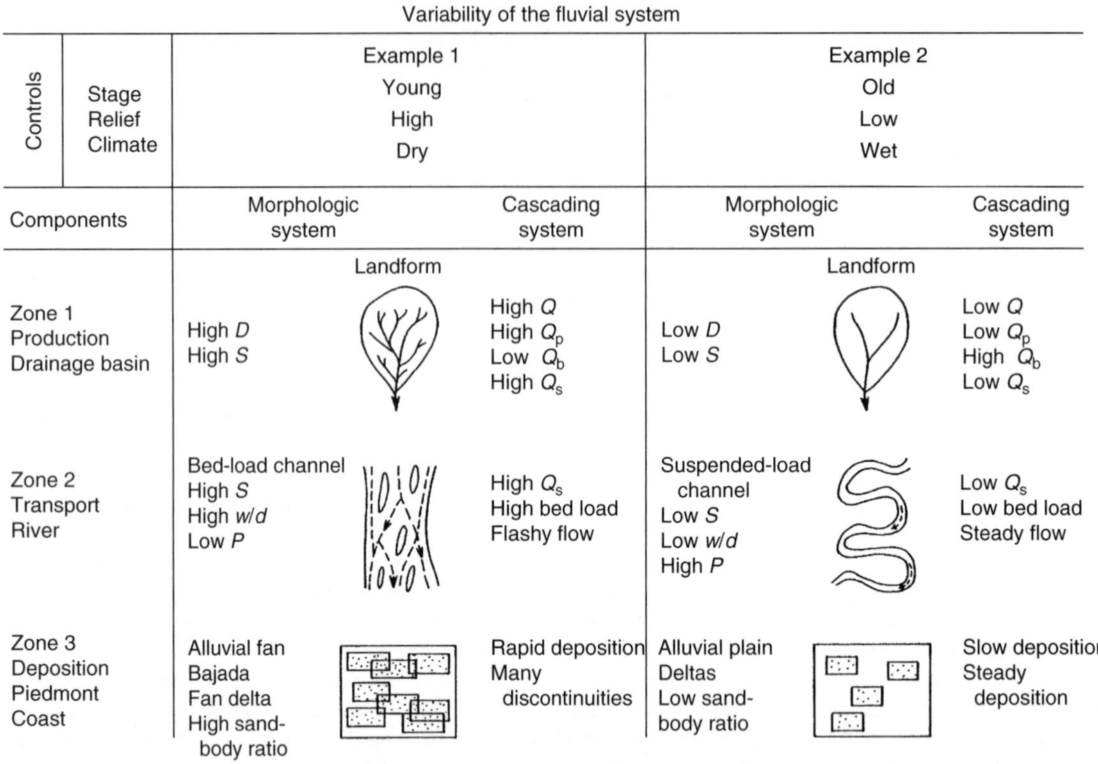

Variability of the fluvial system

	Example 1		Example 2	
Controls: Stage / Relief / Climate	Young High Dry		Old Low Wet	
Components	Morphologic system	Cascading system	Morphologic system	Cascading system
	Landform		Landform	
Zone 1 Production Drainage basin	High D High S	High Q High Q_p Low Q_b High Q_s	Low D Low S	Low Q Low Q_p High Q_b Low Q_s
Zone 2 Transport River	Bed-load channel High S High w/d Low P	High Q_s High bed load Flashy flow	Suspended-load channel Low S Low w/d High P	Low Q_s Low bed load Steady flow
Zone 3 Deposition Piedmont Coast	Alluvial fan Bajada Fan delta High sand-body ratio	Rapid deposition Many discontinuities	Alluvial plain Deltas Low sand-body ratio	Slow deposition Steady deposition

Figure 5.4 Two examples of very different fluvial systems, showing the variability of the morphologic and cascading components in the three geomorphic zones: D, drainage density; S, gradient; w/d, width–depth ratio; P, sinuosity; Q, water discharge per unit area; Q_b, base flow; Q_p, peak discharge; Q_s, sediment load. *The little boxes in Zone 3 are channels illustrating sandy-body ratios* (from Schumm 1981; reproduced by permission of Geological Society of America)

transports relatively fine sediments (fine sands, silt, and clay) at a low slope in a channel with high sinuosity and a low width-to-depth ratio. Discharge will be relatively steady, although during major precipitation events large floods will move through the valley. The fine sediment and the steady nature of the flow will cause slower rates of deposition in Zone 3, with a few sand bodies and an alluvial plain or birds-foot delta will form.

A change of climate can transform Example 1 to Example 2 or vice versa or to some intermediate stage. The character and volume of the sediments delivered to Zones 2 and 3 will also change, and significant channel adjustments will result. Without tectonic interruptions, the erosional evolution of a landscape should result in a transition from drainage basins and channels like Example 1 to those more similar to Example 2 (Figure 5.4). As the relief of the drainage basin is reduced during the erosional evolution, drainage density will decrease, slopes will decline, and the amount and grain size of the sediment will decrease. The result will be a transition from a braided to a meandering channel in Zone 2 and to finer grained, more uniform deposits in Zone 3.

The relationships displayed in Figure 5.4 are straightforward and well known. They demonstrate that, because of the number of variables acting, the fluvial system will have a complex history, as it adjusts to climatic changes and human influences through time. In addition, at any one time, the range of geology, relief, and climate will guarantee that a great range of morphologic characteristics can exist among drainage basins.

5.3 FLUVIAL SYSTEM APPROACHES

Partial Versus Total-system Approach

There is more than one approach to the fluvial system. One may be ambitious and attempt a total-system analysis integrating information on all aspects of the fluvial system, but usually there will be insufficient information to permit such a total-system approach. Rather, one may chose to investigate only the drainage network of Zone 1 or a reach of channel in Zone 2. This reduced partial-system analysis is usually all that can be attempted, but its importance lies in the value of viewing a limited problem or limited study area in a broader perspective.

For example, in the 1950s the US Geological Survey was collecting sediment yield data for the Bureau of Land Management throughout large areas of New Mexico, Arizona, and Utah. This involved annual surveys of small reservoirs, which provided sediment yield data that could be related to grazing pressures and climatic fluctuations. The second author was involved in this process during one field season and insisted on at least a quick reconnaissance of the drainage basins that were supplying the sediment to the surveyed reservoirs. This reconnaissance revealed that for the majority of the sites there were additional reservoirs upstream of the one that was surveyed annually. Obviously, the data were of little use unless the upstream reservoirs were also surveyed. In this case, by moving away from the data-collection site to another level, albeit qualitative, it was possible to evaluate the usefulness of the data that was being collected, and many data-collection sites were abandoned.

Another example involves bank erosion along the Ohio River (Schumm 1994). Riparian landowners claimed that activities of the US Army Corps of Engineers caused serious erosion of their banks. Inspection of 20 sites revealed that there was, indeed, erosion. This could lead to the conclusion that the landowners' claims were correct, but inspection by helicopter of much of the Ohio River revealed that erosion was occurring where it could be expected along this major river and the claims were rejected. Therefore, without an attempt to view the eroding sites in the perspective of the total river, it might have been recommended that the Corps of Engineers rip-rap the eroding banks. This was not done, and several years later it was satisfying to find that the previously eroding banks were stabilizing and being colonized by vegetation. Again, by viewing the sites in the context of the total river, the validity of the claims could be evaluated. This broader look at a very large system was productive, and the partial-system approach, although qualitative can be very valuable.

If geomorphological studies are characterized by spatial limits, they have also temporal limits. In this context, a system approach is always partial because time in the fluvial system is not bounded like a catchment but may change at the seasonal, decadal, century, or Holocene scales. Geomorphic systems can be studied at different timescales depending on the study objectives, which are to explain present geomorphic features or their sensitivity to changes in run-off and sediment yield.

The Fluvial System, a Model Resulting from the Research Process or a Preliminary Hypothesis

The "fluvial system" can be seen as a conceptual model developed by the researcher based on results from a research program. Complexity is added to the original simple model through regional studies, which show the importance of effects such as riparian vegetation and geomorphic facies, and the cascading effects of geomorphic changes on living communities and human uses (Bornette and Heiler 1994, Bravard *et al.* 1997).

Once the conceptual model has been defined, it can be used as a tool to focus efforts early in research process. It is then a basis to formulate hypotheses in a deductive approach, allowing the researcher to build a preliminary rough architecture of the studied component to test the potential factors controlling it, its sensitivity to changes, the acting range of its processes and forms, the geomorphic thresholds. Thus, the fluvial system provides a simple framework into which complexities of the specific river can be placed in contrast, and within which questions can be posed, such as the potential effects of changes in peak flows or sediment load, or when currently occurring adjustment is likely to be finished (Gregory 1987b) (Figure 5.5).

Case Study Versus Comparative Analysis

The fluvial system concept can serve firstly to integrate case studies in a broader spatial and temporal scale context, considering upstream influences on channel and long-term trends. Many monographic studies, such as those of Bravard (1987) on the upper Rhône River (France), or Agnelli *et al.* (1992) on the Arno River (Italy), have been done in this perspective and brought many useful elements to underline complex history of channels. Although a single case study can facilitate the understanding of channel change and structural complexity, it is often risky to generalize the results to provide a reproducible model of how the river functions and is sensitive to given acting forces. In such context, the fluvial system concept

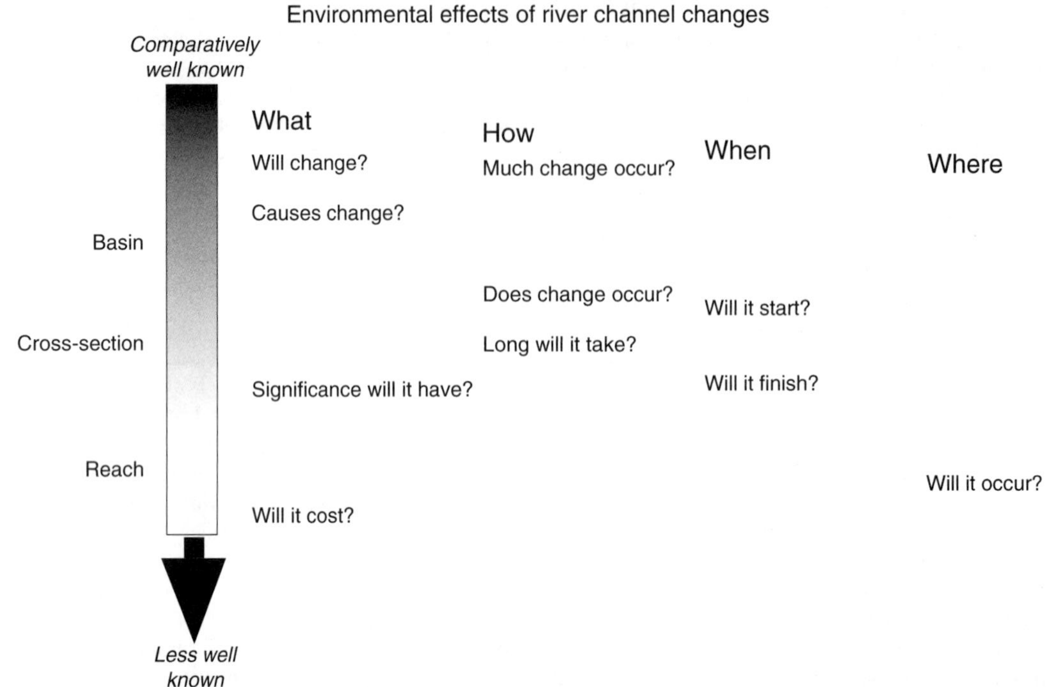

Figure 5.5 Questions posed in the study of river channel changes. Questions are located according to the scale of better known (top) to less well known (bottom) (modified from Gregory 1987b; reproduced by permission of John Wiley and Sons, Ltd.)

facilitates development of comparative studies for generalization purposes and theory.

The analysis of the fluvial system can be then based on comparisons of many spatial objects corresponding to *components* of the system. As noted above, the fluvial system is composed of different open and interacting *components* which are nested (drainage pattern > river reaches > dominant features > sedimentary patches) and described by *attributes* (e.g., a channel reach can be defined by its geometry, water and sediment processes, morphological changes). Each component can be compared to others at a single scale by comparing their attributes or the study can focus on the interactions between the nested components. Two conceptual approaches can be then distinguished.

Similarity analysis focuses on a set of landforms at a single spatial scale for which the comparison of the attributes allows groups to be identified and ordered (Figure 5.6A). This approach is commonly used in partial-system analyses. Similarities or dissimilarities can be assessed at various spatial scales: e.g., a set of channel reaches to study similarities within a set of bars, a set of basins or of reaches within a basin to study similarities within a set of channel cross-sections. This analysis is mostly synchronous in the sense given by Amoros and Bravard (1985), involving essentially simultaneous measurements at numerous sites, results of which can be analyzed statistically.

Connectivity analysis focuses on the links between nested components of the fluvial system (e.g., basin and channels, channel reach and former channels) to better evaluate the sensitivity of the lower, dependant components to changes in processes in upper component (Figure 5.6B). In this context, it is not a comparative analysis of attributes of single scale components, but it is a comparative analysis of changes affecting different sets of single scale components from which causes (agents, chronology) can be determined. Connectivity analysis can also be termed integrated analysis when it concerns links between a catchment and the channel reach (Van Beek 1981, Kirby and White 1994). Connectivity analysis means that elements are interacting in a bounded physical system, here, the watershed. It focuses upon the relationships between morphological units or *components* (e.g., relationships between sub-watersheds, between river reaches, between a

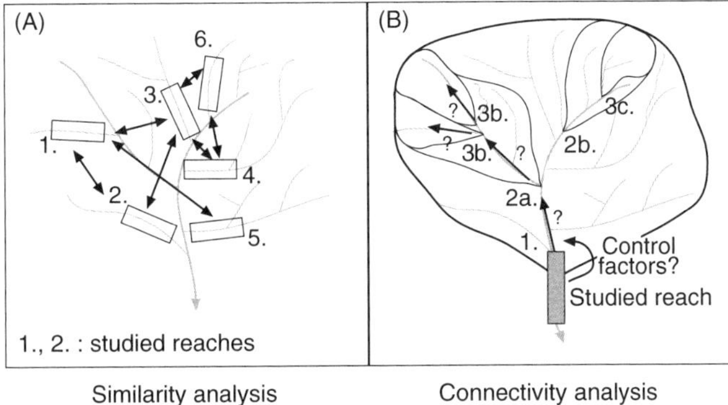

Figure 5.6 Principles of the *similarity analysis* focusing on a set of comparing components (A) and *connectivity analysis* focusing on cascading sediment routing through nested components (B)

watershed and the river channel or between a river channel and its floodplain) integrated within the watershed.

Diachronic or *retrospective analysis* (Amoros and Bravard 1985), involving assessment of changes of a single component over time, using historical sources, is important in this approach. According to the intensity, spatial extent, and chronology of these changes, it is then possible to evaluate the causes of changes of the dependant components, their respective importance, and to assess their present and future effects. Diachronic analysis can identify different adjustments occurring in various components of the system and the timing of those adjustments, especially in relation to changes in land use or other independent watershed variables. The combination of these analyses can provide insights relevant to management questions such as causes of coastal erosion when riverine sand supply has been reduced by dams or in-channel mining.

Thus, *similarity analysis* and *connectivity analysis* (Figure 5.6) are the two ways to study the fluvial system in a comparative manner, one focusing on single scale components, the other focusing on the links between components of different scales. The size and heterogeneity of the study area, the question posed, and the causes of changes should determine which approach (or combination of the two) is chosen. Within a *connectivity approach*, *similarity analysis* can be done at each scale level if a set of components is studied.

In this context, the ways of approaching a river can be summarized by a set of 3D diagrams (Figure 5.7), each axis being respectively the spatial scale (e.g., in-channel feature unit, channel reach unit, floodplain unit, catchment unit), the timescale (season, year, decade, century), the number of spatial objects or *components* being considered. The basic approach focuses on a single spatial unit observed at a single scale and without temporal perspective (upper left diagram). This is the first level of a case study, the description of the geomorphological characters of the spatial unit, which usually is augmented by studying its sub-units in an integrated perspective and its changes over time. The ultimate level of the case study approach considers all the characters of the spatial object: its inner complexity and the relationships between its sub-units (*connectivity analysis*) through time (lowest left diagram). A similar approach can be taken in a comparative perspective (right side of diagram). Rather than describing a single spatial unit, many are described simultaneously to identify differences among them or to order them on a longitudinal or a temporal gradient (similarity analysis). The second level of comparative studies, which is one of the most developed in the geomorphological literature, is to compare nested spatial objects, typically a channel reach, and its catchment. This is the base of hydraulic geometry analysis (Hey 1978, Ferguson 1986) or allometric studies, which are based on empirical power functions, relating catchment size, usually basin area, to channel geometry (e.g., width, depth, cross-sectional area, channel length, area of alluvial fans) (Church and Mark 1980). Here are studies combining both *similarity* and *connectivity* approaches. Comparative studies can also consider the temporal trend of each spatial object to consider differences in adjustment rather than difference in structure. The most advanced approaches compare multiple nested components over time, as done in the French pre-Alps (Liébault *et al.* 1999), and the Bega River in Australia (Brierley and Fryirs 2000). See examples developed later.

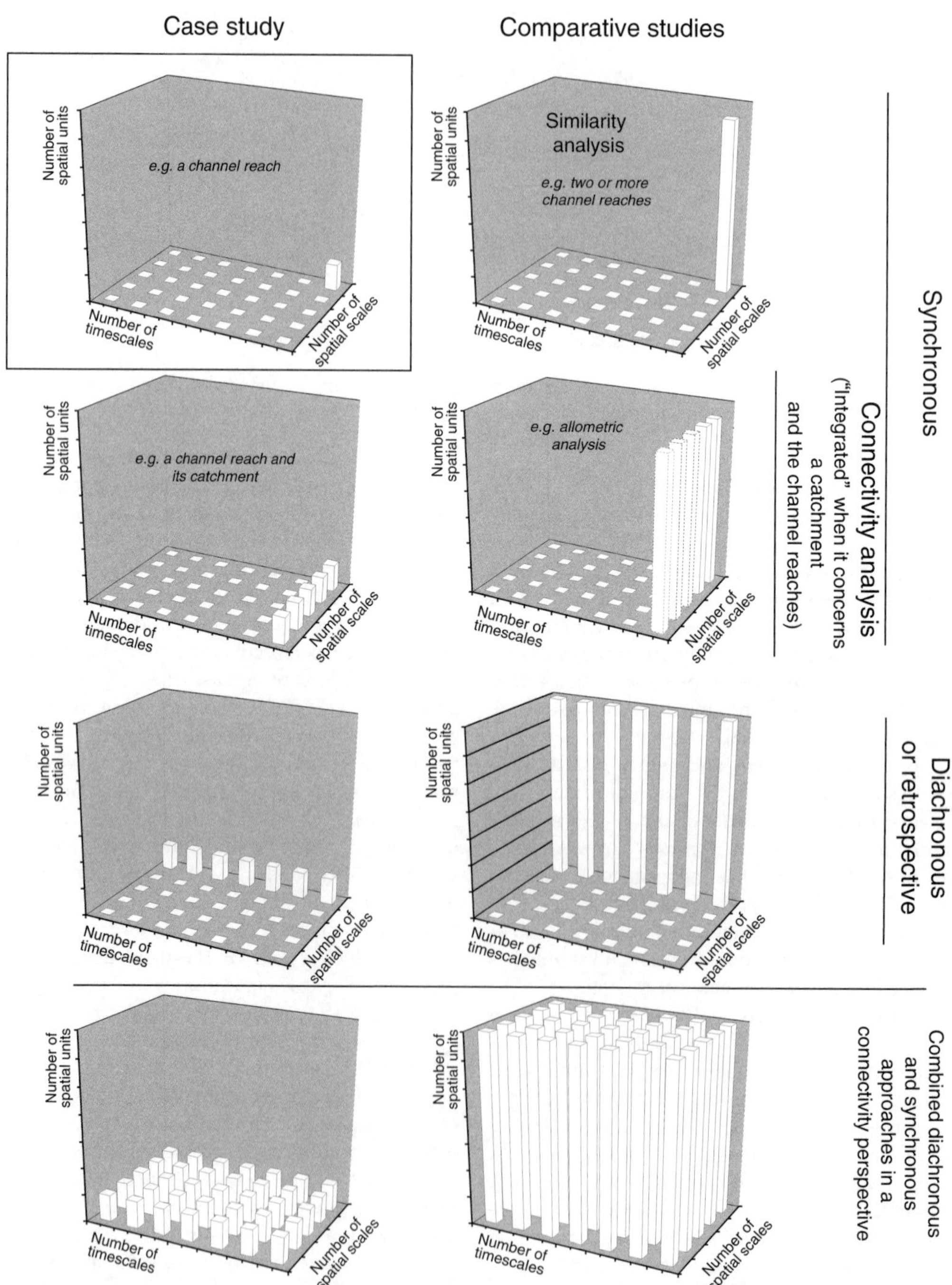

Figure 5.7 Schematic 3D models to summarize how fluvial systems can be studied based on timescale, spatial scale, and number of spatial objects considered. Few different approaches are then possible: monographic or comparative, synchronous or diachronous, based on similarity or connectivity assessment

Quantitative Versus Qualitative Analysis

The system approach is flexible, in that it can be fully qualitative but also very quantitative, supported by experiments and simulations. It can be developed with an increasing precision from expertise to detailed scientific analysis of each of its components, and it can be adapted to the management needs of a particular river. Fully qualitative approaches (e.g., geomorphological expertises) are popular in river management to assess different engineering options.

One can use the system concept to pose preliminary hypotheses, and then as a framework within which to combine other geomorphic tools. Such holistic approaches are best used in conjunction with reductionist's ones, the first providing an understanding of the river functioning at a large spatial and temporal scale while the second can test hypothesize linkages and simulate processes and changes (Richards 1996).

Experimental (flume) studies have been widely used to validate preliminary hypothesis posed by a larger systemic approach (Schumm *et al.* 1987). The approaches are complementary, as field observations can reveal the complexity of the systems without clearly distinguishing the respective importance of controlled factors, which is more accurately done by experiments.

Empirical approaches with large sample sizes can help calibrate models based on physical laws and identify their boundary conditions and the extent of their applicability. This can be done with comparative studies to identify threshold or correlation between attributes of components in similarity or connectivity approaches. Allometric analysis is a well-known quantitative approach based on the fluvial system, which facilitated major developments in geomorphology (Church and Mark 1980). The fluvial system concept also underlined efforts to establish discriminate models of fluvial pattern (Bridge 1993) and regression models linking discharge and channel forms (Hey and Thorne 1986). All these empirical models are based on large samples of spatial objects, each one being a binomial of nested components catchment-channel, to establish similarities.

Moreover, similarity analyses facilitate development of field experimental studies to test hypotheses and identify threshold conditions. Comparative analysis (paired or multi-catchment approach) can then be used to evaluate the effects of human actions on the natural environment, as by Trimble (1997) at a short timescale, and by Brooks *et al.* (2003) at a longer timescale concerning riparian vegetation effects on channel geometry. These approaches can be used in river restoration projects for which experimentation in natural conditions is necessary to improve proposed mitigation measures. Henry and Amoros (1995) proposed to compare a restored reach (geometrical modifications of cut-off channels and their hydrological connections with the main channel) with a control reach unaffected by restoration works but whose functioning is similar to it, to distinguish effects of the intervention from other system-wide influences (Figure 5.8). Such approaches have been used to assess restoration project effects on invertebrate populations (Friberg *et al.* 1994) and fish populations (Shields *et al.* 1997) sensitive to habitat changes. Using similarity analysis (paired or multiple spatial objects) to assess post-project geomorphic changes may require long observation periods (5–10 years) to capture high flow years in which geomorphological changes are more likely to occur at a measurable level.

Basin-scale modeling (Benda and Dunne 1997, Coulthard *et al.* 2000) can use historical data to simulate and predict channel adjustments in response to catchment changes. Geographical information system (GIS) databases can be combined with numerical modeling to reproduce sediment routing and its resultant changes in channel features (Montgomery *et al.* 1998, see Chapter 15), and allowing better predictions of potential channel response, in terms of duration and extent, and to better characterize longitudinal discontinuities and downstream changes in bed elevation, channel geometry, grain size, and habitats.

5.4 DETAILED EXAMPLES

By analyzing a set of components of the fluvial system at any spatial scale, we can distinguish them according to attributes, order them according to key geomorphological questions (such as the stage of evolution and specific process-response), and then build conceptual models (Figures 5.9A and B). *Connectivity analysis* can aid in understanding the cascading factors that control the changes of nested components, and in building causal and chronosequential models (Figure 5.9C).

Applications of Similarity Analysis

Two tools or approaches that are particularly useful in geomorphic investigations are *location for time substitution* (LTS) to develop evolutionary models of landform change, and *location for condition evaluation* (LCE) for assessing landform sensitivities or stability (Schumm 1991).

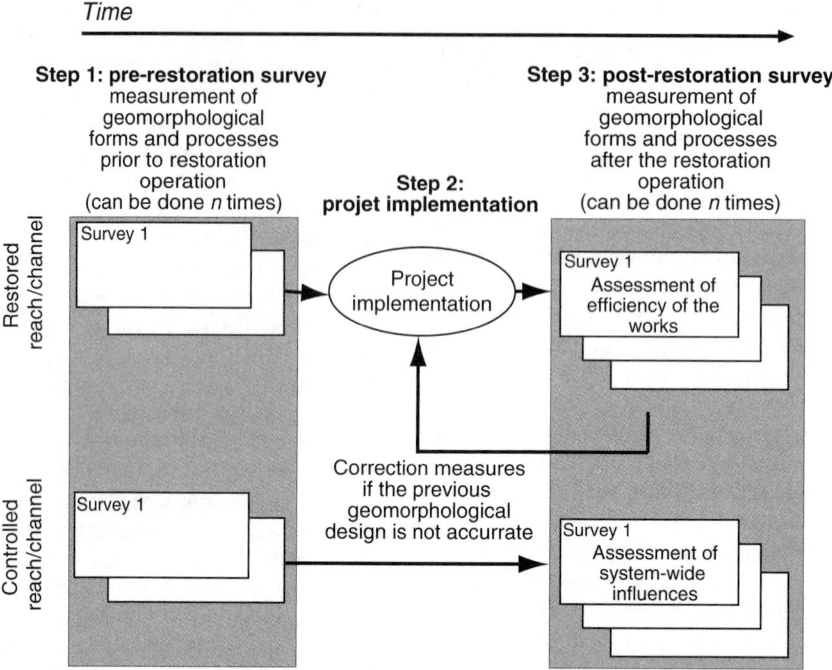

Figure 5.8 Application of similarity analysis in restoration projects. The restored site evolution is compared *n* times, at least one time before the implementation and one time after, to those of a control site unaffected in order to evaluate the efficiency of the measures done, removing the possible effects of factors affecting all the system and proposing corrected measures if the previous project does not reach the objectives (reproduced from Henry and Amoros 1995)

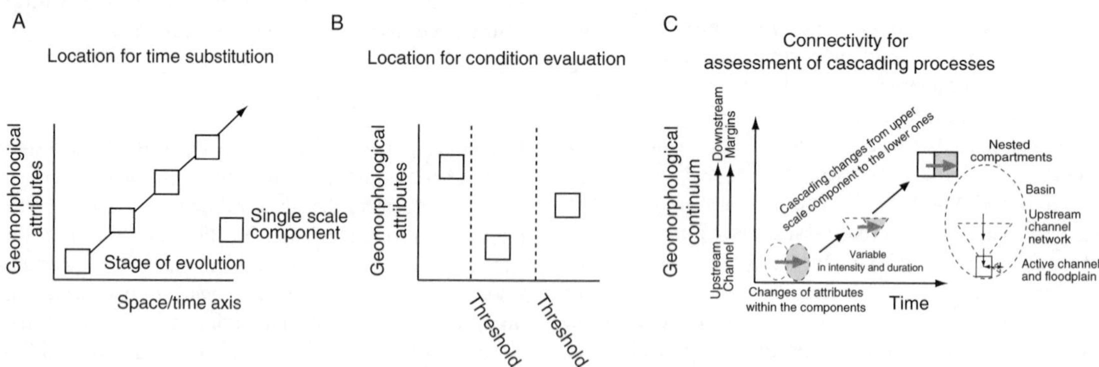

Figure 5.9 Summary of the different conceptual models used in fluvial system analysis: (A) LTS; (B) LCE; (C) connectivity model

Location for Time Substitution

The LTS is a well-known tool to geomorphologists and it is often referred to as the "ergodic" method or "space for time substitution", but following Paine (1985) and Schumm (1991), it will be referred to here as LTS. This involves the selection of a sample of landforms that can be arranged in a sequence that shows change through time (Figure 5.9A). This has been used to show hillslope, drainage network, and

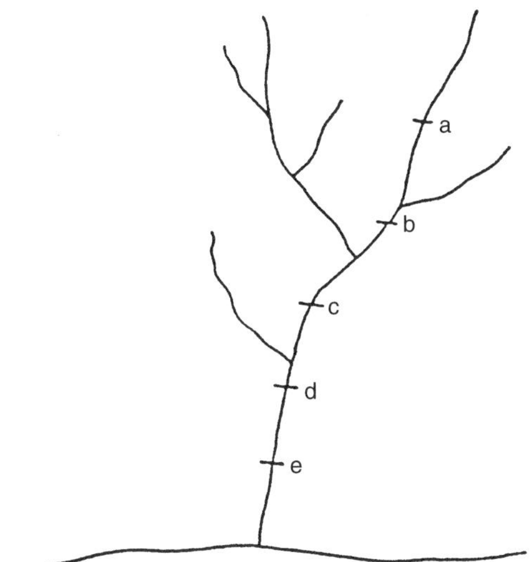

Figure 5.10 Sketch shows method used to obtain data for LTS. Measurements at sites a–e provide information for the development of a model of channelized stream (incised channel) evolution

incised-channel change with time. For example, if a series of cross-sections are surveyed along a channel (Figure 5.10) that has incised, as a result of natural or human-induced changes (e.g., channelization), an evolutionary model of channel adjustment can be developed (Figure 5.11). Glock (1931) used this technique to develop a model of drainage network evolution (Figure 5.12). In spite of severe criticism of his model, subsequent experimental studies support his conclusions (Schumm *et al.* 1987). The model presented in Figure 5.11 was developed for incised channels in northern Mississippi, and it has both academic and practical value because it permits estimation of sediment production and agricultural land loss and the identification of channel reaches that require controls (Schumm *et al.* 1984). The LTS can be an effective means of developing a model of evolving landforms.

In 1962, it was reported that a significant decrease of sediment passing through the Grand Canyon had occurred probably as a result of drought. Other explanations were advanced, but field investigations in the watershed revealed that many of the channel tributaries to the Colorado and Little Colorado River that had incised during the latter part of the 19th century were healing. An LTS study (Gellis *et al.* 1991) revealed that not only was erosion decreasing, but sediment was being stored in newly developing floodplains in the Colorado River Basin. The arroyos of southwestern USA were following the evolutionary sequence developed for the channelized streams of southeastern USA (Figure 5.13), and the LTS technique revealed the reason for decreased sediment yields in the Colorado River.

Recent research in central southern England (Gregory *et al.* 1992), Zimbabwe (Whitlow and Gregory 1989), and Arizona (Chin and Gregory 2001) used an LTS framework and a downstream hydraulic geometry analysis based on empirical relationship between channel cross-sectional area at bankfull and the basin area. Urbanized catchments were characterized by increased flood frequency and consequent channel degradation and widening. As a result the urbanized catchment-channel deviated from the general relationship by being wider and deeper than what the model predicts. Data collected in Fountain Hills Basins (Arizona) showed the expected downstream increase in channel width, depth, and capacity with drainage area while the data collected in reaches disrupted by urbanization yield channel widths up to two times wider than the upstream reaches.

In using LTS it is then important to compare features produced by the same processes that are operating under the same physical conditions. For example, the evolution of an incised channel in alluvium can be determined by surveying cross-sections at several locations where the channel is in alluvium (Figure 5.10), but one cannot combine data or compare channels in weak alluvium with channels in resistant alluvium or bedrock and expect to find meaningful results. Nor can one develop a model of drainage network evolution (Figure 5.12) if the lithology at each site is different. Therefore, if one is asked to evaluate the stability of a site it is wise to search for similar site conditions within the same general area, and to use these to aid in the specific site evaluation. Observed sequences reflect exclusively the temporal evolution step by step which means that we assume that the studied features are in disequilibrium with their controlled factors.

Location for Condition Evaluation

The second technique can be used to identify sensitive landforms in what can be termed the LCE (Figure 5.9B). This involves measuring the characteristics of relatively stable and unstable landforms. A comparison permits the identification of critical threshold conditions and sensitive landforms. Using such a technique, we assume that each case is in dynamic equilibrium with its controlling parameters, which means that threshold conditions depend on the intrinsic characteristics of the system (e.g., geological or climatic setting).

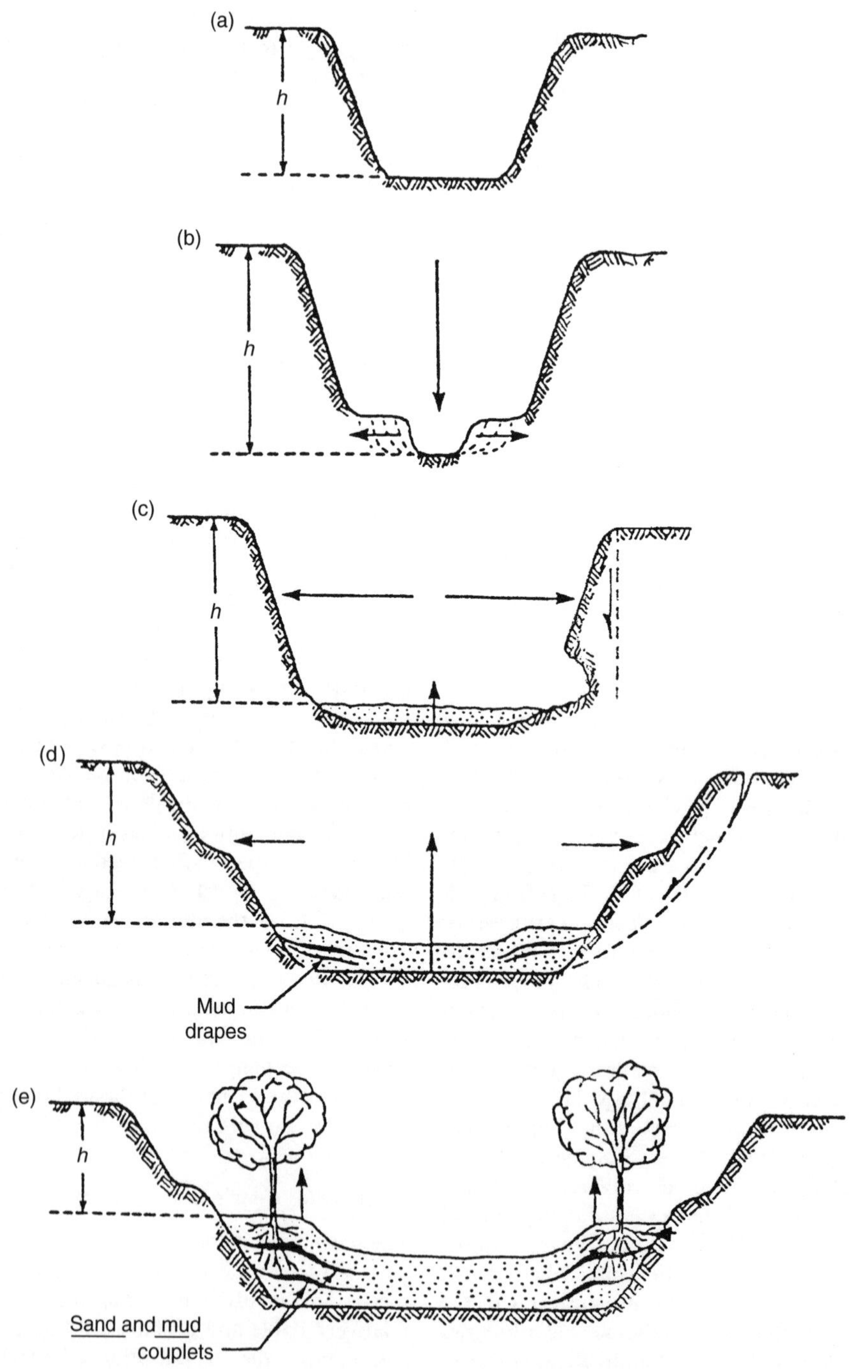

(a)
h

(b)
h

(c)
h

(d)
h
Mud drapes

(e)
h
Sand and mud couplets

Figure 5.11 Evolution of incised channel from initial incision (a, b) and widening (c, d) to aggradation (c, d) and eventual stability (e) (modified from Schumm *et al.* 1984)

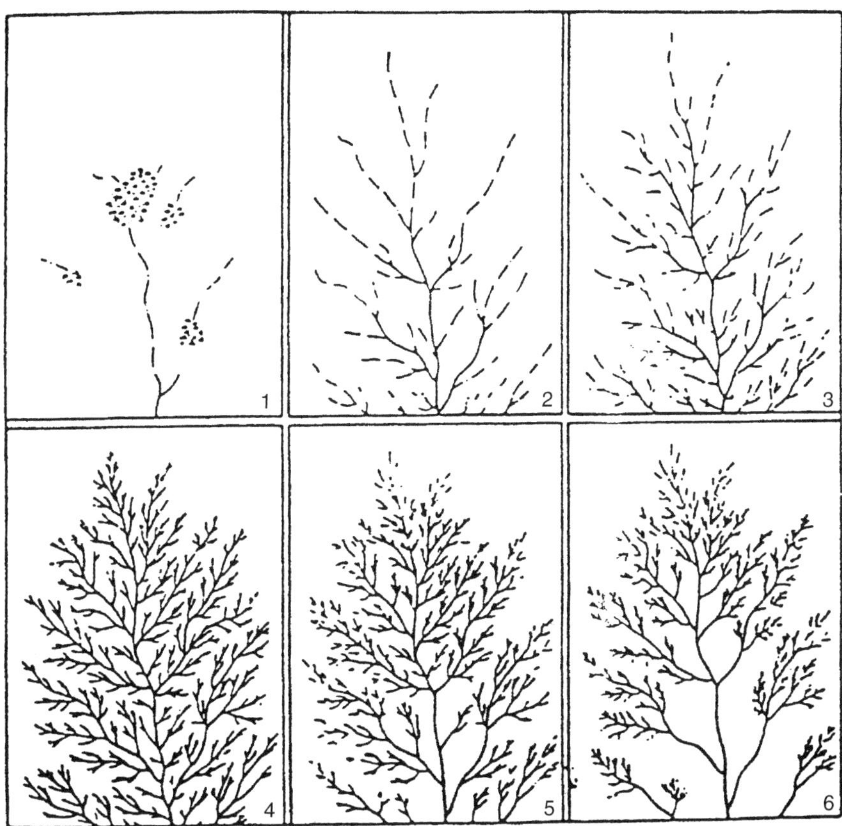

Figure 5.12 Evolution of drainage network based upon LTS by Glock (1931). Each pattern was obtained from a different topographic map and arranged in a sequence from young (1) to old (6)

This technique has been used to identify sensitive valley floors that are likely to gully in Colorado (Figure 5.13) and New Mexico (Patton and Schumm 1975, Wells *et al.* 1983), river reaches that are susceptible to a pattern change from straight to meandering to braided (Figure 5.14) alluvial fans that are susceptible to fan-head incision (Schumm *et al.* 1987), and thresholds of hillslope stability (Carson 1975). This approach is similar to the LTS, as described above, except that it is the present conditions rather than an evolutionary model that need to be evaluated.

In each of these cases, data were collected at a number of locations, and a relation was developed to identify future or threshold conditions. For example, the slope of the line on Figure 5.13 identifies a valley floor slope at a given drainage area at which gullies are likely to form. When a relation such as that of Figure 5.13 is developed between drainage area and alluvial-fan slope, alluvial fans that are susceptible to fan-head trenching can be identified. The curve of Figure 5.14, when developed for a specific river, can be used to identify when a river pattern is susceptible to change from meandering to braided and vice versa. In addition, the vertical position of a point on the plot is an indication of future change. For example, the point with the highest sinuosity (A) represents a river reach where cut-offs will occur, whereas the point that plots very low (B) is adjusting for previous cut-offs, and sinuosity will be increased by meander growth.

Debris flow activity along the western front of the Wasatch Mountains in Utah provides an excellent opportunity for an LCE study. Since settlement in the Salt Lake area, debris flows have caused considerable damage to agricultural lands and buildings. The

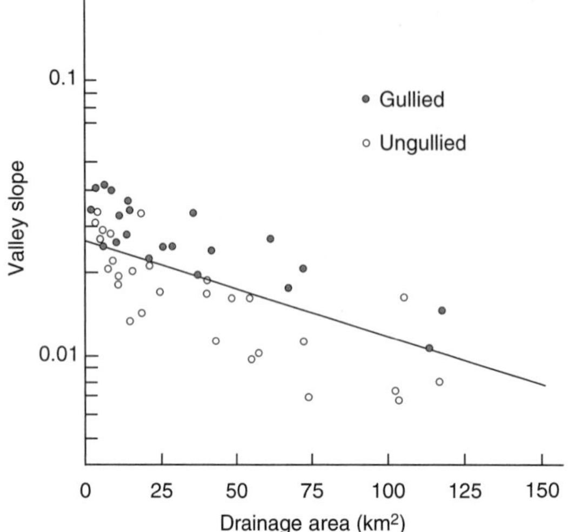

Figure 5.13 *Location for condition substitution* permits identification of threshold valley floor slope at which gullies form (from Patton and Schumm 1975; reproduced by permission of Geological Society of America)

situation is aggravated by increased urban development. In 1983, severe storms triggered debris flows that again caused damage and concern that the future would undoubtedly be plagued by additional debris flows. An engineering solution was suggested; debris basins should be constructed at the mouths of the canyons that produced debris flows in 1983. However, geologists who visited the sediment source areas (Zone 1) discovered that the watersheds that produced debris flows in 1983 had flushed much of their stored sediment, and therefore, they were unlikely to produce debris flows in the foreseeable future (Lowe 1993, Keaton 1995). In fact, if a threat of future debris flows existed, they would more likely be generated in the watersheds that had not produced debris flows in 1983. The geological observations, when related to debris flow locations, explained debris flow occurrences. In addition, an LCE study of the Wasatch Mountains could permit development of a relation between basin characteristics and the likelihood of future debris flows. Measurements of valley-floor gradient in each watershed could yield a relationship similar to that of Figure 5.13 to aid in identification of drainage basins likely to produce debris flows in the future.

Large alluvial rivers are significantly controlled by sediment characteristics, hydrology, and hydraulics. Therefore, one could assume that careful and detailed study of a short reach (10 km) would provide information that could be used to describe the morphology and dynamics of hundreds of kilometers of channel. However, a partial-system study would reveal the error of this assumption and reinforce the conclusion that a narrow approach to a landform investigation can lead to error, which can be avoided by viewing the reach in the context of the total river. The previous Ohio River example provides support for this position.

The lower Mississippi River between the junction of the Ohio River and the junction of the Red River

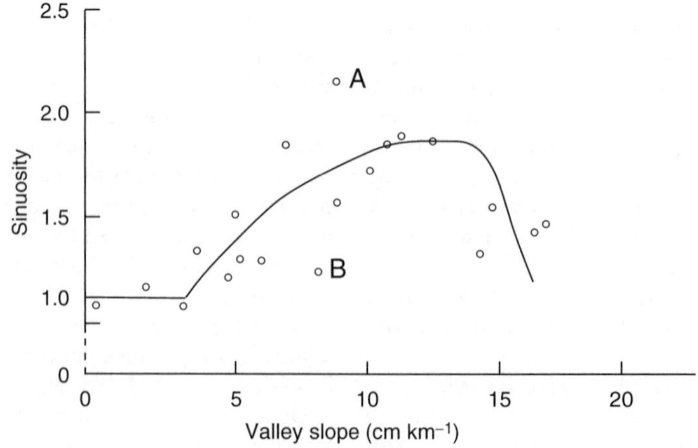

Figure 5.14 *Location for condition substitution* of valley floor slope at which 1890 Mississippi River channel pattern changed from low sinuosity to meandering and from meandering to a transition meandering-braided pattern (high to low sinuosity) as valley slope varied (modified from Schumm *et al.* 1972)

with the Mississippi can be divided into 24 different reaches, as based upon behavior during 150 years and recent morphologic conditions (Schumm *et al.* 1994). Even on a small-scale map it is possible to detect the difference among the reaches. The variability is related to the slope of the water surface or valley floor (Figure 5.15A), which also explains the behavior of the reaches through time (Figure 5.15B). For example, relatively steep reaches have higher sinuosity and great sinuosity variations through time as compared to gentler reaches. The LCE evaluation of these reaches provides a means of predicting reach behavior in the future.

The floodplain of the Ain River (France) has numerous cut-off channels, which range widely in habitat types as they evolve from fully aquatic to terrestrial, as they silt-up through time. This range of habitat types results in high biodiversity. LCE analysis has been conducted on 11 cut-off channels in order to understand their silting dynamics and assessed their sensitivity to terrestrialization (Piégay *et al.* 2000b). The conventional model of sedimentation rate, decreasing as a function of time, as those established by Hooke (1995), has not been observed in these cut-off channels, the youngest forms (20 years old) having thicker overbank sediment than others of 65–80 years old (Figure 5.16A). Retrospective analysis showed that river metamorphosis, from braided to meandering occurred in the last six decades in response to basin land-use changes and flow regulation by upstream dams. As a consequence, the geometry of cut-off channels has also been modified, from mainly straight and steep before 1945, to wandered and meandered after.

Because the geometry of the cut-off channels changed, their sediment trapping efficiency also changed. The sedimentation is then opposite to the expectation that older channels would have accumulated more sediment but it is consistent with the differences in channel geometry. The meander cut-off channels are mainly flooded by backwaters and consequently experience high deposition rates, while abandoned braided channels function as secondary channels during floods and are less susceptible to fine sedimentation due to high flow velocities (Figure 5.16B).

Geomorphologists collect data at many locations to develop evolutionary models (LTS) or to determine the sensitivity of landforms (LCE) for practical purposes of prediction as well as for environmental reconstruction. Of even greater value, both techniques require that the investigator backs away from a single site and looks at many sites, which provides the "big picture" and a basis for generalization.

Applications of Connectivity Analysis

Connectivity analysis is based on a "hydrosystem" framework, which assumes that geomorphological attributes of a component result from multiple adjustments, which have cascading effects on the other attributes (biological ones mostly). The aim is to highlight the cascading causal factors of observed changes and to estimate the relaxation time (Figure 5.9C).

Such models can be time-oriented (Figure 5.17) or component-orientated (Figure 5.18). In the first kind of models, the studied component is the end of a nested system with higher hierarchical levels. When the changes affecting the different hierarchical levels are studied and dated, it is possible to plot them on a temporal axis and identify higher scale changes that explain those observed at a lower scale. The common example concerns the links between a watershed and its channel network, but links between a channel reach and its former channels can also be analyzed, leading to conceptual models of the effects of one level on others, evaluate the intensity and duration of the transmission of the changes downstream or from the channel to its margins. In component-oriented models, the temporal scale and relaxation timescale are less established, but the cascading changes of attributes from one component to another are more clearly modeled.

The East Fork of Pine Creek, Idaho, provides a good example of time-oriented connectivity analysis. Kondolf *et al.* (2002) documented channel widening in the 20th century, and they identified two potential causal factors:

(i) a bed load supply increase from the sub-watersheds caused by mining activities and mining waste inputs;
(ii) an increase of bank sensitivity to erosion from grazing and logging on the floodplain. Detailed study of the chronology of the geomorphological phenomena and their potential causes indicated that the first factor predominated (Figure 5.17).

Good examples of connectivity analysis with component-oriented modeling were given by Bravard *et al.* (1997). These authors described the general trends in river incision in France during the 20th century, underlined the causes and geomorphological consequences, and effects of incision on ecosystems of the alluvial plains, such as riparian vegetation, aquatic vegetation of former channels, benthic and hyporheic macroinvertebrate communities, and fish assemblages. Conceptual models of cascading factors from geomorphological components to biological components

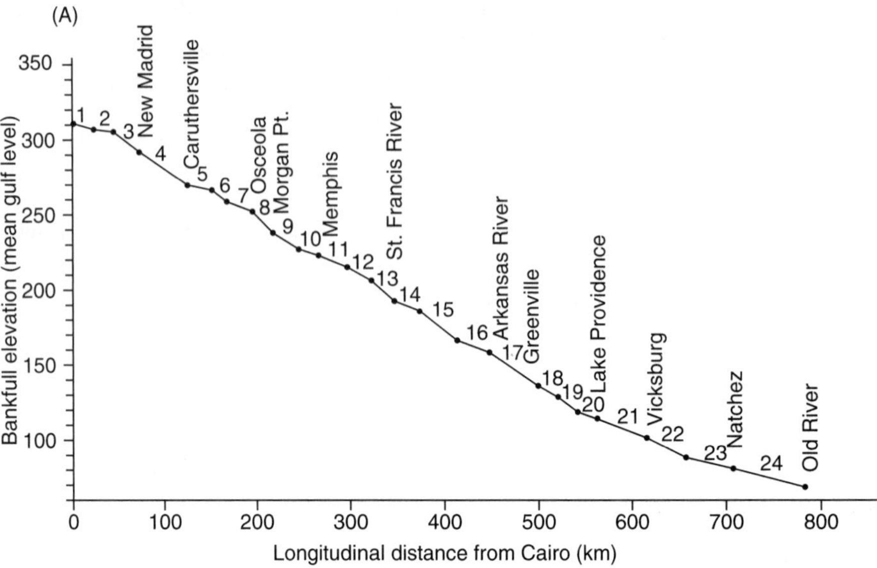

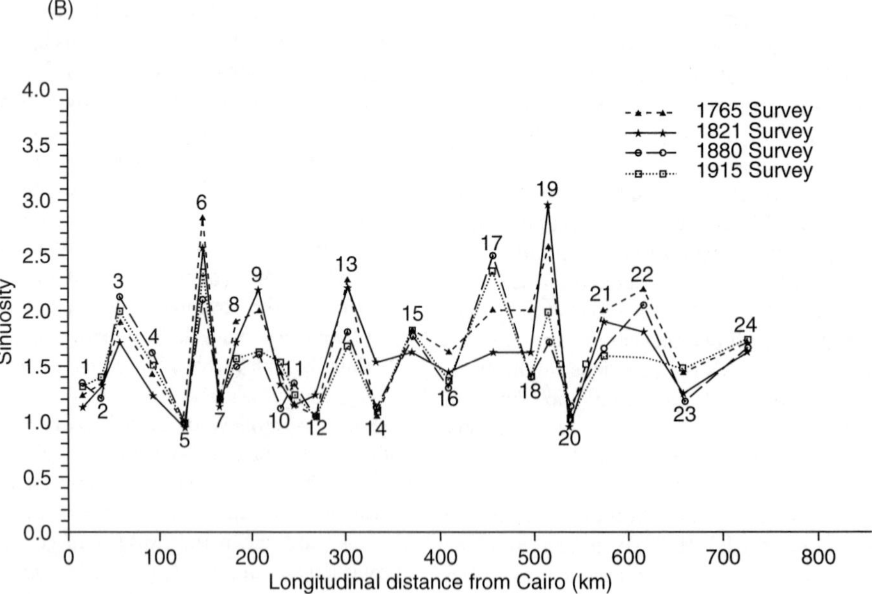

Figure 5.15 Reach of the pre-1930 lower Mississippi River course: (A) projected-channel (floodplain or water surface) longitudinal profile based on 1880 bankfull elevation; (B) average sinuosity for each reach of A, 1765–1915. Numbers identify reaches (from Schumm *et al.* 1994; reproduced by permission of the American Society of Civil Engineers)

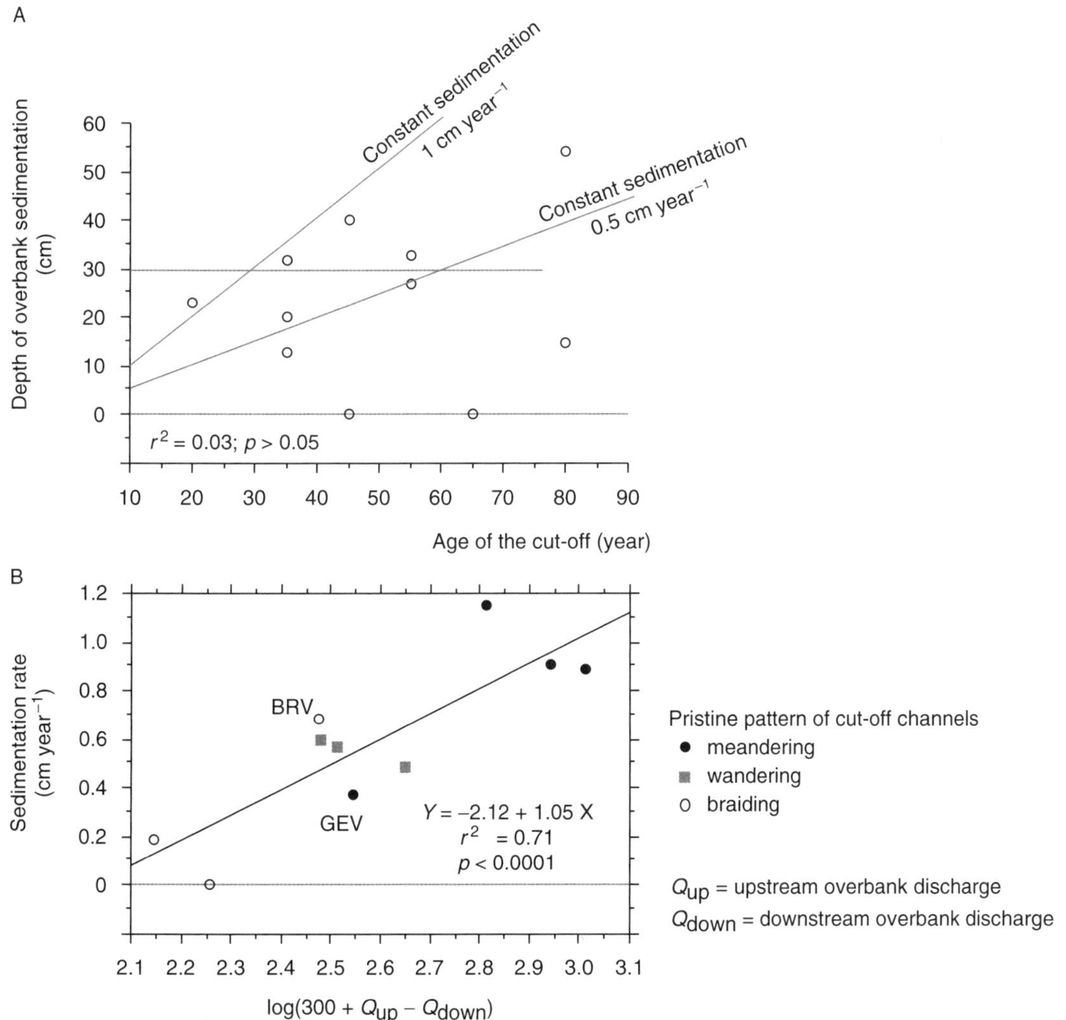

Figure 5.16 *Location for condition substitution* study applied to the former channels of the Ain River. (A) a statistical independence between the channel cut-off age and the thickness of overbank sedimentation; (B) an overbank sedimentation rate controlled by the pristine geometry of the former channels (reproduced by permission of John Wiley and Sons, Ltd.). See text for explanation

show how vertical channel changes (e.g., aggradation, incision) affect interactions between the channel and former channels, notably rates of fine sediment deposition, and rates of vegetation succession (Figure 5.18).

Coupling Similarity and Connectivity Analysis

Research done using a system approach commonly combines multiple approaches such as *connectivity* and *similarity* (mainly LTS) analysis, as illustrated by three case studies.

Ain River, France

Along the Ain River, vegetation encroachment and channel narrowing in the 20th century was initially attributed to decreased peak flows due to upstream dam built in 1968. However, studies of the chronology of vegetation encroachment on the Ain and other Rhône River tributaries (Figure 5.19) showed that

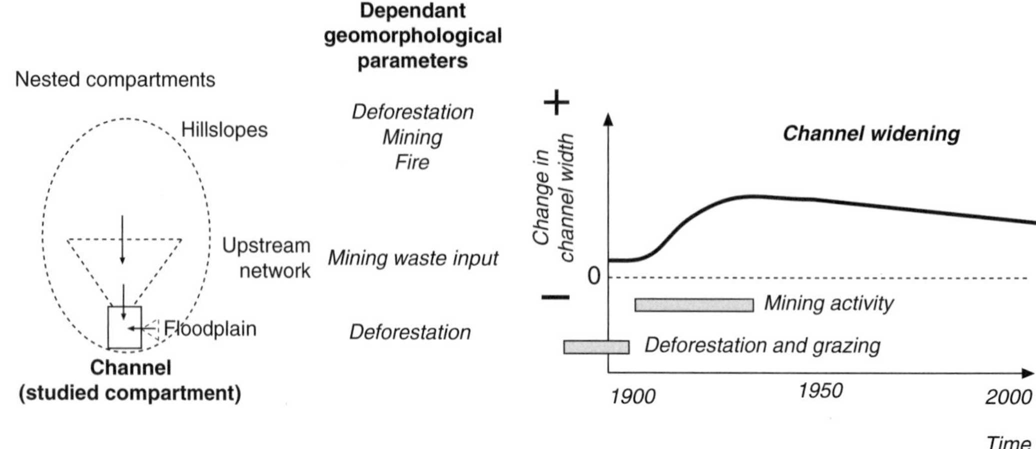

Figure 5.17 Conceptual models of time-oriented connectivity analysis: effects of basin land-use changes on downstream channel morphology (East Fork Pine Creek, Idaho)

the encroachment could not be explained primarily by dam-induced flow changes (Piégay *et al.* 2003a). Vegetation encroachment began on the Ain before the dam, and was observed on other rivers whose peak flows were not disrupted by dams, evidently affected by land-use changes on the floodplain and in the catchment. Two main types of vegetation encroachment were observed:

(i) Early in the 20th century, vegetation encroached along braided mountain reaches (such as along the Ubaye River in the 1920s) in response to decreased bed load supply from afforestation of the catchment and installation of check-dams from 1880 to 1910.
(ii) Vegetation encroached from 1945 to 1970 along the Ain and other rivers in the region (e.g., the Eygues, Roubion, Drôme, Ouvèze, Loire, and Allier Rivers), whether influenced by dams or not.

This encroachment occurred as the floodplain was abandoned, pastures were replaced by forest, and trees colonized gravel bars. In this study, a historical analysis conducted on the Ain showed the vegetation encroachment preceded the dam. This was then confirmed for other rivers in the region. The multiple case studies were the basis for an LTS analysis, from which a conceptual model of channel changes over the 20th century was developed. The chronology of changes was key to understanding the causal relations. Mountain reaches located close to the sediment sources were the first to show vegetation encroachment, due to decrease in sediment delivery, while piedmont reaches downstream experienced encroachment later, probably reflecting both floodplain land-use change as well as lately decreased sediment input.

Bega River, Australia

Brierley and Fryirs (2000) and Fryirs and Brierley (2000) illustrated the importance of considering geomorphological adjustment to human impacts at a broad scale to frame river management and biological restoration. They used a total-system perspective to assess consequences of European settlement on fluvial forms of the Bega River (1040 km^2) on the south coast of the New South Wales. The pre-settlement river was characterized by extensive swamps along the middle and upper reaches and a continuous low capacity channel along the lowest gradient reaches. European settlement strongly modified hydrologic regime, sediment supply and transfer, and bank resistance, producing widespread channel widening and incision. To identify the character, capacity, and stages of river recovery, comparative analysis (LTS) was based on retrospective analysis (e.g., archival plans, explorer's accounts, old ground and air photographs, hydrological data analysis to derive critical discharges) and on current field observations and measurements (long profiles and channel cross-sections, description of valley floor sedimentary structures, valley and channel morphology).

Channel features varied in space because of internal characters of reaches (mainly valley morphology and

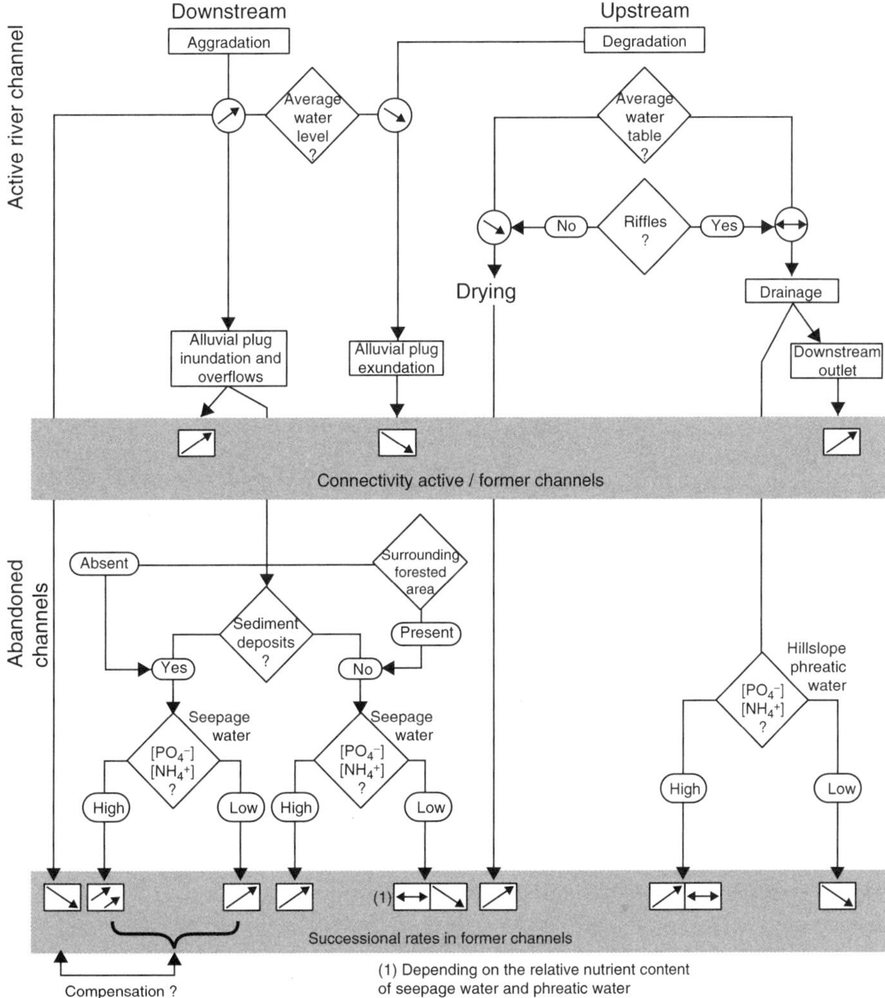

Figure 5.18 Conceptual model predicting effects of vertical main channel changes on vegetation successional rates in former channels (Bravard *et al.* 1997). Two components are distinguished, the main channel (the upper scale level) and the former channels (from the lower scale level, reproduced by permission of John Wiley and Sons, Ltd.), and also the changes of their attributes with vertical main channel change. The long arrows express the cascading effects from one attribute to the other whereas the arrows in squares express the sense of these effects (increase, decrease, or constant)

distance downstream from the sediment sources) and in time because they have not reached the same adjustment stage at time t. Several homogeneous structural reaches were identified, within each of which the LTS model produced a distinct set of evolutionary stages:

(i) the cut and fill river style, in wide, fully shaped valleys with steep slopes,

(ii) the transfer style valley occupying the mid-catchment reaches, bedrock confined with a lower gradient and a valley width up to 200 m, and

(iii) the floodplain accumulation river style in downstream reaches with a wide and low slope valley (Figure 5.20A).

With a detailed and fully documented evolutionary framework of river change and an appreciation of

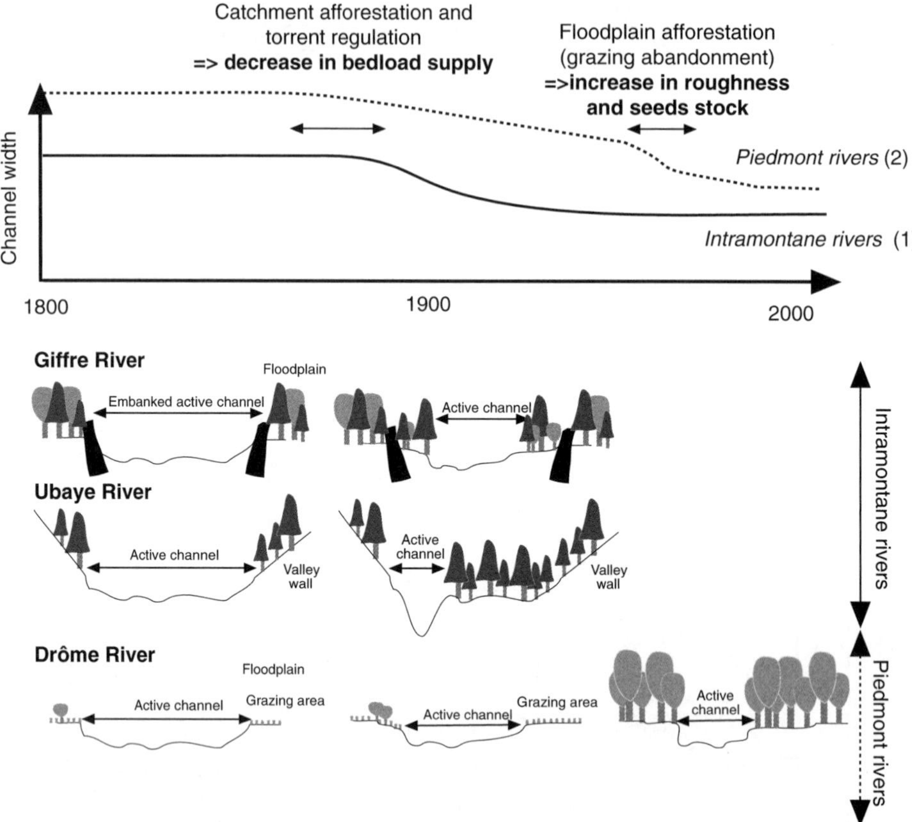

Figure 5.19 Conceptual model summarizing causes and chronology of channel narrowing affecting the alpine and piedmont tributaries of the Rhône River (France) during the contemporary period

geomorphic linkages with a catchment and associated limiting factors that may inhibit recovery potential, five stages of the LTS model were distinguished: the intact stage, the self-restored stage, the turning point stage, the degraded stage, and the created stage for which the character and the behavior of the river reach do not equate to those of the predisturbance conditions (Figure 5.20B; from Fryirs and Brierley 2000). None of the narrow channels documented historically still exist. By 1900, the degradation and widening process was well advanced (real case B) and was still acting in the 1940s (real case C). The turning point occurred in the 1960s with island formations and exotic vegetation establishment (real case D). The authors expect created conditions (predicted case E) with a low flow channel deepened and an increase in floodplain–channel connectivity with sediment removing along the channel bed.

The Drôme, Roubion, and Eygues Rivers

Research on channel incision on the Drôme (1640 km^2), the Roubion (635 km^2), and the Eygues (1100 km^2) also illustrates the application of different conceptual tools presented in this chapter to understand the system evolution and to inform management decisions. Since 1994, an integrated analysis has been conducted on these systems located in the southern French pre-Alps, focusing on multiple temporal and spatial scales, and a wide range of tools (Figure 5.21). The rivers, all drain westward from limestone mountains under 2000 m in elevation, and flow into the middle Rhône River.

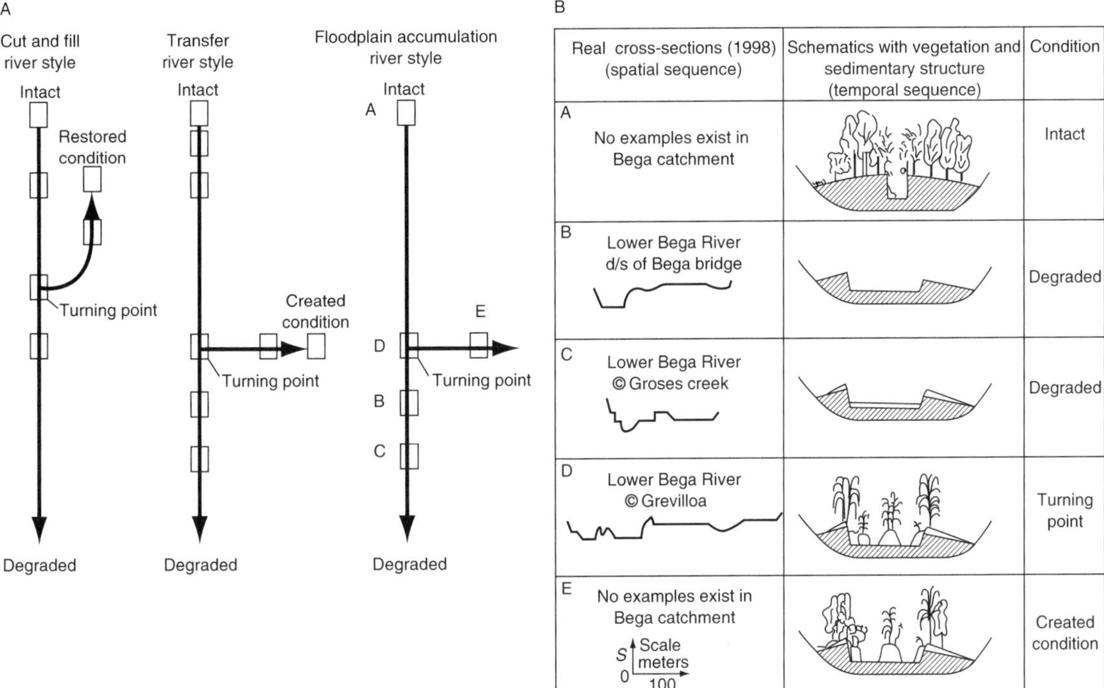

Figure 5.20 Conceptual model illustrating the temporal positions of different reaches of the Bega River system according to their style and the characters of the adjustment following human disturbances. A shows the different potential evolutionary stages according to the style and B gives stage examples concerning the floodplain accumulation river style (Fryirs and Brierley 2000; reprinted with permission from the Physical Geography, Vol. 21, No. 3, pp. 244–277, V.H. Winston & Son, Inc., 360 South Ocean Boulevard, Plan Beach, FL 33480. All rights reserved)

The approach utilized (Figure 5.21) can be considered as an integrated or total-system analysis, as the geomorphological question is posed within basins, but with nesting information from sub-units. In this broad context, at each given spatial scale, comparisons are made between (i) the Drôme, the Eygues, and the Roubion and (ii) the set of sub-basins, the tributary reaches, the main river segments (Figure 5.21). The approach is then based both on *similarity analysis*, as single scale components are compared (e.g., downstream alluvial reaches of the tributaries), and *connectivity analysis,* as the changes occurring on the upper levels are compared to those occurring at a lower level (downstream reach of tributaries in relation to their respective basins).

On the Drôme River, incision averaging 3 m was observed in downstream reaches where gravel mining was concentrated. Channel degradation has caused serious environmental problems, such as reduced channel dynamics and riparian vegetation regeneration, groundwater draw down, and undermining of levees and other infrastructure. River managers have recognized the need to assess the causes of the degradation beyond the reach scale and to manage bed load on longer timescale than previously (e.g., decades instead of years) and over a larger spatial scale (i.e., from upper reaches to the Rhône instead of reach scale).

Enlarging the scope of analysis to the tributaries showed that the bed load supply from the catchment was reducing as a result of afforestation and erosion control since the 19th century (Landon *et al.* 1998). It was then necessary to understand the sediment transfer changes affecting the watershed, to develop precise chronologies and analyze causes, and to identify the active and potential sediment sources.

A *similarity analysis* was conducted on 50 subwatersheds with field measurements (geometry and grain size analysis, scour chains and tracers in order to assess the bed load transport, erosion pins to evaluate the inner bed load input, Cs-137 and Pb-210 profiles but also dendrochronology to precise chronology of

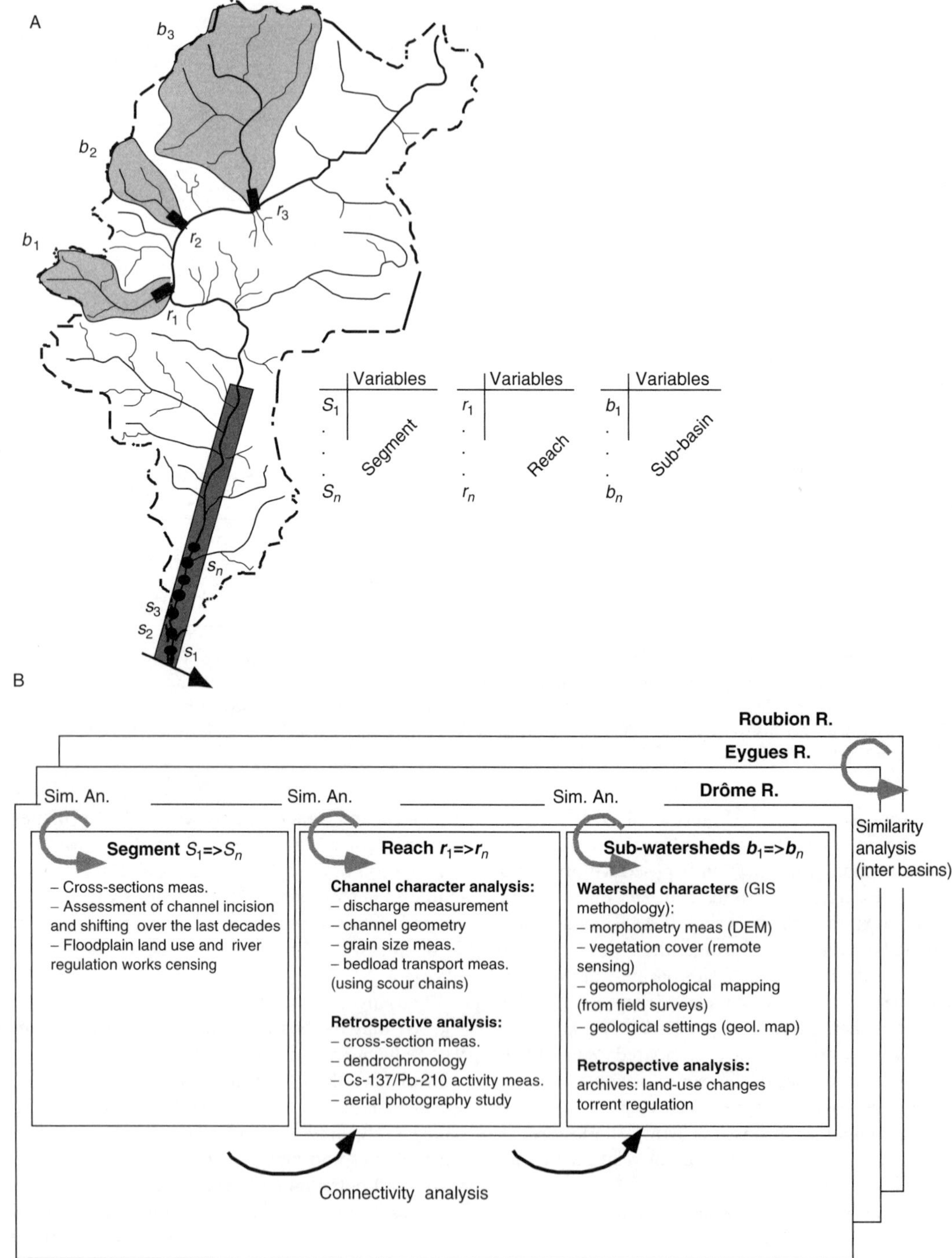

Figure 5.21 General schedule of geomorphological researches based on both similarity and connectivity analyses within the south pre-Alps of France for assessing factors controlling the main stem changes: (A) theoretical basin with nested components: tributary basins, tributary main reaches, main stem segments; (B) data collection

channel narrowing and deepening) in an historical perspective, aerial photographs taken in 1945, 1970, and 1995, and a land-survey map of the middle 19th century being studied. A GIS developed from a digital elevation model (DEM), remote sensing images, air photos, and also archival data (maps, diagrams, written documents) was used to evaluate changes in vegetation and stream regulation, and multivariate statistical analysis was performed to identify similarities among sub-systems (Figure 5.22). Three fundamental phases of the approach are outlined in Table 5.1. Phase 1 is a detailed investigation of the study site in Zone 2. Phase 2 is an expansion of the study to adjacent landforms and to Zone 1. Phase 3 involves collection of historical information and the integration of the results of all three phases.

The connectivity analysis showed that tributaries still actively yielding high sediment loads are typically high-gradient, with well-developed steep headwaters and many contacts between the stream network and highly erodible geomorphological units (Figure 5.22 for the Drôme case; Liébault *et al.* 1999). The results indicated that a self-restoration process following the mining period was possible on the Eygues, but not on the Drôme. Even with a history of gravel mining comparable to the Drôme, the Eygues still experienced high bed load delivery from the watershed, because the basin is more influenced by a Mediterranean climate (the vegetation cover is less extensive and the rainfall is more intense) and because of its geological setting, which resulted in a rapid transfer of sediment from the valley slopes to the channel. Of the three, the Roubion River has experienced the greatest reduction in sediment supply. An LCE analysis then indicated potential threshold factors explaining differences in channel adjustment among the three river systems.

5.5 CONCLUSIONS

A system approach is useful in fluvial geomorphological research, as it can provide an holistic framework within which to organize research, to understand sediment routing, and to integrate sediment sources and their spatial and temporal variability (Table 5.2). The approach is also useful in river management as it permits hydraulic, hydrological, socio-economic and ecological questions to be posed simultaneously, and answered to solve interdisciplinary and applied problems.

Table 5.1 List of elements to be considered in a coupled similarity—connectivity analysis, the example of the Drôme, Roubion, and Eygues catchments

Basic axis		Description
1	Present forms and processes within the study site	Geomorphic description of the reach (channel geometry and grain size), analysis of processes, e.g., assessment of reach conveying and trapping capacity (measurement of velocity, discharge and MES concentration, bed load transport, sedimentation rate within the floodplain, channel shifting, and bank stability)
2	Spatial enlargement considering the floodplain/ valley bottom and/or the catchment	Study of floodplain (stratigraphy and geometry, vegetation cover, land use) and watershed characteristics (hydrographic network, basin morphometry, rainfall distribution, geology and vegetation patterns) from fieldwork and laboratory procedures (remote sensing and GIS analysis)
3	Temporal enlargement considering the channel, the floodplain/valley bottom, and/or the catchment	Study of changes over time in channel form and the variables listed above. At this stage, research not only may consider biological, physico-chemical, geoarcheological, and sedimentary indicators but also archives (stream gauging records, plans for regulation of channel reaches, etc.). Written archives can be useful to understand the character and the chronology of land-use changes and the previous state of the system. The historical analysis should be conducted at an holistic scale, encompassing the nature and timing of changes affecting the neighboring floodplain, the upstream channel network, or the whole catchment

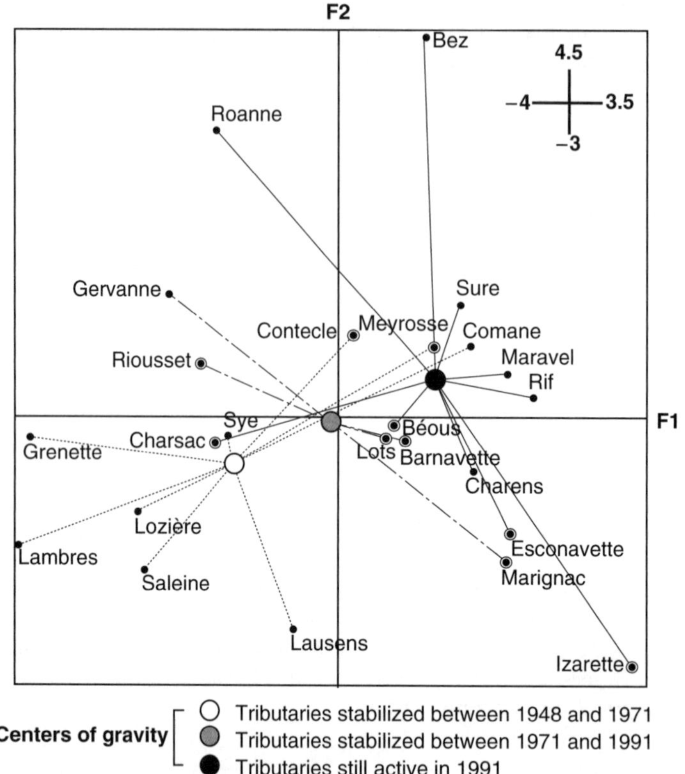

Sub-watersheds which have been strongly restored by the *MLR service* (the restoration area represents more than 30% of the watershed area)

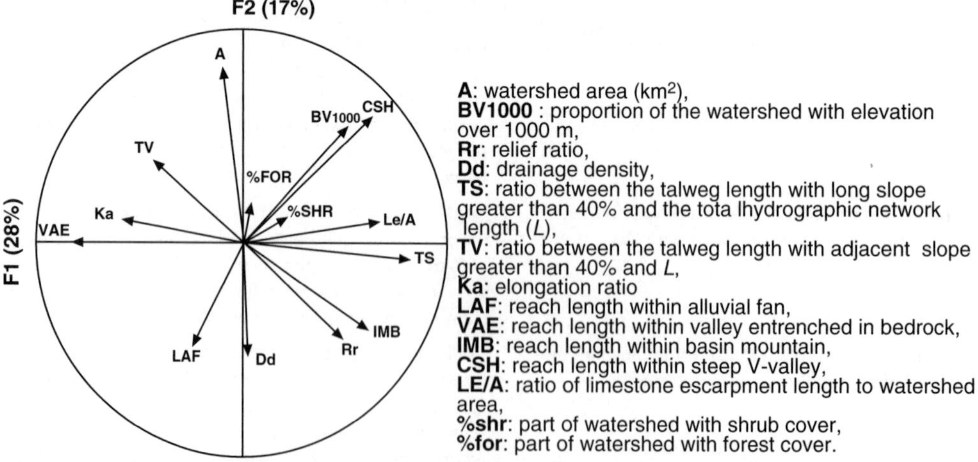

F2 (17%)
F1 (28%)
A
BV1000
CSH
TV
%FOR
%SHR
Ka
VAE
Le/A
TS
LAF
Dd
Rr
IMB
A: watershed area (km2),
BV1000 : proportion of the watershed with elevation over 1000 m,
Rr: relief ratio,
Dd: drainage density,
TS: ratio between the talweg length with long slope greater than 40% and the tota lhydrographic network length (*L*),
TV: ratio between the talweg length with adjacent slope greater than 40% and *L*,
Ka: elongation ratio
LAF: reach length within alluvial fan,
VAE: reach length within valley entrenched in bedrock,
IMB: reach length within basin mountain,
CSH: reach length within steep V-valley,
LE/A: ratio of limestone escarpment length to watershed area,
%shr: part of watershed with shrub cover,
%for: part of watershed with forest cover.

Figure 5.22 An Example of integrated analysis performed on the Drôme Basin to highlight basin characters which control tributary narrowing. The two graphs give the results of a normed principal component analysis performed on 24 sub-watersheds with 14 morphometric, geomorphic, and biogeographic variables. *On the upper graph are projected the positions of the 24 tributaries grouped in three sets according to the chronology of their changes. On the lower graph, "the correlation circle", are projected the variables measured on each tributary. Their directions allow interpretation of the position of the tributaries and the sets of tributaries in the graph above* (from Liébault *et al.* 1999; reproduced by permission of Regents of the University of Colorado)

Table 5.2 Advantages and limitations of fluvial system approaches

Main advantages
Enlarge time and spatial scales when considering channel reach sensitivity allow considering medium and long-term changes and then consequences of human actions on river processes and forms
Underline geographical complexity to understand limitations of reductionist's approaches
Formulate interdisciplinary questions and apply geomorphological knowledge for ecology and engineering purposes
Main limitations
Cannot be used alone but provides a framework for formulating hypothesis, which can be tested by geomorphic tools such as experiments, mathematical modeling or GIS
Time consuming because it must cover a large area and uses multiple methods and materials (field data, documents, archives) to get robust conclusions
Based on empirical laws or expertise judgement, not necessarily on physical laws controlling river system
Strong effects of human experience and data available. Errors in interpretation are common and risks of confounding facts and interpretation of facts are high

The catchment is recognized as the obvious unit for analysis and planning, a well-defined territory. Management at the catchment level is increasingly recognized in the literature as necessary, and increasingly adopted by government agencies, as illustrated by the "watershed"-based planning by the US Environmental Protection Agency and the 1992 French water law. The French law promotes new management strategies, balancing between the human activities and environmental preservation through catchment-level planning procedures called SDAGE and SAGE (regional and catchment water management plans) (Piégay *et al.* 1994, 2002).

In this new context of river management, a geomorphological approach allows better assessment of how and at what rate natural or human changes in a given part of a watershed are likely to influence sedimentary and morphological features upstream and downstream (Newson 1994), and provide guidelines for restoration (Sear 1994, Kondolf and Downs 1996). Several authors have underlined the need to consider geomorphological framework for biological improvement strategies (Sear 1994, Downs 1995, Brierley *et al.* 1999; see also Chapter 21 for examples and detailed references) as well as engineering guidelines. Following Gilvear (1999), there are key contributions that fluvial geomorphology can make to the engineering profession with regard to river and floodplain management such as promoting recognition of connectivity and interrelationships between river planform, profiles and cross-sections, stressing importance of understanding fluvial history and chronology over a range of timescales, highlighting the sensitivity of geomorphic systems to environmental disturbances and changes, especially when close to geomorphic thresholds. Physical habitats are often mapped in detail, but the temporal evolution of habitat mosaics, various spatial scales, and the connectivity between components (nested perspective) must also be addressed to solve interdisciplinary problems (Newson and Newson 2000).

Fluvial system approaches have not only advantages, but can yield questionable conclusions when conducted without sufficient care or without adequate background (Table 5.2). When a good scientific practice is used, it can be very time consuming to collect sufficiently large data sets to describe different components of the system and careful selection of samples is needed to develop inferential statistical approach and support robust conclusions.

ACKNOWLEDGMENTS

The research in the Rhône Basin system used to exemplify the chapter was carried out by a team that included G. Bornette, A. Citterio, F. Liébault, and N. Landon, and also the senior researchers of the former PIREN team of Lyon (particularly C. Amoros, J.P. Bravard, M. Coulet, G. Pautou, A.L. Roux...) who developed the hydro-system concept, which provides the interdisciplinary framework for this work.

The authors would also like to thank G.M. Kondolf, D. Gilvear, and G.E. Petts who reviewed the text and brought forward useful comments to improve it.

REFERENCES

Agnelli, A., Billi, P., Canuti, P. and Rinaldi, M. 1992. Dinamica morphologica recente dell'alveo del fiume Arno, CNR-GNDCI, Volume 1739, 191 p.

Amoros, C. and Bravard, J.P. 1985. L'intégration du temps dans les recherches méthodologiques appliquées à la gestion écologique des vallées fluviales: l'exemple des écosystèmes aquatiques abandonnés par les fleuves. *Revue Française des Sciences de l'Eau* 4: 349–364.

Amoros, C. and Petts, G.E. 1993. *Hydrosystèmes fluviaux*, Paris: Masson, 300 p.

Benda, L.E. and Dunne, T. 1997. Stochastic forcing of sediment routing and storage in channel networks. *Water Resources Research* 33: 2865–2880.

Bennett, R.J. and Chorley, R.J. 1978. *Environmental Systems: Philosophy, Analysis and Control*, London: Methuen and Co. Ltd., 624 p.

Bornette, G. and Heiler, G. 1994. Environmental and biological responses of former channels to river incision: a diachronic study on the upper Rhône River. *Regulated Rivers: Research and Management* 9: 79–92.

Bravard, J.P. 1987. *Le Rhône, du Léman à Lyon*, Lyon: La Manufacture, 450 p.

Bravard, J.P. and Petit, F. 1997. *Les cours d'eau, dynamique du système fluvial*, Paris: Armand Colin, 222 p.

Bravard, J.P., Amoros, C., Pautou, G., Bornette, G., Bournaud, M., Creuzé des Chatelliers, M., Gibert, J., Peiry, J.L., Perrin, J.F. and Tachet, H. 1997. River incision in South-east France: morphological phenomena and ecological effects. *Regulated Rivers: Research and Management* 13: 1–16.

Bridge, J.S. 1993. The interaction between channel geometry, water flow, sediment transport and deposition in braided rivers. In: Best, J.L. and Bristow, C.S., eds., *Braided Rivers*, Special Publication No. 75, Geological Society of London, pp. 13–63.

Brierley, G.J. and Fryirs, K. 2000. River styles, a geomorphic approach to catchment characterization: implications for river rehabilitation in Bega catchment, New South Wales, Australia. *Environmental Management* 25(6): 661–679.

Brierley, G.J., Cohen, T., Fryirs, K. and Brooks, A. 1999. Post-European changes to the fluvial geomorphology of Bega catchement, Australia: implications for river ecology. *Freshwater Biology* 41: 839–848.

Brooks, A.P., Brierley, G.J. and Millar, R.G. 2003. A paired catchment study between a pristine and a disturbed lowland alluvial river in southeastern Australia: channel morphodynamics before and after riparian vegetation clearance and wood removal. *Geomorphology* 51: 7–29.

Carson, M.A. 1975. Threshold and characteristic angles of straight slopes. In: *4th Guelph Symposium on Geomorphology*, Norwich, UK: Geobooks, Volume 3, pp. 19–34.

Chin, A. and Gregory, K.J. 2001, Urbanization and adjustment of ephemeral stream channels. *Annals of the Association of American Geographers* 91(4): 595–608.

Chorley, R.J. and Kennedy, B.A. 1971. *Physical Geography: A Systems Approach*, London: Prentice-Hall.

Church, M. and Mark, D.M. 1980. On size and scale in geomorphology. *Progress in Physical Geography* 4: 342–390.

Coulthard, T.J., Kirkby, M.J. and Macklin, M.G. 2000. Modelling geomorphic response to environmental change in an upland catchment. *Hydrological Processes* 14: 2031–2045.

Downs, P.W. 1995. River channel classification for channel management purposes. In: Gurnell, A.M. and Petts, G.E., eds., *Changing River Channels*, Chichester, UK: John E. Wiley and Sons, pp. 347–365.

Ferguson, R.I. 1986. Hydraulics and hydraulic geometry. *Progress in Physical Geography* 10: 1–31.

Friberg, N., Kronvang, B., Svendsen, L.M., Hansen, H.O., Nielsen, M.B. 1994. Restoration of channelized reach of the river Gelsa, Denmark: effects on the macroinvertebrate community. *Journal of Aquatic Conservation: Marine and Freshwater Ecosystems* 4: 289–296.

Fryirs, K. and Brierley, G. 2000. A geomorphic approach to the identification of river recovery potential. *Physical Geography* 21 (3): 244–277.

Gellis, A., Hereford, R., Schumm, S.A. and Hayes, B.R. 1991. Channel evolution and hydrologic variations in the Colorado River Basin: factors influencing sediment and salt loads. *Journal of Hydrology* 124: 317–344.

Gilvear, D.J. 1999. Fluvial geomorphology and river engineering: future roles utilizing a fluvial hydrosystems framework. *Geomorphology* 31: 229–245.

Glock, W.S. 1931. The development of drainage systems: a synoptic view. *Geographical Review* 21: 475–482.

Gregory, K.J. 1987a. River channels. In: Gregory, K.J. and Walling, D.E., eds., *Human Activity and Environmental Processes*, Chichester, UK: John Wiley and Sons, pp. 207–235 (Chapter 9).

Gregory, K.J. 1987b. Environmental effects of river channel changes. *Regulated Rivers Research and Management* 1: 358–363.

Gregory, K.J., Davis, R.J. and Downs, P.W. 1992. Identification of river channel change due to urbanization. *Applied Geography* 12: 299–318.

Hack, J.T. 1960. Interpretation of erosional topography in human temperate regions. *American Journal of Science* 258: 80–97.

Henry, C.P. and Amoros, C. 1995. Restoration ecology of riverine wetlands. I. A scientific base. *Environmental Management* 19(6): 891–902.

Hey, R.D. 1978. Determinate hydraulic geometry of river channels. *Journal of Hydraulics Division* 104(HY6): 869–885.

Hey, R.D. and Thorne, C.R. 1986. Stable channel with mobile gravel-bed rivers. *Journal of Hydraulic Engineering* 8: 671–689.

Hooke, J.M. 1995. River channel adjustment to meander cutoffs on the river Bollin and river Dane, northwest England. *Geomorphology* 14: 235–253.

Keaton, J.R. 1995. Dilemmas in regulating debris-flow hazards in Davis County, Utah. In: Lund, W.R., ed., *Environmental and Engineering Geology of the Wasatch Front Region: Utah*, Geological Association Publication No. 24, pp. 185–192.

Kirby, C. and White, W.R. 1994. *Integrated River Basin Development*, Chichester, UK: John Wiley and Sons, 537 p.

Kondolf, G.M. and Downs, P.W. 1996. Catchment approach to planning channel restoration. In: Brookes, A. and Shields, F.D., Jr., eds., *River Channel Restoration: Guiding Principles for Sustainable Projects*, Chichester, UK: John Wiley and Sons, pp. 129–148.

Kondolf, G.M., Piégay, H. and Landon, N. 2002. Channel response to increased and decreased bedload supply from land-use change contrasts between two catchments. *Geomorphology* 45: 35–51.

Landon, N., Piégay, H. and Bravard, J.P. 1998. The Drôme River incision (France): from assessment to management. *Landscape and Urban Planning* 43: 119–131.

Liébault, F. and Piégay, H. 2002. Causes of 20th century channel narrowing in mountain and piedmont rivers and streams of Southeastern France. *Earth Surface Processes and Landforms* 27: 425–444.

Liébault, F., Clément, P., Piégay, H. and Landon, N. 1999. Assessment of bedload supply potentiality from the tributary watersheds of a degraded river: the Drôme (France). *Arctic, Antarctic, and Alpine Research* 31(1): 108–111.

Lowe, M. 1993. *Debris-flow Hazards: A Guide for Land-use Planning*, Davis County, Utah: US Geological Survey Prof. Paper 1519.

Montgomery, D.R., Dietrich, W.E. and Sullivan, K. 1998. The role of GIS in watershed analysis. In: Lane, S.N., Richards, K.S. and Chandler, J.H., eds., *Landform, Monitoring, Modelling and Analysis*, Chichester, UK: John Wiley and Sons, pp. 241–261.

Newson, M.D. 1994. Sustainable integrated development and the basin sediment system: guidance from fluvial geomorphology. In: Kirby, C. and White, W.R., eds., *Integrated River Basin Development*, Chichester, UK: John Wiley and Sons, pp. 1–10.

Newson, M.D. and Newson, C.L. 2000. Geomorphology, ecology, and river channel habitat: mesoscale approaches to basin-scale challenges. *Progress in Physical Geography* 24 (2): 195–218.

Orme, A.R. and Bailey, R.G. 1970. Effect of vegetation conversion and flood discharge on stream channel geometry: the case of Southern California watersheds. *Proceedings of the American Association of Geographers* 2: 101–106.

Paine, A.D.M. 1985. Ergodic reasoning in geomorphology: time for a review of the term? *Progress in Physical Geography* 9: 1–15.

Park, C.C. 1981. Man, river systems and environmental impacts. *Progress in Physical Geography* 5: 1–31.

Patton, P.C. and Schumm, S.A. 1975. Gully erosion, northern Colorado: a threshold phenomenon. *Geology* 3: 88–90.

Peiry, J.L. 1987. Channel degradation in the middle Arve River, France. *Regulated Rivers: Research and Management* 1: 183–188.

Peiry, J.L., Salvador, P.G. and Nouguier, F. 1994. L'incision des rivières des Alpes du Nord: état de la question. *Revue de Géographie de Lyon* 69(1): 47–56.

Piégay, H., Bravard, J.P. and Dupont, P. 1994. The French water law: a new approach for alluvial hydro-system management, French Alpine and Perialpine stream examples. In: Marston, R.A. and Hasfurther, V.R., eds., *Effects of Human-induced Changes on Hydrologic Systems, Annual Summer Symposium of the American Water Resources Association, Jackson Hole, Wyoming, USA*: American Water Resources Association, pp. 371–383.

Piégay, H., Dupont, P. and Balland, P. 1996. Intégrer l'espace dans notre mode de gestion des systèmes-rivières. *Bulletin du Conseil Général du GREF* 46: 23–32.

Piégay, H., Salvador, P.G. and Astrade, L. 2000a. Réflexions relatives à la variabilité spatiale de la mosaïque fluviale à l'échelle d'un tronçon. *Zeitschrift für Geomorphologie* 44(3): 317–342.

Piégay, H., Bornette, G., Citterio, A., Hérouin, E., Moulin, B. and Statiotis, C. 2000b. Channel instability as control factor of silting dynamics and vegetation pattern within perifluvial aquatic zones. *Hydrological Processes* 14(16/17): 3011–3029.

Piégay, H., Gazelle, F. and Peiry, J.L. 2003. Les ripisylves, un facteur de contrôle de la géométrie et de la dynamique du lit fluvial et de son aquifère. In: Pautou, G., Piégay, H., Ruffinoni, C., eds., *Les ripisylves dans les hydrosystèmes fluviaux*, Institut pour le Développement Forestier, Paris.

Piégay, H., Dupont, P. and Faby, J.A. 2002. Question of water resources management: feedback on the first implemented plans SAGE and SDAGE. *Water Policy* 4: 239–262.

Richards, K. 1996. Samples and cases: generalisation and explanation in geomorphology. In: Rhoads, B.L. and Thorn, C.E., eds., *The Scientific Nature of Geomorphology*, Chichester, UK: John Wiley and Sons, pp. 171–190.

Roux, A.L. 1982. Cartographie polythématique appliquée à la gestion écologique des eaux; étude d'un hydrosystème fluvial: le Haut Rhône français. Report, CNRS, Lyon, 116 p.

Schumm, S.A. 1977. *The Fluvial System*, Chichester, UK: John Wiley and Sons, 338 p.

Schumm, S.A. 1981. *Evolution and Response of the Fluvial System: Sedimentologic Implications*. Social Economic Paleontologists Mineralogists, Special Publication No. 31, pp. 19–29.

Schumm, S.A. 1991. *To Interpret the Earth: Ten Ways to be Wrong*, Cambridge: Cambridge University Press, 133 p.

Schumm, S.A. 1994. Erroneous perceptions of fluvial hazards. *Geomorphology* 10: 129–138.

Schumm, S.A., Kahn, H.R., Winkley, B.R. and Robbins, L.G. 1972. Variability of river patterns. *Nature* (*Phy. Sci.*) 237: 75–76.

Schumm, S.A., Harvey, M.D. and Watson, C.C. 1984. *Incised Channels: Initiation, Evolution, Dynamics, and Control*, Littleton, CO: Water Resources Publication, 200 p.

Schumm, S.A., Mosley, M.P. and Weaver, W.E. 1987. *Experimental Fluvial Geomorphology*, New York: Wiley, 413 p.

Schumm, S.A., Rutherfurd, I.D. and Brooks, J. 1994. Precutoff morphology of the lower Mississippi River. In: Schumm, S.A. and Winkley, B.R., eds., *The Variability of Large Alluvial Rivers*, New York: American Society of Civil Engineers Press, pp. 13–44.

Sear, D.A. 1994. Viewpoint: river restoration and geomorphology. *Aquatic Conservation: Marine and Freshwater Ecosystems* 4: 169–177.

Shields, F.D., Knight, S.S. and Cooper, C.M. 1997. Rehabilitation of warmwater stream ecosystems following channel incision. *Ecological Engineering* 8: 93–116.

Strahler, A.N. 1950. Equilibrium theory of erosional slopes approached by frequency distribution analysis. *American Journal of Science* 248: 673–698.

Trimble, S.W. 1997. Stream channel erosion and change resulting from riparian forests. *Geology* 25(5): 467–469.

Van Beek, J.L. 1981. Planning for integrated management of the Atchafalaya River Basin: natural system viability and policy constraints. In: North, R.M., Dworsky, L.B. and Allee, D.J., eds., *Unified River Basin Management*, Minneapolis: American Water Resources Association, pp. 328–337.

Wells, S.G., Bullard, T.F., Miller, J. and Gardner, T.W. 1983. Applications of geomorphology to uranium tailings silting and groundwater management. In: Wells, S.G., Love, D.W. and Gardner, T.W., eds., *Chaco Canyon Country*, American Geomorphological Field Group, Guidebook, pp. 51–56.

White, I.D., Mottershead, D.N. and Harrison, S.J. 1992. *Environmental Systems: An Introductory Text*, London: Chapman & Hall, 616 p.

Whitlow, J.R. and Gregory, K.J. 1989. Changes in urban stream channels in Zimbabwe. *Regulated Rivers: Research and Management* 4: 27–42.

6

Analysis of Aerial Photography and Other Remotely Sensed Data

DAVID GILVEAR[1] AND ROBERT BRYANT[2]
[1]*Department of Environmental Sciences, University of Stirling, UK*
[2]*Department of Geography, University of Sheffield, UK*

6.1 INTRODUCTION

Aerial photography and other remotely sensed data have increasingly been used as tools by the geomorphologist. Remote sensing is based upon principles surrounding the transfer of energy from a surface to a sensor. Prior to 1970s the sensor, in the context of geomorphological mapping, was usually black and white photographic film and the platform an aeroplane. Since the early 1970s, however, there has been a huge increase in the number of sensors and platforms (Tables 6.1 and 6.2) offering the geomorphologist enhanced capabilities for interrogating the earth's surface. Moreover, the number of operational remote sensing systems is continually increasing; providing improved spatial and spectral coverage and resolution. Remote sensing compared to traditional cartographic and field-based data collection has several advantages including better spatial and temporal resolution, storage of data in digital format, and interrogation of electromagnetic radiation (EMR), emitted or reflected, from land and water that is not detected by the human eye. A number of recent reviews have thus advocated the potential of using remote sensing as a tool to aid the investigation of rivers (e.g. Muller *et al.* 1993, Malthus *et al.* 1995, Milton *et al.* 1995, Lane 2000). Given the impending launch of higher specification satellite sensors together with improvements in airborne sensors, and digital camera and camcorder technologies, the future appears exciting in terms of gaining panoptic geomorphic coverage of catchments, valley floors, river systems and all but the smallest streams. The smallest of streams may also be interrogated at the reach scale using remotely sensed data acquisition methods via hand-held, tripod, crane or 'blimp' mounted sensors.

This chapter aims to provide a general review of the analysis of aerial photography and other remotely sensed data as a tool for studying fluvial processes and landforms with an emphasis on channel and floodplain environments. In particular, we aim to focus on remote sensing data taken from above-ground, aerial and space-borne remote systems. Techniques such as echo-sounding, use of electrical resistivity and surface-based ground penetrating radar are also forms of remote sensing but are not within the scope of this chapter. It is worth noting that remote sensing does not seek to replace traditional field-based methods of investigation, but rather to complement them by providing greater spatial coverage and in some cases greater temporal resolution; in each case giving access to a larger and more dynamic sample population. Indeed, the real potential of applying remotely sensed data to fluvial research may only be realised if field-based methods are used to support remotely sensed data. For example, morphological data obtained at a cross-section on the ground can be extended to the reach and thence to the channel segment and finally the catchment scale. Overall, therefore, image analysis applied to remotely sensed data can potentially be used to provide information on hydrology, fluvial processes and spatial and temporal variability in land-use at the catchment scale, thus putting riverine data into a landscape context. Moreover, in the case of very large rivers (e.g. Amazon or Brahmaputra) viewing and capturing an image from the air is the only way to observe and quantify the overall morphology of the river.

Table 6.1 Summary of (A) current high and medium (B) forthcoming high-resolution and (C) forthcoming medium-resolution earth resource satellite remote sensing platforms and sensors
(A)

Sensor name and platform	Launch date	Archive	Spectral bandwidths	Spatial resolution (m)	Temporal resolution
SPOT 4 (HRV)	1998	1986–	Pan: 0.51–0.73	10	2–26 days
			1. 0.5–0.59	20	depending on
			2. 0.61–0.68	20	overlap
			3. 0.79–0.89	20	coverage
Landsat 5 (SS)	1986	1972–	1. 0.5–0.6	80	16–18 days
Landsat 4	1982		2. 0.6–0.7		
Landsat 3	1978		3. 0.7–0.8		
Landsat 2	1975		4. 0.8–1.1		
Landsat 1	1972				
Landsat 6 (ET)	1993 (failed)		1. 0.45–0.52	30	16 days
Landsat 5 (T)	1984		2. 0.52–0.60	30	
Landsat 4	1982		3. 0.63–0.69	30	
			4. 0.76–0.90	30	
			5. 1.55–1.75	30	
			6. 10.4–12.5	120	
			7. 2.1–2.35	30	
AVHRR NOAA-15 (K)	1988	1978–	1. 580–680 n	1100	12 h
			2. 725–1100 n		
			3a. 1580–1640 n		
			3b. 3550–3930 n		
			4. 10 300–11 500 n		
			5. 10 500–12 500 n		
IRS-1C/D (LISS III)	1995 (1C)	1988–	Pan: 500–800 n	5.8	5–24 days
	1997 (ID)		Green: 500–600 n	23	depending on
			Red: 600–700 n	23	latitude
			Near-IR: 800–900 n	23	
Ikonos –1/2	1999	1999–	Pan: 450–900 n	1	1–3 days
			Blue: 450–530 n	4	
			Green: 520–610 n	4	
			Red: 640–720 n	4	
			Near-IR: 770–880 n	4	
Landsat 7 (ET+)	1999	1999–	Pan: 520–900 n	15	16 days
			1. 450–515 n	30	
			2. 525–605 n	30	
			3. 630–690 n	30	
			4. 750–900 n	30	
			5. 1550–1750 n	30	
			6. 10 400–12 500 n	60	
			7. 2090–2350 n	30	
SPIN-2	1999	1999–	TK-350 Caera	10	Variable
			Pan: 510–760 n		
			KVR-1000 Caera		
			Pan: 510-760 n	2	
Terra (EOS-A1) (Aster)	2000	2000–	VNIR		Variable
			1. 520–600 n	15	
			2. 630–690 n	15	

(*Continues*)

Table 6.1 (A) (*Continued*)

Sensor name and platform	Launch date	Archive	Spectral bandwidths	Spatial resolution (m)	Temporal resolution
			3. 760–860 n (nadir)	15	
			4. 760–860 n (back)	15	
			SWIR: 1640–2430 n (6 ch)	30	
			TIR: 8125–11 650 n (5 ch)	90	
CERBS (CCD)	1999	1999–	Pan: 510–730 n	20	26 days
			1. 450–520 n		
			2. 520–590 n		
			3. 630–690 n		
			4. 770–890 n		
ADEOS (AVNIR)	1996	1996–	Pan: 520–690 n	8	41 days (3 days sub-cycle)
			u1: 420–500 n	16	
			u2: 520–600 n	16	
			u3: 610–690 n	16	
			u4: 760–890 n	16	

(B)

Sensor and platform	Launch date	Spectral bandwidths	Spatial resolution (m)	Swath width (km)	Temporal resolution
Quickbird 1	2000	Pan: 450–900 n	1	22	1–5 days depending on latitude
		Blue: 450–520 n	4		
		Green: 520–600 n	4		
		Red: 630–690 n	4		
		Near-IR: 760–890 n	4		
Orbview 3	2001	Pan: 450–900 n	1	8	Less than 3 days
		Blue: 450–520 n	4		
		Green: 520–600 n	4		
		Red: 625–695 n	4		
		Near-IR: 760–900 n	4		
Orbview 4	2001	Pan: 450–900 n	1	8	Less than 3 days
		Blue: 450–520 n	4		
		Green: 520–600 n	4		
		Red: 625–695 n	4		
		Near-IR: 760–900 n	4		
		Hyper-spectral: 450–2500 n (200 channels)	8	5	
EROS-A+/B2	2000	Pan: 450–900 n	1 (1.5)	12 (15)	1–15 days
SPOT-5	2002	Pan: 610–680	5	60	26 days
		1. 500–590 n	10		
		2. 610–680 n	10		
		3. 790–890 n	10		
		4. 1580–1750 n	20		
EO-1 (Hyperion)	2000	Hyper-spectral: 400–2500 n		7.5	Variable

(*Continues*)

Table 6.1 (B) (*Continued*)

Sensor and platform	Launch date	Spectral bandwidths	Spatial resolution (m)	Swath width (km)	Temporal resolution
		220 bands with a 10-n resolution	30		
EO-1 (ALI)	2000	Pan: 480–690 n	10	37	Variable
		S-1. 433–453 n	30		
		S-1. 450–510 n	30		
		S-2. 525–605 n	30		
		S-3. 630–690 n	30		
		S-4. 775–805 n	30		
		S-4. 845–890 n	30		
		S-5. 1200–1300 n	30		
		S-6. 1550–1750 n	30		
		S-7. 2080–2350 n	30		
NEO (COIS)	2000	Pan: 450–690 n	10		7 days (2.5 days global)
		Hyper-spectral: 400–2500 n 10 n spectral resolution	30–60		
AERIES	2001	Pan: 480–690	10	15	7 days
		Hyper-spectral: 400–1050 n of 20 n band spacing	30		
		1050–2000 n spectrally binned by two (i.e. 32 n band spacing) and possibly only optionally available	30		
		2000–2500 n of 16 n band spacing	30		

(C)

Sensor and planform	Launch date	Spectral bandwidths	Spatial resolution (m)	Swath width (km)	Temporal resolution
ERIS (ENVISAT)	2001	Hyper-spectral: 390–1040 n 15 programmable bandwidths with a spectral resolution of 2.5 n	300	1150	2–3 days
EO-1 (LAC)	2000	Hyper-spectral: 850–1600 n 256 wavebands with a spectral resolution of 2–6 n	250	185	Variable
ADOS-II (GLI)	2000	Hyper-spectral: 36 channels		1600	4 days
		VNIR: 380–865 n (19 ch)	1000		
		VNIR: 460–825 n (4 ch)	250		
		SWIR: 1050–1380 n (4 ch)	1000		
		SWIR: 1640–2210 n (2 ch)	250		
		TIR: 3715–12 000 n (6 ch)	1000		

Table 6.2 Examples of airborne multi-spectral and hyper-spectral remote sensing systems

Sensor name	Acronym	Spectral coverage (μm)	Number of available wavebands
Airborne visible/infrared imaging spectrometer	AVIRIS	410–2450	224
Reflective optics system imaging spectrometer	ROSIS	430–880	28
Multi-spectral infrared and visible imaging spectrometer	MIVIS	400–2500	92
Modular airborne imaging spectrometer	MAIS	440–2500 7800–11 800	71 7
CCD airborne experimental scanner for applications in remote sensing	CAESAR	520–780	9
Digital airborne imaging spectrometer	DAIS	400–2500	72
Compact airborne spectrographic imager	CASI	410–925	288 or 15
Daedalus 1268 ATM	ATM	420–2350 850–13 000	10 1

Furthermore, seeing the problem from a different viewpoint (literally) can provide new insights and suggest new hypotheses, which can then be tested in the field (Milton *et al.* 1995).

6.2 THE PHYSICAL BASIS

Photogrammetry

The two key geometric properties of an aerial photography are those of angle and scale. According to the angle at which an aerial photography is taken it is referred to as either vertical, high oblique or low oblique. The following discussion relates to vertical aerial photography. Vertical aerial photographs are normally taken in sequences along an aircraft's flight line with an overlap of approximately 60% to allow the photographs to be viewed three-dimensionally or stereoscopically. A small-scale aerial photograph will provide a synoptic, low spatial resolution overview (e.g. 1:50 000) of a large area; such a photograph may be useful for mapping drainage networks but is only appropriate for detailed reach-scale analysis of river morphology on large rivers. A large-scale aerial photograph will provide a high spatial resolution view of a small area; such a photograph will be useful for detailed analysis of a reach but if data for a long length of river was needed it would entail use of a large number of such aerial photographs. The scale of a photograph is determined by the focal length of the camera and the vertical height of the lens above the ground.

Overlapping pairs of aerial photographs can provide a 3D view of the earth's surface by the effect of parallax. Parallax refers to viewing an object from two different angles. Humans use the principle by focusing on an object with their left and right eye. With aerial photographs, optical devices called stereoscopes are used to view a pair of stereo aerial photographs, and the ground appears to the viewer to be in 3D. The phenomena of parallax can be used to measure the height of objects. Parallax results in points of higher elevation having a greater horizontal displacement on successive aerial photographs than a lower elevation feature. The value of parallax displacement is positively related to the distance between the centre of the two photographs and the height of the object of interest and negatively related to the height above the ground from which the photograph was taken. Modern computer-based photogrammetry allows automated production of digital terrain models from stereo aerial photography and such techniques are obviously important for the subject of geomorphology (Lane *et al.* 1993). More detail on the potential of analytical and digital photogrammetry in geomorphological research can be found in Dixon *et al.* (1998), Lane *et al.* (1993) and Chandler (1999). An excellent review of photogrammetric applications for the study of channel morphology and associated data quality issues can also be found in Lane (2000).

Electromagnetic Radiation and Remote Sensing Systems

EMR reflected from the earth's surface can vary with location, time, geometry of observation and waveband (Verstraete and Pinte 1992; Figure 6.1).

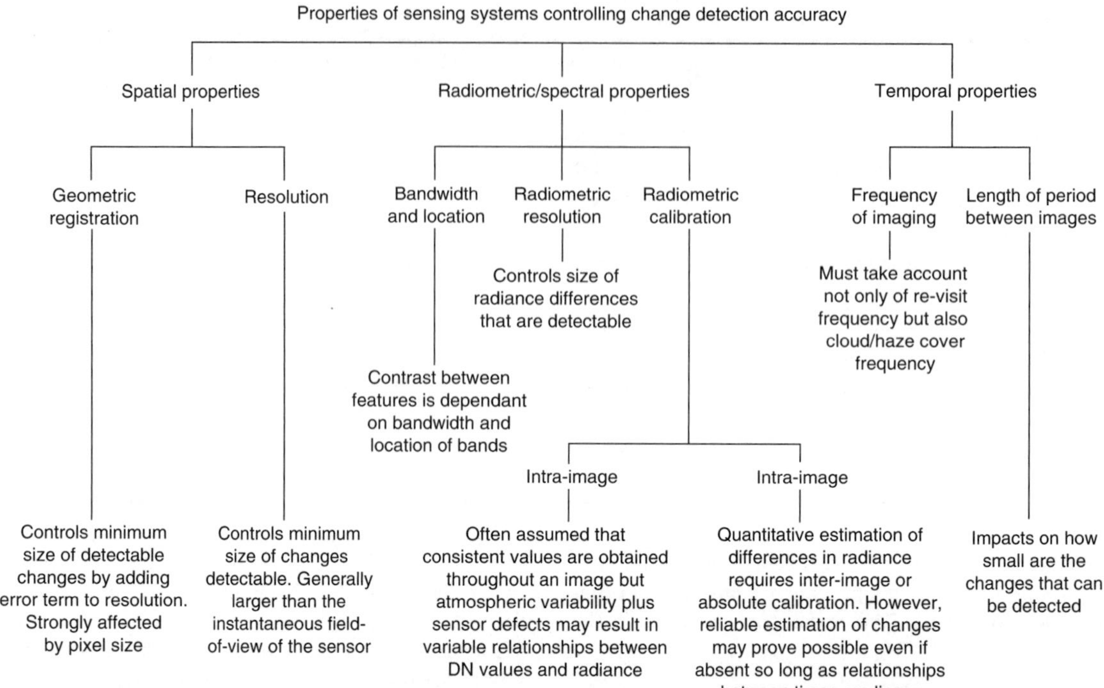

Figure 6.1 The main properties of remote sensing systems controlling the accuracy of temporal change and spatial variability detection (after Townshend and Justice 1988, reproduced by permission of Taylor & Francis, Ltd., http://www.tandf.co.uk/journals)

Consequently, the successful interpretation of remotely sensed data for a particular river will depend upon an understanding or characterisation of these four factors. In particular, an understanding of the way in which EMR interacts with the surface of the earth and what factors affect its capture by a sensor is important. Many good reviews of the nature and interaction of EMR and earth surface features exist (e.g. Asrar 1989). Aspects relevant to fluvial geomorphology are summarised here.

Electromagnetic Radiation

EMR occurs as a continuum of wavelengths. The wavelengths of greatest interest when remotely sensing the earth are the reflected radiation in the visible, near and middle infrared wavebands, emitted radiation in the middle and thermal infrared wavebands and reflected radiation in the microwave wavebands. EMR originates from a source; this is usually the sun's reflected light or the earth's emitted heat but can be man-made as in active microwave radar. Initially, EMR passing through the atmosphere may be distorted and scattered. In general, greater scattering and distortion occurs with greater distance between the earth and sensor and the greater the levels of atmospheric moisture, pollutants and dust. Generally, atmospheric noise is wavelength-specific, and can be easily removed by ignoring those wavelengths that are affected (e.g. for hazy image scenes caused by Rayleigh scattering, short wavelengths can be omitted from the image set). However, some atmospheric effects (e.g. Mie and non-selective scatter of EMR) are more difficult to remedy or take account of (Kaufman 1989, Cracknell and Heyes 1993). Overall, the level of correction undertaken for atmospheric effects can depend on whether qualitative or quantitative data are to be extracted from imagery. For the latter, correction and calibration using in situ (alternatively called ground-truthed) data are commonly necessary.

Once EMR interacts with the surface, one of three processes can occur:

(i) reflection of energy;
(ii) absorption of energy;
(iii) transmission of energy.

In general, the amount and characteristics of each of these three energy interactions will depend upon the inherent characteristics of the earth's surface and the wavelength of EMR that is interacting with it. For example, visible wavelengths are reflected from water in a different way than those wavelengths in the near and middle infrared. Consequently, in order to successfully generate geomorphic information from remote sensing data, the knowledge of how EMR interacts with the specific surfaces is needed. Most objects can only be differentiated if the reflectance from the surface is different from that of the adjacent object in the wavebands being captured by the sensor and above the radiometric precision of the sensor. In most cases, such information can either be obtained from the literature (e.g. Irons *et al.* 1989), spectral libraries (e.g. on-line libraries such as at: http://speclib.jpl.nasa.gov) or from field collection of reflectance spectra co-incident with image acquisition using a spectroradiometer. A brief review of the spectral property of surfaces within the fluvial realm is outlined later.

Sensors and Platforms

Most sensors commonly have several channels capturing information in narrow, broad or continuous bandwidths. Generally, sensors with few channels (1–10) and broad bands (50–100 nm) are referred to as multi-spectral. Sensors with the capability to measure in numerous (up to 250 bands), narrowly defined (to 1–10 nm) bands or continuous parts of the electromagnetic spectrum are referred to as hyper-spectral. Different sensors may also have different radiometric resolutions, which will control the size of radiance differences at the earth's surface that can be detected. Different sensors normally also capture different components of EMR (Tables 6.1 and 6.2). Given the knowledge of the reflectance properties of fluvial surfaces, it is important to understand how these data will be recorded at the sensor (Tables 6.1 and 6.2). Most photographic sensors/cameras differ from hand held cameras only in that they have dedicated film magazines, automated drive mechanisms, and a large supporting lens cone. Similarly, the cameras can record data using common film types. For visible wavelengths, either black and white panchromatic film or true colour film can be used. For other wavelengths, black and white and false-colour near infrared film may also be utilised. Photographic sensors can produce hard-copy images using either: (1) strip, (2) panoramic, or (3) frame formats. Digital sensors can essentially have two different types of design, either:

(i) an optical mechanical scanner (or multi-spectral across-track scanner), or
(ii) a linear array (i.e. an along-track push-broom of charge coupled devices, CCD).

In addition, sensors of each type have a pre-determined spatial resolution (the edge length of a square or rectangular land parcel from which an individual signal can be deduced; see later for more detail) and swath width (visible area on each pass). A useful review of imaging spectrometry and current and forthcoming systems can be found in Curran (1994), and Plummer *et al.* (1995). It should also be noted that recent advances in combining the output of global positioning systems with image capture on a variety of platforms has increased the potential for accurate identification of an absolute location on the earth. Similarly, the use of spatially accurate global positioning systems in fluvial research (e.g. Milne and Sear 1997) allows measurements taken on the ground to be linked to individual pixels on imagery permitting more accurate image calibration and validation. Sensors can be further characterised by their platform, which can range from a satellite to aircraft or even balloons. For most existing and forthcoming satellite platform/sensor combinations, the repeat-period can range from 12 h to 44 days (Table 6.1). In the case of airborne sensors, a greater temporal flexibility can be afforded (Table 6.2). The pros and cons of airborne data versus satellite data for river research are given in Table 6.3.

Considerations

The most important considerations when acquiring imagery for a particular site are whether the radiometric resolution of the sensor, the amount of atmospheric scatter, the surface roughness of the objects and the spatial variability of reflectance within the wider field of view can affect the ability to differentiate between objects. The last factor is important because the radiance recorded from an area of ground also contains radiance from surrounding areas. Another important factor to be aware of is that the raw digital number (DN) values often need to be calibrated to radiance units, and this calibration may not be constant across an image or between images if atmospheric distortion or illumination is variable. Even after calibration some wavebands may have to be discarded due to high noise to signal ratios or

Table 6.3 The pros and cons of airborne data versus space-borne data for fluvial geomorphology (modified from Dekker *et al.* 1995)

	Airborne photography[a]	Airborne imaging spectrometer[b]	Airborne multi-spectral scanner[c]	Space-borne multi-spectral scanner[d]
Resolution				
Spatial range (m)	<0.5	0.5–20	0.5–20	10–80
Spectral bandwidth (nm)	Approx. 10	1.8	5–20	60
Radiometric range (DN)	Hard copy	4092	256–4029	256
Temporal repetition (days)	Upon request	Upon request	Upon request	3–18
Logistics				
Number of spectral bands	1–3	288	1–15	7
Swath width	Small*	Small*	Small*	Large
Mission targeting	Upon request	Upon request	Upon request	None or limited
Flight time	Upon request	Upon request	Upon request	Fixed
Repetitive coverage	Low cost	High cost	High cost	Default (lower, cost)
Cost/km^2	Low	High	High	Lower
State-of-the-art technology	No	Most recent	Most recent	10–15 year lag
Atmospheric influence	Low	Less	Less	Highest
Sky conditions required	Less critical	Critical	Critical	Extremely critical
Sensor platform motions	Roll, pitch and yaw	Roll, pitch and yaw	Roll, pitch and yaw	Negligible
Hands-on repairs/ adjustments	In situ	In situ	In situ	Almost impossible
Standardised products	All	Few	Some	More

[a]For example, Wild RC10 Survey Camera (b/w panchromatic, b/w infrared, colour and colour infrared).
[b]For example, CASI, AVIRIS.
[c]For example, Daedalus 1268 ATM.
[d]For example, SPOT, Landsat TM.
*Depends on the flying height.

simply due to poor calibration (e.g. Bryant and Gilvear 1999). Uneven radiation capture at the sensor, due to variations in scene illumination, equally applies when scanning aerial photographs to allow image processing. Hence, Gilvear *et al.* (1995) needed to apply shade correction to scanned aerial photographs of different parts of Faith Creek in Alaska and for different dates before image analysis to detect meso-scale habitat change. Shade correction was necessary because of differences in natural illumination (i.e. atmospheric conditions) at the time the photographs were taken, uneven illumination when the photographs were scanned (using a video camera system) and differences in photographic processing. The grey-tone in an aerial photograph when captured in digital format is assigned an 8-bit DN value between 0 (black) and 255 (white) according to its grey-tone. This number of grey-tones is much greater than the human eye can detect, allowing image analysis to identify spatial variability that would previously go undetected.

Scale and Spatial Accuracy Issues

Size of River

One of the main considerations in using remote sensing to study the fluvial geomorphology of river systems is the size of river. A number of studies of large rivers (>200 m wide) have been undertaken using satellite remote sensing (e.g. Salo *et al.* 1986, Phillip *et al.* 1989, Ramasamy *et al.* 1991) and more recently during space shuttle missions (see later). Milton *et al.* (1995) suggest that for rivers approximately 20–200 m wide, airborne remote sensing (incorporating high resolution advanced sensors and improved temporal/spatial flexibility) may be a more suitable approach for mapping and monitoring change. For small rivers (<20 m wide), a hand-held helium blimp or model aircraft with remotely operated camera, or oblique imagery, that is subsequently rectified, may be more appropriate in gaining the spatial resolution of imagery required. Thus, Kennedy (1998) working on a small stream of approximately 10 m width in the HJ Andrews Experimental Forest in Oregon used a 'blimp' at 80 m altitude to gain 1:2500 aerial photography from which gravel bar morphology, vegetation patches and coarse woody debris accumulations could be identified.

Image Format

One key characteristic of aerial photography and satellite imagery is that it provides typically a square format (whether digital or hard-copy). Unfortunately, this does not match well with rivers which form linear features in the landscape. When these images are used the length of river shown will not be much greater than the edge length of the image unless it has high sinuosity (Figure 6.2A). Therefore, to cover an appreciable river length a number of images often have to be pieced together to form a mosaic unless very small-scale images (e.g. >1:25 000) are used. Even with large-scale photographs covering a small reach, with small to medium size rivers channel features may be hard to detect given the small area the river covers on the image, especially where riparian woodland obscures some of the channel. This mismatch in geometry results in increased costs in the purchase of imagery, increased time for image rectification, and more ground control points (GCPs). A distinction should also be highlighted here between aerial photography flown specifically for the purpose of a riverine study with systematic coverage of a whole region or country. In the first case, overlapping aerial photographs will be orientated along the direction that the river is flowing, and this will maximise the length of river on each photograph. In the second case the photographs will not be orientated parallelly or focus on the river, and only small lengths of the river will appear on some of the appropriate photographs. Coverage of countries or regions is also normally small scale, as otherwise the number of flight lines needed to permit full coverage is excessively large. Following a watercourse (unless it is very wide and very small-scale photography is needed) requires only one flight line and thus larger-scale photography is more acceptable. In the US regional surveys of 1:20 000 or 1:25 000 scale, for example, might be commissioned by the US Geological Survey or Department of Agriculture, whereas 1:12 000 scale or better photography of the river channel might be commissioned by the US Army Corps of Engineers or Bureau of Reclamation for flood control mapping, for instance, but these flights may not include good coverage of riparian and floodplain areas. With digital data and appropriate image analysis software, one can zoom-in on areas of the channel of interest to gain a large-scale picture but this may not be appropriate if the spatial resolution of the image is too small. However, airborne remote sensing captures digital data with a fixed swath width, for a given flying elevation, but with infinite length. Thus, data can be captured, for example, as a 1-km × 10-km area, covering perhaps a 15-km length of a medium size river together with its floodplain. It is thus ideally suited to collecting data on rivers (Milton *et al.* 1995).

Sensor Resolution

The spatial resolution of a sensor is usually described by distance in metres, which relates to the edge length of a single square or rectangular parcel of land from which a radiation value can be assigned (a pixel). Pixel size relates to sensor type and altitude. The higher the platform, the larger the pixel size for a given sensor, and the wider the swath width (Figure 6.3), although this can vary depending on the sensor optics and the size of the CCD used. In scanning aerial photographs and producing digital imagery one must also calculate the pixel size in relation to the photographic scale and scanning resolution (Figure 6.2B). This will limit the amount of detail and minimum size of object that can be detected. Although one can enlarge a particular area of a photographic print or zoom-in on a digital image to gain greater detail, there will be a point where no further information can be visually obtained, particularly on an image made up of pixels. The grain size of the photographic film may also, but only rarely, limit the minimum size of fluvial features that can be detected (Lane *et al.* 1994). Grain size is the theoretical minimum and scanning density the practical minimum.

A key question in remote sensing is whether the pixel size is smaller than the object (e.g. landform) of interest. If larger, identification of a landform will be difficult and delimitation of its boundaries is impossible. Even if the pixel size is smaller than the landform of interest, problems can still arise because normally a number of pixels are needed to effectively identify a feature. Firstly, pixel edges may not necessarily coincide with the edge of features on the

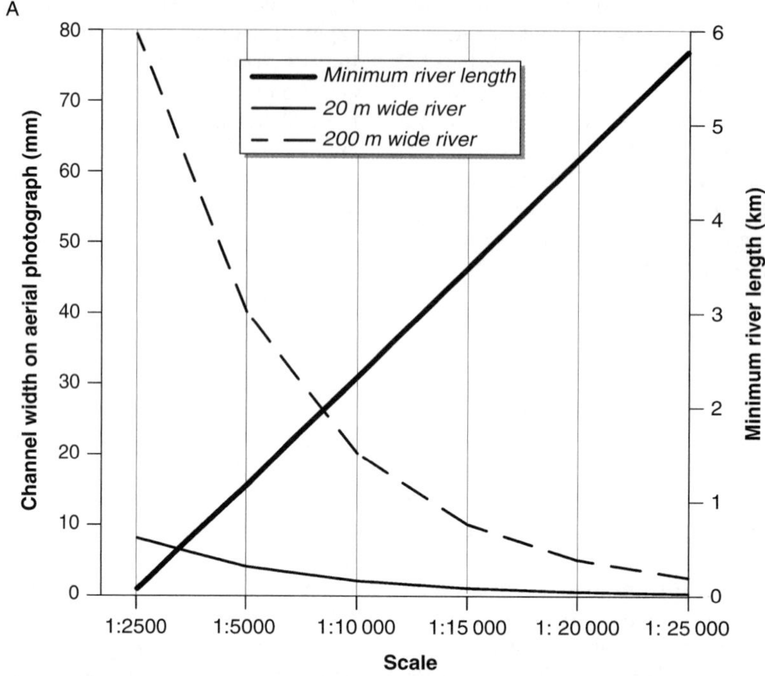

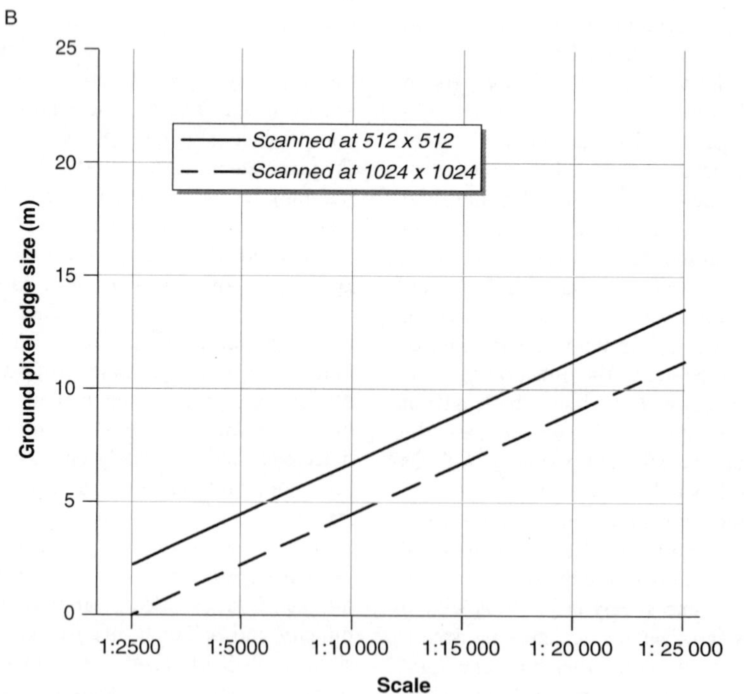

Figure 6.2 The effects of image scale, scanning resolution and river size on spatial resolution. (A) Aerial photograph scale, minimum river length observable and the channel width on the photograph for 20 and 200 m width. (B) Ground pixel resolution on scanned aerial photographs for differing scales and scan resolution

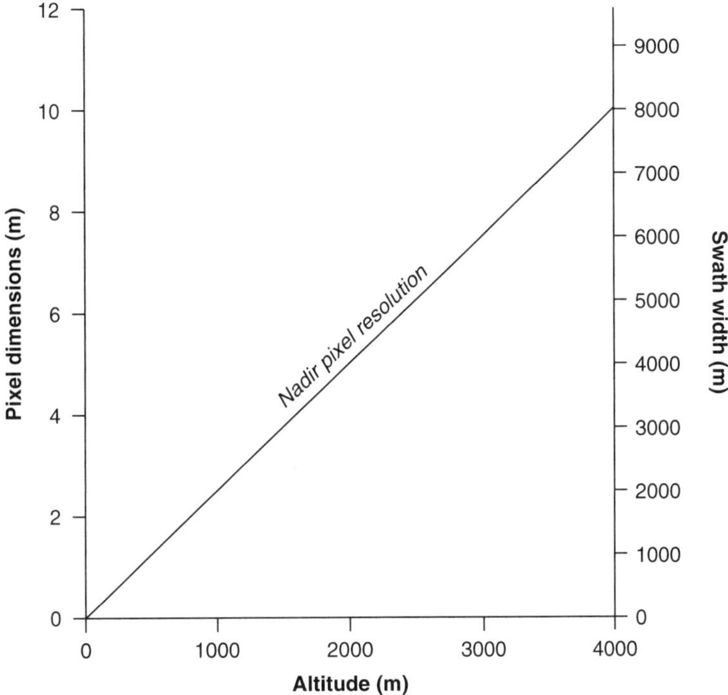

Figure 6.3 Relationship between pixel size and swath width with platform altitude for the Daedalus 1268 ATM scanner (modified from Wilson 1995)

ground. Therefore, many pixels will contain a number of objects (mixed pixels). In the case of the fluvial environment this could be bed material, water and vegetation. The combination of these reflectances may result in the pixel appearing similar to another type of surface and hence be misinterpreted. In the case of multi-spectral imagery, the problems of mixed pixels can be met using mixture modelling to estimate the proportion of each spectral 'end-member' present within a pixel (Mertes *et al.* 1993, Bryant 1996). Secondly, problems arise because a scanner not only receives radiation from the area of ground demarcated by pixel edges (the ground resolution element—GRE) but also surrounding areas (instantaneous working area—IWA). Indeed even within a GRE the scanner will not respond uniformly to radiation from its area because pixel intensities are not independent but auto-correlated. More detail on what a pixel means in reality can be found in Cracknell (1998). On a scanned black and white aerial photograph individual pixels of soil and water can have the same DN value. A density sliced image would therefore assign them to a similar land cover type when from visual observation of the image the difference in surface type would be obvious because of the pixels' location in the larger image and contextual information. In this regard feature integrity needs to be incorporated into the classification procedure used. Image analysis should therefore not always be seen as superior to visual interpretation, but as a complementary approach. With colour aerial photographs, soil, water, and vegetation surfaces are more easily distinguished but distinguishing pure water and areas with submergent and emergent aquatic plants may be difficult.

Geometric Accuracy

Spatial resolution and scale should not be confused with accuracy in that the image may not be geometrically correct and may include tilt and warping. To rectify images, GCPs, for which a relative or absolute location is known, have to be matched with the corresponding feature on the image using a mathematical transformation. In the consideration of temporal change, and an absence of GCPs, image registration to each other using objects that are known not to have moved can be undertaken. In some remote areas, with no man-made objects, such identification can be problematic. If the chosen objects move (e.g. bank lines due to erosion) results can be spurious. Satellite platforms are highly stable (i.e. remain perpendicular to the ground surface and do not suffer roll and pitch as with an aircraft) and often a first-order transformation based on relatively few GCPs is sufficient to gain a high level of accuracy. Airborne platforms are often less stable, but techniques to geometrically correct photographic images from nadir and oblique pointing cameras are well developed (Lane *et al.* 1993, Chandler 1999). Successful rectification of airborne scanner data may require the survey area to be split into smaller sections (e.g. Christensen *et al.* 1988), or the transformation to be localised (e.g. Devereux *et al.* 1990), or the use of a parametric correction procedure based on aircraft altitude measurements. It may benefit from the incorporation of a digital elevation model (DEM) (Cosandier *et al.* 1994) but probably not in level, relatively flat floodplain environments. Townshend and Justice (1988) have reviewed, focussing on spatial aspects, those properties of remote sensing systems that control the accuracy of land cover assessments (Figure 6.1). They emphasise the importance of matching the spectral, radiometric and spatial capabilities of the sensor with the properties of the surfaces being sensed. Moreover, the timing and frequency of sensing must coincide with that of temporal change or events within the fluvial environment, and even then, in the case of many sensors, cloud cover may prevent observation.

Spectral Properties and the Fluvial Environment

Landscape Components

Figure 6.4 shows typical spectral response curves, measured using airborne thematic mapper (ATM) data, for surfaces and features found in the fluvial environment. It is apparent that spectral responses normally fall into three distinct classes (Hooper 1992, Milton *et al.* 1995):

(i) water, shadow and aquatic vegetation;
(ii) trees and other green vegetation;
(iii) exposed sediment.

It should be noted, however, that in the case of black and white aerial photographs and some other spectral wavebands, these classes might not be so clearly differentiated. These three classes may be thought of as the 'spectral end-members' of the riverine environment. In the context of fluvial geomorphology, the two main-end members are water and sediment. Interrogation of subtle differences in the radiation from water and exposed sediment can reveal more about their nature and hence the relationship between these surfaces and EMR is explored further below. Vegetation is another component that is often of interest to the fluvial geomorphologist in that floodplain vegetation mosaics often relate to topography, soils, and channel mobility (Hickin and Nanson 1975).

As mentioned earlier, the majority of visible, near and middle infrared radiation reaching a soil or sediment surface is either reflected or absorbed, and little is transmitted. The five characteristics of sediment, which are inter-related, and which determine its reflectance properties are in order of importance: moisture content, organic content, texture, structure and iron oxide content. Most important is the relationship between soil moisture and reflectance. Reflectance reduces substantially in the visible wavelengths with increasing moisture content until soil reflectance becomes saturated. Reflectance in the near and middle infrared wavelengths is also negatively correlated with soil moisture and an increase in soil moisture will result in rapid falls in reflectance particularly in wavelengths centred at 0.9, 1.4, 1.9, 2.2 and 2.7 nm. Moisture will have a greater effect on the reflectance of clay soils than sandy sediments. Soil organic matter will also decrease reflectance up to a content of 4–5%. The complex relationship between fluvially deposited sediment and spectral characteristics is demonstrated by the results of Bryant *et al.* (1996) and Rainey *et al.* (2000) (Figure 6.5). Figure 6.5 shows that immediately before a high tide there is a positive reflectance between ATM Band 9 reflectance and increasing particle size but immediately after the high tide and thorough wetting of the inter-tidal sediments there is a negative relationship. Such knowledge of changes in spectral reflectance–physical substrate properties with differing degrees of wetness is obviously vital to sound interpretation of remotely sensed data and also illustrates the need for concomitant field-based measurements. It also illustrates that sound knowledge of the spectral characteristics of the features of interest and how they respond to environmental variables can also help in optimising the use of remotely sensed data. The radiance of thermal infrared wavelengths from a soil is primarily determined by its moisture content. The wetter the soil is, the cooler it

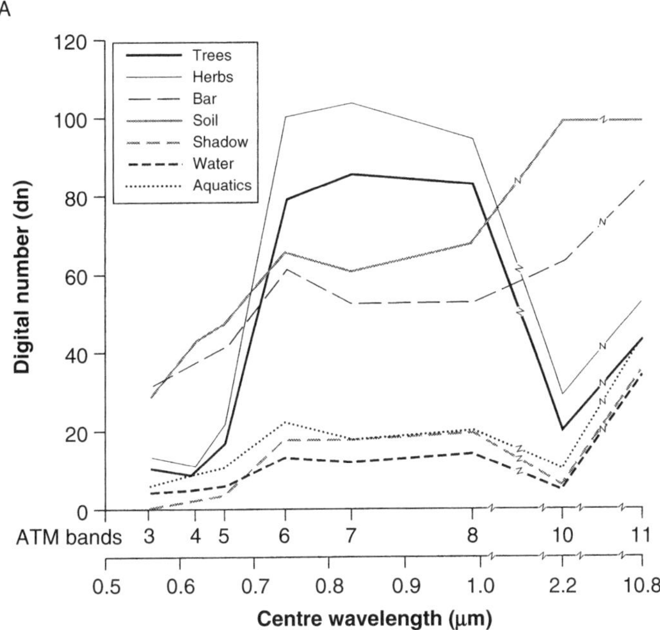

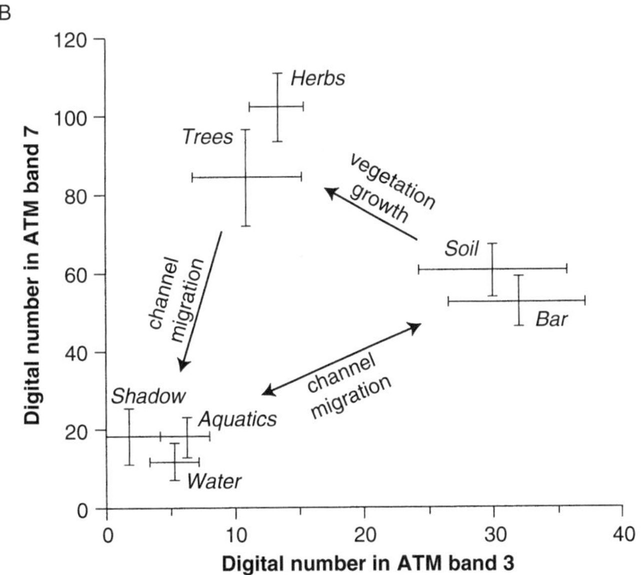

Figure 6.4 Spectral signatures for the fluvial environment based on imagery of the river Teme, UK (modified from Hooper 1992, Milton *et al.* 1995). (A) Representative spectral response curves ascertained from ATM data. (B) Differentiation of fluvial environments using the range of ATM Band 7 and Band 3 values for each surface type and their relationship to channel change

will be during the day and the warmer it will be at night. Soils and sediment in general generate a low radar return and only when they are recorded at moderate to low incidence angles do they generate a moderate return and are sensitive to soil moisture variations.

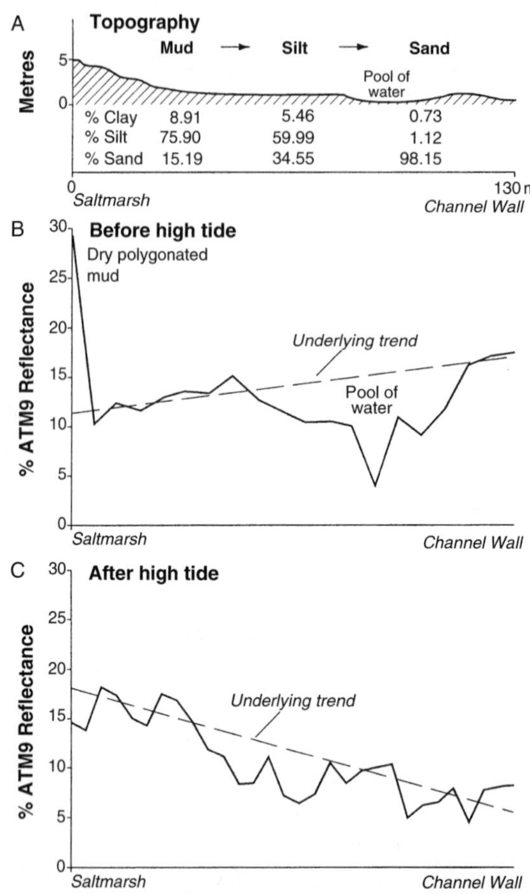

Figure 6.5 Variable ATM reflectance from inter-tidal sediments either side of a high tide on the river Ribble, England, showing that moisture content not only results in a shift in reflectance but also reverses the relationship between reflectance and particle size (after Rainey *et al.* 2000). (A) Inter-tidal morphology and sedimentology. (B) Spatial variation in spectral reflectance of the inter-tidal sediments before the high tide. (C) Spatial variation in spectral reflectance of the inter-tidal sediments after the high tide and re-wetting

Unlike soil and sediment, pure water surfaces absorb or transmit the majority of visible, near and middle infrared radiation. In visible wavelengths, little light is absorbed, a small amount (usually under 5% is reflected) and the majority is transmitted. Water also absorbs near and middle infrared wavelengths strongly. This results in sharp contrasts between water and land boundaries with pure water appearing black, for instance, on infrared aerial photographs. The factors that affect the spatial variability in the reflectance are depth of water, the suspended and solute content of the water and the surface roughness of the water. In shallow water, some radiation is reflected not by the water itself but by the substrate. Therefore, in shallow water it is often the channel bed that determines the water's reflectance properties and colour, in the absence of high-suspended sediment loads or colour levels. The effects of differing water depths and differing substrates on spectral reflectance, as measured using a field spectrometer is shown in Figure 6.6. Water has a similar thermal inertia to soil and yet it has a much smaller diurnal thermal range. It may therefore be differentiated by being warmer at night and cooler during the day with differences most marked early in the morning. Thermal imagery, especially where collected in the early hours of the morning, is therefore often best used in differentiating soil, water boundaries and soil moisture

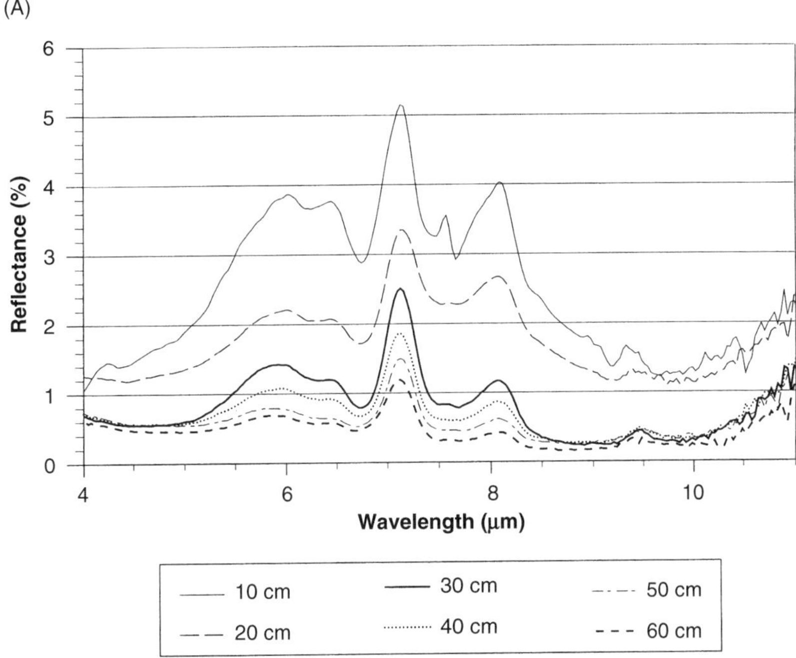

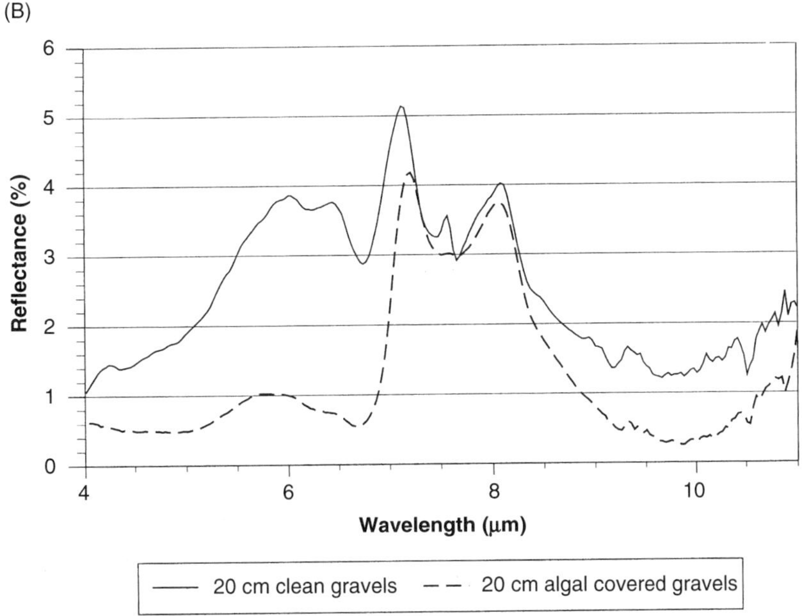

Figure 6.6 Spectral reflectance as recorded on the river Tummel, Scotland, using a field spectrometer showing the effect of (A) water depth and (B) bottom types (after Gilvear *et al.* 1998; reproduced by permission of Taylor & Francis, Ltd., http://www.tandf.co.uk/journals)

variability. A water body is usually an area of low radar return and appears black on radar images although speckling may occur if waves are present and at right angles to the radar pulse.

In the case of the spectral reflectance from an image or part of an image of particular interest not encompassing the full range of DN values, contrast stretching can be applied to heighten differences in the image. Thus, given that the grey-tone variability across the water surface was limited on scanned aerial photographs of Faith Creek, Alaska, a contrast stretch was applied solely to the water surface (Gilvear *et al.* 1995, Winterbottom 1995, Winterbottom and Gilvear 1997) to allow a stronger link to water depth to be extrapolated. Other parts of the image were masked off and the lightest grey-tone (equivalent to the shallowest areas) was assigned a value of 255 and the darkest areas (equivalent to the deepest areas) a value of zero. This allowed the differences in water depth in the image to be more clearly identified. Many other image enhancement techniques are available to improve the 'quality' of images but are outside the scope of this chapter. Classification of imagery into earth surface categories based on their spectral properties can be undertaken using a variety of methods. For more detailed types of classification, and further information on enhancement and classification techniques, a textbook on remote sensing should be consulted (e.g. Sabins 1996). Raw DN values or radiation fluxes can be converted to an environmental variable (e.g. water depth or soil moisture content) using an empirically based relationship derived with ground truth data or an established algorithm. Supervised classification of pixels is based upon assigning pixels with similar spectral characteristics to pixels identified in training areas. Training areas are one or more pixels assigned a land surface category where the cover type for those pixels in known from field survey. Unsupervised classification relies on methods that assign pixels to a pre-determined number of groups according to spectral similarity. This may or may not relate to land cover types or features on the ground.

Summary Overview

The preceding sections have demonstrated that there is a large number of factors that have to be taken into account when selecting a remote sensing approach to interrogate a fluvial system. The key variables that will determine the choice of data and approach taken are:

(i) the length of river being studied;
(ii) the width of the river;
(iii) the spatial resolution required;
(iv) the land cover or water surface or sub-surface properties to be detected;
(v) degree of precision and accuracy acceptable;
(vi) the frequency with which changes might need to be detected.

Following consideration of these variables the fluvial geomorphologist may then be able to determine whether a ground-based, airborne, or space-borne approach will be most appropriate and whether single spectra, multi-spectral or hyper-spectral data are required. Table 6.5 provides a generic protocol for assessing the most suitable remote sensing approach according to the type of fluvial study being undertaken.

6.3 RIVER GEOMORPHOLOGY AND IN-CHANNEL PROCESSES

2D Channel Morphology and Channel Change

2D mapping of river channel morphology and channel change has been the focus of fluvial geomorphology for a number of decades (Table 6.4). A key issue is scale when using remotely sensed data. Large channels are easy to observe on aerial photographs and satellite data using a range of imagery at all scales. The only constraint is restricted river lengths on large-scale images and thus high purchase costs and time involved in producing maps of drainage networks for large areas. For example, the anastomosing channels on the Niger delta has been mapped with satellite data (Diakite *et al.* 1986). As the channel gets smaller, however, spatial resolution becomes critical. France *et al.* (1986) working in Wales concluded that Landsat TM data could record lakes as small as 0.6 ha and streams down to 3–5 m width with acceptable accuracy, and thus detected 33 first-order streams. However, 1:10 000 air-photo interpretation revealed 156 first-order streams many of which were less than 1.0 m wide. For the delineation of ephemeral streams in Nevada 76% of second-order and larger streams could be identified in SPOT panchromatic images (Gardner *et al.* 1987). At a smaller scale and in a more complex situation, Schumann (1989) identified the 'parent' channel and the relative importance of anabranches in an anastomosing reach of Red Creek, Wyoming, using black and white aerial photographs. The parent channel was darkest due to grasses

Table 6.4 Recent examples of the application of remote sensing to fluvial geomorphology

Location	Geomorphic purpose	Imagery type, and scale/pixel size	Reference
(a) *2D channel morphology and channel change*			
Tanana River, Alaska, USA	Braid bar morphology	Synthetic aperture radar (SAR)	Nykanen *et al.* (1998)
Soda Butte Creek, Montana, and Cache Creek Wyoming, USA	In-channel hydrogeomorphic units including coarse woody debris	Airborne multi-spectral (blue, green, red and infrared bandwidths); 1 m resolution	Wright *et al.* (2000)
River Tummel, Scotland	Channel planform change and bank erosion	Aerial photography, 1:10 000 and 1:12 000	Winterbottom (2000)
(b) *3D and quasi-3D channel morphology and channel change*			
North Ashburton River, New Zealand	Below water line morphology and exposed sediments	Aerial photography 1:3000	Westaway *et al.* (2000)
Waimakariri River, New Zealand	Channel morphology and DEM production in large braided systems	Aerial photography; 0.5 m resolution	Lane *et al.* (2000)
River Tummel, Scotland	Below water line morphology and riparian vegetation	Airborne multi-spectral imagery; 1 m resolution	Bryant and Gilvear (1999)
(c) *2D mapping of turbidity suspended solids concentrations and bed material*			
River Ribble, North West England	Percentage clay, silt and sand in inter-tidal sediments	Airborne thematic mapper data; 2 m pixel size	Rainey *et al.* (2000)
River Affric, Scotland	Bed material size	Gantry mounted vertical black and white photographs; 1:24 m scale	Butler *et al.* (in press)
(d) *2D and 3D mapping of floodplain morphology*			
Garonne Valley, France	General floodplain discrimination	Landsat TM; 30 m resolution	Muller (1992)
River Tummel, Scotland	Riparian vegetation changes	Airborne multi-spectral imagery; 1 m resolution (2 m resolution)	Bryant and Gilvear (1999)
(e) *2D mapping of flood inundation*			
Amazon rainforest	Mapping of the extent of inundation below a forested canopy	SAR data on JERS I	Miranda *et al.* (1998)
	Floodplain inundation	SAR and optical imagery	Townsend and Walsh (1998)
River Meuse in The Netherlands	Mapping of the extent of inundation	SAR	Bates and De Roo (2000)
(f) *2D and 3D mapping of overbank sedimentation, deposition and scour*			
River Ob, Siberia	Floodplain deposition and scour	SAR	Smith and Alsdorf (1998)
Mississippi, USA	Overbank sedimentation	Landsat TM and oblique aerial photography	Gomez *et al.* (1995, 1997)

and sagebrush flanking the channel where moisture availability was highest. In the case of the spatial resolution of the imagery being used to its limit, a waveband that achieves the greatest contrast between land and water is most suitable (e.g. near and middle infrared) for detection of channel planform.

Table 6.5 Feasibility, advantages and disadvantages of various remote sensing approaches according to river channel size and type of geomorphic investigation (see also Table 6.3)

(A) Small sized channels (<20 m wide)

Investigation type and remote sensing format	2D channel morphology and channel change	3D and quasi-3D channel morphology and channel change	2D mapping of suspended solids concentrations and bed material	2D and 3D mapping of floodplain morphology	2D mapping of flood inundation	2D and 3D mapping of overbank sedimentation, deposition and scour
Black and white aerial photography						
	Advantages	*Advantages*		*Advantages*	*Advantages*	*Advantages*
(General advantages) Widely available, relatively cheap and easy to commission flights which can coincide with cloudless skies. Historical record as far back as the 1940s and 1950s for most developed countries	1:5000 scale or larger photography sufficient for detailed analysis of planform	Image analysis applied to scanned photographs can be used in clear shallow streams to detect variations in water table depth		1:5000 scale or larger photography sufficient for detailed mapping of floodplain landforms. Photogrammetry can be applied to stereo pairs for detection of relief	Feasible	1:5000 scale or larger photography sufficient for detailed mapping of floodplain landforms. Photogrammetry can be applied to stereo pairs for detection of relief
	Disadvantages	*Disadvantages*	*Disadvantages*	*Disadvantages*	*Disadvantages*	*Disadvantages*
	For long reaches (e.g. >2 km) a large number of individual photographs are needed especially when the photography is not specially commissioned	Not possible in relatively deep or turbid rivers. Need for validation of water depths using ground-based measurements	Not generally possible unless very marked differences in concentrations are apparent. Need for ground-based measurements. Possibility of confusion with variations in stream bed reflectance in shallow streams	For long reaches a large number of individual photographs are needed especially when the photography is not specially commissioned. Experience in air photograph interpretation may be necessary for distinguishing some features	Confusion over classification of water with other land uses possible. For long reaches a large number of individual photographs are needed especially when the photography is not specially commissioned	For long reaches a large number of individual photographs are needed especially when the photography is not specially commissioned. Experience in air photograph interpretation may be necessary for distinguishing some features
Colour and infrared aerial photography						
(General advantages) As above but less widely available and limited historical record. Easier to interpret than above	As above	As above	As above but variability in turbidity often more marked than on black and white aerial photographs	As above although colour aerial photography can aid feature recognition. Infrared also enhances differences in soil moisture, which often aids feature recognition	As above but enhanced capability for water detection. On infrared photographs water shows up as black	As above although colour aerial photography can aid feature recognition. Infrared also enhances differences in soil moisture which often aids feature recognition

Airborne multi- and hyper-spectral imagery						
		Advantages	*Advantages*	*Advantages*	*Advantages*	*Advantages*
(General advantages) Digital format and specific spectral wavelengths can be prescribed according to the purpose of the study. Flights can be planned to coincide with cloudless skies	The scale or such imagery is not generally appropriate. Only on streams greater than 10 m wide information would be useful and then detection of bank position only accurate to approximately the nearest metre. Problems of mixed pixels	Enhanced water depth penetration at certain wavelengths. Laser altimetry provides the opportunity of accurate digital elevation models for exposed sediments (useful in braided rivers at low flows)	Enhanced detection but confusion with bottom reflectance in non-turbid shallow stream	Laser altimetry provides the opportunity of accurate digital elevation models. Possible to map the floodplain surface through wooded canopies. Use of specific wave bands aids feature recognition and soil moisture and particle size variations	Easy detection of water surfaces. Possibility of estimating water depth	Use of specific wavebands aids feature recognition and soil moisture and particle size variations. Repeat flights using laser altimetry provides the opportunity of mapping areas that have undergone significant changes in erosion and deposition. Possible to map the floodplain surface through wooded canopies
		Disadvantages	*Disadvantages*	*Disadvantages*		*Disadvantages*
		The scale or such imagery is not generally appropriate. Problems of mixed pixels. Need for validation of water depths using ground-based measurements	The scale of such imagery is not generally appropriate. Problem of mixed pixels. Need for ground-based measurements	On small floodplains problems of spatial resolution and mixed pixels		On small floodplains problems of spatial resolution and mixed pixels
Satellite and space-borne imagery						
(General advantages) Wide scale coverage and availability and high frequency of over flights. Spatial and spectral resolution becoming higher	Generally not appropriate except for crude channel planform detection	Not appropriate	Not appropriate	Generally not appropriate	Spatial resolution generally not appropriate except on small rivers with wide floodplains. Problems of mixed pixels	Generally not appropriate

(*Continues*)

Table 6.5 (*Continued*)

(B) Medium sized channels (20–200 m wide)

River channel size/ investigation type	2D mapping of channel morphology and channel change	3D mapping of channel morphology (3D)	2D mapping of suspended solids concentrations and bed material	2D and 3D mapping of floodplain morphology	2D mapping of flood inundation	2D and 3D mapping of overbank sedimentation, deposition and scour
Black and white aerial photography						
	Advantages	*Advantages*		*Advantages*	*Advantages*	*Advantages*
	1:20 000 scale or larger photography sufficient for detailed analysis of planform	1:10 000 scale or larger photography sufficient. Image analysis applied to scanned photographs can be used in clear shallow streams to detect variations in water table depth		1:10 000 scale or larger photography sufficient for detailed mapping of floodplain landforms. Photogrammetry can be applied to stereo pairs for detection of relief	Feasible using 1:20 000 scale or larger	1:10 000 scale or larger photography sufficient for detailed mapping of floodplain landforms. Photogrammetry can be applied to stereo pairs for detection of relief
	Disadvantages	*Disadvantages*	*Disadvantages*	*Disadvantages*	*Disadvantages*	*Disadvantages*
	For reaches longer than 5 km, more than one photograph is needed. Number of photographs may be large for long reaches especially when the photography is not specifically commissioned	Not possible in relatively deep or turbid rivers. Need for validation of water depths using ground-based measurements. For reaches longer than 5z km, more than one photograph is needed	Not generally possible unless very marked differences in concentrations are apparent. Need for ground-based measurements. Possibility of confusion with variations in stream bed reflectance in shallow streams	For long reaches a large number of individual photographs are needed especially when the photography is not specially commissioned. Experience in air photograph interpretation may be necessary for distinguishing some features	Confusion over classification of water with other land uses possible. For long reaches a large number of individual photographs are needed especially when the photography is not specially commissioned	For long reaches a large number of individual photographs are needed especially when the photography is not specially commissioned. Experience in air photograph interpretation may be necessary for distinguishing some features
Colour and infrared aerial photography						
	As above	As above	As above but variability in turbidity often more marked than on black and white aerial photographs	As above although colour aerial photography can aid feature recognition. Infrared also enhances differences in soil moisture, which often aids feature recognition	As above but enhanced capability for water detection. On infrared photographs water shows up as black	As above although colour aerial photography can aid feature recognition. Infrared also enhances differences in soil moisture, which often aids feature recognition

Airborne multi- and hyper-spectral imagery						
	Advantages	*Advantages*	*Advantages*	*Advantages*	*Advantages*	*Advantages*
	Spatial resolution is often appropriate except at the lower end of this channel size range where mixed pixels are a problem and reduce the accuracy with which the position of river banks can be mapped	Spatial resolution is often appropriate except at the lower end of this channel size range where mixed pixels are a problem. Enhanced water depth penetration at certain wavelengths. Laser altimetry provides the opportunity of accurate digital elevation models for exposed sediments (useful in braided rivers at low flows)	Enhanced detection but confusion with bottom reflectance in non-turbid shallow stream	Laser altimetry provides the opportunity of accurate digital elevation models. Possible to map the floodplain surface through wooded canopies. Use of specific wavebands aids feature recognition and soil moisture and particle size variations	Easy detection of water surfaces. Possibility of estimating water depth and seeing water through wooded canopies	Use of specific wavebands aids feature recognition and soil moisture and particle size variations. Repeat flights using laser altimetry provides the opportunity of mapping areas that have undergone significant changes in erosion and deposition. Possible to map the floodplain surface through wooded canopies
		Disadvantages	*Disadvantages*	*Disadvantages*		*Disadvantages*
		Need for validation of water depths using ground-based measurements. Not appropriate for depths greater than 2 m or turbid river systems	Need for ground-based measurements. Spatial resolution is often appropriate except at the lower end of this channel size range where mixed pixels are a problem	On small floodplains problems of spatial resolution and mixed pixels		On small floodplains problems of spatial resolution and mixed pixels
Satellite and space-borne imagery						
	Generally not appropriate except for crude channel planform detection	Not appropriate	Not appropriate		Spatial resolution generally not appropriate except with small rivers with wide floodplains. Problems of mixed pixels	Generally not appropriate

(*Continues*)

Table 6.5 (*Continued*)

(C) Large river channels (>200 m wide)

River channel size/ investigation type	2D mapping of channel morphology and channel change	3D mapping of channel morphology (3D)	2D mapping of suspended solids concentrations and bed material	2D and 3D mapping of floodplain morphology	2D mapping of flood inundation	2D and 3D mapping of overbank sedimentation, deposition and scour
Black and white aerial photography						
	Advantages	*Advantages*		*Advantages*	*Advantages*	*Advantages*
	1:50 000 scale or larger photography sufficient for detailed analysis of planform	1:10 000 scale or larger photography sufficient. Image analysis applied to scanned photographs can be used in clear shallow streams to detect variations in water table depth		1:10 000 scale or larger photography sufficient for detailed mapping of floodplain landforms. Photogrammetry can be applied to stereo pairs for detection of relief	Feasible using 1:20 000 scale or larger	1:10 000 scale or larger photography sufficient for detailed mapping of floodplain landforms. Photogrammetry can be applied to stereo pairs for detection of relief
	Disadvantages	*Disadvantages*	*Disadvantages*	*Disadvantages*	*Disadvantages*	*Disadvantages*
	For reaches longer than 10 km more than one photograph is needed	Not possible in relatively deep or turbid rivers. Need for validation of water depths using ground-based measurements	Not generally possible unless very marked differences in concentrations are apparent. Need for ground-based measurements. Possibility of confusion with variations in stream bed reflectance in shallow streams	For long reaches, a large number of individual photographs are needed especially when the photography is not specially commissioned. Experience in air photograph interpretation may be necessary for distinguishing some features	Confusion over classification of water with other land uses possible. For long reaches, a large number of individual photographs are needed especially when the photography is not specially commissioned	For long reaches, a large number of individual photographs are needed especially when the photography is not specially commissioned. Experience in air photograph interpretation may be necessary for distinguishing some features
Colour and infrared aerial photography						
	As above	As above	As above but variability in turbidity often more marked than on black and white aerial photographs	As above although colour aerial photography can aid feature recognition. Infrared also enhances differences in soil moisture which often aids feature recognition	As above but enhanced capability for water detection. On infrared photographs water shows up as black	As above although colour aerial photography can aid feature recognition. Infrared also enhances differences in soil moisture, which often aids feature recognition

Airborne multi- and hyper-spectral imagery						
		Advantages	*Advantages*	*Advantages*	*Advantages*	*Advantages*
	Scale is sufficient for overall planform detection and individual reaches can be zoomed in on for detailed analysis	Spatial resolution is not a problem. Enhanced water depth penetration at certain wavelengths. Laser altimetry provides the opportunity of accurate digital elevation models for exposed sediments (useful in large braided rivers at low flows)	Spatial resolution is not a problem. Enhanced detection but confusion with bottom reflectance in non-turbid shallow stream	Laser altimetry provides the opportunity of accurate digital elevation models. Possible to map the floodplain surface through wooded canopies. Use of specific wavebands aids feature recognition and soil moisture and particle size variations	Easy detection of water surfaces. Possibility of estimating water depth and seeing water through wooded canopies	Use of specific wavebands aids feature recognition and soil moisture and particle size variations. Repeat flights using laser altimetry provides the opportunity of mapping areas that have undergone significant changes in erosion and deposition. Possible to map the floodplain surface through wooded canopies
		Disadvantages	*Disadvantages*			*Disadvantages*
		Need for validation of water depths using ground-based measurements. Not appropriate for depths greater than 2 m or turbid river systems	Need for ground-based measurements			Huge volumes of data are generated even with modest channel lengths (e.g. 5 km)
Satellite and space-borne imagery						
	Advantages					
	Spatial resolution generally appropriate except at the lower end of the size range and with lower resolution imagery	Water depths and turbidity usually too high	Ideal but need ground-based measurements for calibration	Ideal although mixed pixels cause detection of smaller features to be problematic. Use of specific wavebands aids feature recognition and soil moisture and particle size variations. Digital elevation modelling not normally feasible	Ideal	Ideal although mixed pixels cause detection of smaller features to be problematic. Use of specific wavebands aids feature recognition and soil moisture and particle size variations. Sequential synthetic aperture radar

(Continues)

Table 6.5 (c) (*Continued*)

River channel size/ investigation type	2D mapping of channel morphology and channel change	3D mapping of channel morphology (3D)	2D mapping of suspended solids concentrations and bed material	2D and 3D mapping of floodplain morphology	2D mapping of flood inundation	2D and 3D mapping of overbank sedimentation, deposition and scour
						Interferometry allows disturbance mapping by identifying changes in the roughness of surfaces—difficult to use in well-vegetated areas
	Disadvantages					
	More than one image for long river lengths (>100 km) needed					

Increasingly, robust automated classification of channels will become possible (Argialas *et al.* 1988). Many other examples of imagery being used to map channel planform could be cited but this is not necessary given the straightforward and obvious simplicity of the technique. However, problems can sometimes occur, in detection and bankfull definition, and here the expertise of the geomorphologist is of paramount importance.

In-channel features have also been mapped extensively using aerial photographs and satellite imagery. Aerial photography, for example, has been used to map bar forms for over five decades as part of channel planform studies (e.g. Ferguson and Werritty 1983, Warburton *et al.* 1993). Similarly on large rivers, satellite imagery has been used to map bar morphology (Thorne *et al.* 1993). Figure 6.7, for example, clearly shows overall channel planform and bar morphology of the Mississippi river above Vicksburg. The image, covering an area of about 28 km × 21 km was acquired in October 1994 by space-borne imaging radar. Imagery, such as that shown in Figure 6.7, can be used easily to produce quantitative data on geomorphic attributes such as channel width, and size and shape of exposed channel bars. However, the bar size and shape as depicted on the image is stage dependent and successive images cannot always be compared directly unless water levels are known to be the same at each of the epochs. More recently, airborne multi-spectral imagery has been used to try and map a wide-range of geomorphic features with mixed success (e.g. Wright *et al.* 2000). The potential is high but increased knowledge of the spectral characteristics of geomorphic features is still required.

Channel planform change has also been the research focus of a number of fluvial geomorphologists and sequential sets of aerial photographs have commonly been used to detect change (e.g. Ferguson and Werritty 1983, Lapointe and Carson 1986, Table 6.4). Large-scale changes in channel position are often apparent from visual interpretation but geometric rectification is required where accurate measurement of change is necessary or changes in bank position are small. For example, although, Gilvear *et al.* (1999b) were able to identify large-scale changes in channel position on the Luangwa River, Zambia, visually (Figure 6.8), geometric rectification and digitisation of rivers bank lines within a GIS were necessary to detect channel change elsewhere and to measure accurately bank erosion rates. The accuracy of visual comparison, without the use of geometric rectification, or rectification of photographs to each other using fixed ground-control points, will depend upon the degree of tilt and distortion of the aerial photographs. Williams *et al.* (1979) managed to superimpose 39 photogrammetrically recovered bank profiles to measure retreat rate to an accuracy of 0.06 ms/year. Temporal resolution of aerial photography will vary widely between regions. Sixteen dates between 1921 and 1977 were available to Williams *et al.* (1979) for the Ottawa River. In many cases, particularly in the New World, however, the temporal resolution will be much less. If a short-term change, particularly resulting from a single flood event, is of interest, commissioned flights are often necessary to gain temporal resolution (e.g. Winterbottom 2000). When comparing images it is also necessary to remember that gross changes between dates will mask cycles of erosion and accretion, and that channels may ‘flip’ position, returning to their original courses at a later date.

Since 1970s, satellite data have been used to enhance our knowledge of channel planform change of large river systems for which the synoptic view of a space-borne sensor has advantages over the restricted coverage of individual aerial photographs (e.g. Phillip *et al.* 1989, Perez and Muller 1990). For example, Salo *et al.* (1986) used multi-date Landsat MSS images of the meandering and anastomosing stretches of the Ucayili and Amazon in Peru, to quantify lateral migration rates of 200 m/year between 1979 and 1983. A more rudimentary approach based on the visual interpretation of photographic images was used by Ramasamy *et al.* (1991) to identify relict channels in the Yamuna River, western India. Most remarkably, Jacobberger (1988) mapped abandoned river channels that were active in Sahelian Mali 6000–8000 years ago using MSS and TM images.

3D and Quasi-3D Channel Morphology and Channel Change

Increasingly, various remote sensing methods have been used to produce 3D or quasi-3D reach scale channel morphology. The major problem with quantifying channel morphology in three dimensions using remote sensing methods is the fact that a different approach is needed for exposed and submerged areas of the river channel. For exposed channel bars, large-scale aerial photogrammetry can be used to produce accurate elevation data (e.g. Westaway *et al.* 2000; Table 6.4). Laser altimetry is also now being used to map the morphology of exposed channel beds. Sequential sets of laser altimetry data have also been

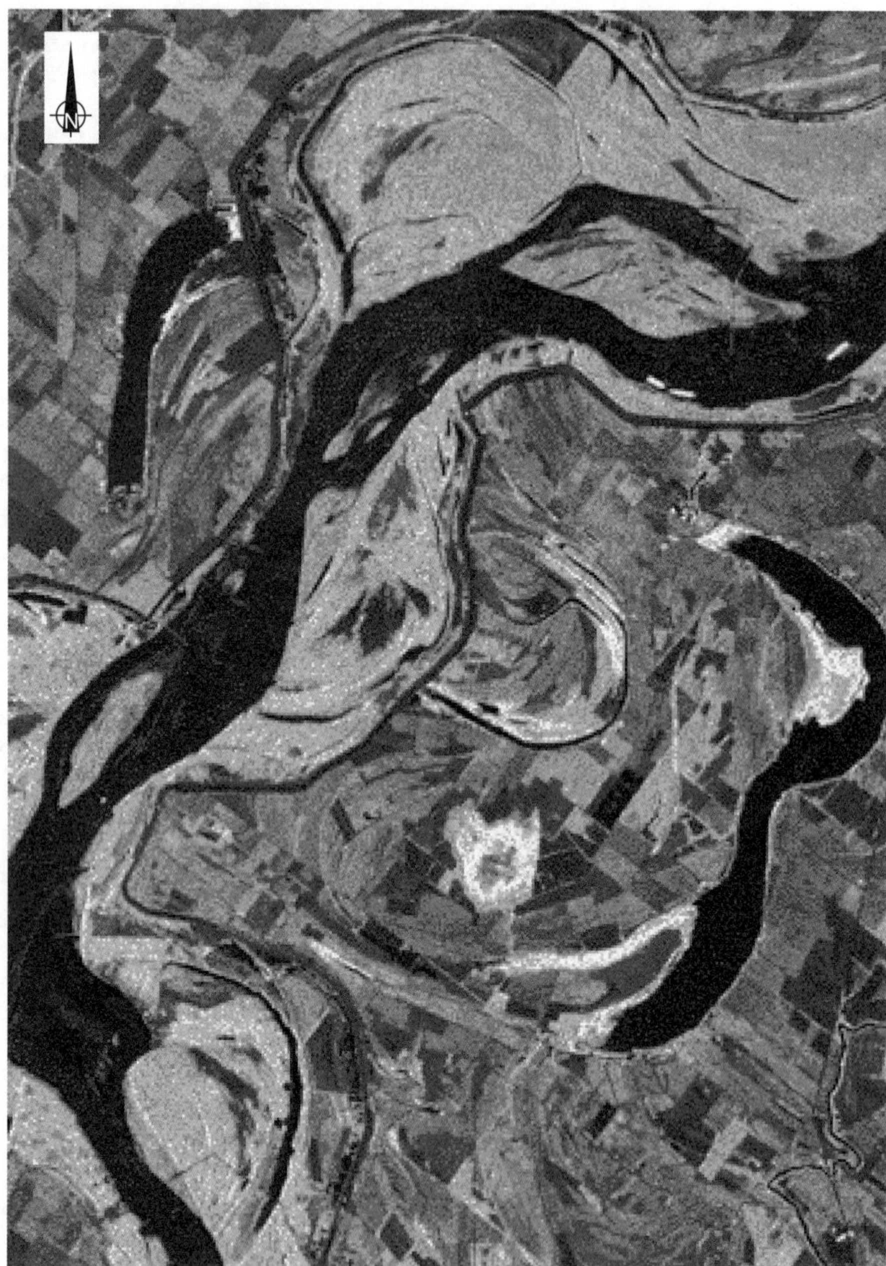

Figure 6.7 An image of approximately 28 km × 21 km of the river Mississippi and its floodplain, north of Vicksburg (http://www.jpl.nasa.gov/radar/sircxsar.html). The image was acquired in October 1994 by the space-borne imaging radar, C/X band synthetic aperture imaging radar system (SIR-C/X-SAR) (reproduced by permission of John Wiley and Sons, Ltd.)

used to quantify channel change for exposed areas of braided rivers on large New Zealand rivers. (Westaway, pers. comm.). Unfortunately, only on large braided river systems are large areas of the channel left exposed under low flow conditions. Thus, a more important concern for the fluvial geomorphologist,

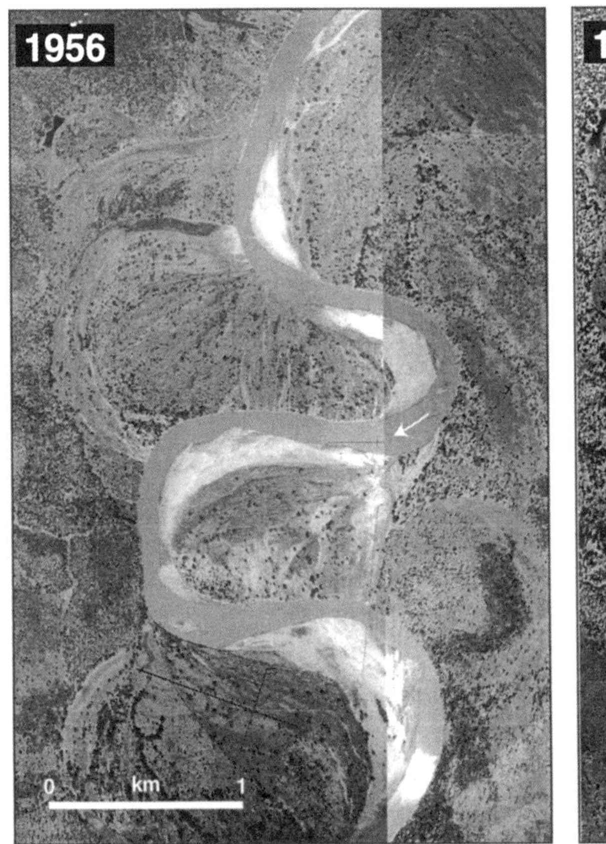

Figure 6.8 Channel change on the Luangwa River, Zambia, identified by comparison of aerial photography for the years 1956 and 1988 (after Gilvear *et al.* 1999b)

and from a remote sensing perspective a more challenging problem, is that of mapping submerged areas.

A relatively robust technique for applying image analysis to aerial photographs to derive water depths for shallow non-turbid rivers has been developed in recent years (Gilvear *et al.* 1995, Winterbottom and Gilvear 1997, Lane *et al.* 2000). The technique relies on there being a good correlation between the grey-tone on aerial photographs and water depth. This is not always visible to the eye but with can be detected image enhancement. The variation in grey-tone in shallow clear water rivers and simple situations relates to variations in the reflectance of light from the river bed. The absorption of light radiation in water increases exponentially with depth and a number of algorithms have been developed to model this relationship (Lyzenga 1981a,b, Clark *et al.* 1987, Bierwirth *et al.* 1993). Lyzenga's (1981a,b) algorithm has been employed in several studies to map the 3D morphology of gravel-bed rivers (e.g. Winterbottom 1995, Gilvear *et al.* 1995). The results of these studies have proved to be relatively accurate in comparison with ground-truth data collected contemporaneously with imagery. Such an approach unless tied in with methods of measuring the elevation of exposed sediments only producing a quasi-3D model in that absolute elevations are not known. Another approach to obtain 3D channel morphology using a combined remote sensing and ground data was the one undertaken by Lane *et al.* (1994). Above water line topography was quantified repeatedly over a 21-day period by rigorous analytical photogrammetry applied to oblique aerial photography. Below water line measurements were undertaken using rapid tacheometric survey and tied into the same ground control network, producing daily 3D images of the stream bed. Differences between these images then formed the basis for calculation of bedload transport rates and zones of aggradation and degradation. For their river, the results suggest that a cross-section spacing of less than 2 m is required to estimate cut or fill to within 20% of the correct value. This demonstrates the need for rapid high spatial resolution techniques for mapping 3D channel form to be developed for further understanding of channel bed dynamics.

Potentially, multi- and hyper-spectral imagery provides the geomorphologist with the ability to detect variations in water depth to deeper and more turbid channels. Using airborne multi-spectral data, Winterbottom and Gilvear (1997) found the best relationship between water depth and radiance in the interval 605–625 nm (Figure 6.9). Comparison of the 3D image produced by Winterbottom and Gilvear (1997) with a later image produced by applying the same technique, and which followed a 1:70-year return period flood also allowed subtle change in bedforms to be identified (Bryant and Gilvear 1999). In contrast to the work of Winterbottom and Gilvear (1997), Acornley *et al.* (1995), using CASI data, found the wavelengths 800–820 nm produced the best results although all bands in the ranges 510–610 and 645–820 gave satisfactory results. Here, the imagery was captured in autumn when aquatic macrophytes were absent, which would have complicated image analysis. Using a multi-spectral video imaging system that could detect in the green (550 nm), red (650 nm) and near infrared (850 nm) of the electromagnetic spectrum, Hardy *et al.* (1994) were also able to classify water depths and features such as runs, pools, and riffles on the Green River, Utah. The accuracy of such depth classifications are greatest in areas of low surface roughness, because a 'broken' water surface can scatter light and cause erroneous values; such phenomena, however, may aid the mapping of hydraulic habitat. Other limitations to the technique may include excessive shading of the bed, presence of submerged, floating and emergent vegetation and high water turbidity. A range of new multi-spectral videography systems are also becoming available which should give good image geometry, a greater number of bands and spectral resolution, and a convenience in deployment and data processing unmatched by more traditional systems (Sun and Anderson 1994, Hardy and Shoemaker 1995, Hardy 1998). On larger rivers, bathymetric mapping has been undertaken using satellite data but high turbidity and deeper water often preclude success. Vinod Kumar *et al.* (1997), however, undertook bathymetric mapping in the vicinity of the Rupnarayan-Hooghly River confluence, India, to depths of 8–10 m using LISS-II data in the wavelengths

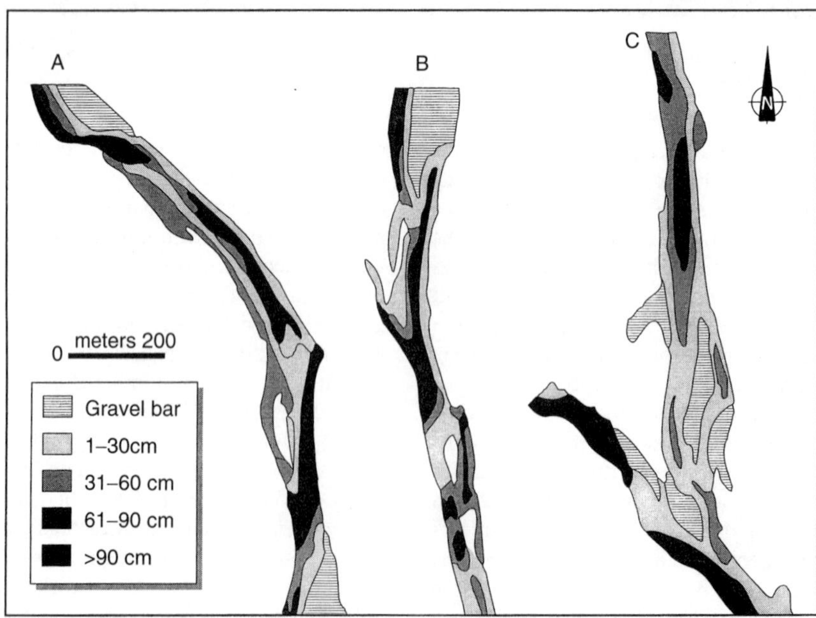

Figure 6.9 Example of a bathymetric map produced by image analysis applied ATM data of the confluence of the rivers Tay and Tummel, Scotland (after Winterbottom and Gilvear 1997). A, B and C are adjoining reaches in the downstream direction

0.77–0.80 nm. This was undertaken to guide dredging operations within the port of Calcutta. Production of 3D images of bed topography for long channel reaches from remotely sensed data offers great potential for linking hydraulic modelling with channel change. The primary assumptions of these techniques are that (a) the attenuation coefficient, and (b) the substrate reflectance, both remain constant over the full length and breadth of the extrapolated area. For the most part, these assumptions will hold true for short river reaches but ground data are really needed to verify the assumption, or produce separate algorithms where differences in the attenuation coefficient and substrate occurs. In time it is likely that dual frequency laser altimately will allow mapping of both exposed and submerged surfaces in one operation.

2D Mapping of Turbidity, Suspended Solid Concentrations and Bed Material

The use of satellite imagery on large river systems, and airborne data on smaller rivers, provides the opportunity to measure spatial and temporal changes in suspended sediment concentrations at the water surface over long reaches. However, because water chemistry, surface water roughness, sediment size, shape and mineralogy, atmospheric conditions, and shadow all also affect the spectral properties of deep water, as well as water depth in shallow areas ground truth data are often needed to allow calibration. Using a field-derived relationship between suspended sediment concentrations and spectral data, Aranuvachapun and Walling (1988) used satellite data to map spatial variability in suspended sediments for the Yellow River. To date, however, nearly all the work on the spectral reflectance–suspended sediment relationships has been undertaken on coastal, estuarine and lake waters (e.g. Lathrop and Lillesand 1986, Novo *et al.* 1989a,b, Xia 1993, Ferrier 1995). For a thorough understanding of the relationship between suspended sediment concentrations and reflectance, scrutiny of the literature relating to these environments is recommended. Few studies have examined the extension of results in estuaries and coastal areas to streams and rivers, and problems may arise in shallow and heterogeneous aquatic environments where mixed pixels are also likely. In coastal environments, Lillesand *et al.* (1983) recommended avoiding the use of MSS data in water depths of less than 2 m. The most appropriate spectra may also vary according to the sediment concentrations present (Liu and Klemes 1988; Table 6.4). However, the use of remotely sensed data in understanding the suspended sediment dynamics of medium sized and large rivers is likely to increase (Figure 6.10). Bale *et al.* (1994) suggest that reflectance in the near infrared is unaffected by other water quality parameters and is also largely independent of the colour or nature of the particles. However, it must be noted that the remotely sensed data will relate to near-surface values in turbid rivers, and not provide depth-integrated information.

Recent success has been obtained with mapping bed material size remotely (Table 6.4). This has been achieved using two different approaches. The first technique relies on sands, silts and gravels having differing spectral characteristics. Bed material size on exposed inter-tidal areas has thus been mapped (Rainey *et al.* 2000; Figure 6.11). In this situation, however, mapping was complicated by differences in soil moisture content because different areas have been subject to differing drying times since the last high tide, and spectral properties of sediments are moisture dependent (see earlier section and Figure 6.5). Image analysis techniques, which identify edges, have also been applied to aerial photography taken from a gantry to determine bed material size for weakly sorted river gravels (e.g. Butler *et al.* in press). However, up-scaling from such localised measurements of bed material size to the reach scale for gravel bed rivers has yet to be achieved.

6.4 FLOODPLAIN GEOMORPHOLOGY AND FLUVIAL PROCESSES

2D and 3D Mapping of Floodplain Morphology

Many landforms, including oxbow lakes, levees and scroll bars are present on floodplains resulting in a complex mosaic of topographical and sedimentological forms often masked by vegetation. Identification of these features may be possible from variations in soil moisture and vegetation. Aerial photography has thus been used extensively to map floodplain features. Colour aerial photographs are particularly useful in that subtle differences in land cover that relate to underlying topography and sedimentology are more easily seen. Lewin and Manton (1975) using 1:5000 stereo pairs to map the floodplain topography of three Welsh rivers to a vertical resolution of 0.10 m and horizontal resolution of 0.3 m. In the Garonne Valley, Muller (1992) found Band 5 of TM imagery best for discriminating the floodplain from adjacent terraces and mapping spatial variability in alluvial surfaces on the floodplain. Davidson and Watson (1995) were able to map spatial variability in soil moisture on the

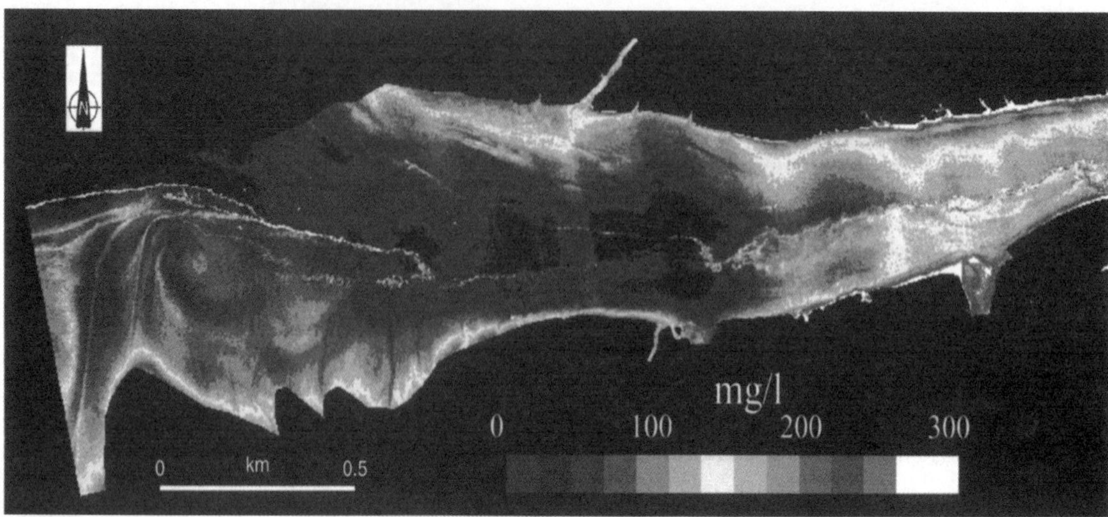

Figure 6.10 Suspended solid concentration on a rising tide within the freshwater reach of the tidal river Ribble, England, as mapped using CASI data (Atkin, pers. comm.). The high degree of spatial variation in the image relates to the leading edge of the saltwater wedge on the incoming tide moving up a freshwater reach of the estuary and shows the potential for improved understanding of suspended sediment dynamics (reproduced by the permission of John Wiley and Sons, Ltd.)

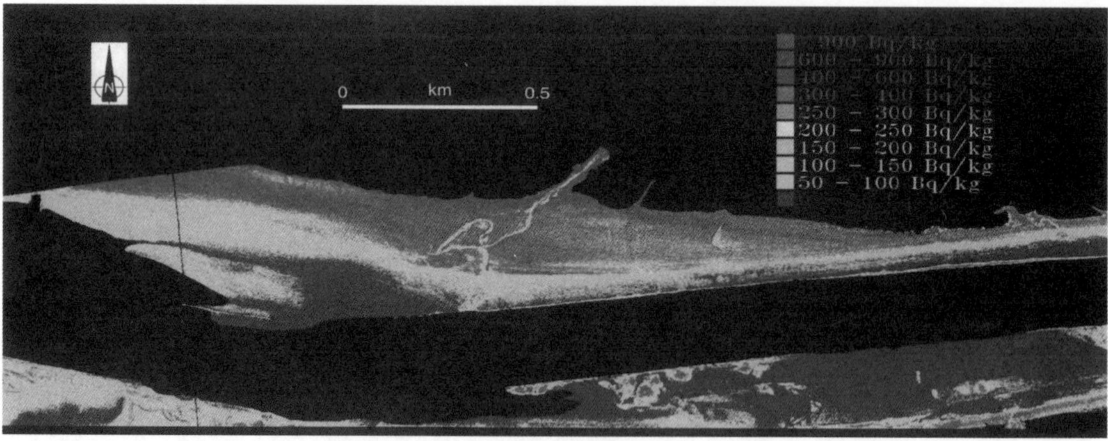

Figure 6.11 Particle size variation for the inter-tidal area of the river Ribble as mapped using ATM data (Rainey 1999). Using a regression relationship between particle size and radionuclide concentrations on the river, the data were used to produce a map of radionuclide concentrations in Bequerals per kilogram (Bq/kg) hence the key classification

floodplain. The areas of highest soil moisture were in topographic hollows left by relic channels. High spatial and vertical resolution topographic data can also be acquired by using scanning aircraft laser altimetry (Ritchie 1996). Laser altimetry is now being used by the UK Environmental Agency to map floodplain topography for flood hazard mapping. Recent advances in the integration of scanning lidar technology with CCD digital imaging technology has produced airborne technology with access to real time orthoimaging systems. The NASA ATM is a conically scanning airborne laser altimeter system capable of acquiring a swath width of 250 m wide with a spot spacing of 1–3 m, and vertical precision of 10–15 cm. The potential of this in geomorphological and floodplain research has been demonstrated by Garvin and

Williams (1993), and Marks and Bates (2000). Changes to the floodplain either side of a 1:65-year flood event were also quantified by Bryant and Gilvear (1999) using ATM data. Flood-induced depositional forms such as gravel lobes and sand splays were mapped.

2D Mapping of Flood Inundation

The use of airborne and satellite imagery to provide a synoptic perspective of flooding is relatively straightforward, except in forested floodplains, and has been extensively reviewed (e.g. Salomonson *et al.* 1983, Barton and Bathols 1989, Smith 1997). Sensor and platform use will depend upon the extent of inundation and spatial resolution, timing of the flood in relation to orbiting satellites or response times of airborne campaigns, the importance of emergent and floating vegetation and weather conditions. On small river systems, the extent of inundation can easily be seen on aerial photographs taken at the time of flooding. However, on such systems inundation is often short-lived and rarely are photographs available, particularly for the time of maximum inundation which is often of greatest interest. Gilvear and Davies (unpublished) were able to reconstruct the maximum extent of inundation using 1:5000 colour aerial photography taken 10 days after a 1:100-year flood event on the river Tay, Scotland, by the location of strand lines (i.e. flood debris).

When the flood coincides with a satellite orbit over head and an absence of clouds, and when large areas of open water exist, inundation mapping can be undertaken simply using satellite imagery (Table 6.4). However, inundation mapping below a forest canopy can be problematic, although Ormsby *et al.* (1985) found that the L-band data from the shuttle imaging radar (SIR-A) was helpful in separating forest vegetation from partially submerged grasses and shrubs and permitted a good definition of the land–water boundary even below a forest canopy. Cloud cover can be a problem in mapping inundation during the height of a flood except in the case of radar. The all-weather capability of radar is thus highly advantageous (Rudant 1994, Wagner 1994). Radar images record differences in roughness that indicate flood conditions. Sippel *et al.* (1994, 1998) thus used the scanning multi-channel microwave radiometer on board the Nimbus 7 satellite to track changes in inundation on the Amazon River near Manaus over a 7-year period. Despite the coarse spatial resolution of the annual inundation area determined using mixing models correlated well with changes in river stage. Similarly, Brakenridge *et al.* (1994) were able to map the extent of flooding during the July 1993 flood on the Mississippi River using a synthetic aperture radar (SAR) image of Iowa from the ERS-1 satellite. Moreover, by coupling SAR imagery with topographic imagery during the same flood but in Wisconsin, Brakenridge *et al.* (1998) were able to measure water surface and hence the flood wave.

2D and 3D Mapping of Overbank Sedimentation, Deposition and Scour

Remotely sensed data also affords the possibility of deriving estimates of suspended sediment concentrations in floodwaters and floodplain deposition. Mertes *et al.* (1993), working within the floodplain wetlands of the Amazon, showed that after nominal calibration to water-surface reflectance, near surface suspended sediment concentrations could be estimated for each 30 m × 30 m pixel using linear spectral mixture analysis. Similarly, Gomez *et al.* (1995) used a Landsat 5 Thematic Mapper image to derive estimates of near surface overbank suspended sediment concentrations in floodwaters during the 1993 Mississippi floods. Gomez *et al.* (1997), in conjunction with field measurements of deposition, were also able to use a TM image to produce a high spatial resolution map of floodplain sedimentation within the vicinity of the 1993 Sny Island levee break on the Mississippi near Canton in Missouri (Figure 6.12). Oblique aerial photography was also used to map scour, topsoil stripping a sand rim and sand sheets close to the levee break but spatial accuracy is compromised in such situations unless rigorous photogrammetric methods are adhered to. Using evidence from field survey and 1:10 000 colour aerial photography (Gilvear and Black 1999), and ATM imagery (Bryant and Gilvear 1999) found similar geomorphological patterns to that of Gomez *et al.* (1997) in relation to flood embankment failures during a large flood in the same year on the river Tay, Scotland.

SAR interferometry also has some potential for assessing widescale floodplain erosion and deposition by allowing sequential construction of high resolution DEMs and disturbance mapping from repeat-pass interferometric phase de-correlation. The latter is based on the fact that interferometric correlation or phase coherence will decrease if the scattering properties of a surface change over time (Smith and Alsdorf 1998). Thus, floodplain scour or

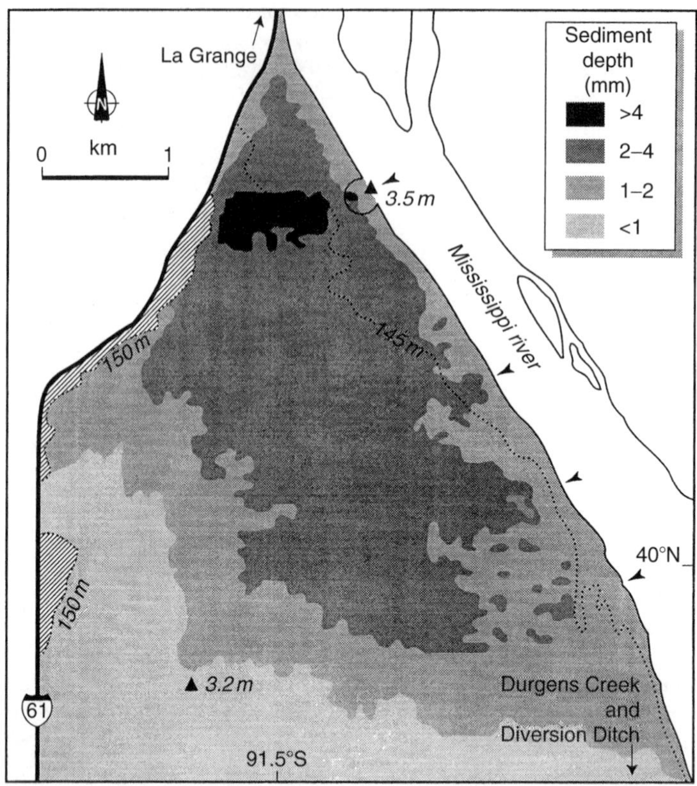

Figure 6.12 Spatial variation of the depth of overbank deposits in the vicinity of the Sny Island and levee break during the 1993 Mississippi floods as derived from a calibrated TM image (Gomez *et al.* 1997)

deposition can create a new scattering surface. While other factors can cause temporal phase de-correlation (e.g. soil moisture differences, vegetation growth), areas that do yield high phase coherence can be assumed to remain stable. Construction of accurate DEMs using SAR can also be problematic in heavily vegetated areas but the method has been used successfully in identifying flood damage (Izenberg *et al.* 1996).

6.5 CONCLUSIONS

Analysis of aerial photography and remotely sensed data has wide applications in detecting and mapping landforms, measuring temporal changes in fluvial landforms and controlling processes. The pros and cons of using differing sensor platforms and imagery are summarised in Table 6.5. For the study of large rivers (approx. 200 m wide or greater), space-borne sensors provide the fluvial geomorphologist with information on channel morphology. For medium sized rivers (approx. 20–200 m wide), data derived from airborne remote sensors or relatively large-scale aerial photography (approx. 1:5000–1:25 000) scale is better suited and can provide specific information. Increasingly, space-borne systems will have the spatial resolution to map features on smaller rivers. Small river systems are more amenable to study by traditional terrestrial techniques and large-scale aerial photography (approx. 1:2500 or better) often taken from 'blimps' or remotely controlled aircraft. On small and medium sized rivers conventional photography with a hand-held camera can sometimes also be analysed to reveal information not otherwise obtainable at high spatial resolution.

Maximising the potential of the analysis of aerial photography and other remotely sensed data as a tool in the study of fluvial systems depends upon a sound understanding of the spatial and temporal capabilities of different sensors, the range and usefulness of various of image analysis techniques, the spectral characteristics of the fluvial environment and the nature and scale of the geomorphic problem under investigation. Not one remote sensing system or type of image analysis provides the panacea, in that rivers varying in size, fluvial landforms and features have markedly different spectral characteristics and sensors vary in their spatial and spectral capability. Nevertheless, analysis of various types of terrestrially based and aerial photography, data from first generation satellite sensors and the latest generation of remote sensing systems offers the fluvial geomorphologist a rich set of tools. They allow the visualisation, description, and classification of a host of geomorphic attributes of rivers over a wide range of spatial scales and the detection and analysis of river channel change over timescales from days to decades. Such information is crucial to planners interested in the stability of rivers before authorising adjacent developments, ecologists interested in fluvial disturbance and engineers concerned with river training or whether bridges, transport networks and flood defences may be threatened by erosion.

ACKNOWLEDGEMENTS

The UK Natural Environment Research Council (NERC) supported some of the investigations described here through the provision of ATM data and instrument loans from the NERC equipment pool for field spectroscopy. The authors' knowledge of the subject area has also been aided by collaborative projects with a number of individuals including Dr. Sandra Winterbottom and Dr. Ted Milton. This chapter was also improved by anonymous peer review comments.

REFERENCES

Acornley, R.M., Cutler, M.E.J., Milton, E.J. and Sear, D.A. 1995. Detection and mapping salmonid spawning habitat in chalk streams using airborne remote sensing. In: *Proceedings of the 21st Annual Conference of the Remote Sensing Society, Southampton, September 1995*, pp. 267–274.

Aranuvachapun, S. and Walling, D.E. 1988. Landsat-MSS radiance as a measure of suspended sediment in the Lower Yellow River (Hwang Ho). *Remote Sensing of Environment* 25: 145–165.

Asrar, G. 1989. *Theory and Applications of Optical Remote Sensing*, New York: John Wiley and Sons, 725 p.

Argialas, D.P., Lyon, J.G. and Mintzer, O.W. 1988. Quantitative description and classification of drainage patterns. *Photogrammetric Engineering and Remote Sensing* 54: 505–509.

Bale, A.J., Toucher, M.D., Weaver, R., Hudson, S.J. and Aiken, J. 1994. Laboratory measurements of the spectral properties of estuarine suspended particles. *Netherlands Journal of Aquatic Ecology* 28: 237–244.

Barton, I.J. and Bathols, J.M. 1989. Monitoring floods with AVHRR. *Remote Sensing of Environment* 30: 27–35.

Bates, P.D. and De Roo, A.P.J. 2000. A simple raster-based model for flood inundation simulation, *Journal of Hydrology* 236(1/2): 54–77.

Bierwirth, P.N., Lee, T.J. and Burne, R.V. 1993. Shallow sea-floor reflectance and water depth derived by unmixing multispectral imagery. *PERS* 59: 331–338.

Brackenridge, G.G., Tracy, B.T. and Know, J.C. 1994. Orbital SAR remote sensing for a flood wave. *International Journal of Remote Sensing* 19: 1439–1445.

Bryant, R.G. 1996. Validated linear mixture modelling of Landsat TM data for mapping evaporite minerals on a playa surface: methods and applications. *International Journal of Remote Sensing* 17: 405–412.

Bryant, R. and Gilvear, D.J. 1999. Quantifying geomorphic and riparian land cover changes either side of a large flood event using airborne remote sensing: river Tay, Scotland, *Geomorphology* 23: 1–15.

Bryant, R., Tyler, A, Gilvear, D., McDonald, P., Teasdale, I., Brown, J. and Ferrier, G. 1996. A preliminary investigation into the spectral characteristics of inter-tidal estuarine sediments. *International Journal of Remote Sensing* 17: 405–412.

Brakenridge, G.R., Tracy, B.T. and Know, J.C. 1998. Orbital SAR remote sensing of a flood wave, *International Journal of Remote Sensing* 19: 1439–1445.

Butler, J.B., Lane S.N. and Chandler, J.H. in press. Automated extraction of grain-size data from gravel surfaces using digital image processing. *Journal of Hydraulics Research.*

Chandler, J.H. 1999. Effective application of automated digital photogrammetry for geomorphological research, *Earth Surface Processes and Landforms* 24: 51–63.

Christensen, E.J., Jensen, J.R., *et al.* 1988. Aircraft MSS data registration and vegetation classification for wetland change detection. *International Journal of Remote Sensing* 9: 23–38.

Clark, R.K., Fay, T.H. and Walker, C.L. 1987. Bathymetry calculations with Landsat 4TM imagery under a generalised ratio assumption. *Appllied Optics* 26: 4036–4038.

Cosandier, D., Ivanco, T.A., *et al.* 1994. The integration of a digital elevation model in CASI image geocorrection. In: *Proceedings of the First International Airborne Remote Sensing Conference and Exhibition, Strasbourg, France*, Ann Arbor, Michigan: Environmental Research Institute of Michigan (ERIM), pp. 515–529.

Cracknell, A.P. 1998. Synergy in remote sensing – what's in a pixel? *International Journal of Remote Sensing* 19: 2025–2047.

Cracknell, A.P. and Heyes L.W.B. 1993. *Introduction to Remote Sensing*, London: Taylor & Francis, 293 p.

Curran, P. 1994. Imaging Spectrometry. *Progress in Physical Geography* 18: 247–266.

Davidson, D.A., Watson, A. 1995. Spatial variability in soil moisture as predicted from airborne thematic mapper (ATM) data. *Earth Surface Processes and Landforms* 20: 219–230.

Dekker, T, Malthus, T. and Hoogenboom, P. 1995. The remote sensing of inland water quality. In: Danson, F.M. and Plummer, S.E., eds., *Advances in Environmental Remote Sensing*, London: Chapman and Hall, pp. 123–143 (298 p.).

Devereux, B.J., Fuller, R.M., Carter, L., Parsell, R.J. 1990. Geometric correction of airborne scanner imagery by delauney triangles, *International Journal of Remote Sensing* 11: 2237–2251.

Diakite, M., Yerheau, M. and Bonn, F. 1986. The utility of Landsat TM imagery in the inland delta cartography of Mali. In: *Proceedings of the 20th International Symposium on Remote Sensing of Environment (ERIM), Nairobi*, pp. 567–574.

Dixon, L.F.J., Barker, R., Bray, M., Farres, P., Hooke, J., Inkpen, R., Merel, A., Payne, D. and Shelford, A. 1998. Analytical photogrammetry for geomorphological research. In: Lane, S., Richards, K. and Chandler, J., eds., *Landform Monitoring, Modelling and Analysis*, Chichester: Wiley, pp. 63–49 (454 p.).

Ferguson, R.I. and Werritty, A. 1983. Bar development and channel changes in the gravelly River Feshie, Scotland. In: Collinson, J. and Lewin, J., eds., *Modern and Ancient Fluvial Systems*, Oxford: Blackwell Scientific Publications, pp. 181–194.

Ferrier, G. 1995. A field study of the variability in the suspended sediment concentration-reflectance relationship, *International Journal of Remote Sensing* 16: 2713–2720.

France, M.J., Collins, W.G. and Chindeley, T.R. 1986. Extraction of hydrological parameters from Landsat Thematic Mapper Imagery. In: *Proceedings of the 20th International Symposium on Remote Sensing of Environment (ERIM), Nairobi*, pp. 1165–1173.

Gardner, T.W., Connors, K.F., Hu, H. and Peterson, G.W. 1987. Delineation of ephemeral fluvial networks on low-relief piedmont surfaces, Plutonium Valley, Nevada, USA. In: *Spot-1, Image Utilisation, Assessment Results*, Paris, CNES, pp. 215–220.

Garvin, J.B. and Williams, R. 1993. Geodetic airborne laser altimetry of Breidamerkurjokull and Skeidararjokull, Iceland and Jakobshavn, Greenland, *Annals of Glaciology* 17: 377–386.

Gilvear, D.J. and Black, A. 1999. Flood induced embankment failures on the river Tay: implications of climatically induced hydrological change in Scotland, *Hydrological Sciences Journal* 43: 1–16.

Gilvear, D.J., Waters, T. and Milner, A. 1995. Image analysis of aerial photography to quantify changes in channel morphology and instream habitat following placer mining in interior Alaska. *Freshwater Biology* 44: 101–111.

Gilvear, D.J., Winterbottom, S.J. and Sichingbula, H. 1999b. Character of meander planform change on the Luangwa River, Zambia, *Earth Surface Processes and Landforms* 16: 1–24.

Gomez, B., Mertes, L.A., Phillips, J.D., Milligan, F.J. and James, L.A. 1995. Sediment characteristics of an extreme flood: 1993 upper Mississippi River Valley. *Geology* 23: 963–966.

Gomez, B., Phillips, J.D., Milligan, F.J. and James, L.A. 1997. Floodplain sedimentation and sensitivity: summer 1993 flood, Upper Mississippi River Valley. *Earth Surface Processes and Landforms* 22: 923–936.

Hardy, T.B. 1998. The future of habitat modelling and instream flow assessment techniques. *Regulated Rivers: Research and Management* 14: 405–420.

Hardy, T.B. and Shoemaker, J.A. 1995. Use of multi-spectral videography for spatial extrapolation of fisheries habitat use in the Comal River. In: *Proceedings of the 15th Workshop on Colour Photography and Videography for Resource Assessment*, American Society of Photogrammetry and Remote Sensing, pp. 154–163.

Hardy, T.B., Anderson, P.C., Neal, C.M.U., Stevens, D.K. 1994. Application of multispectral videography for the delineation of riverine depths and mesoscale hydraulic features. In: *Effects on Human Induced Changes on Hydrologic Systems, Paper and presentation at the American Water Resources Association, June 26–29, 1994, Jackson Hole, Wyoming*.

Hickin, E.J. and Nanson, G.C. 1975. Lateral migration rates of meandering rivers. *Journal of Hydrological Engineering* 110: 1557–1567.

Hooper, I.D. 1992. Relationships between vegetation and hydrogeomorphic characteristics of British riverine environments: a remotely sensed perspective. Unpublished Ph.D. Thesis, University of Southampton.

Irons, J.R., Weismiller, R.A. and Peterson, G.W. 1989. Soil reflectance. In: Asrar, G., ed., *Theory and Applications of Optical Remote Sensing*, New York: John Wiley and Sons, pp. 336–428 (Chapter 3).

Izenberg, N.R., Arvidson, R.E., Bracket, R.A., Saatchi, S.S., Osburn, G.R. and Dohrenwend, J. 1996. Erosional and depositional patterns associated with the 1993 Missouri River floods inferred from SIR-C and TOPSAR radar data. *Journal of Geophysical Research* 101: 23 149–23 167.

Jacobberger, P.A. 1988. Mapping abandoned river channels in Mali through directional filtering of thematic mapper data. *Remote Sensing of the Environment* 26: 161–170.

Kaufman, Y.J. 1989. The atmospheric effect on remote sensing and its correction. In: Asrar, G., ed., *Theory and Applications of Optical Remote Sensing*, London: John Wiley and Sons, pp. 189–211 (349 p.).

Kennedy, G. 1998. Channel migration and floodplain patch dynamics of a montane stream, Orgeon, USA, Unpublished B.Sc. Thesis, University of Stirling.

Lane, S.N. 2000. The measurement of river channel morphology using digital photogrammetry. *Photogrammetric Record* 16(96): 937–957.

Lane, S.N., Richards, K.S. and Chandler, J.H. 1993. Developments in photogrammetry; the geomorphological potential. *Progress in Physical Geography* 17: 306–328.

Lane, S.N., Chandler, J.H. and Richards, K.S. 1994. Developments in monitoring and modelling small-scale river bed topography. *Earth Surface Processes and Landforms* 19: 349–368.

Lane, S.N., Westaway, R.M., Hicks, D.M. and Duncan, M. J. 2000. High resolution digital photogrammetry and image analysis for the measurement of large gravel-bed rivers. In: *Paper Presented at the Annual Conference of the Remote Sensing Society, University of Leicester, September 2000*.

Lapointe, M.F. and Carson, M.A. 1986. Migration patterns of an asymmetric meandering river: the Rouge River, Quebec. *Water Resources Research* 22: 731–743.

Lathrop, R.G. and Lillesand, T.M. 1986. Use of thematic mapper data to assess water quality in Green Bay and Central Lake Michigan. *Photogrammetric Engineering and Remote Sensing* 52: 349–354.

Lewin J. and Manton, M.M. 1975. Welsh floodplain studies: the nature of floodplain geometry. *Journal of Hydrology* 25: 37–50.

Lillesand, T.M., Johnson, W.L., Devell, R.L., Lindstrom, O.M. and Mesisner, D.E. 1983. Use of Landsat data to predict the trophic status of Minnesota lakes. *Photogrammetric Engineering and Remote Sensing* 49: 219–229.

Liu, W.-Y. and Klemes, V. 1988. Quantitative analysis of distributions of suspended sediment in the Yellow River estuary from MSS data. *Geocarto International* 1: 51–62.

Lyzenga, D.R. 1981a. Remote Sensing of bottom reflectance and water attenuation parameters in shallow water using aircraft and Landsat data. *International Journal of Remote Sensing* 2: 71–82.

Lyzenga, D.R. 1981b. Remote sensing of bottom reflectance and water attenuation parameters in shallow water using aircraft and LANDSAT data. *International Journal of Remote Sensing* 2: 71–82.

Malthus, T.J., Place, C.J., Bennet, S., North, S. 1995. An evaluation of the airborne thematic mapper sensor for monitoring inland waters. In: *Proceedings of the Remote Sensing Society Meeting, Southampton, 1995*.

Marks, K. and Bates, P. 2000. Integration of high-resolution topographic data with floodplain flow models. *Hydrological Processes* 14(11/12): 2109–2122.

Mertes, L.A.K., Smith, M.O. and Adams, J.B. 1993. Estimating suspended sediment concentrations in surface waters of the Amazon River wetlands from Landsat images. *Remote Sensing of the Environment* 43: 281–301.

Milne, J.A. and Sear, D.A. 1997. Modelling river channel topography using GIS. *International Journal of GIS* 11: 99–519.

Milton, E.J., Gilvear, D.J., Hooper, I.D. 1995. Investigating river channel changes using remotely sensed data. In: Gurnell, A., Petts, G.E., eds., *Changing River Channels*, Chichester: Wiley, pp. 277–301 (398 p.).

Miranda, F.P., Fonseca, L.E.N. and Carr, J.R. 1998. Semivariogram textual classification of JERS-1 SAR data obtained over a flooded area of the Amazon rainforest. *International Journal of Remote Sensing* 19: 549–556.

Muller, E. 1992. Evaluation de la bande TM5 pour la cartographie morpho-hydrogeologique de la moyenne vallee de la Garonne, Project SPOT4/MIR. CNES, Paris.

Muller, E., Decamps, H., Dobson, M.K. 1993. Contribution of space remote sensing to river studies. *Freshwater Biology* 29: 301–312.

Novo, E.M.M., Hansom, J.D. and Curran, P.J. 1989a. The effect of sediment type on the relationship between reflectance and suspended sediment concentration. *International Journal of Remote Sensing* 10: 1283–1289.

Novo, E.M.M., Hansom, J. and Curran, P.J. 1989b. The effect viewing geometry and wavelength on the relationship between reflectance and suspended sediment concentration. *International Journal of Remote Sensing* 10: 1357–1372.

Nykanen, D.K., Foufoula-Georgiou, E. and Sapozhnikov, V.B. 1998. Study of spatial scaling in braided river patterns using synthetic aperture radar imagery. *Water Resources Research* 34: 1795–1807.

Ormsby, J.P., Blanchard, B.J., *et al.* 1985. Detection of lowland flooding using active micro-wave systems. *Photogrammetric Engineering and Remote Sensing* 51: 317–328.

Perez, J. and Muller, E. 1990. Suivi de l'evolution de la vegetation riveraine du Rio Guanare (Venezuela) a l'aide de donees SPOT XS, LANDSAT MSS et de photos aeriennes. In: *Troisiemes Jounrnees Scientifiques du reseau de Teledetectionion de l'UREF, Novembre 13–16, 1990, Toulouse*.

Phillip, G., Gupta, R.P., Bhattacharya, A. 1989. Channel migration studies in the middle Ganga Basin, India, using remote sensing data. *International Journal of Remote Sensing* 10: 1141–1149.

Plummer, S.E., Danson, F.M. and Wilson, A.K. 1995. Advances in remote sensing technology. In: Danson, F.M. and Plummer, S.E., eds., *Advances in Environmental Remote Sensing*, London: John Wiley and Sons, pp. 1–8.

Ramasamy, S.M., Bakliwal, P.C., Verma, R.P. 1991. Remote sensing and river migration in Western India. *International Journal of Remote Sensing* 12: 2597–2609.

Rainey, M.P. 1999. Airborne remote sensing of estuarine intertidal radionuclide concentrations, Unpublished Ph.D. Thesis, University of Stirling.

Rainey, M.P., Tyler, A.N., Bryant, R.G., Gilvear, D.J. and McDonald, P. 2000. The influence of surface and interstitial moisture on the spectral characteristics of inter-tidal sediments; implications for airborne image acquisition and processing. *International Journal of Remote Sensing* 21: 3025–3038.

Ritchie, J.D. 1996. Airborne laser altimeters, remote sensing applications to hydrology. *Hydrological Sciences Journal* 41: 625–636.

Rudant, J.P. 1994. French Guinana through the clouds: first complete satellite coverage. *Earth Observation Quarterly* 44: 1–6.

Sabins, F.F. 1996. *Remote Sensing: Principles and Interpretation*, Sydney: Freeman, 298 p.

Salo, J., Kalliola, R., Hakkinen, I., Makinen, Y., Niemala, P., Puhakka, M., Coley, P.D. 1986. River Dynamics and the diversity of Amazon lowland forest. *Nature* 322: 254–258.

Salomonson, V.V., Jackson, T.J., *et al.* 1983. Water resources assessment. In: *Manual of Remote Sensing*, Falls Church, Virginia: American Society of Photogrammetry, pp. 1497–1570.

Schumann, R.R. 1989. Morphology of Red Creek, Wyoming, an arid-region anastomosing channel system. *Earth Surface Processes and Landforms* 14: 277–288.

Sippel, S.J., Hamilton, S.K., Melack, J.M. and Choudhury, B.J. 1994. Determination of inundation area in the Amazon River floodplain using the SMMR 37 GHz polarization difference. *Remote Sensing of the Environment* 48: 70–76.

Sippel, S.J., Hamilton, S.K., Melacks, J.M. and Novo, E.M.M. 1998. *International Journal of Remote Sensing* 19: 1143–1161.

Smith, L.C. 1997. Satellite remote sensing of river inundation area, stage, and discharge: a review. *Hydrological Processes* 11(10): 1427–1439.

Smith, L.C. and Alsdorf, D.E. 1998. A control on sediment and organic carbon delivery to the Arctic Ocean revealed with satellite SAR: Ob River, Siberia. *Geology* 26: 395–398.

Sun, K. and Anderson, J.M. 1994. An easily deployable miniature, airborne imaging spectrometer. In: *Proceedings of the First International Airborne Remote Sensing Conference and Exhibition, Strasbourg, France*, Ann Arbor: Environmental Research Institute of Michigan (ERIM), pp. 178–189.

Thorne, C.R., Russell, A.P.G. and Alam, M.K. 1993. Planform pattern and channel evolution of the Brahmaputra River, Bangladesh. In: Best, J.L. and Bristow, C.S., eds., *Braided Rivers*, The Geological Society Special Publication No. 75, London: Geological Society, pp. 257–277 (417 p.).

Townsend, P.A. and Walsh, S.J. 1998. Modelling floodplain inundation using an integrated GIS with radar and optical remote sensing. *Geomorphology* 21: 295–312.

Townshend, J.R.G. and Justice, C.O. 1988. Selecting the spatial resolution of satellite sensors required for global monitoring of land transformations. *International Journal of Remote Sensing* 9: 187–236.

Verstraete, M.M. and Pinte, B. 1992. Extracting surface properties from satellite data in the visible and near-infrared wavelengths. In: Mather, P.M., ed., *Terra-1 Understanding the Terrestrial Environment*, London: Taylor & Francis, pp. 203–209.

Vinod Kumar, K., Palit, A. and Bhan, S.K. 1997. Bathymetric mapping in Rupnarayan-Hooghly confluence using Indian remote sensing satellite data. *International Journal of Remote Sensing* 18: 2269–2270.

Wagner, M.J. 1994. ERS-1 images of the Christmas flood over Europe. *Earth Observation Quarterly* 43: 12–13.

Warburton, J., Davies, T.R.H. and Mandl, M.G. 1993. A meso-scale field investigation of channel change and floodplain characteristics in an upland braided gravel bed river, New Zealand. In: Best, J.L. and Bristow, C.S., eds., *Braided Rivers*, The Geological Society Special Publication No. 75, London: Geological Society, pp. 241–257 (417 p.).

Westaway, R.M., Lane, S.N. and Hicks, D.M. 2000. The development of an automated correction procedure for digital photogrammetry for the study of wide, shallow, gravel bed rivers. *Earth Surface Processes and Landforms* 25: 209–226.

Wilson, A.K. 1995. An integrated data system for airborne remote sensing. *International Journal of Remote Sensing* 18: 1889–1901.

Williams, D.R., Romeril, P.M. and Mitchell, R.J. 1979. Riverbank erosion and recession in the Ottawa area. *Canadian Geotechnical Journal* 16: 641–650.

Winterbottom, S. 1995. An analysis of channel change on the rivers Tay and Tummel Scotland, using GIS and remote sensing techniques, Unpublished Ph.D. Thesis, University of Stirling.

Winterbottom, S.J. 2000. Medium and short-term channel planform changes on the Rivers Tay and Tummel, Scotland. *Geomorphology* 34: 195–208.

Winterbottom, S.J. and Gilvear, D.J. 1997. Quantification of channel bed morphology in gravel-bed rivers using airborne multispectral imagery and aerial photography. *Regulated Rivers: Research and Management* 13: 489–499.

Wright, A., Marcus, W.M. and Aspinall, R. 2000. Evaluation of multi-spectral imagery as a tool for mapping stream morphology. *Geomorphology* 33: 107–120.

Xia, L. 1993. A united model for quantitative remote sensing of suspended sediment Concentration. *International Journal of Remote Sensing* 14: 2665–2676.

7

Geomorphic Classification of Rivers and Streams

G. MATHIAS KONDOLF[1], DAVID R. MONTGOMERY[2], HERVÉ PIÉGAY[3] AND LAURENT SCHMITT[4]

[1]*Department of Landscape Architecture and Environmental Planning, University of California, USA*

[2]*Department of Geological Sciences, University of Washington, Seattle, USA*

[3]*UMR 5600, CNRS, Lyon, France*

[4]*UMR 5600, Université Lyon 2, Bron, France*

"You cannot step in the same river twice,
for the second time it is not the same river."
—Heraclitus

7.1 INTRODUCTION

Rivers range widely in size, in channel form, and in their degree of dynamism. Regional variability in river processes and river characteristics imparts a fundamental tension between development of generalizable and regional characterizations of river systems. It is not surprising, therefore, that attempts to classify rivers have resulted in a wide variety of classification schemes, serving a wide range of purposes from typologies for interpreting and understanding landscape evolution over geologic time to those attempting to aid in the development of engineering designs for channel restoration projects. As with any tool, classification can be useful if applied properly to the appropriate problem. But classification schemes are at best limited tools, whose capabilities are often overestimated by users lacking sound technical training in geomorphology. Moreover, reliance on classification systems can lead to serious problems, such as unnecessary and unwise interventions when misapplied or used by unskilled or inexperienced hands. This chapter discusses general philosophies of channel classification in fluvial geomorphology, reviews examples of geomorphic classification systems, and explores uses and limitations of classifications as a tool in fluvial geomorphology and river management.

Classification Defined

Classification is ordering of objects into groups based on common characteristics and attaching labels to the groups. Classification permits objects to be inventoried, so as to tally the number falling into various classes. If subgroups of a collection of objects can be identified with common characteristics and behavior patterns, distinct from other subgroups, then a set of traits can be ascribed to the object (based on detailed study of other members of that class), which may then allow prediction of the behavior of the object under new circumstances. Classification allows scientists to stratify an otherwise confusing universe into sets of similar objects, study representative objects, and extrapolate results to other, similar objects.

Classification refers both to the process of ordering objects in groups (the activity) and the resulting system of groups (the result). In common usage, the term is also used for the application of the resulting system, i.e., encountering new objects and placing them in the predetermined classes, a step referred to in the taxonomic literature as *identification* (Sneath and Snokal 1973). Taxonomists distinguish between *natural* classifications, a codification of natural clustering of objects with similar characteristics, and *special* classifications, which involve arbitrary distinctions drawn across a natural continuum (Sneath and Snokal 1973). Classifications of animals into species are considered natural classifications. Despite disagreements over details, most independent workers would reach similar classification decisions for major taxa, because evolution has provided a natural nested clustering. But animals also can be organized into useful, albeit arbitrary, special classifications such as all carnivores, or all aquatic invertebrates that cannot tolerate water temperatures exceeding a given level.

The process of classification development and application can be broken down into discrete steps (Figure 7.1). Based either on an a priori understanding of the system or cluster analysis on large data sets, a set of categories is proposed. The definition of categories depends, in part, upon the purpose of the classification, as a given set of objects can be classified in many different ways. The variables determined to be particularly diagnostic under the classification scheme are then emphasized in the subsequent collection of data. As additional objects are encountered they are assigned to categories in the existing classification scheme (identification), or recognized as not fitting within pre-existing categories. In the latter case, objects that do not fit into the classification can both indicate the limits of the system, and thereby provide feedback for revising the system or abandoning it for a new approach, and help identify special or unique objects or systems that differ from those encountered previously or considered representative.

Purposes of Classification

A wide range of classification schemes have been developed for fluvial systems, reflecting the intended purpose of the classification, different disciplines involved, and the characteristics of the systems being classified (i.e., the studied environment, as per Gurnell *et al.* 1994). For example, Bavarian Water Law uses a classification system to assign responsibility for river maintenance and flood control of large rivers (Class 1) to the state, medium-sized rivers (Class 2) to the seven districts within Bavaria, and smaller watercourses (Class 3) to local communities (W. Binder, Bayerisches Landesamt für Wasserwirtschaft, Munchen, Germany, personal communication, 1991). The classification system serves admirably for this administrative purpose, but it may not serve for other purposes, such as distinguishing among rivers with different ecological characteristics. Similarly, the US Forest Service incorporates fishery and water supply values in a classification system used to determine the degree of protection from timber harvest afforded to a reach. The system is based on presence of perennial flow, presence of resident or anadromous fish, use for municipal water supply, and relative size of stream (Gregory and Ashkenas 1990).

A fundamental motivation for using classification systems is to improve communication, especially in interdisciplinary settings. As biologists sought to classify aquatic habitat components to provide a common framework for the input of diverse disciplines and sites (Platts 1980, Hawkins *et al.* 1993), a number of authors recognized the need for a classification system for stream channels (Pennak 1971, Hawkes 1975, Terrell and McConnell 1978, Newson and Newson 2000).

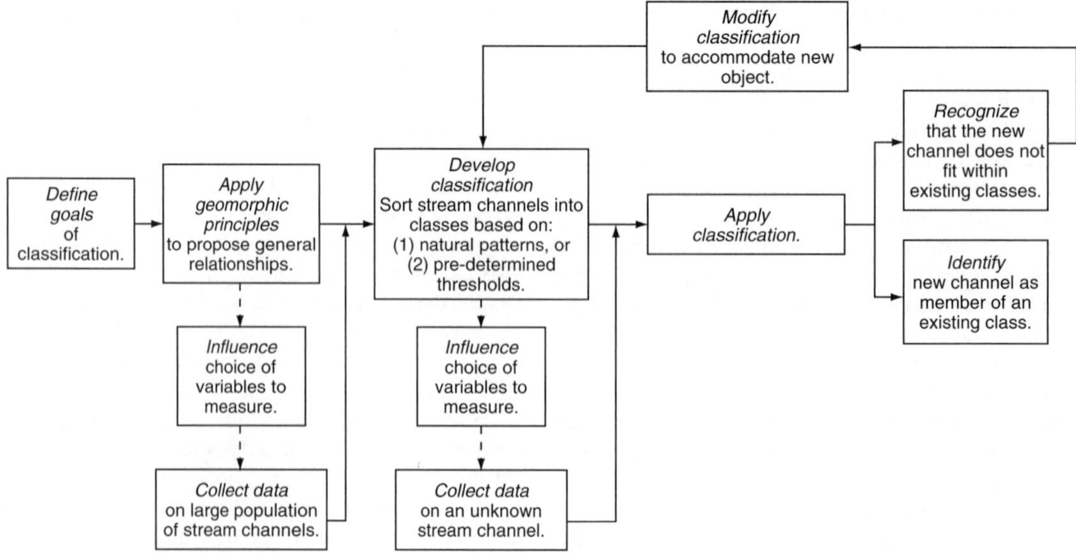

Figure 7.1 Flow chart of the process of development, application, and revision of a classification scheme for stream channels (from Kondolf 1995, reproduced by permission of John Wiley and Sons, Ltd.)

Classification systems intended to improve communication should be objective, so that operators from different disciplines and in different regions will reach the same classification decisions.

The influence of the classifier's discipline is readily apparent in the diverse classification systems proposed for South African rivers: geographic, limnological, chemical, and biological criteria result in different groupings (King *et al.* 1992). Classification systems based on bed material size (ASCE 1992), water quality (BES 1990), macrophytes (Holmes 1989), invertebrates and fish (Pennak 1971, Furse *et al.* 1984), recreational potential (Zachman 1984), restoration potential (NRA 1992), stability characteristics for engineering works (Simons 1978), or based on a mixture of some disciplines (Otto and Braukmann 1983, Biggs *et al.* 1990) may produce entirely different groupings.

Finally, the geomorphic characteristics of fluvial systems in the region under study influence the resultant classification. For example, many channel classes used by the Tsongas National Forest in southeast Alaska, such as "beaver dam/pond" channels and "deeply incised muskeg" channels (Paustian *et al.* 1992) would have little relevance in England, where basin lithology (e.g., chalk or clay) is a principal determinant of channel form (Holmes 1989), or in the Great Plains of North America, where the percent silt and clay in the river bed and banks is a good predictor of channel processes and morphology (Schumm 1963). Table 7.1 summarizes a few classifications based on geomorphic criteria, organized by objectives.

Hierarchy in Fluvial Geomorphic Classification

Fluvial systems can be viewed as inherently hierarchical, with smaller units nested within larger ones. In decreasing scale these could include landscape/ecoregion, floodplain/corridor (valley segments), channel reach, specific channel units (e.g., pools and riffles), and microhabitats (Lotspeich 1980, Amoros *et al.* 1982, Frissell *et al.* 1986) (Figures 7.2 and 7.3). Lower hierarchical levels are controlled asymmetrically by the upper levels, i.e., upper levels control lower but not vice versa (Naiman *et al.* 1992, Amoros and Petts 1993), because within a given landscape ecoregion, similar lithology, climate, geomorphology, and land-use history would tend to give rise to similar stream characteristics and thus constitute a stream system class, within which classes could be defined for progressively smaller features. The asymmetrical control of small scale features by larger scale characteristics implies that one must see beyond local site conditions to understand catchment controls, to view streams in a watershed context (Hynes 1975, Frissell *et al.* 1986). Moreover, it implies that stream classes developed for one region need not be applicable elsewhere.

Because rivers typically undergo profound changes along their length, each level of a classification system must either limit itself to homogeneous sections of channel (Kellerhals *et al.* 1976, Brice 1982, Rosgen 1994, Montgomery and Buffington 1997, Montgomery *et al.* 1998) or address the nature of longitudinal change as a basis for classifying different regions (Brussock *et al.* 1985, Frissell *et al.* 1986, Bethemont *et al.* 1996, Montgomery 1999).

Underlying Philosophies: Rivers as a Continuum or Discrete Types

One interesting aspect of the wide range of views on classification is an often-unstated difference in underlying philosophy, with roots in ancient philosophical debates. As applied to river classification, the issue boils down to whether river systems are composed of a continuum of channel morphology or discrete types of channels either bounded by geomorphic thresholds or controlled by local influences such as a flow constriction imposed by a landslide deposit or differences in bed or clast lithology. In the latter case, it may be possible to develop a natural classification, while in the former case, all channel classification schemes are perforce arbitrary, special classifications.

In many cases, efforts to objectively define discrete classes from large data sets drawn from diverse regions (as discussed below) have not yielded natural classifications based on geomorphic (Mosley 1987), ecological (Wright *et al.* 1984, Furse *et al.* 1984), or water quality (Cushing *et al.* 1980) variables. Thus, Cushing *et al.* (1983) concluded, "Streams are best viewed as gradients, or continua" and attempts to define discrete classes have "little ecological value". Wright *et al.* (1984) concluded, "any classification is bound to be arbitrary". From a geomorphic point of view, depending on the physiographic environment, the boundaries between adjacent stream types can be abrupt or gradual.

Recognizing that fluvial forms vary in a longitudinal direction, longitudinal zonations for rivers (from headwaters to the sea) have been proposed for New Zealand rivers (Nevins 1965), and for Washington state (Palmer 1976). Both of these approaches identified four "geohydraulic river zones", each with distinct channel gradient, channel pattern, valley cross-section, bed

Table 7.1 Examples of geomorphic-based river classification and sectorization objectives

Objective	Scales	References
Describe valley geomorphology, quantify drainage network	Basin, valley, drainage network	Davis (1899), Strahler (1957)
Classify and characterize hydrologic regimes	Basin	Pardé (1968), Gustard (1992)
Provide a theoretic hierarchical framework for river classification	All scales	Hynes (1975), Schumm (1977), Lotspeich (1980), Brussock *et al.* (1985), Frissell *et al.* (1986), Kern (1994)
Elaborate hierarchical typologies and/or ecoregional studies	All scales	Rohm *et al.* (1987), Cupp (1989a), Hugues *et al.* (1993), Omernik (1987), Wasson *et al.* (1993), Imhof *et al.* (1996), Allan and Johnson (1997), Heritage *et al.* (1997), Souchon *et al.* (2000)
Characterize valley bottom or floodplain dynamics	Valley bottom, floodplain	Galay *et al.* (1973), Cupp (1989b), Nanson and Croke (1992), Bravard and Peiry (1999), Ferguson and Brierley (1999)
Describe (or predict) alluvial channel patterns	Channel pattern	Leopold and Wolman (1957), Galay *et al.* (1973), Rust (1978), Paustian *et al.* (1984), Schumm (1985), Van den Berg (1995), Nanson and Knighton (1996), Alabyan and Chalov (1998)
Regionalize channel morphology and dynamics	Channel reach, often viewed in the basin context	Petit (1995), Rosgen (1996)
Sectorize streams in reach having homogeneous geomorphic functioning for management purposes	Channel reach, often viewed in the basin context	Maire and Wilms (1984), Cupp (1989b), Agence de l'Eau Rhin-Meuse *et al.* (1991), Orlowski *et al.* (1995), Van Niekerk *et al.* (1995), Bernot *et al.* (1996), Heritage *et al.* (1997)
Classify streams for management purposes	Channel reach, often viewed in the basin context	NRA (1992), Corbonnois and Zumstein (1994), Rosgen (1994, 1996), Zumstein and Goetghebeur (1994), Bernot and Creuzé des Châtelliers (1998), Schmitt (2001b)
Classify streams on the basis of their morphodynamic processes and adjustments	Channel reach, often viewed in the basin context	Kellerhals *et al.* (1976), Schumm (1963, 1977), Tricart (1977), Brookes (1987), Whiting and Bradley (1993), Downs (1994, 1995), Montgomery and Buffington (1997), Schmitt (2001a)
Classify reference *natural* states of streams ("Leitbild"; German approaches)	Channel reach, often viewed in the basin context	Otto and Braukmann (1983), Otto (1991), Müller *et al.* (1996), Bostelmann *et al.* (1998a,b), Tölk (1998)

(Continues)

Table 7.1 (*Continued*)

Objective	Scales	References
Identify reaches sensitive to erosion	Channel reach	Piégay *et al.* (1997)
Identify reaches producing/storing LWD	Channel reach	Piégay *et al.* (1996)
Stratify a river quality index	Channel reach, often viewed in the basin context	AQUASCOP (1997), Raven *et al.* (1997), Malavoi (2000)
Identify reaches for rehabilitation purposes	Channel reach, often viewed in the basin context	NRA (1992), Bostelmann *et al.* (1998a,b), Brierley and Fryirs (2000)
Manage biological resources	All scales	Otto and Braukmann (1983), Wright *et al.* (1984), Cupp (1989a), Biggs *et al.* (1990), Souchon *et al.* (2000)
Identify aquatic habitats/make biotic typologies (fish, macro-invertebrate, macrophytes)	Channel reach, morphodynamic unit and microhabitat, often viewed in the basin context	Huet (1949), Pennak (1971), Vannote *et al.* (1980), Wright *et al.* (1984), Cupp (1989a), Holmes (1989), Malavoi (1989), Biggs *et al.* (1990), Hawkins *et al.* (1993), Robach *et al.* (1996), Allan and Johnson (1997), Nicolas and Pont (1997), Montgomery *et al.* (1998)

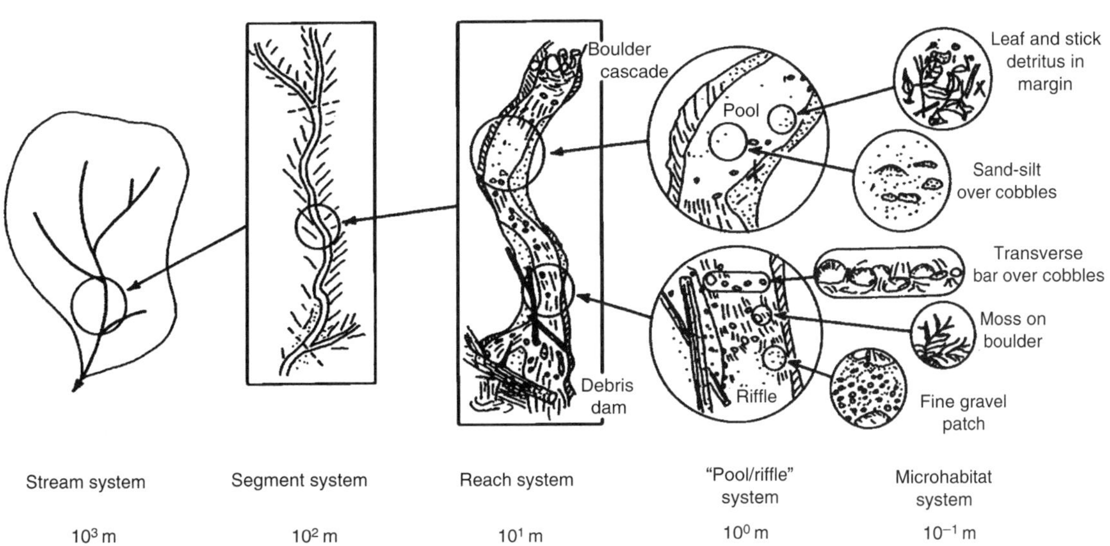

Figure 7.2 Hierarchical organization of a stream system and its habitat subsystems for second- or third-order mountain streams (reproduced from Frissell *et al.* 1986)

material size, and "material budget" (whether the bed is eroding, depositing, or stable) (Figure 7.4). Likewise, Schumm (1977) proposed the concepts of fluvial systems (see Chapter 5) involving a general division of river systems into erosional headwater reaches, connected by transport reaches, ending in depositional zones. Analogous longitudinal zonations in biological characteristics have been proposed in general or for

	Variables										Spatial scales (km^2)					
	Geographical unit (geology, landform...)	Valley or valley bottom morphology	Channel pattern, sinuosity	Channel morphology	Gradient	Stream power	Channel bottom grain size	Sediment load (type and/or intensity)	Morphodynamic units	Morphodynamic adjustments	10^3 and more	10^2	10^1	10^0	10^{-1}	10^{-2}
Leopold and Wolman (1957)			●										●	●	●	
Rust (1978)			●										●	●	●	
Lotspeich (1980)	●										●	●	●	●		
Cloots and Maire (1980)	●	●	●	●								●	●	●		
Ferguson (1981)			●			●							●	●		
Maire and Wilms (1984)		●	●	●	●		●			●				●	●	
Brussock *et al.* (1985)	●										●					
Frissel *et al.* (1986)	●	●	●	●	●		●	●	●	●	●	●	●	●	●	●
Hugues *et al.* (1993)	●										●					
Cupp (1989a, b)		●	●										●	●	●	
Agence de l'Eau Rhin-Meuse *et al.* (1991)	●			●	●								●	●	●	
Otto (1991)		●	●		●		●						●	●	●	
Nanson and Croke (1992)		●	●	•		●	●	●		●			●	●		
Corbonnois and Zumstein (1994)	●	●			●								●	●		
Downs (1994, 1995)	●				●					●			●	●	●	
Rosgen (1994, 1996)		●	●	●	●		●							●	●	●
Petit (1995)			●			●				●			●	●		
Nanson and Knighton (1996)	●	●	●	●	●	●	●	●		●			●	●		
Bethemont *et al.* (1996)	●	●	●	●	●				●		●	●	●	●	●	●
Bernot *et al.* (1996)	●			●	●	●	●						●	●		
AQUASCOP (1997)	●	●	•	•	•	●	•	●				●	●			

•: Variable partially taken into account

Figure 7.3 Variables and spatial scales taken into account for 21 geomorphic channel classification schemes

other regions (Carpenter 1928, Huet 1949, Vannote *et al.* 1980).

Local thresholds are important in explaining the complex succession of fluvial forms along the river system, and the general zonations described above become complicated in rivers with complex geology and anthropic modification. On the Ubaye River, an Alpine tributary to Durance River in Southern France, the longitudinal succession of channel form does not follow the conventional pattern because of its particular geological setting: a braided pattern occurs upstream in a wide marly basin, while downstream it transitions from wandering to meandering and then to straight, with increasing slope and grain size, and decreasing valley bottom width as it traverses more competent lithologies (Piégay *et al.* 2000).

River channels exhibit characteristics of both a downstream continuum (Vannote *et al.* 1980) and locally controlled systems (Montgomery 1999). For example, channels generally widen downstream as a power function of drainage area in both alluvial and bedrock channels (Leopold and Maddock 1953), but local differences in lithology can affect the width of bedrock channels (Montgomery and Gran 2001). Bed material gradually changes from cobbles and boulders in steep mountain channels to sand and gravel in lowland rivers, but local tributary inputs that serve as sources for large boulders can impart substantial local variability—such as forming the rapids along the Colorado River through the Grand Canyon. Similarly, flow obstructions such as logs and log jams can trigger pool scour and bed fining that

Figure 7.4 Oblique view of an idealized river system from headwaters to sea, illustrating the geo-hydraulic zones of Bauer (adapted from Palmer 1976) and corresponding types of Rosgen (1994) and Nevins (1965) (partly after Flosi and Reynolds 1991, from Kondolf 1995, reproduced by permission of John Wiley and Sons, Ltd.)

impart local variability to general downstream changes in channel geometry. Consequently, the appropriate philosophical underpinnings for channel classification inherently depend on the purpose to which it is to be applied.

7.2 OVERVIEW OF CLASSIFICATIONS IN FLUVIAL GEOMORPHOLOGY

A wide range of geomorphic river classification schemes have been proposed since the late 19th century, reflecting the diversity of environmental settings, the variety of potential approaches to ordering complex natural systems, the intellectual framework of the field, and the diverse purposes for which the systems were developed. We can distinguish two main objectives for river classification:

1. scientific understanding of how rivers function, existence of natural thresholds that produce longitudinal complexity on different spatial scales, and whether channels can be clustered in "homogeneous" classes, and
2. geomorphically based management guidance to inform decisions about channel maintenance, improvement, restoration or conservation. In this case, geomorphic criteria may be mixed with criteria of other disciplines (e.g., ecology, water chemistry).

Early Classifications

Distinctions between mountain torrents and lowland rivers are perhaps the oldest form of river classification. The mineralogist Dana (1850) offered an elegant description of the difference between mountain streams and lowland alluvial channels based on his experiences scaling the interior of islands in the South Pacific. Powell (1875) proposed a classification of rivers based on their genetic relation to geologic structure. At the close of the 19th century, the *geographic cycle* of Davis (1899) fitted neatly into the philosophical notions derived from evolutionary theory then in vogue by fitting landscapes and rivers into stages of an evolutionary cycle. Early in the 20th century, the relationship of channel network form to geologic history, lithology, and structure was recognized. Zernitz (1932) classified channel network forms based on branching angles into classes such as the now familiar dendritic, trellis, and radial channel network forms.

Process-based Classifications

Leopold and Wolman (1957) presented a quantitative basis for differentiating straight, meandering, and braided channel patterns based on relationships between slope and discharge. This early pattern-based classification has been revisited and enlarged (Rust 1978, Schumm 1985, Ferguson 1987, Church 1992, Thorne 1997, Alabyan and Chalov 1998), in particular with the inclusion of additional patterns such as anastomosing (Smith and Smith 1980, Knighton and Nanson 1993, Makaske 2001), and more generally anabranching rivers (Nanson and Knighton 1996). Schumm (1963, 1977) classified alluvial rivers on the basis of whether their beds are stable, eroding, or aggrading, and further differentiated them through the dominance of suspended load, mixed load, or bedload sediment transport (Table 7.2). Nanson and Croke (1992) took account of the strong dependence between channel and floodplain to propose a detailed genetic floodplain classification. All of the above systems are oriented to alluvial channels with floodplains and are generally not appropriate for classifying steep channels in relatively small drainage basins. Consequently, they provide only a partial context for channel classification.

Other classification systems are based on channel characteristics. Howard (1980, 1987) described channels as alluvial or bedrock, the former having a layer of active alluvium. He further subdivided alluvial channels into sand-bed and gravel-bed corresponding to regime and threshold channels, respectively. The bed of a regime channel is highly mobile even at low flow and both sediment transport and bedform roughness change with increasing discharge, whereas the bed of a threshold channel typically is mobile only beyond some critical discharge. As fundamental differences in the nature of bed material reflects thresholds in chan-

Table 7.2 Classification of alluvial channels (from Schumm 1963, 1977)

Mode of Sediment transport and type of channel	Channel sediment (*M*) (%)	Bedload (percentage of total load)	Channel stability		
			Stable (graded stream)	Depositing (excess load)	Eroding (deficiency of load)
Suspended load	>20	<3	Stable suspended-load channel. Width/depth ratio < 10; sinuosity usually > 2.0; gradient, relatively gentle	Depositing suspended load channel. Major deposition on banks cause narrowing of channel; initial streambed deposition minor	Eroding suspended-load channel. Streambed erosion predominant; initial channel widening minor
Mixed load	5–20	3–11	Stable mixed-load channel. Width/depth ratio > 10, < 40; sinuosity usually < 2.0, > 1.3; gradient, moderate	Depositing mixed-load channel. Initial major deposition on banks followed by streambed deposition	Eroding mixed-load channel. Initial streambed erosion followed by channel widening
Bed load	<5	>11	Stable bed-load channel. Width/depth ratio > 40; sinuosity usually < 1.3; gradient, relatively steep	Depositing bed-load channel. Streambed deposition and island formation	Eroding bed-load channel. Little streambed erosion; channel widening predominant

nel processes, Howard's (1980) classification provides insight into the potential response of different channel types, but is too broad to distinguish among many channels. Classifications based on channel adjustments were developed by Brookes (1987) and Downs (1994, 1995).

Church (1992) classified channels by scale: small channels are those where the scale of the channel is only 1–10 particle diameters, in which the relative roughness is large and single clast define significant elements of overall channel form. By contrast, in large channels flow depths are typically greater than 10 particle diameters. Whereas small channels are closely coupled to hillslope processes, large channels are those wherein morphology is determined by fluvial processes (i.e., discharge, gradient, and the sediment size and supply) and geological constraints (Church 1992).

For mountain channels, Grant *et al.* (1990) and Whiting and Bradley (1993) elaborated classifications of headwater in-channel features and stream reaches based primarily on the interaction between hillslope processes and channel processes. Montgomery and Buffington (1997) found that different mountain channel reach morphologies had different relative transport capacity as expressed in terms of stream power, or drainage area and reach slope. Although such distinctions provide for a natural classification of channel types, they represent stratification of a continuum of natural channel morphologies.

Stream Power-based Classifications

With recognition of stream power as a key variable in fluvial geomorphology, an increasing number of classifications have been based on this parameter (Table 7.3; Schmitt *et al.* 2001b). The concept of stream power was first used in fluvial geomorphology to study sediment transport in river channels (Knapp 1938, Bagnold 1966) and was later applied for different purposes: bedload and suspended load (Bagnold 1977), channel instability and bank erosion (Brookes 1987, Lawler 1992), thresholds in channel patterns (Ferguson 1981, Van den Berg 1995) and floodplain dynamics (Nanson and Croke 1992, Bravard and Peiry 1999). Stream power-based classifications have been applied at finer spatial resolutions than the common resolution of channel patterns (Newson *et al.* 1998, Schmitt 2001a).

Stream reaches and classes can also be defined based on the downstream variation in stream power (Knighton 1999), using discontinuities in specific stream power to draw boundaries between reaches (Bernot *et al.* 1996, Astrade and Bravard 1999, Schmitt *et al.* 2000). By comparing different stream zonations, thus obtained in a given region or management unit, such as those defined for rivers in the Rhône-Mediterranean region of France (Bernot and Creuzé des Châtelliers 1998), regional classification can be developed (with other geomorphic variables taken into account as well).

When different stream power-based classifications are compared, overlaps of the specific stream power classes are frequently observed (Table 7.3). This imprecision can be due to estimations of basic parameters (Schmitt *et al.* 2001), the lack of clear stream power thresholds between channel patterns (Ferguson 1987), and the effect of the geographic setting. In most cases, stream power-based classifications are supplemented by geomorphic variables at the levels of valley bottom, floodplain or channel. Moreover, specific stream power (stream power per unit channel width) is not an independent variable, as channel slope depends in part on sinuosity, and width depends on channel geometry, which are two dependent variables (Van den Berg 1995, Schmitt et al. 2001). Provided one takes into account these variables, specific stream power appears to be a useful variable to construct geomorphic classifications at different spatial resolutions, and has the potential to take into account channel processes and adjustments (NRA 1992, Kondolf 1995, Newson *et al.* 1998).

Hierarchical Classifications

Classification models as determined above lead generally to the definition of *interlocked spatial units* within which the variability of each smaller hierarchical level is constrained by that of the higher hierarchical level (see Chapter 5). At the broadest scale, differences in styles of precipitation and vegetation lead to differences in river processes and characteristics in major climate zones (e.g., alpine, tropical, temperate, arid, and polar regions). Ecoregions defined by areas of similar climate, vegetation, and topography (Omernik 1987) can be related to the characteristics of aquatic habitats (Rohm *et al.* 1987, Imhof *et al.* 1996, Allan and Johnson 1997). Just as there are a number of ways to broadly stratify general environmental influences on river systems, there are many ways to address channel classification at finer spatial scales.

In the Loire River Basin (100 000 km^2), characteristics of river corridors depended largely on the morphoregion (>100 km^2) drained (Figure 7.5). For example,

Table 7.3 Synthetic and comparative representation of some specific stream power-based river classifications. The correlation between the specific stream power classes is rough (from Schmitt *et al.* (2001), used by permission of Gebrüder Borntraeger, Stuttgart)

Specific stream power W/m^2 (+) ↑	Ferguson (1981, 1987) (channel pattern)	Nanson and Croke (1992) (floodplains) (classification 1st level)	Petit (1995)	Nanson and Knighton (1996) (anabranches)*	Bernot and Creuzé des Châtelliers (1998) (typology 2nd level)
					$+100 < \omega < +1000$; V-shaped valley confined
		>300; non-cohesive floodplains high energy			$+100 < \omega < +1000$; U-shaped valley deep
	120–300; active low sinuosity		>100; frequent channel shifting (braiding is possible)	100–300; Type 6	$50 < \omega < 500$; V-shaped valley widen
				30–100; Type 5	$30 < \omega < 300$; presence of a floodplain
	20–350; confined	10–300; non-cohesive floodplain medium energy		50 to 5–10; Type 3	$30 < \omega < 700$; channel limited by incision
	5–350; active meandering		<35; no self-adjustment after regulation	15–35; Type 4 ≤ 8; Type 1	$30 < \omega < 120$ large floodplain
	1–60; inactive unconfined	<10 cohesive floodplains low energy	<15; inactive channels	4–8; Type 2	$\omega < 30$; littoral floodplain

**Signification of Nanson and Knighton's anabranch types*:
Type 1: cohesive sediment anabranching rivers.
Type 2: sand-dominated, island-forming anabranching rivers.
Type 3: mixed-load, laterally active anabranching rivers.
Type 4: sand-dominated, ridge-forming anabranching rivers.
Type 5: gravel-dominated, laterally active anabranching rivers.
Type 6: gravel-dominated, stable anabranching rivers.

in the highlands of the Massif Central (Upper Massif Central and Granitic plateau of Massif Central) River corridors consist of steep narrow valleys, while in the granitic Armoricain region (Armoricain Massif), they consist of wide, gently sloped valleys (Figure 7.6; Bethemont *et al.* 1996). Each morpho-region has a characteristic longitudinal distribution of valley morphology, channel morphology, and in-channel features (Figure 7.7; Souchon *et al.* 2000). In the granitic Armoricain region, the second-order streams are mainly characterized by shallow water with moderate velocity (also known variously as runs or plane-bed reaches), whereas the third- to fifth-order streams were dominated by lentic channels and deep waters with low velocity. In the "sedimentary" region, the modification of channel features is structured by the stream size and slope gradient. From second to fifth-order, the slope decreases and the geomorphological features change from pool-riffle sequences (MOU, RAP) or runs (PLA) to homogeneous deeper low velocity channels (CLE). This classification reduce yields classes within which channel variability is relatively consistent in each

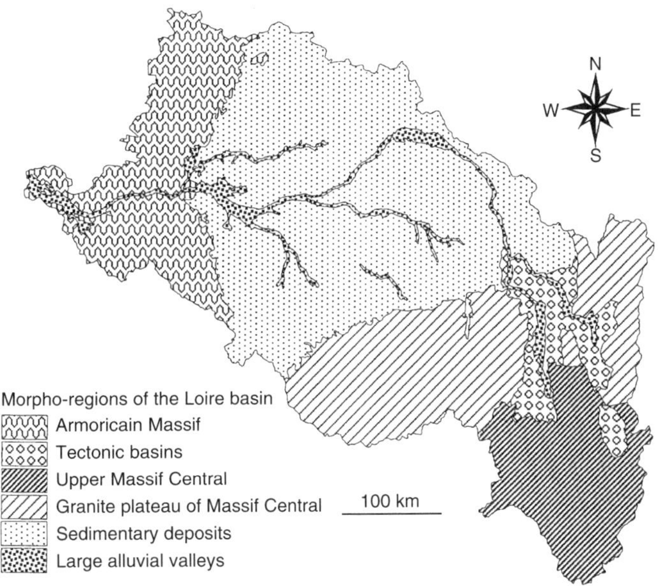

Figure 7.5 Distinct morpho-regions of the Loire River Basin (reproduced from Bethemont *et al.* 1996)

morpho-region of the Loire, and provides a tool to manage aquatic ecosystems at the watershed scale. Cupp's (1989b) valley segment-level classification scheme developed for Western Washington and Paustian *et al.*'s (1992) channel reach-level classification of southeast Alaskan channels are other examples of such regionalized modular components of a general hierarchical fluvial classification system.

Montgomery and Buffington (1997, 1998) proposed a hierarchical valley segment and reach-level classification of mountain channel networks based on morphologic attributes related to relative sediment supply and the ratio of sediment supply to transport capacity. They recognized colluvial, alluvial, and bedrock valley segment types. Colluvial valleys are headwater valley segments with relatively ineffective fluvial sediment transport and in which colluvium delivered from hillslopes accumulates as colluvial valley fills. Bedrock valley segments are those in which little material is stored in the valley bottom, whereas alluvial valley segments are those with thick alluvial valley fills. Montgomery and Buffington (1997, 1998) also recognized eight distinct channel reach types that can be used to characterize a continuum of natural channel reach morphologies (Figure 7.8). These reach types are defined by discrete bed morphologies interpreted to generally reflect relative transport capacity over shorter timescales than the valley morphologies described above. These channel types are intended to allow comparison of comparable reaches, and although these reach-level channel types are generally correlated with reach slopes, they also reflect local conditions and disturbance history.

The concepts of process domains and lithotopo units implicit in this approach provide for classification at spatial scales greater than individual channel reaches (Montgomery 1999). Process domains are areas of a watershed that are dominated by comparable geomorphological processes and therefore with similar sediment transport dynamics and comparable disturbance regimes. Channels within a process domain would be expected to experience similar disturbance processes, and different process domains roughly delineate a longitudinal channel classification (Figure 7.9). The parallel concept of lithotopo units is intended to define areas with similar lithology and topography and within which channels would share common characteristics.

Geomorphic Classification for Management

Classifications have been devised explicitly for management in the UK (NRA 1992, Downs 1994, Newson *et al.* 1998), France (Zumstein and Goetghebeur 1994, AQUASCOP 1997, Schmitt 2001b), Germany (Otto 1991, Bostelmann *et al.* 1998a,b), the US (Cupp

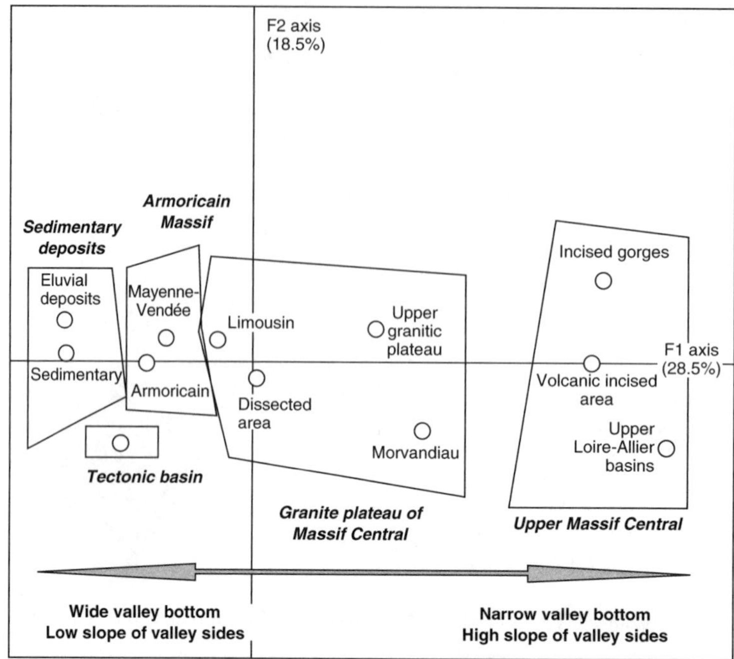

Figure 7.6 Distribution of the 12 elementary morpho-regions of the Loire River Basin on the first factorial map F1–F2 of a Multiple Correspondence Analysis using five morphological variables (stream order, valley slope, valley side slopes, channel sinuosity, low flow channel width). The regions line up along the F1 axis in five main regions, from the sedimentary region near Nantes with wide alluvial valleys and gentle side slopes to the incised volcanic region of the Massif Central with narrow valleys and steep side slopes (see Figure 7.5 for location of the regions) (reproduced from Bethemont *et al.* 1996)

1989b, Rosgen 1994, 1996, Montgomery and Buffington 1997, 1998) and South Africa (King *et al.* 1992, Van Niekerk *et al.* 1995, Heritage *et al.* 1997, Dollar 2000). These management-oriented classifications are mainly single-scaled, focusing on channel reaches along a river continuum or sampled randomly in a given area (with natural but also political boundaries).

Reach Delineation

Longitudinal zonation is an increasingly important topic in applied fluvial geomorphology with the development of integrated management schemes. Increasingly, river management is undertaken at the catchment scale, or at least over river reaches of several kilometers length, with recognition of upstream–downstream interdependence. In such a context, geomorphic classification schemes have been used to give insight into the response potential of some river systems, such as identifying unstable reaches (Brookes 1987, Downs 1994), reaches prone to channel migration (Piégay *et al.* 1996), or reaches more sensitive to effects of upstream land-use changes (Downs 1995, Montgomery and Buffington 1998). Such schemes can also identify homogeneous spatial patterns for river management purposes, such as zones of different erosion hazard (the so-called streamway concept) assessed from historical channel shifting (Piégay *et al.* 1996, 1997), and also a framework to prioritize river restoration based on a mapping of historical changes (see the example of the Bega River, Australia, Brierley and Fryirs 2000) (Figure 7.10).

To define homogeneous reaches for formulating specific management recommendations, sectorization methods based on geomorphological parameters have been developed. For example, Bernot *et al.* (1996) distinguished control (independent) variables such as the valley morphology, the discharge and the geological setting, from adjusting (dependent) variables such as specific stream power and channel pattern. Each identified geomorphic reach was then described

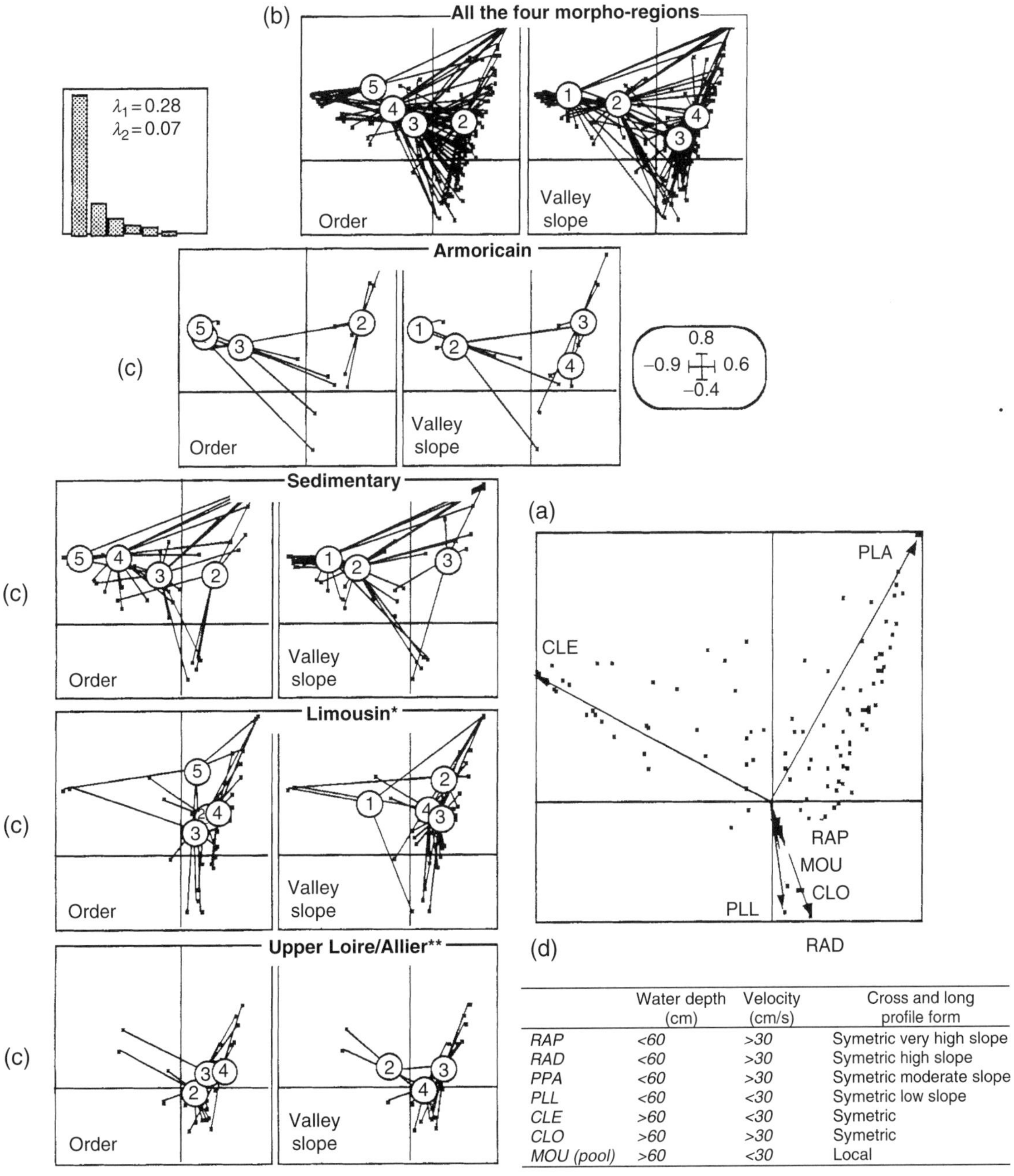

	Water depth (cm)	Velocity (cm/s)	Cross and long profile form
RAP	*<60*	*>30*	Symetric very high slope
RAD	*<60*	*>30*	Symetric high slope
PPA	*<60*	*>30*	Symetric moderate slope
PLL	*<60*	*<30*	Symetric low slope
CLE	*>60*	*<30*	Symetric
CLO	*>60*	*>30*	Symetric
MOU (pool)	*>60*	*<30*	Local

Figure 7.7 Factorial maps of a PCA of the morpho-regions of the Loire Basin (France) based on seven types of geomorphological features, shared by stream orders and valley slope classes (from 1, low slope, to 5, high slope). (a) Whole data set (196 sample reaches by seven types of geomorphological features). (b) Same as (a), with the reaches shared by size classes and valley slope classes. (c) Same as (b) with the reaches shared by morpho-regions. (d) Main characters of the seven types of geomorphological features. *Limousin is a sub-region of the Granite Plateau of Massif Central. **Upper Loire/Allier is a sub-region of the Upper Massif Central (see Figures 7.5 and 7.6) (modified from Souchon *et al.* 2000)

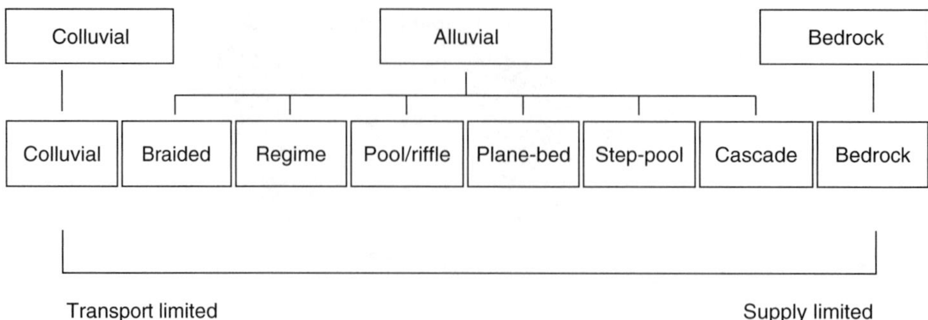

Figure 7.8 Distinct channel reach types of Montgomery and Buffington (1997, 1998) shown as a function of transport or supply limitation (reproduced from Montgomery and Buffington 1997)

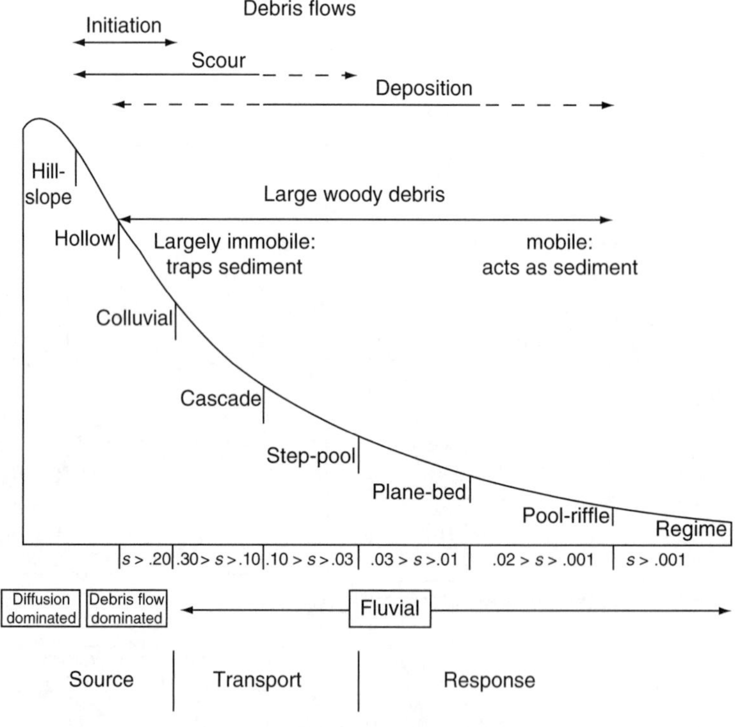

Figure 7.9 Process domains of Montgomery and Buffington (1997) arranged along a longitudinal gradient (reproduced from Montgomery and Buffington 1997)

by other parameters such as channel morphology and dynamics, anthropogenic alterations, water quality indices, characteristics of riparian and aquatic communities. Although not a sophisticated tool from a technical point of view, the approach is easy to develop and reproduce, and can be useful as a preliminary analysis for planning interventions.

Reference Conditions and Stream Power

Classification can help to stratify channels that should have similar reference conditions against which to assess the degree of degradation, and provide an indication of the historical (pre-disturbance) condition at the site to inform restoration designs. However, the

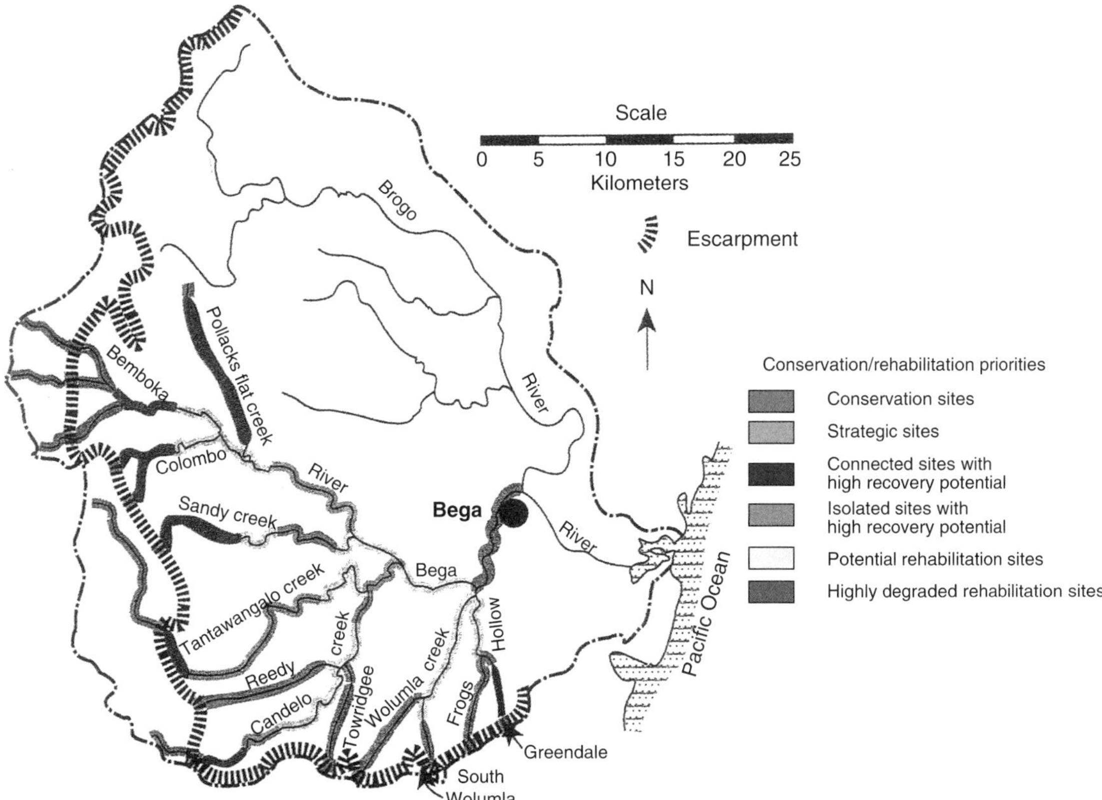

Figure 7.10 Prioritization of river reaches for rehabilitation in the Bega River catchment, building out from nearly intact reaches to those most degraded, and targeting strategic sites (reproduced from Brierly and Fryirs 2000)

historical condition is not necessarily a suitable goal for restoration design, as any such ideal must be adjusted to account for irreversible changes in controlling factors (such as runoff regime) and for considerations based on cultural ecology at the site, such as preservation of historical land uses or creation of habitat for endangered species. Classification can also provide a preliminary indication about whether various bank protection works and habitat enhancement structures are likely to prove successful. Based on Brookes' (1987, 1990) finding that habitat enhancement structures in Denmark and Britain typically did not survive in channels subject to a unit stream power exceeding 35 W/m^2, the National Rivers Authority (now the Environment Agency) included an estimate of stream power in its classification system as a guide to the likely impact of channel modification (NRA 1992).

Channel Restoration

Classification can be a useful tool in stream restoration in at least three ways:

1. surveying existing conditions and setting priorities for restoration,
2. envisioning an end state towards which restoration should proceed, and
3. providing initial indications about restoration measures likely to succeed in a given channel (Kondolf 1995).

Classification has been heavily used as a basis for restoration, despite the limited scientific basis of some systems and disappointing performance of many such projects (see case studies).

7.3 APPLYING GEOMORPHIC CLASSIFICATION SCHEMES TO FLUVIAL SYSTEMS

Before applying classification schemes, several issues must be addressed. Foremost is whether to use an existing classification (and if so, which one) or to develop a new system for the region and/or problem at hand. This is not a trivial question, as the selection of variables as the basis for a classification implicitly assigns greater importance or significance to the variables used than to the variables not used. The selection of variables is also influenced by the degree to which they lend themselves to measurement, and some geomorphically significant variables may not lend themselves readily to quantification.

Procedures for geomorphological data collection in stream channels have been developed independently of stream classification programs (e.g., Mosley 1982; Hicks and Mason 1991), as part of classification programs (Kellerhals *et al.* 1976, Jowett and Duncan 1990, Downs and Brookes 1994, Thorne 1998), and incorporated into some ecological data collection protocols (e.g., Platts *et al.* 1983, 1987, Biggs *et al.* 1990). In general, stream inventory procedures include variables such as channel pattern and sinuosity, channel dimensions, bed material size, channel gradient, whether the channel is alluvial or bedrock-controlled, degree of entrenchment, and catchment variables such as drainage area, basin relief, valley gradient, annual rainfall, and lithology of the basin. These variables, along with indices of channel entrenchment and valley confinement, are readily measured in the field, and some can be measured from maps and aerial photographs. If an existing classification is to be applied, then pre-existing classes can be used. If a new classification is being developed, then the number of classes may be determined by the scale of the river system, considerations such as the need to divide responsibility for rivers among different levels of government or the desire to keep the number of classes small enough that the classification system is practical and useful.

In any system designed for broad application by users besides the scheme's author, selection of variables will be influenced by the availability of data, or at least the degree to which certain variables lend themselves to measurement and quantification. For example, bedrock lithology underlying the catchment and the reach can exert a profound control on river form, not only through direct effects in bedrock-controlled reaches, but also by controlling the valley walls bounding alluvial reaches, and influencing the runoff and sediment load delivered to these reaches (Montgomery 1999). However, there is no simple way to incorporate lithology into most channel classification schemes. Unlike channel width, for example, there is no single number that can represent the range of physical attributes associated with different rock types and structure, such as hardness, permeability, stratification, foliation, and fracturing. More feasibly, one can use underlying rock type as one basis for defining homogeneous regions for which a classification scheme can be developed (Kern 1992, Bethemont *et al.* 1996, Schmitt 2001a), or within which one would expect similar channel types to exhibit similar finer-scale characteristics (Montgomery 1999). But even with expert schemes it is possible to learn new things about the behavior of the rivers in question (and thereby revise the classification scheme) if the scheme's performance is objectively analyzed (Figure 7.1).

Data Collection as Distinct from Identifying Channel Type

Data collecting and recording should be distinct steps from identifying channels as belonging to a particular class. If not, the observer's perceptions may be influenced by expectations that a channel will fit into a particular class. Essentially the same problem in the context of correlating river or marine terrace remnants based on elevation was discussed by Johnson (1944), who noted a tendency for workers to reach premature conclusions about the suite of terraces present in a given area and rounding elevations of subsequently observed terraces to the nearest preconceived terrace elevation. This resulted in spurious correlations and loss of the real data potentially available if actual values of terrace remnant elevation had been recorded.

> A study on the Atlantic coast reports terraces at 100-foot intervals . . . [and a subsequent] French investigator has reported successive terraces at intervals of 100 meters. We must conclude either that a wise Providence not only pays remarkable attention to the magic number 100, but also nicely adjusts uplifts of the land and lowering of sea level to the particular system of measure prevailing in each country; or else that actual elevations of very different character are by observers so roughly approximated to the nearest even figure as to make them valueless for correlation purposes. The latter interpretation seems the more reasonable. [Johnson (1944: 806).]

In using any channel classification, the performance of the classification system should be assessed once the streams are identified as belonging to specific classes.

How many channels were not accommodated by the pre-existing categories? Does the taxonomy need to be modified to accommodate local/regional conditions or the needs/purposes of the project? In other words, channel classification should be used as a flexible tool to help organize understanding, but it should not be blindly relied upon to generate insights that may or may not be appropriate to the specific local or watershed context of the channels in question.

Tools Used to Classify Spatial Units from Data

Most geomorphic classification systems originated as expert systems based on general principles and experience with rivers in a given region. Such classifications "provide a weak form of explanation because all schemes involve a set of criteria which relate to an a priori expectation of the way in which researchers believe their river channels to be distinguished" (Downs 1995: 348). Such a priori classification schemes reflect the training and experience of the scheme's author, both in the selection of variables to include and classes proposed. An alternative approach is to collect large data sets and employ statistical methods (e.g., cluster analysis) to objectively define patterns or groupings of similar channels. In the latter approach, expert judgement is still involved in selecting the variables to use as the basis for the classification scheme, but identification of distinct groupings is left to an objective procedure, although the boundaries may still be subjectively drawn.

Because large sets of sites and variables used to describe them are needed at a regional or nation-wide scale to perform a typology using statistical tools, multivariate methods are becoming popular. It is often efficient to summarize the data set variability into main factors (e.g., components of a principal component analysis) and, then achieve the cluster analysis (e.g., hierarchical ascendant classification, k-means algorithm or other techniques) on these factors (see Chapter 20). Finally, the obtained classification can be validated by discriminant analysis (quantitative variables) or segmentation (qualitative variables). These methods can find the best way to split a set of river reaches or in channel features, whether the descriptive variables cluster into natural classes or form a continuous gradient (in which case the classes obtained are special classes).

Newson *et al.* (1998) applied a twin-span analysis to geomorphic and ecological data from 432 sites around the UK and found no "objective taxomony", but noted that the river types indicated were intuitively realistic and might prove to be statistically definable with a larger data set. In New Zealand, Mosley (1982, 1987) conducted a cluster analysis on geomorphological data from 190 river reaches and found only four clear interpretable clusters. Nor were clusters evident in a three-dimensional plot of width/depth ratio, channel slope, and mean bed material size from 100 rivers measured by Jowett and Duncan (1990). Mosley concluded that a multivariate approach to characterizing rivers was more useful than a classification system for predicting ecological communities supported by the stream and for predicting environmental impacts. While it may not be possible to identify discrete river types in data sets drawn from many landscape provinces, the approach is likely to be more successful when applied within a given physiographic unit, where multivariate approaches can provide results with practical value for management, as done on Upper Rhine floodplain anabranches (Figure 7.11; Schmitt *et al.* 2000). The dendrogram separates clearly "anabranches with moderated or no dynamic" from "dynamic anabranches". The first group corresponds exclusively to groundwater-dominated channels, which do not experience floods and in which morphodynamic activity is weak. The channels in the second group also receive floodwaters from the Ill River, and thus have morphodynamic activity.

Utility and Limitations of Channel Classification

Many classification schemes have been recently commissioned or adopted by river management agencies with the aim of simplifying geomorphological analysis to assist in management, to the extent that they foster understanding and assist in achieving these goals. Classification schemes can be seductive, especially for non-geomorphologists, who may not appreciate the complex nature of geomorphological processes, and who may feel that the channel is completely described once it has been "classified". Fitting nature into the classification system may become the objective, and important information about the channel may be missed. As the users of a classification system may lack background in fluvial geomorphology, the idealizations of the classification scheme can be more "real" to them than the evidence presented in the field or the implications that might follow from an understanding of geomorphic process. For example, a stream restoration design in California based on the Rosgen classification system (described below) recently called for filling of a large pool that had persisted in the channel for many years because it did "not belong in a B stream type". In this case, classification could be

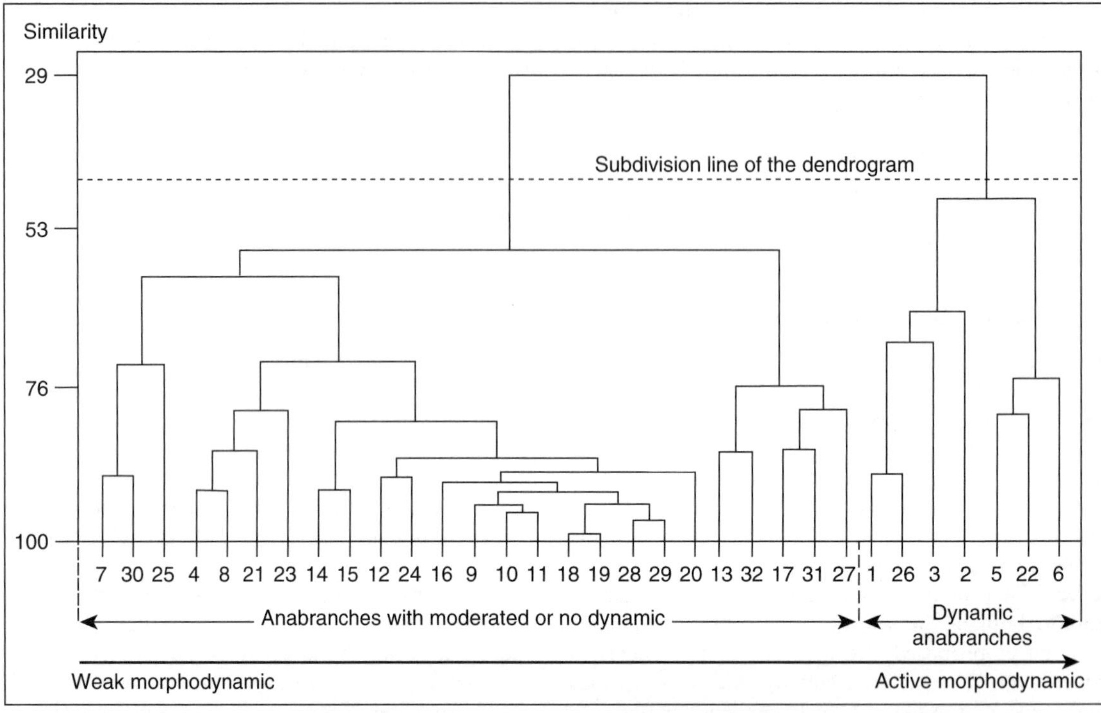

Figure 7.11 Classification of anabranching rivers in the French Rhine floodplain based on a cluster analysis of 32 sites (numbers). Seven hydro-geomorphic variables are taken into account: sinuosity, width/depth ratio, mean and variation coefficient of in-channel coarse sediments *b*-axis, percent of lotic morphodynamic units and two lateral mobility indices. The length of a reach corresponds to 36 bankfull width (reproduced from Schmitt *et al.* 2000)

viewed as impeding rather than fostering understanding of the channel. In addition, a classification may simply provide a snapshot of channel condition that does not reflect temporal variability, disturbance history, or the potential for channel change due to such variability. Perhaps the most important limitation of channel classification is that once put into a class, the channel may be viewed as "known", and critical thinking abandoned in favor of pre-existing assumptions about this class of channel. The focus of many classifications is on the channel form, not geomorphic processes that control it, largely because the former are easier to quantify.

By focusing on channel form in a reach (at the point in time when the classification is applied), channel classifications cannot capture the dynamic behavior of river systems or effects of changes upstream and downstream. Notions of "natural" or reference conditions can be difficult to apply in areas with a long history of human occupation. In small basins, channel form is often linked to nearby geology, land use, and basin characteristics, but with increasing drainage area and heterogeneity in the catchment, upstream effects combine with downstream influences.

7.4 CASE STUDIES

We illustrate the use of channel classification with four case studies presented below. The first three report classification schemes developed several years ago in Germany, the UK, and France, and which are more widely applied since adoption of the EU Water Framework Directive in October 2000 (European Parliament and Council of Europe 2000). This Directive aims to achieve minimum water quality standards for surface and groundwater resources of all member-states within 15 years. The approach is to implement detailed action programs (of regulatory and economic measures) within hydrographic basins. The Directive will profoundly modify national water policies in the 15 states by setting up precise methods for measuring and characterizing each category of surface water body (rivers, lakes, transitional waters, and coastal waters) (Raven *et al.* 2002). For rivers, a network of reference sites of high ecological status for each river type is to be established. On this basis, the member-states shall establish programs for the monitoring of river ecological quality (biological, hydromorphological and physico-chemical quality) in order to establish restoration programs. The last case study describes an American approach to classification widely used to support the design of river restoration projects.

Leitbild Approach to Setting Goals for Restoration

Channel classification can provide a "guiding image", or *Leitbild*, of the channel form that would naturally occur on the site, adjusted to account for irreversible changes in controlling factors (such as runoff regime) and for considerations based on cultural ecology (such as preservation of historical land uses or creation of habitat for endangered species) (Kern 1992, 1994). Attributes of this ideal channel form can be adopted as goals for restoration projects. Thus, the *Leitbild* is a model of the ideal channel design for a site based on physical and ecological considerations, including historical changes to runoff and sediment yield. Based on constraints such as flood control, pre-existing water rights and budget limitations, planners propose an optimal design for review by resource agencies and the public, and ultimately modification into a feasible design for the site (Kern 1992).

The *Leitbild* concept is being applied extensively in western Germany as a basis for assessing existing channel conditions and to provide guidance for restoration activities. Detailed channel *Leitbild* classifications have been developed for Baden-Wurttemberg (Bostelmann *et al.* 1998a,b), Kraichgau (Tölk 1998), Nordrhein-Westfalen (Müller *et al.* 1996, Landesumweltamt Nordrhein-Westfalen 1996), Rhineland-Pfalz (Otto 1991), etc. The classifications are developed only for regions with similar geology, climate, etc., in which a consistent set of valley and channel forms could be expected.

The characteristics of various *Leitbilds* have been defined through detailed field study, and practitioners are encouraged to visit illustrative reaches displaying properties of the *Leitbild* to aid development of more compelling conceptual models for restoration efforts than can be gleaned from diagrams and statistics alone. The *Leitbild* approach requires some judgement (and thus professional background) to apply, largely because historical changes in basin and channel conditions must be understood and considered in developing the *Leitbild* for a given site. The process is more than simple mimicry of remnant natural channels found in undisturbed drainage basins. Remnant natural channels provide an indication of the potential natural state of a given class of channel and attributes that might be considered for restoration objectives, but cannot indicate how to address constraints such as altered runoff patterns when developing a *Leitbild* for the project reach. Similarly, historical reconstructions of former conditions in the subject reach provide useful insights into the ecological potential of the site, but historical conditions may be impossible to recreate and maintain because of changes in the catchment. These *Leitbilds* can provide a basis for establishing the "reference network of sites of high ecological status for each river type" required by the EU Water Framework Directive.

SEQ Physique (Evaluation System of River Physical Quality)

Since 1964, France has had an administrative structure to evaluate and improve water quality, but nonetheless, rivers have continued to be negatively affected by flow regulation, canalization, and in-channel aggregate mining, which have reduced flows, altered channel form, and degraded aquatic ecosystems. Recognizing that water quality improvement must be coupled with physical and biological integrity, the Water Law of 1992 created three SEQs (Quality Evaluation System), respectively, SEQ_{eau} (for water quality indicators), SEQ_{bio} (for biological indicators), and SEQ_{phy} (for physical indicators) (Agences de l'Eau 1998). Each aims to evaluate the state of degradation, from which quality objectives are defined and planned and coherent restoration strategies at the national scale are promoted with priorities. The "SEQ_{phy}" is based on scores of degradation state based on 30 variables collected on homogeneous river reaches of several kilometers. Within each river reach, four components are considered separately. Three are spatial units: the channel, the bank, and the floodplain; the fourth is a dynamic component, the hydrological regime (Figure 7.12; Table 7.4).

To control for variation in rivers from one geological/climatic region to another and from upstream to downstream, channel reaches were classified so scoring could be done by river type. For example, the ecological, hydrological, and morphological functions of floodplains are greater in lowland large rivers than in Alpine torrents, and thus should be scored differently

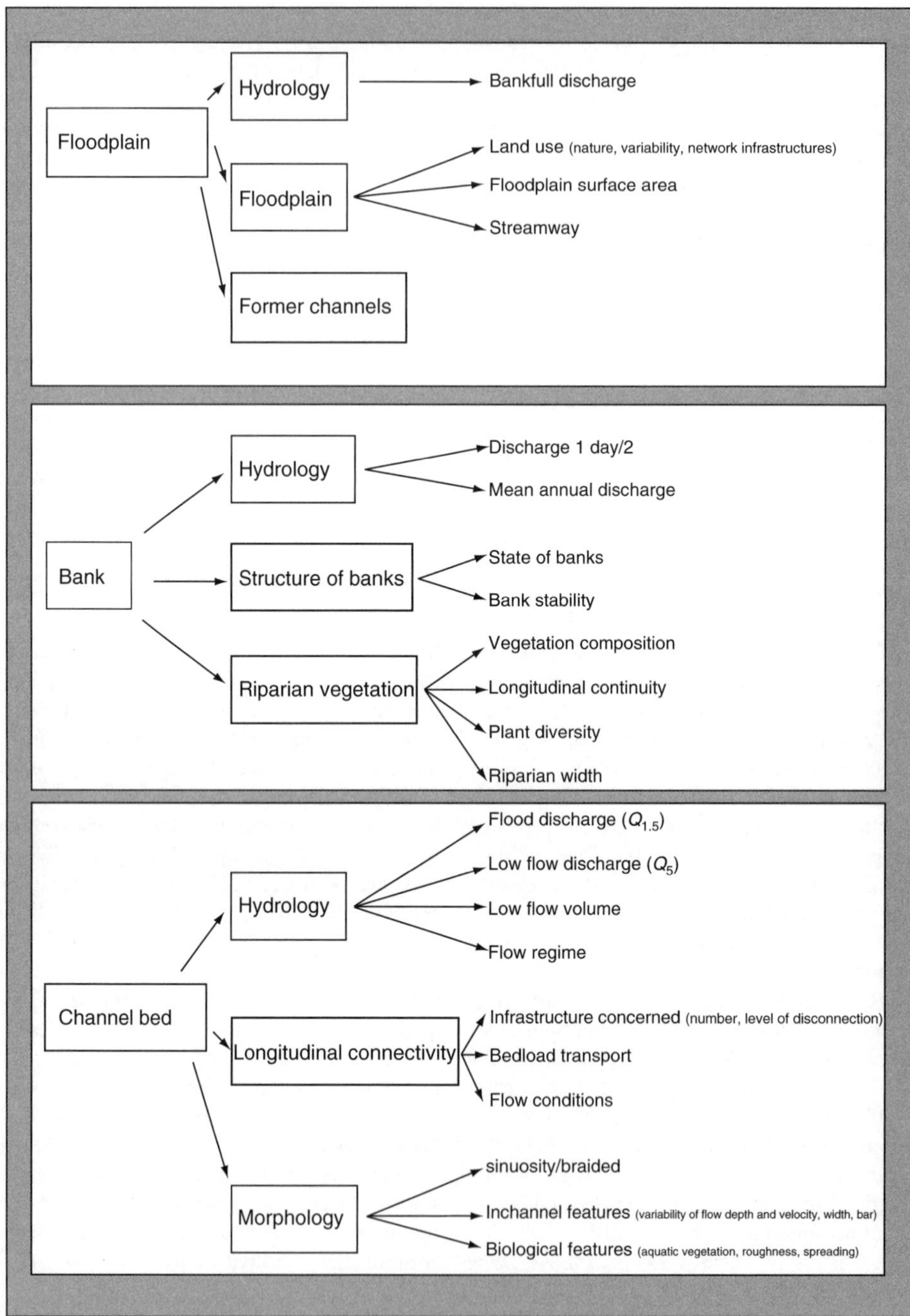

Figure 7.12 Criteria used to classify channels in the SEQ_{phy} system at the nation-wide scale, as now applied across France (adapted from Agence de l'Eau 1998)

when assessing degradation. Rivers were classified by expert-based systems (AQUASCOP 1997) and have since been quantitatively assessed by Schmitt (2001b) for northeastern France. The classification approach at a national scale is based on processes rather than static fluvial forms, variables such as stream power, bedload supply, geological settings, floodplain extension, and hydrological regime being considered. The channelized reaches have been classified in their predisturbance (1850) taxa. In this system, engineering works are considered as degrading elements and integrated into the scoring process. The national classification system, developed by experts, initially distinguished 30 taxa, but only 15 were ultimately incorporated because the scoring procedures were unable to distinguish some of them. For example, shifting large rivers were initially separated according to the climatic character of their drainage areas (Alpine or Mediterranean), but the scores proposed by the set of experts for scoring importance in relative degradation of hydrological regime, channel, bank, and floodplain natural characters were not different.

The classification system supporting the SEQ_{phy} is process-based, and the scoring system takes into account the importance of dynamic physical processes controlling biodiversity (e.g., conservation of shifting, flooding, connection between channel and abandoned channel). However, the temporal perception of the reach is static. Because scoring for each taxon is based on field or map descriptions of biogeomorphic characters (land-use type, riparian forest width and structure, width of active floodplain, channel pattern, grain size, in-channel features), it cannot take into account historical trends that could provide insight into future geomorphic changes. Land-use change on floodplains has a stronger effect on the scoring than do channel changes, in part because the former are often more visible and measurable from aerial photographs and topographic maps. To the extent that the system is based on periodic surveys, improved data sets should be available to measure future alterations. Another limitation is the low spatial resolution of the cartographic analysis, done at a scale of 1:1 000 000, often resulting in significant differences between mapped river classes and field evidence.

The River Habitat Survey (RHS) Approach of the UK

The RHS system was developed in 1994 to provide a unifying basis for river classification and evaluation (Raven *et al.* 1998). RHS comprises four related outputs (Raven *et al.* 1997):

1. a standard field survey method,
2. a large computer database,
3. a classification of unmodified rivers based on a physical predictive model, and
4. a technique for assessing river habitat quality.

The RHS data set comprises 17 000 sites in England, Wales, Scotland, and Northern Ireland (Raven *et al.* 1997), located on the basis of a random stratification sample. Each site is 500 m long and extends 50 m either side of the channel (Raven *et al.* 1997). The collected variables concern essentially the physical structure of rivers, using the habitat level as the basic element for river management (Harper and Everard 1998). The data are obtained from maps and stream gauging records (e.g., altitude, slope, geology, distance from source, mean annual flow) and field surveys (e.g., width, depth, channel substrate and geomorphological units, bank vegetation structure and artificial modifications such as weirs, dams, bank reinforcement, channel deepening or re-alignment). These variables are noted at each of 10 equidistant points along the 500-m reach. After collection, the data are integrated in a computer database to establish a reference network of relatively undisturbed river sites (Raven *et al.* 1997). A key-element of RHS is the elaboration of a semi-natural hydromorphological river typology derived from this subset of reference sites, which support comparison of a site to "reference" conditions and allow a given site to be assessed in the context of all sites of the same river type (Environment Agency 2002). This assessment comprises a simple notation in five classes (excellent, good, fair, poor, bad), which reflects the deviation from the reference state given by the river typology (Raven *et al.* 1997, 1998).

The classification scheme was developed in three stages (Environment Agency 2002). First, a classification of 11 types resulted from statistical analysis, distinguishing sites according to geology, altitude, slope, and mean annual discharge (Raven *et al.* 1997, Environment Agency 2002). This scheme was abandoned because intra-type variability equaled or exceeded the variability between the types. So, a second classification of nine types was elaborated, but this did not adequately predict different habitat features. A third iteration of the classification was based on the observation that most geomorphological features were correlated with map-based variables like altitude, slope, distance from the source, altitude of the source, and geology. A principal component analysis (PCA) on the table "individuals – variables" (individuals being the surveyed stations and the variables being the field measures) simplified the initially large set of variables into a smaller set of synthetic variables, called the principal components. In this case, the first two

Table 7.4 Indicators of the physical alteration of the river channel; the alteration is evaluated from scores which vary according to the river types (example of three types within a classification of 15 types). See Figure 7.13 for details about the indicators (from AQUASCOP 1997)

			Mountain streams	Mountain rivers	Rivers of sandy plains
Floodplain			10	33	20
Bank			30	33	40
Channel bed			60	33	40
		Total	100	100	100
Floodplain indicators					
Hydrology			30	30	30
Former channels			0	30	0
Floodplain features			70	40	70
		Subtotal	100	100	100
	Floodplain feature indicators				
	Land use		100	33	50
	Flooding surface area		0	33	50
	Streamway (meandering band)		0	33	0
		Subtotal	100	100	100
Bank indicators					
Hydrology			30	30	30
Riparian vegetation			0	30	30
Bank character			70	40	40
		Subtotal	100	100	100
	Bank characters				
	State of banks		100	50	100
	Stability of banks		0	50	0
		Subtotal	100	100	100
Channel bed indicators					
Hydrology			30	30	30
Longitudinal connectivity			40	40	40
Morphology			30	30	30
		Subtotal	30	30	30

(*Continues*)

Table 7.4 (*Continued*)

		Mountain streams	Mountain rivers	Rivers of sandy plains
Morphology indicators				
Sinuosity/ braiding		0	40	60
In-channel features		80	40	10
Biological features		20	20	30
	Subtotal	100	100	100

components (called F1 or factor 1, and F2 or factor 2) explained 90% of the total variance (or inertia). The first component represented a gradient of altitude and slope, while the second component was correlated with discharge and reflected a potential "energy" gradient (Jeffers 1998, Environment Agency 2002). A summary of different river characteristics can be viewed on the first factorial map, which shows that the limits between groups were not defined clearly (Figure 7.13). To facilitate understanding, the biplot was divided by lines drawn arbitrarily across a continuum to define eight named "types". The semi-natural features could be predicted by the four map-based variables, or by scores on the two principal components (Jeffers 1998, Environment Agency 2002). In the future, the integration of variables such as geology should still improve the models, and many of the variables could be extracted automatically by GIS (Dawson *et al.* 2002). In a complementary approach, using RHS data and other process-related variables, Newson *et al.* (1998) highlighted the basis of a dynamic river classification and emphasized the importance of specific stream power as a variable for classifying rivers, although this improved the prediction of class type by only 17%.

RHS is presently being revised to include further geomorphological data in support of river restoration. RHS is applied in support of both national and regional assessments of state of the environment, and also to provide longitudinal information on channel typology and physical quality.

Rosgen's Classification and Restoration Projects

Rosgen (1985, 1994, 1996) developed a classification system based in part on application of hydraulic geometry concepts of Leopold and Maddock (1953) and Leopold *et al.* (1964) and field experience in the Rocky Mountain region. Rosgen's (1985) scheme recognized 25 distinct stream "types", based on gradient, sinuosity, width/depth ratio, bed material size, and degree of valley confinement, in four major groups along the longitudinal gradient of the river, from steep mountain streams designated "A" to estuarine channels designated "D" (Figure 7.3). Within these four broad groups, specific classes were defined and designated by an alpha-numeric code as "A1", "A2", "B1", "B2", etc. (analogous to the climates "Af", "Am", "Aw", etc., of Koppen 1936). Rosgen modified the channel characteristics attached to each alpha-numeric designation several times, recognizing 94 distinct stream classes as of 1996 (Rosgen 1994, 1996). An alpha-numeric code can accommodate more classes than the more descriptive designations employed by other classification systems, but an alpha-numeric such as "C1-1" lacks the descriptive power of a verbal designation such as a "braided, point-bar" channel (Brice 1982).

The Rosgen system has proved extremely popular with land managers, has been widely employed as a basis for stream management and restoration, and has been incorporated by reference into guidance documents of a number of public agencies, such as the US Forest Service (Clary and Webster 1989, McCain *et al.* 1990), the California Department of Fish and Game (Flosi and Reynolds 1991), the Nevada Department of Wildlife (Myers and Swanson 1991), and the US Fish and Wildlife Service (McCandless and Everett 2002).

In its report entitled *Restoration of Aquatic Ecosystems*, a committee of the US National Research Council (NRC), which did not include a geomorphologist,

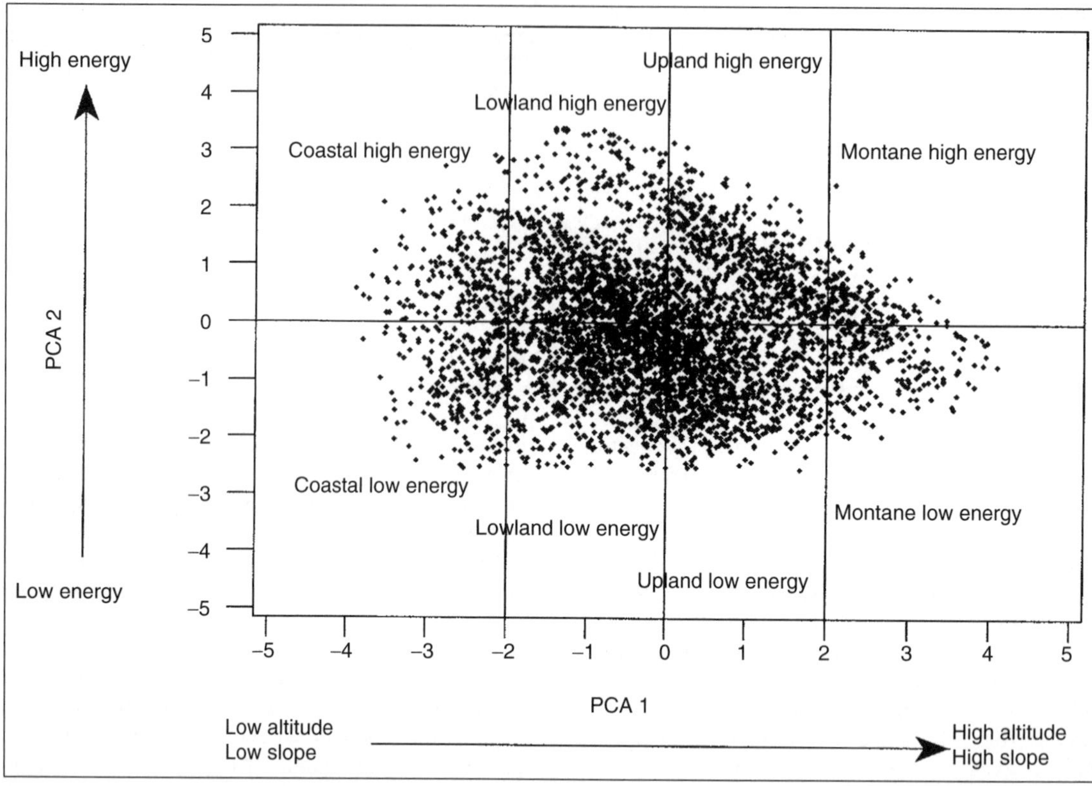

Figure 7.13 Principal Component Analysis performed on 4569 English and Welsh sites described by their altitude, slope, distance from source and altitude of the source, highlighting the semi-natural river typology of the RHS approach (reproduced from Environment Agency 2002)

noted that river and stream restoration projects frequently failed, and concluded that these failures commonly resulted from a failure to take hydrology and natural processes into account. Remarkably, the report then recommended what was essentially a "cookbook approach" to taking geomorphology into account: applying the Rosgen classification scheme and then using a table indicating the types of artificial habitat enhancement structures suitable in different Rosgen "stream types" (NRC 1992: 236–243). The Rosgen scheme has indeed been widely applied in design of stream restoration projects. The procedure for applying the system, as indicated by design documents of several such projects, is a determination of the "proper" stream type for a site based on data collected on site and expectations of transitions from one type to another. The proper stream type is then constructed with heavy equipment with expectation that it will be inherently stable. However, these projects are typically heavily reinforced on the outside of meander bends with large boulders, rootwads, and timbers, and the beds are stabilized with boulder weirs termed "vortex rock weirs".

The NRC's ill-advised recommendation illustrates the need for expertise in fluvial geomorphology to be brought to bear on channel restoration designs to diagnose and integrate various regional and local controls and influences on channel condition and response potential (Montgomery and MacDonald 2002). A number of issues arise in the widespread application of the Rosgen scheme. As acknowledged by Rosgen (1994: 181), many channels do not fit in categories of the classification scheme, "where values of one variable may be outside of the range for a given stream type". To readers with scientific training, this may suggest that the classification system is not as universally applicable as widely believed, and is consistent with the notion that river channels are distributed along a continuum of channel forms rather than in bins of discrete "types" as assumed by the classification. Rosgen (1994) suggested modifying the classes with an alpha-numeric suffix, such as "C4c-", which would indicate that the slope value for a given channel fell out of the prescribed range for C4. In practice, those applying the system have commonly assigned channels to the classification category to which it seems they are most similar, in some cases recording only the ultimate classification category and not recording the ways in which the channel did not fit the criteria for the class assigned.

More fundamentally, this classification system purports to "predict a river's behavior from its appearance" (Rosgen 1994: 170). As noted by Miller and Ritter (1996), to make valid predictions, the criteria on which the system is based (such as the boundaries separating one "type" from another) should have geomorphic significance, which was not demonstrated by Rosgen (1994). Miller and Ritter (1996) noted that the Rosgen system stream "types" were not necessarily linked to the current equilibrium state of the channel; climatic and hydrologic regime were not considered; and that substituting predictions from a classification system for real fluvial investigations could lead to counterproductive management schemes.

The Rosgen system has proved popular among non-geomorphologists as a communication tool. For this purpose, it is evidently helpful to many practitioners, and its drawbacks as a basis for restoration design could be avoided (Miller and Ritter 1996). Perhaps it can best be used as a slope/grain size matrix, with the letter indicating the valley slope and the number indicating grain size. Such shorthand to describe the slope–grain size combination of a reach would be useful in itself.

Despite the widespread popularity among practitioners of the Rosgen classification scheme as a basis for restoration design, the actual performance of the projects designed using this approach has rarely been evaluated. In fact, stream and river restoration projects, whatever their basis, have rarely been subjected to objective post-project evaluation (Kondolf 1995, 1998, Kondolf and Micheli 1995). However, recent post-project evaluations in Maryland and California have shown that a number of projects designed with this scheme have failed within a few months or years of construction. Two examples are presented below.

Deep Run, Maryland

Smith (1997) evaluated the performance of a project constructed in 1995 on Deep Run, southwest of Baltimore, Maryland. The stated rationale for the project was to reduce bank erosion by excavating the existing channel and replacing it with a channel whose dimensions would be stable at the site, according to the Rosgen classification scheme. As observed by Downs and Kondolf (2002), projects like this have been constructed throughout Maryland since the early 1990s, but Deep Run is the only one to be subjected to a thorough post-project appraisal, in which pre-project baseline conditions were documented, as well as as-built project conditions, and performance of the project over subsequent years. Smith (1997) surveyed pre-project channel geometry and measured flow velocities in the channel and on the floodplain during floods, and he repeated the surveys and velocity measurements after the project was built. The channel reconstruction project entailed removal of most existing riparian vegetation and heavy earth-moving so that idealized channel form of symmetrical meander bends could be imposed on the site. The constructed channel was narrower than the original channel, based on the designer's estimate of the "bankfull discharge", and the amplitude and wavelength of the meander bends were sized as a multiple of this bankfull channel width.

Smith's (1997) measurements showed that overbank velocities were considerably higher after the project was constructed, evidently due to the removal of the hydraulically rough riparian vegetation, and probably also because the smaller, constructed channel put more water overbank for a given discharge than did the original channel. In 1996, Deep Run largely abandoned the designed channel, cutting a new channel through the constructed floodplain, abandoning some of the bank revetments, outflanking and eroding others. The project was repaired and failed again in 1997. The resulting channel is wider than the design channel and less stable than the pre-project condition (Smith 1997).

Uvas Creek

A 0.9-km reach of Uvas Creek, near Gilroy, California, was reconstructed as a "C4" channel under the Rosgen (1994) classification system to increase channel stability and improve habitat for steelhead trout (*Oncorhynchus mykiss*). The project plan stated that the channel had formerly been a stable, C4 channel, but presented no historical evidence to support this assertion, which was based on application of the Rosgen classification system. The project involved very much the same sort of earth-moving as undertaken at Deep Run to produce a series of symmetrical meander bends scaled to the bankfull width estimated by the designers. The project was completed in November 1995, only to fail in a flood with a return interval of 6 years in February 1996 (Figure 7.14).

A post-project appraisal by Kondolf *et al.* (2001) showed that the post-1996 channel form, of a wide, braided sand-and-gravel channel, was typical for streams draining this part of the California Coast Ranges, with high uplift rates, a Mediterranean climate, erodible lithologies, and consequent high loads of gravel and sand, and dynamic, episodic channel processes. There was no historical evidence to support the prediction of the Rosgen classification that a single-thread, meandering channel would be stable at this site (Figure 7.15).

The performance of these projects calls into question the increasingly popular approach of using estimates of bankfull discharge (usually defined as the $Q_{1.5}$) and the Rosgen classification system to design projects that involve massive regrading of the valley bottom to achieve a desired, idealized channel form, but without conducting real analysis of the historical and current geomorphic processes in the catchment. Moreover, the notion that there is a "stable" channel form for any given channel fails to acknowledge the important role of channel dynamics and channel change in many rivers, especially semi-arid systems that tend to be dominated more by infrequent floods and periodic disturbance (Wolman and Gerson 1978), which in turn support a diverse range of native species adapted to the dynamic environment (Gasith and Resh 1999). These issues highlight some fundamental shortcomings with using a form-based classification system to design channel restoration projects, as opposed to conducting scientifically sound (albeit longer and more costly) geomorphic and ecological studies to understand the current and historical processes in the channel and the catchment, to develop a real understanding of the factors leading to channel change and thus a more sophisticated assessment of opportunities and constraints for restoration for the specific river. A process-based geomorphic classification system might perform better, but in the end it is unlikely that any classification system can be an adequate basis for design of restoration projects.

7.5 CHANNEL CLASSIFICATION: TOOL OR CRUTCH?

Heraclitus' elegant description of the dynamic nature of river channels masks the fact that river channel change may be either perceptible only over centuries, as illustrated by many lowland English rivers (Brookes 1992), or unnervingly rapid like in the dramatic channel shifts of the Carmel River, California, during a single flood in 1911 (Kondolf and Curry 1986). Since the dependent variables of channel geometry (width, depth, slope, velocity, and bedform roughness) can adjust in a variety of ways to imposed changes in the independent variables of flow and sediment load, it is impossible to predict with certainty the channel's response to a given perturbation in the system from existing formulae (Maddock 1970, Montgomery and Buffington 1998). Each channel is unique, and assessment of its present condition and likely future behavior require understanding of both its current condition and past behavior, as documented in historical maps, aerial photographs, surveys, and archival sources (Hooke and Kain 1982, Kondolf and Larson 1995, Collins and Montgomery 2001, Collins *et al.* 2002). A historical study should be conducted to determine the stream's characteristic behavior, especially its response to (and relaxation time from) perturbations such as large floods, changes in sediment supply, or engineering works.

Systematic description of the existing state of a stream channel can be instructive because the existing channel form integrates the many factors influencing river form and process, and a classification may help some managers understand how to describe these existing conditions. However, when using a classification system, it is important to avoid over-emphasizing the categories, lest the user view the stream system as a series of "snapshots". Ideally, stream classification should permit the comparison of observations from diverse sites and the application of insights developed in one drainage to another. When used appropriately channel classification can provide a powerful and flexible tool for fluvial geomorphologists, but the potential to rely on channel classifications as a crutch should warrant substantial care to ensure the judicious use of this tool. It is only in this condition that it will be useful for restoration and sustainable management of fluvial systems.

Channel classifications are evolving from a priori classifications to large geomorphological databases, which can be used flexibly, allowing the end user to build his own classification to answer specific questions. With such systems, it is possible to add progressively more data, as well as historical data, to better integrate different timescales and consider retrospective analysis, and to combine corridor and channel information. These databases should integrate GIS software (Gurnell *et al.* 1994) and be intelligently

(a)

(b)

Figure 7.14 Uvas Creek channel viewed downstream from the Santa Teresa Rd bridge, the upstream end of the channel reconstruction project. (a) View in January 1996, shortly after November 1995 project construction. (b) View in July 1997, after the designed channel washed out in February 1996 and high flows in winter 1997 (reproduced from Kondolf *et al.* 2001)

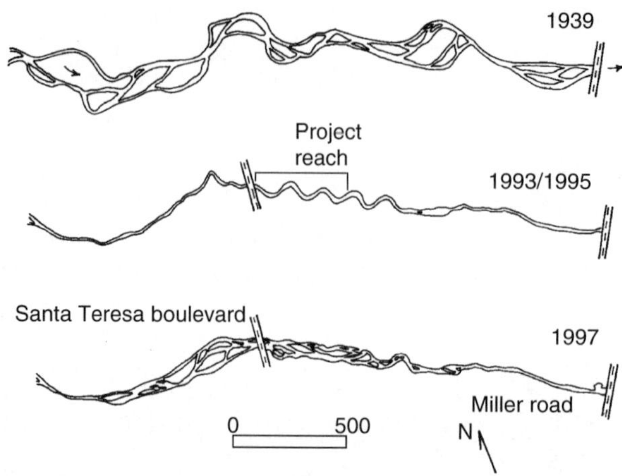

Figure 7.15 The active channel of Uvas Creek as mapped from aerial photographs taken in 1939, 1993, and 1997. The 1939 channel was braided and wide. The reconstruction project completed in November 1995 did not last long enough to be captured on aerial photographs, but is shown here as mapped on a 1993 aerial photograph from design drawings. The 1997 aerial photograph shows the channel had abandoned the design channel project (reproduced from Kondolf *et al.* 2001)

used by well-trained geomorphologists. "Universal" classifications schemes have only limited usefulness (Kondolf 1995; Heritage *et al.* 1997, Brierley and Fryirs 2000) and are not, as evidently believed by many managers and practitioners, "a panacea" that can replace understanding the channel(s) in question (Montgomery and Buffington 1998, Goodwin 1999). On the other hand, pragmatic applications of channel classification can provide useful tools for understanding smaller geographic units within regional frameworks and broader hierarchical levels (Mosley 1987, Bryce and Clarke 1996, Brierley and Fryirs 2000, Schmitt 2001a).

ACKNOWLEDGMENTS

This draft benefited from review comments by David Sear and Angela Gurnell. Manuscript preparation was partially supported by the Beatrix Farrand Fund of the Department of Landscape Architecture and Environmental Planning of the University of California Berkeley.

REFERENCES

Agence de l'Eau Rhin-Meuse, Laboratoire d'Ecologie de l' Université de Metz, Ecolor, Gereea, Loisirs et détente et INRA Rennes. 1991, Etude des végétaux fixes en relation avec la qualité du milieu (méthodologie MEV). *Etude inter-agences, hors série*, 94 p. + ann.

Agences de l'Eau. 1998. SEQ physique: système d'évaluation de la qualité physique des cours d'eau, rapport de présentation. *Rapport Inter-Agences, Ministère de l'Environnement*, 31 p.

Alabyan, A.M. and Chalov, R.S. 1998. Types of river patterns and their natural controls. *Earth Surfaces. Processess and Landforms* 23: 467–474.

Allan, J.D. and Johnson, L.B. 1997. Catchment-scale analysis of aquatic ecosystems. *Freshwater Biology* 37(1): 107–111.

Amoros, C., Richardot-Coulet, M. and Pautou, G. 1982. Les "ensembles fonctionnels": des entités écologiques qui traduisent l'évolution de l'hydrosystème en intégrant la géomorphologie et l'anthropisation (exemple du Haut-Rhône français). *Revue de Géographie de Lyon* 57 (1): 49–62.

Amoros, C. and Petts, G.E., eds., 1993. *Hydrosystèmes fluviaux*. Masson, Paris, 300 p.

AQUASCOP. 1997. Système d'évaluation de la qualité du milieu physique. 1. Typologie physique simplifiée des cours d'eau français. *Rapport, Agences de l'eau*, 55 p.

ASCE (American Society of Civil Engineers, Task Committee on Sediment Transport and Aquatic Habitats). 1992. Sediment and aquatic habitat in river systems. *Journal of Hydraulic Engineering* 118: 669–687.

Astrade, L. and Bravard, J.P. 1999. Energy gradient and geomorphological processes along a river influenced by neotectonics (the Saône River, France). *Geodinamica Acta* 12(1): 1–10.

Bagnold, R.A. 1966. An approach to the sediment transport problem from general physics. *US* Geological Survey Professionnal Paper 422 (I): 37 p.

Bagnold, R.A. 1977. Bed load transport by natural rivers. *Water Resources Research* 13(2): 303–312.

Bernot, V. and Creuzé des Châtelliers, D. 1998. Etude de l'adaptation de la méthode de sectorisation mise au point sur le Vidourle à d'autres cours d'eau du bassin Rhône-Méditerranée-Corse. *Rapport final à l'Agence de l'Eau Rhône-Méditerranée-Corse, ARALEPBP*, 63 p. + ann.

Bernot, V., Calland, V., Bravard, J.P. and Foussadier, R. 1996. La sectorisation longitudinale du Vidourle (Gard-Hérault): une méthode appliquée à la gestion environnementale de l'espace fluvial méditerranéen. *Revue de Géographie de Lyon* 4: 323–339.

Bethemont, J., Andriamahefa, H., Rogers, C. and Wasson, J.G. 1996. Une approche régionale de la typologie morphologique des cours d'eau. Application de la méthode "morphorégions" au bassin de la Loire et perspectives pour le bassin du Rhône (France). *Revue de Géographie de Lyon* 71(4): 311–322.

Biggs, B.J.F., Duncan, M.J., Jowett, I.G., Quinn, J.M., Hickey, C.W., Davies-Colley, R.J. and Close, M.E. 1990. Ecological characterisation, classification, and modelling of New Zealand rivers: an introduction and synthesis. *New Zealand Journal of Marine and Freshwater Research* 24: 277–304.

Bostelmann, R., Braukmann, U., Briem, E.U., Drehwald, U.E., Fleischhacker, T.U., Humborg, G.T., Kübler, P.G., Nadolny, I. and Scheurlen, K. 1998a. Regionale Bachtypen in Baden-Würtemberg. Arbeitsweisen und exemplarische Ergebnisse an Keuper- und Gneisbächen. *LFU, Ministerium für Umwelt und Verkehr Baden-Würtemberg, Handbuch Wasser 2, Karlsruhe*, 268 p.

Bostelmann, R., Braukmann, U., Briem, E.U., Fleischhacker, T.E., Humborg, G.T., Nadolny, I.G., Scheurlen, K. and Weibel, U. 1998b. An approach to classification of natural streams and floodplains in South-west Germany. In: Waal, L.C., Large, A.R.G. and Wade, P.M., eds., *Rehabilitation of Rivers: Principles and Implementation*, Chichester: John Wiley and Sons Ltd., pp. 31–55.

Bravard, J.P. and Peiry, J.L. 1999. The *CM* pattern as a tool for the classification of alluvial suites and floodplains along the river continuum. In: Marriott, S.B. and Alexander, J., eds., *Floodplains: Interdisciplinary Approaches*, Special Publication 163, London: Geological Society, pp. 259–268.

Brice, J.C. 1982. Stream Channel Stability Assessment, Technical Report No. FHWA/RD-82/021, Federal Highways Administration, US Department of Transportation, Washington, DC.

Brierley, G.J. and Fryirs, K. 2000. River styles, a geomorphic approach to catchment characterization: implications for river rehabilitation in Bega catchment, New South Wales, Australia. *Environmental Management* 25(6): 661–679.

BES (British Ecological Society). 1990. *Ecological Issues No. 1, River Water Quality*, British Ecological Society, Field Studies Council.

Brookes, A. 1987. The distribution and management of channelized streams in Denmark. *Regulated Rivers: Research and Management* 1: 3–16.

Brookes, A. 1990. Restoration and enhancement of engineered river channels: some European experiences. *Regulated Rivers: Research and Management* 5: 45–56.

Brookes, A. 1992. Recovery and restoration of some engineered British river channels. In: Boon, P.J., Calow, P. and Petts, G.E., eds., *River Conservation and Management*. Chichester: John Wiley and Sons Ltd., pp. 337–352.

Brussock, P.P., Brown, A.V. and Dixon, J.C. 1985. Channel form and stream ecosystem models. *Water Resources Bulletin* 21: 859–866.

Bryce, S.A. and Clarke, S.E. 1996. Landscape-level ecological regions: linking state-level ecoregion frameworks with stream habitat classifications. *Environmental Management* 20(3): 297–311.

Carpenter, K.E. 1928. *Life in Inland Waters with Especial Reference to Animals*, London: Sidgwick and Jackson (as cited by Hawkes 1969).

Church, M. 1992. Channel morphology and typology. In: Calow, P. and Petts, G., eds., *The Rivers Handbook: Hydrological and Ecological Principles*, Oxford: Blackwell Scientific Publications, pp. 126–143.

Clary, W.P. and Webster, B.F. 1989. Managing grazing of riparian areas in the intermountain region, General Technical Report INT-263, USDA Forest Service, Intermountain Research Station, Ogden, Utah.

Collins, B.D. and Montgomery, D.R. 2001. Importance of archival and process studies to characterizing presettlement riverine geomorphic processes and habitat in the Puget Lowland. In: Dorava, J.B., Montgomery, D.R., Palcsak, B. and Fitzpatrick, F., eds., *Geomorphic Processes and Riverine Habitat*, Washington, DC: American Geophysical Union, pp. 227–243.

Collins, B.D., Montgomery, D.R. and Haas, A. 2002. Historic changes in the distribution and functions of large woody debris in Puget Lowland rivers. *Canadian Journal of Fisheries and Aquatic Sciences* 59: 66–76.

Corbonnois, J. and Zumstein, J.F. 1994. Proposition de typologie des cours d'eau. Application au réseau hydrographique du Nord-Est de la France (bassin de la Moselle). *Revue de Géographie Alpine* 2: 15–24.

Cupp, C.E. 1989a. Identifying spatial variability of stream characteristics through classification, MS Thesis, University of Washington, Seattle.

Cupp, C.E. 1989b. Stream corridor classification for forested lands of Washington. Washington Forest Protection Association, Olympia, Washington, USA., 24 p. + ann.

Cushing, C.E., McIntire, C.D., Sedell, J.R., Cummins, K.W., Minshall, G.W., Petersen, R.C. and Vannote, R.L. 1980. Comparative study of physical variables of streams using multivariate analysis. *Archive fur Hydrobiologie* 89: 343–352.

Cushing, C.E., McIntire, C.D., Cummins, K.W., Minshall, G. W., Petersen, R.C., Sedell, J.R. and Vannote, R.L. 1983. Relationships among chemical, physical, and biological indices along river continua based on multivariate analyses. *Archive für Hydrobiologie* 98: 317–326.

Dana, J. 1850. On denudation in the Pacific. *American Journal of Science, Series 2*, 9: 48–62.

Davis, W.M. 1899. The geographical cycle. *Geographical Journal* 14: 481–504.

Dawson, F.H., Hornby, D.D. and Hilton, J. 2002. A method for the automated extraction of environmental variables to help the classification of rivers in Britain. *Aquatic Conservation: Marine and Freshwater Ecosystems* 12: 391–403.

Dollar, E.S.J. 2000. Fluvial geomorphology. *Progress in Physical Geography* 24(3): 385–406.

Downs, P.W. 1994. Characterization of river channel adjustments in the Thames Basin, South-East England. *Regulated Rivers: Research and Management* 9: 151–175.

Downs, P.W. 1995. Estimating the probability of river channel adjustment. *Earth Surface Processes and Landforms* 20: 687–705.

Downs, P.W. and Brookes, A. 1994. Developing a standard geomorphological approach for the appraisal of river projects. In: Kirby, C. and White, W.R., eds., *Integrated River Basin Development*, Chichester: John Wiley and Sons Ltd.

Downs, P.W. and Kondolf, G.M. 2002. Post-project appraisal in adaptive management of river channel restoration. *Environmental Management* 29(4): 477–496.

Environment Agency. 2002. *RHS Applications Manual*, Version 1.1, 71 p.

European Parliament and Council of Europe. 2000. Directive 2000/60/EC of the European Parliament and of the Council of 23 October 2000 establishing a framework for Community action in the field of water policy. *Official Journal* L 327(22/12/2000): 73 p.

Ferguson, R. 1981. Channel form and channel changes. In: Lewin, J., ed., *British Rivers*, Boston: Allen & Unwin, pp. 90–125.

Ferguson, R. 1987. Hydraulic and sedimentary controls of channel pattern. In: Richards, K., ed., *River Channels. Environment and Process,* Basil: Blackwell, pp. 129–158.

Ferguson, R.J. and Brierley, G.J. 1999. Downstream changes in valley confinement as a control on floodplain morphology, Lower Tuross River, New South Wales, Australia: a constructivist approach to floodplain analysis. In: Miller, A.J. and Gupta, A., eds., *Varieties of Fluvial Form*, Chichester: John Wiley and Sons Ltd., pp. 377–407.

Flosi, G. and Reynolds, F.L. 1991. *California Salmonid Stream Habitat Restoration Manual*, Sacramento: California Department of Fish and Game.

Frissell, C.A., Liss, W.J., Warren, C.E. and Hurley, M.D. 1986. A hierarchical framework for stream habitat classification: viewing streams in a watershed context. *Environmental Management* 10: 199–214.

Furse, M.T., Moss, D., Wright, J.F. and Armitage, P.D. 1984. The influence of seasonal and taxonomic factors on the ordination and classification of running-water sites in Great Britain and on the prediction of their macroinvertebrate communities. *Freshwater Biology* 14: 257–280.

Galay, V.J., Kellerhals, R. and Bray, D.I. 1973. Diversity of river types in Canada. In: *Fluvial Processes and Sedimentation, Proceedings of Hydrology Symposium No. 9*, Edmonton, Alberta: National Research Council of Canada, pp. 216–250.

Gasith, A. and Resh, V.H. 1999. Streams in Mediterranean climate regions: abiotic influences and biotic responses to predictable seasonal events. *Annual Review of Ecology and Systematics* 30: 51–81.

Goodwin C.N. 1999. Fluvial classfication: Neanderthal necessity or needless normalcy? In: Olson, D.S. and Potyondy, J.P., eds., *Wildland Hydrology*, Herndon, Virginia: American Water Resources Association, TPS-99-3.

Grant, G.E., Swanson, F.J. and Wolman, M.G. 1990. Pattern and origin of stepped-bed morphology in high-gradient streams, Western Cascades, Oregon. *Geological Society of America Bulletin* 102: 340–352.

Gregory, S. and Ashkenas, L. 1990. *Riparian Management Guide, Willamette National Forest*, Oregon: Eugene.

Gurnell, A.M., Angold, P. and Gregory, K.J. 1994. Classification of river corridors: issues to be addressed in developing an operational methodology. *Aquatic Conservation: Marine and Freshwater Ecosystems* 4: 219–231.

Gustard, A. 1992. Analysis of river regimes. In: Calow, P. and Petts, G.E., eds., *The Rivers Handbook. Hydrological and Ecological Principles*, Volume 1, Oxford: Blackwell Scientific Publications, pp. 29–47.

Harper, D. and Everard, M. 1998. Why should the habitat-level approach underpin holistic river survey and management? *Aquatic Conservation: Marine and Freshwater Ecosystems* 8: 395–413.

Hawkes, H.A. 1975. River zonation and classification. In: Whitton, B., ed., *River Ecology*. Berkeley: University of California Press, pp. 312–274.

Hawkins, C.P., Kershner, J.L., Bisson, P.A., Bryant, M.D., Decker, L.M., Gregory, S.V., McCullough, D.A., Overton, C.K., Reeves, G.H., Steedman, R.J. and Young, M.K. 1993. A hierarchical approach to classifying stream habitat features. *Fisheries* 18: 3–12.

Heritage, G.L., Van Niekerk, A.W. and Moon, B.P. 1997. A comprehensive hierarchical river classification system. *Geoökoplus* 4: 75–84.

Hicks, D.M. and Mason, P.D. 1991. *Roughness Characteristics of New Zealand Rivers*, Wellington: New Zealand Department of Scientific and Industrial Research, Marine and Freshwater, Water Resources Survey.

Holmes, N. 1989. British rivers: a working classification. *British Wildlife* 1: 20–36.

Hooke, J.M. and Kain, R.J.P. 1982. *Historical Change in the Physical Environment: A Guide to Sources and Techniques*. London: Butterworth Scientific.

Howard, A.D. 1980. Thresholds in river regimes: In: Coates, D.R. and Vitek, J.D., eds., *Thresholds in Geomorphology*, London: Allen & Unwin, pp. 227–258.

Howard, A.D. 1987. Modelling fluvial systems: rock-, gravel- and sand-bed channels. In: Richards, K., ed., *River Channels. Environment and Process.* Basil: Blackwell, pp. 69–94.

Huet, M. 1949. Aperçu des relations entre la pente et les populations piscicoles des eaux courantes. *Schweizerische Zeitschrift fuer Hydrologie* 11(3/4): 332–351.

Hugues, R.M., Heiskary, S.A., Matthews, W.J. and Yoder, C.O. 1993. Use of ecoregions in biological monitoring. In: Loeb, S.L. and Spacie, A., eds., *Biological Monitoring of Aquatic Systems*, USA: Lewis Publishers, pp. 125–151.

Hynes, H.B.N. 1975. Edgardo Baldi Memorial Lecture. The stream and its valley. *Verh. Internat. Verein. Limnol.* 19: 1–15.

Imhof, J.G., Fitzgibbon, J. and Annable, W.K. 1996. A hierarchical evaluation system for characterizing watershed ecosystems for fish habitat. *Canadian Journal of Fisheries and Aquatic Sciences* 53(Suppl. 1): 312–326.

Jeffers, J.N.R. 1998. Characterization of river habitats and prediction of habitat features using ordination techniques. *Aquatic Conservation: Marine and Freshwater Ecosystems* 8: 529–540.

Johnson, D.W. 1944. Problems of terrace correlation. *Bulletin of the Geological Society of America* 55: 793–818.

Jowett, I.G. and Duncan, M.J. 1990. Flow variability in New Zealand rivers and its relationship to in-stream habitat and biota. *New Zealand Journal of Marine and Freshwater Research* 24: 305–317.

Kellerhals, R., Church, M. and Bray, D.I. 1976. Classification and analysis of river processes. *Journal of the Hydraulics Division, American Society of Civil Engineers* 102: 813–829.

Kern, K. 1992. Restoration of lowland rivers: the German experience. In: Carling, P.A. and Petts, G.E., eds., *Lowland Floodplain Rivers: Geomorphological Perspectives*, Chichester: John Wiley and Sons Ltd., pp. 279–297.

Kern, K. 1994. *Grundlagen naturnaher Gewassergestaltung-Geomorphologische Entwicklung von Fliessgewassern*, Berlin: Springer, 256 pp.

King, J.M., DeMoor, F.C. and Chutter, F.M. 1992. Alternative ways of classifying rivers in southern Africa. In: Boon, P.J., Calow, P. and Petts, G.E., eds., *River Conservation and Management*, Chichester: John Wiley and Sons Ltd., pp. 213–230.

Knapp, R.T. 1938. Energy-balance in stream-flows carrying suspended load. *American Geophysical Union Transactions* 19: 501–505.

Knighton, A.D. 1999. Downstream variation in stream power. *Geomorphology* 29(3/4): 293–306.

Knighton, A.D. and Nanson, G.C. 1993. Anastomosis and the continuum of channel pattern. *Earth Surface Processes and Landforms* 18: 613–625.

Kondolf, G.M. 1995. Geomorphological stream channel classification in aquatic habitat restoration: uses and limitations. *Aquatic Conservation* 5: 127–141.

Kondolf, G.M. 1998. Lessons learned from river restoration projects in California. *Aquatic Conservation* 8: 39–52.

Kondolf, G.M. and Larson, M. 1995. Historical channel analysis and its application to riparian and aquatic habitat restoration. *Aquatic Conservation* 5: 109–126.

Kondolf, G.M. and Curry, R.R. 1986. Channel erosion along the Carmel River, Monterey County, California. *Earth Surface Processes and Landforms* 11: 307–319.

Kondolf, G.M. and Micheli, E.M. 1995. Evaluating stream restoration projects. *Environmental Management* 19: 1–15.

Kondolf, G.M., Smeltzer, M.W. and Railsback, S. 2001. Design and performance of a channel reconstruction project in a coastal California gravel-bed stream. *Environmental Management* 28(6): 761–776.

Koppen, W. 1936. Das geographische System der Klimate. In: Koppen, W. and Geiger, R., eds., *Handbuch der Klimatologie*, Volume 1, Part C, Berlin: Borntraeger.

Landesumweltamt Nordrhein-Westfalen. 1996. Naturraumspezifische Leitbilder für kleine und mittelgroße Fließgewässer in der freien Landschaft. Eine vorläufige Zusammenstellung von Referenzbach- und Leitbildbeschreibungen für die Durchführung von Gewässerstrukturgütekartierung in Nordrhein-Westfalen, 127 p.

Lawler, D.M. 1992. Process dominance in bank erosion systems. In: Carling, P.A. and Petts, G.E., eds., *Lowland Floodplain Rivers: Geomorphological Perspectives*, Chichester: John Wiley and Sons Ltd., p. 117–143.

Leopold, L.B. and Maddock, T. 1953. *The Hydraulic Geometry of Stream Channels and Some Physiographic Implications*, US Geological Survey Professional Paper 252, 57 p.

Leopold, L.B. and Wolman, M.G. 1957. *River Channel Patterns – Braided, Meandering, and Straight*, US Geological Survey Professional Paper 282(B), pp. 39–85.

Leopold, L.B., Wolman, M.G. and Miller, J.P. 1964. *Fluvial Processes in Geomorphology*. San Francisco: W.H. Freeman.

Lotspeich, F.B. 1980. Watersheds as the basic ecosystem: this conceptual framework provides a basis for a natural classification system. *Water Resources Bulletin* 16(4): 581–586.

Maddock, T., Jr. 1970. Indeterminate hydraulics of alluvial channels. *Journal of the Hydraulics Division, Proceedings of the American Society of Civil Engineers* 96: 2309–2323.

Maire, G. and Wilms, P. 1984. Etude hydro-géomorphologique du Giessen. Détermination de secteurs homogènes. *Région Alsace, Université Louis Pasteur, Centre de Géographie Appliquée, UA 95 du CNRS, Strasbourg*, 78 p. + ann.

Makaske, B. 2001. Anastomosing rivers: a review of their classification, origin and sedimentary products. *Earth-Sciences Reviews* 53: 149–196.

Malavoi, J.R. 1989. Typologie des faciès d'écoulement ou unités morphodynamiques des cours d'eau à haute énergie. *Bulletin français de pêche et de pisciculture* 315: 189–210.

Malavoi, J.R. 2000. Typologie et sectorisation des cours d'eau du bassin Loire-Bretagne. *Agence de l'Eau Loire-Bretagne* 79 p.

McCain, M., Fuller, D., Decker, L. and Overton, K. 1990. *Stream Habitat Classification and Inventory Procedures for Northern California*, San Francisco, California: FHR Currents No. 1, USDA Forest Service Region 5.

McCandless, T.L. and Everett, R.A. 2002. Maryland stream survey: bankfull discharge and channel characteristics of streams in the Piedmont hydrologic region, Report CBFO-802-01 US Fish and Wildlife Service, Chesapeake Bay Field Office, Annapolis, MD (available at: http://www.fws.gov/r5cbfo/Piedmont.pdf).

Miller, J.R. and Ritter, J.B. 1996. Discussion. An examination of the Rosgen classification of natural rivers. *Catena* 27: 295–299.

Montgomery, D.R. 1999. Process domains and the river continuum. *Journal of the American Water Resources Association* 35: 397–410.

Montgomery, D.R. and Buffington, J.M. 1997. Channel reach morphology in mountain drainage basins. *Geological Society of America Bulletin* 109: 596–611.

Montgomery, D.R. and Buffington, J.M. 1998. Channel processes, classification, and response potential. In: Naiman, R.J. and Bilby, R.E., eds., *River Ecology and Management*, New York: Springer-Verlag Inc., pp. 13–42.

Montgomery, D.R. and Gran, K.B. 2001. Downstream hydraulic geometry of bedrock channels, *Water Resources Research* 37: 1841–1846.

Montgomery, D.R. and MacDonald, L.H. 2002. Diagnostic approach to stream channel assessment and monitoring, *Journal of the American Water Resources Association* 38: 1–16.

Montgomery, D.R., Dietrich, W.E. and Sullivan, K. 1998. The role of GIS in watershed analysis. In: Lane, S.N., Richards, K.S. and Chandler, J.H., eds., *Landform Monitoring, Modelling and Analysis*, Chichester: John Wiley and Sons Ltd., pp. 241–261.

Mosley, M.P. 1982. *A Procedure for Characterising River Channels*, Water and Soil Misc. Publication No. 32, Christchurch: New Zealand Ministry of Works and Development.

Mosley, M.P. 1987. The classification and characterization of rivers. In: Richards, K., ed., *River Channels, Environment and Process*, Oxford: Basil Blackwell, pp. 295–320.

Müller, A., Glacer, D., Sommerhäuser, M. and Timm, T. 1996. Leitbilder für die Gewässerstrukturgütekartierung in Nordrhein-Westfalen. *Kasseler Wasserbau-Mitteilungen Heft* 6: 95–105.

Myers, T.J. and Swanson, S. 1991. Aquatic habitat condition index, stream type, and livestock bank damage in northern Nevada. *Water Resources Bulletin* 27: 667–677.

Naiman, R.J., Lonzarich, D.G., Beechie, T.J. and Ralph, S.C. 1992. General principles of classification and the assessment of conservation potential in rivers. In: Boon, P.J., Calow, P. and Petts, G.E., eds., *River Conservation and Management*, Chichester: John Wiley and Sons Ltd., p. 93–124.

Nanson, G.C. and Croke, J.C. 1992. A genetic classification of floodplains. *Geomorphology* 4: 459–486.

Nanson, G.C. and Knighton, A.D. 1996. Anabranching rivers: their causes, character and classification. *Earth Surface Processes and Landforms* 21(3): 217–239.

Nevins, T.H.F. 1965. River classification with particular respect to New Zealand. In: *Proceedings of the 4th New Zealand Geography Conference, Dunedin*, New Zealand Geographical Society Conference Series No. 4, pp. 83–90.

Newson, M.D., Clark, M.J., Sear, D.A. and Brookes, A. 1998. The geomorphological basis for classifying rivers. *Aquatic Conservation* 8: 415–430.

Newson, M.D. and Newson, C.L. 2000. Geomorphology, ecology and river channel habitat: mesoscale approaches to basin-scale challenges. *Progress in Physical Geography* 24(2): 195–217.

Nicolas, Y. and Pont, D. 1997. Hydrosedimentary classification of natural and engineered backwaters of a large river, the lower Rhône: possible applications for the maintenance of high fish biodiversity. *Regulated Rivers: Research and Management* 13: 417–431.

NRA (National Rivers Authority). 1992. River channel typology for river planning and management, Internal report to NRA Thames region, prepared by Geodata Unit, Southampton University, England.

NRC (National Research Council). 1992. *Restoration of Aquatic Ecosystems: Science, Technology, and Public Policy*, Washington, DC: NRC Committee on Restoration of Aquatic Ecosystems.

Omernik, J.M. 1987. Ecoregions of the coterminous United States. *Annals of the Association of American Geographers* 77: 118–125.

Orlowski, L.A., Schumm, S.A. and Mielke, P.W., Jr. 1995. Reach classifications of the lower Mississippi River. *Geomorphology* 14: 221–234.

Otto, A. 1991. Grundlagen einer morphologischen Typologie der Bache. In: Larsen, P., ed., *Mitteilungen 180*, University of Karlsruhe: Institute for Hydraulic Structures and Agricultural Engineering, pp. 1–94.

Otto, A. and Braukmann, U. 1983. Gewässertypologie im ländlichen Raum. In: *Schriftenreihe des Bundesministers für Ernährung, Landwirtschaft und Forsten. Reihe A: Angewandte Wissenschaft, Heft 288, Landwirtschaftsverlag GmbH*, 61 p.

Palmer, L. 1976. River management criteria for Oregon and Washington. In: Coates, D.R., ed., *Geomorphology and Engineering*, Stroudsburg, Pennsylvania: Dowden, Hutchinson, and Ross, pp. 329–346.

Pardé, M. 1933. *Fleuves et rivières*, Armand Colin, 241 p.

Paustian, S.J., Marion, D.A. and Kelliher, D.F. 1984. Stream channel classification using large scale aerial photography for Southeast Alaska watershed management. In: Murtha, P.A. and Harding, R.A., eds., *Renewable Resources Management; Application of Remote Sensing. Am. Soc. Photogramm., Symposium on the Application of Remote Sensing to Resource Management, Seattle, WA, USA*, pp. 670–677.

Paustian, S.J. (and 12 others), *et al.* 1992. A channel type users guide for the Tsongas National Forest, Southeast Alaska, USDA Forest Service, Alaska Region, R 10 Technical Paper 26.

Pennak, R.W. 1971. Towards a classification of lotic habitats. *Hydrobiologia* 38: 321–334.

Petit, F. 1995. Régime hydrologique et dynamique fluviale des rivières ardennaises. In: Demoulin, A., ed., *L'Ardenne, Essai de Géographie Physique, Livre en hommage au Professeur A. Pissart*, Université de Liège: Laboratoire d'Hydrographie et de Géomorphologie fluviatile, pp. 194–223.

Piégay, H., Barge, O. and Landon, N. 1996. Streamway concept applied to river mobility/human use conflict management. In: *International Water Resources Association: Rivertech '96: New/emerging Concept for Rivers, Chicago, USA, September 1996*, pp. 681–688.

Piégay, H., Cuaz, M., Javelle, E. and Mandier, P. 1997. A new approach to bank erosion management: the case of the Galaure River, France. *Regulated Rivers: Research and Management* 13: 433–448.

Piégay, H., Salvador, P.G. and Astrade, L. 2000. Réflexions relatives à la variabilité spatiale de la mosaïque fluviale à l'échelle d'un tronçon. *Zeitschrift für Geomorphologie* 44(3): 317–342.

Platts, W.S. 1980. A plea for fishery habitat classification. *Fisheries* 5: 2–6.

Platts, W.S., Megahan, W.F. and Minshall, G.W. 1983. Methods for evaluating stream, riparian, and biotic conditions, USDA Forest Service General Technical Report INT-138, USDA Forest Service Intermountain Research Station, Ogden, Utah.

Platts, W.S., Armour, C., Booth, G.D., Bryant, M., Cuplin, P., Jensen, S., Lienkaemper, G.W., Minshall, G.W., Monsen, S.B., Nelson, R.L., Sedell, J.R. and Tuhy, J.S. 1987. Methods for evaluating riparian habitats with applications to management, USDA Forest Service General Technical Report INT-221, USDA Forest Service Intermountain Research Station, Ogden, Utah.

Powell, J.W. 1875. *Exploration of the Colorado River of the West and its Tributaries*, Washington, DC: Government Printing Office, 291 p.

Raven, P.J., Fox, P., Everard, M., Holmes, N.T.H. and Dawson, F.H. 1997. River habitat survey: a new system for classifying rivers according to their habitat quality. In: Boon, P.J. and Howell, D.L., eds., *Freshwater Quality: Defining the Indefinable*? *Scottish Natural Heritage*, pp. 215–234.

Raven, P.J., Boon, P.J., Dawson, F.H. and Ferguson, A.J.D. 1998. Towards an integrated approach to classifying and evaluating rivers in the UK. *Aquatic Conservation: Marine and Freshwater Ecosystems* 8: 383–393.

Raven, P.J., Holmes, N.T.H., Charrier, P., Dawson, F.H., Naura, M. and Boon, P.J. 2002. Towards a harmonized approach for hydromorphological assessment of rivers in Europe: a qualitative comparison of three survey methods. *Aquatic Conservation: Marine and Freshwater Ecosystems* 12: 405–424.

Robach, F., Thiébaut, G., Trémolières, M. and Muller, S. 1996. A reference system for continental running waters: plant communities as bioindicators of increasing eutrophication in alkaline and acidic waters in north-east France. *Hydrobiologia* 340: 67–76.

Rohm, C.M., Giese, J.W. and Bennett, C.C. 1987. Evaluation of an aquatic ecoregion classification of streams in Arkansas. *Journal of Freschwater Ecology* 4(1): 127–140.

Rosgen, D.L. 1985. A stream classification system. In: *Riparian Ecosystems and Their Management, Proceedings of the First North American Riparian Conference, April 16–18, Tucson, Arizona*, USDA Forest Service General Technical Report No. RM-120.

Rosgen, D.L. 1994. A classification of natural rivers. *Catena* 22(3): 169–199.

Rosgen, D.L. 1996. *Applied River Morphology*, Wildland Hydrology, Pagosa Springs, Colorado, 390 p.

Rust, B.R. 1978. A classification of alluvial channel systems. In: Miall, A.D., ed., *Fluvial Sedimentology*, Canadian Society of Petroleum Geologist Memoir 5, pp. 187–198.

Schmitt, L. 2001a. Typologie hydro-géomorphologique fonctionnelle de cours d'eau, Recherche méthodologique appliquée aux systèmes fluviaux d'Alsace, Ph.D. Thesis, University Louis Pasteur of Strasbourg, 217 p. + ann.

Schmitt, L. 2001b. *Etude de la typologie des cours d'eau alsaciens : expertise scientifique et consolidation des acquis. Rapport final*. ULP, Centre d'Etudes et de Recherches Eco-Géographiques Fre 2399, Agence de l'Eau Rhin-Meuse, 166 p. + ann.

Schmitt, L., Maire G. and J. Humbert 2000. Typologie hydro-géomorphologique des cours d'eau : vers un modèle adapté à la gestion du milieu physique des rivières du versant sud-occidental du fossé rhénan. *GéoCarrefour* 75: 347–363.

Schmitt, L., Maire G. and Humbert, J. 2001. La puissance fluviale : définition, intérêt et limites pour une typologie hydro-géomorphologique de rivières. *Zeitschrift für Geomorphologie* 45: 201–224.

Schumm, S.A. 1963. *A Tentative Classification of Alluvial River Channels*, US Geological Survey Circular 477, Washington, DC.

Schumm, S.A. 1977. *The Fluvial System*, New York: John Wiley and Sons Ltd.

Schumm, S.A. 1985. Patterns of alluvial rivers. *Annual Review of Earth and Planetary Sciences* 13: 5-27.

Simons, P.K. 1978. The river scene: comments on North Island rivers. In: *Proceedings of Erosion Assessment and Control conference, Christchurch, New Zealand*, New Zealand Association of Soil Conservators, pp. 156–158.

Smith, D.G. and Smith, N.D. 1980. Sedimentation in anastomosed river systems: examples from alluvial valleys near Banff, Alberta. *Journal of Sedimentary Petrology* 50(1): 157–164.

Smith, S. 1997. Changes in the hydraulic and morphological characteristics of a relocated stream channel, Unpublished Master of Science Thesis, Department of Geology University of Maryland, College Park, 148 p.

Sneath, P.H.A. and Snokal, R.R. 1973. *Numerical Taxonomy: The Principles and Practice of Numerical Classification*, San Francisco: W.H. Freeman.

Souchon, Y., Andriamahefa, H., Cohen, P., Breil, P., Pella, H., Lamouroux, N., Malavoi, J.R. and Wasson, J.G. 2000. Régionalisation de l'habitat dans le bassin de la Loire, Unpublished report, Cemagref, Agence de l'eau Loire Bretagne, p. 291.

Strahler, A.N. 1957. Quantitative analysis of watershed geomorphology. *Transactions, American Geophysical Union* 38 (6): 913–920.

Terrell, T.T. and McConnell, W.J. 1978. *Stream Classification 1977: Proceedings of the Workshop Held at Pingree Park, Colorado, October 1977*, US Fish and Wildlife Service Publication No. FWS/OBS-78/23, USFWS Instream Flow Group, Fort Collins, Colorado, pp. 1–38.

Thorne, C.R. 1997. Channel types and morphological classification. In: Thorne, C.R., Hey, R.D. and Newson, M.D., eds., *Applied Fluvial Geomorphology for River Engineering and Management*, Chichester: John Wiley and Sons Ltd., pp. 175–222.

Thorne, C.R. 1998. *Stream Reconnaissance Handbook*. Chichester: John Wiley and Sons Ltd., 133 p.

Tölk, J. 1998. Leitbilderstellung für fliessgewässer im Kraichgau, Diplomarbeit, Institut für Geographie und Geoökologie der Universität Karlsruhe, Gewässerdirektion Nördlicher Oberrhein, 138 p. + ann.

Tricart, J. 1977. *Précis de Géomorphologie. Tome 2, Géomorphologie dynamique générale*. Sedes, 345 p.

Van Den Berg, J.H. 1995. Prediction of alluvial channel pattern of perennial rivers. *Geomorphology* 12: 259–279.

Van Niekerk, A.W., Heritage, G.L. and Moon, B.P. 1995. River classification for management: the geomorphology of the Sabie River in the Eastern Transvaal. *South Africa Geographical Journal* 77(2): 68–76.

Vannote, R.L., Minshall, G.W., Cummins, K.W., Sedell, J.R. and Cushing, C.E. 1980. The river continuum concept. *Canadian Journal of Fisheries and Aquatic Sciences* 37: 130–137.

Wasson, J.G., Bethemont, J., Degorce, J.N., Dupuis, B. and Joliveau, T. 1993. Approche écosystèmique du bassin de la Loire: Eléments pour l'élaboration des orientations fondamentales de gestion. Phase 1: Etat initial – Problématique. Rapport d'étape. CEMAGREF, Groupement de Lyon et Université Jean Monnet de St Etienne, CRENAM, CNRS U.R.A. 260, 102 p.

Whiting, P.J. and Bradley, J.B. 1993. A process-based classification system for headwater streams. *Earth Surface Processes and Landforms* 18: 603–612.

Wolman and Gerson. 1978. Relative scales of time and effectiveness of climate in watershed geomorphology. *Earth Surface processes and Landforms* 3: 189–208.

Wright, J.F., Moss, D., Armitage, P.D. and Furse, M.T. 1984. A preliminary classification of running water sites in Great Britain based on macro-invertebrate species and prediction of community type using environmental data. *Freshwater Biology* 14: 221–256.

Zernitz, E.R. 1932. Drainage patterns and their significance. *Journal of Geology* 40: 498–521.

Zumstein, J.F. and Goetghebeur, P. 1994. Typologie des rivières du bassin Rhin-Meuse, *Agence de l'Eau Rhin-Meuse*, 6 p. + 1 carte.

Zachman, W.R. 1984. A river classification system: management for Minnesota's rivers. In: *Proceedings of the 1984 River Recreation Symposium, October 31–November 3*, Baton Rouge, Louisiana: Louisiana State University, pp. 619–627.

8

Modelling Catchment Processes

PETER W. DOWNS[1,2] AND GARY PRIESTNALL[2]
[1]Stillwater Sciences, Berkeley, CA, USA
[2]Department of Geography, University of Nottingham, Nottingham, UK

8.1 INTRODUCTION

Modelling the geomorphology of fluvial landforms is dependent upon understanding a variety of processes that derive from fundamental forces that act as process 'drivers'. The majority of these forces are ultimately controlled by catchment-related parameters, including human action, and this provides the justification for developing catchment-based models but explains also the extreme complexity of achieving a detailed and robust output. For example, at the scale of the channel cross-section, geomorphological processes can be deduced through measurement and analysis of the channel form and its geotechnical properties (see Chapter 11). However, processes occurring at a channel cross-section are controlled by the processes and channel dynamics of the river reach, resulting from interactions between flow hydraulics, the sediment composition of the channel bed and banks, the channel morphology and its in-stream and riparian vegetation (Chapter 19). Many of these factors are, in turn, controlled by the dynamics of flow and sediment over an extended reach basis, and can be approximated through modelling flow hydraulics and sediment transport (see Chapter 18).

Process-mechanical simulation in fluvial geomorphology rarely extends beyond the river reach scale. However, the boundary conditions governing reach-scale flow hydraulics and dominant sediment transport processes are determined by the channel network hydrology and the channel network sediment budget, respectively (see Chapter 16). These network controls combine to determine the frequency and distribution of geomorphologically effective flow events and the dynamics of sediment supply, transport and deposition. Ultimately, these features are themselves functions of the hydrology and sediment supply from the terrestrial surfaces within the catchment. Therefore, the fundamental process drivers of fluvial geomorphology become factors such as the contemporary regional climate, topography and vegetation, set within the context of recent weather events and allied to a set of individual system history features that define the 'uniqueness' (Schumm 1991) of the individual catchment (Haff 1996). These may include the catchment's climate history since the beginning of Holocene period or longer, the geological structure and lithology of the catchment, and (not least) the activities of humans within the catchment.

This chapter examines the data challenges facing geomorphologists intent on catchment-scale modelling of fluvial geomorphological processes, and reviews the contemporary responses to three types of catchment-scale model, namely, integrated component process models, watershed analysis and conceptual models. It outlines tools available within geographical information systems (GIS) to initiate catchment-scale modelling and concludes with prospects for would-be modellers of catchment-scale fluvial geomorphology processes.

8.2 INPUT DATA REPRESENTATION

Modelling catchment-scale influences on fluvial geomorphology is inherently data demanding. Data may be required to represent the river channel network and the catchment surface including parameters spatially distributed over (and beneath) the catchment surface. Information relating to the river channel itself is typically surveyed directly although some characteristics of the channel can be extracted from large-scale aerial photography and from very high resolution remote sensing imagery. However, the impracticability of directly surveying spatially extensive parameters of the catchment topography and its surface features lead usually to a requirement for secondary data sources such as maps and remote sensing by aircraft or satellite imagery (see Chapter 6). Given also the importance of different spatial scales of data and the need to represent the relevant parameters through time as well as in space, careful management and manipulation of this data becomes vital. Overall, demands on data input and management have led to an increased awareness and use of remote sensing data and GIS within geomorphology (Walsh *et al.* 1998).

One particular challenge in catchment-scale process modelling is to capture and represent the infinite complexity of the real world in sufficient detail to offer meaningful abstractions of processes and morphologies, whilst maintaining a manageable model. Achieving acceptable data accuracy within a suitable temporal and spatial framework is paramount in enabling the necessary model flexibility in data representation. The greater the reliance on digital data sets as input parameters, the greater the care that must be taken in understanding errors present in these data sets. Burrough and McDonnell (1998) outline the various errors that can occur in capturing and storing geographical information in digital form. There is a real inherent danger that uncertainties in digital data representation will mask uncertainties in the fluvial geomorphological process modelling, and such uncertainty can potentially propagate throughout the entire analysis.

Catchment-scale studies often involve the integration of data derived from a wide range of sources, at a range of scales, and over a period of time. Consequently, understanding and management of the uncertainty in data is very important, and information describing the nature and history of the data is required. Metadata, the detailed information describing the actual data themselves, are vital if researchers are to be able to make full and appropriate use of digital data sets. Every effort should be made to obtain and use metadata, and any new metadata developed should follow an established standard such as the Content Standard for Digital Geospatial Metadata endorsed by the Federal Geographic Data Committee (www.fgdc.gov/metadata/metadata.html, accessed 15th June 2001).

GIS offers several ways of representing spatial data in digital form, for example, discrete objects such as point, line or areal features are often best stored in vector form with their co-ordinates stored to record the exact locations and extent of the features. Alternatively, the raster method can be employed with a value allocated to each cell in a regular grid. The vector–raster distinction can be exemplified using elevation data. Where contours are stored in vector form, each contour is held as a string of co-ordinates that define its location, with a value for height allocated to each co-ordinate pairing. Alternatively, in raster format, every cell on the ground is allocated a height providing a partial representation of the continuously varying surface morphology. Although most GIS will now allow conversion between these two storage conventions, it is important to be aware of the advantages and disadvantages of each method and to recognise the analytical techniques that are appropriate or possible with each. Reference can be made to core GIS texts such as Longley *et al.* (2001).

For catchment-scale studies, three facets of data representation are now considered, identifying the particular issues regarding spatial data representation in each case. These facets are catchment surface representation, land cover and land use estimation, and the representation of channel networks.

Catchment Surface Representation

The importance of topography in a review dedicated ostensibly to process modelling stems from the argument that, at the catchment-scale, the 'process' label infers a dominant concern for process over landform that may not be justified. As Lane *et al.* (1998a) note, although there is an interdependent relationship between form and process in geomorphology, for most models of 'contemporary processes', process rates *and types* will actually be controlled by topography. If topographic representation provides the boundary conditions under which processes operate (for instance, in defining the locations of surface and subsurface flow convergence and divergence), it may also, to a large extent, determine model output (Moore *et al.* 1994). This is especially likely at the catchment-scale where necessary simplifications to process models and terrain representation are inevitable in order to achieve a working model.

The representation of earth surface elevations in digital form, often as a regular grid of elevation values, is termed a digital elevation model (DEM). The density and distribution of surveyed height information is fundamental in ensuring control over the resulting DEM (McCullagh 1998). Errors introduced in the

DEM can have major repercussions during analysis due to the number of parameters that are derived from the original elevation data. The initial elevation data is often compromised because, for reasons of data availability, time and cost constraints and ease of data management, catchment-scale studies incorporate medium to low resolution DEMs (from 50 m up to 1 km for large catchments). Unfortunately, the resulting terrain surface from low resolution DEMs may 'smooth out' certain critical topological landscape elements and therefore derived parameters such as slope or flow direction may be grossly misleading (Evans 1998). Even with a 25-m resolution DEM derived from air photo interpretation, significant smoothing effects can be observed particularly in valley bottoms, and this may be influential in derived parameters such as soil moisture (Watson *et al.* 1996). Because data 'smoothing' can be responsible for uncertainties in flow direction, catchments delineated by automated techniques are also at risk (Miller and Morrice 1996). Outlet points placed in gently undulating low lying areas are particularly prone to misrepresentation in the DEM due in part to the scarcity of surveyed elevation data, and small errors can cause erratic flow directions and result in the derivation of highly misleading upstream area calculations. One solution is to use the known river network (held as a topologically structured form) to guide the extraction of upstream catchment areas (Priestnall and Downs 1996).

Elevation surfaces can be represented either by triangular irregular networks (TINs) or by a regular grid of elevation values (a *raster* representation). TINs utilise a triangulation between heighted data points such as contour vertices or spot heights. Differing complexities of landform can be represented more efficiently through the use of a fine mesh of triangles in areas of greatly varying terrain and fewer, larger, triangles where the terrain surface is less variable and surface height information is sparse. Although many derivatives, such as water flow direction, can be obtained from TINs (Maidment *et al.* 1989), the computational simplicity of a regular grid makes it a more common choice of surface representation. Figure 8.1 illustrates the creation of a TIN, and then a gridded DEM, from input contour data. This DEM can become a crucial layer in overlay procedures involving both vector layers such as the river channel and raster layers such as land cover or geology. The relative merits of gridded and triangulated terrain models are considered by McCullagh (1998) but, prior to concerns about DEM construction, suitable input data are required.

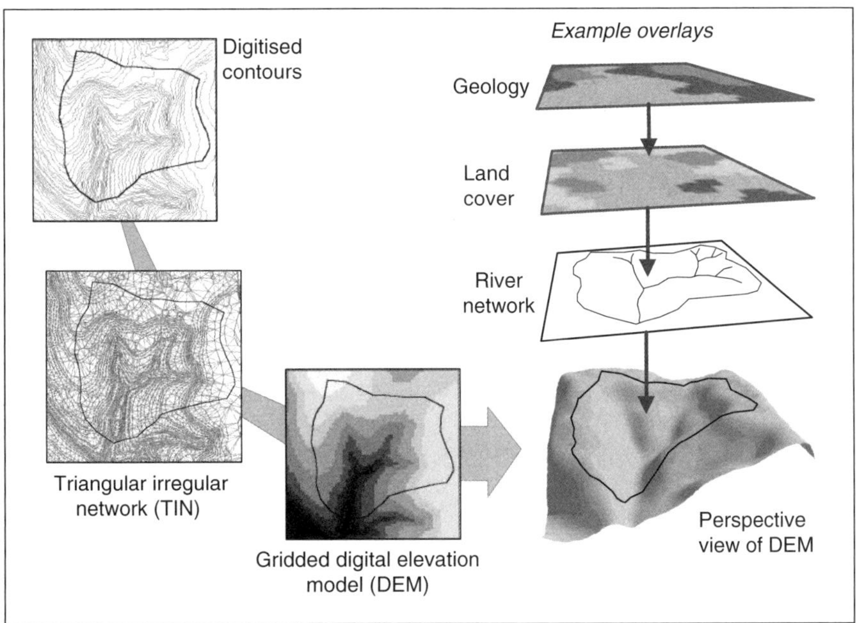

Figure 8.1 Example creation of a DEM from a triangulated network of digitised contours, and a simplification of subsequent overlay possibilities

For many catchments, digital data may exist already either as vector contours or gridded raster DEMs. In the UK, for example, the Ordnance Survey (OS) market several digital height products including PROFILE that offers 5–10 m interval terrain contours, or gridded DEMs, covering 25 km^2 'tiles' at around £100 per tile (http://www.ordsvy.gov.uk, accessed 3rd July 2001). Digital contour data sets such as PROFILE also contain selected rivers heighted in discrete sections and these allow some degree of control over the creation of catchment DEMs. Where gridded terrain models are produced from digital contours, the effects of interpolation errors on the accuracy of the model and on derived parameters such as slope, aspect and flow direction should be noted (Wise 1998). Wherever possible, vector lines should be incorporated into the triangulation process in addition to contours and spot heights (Priestnall and Downs 1996). These heighted vector strings constrain the triangulation and preserve important valley detail. In addition to creating DEMs from digital contours, whereupon the researcher can define the output grid resolution, ready-made gridded DEM data can be obtained directly from some government agencies. For instance, in the USA, a national coverage of DEM data (7.5-min series, 30 m DEM) is available through the United States Geological Survey (USGS, http://mcmcweb.er.usgs.gov/status/dem_stat.html, accessed 3rd July 2001) and offers a base data set suitable for studies in large catchments (Hutchinson 1993).

DEMs can also be derived from photogrammetric analysis of elevation data from stereo pairs of remotely sensed images, either from airborne or satellite platforms (Lillesand and Kiefer 1994). Dixon *et al.* (1998) argued that aerial photography offers accurate, non-evasive data capture which is repeatable and can, therefore, allow landform change monitoring. Disadvantages of this technique include potentially large data volumes and an overall procedure that can be expensive and time consuming, which often prohibits the use of aerial stereo pairs in catchment-scale studies. Satellite images offer a wider geographical coverage but satellite-derived DEMs generally have low vertical and horizontal accuracy compared to other sources of DEM (McCullagh 1998).

Parameterising Land Cover and Obtaining Land Use

In addition to terrain surface errors, there also exist inherent difficulties in delineating some thematic parameters. For instance, at the catchment-scale, information on soil type and land cover types are often derived directly from remotely sensed data and stored in terms of discrete areas or classified pixels. In reality, the boundaries between soil type and land cover may be neither rigid nor necessarily definable (see Burrough and Frank 1996). This concern in classifying 'continuously varying' data extends from terrestrial surfaces to river channels (Mosley 1987, Downs 1995, Kondolf 1995).

It is possible to distinguish data sets representing *land cover* from those representing *land use*. *Land cover* describes the type of surface material and may include, for example, roughness information that influences surface water flows. *Land use* describes the nature of the human activities associated with a particular area of land. Typically, land cover is classified for use in hydrological modelling but it is likely to be a benefit for catchment models of geomorphology only if land use information is also included. One such example is provided by the 'catchment characteristics' hydrological equation contained in the *Flood Studies Report* (Institute of Hydrology 1993) wherein the extent of urban area can be incorporated alongside land cover characteristics.

Data sources of land cover and land use information include a wide variety of map-based and image-based products, in addition to any directly sampled data. Map-based data can be digitised, either as discrete areas or as lines of equal value, and then converted to form a spatially referenced data 'layer' within a GIS-based catchment-process model which includes the DEM (Downs and Priestnall 1999). Alternatively, relevant digital data may already exist in the region such as, in the USA, the STATSGO soil survey data discussed by Lytle *et al.* (1996) and, from the Soil Survey in the UK, the HOST data comprising soil data reclassified according to hydrological behaviour.

Remotely sensed data can provide frequent coverage of earth surface properties covering wide areas and have been invaluable for resource management, monitoring and mapping for many years, and are now becoming increasing important as a data source for geomorphologists (Walsh *et al.* 1998). Gilvear and Bryant (Chapter 6, this volume) discuss the importance of remote sensing for fluvial geomorphologists. Issues of particular importance when defining catchment-scale land cover and/or land use include:

- *Spatial resolution*. Finer image resolutions distinguish critical land covers such as sediment sources but make for large and unwieldy data sets.

- *Spectral resolution.* Sensors should be able to distinguish land covers that are meaningful in terms of sediment supply.
- *Geographical coverage.* Catchment studies often involve large areas, therefore, many images.
- *Temporal coverage.* Frequency of coverage has direct implication for incorporating temporal process analyses into catchment research but may also influence the time required to obtain cloud free coverage of large areas, and therefore the time taken to generate land cover databases.

In addition to sensors that passively collect reflected or emitted energy, active sensors such as synthetic aperture radar (SAR) have been used to map drainage patterns, soil texture and soil moisture (Beaudoin *et al.* 1990) in addition to linear geomorphological features (Vencatasawmy *et al.* 1998). Also, airborne passive microwave radiometers have been employed successfully in monitoring spatial and temporal variability in the soil moisture content of the top 5 cm of the soil profile (Mattikalli *et al.* 1996).

Extraction of Channel Planform and Networks

Studies that consider river channel planform in addition to processes encounter problems of real channel representation. Natural systems are generally more difficult to represent digitally than phenomena observed, for example, in the urban environment, but river channels represent a particularly challenging case. GIS data models generally provide a choice between vector lines or raster cell representation. In terms of representing form, raster-based data offers a poor mechanism for storing linear features and although vector-based representation is a more logical choice for river networks, the question of scale is critical. The point at which representation of the river according to its centre line is replaced by a surface linking both banks is a matter for careful consideration. The meaningful representation of the river bank tops themselves are of particular importance given the numerous fluvial geomorphology relationships based on 'bankfull' channel width.

Creating a structured river channel network is critical to catchment process modelling as it enables water and sediment entering the channel reach (from upstream or the adjoining side slope) to be routed downstream. A structured network can also enable upstream channels to be identified and from these the contributing area can be derived (Priestnall and Downs 1996). Channel networks are also likely be represented by data sources other than direct survey. Maps are often utilised either by digitising paper copies (see Downward 1995 for cautions) or by purchasing digital maps from mapping agencies or from research institutions (e.g., UK Institute of Hydrology). Maps offer an approximation of river channel locations that can be digitised and processed to form topologically structured networks using GIS software such as ArcInfo. They do, however, suffer from the problems including:

- channel networks becoming inaccurate where planform shifts have occurred between map editions;
- selective representation of the upper reaches of channel networks;
- under-representation of channel heads which may be critical elements in generating sediment for the catchment (Kirkby 1980, Montgomery and Dietrich 1989, Dietrich and Dunne 1993);
- cartographic generalisation of networks in small-scale maps effectively reducing the channel length. Using a fractal-based approach, Wilby (1996) demonstrates that the timing of a flood hazard is altered according to the scale over which the channel length is measured.

These issues are reviewed in more detail in Chapter 4 (Gurnell *et al.* this volume).

Increasingly, digital imagery may also be used to extract channel networks, including river bank boundaries. This is an attractive prospect as satellite sensors hold the potential that semi-automated extraction of 'river lines' could be achieved over wide spatial extents and at regular temporal intervals (see Chapter 5, this volume). To date, however, the automated extraction of linear features such as roads (Gruen and Li 1995) and rivers (Haala and Vosselman 1992) has received some attention, but most research effort in terms of automated extraction of geographical features from imagery has been concerned with buildings. With the exception of the world's very largest rivers, using satellite imagery for river representation in geomorphological research will require:

- *Improved spatial resolution.* Most satellite sensors have spatial resolutions that are too coarse to represent the majority of river channels. For an accurate depiction of river channel planforms, the channel width must be represented by several pixels (at a minimum) rather than just one. As Figure 8.2 shows, a 6–8-m wide river channel is evident to the human eye once the spatial resolution increases

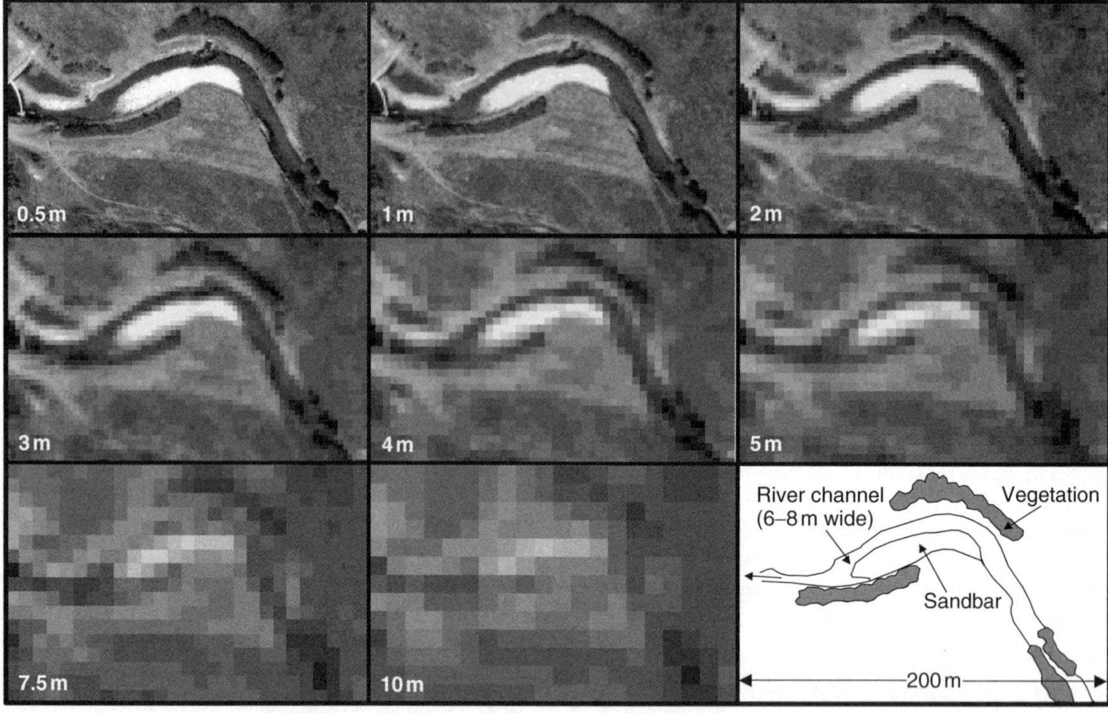

Figure 8.2 River recognition at different spatial resolutions: the implications of using different satellite sensors

below 5 m. Therefore, images from very high resolution satellite sensors offer improved channel network representation but may be extremely costly in large catchments. Detailed imagery such as digital aerial orthophotographs have many advantages for in-channel modelling applications and for identifying catchment features such as landslide scars, but may have only limited geographical coverage in individual catchments.

- *Improved spectral contrast.* Contrast between the river and the surrounding land is often poor and cannot always be relied upon to allow extraction of morphological features even by the human eye.
- *Compensation for obstruction by vegetation.* Small river channels are often partially obscured by overhanging trees resulting in the linear feature being broken up, if not by the trees themselves then by shadows cast by these trees. More sophisticated, automated, feature extraction techniques allowing linear features to be built-up from broken segments are required but are currently still a subject for research.
- *Improved automated representation.* Automated channel recognition procedures may produce a totally different representation of the channel than previous imagery.

8.3 REPRESENTING DYNAMIC CATCHMENT FEATURES

In addition to issues faced in representing 'static' phenomena such as the terrain surface in an adequate data framework that minimises data error and generalisation, catchment models require the ability to model data dynamically, as it changes in time. Catchment-scale, time-series data are often suited to a raster representation with each time interval being represented by a separate grid. GISs have not represented 'time' easily in the past but accessible tools are becoming available to simulate the change in time of spatially distributed processes such as the routing of water flows. In this case, directional linkages between cells are coded into the flow matrix and, in addition, each cell can possess finite element process models that determine flow characteristics and sediment transport. The water and/or sediment output from each cell represents the input value to the next cell. The process is repeated until a new grid representing the next time interval is completed. Several models reviewed later in this chapter have used this basis for modelling 'flows' in channels (SHETRAN, Bathurst *et al.* 1996) and on hillslopes (Montgomery *et al.* 1998). Basic dynamic flow modelling capabilities are becoming increasingly common in generic GIS such as ArcInfo and PCRaster. In addition, improved animation facilities within GIS, or using dedicated visualisation software, allows parameter variations through time to be presented effectively.

Conversely, network-scale modelling of changes in river channel morphology is far less advanced (Downs and Priestnall 1999). Channel morphology change cannot be discerned directly from finite-element sediment transport models because numerous parameters other than sediment transport capacity are required because the dominant processes are scale-dependent (e.g., Lawler 1992) and related to antecedent conditions and human management actions, and can involve significant time-lags between cause and resultant effect. As a result, non-contiguous changes are possible that cannot be modelled on a simple input–output basis.

8.4 NEW DATA SOURCES

With increasing demand for catchment models, the role of remotely sensed imagery and more advanced GIS analytical capabilities is set to become increasingly important. Catchment modelling capabilities have traditionally suffered because covering spatially extensive areas has necessitated low resolution data. High resolution multi-spectral imaging systems for airborne platforms have existed for some time such as the compact airborne spectrographic imager (CASI). CASI offers 15 channels between the blue visible and the near infrared wavelengths at spatial resolutions of 2–4 m allowing reclassification into 'standard' land cover classes such as, in the UK, the Land Cover Map of Great Britain (Fuller *et al.* 1994).

The launch of the first 1-m resolution commercial satellite IKONOS in 1999 (http://www.spaceimaging.com, accessed 3rd July 2001) heralded the start of the new era in very high resolution satellite data capture (previewed by Ridley *et al.* 1997) which will offer, in addition to 1 m panchromatic images, multi-spectral images of around 3 m including stereo coverage. These sensors may well provide the necessary balance between geographical coverage and spatial and spectral resolution acceptable for catchment geomorphological studies. The cost of such image products may, however, still prove prohibitively expensive for many catchment-scale research studies.

Advances in feature extraction using these data sources coupled with increasingly 'knowledge-rich' digital map products will mean that useful land cover and land use data sets will become more readily accessible in the near future, and with a reduced commitment for processing the raw data into a useable form. Certainly, it seems that these advances will improve the accuracy of hydrological modelling as a key driver for geomorphological processes.

A relatively new technology that can provide elevation data to decimetre accuracy is light detection and ranging (LiDAR) (Hug 1997, Lohr 1998). The digital surface models produced are free from many of the problems associated with digital photogrammetry, such as the complexities associated with image matching techniques (Smith and Smith 1996, Smith *et al.* 1997). The surface elevations of all features on the ground are captured and these surface objects such as buildings and trees can be separated from the ground leaving a DEM (Jaafar *et al.* 1999). LiDAR is often used for mapping large areas of river floodplains and coastal zones for flood inundation studies, for example, by the Environment Agency in the UK (Holden and Butcher 1998). As programmes for LiDAR data capture continue, a greater geographical extent of accurate elevation data may become available for import into catchment studies. Ultimately, it may be possible to distinguish automatically individual features created by humans by using CASI imagery in combination with LiDAR elevation data (Priestnall *et al.* 2000).

8.5 INTEGRATED COMPONENT PROCESS MODELS

The data-demanding nature of catchment-scale modelling in fluvial geomorphology requires models capable of accommodating a wide variety of data sources measured over different spatial and temporal scales, integrated according to numerous process-mechanical sub-models also operating at a variety of different spatial and temporal scales. This ideal probably explains why comprehensive catchment-level models of fluvial geomorphological processes do not exist yet, and are unlikely to in the near future (cf. Howard 1996 on floodplain evolution modelling). Progress since one of the last general reviews of geomorphological process models (Anderson 1988) has resulted in the initiation of a number of process-based models that may be applied to fluvial geomorphology, broadly defined. These models, termed here *integrated contemporary process models* (ICP models), may include sub-models of hillslope hydrology and soil erosion combined with hydrological and hydraulic routing models capable of simulating in-channel sediment transport, although all fall short of simulating resultant river channel adjustment. Development of these models is a long-term project calling upon teams of researchers whose process sub-models will be at different stages of development or revision. As Anderson and Sambles (1988) foresaw, one constraint in disseminating geomorphological modelling activity is the time taken to understand detailed, spatially extensive models. Therefore, we simply outline the facilities and performance offered by each of three examples: there are no simple 'methods in fluvial geomorphology' at this level.

Each model calculates and represents catchment processes in different ways. The Water Erosion Prediction Project (WEPP) watershed model was developed by the United States Department of Agriculture as a process-based successor to Wischmeier and Smith's (1978) empirically based universal soil loss equation (Ascough *et al.* 1997). Intended for application in agricultural field management, it has been developed extensively from its first conceptions (Foster and Lane 1987, Lane and Nearing 1989) to the official release of a comprehensive software package in 1995 ('WEPP95', Flanagan *et al.* 1995). WEPP95 (Figure 8.3) is based on "fundamentals of stochastic weather generation, infiltration theory, hydrology, soil physics, plant science, hydraulics and erosion mechanics" (Ascough *et al.* 1997, p. 921) calculated initially on a field-by-field basis. The model combines process models and physically based empirical relationships to simulate the following components: climate, winter weather processes, sprinkler and furrow irrigation systems, water balance, plant growth, residue decomposition and tracking, soil parameters and their effect on hydrology and erosion,

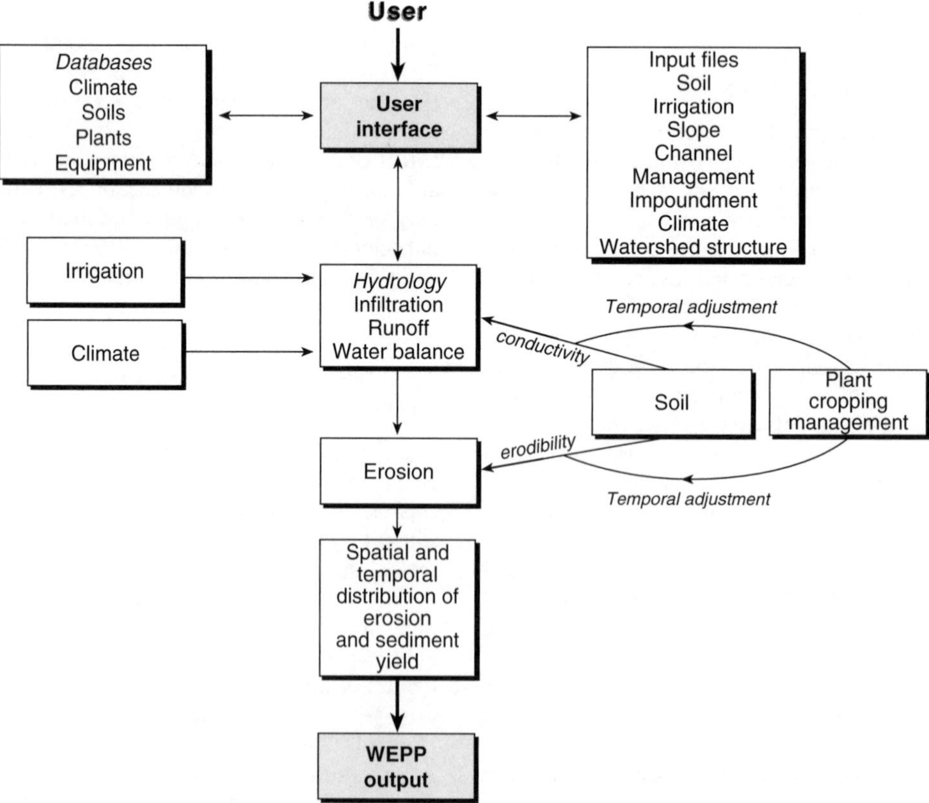

Figure 8.3 Input parameters and flow chart of major operations modelled within the WEPP95 hillslope erosion model (redrawn from Ascough *et al.* 1997)

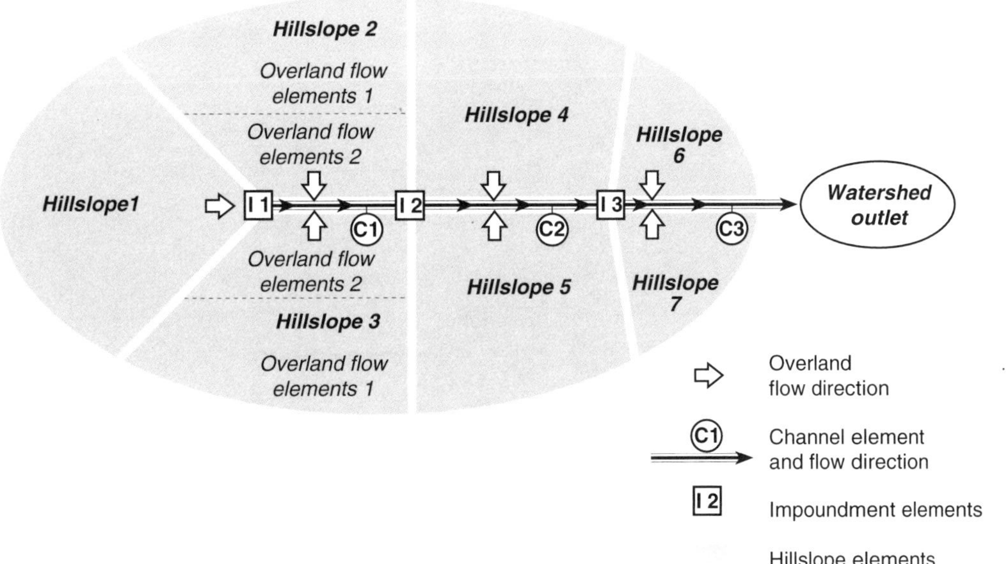

Figure 8.4 Schematic representation of the WEPP95 model over a small watershed to include the integrated components of multiple hillslopes, channel processes and impoundment structures (redrawn from Ascough *et al.* 1997)

channel hydrology and water balance, channel erosion and impoundment trapping of incoming sediment. Routing through the catchment is achieved by assigning a topological format to each slope relative to 'channel elements' and 'impoundment elements' and invoking hydraulic process models (Figure 8.4).

The MEDALUS model developed from the Mediterranean Desertification and Land Use programme ('MEDALUS', Brandt and Thornes 1996). A process model of hillslope erosion (Kirkby 1993a) is scaled-up to the catchment scale using the MEDRUSH model (Kirkby 1993b). The MEDALUS hillslope component operates at a 'field-scale' initially, calculating erosion according to fundamental principles associated with four sub-system components (atmosphere, vegetation, surface and soil) at three points down the slope (Figure 8.5). Thornes *et al.* (1996, p. 138) argue that the novelty of the hillslope model includes its accommodation of macro-scale roughness elements (rills and gullies), its selective patterning of infiltration according to topographic, vegetation and soil characteristics and its subsurface lateral re-distribution of soil moisture. Kirkby *et al.* (1998) note three additional novel aspects including the ability to simulate long-term model behaviour (a critical concern for geomorphological modelling; Brooks and Anderson 1998) and the ability to accommodate the micro-topography of surface roughness. The third aspect relates to flow routing via 'MEDRUSH', which subdivides a catchment into representative 'flow strips' with sizes that vary according to their likely importance for sediment generation (Figure 8.6). Flow and sediment are routed through sub-basins to the catchment outlet via the channel network using linear transfer functions (Abrahart *et al.* 1996).

The third model is SHETRAN (Bathurst *et al.* 1996), a derivative of the Système Hydrologique Européen distributed hydrological model ('SHE', Abbott *et al.* 1986a,b) wherein algorithms for the generation and transport of sediment are laid over SHE's hydrological capabilities. SHETRAN is a cell-based distributed model in which surface/subsurface relations, including those of soil erosion (based on a sub-model, 'SHESED', Wicks and Bathurst 1996), are calculated as finite differences according to the cell routing network (Figure 8.7). Each physical variable is represented by one parameter in each cell and the processes are "modelled either by finite difference representations of the partial differential equations of mass and energy conservation or by empirical equations derived from independent experimental research" (Bathurst *et al.* 1996, p. 356). There are parameters for the hydrological functions of the soil, the vegetation, overland flow, the ease with which soil can

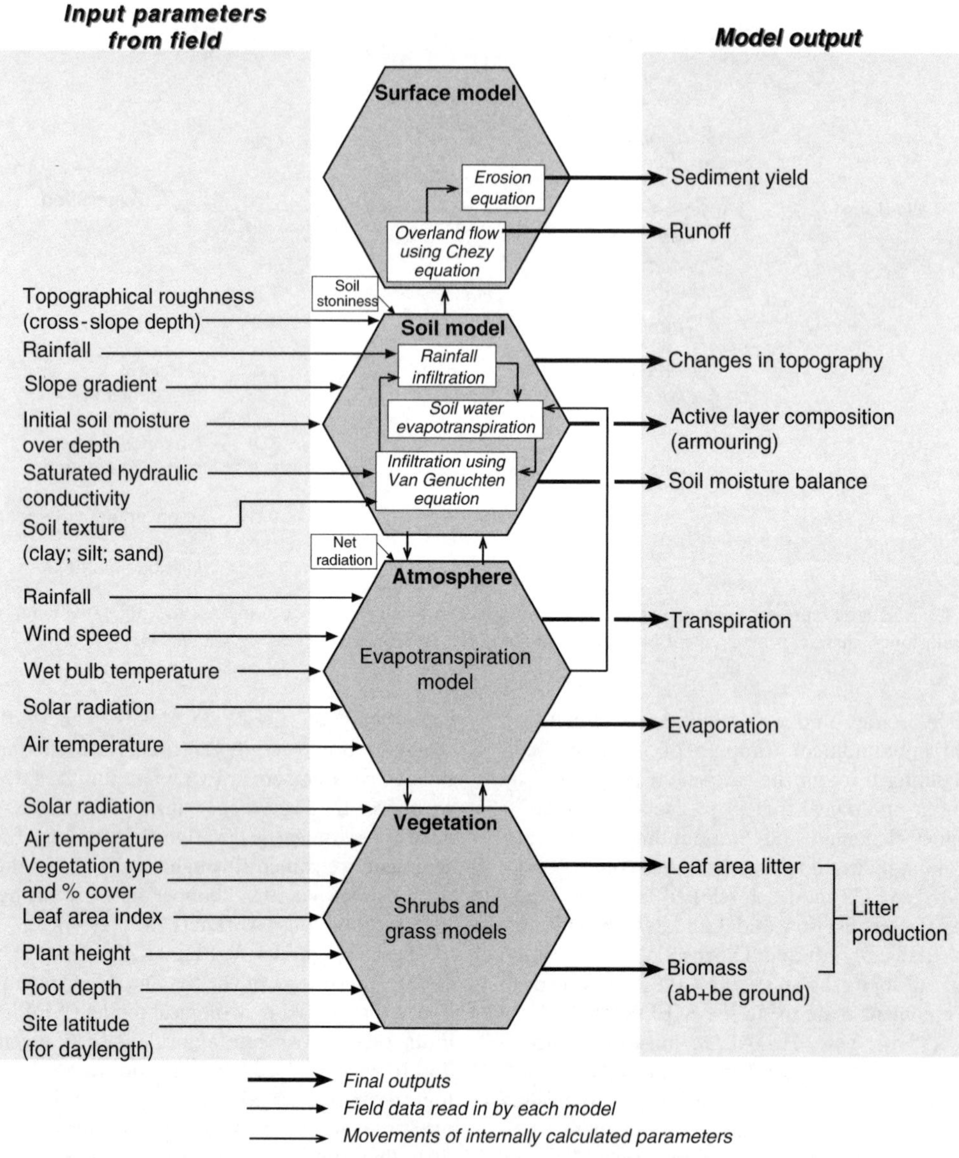

Figure 8.5 Input parameters and flow chart of major operations within the MEDALUS hillslope erosion model (redrawn from Thornes *et al.* 1996)

be eroded, topography and channel characteristics (Bathurst *et al.* 1996). River channel representation is also cell-based. SHETRAN is intended primarily for environmental management applications including issues concerning surface water and groundwater resources management, pollutant movement and land erosion.

Validation

A fundamental requirement of each model is an ability to model accurately the processes taking place, both over periods of individual events and for extended timescales. As Thornes *et al.* (1996, p. 137) state "...models are only as good as their capacity to

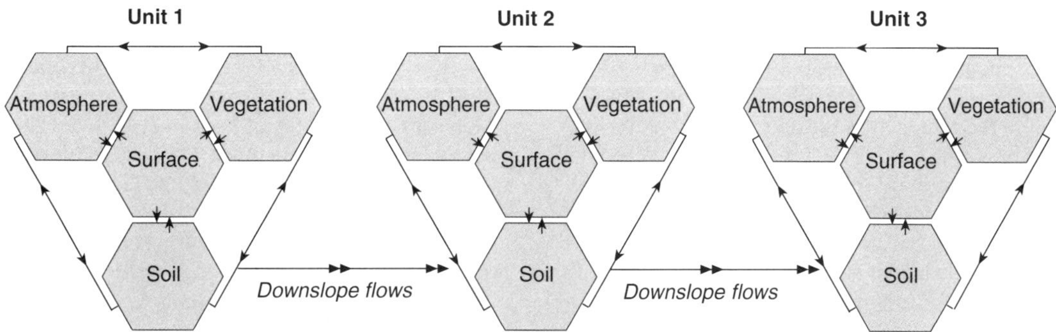

Figure 8.6 Schematic representation of sub-model coupling within a hillslope catena in MEDALUS (redrawn from Kirkby *et al.* 1998) (reproduced by permission of John Wiley and Sons, Ltd.). Unit numbers refer to sequential downslope components

Evapotranspiration loss model

Rain and snow output

Canopy interception model

Snowmelt model

Root zone model

Overland and channel flow model

Saturated flow model

One-dimensional unsaturated zone model

Stream/aquifer interactions

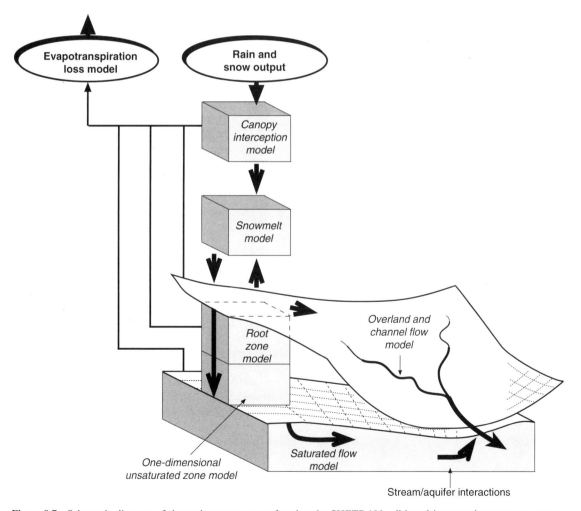

Figure 8.7 Schematic diagram of the major components forming the SHETRAN cell-based integrated component process model (redrawn from Dunn *et al.* 1996)

replicate, to an acceptable level, the magnitude, pattern in space and time and character of real world processes". In geomorphological applications where processes deriving from the topography are routed back to measure their impact on future topography, this may encompass periods exceeding 100 years. The ICP models have been mostly tested using a 'split sample' basis (Brooks and Anderson 1998), where the model is calibrated using one data set and the success is judged against a single model output using separate test data. For instance, in tests against plot data from South-eastern Spain and with minor calibration of empirical coefficients, the MEDALUS hillslope component predicted overland flow from single events reasonably well (explained variance, $r^2 = 0.64$) if large rainfall events on bare soils were excluded from analysis (Thornes *et al.* 1996). Likewise, with large events excluded, the model predicts sediment yield quite well ($r^2 = 0.83$).

However, the validation of geomorphological models is in its infancy (Kirkby 1996). One problem when validating catchment models is the limited availability of high quality, catchment-scale sediment data over extended time periods. For instance, in Bathurst *et al.*'s (1996) test of SHETRAN at their Spanish test catchment, sediment production was compared to a long-term average provided by the bathymetry of reservoir sediment deposits rather than directly to sediment yield data. Also, in testing the hillslope component of MEDALUS, Thornes *et al.* (1996) relied on a *simulated* 15-year period consisting of eight 'wet' and seven 'dry' years with the results compared to sediment yields noted in the published literature (Thornes *et al.* 1996).

As discussed in Chapter 17, another problem in validating catchment process models is *equifinality*, where the same model output can result from many combinations of internal processes (see Beven 1996). Therefore, the right prediction may be obtained without the model's internal mechanisms accurately representing the catchment processes. This problem is compounded when models are tested only against one or two output parameters, such as discharge and sediment yield. This has led to calls for new modes of validation (e.g., Brooks and Anderson 1998) such as multiple response validation, where the model is optimised against multiple outputs (Mroczkowski *et al.* 1997). Another approach is internal validation (Fawcett *et al.* 1995) where assessment of the model component parts is seen to be as important as predicting the output. This latter technique is a logical approach for catchment process models given their internal complexity and numerous sub-models.

These problems may be avoided where sufficient test data exist. For example, sediment yield predictions from WEPP95 have been tested against real data from 15 small (0.34–5.14 ha) experimental agricultural watersheds in the USA (Liu *et al.* 1997). Sensitivity tests for an earlier version of WEPP's hillslope and watershed models are reported in Tiscareno-Lopez *et al.* (1993, 1994). The watersheds are under different agricultural practices, and simulations were completed for individual events and for longer-term performance against combined individual events from each catchment (events at each station ranged in number from 3 to 241). For the combined events, the ability of the model to simulate runoff and sediment yield is good (runoff $r^2 = 0.86$, sediment yield $r^2 = 0.91$). However, for individual watersheds, the results from individual events are far more variable with explained variance for runoff varying between 0.01 and 0.85, and for sediment yield, from 0.02 to 0.90. Clearly, this suggests that the model's internal components simulate some agricultural practices better than others. Additionally, it illustrates the difficulty that catchment models face in representing antecedent conditions prior to individual events. Experimental plot tests of SHETRAN in Portugal showed similar results: one event, producing extreme sediment production, was especially difficult to model even though its runoff parameters were similar to other events (Bathurst *et al.* 1996).

Scale Issues

Representing catchments of different sizes involves more than simply scaling-up the fundamental units over which the model is measured because, as size changes, so too does the dominant group of processes relevant to the model (Kirkby *et al.* 1998). Emphasis may shift from concern with slope micro-topography and individual flow hydrographs at the hillslope scale, to topography, soil and vegetation patterns over longer periods at the catchment scale, and to lithology and climate extending back over thousands of year at the regional and global scale (Kirkby *et al.* 1998). Therefore, as catchment size increases, so its component parts should alter. In SHETRAN, for instance, the user can increase cell size according to the catchment size. However, as process-based equations are used to calculate one value for each parameter in each cell, this capability operates effectively only when the cell size is small relative to variations in the local hydrological controls and responses (Bathurst *et al.* 1996). Once the grid cells are larger than the distance over which parameters show significant variation, then model accuracy is likely to decrease rapidly as output errors in one cell generate input errors in the next. To tackle this concern, spatially integrated 'effective' parameter calculations have been developed and applied with reasonable success the 701 km^2 Cobres catchment in Portugal using a 2 km × 2 km grid cell (Bathurst *et al.* 1996). On this basis, an upper limit catchment size of 2500 km^2 is advised (http://www.ncl.ac.uk/wrgi/wrsrl/rbms/rbm.html SHETRAN, accessed 4th July 2001).

Alternatively, the 'flow strip' approach being developed for MEDRUSH may provide a solution by sizing cells and sub-catchments according to their importance to sediment generation. Kirkby *et al.* (1998) propose that, for a basin of 5000 km^2, cells in the headwater areas (providing a relatively large amount of eroded sediment) should be modelled at 1–5 km^2 resolution whereas in the downstream reaches, cells of 10–50 km^2 may be feasible. In addition, flow strips are grouped according to their slope: catchment area relation (Figure 8.6). Processes calculated using one representative slope from each slope/area group are applied to the similar strips, reducing data requirements and computing complexity. This approach is supported by tests to find representative hillslope components (discretization) in the WEPP95 model. Using invariant process-based equations, sensitivity testing was undertaken for progressively longer hillslopes by Baffaut *et al.* (1997). They conclude that hillslope length is a major constraint on WEPP95 accuracy and that the model becomes less reliable over longer hillslopes as a consequence of changes in the dominant controls on processes. They recommend that a maximum hillslope length of 100 m be used, thus restricting the permissible size of an individual catchment to approximately 0.4 km^2.

Simulation of Channel Adjustments

For catchment-process models in fluvial geomorphology, one potential endpoint regarding landscape change is for the model to be capable of simulating river channel adjustments, requiring both highly accurate predictions of terrestrial sediment production and an equally accurate in-channel sediment erosion, transport and deposition modelling capability. The channel components of MEDRUSH are under development and will incorporate process mechanisms that are scaled up from the MEDALUS hillslope model (Kirkby *et al.* 1998). SHETRAN facilitates channel sediment routing and bed armouring processes (Wicks and Bathurst 1996) but as the channel is represented topologically rather than topographically, and because bank erosion is not modelled, full channel perimeter interactions are not possible. Testing of changes in the channel bed elevation in comparison to measured data from the 1979 snowmelt season on the East Fork River, Wyoming revealed new bed erosion to be simulated realistically, but the spatial locations of erosion and deposition were not well predicted (Wicks and Bathurst 1996). Modelling of river channel dynamics has been attempted in WEPP95 to obtain accurate sediment yield predictions at the watershed mouth. The model incorporates channel morphology components in addition to hydraulic sediment routing capabilities and calculates sediment detachment and deposition by solving for sediment continuity. However, overall, only a few of the process interactions taking place within real river channels are incorporated. Ascough *et al.* (1997, p. 922) suggest that the model is suited to application in constructed waterways and concentrated flow in gullies, but not to perennial stream channels, locations with dynamic contributing areas, channels with mass bank failures or headcut erosion processes or where erosion is generated by seepage effects. Therefore, despite real progress in ICP modelling capability, where river channel dynamics are an issue, an alternative approach is required for now.

8.6 WATERSHED ANALYSIS

ICP models are physically *based* and strive for a full process-mechanical basis to offer novel system insights, but the complexity and scale diversity of their component sub-systems means that each requires, to some degree, the use of empirical coefficients and/or site-specific calibration in order to function. Another class of models has attempted to use a combination of physical and empirical elements more explicitly to over-ride the problems of accommodating the wide variety of data sources needed for catchment modelling. These *watershed analysis models* (Montgomery *et al.* 1995) have been developed in response to 'cause and effect' environmental management concerns and offer, in potential at least, the ability to incorporate historical factors. The basic goal of watershed analysis is "... to generate a spatially explicit understanding of a landscape and its ecosystems at a resolution sufficient to allow assessment of the integrated environmental consequences of inherently local land use practices" (Montgomery *et al.* 1998, p. 244). Watershed analysis models differ from the ICP models primarily on the basis of their greater concern with predicting and forecasting being problem-led rather than research-led. While watershed analysis may appear to be less based on physical processes than the ICP models, in reality, distinguishing between geomorphological models on this basis can be rather misleading, for as Kirkby (1996, p. 263) states, "In geomorphology, few models rise far above empiricism, and most 'physically based' models are simply pushing the level of empiricism one level further down".

Montgomery *et al.* (1995) argue that the fundamental basis of watershed analysis is fourfold (Figure 8.8). First, an initial stratification of the watershed (i.e., catchment) is made to sub-divide 'landscape-level' functions and processes according to dominant processes. Next, archive data are collected in order to reconstruct historical conditions prior to, thirdly, collecting data and observations related to contemporary conditions, a phase that will include field analysis. Finally, following the application of process-based models of the dominant processes, in their historical context, the landscape is re-classified according to the management questions. Each watershed analysis will vary in detail according to the questions asked of the model, with the level of analysis being determined by the level of confidence required to determine acceptable risk in the outcome (Montgomery *et al.* 1995). Watershed analyses are, therefore, highly suited as a form of exploratory analysis that prompts catchment-scale questions for more detailed analysis, rather than a device for deriving fundamental process insight.

Because watershed analysis is critically dependent on the initial stratification of catchment-level processes, it is applicable to catchments with well-defined breaks in geomorphological processes and dominant land uses, such as the steep, forested, watersheds of the Pacific North-west of the US. For instance, Montgomery *et al.* (1998) illustrate a watershed analysis centred on identifying landslide-derived sediment generation predicted from a physically based model of potential slope instability incorporating topographic controls, soil type, hydrology and rainfall data developed from digital terrain data to offset the incomplete record obtained from aerial photographs and maps. Stream channels were also derived from terrain data using a process-derived indication of headwater extent to define a channel network that is classified according to slope threshold criteria.

Conversely, Downs and Priestnall's (1999) watershed analysis is set in an English lowland catchment

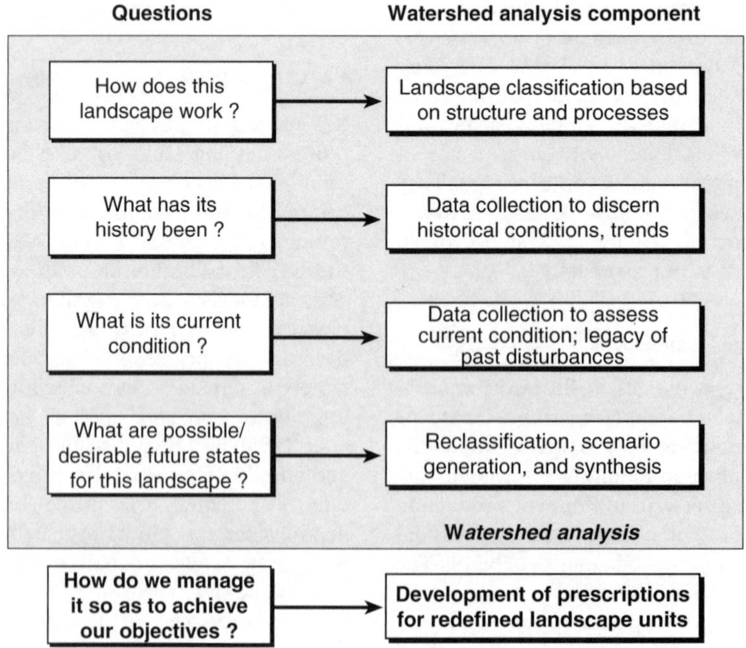

Figure 8.8 Overview of the watershed analysis procedure illustrating major questions asked of the modeller at each stage (redrawn from Montgomery *et al.* 1995)

with a much greater heterogeneity of land use types and histories, and without direct coupling of hillslope and channel processes. In this analysis, identifying the dominant process regimes (the initial stratification) becomes a central question. As such, Downs and Priestnall (1999) develop a model structure that focuses directly upon the river channel, asking questions about its ability to adjust its form according to in-channel, riparian and catchment conditions. As expected, the discriminator of stream power (as energy available to do work) is, per se, insufficient to distinguish types of channel adjustments observed using field survey. In this setting, process dominance within the channel cannot be stratified without knowledge of human activities in the catchment and model development requires that human impacts are accommodated according to their process implications.

Watershed analyses are developing alongside catchment-based concerns for environmental management, and are often closely associated with cumulative watershed effects (Reid 1993) and estimation of catchment sediment budgets (e.g., Reid and Dunne 1996). Although, scientifically, they may offer less long-term insight than do the ICP models, they are more suited for development by small teams of researchers to address specific issues. Overall, because landscape-level stratification is not constrained by a pre-determined scale, watershed analyses will need to be extremely flexible in order to be updated and revised as new material becomes available (Montgomery *et al.* 1995). They are, therefore, likely to utilise digital data as model inputs and to use GIS as the analytical platform from which to develop flexible models. The input data types and analytical tools developed for two watershed analyses are detailed later in this chapter.

8.7 INTERPRETATIVE MODELS

An additional, practical, approach to catchment modelling is provided by *interpretative models* wherein empirical observations of catchment processes are integrated using expert analysis to overcome the spatial and temporal scale problems common to other forms of catchment modelling. Model construction can be achieved rapidly and, with explicit guidelines for field interpretation, replicable results can be achieved (Gregory *et al.* 1992, Downs and Thorne 1996, Thorne 1998). Such models are being developed world-wide, in response to practical concerns in river management (Environment Agency 1998). One example is provided by the 'Fluvial Audit' (Sear *et al.* 1995). This model (Figure 8.9) relies upon a conceptual understanding of the processes linking sediment generation, transport and deposition across a catchment. Field and archive data are gathered and processed to define catchment and reach geomorphological maps and a time-chart of impacts with potential implication for the 'problem' reach. From these outputs, the most likely potentially destabilising phenomena (PDP) are defined (the factors that are the most likely causes of the management problem, Sear *et al.* 1995). The PDP are the issues that have to be tackled or accommodated in developing sustainable river channel management solutions.

Interpretative models are not fully quantitative and stem from the geomorphological tradition of using morphological data to establish process existence by association. However, they are a viable and practicable model technology which over-ride problems of scale and data accuracy, allowing the integration of historical data and facilitating conclusions related directly to the observed river channel geomorphology. The model also provides considerable 'richness' (Kirkby 1996) in terms of the output understanding achieved relative to the data input requirements. While this is not to denigrate the importance of developing fully parameterised catchment-process models in fluvial geomorphology, interpretative models provide a worthy reminder of our current ability to understand highly complex catchment systems.

8.8 EXPLORATORY CATCHMENT MODELLING USING GIS

The factors identified earlier in this chapter as forming the main drivers for contemporary fluvial geomorphological processes at the catchment scale include topography, features spatially distributed over the surface and subsurface, and the channel network. Catchment-scale modelling of fluvial processes therefore necessitates the integration of disparate data sources and the flexibility to combine these data 'layers' according to various sub-models. This general requirement can be met using a GIS as the base on which data sets are overlaid on a common co-ordinate system, allowing spatial interrelationships to be explored (Goodchild 1993). Many of the generic features of GIS that have been considered relevant to hydrological modelling (Kovar and Nachtnebel 1993, 1996) are likely to be of direct relevance to catchment-scale modelling in fluvial geomorphology. These may include the provision of:

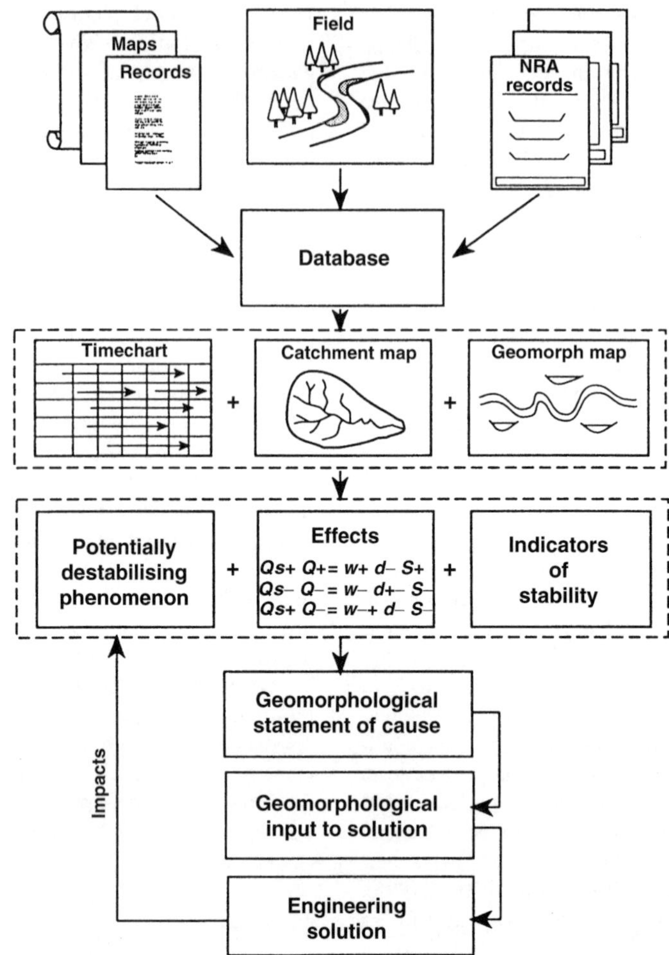

Figure 8.9 Procedure for undertaking a fluvial audit, using expert understanding to guide the interpretation of effects and the identification of potentially destabilising phenomenon (PDP). From Sear *et al.* (1995), reproduced with permission of John Wiley and Sons

- complex map overlays and spatial analysis to derive input data for models;
- a linkage mechanism between models of different spatial resolutions;
- digital landforms of different projections and scales to a standardised format (georeferencing);
- post-simulation graphics for evaluating results (Ross and Tara 1993).

Burrough and McDonnell (1998) offer an introduction to generic GIS tools for the input, verification, storage, analysis and output of geographical information. These functions can be considered to represent the standard functions of a GIS as obtained from the supplier. These tools allow basic manipulation and exploration of catchment-scale data sets and provide a central store for data relating to catchment modelling. In addition to the standard specification many systems contain specific functions designed to assist the study of fluvial systems. For instance, basic functions can be combined using macro-languages such as Arc View 'Avenue'. Complete customisation and control over the interactions between GIS and modelling procedures is possible using a programming language, but at the expense of system complexity (Downs and Priestnall 1999). The following sections offer an introduction to GIS functions that may be useful for the study of fluvial systems at a catchment scale. Table 8.1 describes the data sources and GIS-related components of two watershed analysis models as specific

Table 8.1 Data input sources and GIS-related components of the watershed analysis models of Montgomery *et al.* (1995, 1998) and Downs and Priestnall (1999) to illustrate their specific functions

Data	Watershed analysis model			
Application area	Piedmont/lowland England (Downs and Priestnall 1999)		Pacific North-west USA (Montgomery *et al.* 1998)	
	Method	Comments	Method	Comments
Inauguration of model	Late 1993	Model pre-dates availability of many digital data sources outlined in this chapter	1992	Progressive development of process-state models
Input data representation (see Section 8.2)				
Catchment surface representation	Digitised from 1:25 000 maps (5 m contour interval). Converted to raster DEM	Heighted vector river network used to constrain triangulation	Arbitrary	Applications have used 2–30 m grid DEMs
Parameterisation of land cover/land use	Digitised paper maps and digital sources; land cover classes derived from remote sensing (Fuller *et al.* 1994) and land use change 1985–1990 maps (Department of the Environment 1991)	Sources guided by discharge calculation requirements and objectives of research	Land use effects parameterised through influence on soil and vegetation properties	Requires translation of land cover classes into effect on soil properties
Extraction of channel planform and network	Digitised from 1:25 000 river network into vectors. Supplemented where possible from large-scale mapping where known planform changes had occurred	Planform changes updated with reference to aerial photographs. No checks into headwaters	Either determined from drainage area and slope (Montgomery and Foufoula-Georgiou 1993) or digitised channel vectors	Flexible methodology tailored to data source
Exploratory modelling using GIS (see Section 8.8)				
Core functions				
Capture and storage	Stored within laserscan automated map production system, LAMPS (Laser-Scan Laboratories Ltd. 1992) and later converted to ArcView 3	Guided by resource availability. Mapping system supplemented by 'C' programs	Data integrated into ArcInfo	Guided by data and resource availability
Organisation and retrieval	Customised menu presented user with choice of data layers and guided user through sequence of operations	Macros written to simplify choice of available raster and vector layers using mapping system as the 'graphical system'	ArcInfo and customised menus	
Data modification	Powerful vector editing and topological structuring functions available within LAMPS	These functions facilitated the creation of the 3-D structured river network	ArcInfo	

(*Continues*)

Table 8.1 (*Continued*)

Data	Watershed analysis model			
Application area	Piedmont/lowland England (Downs and Priestnall 1999)		Pacific North-west USA (Montgomery *et al.* 1998)	
	Method	Comments	Method	Comments
Exploratory data analysis				
Derived parameter estimation	Custom spatial units for local area, river corridor and catchment area upstream calculated using 'C' programs working on gridded DEM	Error checking system to eliminate error in upstream area calculation in areas of low relief (Priestnall and Downs 1996)	Custom programs to calculate relative slope stability, reference grain size and channel each slope (Montgomery *et al.* 1998)	Programs combine soil properties with terrain attributes to map relative patterns of process of interest
Terrain indices and associated functions	ArcInfo used for sink-filling and flow direction calculations	Flow direction grid re-imported into LAMPS	Custom macros used for DEM pre-processing, drainage area and slope determination	
Visualisation	Custom function written to enable any combination of surface/network data overlay (see Figure 5, Downs and Priestnall 1999)	ArcView 3 used in later stages to visualise results of observed versus predicted stream power	Customised to application	
Model implementation				
Coupling of models and GIS	FORTRAN export files written for export to discharge calculation software and for exploratory analysis achieved in Excel/SPSS	Very loose coupling. Macros controlled 'C' programs which in turn used text files to pass information around system	Models coded into custom macros, others function as 'C' programs	
Custom functions	Whole system design based around custom macros		Whole system design based around custom macros	

examples illustrating the functions outlined below. Even from two examples it becomes apparent that a standard system design for catchment modelling is unlikely to exist ever.

Core Functions

Capture and storage. An important aspect of GIS for catchment modelling is the ease with which data sets (as described in earlier sections) can be imported and integrated onto a single platform. Most GIS can easily map point data sampled directly in the field and, increasingly, systems are facilitating the integration of data captured using global positioning systems (GPS), data loggers and other surveying tools. Data should ideally be obtained in a format that is easily imported into the chosen GIS, but often the data have already been surveyed. This information, such as river soundings, for example, may have already been imported or digitised into computer aided design (CAD) systems, but can be routinely imported into most GISs via transfer formats such as DXF or directly from a CAD design file.

Organisation and retrieval. GIS should allow data sets to be organised for easy retrieval and display of their information either individually or in combination with other data sets. Multiple data set overlays are facilitated because GIS data are geo-referenced to 'real world' spatial co-ordinates. This ability to organise, retrieve and display geo-referenced information covering a whole catchment area should accelerate the process of data exploration and analysis. GIS should also facilitate the management of attribute information associated with features or whole data sets. For example, if point features representing land use change are mapped and their associated attribute information stored, any query on this data set (such as 'show me all incidences of rural to urban land use change') should highlight the locations satisfying this query. If desired, a new layer may be derived and stored from this result.

Data modification. The ability to edit data is important for keeping data sets up-to-date. Most GIS include tools for editing data in both raster and vector forms, including the ability to change or supplement both the spatial information and the attribute data associated with the graphical features. Certain analytical operations in GIS may require data to be structured in such a way that the spatial relationship between features is stored, in addition to data itself. This *topological structuring* can, for example, allow the formation of structured river networks with features that direct routing operations through that network.

Exploratory Data Analysis

Derived parameter estimation. Tools to manipulate and analyse spatial information are of particular importance when exploring catchment influences on river channels. This may include the use of distance operators such as *buffering* to derive zones of influence around river channel networks, for example. The *overlay* operation is fundamental to GIS and allows parameter estimation over defined spatial units, for example, summarising the extent of a single land cover type over a catchment extent. The overlay procedure illustrated in Figure 8.1 could involve summarising geology or land cover information according to sub-basins derived from the DEM, or according to buffer zones around the vector river channel network. Extending these basic exploratory functions to *matrix algebra* can produce various derived parameters: raster-based representations from numerous data layers are overlaid according to mathematical criteria to produce a new map of previously unknown information. For example, an index representing susceptibility to erosion could be estimated using an index of gradient and surface material characteristics. Such examples are covered in the core GIS text by Burrough and McDonnell (1998) and a review of cell-based modelling is presented by Gao *et al.* (1996).

Terrain indices and associated functions. Terrain data are also central to the exploratory stages of catchment modelling, allowing the production of relief maps and three-dimensional views to give valuable clues to the nature of the catchment and the parameters distributed over that catchment. However, in addition to general arithmetic and statistical functions, terrain data may now be analysed using several topographic and hydrological functions commonly available in 'generic' GISs such as Idrisi, GRASS, ArcInfo or ArcView. Functions for deriving topographic or hydrological parameters, many of which may be of use in catchment-process modelling, are summarised in Table 8.2. Several systems, targeted at the hydrological modeller or catchment manager, are available which offer a subset of the generic GIS functions alongside a wide range of hydrological functions. Examples of these systems include ILWIS (http://www.itc.nl/ilwis/, accessed 5th July 2001) and RIVERTOOLS (via http://www.rsinc.com/rivertools/, accessed 5th July 2001) and the more general modelling system PCRaster (http://www.geog.uu.nl/pcraster/, accessed 5th July 2001).

Visualisation. A feature of most GISs is the ability to shuffle data layers to display the desired layer on the top in a two- or three-dimensional visualisation. Also, surveyed points or lines can be displayed within the context of other parameters distributed across the catchment (including those derived from the terrain). This perspective may allow the recognition of unforeseen patterns or relationships that warrant further, more rigorous, exploration. Similarly, the selective display of features according to their attributes can be of value in exploratory data analysis. For example, displaying all sample points possessing an infiltration rate above a certain threshold value may provide vital clues for establishing the initial process-based subdivision of 'landscape-level' functions and processes in watershed analysis.

Exploratory data analysis may also be assisted using simulated three-dimensional (3-D) visualisation capabilities now common in desktop GISs to allow the terrain surface (or any chosen surface) to be displayed in perspective from any given point. Other features or imagery can be draped over surfaces, including the results of analytical queries based upon feature attributes that may not normally be visible, for example, error or uncertainty. Figure 8.10 shows a perspective view of a digital surface model produced by an airborne LiDAR sensor draped with a true colour composite image from a CASI sensor

Table 8.2 Summary of parameters that can be derived from digital terrain models (DTMs)

Operation	Description
Relief (including 3-D visualisation)	Digital representations of the terrain surface in a variety of display formats to explore the nature of the catchment. Colouring by height and hillshading may be particularly effective in highlighting subtle undulations in the terrain (hillshading especially). Three-dimensional views from any angle or altitude can be produced, either colouring the terrain by height or draping in image over the surface. Image drapes can include aerial photographs, remotely sensed images or gridded representations of any of the data sets for the catchment, including the derived parameters described below
Slope	A variety of algorithms allow the calculation of maximum slope at any cell in the terrain model. Slope may be an important constituent of indices representing geomorphological dynamics
Sinks/depressions	Depressions or sinks in the terrain model often result from errors in input survey data or through the triangulation process. These can be automatically identified and filled to prevent them being treated as areas of internal drainage
Flow direction	To model water flow through the catchment the direction of flow from each cell can be calculated resulting in a 'flow matrix'. In many flow matrices, several flow directions are possible from each cell so unique directional 'values' are required to result in a realistic flow matrix
Catchment area upstream	From the flow matrix, cells contributing flow to any given 'outlet' cell (representing a point on a river, for example) can be automatically calculated (one approach is detailed in Priestnall and Downs 1996). The resulting cells can form an 'area of interest' when overlaying other data sets to explore catchment-scale influences (Downs and Priestnall 1999). With programming or customisation, other spatial units of study such as flow strips on valley sides can be defined from the flow matrix
Flow accumulation	For each cell the number of other cells which contribute flow to that point can be calculated and allocated to a new grid. One of the main uses of this function is to define a channel network along cells of high flow accumulation and thence to use this network in routing operations. The channel network is defined according to cells exceeding a threshold value of flow accumulation. Conversely, cells having very low flow accumulation values can be taken to represent ridges or watersheds
Aspect	Aspect, derived from the direction of steepest slope at each cell can be used, for example, to estimate the effectiveness of sunlight in increasing the rate of evaporation or snowmelt
Curvature	The terrain curvature in both plan and profile about a cell can be calculated and can contribute to calculations of hillslope hydrology parameters such as flow convergence

on board the aircraft. The sophistication of most visualisation systems is such that the impression of distance and perspective is sufficiently 'believable' to allow catchment relief features to be recognisable and offers a more intuitive view of the data sets than the two-dimensional equivalent. The size, extent, and juxtaposition of features of potential geomorphic interest such as river embankments or sub-basin units can be seen clearly both with a 'realistic' drape as in Figure 8.10, or with any parameter thought to be influential at the catchment scale. The example shown was created using 'Vistapro 4' (approximate cost $80) and demonstrates that effective visualisation can be a relatively inexpensive tool.

Model Implementation

Coupling of models and GIS. Beyond exploratory analyses, the degree to which environmental models can be created directly within GIS varies considerably (Goodchild *et al.* 1996), usually according to an inverse relation with model complexity. Whereas the generic GIS functions described earlier can support the development of simple models within the GIS itself, for models of greater complexity it can be desirable to maintain a loose coupling between the modelling software and the GIS. Continued improvements in programming interfaces described by Kopp (1996) are allowing better integration of models and GIS. However, as yet, there are fewer examples of GIS coupled

Figure 8.10 LiDAR digital surface model draped with imagery from CASI airborne sensor: data illustrate the river Arun, England. Courtesy of the Centre for Environmental Data and Surveillance, The Environment Agency, Bath

with catchment-scale fluvial process models than in other areas of environmental modelling, such as the study of atmospheric systems. Eventually, the innovative use of generic GIS tools may offer experimental catchment-scale analysis and simple modelling without the need for highly complex system designs (see Downs and Priestnall 1999). From specific innovations in the field of geocomputation, including models utilising neural computing techniques for hydrological applications, more generic 'toolkits' should develop, which may offer much to the development of catchment-scale models (Fischer and Abrahart 2000).

Custom functions. Many GIS offer the ability for custom functions to be written using the macro-language associated with that system. However, it is becoming increasingly common that the flexibility of GISs is being improved through custom functions written in a conventional programming language. For instance, Batelaan *et al.* (1996) present a set of custom functions for the GRASS GIS written in the C++ programming language. These 'off-the-shelf' custom functions can be of great value in preparing and manipulating distributed parameters at a range of spatial scales and over any number of different spatial units. One possible disadvantage of the use of custom functions is that a system may become more complex and difficult to maintain than one based upon a set of generic GIS operations. However, as GIS developers increase their support of conventional languages such as Visual Basic, libraries of code should be developed which, if well documented, will encourage easier sharing and further development of more specific functions than can be offered by a generic GIS.

8.9 CONCLUSIONS AND PROSPECTS

This chapter has highlighted the fact that progress in catchment modelling of fluvial geomorphological processes is determined not only by contemporary process understanding and scale-related issues regarding the significant contributory factors, but by a wide variety of issues related to terrain data acquisition and representation. In this regard, GISs are offering ever-improving integral analytical routines suitable for catchment-process analyses, greater capacity for developing customised functions using a conventional programming language, more flexible storage facilities and enhanced visualisation capabilities for graphically exploring base data relationships. While Brooks and Anderson (1998) rightly warn that a search for ever improving model resolution can be a 'dangerous distraction' from improving the process basis of such models, a basic concern for input data quality to catchment models should figure highly in the geomorphological modeller's mind, particularly given the potential impact of terrain on geomorphological process (Lane *et al.* 1998a). Decisions are required over the sampling strategy and the method for estimating the model input parameters. Large areas such as catchments will require data interpolation between sample points so that the spatial unit of study is critical. Furthermore, creating digital data sets from the input parameters provides researchers with a potentially wide range of analytical possibilities but introduces new data quality uncertainties related to, for instance, methods of data abstraction from raw field measurements, choices regarding storage method, types of algorithm used to derive new parameters and suitable levels of data resolution. These issues will, to a large extent, determine the errors inherent to the model building process (see Lane *et al.* 1998b). Inevitably, catchment-scale process models in fluvial geomorphology will require more accurate digital data at appropriate combinations of data resolution and geographical coverage, and improved algorithms for model construction that result in a more faithful representation of catchment surfaces.

The reduction of dynamic environmental systems such as catchment fluvial geomorphology into simplified mathematical or statistical models allows researchers to comprehend the complexity of the real world (Goodchild *et al.* 1996) but brings with it uncertainty at every level of the procedure (Burrough *et al.* 1996). Despite the numerous uncertainties involved in modelling the interaction of multiple geomorphological processes with extensive and varied catchment topography, the 1990s witnessed construction of the first process-based models of fluvial geomorphology at the catchment-scale. However, because of our imperfect understanding of fluvial geomorphological process, all models require empirical data in order to function. As a result, catchment-process modelling can be approached at numerous levels of 'sophistication', including watershed analysis and interpretative catchment models in addition to the ICP models. The ability to model individual contributory processes, such as those outlined in the introduction to this chapter, is determined by the contemporary state of understanding in the specific sub-discipline (see appropriate chapters in this volume). For catchment-scale process models in fluvial geomorphology to obtain a high 'richness' in terms of net information gain (Kirkby 1996), we suggest that they should provide both transferable process insights and reliable predictive capabilities for the modelled catchment. In addition to accurate terrain representation, this requires enhanced internal process models of geomorphological systems. Generic issues in this regard include the need to derive consistent parameter identification (Kirkby 1996), to tackle 'upscaling' problems associated with process representation (Haff 1996, Kirkby 1996, Brooks and Anderson 1998, Kirkby *et al.* 1998) and to resolve the 'inverse problem' of increasing parameter uncertainty away from the present day (Yeh 1986, Brooks and Anderson 1998). Only in this way can models that tackle long-term geomorphological change be constructed.

Advances are also required in the validation of input data and model response. Increasing reliance on remotely gathered information will place a premium on field-based verification of input data to reduce uncertainty (Carver *et al.* 1996, Watson *et al.* 1996). Ideally, tests of variations in individual parameters should include a sensitivity analysis designed to direct fieldwork towards obtaining improved data for those parameters suspected of being most error prone. Review of the ICP models showed that rarely do data of sufficient quality exist to test the model outputs. Model response validation requires high quality field data that are spatially extensive and are maintained over extended time periods. Therefore, resources must be provided for field data collection alongside those for model construction. Because there is little routine sediment monitoring world-wide, the most feasible source of data for validation may be from new technologies such as LiDAR and differential GPS surveys which allow rapid, repeat surveying of catchment terrain and/or the river channel (Higgitt and Warburton 1999). From this data, volumetric analyses of landform change are possible from sequential DEMs.

Technological advances in digital data acquisition and greater analytical capabilities within GIS seem likely to allow rapid advances in our ability to construct catchment models of fluvial geomorphology processes. Exploratory analyses, of the type exemplified by watershed analysis, are especially likely to benefit. However, in the short term, it is possible that enhanced catchment-scale analytical capability brought about by new technology will be more than offset by additional sources of data inaccuracy and generalisation. This may result in better model 'conceptualisation', but less accurate models in terms of their output. Therefore, it is an attention to error sources and the collection of suitable field data for validation purposes that will determine whether the models as constructed truly represent improvements in understanding of catchment-scale fluvial geomorphology processes.

ACKNOWLEDGEMENTS

We would like to thank Karen Kemp for helpful discussions in formulating this paper, Dave Montgomery for providing the information to complete Table 8.1, and Elaine Watts of the Cartographic Unit, School of Geography, University of Nottingham for drafting Figures 8.3–8.8. Research into the Bollin catchment model was facilitated by University of Nottingham Grant NLRG020.

REFERENCES

Abbott, M.B., Bathurst, J.C., Cunge, J.A, O'Connell, P.E. and Rasmussen, J. 1986a. An introduction to the European Hydrological System – Système Hydrologique Européen, 'SHE'. 1. History and philosophy of a physically-based, distributed modelling system. *Journal of Hydrology* 87: 45–59.

Abbott, M.B., Bathurst, J.C., Cunge, J.A, O'Connell, P.E. and Rasmussen, J. 1986b. An introduction to the European Hydrological System – Système Hydrologique Européen, 'SHE'. 2. Structure of a physically-based, distributed modelling system. *Journal of Hydrology* 87: 61–77.

Abrahart, R.J., Kirkby, M.J., McMahon, M.L., Bathurst, J.C., Ewen, J., Kilsby, C.G., White, S.M., Diamond, S., Woodward, I., Hawkes, J.C., Shao, J. and Thornes, J.B. 1996. MEDRUSH – spatial and temporal river-basin modelling at scales commensurate with global environmental change. In: Kovar, K. and Nachtnebel, H.P., eds., *Application of Geographic Information Systems in Hydrology and Water Resources Management*, International Association of Hydrological Sciences Publication 235, Wallingford: IAHS, pp. 47–54.

Anderson, M.G., ed. 1988. *Modelling Geomorphological Systems*, Chichester: John Wiley and Sons.

Anderson, M.G. and Sambles, K.M. 1988. A review of the bases of geomorphological modeling. In: Anderson, M.G., ed., *Modelling Geomorphological Systems*, Chichester: John Wiley and Sons, pp. 1–32.

Ascough, J.C., II., Baffaut, C., Nearing, M.A. and Liu, B.Y. 1997. The WEPP watershed model. I. Hydrology and erosion. *Transactions of the American Society of Agricultural Engineers* 40: 921–933.

Batelaan, O., Wang, Z. and De Smedt, F. 1996. An adaptive GIS toolbox for hydrological modelling. In: Kovar, K. and Nachtnebel, H.P., eds., *Application of Geographic Information Systems in Hydrology and Water Resources Management*, International Association of Hydrological Sciences Publication 235, Wallingford: IAHS, pp. 3–10.

Bathurst, J.C., Kilsby, C. and White, S. 1996. Modelling the impacts of climate change and land-use change on basin hydrology and soil erosion in Mediterranean Europe. In: Brandt, C.J. and Thornes, J.B., eds., *Mediterranean Desertification and Land Use*, Chichester: John Wiley and Sons, pp. 355–387.

Baffaut, C., Nearing, M.A., Ascough, J.C., II. and Liu, B.Y. 1997. The WEPP watershed model. II. Sensitivity analysis and discretization on small watersheds. *Transactions of the American Society of Agricultural Engineers* 40: 935–943.

Beaudoin, A., Le Toan, T., Gwyn, Q.H.W. 1990. SAR observations and modelling of the C-Band backscatter variability due to multi-scale geometry and soil moisture. *IEEE Transactions on Geoscience and Remote Sensing* 28: 886–895.

Beven, K.J. 1996. Equifinality and uncertainty in geomorphological modelling. In: Rhoads, B.L. and Thorn, C.E., eds., *The Scientific Nature of Geomorphology*, Chichester: John Wiley and Sons, pp. 289–313.

Brandt, C.J. and Thornes, J.B., eds. 1996. *Mediterranean Desertification and Land Use*, Chichester: John Wiley and Sons.

Brooks, S.M. and Anderson, M.G. 1998. On the status and opportunities for physical process modelling. In: Longley, P.A., Brooks, S.M., McDonnell, R. and Macmillan, B., eds., *Geocomputation: A Primer*, Chichester: John Wiley and Sons, pp. 193–230.

Burrough, P.A., van Rijn, R. and Rikken, M. 1996. Spatial data quality and error analysis issues: GIS functions and environmental modelling. In: Goodchild, M.F., Steyaert, L.T., Parks, B.O., Johnson, C., Maidment, D., Crane, M. and Glendinning, S., eds., *GIS and Environmental Modeling: Progress and Research Issues*, Fort Collins: GIS World Books, pp. 29–34.

Burrough, P.A. and Frank, A.U. 1996. *Geographic Objects with Indeterminate Boundaries*, London: Taylor & Francis.

Burrough, P.A. and McDonnell, R.A. 1998. *Principles of Geographical Information Systems*, Oxford: Oxford University Press.

Carver, S., Heywood, I., Cornelius, S. and Sear, D. 1996. Evaluating field-based GIS for environmental characterization, modeling, and decision support. In: Goodchild, M.F., Steyaert, L.T., Parks, B.O., Johnson, C., Maidment, D., Crane, M. and Glendinning, S., eds., *GIS and Environmental Modeling: Progress and Research Issues*. GIS World Books, Fort Collins, pp. 43–47.

Department of the Environment. 1991. Land use change in England. *Statistical Bulletin* 90(5) (Department of the Environment, London).

Dietrich, W.E. and Dunne, T. 1993. The channel head. In: Beven, K.J. and Kirkby, M.J., eds., *Channel Network Hydrology*, Chichester: John Wiley and Sons, pp. 175–219.

Dixon, L.F.J., Barker, R., Bray, M., Farres, P., Hooke, J., Inkpen, R., Merel, A., Payne, D. and Shelford, A. 1998. Analytical photogrammetry for geomorphological research. In: Lane, S.N., Richards, K.S. and Chandler, J.H., eds., *Landform Monitoring, Modelling and Analysis*, Chichester: John Wiley and Sons, pp. 63–94.

Downs, P.W. 1995. River channel classification for channel management purposes. In: Gurnell, A.M. and Petts, G.E., eds., *Changing River Channels*, Chichester: John Wiley and Sons, pp. 347–365.

Downs, P.W. and Priestnall, G. 1999. System design for catchment-scale approaches to studying river channel adjustments using a GIS. *International Journal of Geographical Information Systems* 13: 247–266.

Downs, P.W. and Thorne, C.R. 1996. The utility and justification of river reconnaissance surveys. *Transactions of the Institute of British Geographers* 21: 455–468.

Downward, S.R. 1995. Information from topographic survey. In: Gurnell, A.M. and Petts, G.E., eds., *Changing River Channels*, Chichester: John Wiley and Sons, pp. 303–323.

Dunn, S.M., Mackay, R., Adams, R. and Oglethorpe, D.R. 1996. The hydrological component of the NELUP decision-support system: an appraisal. *Journal of Hydrology* 177: 213–235.

Evans, I.S. 1998. What do terrain statistics really mean? In: Lane, S.N., Richards, K.S. and Chandler, J.H., eds., *Landform Monitoring, Modelling and Analysis*, Chichester: John Wiley and Sons, pp. 119–138.

Environment Agency. 1998. *River Geomorphology: A Practical Guide*, National Centre for Risk Analysis and Options Appraisal Guidance Note 18, prepared by Thorne, C.R., Downs, P.W., Newson, M.D. Clark, M.J. and Sear, D.A., London: Environment Agency.

Fawcett, K.R., Anderson, M.G., Bates, P.D., Jordan, J.-P. and Bathurst, J.C. 1995. The importance of internal validation in the assessment of physically-based distributed models. *Transactions of the Institute of British Geographers* 20: 248–265.

Fischer, M.M. and Abrahart, R.J. 2000. Neurocomputing – tools for geographers. In: Openshaw, S. and Abrahart, R.J., eds., *GeoComputation*, London: Taylor & Francis, pp. 187–217.

Flanagan, D.C., Nearing, M.A. and Laflen, J.M., eds. 1995. *USDA – Water Erosion Prediction Project: Hillslope Profile and Watershed Model Documentation*, United States Department of Agriculture-Agricultural Research Service National Soil Erosion Research Laboratory Report 10, West Lafayette: USDA.

Foster, G.R. and Lane, L.J. 1987. *User Requirements: USDA – Water Erosion Prediction Project WEPP*, United States Department of Agriculture-Agricultural Research Service National Soil Erosion Research Laboratory Report 1, West Lafayette: USDA.

Fuller, R.M., Groom, G.B. and Jones, A.R. 1994. The land cover map of Great Britain: an automated classification of Landsat Thematic Mapper data. *Photogrammetric Engineering and Remote Sensing* 60: 553–562.

Gao, P., Zhan, C. and Menon, S. 1996. An overview of cell-based modeling with GIS. In: Goodchild, M.F., Steyaert, L.T., Parks, B.O., Johnson, C., Maidment, D., Crane, M. and Glendinning, S., eds., *GIS and Environmental Modeling: Progress and Research Issues*, Fort Collins: GIS World Books, pp. 325–331.

Goodchild, M.F.1993. The state of GIS for environmental problem-solving. In: Goodchild, M.F., Parks, B. and Steyaert, L., eds., *Environmental Modelling with GIS*, Oxford: Oxford University Press, pp. 8–15.

Goodchild, M.F., Steyaert, L.T., Parks, B.O., Johnson, C., Maidment, D., Crane, M. and Glendinning, S., eds. 1996. *GIS and Environmental Modeling: Progress and Research Issues*, Fort Collins: GIS World Books.

Gregory, K.J., Davis, R.J. and Downs, P.W. 1992. Identification of river channel change due to urbanisation. *Applied Geography* 12: 299–318.

Gruen, A., and Li, H. 1995. Semi-automatic road extraction by dynamic programming. *ISPRS Journal of Photogrammetry and Remote Sensing* 50: 11–20.

Haala, N. and Vosselman, G. 1992. Recognition of road and river patterns by relational matching. *ISPRS Journal of Photogrammetry and Remote Sensing* 29: 969–975.

Haff, P.K. 1996. Limitations on predictive modelling in geomorphology. In: Rhoads, B.L. and Thorn, C.E., eds., *The Scientific Nature of Geomorphology*, Chichester: John Wiley and Sons, pp. 337–360.

Higgitt, D.L. and Warburton, J. 1999. Applications of differential GPS in upland fluvial geomorphology. *Geomorphology* 29: 121–134.

Holden, N. and Butcher, P. 1998. Catchment mapping and monitoring. In: *Profiting from Collaboration. AGI Proceedings at GIS'98, Birmingham, England, October 13–15, 1998*, London: AGI/Miller Freeman, pp. 4.10.1–4.10.3.

Howard, A.D. 1996. Modelling channel evolution and floodplain morphology. In: Anderson, M.G., Walling, D.E. and Bates, P.D., eds., *Floodplain Processes*, Chichester: John Wiley and Sons, pp. 15–62.

Hug, C. 1997. Extracting artificial surface objects from airborne laser scanner data. In: Gruen, A., Baltsavias, E.P. and Henricson, O., eds., *Automatic Extraction of Man-made Objects From Aerial and Space Images II*, Basel: Birkhauser Verlag, pp. 203–212.

Hutchinson, M. 1993. Development of a continent-wide DEM with applications to terrain and climate analysis. In: Goodchild, M.F., Parks, B. and Steyaert, L., eds., *Environmental Modelling with GIS*, Oxford: Oxford University Press, pp. 392–399.

Institute of Hydrology. 1993. *Flood Studies Report*. Wallingford: Institute of Hydrology.

Jaafar, J., Priestnall, G., Mather, P.M. and Vieira, C.A. 1999. Construction of DEM from LiDAR DSM using morphological filtering, conventional statistical approaches and artificial neural networks. In: *Earth Observation: From Data to Information, Proceedings of the 25th International Conference of the Remote Sensing Society (RSS'99), University of Wales at Cardiff and Swansea*, Nottingham: The Remote Sensing Society, pp. 299–306.

Kirkby, M.J. 1980. The stream head as a significant geomorphic threshold. In: Coates, D.R. and Vitel, A.D., eds. *Thresholds in Geomorphology*, London: Allen and Unwin, pp. 53–73.

Kirkby, M.J. 1993a. *Physically-based Process Model for Hydrology, Ecology and Land Degradation*, MEDALUS I Final Report, Thatcham, Berkshire: MEDALUS.

Kirkby, M.J. 1993b. *MEDRUSH Model*, MEDALUS II First Interim Report, Thatcham, Berkshire: MEDALUS.

Kirkby, M.J. 1996. A role for theoretical models in geomorphology? In: Rhoads, B.L. and Thorn, C.E., eds., *The Scientific Nature of Geomorphology*, Chichester: John Wiley and Sons, pp. 257–272.

Kirkby, M.J., Abrahart, R., McMahon, M.D., Shao, J. and Thornes, J.B. 1998. MEDALUS soil erosion models for global change. *Geomorphology* 24: 35–49.

Kondolf, G.M. 1995. Geomorphological stream channel classification in aquatic habitat restoration: uses and limitations. *Aquatic Conservation: Marine and Freshwater Ecosystems* 5: 127–141.

Kopp, S.M. 1996. Linking GIS and hydrological models: where we have been, where we are going. In: Kovar, K. and Nachtnebel, H.P., eds., *Application of Geographic Information Systems in Hydrology and Water Resources Management*, International Association of Hydrological Sciences Publication 235, Wallingford: IAHS, pp. 133–139.

Kovar, K. and Nachtnebel, H.P., eds. 1993. *Application of Geographic Information Systems in Hydrology and Water Resources Management*. International Association of Hydrological Sciences Publication 211, Wallingford: IAHS.

Kovar, K. and Nachtnebel, H.P., eds. 1996. *Application of Geographic Information Systems in Hydrology and Water Resources Management*, International Association of Hydrological Sciences Publication 235, Wallingford: IAHS.

Lane, L.J. and Nearing, M.A., eds. 1989. USDA – *Water Erosion Prediction Project: Hillslope Profile Model Documentation*, United States Department of Agriculture-Agricultural Research Service National Soil Erosion Research Laboratory Report 2, West Lafayette: USDA.

Lane, S.N., Richards, K.S. and Chandler, J.H. 1998a. Landform monitoring, modelling and analysis: landform in geomorphological research. In: Lane, S.N., Richards, K.S. and Chandler, J.H., eds., *Landform Monitoring, Modelling and Analysis*, Chichester: John Wiley and Sons, pp. 1–18.

Lane, S.N., Richards, K.S. and Chandler, J.H., eds. 1998b. *Landform Monitoring, Modelling and Analysis*, Chichester: John Wiley and Sons.

Laser-Scan Laboratories Ltd. 1992. *DTMPREPARE and DTMCREATE Reference*, Cambridge: Laser-Scan Laboratories Ltd.

Lawler, D.M. 1992. Process dominance in bank erosion systems. In: Carling, P.A. and Petts, G.E., eds., *Lowland Floodplain Rivers: Geomorphological Perspectives*, Chichester: John Wiley and Sons, pp. 117–143.

Liu, B.Y., Nearing, M.A., Baffaut, C. and Ascough, J.C., II. 1997. The WEPP watershed model. III. Comparisons to measured data from small watersheds. *Transactions of the American Society of Agricultural Engineers* 40: 945–952.

Lillesand, T.M. and Kiefer, R.W. 1994. *Remote Sensing and Image Interpretation*, New York: John Wiley and Sons.

Lohr, U. 1998. Laserscanning for DEM generation. In: Brebbia, C.A. and Pascolo, P., eds., *GIS Technologies and Their Environmental Applications*, Southampton: Computational Mechanics Publications, pp. 243–249.

Longley, P.A., Goodchild, M.F., Maguire, D.J. and Rhind, D.W., eds. 2001. *Geographic Information Systems and Science*, Chichester: John Wiley and Sons.

Lytle, D.J., Bliss, N.B. and Waltman, S.W. 1996. Interpreting the state soil geographic database (STATSCO). In: Goodchild, M.F., Steyaert, L.T., Parks, B.O., Johnson, C., Maidment, D., Crane, M. and Glendinning, S., eds., *GIS and Environmental Modeling: Progress and Research Issues*, Fort Collins: GIS World Books, pp. 49–52.

McCullagh, M.J. 1998. Quality, use and visualisation in terrain modelling. In: Lane, S.N., Richards, K.S. and Chandler, J.H., eds., *Landform Monitoring, Modelling and Analysis*, Chichester: John Wiley and Sons, pp. 95–118.

Maidment, D.R., Djokic, D. and Lawrence, K.G. 1989. Hydrologic modelling on a triangulated irregular network. *Transactions American Geophysical Union* 70: 1091.

Mattikalli, N.M., Engman, E.T., Ahuja, L.A. and Jackson, T.J. 1996. A GIS for spatial and temporal monitoring of microwave remotely sensed soil moisture and estimation of soil properties. In: Kovar, K. and Nachtnebel, H.P., eds., *Application of Geographic Information Systems in Hydrology and Water Resources Management*, International Association of Hydrological Sciences Publication 235, Wallingford: IAHS, pp. 621–628.

Miller, D.R. and Morrice, J.G. 1996. An assessment of the uncertainty of delimited catchment boundaries. In: Kovar, K. and Nachtnebel, H.P., eds., *Application of Geographic Information Systems in Hydrology and Water Resources Management*, International Association of Hydrological Sciences Publication 235, Wallingford: IAHS, pp. 445–451.

Montgomery, D.R. and Dietrich, W.E. 1989. Source areas, drainage density, and channel initiation. *Water Resources Research* 25: 1907–1918.

Montgomery, D.R. and Foufoula-Georgiou, E. 1993. Channel network source representation used digital elevation models. *Water Resources Research* 29: 3925–3934.

Montgomery, D.R., Dietrich, W.E. and Sullivan, K. 1998. The role of GIS in watershed analysis. In: Lane, S.N., Richards, K.S. and Chandler, J.H., eds., *Landform Monitoring, Modelling and Analysis*, Chichester: John Wiley and Sons, pp. 241–261.

Montgomery, D.R., Grant, G.E. and Sullivan, K. 1995. Watershed analysis as a framework for implementing ecosystem management. *Water Resources Bulletin* 31: 369–386.

Moore, I.D, Grayson, R.B, Ladson, A.R. 1994. Digital terrain modelling: a review of hydrological, geomorphological and biological applications. In: Beven, K.J. and Moore, I.D., eds., *Terrain Analysis and Distributed Modelling in Hydrology: Advances in Hydrological Processes*, Chichester: John Wiley and Sons, pp. 7–34.

Mosley, M.P. 1987. The classification and characterisation of rivers. In: Richards, K.S., ed., *River Channels: Environment and Process*, Oxford: Blackwell, pp. 295–320.

Mroczkowski, M., Raper, G.P. and Kuczera, G. 1997. The quest for a more powerful validation of conceptual catchment models. *Water Resources Research* 33: 2325–2336.

Priestnall, G. and Downs, P.W. 1996. Automated parameter estimation for catchment scale river channel studies: the benefits of raster-vector integration. In: Kovar, K. and Nachtnebel, H.P., eds., *Application of Geographic Information Systems in Hydrology and Water Resources Management*, International Association of Hydrological Sciences Publication 235, Wallingford: IAHS, pp. 215–223.

Priestnall, G., Jaafar, J. and Duncan, A. 2000. Extracting urban features from LiDAR digital surface models. *Computers, Environment and Urban Systems* 24: 1–14.

Reid, L.M. 1993. *Research and Cumulative Watershed Effects*, General Technical Report PSW-GTR-141, Albany, CA: Pacific Southwest Research Station, Forest Service, US Department of Agriculture, 118 p.

Reid, L.M. and Dunne, T. 1996. *Rapid Evaluation of Sediment Budgets*, Reiskirchen, Germany: Catena, 164 p.

Ridley, H.M., Atkinson, P.M., Aplin, P., Muller, J.P. and Dowman, I. 1997. Evaluating the potential of the forthcoming commercial US high-resolution satellite sensor imagery at the Ordnance Survey. *Photogrammetric Engineering and Remote Sensing* 63: 997–1005.

Ross, M.A. and Tara, P.D. 1993. Integrated hydrologic modelling with geographic information systems. *Journal of Water Resources Planning and Management, Proceedings of the American Society of Civil Engineers* 119: 129–139.

Schumm, S.A. 1991. *To Interpret the Earth: Ten Ways to be Wrong*, Cambridge: Cambridge University Press.

Sear, D.A., Newson, M.D. and Brookes, A. 1995. Sediment-related river maintenance: the role of fluvial geomorphology. *Earth Surface Processes and Landforms* 20: 629–647.

Smith, J.S. and Smith, D.G. 1996. Operational experiences of digital photogrammetric systems. *International Archives of Photogrammetry and Remote Sensing* XXXI(Part B2): 357–362.

Smith, M.J., Smith, D.G., Tragheim, D.G. and Holt, M. 1997. DEMs and ortho-images from aerial photographs. *Photogrammetric Record* 15: 945–950.

Thorne, C.R. 1998. *Stream Reconnaissance Handbook*, Chichester: John Wiley and Sons.

Thornes, J.B., Shao, J.X., Diaz, E., Roldan, A., McMahon, M. and Hawkes, J.C. 1996. Testing the MEDALUS hillslope model. *Catena* 26: 137–160.

Tiscareno-Lopez, M., Lopes, V.L., Stone, J.J. and Lane, L.J. 1993. Sensitivity analysis of the WEPP watershed model for rangeland applications. I. Hillslope processes. *Transactions of the American Society of Agricultural Engineers* 36: 1659–1672.

Tiscareno-Lopez, M., Lopes, V.L., Stone, J.J. and Lane, L.J. 1994. Sensitivity analysis of the WEPP watershed model for rangeland applications. II. Channel processes. *Transactions of the American Society of Agricultural Engineers* 37: 151–158.

Vencatasawmy, C.P., Clark, C.D. and Martin, R.J. 1998. Landform and lineament mapping using radar remote sensing. In: Lane, S.N., Richards, K.S. and Chandler, J.H., eds., *Landform Monitoring, Modelling and Analysis*, Chichester: John Wiley and Sons, pp. 165–194.

Walsh, S.J., Butler, D.R. and Malanson, G.P. 1998. An overview of scale, pattern, process relationships in geomorphology: a remote sensing and GIS perspective. *Geomorphology* 21: 183–205.

Watson, F.G.R., Vertessy, R.A. and Band, L.E. 1996. Distributed parameterization of a large scale water balance model for an Australian forested region. In: Kovar, K. and Nachtnebel, H.P., eds., *Application of Geographic Information Systems in Hydrology and Water Resources Management*, International Association of Hydrological Sciences Publication 235, Wallingford: IAHS, pp. 215–223.

Wicks, J.M. and Bathurst, J.C. 1996. SHESED: a physically-based, distributed erosion and sediment yield component for the SHE hydrological modelling system. *Journal of Hydrology* 175: 213–238.

Wilby, R.L. 1996. The fractal nature of river channel networks: applications to flood forecasting in the river Soar, Leicestershire. *The East Midland Geographer* 19: 59–72.

Wischmeier, W.H. and Smith, D.D. 1978. Predicting Rainfall Erosion Losses – A Guide to Conservation Planning, United States Department of Agriculture Agricultural Handbook 537, Washington, DC: USDA.

Wise, S.M. 1998. The effect of GIS interpolation errors on the use of digital elevation models in geomorphology. In: Lane, S.J., Richards, K.S. and Chandler, J.H., eds., *Landform Monitoring, Modelling and Analysis*, Chichester: John Wiley and Sons, pp. 139–164.

Yeh, W.W.-G. 1986. Review of parameter identification procedures in groundwater hydrology: the inverse problem, *Water Resource Research* 22: 95–108.

Part IV

Chemical, Physical and Biological Evidence: Dating, Emphasizing Spatial Structure and Fluvial Processes

9

Radiogenic and Isotopic Methods for the Direct Dating of Fluvial Sediments

STEPHEN STOKES[1] AND DES E. WALLING[2]

[1]*School of Geography and the Environment, University of Oxford, Oxford, UK*

[2]*Department of Geography, University of Exeter, Exeter, UK*

9.1 INTRODUCTION

Rivers are a ubiquitous component of the continental landscape and are sensitive to change, adjusting their form over time as a result of a wide range of internal and external factors (e.g. Schumm 1981, Schumm and Winkely 1994). Their deposits potentially constitute a long, albeit discontinuous, record of terrestrial landscape evolution, and in many instances may represent the only available record of long-term terrestrial environmental responses to climate change. As such, fluvial deposits are a critical archive of river response and form an important part of many Quaternary sedimentary sequences. It is essential to date such fluvial sediments in order to fully exploit their preserved record of environmental changes and responses to anthropogenic landscape modification.

There are a wide variety of techniques available for the age evaluation of later Quaternary sediments. However, determining the fluvial response to environmental change is, in many cases, still limited by a lack of suitable, widely applicable sediment tracing and dating techniques, which can be applied directly to fluvial material of Pleistocene, Holocene and historical age (Fuller *et al.* 1996). The purpose of this chapter is to discuss contemporary approaches to the dating of fluvial, colluvial and related sediments. We do this by first briefly overviewing some of the common methods available for the age determination of Quaternary sediments, and secondly by outlining the principles, developments and some case example applications of two recently developed groups of methods which hold great potential for future research in fluvial and colluvial environmental archives.

9.2 A CLASSIFICATION OF DATING TECHNIQUES

There are numerous routinely employed chemical and physical techniques which provide qualitative and quantitative insights into both contemporary and ancient fluvial systems. These methods have undergone rapid progress in the past few decades and, in addition, whole new suites of techniques relating to cosmogenic and short-lived isotopes have been developed in response to improvements in both theory and instrumentation. A summary of some of the key techniques employed in the dating of fluvial sediments is presented in Figure 9.1. These approaches vary, in addition to other factors, in the time range applicable, the precision of the resulting age estimates, and the nature of the event being dated. Classification of such a broad range of methods is in itself complex. Colman and Pierce (2000) propose a two-stage classification whose basic structure is encapsulated within Figure 9.1. They propose the classification of dating methods in terms of both broad type of methodologies (i.e., assumptions and mechanisms which form the basis of the technique) and on the type of results which the methods can produce. The methodological classification is sixfold (Colman and Pierce 2000):

1. *Sidereal* (calendar or annual) methods allow direct age evaluation based on annual cycles or rhythms.
2. *Isotopic* methods measure changes in isotopes or isotopic ratios due to radioactive decay. The most common of these methods is radiocarbon dating.

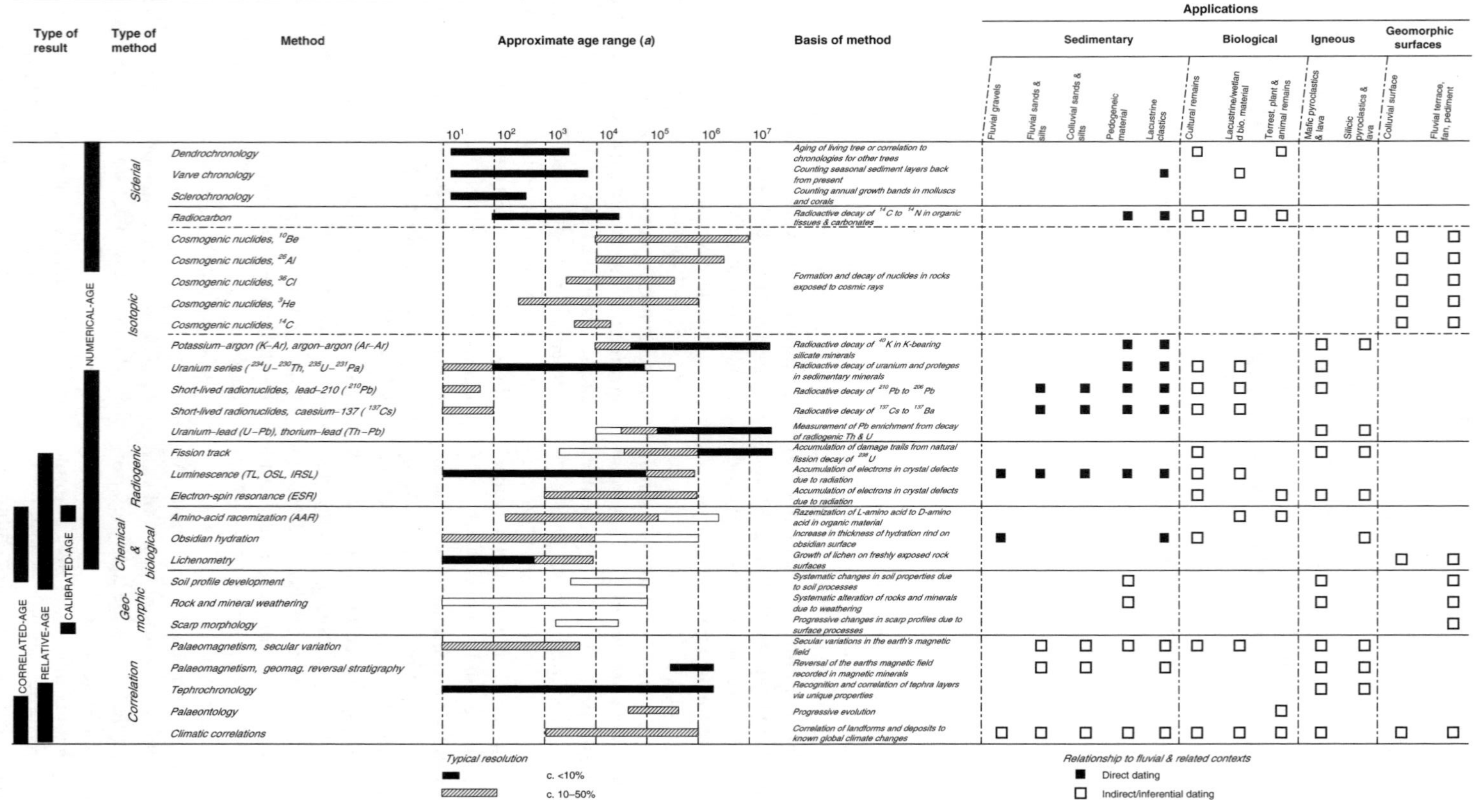

Figure 9.1 Summary classification of dating methods capable of providing age-related information for Quaternary fluvial and associated deposits (adapted from Sowers *et al.* 2000)

3. *Radiogenic* methods measure the cumulative effects of radioactive decay, including crystal damage and electron dislocation/trapping.
4. *Chemical and biological* methods are based on the measurement of time-dependent chemical or biological processes.
5. *Geomorphic* methods are based on the time-dependent geomorphic response to environmental factors. These include a wide cross-section of biological, physical, chemical and inter-related processes which are typically complex and the resulting environmental response is frequently non-linear.
6. *Correlation* methods are techniques which do not directly relate to an age-dependent process. Instead, they are exploited based on correlation with other classes of dating method via time-independent properties (e.g. palaeomagnetism).

While a practitioner is primarily interested in the methodological classification of dating techniques, the applied fluvial geomorphologist/sedimentologist may be more concerned with the results that the methods produce. The classes of techniques based on the results they produce are not mutually exclusive and include (Colman and Pierce 2000):

1. *Numerical-age* methods are those that provide results on an absolute timescale (i.e., they produce estimates of age and uncertainty on a comparable and absolute timescale).
2. *Calibrated-age* methods provide approximate numerical age estimates based on systematic changes resulting from environmentally modulated individual processes or groups of processes.
3. *Relative-age* methods provide only an order or relative age sequences (i.e., older to younger, less developed to most developed).
4. *Correlated-age* methods produce ages only by demonstrating equivalence to independently dated deposits or events.

In addition to the clear desirability of establishing an absolute (numerical-age) timescale, it is preferred to employ methods which can directly date sedimentary depositional and erosional events. For this reason, we subdivide the methods once more into those which can provide direct age estimates for sediments and those which can provide an indirect estimate or inference (Figure 9.1). We define *indirect age* estimation methods as those approaches that provide 'bracketing' ages, which define maximum or minimum time spans of fluvial events, or dates materials associated with fluvial deposits. As such, while they are still useful, they may not categorically describe the timing of the fluvial activity itself. *Relative dating* approaches, such as soil profile development (e.g. Rockwell 2000) or mineral weathering indexes (e.g. Birkeland and Noller 2000), provide approximate geomorphological indexes of the time elapsed since the cessation of fluvial deposition or erosion are likewise indirect. A scarcity of materials which can be dated by other indirect numerical-age or calibrated-age techniques (such as large mammal teeth for electron–spin resonance (ESR); shells for amino acid racemisation (AAR) dating) means that in many cases, chronological control of fluvial deposits relies entirely on (indirect) faunal or floral correlation's (Perkins and Rhodes 1994) or relative-age approaches. Development of an absolute timescale from such methods requires calibration of the relative indexes via cross-referencing to other methods.

Thus, for direct and absolute dating of fluvial sediments we are left with a limited subset of chemical, isotopic and radiogenic methods (Figure 9.1). These include radiocarbon dating, luminescence-based techniques, and dating via incorporated short-lived radioisotopes. In fluvial systems, there is often a complete lack of material suitable for radiocarbon dating; where it is possible, large uncertainties for young samples (<500 years), and the limitations of a 35–50-ka maximum age limit its applicability to latest Pleistocene and Holocene fluvial deposits. Fortunately, over the past few decades there have been considerable advances in short-lived isotopic and radiogenic, particularly luminescence dating, methods. These relatively new and developmental methods provide a great potential for the future age evaluation of fluvial depositional systems and they are the focus of this chapter. This in part reflects the expertise of the authors, but also underlies the great importance that researchers continue to place on the development of robust, direct chronologies for depositional fluvial systems, and on the timing of fluvial processes over timescales, which extend beyond those of observational records. The techniques are described in some detail in separate sections that follow.

9.3 LUMINESCENCE DATING OF FLUVIAL SEDIMENTS

General Dating Principles

Luminescence dating methods encompass a range of techniques which can determine the period of time elapsed since a sediment was exposed to daylight. They are based on the accumulation of charge populations trapped within crystalline sedimentary materials such as quartz and feldspar. These charge populations build up in response to the dissipation of energy within the crystal structure during the radioactive decay of long-lived radionuclides and cosmic rays. The method estimates the last time the accumulated charge population was reset. Electrons may be evicted from their ground state by an addition of energy, as is the case when they are exposed to ionising radiation emitted during radioactive decay of uranium, thorium and radioactive potassium (^{40}K). An additional, typically small, source of energy is derived from cosmic rays. Once trapped, electrons may be held at some such localities for extended periods of time ($>10^5$ years), or until a further amount of energy is introduced via thermal (phonon) or optical (photon) excitation. The additional energy overcomes an activation potential and allows electrons to return to their ground state. In doing so they emit excess energy in the form of a photon, termed luminescence. For sedimentary materials, the resetting event is the last exposure to daylight prior to deposition. As such, the method uniquely allows the age of sediment deposition (as opposed to formation) to be established.

Luminescence emission following a thermal stimulation is termed thermoluminescence (TL); luminescence emission following an optical stimulation is termed *optically stimulated luminescence* (OSL). Stimulation via infrared is *termed infrared stimulated luminescence* (IRSL). While there are other terms in usage (see Duller 1996), the practitioner community is moving towards adopting only these terms and at the same time providing information relating to the wavelength and intensity of the photon flux. Detection of the luminescence emission requires the use of a high sensitivity and, low background photomultiplier (PMT). The family of techniques has undergone rapid developments and enhancements in the 1980s and 1990s. Aitken (1985, 1998), Prescott and Robertson (1997) and Wintle (1997) overview these developments in considerably more detail than is possible here.

Luminescence ages are calculated using the general equation

$$\text{age} = \text{equivalent dose (Gy)/dose rate} \quad (1)$$

The equivalent dose (D_e) (measured in grays, Gy), relates to the ionising radiation from decay of the U and Th decay series, ^{40}K and a minor contribution from cosmic radiation, which the sample has absorbed since burial. Alternative names for this quantity that have previously been used in the literature include palaeodose (*P*) and accumulated dose (AD). The dose rate corresponds to the rate at which the sample was exposed to ionising radiation, and hence the rate at which the trapped charge population was deposited. Where the time interval concerned is a year, the dose rate is known as the annual dose.

The datable range of the luminescence techniques relates both to the capacity of the dosimeter material in question to take up charge and the rate at which trapped electrons are created within the lattice (which is directly related to ambient environmental radiation levels). For typical levels of environmental radiation, quartz provides a dosimeter which can be routinely used for samples ranging in ages up to 150 ka, and ages as old as 800 ka have been reliably generated (e.g. Huntley *et al.* 1993). In the case of feldspathic minerals, the saturation dose level is typically much higher and ages of the order of hundreds of thousands of years have been reported (e.g. Berger *et al.* 1992). In any case, use of either mineral types provides a means of obtaining ages from sedimentary contexts over fourfold greater in time span than the radiocarbon method, and spanning at least the full last interglacial–glacial cycle.

Equivalent Dose Determination

The determination of an equivalent dose requires measurement of either TL or OSL, in order to compare the abundance of naturally accumulated charge to that corresponding to known doses of radiation in the laboratory. TL measurements involve heating a small (typically 4–6 mg) aliquot of refined quartz or feldspar resulting in a typical glow curve of temperature versus luminescence in which various peaks in the curve correspond to the thermal activation temperatures of differing electron traps. Artificial irradiation of aliquots in the laboratory may result in additional lower temperature peaks which are not seen in natural samples collected from a depositional context. Their absence from geological samples demonstrates their instability over long timescale and their unsuitability for dating. It is necessary to pre-heat artificially irradiated aliquots to remove these components.

By contrast, OSL measurements are achieved by exposing the sample aliquot, prepared in an identical fashion as for TL measurements, to light of a fixed wavelength for a set duration of time, at a fixed intensity. Generally, quartz is stimulated with blue or green ($\lambda = 470$–514 nm) photons and feldspar by infrared ($\lambda = 830$–900 nm) photons (Wintle 1997). A plot of light exposure versus OSL reveals a rapidly depleting signal, which may be mathematically modelled by a complex exponential function.

There are two widely used strategies for harnessing such light exposure–decay curves to obtain an equivalent dose (Figure 9.2). The *additive dose method* involves measurement of both natural luminescence intensity and the measurement of the luminescence from other aliquots which have been subjected to additional amounts of ionising radiation. As the addition of radiation populates low-temperature traps, it is necessary to pre-heat all aliquots prior to measurement. This is thought to simulate extended burial periods at ambient temperatures, and various strategies have been used ranging from short duration (e.g., 10 s), high temperature (e.g. 280 °C) to long duration (e.g. 16 h) low-temperature (e.g. 160 °C) treatments (Aitken 1998). The resulting luminescence versus added radiation dose growth curve may then be used to define the trend of signal growth with dose, and with this information the equivalent dose may be estimated by extrapolating the trend line back to its x-axis intercept (Figure 9.3a). The *regeneration* method involves the measurement of the natural aliquot luminescence intensity and the measurement of the luminescence intensity of aliquots whose natural signal was first removed by an optical bleaching. In this case, D_e is estimated by interpolation of the natural aliquot intensity level to the regenerated growth curve (Figure 9.2b). Pre-heating is again required.

Traditionally, a large (c. up to 80) number of aliquots are required for the generation of a D_e. The resulting multiple aliquot 'growth curves' tend to exhibit a linear trend of signal growth at low (c. 0–20 Gy) doses, while at higher doses growth curves generally exhibit a saturating exponential growth trend. Extrapolation of growth curves for data which are close to saturation is complex and typically results in large uncertainties in the resulting D_e estimates.

Single Aliquot and Single Grain Techniques

In many instances, application of multiple aliquot procedures may be complicated, either by excessive scatter in multiple aliquots, or by a lack of sufficiently

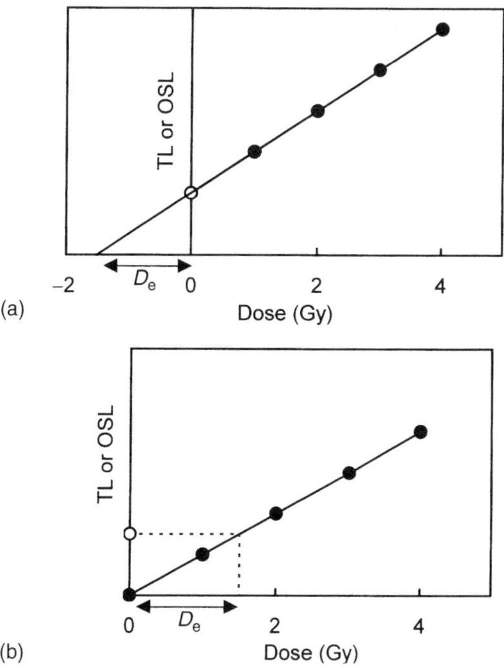

Figure 9.2 Methods of D_e determination: (a) additive dose method; (b) regeneration method

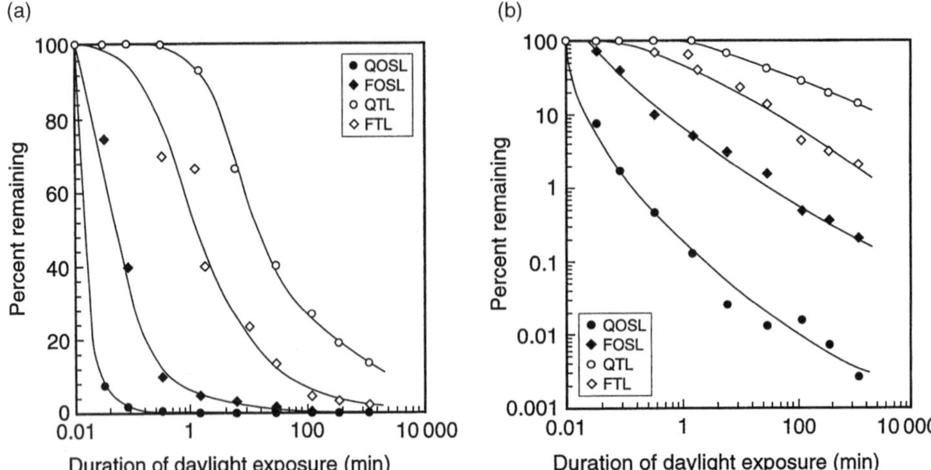

Figure 9.3 Depletion of luminescence due to daylight exposure. Filled circles: quartz optically stimulated luminescence (QOSL) response. Filled diamonds: feldspar optically stimulated luminescence (FOSL) response. Open circles: Quartz thermoluminescence (QTL) response. Open diamonds: feldspar thermoluminescence (FTL) response (a, linear scale; b, log scale)

large quantities of material for analysis. Over the past decades a series of important developments has been made which have provided single aliquot alternatives to the traditional multiple aliquot approach.

Duller (1991) was the first to propose procedures for single aliquot dating, developing an additive dose-based method for potassium feldspar grains, which involved repeatedly sampling a portion of the total IRSL by a brief (0.1 s) IR exposure, followed by a cycle of artificial irradiation and pre-heating. A further sample aliquot is subjected to the sample IR stimulation and pre-heating cycle without added dosing to quantify the combined effect (signal depiction) of the two experimental steps. The data measured for the first aliquot are then proportionately corrected for the depletion observed in the second. Key advantages of this approach are:

1. A population of D_e's may be generated for a sample instead of a single multiple aliquot D_e, and for well-bleached samples the standard error of the mean of such a population provides a more precise D_e assessment.
2. It is possible to test for homogeneity of luminescence behaviour from aliquot-to-aliquot which may reflect incomplete bleaching histories or other factors. These are discussed in detail later.
3. The method avoids the aliquot-to-aliquot scatter frequently noted in multiple aliquot growth curves.
4. The method is highly suited to automation (Duller 1996).

Many single aliquot D_e determination strategies have evolved in the ensuing period, employing both additive dose and regenerative approaches. The utility of the single aliquot additive dose method, and variations of it, for application to quartz sands using green light has been demonstrated widely (e.g. Mejdahl and Botter-Jensen 1994, 1997, Stokes 1994, Murray and Mejdahl 1998, Stokes *et al.* 2001). Much of this research has focussed, though not exclusively, on fluvial and aeolian-associated sediments from Australia (e.g. Murray and Roberts 1998, Olley *et al.* 1998, 1999), where intercomparisons between single aliquot and multiple aliquot procedures have indicated good agreement over a wide range of D_e (0.1–100 Gy). Since the initial single aliquot applications developed by Duller, there has been a near ubiquitous shift towards the exclusive use of quartz as the dosimeter.

Single Aliquot Regeneration Procedures

The single aliquot regeneration (SAR) procedure (Murray and Roberts 1998, Murray and Wintle 2000) is an extremely promising method of D_e determination for a wide range of luminescence dating application. In the SAR approach, D_e's are calculated for each aliquot by an iterative procedure involving the measurement of the natural OSL signal, and then regenerating the signal via a series of irradiation—pre-heat—stimulation cycles. The great advance in the SAR methodology is that additional steps are incorporated to directly test and correct for any changes in the response of the sample during the measurement steps. The specific advantages of the SAR procedure include:

1. The use of a growth curve interpolation dramatically reduces uncertainties (typically <4%) in D_e estimates in comparison to additive dose methods (uncertainties typically >15%).
2. A range of internal checks on sample behaviour is possible, which provide added robustness to resulting age estimates.
3. Populations of D_e's produced for samples allow data interrogation, in particular for partial bleaching.
4. The procedure is more rapid than multiple aliquot or single grain additive dose approaches.

Although recently developed, these methods have already made considerable impact in palaeohydrological investigations and a number of these are discussed later.

Analyses of Individual Mineral Grains

Single aliquot procedures have also been used to generate D_e's for individual grains. Age estimation on a grain-by-grain basis is highly desirable as construction of sample D_e or age populations provides a means by which the homogeneity of a given sample, and hence its suitability for dating, may be assessed. Sample heterogeneity can be introduced by either incomplete resetting of previously accumulated charge prior to deposition (see below) or by post-depositional modification (e.g. pedoturbation, bioturbation).

Lamothe *et al.* (1994) first demonstrated this potential by hand picking potassium feldspar grains from Late Glacial age marine sediments from Southern Quebec. Subsequently, Murray and Roberts (1998) applied a single aliquot additive dose approach to date single grains of Australian quartz. Their studies have been followed up by a range of single grain applications to Australian archaeological, aeolian and fluvial (see later) quartz (e.g. Roberts *et al.* 1997, 1998, Olley *et al.* 1998, 1999). Equipment is now available that allows the analysis of up to thousands of individual grains simultaneously using a computer controlled, finely focussed high power laser (which increases luminescence output considerably) (e.g. Duller *et al.* 1999).

It is likely that single grain approaches will shortly be adopted as the technique of preference in many optical dating applications.

Rates of Zeroing of Previously Accumulated Charge

It is an implicit assumption of the method that during sediment erosion, transportation and deposition any previously resident (latent) charge population within the crystal lattice is substantially reduced or removed. This has been elegantly demonstrated for both OSL and TL in quartz and feldspar grains by direct measurements of signal reduction versus daylight exposure time (e.g. Godfrey-Smith *et al.* 1988, Berger 1990, Stokes and Gaylord 1993, Rendell *et al.* 1994). As an illustrative example, Godfrey-Smith *et al.* (1988) calculated quartz and feldspar OSL reductions to levels below 5% of their initial values following exposure to 1 min of daylight, the TL signal from the same samples reduced considerably more slowly, falling to levels below 20% of the initial signal after 35 (feldspar) and 200 (quartz) min of daylight exposure, respectively (Figure 9.3). Rendell *et al.* (1994) used a similar approach to demonstrate that resetting of the OSL signal in quartz and feldspar following a 3 h light exposure occurs efficiently in water depths of over 12 m despite the solar spectrum being dramatically attenuated in water (Figure 9.4). As with the results for surface bleaching, they found that the TL signal was considerably less efficiently, and ultimately incompletely, reset. While representing 'ideal' circumstances for bleaching, which may not be realised in many fluvial settings, the results clearly emphasise the advantage which OSL-based techniques have over TL, namely, that resetting of latent luminescence signals will always be greater for OSL than for TL.

Such studies not only confirm one of the key assumptions of the methods, but also demonstrate a critical advantage of techniques which employ OSL methods; namely, that for many sedimentary environments it can be assumed that the latent charge population at the time of deposition can be assumed to be negligible in comparison to latent TL, and that generally the level of OSL measured in a sediment sample relates to charge redistributed only since the last depositional cycle. In the case of TL methods, in the absence of strong independent evidence to suggest that samples were subjected to prolonged bleaching during their transport and deposition, it is necessary

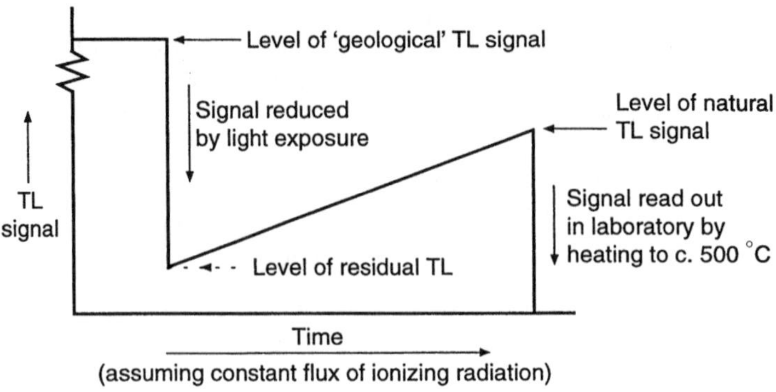

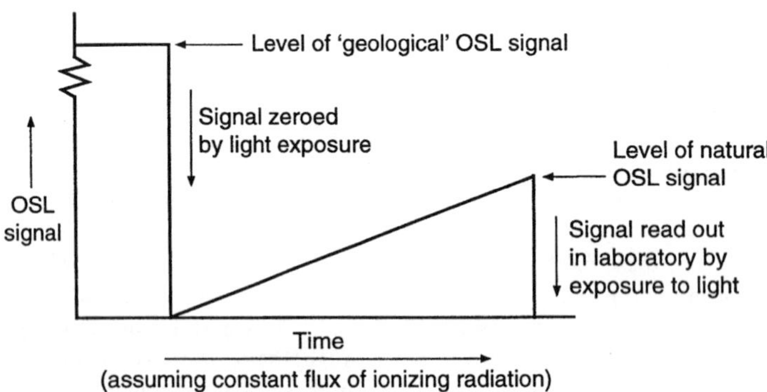

Figure 9.4 Schematic diagram depicting the process of accumulation and resetting of trapped charge in minerals due to radiation dose from long-lived radionuclides (accumulation) and day light exposure during erosion, transportation and surface exposure (resetting)

to make some estimate of the contribution which latent TL makes to the total TL measured in the laboratory. Central to an understanding of the applicability and suitability of fluvial sediments for optical dating is the concept of partial or incomplete bleaching. This is discussed in some detail in later sections.

Resetting Underwater

Underwater, light is transmitted with a severely restricted bleaching spectrum, which results from the attenuation of the UV component of the solar spectrum as it passes through water (Fuller *et al.* 1994). This will occur in still water and there will be further attenuation in turbid water, where fine-grained particles scatter light. Berger and Luternauer (1987) have demonstrated (Figure 9.5) that the light intensities at 4 m depth in a turbulent river are about 10^4 times less than at the surface, and are severely attenuated below 500 nm and above 690 nm. Such conditions may partially or completely inhibit the bleaching process and result in fluvial deposits consisting of mixed populations of well and poorly bleached grains and failing the central assumption.

Since grains incorporated in a fluvial deposit may have originated in any part of the river system, any particular grain will have an unknown light exposure before deposition. In addition, different grain sizes are likely to have different histories of light exposure, with fine grains being nearer to the top of the water surface and coarse grains being moved by saltation close to the river-bed (Fuller *et al.* 1994). In contrast to the above inference, Olley *et al.* (1998) noted that coarse grains were marginally better bleached than finer sands in their study of historical Australian flood deposits. The extent of bleaching within fluvial sediment populations is the critical issue for dating applications. Where deposits consist of mixed populations of bleached and unbleached grains, an age excess would be anticipated and much research has been

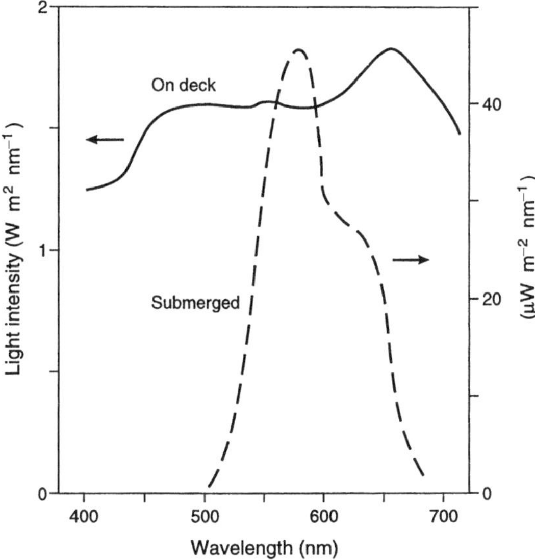

Figure 9.5 The solar spectrum in daylight and underwater (after Berger and Luternauer 1987)

expended on known age fluvial sediments to test for excess.

Tests of the assumed completeness of bleaching can be achieved by a range of laboratory simulation and facies analogue approaches. For example, Ditlefsen (1992) investigated the bleaching of potassium feldspars through a 75-cm column of different water suspensions and compared the residual TL and IRSL signals after light exposure. Ditlefsen found that clear, shallow water has a minimal impact on bleaching levels. He demonstrated that for dilute suspensions ($<0.02\,g\,l^{-1}$), with an air-generated turbulence introduced to keep material in suspension, the IRSL signal was reduced by 95% within 20 h whereas the TL was only reduced by 25–50%. In more dense suspensions ($>0.05\,g\,l^{-1}$), however, there was little bleaching of any luminescence signal and it was concluded that bleaching times well above 20 h are needed to reset any of the luminescence signals to levels tolerable to dating.

Another approach is to determine effective ages for contemporary or historically dated fluvial and other sediments. The presence of a significant non-zero residual signal would provide evidence for incomplete bleaching at deposition. This has obvious limitations but does serve to provide a guideline for the bleaching conditions of a particular fluvial environment. Stokes (1994) combined such an approach with multiple aliquot methods to date samples from a range of depositional environments. He found that while ages estimated for aeolian sediments indicated complete bleaching, analysis of selected fluvial sediments indicated a depositional residual of up to 1 Gy (or equivalent to between 500 and 1000 years assuming 'typical' dose rates).

In a further study, Stokes *et al.* (2001) collected samples from two large (catchment area $> 10^5\,km^2$) drainage basins to evaluate lateral and downstream variations in the extent of bleaching in modem river systems. SAR data from various facies within a point bar sequence on the Colorado River indicate that channel bar and upper flow regime point bar sediments deposited during high flow conditions exhibit evidence of partial bleaching. In contrast, floodplain and lower point bar silty sands did not indicate evidence of widespread partial bleaching. Depositional residual ages of c. 70–300 years were estimated based primarily on minimum D_e. The analysis demonstrates that it is possible to identify and largely circumvent problems associated with this form of partial bleaching by SAR analysis.

In the same study, SAR data for Loire River bed material samples collected along the full (source–mouth) distance of the river produced wide (up to 40 Gy) ranges of values which generally reduced downstream (Figure 9.6). All D_e distributions were positively skewed and mean, median or minimum D_e estimates all exhibited a distinctive downstream trend

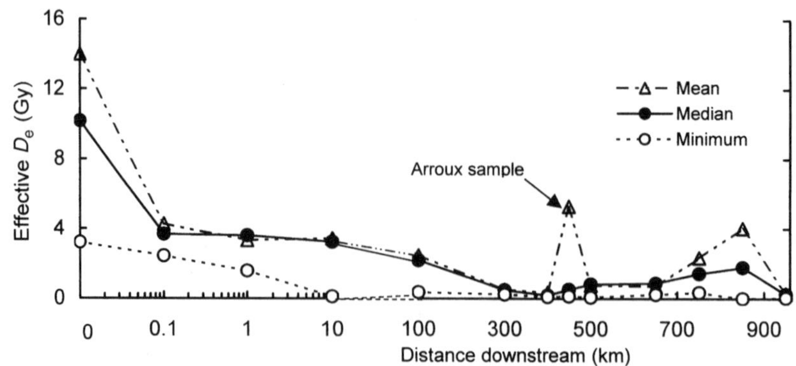

Figure 9.6 Reduction of apparent equivalent dose with distance downstream for sand-sized quartz extracted from bedload samples from the Loire River (modified from Stokes *et al.* 2001)

which initially reduces dramatically (indicating rapid bleaching during transport) over the first 10–100 km and then varied relatively little down the remaining c. 900 km. D_e versus OSL intensity relationships implied that partial bleaching was not present. Median D_e estimates imply a monotonic downstream reduction in residual age from c. 3.3 ka to 300–80 a, while minimum D_e estimates as recommended by Olley *et al.* (1998) (see below) imply downstream residual ages of 1 ka at source, down to a few decades over as little as 10 km.

In addition, it is possible to evaluate to some degree the extent of partial bleaching within a sample by examining the nature of single aliquot or single grain D_e distributions (e.g. Duller 1996, Olley *et al.* 1998, 1999, Colls *et al.* 2001). These methodologies focus either on the relationship between aliquot luminescence intensity and equivalent dose (D_e) or the actual distribution of D_e estimates. In the first strategy, a positive correlation between aliquot intensity and D_e is considered to imply that the sediment population consists of a heterogeneous mixture of grains which were variably (i.e., incompletely) bleached prior to their ultimate deposition. If such a relationship exists, it has been assumed that the lowest D_e estimate is likely to reflect the aliquot which comprises the most bleached grain population and therefore the best estimate of the true D_e, although this may still incorporate grains possessing a pre-depositional residual charge population and therefore be an overestimate (Olley *et al.* 1998). Wintle *et al.* (1993, 1995) used this approach to date colluvial deposits from QuaZulu Natal which were independently dated by ^{14}C. The other approach, requiring the analysis of many (c. $n > 50$) aliquots involves the definition of D_e variability by the construction of a D_e frequency histogram (e.g. Olley *et al.* 1998, 1999). In this approach, the limited variation of a D_e frequency ogive constructed for a homogeneously well-bleached deposit is readily contrasted against the wide and asymmetrically (positive skewed) distributed histograms of poorly bleached sediments.

Thus, it is possible to assess the suitability of a sample for dating by its luminescence properties alone, without reference to the geological interpretation of the depositional environment. In situations where the single aliquot data indicate a mixture of well-bleached and poorly bleached grains, the true D_e probably lies towards the bottom end of the range of D_e's derived from the single aliquot analysis. This does, however, assume that at least a proportion of the grains have been well-bleached and that post-depositional processes have not introduced 'young' grains.

Olley *et al.* (1999) have taken this approach further by testing the significance of sample mass, assuming that larger aliquot sizes increase the probability of incorporating unbleached grains. They have demonstrated that single aliquot D_e frequency histogram asymmetry is, at least in part, related to mixed grain populations, that minimum D_e estimates are likely to reflect the best estimate of the time elapsed since burial, and that small aliquot (60–100 grains) or single grain analysis are the best approach to ensuring accurate optical dating results from fluvial deposits.

Strategies for Future Luminescence Dating of Fluvial Depositional Environments

Fluvial sequences remain the most enigmatic for luminescence dating application and have produced results which could be described as ranging from the ecstatic to agonised, the latter being more prevalent. Some of these results are described in Section 9.5. The likelihood of successful future luminescence dating applications lies in a combination of the utilisation of appropriate dating strategies and careful attention to sedimentological and geomorphological factors. Perhaps the most significant geomorphological indicator, which might be used to predict the suitability of fluvial samples for dating, is the estimated concentration of suspended sediment within the water column during transportation; high concentrations such as in proximal glaciofluvial sequences being problematic for dating (e.g. Gemmell 1988, 1997, Rhodes and Pownell 1994), and low concentrations, such as within sandy fluvial settings, most likely resulting in complete bleaching at deposition. Some further factors which may control optical resetting and correspondingly the suitability of fluvial sedimentary deposits for optical dating are summarised in Figure 9.7.

It is clear that not all fluvial settings are presently capable of being routinely dated by luminescence techniques. The majority of larger catchment systems should, however, be readily datable provided an appropriate methodology is employed. Such a methodology must combine geomorphological considerations and dating practices. Systematic dating of fluvial sediments is presently undergoing rapid development (e.g. Olley *et al.* 1998, 1999, Colls *et al.* 2001, Stokes *et al.* 2000). A series of important factors are emerging from such studies which should be borne in mind when planning a dating programme:

1. A consideration of the fluvial sedimentary facies, transportation regime and distal–proximal relationships allows a degree of anticipation of the suitability of the sediments for dating.
2. Sampling of equivalent modern sedimentary facies deposited under similar hydrological regimes may provide insights into the extent of resetting.
3. Use optical dating methods. In fluvial applications it is now clear that there is no advantage in using TL methods.
4. Employ small aliquot and single grain approaches to minimise contamination from poorly bleached grains. These techniques can assist in identifying the existence of heterogeneously bleached samples. Wherever there is some uncertainty about the extent of bleaching, a combination of small single aliquots and single grain analyses should provide a quantitative estimate of the depositional event (Olley *et al.* 1998, 1999).

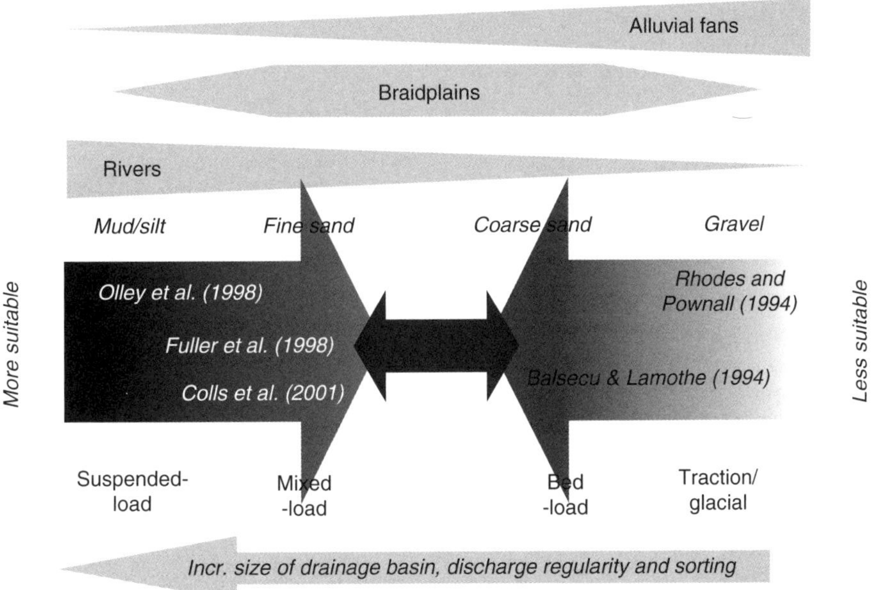

Figure 9.7 Schematic diagram depicting a selection of significant parameters of fluvial depositional environments and the likely suitability of a selection of fluvial sub-environments for optical dating

5. Collaborative research programmes, which harness the combined expertise of both dating practitioners and fluvial geomorphologists or sedimentologists, are the best means by which a problem can be fully interrogated.

9.4 USING SHORT-LIVED RADIOISOTOPES TO ESTABLISH CHRONOLOGIES FOR RECENT FLUVIAL SEDIMENTS

The Context

The application of luminescence dating, as described above, focuses on timescales that are typically measured in thousands of years. Alternative approaches must be sought when attention is directed to contemporary fluvial processes and their associated deposits and when, for example, information on contemporary or recent deposition rates is required. In this case, there is a need to consider timescales of the order of 10^1–10^2 years. The assessment of recent rates of overbank sedimentation on river floodplains is likely to represent a key requirement in a range of investigations and this will provide the focus of the following discussion. Information on overbank sedimentation rates will frequently be required, both as measure of the rate of development of contemporary fluvial landforms and as a means of quantifying the role of overbank sedimentation as a conveyance loss within the sediment budget of a catchment. In order to establish the accretion rate at a particular site, there is a need to define the age/depth curve and thus to date specific horizons.

Traditional approaches to dating specific levels or horizons, and thereby calculating the rate of deposition from the depth of sediment overlying this level, have included the use of prior benchmark surveys (cf. Happ 1968, Leopold 1973, Trimble 1983); the existence of datable surfaces or materials (cf. Costa 1975, Hupp 1988); and relating trace metal concentrations in the sediment profile to the known history of mining activity in the upstream catchment (e.g. Knox 1987, Lewin and Macklin 1987, Popp *et al.* 1988, Macklin and Dowsett 1989). All of these approaches are likely to have constraints on their application by virtue of the need for pre-existing surveys or for specific local conditions or information. Recent work on the application of environmental or fallout radionuclides, and more particularly caesium-137 (^{137}Cs) and unsupported Pb-210 (unsupported ^{210}Pb) (e.g. Ritchie *et al.* 1975, McHenry *et al.* 1976, Popp *et al.* 1988, Walling and Bradley 1989, Bishop *et al.* 1991, Walling *et al.* 1992, Walling and He 1993, He and Walling 1996a, Allison *et al.* 1998, Goodbred and Kuehl 1998) has demonstrated the potential of an approach which is capable of being applied more generally across a wide range of environments and locations. The basis for using fallout radionuclides to establish a chronology or age/depth relationship for overbank sediment deposits is discussed below.

The ^{137}Cs Technique

Fallout ^{137}Cs is an artificial radionuclide with a half-life of 30.17 years that was introduced into the environment by the atmospheric testing of thermonuclear weapons during a period extending from the mid-1950s through to the 1970s (see Cambray *et al.* 1989). The radiocaesium released into the stratosphere by these weapons tests was dispersed globally and subsequently deposited as fallout on the land surface. Significant levels of ^{137}Cs fallout were first recorded in 1954 and most of the fallout occurred in the period between 1956 and the early 1970s. In the northern hemisphere, there was a high input in 1959 and maximum deposition in 1963, the year of the Nuclear Test Ban Treaty (cf. Figure 9.8). In the southern hemisphere, fallout amounts were significantly lower, due to the greater distance from the main bomb-test sites, and, in some locations, maximum fallout was delayed until 1965. Since the early 1980s, rates of ^{137}Cs fallout have been very low, although in many parts of Europe and in some adjacent regions, an additional short-term input was received in 1986 as a result of the Chernobyl accident, which also released radiocaesium into the atmosphere (cf. Cambray *et al.* 1989). Deposition fluxes of ^{137}Cs from the atmosphere have been documented by monitoring stations in several areas of the world and the pattern of fallout shows considerable global variability, with substantially higher values in the northern hemisphere than in the southern hemisphere and with higher values in the mid-latitudes and a decrease towards the equator. This pattern parallels that of strontium-90 (^{90}Sr) fallout, another product of weapons testing and, since data for ^{90}Sr fallout inventories are available for many more sites around the world, reference is usually made to the pattern of ^{90}Sr fallout to demonstrate the main features of the global pattern of ^{137}Cs fallout (cf. Table 9.1). In most environments, ^{137}Cs deposited as fallout is rapidly and strongly fixed by clay particles in the surface soil or sediment (cf. Frissel and Pennders 1983, Livens and Rimmer 1988, Ritchie and McHenry 1990, He and Walling 1996b), and its subsequent

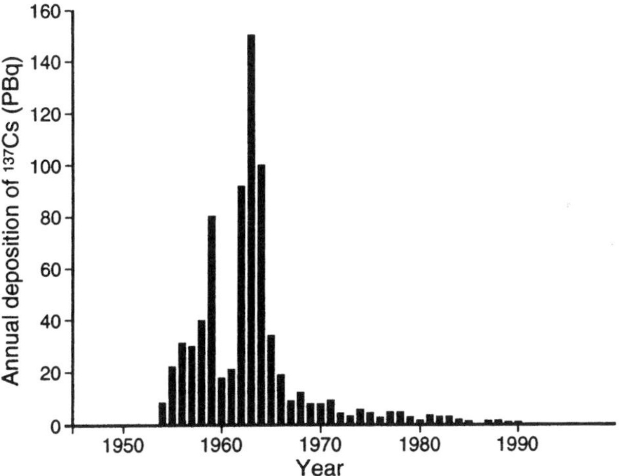

Figure 9.8 The temporal pattern of bomb-derived ^{137}Cs fallout in the northern hemisphere (based on Cambray *et al.* 1989)

Table 9.1 Latitudinal variation of mean bomb-derived ^{90}Sr inventories at the end of 1983, based on data presented by Larsen (1985)

Latitude band (degree)	Relative magnitude of the mean ^{90}Sr inventory[a]	
	Northern hemisphere	Southern hemisphere
0–10	0.25	0.14
10–20	0.35	0.13
20–30	0.54	0.22
30–40	0.74	0.25
40–50	1.00	0.29
50–60	0.93	0.17
60–70	0.54	0.12
70–80	0.23	0.07
80–90	0.11	0.04

[a]Inventory values have been estimated from the cumulative ^{90}Sr deposition on the land and ocean surface of individual latitudinal belts and have been expressed as a ratio to the value for the 40–50° N belt, which has the greatest inventory.

redistribution occurs in association with soil and sediment particles.

The basis for using ^{137}Cs to estimate rates of accretion of overbank floodplain sediments reflects its accumulation in these sediments as a result of inputs from two primary sources. These sources are, firstly, direct atmospheric fallout to the floodplain surface and, secondly, the deposition of sediment-associated ^{137}Cs during the process of sediment accretion. In the latter case, the sediment-associated ^{137}Cs represents radiocaesium originating as fallout over the surface of the upstream catchment, which has been adsorbed by soil and sediment particles and subsequently mobilised by erosion and transported downstream as part of the suspended sediment load of the river. Some of this suspended sediment load and its associated ^{137}Cs will be deposited on the floodplain during overbank flooding. Both the vertical distribution and the total inventory (Bq m^{-2}) of ^{137}Cs in floodplain sediments will therefore commonly differ from those in the soils of adjacent undisturbed areas above the level of floodplain inundation, since the latter will have received ^{137}Cs only from direct fallout. Figure 9.9 compares the ^{137}Cs profile recorded for a sediment core collected from the floodplain of the river Stour in Dorset, UK, with that for a soil core collected from an adjacent area of undisturbed permanent pasture above the level of flood inundation. The substantially greater inventory of the floodplain core reflects the accumulation of additional ^{137}Cs associated with the deposition of fine sediment during overbank flood events. The progressive accumulation of fine sediment on the floodplain is reflected by the occurrence of ^{137}Cs to much greater depths in the floodplain core and the similarity between the ^{137}Cs depth distribution and the temporal pattern of fallout shown in Figure 9.8. The peak level of ^{137}Cs activity found at

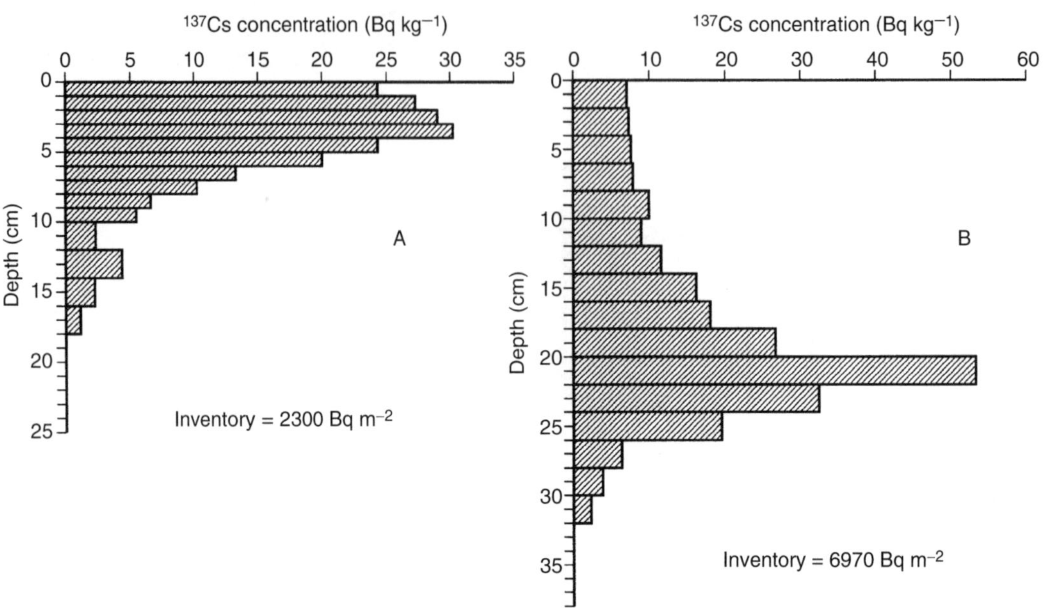

Figure 9.9 Comparison of ^{137}Cs profiles associated with a soil core from undisturbed pasture (A) and a sediment core from an adjacent site on the floodplain of the river Stour, Dorset, UK (B)

c. 21 cm can be ascribed to the surface of the floodplain in 1963, since this was the year of peak fallout and maximum activity in the sediment eroded from the upstream catchment (cf. Walling and He 1992).

Existing procedures for using ^{137}Cs measurements to estimate the age/depth relationship, and thus the sedimentation rate associated with a floodplain sediment core have used two approaches (cf. Walling and He 1997a). In the first, the shape of the ^{137}Cs depth profile (e.g. Figure 9.9B) is used in conjunction with information on the temporal pattern of fallout input, whilst in the second the sedimentation rate is estimated from a single value for the total inventory of the core. It is important to note that in both cases it is only possible to derive an estimate of the mean sedimentation rate for the period extending from the time of the main bomb fallout input to the time of collection of the sediment core and this in turn assumes that sedimentation is an essentially continuous, rather than more intermittent, process. Both approaches will be considered in turn.

In analysing the shape of the ^{137}Cs depth profile in relation to the record of fallout input, the absolute magnitude of the annual fallout inputs may not be known. However, it can generally be assumed that the relative magnitude of these values will follow the pattern shown by monitoring stations in the same region of the world. The record for Milford Haven (cf. Cambray *et al.* 1989) is thus widely used in Europe, whilst those for New York and Adelaide/Brisbane have been frequently used in North America and Australia, respectively. In the simplest application of this approach, either the depth of the level with peak activity is equated with the floodplain surface in the year of peak fallout input, which is commonly 1963 (cf. Ritchie *et al.* 1975, McHenry *et al.* 1976, Walling and He 1992), or the depth at which significant ^{137}Cs concentrations first appear is assumed to indicate the position of the surface in the mid-1950s (cf. Popp *et al.* 1988, Ely *et al.* 1992). In both cases, it is necessary to take account of possible post-depositional redistribution of ^{137}Cs in the sediment profile. For this reason the age/depth relationship is commonly established using the depth of the peak activity, since the position of the peak is less likely to be influenced by post-depositional redistribution than the depth at which ^{137}Cs first appears. An estimate of the downward migration rate associated with a ^{137}Cs profile can be obtained by examining the shape of the ^{137}Cs profile for an adjacent undisturbed soil

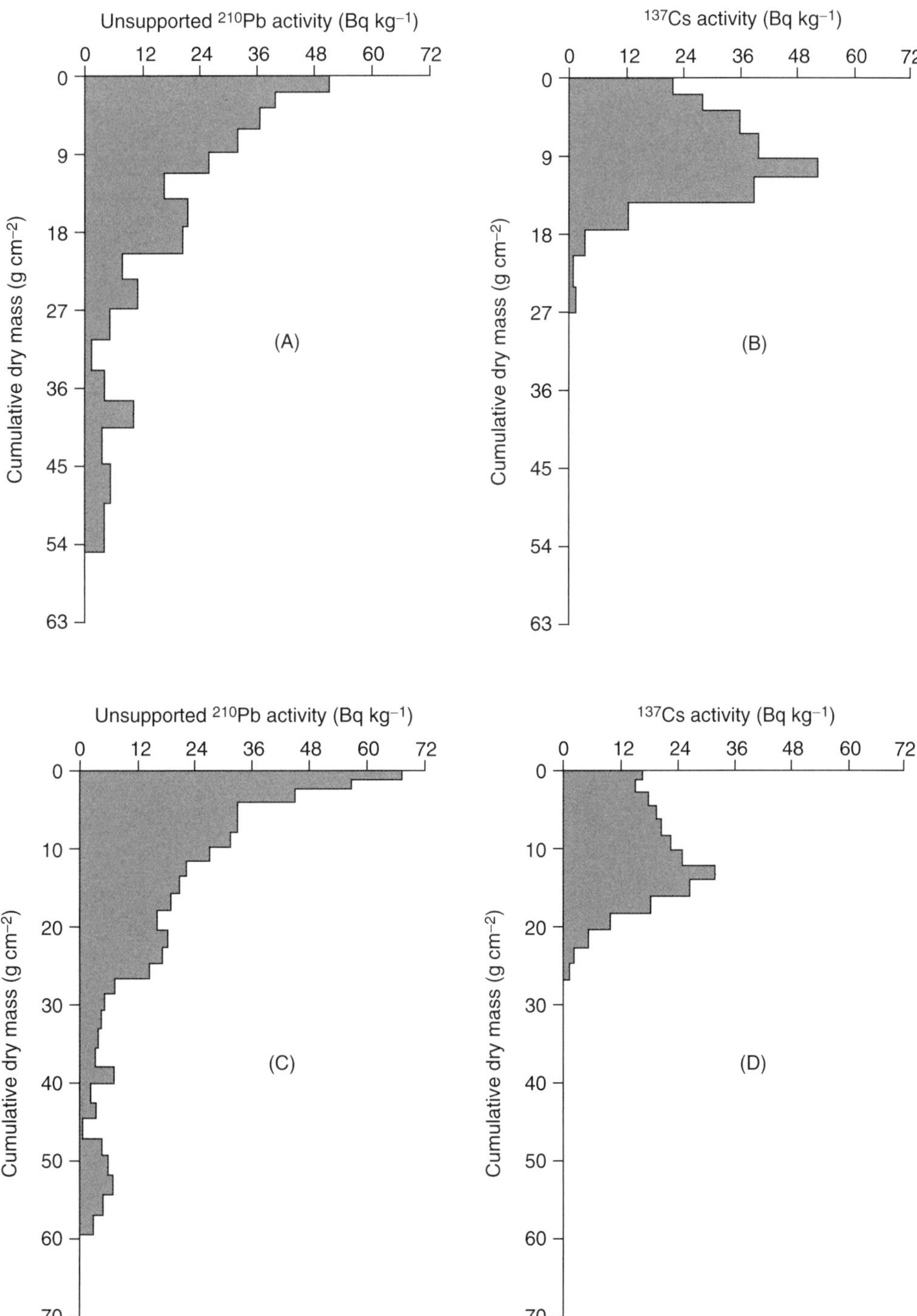

Figure 9.10 Unsupported ^{210}Pb and ^{137}Cs profiles for two sediment cores collected from the floodplains of the river Culm (A and B) and the river Exe (C and D)

above the level of floodplain inundation (e.g. Figure 9.9A). In the case of an undisturbed soil not subject to sediment accumulation, the peak ^{137}Cs activity should remain at the surface in the absence of downward migration. In most cases, however, the peak will be displaced slightly below the surface and the magnitude of the displacement affords a means of estimating the downward migration rate for the period between the year of peak fallout (commonly 1963) and the time of collection of the sample. In order to avoid complications in estimating sedimentation rates associated with downcore changes in bulk density, the ^{137}Cs depth profile is frequently plotted with depth expressed as a cumulative mass depth (g cm^{-2}) below the surface. Thus, if both the cumulative mass depth at which the peak ^{137}Cs activity occurs D_{pk}, (g cm^{-2}) and the downward migration rate of the ^{137}Cs peak R_m (g cm^{-2} year^{-1}) are known, the mean sedimentation rate R (g cm^{-2} year^{-1}) can be estimated as

$$R = (D_{pk}/T_p) - R_m \quad (2)$$

where T_p is the time (year) elapsed between the year of peak fallout and the time the core was collected. In a more complex procedure described by Walling and He (1997a), a model describing the key processes responsible for the development of the ^{137}Cs depth profile (e.g. fallout inputs, sediment deposition, and downward migration) was fitted to the measured depth distribution, to establish the sedimentation rate, which was treated as an unknown.

In the above discussion of the use of the ^{137}Cs depth profile to establish the age/depth relationship for a floodplain sediment core, it is implicitly assumed that the floodplain sediments are undisturbed and that the ^{137}Cs depth profile reflects the interaction of the progressive accumulation of sediment with the fallout input. In some situations, however, the floodplain will be cultivated and the sediment will be disturbed and in many cases mixed within the plough depth. Although it will no longer be possible to use the downcore ^{137}Cs profile, Allison *et al.* (1998) and Walling *et al.* (1998, 1999a) have shown that it is still possible to estimate the sedimentation rate, if the floodplain has been regularly cultivated since the inception of fallout. If no sedimentation has occurred, ^{137}Cs will only be found to a depth equivalent to the plough depth. If, however, sedimentation has occurred, its depth will be represented by the depth to which ^{137}Cs is found below the plough depth, due to the progressive accretion of the surface. Comparison of the plough depth D_{pl} (g cm^{-2}) with the depth to the base of the ^{137}Cs profile D_b (g cm^{-2}) therefore affords a means of estimating the sedimentation rate, namely,

$$R = \frac{D_b - D_{pl}}{T_s} \quad (3)$$

where T_s is the time elapsed (year) since the beginning of significant fallout and the collection of the sediment core.

The second approach to estimating the mean sedimentation rate for a sediment core is based on a single measurement of the total inventory I_{tot} (Bq m^{-2}) for the core. This value is, in turn, used to estimate the excess inventory I_{ex} (Bq m^{-2}) defined as the total inventory less the estimated local direct atmospheric fallout inventory I_{at} (Bq m^{-2}). The latter can be established using cores collected from adjacent areas of stable undisturbed grassland above the level of inundating floodwater. The excess inventory I_{ex} represents the ^{137}Cs input associated with deposited sediment and its magnitude will therefore reflect the rate of sediment deposition and the ^{137}Cs concentration in the deposited sediment. The latter will have varied through time in response to the temporal pattern of ^{137}Cs fallout, its accumulation within catchment soils and its subsequent remobilisation by erosion, and Walling and He (1997a) report the use of a model to represent this value and thereby estimate the mean sedimentation rate. A simpler approach invokes the assumption that, in relative terms, temporal variation of the ^{137}Cs concentration in deposited sediment will have been similar across the floodplain surface and that the main factor giving rise to spatial variation in the ^{137}Cs content of deposited sediment will be its grain size composition. If the sedimentation rate R_o (g cm^{-2} year^{-1}) and excess ^{137}Cs inventory $I_{ex,o}$ (Bq m^{-2}) are known for a specific reference point on the floodplain (for example, from a well-defined ^{137}Cs profile obtained from a sectioned sediment core), the sedimentation rate R at other points on the floodplain can be estimated from a comparison of the values of excess inventory I_{ex} and the grain size composition of the sediment at those points with that of sediment from the reference point, i.e.,

$$R = R_o \frac{I_{ex}}{I_{ex,o}} \left(\frac{S_{d,o}}{S_d} \right)^v \quad (4)$$

where $S_{d,o}$ and S_d are the specific surface area (m^2 g^{-1}) of surface sediment from the reference point and the measurement point, respectively, and v is a constant reflecting the relationship between ^{137}Cs concentration and specific surface area, which has been shown by He and Walling (1996b) to have a value of ca. 0.7. It is important to include the second term in Equation (4) and therefore the comparison of the specific surface area of surface sediment from the two sampling points, since, in addition to the deposition rate, the magnitude of the ^{137}Cs inventory will also be influenced by the grain size composition of the sediment. It is well known that the ^{137}Cs content of sediment from a common source will commonly increase as its grain size decreases (cf. He and Walling 1996b). Use of the specific surface area for surface sediment assumes that this is representative of sediment from the entire depth of the core and this is likely to be valid for most overbank deposits. This second approach is also applicable to cultivated floodplain surfaces, in that cultivation should not greatly influence the magnitude of the ^{137}Cs inventory, although tillage will itself serve to redistribute the floodplain sediment horizontally and may therefore smooth any spatial variation in sedimentation rate. However, the potential for winnowing of ^{137}Cs-rich fines from cultivated areas by wind erosion must also be recognised for cultivated floodplains, the value of R_o for the reference point will be based on Equation (3) rather than Equation (2).

A key factor in assessing the relative merits of the two approaches for estimating floodplain sedimentation rates from ^{137}Cs measurements outlined above is the number of laboratory measurements required. In order to define the ^{137}Cs profile, it is necessary to section the core into depth increments and to analyse each increment for its ^{137}Cs content. In view of the lengthy count times involved, it may therefore only be possible to analyse a small number of cores and this will necessarily limit the potential for studying spatial patterns of floodplain sedimentation. In contrast, the use of total inventory values involves only a single analysis of the bulk core and it is therefore possible to assemble information for a greater number of points on a floodplain and to document the spatial variability of sedimentation rates across a floodplain. In practice, a combination of the two approaches will frequently be used, since the deposition rate derived from a sectioned core can be used as a reference value to estimate the deposition rates for a large number of bulk cores collected from the adjacent area using Equation (4).

The above discussion of the use of ^{137}Cs measurements to estimate rates of overbank deposition on river floodplains relates to situations where the presence of radiocaesium in floodplain sediments reflects the input of bomb-derived ^{137}Cs during the period extending from the mid-1950s through to the 1970s. In some areas of Europe and adjacent regions, the Chernobyl disaster of 1986 will have provided additional inputs of ^{137}Cs fallout at that time (see De Cort *et al.* 1998). In most areas, the additional ^{137}Cs inventory associated with the Chernobyl input will be significantly less than the pre-existing bomb-derived inventory, but in some locations relatively close to Chernobyl, the Chernobyl input could be two orders of magnitude or more greater than the bomb fallout input (cf. Walling *et al.* 2000). Although the presence of Chernobyl-derived ^{137}Cs will complicate the interpretation of ^{137}Cs depth profiles and inventories, it can also afford increased potential for establishing the chronology of floodplain deposits. In the case of the ^{137}Cs depth profile, the existence of a second well-defined fallout input in 1986 will frequently be clearly apparent in the profile and this can afford a basis for establishing the depth of both the 1963 and the 1986 surfaces and thus deposition rates over the period 1963–1986 and 1986 to the present. Comparison of the two deposition rates could afford valuable information on changes in sedimentation rates over the recent past. The values of total ^{137}Cs inventory obtained for individual cores containing both bomb- and Chernobyl-derived radiocaesium will reflect both sources of radiocaesium, but, since the basic positive relationship between excess ^{137}Cs inventory and deposition rate will apply to both bomb-derived and Chernobyl fallout, it should again be possible to estimate deposition rates by relating the excess ^{137}Cs inventories for individual cores to that for a core where the total depth of sedimentation since 1963 has been established from the ^{137}Cs depth profile.

Using Unsupported ^{210}Pb Measurements

The use of unsupported ^{210}Pb measurements offers an alternative to the use of the ^{137}Cs technique for estimating the age/depth relationship for overbank floodplain sediments and thus deposition rates. Although the principles involved are essentially similar, the potential advantages of unsupported ^{210}Pb measurements lie, firstly, in the ability to cover somewhat longer timescales and, secondly, in providing an alternative in areas of the world where ^{137}Cs inventories are either very low due to limited bomb-derived fallout or unduly complicated by additional inputs of Chernobyl fallout. Pb-210 is a product of the U-238 (^{238}U) decay series with a half-life of 22.2 years and its presence in the atmosphere is due to the diffusion of Rn-222 (^{222}Rn), the daughter of Ra-226 (^{226}Ra), from the lithosphere into the atmosphere and its subsequent decay. As in the case of ^{137}Cs, deposition of ^{210}Pb from the atmosphere to the land surface occurs primarily as wet fallout in association with precipitation (cf. Nozaki *et al.* 1978, Nevissi 1985). However, whereas ^{137}Cs fallout has varied through time and, in the case of bomb fallout, was effectively restricted to the period extending from the mid-1950s to the 1970s, ^{210}Pb fallout represents a natural process and can be viewed as having been essentially constant through time. As with ^{137}Cs, the ^{210}Pb fallout flux to the land surface is rapidly and strongly adsorbed by soil and sediment particles and its subsequent redistribution reflects the movement of those particles. Fallout ^{210}Pb is commonly designated as excess or unsupported ^{210}Pb, when incorporated into soils or sediments, in order to distinguish it from the in situ ^{210}Pb produced by the decay of ^{226}Ra. It is this unsupported component of ^{210}Pb in sediment that offers potential for use in estimating floodplain sedimentation rates. As in the case of ^{137}Cs, the total inventory of excess ^{210}Pb in a floodplain core reflects both the direct atmospheric fallout flux to the floodplain surface and its deposition during overbank flood events in association with suspended sediment mobilised from the upstream catchment by erosion. Because of the essentially constant fallout flux and its half-life of 22.2 years, unsupported ^{210}Pb can be used to derive age/depth relationships for floodplain sediments and estimates of floodplain sedimentation rates extending back over about 100 years (i.e., 4–5 half-lives).

Figure 9.10 presents typical examples of the depth distribution of unsupported ^{210}Pb in floodplain sediment cores collected from the river Culm and the river Exe in Devon, UK The ^{137}Cs profiles for the same cores are shown for comparative purposes. Whereas with ^{137}Cs, establishment of the age/depth relationship depends on the depth of the peak concentration, which can be linked to the period of peak fallout in 1963, the essentially constant fallout input for ^{210}Pb means that there is no specific chronological marker and information on the age/depth relationship must be derived from the rate of decrease of the unsupported ^{210}Pb activity down the core in relation to the half-life of the radionuclide (22.2 years). Unsupported ^{210}Pb has been successfully used as a basis for establishing lake sediment chronologies over the past 100–150 years, through application of the constant flux and constant sedimentation rate (CFCS), constant initial concentration (CIC) and constant rate of supply (CRS) ^{210}Pb dating models (Appleby and Oldfield 1978, Robbins 1978, Oldfield and Appleby 1984a,b). However, He and Walling (1996a) have indicated that these models are not directly applicable to floodplain situations, due to the need to take account of essentially separate inputs of unsupported ^{210}Pb associated with fallout to the floodplain surface and deposition of sediment during infrequent overbank flood events. The latter will vary through time in response to variations in the magnitude and frequency of such events. In addition, post-depositional remobilisation of sediment and/or of ^{210}Pb itself may further complicate the depth distribution of fallout ^{210}Pb in the floodplain sediment profile. In view of these constraints, any attempt to date specific horizons in a sediment core and thereby decipher downcore changes in sedimentation rate is likely to be unsuccessful. Nevertheless, He and Walling (1996a) have demonstrated that it is possible to obtain an estimate of the long-term average sedimentation rate for a core using a constant initial concentration, constant sedimentation rate (CICCS) model. This also has the important advantage that it requires measurement of only the total unsupported ^{210}Pb inventory of a core rather than the sectioning of the core and measurement of the unsupported ^{210}Pb content of the individual depth increments. The basis of the approach is outlined below.

As noted above, the input of unsupported ^{210}Pb to a depositional location on a floodplain will commonly comprise contributions from two sources, namely, direct atmospheric fallout associated with precipitation inputs and an excess or catchment-derived input associated with suspended sediment deposited during overbank flood events. The total unsupported ^{210}Pb inventory I_{tot} (Bq m^{-2}) therefore comprises two components, namely, the local fallout input I_{at} (Bq m^{-2}) and the excess or catchment-derived inventory I_{ex} (Bq m^{-2}).

Since it is possible to establish I_{at} by measuring the unsupported ^{210}Pb inventory of an adjacent undisturbed stable site above the level of floodplain inundation, the catchment-derived inventory I_{ex} for a specific core can be estimated by subtracting I_{at} from I_{tot}. The magnitude of the value for the excess or catchment-derived inventory will reflect the sedimentation rate R (g cm^{-2} year^{-1}), the unsupported ^{210}Pb content of the sediment when originally deposited on the floodplain C_i (Bq kg^{-1}) and the decay constant of ^{210}Pb, λ_{Pb} (year^{-1}). If C_i can be estimated and can be assumed to be constant through time (CIC), the sedimentation rate R can be estimated as

$$R = \lambda_{Pb} \frac{I_{ex}}{C_i} \quad (5)$$

This relationship can be applied to both undisturbed and cultivated floodplain sediments, since the value of I_{ex} will not be influenced by cultivation.

In order to use Equation (5) to determine the sedimentation rate, the initial concentration of catchment-derived unsupported ^{210}Pb in deposited sediment, C_i, must be known. The value of C_i can be determined directly, for example, by deploying a sedimentation trap on the floodplain close to the core position and analysing the sediment collected. Alternatively, C_i may be estimated indirectly by measuring the unsupported ^{210}Pb content of suspended sediment transported by the river C_f (Bq kg^{-1}), at a location close to the coring site, and taking account of differences in grain size composition between the suspended and deposited sediment from the core by use of the relationship proposed by He and Walling (1996b), i.e.,

$$C_i = \left(\frac{S_c}{S_f}\right)^v C_f \quad (6)$$

where S_f and S_c represent the mean specific surface area (m^2 g^{-1}) of suspended sediment and deposited sediment from the floodplain core, respectively, and v is a constant with a value of about 0.7, reflecting the partitioning of unsupported ^{210}Pb between different size fractions. It is also possible to obtain an approximate estimate of the concentration of unsupported ^{210}Pb in deposited sediment contributed from the catchment, by subtracting the direct fallout contribution from the total unsupported ^{210}Pb concentration in recently deposited sediment, if the atmospheric-derived contribution can be estimated. Goodbred and Kuehl (1998) report the use of a similar relationship to those represented in Equations (5) and (6) to estimate sedimentation rates on the floodplain of the Ganges-Brahmaputra River within the Bengal Basin in India, although in their case the influence of particle size was incorporated by expressing C_i in terms of the ^{210}Pb content of the clay fraction of the catchment-derived sediment and including a factor to take account of the proportion of clay in the sediment core. In addition, account was taken of the potential for focussing of the atmospheric fallout reaching the floodplain surface, through concentration of surface runoff into the low-lying depressions or bils, by including a factor that increases the value of I_{at} for these areas.

Unlike the CIC and CRS models used to derive chronologies for lake sediment cores, the CICCS model described above can only be used to establish a linear age/depth relationship for a depositional floodplain site. The deposition rate given by Equation (5) assumes quasi-continuous sedimentation and therefore represents the mean value over approximately the last 100 years. Inspection of a representative unsupported ^{210}Pb depth profile (e.g. Figure 9.10) can be used to check the assumption of quasi-continuous sedimentation, since this should be reflected by an essentially monotonic reduction in the unsupported ^{210}Pb concentration with depth. The assumption of a constant C_i also implies that sediment sources and sediment transport and deposition processes have not shown major changes over the past 100 years. Again, this assumption can be verified in general terms by inspection of a representative unsupported ^{210}Pb depth profile. An essentially monotonic reduction in concentration with depth should be evident. The provision of a linear age/depth relationship and thus a mean sedimentation rate is also essentially similar to the result provided by the ^{137}Cs technique.

Field and Laboratory Techniques

The collection of sediment cores from the floodplain under investigation, for sectioning or analysis as a bulk core, is a key requirement of the ^{137}Cs and unsupported ^{210}Pb techniques outlined above. Core depth (i.e., length) will depend on the sedimentation rate, since it is clearly important that the core should extend to the base of the ^{137}Cs or unsupported ^{210}Pb activity. This is particularly the case for bulk cores, since it will not be possible to confirm the position of the base from the depth profile, and it may be advisable to analyse a slice from the base of the core to confirm that this slice contains no ^{137}Cs or unsupported ^{210}Pb. Motorised percussion corers are frequently used to obtain deeper cores and the choice of an appropriate core tube diameter should reflect both the potential heterogeneity of the sediment deposit and the sample mass required for analysis. In the former case, fine-grained overbank deposits can commonly be assumed to be relatively homogeneous, but if heterogeneity is likely to exist as a result of desiccation cracks, plant roots or similar causes, a larger diameter core is likely to provide more representative values for both the bulk inventory and also the inventory of individual depth increments. When establishing the reference or local fallout inventory for a study site, by collecting cores from a location above the level of inundating floodwater, it will frequently prove useful to collect multiple cores from a small area, in order to determine the mean inventory. A core tube with a diameter of ca. 6 cm is frequently employed for bulk cores. Sample mass is often an important consideration for sectioned cores, since the mass available for analysis will decrease as the thickness of the slice is decreased. A 12-cm diameter core tube should prove appropriate, where the core is sliced into 1 or 2 cm depth increments.

Laboratory analysis of core samples for ^{137}Cs content is commonly undertaken by gamma spectrometry using a p-type, coaxial, high purity germanium (HPGe) detector. Count times will depend on both the detector efficiency and on sample mass and activity, and count times of ca. 30 000 s or more are frequently required, in order to obtain a precision of ca. ±10% at the 95% level of confidence. As noted above, such extended count times can impose an important limitation on the number of sectioned cores that it is possible to analyse and it is for this reason that the potential for estimating sedimentation rates from a single measurement of the inventory of a bulk core is frequently seen as a major advantage.

In the case of unsupported ^{210}Pb, a number of analytical procedures, involving alpha, beta and gamma spectrometry can be employed (cf. Crickmore *et al.* 1990). Most early applications of ^{210}Pb dating for lake sediments employed alpha spectrometry to determine ^{210}Pb activities by measuring the alpha emissions of its grand-daughter polonium-210 (^{210}Po). Such measurements require chemical digestion and concentration of the sample prior to alpha counting (cf. Eakins 1983) and this requirement is one reason why recent workers have favoured the use of gamma spectrometry (cf. Joshi 1987). In this case, samples can be counted directly, without the need for extraction and concentration. A low-background, low-energy, n-type, HPGe detector can be used to measure both ^{210}Pb and ^{226}Ra at the same time, with the value for the latter commonly being derived through measurement of the activity of its short-lived daughter ^{214}Pb. The unsupported ^{210}Pb concentration is calculated as the difference between the total ^{210}Pb activity and the ^{226}Ra-supported ^{210}Pb activity. The ability to measure ^{226}Ra, and therefore calculate the ^{226}Ra-supported ^{210}Pb, is an important advantage of gamma spectrometry. This is not possible with alpha counting and in that case the ^{226}Ra-supported ^{210}Pb is generally estimated by documenting the near-constant ^{210}Pb activity at the base of a core, where the unsupported ^{210}Pb activity can be assumed to be minimal. This can, however, introduce uncertainties due to the need for the core to extend to sufficient depth to ensure that no unsupported ^{210}Pb is present. Samples analysed by gamma spectrometry must be sealed for ca. 20 days prior to counting to ensure equilibrium between ^{226}Ra and its daughter ^{222}Rn. As with ^{137}Cs, extended count times (ca. 80 000 s) are generally required in order to obtain a precision of ±10% at the 95% level of confidence. If, as indicated above, a suitable detector is available, it is possible to measure both ^{210}Pb and ^{137}Cs at the same time, although it must be recognised that the need to count the low energy (46 keV) ^{210}Pb peak may compromise the ability to measure the higher energy (662 keV) ^{137}Cs peak, if a planar detector is used. Reduced thickness, n-type low-background coaxial detectors are, however, generally suitable for this purpose. Where only a small sample mass is available (e.g. <20 g), a well detector will be required.

The Use of Other Environmental Radionuclides

The above discussion has focussed on the use of ^{137}Cs and unsupported ^{210}Pb to establish chronologies for recent fluvial sediments, since most such work has involved these two radionuclides. Scope also undoubtedly exists to make use of other environmental radionuclides, but further work is required to test their potential. Silicon-32 (^{32}Si), a naturally occurring cosmogenic fallout radionuclide with a half-life of ca. 140 years could, for example, provide a basis for extending chronologies back beyond the ca. 100 years offered by unsupported ^{210}Pb. Recent work in using beryllium-7 (^{7}Be), another naturally occurring cosmogenic fallout radionuclide, as a tracer in soil erosion investigations (Walling *et al.* 1999b) could also undoubtedly be extended to floodplain sediments. In the case of ^{7}Be, its very short half-life (53 days) can provide the basis for documenting sedimentation rates associated with individual events, rather than longer-term average rates of sediment accretion. If an isolated overbank flood event occurs, measurements of the excess ^{7}Be inventories (i.e., $I_{ex} = I_{tot} - I_{at}$) associated with a network of shallow cores collected shortly after the event, coupled with information on the ^{7}Be content of the sediment deposited on the floodplain, can be used to establish both the magnitude (g cm^{-2}) and the spatial pattern of overbank deposition rates for the event. In this application, however, it is important that the overbank flood event should be separated in time from any preceding overbank flood by several half-lives, so that the spatially variable I_{ex}

associated with sediment deposition during the preceding event will have effectively disappeared through radioactive decay. Immediately prior to the flood event I_{tot} will therefore be effectively equal to I_{at}, which in turn will be spatially uniform.

9.5 CASE STUDIES

OSL Dating of Fluvial Sediments

There have been several studies which have successfully dated fluvial deposits using TL and optical dating techniques (Table 9.2). Given the developmental status of the methods it remains important, where possible, to evaluate the quality of the results by comparison with other (where available particularly ^{14}C) methods, and to seek parameters obtained in the course of the dating exercises which provide a means of independently testing the robustness of the age-related information. Three case study examples are provided below which exemplify some of the many approaches which are currently being adopted.

Dating Colluvium from South Africa

For a considerable time, there has been uncertainty as to whether it is possible to date colluvial deposits given the anticipated low levels of bleaching that such samples are likely to have been subjected during transport and deposition (Stokes 1999). Wintle *et al.* (1995) demonstrated that by a careful interrogation of a combination of measurement procedures it is possible both to date some horizons and to evaluate the accuracy of other results from less well-bleached deposits. They achieved this by applying both TL and optical dating to two colluvial samples from St Paul's Mission, Natal, South Africa, which were independently indirectly dated via ^{14}C ages on an intercalated organic-rich palaeosol horizon (c. 1420 ± 60 BP). All equivalent dose estimates were based on either sand-sized (125–150 μm) potassium feldspar grains or fine (4–11 μm) grains of a mixed (carbonate-free) composition. Their optical dating strategy incorporated stimulation via infrared diodes (i.e., IRSL), and both multiple aliquot and single aliquot, and additive dose and regenerative D_e determination approaches. In the case of finer grain, overlying colluvial unit they found a wide difference in age between the TL (mean 25.7 ka) and the IRSL (mean c. 7 ka) ages and a considerable spread in the distribution of single aliquot ages. All ages significantly overestimated the maximum possible age of the unit. The underlying other sample was more sandy in texture and produced TL and IRSL ages which were comparable and a smaller range of D_e values. The mean age of the single aliquot-based analyses produced an age of 1770 ± 250 years, in good agreement with the radiocarbon age on the soil. Further promising research has recently been described using refinements of these basic approaches to fine grains from European colluvial deposits by Lang and Nolte (1999) and Banerjee *et al.* (2001).

Long-term Environmental Histories of Drainage Basins

A wide-ranging study of fluvial responses to environmental changes within the middle reaches of the Loire River, France, combined analysis of remotely sensed data, sedimentological analysis and optical dating of sand-sized quartz from within the fluvial sequences (Straffin *et al.* 1999, Colls *et al.* 2001). A significant advantage of the project was the presence of a suite of six radiocarbon dates which provided a basis for cross-checking of the younger (mid-Late Holocene) allostratigraphic units. This study produced a total of 19 single aliquot (SAR) dates which were demonstrated to be in agreement with the available (indirect) radiocarbon chronology, and that the eight widely recognisable terrace levels within the middle Loire reflected punctuated accumulation of sediments over the past 120 000 years (Figure 9.11). A significant novelty of this study was the incorporation of non-subjective tests for partial bleaching by comparison of SAR single aliquot luminescence intensities and D_e estimates.

Linking Monsoonal Variability and Palaeofloods in Low Latitudes

Well-integrated drainage systems are often rare within desert basins. However, given the extreme levels of inter-annual variability within such systems, large flood events may occur and produce characteristic slack water flood deposits within back-flooded tributary channel. Such deposits might be inferred to be only partially bleached during their transport within highly turbid water. On the other hand, the sediment transported has a high likelihood of being exposed at or near the desert surface for a considerable period. Kale *et al.* (2000) have demonstrated that it is possible to use optical dating approaches to exploit such sediments and develop high precision chronologies for multiple latest pre-historical flood events from the Thar Desert.

Table 9.2 Selected examples of previous studies which have used luminescence dating methods for the age evaluation of fluvial sediments

Study	Location	Fluvial/ colluvial deposit	Dosimeter	Size fraction (μm)	TL/ OSL[a]/ IRSL[a]	PMT[a]	Filter	Excitation source	Equivalent dose determination method	Age range (ka)	Independent age control?	General success?
Berger (1990)	Various	Various	Polymineral (predom. feldspars)	2–11	TL				Partial bleach	<0.1	N/a	Dependent upon depositional context
Page *et al.* (1991)	S.E. Australia	Riverine plain deposits	Unclear (Kfeldspar?)	Unknown	TL				Regenerative	27–105	U/Th (indirectly)	Yes
Ditlefsen (1992)	Laboratory study	Danish fluvioglacial sand	Kfeldspar	100–200	TL/IRSL		Scott B39 and HA3	IR diode	Additive dose	N/a	N/a	Yes for IRSL
Forman and Ennis (1992)	Svalbard	Glacial waterlain sediments	Polymineral	4–11	TL				Modified total and partial bleach method	Poor results	No?	No
Kamaludin *et al.* (1993)	Malaysia	Old alluvium	Quartz	90–125	TL				Combined regenerative and additive method	30–65	Study compared ^{14}C with TL	Yes (in that shows potential)
Murray *et al.* (1992)	Australia	Fluvial sands	Quartz	90–125	TL				Total bleach and regenerative methods	0–6	Excess 226RA decay; ^{14}C	Limited value for Holocene deposits
Proszynska-Bordas *et al.* (1992)	Spain	River terraces and fossil soils	Polymineral	4–11	TL	M12Fs52A (E. German)	Scott B12 filter		Partial bleach and regeneration	20–110	^{14}C; archaeological	Yes
Balescu and Lamothe (1994)	S. Italy, N. France, Quebec	Shallow marine, estuarine, lacustrine, sublittoral	Alkali feldspars	150–250	TL/IRSL	EMI 9635QA	Blue-transmitting Corning 7-59/ Scott BG39	30 IR emitting diodes	Additive gamma dose	10–177	^{14}C, amino acid and ESR dates	Yes
Duller (1994)	Scotland	Glacial outwash gravels/ glaciofluvial outwash	Kfeldspar	180–211	TL/IRSL		Corning 5–58 and Scott BG39 combination	13 TSHA 6203 IR diodes	Single aliquot/ partial bleach	c. <15	^{14}C and others not mentioned	Mixed results; some close agreement/some signif. overest.
Fuller *et al.* (1994)	Danube	Alluvial	Kfeldspar	4–11	IRSL		Schott BG39 and Corning 3-67		Partial bleach	various	Skeleton (Mesolithic buriel site)	18% errors in simulation experiment
Lamothe *et al.* (1994)	Quebec	Fluvial sands	Kfeldspar	0.5–1.0 mm	IRSL				Single grain	10	^{14}C	Potential; 30% underest. anom. fading

Lang (1994)	S. Germany	Colluvial; alluvial; lacustrine	Polymineral	4–11	IRSL	EMI 9635	Scott BG39 and Corning 3-67		Multiple aliquot	0–7	1990 flood deposit	No, big age overest.
Perkins and Rhodes (1994)	UK	Cold stage fluvial sands and gravels	Quartz	90–125	OSL		2 U-340 filters	3 GG420	Multiple aliquot additive growth	21–148	ESR, ^{14}C	Yes
Wintle *et al.* (1993)	South Africa	Colluvial	Polymineral	4–11	IRSL, TL	EMI 9635	BG 39	880 nm	Single and multiple aliquot	0–110	^{14}C	No
Wintle et al. (1995)	South Africa	Colluvial	Kfeldspar	180–212	IRSL	EMI 9635	BG 39	880 nm	Single aliquot	0–110	^{14}C	Yes
Fuller *et al.* (1996)	N.E. Spain	Quaternary fluvial deposits	Kfeldspars	>100	IRSL	EMI 9635			Partial bleach	13–250	No	Yes
Gemmell (1997)	Italian Alps	Suspended glacial meltwater	Polymineral	4–11	TL	EMI 9135	Schott UG-11 with a Chance-Pilkington HA3			N/a	N/a	No
Shaw *et al.* (1998)	Botswana	Lacustrine (diatomite)	Quartz	90–125	OSL	EMI 9135	2 7-51 and 1 BG-39	Ar^+ laser (λ = 514.5 nm)	Single aliquot additive dose	20–30	N/a	Yes
Olley *et al.* (1997)	Australia	Fluvial sands	Quartz	63–250	OSL	EMI 9635QA	3 U-340 filters	420–575 nm green light source	SARA, a basic SAR[a] method	<0.5	Geomorphic evidence	Yes
Colls *et al.* (2001)	France	Fluvial sands and gravelly sands	Quartz	90–600	OSL	EMI 9635QA	2 U-340 filters	420–575 nm green light and blue (470 nm) LED[a] sources	(SAAD[a] and SAR[a]	0–140	^{14}C	SAAD[a], No, method problematic SAR[a], yes
Olley *et al.* (1998)	Australia	Fluvial sands	Quartz	63–250	OSL	EMI 9635QA	3 U-340 filters	420–575 nm green light source	Single grain	<0.5	Geomorphic evidence	Yes
Straffin *et al.* (1999)	France	Fluvial sands and gravelly sands	Quartz	90–600	OSL	EMI 9635QA	2 U-340 filters	420–575 nm green light and blue (470 nm) LED[a] sources	SAAD[a] and SAR[a]	0–140	^{14}C	SAR[a], yes
Kale *et al.* (2000)	India	Fluvial sands	Quartz and feldspar	90–150	OSL/ IRSL	EMI 9635QA	2 U-340 filters	420–575 nm green light source	SAR[a] and differential partial bleach	0.1–1.0	No	Yes
Banerjee *et al.* (2001)	Germany	Colluvium	Polymineral fine grains	4–11	OSL/ IRSL	EMI 9635QA	2 U-340 filters	420–575 nm green light source	Modified SAR[a] using both IRSL[a] and OSL[a]	1–3	^{14}C	Yes

[a]PMT, photomultiplier; OSL, optically stimulated luminescence; IRSL, infrared stimulated luminescence; SAAD, single aliquot additive dose; SAR, single aliquot regeneration; LED, light emitting diode.

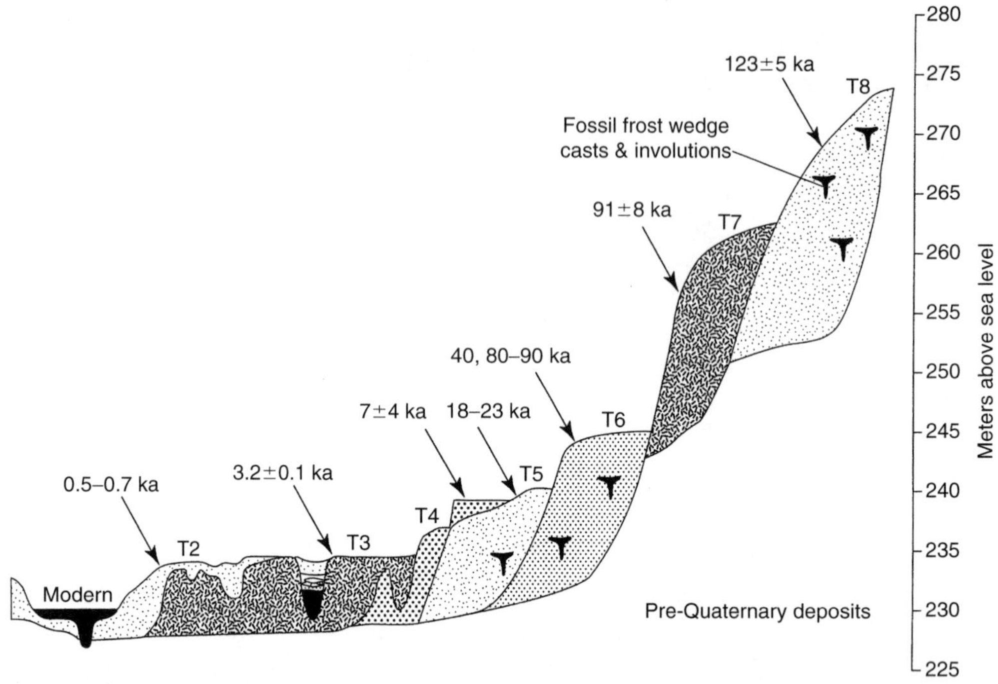

Figure 9.11 Summary of chronostratigraphy developed for fluvial terraces of the Middle Loire (after Colls *et al.* 2001)

Their approach was to apply a combination of additive dose and regeneration D_e determination procedures to both quartz and feldspar isolated from within the flood deposits. They produced a series of 12 age estimates ranging from 190 to 990 years which were internally consistent and indicated a flood recurrence interval of 40 years, with large (megaflood) events occurring ever c. 200 years. The data support the view that the frequency of flood events increased during the Medieval Warm Period, further suggesting regionally increased moisture level during that time.

^{137}Cs and ^{210}Pb Dating of Fluvial Sediments

Provision of a Chronology

Both ^{137}Cs and unsupported ^{210}Pb profiles for three sediment cores collected from the floodplain of the river Ouse in Yorkshire, UK, were used to establish chronologies to assist in the interpretation of downcore changes in sediment properties, which were related to changes in the relative importance of key sediment sources in the upstream catchment (cf. Owens *et al.* 1999). Measurements of ^{137}Cs and unsupported ^{210}Pb activities were undertaken on the sectioned cores and the downcore profiles were used to identify the depths associated with sediment deposited in 1963 (^{137}Cs) and 1900 (unsupported ^{210}Pb). Sediment fingerprinting procedures (cf. Walling and Woodward 1995, Collins *et al.* 1997, Walling *et al.* 1999c) were used to investigate downcore changes in the source of the suspended sediment deposited on the floodplain over the last 100 years or more. Multicomponent source fingerprints, involving a range of sediment properties, and a multivariate mixing model were employed in association with measurements of the downcore variation in the same sediment properties to establish the relative importance of topsoil and subsoil/channel bank sources within the upstream catchment in contributing to the sediment deposits represented by the cores (cf. Owens *et al.* 1999). The results are presented in Figure 9.12. In the case of core 1, which was collected from the floodplain in the lower reaches of the catchment, Figure 9.12 indicates that topsoil represents the dominant sediment source and suggests that there has been little change in the relative importance of the two sediment sources over the past 100 years represented by the core, although there is some evidence that topsoil sources increased in relative importance in the early part of the 20th century. Cores 2 and 3 were collected from floodplain sites further upstream

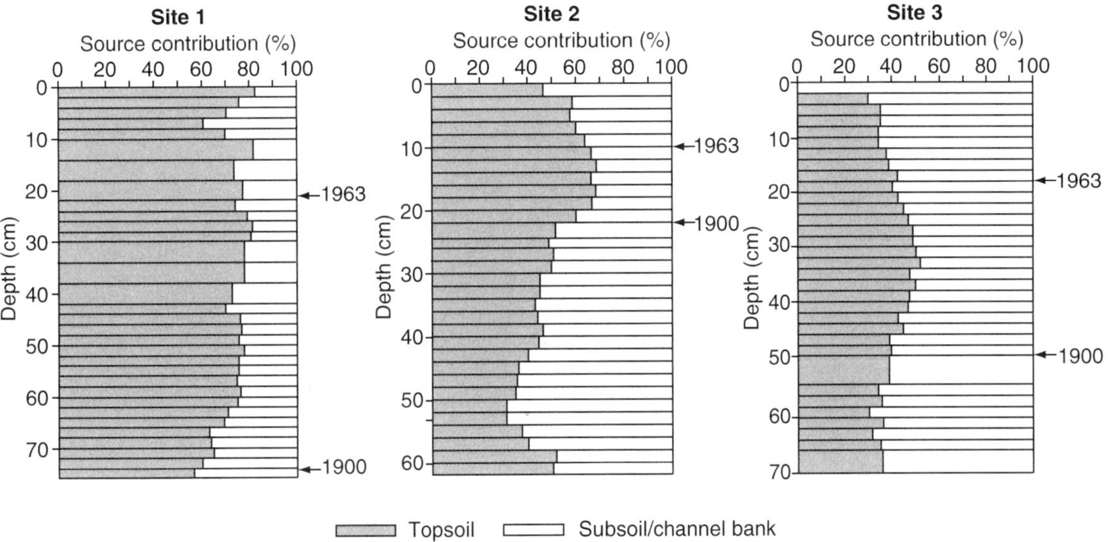

Figure 9.12 Downcore changes in the relative contribution of topsoil and subsoil/channel bank sources for the three cores collected from the floodplain of the river Ouse, Yorkshire, UK

on the river Ouse and these two cores show more evidence of changing sediment sources. More particularly, there is evidence that contributions from subsoil/channel bank material are dominant in the lower parts of the cores, which predate 1900, whereas topsoil sources can be seen to have increased in importance between 1900 and the 1950s. Owens *et al.* (1999) have related this increase to changes in land use within the upstream catchment areas during the early and middle parts of the 20th century.

Documenting Spatial Patterns of Overbank Sedimentation Rates of River Floodplains

The use of ^{137}Cs and unsupported ^{210}Pb measurements on bulk cores collected from river floodplains, in order to estimate the overbank sedimentation rate associated with the individual cores, affords a means of assembling data for a large number of points on a floodplain and thereby documenting the detailed spatial patterns involved. The potential of this approach can be demonstrated by considering some results from a study of a portion of the floodplain of the river Severn near Buildwas, Shropshire, UK. The study site occupies the inside of a meander bend and 124 cores (bulk and sectional cores) were collected from this area at the intersections of a 25 m × 25 m grid. The floodplain is occupied by permanent pasture. Additional reference cores were collected from adjacent areas of undisturbed pasture above the level of inundation, in order to establish the local ^{137}Cs and unsupported ^{210}Pb fallout inventories. In the case of ^{137}Cs, the sedimentation rates estimated for the sectioned cores were extrapolated to the bulk cores using the procedure represented by Equation (4). For the unsupported ^{210}Pb measurements, the average sedimentation rates over the past 100 years were estimated for all cores using the CICCS model and the procedure represented by Equation (5). In both cases, information on the particle size composition of sediment representative of the individual coring points was incorporated into the calculations to take account of spatial variability of sediment grain size across the study site.

The point estimates of sedimentation rate relating to the past 40 and 100 years provided by the ^{137}Cs and unsupported ^{210}Pb measurements, respectively, have been used to derive the maps of sedimentation rates presented in Figure 9.13. The average sedimentation rate for the set of 124 cores, based on the ^{137}Cs measurements, was 0.28 g cm^{-2} year^{-1}, whereas that based on the unsupported ^{210}Pb measurements was 0.33 g cm^{-2} year^{-1}. Sedimentation rates at this site therefore show little evidence of change between the two periods and their overall range (0.15–0.50 g cm^{-2} year^{-1}) is similar to that reported for other British rivers (e.g. Walling *et al.* 1996, He and Walling 1997, Walling and He 1998). The detail of the patterns

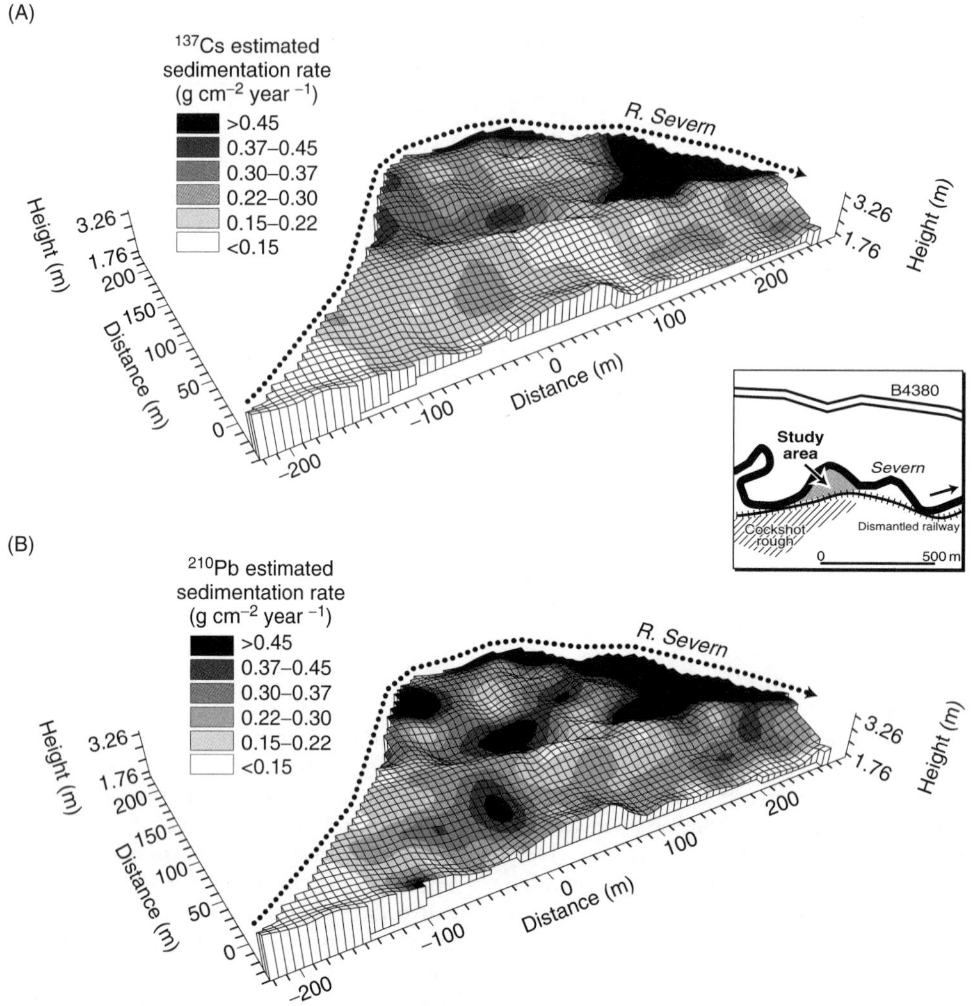

Figure 9.13 Deposition rates across the floodplain study site bordering the river Severn near Buildwas, estimated from ^{137}Cs (A) and unsupported ^{210}Pb (B) measurements

shown in Figure 9.13 reflects both the microtopography of the site, and more particularly the series of ridges and swales associated with meander development, and also a general trend of decreasing sedimentation with increasing distance from the channel and of increased sedimentation in areas occupied by deeper water during periods of flood inundation.

Quantifying Conveyance Losses Associated with Overbank Sedimentation

The results presented in Figure 9.13 relate to a detailed investigation of sedimentation patterns in a small area of floodplain. The use of ^{137}Cs and unsupported ^{210}Pb measurements can also be extended to larger scale investigations as demonstrated by Walling *et al.* (1998, 1999b) in their study of conveyance losses associated with overbank sedimentation on the floodplains bordering the main channel systems of the river Ouse and the river Wharfe, which together drain a catchment area of ca. 4100 km^2 in Yorkshire, UK (Figure 9.14). In this case, there was a need to estimate the total mass of sediment deposited on the floodplains bordering the main channels of the river system and thus the magnitude of the conveyance loss. By relating this loss to the suspended sediment output

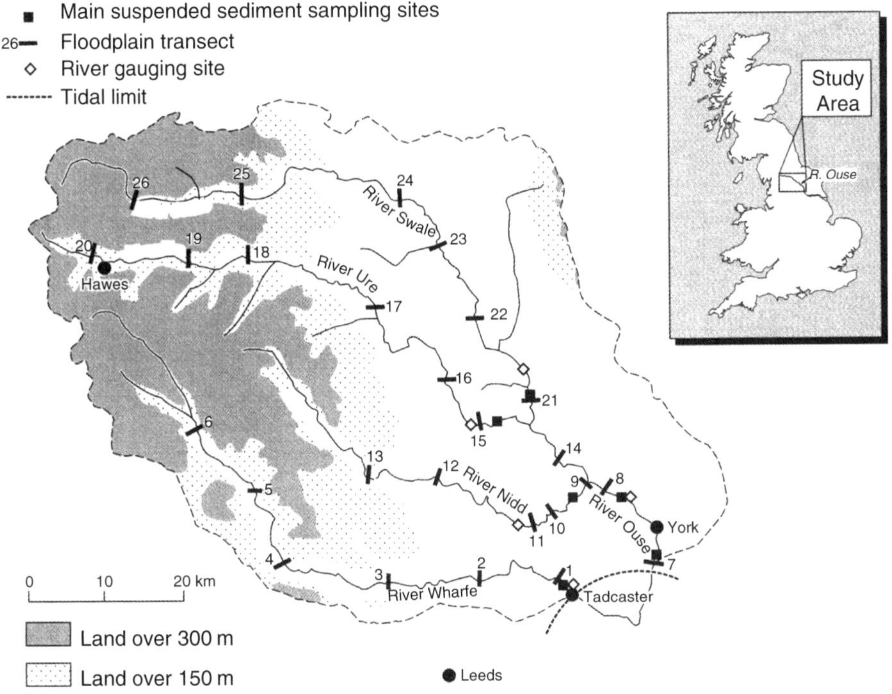

Figure 9.14 The catchment of the rivers Ouse and Wharfe, Yorkshire, UK, and the location of the floodplain transects used to estimate sediment deposition rates

from the river system, it was possible to establish the importance of the conveyance loss relative to the total suspended sediment input to the main channel system.

This application necessitated scaling-up sedimentation rate estimates obtained for individual points on the floodplain to the entire floodplain area bordering the main river system. This was achieved through the use of transects across the floodplain at representative sites along the main channels and extrapolation of the resulting estimates of the mean sedimentation rates for the individual transects to the adjacent floodplain reaches. Twenty six representative transects (Figure 9.14), involving a total of more than 250 sediment cores, were employed and emphasis was placed on the use of ^{137}Cs measurements to obtain estimates of average sedimentation rates over the past ca. 40 years. The cores collected from each transect included a single sectioned core taken from a representative location and a series of bulk cores taken from other representative points along the transect. Reference cores were also collected at a number of stable (non-eroded) undisturbed sites located along the river system, but above the level of flood inundation, in order to establish the local fallout inventory for individual transects. The value of sedimentation rate obtained for the sectioned core was used to derive estimates of the sedimentation rate associated with the bulk cores, based on their values of excess inventory and information on their grain size composition, using the procedure contained in Equation (4). Values of mean sedimentation rate (g cm^{-2} year^{-1}) were derived for each transect and extrapolation of these values to the individual reaches between adjacent transects provided estimates of the mean annual conveyance losses associated with overbank sedimentation on the floodplains bordering the main channel system of the river Ouse and its major tributaries and the river Wharfe (Table 9.3). By relating these losses to estimates of the mean annual suspended sediment load of the rivers (Table 9.3), it was possible to establish the relative importance of floodplain storage in the sediment budget of the main channel system. In the case of the main river Ouse system, floodplain deposition was estimated to account for ca. 40% of the total mass of sediment delivered to the main channel system over the past 40 years. For the river Wharfe, the equivalent value was 49%.

Table 9.3 A comparison of the estimates of total sediment storage on the floodplains bordering the main channel systems of the river Ouse and its primary tributaries and the river Wharfe, with the estimated suspended sediment loads for the study rivers (based on Walling *et al.* 1998, 1999a)

River	Floodplain storage ($t\,year^{-1}$)	Mean annual suspended sediment load ($t\,year^{-1}$)	Total sediment delivered to channel ($t\,year^{-1}$)	Floodplain storage a percentage of sediment input
Swale	19 214	42 352	61 566	31.2
Nidd	7 573	7 719	15 292	49.5
Ure[a]	15 125	28 887	44 012	34.3
Ouse[a]	18 733			
Total to Ouse gauging station	49 041	75 111	124 152	39.5
Total to tidal limit	60 645			
Wharfe	10 325	10 816	21 141	48.8

[a]Ure is the river Ure to its confluence with the river Swale, while Ouse refers to the river Ure/Ouse from below this point to the tidal limit.

Assessing Changing Rates of Floodplain Sedimentation over the Past 100 years

As indicated above, measurements of both ^{137}Cs and unsupported ^{210}Pb can be undertaken simultaneously on the same core and this affords a means of investigating changes in overbank sedimentation rates over the past 100 years, in that estimates of sedimentation rates relating to the past ca. 35 years derived from the ^{137}Cs measurements can be compared with equivalent values for the past ca. 100 years derived from unsupported ^{210}Pb measurements. The overlap between the two periods necessarily reduces the temporal resolution of the results, but they can, nevertheless, provide valuable evidence as to whether sedimentation rates have increased, decreased or remained essentially stationary in recent years (cf. Walling and He 1994, 1999a).

As an example of this approach, Figure 9.15 presents information on the depth distributions of ^{137}Cs and unsupported ^{210}Pb in cores collected from representative sites on the floodplains of two British rivers, namely, the river Tone in Somerset and the river Arun in Sussex. The cores exhibit well-developed ^{137}Cs peaks at about 180 and 130 mm below the surface and the average sedimentation rate over the period extending from 1963 to the time of collection of the cores in 1996 has been estimated using Equation (2). The resulting values of mean annual sedimentation rate are 0.56 and 0.39 $g\,cm^{-2}\,year^{-1}$ for the rivers Tone and Arun, respectively. The equivalent estimates of mean annual sedimentation rate for the past ca. 100 years obtained from the unsupported ^{210}Pb measurements are 0.43 and 0.48 $g\,cm^{-2}\,year^{-1}$, respectively. These results indicate that the average sedimentation rate has increased towards the present in the case of the site on the floodplain of the river Tone, whereas the core from the river Arun shows evidence of a decreasing sedimentation rate towards the present.

Extension to Other Fluvial Sediments

The use of ^{137}Cs and unsupported ^{210}Pb as a means of dating recent fluvial sediments is primarily restricted to fine (i.e., $<$0.063 mm) sediment, due to the preferential association of radionuclides with fine-grained material. The approach therefore mainly has been applied to overbank sediments on river floodplains. Scope may, however, exist to exploit the time-marker provided by the peak ^{137}Cs fallout rate in the early 1960s or by the input of Chernobyl fallout in 1986 to identify a datable horizon in other fluvial sediments containing at least some fine material. Figure 9.16 provides an example of one such application. This relates to the accretion of the braided channel system of a small proglacial stream draining from the Mitdluagkat glacier in eastern Greenland (cf. Hasholt and Walling 1992). This area had been recently deglaciated and interest focussed on the development of the channel system and more particularly the accretion of fine sediment (i.e., sand and silt) within the channel and on adjacent bars. By documenting the vertical distribution of ^{137}Cs in the sediment deposits (cf. Figure 9.16), it was possible to confirm that much

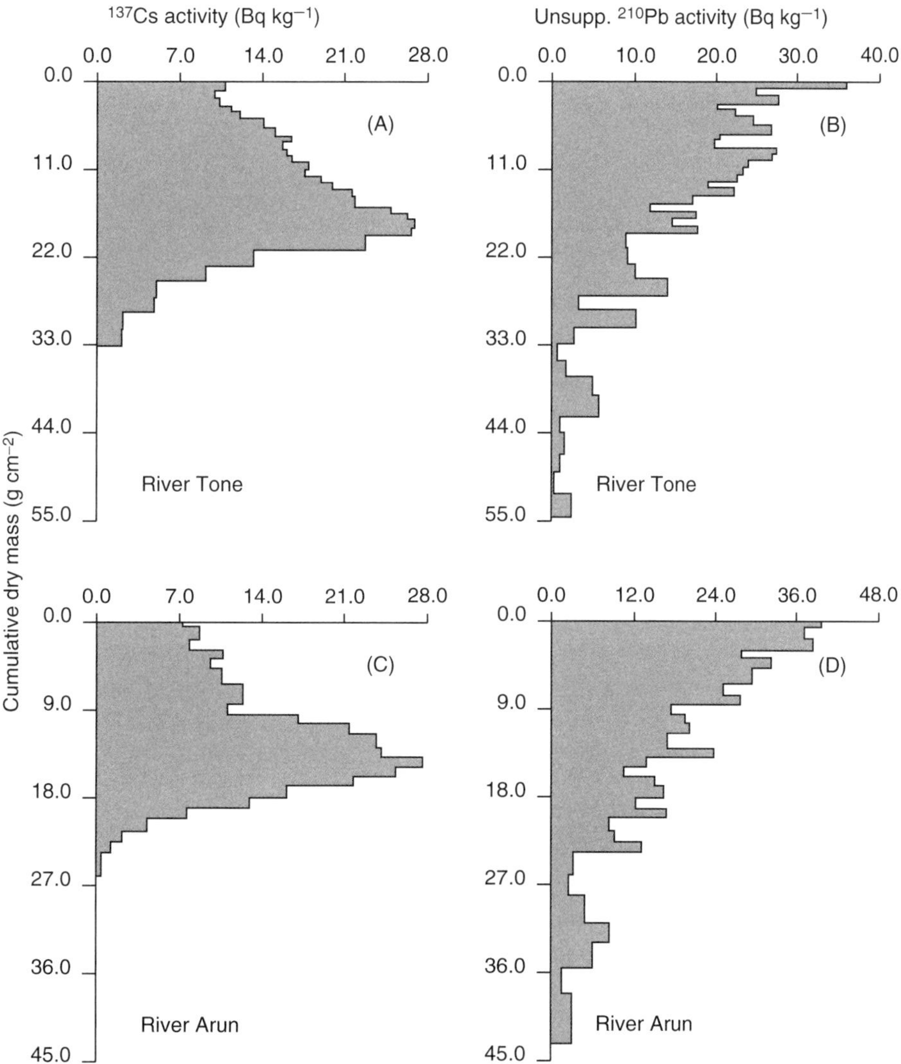

Figure 9.15 ^{137}Cs and unsupported ^{210}Pb profiles associated with sediment cores collected from the floodplains of the river Tone (A and B) and the river Arun (C and D), UK

of the sediment was deposited after the time of peak fallout in 1963 and to estimate the depth of accretion since 1963. Because of the heterogeneity of such sediment deposits, it is important to ensure that vertical variations in ^{137}Cs activity are not simply a reflection of variations in the grain size composition of the sediment. Measurements of unsupported ^{210}Pb activity are likely to be of little or no value in such situations, since they are unable to identify a specific datable horizon.

9.6 CONCLUSIONS

This chapter has primarily sought to review applications and recent developments in two groups of techniques for dating fluvial sediments. The nature of optical dating and its potential for dating fluvial deposits has evolved radically in the past decade. While early studies based on TL or multiple aliquot optical dating demonstrated their capacity to provide good relative and absolute age control in ideal situations, high levels of suspended sediment or short transport distances negated their

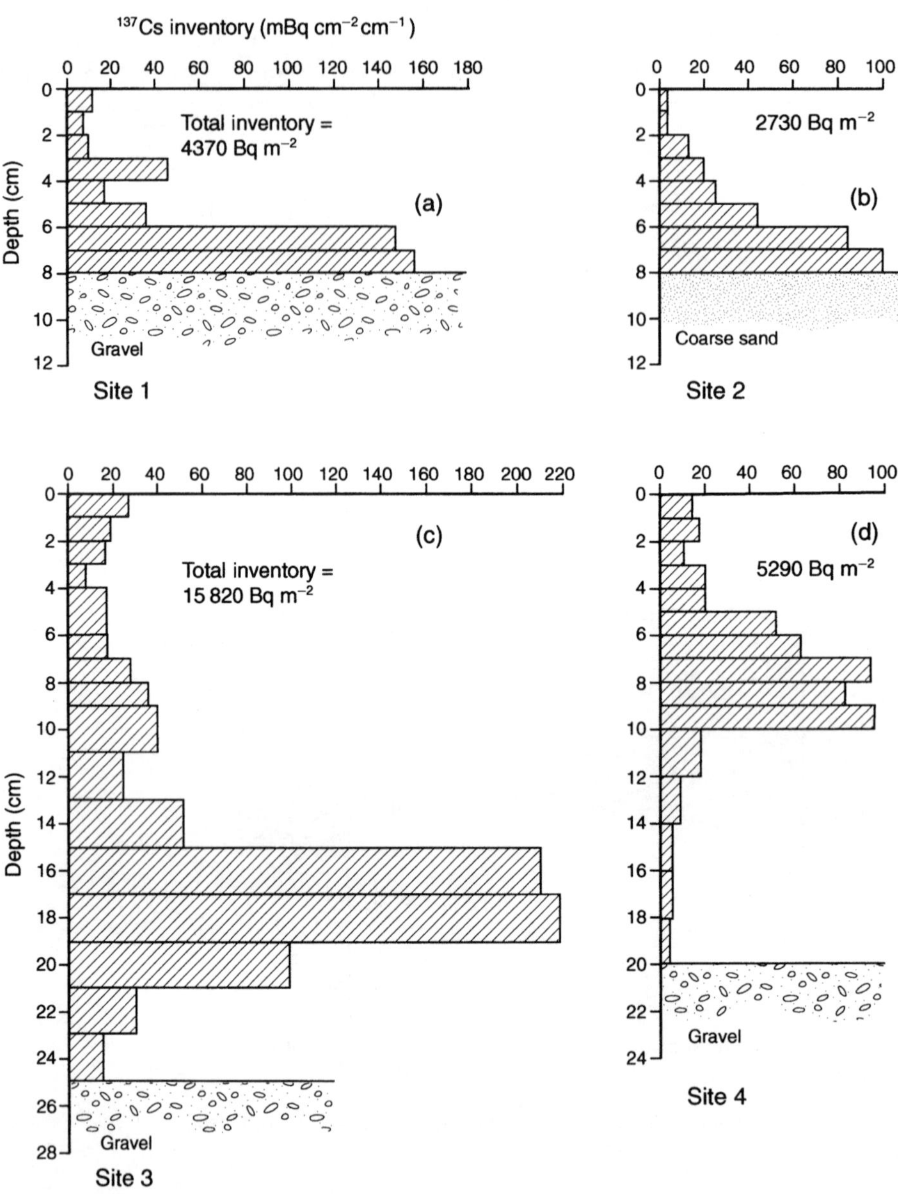

Figure 9.16 ^{137}Cs profiles associated with four depositional sites within the valley of the Mitdluagkat stream, eastern Greenland (based on Hasholt and Walling 1992)

universal applicability, particularly over Holocene and historical time ranges. The latest methodological developments have incorporated single aliquot and single grain techniques for D_e evaluation. Such methods allow a much higher precision than multiple aliquot procedures and additionally provide a range of measures by which the accuracy of the date, and the presence of poorly bleached grains, can be evaluated. They represent a powerful, yet not fully utilised, means by which chronologies may be constructed for many fluvial deposits over timescales ranging from decades to hundreds of thousands of years. The extent to which assumptions relating to the completeness of bleaching can be confirmed are central to the accuracy of luminescence dates for fluvial sediments. Techniques for the assessment of sediment population bleaching have evolved radically in the past few years and their use, in combination with the SAR method, mean that high precision, accurate chronologies for fluvial deposits can be constructed.

In the case of the use of short-lived environmental radionuclides, and more particularly ^{137}Cs and unsupported ^{210}Pb, applications to fluvial sediments have largely developed from earlier works on dating lake sediments. Their value in studies of fluvial sediments has been demonstrated clearly now. Limitations inevitably exist in terms of the preferential association of the radionuclides with the fine fraction (clay and silt) and this has favoured their application in studies of overbank sedimentation on river floodplains. However, considerable scope undoubtedly exists to extend the use of short-lived radionuclides to other fluvial sediments. Furthermore, an important characteristic of such fallout radionuclides is their ubiquitous input to the landscape and thus the potential to couple studies of fluvial deposits with complementary investigations of colluvial deposits. Both ^{137}Cs and unsupported ^{210}Pb have also been applied very successfully as tracers in soil erosion investigations (cf. Ritchie and McHenry 1990, Walling and He 1999b) and there is a considerable scope to use these radionuclides as a tool for studying both erosion and deposition rates and therefore for undertaking integrated assessments of the various components of catchment sediment budgets (cf. Owens *et al.* 1997, Walling 1998).

REFERENCES

Aitken, M.J. 1985. *Thermoluminescence Dating*, London: Academic Press.

Aitken, M.J. 1988. *Introduction to Optical Dating*, Oxford: Oxford University Press.

Aitken, M.J. 1998. *An Introduction to Optical Dating: The Dating of Quaternary Sediment by Use of Photon-stimulated Luminescence*, Oxford: Oxford University Press.

Allison, M.A., Kuehl, S.A., Martin, T.C. and Hassan, A. 1998. Importance of flood-plain sedimentation for river sediment budgets and terrigenous input to the oceans: insights from the Brahmaputra-Jamuna River. *Geology* 26: 175–178.

Appleby, P.G. and Oldfield, F. 1978. The calculation of lead-210 dates assuming a constant rate of supply of unsupported ^{210}Pb to the sediment. *Catena* 5: 1–8.

Balescu, S. and Lamothe, M. 1994. Comparison of TL and IRSL age estimates of feldspar coarse grains from waterlain sediments. *Quaternary Geochronology* 18: 437–444.

Banerjee, D., Murray, A.S., Botter-Jensen, L. and Lang, A. 2001. Equivalent dose estimation using a single aliquot of polymineral fine grains. *Radiation Measurements* 33: 73–94.

Berger, G.W. 1990. Effectiveness of natural zeroing of the thermoluminescence in sediments. *Journal of Geophysical Research* 95: 12 375–12 397.

Berger, G.W. and Luternauer, J.J. 1987. Preliminary fieldwork for thermoluminescence dating studies at the Fraser River delta, British Colombia. *Geological Survey of Canada Paper* 87: 901–904.

Berger, G.W., Pillans, B.J. and Palmer, A.S. 1992. Dating loess up to 800 ka by thermoluminescence. *Geology* 20: 403–406.

Birkeland, P.W. and Noller, J.S. 2000. Rock and mineral weathering. In: Stratton, J.S., Sowers, J.M. and Lettis, W.R., eds., *Quaternary Geochronology: Methods and Applications*, AGU Reference Shelf 4, Washington: American Geophysical Union, pp. 273–292.

Bishop, P., Campbell, B. and McFadden, C. 1991. Absence of caesium-137 from recent sediments in eastern Australia: indication of catchment processes? *Catena* 18: 61–69.

Cambray, R.S., Playford, K. and Carpenter, R.C. 1989. Radioactive fallout in air and rain: results to the end of 1988, U.K. Atomic Energy Authority Report AERE-R 13575, HMSO, London.

Collins, A.L., Walling, D.E. and Leeks, G.J.L. 1997. Sediment sources in the Upper Severn catchment: a fingerprinting approach. *Hydrology and Earth System Sciences* 1: 509–521.

Colls, A.E.L., Stokes, S., Blum, M.D. and Straffin, E. 2001. Age limits on the Late Quaternary evolution of the upper Loire River. *Quaternary Science Reviews* (*Quaternary Geochronology*) 20: 743–750.

Colman, S.M. and Pierce, K.L. 2000. Classification of Quaternary geochronologic methods. In: Noller, J.S., Sowers, S. and Lettis, W.R., eds., *Quaternary Geochronology: Methods and Applications*, Washington, DC: American Geophysical Union.

Costa, J.E. 1975. Effects of agriculture on erosion and sedimentation in the Piedmont Province, Maryland. *Geological Society of America Bulletin* 86: 1281–1286.

Crickmore, M.J., Tazioli, G.S., Appleby, P.G. and Oldfield, F. 1990. *The Use of Nuclear Techniques in Sediment Transport and Sedimentation Problems*, UNESCO Technical Documents in Hydrology, Paris: UNESCO.

De Cort, M., *et al.* 1998. *Atlas of Caesium Deposition on Europe After the Chernobyl Accident*, ECSC-EEC-EAEC, Brussels-Luxembourg.

Ditlefsen, C. 1992. Bleaching of K-feldspars in turbid water suspensions: a comparison of photo- and thermoluminescence signals. *Quaternary Science Reviews* 11: 33–38.

Duller, G.A.T. 1991. Equivalent dose determinations using single aliquots. *Nuclear Tracks and Radiation Measurements* 18: 371–378.

Duller, G.A.T. 1994. Luminescence dating of poorly bleached sediments from Scotland. *Quaternary Geochronology* (*Quaternary Science Reviews*) 13: 521–524.

Duller, G.A.T. 1996. Recent developments in luminescence dating of Quaternary sediments. *Progress in Physical Geography* 20: 127–145.

Duller, G., Boetter-Jensen, L., Murray, A. and Truscott, A. 1999. Single grain laser luminescence (SGLL) measurements using a novel automated reader. *Nuclear Instruments and Methods in Physics Research B* 155: 506–514.

Eakins, J.D. 1983. The ^{210}Pb techniques for dating sediments and some applications, Report AERE-R-10821, UK AEARE, Harwell.

Ely, L.L., Webb, R.H. and Enzel, Y. 1992. Accuracy of post-bomb ^{137}Cs and ^{14}C in dating fluvial deposits. *Quaternary Research* 38: 196–204.

Forman, S.L. and Ennis, G. 1992. Limitations of thermoluminescence to date waterlain sediments from glaciated fjord environments of western Spitsbergen, Svalbard. *Quaternary Science Reviews* 11: 61–70.

Frissel, M.J. and Pennders, R. 1983. Models for the accumulation and migration of ^{90}Sr, ^{137}Cs, ^{239,240}Pu and ^{241}Am in the upper layers of soils. In: Coughtrey, P.J., ed., *Ecological Aspects of Radionuclide Release*, Oxford: Blackwell, pp. 63–72.

Fuller, I.C., Wintle, A.G. and Duller, G.A.T. 1994. Test of the partial bleach methodology as applied to the infra-red stimulated luminescence of an alluvial sediment from the Danube. *Quaternary Geochronology* (*Quaternary Science Reviews*) 13: 539–543.

Fuller, I.C., Macklin, M.G., Passmore, D.G., Brewer, P.A., Lewin, J. and Wintle, A.G. 1996. Geochronologies and environmental records of Quaternary fluvial sequences in the Guadalope Basin, northeast Spain, based on luminescence dating. In: Branson, J., Brown, A.G. and Gregory, K., eds., *Global Continental Changes: The Context of Palaeohydrology*, Geological Society Special Publication, London: Geological Society, pp. 99–120.

Fuller, I.C., Macklin, M.G., Lewin, J., Passmore, D.G. and Wintle, A.G. 1998. River response to high-frequency climate oscillations in southern Europe over the past 200 ky. *Geology* 26: 275–278.

Gemmell, A.M.D. 1988. Zeroing of the TL signal in sediment undergoing fluvio-glacialtransport: an example from Austerdalen, Western Norway. *Quaternary Science Review* 7: 339–345.

Gemmell, A.M.D. 1997. Fluctuations in the thermoluminescence signal of suspended sediment in an Alpine glacial meltwater stream. *Quaternary Geochronology* (*Quaternary Science Reviews*) 16: 281–290.

Godfrey-Smith, D.I., Huntley, D.J. and Chen, W.-H. 1988. Optical dating studies of quartz and feldspar sediment extracts. *Quaternary Science Reviews* 7: 373–380.

Goodbred, S.L. and Kuehl, S.A. 1998. Floodplain processes in the Bengal Basin and the storage of Ganges-Brahmaputra River sediment: an accretion study using ^{137}Cs and ^{210}Pb geochronology. *Sedimentary Geology* 121: 239–258.

Happ, S.C. 1968. Valley sedimentation in north-central Mississippi. In: *Proceedings of the Mississippi Water Resources Conference*, pp. 1–8.

Hasholt, B. and Walling, D.E. 1992. Use of caesium-137 to investigate sediment sources and sediment delivery in a small glacierized mountain drainage basin in eastern Greenland. In: *Erosion, Debris Flows and Environment in Mountain Regions*, IAHS Publication No. 209, Wallingford, UK: IAHS Press, pp. 87–100.

He, Q. and Walling, D.E. 1996a. Use of fallout Pb-210 measurements to investigate longer-term rates and patterns of overbank sediment deposition on the floodplains of lowland rivers. *Earth Surface Processes and Landforms* 21: 141–154.

He, Q. and Walling, D.E. 1996b. Interpreting particle size effects on the adsorption of ^{137}Cs and unsupported ^{210}Pb by mineral soils and sediments. *Journal of Environmental Radioactivity* 30: 117–137.

He, Q. and Walling, D.E. 1997. Rates of overbank deposition on the floodplains of British lowland rivers documented using fallout ^{137}Cs. *Geografiska Annaler* 78A: 223–234.

Huntley, D.J., Hutton J.T. and Prescott, J.R. 1993. Optical dating using inclusions within quartz grains. *Geology* 21: 1087–1090.

Hupp, C.R. 1988. Plant ecological aspects of flood geomorphology and palaeoflood history. In: Baker, V.R., Kochel, R.C. and Patton, P.C., eds., *Flood Geomorphology*, Chichester: Wiley, pp. 330–356.

Joshi, S.R. 1987. Nondestructive determination of lead-210 and radium-226 in sediments by direct photon analysis. *Journal of Radioanalysis and Nuclear Chemistry, Articles*, 116: 169–182.

Kale, V.S., Singhvi, A.K. Mishra, P.K. and Banerjee, D. 2000. Sedimentary records and luminescence chronology of Late Holocene palaeofloods in the Luni River. *Catena* 40: 337–358.

Kamaludin, B.H., Nakamura, T., Price, D.M., Woodroffe, C.D. and Fujii, S. 1993. Radiocarbon and thermoluminescence dating of the Old Alluvium from a coastal site in Perak, Malaysia. *Sedimentary Geology* 88: 199–210.

Knox, J.C. 1987. Historical valley floor sedimentation in the Upper Mississippi Valley. *Annals Association of American Geographers* 77: 224–244.

Lamothe, M., Balescu, S. and Auclair, M. 1994. Natural IRSL intensities and apparent luminescence ages of single feldspar grains extracted from partially bleachedsediments. *Radiation Measurements* 23: 555–562.

Lang, A. 1994. Infra-red stimulated luminescence dating of Holocene reworked silty sediments. *Quaternary Geochronology* (*Quaternary Science Reviews*) 13: 525–528.

Lang, A. and Nolte, S. 1999. The chronology of Holocene alluvial sediments from the Wetterau, Germany, provided by optical and C-14 dating. *The Holocene* 9: 207–214.

Larsen, R.J. 1985. *Worldwide deposition of* ^{90}Sr *through 1983*, New York: United States Department of Energy.

Leopold, L.B. 1973. River channel change with time: an example. *Geological Society of America Bulletin* 84: 1845–1860.

Lewin, J. and Macklin, M.G. 1987. Metal mining and floodplain sedimentation in Britain. In: Gardiner, V., ed., *Inter-*

national Geomorphology, Part 1, Chichester: Wiley, pp. 1009–1027.

Livens, F.R. and Rimmer, D.L. 1988. Physico-chemical controls on artificial radionuclides in soils. *Soil Use and Management* 4: 63–69.

Mackey, S.D. and Bridge, J.S. 1995. Three-dimensional model of alluvial stratigraphy: theory and application. *Journal of Sedimentary Research* B 65: 7–31.

Macklin, M.G. and Dowsett, R.B. 1989. The chemical and physical speciation of trace metals in fine-grained overbank flood sediments in the Tyne Basin, north-east England. *Catena* 16: 135–151.

McHenry, J.R., Ritchie, J.C. and Verdon, J. 1976. Sedimentation rates in the Upper Mississippi River, In: *Rivers '76*, Volume II, New York: ASCE, pp. 1339–1349.

Mejdahl, V. and Botter-Jensen, L. 1994. Luminescence dating of archaeological sediments using a new techniques based on single aliquot measurements. *Quaternary Geochronology* 7: 551–554.

Mejdahl, V. and Botter-Jensen, L. 1997. Experience with the SARA OSL method. *Radiation Measurements* 27: 291–294.

Murray, A.S. and Mejdahl, V. 1998. Comparison of regenerative-dose single-aliquot and multiple-aliquot (SARA) protocols using heated quartz from archaeological sites. *Quaternary Science Reviews* (*Quaternary Geochronology*) 18: 223–229.

Murray, A.S. and Roberts, R.G. 1998. Measurement of the equivalent dose in quartz using a regenerative-dose single-aliquot protocol. *Radiation Measurements* 29: 503–515.

Murray A.S. and Wintle A.G. 2000. Luminescence dating of quartz using an improved single aliquot regenerative-dose protocol. *Radiation Measurements* 32: 207–217.

Murray, A.S., Wohl, E. and East, J. 1992. Thermoluminescence and excess Ra decay dating of Late Quaternary fluvial sands, East Alligator River, Australia. *Quaternary Research* 87: 29–41.

Nevissi, A.E. 1985. Measurement of ^{210}Pb atmospheric flux in the Pacific northwest. *Health Physics* 48: 169–174.

Nozaki, Y., DeMaster, D.J. Lewis, D.M. and Turekian, K.K. 1978. Atmospheric ^{210}Pb fluxes determined from soil profiles. *Journal of Geophysical Research* 83: 4047–4051.

Oldfield, F. and Appleby, P.G. 1984a. Empirical testing of ^{210}Pb dating models for lake sediments. In: Haworth, E.Y. and Lund, J.W.G., eds., *Lake Sediments and Environmental History*, Leicester, UK: Leicester University Press, pp. 93–114.

Oldfield, F. and Appleby, P.G. 1984b. A combined radiometric and mineral magnetic approach to recent geochronology in lakes affected by catchment disturbance and sediment redistribution. *Chemical Geology* 44: 67–83.

Owens, P.N., Walling, D.E., He, Q, Shanahan, J. and Foster, I.D.L. 1997. The use of caesium-137 measurements to establish a sediment budget for the Start catchment, Devon, UK. *Hydrological Sciences Journal* 42: 405–423.

Owens, P.N., Walling, D.E. and Leeks, G.J.L.1999. Use of floodplain sediment cores to investigate recent historical changes in overbank sedimentation rates and sediment sources in the catchment of the River Ouse, Yorkshire, UK. *Catena* 36: 21–47.

Olley, J., Caitcheon, G.G. and Murray, A. 1998. The distribution of apparent dose as determined by optically stimulated luminescence in small aliquots of fluvial sediments: implications for dating young sediments. *Quaternary Science Reviews* (*Quaternary Geochronology*) 17: 1033–1040.

Olley, J.M., Caicheon, G.G. and Roberts, R.G. 1999. The origin of dose distribution in fluvial sediments, and the prospect of dating single grains from fluvial deposits using optically stimulated luminescence. *Radiation Measurements* 30: 207–217.

Page, K.J., Nanson, G.C. and Price, D.M. 1991. Thermoluminescence chronology of Late Quaternary deposition on the riverine plain of south-eastern Australia. *Australian Geographer* 22: 14–23.

Perkins, N.K. and Rhodes, E.J. 1994. Optical dating of fluvial sediments from Tattershall, U.K. *Quaternary Geochronology* (*Quaternary Science Reviews*) 18: 517–520.

Popp, C.L., Hawley, J.W., Love, D.W. and Dehn, M. 1988. Use of radiometric (Cs-137, Pb-210), geomorphic and stratigraphic techniques to date recent oxbow sediments in the Rio Puerco drainage, Grants Uranium region, New Mexico. *Environmental Geology and Water Science* 11: 253–269.

Prescott, J.R. and Robertson, G.B. 1997. Sediment dating by luminescence: a review. *Radiation Measurements* 27: 893–922.

Proszynska-Bordas, H., Stanska-Proszynska, W. and Proszynski, M. 1992. TL dating of river terraces with fossil soils in the Mediterranean region. *Quaternary Science Reviews* 11: 53–60.

Rendell, H.M., Webster, S.E. and Sheffer, N.L. 1994. Underwater bleaching of signals from sediment grains – new experimental-data. *Quaternary Science Reviews* (*Quaternary Geochronology*) 13: 433–435.

Rhodes, E.J. and Pownell, L. 1994. Zeroing of the OSL signal in quartz from young glaciofluvial sediments. *Radiation Measurements* 23(2/3): 581–585.

Ritchie, J.C., Hawks, P.H. and McHenry, J.R. 1975. Deposition rates in valleys determined using fallout cesium-137. *Geological Society of America Bulletin* 86: 1128–1130.

Ritchie, J.C. and McHenry, J.R. 1990. Application of radioactive fallout cesium-137 for measuring soil erosion and sediment accumulation rates and patterns: a review. *Journal of Environmental Quality* 19: 215–233.

Rockwell, T.K. 2000. Use of soil geomorphology in fault studies. In: Stratton, J.S., Sowers, J.M. and Lettis, W.R., eds., *Quaternary Geochronology: Methods and Applications*. AGU Reference Shelf 4, Washington: American Geophysical Union, pp. 273–292.

Robbins, R.A. 1978. Geochemical and geophysical applications of radioactive lead. In: Nriagu, J.O., ed., *The Biogeochemistry of Lead in the Environment*, Amsterdam: Elsevier, pp. 286–283.

Roberts, R.G., Bird, M., Olley, J., Galbraith, R.F., Lawson, E., Laslett, G.M., Yoshida, H., Jones, R., Fullagar, R., Jacobsen, G. and Hua, Q. 1998. Optical and radiocarbon dating at Jinmium rock shelter in northern Australia. *Nature* 393: 358–362.

Roberts, R., Walsh, G., Murray, A., Olley, J., Jones, R., Morwood, M., Tuniz, C., Lawson, E., MacPhail, M., Bowdery, D. and Naumann, I. 1997. Luminescence dating of rock art and past environments using mud-wasp nests in northern Australia. *Nature* 387: 696–699.

Schumm, A.S. 1981. Evolution and response of the fluvial system, sedimentologic implications. In: Ethridge, F.G. and Flores, R.M., eds., *Recent and Ancient Nonmarine Depositional Environments: Models for Exploration*, Special Publication 31, Tulsa: Society of Economic Paleontologists and Mineralogists, pp. 19–29.

Schumm, S.A. and Winkely, B.R., eds., 1994. *The Variability of Large Alluvial Rivers*, New York: American Society of Civil Engineers.

Shaw, P.A., Stokes, S., Thomas, D.S.G., Davies, F.B.M. and Holmgren, K. 1997. Palaeoecology and age of a Quaternary high lake level in the Makgadikgadi Basin of the Middle Kalahari, Botswana. *South African Journal of Science* 93: 273–276.

Sowers, S., Noller, J.S. and Lettis, W.R. 2000. Methods for dating Quaternary sediment. In: Noller, J.S., Sowers, S. and Lettis, W.R., eds., *Quaternary Geochronology: Methods and Applications*. Washington, DC: American Geophysical Union.

Stokes, S. 1994. The timing of OSL sensitivity changes in a natural quartz. *Radiation Measurements* 23: 601–605.

Stokes, S. and Gaylord, D.R. 1993. Optical dating of Holocene dune sands in the Ferris Dune Field, Wyoming. *Quaternary Research* 39: 274–281.

Stokes, S. 1999. Luminescence dating applications in geomorphological research. *Geomorphology* 29: 153–171.

Stokes, S., Bray H.M. and Blum, M.D. 2000. Optical resetting in fluvial depositional systems. *Quaternary Science Reviews* (*Quaternary Geochronology*) 20: 879–885.

Stokes, S., Colls, A.E.L., Fattahi, M. and Rich, J. 2001. Investigation of the performance of quartz single aliquot D_e determination procedures. *Radiation Measurements* 32: 585–594.

Straffin, E.C., Blum, M.D., Colls, S. and Stokes, S. 1999. Alluvial stratigraphy of the Loire and Arroux Rivers. *Quaternaire* 10: 271–282.

Trimble, S.W. 1983. A sediment budget for Coon Creek Basin in the Driftless Area, Wisconsin, 1853–1977. *American Journal of Science* 283: 454–474.

Walling, D.E. 1998. Opportunities for using environmental radionuclides in the study of watershed sediment budgets. In: *Proceedings International Symposium on Comprehensive Watershed Management, Beijing, China, September 1998*, pp. 3–16.

Walling, D.E. and Bradley, S.B. 1989. Rates and patterns of contemporary floodplain sedimentation: a case study of the river Culm, Devon, UK. *Geojournal* 19: 53–62.

Walling, D.E. and He, Q. 1992. Interpretation of caesium-137 profiles in lacustrine and other sediments: the role of catchment-derived inputs. *Hydrobiologia* 235/236: 219–230.

Walling, D.E. and He, Q. 1993. Use of caesium-137 as a tracer in a study of rates and patterns of floodplain sedimentation. In: *Tracers in Hydrology, Proceedings of the Yokohama symposium*, IAHS Publication No. 215, Wallingford, UK: IAHS Press, pp. 319–328.

Walling, D.E. and He, Q. 1994. Rates of overbank sedimentation on the flood plains of several British rivers during the past 10 years. In: *Variability in Stream Erosion and Sediment Transport, Proceedings of the Canberra symposium*, IAHS Publication No. 224, Wallingford, UK: IAHS Press, pp. 203–210.

Walling, D.E. and He, Q. 1997a. Use of fallout ^{137}Cs in investigations of overbank sediment deposition on river floodplains. *Catena* 29: 263–282.

Walling, D.E. and He, Q. 1997b. Investigating spatial patterns of overbank sedimentation on river floodplains. *Water, Air and Soil Pollution* 99: 9–20.

Walling, D.E. and He, Q. 1998. The spatial variability of overbank sedimentation on river floodplains. *Geomorphology* 24: 209–223.

Walling, D.E. and He, Q. 1999a. Changing rates of overbank sedimentation on the floodplains of British rivers during the past 100 years. In: Brown, A.G. and Quine, T.A., eds., *Fluvial Processes and Environmental Change*, Chichester: Wiley, pp. 207–222.

Walling, D.E. and He, Q. 1999b. Use of fallout lead-210 measurements to estimate soil erosion on cultivated land. *Soil Science Society of America Journal* 63: 1404–1412.

Walling, D.E. and Woodward, J. C. 1995. Tracing sources of suspended sediment in river basins. A case study of the river Culm, Devon, UK. *Marine and Freshwater Research* 46: 327–336.

Walling, D.E., Golosov, V.N., Panin, A.V. and He, Q. 2000. Use of radiocaesium to investigate erosion and sedimentation in areas with high levels of Chernobyl fallout. In: Foster, I.D.L., ed., *Tracers in Geomorphology*, Chichester: Wiley, pp. 291–308.

Walling, D.E., He, Q. and Nicholas, A.P. 1996. Floodplains as suspended sediment sinks. In: Anderson, M.G., Walling, D.E. and Bates, P.D., eds., *Floodplain Processes*, Chichester: Wiley, pp. 399–440.

Walling, D.E., He, Q. and Blake, W. 1999b. Use of ^{7}Be and ^{137}Cs measurements to document short- and medium-term rates of water-induced soil erosion on agricultural land. *Water Resources Research* 35: 3865–3874.

Walling, D.E., Owens, P.N. and Leeks, G.J.L. 1998. The role of channel and floodplain storage in the suspended sediment budget of the river Ouse, Yorkshire, UK. *Geomorphology* 22: 225–242.

Walling, D.E., Owens, P.N. and Leeks, G.J.L. 1999a. Rates of contemporary overbank sedimentation and sediment storage on the floodplains of the main channel systems of the Yorkshire Ouse and the river Tweed, UK. *Hydrological Processes* 13: 993–1009.

Walling, D.E., Owens, P.N. and Leeks, G.J.L. 1999c. Fingerprinting suspended sediment sources in the catchment of the river Ouse, Yorkshire, UK. *Hydrological Processes* 13: 995–975.

Walling, D.E., Quine, T.A. and He, Q. 1992. Investigating contemporary rates of floodplain sedimentation. In: Petts, G. and Carling, P.A., eds., *Lowland Floodplain Rivers*, Chichester: Wiley, pp. 154–184.

Wintle, A.G. 1997. Luminescence dating: laboratory procedures and protocols. *Radiation Measurements* 27: 769–817.

Wintle, A.G., Li, S.-H. and Botha, G.A. 1993. Luminescence dating of colluvial deposits from Natal, South Africa. *South African Journal of Science* 89: 77–82.

Wintle, A.G., Li, S.H., Botha, G.A. and Vogel, J.C. 1995. Evaluation of luminescence-dating procedures applied to late-Holocene colluvium near St Paul's Mission, Natal, South Africa. *The Holocene* 10: 97–102.

10

Vegetation as a Tool in the Interpretation of Fluvial Geomorphic Processes and Landforms in Humid Temperate Areas

CLIFF R. HUPP[1] AND GUDRUN BORNETTE[2]
[1] *US Geological Survey, Virgina, USA*
[2] *UMR 5023 CNRS, Université Claude Bernard Lyon I, Villeurbanne, France*

10.1 INTRODUCTION

Biota may interact with geomorphic processes in a variety of ways. Recent appreciation for this interaction has led to the development of a field of endeavor termed "biogeomorphology" (Viles 1988). Organisms from bacteria and their effects in weathering (Viles 1995) to beaver (*Castor*) and their effects on sedimentation and channel dynamics (Rostan *et al.* 1987, Butler and Malanson 1995, Schwarz *et al.* 1996) may play a profound role in geomorphic processes. Early ecologists understood the need to document geomorphic form and process to explain plant species distributions (Cowles 1901). Thus, the relationship between vegetation and geomorphic processes has been acknowledged for at least a century. Recent reviews on the subject of biogeomorphology include Viles (1988), Thornes (1990), Hupp *et al.* (1995). The field has expanded at such a rapid rate in the past 20 years that it would be impossible to treat fairly all the forms of biota in various physiographic and climatic settings within the confines of a single book chapter.

The present chapter will largely confine discussions to vegetation and fluvial geomorphic relations in humid, temperate regions of eastern North America and Western Europe. Vegetation can be used as a tool for geomorphic interpretation in at least two major ways:

1. through dendrogeomorphic analyses (tree-rings) to estimate the timing of important geomorphic events including floods, mass wasting, and to estimate rates of erosion and sedimentation; and
2. through the documentation and interpretation of species distributional patterns that establish in response to prevailing hydro-geomorphic conditions.

One non-fluvial subject where vegetation may be an important tool is the variable source or partial area concept in relation stream flow. Not all areas in a catchment contribute equally to stream flow (Dunne and Black 1970). In a classic paper on the relations between plant ecology and geomorphic form and process, Hack and Goodlett (1960) showed that forest types could be related to slope types where convex upward slopes contributed mostly to runoff; concave slopes contributed mostly to groundwater recharge; and linear slopes were intermediate. They produced a map of vegetation types using key species that clearly delimited slope declivity and laid the groundwork for future studies. The possible importance of vegetation mapping in delineating variable source areas was explicitly stated in Dunne and Black (1970). In the UK, especially, upland vegetation types have been successfully related to runoff in studies by Gurnell and Gregory (1987, 1995) and Thornes (1988, 1990). However, other research in this important area has been slow in forthcoming, and the subject justly warrants further study worldwide.

10.2 DENDROGEOMORPHOLOGY IN FLUVIAL SYSTEMS

The use of tree-ring dating for the interpretation of geomorphic processes has become an increasingly common technique largely as a result of several important papers including Sigafoos (1964), Everitt (1968), Alestalo (1971), Helley and LaMarch (1973) and Schweingruber (1988). Shroder (1978) coined the term dendrogeomorphology, which now describes this methodology; it is beyond the scope of the present chapter to provide an in-depth review. Rather extensive reviews can be found in several chapters of Jacoby and Hornbeck (1987). Dendrogeomorphology has been used in a wide range fluvial applications, including floods (Sigafoos 1964, Yanosky 1982, Hupp 1988), floodplain deposition (Sigafoos 1964, Hupp 1988, Hupp and Bazemore 1993), channel dynamics (Hupp 1992, Friedman *et al.* 1996, Scott *et al.* 1997), lake shoreline dynamics (Begin and Filion 1995), saltwater intrusion along streams (Yanosky *et al.* 1995), and mountain glacier activity and debris flows (Sigafoos and Hendricks 1961, Hupp *et al.* 1987). Standard dendrochronological techniques that incorporate ring-width measurement to interpret geomorphic processes may be quite useful.

Floods and Inundation

Floods, from prolonged inundation characteristic of relatively large, low-gradient basins to high-gradient, short-period (flashy), destructive events are, perhaps, the most important extrinsic factor in bottomland systems. Thus, the knowledge of flooding characteristics and magnitude/frequency information is of great utility to students of fluvial geomorphology. Dates of past floods may be determined from ages of trees on fluvial landforms, from scars and sprouts from flood-damaged stems (Sigafoos 1964, Hupp 1988), and from differences in properties of wood anatomy related to flooding (Yanosky 1982).

Four basic types of botanical evidence of geomorphic events, floods in this case, are routinely used (Hupp 1988):

1. corrasion scars,
2. adventitious sprouts,
3. tree age, and
4. ring anomalies (Figure 10.1).

A description of each follows. It is assumed that all samples are cross-dated, which reduces problems associated with false or missing rings (Cleaveland 1980, Cleaveland and Stahle 1989).

Corrasion scars. Scars may be the most conspicuous evidence of past flooding on riparian trees and shrubs (Figure 10.1A). Currently, the most reliable and accurate method of tree-ring-determined dating of floods is the analysis of increment cores or cross-sections through scars. These samples may yield the exact date (year) of an event and because of differences in wood produced during the growing season, the season of the event also may be inferred. Corrasion destroys the cambium (wood-producing tissue) in the area of impact, thus growth is stopped in the damaged area, the event is recorded as undamaged tissue grows in annual increments around and over the scar. Maximum scar heights may also allow for the estimation of the peak stage of a flood or elevation of a debris flow.

Adventitious sprouts. Sprouts from inclined or broken stems are easily determined in the field and can date an event by coring at the base of the sprout. This type of evidence has the appearance of vertically growing sprouts from a tilted main stem (Figure 10.1B) that is usually trained in the downstream direction or has two or more sprouts growing from a split base (Figure 10.1C). Flood training of riparian trees is a highly visible feature along streams subjected to periodic high-velocity floods. Use of the term sprout does not imply youth, as some sprouts may be quite old. The age of sprouts is usually within 1 year of the age of the event; obviously, the first year (center ring) of the sprout must be included in the core or cross-section. Some trees may bear evidence of several floods. The accuracy obtained from adventitious sprouts is exceeded only by corrasion scar analysis, which, because of healing, may have limited use in dating older events.

Ring anomalies. Geomorphic events may affect trees without leaving obvious evidence as described above. This may occur as changes in ring width or as various alterations in intra-ring tissue (Yanosky 1984). Cores from trees must be analyzed, usually microscopically, to detect anomalous rings or ring patterns. Eccentric ring patterns (Figure 10.1D) occur when a tree is tilted from vertical. Slight inclinations can induce eccentric rings without causing the formation of adventitious sprouts. Abrupt tilting of a tree (as is typical of some geomorphic events) results in subsequent rings that are wide on one side of the trunk while on the opposite side the same rings are relatively narrow. When

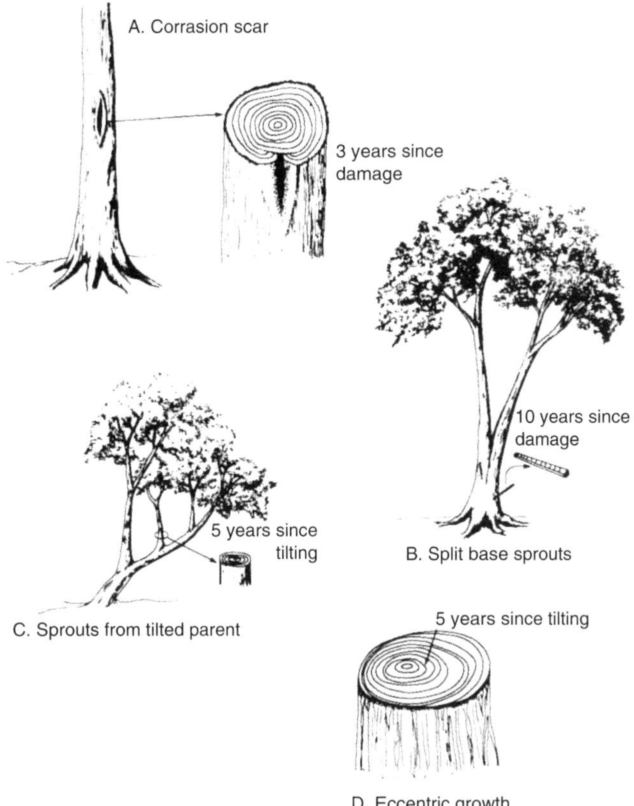

Figure 10.1 Types of botanical evidence of geomorphic processes (Hupp 1988)

this pattern occurs after concentric ring production, the date of the onset of eccentric growth is usually within one year of the event. This line of evidence is particularly useful in areas dominated by conifers, which do not typically form adventitious sprouts regardless of degree of inclination.

Tree age. Flood-deposited sediment or flood-scoured areas provide "new" sites for vegetation establishment. For example, flood-deposited longitudinal bars or point bars may be rapidly colonized by certain woody species (Everitt 1968, Johnson 1994, McKenney *et al.* 1995, Scott *et al.* 1997). The age of trees and shrubs growing on these new surfaces indicate a minimum time since initial deposition or scour. When all the oldest trees on a geomorphically delineated surface are nearly the same age, it is reasonable to assume the age of the oldest is close to the age of the surface, barring subsequent, unrelated, events such as lumbering or fire. Along actively migrating channels, tree ages can be used to estimate the timing and rate of channel shifts. Point bars, in particular, may have several bands of woody plants whose age is progressively younger toward the channel. This technique is useful for dating or at least obtaining minimum ages for surfaces created by large, exceptionally infrequent (hundreds of years) events.

Sediment Deposition and Erosion

Substantial amounts of sediment can be deposited on or eroded from alluvial areas during flooding events. Rates and amount of sediment deposition, scour, and associated channel shifting can be estimated using woody vegetation (e.g. Sigafoos 1964, Hupp 1988, 2000). Suspended sediment is an especially important environmental concern because of problems associated solely with sediment and because of problems associated with the adherence of hydrophobic contaminants to fine particles. Unfortunately, sedimentation trapping rates and retention time in fluvial systems is poorly understood. Indeed, as of the early 1990s only four accounts of vertical sediment accretion rates in any type of wetland in the United States had been published (Johnston 1991). Since then, other streams, particularly in southeastern United States (Hupp 2000) have been studied for sediment trapping rates (Table 10.1) but worldwide the number of riparian areas with documented sedimentation rates is incredibly low. Woody vegetation analyses offer an inexpensive and relatively accurate method for obtaining sedimentation rate data along many streams.

Buried stems are the principal form of botanical evidence used to estimate sedimentation rates. Initial tree roots (at germination) grow just below the ground surface and eventually form the major root trunks that radiate horizontally from the germination point. The basal flare, or root collar, and initial root zone are thus a distinctive marker of the original ground surface at the time of germination (Figure 10.2). Trees subjected to substantial sediment deposition typically have the appearance of a telephone pole because burial may obscure the normal flare at the base of the tree (Figure 10.2). Net sedimentation rates can be estimated by determining the depth of root burial some distance from the exposed trunk (usually one or two tree diameters), then extracting a tree core near the ground surface and determining the age of the tree, and finally dividing the depth of burial by the age of the tree. Several trees at a location should be analyzed similarly to obtain an average net rate to reduce possible micro-site variation. This technique yields sedimentation rate estimation over the average age of the sampled trees (Hupp and Bazemore 1993) for specific areas over a bottomland and has been shown to be usefully accurate when compared to repeat cross-sectional analyses (Hupp and Simon 1991, Hupp *et al.* 1993). However, the estimation of short-term (past 1–5 years) sedimentation rates, may be measured more accurately using artificial marker horizons such as white feldspar clay pads (Hupp and Bazemore 1993, Kleiss 1993, 1996).

The use of dendrogeomorphic techniques to document erosion rates has largely been confined to hillslope processes (LaMarche 1968, Shroder 1978, Hupp and Carey 1990). In eroding areas the lateral roots become exposed. The technique for estimating erosion is similar to that used in estimating deposition; measurements are made from the present ground surface to the top of the exposed roots to obtain a distance. The tree is then cored near its base to determine the age, which is divided into the erosion distance to obtain a net erosion rate over the life span of the tree. Sigafoos (1964) used this technique to document flood-related cycles of deposition and erosion along a large river floodplain. Similarly, rates of bank retreat have been estimated using this technique (Simon and Hupp 1987).

Table 10.1 Mean sediment deposition rates (mm year^{-1}) for coastal plain rivers; data from dendrogeomorphic analyses. The Cache River was investigated twice in different studies and locations

River	Type	Rate	Authorship	Date
Hatchie, TN	Alluvial	5.4	Bazemore *et al.*	1991
Forked Deer, TN	Alluvial	3.5	Bazemore *et al.*	1991
Chicahominy, VA	Alluvial	3.0	Hupp *et al.*	1993
Obion, TN	Alluvial	3.0	Bazemore *et al.*	1991
Patuxent, MD	Alluvial	2.9	Schening *et al.*	1999
Cache, AR	Alluvial	2.7	Hupp and Schening	1997
Roanoke, NC	Alluvial	2.3	Hupp *et al.*	1999
Cache, AR	Alluvial	1.8	Hupp and Morris	1990
Wolf, TN	Alluvial	1.8	Bazemore *et al.*	1991
Mattaponi/Pamunkey, VA	Alluvial	1.7	Schening *et al.*	1999
Coosawhatchie, SC	Blackwater	1.6	Hupp and Schening	1997
Choptank, MD	Blackwater	1.5	Schening *et al.*	1999
Pocomoke, MD	Blackwater	1.5	Hupp *et al.*	1999

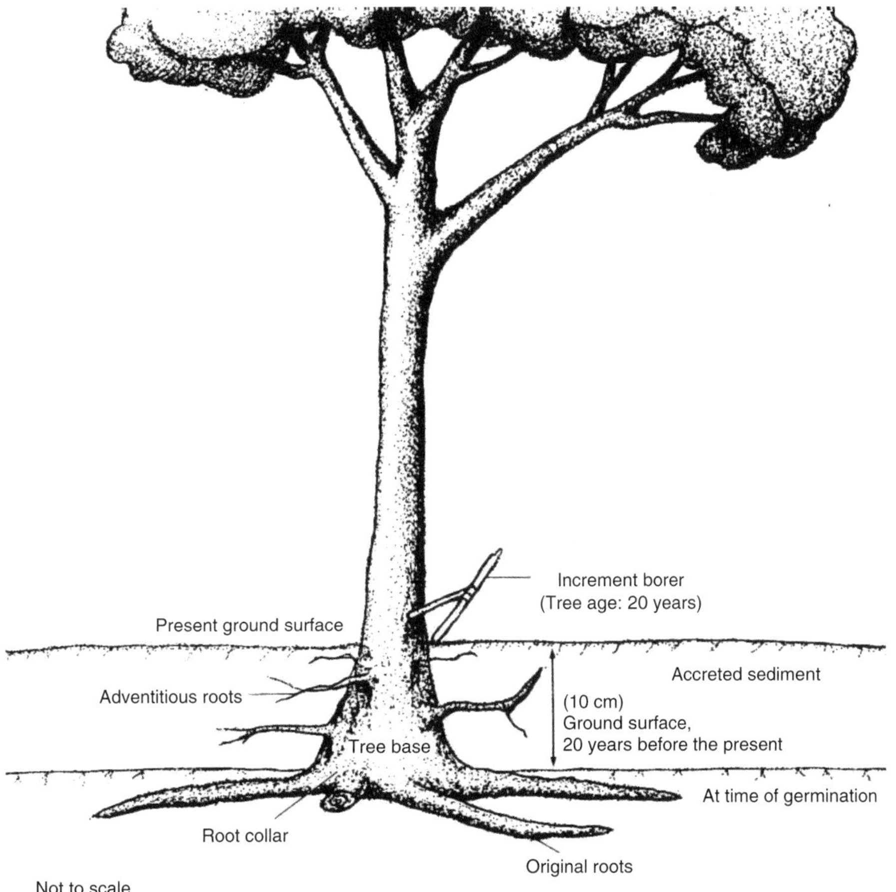

Figure 10.2 Generalized "buried" floodplain tree. Determination of depth to root collar at time of germination (just below original ground surface) shows a net accretion of 10 cm. An increment core taken at the base of the tree indicates the tree is 20 years old. Thus, a net sedimentation rate is conservatively estimated to be 0.5 cm $year^{-1}$ for the past 20 years (Hupp and Morris 1990)

Temporal Trends

One of the more unique but underutilized applications of dendrogeomorphology is the detection of changes in geomorphic processes, particularly deposition and erosion over time that may be related to environmental shifts. It is possible to infer changes in these processes by organizing the dendrogeomorphic data into tree-age classes (cohorts) over the life span of most of the samples. Intentionally sampling many trees over a wide range of ages allows for the calculation of mean rates of deposition over, say decadal, time periods (Figure 10.3, Hupp and Bazemore 1993). However, there are several caveats that must be assumed: all rates are net rates, compaction is not normally taken into account, and accuracy of early rates of deposition is dampened because subsequent deposition affects calculation of any prior time period. Subtracting out subsequent deposition is usually subjective, provides additional error, and does not typically improve the overall interpretative ability (Hupp and Bazemore 1993).

Temporal dendrogeomorphic techniques have been used successfully in coastal plain bottomlands of Arkansas (Hupp and Morris 1990) and Tennessee (Hupp and Bazemore 1993) to interpret the effects of deforestation for agriculture and channelization. For example, Hupp and Bazemore (1993) found that channelized rivers in West Tennessee consistently through time have lower floodplain sedimentation

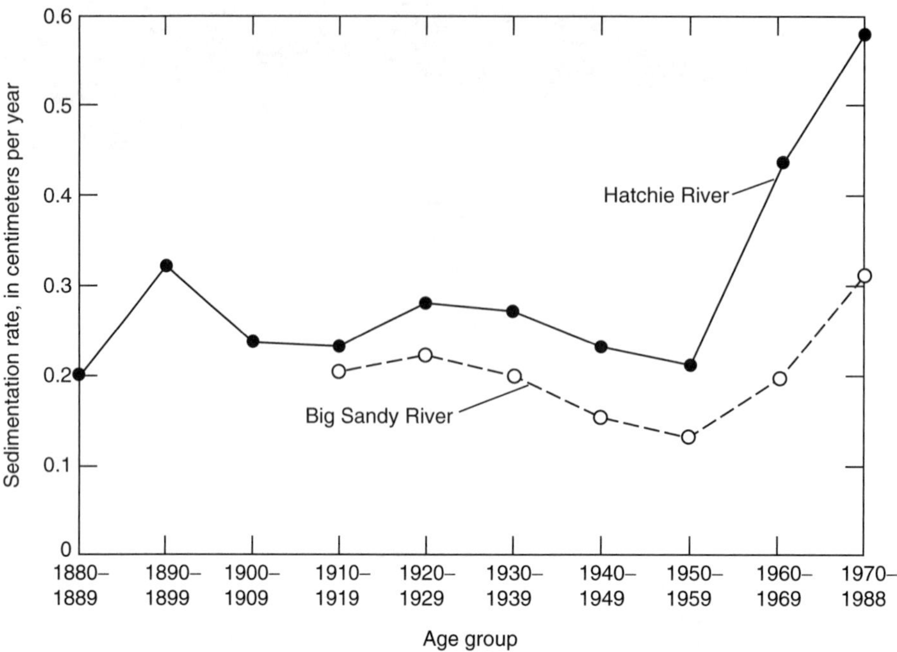

Figure 10.3 Sedimentation rates by age class of trees sampled for rate determination along the Hatchie and Big Sandy Rivers, West Tennessee. Sedimentation rates are consistently higher along the unchannelized Hatchie River; note a simultaneous rise in sedimentation rates beginning about 1950 (Hupp and Bazemore 1993)

rates than unchannelized rivers (Figure 10.3), not unexpected given that channelization intentionally and typically severely reduces overbank flooding. Further, both systems simultaneously responded with increased sedimentation rates after substantial bottomland land clearing for agriculture following the close of World War II (Figure 10.3). Although the sedimentation rate for an historic decade is imprecise, it is nevertheless possible to make relative comparisons that may yield valid environmental interpretations. Few, other relatively inexpensive techniques may provide long-term sedimentation rates in this detail.

10.3 PLANT ECOLOGICAL–FLUVIAL GEOMORPHIC RELATIONS

The community organization and dynamics of vegetation on the bottomlands of large rivers are strongly governed by fluvial geomorphic processes and landforms, which are largely created and maintained by fluctuations of water discharge. The likelihood of a given species vigorously growing on a particular landform is a function of (1) the suitability of the site for germination and establishment (ecesis), and (2) the ambient environmental conditions at the site that permit persistence at least until reproductive age (Grubb 1977, Zimmermann and Thom 1982, Hupp and Osterkamp 1996). Thus, the presence of a given species on a particular landform may allow for considerable inference regarding the hydrogeomorphic conditions of the landform. The distributional pattern may be limited by the tolerance of a species for specific disturbance regimes or stress, and consequently by tolerance for biotic interactions that prevail at this disturbance or stress level.

The bottomland landscape can be characterized as a shifting mosaic of landforms adjacent to stream channels (Bravard *et al.* 1986, Swanson *et al.* 1988) and/or a relatively predictable, largely linear array of landforms (Osterkamp and Hupp 1984, Hupp and Osterkamp 1996). Regardless, considerable variation in hydrogeomorphic processes can occur over short distances across this landscape and indeed across a single fluvial landform such as a floodplain. This complexity of physical form and process within the riparian zone has led to considerable diversity within vegetative communities supported by fluvial systems (Nilsson *et al.* 1989, Naiman *et al.* 1993). Periodic disturbance by relatively frequent floods (that may scour or aggrade) and/or inundation duration (hydroperiod) has been cited as the principal fluvial geomorphic process responsible for the high biodiversity in riparian ecosystems (Vannote *et al.* 1980, Hupp and Osterkamp 1985, 1996, Nilsson *et al.* 1989, Gregory 1992, Sharitz and Mitsch 1993, Bornette *et al.* 1994).

The term bottomland, as used here, refers to all fluvially generated landforms and the vegetation they support. These landforms occur as terraces high in the valley section and in descending order, proceed through the floodplain and various riparian features, channel bars, to the channel bed. The riparian zone is here defined as those surfaces that are inundated or saturated by the dominant discharge (typically the bankfull discharge, occurring at least once every two years on the average). This would include most modern floodplains and all surfaces lower in the valley section. It does not include terraces with long flood-return intervals (Osterkamp and Hupp 1984). Although this definition may not be as inclusive as others (e.g. Malanson 1993), it is quantitative and tied to the geomorphic concept of dominant discharge. We will limit most of our discussion to the floodplain, a horizontal to gently sloping surface that may be aerially extensive along many streams and is periodically inundated (Osterkamp and Hupp 1984). Some of the most influential effects include:

1. the creation of new areas for establishment such as point bars, depositional islands, abandoned channels, and scour pools;
2. regime-related (normal physical characteristics for various stream types under given climatic conditions) variation in bank stability;
3. the formation of flood intensity gradients normal to the low-flow channel;
4. spatial variation in sediment deposition and erosion rate, size clast, and nutrient availability;
5. gradients related to water availability or inundation duration (hydroperiod).

Separating these and other factors that influence bottomland-vegetation patterns is difficult because most are distinctly interdependent and a general lack of consistent or accepted landform and process definitions exists within the geomorphic sciences, and especially between the geomorphic and ecologic sciences.

Fluvial Landforms and Floods

Fluvial geomorphic processes via the action of flowing water create a number of widely recognized landforms from small channel bedforms to extensive floodplains. The latter typically support diverse forested ecosystems in humid temperate regions of the world. Floodplains, like most fluvial landforms, are dynamic features almost constantly eroding in places while aggrading in others. Channel dynamics, especially those of meandering, pseudo-meandering, and wandering streams, provide the energy necessary to erode and transport floodplain sediments. Meanders typically extend, eroding accreted sediments until they are cut off by an avulsion (channel cutoff) leaving an oxbow lake and a new channel (Figure 10.4). Whole meander loops, additionally, tend to migrate downstream. Thus, over geomorphic time, nearly all floodplain alluvium is in a state of flux. Large rivers that drain Alpine areas and their piedmonts (e.g. Europe and western North America) tend to have relatively high energy, abundant gravel bedload, and straight, braided or wandering channel patterns (Bravard *et al.* 1986, Kellerhals and Church 1989, Ritter *et al.* 1995, Müller 1995). Lateral instability and avulsion in these systems during even moderate flows promote the development of abandoned channels such as chute cutoffs, oxbow lakes, and dead arms (Bravard and Gilvear 1996). This lateral mobility of the main channel increases the biodiversity on these floodplains by creating coincident patches with different ages and hydrologic characteristics (Pautou 1984, Amoros *et al.* 1987, Amoros and Roux 1988). Large upland rivers draining relatively moderate gradient areas (e.g. eastern North America) tend to carry considerable silt/clay sediment loads and may be relatively stable with large, fine-grained floodplains. Typically high-magnitude low-frequency flow events are necessary for significant avulsion activity. However, consistent variation in hydrogeomorphic conditions with surface elevation creates long linear patches analogous to the mosaic on the more dynamic Alpine rivers. Floodplains of large lowland rivers (e.g. the coastal plain of eastern United States) tend to have net sediment storage during high or rising sea level such as the conditions over the past several thousand years. These lowland rivers, with active meandering, tend to be more laterally dynamic than stable moderate gradient upland rivers and share many of the hydrogeomorphic characteristics of the high-energy Alpine streams.

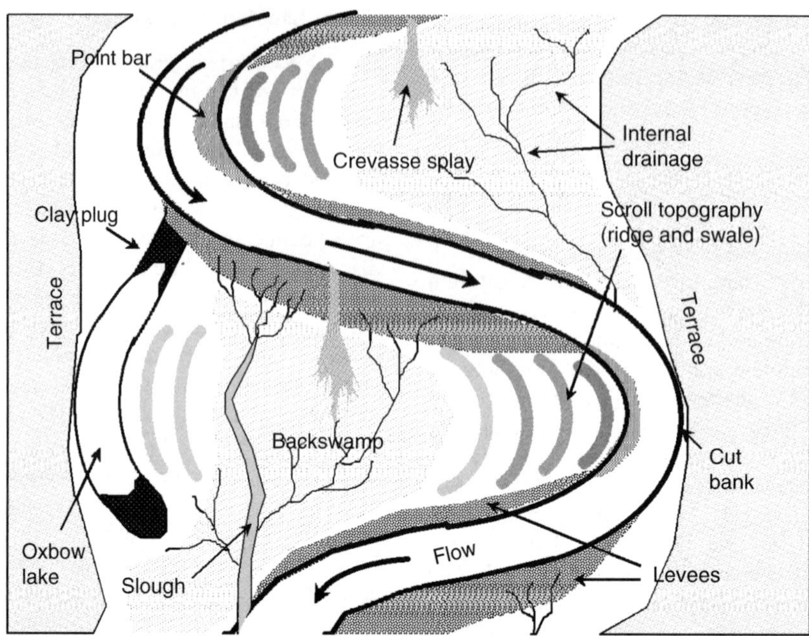

Figure 10.4 Generalized fluvial landforms on a coastal plain bottomland. Note greater levee development along straight reaches and on the downvalley side of the stream (Hupp 2000)

Lowland floodplains aggrade in two ways. First, by lateral accretion or point-bar extension where coarse material is deposited on the inside bank of channel bends; a corresponding volume is typically eroded on the opposite or cut bank (Figures 10.4 and 10.5). Second, by vertical accretion where suspended sediment (typically fines) is deposited over the floodplain during overbank flows. Lateral accretion is an episodic process that occurs during high flows, building the point bar into an often crescent-shaped ridge. Over time, a series of high-flow events produce the ridge and swale topography (Figure 10.4) associated with meander scrolls. The establishment of ruderal woody vegetation during intervening low-flow periods on fresh scroll surfaces creates bands of increasingly younger vegetation toward the main channel (McKenney *et al.* 1995). These bands of vegetation may accentuate the ridge and swale topography by creating contrasting depositional environments during subsequent high flows; the hydraulics necessary to produce meander scroll topography and the role of vegetation in its development are poorly understood.

The often drastic and sudden reduction in flow velocity after leaving the main channel and entering the hydraulically rough floodplain environment facilitates fine-sediment deposition (Figure 10.5). As rising floodwaters overtop the bank, the coarser (or heavier) sediment is deposited first, creating natural levees along the floodplain margin. Levees tend to be most pronounced along relatively straight reaches between meanders and are often the highest ground on the floodplain. Levees are sometimes breached by stream flow resulting in a crevasse splay that may insert coarse material deep into the otherwise fine-grained bottom (Figure 10.4). Levee development and the breaches that form are poorly documented in the literature, yet are critical in the understanding of the surface-water hydrology of most bottomlands. Levee height and breaches strongly affect the hydroperiod (and thus, sedimentation dynamics) in systems dominated by surface-water flow (Patterson *et al.* 1985). Natural levee development and dynamics has received relatively little fluvial geomorphic study.

River overflows can result in flooding over large areas of the floodplain that may have only minor erosion, or, in the case of spates (flash floods), considerable scouring, massive slope failure, and bank erosion may occur (Resh *et al.* 1988, Ward *et al.* 1999). In the latter, which may be more typical of Alpine braided rivers, erosion occurs with aggradation, leading to a shifting mosaic of scoured and aggraded patches, depending on the local slope and coarse

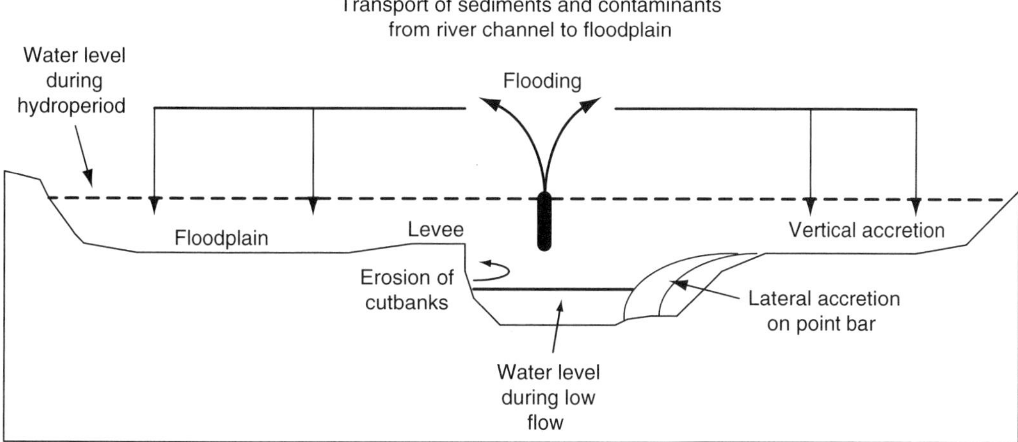

Figure 10.5 Transport paths of sediment (and any associated contaminants) for both lateral and vertical accretion. Annual hydroperiod is indicated (Hupp 2000)

material supply (Kalliola *et al.* 1991, Bravard and Gilvear 1996). Similarly, at the scale of river reaches, scouring occurs when sufficient slope is combined with low sediment supply from upstream; erosion processes depend on sediment supplies from the channel sector located just upstream (Petts and Bravard 1996). Conversely, when the gradient decreases, or the bedload increases, scouring efficiency of the river decreases, and the reach tends to aggrade. Non-equilibrium situations, such as chronic incision or aggradation, usually result from anthropogenic impacts (Galay 1983, Bravard *et al.* 1997). Any decrease in coarse material supply, or any alteration of the river transport ability can lead to the incision of the riverbed. The main channel incision triggers a progressive disconnection of the riparian zone and riverine wetlands from the river, and thus to a decrease in scouring frequency (Bornette and Heiler 1994, Bravard *et al.* 1997) and floodplain inundation (Hupp 1999).

The formation of a cutoff channel commonly occurs on meandering, braiding, and wandering river systems. At the cutoff channel scale, natural disconnection (plug formation, Figure 10.4) proceeds between the upstream end of the cutoff channel and the main river as a result of sedimentation and a decrease in erosion (scouring) efficiency. This plug impedes river overflows for small floods. For higher floods, the river may overflow into the cutoff channel with an increased slope from the plug to the aquatic zone, which may then increase scouring efficiency in the downstream aquatic zone at a local scale. Along aggrading river reaches, sediment deposition, beginning upstream on a cutoff, may proceed throughout the cutoff and eventually fill (or nearly so) the entire former channel.

Large Woody Debris—Channel Morphology

Research in forested areas over the past two decades (Bilby and Likens 1980) has shown that large woody debris (LWD), or drift, in along stream channels to be an important element of fluvial geomorphic form and process (Gregory 1992, Gurnell and Gregory 1995). Although not a tool in the same sense as dendrogeomorphic techniques, the knowledge and assessment LWD is an important aspect in fluvial geomorphic understanding. Piégay and Gurnell (1997), in a review of LWD, found 104 papers on the subject published since 1975. Summary content of these papers suggests three major LWD influences on the riverine system:

1. sediment storage, by influencing the distribution of stream power;
2. flow hydraulics and the average condition and variance of channel dimensions and stability;
3. ecological habitat, the sustained maintenance of sediment and organic material transport dynamics and structural complexity of channels.

Obviously, this is an important interaction between vegetation and fluvial geomorphology. However, it is beyond the scope of the present paper to provide a lengthy discussion on this topic; the reader is directed to the citations in this section for further information.

Most streams, even those in semi-arid environments, naturally support woody riparian vegetation. Meandering, avulsion, braiding, sediment deposition, and changes in channel width along streams with basins and/or forested riparian areas may be controlled at least in part by LWD (Zimmermann *et al.* 1967, Keller and Swanson 1979, Bilby and Ward 1989, Robison and Beschta 1990, Fetherson *et al.* 1995, Wood-Smith and Swanson 1997). Accumulations of LWD may armor the channel bed and banks in some places, while in others it may scour the channel through flow concentration and deflection; thus, simultaneously contributing sediment to the channel and, in some places, eroding pools (Wood-Smith and Swanson 1997). In coastal plain systems, Simon and Hupp (1992) showed that in streams lacking coarse bed material, LWD may ultimately control depth of degradation (following channelization) and timing of the onset of channel recovery.

The amount of LWD along most streams is related to disturbance history, forest type, successional stage, decomposition rate, and channel size (Sedell *et al.* 1988). Models of theoretical LWD loading have been developed for several types of streams, notably the Lymington River Basin in the UK (Gregory and Davis 1992, Gregory *et al.* 1993). Management practices (debris removal for example) may be an important influence in LWD loading (Gregory *et al.* 1993).

10.4 PLANT COMMUNITIES AND DYNAMICS IN BOTTOMLANDS

Disturbance characteristics play a major role in the development of many vegetation patterns (Johnson *et al.* 1985, Day *et al.* 1988, Kirkman and Sharitz 1994). Periodic flood disturbances control certain communities that persist at a dynamic equilibrium stage (Pickett 1980, Hupp and Osterkamp 1985, Bornette and Amoros 1991, 1996, Bendix and Hupp 2000) with no loss of species compositional integrity through time. Along low-gradient systems, with extensive forested wetlands, the tight relation between vegetation type and annual length of inundation (hydroperiod) is well documented (Wharton *et al.* 1982, Sharitz and Mitsch 1993). Moderate gradient streams of temperate eastern North America typically develop consistent and persistent linear fluvial landforms that are maintained by predictable variation in discharge. Coincident hydrogeomorphic analyses (Osterkamp and Hupp 1984, Hupp and Osterkamp 1985, Bendix and Hupp 2000) along several of these streams suggest that the overriding influence on the distributional patterns of species is the frequency of inundation and the susceptibility of plants to destructive flooding. Thus, the hydrogeomorphic processes operating differently on the different landforms affect the plant patterns, not the landforms per se (Table 10.2). As examples, two shrubs common on the channel shelf (a bank feature), *Alnus serrulata* and *Cornus amomum* (riparian shrub community, Table 10.2), are relatively resistant to destruction by floods because of small, highly resilient stems, and the ability to sprout rapidly from damaged stumps. Conversely, *Cornus florida* and some species of *Quercus* and *Carya*, which commonly grow on terraces but rarely on lower surfaces, may be intolerant to repeated flood damage or inundation. Floodplain species, such as *Carya cordiformis*, *Juglans nigra*, and *Ulmus americana*, are less tolerant to destructive (scouring) flooding than are channel-shelf species, but more tolerant to periodic inundation than are terrace species (Table 10.2). Depositional bars (excepting point bars) rarely support woody species, however, several perennial herbaceous species survive destruction through deeply anchored perenating rootstocks.

Across most low-gradient lowland floodplains striking vegetation zonation is displayed (coastal

Table 10.2 Summary of fluvial–landform relations with vegetation type, flow duration (percent of time a flow elevation is equaled or exceeded), and flood frequency (after Hupp and Osterkamp 1996)

Fluvial landform	Vegetation type	Flow duration	Flood frequency
Depositional bar	Herbaceous	About 40%	–
Channel shelf	Riparian shrubs	5–25%	–
Floodplain	Floodplain forest	–	1–3 years
Terraces	Terrace assemblage	–	>3 years

plain of southeast USA). Small differences in elevation, often measured in centimeters, may lead to pronounced differences in length of inundation (hydroperiod) and thus community composition (Mitsch and Gosselink 1993). Many vegetation classification systems infer that the length of hydroperiod is the most influential factor in affecting lowland floodplain species patterns. It should be noted that flooding, per se, may not limit species distribution but rather the anaerobic conditions associated with flooding (Wharton *et al.* 1982). Anaerobic respiration within the roots of plants leads to the production of toxic by-products and limits the plants' ability to take up nutrients and water. Plants tolerant of varying degrees of flooding have developed physical and/or metabolic adaptations to deal with inundation and anoxia (Wharton *et al.* 1982). Presumably, it is the degree to which individual species have adapted to anoxia-related stresses that has led to the distinct and drastic changes in vegetation composition over very short (meters) distances across many lowland floodplains (Huffman and Forsythe 1981). Sloughs and other floodplain depressions along southeastern US bottomlands characteristically support stands of *Taxodium distichum* and *Nyssa aquatica*. Just outside these wettest areas, *Quercus lyrata*, *Gleditsia aquatica*, and *Carya aquatica* may dominate. Middle floodplain elevations are characterized by bottomland hardwood stands (*Quercus*, *Fraxinus*, *Ulmus*); while well-drained locations may support *Pinus* species and *Fagus*.

In aquatic ecosystems, scouring as well as flooding affect vegetation patterns. Scouring during river overflow is considered a disturbance, that may uproot plants and disrupt communities (Jones 1956, van der Valk and Bliss 1971, Bilby 1977), and in some cases, may completely remove fine-sediment deposits that had accumulated since the last scouring event. Natural successional processes that occur in riverine ecosystems can be slowed or stopped, depending on the intensity and frequency of flood scouring (Sparks *et al.* 1990, Foeckler *et al.* 1991, 1994, Müller 1995, Bornette and Amoros 1996). In accordance with the intermediate disturbance hypothesis (Connell 1978, Sousa 1984), intermediate frequency and intensity of flood scouring sustains a dynamic equilibrium in such ecosystems. It allows for the maintenance of a highly diverse, shifting mosaic of plant species (Resh *et al.* 1988, Bornette and Amoros 1991, Roberts and Ludwig 1991, Bornette *et al.* 1998). High flood scouring frequency and/or intensity usually leads to a decrease in biodiversity, because plants communities are unable to recover before the next disturbance, or because of the lack of a substrate favorable for establishment (Kohler and Schiele 1985, Resh *et al.* 1988). Conversely, low frequency and/or intensity of flood scouring is unable to impede completely successional processes and substrate accumulation that may occur, leading to the terrestrialization of aquatic ecosystems, and a progressive reduction in flood frequency. Large-sized plants (*Nuphar lutea*, *Nymphea alba*) unable to resprout from above-ground parts typically colonize undisturbed or rarely disturbed channels (Bornette *et al.* 1998, Amoros and Bornette 1999). Conversely, intermediately scoured habitats, maintained at a dynamic equilibrium stage by disturbance frequency and intensity typically support highly diverse communities of evergreen perennial species (Greulich and Bornette 1999) able to resprout efficiently from any broken part of their individual (Barrat-Segretain *et al.* 1998, 1999). In lowland floodplains, or in locally aggradational areas, rooted plants may ultimately be eliminated and replaced by not anchored plants (*Ceratophyllum demersum*, Lemnaceae), but moderate silting may allow the persistence of some species (as *Vallisneria* sp., *Najas* sp., Haslam 1978, Ribicki and Carter 1986, Amoros *et al.* 2000).

10.5 THE EROSIONAL–DEPOSITIONAL ENVIRONMENT AND EQUILIBRIUM

Most bottomland substrates available for vegetation establishment are alluvial, thus hydraulic scour and size-clast sorting may be important limiting factors. Erosion and channel cutoffs may lead to almost permanently inundated pools that support aquatic vegetation. Terrestrial wetland vegetation may be restricted to narrow ranges of sediment size. Most woody plants must maintain their actively absorbing root zones in the upper 15 cm of soil (Kramer and Kozlowski 1979). Thus, only species capable of rapid root growth along newly buried stems (e.g. species of *Salix* and *Populus*) may occupy bottomland areas periodically affected by frequent or large amounts of sediment deposition (Figure 10.6) or active channel migration such as along point bars (Hupp 1988). Obviously, where erosion removes all part of the soil in the root zone of trees, they will be killed or severely damaged (Figure 10.6). Flooding regime is intimately associated with the depositional environment along most streams with adequate sediment supply.

Aggradation along stream reaches, natural or otherwise, provides new sites for vegetation colonization and may initiate a succession of vegetation stages indicative of progressively changing hydrogeomorphic conditions. Hupp (1992) and Johnson (1994) found that dominance by pioneer species can be explained by life history characteristics including: a large, dependable seed crop, seeds effectively dispersed by wind or water to suitable riparian sites, rapid germination, and rapid shoot and root development (to withstand flooding and desiccation). Initial vegetation establishment increases hydraulic roughness facilitating further sedimentation and ultimately modifies the relatively unstable initial surfaces into relatively stable surfaces favorable for the recruitment of later, stable-site species (Hupp 1992, Johnson 1994, Friedman *et al.* 1996).

Along streams dominated by aggradation, extensive depositional areas may be generated during each flood. Such floodplains are common along many southeastern US streams where upstream agriculture practices and channelization generate large amounts of suspended fines (Bazemore *et al.* 1991, Hupp 1992). In the degradation situation, large eroded zones are generated, and the river is decreasingly connected to its floodplain. Such conditions are common in rivers subjected to incision, e.g. as a consequence of flow regulation (Galay 1983, Babinsky 1992, Bravard 1994, Peiry *et al.* 1994, Hupp 1999).

It is possible to plot each floodplain or floodplain reach on a curve that represents erosion and deposition processes (Figure 10.6). The position along this curve depends on the dominant process (vertical axis) and on the size of the patches (horizontal axis). For each river system or river reach, the amount of net aggradation vs. erosion determines the position on the hypothetical curve simulating these two processes. At the ends of the gradient (Figure 10.6), the highest patch sizes occur. Near the center (equilibrium) of the gradient (Figure 10.6), the number and size of eroded vs. deposited patches tend to be equal, the turnover rate of these patches remain the same along the curve. Most temperate river floodplains can be plotted somewhere between these two polar situations, and the size of the deposition patches vs. erosion patches is related to the location of the river floodplain along the curve. A central equilibrium point corresponds to the situation where silted patches are as large as eroded patches (Figure 10.6). This point reflects the highest habitat heterogeneity. In this case, a dynamic equilibrium is reached, corresponding to a shifting mosaic of habitats at the floodplain scale. Such situations are increasingly less common in temperate areas, due to the frequent and heavy impacts of human activities that disrupt and drive fluvial processes toward one end or the other of the gradient. At the scale of an entire river, some reaches may be subjected to migrating aggradation

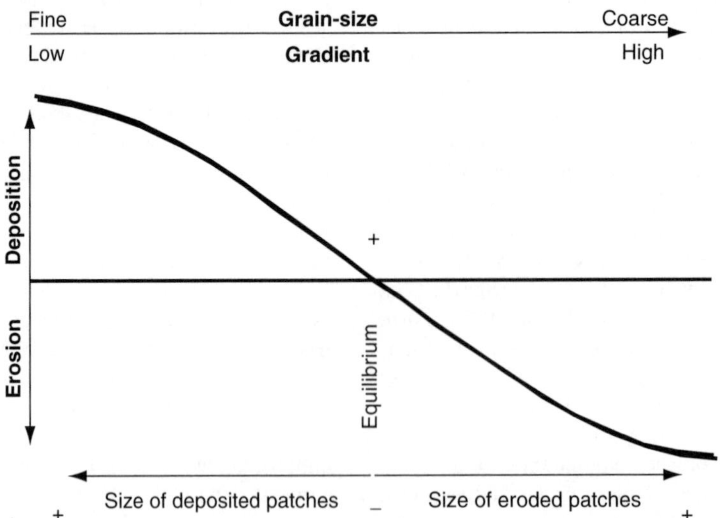

Figure 10.6 Pattern of sediment erosion vs. deposition on river floodplains. Each river floodplain or river reach is characterized by deposition and/or erosion processes, and the partitioning devoted to the two processes determines the position of the river reach along the curve. *Note*: depending on stream regime equilibrium conditions may be on either side of the equal deposition/erosion point

while others to degradation. Thus, it is at the reach scale that the disruption of equilibrium may be most easily observed.

Stream gradient (or power) also varies systematically along this conceptual gradient (Figure 10.6). Where stream gradients are high, erosional processes will dominate (e.g. low-order high mountain cascades), conversely along low-gradient reaches, depositional processes usually dominate (e.g. coastal plain rivers during periods of rising sea level). Thus, equilibrium conditions and highest biodiversity may also shift along this conceptual gradient depending on the flow regime. The turnover rate of the floodplain patches also depends on the flow regime of the river, or of the river reach, and may vary greatly from one to another.

Most streams in the West Tennessee portion of the southeastern US coastal plain have been channelized (Simon and Hupp 1992). This channelization led to severe degradation or erosion of the affected and upstream reaches of the streams. Degradation occurs on both the channel bed and banks until some quasi-equilibrium is attained and aggradation begins low in the channel section (Simon and Hupp 1992). A cycle of erosion, accretion, and return to equilibrium is described in a six-stage model that incorporates vegetation in the geomorphic processes (Simon and Hupp 1987, 1992, Hupp 1992).

The erosional phase of this cycle often completely removes all woody vegetation. Late in this phase refugia occur in protected areas, usually downstream of slump blocks from mass wasting on the banks (Simon and Hupp 1992). These refugia offer enough stability for ruderal riparian vegetation to establish (Figure 10.7A). However, upland species (*Rhus glabra*, *Ulmus alata*, *Gleditsia tricanthos*, and *Robinia pseudoacacia*) may occur high on the banks and represent species especially tolerant of erosive conditions. During the highly aggradational phase of this cycle, species found in refugia of the late erosional phase were tolerant of high sediment accretion rates (*Salix nigra*, *Betula nigra*, *Acer saccharinum*, *Acer negundo*, *Populus deltoides*, and *Platanus occidentalis*) become dominant (Figures 10.7A and C). Note that there is a twofold increase in percent occurrence for most species and that there is a nearly fourfold decrease in un-vegetated sites in the aggradational phase (Figures 10.7A and C); excepting a small percent of *Rhus glabra* the erosional indicator species are lacking.

The equilibrium phase of this cycle is, by far, the most diverse and is always vegetated (Figure 10.7B). In West Tennessee, the equilibrium sites were still young enough to support high percentages of some aggradational species (*Betula nigra*, *Platanus occidentalis*, and *Acer saccharinum*); however *Salix nigra* and *Populus deltoides*, perhaps the most indicative species of aggradation, experienced substantial declines in percent occurrence (Figures 10.7B and C). Many species normally found in equilibrated bottomland hardwood forest of the region are present in the equilibrium phase including: *Fraxinus pennsylvanica*, *Ulmus americana*, *Liquidambar styracilflua*, *Taxodium distichum*, *Nyssa aquatica*, and three oak (*Quercus*) species.

The floodplain submersion during flooding is a major limiting factor for aquatic vegetation, particularly for long-duration or high-discharge floods. Sediment deposition usually increases the terrestrialization rate not only through increasing aggradation, but also through eutrophication, as suspended sediment is usually nutrient-rich (Shelford 1954, Peck and Smart 1986, Bhowmik and Adams 1989, Sparks *et al.* 1990). Plants do not succeed if they are unable to tolerate these submersion periods (primarily helophytes and hydrophytes with low reserves and growth rates; van der Valk and Bliss 1971, Hamel and Bhéreur 1982, Brock *et al.* 1987, Sparks *et al.* 1990, van den Brink *et al.* 1995). Rooted submerged plants are particularly disfavored, due to the high turbidity of the water and consecutive high growth of phytoplankton in the more eutrophic situations (Sparks *et al.* 1990, van den Brink *et al.* 1993). Nymphaeides are also disfavored because deposition ultimately impedes the regrowth of the plant from rhizomes, and because the recruitment of such species from seeds requires an early supply of light to hypocotyles (Smits *et al.* 1990). Only species that are able to develop in very turbid water such as floating plants (e.g. *Lemna*) succeed in the most eutrophic ecosystems, such as not-anchored species that require nutrients in water, assuming the water velocity during floods is not a detriment for such species. Commonly, in permanently connected waterbodies, floating plants are easily washed away by water even with low velocity (Bornette *et al.* 1998). Thus, plants able to resprout after burial and annual fast growing species are usually favored in such situations (Haslam 1978, Kalliola *et al.* 1991, Amoros and Bornette 1999). Ultimately, sediment deposition can completely impede the growth of aquatic plants.

Deposition of fines may progressively isolate aquatic ecosystems from shallow aquifers by clogging substrate interstices (Foeckler *et al.* 1991), and consequently eliminate plant species that require water

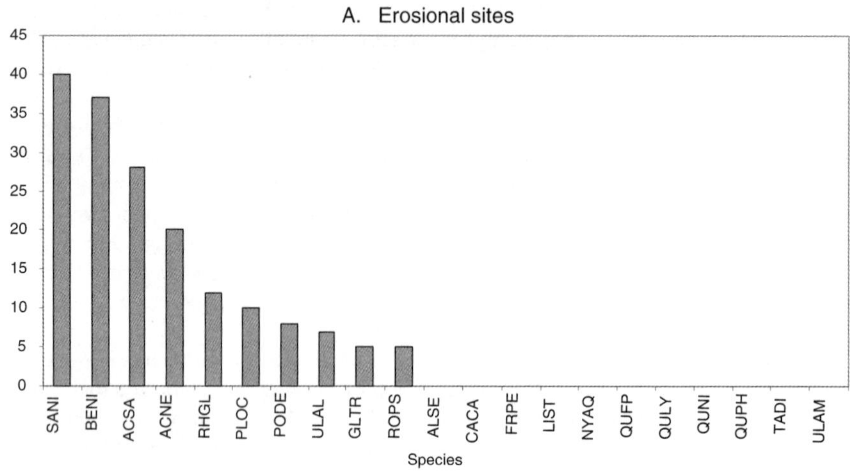

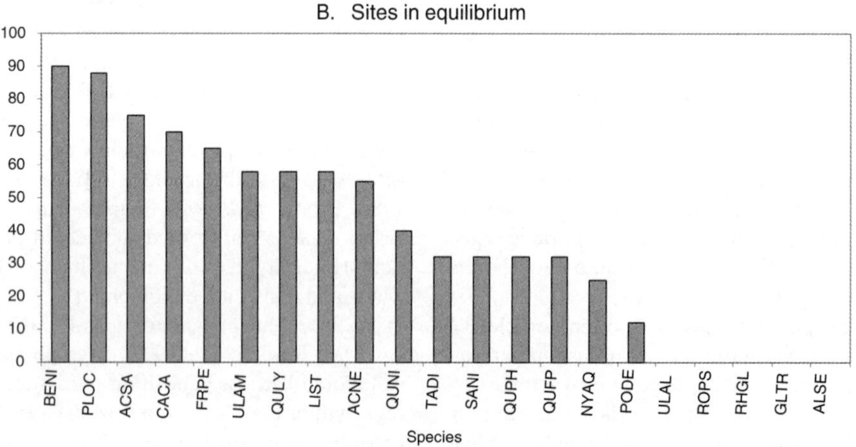

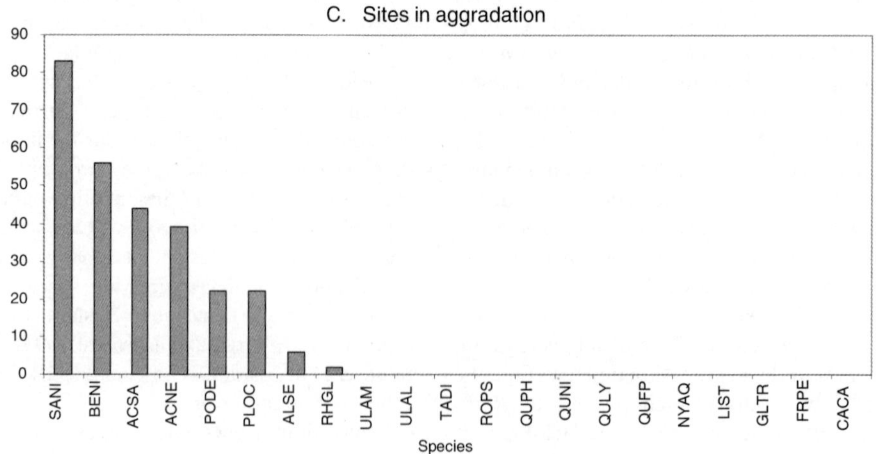

Figure 10.7 Percent of sites (>100) present for selected species during channel evolution following channelization: (A) erosional sites; (B) equilibrium sites, (C) aggradational sites. From Simon and Hupp's (1992) data

renewal (Amoros *et al.* 2000). High water turnover rate can limit the siltation effect, consequent eutrophication, and terrestrialization. Indeed, many coarse gravel cutoff channels are supplied by groundwater (river seepage or hillslope aquifer). If the slope is sufficiently high and/or the hydraulic capacity low, the exportation of fine sediment is facilitated and sediment accretion (siltation) is slowed down, and sometimes stopped (Bornette *et al.* 1994). In these situations, groundwater is able to rapidly restore low turbidity levels after each flood event (Reygrobellet and Castella 1987, Trémolières *et al.* 1991, Malanson 1993). In a cutoff channel of the Ain river (France), contrasted plant communities corresponding to contrasted trophic levels occur with the same cutoff channel due to strong difference in hydraulic capacity, and despite no significant difference in water physico-chemistry in the two zones ($[N\text{-}NH_4^+] = 0.035 \pm 0.03\ mg\,l^{-1}$ and 0.046 ± 0.03, respectively, in the upstream and downstream zones and $[P\text{-}PO_4^{2-}] = 0$ in the two zones, averaged values calculated from 12 samples collected monthly). In the downstream part of the channel (in black in Figure 8), water quality is restored after each flood event by sufficiently high groundwater supplies, but the discharge remains insufficient to export fine nutrient-rich sediment responsible to eutrophication, as indicated by the floristic content (abundance of *Elodea nutallii*, *Sparganium emersum*, *Callitriche platycarpa* in the downstream zone). In this zone, an apparent contradiction appears between the water quality and the composition of plant communities. In the upper zone, groundwater supplies impede silting, and oligotraphent species dominate (*Potamogeton coloratus*, *Juncus subnodulosus*, *Hydrocotyle vulgaris*). Scouring typically breaks aquatic plant stems, especially those with large leaves (e.g. *Nuphar lutea*; Brierley *et al.* 1989), and favors plants with an ability to resist uprooting and breakage. Plants with flattened and linear leaves, or those with strong anchorage

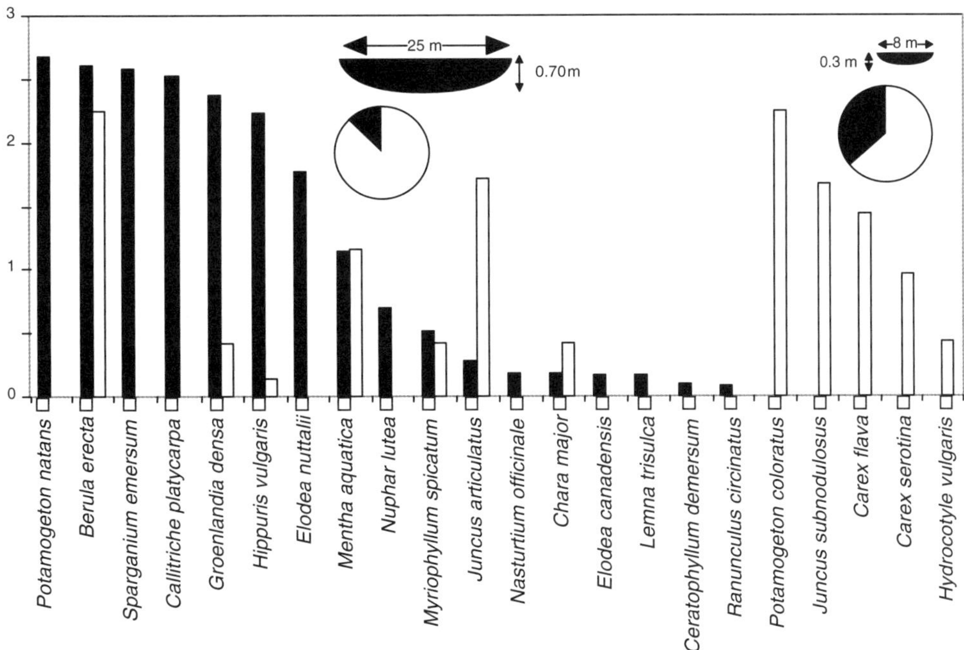

Figure 10.8 Abundance of aquatic plants in a cutoff channel submitted to river backflows and subsequent siltation processes and eutrophication. The abundance of plant species in the upstream (white bars) and downstream zone (black bars) is calculated as the average Braun-Blanket cover indices of the species in the zones. The circles in the top of the figure indicate the part of the coarse substrate (in black) compared to the fine substrate (silt + organic matter, in white) in the downstream zone (circle in the left) and in the upstream zone (circle in the right), and the black half circles represent the average cross-profiles of the two zones

are thus usually favored, as well as those able to regrow from fragments or parts preserved in the sediment (Kalliola *et al.* 1991, Barrat-Segretain 1996, Barrat-Segretain *et al.* 1998 and 1999). Such disturbances create gaps that can be rapidly colonized by propagules (Barrat-Segretain and Amoros 1996). Similarly, scouring eliminates at least a part of the fine sediment, and consequently tends to disfavor plants propagating with buried seeds (Henry *et al.* 1996). The unpredictable nature of scouring events may favor vegetative reproduction over seed recruitment, because of the usually short seasonal period associated with germination (Amoros and Bornette 1999). Conversely, fragments produced during the scouring event are immediately available for the colonization of sites newly opened by the disturbance (Kalliola *et al.* 1991, Barrat-Segretain *et al.* 1998 and 1999). Plants that are deeply anchored may use sediment as refugia and may be also at least partly preserved (Henry *et al.* 1996).

10.6 CONCLUSIONS

Fluvial geomorphic processes and landforms normally exert a profound influence on the vegetation patterns of temperate bottomlands. Thus, inference may be drawn on hydro-geomorphic conditions through the analysis of dendrogeomorphic evidence, riparian species composition, diversity, and life-history characteristics. Predictable patterns of species composition and community structure occur along high-energy Alpine rivers, medium-energy stable piedmont rivers, and low-energy lowland rivers. Also, a similar pattern may occur along a flow regime gradient, and the vegetation response may be directly related to specific hydro-geomorphic conditions. The highly dynamic rivers draining Alpine areas of western Europe offer an exceptional opportunity to investigate in detail vegetation responses (diversity, strategies, community composition) across a wide range of hydro-geomorphic conditions. Additionally, we believe that eco-geomorphic investigations of these areas may yield interpretations applicable across a broad range of spatial and temporal scales.

Vegetation organization, composition, and plant community dynamics on river floodplains are controlled by:

1. disturbance type (erosion, deposition) and scale (frequency and intensity of erosion vs. deposition, and flood duration);
2. biological characteristics of plants linked to resistance to disturbance, resilience (diaspore type, abundance and dispersability, recruitment efficiency), and competitive ability.

REFERENCES

Alestalo, J. 1971. Dendrochronological interpretation of geomorphic process. *Fennia* 105: 1–140.

Amoros, C. and Roux, A.L. 1988. Interactions between water bodies within the floodplains of large rivers: function and development of connectivity. *Münstersche Geogr. Arbeiten* 29: 125–130.

Amoros, C. and Bornette, G. 1999. Antagonist and cumulative effects of connectivity: a predictive model based on aquatic vegetation in riverine wetlands. *Arch. Hydrobiol.* 115(3; Suppl.): 311–3210.

Amoros, C., Bornette, G. and Henry, C.P. 2000. A vegetation-based method for the ecological diagnosis of riverine wetlands. *Environ. Manage.* 25: 211–227.

Amoros, C., Rostan, J.C., Pautou, G. and Bravard, J.P. 1987. The reversible process concept applied to the environmental management of large rivers systems. *Environ. Manage.* 11: 607–617.

Babinsky, Z. 1992. Hydromorphological consequences of regulating the Lower Vistula, Poland. *Regul. Riv.* 7: 337–348.

Barrat-Segretain, M.H. 1996. Germination and colonisation dynamics of *Nuphar lutea* (L.) Sm. In a former river channel. *Aquat. Bot.* 55: 31–38.

Barrat-Segretain, M.H. and Amoros, C. 1996. Recovery of riverine vegetation after experimental disturbance: a field test of the patch dynamics concept. *Hydrobiologia* 321: 53–68.

Barrat-Segretain, M.H., Bornette, G. and Vilas-Boas, A. 1998. Comparative abilities of vegetative regeneration of aquatic plants growing in disturbed habitats. *Aquat. Bot.* 60: 201–211.

Barrat-Segretain, M.H., Henry, C. and Bornette, G. 1999. Regeneration and colonization of aquatic plant fragments in relation with the disturbance frequency of their habitats. *Arch. Hydrobiol.* 145: 111–127.

Bazemore, D.E., Hupp, C.R. and Diehl, T.M. 1991.Wetland sediment deposition and vegetation patterns near selected highway crossings in West Tennessee. US Geological Survey, Water Resources Investigations Report 91-4106, 46 p.

Begin, Y. and Filion, L. 1995. A recent downward expansion of shoreline shrubs at Lake Bienville (subarctic Quebec). *Geomorphology* 13: 271–282.

Bendix, J. and Hupp, C.R. 2000. Hydrologic and Geomorphic Impacts on Riparian Plant Communities. *Hydrologic Processes* 14: 2977–2990 (special issue).

Bhowmik, N.G. and Adams, J.R. 1989. Successional changes in habitat caused by sedimentation in navigation pools. *Hydrobiologia* 176/177: 17–27.

Bilby, R. 1977. Effects of a spate on the macrophyte vegetation of a stream pool. *Hydrobiologia* 56: 109–112.

Bilby, R.E. and Likens, G.E. 1980. The importance of organic debris dams in the structure and function of stream ecosystems. *Ecology* 61: 1107–1113.

Bilby, R.E. and Ward, J.W. 1989. Changes in characteristics and function of woody debris with increasing size of stream in western Washington. *Trans. Am. Fish. Soc.* 118: 368–278.

Bornette, G. and Amoros, C. 1991. Aquatic vegetation and hydrology of a braided river floodplain. *J. Veg. Sci.* 2: 497–512.

Bornette, G. and Amoros, C. 1996. Disturbance regimes and vegetation dynamics: role of floods in riverine wetlands. *J. Veg. Sci.* 7: 615–622.

Bornette, G. and Heiler, G. 1994. Environmental and biological responses of former channels to river incision: a diachronic study on the Upper Rhône River. *Regul. Riv.* 9: 79–92.

Bornette, G., Amoros, C. and Chessel, D. 1994. Effect of allogenic processes on successional rates in former river channels. *J. Veg. Sci.* 5: 237–246.

Bornette, G., Amoros, C. and Lamouroux, N. 1998. Aquatic plant diversity in riverine wetlands: the role of connectivity. *Freshwat. Biol.* 39: 267–283.

Bravard, J.P. 1994. L'incision des lits fluviaux: du phénomène morphodynamique naturel et réversible aux impacts irréversibles. *Rev. Geogr. Lyon* 69: 5–10.

Bravard, J.P. and Gilvear, D.J. 1996. Hydrological and geomophological structure of hydrosystems. In: Petts, G. and Amoros, C., eds., *Fluvial Hydrosystems*, London, UK: Chapman & Hall, pp. 98–116.

Bravard, J.P., Amoros, C., Pautou, G., Bornette, G., Bournaud, M., Creuzé des Châtelliers, M., Gibert, J., Peiry, J.L., Perrin, J.F. and Tachet, H. 1997. Stream incision in Southeast France: morphological phenomena and impacts upon biocenoses. *Regul. Riv.* 13: 75–90.

Bravard, J.P., Amoros, C. and Pautou, G. 1986. Impact of civil engineering works on the successions of communities in a fluvial system. *Oikos* 47: 92–111.

Brierley, S.J., Harper, D.M. and Barham, P.J. 1989. Factors affecting the distribution and abundance of aquatic plants in a navigable lowland river; the River Nene, England. *Regul. Riv.* 4: 263–274.

Brock, C.M., van der Velde, G. and Van de Steeg, H.M. 1987. The effects of extreme water level fluctuations on the wetland vegetation of a Nymphaeid-dominated oxbow lake in the Netherlands. *Arch. Hydrobiol. Beih. Ergebn. Limnol.* 27: 57–73.

Butler, D.R. and Malanson, G.P. 1995. Sedimentation rates and patterns in beaver ponds in a mountain environment. *Geomorphology* 13: 255–269.

Cleaveland, M.K. 1980. Dating tree rings in Eastern United States. Va. Polytech. Inst. State Univ. FWS-2-80, pp. 110–124.

Cleaveland, M.K. and Stahle, D.W. 1989. Tree ring analysis of surplus and deficit runoff in the White River, Arkansas. *Water Resources Res.* 25: 1391–1401.

Connell, J.H. 1978. Diversity in tropical rain forests and coral reefs. *Science* 199: 1302–1310.

Cowles, H.C. 1901. The physiographic ecology of Chicago and vicinity: a study of the origin, development, and classification of plant societies. *Botanical Gazette Chicago* 31: 73–182.

Day, R.T., Keddy, P.A. and McNeill, J. 1988. Fertility and disturbance gradients: a summary model for riverine marsh vegetation. *Ecology* 69(4): 1044–1054.

Dunne, T. and Black, R.D. 1970. An experimental investigation of runoff production in permeable soils. *Water Resources Res.* 6: 478–490.

Everitt, B.L. 1968. Use of cottonwood in an investigation of the recent history of a floodplain. *Am. J. Sci.* 266: 417–439.

Fetherson, K.L., Naiman, R.J. and Bilby, R.E. 1995. Large woody debris, physical processes, and riparian forest development in montane river networks of the Pacific Northwest. *Geomorphology* 13: 133–144.

Foeckler, F., Diepolder, U. and Deichner, O. 1991. Water Mollusk communities and bioindication of Lower Salzach floodplain waters. *Regul. Riv.* 6: 301–312.

Foeckler, F., Kretschmer, W., Deichner, O. and Schmidt, H. 1994. Les communautés de macroinvertébrés dans les chenaux abandonnés par une rivière en cours d'incision, la basse Salzach (Bavière, Allemagne). *Rev. Geogr. Lyon* 69: 31–40.

Friedman, J.M., Osterkamp, W.R. and Lewis, W.M. 1996. The role of vegetation and bed-level fluctuations in the process of channel narrowing. *Geomorphology* 14: 341–351.

Galay, V.J. 1983. Causes of river bed degradation. *Water Resource Res.* 19: 1057–1090.

Gregory, K.J. 1992. Vegetation and river channel processes. In: Boon, P.J., Calow, P. and Petts, G.E., eds., *River Conservation and Management*, Chichester, UK: John Wiley and Sons, pp. 255–269.

Gregory, K.J. and Davis, R.J. 1992. Coarse woody debris in stream channels in relation to river channel management in woodland areas. *Regul. Riv.: Res. Manage.* 7: 237–260.

Gregory, K.J., Davis, R.J. and Tooth, S. 1993. Spatial distribution of coarse woody debris in the Lymington Basin, Hampshire, UK. *Geomorphology* 6: 207–224.

Greulich, S. and Bornette, G. 1999. Competitive abilities and related strategies in four aquatic plant species from an intermediately disturbed habitat. *Freshwat. Biol.* 41: 493–506.

Grubb, P.J. 1977. The maintenance of species-richness in plant communities: the importance of the regeneration niche. *Biol. Rev.* 52: 107–145.

Gurnell, A.M. and Gregory, K.J. 1987. Vegetation characteristics and the prediction of runoff: analysis of an experiment in the New Forest, Hampshire, UK. *Hydrol. Processes* 1: 125–142.

Gurnell, A.M. and Gregory, K.J. 1995. Interactions between semi-natural vegetation and hydrogeomorphic processes. *Geomorphology* 13: 49–69.

Hack, J.T. and Goodlett, J.C. 1960. Geomorphology and forest ecology of a mountain region in the central Appalachians, US Geological Survey, Professional Paper 347.

Hamel, C. and Bhéreur, P. 1982. Méthodes d'interprétation de l'évolution spatiale et temporelle des hydrophytes vasculaires. In: Symoens, J.J., Hooper, S.S. and Compère, P., eds., *Studies on Aquatic Vascular Plants*, Brussels: Royal Botanical Society of Belgium, pp. 294–303.

Haslam, S.M. 1978. *River Plants*. Cambridge, UK: Cambridge University Press.

Helley, E.J. and LaMarch, V.C., Jr. 1973. Historic flood information for northern California streams from geological and botanical evidence, US Geological Survey, Professional Paper 485-E, pp. 1–16.

Henry, C., Amoros, C. and Bornette, G. 1996. Species traits and recolonization processes after flood disturbance in riverine macrophytes. *Vegetatio* 122: 13–27.

Huffman, R.R. and Forsythe, S.W. 1981. Bottomland hardwood forest communities and their relation to anaerobic soil conditions. In: Clark, J.R. and Benforado, J., eds., *Wetlands of Bottomland Hardwoods*, New York, USA: Elsevier Scientific Publications, pp. 187–196.

Hupp, C.R. 1988. Plant ecological aspects of flood geomorphology and paleoflood history. In: Baker, V.R., Kochel, R.C. and Patton, P.C., eds., *Flood Geomorphology*, New York, USA: Wiley and Sons, pp. 335–356.

Hupp, C.R. 1992. Riparian vegetation recovery patterns following stream channelization: a geomorphic perspective. *Ecology* 73: 1209–1226.

Hupp, C.R. 1999. Relations among riparian vegetation, channel incision processes and forms, and large woody debris. In: Darby, S., Simon, A., eds., *Incised River Channels*, Chichester, UK: John Wiley and Sons, pp. 217–245.

Hupp, C.R. 2000. Hydrology, Geomorphology, and Vegetation of Coastal Plain Rivers in the Southeastern United States. *Hydrologic Processes* 14: 2991–3010 (special issue).

Hupp, C.R. and Bazemore, D.E. 1993. Spatial and temporal aspects of sediment deposition in West Tennessee forested wetlands. *J. Hydrol.* 141: 179–196.

Hupp, C.R. and Carey, W.P. 1990. Dendrogeomorphic approach to slope retreat, Maxey Flats, Kentucky. *Geology* 18: 658–661.

Hupp, C.R. and Morris, E.E. 1990. A dendrogeomorphic approach to measurement of sedimentation in a forested wetland, Black Swamp, Arkansas. *Wetlands* 10: 107–124.

Hupp, C.R. and Osterkamp, W.R. 1985. Bottomland vegetation distribution along Passage Creek, Virginia, in relation to fluvial landforms. *Ecology* 66: 670–681.

Hupp, C.R. and Osterkamp, W.R. 1996. Riparian vegetation and fluvial geomorphic processes. *Geomorphology* 14: 277–295.

Hupp, C.R. and Schening, M.R. 1997. Patterns of sedimentation and woody vegetation along black- and brownwater riverine forested wetlands. *Bull. Assoc. Southeastern Biologists* 44: 175.

Hupp, C.R. and Simon, A. 1991. Bank accretion and the development of vegetated depositional surfaces along modified alluvial channels. *Geomorphology* 4: 111–124.

Hupp, C.R., Osterkamp, W.R. and Howard, A.D., eds., 1995. Biogeomorphology—*Terrestrial and Freshwater Systems*, Amsterdam: Elsevier Science, 347 p.

Hupp, C.R., Osterkamp, W.R. and Thornton, J.L. 1987. Dendrogeomorphic evidence and dating of recent debris flows on Mount Shasta, Northern California, US Geological Survey, Professional Paper 1396-B, 39 p.

Hupp, C.R., Walbridge, M.R. and Lockaby, B.G. 1999. Fluvial geomorphic processes, water quality, and nutrients of Bottomland Hardwood Systems. In: *Proceedings of the Conference On Bottomland Hardwood systems, Memphis, Tennessee.*

Hupp, C.R., Woodside, M.D. and Yanosky, T.M. 1993. Sediment and trace element trapping in a forested wetland, Chickahominy River, Virginia. *Wetlands* 13: 95–104.

Jacoby, G.C. and Hornbeck, J.W., eds., 1987. *Proceedings of the International Symposium on Ecological Aspects of Tree-Ring Analysis*, Publications CONF 8608144, Washington, DC: US Department of Energy, 716 p.

Johnson, W.C. 1994. Woodland expansion in the Platte River, Nebraska: patterns and causes. *Ecol. Monogr.* 64: 45–84.

Johnson, W.B., Sasser, C.E. and Gosselink, J.G. 1985. Succession of vegetation in an evolving river delta, Atchafalaya Bay, Louisiana. *J. Ecol.* 73: 973–986.

Jones, H. 1956. Studies on the ecology of the river Rheidol. II. An ox-bow of the lower Rheidol. *J. Ecol.* 44: 12–27.

Johnston, C.A. 1991. Sediment and nutrient retention by freshwater wetlands: effects on surface water quality. *Crit. Rev. Environ. Control* 21: 491–565.

Kalliola, R., Salo, J., Puhakka, M. and Rajasilta, M. 1991. New site formation and colonizing vegetation in primary succession on the western Amazon floodplains. *J. Ecol.* 79: 877–901.

Keller, E.A. and Swanson, F.J. 1979. Effects of large organic debris on channel form and fluvial process. *Earth Surf. Processes* 4: 361–380.

Kellerhals, R. and Church, M. 1989. The morphology of large rivers: characterization and management. In: Dodge, D.P., eds., *Proceedings of the International Large River Symposium, Can. Spec. Publ. Fish. Aquat. Sci.* 106, pp. 13–30 (629 p.).

Kleiss, B.A. 1993. *Cache River, Arkansas: Studying a Bottomland Hardwood (BLH) Wetland Ecosystem*, The Wetlands Research Program, Volume 3, Number 1, Vicksburg, Mississippi, USA: US Army Waterways Experiment Station.

Kleiss, B.A. 1996. Sediment retention in a bottomland hardwood wetland in eastern Arkansas. *Wetlands* 16: 321–333.

Kirkman, L.K. and Sharitz, R.R. 1994. Vegetation disturbance and maintenance of diversity in intermittently flooded Carolina bays in South Carolina. *Ecological Applications* 41: 177–188.

Kohler, A. and Schiele, S. 1985. Veränderungen von Flora und Vegetation in den kalkreichen Fließgewässern der Friedberger Au (bei Augsburg) von 1972 bis 1982 unter veränderten Belastungbedingungen. *Arch. Hydrobiol.* 103: 137–199.

Kramer, P.J. and Kozlowski, T.T. 1979. *Physiology of Woody Plants*, New York: Academic Press.

LaMarche, V.C., Jr. 1968. Rates of slope degradation as determined from botanical evidence, White Mountains, California, US Geological Survey, Professional Paper 352-I, pp. 1341–1377.

Malanson, G.P. 1993. *Riparian Landscapes*. Cambridge: Cambridge University Press.

McKenney, R., Jacobson, R.B. and Wertheimer, R.C. 1995. Woody vegetation and channel morphogenesis in low gradient, gravel-bed streams in the Ozark Plateaus, Missouri and Arkansas. *Geomorphology* 13: 175–198.

Mitsch, W.J. and Gosselink, J.G. 1993. *Wetlands*, Second Edition, New York, USA: Van Nostrand Reinhold Co.

Müller, N. 1995. River dynamics and floodplain vegetation and their alterations due to human impact. *Arch. Hydrobiol.* 9(3/4; Suppl. 101, Large Rivers): 477–512.

Naiman, R.J., Décamps, H. and Pollock, M. 1993. The role of riparian corridors in maintaining regional diversity. *Ecol. Appl.* 3: 209–212.

Nilsson, C., Grelsson, G., Johansson, M. and Sperens, U. 1989. Patterns of plants species richness along riverbanks. *Ecology* 70: 77–84.

Osterkamp, W.R. and Hupp, C.R. 1984. Geomorphic and vegetative characteristics along three Northern Virginian streams. *Geol. Soc. Am. Bull.* 95: 1093–1101.

Patterson, G.G., Speiran, G.K. and Whetstone, B.H. 1985. Hydrology and its effects on distribution of vegetation in Congaree National Monument, South Carolina, US Geological Survey, Water-Resources Investigations Report 85-4256.

Pautou, G. 1984. L'organisation des forêts alluviales dans l'axe rhodanien entre Genève et Lyon; comparaison avec d'autres systèmes fluviaux. *Doc. Cartogr. Ecol.* 27: 43–64.

Peck, J.H. and Smart, M.M. 1986. An assessment of the aquatic and wetland vegetation of the upper Mississippi River. *Hydrobiologia* 136: 57–76.

Peiry, J.-L., Salvador, P.-G., and Nouguier, F. 1994. L'incision des rivières dans les Alpes du Nord: état de la question. *Rev. Geogr. Lyon* 69, 47-56.

Petts, G.E. and Bravard, J.P. 1996. A drainage basin perspective. In: Petts, G.E. and Amoros, C., eds., *Fluvial Hydrosystems*, London, UK: Chapman & Hall, pp. 13–35.

Pickett, S.T.A. 1980. Non-equilibrium coexistence of plants. *Bull. Torrey Bot. Club* 107: 238–248.

Piégay, H. and Gurnell, A.M. 1997. Large woody debris and river geomorphological pattern: examples from S.E. France and S. England. *Geomorphology* 19: 99–116.

Resh, V.H., Brown, A.V., Covich, A.P., Sheldon, A.L., Wallace, J.B. and Wissman, R.C. 1988. The role of disturbance in stream hydrology. *J. N. Am. Benthol. Soc.* 7: 433–455.

Reygrobellet, J.L. and Castella, E. 1987. Some observations on the utilization of groundwater habitats by Odonata larvae in an astatic pool of the Rhône alluvial plain (France). *Adv. Odonatol.* 3: 127–134.

Ribicki, N.B. and Carter, V. 1986. Effect of sediment depth and sediment type on the survival of *Vallisneria americana* Michx grown from tubers. *Aquat. Bot.* 24: 233–240.

Ritter, D.F., Kochel, R.C. and Miller, J.R. 1995. *Process Geomorphology*, Third Edition, Chicago, USA: W.C. Brown Publishers.

Roberts, J. and Ludwig, J.A. 1991. Riparian vegetation along current-exposure gradients in floodplain wetlands of the River Murray, Australia. *J. Ecol.* 79: 117–127.

Robison, E.G. and Beschta, R.L. 1990. Coarse woody debris and channel morphology interactions for undisturbed streams in southeast Alaska, USA. *Earth Surf. Processes Landforms* 15: 149–156.

Rostan, J.C., Amoros, C. and Juget, J. 1987. The organic content of the surficial sediment: a method for the study of ecosystems development in abandoned river channels. *Hydrobiologia* 148: 45–62.

Schening, M.R., Hupp, C.R. and Herbst, A.R. 1999. Sediment transport and storage in forested wetlands along the Chesapeake Bay tributaries. *Bull. Soc. Wetland Sci.* 16: 40.

Schwarz, W.L., Malanson, G.P. and Weirich, F.H. 1996. Effect of landscape position on the sediment chemistry of abandoned-channel wetlands. *Landscape Ecol.* 11: 27–38.

Schweingruber, F.H. 1988. *Tree Rings: Basics and Applications of Dendrochronology*, Boston, MA: Kluwer Academic Publishers.

Scott, M.L., Auble, G.T. and Friedman, J.M. 1997. Flood dependency of cottonwood establishment along the Missouri River, Montana, USA. *Ecol. App.* 7: 677–690.

Sedell, J.R., Bissson, P.A., Swanson, F.T. and Gregory, S.V. 1988. What we know about large trees that fall into streams and rivers. In: Masser, G., Tarrant, R.F. and Franklin, J.F., eds., *From the Forest to the Sea: A Story of Fallen Trees*, General Technical Report No. 229, USDA Forest Service, pp. 47–81.

Sharitz, R.R. and Mitsch, W.J. 1993. Southern floodplain forests. In: Martin, W.G., Boyce, S.G. and Echternacht, A.C., eds., *Biodiversity of the Southeastern United States, Lowland Terrestrial Communities*, New York, USA: John Wiley and Sons, pp. 311–372.

Shelford, V.E. 1954. Some lower Mississippi Valley flood plain biotic communities; their age and elevation. *Ecology* 35: 126–142.

Shroder, J.F. 1978. Dendrogeomorphological analysis of mass movement of Table Cliffs Plateau, Utah. *Quaternary Res.* 9: 168–185.

Sigafoos, R.S. 1964. Botanical evidence of floods and floodplain deposition, US Geological Survey, Professional Paper 485-A, pp. 1–35.

Sigafoos, R.S. and Hendricks, E.L. 1961. Botanical evidence of the modern history of Nisqually Glacier, Washington, US Geological Survey, Professional Paper 387-A, pp. 1–20.

Simon, A. and Hupp, C.R. 1987. Geomorphic and vegetative recovery processes along modified Tennessee

Streams: an interdisciplinary approach to disturbed fluvial systems. *Intl. Assoc. Hydro. Sci., IAHS-AISH* 167: 251–262.

Simon, A. and Hupp, C.R. 1992. Geomorphic and vegetative recovery processes along modified stream channels of West Tennessee, US Geological Survey, Open File Report 91-502.

Smits, A.J.M., van Avesaath, P.H. and van der Velde, G. 1990. Germination requirements and seed-banks of some nymphaeid macrophytes: *Nymphaea alba* L., *Nuphar lutea* (L.) Sm. and *Nymphoides peltata* (Gmel.) O. Kuntze. *Freshwat. Biol.* 24: 315–326.

Sousa, W.P. 1984. The role of disturbance in natural communities. *Ann Rev. Ecol. Syst.* 15: 353–391.

Sparks, R.E., Bayley, P.B., Kohler, S.L. and Osborne, L.L. 1990. Disturbance and recovery of large floodplain rivers. *Environ. Manage.* 14: 699–709.

Swanson, F.J., Kratz, T.K., Caine, N. and Woodmansee, R.G. 1988. Landform effects on ecosystem patterns and processes. *BioScience*: 92–98.

Thornes, J.B. 1988. Erosional equilibria under grazing.In: Bintliff, J., Davidson, D. and Grant, E., eds., *Conceptual Issues in Environment Archaeology*, Edinburgh, UK: Edinburgh University Press, pp. 193–210.

Thornes, J.B., ed., 1990. *Vegetation and Erosion: Processes and Environments*, British Geomophological Research Group Symposia Series, Chichester, UK: John Wiley and Sons, 518 p.

Trémolières, M., Carbiener, D., Carbiener, R., Eglin, I., Robach, F., Sanchez-Pérez, J.M., Schnitzler, A. and Weiss, D. 1991. Zones inondables, végétation et qualité de l'eau en milieu alluvial rhénan: l'île de Rhinau, un site de recherches intégré. *Bull. Ecol.* 22: 317–336.

van den Brink, F.W.B., De Leuw, J.P.H.M., Van der Velde, G. and Verheggen, G.M. 1993. Impact of hydrology on the chemistry and phytoplankton development in floodplain lakes along the lower Rhine and Meuse. *Biogeochemistry* 19: 103–128.

van den Brink, F.W.B., van der Velde, G., Bosman, W.W. and Coops, H. 1995. Effect of substrate parameters on growth responses of eight helophyte species in relation to flooding. *Aquat. Bot.* 50: 79–87.

van der Valk, A.G. and Bliss, L.C. 1971. Hydrarch succession and net primary production of oxbow lakes in central Alberta. *Can. J. Bot.* 49: 1177–1199.

Vannote, R.L., Minshall, G.W., Cummins, K.W., Sedell, J.R. and Cushing, C.E. 1980. The river continuum concept. *Can. J. Fish. Aquat. Sci.* 37: 130–137.

Viles, H.A., ed., 1988. *Biogeomorphology*, Oxford, UK: Basil Blackwell.

Viles, H.A. 1995. Ecological perspectives on rock surface weathering: towards a conceptual model. *Geomorphology* 13: 21–35.

Ward, J.V., Tockner, K., Edwards, P.J., Kollman, J., Bretschko, G., Gurnell, A.M., Petts, G.E. and Rossaro, B. 1999. A reference river system for the Alps: the "Fiume Tagliamento". *Regul. Riv.* 15: 63–75.

Wharton, C.H., Kitchens, W.M., Pendleton, E.C. and Sipe, T.W. 1982. The ecology of bottomland hardwood swamps of the southeast: a community profile, US Fish and Wildlife Service FWS/OBS-81-37.

Wood-Smith, R.D. and Swanson, F.J. 1997. The influence of large woody debris on forest stream geomorphology. In: Wang, S.S.Y., Langendoen, E.J. and Shields, F.D., eds., *Management of Landscapes Disturbed by Channel Incision*, Oxford, Mississippi: University of Mississippi, pp. 133–138.

Yanosky, T.M. 1982. Hydrologica inferences from ring widths of flood-damaged trees, Potomac River, Maryland. *Environ. Geol. Water Sci.* 4: 43–52.

Yanosky, T.M. 1984. Documentation of high summer flows on the Potomac River from the wood anatomy of ash trees. *Water Resources Bull.* 20: 241–250.

Yanosky, T.M., Hupp, C.R. and Hackney, C.T. 1995. Chloride concentrations in growth rings of *Taxodium distichum* in a saltwater-intruded estuary. *Ecol. Appl.* 5: 785–792.

Zimmermann, R.C. and Thom, B.G. 1982. Physiographic plant geography. *Prog. Phys. Geogr.* 6: 45–59.

Zimmermann, R.C., Goodlett, J.C. and Comer, G.H. 1967. The influence of vegetation on channel form of small streams. *Intl. Assoc. Sci. Hydrol.* 75: 255–275.

Part V

Analysis of Processes and Forms: Water and Sediment Interactions

11

Measurement and Analysis of Alluvial Channel Form

ANDREW SIMON[1] AND JANINE CASTRO[2]
[1]*USDA Agricultural Research Service, Oxford, MS, USA*
[2]*US Fish and Wildlife Service, Portland, OR, USA*

11.1 INTRODUCTION

Channel form, or morphology, has long been recognized as a diagnostic tool in evaluating fluvial landforms. Since Davis (1909) conceptualized the temporal aspect of channel and drainage basin evolution in the "cycle of erosion", geographers, geologists, and geomorphologists have used channel form as a parameter in classification, analysis, and prediction of fluvial response. Davis' view of fluvial landscapes was simplistic, but in combination with the detailed measurements of channel forms and processes in the studies by Gilbert (1914), we can envision these works as representing opposite approaches by which to direct future work. Davis' work represents larger-scale, qualitative assessments of channel form by which inferences about smaller-scale processes were advanced. Conversely, Gilbert's work represents the use of quantitative measurements by which inferences about larger-scale processes were advanced. The links implied here between channel form and process have been central in understanding fluvial geomorphology and, as such, have been the topic of many textbooks (e.g. Leopold *et al.* 1964, Morisawa 1968, Gregory and Walling 1973, Schumm 1977, Richards 1982, Knighton 1998).

Channel form can be considered to include aspects of the shape of a river channel in profile, in cross-section, and aerially, which includes channel planform and channel pattern. Profile characteristics include parameters such as channel-bed gradient and valley slope, and features such as pools, riffles, and cascades. Cross-sectional characteristics include channel width and depth, and features such as the bed, bars, banks, floodplains, and terraces. Planform characteristics include sinuosity, meander wavelength, belt width, and features such as meanders, braids, and abandoned channels.

Channel form, which includes measurements and descriptions of the shape of channel profiles, cross-sections, and planforms, can be used in combination with other attributes of a stream system such as riparian vegetation and character of the boundary sediments, to infer dominant trends in channel processes and response (Simon and Hupp 1986, Montgomery and Buffington 1997, Elliott *et al.* 1999). However, using gross channel form to quantitatively predict channel behavior, such as channel adjustments, system disturbances or rates of sediment transport, without rigorous analysis of channel processes is flawed (Miller and Ritter 1996). Thus, the key to using channel form in the analysis of fluvial landforms must be based on either (1) measurements of parameters that aid in quantifying channel processes such as flow hydraulics, sediment transport, and bank stability or (2) observations of diagnostic characteristics that provide information on active channel processes. In turn, measurements should either directly or indirectly lead to analysis of those forces acting on the channel boundary and those forces resisting entrainment of boundary sediments. Change in channel form is a matter of the applied forces overcoming resistance.

The overall purpose of this chapter is to provide a guide for field measurements and analytic techniques related to alluvial channel form that are central to understanding aspects of alluvial channel behavior. This is accomplished by concentrating on those factors that directly control the balance or imbalance between applied forces and boundary resistance. Generally, if force and resistance are balanced, the stream neither erodes nor deposits, and is capable of transporting the sediment load delivered from upstream reaches. This balance indicates a stability of channel dimensions and can be mathematically expressed as the stream power proportionality (Lane 1955):

$$QS_b \propto Q_s D_{50} \qquad (1)$$

where Q is the discharge, S_b the bed slope, Q_s the bed-material discharge, and D_{50} is the median grain size of bed material, indicating that 50% of the bed material is finer.

Equation (1) indicates that if available stream power were augmented by an increase in the discharge or the gradient of the stream, there would be an excess amount of stream power relative to the discharge of bed-material sediment whose resistance is a measure of particle diameter. Additional sediment would be eroded from the channel boundary resulting in (1) an increase in bed-material discharge to an amount commensurate with the heightened stream power, and (2) a decrease in channel gradient and, consequently, stream power as the elevation of the channel bed is lowered. A similar response would be expected from a decrease in the erosional resistance of the channel boundary or a decrease in the size of bed-material sediment (assuming the bed is not cohesive). In contrast, a decrease in available stream power or an increase in the size or discharge of bed-material sediment would lead to aggradation on the channel bed. Aggrading or degrading channels represent end members on a continuum where vertical stability is represented at the center point.

The conceptual and semi-quantitative relation provided by Lane (1955) provides only limited insight into the type and hierarchy of adjustment processes. Excess stream power can erode additional sediment from the channel boundary, however, Equation (1) does not indicate where the erosion will occur and, therefore, how channel form might change. Identifying instream sediment sources in this case becomes a matter of determining the relative resistance of the bed and bank material to the applied forces imposed by the flow and/or by gravity. For a sand-bedded stream with cohesive banks, an initial adjustment might involve streambed incision because of low critical shear stresses, higher applied shear stresses on the bed than on the bank-toe, and more frequent exposure to hydraulic shear than adjacent streambanks. Conversely, if we assume that the streambed is highly resistant, composed of cohesive clays, bedrock, or large particles such as cobbles or boulders, and that the bank-toe is composed of significantly weaker materials, we could expect bank erosion to be the initial adjustment.

The organization of this chapter is based on the simple question of: What can channel form tell us about channel processes and channel stability? Measurements that play an important role in analyzing channel processes are presented first, followed by a review of some of the techniques that have proven useful in analyzing these processes. General and empirical techniques are addressed before the more detailed deterministic approaches. Tools are provided by which to (1) diagnostically evaluate channel conditions and processes, and (2) select appropriate analytic techniques to estimate future channel changes or stable-channel dimensions.

Because of the broad scope of this chapter, it is not possible to specifically address all channel form measurement and analysis techniques. However, there are several other chapters that focus on specific components of channel form measurement and analysis including: Chapter 4, Using Historical Data in Fluvial Geomorphology; Chapter 6, Analysis of Aerial Photography and Other Remotely Sensed Data; Chapter 13, Bed Sediment Measurement; and Chapter 15, Sediment Transport.

11.2 FLUVIAL ADJUSTMENTS OF CHANNEL FORM

Measurement and analysis of channel form provide a context for interpretation of present and future channel morphologies and are often a central theme in studies which strive to identify and determine the magnitude and extent of channel change. Because streams are open systems, an alluvial channel has the ability to adjust to altered environmental conditions.

Scour and/or fill in a streambed over the course of a storm hydrograph, although representing streambed mobility, do not necessarily indicate instability because the short time period of the event is not indicative of rapid, progressive change. In fact, the important distinctions between the processes of scour and degradation, and fill and aggradation are a temporal issue. Slow, progressive erosion of a meander bend with concomitant deposition on the opposite point bar, which maintains an average channel width over time, also does not indicate instability. Meander migration of an alluvial stream is expected over long periods of time; again, potential instability is not inherent in the change, but rather a result of the rate of change.

The previous examples highlight the importance of timescales in interpretation and analysis of channel form. Even the dependency of variables can change as a function of the timescale applied (Schumm and Lichty 1965). Using valley slope as an example, over geologic timescales (long-term) there will be a progressive flattening with time as upstream areas degrade and downstream areas aggrade. However, if we were to take a "slice" of that period representing several years (medium-term) we would find that channel gradient tends to vary within a small range about some mean value. By taking another "slice", this time from the medium-term representing an almost instantaneous time period, channel gradient is constant. Thus, channel gradient is a dependent variable over the long-term and an independent variable over the medium- and short-terms. Variables describing channel form are indeterminate over geologic time, dependent over medium timescales and independent over short timescales (Schumm and Lichty 1965).

Longitudinal Profile

Initial modifications to channel form are often manifest by adjustments in profile through erosion or deposition of streambed materials. In simplest terms, we can conceptualize vertical stability as defined by Mackin (1948), where aggradation or degradation does not occur over a period of years. In physical terms, this indicates that just enough bedload is supplied from upstream relative to the available stream power [Equation (1)] or shear stress over a range of flows. Average boundary-shear stress (τ_0) is the drag exerted by the flow on the bed and is defined as:

$$\tau_0 = \gamma_w R\, S_b \tag{2}$$

where γ_w is the unit weight of water (N/m^3) and R is the hydraulic radius (area/wetted perimeter) (m). Resistance of non-cohesive materials is a function of bed roughness and particle size (weight), and is expressed in terms of a dimensionless critical shear stress (Shields 1936):

$$\tau^* = \tau_0/(\rho_s - \rho_w)g\, D \tag{3}$$

where τ^* is the critical dimensionless shear stress, ρ_s the sediment density (kg/m^3), ρ_w the water density (kg/m^3), g the gravitational acceleration (m/s^2), and D is the characteristic particle diameter (m). Critical shear stress (τ_c) in dimensional form can be obtained by invoking the Shields criterion and, for hydrodynamically rough beds, utilizing a value of 0.06 for τ^*:

$$\tau_c = 0.06(\rho_s - \rho_w)g\, D \tag{4}$$

Thus, the shear stress required to entrain a grain of diameter D can be estimated. Other commonly used values of τ^* are 0.03 and 0.047 (Vanoni 1975). The general criteria for incipient motion of cohesionless particles based on shear stress is similar to one developed by Hjulstrom (1935) using flow velocity and should be considered approximate because of heterogeneous bed-material sizes, imbrication, grain exposure, and turbulent velocity fluctuations near the streambed.

Cross-sectional Form

Adjustment of cross-sectional form is primarily by bank erosion and lateral channel migration, which represents a dominant means of channel response (Simon 1992, Simon and Darby 1997). Because channel banks generally erode by mass failure (enhanced by fluvial erosion at the bank-toe), the shear strength of the bank material represents the resistance of the boundary to erosion. The height and angle of the bank along with the unit weight of the bank material influence the gravitational driving force. The relative stability of the bank, therefore, becomes a matter of defining the ratio of resisting forces (shear strength) to driving forces (gravity), which is known as the factor of safety (F_s). A value of unity indicates the critical case and imminent failure.

Ignoring vegetative effects, shear strength (S_r) is comprised of two components: cohesive and frictional strengths. For the simple case of a planar failure of unit width and length, bank resistance is represented by the Coulomb equation:

$$S_r = c' + (\sigma - \mu_w)\tan\phi' \tag{5}$$

where S_r is the shear strength (kPa), c' the effective cohesion (kPa), σ the normal stress (kPa), μ_w the pore-water pressure (kPa), and ϕ' is the effective friction angle (degrees). The normal stress is given by:

$$\sigma = W\cos\beta \tag{6}$$

where W is the weight of the failure block (kg) and β is the angle of the failure plane (degrees). The driving (gravitational) force for this example is given by:

$$W \sin \beta \quad (7)$$

The factor of safety (F_s) can then be expressed by dividing Equation (5) by Equation (7). Factors that decrease the erosional resistance (S_r) such as excess pore-water pressure from saturation and the development of vertical tension cracks favor bank instabilities. Similarly, increases in either bank height (by bed degradation) or bank angle (by undercutting) favor bank failure by causing the gravitational component to increase. In contrast, vegetated banks are generally drier and provide improved bank drainage, which enhances bank stability (Thorne 1990, Simon and Collison 2002). Plant roots provide tensile strength to the soil, which is generally strong in compression, resulting in reinforced earth (Vidal 1969) that resists mass failure, at least to the depth of vegetation roots, which is often not greater than 1 m (Simon and Collison 2002).

Streams, particularly those that are incised, generally have a substantial proportion of their banks above the level of the phreatic surface during low and even moderate river stages. Above the water table, the soil has a degree of saturation less than 100% and negative pore-water pressures (matric suction) develop. For circumstances where the failure surface passes through unsaturated soil, slope-stability analyses need to evaluate matric suction (ψ): the difference between the air pressure (μ_a) and the water pressure in the pores (μ_w). The increase in shear strength due to an increase in matric suction is described by the angle ϕ^b. Incorporating this effect into the standard Mohr–Coulomb equation produces (Fredlund *et al.* 1978):

$$S_r = c' + (\sigma - \mu_a) \tan \phi' + (\mu_a - \mu_w) \tan \phi^b \quad (8)$$

where $(\sigma - \mu_a)$ is the net normal stress on the failure plane at failure, and μ_w is the pore-water pressure on the failure plane at failure. The value of ϕ^b is generally between 10° and 20°, with a maximum value of ϕ' under saturated conditions (Fredlund and Rahardjo 1993, Simon *et al.* 1999a).

The effects of matric suction on shear strength are reflected in the apparent or total cohesion (c_a) term:

$$c_a = c' + (\mu_a - \mu_w) \tan \phi^b = c' + \psi \tan \phi^b \quad (9)$$

Negative pore-water pressures (positive matric suction, ψ) in the unsaturated zone provide for cohesion greater than the effective cohesion, and thus, greater shearing resistance. This is often manifest in steeper bank slopes than would be indicated by ϕ'. To quantify the magnitude of this effect, data are required on pore-pressure distributions in the bank or can be estimated by assuming various levels of the phreatic surface.

Channel Planform

The courses of rivers and streams have historically been among the first geographic features mapped. The basic tri-partite division of channel patterns proposed by Leopold and Wolman (1957) of braided, meandering, and straight is widely accepted, but these broad categories are inadequate for many purposes, so additional channel pattern classes have been proposed by other authors (see Chapter 7). Leopold and Wolman (1957) proposed a relation to predict whether a channel would be braided (Figure 11.1), meandering (Figure 11.2), or straight, based on channel slope and bankfull discharge (Figure 11.3). The range of channel planforms is probably better described as a continuum having a qualitative range of typical characteristics (Schumm 1981) (Figure 11.4).

A channel that is constantly shifting and avulsing over relatively short timescales is considered unstable. Planform stability is a culmination of profile and cross-sectional stability and as such it is not always clear what process initiated the primary instability. Planform instability is closely linked with both floodplain and channel alterations. Significant changes of floodplain roughness can initiate planform instability, especially when channel roughness exceeds floodplain roughness creating a preferential flow path across the floodplain and setting the stage for a channel avulsion (Figure 11.5).

The Manning (1891) equation relates average flow velocity to variables of channel form and flow resistance by:

$$v = (1/n)\ R^{2/3} S_b^{1/2} \quad (10)$$

where v is the velocity (m/s), n the a roughness coefficient, and R is the hydraulic radius or hydraulic mean depth (m). Thus, flow velocity is directly related to hydraulic radius and slope and inversely related to flow resistance.

Valley slope, and hence floodplain slope (S), is generally much greater than channel slope. Holding all other variables equal, this would indicate that

Figure 11.1 The Rakaia River, a classic braided river on the south island of New Zealand (photograph by G. Mathias Kondolf)

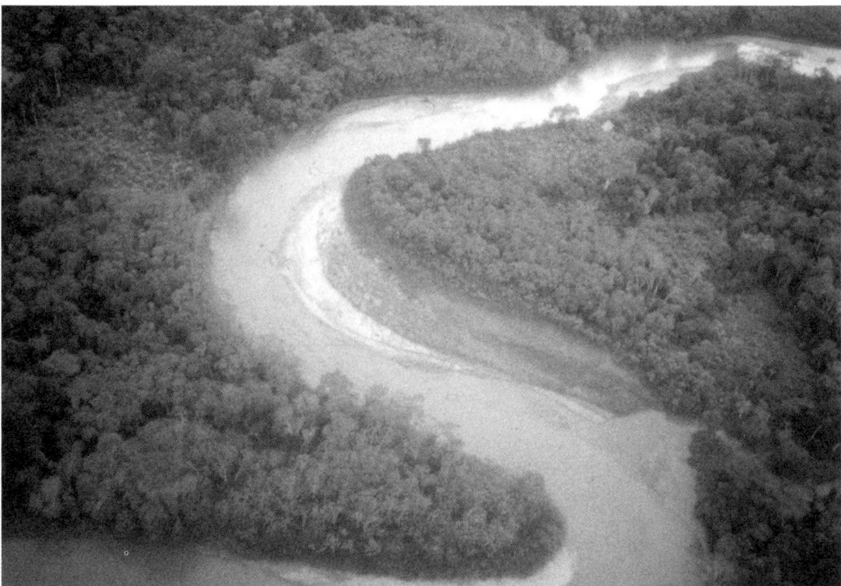

Figure 11.2 The Caraway River, Equador, a typical meandering stream

flows over the floodplain should have higher velocities than in the stream channel. This does not typically occur because (1) roughness (n) is much greater on the floodplain than in the channel, and (2) hydaulic radius (R) over the floodplain is much less. If, however, the primary channel is wide and shallow, and the floodplain has been cleared of vegetation, velocities over the floodplain can approach in-channel velocities.

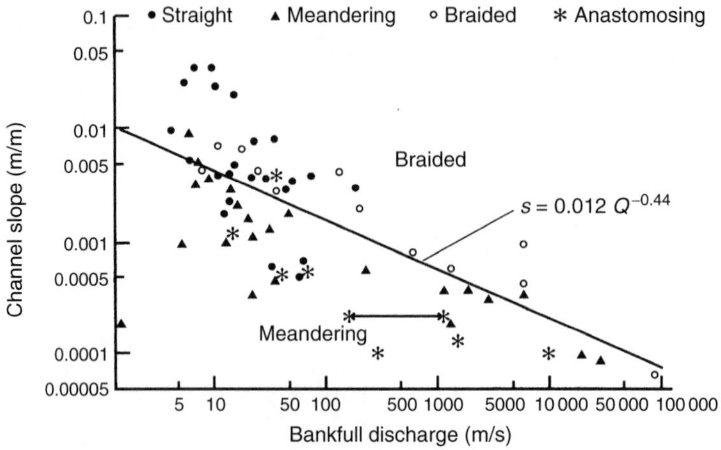

Figure 11.3 Threshold relation between braided and meandering channel patterns as defined by Schumm (1981) (from Knighton 1998, Figure 5.16) (reproduced by permission of John Wiley and Sons, Ltd.)

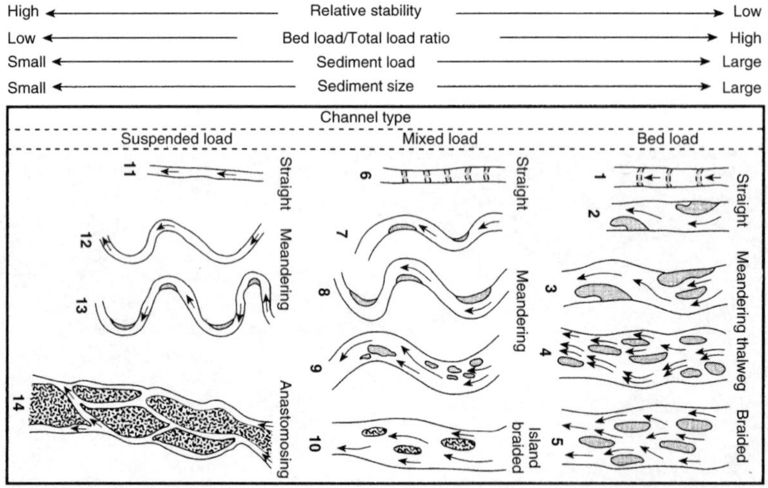

Figure 11.4 Classification of channel patterns by Schumm (1981) (from Knighton 1998, Figure 5.17) (reproduced by permission of John Wiley and Sons, Ltd.)

Given this condition, the threshold for channel avulsion is approached or exceeded. In incising channels shear stresses are particularly heightened on outside bends, resulting in accelerating cutting, upper bank failure and meander migration.

11.3 MEASUREMENTS OF CHANNEL FORM

Adjustment processes that can affect entire fluvial systems include channel degradation and aggradation, lateral channel migration, channel widening or narrowing, channel avulsion, and changes in the quantity and character of the sediment load. These processes differ from short-term, event-related localized processes such as scour and fill, which can be limited in magnitude as well as in spatial and temporal scale. Analysis of channel form requires that the investigator determine whether processes are localized disturbances or system-wide adjustments.

It is often difficult to differentiate between localized and system-wide processes without extending the investigation upstream and downstream of a particular

Figure 11.5 Photograph of channel avulsion showing new and original course (photograph by Janine Castro)

stream reach. Similar channel forms can be the result of dissimilar causes, and because channel adjustments migrate over time and space and may affect previously undisturbed reaches, it is essential to properly identify the cause of channel change and not just the local symptoms. To determine channel stability or to simply quantify channel processes, direct measurements of channel form are necessary.

Longitudinal Profile

Longitudinal profiles are one of the most useful pieces of field information that is gathered. From these profiles, thalweg slope, energy-grade lines, bed features, residual pools, flow obstructions, knick points, and gradient changes can be identified. This information can then be used to estimate flow velocity, discharge, stream power, flow resistance, shear stress and sediment transport.

Generally, a longitudinal survey will start at a stable point within the stream channel. The type of geomorphic surfaces (e.g. pools, riffles, etc.) that are encountered during survey should be noted. Sufficient detail should be provided to clearly define bed features such as steps, pools, riffles, runs, and any unique conditions. Surveys of a thalweg profile should ideally encompass a reach at least 6–30 channel widths in length, especially if the purpose is to determine channel gradient, although this is not always practical. Surveys of this length will generally include a series of pool–riffle sequences. At least two complete sequences should be included with the final calculation of slope comprising a linear regression between distance and elevation. Site conditions may limit the extent of the survey, or unique conditions may require a more expansive survey. Where flows are too deep to obtain measurements of bed elevation, survey shots of the water surface (edge of water) can be used.

A total station or laser level is preferred for longitudinal profiles because of the great distances that can be obtained for individual shots and because of the sensitivity of channel slope in various hydraulic models. Using the appropriate equipment is especially critical in areas that are flat. In steep terrain, a hand level that provides a derived accuracy of 0.25% for channel gradient may be sufficient; however, this is unacceptable for a stream channel with a gradient of 0.05%. See Harrelson *et al.* (1994) for more specific guidelines on survey techniques.

To measure short-term changes in the elevation of non-cohesive channel beds, scour chains can be installed. Chains are anchored to a pin placed horizontally below the estimated maximum depth of scour, extended vertically upwards and draped over the bed surface. The elevation of the bottom of the chain and the bed surface are surveyed and the length of chain exposed is measured. Scour chains are inspected and measured after peak-flow events. The exact location of the scour chain should be carefully noted, since locating the scour chain after a large flow event is very labor intensive and can be extremely difficult. During a sediment-transporting event, the material around the chain may be scoured causing the chain to lie over against the remaining bed material. This level indicates the depth of scour for the event (Lisle and Eads 1991). If a scour chain is left in place through numerous events, it is not possible to determine which event caused the greatest degree of scour, so it is essential to locate, measure, and replace the scour chain after every significant event if the purpose is to determine average scour conditions for various size events.

Cross-sectional Form

Measurements of channel cross-sections are probably the most common channel data collected, however, surveys of the floodplain surface are often neglected. Surveys provide the needed data for a number of uses, including active channel width and depth, wetted perimeter, bank height and angle, and the presence, elevation, and extent of floodplain and adjacent terraces. Derived attributes for the active channel include cross-sectional area, average depth, hydraulic radius, and width to depth ratio. These values can be combined with longitudinal and hydraulic data to calculate channel velocity and discharge at various stages, stream power, shear stress, and other parameters that begin to quantify channel processes.

Generally a cross-sectional survey should start on the floodplain/low terrace interface (in non-incised streams) or higher terrace surface and proceed across the floodplain, down the bank, and across the channel, finishing on the opposite side of the valley. Sufficient detail should be provided to clearly define floodplain topography, natural levee deposits, bank form such as the vertical face and any failed debris, the bank-toe, the edge of water, thalweg, and any bar surfaces. Notations should also be made of the existence, type, and abundance of riparian vegetation on floodplain, bank, and bar surfaces. Transitions between upland and riparian species should be clearly indicated as this marks the zone of relative inundation frequency.

Depending upon the level of accuracy desired and the site conditions, total stations, laser levels, transits, hand levels, or level lines can be used to survey. Laser levels with remote sensors allow a single surveyor to collect cross-sectional data, or several surveyors to collect data on various cross-sections concurrently. Transits provide the same level of accuracy as a laser level, but require at least two surveyors on site. Hand levels have reduced accuracy and require at least two surveyors, but they are very transportable. For surveys in remote, rugged terrain, hand levels are often the preferred surveying equipment. Stretching a level line across a stream channel and directly measuring vertical distance with a rod is commonly used for cross-section measurement. Errors occur when the line is not level from left to right bank, or when the line sags in the middle. Small gage cable marked at regular intervals with beads eliminates much of the sagging and also eliminates fluttering of tapes due to wind.

Active Channel

The overall method for cross-sectional surveys is to define the length of the study reach and to then separate it into sub-reaches/cross-sections. A reach should represent a homogeneous length of stream including similar slope, bed and bank material, bed forms, floodplain/terraces, vegetation, and dominant channel processes, so that measurement variance is minimized. A reach length of approximately 30 channel widths provides a relatively homogeneous sample unit in many alluvial channels (Myers and Swanson 1997). Selection of the number of sub-reaches or cross-sections of a given type (pool or riffle; convergent or divergent flow; meander bend or straight) is a function of the percentage of these sub-reach types over the length of the entire reach, however, cross-sections that are spaced closer than 3 channel widths can result in autocorrelation unless there is a significant, abrupt change in morphology (Myers and Swanson 1997). Systematic sampling of sub-reaches based strictly on multiples of channel width rather than on morphologic types may result in over- or under-sampling of specific channel types because of cyclic spacing especially in alluvial channels (e.g. pool/riffle spacing of 7 channel widths). Once reaches and sub-reaches have been defined, data collection proceeds according to standard surveying techniques (see Harrelson *et al.* 1994). Reach-average values automatically represent a weighted average of the conditions at the surveyed cross-sections.

Bed material. Methods to sample boundary sediments vary by the size class of sediments and their position along the channel boundary. Particle-size characteristics of bed-material samples are used to calculate the shear stress required to erode non-cohesive beds. Bed-material samples (both armor and sub-armor if present) should be obtained from each of the cross-sections selected in the reach and should represent the morphologic features present. Sampling of coarse-grained bed material is covered in detail in Chapter 13.

For cohesive (silt/clay) streambeds, different sampling techniques are often required because the electro-chemical bonding between particles determines resistance to erosion rather than particle size and weight. Critical shear stresses and erodibility coefficients can be obtained on in situ materials in the field with a submerged jet-test device (Figure 11.6) (Hanson

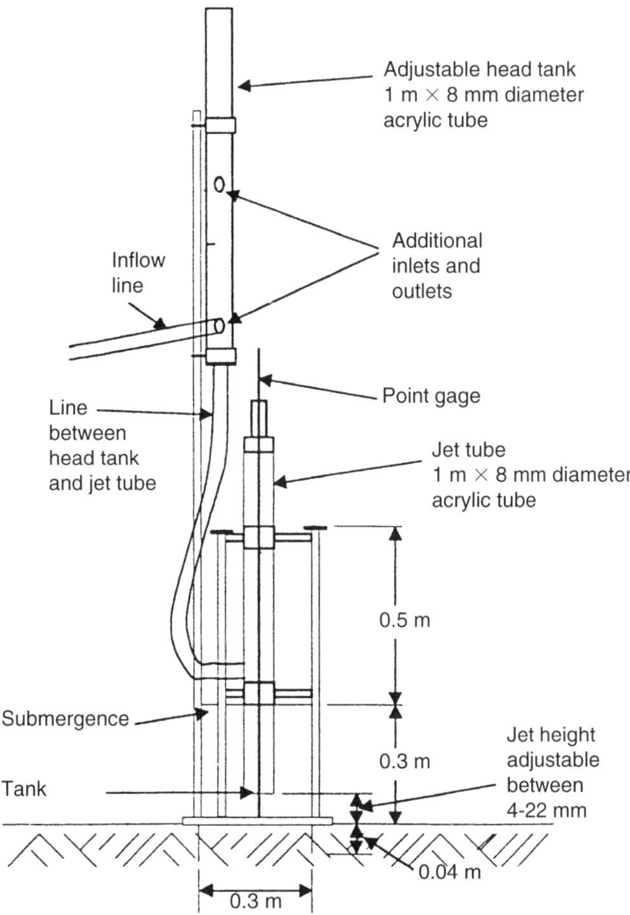

Figure 11.6 Schematic of jet-test device used for in situ measurement of the erodibility of cohesive materials

1990, 1991). The circular mounting plate is driven into the channel bed and water at a given head is pumped through a submerged nozzle. Measurements are made at regular intervals of the distance between the nozzle and the eroding bed. As the bed erodes beneath the nozzle the applied shear stress is reduced. From these data a relation between erosion rate and shear stress is developed for the in situ material and a critical shear stress (τ_c in Pa) and erodibility coefficient (k in cm^3/N s) are calculated. Jet-test results have been used to develop a relation between critical shear stress (τ_c) and the erodibility coefficient (k) for cohesive streambeds (Hanson and Simon 2001). In the absence of jet-test measurements, the percent clay or plasticity index of cohesive streambeds can be used as a measure of the relative resistance to erosion with higher values indicating greater resistance.

Bank material. Samples of bank material are collected to discern the resistance of the channel banks to erosion by hydraulic forces. These hydaulic forces edode by both the shear of flowing water and gravitational forces, which then cause additional erosion due to mass-failure mechanisms.

Bank-material samples should be collected by bulk sampling if possible and analyzed to determine the percentage of major size classes (i.e. sand, silt, and clay). The particle-size distribution of non-cohesive bank material can be used to calculate critical shear stresses for entrainment and compared to likely shear stresses to determine relative erodibility due to hydraulic shear. To determine the erodibility of fine-grained materials (silt and clay) it is recommended that jet-test measurements be made to calculate the potential for bank-toe erosion and steepening during flow events.

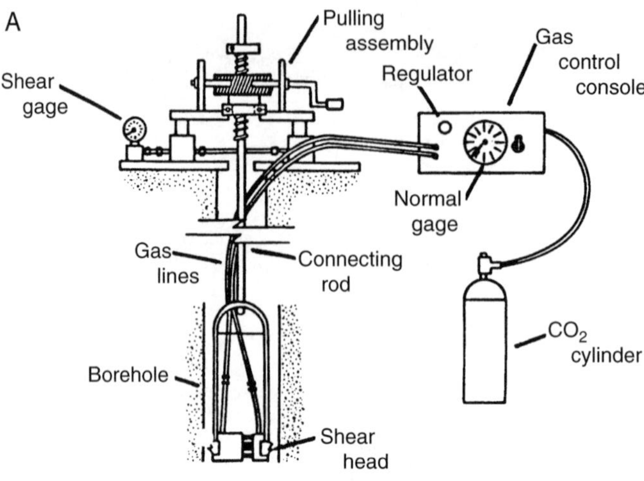

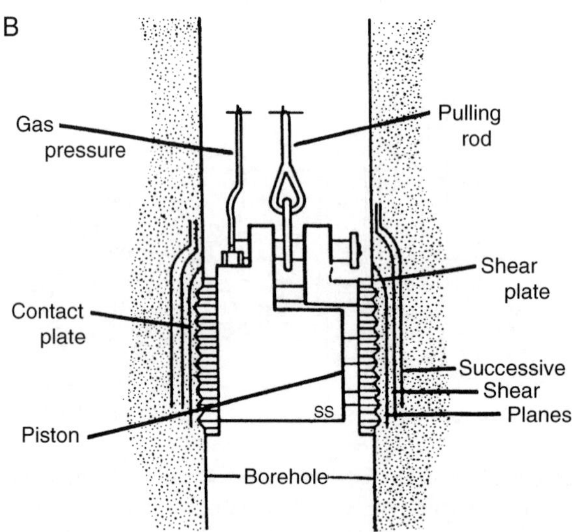

Figure 11.7 Schematic of (A) borehole shear test device for measuring geotechnical properties of streambanks and (B) detail of shear head in borehole (reproduced from Thorne *et al.* 1981)

Bank materials that erode by mass failure need to be sampled for their geotechnical properties. To properly determine the resistance of cohesive materials to erosion by mass movement, data must be acquired on those characteristics that control shear strength; that is cohesion, angle of internal friction, pore-water pressure, and bulk unit weight. At zero normal stress, the magnitude of this bonding (resisting) force is termed cohesion. Cohesion and friction-angle data can be obtained from standard laboratory testing (triaxial shear or unconfined compression tests), or by in situ testing with a borehole shear-test (BST) device (Lohnes and Handy 1968, Lutenegger and Hallberg 1981, Thorne *et al.* 1981, Little *et al.* 1982, Simon 1989a) (Figure 11.7). The BST provides direct, drained shear-strength tests on the walls of a borehole.

Another means of obtaining shear-strength data in the field is with a portable vane shear device. This is an inexpensive device and allows for rapid testing. However, the experience of the authors with these devices is that the data are (1) not reproducible and (2) difficult to use in stability analyses because the friction and cohesion terms are combined into a single value.

Since shear strength varies with moisture content, measurements of pore-water pressure are made in conjunction with shear-strength testing at all test depths. These can be obtained by extracting a core from the appropriate depth with a hammer sampler. A portable piezometer/tensiometer is then inserted into the core to determine the magnitude of pore-water pressure or matric suction (negative pore-water pressure) (Simon *et al.* 1999). Effective cohesion (c') is then calculated as the difference between total cohesion and cohesion due to matric suction and is accomplished by re-arranging Equation (9).

Bulk unit weight of the bank material must also be obtained because it determines, in part, the magnitude of both the driving and resisting gravitational forces. Samples of a known volume are taken within the borehole by means of hammer sampler. These samples are sealed and then returned to the laboratory for weighing, drying and re-weighing to calculate ambient, and dry unit weights. Values of saturated unit weight, required for determining worst-case conditions can be obtained by assuming a value for specific weight of the material. For coarse-grained cohesionless soils, estimates of friction angles can be obtained from reference manuals (i.e. Selby 1982).

Bank-toe erosion. Fluvial erosion at the bank-toe and of in situ materials is important in assessing undercutting of streambanks as a pre-cursor to top-bank gravitational failure. This process can be monitored using a network of erosion pins spaced longitudinally along the bank. Each set of pins is made up of pieces of rebar inserted horizontally into the bank-toe region and displaced vertically. It should be noted whether the pin is initially placed in failed, reworked, or in situ materials. Each pin is measured after runoff events or at a frequency conducive to the temporal scope of the study. Based on measurements of protrusion lengths made by different operators on the same day, erosion-pin data are accurate to within ± 5 mm (Simon *et al.* 1999). Estimates of the change in length between visits are accurate, therefore, to within 1 cm. However, additional systematic errors may be introduced through the effects of (1) turbulent scour around the tip of the pins and (2) disturbance of the bank-material fabric during insertion of the pins. Lawler (1993) has designed Plexiglas, photoelectric erosion pins that provide a continuous digital record of pin exposure and bank-toe erosion. Experience has shown, however, that it is susceptible to breakage from the impact of sediment or debris, mass wasting events and cannot operate at night.

Floodplain

The geomorphic floodplain is defined as "the flat area adjacent to the river channel, constructed by the present river in the present climate and frequently subject to overflow" (Leopold *et al.* 1964). Once a stream channel and floodplain have become decoupled, the feature is considered a terrace, and may interact with stream flows only during moderate to extreme flow events. Determining the recurrence interval of the channel-forming discharge and comparing this to inundation periods of the floodplain will help define the connection between these geomorphic elements. Using Equation (10), and estimating floodplain roughness, it is possible to calculate flow velocity, depth, and with measurements of floodplain slope, shear stress for various stages.

Field evidence of floodplain inundation include recent sediment deposits, rafted wood and debris wrapped around trees and fences, scour around existing trees and other flow deflectors, stain lines, relatively open high flow channels, and general vertical accretion. Flow direction can be determined by orientation of organic material, particle sorting in sediment deposits, imbrication in high flow channels, rafted debris at flood entry points, and potential scour where flood flows reenter the main channel.

Floodplain surfaces commonly originate as depositional bars within active stream channels. These incipient floodplains build vertically over time as additional sediment is deposited. Vegetation and downed wood provide increased hydraulic roughness, reducing velocities and therefore encouraging more deposition. Areas of deposition can be identified by partial burial of the trunks of woody-riparian vegetation. More reliable estimates of deposition rates can be obtained by dating trees using an increment borer and then excavating the tree to the depth of the root collar or root flare (Simon and Hupp 1992). This represents the germination point of the tree. An average deposition rate can then be calculated by dividing the amount of burial by the age of the tree. Event-related deposition can be identified and dated based on even-age-class stands of woody-riparian vegetation. More detailed data can be obtained by placing clay pads in various locations on the floodplain surface and measuring the amount of deposition that occurs on these pads following floodplain inundation.

Terraces

Terraces represent abandoned floodplains and are found at various elevations, which are generally controlled by either the amount of channel incision that has subsequently occurred, or by geologic events such as uplift (Figure 11.8). Terms such as the "100-year floodplain" refer to terrace surfaces that are inundated by a flow event with a return period of 100 years (or 1% probability in any given year), and are not true floodplains (Leopold 1994). Recently formed terraces are common and are typically the result of channel incision due to land management practices or channelization projects.

Terrace surfaces can be dated based on their soil profile and depositional characteristics. Soil type, genesis, and approximate age can be derived from soils maps and in conjunction with terrace elevations can be used to reconstruct paleoenvironments and potentially shifts in vegetation and climate (Balster and Parsons 1968). These maps can also be used to determine sediment sources for specific particle-size classes or mineralogy/lithology.

Terrace boundaries can be a tremendous source of sediment to a stream channel, especially when the stream has impinged upon this geomorphic surface (Figure 11.9). Slope instability combined with basal clean-out during sediment-transporting flow events may result in mass failures which deliver sediment directly to the stream channel.

Channel Planform

Measurements of channel planform include number of active channels, sinuosity, meander belt width, meander amplitude and wavelength, radius of curvature, and valley width (Figure 11.10). From these measurements, various parameters related to flood and sediment routing can be calculated and the relative type and abundance of habitat types can be inferred.

Measuring planform changes over time can provide an accurate estimate of system variability and valuable information for potential rates and magnitude of

Figure 11.8 Photograph of paired and unpaired terraces and floodplain from the south island of New Zealand (photograph by G. Mathias Kondolf)

Figure 11.9 Photograph of Rio Puerco, New Mexico, USA, showing large contribution of sediment from terrace slope (photograph by US Geological Survey)

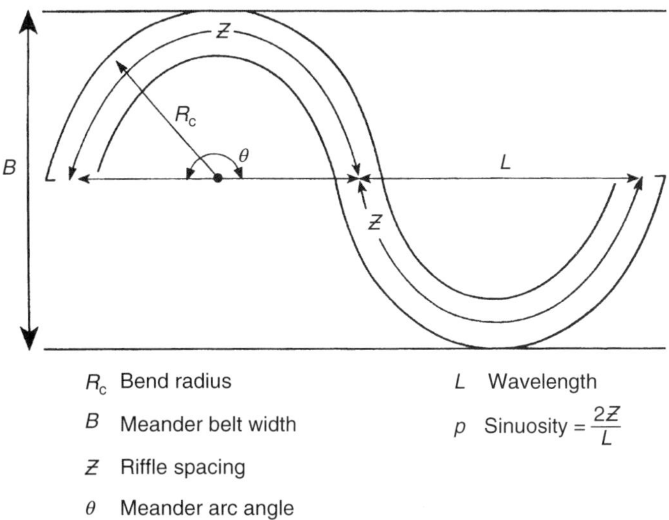

Figure 11.10 Parameters used to characterize meander geometry

future channel changes. Evidence of meander translation and compression may indicate variable boundary resistance, while braiding indicates high sediment loads or sediment pulses. Consistent features that have not migrated upstream or downstream over time may represent significant valley controls or gradient breaks.

Maps and Aerial Photos

One of the simplest methods for quantifying changes in channel planform with time is the use of maps and aerial photos. Aerial photos are generally superior to maps because there is no filter of interpretation, however, historic maps provide important information regarding channel position, complexity, and simplification prior to the advent of aerial photography (see Chapters 4 and 6).

Aerial photos with decadal intervals are ideal for determining planform changes over time. Streams represented on photos with various scales can be digitized and overlain in a GIS-based system, which allows for direct comparison and analysis.

Periods of rapid channel change can be narrowed to specific intervals and can potentially be linked to changes in the watershed or direct changes to the stream channel. Watershed changes can be further quantified if a stream gauging station is located within the watershed.

Field Surveys

Direct field surveys utilizing distance, bearing, and elevation will result in a current representation of channel sinuosity and planform and may be preferable to aerial photos for small streams. Field surveys using total stations are very effective and efficient for stream systems that have minimal vegetative cover, or for areas that lose their foliage during the winter months. In heavily vegetated areas, field surveys utilizing hand levels, compasses, and hip chains or tapes become a more efficient alternative, although the accuracy is greatly reduced.

11.4 MEASUREMENT AND DERIVATION OF CHANNEL HYDRAULIC PARAMETERS

Design Discharge

To estimate fluvial responses for various flow events, it may be necessary to determine which events are significant in the context of sediment transport for a specific study, project or design. For flood-control works, discharges such as the Q_{10}, Q_{25}, Q_{50}, and Q_{100} are typically evaluated. In the realm of fluvial geomorphology, more frequent discharges are often of greater interest, although it is still important to evaluate the potential geomorphic response for a large event. In recent years, the flow or range of flows that transports the most sediment over the long-term and, defined as the "effective discharge" has received renewed attention. It should be stressed that the effective discharge is a concept that most likely represents a range of flows and is not to be confused with "bankfull discharge" and "dominant discharge". So as to avoid confusion, definitions are given:

1. *Bankfull discharge*. The maximum discharge which can be contained within the channel without overtopping the banks (Leopold *et al.* 1964) and generally accepted to represent the flow that occurs, on average, every 1–2.3 years (Williams 1978a).
2. *Dominant discharge*. The single steady discharge that would produce the same morphology and dimensions as produced by the actual flow regime (Inglis 1949). This flow, however, is difficult if not impossible to define.
3. *Effective discharge*. The discharge or range of discharges that transports the largest proportion of the annual sediment load over the long-term (Wolman and Miller 1960, Andrews 1980). Although originally defined for suspended-sediment load, subsequent workers have applied it to bed load, bed-material load, and total load.

Bankfull Discharge

Indicators of bankfull stage can be estimated analytically or based on field observations. In stable, "natural" streams, the best indicator of bankfull stage is often the active floodplain surface. This surface, however, is not always easy to identify, particularly in steep, cobble–boulder streams, and along braided, incised, or aggraded channels. In the absence of a well-defined floodplain surface, other indicators, whose importance will depend on the specific fluvial environment, are useful. In any case, parallel lines of evidence should be used.

Along braided streams or streams containing bars, the top of the bar surface (the proto-floodplain), particularly if it supports woody vegetation is often a good indicator of bankfull stage, although it represents a minimum level for bankfull. In incised channels, where the previous floodplain surface has become a terrace, the bankfull stage can be identified as the lowermost limit of establishing woody-riparian vegetation. Williams (1978a) and Harrelson *et al.* (1994) list additional useful indicators although they should be used with caution and accepted only if there are other lines of evidence.

Field-based observations should, if possible, be verified with stream-gauge data. Stage–discharge relations established at gauging stations are available in the United States from the US Geological Survey and in other countries from those agencies responsible for monitoring water resources. The stage–discharge relation will show a bend to a flatter slope, representing the stage at which flow spreads out across the floodplain or braid-plain surface. If the elevation of this stage is in agreement with surveyed, field-based observations of bankfull indicators, one can be reasonably confident in the selection of the bankfull stage.

Effective Discharge

A convenient means of estimating the long-term potential for sediment transport is to compare critical and available shear stresses for the effective discharge. The effective discharge is generally accepted to occur, on average, about every 1–2 years (Simon *et al.* in press). To verify this, an established sediment-transport relation (concentration versus flow) and a flow-frequency distribution are required. This involves a three-step process:

1. Construct a frequency distribution (histogram for discharge).
2. Construct a sediment-transport rating relation.
3. Integrate the two relations by multiplying the sediment-transport rate for a specific discharge class by the frequency of occurrence for that discharge, with the maximum product being the effective discharge.

Mean-daily flows are often used with instantaneous values of sediment concentration because these data are readily available. The minimum period of record should be at least 10 years. Except for large rivers this approach is biased towards low-flow conditions because short-duration peak discharges are neglected. A better approach is to use shorter-termed flow data, such as those corresponding to the 15-minute stage data collected by various agencies. If sediment data are not available, bed-material load transport rates can be derived from a variety of transport functions. Procedures and examples are provided in Stevens and Yang (1989) and Andrews and Nankervis (1995). Regardless of the type of discharge data used, data are ranked and then subdivided into 25–33 classes (Yevjevich 1972). Subdividing classes using an arithmetic distribution often results in the majority of flows falling into the lowest discharge class. To overcome this problem a logarithmic distribution is used. The effective discharge can then be calculated as the discharge class that has the maximum sediment concentration/discharge product for the classed flow-frequency data.

Energy-grade Line

In many hydraulic models, the independent variable, stream slope is better defined as the energy slope or energy-grade line. Flow energy (or total mechanical energy) at a cross-section expressed as head is comprised of three components: pressure head, velocity head, and datum or potential-energy head. This is expressed by the Bernoulli equation:

$$h_1 = \gamma y + z + v^2/2g \qquad (11)$$

where h_1 is the total head at a point (m), y the flow depth (m), and z is the elevation (m). The energy slope (S_e) is the difference in total head at two points along the channel divided by the reach length:

$$S_e = (h_1 - h_2)/L = h_f/L \qquad (12)$$

where h_f is the head loss due to friction (m), h_2 the total head at point two (m), and L is the reach length (m). This head loss, or dissipation of energy is merely a conversion to heat energy and is not truly a "loss". In practice, water-surface slope or thalweg slope are commonly substituted into equations for the energy gradient because they are simpler to understand and easier to measure. Changes in channel morphology during fluvial adjustment expressed as variations in these flow-energy variables [Equations (11) and (12)] have been used to explain dominant modes of adjustment in diverse environments (Simon 1992).

Hydraulic Roughness

Hydraulic roughness is defined as friction-inducing elements that cause energy dissipation [Equation (12)] and it is extremely important in fluvial geomorphology because it is the basis for calculating average velocity and, therefore, discharge. General rules and assumptions are often used to determine roughness values, however, small changes in roughness can cause dramatic changes in the calculated average velocity.

Roughness is typically represented by a single coefficient, such as Manning's n. This coefficient can represent every form of hydraulic resistance in the channel including particle protrusion, bedforms, curvature, vegetation, obstructions, bedrock outcrops, and skin friction. Roughness varies with increasing stage as certain elements are muted by the flow (particle roughness) and other elements are introduced to the flow (vegetation, obstructions and floodplains). Vegetation can contribute greatly to channel roughness, however, defining the effects of vegetation is extremely complicated because it varies with species, age, abundance, distribution, and season (Watts and Watts 1990).

Chézy

The Chézy equation was developed in 1769 and represents the first open channel, uniform flow formula (Chow 1959). The Chézy equation assumes steady, uniform flow and was developed to calculate velocity and discharge for canals in France. Most flow resistance equations can be traced back to Chézy's original work (Chow 1959):

$$v = C(RS_e)^{1/2} \tag{13}$$

where C is the roughness coefficient.

The Chézy equation is not typically used for practical application due to the lack of acceptable C values for natural stream channels, although historically, the equation was used extensively and is well represented within the literature.

Manning

The Manning equation (10) was first presented in the late 1880s (Manning 1891), and remains one of the most commonly used tools by stream hydraulic engineers. Manning developed the equation by utilizing seven different open channel flow equations (Du Buat 1786, Eytelwein 1814, Weisbach 1845, St. Venant 1851, Neville 1860, Darcy and Bazin 1865, Ganguillet and Kutter 1869) averaging his results, and field-testing his findings (Chow 1959; Fischenich 2000).

The Manning equation assumes steady, uniform flow, which rarely exists in nature. The equation, therefore, is most applicable for relatively straight stream reaches with sub-critical, relatively non-turbulent flow. Velocity derived from Manning's equation represents an average for a particular cross-section at a specific stage.

Several visual methods for determining Manning's n value are available. Pictures of channels are presented with cross-section data and corresponding n values calculated at gauging sites where the other variables have been measured (Fasken 1963, Barnes 1967).

Estimating a composite Manning's n for a stream channel and floodplain rather than using an average provides a more realistic representation of roughness. There are at least two approaches for composite calculations. The first breaks down a cross-section into component parts. Typically the active channel, banks, and floodplain surfaces are treated as individual components. The area of each "roughness zone" is calculated and multiplied by the roughness coefficient assigned to each zone based on its characteristics. The final n value is determined by the sum of the zones divided by the total area.

The second method for composite calculation evaluates the cross-section as a whole, but uses modifying values from a base n value. A base value of n is assigned contingent upon bed material (0.02 for cohesive soils, for instance); various values are added to the base value dependent on surface irregularity, size and shape of the cross-section, obstructions, vegetation, meandering, and floodplains. The final n value is a summation of the base value and modifying values (Fasken 1963).

Darcy–Weisbach

The Darcy–Weisbach equation was developed during the mid-1800s to calculate friction losses for pipe flow. The equation was not used extensively until the development of the Moody diagram, which provided representative friction factors for various boundary materials such as polished brass, smooth wood, and structural steel. Because the equation was developed for pipe flow, certain assumptions, such as an immobile boundary and uniform flow, are inherent in the relation. The Darcy–Weisbach equation generally takes the following form:

$$f = 8gRS_e/v^2 \tag{14}$$

where f is the friction factor.

Equation (14) is typically reorganized to solve for velocity:

$$v = (8gRS_e/f)^{1/2} \tag{15}$$

Rather than relying on qualitative approximations, f can be calculated using the Keulegan equation:

$$1/\sqrt{f} = 2.03 \log(12.2R/k_s) \tag{16}$$

where k_s is the equivalent roughness (m) (Millar 2000),

$$k_s = 6.8\ D_{50} \tag{17}$$

Shear Stress

A measure of the erosional force provided by flows can be represented by average boundary-shear stress [Equation (2)]. To enable calculations of shear stress, measurements of flow depth (or a survey of the channel cross-section) and slope are required. Because bed slope for moderate and high flows is difficult to obtain, water-surface slope is often used as a surrogate. An inexpensive way of obtaining peak-flow data is to install crest-stage gages along the study reach. A crest-stage is simply a stick placed inside a capped PVC or metal pipe with a slotted, basal cap containing cork. The pipe is anchored to the channel bank or to a bridge abutment. When stage rises and submerges the basal cap, cork floats up within the pipe. Upon recession of stage, the cork adheres to the stick at the peak stage and the distance from the top of the stick to the cork mark is measured. By surveying the elevations of the mark made by the cork on the crest-stage gages and the longitudinal distance between them, flow depth at each gage and water-surface slope can be obtained. Average boundary-shear stress for the peak flow can then be calculated by:

$$\tau_0 = \gamma_w (E_w - E_b) S_{wp} \quad (18)$$

where E_w is the elevation of water surface at peak flow (m), E_b the elevation of the bed (m), and S_{wp} is the peak-flow water-surface slope (m/m). Pressure transducers can be installed within the PVC pipes to monitor stage continuously. These data can then be used to calculate shear stresses for the range of measured stages, including the channel-forming discharge.

11.5 FIELD METHODS TO EVALUATE FLUVIAL ADJUSTMENTS: CHANNEL-EVOLUTION MODELS

Initial evaluation of a field site should always include an investigation of the dominant processes, not simply a symptomatic inventory. Determining whether the processes are localized or systemic, short-term or long-term, gradual or catastrophic will provide a context in which analyses can be focused and management decisions can be made.

Researchers in fluvial geomorphology have noted that alluvial channels in different environments, destabilized by different natural and human-induced disturbances, pass through a sequence of channel forms with time (Davis 1909, Ireland *et al.* 1939, Schumm and Hadley 1957, Daniels 1960, Emerson 1971, Keller 1972, Elliot 1979, Schumm *et al.* 1984, Simon and Hupp 1986, Simon 1989a). These systematic temporal adjustments are collectively termed "channel evolution" and permit interpretation of past, present, and future channel processes.

The channel-evolution model developed by Simon and Hupp (1986) and Simon (1989a; 1994) is presented as an example of an effective field reconnaissance methodology (Figure 11.11; Table 11.1). The model has six stages, which is based on shifts in dominant adjustment processes and has been linked to rates of sediment transport (Simon 1989b), bank stability, sediment accretion, and ecologic recovery (Hupp 1992, Simon and Hupp 1992). The model has been successfully used to rapidly identify dominant, system-wide channel processes in watersheds impacted by various human and natural disturbances via aerial reconnaissance or ground observations in diverse regions of the United States and Europe. Identification of stages is accomplished by utilizing diagnostic attributes of channel form, such as the presence or absence of bank failures and accreted bank sediments, and the lowermost limit of woody vegetation relative to mean low-water levels to infer dominant channel processes. Because stages of evolution are systematic over time and space, the spatial distribution of stages within a watershed can be used to infer past, present, and future channel processes and aid in determining the appropriate mitigation measures that may be required in unstable systems.

The timing associated with each stage of channel evolution varies greatly between different stream systems and may vary within a single system. Controlling factors are again related to relations between force and resistance, expressed as the magnitude and scope of the disturbance, geology, bed and bank material, event history, infrastructure (bed and bank armoring), and riparian vegetation. In severely disturbed sand-bed systems it is not uncommon for stages III and IV to last 10–20 years (Simon 1994) and 40–100 years in silt-bed systems (Simon and Rinaldi 2000). Symptoms of re-stabilization (stages V and VI) such as infilling, deposition of bars and berms, and encroachment of woody-riparian vegetation occurs over periods ranging from 1 to about 100 years. Complete re-stabilization including formation of a new floodplain may take hundreds to thousands of years. Re-initiation of channel evolution can occur at any point. It is not uncommon for a stream in the widening stage to begin actively incising if another disturbance affects the reach.

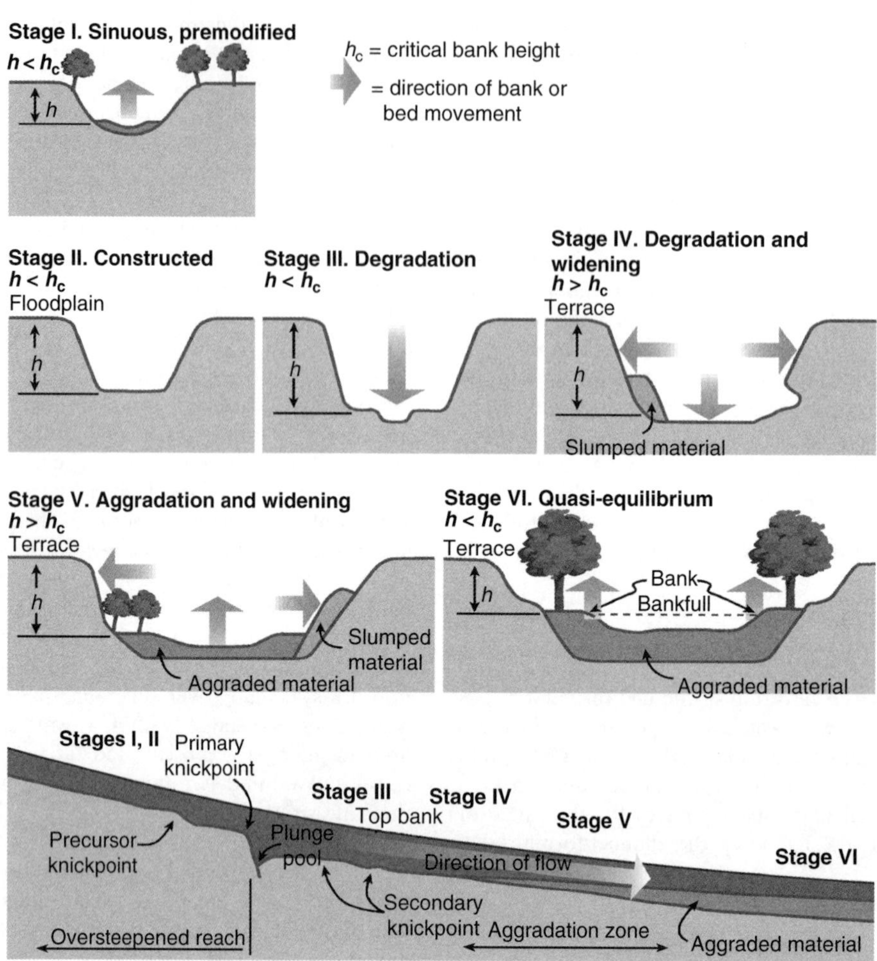

Figure 11.11 Six-stage model of channel evolution (modified from: Simon and Hupp 1986, Simon 1989a)

Table 11.1 Six stages of channel evolution showing dominant processes, characteristic forms and diagnostic criteria (modified from Simon 1989a)

Stage		Dominant processes			
Number	Name	Fluvial	Gravitational	Characteristic forms	Geobotanical evidence
I	Premodified	Sediment transport; mild aggradation; basal erosion on outside bends; deposition on inside bends	–	Stable, alternate channel bars; convex top bank shape; flow line high relative to top bank; channel straight or meandering	Vegetated banks to low-flow line

(*Continues*)

Table 11.1 (*Continued*)

Stage		Dominant processes			
Number	Name	Fluvial	Gravitational	Characteristic forms	Geobotanical evidence
II	Constructed	–	–	Trapezoidal cross-section; linear bank surfaces; flow line lower relative to top bank	Removal of vegetation (?)
III	Degradation	Degradation; basal erosion on banks	Pop-out failures	Heightening and steepening of banks; alternate bars eroded; flow line lower relative to top bank	Riparian vegetation high relative to flow line and may lean towards channel
IV	Threshold	Degradation; basal erosion on banks	Slab, rotational and pop-out failures	Large scallops and bank retreat; vertical face and upper bank surfaces; failure blocks on upper bank; some reduction in bank angles; flow line very low relative to top bank	Tilted and fallen riparian vegetation
V	Aggradation	Aggradation; development of meandering thalweg; initial deposition of alternate bars; reworking of failed material on lower banks	Slab, rotational and pop-out failures; low-angle slides of previously failed material	Large scallops, bank retreat; vertical face, upper bank and slough line; flattening of bank angles; flow line low relative to top bank; development of new flood plain (?)	Tilted and fallen riparian vegetation; re-establishing vegetation on slough line; deposition of material above root collars of slough line vegetation
VI	Restabilization	Aggradation; further development of meandering thalweg; further deposition of alternate bars; reworking of failed material; some basal erosion on outside bends; deposition of flood plain and bank surfaces	Low-angle slides; some pop-out failures near flow line	Stable, alternate bars; convex short vertical face on top bank; flatten of bank angles; development of new flood plain (?); flow line higher relative to top bank	Re-establishing vegetation extends up slough line and upper bank; deposition of material above root collars of slough line and upper bank vegetation; vegetation establishing on bars

11.6 METHODS TO ANALYZE CHANNEL FORM

Once the dominant processes have been identified and the appropriate physical data collected, various empirical and numerical methods are available to quantify subsequent changes in channel form and to estimate future channel configurations (Thorne *et al.* 1981, Schumm *et al.* 1984, Hey and Thorne 1986, Simon and

Hupp 1986, Chang 1988, Thorne and Osman 1988, Molinas 1989, Lohnes 1991, Simon and Hupp 1992, Simon and Downs 1995, Langendoen 2000). These studies include "regime" and other empirical methods as well as numerical-simulation models. There is insufficient space to review all of these in detail and the reader is directed, therefore, to these publications.

Empirical Methods

Empirical methods refer to techniques that rely on relations developed from measurements and observations in the field or laboratory, which may or may not be physically based.

Bed Level Changes

In unstable channels, bed elevation with time (years) can be described by non-linear functions, where change, or response to a disturbance, occurs rapidly at first and then slows and becomes asymptotic. Plotting of bed elevations with time permits evaluation of bed-level adjustment trends and indicates whether the major phase of degradation or aggradation has passed or is ongoing. Various mathematical forms of this function, including exponential, power, and hyperbolic, have been used to characterize bed level adjustment at a site with time, and to predict future bed elevations (Graf 1977, Williams and Wolman 1984, Simon and Hupp 1986, Simon 1989a, 1992, Wilson and Turnipseed 1993, 1994).

Extensive study of bed level adjustment in streams representing a wide range of bed-material sizes showed that the power and exponential functions accurately described upstream degradation and downstream aggradation with time (Simon 1989a, 1992). An exponential function converges to an asymptote and is preferred (H. Jobson, US Geological Survey, written communication, 1992); however, the power function is easier to use:

$$E = at^b \tag{19}$$

where a is coefficient, determined by regression, representing the premodified elevation of the channel bed, t the time since beginning of the adjustment process (years), where $t_0 = 1.0$ (year prior to onset of the adjustment process), and b is the dimensionless exponent, determined by regression and indicative of the non-linear rate of channel-bed change (negative for degradation and positive for aggradation).

The dimensionless form of the exponential equation is (Simon 1992, Simon and Rinaldi 2000):

$$z/z_0 = a + b\,e^{(-kt)} \tag{20}$$

where z is the elevation of the channel bed (at time t), z_0 the elevation of the channel bed at t_0, a the dimensionless coefficient, determined by regression and equal to the dimensionless elevation (z/z_0) when Equation (20) becomes asymptotic, $a > 1$ the aggradation, $a < 1$ the degradation, b the dimensionless coefficient, determined by regression and equal to the total change in the dimensionless elevation (z/z_0) when Equation (20) becomes asymptotic, k the coefficient determined by regression, indicative of the rate of change on the channel bed per unit time, and t is the time since the year prior to the onset of the adjustment process (years, $t_0 = 0$).

Future elevations of the channel bed can be estimated by fitting Equations (19) or (20) to bed elevations and by solving for the time period of interest. Either equation provides acceptable results, depending on the statistical significance of the fitted relation. The predisturbed bed elevation, obtained from field survey, is required along with at least one other bed elevation from a different time period. Degradation and aggradation curves for the same site are fit separately (Figure 11.12). For degrading sites, this method will provide projected minimum channel elevations when the value of t becomes large and, by subtracting this result from the floodplain elevation, will provide projected maximum bank heights. A range of bed adjustment trends can be estimated by using different starting dates when the initial timing of bed level change is unknown (Figure 11.13).

Abundant data are available at stream gauging stations; mean channel bed elevations can be obtained by subtracting mean flow depth from water-surface elevation for each discharge measurement where flows are at bankfull stage or below. This method is described in detail by Jacobson (1995). If this procedure is used for at least several years of record, it should be possible to determine if bed-level adjustment is ongoing. A similar method is employed at gauging stations by Wilson and Turnipseed (1994) to obtain minimum bed elevations. Maximum flow depth of the annual minimum daily stage is subtracted from the water-surface elevation of that stage resulting in minimum thalweg elevations with time.

The longitudinal distribution of b-values [from Equation (19)] or a-values [from Equation (20)] can be used as an empirical model of bed-level adjustment providing there are data from enough sites to establish a relation with distance along the channel or river system. An example using Equation (19) is provided

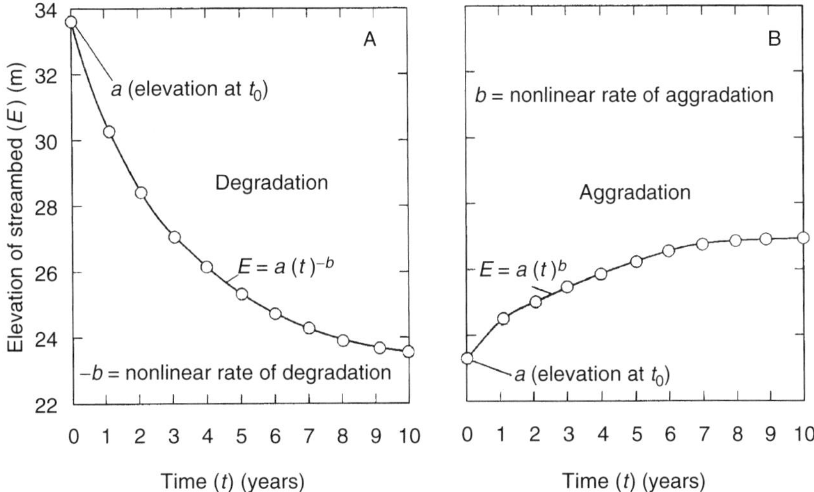

Figure 11.12 Idealized graphs of fitting power functions to (A) degradation and (B) aggradation trends

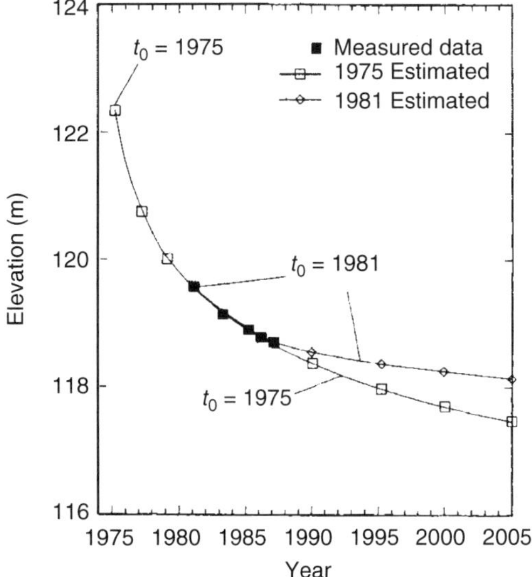

Figure 11.13 Method to predict future bed-level elevations using power function [Equation (19)] and if the timing of the start of the adjustment process (t_0) is unknown

for the Obion River System, Western Tennessee (Figure 11.14) and using Equation (20) for West Tarkio Creek, Iowa and Missouri (Figure 11.15). With knowledge of t_0, b-values can be interpolated for unsurveyed sites that can be used to obtain estimates of bed-level change with time. For channels downstream from dams, the shape of the curve in Figure 11.14 would be similar but reversed; maximum amounts of degradation (minimum b-values) occur immediately downstream of the dam and attenuate non-linearly with distance further downstream (Williams and Wolman 1984). Once the minimum bed elevation has been obtained using Equations (19) or (20), that elevation can be substituted back into the equation and used as the starting elevation at a new t_0 for the "secondary" aggradation phase (Figure 11.12B).

A procedure termed the "geomorphic method" (Lohnes 1994) uses the assertion that the longitudinal

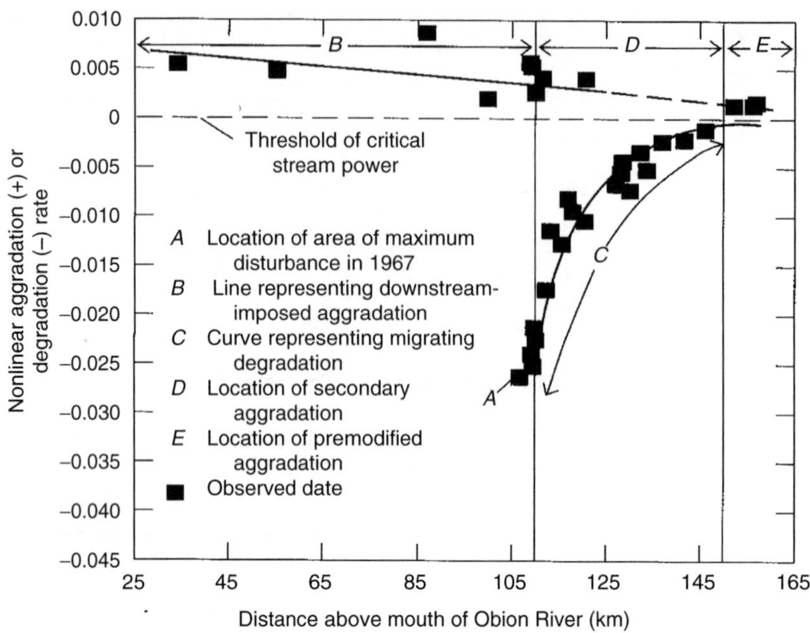

Figure 11.14 Empirical model of bed-level response for the Obion River System, western Tennessee, based on fitting power functions [Equation (19)] to time series bed-level data. Positive *b*-values represent aggradation; negative *b*-values represent degradation

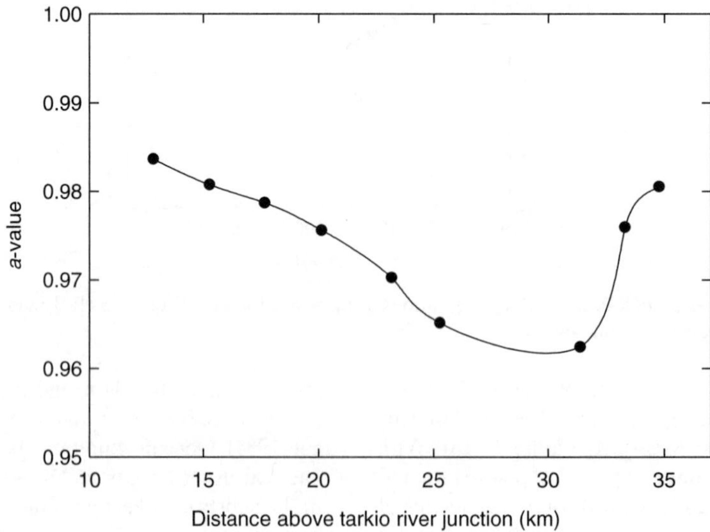

Figure 11.15 Empirical model of bed-level response for West Tarkio Creek, Iowa and Missouri, based on fitting dimensionless exponential functions [Equation (20)] to time series bed-level data. *a*-values less than 1.0 represent degradation with the smaller absolute values representing more severe degradation

profile of a stream will plot as a straight line on a semi-logarithmic graph (Hack 1957), where bed elevation (y-axis) is plotted on the logarithmic scale. Assuming that the downstream reaches are stable (have re-stabilized) and that there is an upstream length of channel that has not been completely incised, the profile of a degrading stream should plot as two linear segments, separated by a knick point or other locally steep area. The lower profile can then be projected upstream beneath the actively degrading reach. The difference in elevation between the upstream profile and the extrapolated line represents the maximum amount of incision that is likely to occur under the assumption that no grade control will be encountered. Daniels (1960) used this method to predict amounts of degradation on Willow Creek, western Iowa. Those predictions were verified by Lohnes *et al.* (1980). Although this method does not contain a time component, it is simple to use.

Channel Width Changes

Estimates of potential channel widening on streams with unstable banks can be obtained by noting the angle of the low-bank surface, which indicates renewed stability because of the establishment of supporting woody vegetation. This stable low-bank surface, termed the "slough line" (Simon and Hupp 1986) is formed as bank angles recede through successive failures and is capped with fluvially reworked material. The horizontal distance between the intersection of the projected slough-line angle with the floodplain surface and the present top bank is the minimum estimated widening for one side of the channel (Simon and Hupp 1992).

Various geotechnically based methods of estimating a "temporary" or "ultimate" angle of stability have been advanced; however, these methods ignore the cohesion component of shear strength because, in many cases, it is assumed to be zero (Skempton 1953, Carson and Kirkby 1972). For thick soil masses, a temporary angle of stability (θ) is defined as (Carson and Kirkby 1972):

$$\tan\theta = \tan\phi \tag{21}$$

where ϕ is the friction angle (degrees). For residual soil mantles, an ultimate stability angle θ_u, can be defined as (Carson and Kirkby 1972):

$$\tan\theta_u = 0.5\tan\phi \tag{22}$$

Simon and Hupp (1992) reported that use of Equation (22) for estimating θ_u at stabilizing reaches in West Tennessee produced values lower than those observed in the field. Instead, to obtain estimates of θ and potential channel widening, Equation (23) was utilized to calculate the failure plane angle (Lohnes and Handy 1968, Thorne *et al.* 1981):

$$\tan\theta = \tan[(\phi+\alpha)/2] \tag{23}$$

where α is the bank angle. The temporary stability angle obtained is then projected to the floodplain surface as previously described.

Estimates of potential channel widening can be made using measured channel width data over a period of years and then fitting a non-linear function to the data. Williams and Wolman (1984) used a dimensionless hyperbolic function of the form to estimate channel widening downstream from dams:

$$(w_1/w_t) = j_1 + j_2(1/t) \tag{24}$$

where w_1 is the initial channel width, w_t the channel width at t years after w_1, j_1 the intercept, and j_2 is the slope of the fitted straight line on a plot of w_1/w_t versus $1/t$. Wilson and Turnipseed (1994) used a power function to describe channel widening after channelization and to estimate potential channel widening in the loess area of northern Mississippi, USA:

$$w = xt^d \tag{25}$$

where w is the channel width, x the coefficient determined by regression indicative of the initial channel width, and d is the coefficient, determined by regression indicative of the rate of channel widening.

Stable-channel Dimensions

Numerous empirical methods are available by which to estimate "stable" channel dimensions. Leopold and Maddock (1953) derived relations between mean velocity (v), mean depth (d), and water-surface width (w) as a function of discharge (Q). Collectively termed "hydraulic geometry", these relations can be expressed as "at-a-station" (for a single cross-section) or as "downstream" relations. For the "downstream" relations, the bankfull or mean-annual discharge is used (Leopold *et al.* 1964). Graphs of these variables often plot as a straight line on logarithmic paper, indicating power functions:

$$w = aQ^b \tag{26}$$

$$d = cQ^f \tag{27}$$

$$v = kQ^m \tag{28}$$

Because $w \times d \times v = Q$ (the continuity equation), it follows then that $b + f + m = 1$. Average values of the downstream hydraulic geometry exponents b, f, and m are, respectively: 0.26, 0.40, and 0.34 for the Midwestern United States (Leopold and Maddock 1953); 0.49, 0.38, and 0.13 for the Pacific Northwest United States (Castro and Jackson 2001); and 0.12, 0.45, and 0.43 for 158 American gauging stations (Leopold *et al.* 1964). Thus, exponents vary by region due to differences in climate, rainfall–runoff relations and specifically, by the type and resistance of boundary sediments (Castro and Jackson 2001, Williams 1978b). Results from these equations should be used with extreme caution for channel design because of the uncertainty in regression estimates even with high r^2 values.

Similar empirical procedures termed "regime methods" developed by engineers in studies of irrigation channels generally rely on three formulas to describe a stable width, depth, and slope (Lacey 1930, 1935, 1958). Blench (1952, 1970) modified Lacey's approach by accounting for differences due to the variability in bank materials, and Simons and Albertson (1960) and Hey and Thorne (1986) allowed for channels other than those composed of sand beds and cohesive banks.

Deterministic Methods

Deterministic methods refer to numerical techniques that rely on physically based field and laboratory measurements of variables and processes. The methods require appropriate identification of the dominant processes, and application of those equations that describe the force and resistance mechanisms for the particular process.

Bed Level Changes

Non-cohesive materials. Average boundary-shear stress [Equation (2)] for a range of flows can be compared to a calculated critical boundary-shear stress [Equation (4)] to identify those flows where excess shear stress (erosion) is likely to occur. The Shields criterion is then invoked to calculate the equivalent particle diameter for the measured critical shear stresses. For uniform, non-cohesive sediments, τ^* can be obtained from the Shields (1936) diagram. Typical values are 0.03, 0.047, and 0.06 [see Equation (4)]. Heterogeneous sediments present additional complications because of hiding and protrusion factors. This issue has been addressed by several researchers, notably Wiberg and Smith (1987) who developed a method to be used for poorly sorted (mixed-size), non-cohesive sediment particles that can also account for variations in particle density. τ^*_{cr} is obtained graphically using a non-dimensional particle diameter (K_*) as the abscissa in a Shields-type entrainment curve, where τ^*_{cr} is the ordinate (Wiberg and Smith 1987):

$$K_* = 0.0047\,(\zeta_*)^{1/3} \tag{29}$$

where K_* is the bed roughness length (k_s) for sediment of specific gravity 2.65 and fluid temperature of 10 °C, and

$$\zeta_* = \frac{D^3(\rho_s - \rho_w)}{v^2 \rho_w} \tag{30}$$

The non-dimensional parameter (K_*) is a function of grain size, particle density, fluid density, and viscosity such that any grain will have a unique value of ζ_* in a particular fluid environment. The factor $3.5 * D_{84}$ of the substrate is used for k_s in calculations of D/k_s to select the appropriate Shields-type curve for obtaining τ^*_{cr} (Wiberg and Smith 1987). Thus, a series of Shields-type entrainment functions are available based on the values of D/k_s. The conventional Shields entrainment function for uniform-sized particles is expressed as a curve with a value of $D/k_s = 1.0$. Other tractive-force methods are also available (Einstein 1950, Lane 1955, Vanoni and Brooks 1957). A good summary of these methods is provided by Vanoni (1975).

Cohesive materials. For cohesive streambeds, data obtained with a jet-test device can be used to estimate erosion rates due to hydraulic forces (Hanson 1990, 1991, Hanson and Simon 2001). The rate of erosion ε (m/s) is assumed to be proportional to the excess shear stress (Foster *et al.* 1977):

$$\varepsilon = k(\tau_0 - \tau_c)^a = k(\tau_e) \tag{31}$$

where k is the erodibility coefficient (m^3/N s), a the exponent assumed to be 1.0, and τ_e is the excess shear stress (Pa).

An inverse relationship between τ_c and k occurs when soils exhibiting a low τ_c have a high k or when soils having a high τ_c have a low k. Similar trends were observed by Arulanandan *et al.* (1980) during laboratory flume testing of soil samples from cohesive streambed materials obtained across the United States. Based on observations from across the United

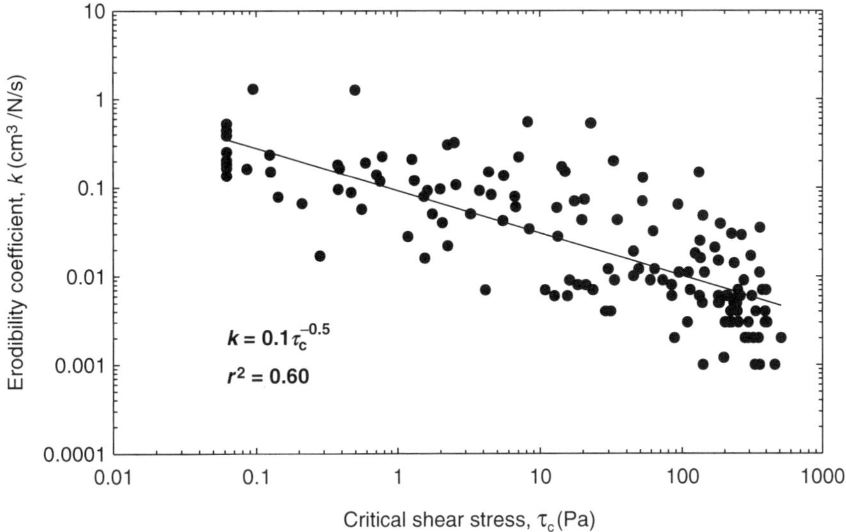

Figure 11.16 Relation between erodibility coefficient (k) and critical shear stress based on jet tests in cohesive materials from sites across the United States (τ_c)

States, Hanson and Simon (2001), estimated k as a function of τ_c (Figure 11.16; $r^2 = 0.64$). Here, k is expressed in cm^3/N s:

$$k = 0.1\tau_c^{-0.5} \qquad (32)$$

To relate these values to the relative potential for flows to erode cohesive beds and to compare cohesive resistance to the resistance of a non-cohesive particle or aggregate (equivalent diameter), an average boundary-shear stress is calculated from Equation (2). The Shields criterion is then invoked to calculate an equivalent particle diameter for the measured critical shear stresses using Equation (4). To calculate erosion rates (ε), values of average boundary or local shear stress are used in conjunction with values of τ_c and k using Equations (31) and (32).

Erosion at the bank toe. The tractive-force method can be modified to evaluate bank-toe erosion by accounting for the fact that the particles rest on an inclined surface. The critical, dimensionless shear stress is adjusted such that (Lane 1953):

$$\tau^*_{cb} = \tau^*_{cr} \cos\varepsilon \sqrt{(1 - \tan^2\varepsilon/\mu^2)} \qquad (33)$$

where τ^*_{cr} is the dimensionless critical shear stress on the side slope, ε the side-slope angle of the bank on which the failed material is deposited, and μ is the Coulomb coefficient of friction based on the assumption that it is equivalent to the tangent of the friction angle of the sediment ($\tan\phi'$) (Bagnold 1953, 1966, Francis 1973). Dimensionalized critical shear-stress values are obtained using:

$$\tau_c = \tau^*_{cb}((\rho_s - \rho)gd) \qquad (34)$$

Equation (33) is only applicable where the side-slope angle (α) is less than the friction angle (ϕ'). However, packing, matric suction, and cementation can result in side-slope angles being steeper than the friction angle. This is often the case with in situ bank-toe material in incised channels. In this case, a modification proposed by Millar and Quick (1993) and based on empirical evidence of the steepest angles measured in the field (ϕ^*) is applicable:

$$\tau_{cb} = 0.067\tan\phi^* \sqrt{[1 - (\sin\varepsilon/\sin\phi^*)^2][(\rho_s - \rho)g]D_{50}} \qquad (35)$$

The value of this equation, however, is limited due to the numerical problems when ϕ^* approaches the vertical because $\tan\phi^*$ becomes indeterminate. This is in the very range of ϕ^* that is common in the bank/bed interface region and the most critical for estimating the stability of the bank-toe.

Critical shear stresses and erodibility coefficients for bank-toe material can be obtained with the jet-test device as described earlier. Rates of erosion can then be calculated using the same procedure as described for cohesive beds.

Channel Width Changes

Culman method. The critical height (H_c) and angle for a wedge-type bank failure can be estimated by solving the following equation (Selby 1982):

$$H_c = [4c' \sin\alpha \cos\phi']/[\gamma(1 - \cos(\alpha - \phi'))] \quad (36)$$

where α is the bank angle (degrees) and γ is the soil unit weight (kg). H_c can be reduced by the tension-crack depth (z_c):

$$H_{cz} = H_c - z_c \quad (37)$$

where H_{cz} is the critical bank height with tension crack (m) and z_c is the tension-crack depth (m) (Selby 1982). This analysis does not provide for layered banks of variable strength. Weighted-average values of bank-material characteristics are used for layered banks.

Equation (36) is solved for ambient and worst-case (saturated) conditions over a range of bank angles (generally 40–90°) and plotted against H_c to provide a bank-stability chart for critical conditions (Figure 11.17). Critical-bank heights for saturated conditions can be assessed using unconsolidated, undrained-strength parameters (ϕ_u = friction angle = 0° and c_u = cohesion) and the saturated unit weight (γ_s). However, without undrained-strength data one can assume that frictional strength is reduced to zero under saturated conditions because of excess pore-water pressures, leaving only effective cohesion (c') to resist mass failure. This is a static-type analysis in that the frequency of occurrence of "worst-case" or "ambient" conditions is not considered.

Two shortcomings of the Culman approach are:

1. only a single, average bank angle can be incorporated, thereby negating the affects of bank-toe erosion;
2. layered banks cannot be accommodated, thereby requiring only single, albeit weighted values of the geotechnical parameters.

Process-based bank-stability model. An analytical method that accounts for variations in bank material

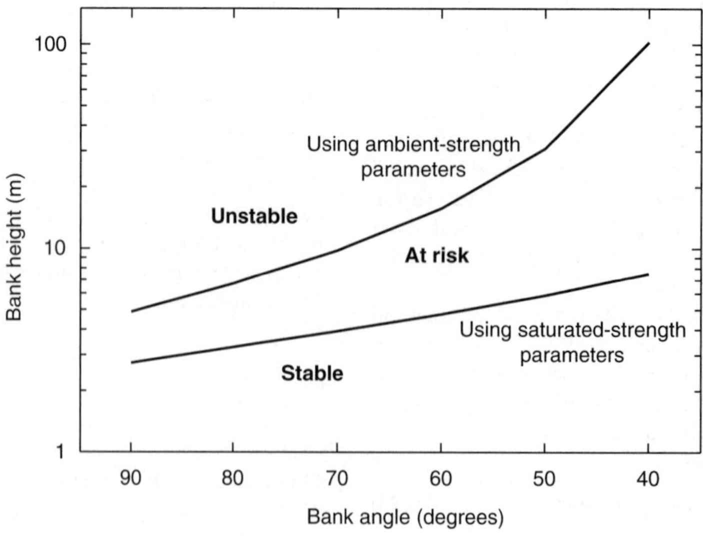

Figure 11.17 Bank-stability chart developed by using the Culman equation [Equation (36)] to solve for a range of bank angles under ambient and saturated conditions

as well as the effects of pore-water and confining pressures is required to obtain a better understanding of bank-failure conditions. A bank-stability algorithm for cohesive, layered banks has been developed by Simon and Curini (1998) and Simon *et al.* (1999, 2000), incorporating both the failure criterion of Mohr-Coulomb for the saturated part of the failure surface [Equation (5)], and the failure criterion modified by Fredlund *et al.* (1978) for the unsaturated part of the failure surface [Equation (8)].

The algorithm is based on the limit equilibrium method, and accounts explicitly for several additional forces acting on a planar failure surface that are not included in earlier bank-stability models (Osman and Thorne 1988). These include the force produced by matric suction on the unsaturated part of the failure plane (S), hydrostatic-uplift force due to positive pore-water pressures on the saturated part of the failure plane (U), and the hydrostatic-confining force provided by the water in the channel and acting on the bank surface (P) (Casagli *et al.* 1997, Curini 1998, Simon and Curini 1998, Rinaldi and Casagli, 1999, Simon *et al.* 1999a, 2000).

The effect of matric suction on shear strength is reflected in the apparent or total cohesion (c_a) term [Equation (9)]. The hydrostatic-uplift (U) and confining (P) forces are calculated from the area of the pressure distribution of pore-water ($h_u * \gamma_w$) and confining ($h_{cp} * \gamma_w$) pressures (μ_w) by:

$$U = \gamma_w h_u^2 / 2 \sin\alpha \tag{38}$$

$$P = \gamma_w h_{cp}^2 / 2 \sin\beta \tag{39}$$

where $\gamma_w = 9.81\,\text{kN/m}^3$, h_u the pore-water head (m), h_{cp} the confining-water head (m), and α and β are the bank-slope and failure-plane angles, respectively (degrees). The loss of the hydrostatic-confining force (P) provided by the water in the channel is the primary reason bank failures often occur on the recession of stormflow hydrographs.

Multiple layers are incorporated through summation of forces in a specific (ith) layer acting on the failure plane. The factor of safety (F_s) is (Simon *et al.* 1999a, 2000):

$$F_s = \frac{\sum c_i' L_i + (S_i \tan\phi_i^b) + [W_i \cos\beta - U_i + P_i \cos(\alpha - \beta)] \tan\phi_i'}{\sum W_i \sin\beta - P_i \sin(\alpha - \beta)} \tag{40}$$

where W_i is the weight of the failure block (kg); L_i the length of the failure plane incorporated within the ith layer, and S is the force produced by matric suction on the unsaturated part of the failure surface (kPa). Equation (40) represents the continued refinement of bank-failure analyses by incorporating additional forces and soil variability (Osman and Thorne 1988, Simon *et al.* 1991, Casagli 1994, Casagli *et al.* 1997, Curini 1998, Simon and Curini 1998, Rinaldi and Casagli 1999, Simon *et al.* 1999a, 2000). In its present form (2003) the model can handle complex bank geometries and is available on the worldwide web at http://www.sedlab.olemiss.edu/cwp_unit/bank.html.

Worst-case conditions are represented by high levels of the phreatic surface and low-flow levels of river stage. This is the classic "drawdown" condition where the counteracting effects of confining pressure are minimized. For the static case, levels of the phreatic surface and river stage are assumed and the model is run based on the pressure distributions represented by those water levels. The effects of varying the levels of these surfaces can be seen for sites along the Missouri River, eastern Montana (Table 11.2). Here, the most critical case is represented by a phreatic-surface level at the elevation of the 736 m^3/s discharge and the river stage for the 283 m^3/s discharge (Simon *et al.* 2002). The model can also be used with dynamic pore-water pressure and stage data to temporally evaluate the conditions leading to bank instability along stream reaches (Figure 11.18) (Simon *et al.* 2000).

11.7 CONCLUSIONS

There are many tools available for measuring and analyzing channel form yet space constraints allows us to provide only a limited overview of techniques that we have found to be useful. Quantitative analysis of channel processes, however, is often given only a cursory review by practitioners who are implementing stream-related projects. A few common reasons that are given for the lack of adequate analyses include: (1) lack of data, (2) lack of time, (3) budget constraints, or (4) personnel limitations. In the present climate of computer modeling, it is very easy to ignore real data by making broad assumptions and/or creating synthetic data—but collecting field data is at the very core of a usable, reliable model. Collecting basic stream data should be one of the primary goals for anyone planning, designing, or implementing projects that are related to the stream corridor.

Table 11.2 Factor of safety data from Simon *et al.* (2002) for sites along the Missouri River, eastern Montana, using a process-based model of bank stability [Equation (40)]. Last column represents worst-case condition with high level of phreatic surface (GW) and low level of river stage (RS)

Site	Planar failures	
	Minimum factor of safety	
	RS-736 GW-736	RS-283 GW-736
1589	1.36	0.85
1604	1.00	0.80
1621	1.64	1.41
1624	1.23	0.83
1630	1.49	0.61
1631	0.77	0.11
1646	1.05	0.80
1676	1.20	0.95
1682	1.52	1.52
1701	1.23	0.99
1716	0.94	0.57
1728	0.81	0.60
1737	1.63	0.97
1744	1.92	1.57
1762	2.85	2.41
1765	0.99	0.64

RS, elevation of river at given discharge; GW elevation of groundwater equal to specified river discharge.

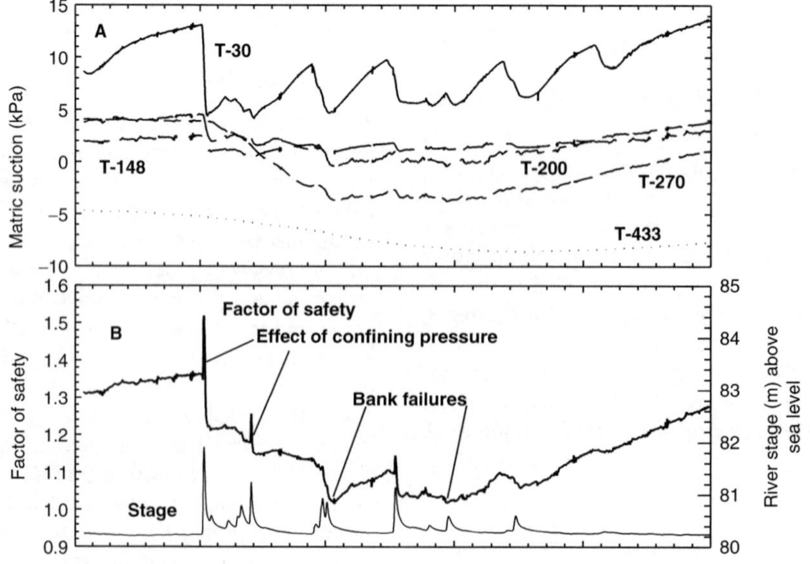

Figure 11.18 Variation in (A) matric suction at five depths in a layered, cohesive streambank (Goodwin Creek, Mississippi, USA) and its effects on the (B) factor of safety [Equation (40)] and bank stability

REFERENCES

Andrews, E.D. 1980. Effective and bankfull discharge of streams in the Yampa River Basin, Colorado and Wyoming. *Journal of Hydrology* 46: 311–330.

Andrews, E.D. and Nankervis, J.M. 1995. Effective discharge and the design of channel maintenance flows for gravel-bed rivers. In: Costa, J.E., Miller, A.J., Potter, K.W. and Wilcock, P.R., eds., *Natural and Anthropogenic Influences in Fluvial Geomorphology*, Geophysical Monograph 89, American Geophysical Union, pp. 151–164.

Arulanandan, K., Gillogley, E. and Tully, R. 1980. Development of a quantitative method to predict critical shear stress and rate of erosion of natural undisturbed cohesive soils, Waterways Experiment Station Technical Report GL-80-5, University of California, Davis.

Bagnold, R.A. 1953. Experiments on a gravity free dispersion of large solid spheres in a Newtonian fluid under shear. *Proceedings of the Royal Society of London, Series A* 255: 49–63.

Bagnold, R.A. 1966. An approach to the sediment transport problem from general physics. *US Geological Survey Professional Paper* 422-I: 1–37.

Balster, C.A. and Parsons, R.B. 1968. Geomorphology and soils: Willamette Valley, Oregon, Special Report 265, Agricultural Experiment Station, Oregon State University.

Barnes, H.H. 1967. *Roughness Characteristics of Natural Channels*, Water Supply Paper 1849, Washington, DC: US Geological Survey, 213 p.

Blench, T. 1952. Regime theory of self-formed sediment-bearing channels. *Transactions of the American Society of Civil Engineers* 117: 383–408.

Blench, T. 1970. Regime theory design of canals with sand beds. *Journal of the Irrigation Division, ASCE* 96(IR2): 205–213.

Carson, M.A. and Kirkby, M.J. 1972. *Hillslope Form and Process*. Cambridge, London: Cambridge University Press, 475 p.

Casagli, N. 1994. Determinazione del fattore di sicurezza per scivolamenti planari nelle sponde fluviali, Studi di Geologia Applicata e Geologia dell'Ambiente, N. 30, Dipartimento Scienze della Terra, Universita degli Studi di Firenze.

Casagli, N., Curini, A., Gargini, A., Rinaldi, M., Simon, A. 1997. Effects of pore pressure on the stability of streambanks: preliminary results from the Sieve River, Italy. In: Wang, S.S.Y., Langendoen, E.J. and Shields, F.D., Jr., eds., *Management of Landscapes Disturbed by Channel Incision*, pp. 243–248.

Castro, J.M. and Jackson, P.L. 2001. Bankfull discharge recurrence intervals and regional hydraulic geometry relationships: patterns in the Pacific Northwest, USA. *Journal of the American Water Resources Association* 37(5): 1249–1262.

Chang, H.H. 1988. *Fluvial Processes in River Engineering*, New York: John Wiley and Sons, 432 p.

Chow, V.T. 1959. *Open Channel Hydraulics*, New York: McGraw-Hill, 680 p.

Curini, A. 1998. *Analisi dei processi di erosione di sponda nei corsi d'acqua*, Universita degli Studi di Firenze, Dipartimento di Scienze della Terra, 147, Unpublished thesis.

Daniels, R.B. 1960. Entrenchment of the Willow Drainage Ditch, Harrison County, Iowa. *American Journal of Science* 258: 161–176.

Davis, W.M. 1909. Geographical Essays, Boston: Ginn (republished 1954, New York: Dover).

Einstein, H.A. 1950. The bedload function for sediment transportation in open-channel flow. *US Department of Agriculture Technical Bulletin*, 1026.

Elliott, J.G. 1979. Evolution of large arroyos, the Rio Puerco of New Mexico, Master of Science Thesis, Colorado State University, Fort Collins, Colorado, USA, 106 p.

Elliott, J.G., Gellis, A.C. and Aby, S.B. 1999. Evolution of arroyos: incised channels of the southwestern United States. In: Darby, S.E. and Simon, A., eds., *Incised Channels: Processes, Forms, Engineering, and Management*, Chichester, England: John Wiley and Sons, pp. 153–186.

Emerson, J.W. 1971. Channelization: a case study. *Science* 172: 325–326.

Fasken, G.B. 1963. *Guide for Selecting Roughness Coefficient 'n' Values for Channels*, Milwaukee, Wisconsin: USDA, Soil Conservation Service.

Fischenich, C. 2000. Robert manning: a historical perspective, TN-EMRRP-SR-10, US Army Corps of Engineers Research and Development Center, Environmental Laboratory.

Foster, G.R., Meyer, L.D. and Onstad, C.A. 1977. An erosion equation derived from basic erosion principles. *Transactions of the ASAE*: 678–682.

Francis, J.R.D. 1973. Experiments on the motion of solitary grains along the bed of a water stream. *Proceedings of the Royal Society London, Series A* 332: 443–471.

Fredlund, D.G. and Rahardjo, H. 1993. *Soil Mechanics of Unsaturated Soils*, New York: John Wiley and Sons, 517 p.

Fredlund, D.G., Morgenstern, N.R. and Widger R.A. 1978. The shear strength of unsaturated soils. *Canadian Geotechnical Journal* 15: 313–321.

Gilbert, G.K. 1914. The transportation of debris by running water. *US Geological Survey Professional Paper* 86.

Graf, W.L. 1977. The rate law in fluvial geomorphology. *American Journal of Science* 277: 178–191.

Gregory, K.J. and Walling, D.E. 1973. *Drainage Basin Form and Process: A Geomorphological Approach*, London: Edward Arnold.

Hack, J.T. 1957. Studies of longitudinal stream profiles in Virginia and Maryland. *US Geological Survey Professional Paper* 294B.

Hanson, G.J. 1990. Surface erodibility of earthen channels at high stresses. Part II – Developing an in-situ testing device. *Transactions of the ASAE* 33(1): 132–137.

Hanson, G.J. 1991. Development of a jet index to characterize erosion resistance of soils in earthen spillways. *Transactions of the ASAE* 36(5).

Hanson, G.J. and Simon, A. 2001. Erodibility of cohesive streambeds in the loess area of the midwestern USA. *Hydrological Processes* 15(1): 23–38.

Harrelson, C.C., Rawlins, C.L. and Potyondy, J.P. 1994. Stream channel reference sites: an illustrated guide to field technique, General Technical Report RM-245, USDA, Forest Service, Fort Collins, CO, 61 p.

Hey, R.D. and Thorne, C.R. 1986. Stable channels with mobile gravel beds. *Journal of the Hydraulic Research, ASCE* 112(8): 671–689.

Hjulstrom, F. 1935. Studies of the morphological activity of rivers as illustrated by the river Fyris. *Bulletin of the Geological Institute, University of Uppsala* 25: 221–527.

Hupp, C.R. 1992. Riparian vegetation recovery patterns following stream channelization: a geomorphic perspective. *Ecology* 73: 1209–1226.

Inglis, C.C. 1949. The behaviour and control of rivers and canals. *Central Waterpower Irrigation and Navigation Research Station, Poona, Research Publication* 13.

Ireland, H.A., Sharpe, C.F.S. and Eargle, D.H. 1939. Principles of gully erosion in the Piedmont of South Carolina. *US Department of Agriculture Technical Bulletin* 633: 142 p.

Jacobson, R.B. 1995. Spatial controls on patterns of land-use induced stream disturbance at the drainage-basin scale—an example from gravel-bed streams of the Ozark Plateaus, Missouri. In: Costa, J.E., Miller, A.J., Potter, K.W. and Wilcock, P.R., eds., *Natural and Anthropogenic Influences in Fluvial Geomorphology*, Geophysical Monograph 89, American Geophysical Union, pp. 219–239.

Keller, E.A. 1972. Development of alluvial stream channels: a five-stage model. *Geological Society of America Bulletin* 83: 1531–1536.

Knighton, D. 1998. *Fluvial Forms and Processes*, New York: John Wiley and Sons, 383 p.

Lacey, G. 1930. Stable channels in alluvium. *Proceedings of the Institute of the Civil Engineers, London*, 229.

Lacey, G. 1935. Uniform flow in alluvial rivers and canals. *Proceedings of the Institute of the Civil Engineers, London*, 237.

Lacey, G. 1958. Flow in alluvial channels with sand mobile beds. *Proceedings of the Institute of the Civil Engineers, London*, 9 (discussion 11).

Lane, E.W. 1953. Progress report on studies on the design of stable channels of the Bureau of Reclamation. *Proceedings of the American Society of Civil Engineers* 79: 246–261.

Lane, E.W. 1955. Design of stable alluvial channels. *Transactions of the American Society of Civil Engineers* 120 (2776): 1234–1260.

Langendoen, E. J. 2000. CONCEPTS – CONservational Channel Evolution and Pollutant Transport System, Report, US Department of Agriculture, Agricultural Research Service, National Sedimentation Laboratory, Oxford, MS.

Lawler, D.M. 1993. The measurement of river bank erosion and lateral channel change: a review. *Earth Surface Processes and Landforms* 18: 777–821.

Leopold, L.B. 1994. *A View of the River*, Cambridge, Mass.: Harvard University Press.

Leopold, L.B. and Maddock, T., Jr. 1953. The hydraulic geometry of stream channels and some physiographic implications. *US Geological Survey Professional Paper* 252.

Leopold, L.B. and Wolman, M.G. 1957. River channel patterns – braided, meandering and straight. *US Geological Survey Professional Paper* 282B: 39–85.

Leopold, L.B., Wolman, M.G. and Miller, J.P. 1964. *Fluvial Processes in Geomorphology*, New York: W.H. Freeman and Co., 522 p.

Lisle, T.E. and Eads, R. 1991. Methods to measure sedimentation of spawning gravels, Research Note PSW-411, USDA Forest Service, Pacific Southwest Research Station, Berkeley, CA.

Little, W.C., Thorne, C.R. and Murphy, J.B. 1982. Mass bank failure analysis of selected Yazoo Basin streams. *Transactions of the American Society of the Agricultural Engineers* 25: 1321–1328.

Lohnes, R.L. 1991. A method for estimating land loss associated with stream channel degradation. *Engineering Geology* 31: 115–130.

Lohnes, R.A. 1994. Stream stabilization research. In: Hadish, G.A., ed., *Stream Stabilization in Western Iowa*, Iowa DOT HR-352, Golden Hills Resource Conservation and Development, Oakland, Iowa, pp. 3.1–3.54.

Lohnes, R.A. and Handy, R.L. 1968. Slope angles in friable loess. *Journal of Geology* 76(3): 247–258.

Lohnes, R.A., Klaiber, F.W. and Dougal, M.D. 1980. Alternate methods of stabilizing degrading stream channels in western Iowa, Iowa DOT HR-208, Department of Civil Engineering, Engineering Research Institute, Iowa State *University*, Ames, Iowa, 132 p.

Lutenegger, J.A. and Hallberg, B.R. 1981. Borehole shear test in geotechnical investigations. *ASTM Special Publication* 740: 566–578.

Mackin, J.H. 1948. Concept of a graded river. *Geological Society of America Bulletin* 59: 463–511.

Manning, R. 1891. On the flow of water in open channels and pipes. *Transactions of the Institution of Civil Engineers of Ireland.*

Millar, R.G. 2000. Influence of bank vegetation on alluvial channel patterns. *Water Resources Research* 36(4): 1109–1118.

Millar, R.G. and Quick, M.C. 1993. Effect of bank stability on geometry of gravel rivers. *Journal of Hydraulic Engineering* 119: 1343–1363.

Miller, J.R. and Ritter, J.B. 1996. An examination of the Rosgen classification of natural rivers, discussion. *Catena* 27: 295–299.

Molinas, A. 1989. *User's Manual for BRI-STARS*, National Cooperative Highway Research Program, Project No. HR-15-11, Hydrau-Tech Engineering and Software, Fort Collins, Colorado, USA, 156 p.

Montgomery, D.R. and Buffington, J.M. 1997. Channel-reach morphology in mountain drainage basins. *Geological Society of America Bulletin* 109(5): 596–611.

Moody, L.F. 1944. Friction factors for pipe flow. *Transactions of ASME* 66.

Morisawa, M. 1968. *Streams: Their Dynamics and Morphology*. New York: McGraw-Hill.

Myers, T.J. and Swanson, S. 1997. Precision of channel width and pool area measurements. *Journal of the American Water Resources Association* 33(3): 647–659.

Osman, A.M. and Thorne, C.R. 1988. Riverbank stability analysis. I. Theory. *Journal of Hydraulic Engineering* 114 (2): 134–150.

Richards, K. 1982. *Rivers: Form and Process in Alluvial Channels*, London: Methuen, 358 p.

Rinaldi, M. and Casagli, N. 1999. Stability of streambanks formed in partially saturated soils and the effects of negative pore water pressures: the Sieve River (Italy). *Geomorphology* 26: 253–277.

Schumm, S.A. 1977. *The Fluvial System*, New York: John Wiley and Sons, 338 p.

Schumm, S.A. 1981. Evolution and response of the fluvial system, sedimentologic implications. *Society of Economic Paleontologists and Mineralogists Special Publication* 31: 19–29.

Schumm, S.A. and Hadley, R.F. 1957. Arroyos and the semiarid cycle of erosion. *American Journal of Science* 225: 161–174.

Schumm, S.A. and Lichty, R.W. 1965. Time, space, and causality in geomorphology. *American Journal of Science* 263: 110–119.

Schumm, S.A., Harvey, M.D. and Watson, C.C. 1984. *Incised Channels: Morphology, Dynamics and Control*, Littleton, Colorado: Water Resources Publications, 200 p.

Selby, M.J. 1982. *Hillslope Materials and Processes*, Oxford, UK: Oxford University Press, 264 p.

Shields, A. 1936. Anwendung der ahnlichkeitsmechanik und turbulenz forschung auf die geschiebebewegung. *Mitteil. Preuss. Versuchsanst. Wasser*, Erd, Schiffsbau, Berlin, Nr. 26.

Simon, A. 1989a. A model of channel response in disturbed alluvial channels. *Earth Surface Processes and Landforms* 14: 11–26.

Simon, A. 1989b. The discharge of sediment in channelized alluvial streams. *Water Resources Bulletin* 25(6): 1177–1188.

Simon, A. 1992. Energy, time, and channel evolution in catastrophically disturbed fluvial systems. In: Phillips, J.D. and Renwick, W.H., eds., *Geomorphic Systems: Geomorphology*, pp. 5:345–372.

Simon, A. 1994. Gradation processes and channel evolution in modified West Tennessee streams: process, response, and form. *US Geological Survey Professional Paper* 1470: 84 p.

Simon, A. and Collison, A.J.C. 2002. Quantifying the mechanical and hydrologic effects of riparian vegetation on streambank stability. *Earth Surface Processes and Landforms* 27: 527–546.

Simon, A. and Curini, A. 1998. Pore pressure and bank stability: the influence of matric suction. In: Abt, S.R., Young-Pezeshk, J. and Watson, C.C., eds., *Water Resources Engineering '98*, New York: ASCE, pp. 358–363.

Simon, A. and Darby, S.E. 1997. Process-form interactions in unstable sand-bed river channels: a numerical modeling approach. *Geomorphology* 21: 85–106.

Simon, A., Curini, A., Darby, S.E. and Langendoen, E. 1999. Streambank mechanics and the role of bank and near-bank processes in incised channels. In: Darby, S.E. and Simon, A., eds., *Incised Channels: Processes, Forms, Engineering, and Management*, Chichester, England: John Wiley and Sons, pp. 123–152.

Simon, A., Curini, A., Darby, S.E. and Langendoen, E.J. 2000. Bank and near-bank processes in an incised channel. *Geomorphology* 35: 193–217.

Simon, A. and Downs, P.W. 1995. An interdisciplinary approach to evaluation of potential instability in alluvial channels. *Geomorphology* 12(3): 215–232.

Simon, A., Dickerson, W. and Heins, A. in press. Suspended-sediment transport rates at the 1.5-year recurrence interval: transport conditions at the bankfull and effective discharge? *Geomorphology*.

Simon, A. and Hupp, C.R. 1986. Channel evolution in modified Tennessee channels. In: *Proceedings, Fourth Federal Interagency Sedimentation Conference, Las Vegas, March 24–27, 1986*, 2, pp. 5-71–5-82.

Simon, A. and Hupp, C.R. 1992. Geomorphic and vegetative recovery processes along modified stream channels of West Tennessee, US Geological Survey Open-file Report 91-502, 142 p.

Simon, A. and Rinaldi, M. 2000. Channel instability in the loess area of the Midwestern United States. *Journal of the American Water Resources Association* 36(1): 133–150.

Simon, A., Thomas, R.E., Curini, A. and Shields, F.D., Jr. 2002. Case Study: channel stability of the Missouri River, eastern Montana. *Journal of Hydraulic Engineering, ASCE* 128(10): 880–890.

Simon, A., Wolfe, W.J. and Molinas, A. 1991. Mass wasting algorithms in an alluvial channel model. In: *Proceedings of the 5th Federal Interagency Sedimentation Conference, Las Vegas, Nevada*, 2, pp. 8-22–8-29.

Simons, D.B. and Albertson, M.L. 1960. Uniform water conveyance channels in alluvial material. *Journal of the Hydraulics Division, ASCE* 86(HY5): 33–71.

Skempton, A.W. 1953. Soil mechanics in relation to geology. *Proceedings, Yorkshire Geological Society* 29: 33–62.

Stevens, H.H., Jr. and Yang, C.T. 1989. Summary and use of selected fluvial sediment-discharge formulas, US Geological Survey Water Resources Investigations Report 89-4026, 62 p.

Thorne, C.R. 1990. Effects of vegetation on riverbank erosion and stability. In: Thornes, J.B., ed., *Vegetation and Erosion*, Chichester, England: John Wiley and Sons, pp. 125–144.

Thorne, C.R., Murphey, J.B. and Little, W.C. 1981. *Stream Channel Stability. Appendix D: Bank Stability and Bank Material Properties in the Bluffline Streams of Northwest*

Mississippi, Oxford, Mississippi: US Department of Agriculture Sedimentation Laboratory, 257 p.

Thorne, C.R. and Osman, M.A. 1988. Riverbank stability analysis. Part II – Application. *Journal of the Hydraulics Division*, ASCE: 114 p.

Vanoni, V.A. 1975. *Sedimentation Engineering*, ASCE Manuals and Reports on Engineering Practice – No. 54, 745 p.

Vanoni, V.A. and Brooks, N.H. 1957. Laboratory Studies of the Roughness and Suspended Load of Alluvial Streams, Missouri River Division Sediment Series, No. 11, US Army Corps of Engineers.

Vidal, H. 1969. The principle of reinforced earth. *Highway Research Record* 282: 1–16.

Watts, J.F. and Watts, G.D. 1990. Seasonal change in aquatic vegetation and its effect on river channel flow. In: Thornes, J.B., ed., *Vegetation and Erosion*, Chichester, England: John Wiley and Sons, pp. 257–257.

Wiberg, P.L. and Smith, J.D. 1987. Calculations of the critical shear stress for motion of uniform and heterogeneous sediments. *Water Resources Research* 23: 1471–1480.

Williams, G.P. 1978a. Bankfull discharge of rivers. *Water Resources Research* 14: 1141–1154.

Williams, G.P. 1978b. Hydraulic geometry of river cross-sections – theory of minimum variance. *US Geological Survey Professional Paper* 1029.

Williams, G.P. and Wolman, M.G. 1984. Downstream effects of dams on alluvial rivers. *US Geological Survey Professional Paper* 1286: 83 p.

Wilson, K.V., Jr. and Turnipseed, D.P. 1993. Channel-bed and channel-bank stability of Standing Pine Creek tributary at State Highway 488 at Free Trade, Leake County, Mississippi. *US Geological Survey Open-File Report* 93–37: 20 p.

Wilson, K.V., Jr. and Turnipseed, D.P. 1994. Geomorphic response to channel modifications of Skuna River at the State Highway 9 crossing at Bruce, Calhoun County, Mississippi. *US Geological Survey Water-Resources Investigations Report* 94-4000: 43 p.

Wolman, M.G. and Miller, J.P. 1960. Magnitude and frequency of forces in geomorphic processes. *Journal of Geology* 68: 58–74.

Yevjevich, V. 1972. *Probability and Statistics in Hydrology*, Fort Collins, Colorado, USA: Water Resources Publications.

12

Flow Measurement and Characterization

PETER J. WHITING

Department of Geological Sciences, Case Western Reserve University, Cleveland, Ohio, USA

12.1 INTRODUCTION

A basic tool of the geomorphologist and hydrologist is the measurement of streamflow; to determine stream discharge; to estimate the flow resistance and boundary shear stress; and to characterize the turbulence and other flow attributes. There are many examples of the need for such information. The discharge at upstream locations is usually needed to anticipate downstream flood levels and to decide whether to open or close dam spillways. These decisions can affect public safety and have significant societal, environmental, and monetary costs. Water allocation and administration requires accurate discharge measurement as does construction of a basic water budget, for example, to determine the effect of silvicultural activities on the volume and timing of runoff. Sediment discharge is often related to flow discharge thus sediment budgets often depend upon flow discharge estimation. The characterization of local flow velocity or turbulence is often needed in process-based studies. Moreover, discharge is determined often from local measurements of flow velocity; thus the ability to characterize the magnitude and direction of flow through the water column is critical. Velocity profiles can be used to estimate the magnitude of the local boundary shear stress from the law-of-the-wall. Very detailed profiles can reveal the magnitude of the resistance to flow associated with bedforms and banks. Faithful characterization of the flow field is one test of the ability of two- and three-dimensional models to predict flow patterns. Fisheries biologists often estimate the potential utilization of habitat based upon the availability of areas with characteristic ranges of velocity. Efforts to better understand sediment transport processes often link sediment particles to boundary shear stress and turbulence.

A variety of equipment and techniques are available to measure flow and discharge in the field and the range of technologies is expanding. These new technologies can improve the accuracy and precision of measurements, speed collection of data, and/or provide new information on the flow. This chapter describes and compares available methods to measure velocity and flow in some detail and explores issues that should be considered in selecting a method for measuring flow in the context of fluvial geomorphology. More comprehensive treatments of the subject, at least for standard methods, can be found in Herschy (1985), Bureau of Reclamation (1984), and ISO (1983) among other books. This chapter will be most useful in helping scientists and decisionmakers select the most appropriate methods for their specific problems. Details about implementing specific methods are presented in the references cited in the text.

12.2 VELOCITY MEASUREMENT

In this section, the techniques and equipment for measuring flow velocity are described; the principles underlying the approaches explained; and the accuracy and appropriateness of the approaches discussed.

An example of the measurement of velocity comes from my work at Solfatara Creek, Wyoming (USA) where I sought to understand the downstream and cross-stream accelerations of flow induced by rapid shoaling associated with a mid-channel bar (Whiting and Dietrich 1991). A key part of examining the magnitude of the accelerations was the characterization of the flow field at many points in the water column, at multiple points across the channel, and at multiple closely spaced cross-sections. Both the downstream and cross-stream components of velocity were measured at each point. Multiple points had to be measured in the water column for several reasons. One reason was to define the lateral and downstream fluxes that varied with height in the water column. Near bed measurements were also needed to provide an estimate of local boundary shear stress at each vertical across each section. Finally, the shoaling was dramatic enough that fine spatial resolution was required to investigate the phenomena.

Floats

Water velocity can be estimated from floats (Dunne and Leopold 1978, Herschy 1985). In situations where another technique is inappropriate or too hazardous, or the proper equipment is unavailable, the downstream displacement over time of buoyant objects such as sticks or oranges can be measured. It is recommended that travel time be at least 20 s. Floats provide a quick estimate of surface velocity, but the accuracy of such an estimate of velocity is less than other methods; probably no better than ±10–20%. The only equipment required for this method is a watch and a tape measure.

An alternative use of floats is to determine gross patterns of flow or the location of eddies. Thus, floats can be helpful in the selection of cross-sections for measurement or in the selection of the measurement approach.

Mechanical Current Meters

Current meters, especially mechanical current meters, are the most commonly used equipment for measuring flow velocity. Mechanical current meters measure flow velocity from the rotation of a vertical-axis bucket wheel with cups or a horizontal-axis impeller. In standard operation, the rotational speed of the cups or impellor is determined by the number of revolutions per time as measured optically, magnetically, or electrically. Determining the number of rotations by counting 'clicks', as a circuit is completed, is rarely done any longer.

The vertical-axis meters include the larger Price AA and smaller Price mini-meter (or 'pygmy' meter) (Figures 12.1a and b). The diameter of the bucket wheel of

Figure 12.1 Mechanical current meters: (a) standard Price AA current meter—an example of a vertical-axis meter; (b) Price mini-meter—a smaller version of the Price AA meter; (c) Ott current meter—an example of a horizontal-axis meter (photo courtesy of USGS)

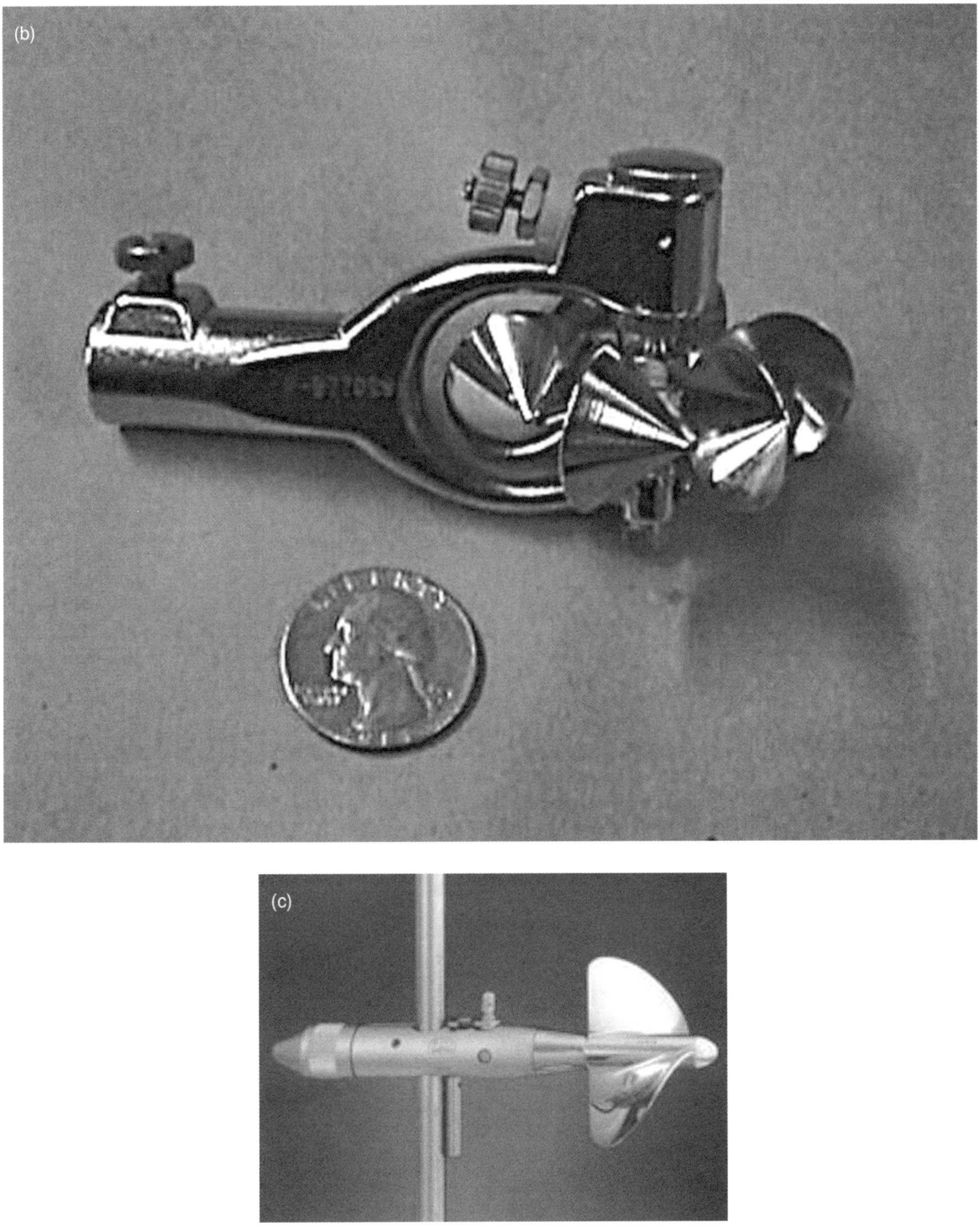

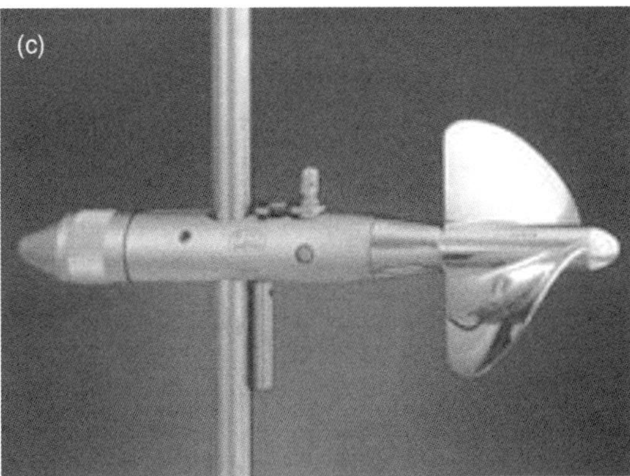

Figure 12.1 (b) and (c)

the meters is 13 and 5 cm, respectively. The bucket wheels should be constructed of metal rather than plastic (Jarrett 1992). The Price AA can be used in water as shallow as 0.15 m and to measure flow velocities from 0.06 to almost 8 m/s. The smaller minimeter was designed for use in shallower and/or slower flows; water depths from 0.08 to 0.45 m and water velocities from 0.02 up to 0.9 m/s. In principle, there is no problem using the mini-meter in deeper flow if flow velocity is in the appropriate range. The Price and mini-meters are often deployed with a vane serving to orient the meter in the flow. The calibration of these devices for flow approaching the meter at an angle has been determined (Fulford *et al.* 1994). The accuracy of the magnitude of the flow velocity of Price AA and mini-meters is about 0.5% (Fulford *et al.* 1994). While Price meters have been used to measure turbulence, the frequency response of both meters is less than 1 Hz thus they are not capable of quantifying the higher frequency part of the turbulence spectrum.

The horizontal-axis meters include the Ott meter (Figure 12.1c) and Smith meter (Smith 1978). These meters have a screw-type impeller which are typically 5–8 cm across and are capable of measuring flow from 0.1 to 8 m/s. Other miniature impeller current meters appropriate for the field, nonetheless fragile, can be as small as 1.2 cm in diameter. They are capable of measuring velocities from 0.03 to 3 m/s. The calibration of these devices for flow approaching the meter at other than parallel to the spindle of the impeller varies by model if it is known at all. The accuracy of flow velocity varies widely by model and manufacturer; the range is about 0.75–2.0% (Fulford *et al.* 1994).

Comparing the two types of mechanical current meters, the vertical-axis meters can be used at lower flow velocity (except perhaps the fragile miniature screw-type meters), but disturb the flow more and are more prone to becoming tangled by debris or growing vegetation than horizontal-axis meters (Fulford *et al.* 1994). The uncertainty associated with vertical- and horizontal-axis meters is similar but in general the accuracy of the vertical-axis meters is higher (Fulford *et al.* 1994).

The mechanical current meters, with the exception of the miniature meters, are very robust in the field. Even if damaged, many repairs are possible in the field. The maintenance requirements are modest, particularly for the vertical-axis meters. The equipment should be cleaned daily after use. Prior to use, the spin of the rotor should be checked. It should take over 30 s for the rotor to stop spinning after it is spun by hand in air (Rantz *et al.* 1982).

Current meters (mechanical and other types) are deployed from wading rods, cables, bridges or other structures, and boats. Top-set wading rods allow the hydrographer to stand in the flow and to re-position the meter in the water column without reattaching the meter to the support and without removing the wading rod from the water or alternatively getting her hands wet (Figure 12.2a). When water depth or velocity is too large for wading, meters can be suspended from a cableway (Figure 12.2b) or bridge, or boat (Figure 12.2c). A weight can be used to submerge the suspended current meter. A weight is attached to the base of a cable to maintain its position in the streaming flow deflecting the cable downstream. Various equipment for use in suspending current meters (reels, weights, cables, bridge boards, etc.) is described in Buchanan and Somers (1969) and Herschy (1985). If flow and depth permit, measurement by wading is often preferred because of the greater control the hydrographer can employ in the holding and placement of the current meter and wading rod. For very detailed small-scale studies of flow structure, current meters can be lowered from portable bridges. To determine the magnitude of the two horizontal components of velocity, two meters can be positioned at 30–90° to one another. An alternative to deploying two meters is to measure one component of flow with one meter and to use a piece of flexible flagging to determine the net direction of flow and the angle between the meter and the flow.

In addition to the current meter itself, the following equipment is needed for velocity measurement: a support rod or cable, an output device to convert revolutions to velocity, and power (usually battery). A portable computer is often useful for storing information particularly time series and turbulence information.

Electromagnetic Current Meters

Electromagnetic current meters measure flow velocity based upon the principle that voltage is produced when a conductor (water) moves through a magnetic field produced by the probe. Electrodes on the surface of the probe measure the resulting voltage and the voltage is linearly proportional to the flow velocity.

The best known of the electromagnetic meters in North America are those manufactured by Marsh-McBirney, such as the Flo-Mate 2000™, a 5-cm diameter tear-drop shaped probe with three electrodes (Figure 12.3). This model provides a measurement of the magnitude of the downstream component of flow over a range of –0.15 to 6 m/s with an uncertainty of

±2% (Marsh-McBirney 1995). This device attaches to wading rods or cables with the same connector as the Price and mini-current meters. Other models (typically used in oceanographic settings) have 3.7 and 1.3 cm-in-diameter spherical sensors with four electrodes in the horizontal plane thus allowing for the determination of both the downstream and cross-stream components of flow. The current meters have a cosine-response to velocity components that are at an angle to the plane of the four electrodes. Both meters can measure bi-directional flow velocities up to ±3 m/s with an uncertainty of ±2%. In situations where the flow direction may reverse (in separated flow in the lee of bedforms or other obstacles or at depth with stratified flow), the ability of the electromagnetic current meter to measure bi-directional flow can be a distinct advantage over mechanical current meters. This potential limitation in the use of mechanical current meters can be circumvented in clear water with a flexible flag to indicate flow direction. Dinehart (1999) reports that the frequency response of the Marsh McBirney Flo-Mate 2000™ is ~1 Hz or less and therefore does not resolve finer turbulent fluctuations (i.e. >2 Hz).

Electromagnetic meters may be affected by strong electrical and magnetic fields and by other electromagnetic current meters placed less than 0.6 m apart, depending upon the model (Marsh-McBirney 1995). The proximity of meters is less of a problem with mechanical current meters.

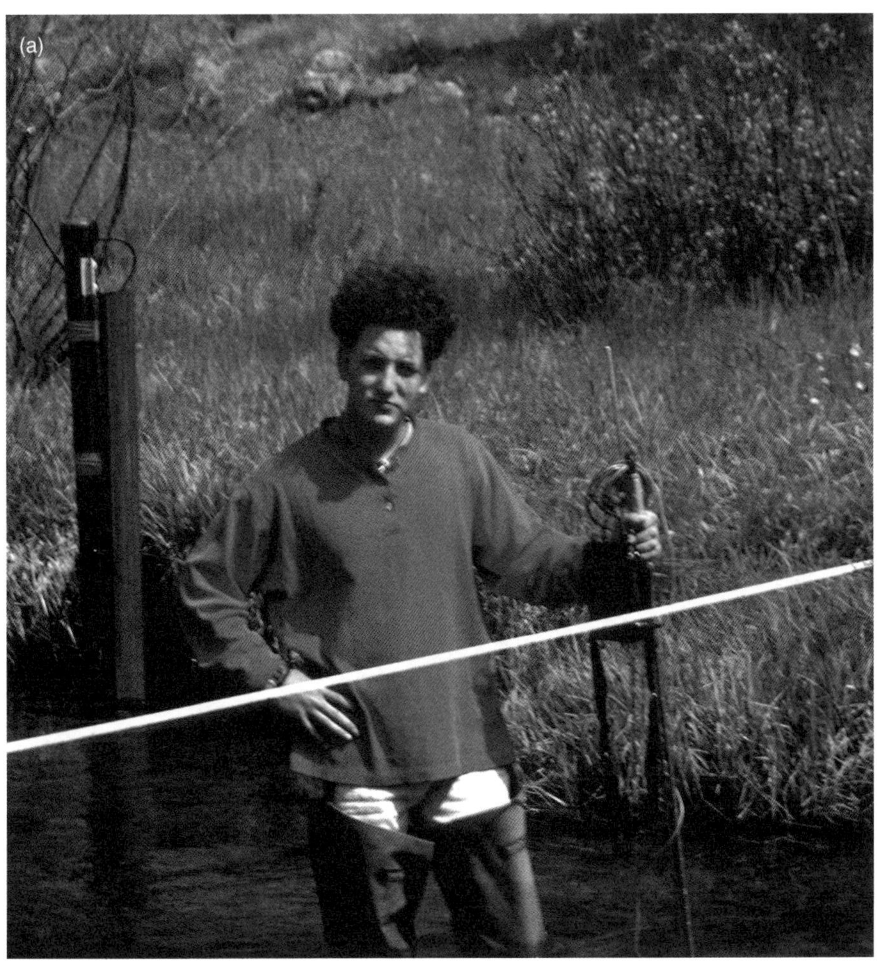

Figure 12.2 Modes of deploying current meters: (a) by wading in the stream

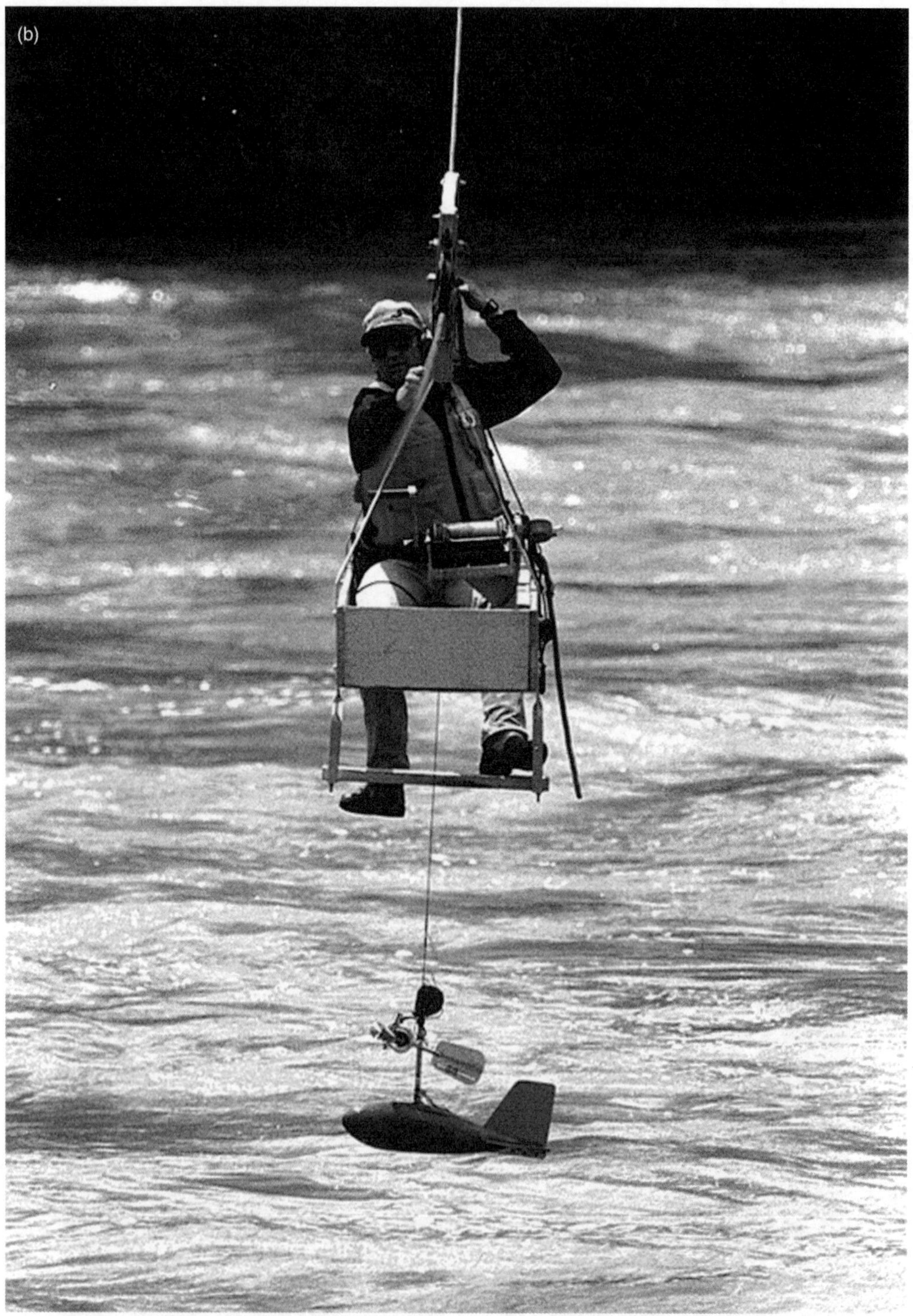

Figure 12.2 (b) Suspended from a cableway (photo courtesy of Post Register Newspaper, Idaho Fall, Idaho) (reproduced by permission of Randy Hayes)

Figure 12.2 (c) Attached to a boat (photo courtesy of Robb Jacobson)

The electromagnetic meter has no mechanical parts thus it is fairly rugged and is not prone to fouling by debris. Dirt and non-conductive grease and oil should be rinsed from the surface of the probe before storage. Periodically, the zero reading of the meter should be checked. Like the mechanical meters, these meters are robust.

In addition to the current meter itself, the following equipment is needed: a support rod or cable, an output device to convert voltage to velocity, and power (usually battery). It is common to import the velocity data directly to a portable computer especially if turbulence is being characterized.

Acoustic Doppler Velocimeters

Acoustic Doppler velocimeters (ADV) measure the 3-components of velocity by the emission of acoustic signals, their reflection by particles suspended in the flow, and the reception of the reflected signal. With the ADV, regions near the sensor are not measurable because there is not enough separation of time between signal emission and reception. Similarly, regions near boundaries are not measurable because the strength of reflections from nearby boundaries swamps the signal of small particles in the flow. A major benefit of such equipment is that there is no device in the sampling volume to distort the flow. Acoustic Doppler velocimetry is useful for characterizing turbulence and turbulent parameters such as the Reynolds stress.

There exist many types of ADV. Some are small, mounted on a wand with a measuring volume that is relatively small (0.1–0.5 cm^3) and close to the device (5–10 cm) (Figure 12.4). Such devices are very useful for measuring complex flow fields especially near boundaries. Other devices are much larger with a sampling volume that is large and somewhat distant from the device (programmable from 0.5–15 m). The exact specifications vary among manufacturers. The smaller devices can measure velocities from 0.001 up to 5 m/s while some of the larger devices can measure velocities up to 10 m/s. The uncertainty of measurements is about 0.5–1%. Sampling rates are typically up to 25 or 50 Hz. The devices must be moved to get information at other locations.

After use, the ADV should be inspected, washed, and dried before being stored. Prior to use, all connections should be checked. Otherwise maintenance of the ADV requires minor effort.

Power (usually a 12-V battery) and a computer to run software and record data are needed to employ the ADV. Some support, rod or structure, is needed to hold the ADV steady in the flow.

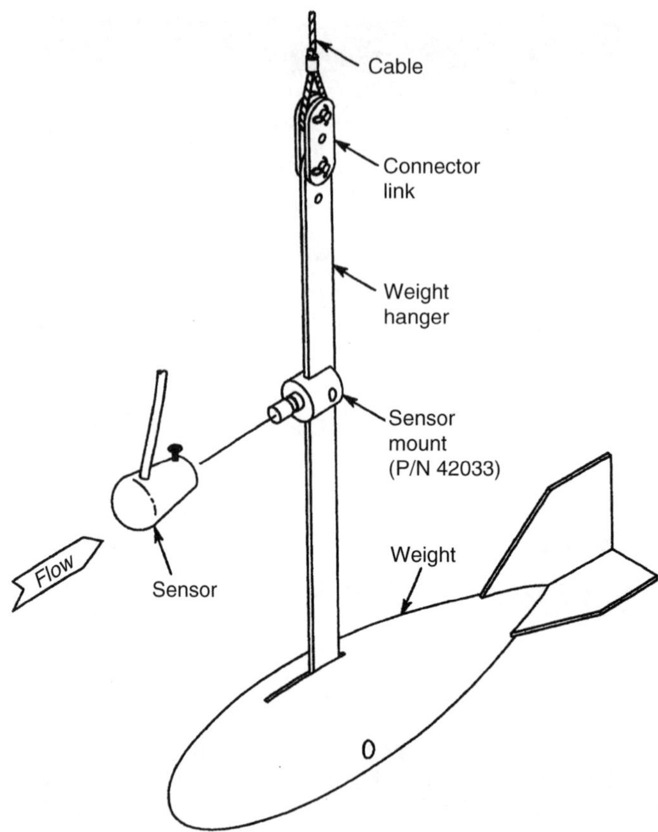

Figure 12.3 Electromagnetic current meter—Flo-Mate 2000 as it would attach to a cable set-up. Diagram courtesy of Marsh McBirney

Acoustic Doppler Current Profilers

Acoustic Doppler current profilers (ADCP) use the same basic approach as ADV, however, the velocity at multiple points is characterized rather than at a single point. ADCP measures the 3-components of velocity at multiple points by the emission of acoustic signals, their reflection by particles suspended in the flow at various distances from the transducer, and the reception of the reflected signals from the various particles. Figure 12.5 shows a picture of such a device. The reflections are separated by time of arrival into uniformly spaced cells for which an average velocity is calculated. The device is deployed at the surface such that signals are sent toward the bed. As with ADV, the ADCP cannot collect data very close to the device—usually within about 0.2 m of the device—in a layer called the blanking distance. Also, the very strong reflection from the stream bottom, called side-lobe interference, precludes measurement in a region near the bed generally about 0.05–0.1 m thick. As a consequence of these limitations, the device is generally limited to water depths greater than about 0.5 m; even then, only the velocity in cells away from either boundary is determinable. These devices measure velocity in the range of 0.00–10 m/s. RD Instruments (1998) reports an accuracy of 0.25%. The sampling rate can be as high as 4 Hz.

Meters can be deployed at a single site to provide continuous flow information through the depth profile or towed across the channel to provide a transverse characterization of velocity in addition to the vertical characterization as appropriate for discharge determination. The ADCP provides a great deal of information about the interior of the flow. If detailed information is needed near a boundary, ADV may be more appropriate. Barua and Rahman (1998) provide an example of the use of this equipment to study a large river.

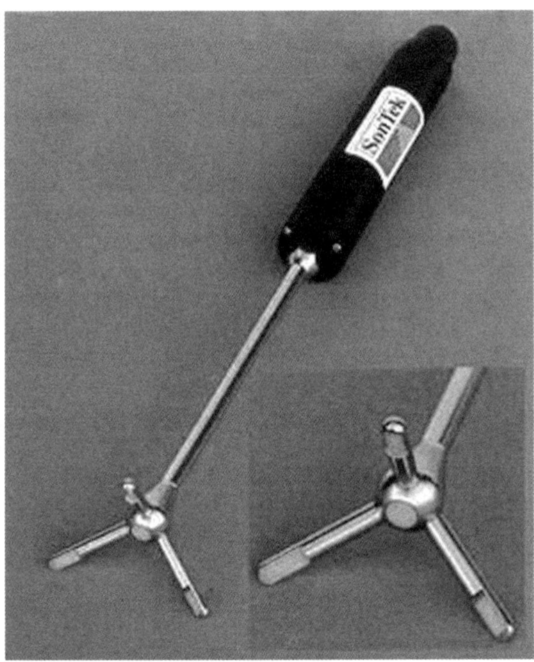

Figure 12.4 Two views of an ADV. The larger view shows the probe and wand and the inset provides a close-up of the wand. Signals are emitted at the base of wand and are received at the tips of each arm. The total spread of the arms is 77 mm (photo courtesy of SonTek/YSI, Inc.)

After use, the ADCP should be inspected, washed, and dried before being stored. Prior to use, all connections should be checked. Otherwise maintenance of the ADCP requires minimal effort. The ADCP has a fairly rugged housing, nonetheless, the circuitry means that the equipment should handled carefully. Repair in the field is unlikely.

In addition to the device, the following equipment is needed: a boat (Figure 12.2c) or support to suspend the ADCP, a computer to run software operating the device and processing data and to record data, and a power source. A 12-V battery is usually sufficient.

Laser Doppler Velocimetry (LDV)

Laser light scattered by small particles during their movement through the sampling volume created by the intersection of beams of laser light can be used to determine the velocity. Light is scattered at frequencies related to the velocity components—the well-known Doppler shift. Two beams of laser light are used to measure a single component of flow and multiple components of the flow can be measured with additional beams. A major benefit of such equipment is that there is no device in the sampling volume to distort the flow. The uncertainty of the measurements is about 0.1%. The sampling volume is smaller than that of ADV or ADCP: a few tenths of a millimeter on a side. The device must be repositioned to measure velocity at a different location in the flow. There are usually sufficient particles in streamwater to serve as scatterers. The measurement of vertical and downstream components at high frequencies (at least 50 Hz) permits determination of the Reynolds stress.

This device has seen very little use in rivers and streams although it has seen some use in oceanographic settings. Given the complexity of the equipment, it is probably most appropriately used in the laboratory.

Other Velocity Measurements

Hot-wire or hot-film anemometry uses the fact that the rate of heat transfer from a solid object is related to the velocity of flow past the object (McQuivey 1973). Though used in the controlled laboratory setting (e.g. Richardson and McQuivey 1968), hot-film anemometry has been used rarely in the field (Grant *et al.* 1968). The calibration is very sensitive to temperature shifts, and the probe is prone to breakage and to react with dissolved constituents in the water thus producing a scale that changes the calibration. With hot-film anemometry, one can measure environmentally common flow velocities of 1–400 cm/s. Similarly, pitot tubes could be used in the field but are used rarely.

The velocity field at the water surface can be estimated by filming or videotaping the stream (Meselhe *et al.* 1998). Substantial processing and image analysis is required. Groundlevel oblique photography will work, but overhead views from vantage points, tethered balloons, or planes are likely to be superior. If flow is relatively clear, cameras can be placed in clear housings to film buoyant particles in the flow. These provide two-dimensional velocity information through the water column (Drake *et al.* 1988). In conjunction with another camera, a fully three-dimensional description of flow fields is possible. Another technique uses thin wire to generate hydrogen bubbles by passing a current through the wire (Schraub *et al.* 1965). These bubbles are then filmed or photographed.

12.3 DISCHARGE MEASUREMENTS

There exists a variety of techniques and equipment for measuring flow discharge. In this section, the

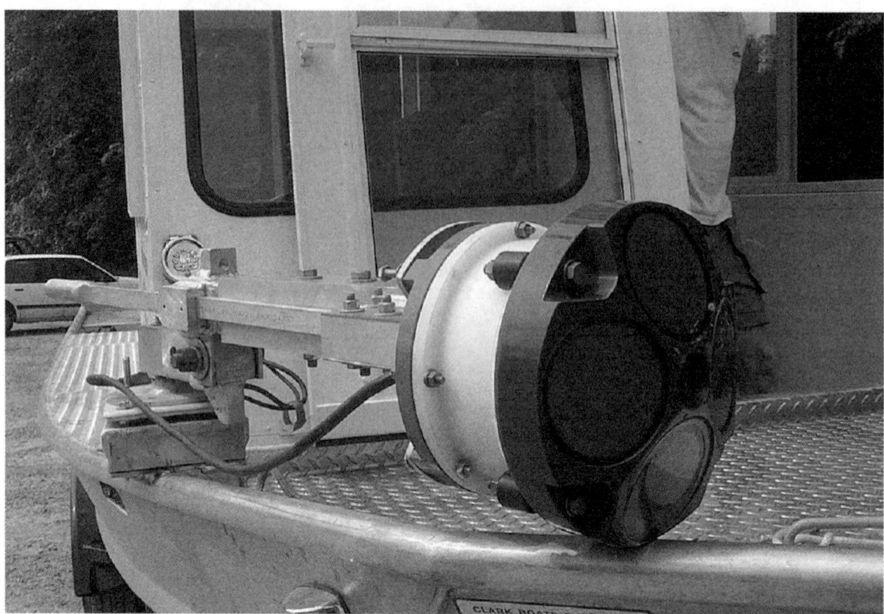

Figure 12.5 A side and bottom view of an ADCP (photo courtesy of Robb Jacobson)

approaches are described; the principles underlying the approaches are explained; and the accuracy and appropriateness of the approaches are discussed.

An example of the use of discharge measurements comes from my work in gravel-bed streams in Idaho (Whiting *et al.* 1999). I was involved in a water rights case that required knowledge of the streamflow in reaches of channel where the US Forest Service was claiming water. The theory of the case was that sufficient water must flow through the channels to preserve the ability of the channel to move all the bedload over the long term. Knowledge of the streamflow was necessary for several reasons and various descriptors of the streamflow were required. We needed stream discharge because the currency of water rights is the volume of water per unit time. Instantaneous measurements were integrated over time to provide estimates of mean daily flow. We also used instantaneous estimates of flow to associate with concurrent sediment transport measurements and to build a bedload rating curve. The sediment rating curve multiplied by each daily value of flow over the period of record to give the total flux of sediment was required in the analysis. We also had to determine the annual instantaneous peak flow for flood frequency analysis (Section 12.5) because the policy of the US Forest Service was that no flow above the 25-year flood would be claimed.

Integration of Point Measurements

The most common method for measuring flow discharge is the summation across the channel width (w) of the local products of subsection area (a) and mean flow velocity (u):

$$Q = \sum_{0}^{w} ua$$

The mid-section method consists of using the mean velocity as representative of a rectangular area with the dimensions of the measured depth at the point of the velocity measurement and the distance between the two adjacent measurements divided by two (Figure 12.6). Other methods include averaging adjacent velocity and depth measurements to calculate subsection discharges or accounting for the cross-sectional area enclosed by various velocity contours. Hipolito and Leoureiro (1988) report that the mid-section method gives the most precise measurements of total discharge through a section. Typically 20–30 verticals are required across the channel and no subsection should have more than 10% of the flow. A more stringent criterion is that no more than 5% of the discharge is in any vertical. The spacing between verticals is commonly equidistant, but it

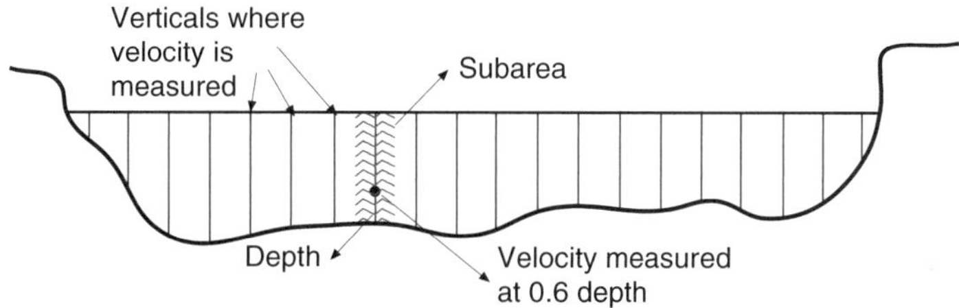

Figure 12.6 Discharge can be determined from the integration the local product of point measurements of velocity and associated area of flow

may be advisable to space verticals to produce segments of equal discharge or to place verticals at breaks in the profile. It is usually necessary to estimate the velocity in the end sections as some fraction of the nearest-bank measurement. Cross-sections where measurements are made should be relatively uniform, and in straight reaches without eddies. If possible, sections should not contain vegetation.

The determination of the mean velocity of the subsection is typically with a current meter. There are several approaches to determining the mean velocity in a vertical. Where roughness in the channel is very large relative to water depth and a logarithmic velocity profile may not be typical, velocity can be measured at evenly spaced multiple points in the vertical (0.1 depth intervals) and the velocities averaged. Where the depths are larger (typically 0.75 m), velocity at 0.2 and 0.8 of the depth below the surface can be averaged. Measurement of velocity at 0.6 of the depth from the surface often gives a good estimate of the mean velocity. This depth corresponds approximately to the elevation of mean velocity given a logarithmic velocity profile. Alternative methods include: measurement of the velocity at 0.2 of the flow depth and multiplication of the observation by a coefficient (usually 0.87); and measurement at three points (averaging the mean of measurements at 0.2 and 0.8 of the flow depth with the measurement at 0.6 of the flow depth). Accuracy can be improved by using multiple measurements in the profile especially near the bed where there often exists the largest gradient of velocity. Velocity should be measured for at least 30 s to account for at least the more frequent pulsation in the flow, especially if the velocity is low.

Carter and Anderson (1963) estimated that instrumental and sampling error was 4% using a single measurement of velocity at 0.6 of flow depth whereas instrumental and sampling error was 2.5% using the average of the measurement of velocity at 0.2 and 0.8 of the flow depth: both of these were for a 25-vertical transect. Fulford *et al.* (1994) observed about 2% differences between stream gagers. The magnitude of the error of the other approaches is unknown.

Earlier in the paper, the estimation of velocity from floats was described. Ideally the flow is uniform and the reach straight. Multiplication of the average surface velocity of several floats by a coefficient gives an approximation of the average velocity. The value of the coefficient is typically 0.8–0.9 depending on the resistance to flow. Mosley and McKerchar (1992) suggested a value of 0.86 for the coefficient. Alternatively, the coefficient can be estimated from the Chezy or Manning equations (Herschy 1985). Multiple floats should be used at intervals across the channel to describe the flow field more fully. While the average velocity determined from several floats can be used to estimate discharge, a better approach is to divide the cross-section into subsections and calculate the discharge through each subsection based upon the width, average depth, and average velocity in each subsection. These subsection discharges are summed to give the total discharge. The accuracy of such an estimate of discharge is less than other methods: Herschy (1985) estimates it to be ±10–20%. Such an approach for estimating discharge might be appropriate when other means are too hazardous or when other equipment is not available. If this method is used, the quality of the estimate can be improved with a straight uniform section free of obvious eddies or secondary currents.

Acoustic Doppler Current Profiling

The ADCP uses acoustic signals to measure water velocity and depth. The ADCP transmits pairs of short acoustic signals from four transducers. These pulses travel through the water column, strike suspended particles, and are reflected back to the

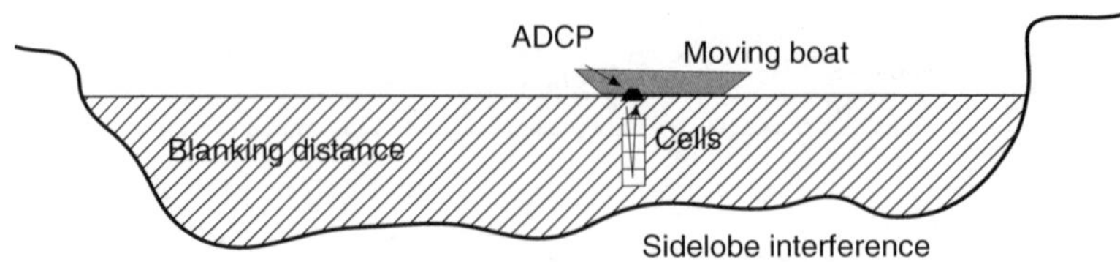

Figure 12.7 Sketch of regions in a channel cross-section analyzed by Acoustic Doppler current profiling (ADCP). In the hatchured areas, velocity measurements are not possible. Near the surface is the blanking distance. Near the bed and banks, this is due to sidelobe interference

receiver. The reflected pulses are separated into stacked 'depth cells' based upon travel time.

For discharge measurement, the ADCP is deployed from a boat (Figure 12.7). The head of the transducer is submerged just below the water surface and discharge is measured by moving the ADCP across the channel to measure vertical velocity profiles and flow depth. Usually multiple passes across the channel (4–6) are averaged. The vertical velocity profiles generated by the ADCP will not include measurements from near the surface or the bottom. In these regions, acoustic signals have not had a chance to travel a sufficient distance before reflection or the bottom echo makes reflections from small scatters in the flow unrecognizable. Velocities in the unmeasured portions of the profile are estimated using a power-law approximation. Proprietary software to track the bottom, estimate cross-stream position, and process the data to yield a discharge value usually comes with the equipment. Additional information on measurement from a boat can be found later in this section.

Morlock (1996) evaluated the device at 12 US Geological Survey gaging stations. The discharge estimated by the profiler was usually close to the conventional estimate; the maximum difference was 8%. The standard deviations of ADCP measurements ranged from about 1% to 6% and were for the most part larger than would be expected from the propagation of errors. Uncertainty in the estimate of velocity is probably about 5%. The device is limited to flow depths greater than 0.5 m. In channels that are 0.5–4 m deep, velocities in excess of 2 m/s are difficult to measure. The ADCP can be very useful in speeding measurements in large rivers.

Rating Curves

A common means for estimating discharge is with a simple streamflow rating, typically built from multiple simultaneous measurements of stage and flow discharge (Figure 12.8). When there is a hydraulic control, there will be a unique relationship between the depth of flow and the discharge. The simple rating can be a single curve or a compound curve (set of intersecting single curves) to account for low, moderate, and high flows. The rating can be in the form of a table, but more typically is an equation. The general form of most equations is

$$Q = a(G - e)^b$$

where Q is the flow discharge, G is the stage, e is typically the height of zero flow, and a and b are constants. Common values of b range from 1.5 to 2.5 (Herschy 1985). It is usually unnecessary to account for retransformation bias (e.g. Duan 1983). The stage

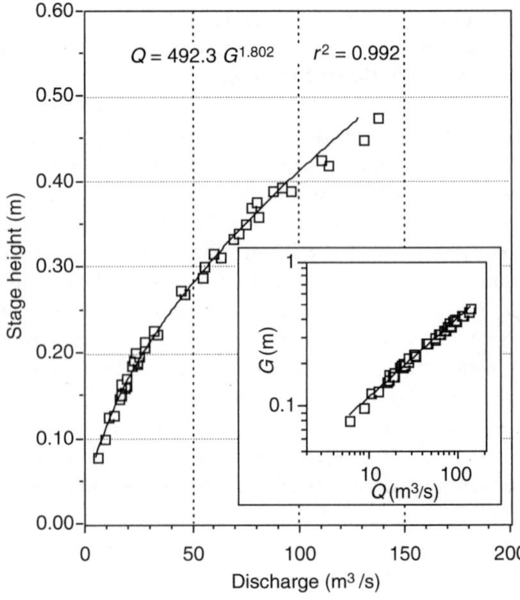

Figure 12.8 An example rating curve from Fourth of July Creek, Idaho. The value of e, the zero-offset, is 0.0 m

of zero flow can be estimated from the thalweg or by the value that makes the log–log plot linear. The stage is noted visually or monitored by float, bubble gage (becoming rare), pressure transducer, or ultrasonic sensor (described by Herschy 1985) and recorded by paper trace (becoming rare), punch tape (less rare), or digitally. The accuracy of different sensors vary: ultrasonic [±0.03 m (0.3%; Latkovich and Leavesley 1992)], pressure transducers (0.1–1.0%); and float (0.5–1.0%). A stilling well acts to protect the float, pressure transducer, or sensor, but more importantly dampens vertical fluctuations in the water level associated with wind and flow turbulence. Errors associated with simple ratings are usually less than 5% except at higher flows.

The placement of stream gages should be done carefully. The best sites are straight reaches of channel without vegetation where flow is confined to a single channel that is not prone to lateral migration or scour and fill. Ideally, a pool exists above the hydraulic control to allow measurement of stage even at low flow. Discharge ratings should be checked at regular intervals—Carter and Davidian (1968) suggest monthly measurement, but more frequent measurement is necessary in streams prone to scour and fill, changing bedforms, and vegetative growth. Ratings should be checked after major floods. Discrepancies between the discharge measured and the discharge expected from the stage, if greater than 5%, are usually addressed by subtracting, from the gage height, a vertical shift that compensates for the discrepancy. Shifts are usually necessary when the bed scours or fills or as other changes in the channel control occur. The data of all gages should be checked periodically against several pre-established reference points outside the channel on stable ground.

Complex curves relate discharge to stage and to other variables—such as velocity, or the rate of rise or fall of stage—and are used where the stage–discharge relation is complicated by storage or where tidal cycles create variable backwater or reverse flow. Index–velocity ratings can be built by measuring velocity continuously at some specific point in the cross-section using a current meter or across the section using an array(s) of equipment emitting and recording acoustic signals (see the subsection *Ultrasonic Methods*). Index–slope ratings can be built by a synchronous array of gages measuring stage along a reach.

It is often necessary to extrapolate rating curves; this is often less of an issue downward, but can be problematic upward. Herschy (1985) suggested using the Manning equation to estimate discharge by establishing how the quantity $nS^{0.5}$ varied with stage, where n is Mannings' roughness and S is the stream gradient. Alternatively, the rate of increase in average velocity with stage often is very small at high discharge and may approach a constant value. This constant value multiplied by the area of flow provides an estimate of discharge. Nonetheless, extrapolation of the rating curve may provide a reasonable estimate.

Kennedy (1984) provides a detailed description of methods and other issues in discharge rating. Buchanan and Somers (1968) and Herschy (1985) describe methods for stage measurement.

Flumes

Flumes have been used to measure discharge in situations where stream characteristics are such that the stage–discharge relationship is prone to shifts due to scour or fill and where the stream is sufficiently small or flashy that other means of gaging are impractical. Most flumes rely on a contraction in the width or a drop in the bed profile to induce critical or supercritical flow in the throat of the flume (Figures 12.9a and b). Under such conditions, the discharge can be determined by a single depth measurement usually in the throat of the flume because of the unique relation between depth and velocity at critical discharge. Moreover, if the flume is built to specifications, the discharge–head (depth) relation is pre-calibrated and need only be verified by occasional discharge measurements. Other advantages of flumes are that they can operate when head loss is small, and over a range of approach velocities.

Flumes can be characterized as sub-critical, critical, or super-critical. Sub-critical flumes are little used at present because flow must be measured in both the approach and the throat of the flume. Critical flow flumes include the Parshall flume and its variants. The Parshall flume is capable of passing efficiently small sediment thus avoiding clogging the structure and shifting the rating curve. Flumes can be modified with a V-notch. The portable Parshall flume, with a standard throat of 3 in., can be used in settings where discharge is too low for current-meter measurements. Also used for measuring discharge on small watersheds are the HS, H, and HL trapezoidal flumes developed by the US Soil Conservation Service. The flumes differ primarily in their capacity. The supercritical flumes are installed typically where sediment is likely to accumulate in the structure unless very high velocities can be used to preclude sedimentation. All flumes can be equipped with a stage recorder to

Figure 12.9 Flumes may be used in various settings and constructed of various materials: (a) a concrete flume associated with a bridge and a pump house; (b) a flume constructed of sheet metal and wood. Both flumes are in the Goodwin Creek watershed in Mississippi

provide a continuous record of stream discharge. Super-critical flumes can have vertical sides (i.e. San Dimas flumes) or trapezoidal sides. Herschy (1985) suggested that the error in estimates of discharge using a flume is 2–4%.

Weirs

Weirs are amongst the oldest, simplest, and most reliable means for measuring discharge. A weir temporarily ponds streamflow prior to its spilling in freefall over a controlled outlet (Figure 12.10). If a standard geometry is used, the weir is pre-calibrated for discharge. Weir blades are either sharp- or broad-crested. Weirs can be rectangular, trapezoidal, or have a 90° V-notch. The height of water over the weir blade (the height and the length of the weir crest in the case of rectangular weirs) is measured to estimate the discharge. There is a trade-off between accuracy and maintenance between the two types of crests. The sharp-crested weir is more prone to clogging by floating debris, but is more accurate. The elevation of the water in the pond can be measured continuously and, in conjunction with a rating curve, provides a continuous record of discharge. While debris screens can be installed upstream of weirs to minimize accumulation of floating debris or sediment in the weir, a flume, in particular a super-critical flume, should be used if substantial amounts of sediment or debris are expected. Herschy (1985) suggested that the error in estimates of discharge using a weir is up to 3%. A fairly complete summary of the use of flumes (and weirs) in measuring discharge is provided by Kilpatrick and Schneider (1983).

Ultrasonic Methods

This less common technique is based upon the continuous measurement of the time of travel of sound waves emitted from transmitters deployed below the water surface. The time of travel over the distance between transmitter and receiver is related to the

Figure 12.10 V-notch weir (photo courtesy of USGS)

average flow velocity, which multiplied by the flow area associated with the velocity gives the discharge. The elevation in the water column at which the velocity is measured is fixed so the relation between velocity at the measurement height and stage must be established. An array of transducers can be deployed to measure the velocity at various heights in the water column (e.g. Genthe and Yamamoto 1974). Total discharge is calculated as the sum of the product of average velocity at a given height and the cross-sectional area associated with the velocity measurement. Such an arrangement may be useful where the range in stage is large or greater precision in discharge estimates is needed. The velocity may be determined to 0.1% (Herschy 1985) but the uncertainty of a discharge estimate based upon such a measurement is approximately 5% (Herschy 1985).

The method may be useful where no stable stage–discharge relationship exists and the construction of a weir or flume is not possible or feasible. Such an approach can be used in backwater from tides, downstream tributaries, or dams.

Dilution and Tracer Gaging

The dilution method for determining flow discharge involves the addition of a conservative tracer to the flow and the determination of the concentration of the tracer downstream after the tracer has been well mixed throughout the flow. In the constant rate injection method, the tracer of concentration, C_1, is applied at a constant measured rate, q. Downstream, where the concentration is C_2, the discharge, Q, is

$$Q = \frac{C_1}{C_2} q$$

When using continuous tracer injection, the downstream measurement point must be sufficiently far downstream that the tracer is well mixed across the channel and at depth. An alternative approach is to introduce a slug of tracer and to monitor the concentration at a downstream location. The discharge is

$$Q = \frac{V}{T} \frac{C_1}{\bar{C}_2}$$

where V is the initial volume of the tracer, T the time of passage for the tracer (first arrival to last arrival), and $\bar{C}_2$ is the average concentration over the time of passage. Hubbard *et al.* (1982) measured the discharge by tracking a slug of tracer. The average velocity of the centroid of the dye cloud should approximate the mean flow velocity which when multiplied by the cross-sectional area gives the discharge.

The tracer for use in dilution gaging can be a dye or solute. Radioactive tracers were used in the past but are generally not used at present. Fluorescent dyes have been used as tracers to determine discharge (Wilson 1968). Fluorescence varies with concentration, which in turn varies with discharge. Factors affecting the relationship between fluorescence and concentration and in turn discharge include temperature, pH, reactions with other constituents, and photochemical degradation of the dye. Concentrations can be detected with a flourometer to below 100 ng/l. When solutes are used, sodium chloride is the most common. It is not generally a problem if the tracer already exists in the stream; if the background concentration remains steady, the dilution method can still be used. If salts are used in the tracing, the concentration of dissolved salts is linearly related to electrical conductivity that is relatively easy to monitor.

Tracer dilution may be useful in various situations, but is particularly useful where flow is shallow or clogged with vegetation or other debris, or where the lateral input of water is large (many tributaries and/or or seepage) and the change in discharge must be known through the reach. Herschy (1985) suggests dilution gaging has been used at discharges as large as 2000 m^3/s. As noted in the introduction, the technique is predicated upon the flow being well mixed at the sampling location. Uncertainty is about 5%.

More detailed discussions of the use of dye or salts in dilution and tracer gaging is available in Wilson (1968), Hubbard *et al.* (1982), and US Bureau of Reclamation (1984).

'Moving-boat' Method

The measurement of discharge by the moving-boat method may be practical for wide rivers, remote locations or when conditions are unsteady (e.g. tidally influenced) or hazardous such that measurements need to be made rapidly. Discharge is measured by equipping a boat with a depth sounder and a continuously operated current meter. While traversing the stream, the depth and combined stream and boat velocities, and angle of the flow with respect to the traverse are measured. The current meter is set at some characteristic depth below the surface and the measured velocity correlated to the average velocity. The value of the coefficient is typically 0.87–0.96 for velocities measured 1 m below the surface (Herschy 1985). The minimum speed of the boat should be such that the boat traverses the river in a straight line roughly orthogonal to the flow; this will require that the boat point at an angle 20–60° (upstream) to the cross-stream direction (Herschy 1985). The mid-section method for integrating subsection measurements to yield total discharge is recommended. For the best measurements, traverses should contain 30–40 observation points and the discharge from six or more traverses should be averaged. Smoot and Novak (1969) estimate that discharges by this technique are within 5% of measurements by conventional techniques.

The ACDP approach described earlier relies upon the moving-boat method. Herschy (1985) includes a chapter devoted to moving-boat measurements.

Electromagnetic Method

Electromagnetic determination of streamflow discharge has been accomplished in some settings. The flow of water in a stream through an electromagnetic field induces an electromotive force that is measured and is directly related to the average flow velocity in the cross-section. The earth's electromagnetic field is useful in principle for such measurements, but electrical interference is a problem. Typically, a coil is buried in the streambed through which electric current is passed and the resulting electromagnetic field used to measure discharge. The equipment is relatively expensive to install but in small streams where the flow can be passed through a pipe, the costs can be more reasonable. Usually AC current is needed. Electromagnetic discharge measurement can be used to determine discharge when average velocities are as low as 0.2 cm/s (Herschy 1985).

Correlation of Point Measurements to Discharge

Point measurements of flow velocity or average velocity in a vertical have been used to estimate discharge. For instance, ADV can be placed on the stream bottom looking upward, moored looking downward, or mounted on some structure looking sideways to provide a measure of velocity in some poorly defined region at a distance 0.5–2 m from the device. Alternatively, ADCP can be attached to the bottom or some other structure, or moored and the average velocity determined for the vertical (Williams 1996). These local measurements of velocity are correlated to measured stream flow, much like a rating curve, to estimate flow discharge. The accuracy of such correlation methods is probably no better than 5%.

Other Techniques for Discharge Determination

In some settings, discharge can be determined by measuring the volume or mass of water collected over a specific time interval. Volumetric measurement of the freefall of water is a convenient way to verify flume and weir calibration. These methods are most useful for small discharges (up to few liters per second).

12.4 INDIRECT METHODS OF DISCHARGE ESTIMATION

There exist several indirect approaches for estimating flows when other techniques are not suitable or available. For instance, it is often necessary to estimate the magnitude of streamflow where the best evidence is high water marks—mudlines, deposited sediment, and/or debris lines. Other records may not exist because there was no gage in the reach of channel, existing equipment was inoperable or destroyed by the flows, or access by personnel to the sites during the high flow was impossible or unsafe.

Slope-area Method

One such indirect method for discharge estimation is the slope-area method. It is based upon resistance formulae and the cross-sectional area and slope of the channel. The best known of these methods, at least in the USA, is the Manning equation

$$Q = AR^{2/3} S^{1/2}/n$$

which is similar in form to the Chezy equation

$$Q = cA(RS)^{1/2}$$

The slope (S) and the hydraulic radius (R) of the channel should be measured in as uniform a reach as is possible. The area of flow is A and n and c are roughness and conveyance factors. The slope should be measured over a length of 20 or more channel widths unless such a distance would include major discontinuities in width or depth, or falls. The value of the roughness factor, n, can be determined by calibration, from empirical relations, or by comparison with descriptive tables or to a viewbook (Barnes 1967). The roughness factor depends upon the size of the bed material, the presence of bedforms, and the amount of vegetation. For most channels, it varies from about 0.01 to 0.06 but values of 0.1 and greater are observed in channels with boulders and other large roughness elements (Hicks and Mason 1991). In compound channels or with flow over a floodplain, different roughness values can be ascribed to various subsections of the cross-section. Discharge estimated by the slope-area method has an uncertainty of at least 10–20% (Herschy 1985).

Dalrymple and Benson (1967) explain the use of high water marks for estimating peak flows by the slope-area method.

Contraction Method

Another indirect method includes the use of the energy equation. At channel width contractions, such as those created by bridge abutments, the change in the water surface elevation through the contraction can be used to estimate peak discharge. The water surface elevation could be measured during high flow but is more often taken from high water marks. One issue that needs to be considered is the possibility of scour and fill of the bed during high flow (Matthai 1967). Peak discharges can be determined also from high water marks of the headwater and tailwaters above and below culverts. The approach can be used for subcritical and critical flows, transitions between such flows, and for submerged outlets (Bodhaine 1968). Uncertainties associated with estimates of peak flow by these methods are probably 20% or greater.

Step-backwater Modeling

A final suggestion for estimating discharge in the absence of measurement is the use of step-backwater models for computing the discharge associated with observed water surface profiles as summarized by Miller and Cluer (1998). Various hydraulic models based upon the one-dimensional energy equation are available that use an iterative solution technique known as the step-backwater method. The hydraulic package HEC-2 (Hydrologic Engineering Center 1982) is an example of such a model. To estimate flows, bathymetry and estimate of roughness are necessary. The discharge that best matches the observed water surface elevations (often highwater marks) along the channel is the estimated streamflow. Uncertainties are relatively large—probably at least 10–20%.

12.5 FLOW HYDROGRAPHS AND ANALYSIS OF FLOW RECORDS

The measurement of velocity and discharge is important in its own right, but these measurements are often used to develop a hydrograph describing the flow rate over time (Figure 12.11). Hydrographs can be used to correlate runoff timing and volume to precipitation timing, intensity, and duration or to determine 1-day, 7-day, 28-day, etc., high or low flows; among a few examples. Instead of the history of flows, the distribution of flows can be analyzed (Figure 12.12).

At the beginning of the section on discharge measurement (Section 12.3), the example of a water rights case was presented. The history of flows was used to

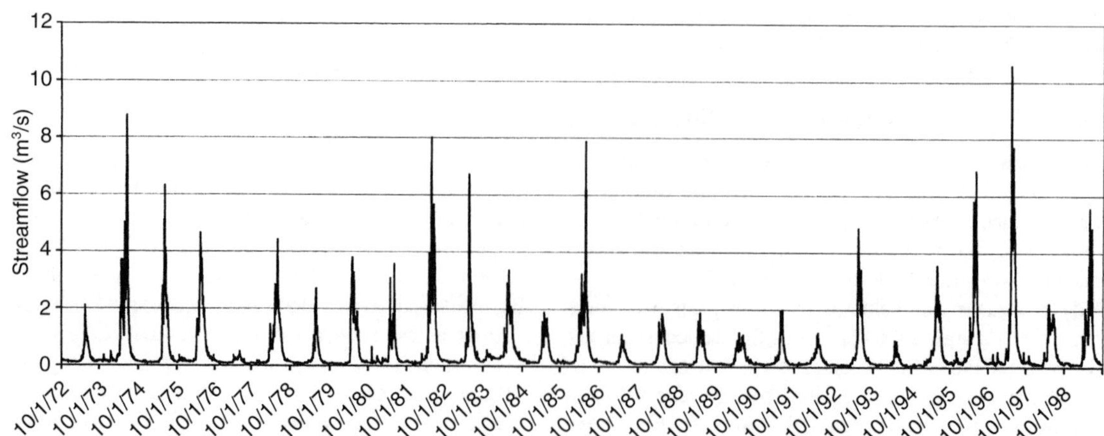

Figure 12.11 Flow hydrograph for Thompson Creek, Idaho, USA, for the period October 1, 1972 to September 30, 1999

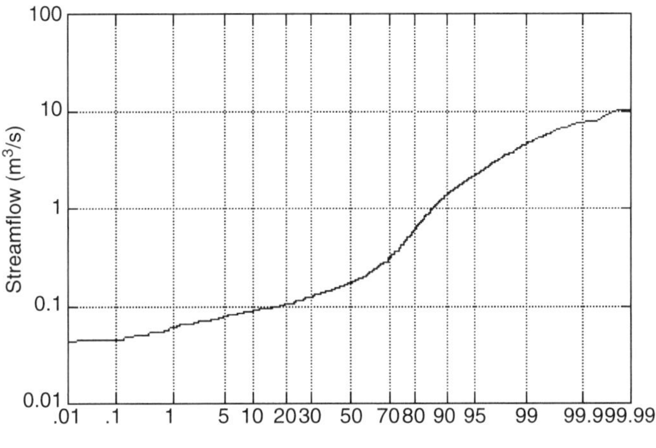

Figure 12.12 Flow duration curve for Thompson Creek, Idaho, USA, for the period October 1, 1972 to September 30, 1999. The discretization period is daily

answer various questions about the effect of the US Forest Services claims for water—How much water would be claimed? During which months? How many years would there be no Forest Service claim because flow levels did not rise high enough?

In the next few subsections, the presentation and analysis of flow records are discussed.

Flow Hydrograph

In Section 12.3, various methods were suggested for determining discharge. While in principle, almost any of these methods could be repeated with great frequency to provide hydrograph of streamflows, it is most common to use a stage–discharge rating curve and the history of stage to produce the hydrograph (Figure 12.11). Stage is commonly measured at 15-min intervals and recorded, the stage converted to discharge, and then the discharge recorded. Typically only daily discretizations of flow are published by governmental agencies but finer resolution data are sometimes archived.

Flow Duration Curves

Flow duration curves (Figure 12.12) describe the percent of the time flow is greater than a specified value (percent exceedence). An important consideration in such cases is the time interval of measurement (discretization). In basins with short lag times (small, urban, and/or extensive bare rock), the mean daily flow is a poor descriptor of the observed flow. For such basins, the appropriate interval with which to build flow duration curves may be 15 min or shorter. In many cases, shorter intervals were used to determine the mean daily flow, but it may be necessary to return to primary records (digital files, punch tapes, hydrograph traces) to recover the finer time resolution. The flow duration curve is developed by sorting the average discharge over the selected discretization interval and assigning the appropriate probability of exceedence based upon the total length of the record. It is critical that there be no missing values over the period of record. If data are missing, they must be estimated. Frequency analyses should be avoided with records shorter than 10 years or for estimating the frequency of events greater than twice the record length (Viessman *et al.* 1977).

Extreme Value Plots

For many purposes, extremes of streamflow (high and low) are critical information. Peak instantaneous discharge is typically determined for a gage to characterize the recurrence interval of floods (Figure 12.13). Usually the annual peak over the period of record is determined, ranked by magnitude, and depending upon the record length, the probability of exceedence is determined. The inverse of the probability of exceedence is the recurrence interval. Instead of analyzing the peak flow of each year, which is called an annual series, a partial series including all peaks above some

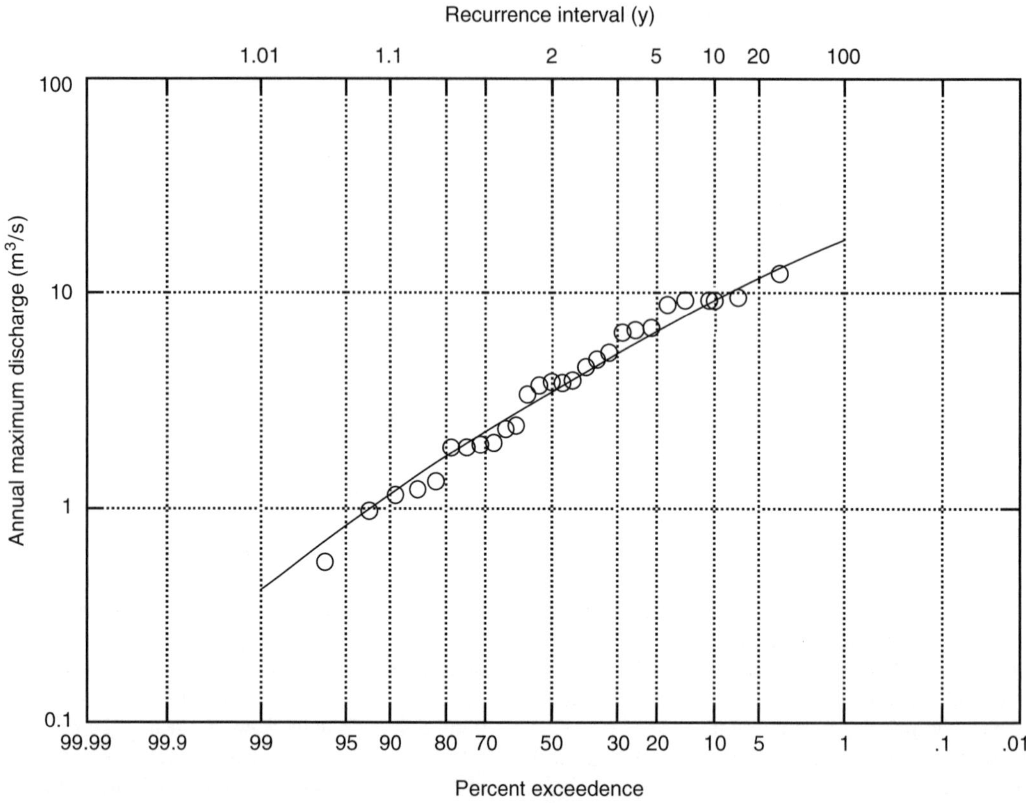

Figure 12.13 The frequency and recurrence interval of annual peak floods for Thompson Creek, Idaho, USA, for the period October 1, 1972 to September 30, 1999

threshold can be analyzed. Annual low flows (e.g. 1-day, 7-day, 28-day) are very important for water supply and environmental studies and can be analyzed similarly.

These tools are treated extensively in hydrology textbooks such as Dunne and Leopold (1978) and Dingman (1994).

12.6 ISSUES IN SELECTING METHODS

As shown above, there are many approaches and technologies that may be appropriate for characterizing flow and flow velocity and determining discharge. The issues to consider in the selection of an approach or equipment are numerous but might be categorized as follows:

- the purpose of the measurements;
- appropriateness of pre-existing data;
- the required precision and accuracy;
- the channel attributes (size and geometry of the channel, stability of the reach);
- the hydrologic attributes (steadiness of flow, unidirectional or reversing flow);
- site accessibility and infrastructure for making measurements;
- the equipment available;
- time available to make measurements;
- cost (equipment and personnel time).

The following sections elaborate briefly upon these issues.

Purpose of Measurements

There exists many reasons for collecting flow information, a number of which were posed in the introduction and include monitoring (e.g. the amount of water), basic research (e.g. the turbulence associated with sediment motion) and applied research (e.g. the amount of habitat available at different flow levels). Depending upon the purpose for which flow information is collected, various methods may or may not be suitable. An important question to consider is the level of spatial and temporal details required. If spatial detail is required, near steady flow for long periods is helpful. Snowmelt-driven systems, spring-fed streams, and streams with reservoir releases are likely to maintain relatively high flows near formative conditions which is especially important for detailed sediment transport studies. Measurements near base flow may provide opportunities for detailed measurement but at stages far different than the flows that have the most influence on channel form.

Pre-existing Data

A number of governmental and non-governmental agencies collect flow information that may be suitable for the problem at hand. In other situations, pre-existing data may be a useful starting point. For example, a discontinued gaging station could be re-occupied. If the rating curve could be shown to be still valid with a few measurements of stage and discharge, a great deal of effort could be avoided. If nothing else, pre-existing data on the stream or nearby streams may suggest the sort of flows expected or the timing of flows thus aiding in the experimental design.

It is often the case that only the processed mean daily discharge values and instantaneous peaks are published. Sometimes it is possible to retrieve more detailed flow information (cross-section data, hydrograph chart traces (digital or paper)) by contacting the agency that collected the information.

Precision and Accuracy

Depending in part on the purpose of the measurement, various levels of precision and accuracy may be required. For instance, if the question being addressed is the change in runoff volume associated with a small change in impervious area, and the expected change is about 10%, it makes sense to use a methodology with an uncertainty substantially less than 10%. While many sets of discharge measurement using a technique with larger uncertainty have the potential to be suitable, the effort required is greater. In other situations, the flow or discharge may need to be known very accurately because of the importance or costs of decisions based upon such information.

Table 12.1 Uncertainty in velocity measurement

Method	Uncertainty (± percentage)
Floats	10–20
Mechanical current meters	0.5–2
Electromagnetic current meters	~2
Acoustic Doppler velocimeters	0.5–1
Laser Doppler velocimeters	~0.1

Table 12.2 Uncertainty in discharge measurement

Method	Uncertainty (± percentage)
Floats	10–20
Integration of point measurements	~5
ADCP	5
Rating curve	<5
Flume	2–4
Weir	1–3
Ultrasonic	~5
Dilution	5
Moving boat	~5

The uncertainties in velocity measurement as summarized from the earlier discussion are shown in Table 12.1. The uncertainties in discharge measurement (following Herschy 1985) for moderate flow conditions are shown in Table 12.2. Uncertainties are likely to be larger at very high and very low flows. Uncertainties in mean daily, mean monthly, and mean annual discharge will be lower.

The uncertainties outlined in Tables 12.1 and 12.2 should be taken as approximate estimates. Factors that can affect the magnitude of the uncertainty are the training and care of the operator, the condition of the equipment, the precision and accuracy of the instrumentation, the number of verticals in the section, the precision of depth measurement, the measurement of stage, the stage–discharge relation, the steadiness and uniformity of flow in the measurement reach, the geometry in the measurement reach, the relative amount of unmeasured flow, and other factors.

While not universally true, there is some truth to the generalization that more accurate measurements require more expensive equipment, more time, and more personnel.

Channel Attributes

Channel attributes include the size and geometry of the channel, the nature and size of the substrate, and the stability of the reach. The size of a system can influence the type of equipment and methods appropriate for use. Discharge from small streams may be measured best with weirs or flumes whereas the largest rivers may be measured best with a moving boat—equipped with current meters or an ADCP. The nature and size of the substrate can be important. The size of the sediment on the stream bottom can influence the measurement approach in several ways. If the size of the particles on the bed is larger relative to the flow depth, the velocity profile may be non-logarithmic thus measuring velocity at a single elevation at 6/10 the depth will not be appropriate. If the relative roughness is large, techniques relying upon the transmission of a signal and its reflectance may give spurious results. On the other hand, fine beds will often develop bedforms potentially requiring longer averaging periods to account for migration of bed features. Where flow accelerations are large, for example, Solfatara Creek mentioned earlier, the measurement of discharge will require multiple measurement points in the vertical. The presence of in-channel vegetation can affect measurement. Vegetation can clog mechanical current meters whereas the vegetation does not affect electromagnetic current meters.

The stability of the reach can be critical for certain techniques of discharge determination. Discharge, as determined by the building of a rating curve, requires that there is a consistent relationship between stage and volume of flow. If the bed is aggrading or degrading, this requirement is not met. One of the other techniques would have to be used in such situations.

Hydrologic Attributes

Hydrologic attributes to consider in the selection of methods include whether flow varies periodically (with diurnal snowmelt), whether the response is rapid (as with a small urbanized basin), and whether flow is unidirectional or reverses (as with tides). Systems prone to rapid changes in flow generally require equipment that collects information automatically. For example, where there is a diurnal signal, it may be warranted to measure flow at a consistent time; where runoff occurs rapidly, stage could be recorded by a pressure transducer and converted to discharge; and in reaches affected by tides, flow reversals may require that meters are capable of indicating the direction of flow.

The range of flows to be measured may also affect the selection of equipment and methods. For instance, there are situations where discharge measurement by wading at low flow is possible (even preferable) but at high flow suspension of equipment from a bridge or cables is required. Should a single approach be used at all flows or should different measurement approaches be used at different stages? There is not a simple answer to the question: the relative merits of collecting the best information over a particular range of flows and using a consistent approach must be balanced.

Site Accessibility and Infrastructure for Making Measurements

The accessibility and remoteness of sites may influence the selection of methods and equipment. If a site is remote, it may be impractical to rely primarily upon personnel to collect flow measurements or to reach the site in a timely manner. If a site is not accessible by a road or easily accessible by boat, measurements may need to rely upon equipment that collects data automatically. If equipment must be carried overland, the weight of equipment and peripherals becomes a consideration.

The infrastructure at a site or site conditions can affect the approach selected. The presence of electric power makes it easier to use electromagnetic discharge determination techniques and ultrasonic techniques. A bridge spanning the river may make a site suitable for suspending current meters (even though they may affect flow) whereas the absence of the bridge may require working from a boat.

Some sites may be inaccessible at certain times of the year. Breeding grounds or spawning habitat of endangered or threatened species often prevent personnel from visiting sites to collect data or download automatically collected data. In these situations, automatic data collection and either large data storage ability or data transmission capability are necessary.

Equipment

It is not unusual for the availability to equipment or familiarity with a particular type of equipment to affect the design of the data collection plan. The robustness of equipment and ease of operation is important to consider as is the weight and the amount of peripheral equipment especially if the site is remote and walking into the site or boat access is required.

Time

The time required to make flow and discharge measurements varies appreciably by method and by equipment. In addition, time for data processing can be appreciable as well.

In some situations, the speed at which measurements can be made is important. If debris is in the channel or if there is boat traffic, time can be of the essence. In situations where flow velocity or discharge is changing rapidly, the ability to make a measurement in a short time interval is important. As an example, characterizing the flow field over a bedform from the bed to the water surface should be accomplished very rapidly—before the bedform migrates any appreciable distance. The time required to make a measurement and process data will probably be considered primarily in the context of the purpose of the measurement, the required accuracy, and the cost.

Cost

The expenditure of funds in the purchase of equipment for flow measurement can vary by an order of magnitude at least. Some variation in cost (and quality) of largely similar equipment exists between manufacturers. Additional costs are associated with the collection and with the processing of data. A modest sum should be reserved for the maintenance of equipment.

12.7 SUMMARY

As laid out in earlier parts of this chapter, there are various reasons for making measurements of flow, various equipment and methods available, and a variety of reasons for selecting one particular approach. This primer on flow measurement should not be taken as sufficiently detailed to serve as a stand-alone handbook on any particular method or equipment. Perhaps, the most appropriate use of the primer is in the initial phases of investigative design. It is recommended that the hydrographer who is considering one of the methods outlined herein read more detailed descriptions as suggested at the end of Section 12.1 or listed in the references.

REFERENCES

Barnes, H.H. 1967. Roughness characteristics of natural channels, US Geological Survey Water Supply Paper 1849, 213 p.

Barua, D.K. and Rahman, K.H. 1998. Some aspects of turbulent flow structure in large alluvial rivers. *Journal of Hydraulic Research* 36: 235–252.

Bodhaine, G.L. 1968. Measurement of peak discharge at culverts by indirect methods. In: *Techniques of Water Resource Investigations of the United States Geological Survey*, Washington: US Government Printing Office, 60 p. (Chapter A3).

Buchanan, T.J. and Somers, W.P. 1968. Stage measurement at gaging stations. In: *Techniques of Water Resource Investigations of the United States Geological Survey*, Washington: US Government Printing Office, 28 p. (Chapter A7).

Buchanan, T.J. and Somers, W.P. 1969. Discharge measurements at gaging stations. In: *Techniques of Water Resource Investigations of the United States Geological Survey*, Washington: US Government Printing Office, 65 p. (Chapter A8).

Bureau of Reclamation. 1984. *Water Measurement Manual*, Denver, CO: US Government Printing Office, 327 p.

Carter, R.W. and Anderson, I.E. 1963. Accuracy of current-meter measurements. *American Society of Civil Engineers Journal* 89: 105–115.

Carter, R.W. and Davidian, J. 1968. General procedure for gaging streams. In: *Techniques of Water Resource Investigations of the United States Geological Survey*, Washington: US Government Printing Office, 13 p. (Chapter A6).

Dalrymple, T. and Benson, M.A. 1967. Measurement of peak discharge by the slope-area method. In: *Techniques of Water Resource Investigations of the United States Geological Survey*, Washington: US Government Printing Office, 44 p. (Chapter A2).

Dingman, S.L. 1994. *Physical Hydrology*, New York: Macmillan Publishing Co., 575 p.

Dinehart, R.L. 1999. Correlative velocity fluctuations over a gravel bed river. *Water Resources Research* 35: 569–582.

Drake, T.G., Shreve, R.L., Dietrich, W.E., Whiting, P.J. and Leopold, L.B. 1988. Bedload transport of fine gravel observed by motion-picture photography. *Journal of Fluid Mechanics* 192: 193–217.

Duan, N., 1983. Smearing estimate: a nonparametric retransformation method. *Journal of American Statistical Association* 78: 605–610.

Dunne, T. and Leopold, L.B. 1978. *Water in Environmental Planning*, New York: W.H. Freeman and Co., 818 p.

Fulford, J.M., Thibodeaux, K.G. and Kaehrle, W.R. 1994. Comparison of current meters used for stream gaging. In: Pugh, C.A., ed., *Fundamentals and Advancements in Hydraulic Measurements and Experimentation*, Am. Soc. Civil Engr., pp. 376–385.

Genthe, W.K. and Yamamoto, M. 1974. A new ultrasonic flowmeter for flows in large conduits and open channels. In: Wendt, W.E., ed., *Flow: Its Measurement and Control in Science and Industry*, Pittsburgh: Instrument Society of America, pp. 947–955.

Grant, H.L., Stewart, R.W. and Moilliet, A. 1968. The spectrum of temperature fluctuations in turbulent flow. *Journal of Fluid Mechanics* 34: 423–442.

Herschy, R.W. 1985. *Streamflow Measurement*, London: Elsevier, 553 p.

Hicks, D.M. and Mason, P.D. 1991. *Roughness Characteristics of New Zealand Rivers*, Christchurch, New Zealand: Water Resources Survey, DSIR Marine and Freshwater.

Hipolito, J.N. and Leoureiro, J.M. 1988. Analysis of some velocity-area methods for calculating open-channel flow. *Hydrological Sciences Journal* 33: 311–320.

Hubbard, E.F., Kilpatrick, F.A., Martens, L.A. and Wilson, J.F. 1982. Measurement of time of travel and dispersion in streams by dye tracing. In: *Techniques of Water Resource Investigations of the United States Geological Survey*, Washington: US Government Printing Office, 44 p. (Chapter A9).

Hydrologic Engineering Center. 1982. *HEC-2 Water Surface Profiles – Users Manual*, Davis, CA: US Army Corps of Engineers.

ISO. 1983. *Measurement of Liquid Flow in Open Channels. ISO Standards Handbook*, 518 p.

Jarrett, R.D. 1992. Hydraulics of mountain rivers. In: Yen, B.C., ed., *Channel Flow Resistance: Centennial of Manning's Formula*, Littleton, CO: Water Resources Publications, pp. 287–298.

Kennedy, E.J. 1984. Discharge ratings at gaging stations. In: *Techniques of Water Resource Investigations of the United States Geological Survey*, Washington: US Government Printing Office, 59 p. (Chapter A10).

Kilpatrick, F.A. and Schneider, V.R. 1983. Use of flumes in measuring discharge. In: *Techniques of Water Resource Investigations of the United States Geological Survey*, Washington: US Government Printing Office, 46 p. (Chapter A14).

Latkovich, V.J. and Leavesley, G.H. 1992. Automated data acquisition and transmission. In: Maidmont, D.R., ed., *Streamflow in Handbook of Hydrology*, New York: McGraw-Hill, pp. 25.1–25.21.

Marsh-McBirney, Inc. 1995. *Operation and Maintenance – Electromagnetic Current Meters*, 116 p.

Matthai, H.F. 1967. Measurement of peak discharge at width contractions by indirect methods. In: *Techniques of Water Resource Investigations of the United States Geological Survey*, Washington: US Government Printing Office, 44 p. (Chapter A4).

Meselhe, E.A., Bradley, A.A., Kruger, A. and Muste, M.V.I. 1998. PIV and numerical modeling for flow analysis. In: *ASCE International Water Resources Engineering Conference Proceedings*, Volume 1, pp. 526–531.

McQuivey, R.S. 1973. *Principles and Measuring Techniques of Turbulence Characteristics in Open-channel Flows*, Geological Survey Professional Paper 802-A, Washington: US Government Printing Office, 82 p.

Miller, A.J. and Cluer, B.L. 1998. Modeling considerations for simulation of flow in bedrock channels. In: Tinkler, K.J. and Wohl, E.E., eds., *Rivers Over Rock: Fluvial Processes in Bedrock Channels*, AGU Geophysical Monograph 107, pp. 61–104.

Morlock, S.E. 1996. *Evaluation of Acoustic Doppler Current Profiler measurements of river discharge*, US Geological Survey Water Resources Investigations Report 95-4218, Washington: US Government Printing Office, 37 p.

Mosley, M.P. and McKerchar, A.I. 1992. In: Maidmont, D.R., ed., Streamflow in Handbook of Hydrology, New York: McGraw-Hill, pp. 8.1–8.39 (Chapter 8).

Rantz, S.E., *et al.* 1982. *Measurement and Computation of Streamflow. Volume 1. Measurement of Stage and Discharge*, US Geological Survey Water Supply Paper 2175, Washington: US Government Printing Office, 631 p.

RD Instruments, Inc. 1998. *Rio Grande User's Guide*, San Diego, CA: RD Instruments.

Richardson, E.V. and McQuivey, R.S. 1968. Measurement of turbulence in water. *Proceedings of the ASCE Hydraulics Division* 94: 411–430.

Schraub, F.A., Kline, S.J., Henry, J., Runstadler, P.W. and Littell, A. 1965. Use of hydrogen bubbles for quantitative determination of time dependent velocity fields in low speed water flows. *Journal of Basic Engineering* 87: 429–444.

Smith, J.D. 1978. Measurement of turbulence in ocean boundary layers. In: *Working Conference on Current Measurement, Office of Ocean Eng., Natl. Oceanic and Atmos. Admin., Univ. Del., Newark, Del, January 11–13.*

Smoot, G.F. and Novak, C.E. 1969. Measurement of discharge by the moving boat method. In: *Techniques of Water Resource Investigations of the United States Geological Survey*, Washington: US Government Printing Office, 22 p. (Chapter A11).

Viessman W., Knapp, J.W., Lewis, G.L. and Harbaugh, T.E. 1977. *Introduction to Hydrology*, New York: Harper and Row, 704 p.

Whiting, P.J. and Dietrich, W.E. 1991. Convective accelerations and boundary shear stress over a channel bar. *Water Resources Research* 27: 783–796.

Whiting, P.J., Stamm, J.F., Moog, D.B. and Orndorff, R.L. 1999. Sediment transporting flows in headwater streams. *Bulletin Geological Society America* 111: 450–466.

Williams, A.J. 1996. Current measurement technology development in the '90s, a review. In: *Proceedings of the 1996 MTS/IEEE Oceans Conference, Ft. Lauderdale, FL*, pp. 105–109.

Wilson, J.F. 1968. Flourometric procedures for dye tracing. In: *Techniques of Water Resource Investigations of the United States Geological Survey*, Washington: US Government Printing Office, 31 p. (Chapter A12).

13

Bed Sediment Measurement

G. MATHIAS KONDOLF[1], THOMAS E. LISLE[2] AND GORDON M. WOLMAN[3]
[1]*Department of Landscape Architecture and Environmental Planning and Department of Geography, University of California, CA, USA*
[2]*US Forest Service Redwood Sciences Laboratory, Arcata CA, USA*
[3]*Department of Geography and Environmental Engineering, Johns Hopkins University, Baltimore MD, USA*

13.1 INTRODUCTION

Bed material is sampled for a range of purposes, including measurement of the suitability of substrate for spawning fish and other aquatic organisms, as input for equations to calculate bed mobilization, bedload transport rates, and likelihood of scour, and as a measure of grain roughness in the channel. Particularly on gravel bed rivers, a variety of techniques have been used to sample gravels, ranging widely in effort and cost, mostly obtaining a gravel sample and passing it through a series of sieves to determine the proportions of various sizes, or measuring particles under randomly located sample points on the bed. Although it may seem obvious, our principal message in this chapter is that selection of sampling technique and analytical approach should be driven by the purpose of the study, i.e., the questions posed, the type of data needed to answer the question posed, the level of confidence needed in the result, and consequently the requisite sample size. Many well-intentioned sampling programs have produced data sets of ultimately questionable value because the purpose of the field data collection effort was not clearly thought out or sample size requirements not recognized.

Theoretical and practical considerations for sampling gravel beds have been thoroughly reviewed in excellent works by Kellerhals and Bray (1971) and Church *et al.* (1987). Church *et al.* (1987) is a classic, dealing with fundamental issues of sample size, comparisons of different sampling methods, and underlying study design issues. In this chapter, we review these considerations and specifically consider issues that arise in sampling for purposes such as assessing the quality of aquatic habitat or effects of upstream land use activities (Lisle and Eads 1991, Young *et al.* 1991). Sand and finer-grained sediments can be adequately analyzed with relatively small samples, but gravels require large samples and thus pose greater challenges in sampling. A considerable literature in fluvial geomorphology concerns sampling of gravels, thus much of this chapter relates to sampling gravel bed rivers, sampling methods most appropriate for various objectives, and advantages and disadvantages of various methods. The reader is also referred to up-to-date and thorough treatment of sampling coarse-grained rivers by Bunte and Abt (2001b).

13.2 ATTRIBUTES AND REPORTING OF SEDIMENT SIZE DISTRIBUTIONS

Natural streambed gravels consist of a mixture of sizes, commonly ranging from clay (<0.004 mm) to boulders (>256 mm). If silt and clay are present in the mixture, particle size may range over five orders of magnitude. Many sediments (and sedimentary rocks) are characterized by larger particles that make up the structure of the deposit (the framework grains) with finer sediments filling the pore spaces between the framework grains (the matrix) (Carling and Reader 1982). Some sediments contain so much matrix that most framework grains are not touching and thus not carrying the weight of the deposit; these are termed 'matrix-supported' deposits (Williams *et al.* 1982). The threshold size between framework gravel and matrix sediment should be a function of the pore sizes in the framework. In a bimodal distribution, the distinction between framework and matrix may be straightforward. Otherwise, defining the upper limit of matrix sediment may be arbitrary.

For each grain, three perpendicular axes can be identified: a long axis, or '*a*-axis', a short axis, or '*c*-axis', and an intermediate axis, or '*b*-axis' (Krumbein 1941) for length, thickness, and breadth, respectively. Grain diameter is usually measured by the intermediate axis because it is the dimension that controls whether a particle can pass through a sieve.

The range in particle size of natural sediments is continuous, but we customarily subdivide the range into size classes for standard terminology (e.g., sand, silt) and to yield sufficient classes for analysis (Pettijohn 1975). Because the range of sizes in natural sediments is so great, it cannot be captured effectively with a linear scale. Instead a geometric scale is used, giving:

> ... larger classes ... for larger sizes and smaller classes for smaller sizes. As Bagnold (1941: p. 2) puts it, linear scales are seldom acceptable to Nature. Nature, if she has any preference, probably takes more interest in the ratios between quantities; she is rarely concerned with size for the sake of size. A millimeter difference between the diameters of two boulders is insignificant, but a millimeter difference between one sand grain and the next is a large and important inequality. (Pettijohn 1975: 34–35)

The most commonly used size scale in fluvial geomorphology and engineering is the Wentworth scale, which defines size classes in millimeters, and with intervals that increase by powers of 2 (Figure 13.1). There is also a considerable literature (especially in engineering and biology) that has reported sizes in inches, and many commonly used sieves are sized in fractions of inches. Fluvial gravels span such a wide range of grain sizes that their distributions are usually plotted either with log-transformed axes, or the size data themselves are log-transformed to the so-called *phi* scale (Krumbein 1934, Inman 1952). The phi scale consists of size units corresponding to powers of 2 (in mm), but with the phi values increasing with decreasing size such that phi = 0 is equivalent to 1 mm, phi = 1 is equivalent to 0.5 mm, phi = 2 is equivalent to 0.25 mm, phi = –1 is equivalent to 2 mm, etc. Within approximately the past decade, following the example of Gary Parker, some geomorphologists have used the negative of the phi scale (the *psi* scale), thereby avoiding the counterintuitive decrease in the scale value with increasing particle size (Parker and Andrews 1985, Bunte and Abt 2001b). The log-transformation to the phi scale provided a computational advantage when it was introduced in 1934, allowing computations to be easily made despite a wide range of grain sizes in natural sediments. However, this computational advantage is no longer meaningful with current computational capabilities. Because the phi size classes are less readily comprehended than actual grain sizes values expressed in millimeters, we prefer to report and plot actual grain sizes (in mm), thereby reducing jargon and increasing clarity of communication. With log-transformed scales, actual grain size values for different percentiles can be easily read from size distribution curves.

Presenting Particle Size Distribution Curves

Particle size distributions can be presented as histograms of the percentage of particle (or sample weight) occurring in each size class, as cumulative size distribution curves, or as box-and-whisker plots. Unless the

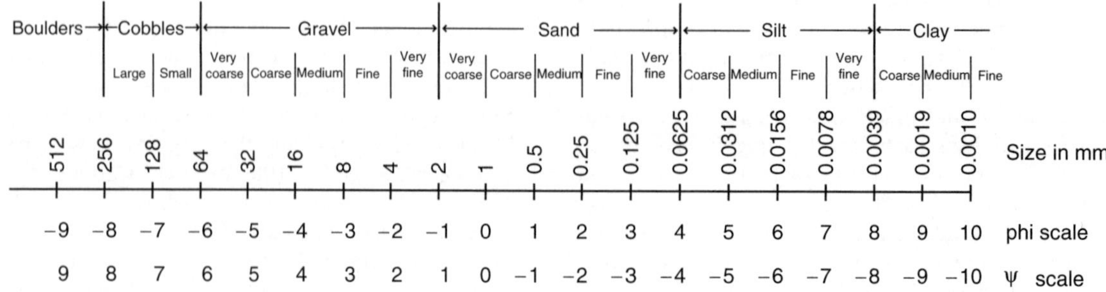

Figure 13.1 The Wentworth grain size scale and equivalent phi values of Krumbein (1934) (modified from Pettijohn 1975)

intervals between sieve sizes follow a geometric progression such as the Wentworth classes (and in many studies the sieves do not), plotted non-cumulative particle size distributions can be misleading. One bar on the histogram may appear larger than the next only because the bar includes particles from a wider range of sizes. Thus, the range of sizes present in a natural sediment is typically presented in cumulative size distribution curves (Figure 13.2). Grain diameters corresponding to specific percentile values can be read directly from the curves plotted on semilogarithmic paper, or by linear interpolation (Bunte and Abt 2001b). D_{16} is the size (in mm), at which 16% of the sample is finer, D_{25} the size at which 25% is finer, etc. Probably, the most widely used percentile value is D_{50}, the median diameter.

While these cumulative size distribution curves, if adequately sampled, can provide complete information on the range of sizes present in a given gravel, it is unwieldy to use these complete pictures for comparison among gravels, and it is impossible to present more than a few similar distributions on the same graph because the lines overlap and characteristics of individual size distributions are obscured (Figure 13.2).

Size distributions can also be presented as modified box-and-whisker plots (Tukey 1977, Kondolf and Wolman 1993), which permit multiple distributions to be presented on the same graph without overlap (Figure 13.3). In the box-and-whisker plots, the rectangle (box) encompasses the middle 50% of the sample, from the D_{25} to D_{75} values, termed the 'hinges'. The median diameter, D_{50}, is represented by a horizontal line through the box. Above and below the box are lines (whiskers) extending to the D_{90} and D_{10} values, a modification from the standard box-and-whisker plot of Tukey (1977), in which the

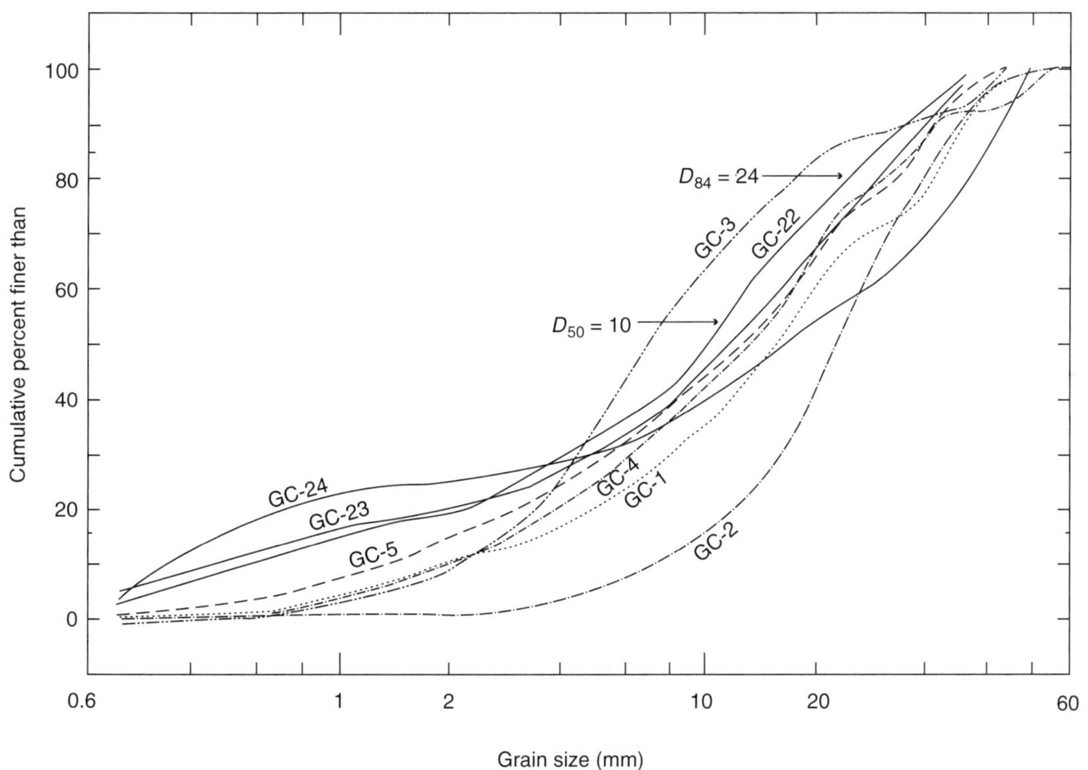

Figure 13.2 Illustrative cumulative size distribution curves for rainbow trout (*Oncorhynchus mykiss*) spawning gravels drawn from the case study in the mainstem Colorado River (solid lines, 3 samples) and Nankoweap Creek (dashed-dotted lines, 5 samples), a tributary downstream of Glen Canyon Dam. In the example shown, a redd gravel from the mainstem Colorado has a D_{84} of 24 mm and a D_{50} of 10 mm. Curves are identified by sample numbers (same as in Figure 13.3) (adapted from Kondolf 2000)

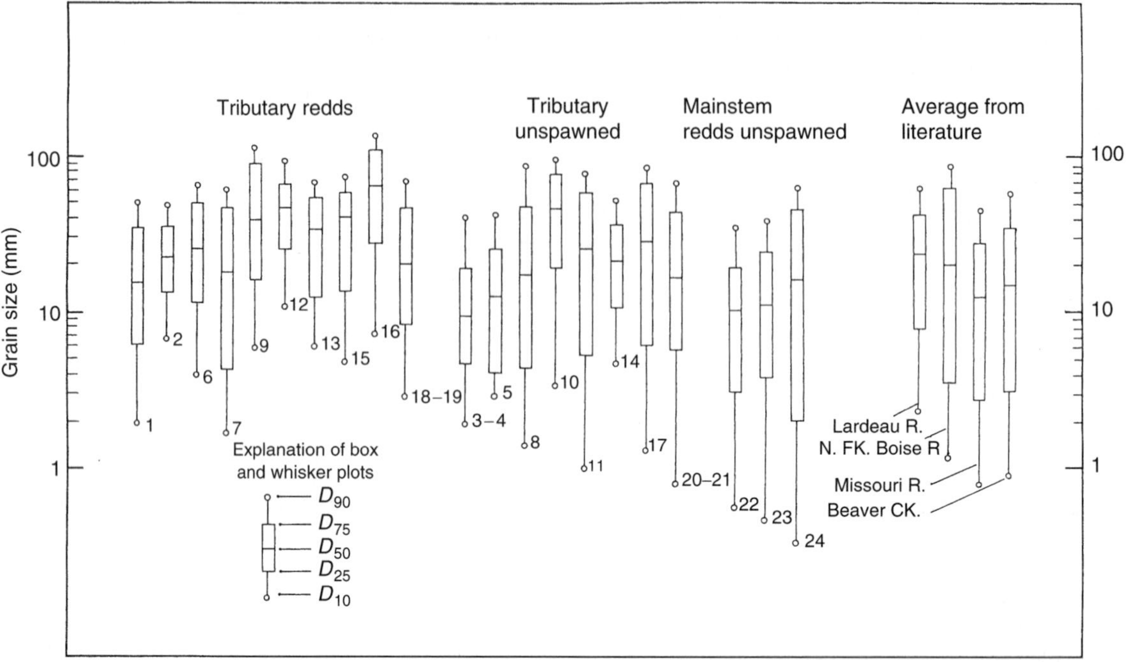

Figure 13.3 Box-and-whisker plots for rainbow trout (*Oncorhynchus mykiss*) spawning gravels from the case study in the Colorado River and tributaries downstream of Glen Canyon Dam, and averages for other rainbow trout spawning gravels (Table 13.5). For each sample, the rectangle (box) encompasses the middle 50% of the sample, from the D_{25} to D_{75} values, termed the 'hinges'. The median diameter, D_{50}, is represented by a horizontal line through the box. Above and below the box are lines (whiskers) extending to the D_{90} and D_{10} values. Numbers refer to samples in Kondolf *et al.* (1989). Box-and-whisker plots are easier to read than cumulative size distribution curves when multiple, similar distributions are plotted on the same graph (reproduced from Kondolf 2000)

whiskers extend out to extreme values. In the case of sediment size distributions, the range of sizes from gravel to clays is so great that it is not practical to plot whiskers to the extremes, so the D_{90} and D_{10} values are often used to capture the range of most (80%) of the size distribution.

Three-component mixtures of sediment "... can be represented by a equilateral triangle diagram in which a single point is a graphic representation of the proportions of the three components" (Pettijohn 1975: p. 32). These ternary diagrams have been most commonly used by sedimentologists and soil scientists to represent mixtures of sand, silt, and clay.

Statistical Descriptors

To facilitate comparison among size distributions, statistics are commonly drawn from the curves for comparison. For example, the median particle diameter, D_{50}, is commonly used in hydrology, geomorphology, and engineering as a measure of central tendency of the distribution because it is easily read and unambiguously interpreted (Inman 1952, Vanoni 1975). Gravel size distributions tend to resemble lognormal, gamma, or Weibull distributions rather than normal distributions (Kondolf and Adhikari 2000). Otto (1939) noted the resemblance of grain size distributions of sediments to the lognormal distribution and as a measure of central tendency, proposed the geometric mean in lieu of an arithmetic mean. The geometric mean can be calculated using D_{16} and D_{84}, based on an analogy to the normal distribution, in which the 16th and 84th percentiles lie 1 standard deviation from the mean (Table 13.1). (Formulae for geometric mean, sorting, skewness, kurtosis based on more values from the size distribution are presented in Folk (1968) and Griffiths (1967).) The geometric mean is thus a measure of central tendency complementary to the median diameter and more influenced by extremes of the distribution.

Table 13.1 Size descriptors commonly drawn from sediment size distributions (*Source*: Kondolf and Wolman 1993)

Measure of:	Quartile-based descriptors	Descriptors based on D_{16}, D_{84}
Central tendency	Median, i.e., D_{50}	Median, i.e., D_{50} Geometric mean (Otto 1939) $D_g = [(D_{84})(D_{16})]^{0.5}$
Dispersion	Trask sorting coefficient (Trask 1932) $\mathrm{si} = (D_{75}/D_{25})^{0.5}$	Geometric sorting coefficient (Otto 1939) $\mathrm{sg} = [(D_{84})(D_{16})]^{0.5}$
Skewness	Quartile coefficient of skewness (Krumbein and Pettijohn 1938) $\mathrm{SK} = [(D_{75}D_{25})/(D_{50})^2]^{0.5}$	Geometric skewness coefficient (Inman 1952) $\mathrm{sk}{=}\log(D_g/D_{50})/\log(\mathrm{sg})$
Peakedness	Kelley's quartile kurtosis (Krumbein and Pettijohn 1938) $\mathrm{KR} = (D_{75} - D_{25})/2\,(D_{90} - D_{10})$	Geometric kurtosis (Inman 1952) $\mathrm{Kr} = \log\,[(D_{16}D_{95})/(D_{05}D_{84})]\ \log(\mathrm{sg})$

Other commonly reported attributes of size distributions are sorting and skewness. Sorting, or dispersion, refers to the degree of concentration or dispersion among the particles. Sorting in fluvial gravels is the process by which particles of a given size are concentrated. In geologic parlance (as followed here), 'well-sorted' means of similar size. In engineering usage, the same term may be used for a well-dispersed size distribution. In downstream reaches of larger river systems, currents may deposit bars composed entirely of gravel, other bars entirely of sand. These deposits would be considered well sorted, or having low dispersion. The Trask (1932) sorting coefficient is based on quartile values drawn from the size distribution and has been used widely in geological studies, but has been largely replaced by the geometric sorting coefficient, sg of Otto (1939) (Table 13.1), based on D_{16} and D_{84}. Both of these sorting coefficients increase with dispersion (and thus decrease with sorting).

Skewness (sk) refers the degree to which the distribution is skewed from a normal or lognormal distribution. Again, there are skewness coefficients based on quartiles and based on D_{16} and D_{84} (Table 13.1). When plotted on an arithmetic scale, gravel size distributions tend to be positively skewed (i.e., the coarse tails extend farther than the fine). However, log-transformed distributions (which are most commonly seen) tend to be negatively skewed, so the geometric mean diameters tend to be less than median diameters (Kondolf and Wolman 1993). Kurtosis refers to the peakedness of the distribution, and can be calculated using D_{10}, D_{25}, D_{75}, and D_{90} (Kelley's quartile kurtosis in Krumbein and Pettijohn 1938) or using the D_5, D_{16}, D_{84}, and D_{95} (Inman 1952) (Table 13.1).

Arguing that fluvial gravels were not lognormally distributed (and certainly many are not), Beschta (1982) suggested that use of measures derived by analogy to the lognormal distribution (D_{16}, D_{84}) were not valid and (with Lotspeich and Everest 1981, Shirazi *et al.* 1981) proposed calculating moment measures as an alternative. The moment measures of traditional statistics are defined in terms of moments about the origin (for the first moment, the mean) and moments about the mean (for higher moments) of individual observations. Grain size distributions are continuous curves, and formulae for computing moment measures for continuous distributions are well known. But these formulae are not directly applicable to sediment size distribution from bulk samples, because, in practice, data on grain size distributions are obtained from a limited number of observations (sieves) and reported in percent weight, and thus, for example, the extreme values are unknown.

The first moment, the mean, cannot be computed for sediments as the simple average of *n* observations. The most closely analogous approach is to compute a weighted average from the percentage in each size class and the size at the midpoint of that class. If an arithmetic mean is desired, the arithmetic midpoint would be used; if the geometric mean is desired, the geometric midpoint is used (e.g., Shirazi *et al.* 1981). This approach is somewhat cumbersome, especially if applied to data collected in different studies, as the selection of sieve sizes has varied enormously among investigators, and the intervals between sieves have also not been consistent across studies. Moreover, as noted by Inman (1952: p. 126), "... the dependence of moment measures on the entire size distribution is a limitation on their practical application to sediments, since mechanical analyses of sediments frequently result in open-ended curves, and thus do not give the coarse and fine limits of the distribution...These facts have resulted in very limited use of these parameters in the description of sediments". Thus, unless size distribution curves are precisely defined at the tails, moment measures cannot be computed, strictly speaking.

Accordingly, the usual practice is to draw values from the cumulative size distribution and compute measures from them (Table 13.1). The Trask (1932) measures are based on the central 50% of the distribution only, and thus they fail to capture important features of many distributions (Inman 1952). The geometric mean, sorting index, skewness, and kurtosis measures proposed by Otto (1939) and Inman (1952) are based on percentile values that encompass more of the distribution, so they can be regarded as providing better measures of many attributes of the size distributions. As a measure of central tendency, however, the median size, D_{50}, is arguably the best, as it is least affected by the tails. Given that some measures of mean, sorting, and skewness must be used even when the distributions are not lognormal, these descriptors still serve as useful measures of attributes of the grain size distribution, they are widely used, and they were recommended as standard measures by Vanoni (1975).

The oft-debated question of whether fluvial gravels are lognormal or 'Rosin' distributed may not be as important as implied by the literature devoted to it (e.g., Ibbeken 1983). The Rosin distribution is actually the same as the Weibull distribution, well studied in the mathematics and statistics literature. The lognormal, Weibull, and gamma distribution are quite similar, and it can be difficult to distinguish which is the best fit to a given gravel distribution (Kondolf and Adhikari 2000). One test that has been used in the literature is to plot the cumulative curve of a sample on a lognormal or Weibull-transformed scale and determine the r^2 value of the fit (e.g., Beschta 1982, Ibbeken 1983). However, because these distributions are so similar, high r^2 values are obtained even when an ideal Weibull (Rosin) distribution is plotted on a lognormally transformed axis (Kondolf and Adhikari 2000). More fundamentally, the relation between the underlying mathematical properties of the different distributions and the physical mechanisms giving rise to them is not clear.

The choice of descriptor to use depends largely on the purpose of the study. For example, D_{50} seems useful in transport relations because its entrainment threshold appears to be constant, for a given range of slope, for varying degrees of sorting. D_g was used by Parker (1990) for the same purpose. Note that while D_g is usually close to D_{50}, the difference between these two measures increases with skewness. As discussed in Section 13.10, both D_{50} and percent finer than 1 mm are useful descriptors to assess the quality of salmonid spawning gravels.

13.3 PARTICLE SHAPE AND ROUNDNESS

Grain shape varies widely, reflecting properties of the source rock and subsequent weathering and abrasion. Zingg (1935) recognized four basic shape classes, each detailed by ratios of the principal axes as shown in Figure 13.4(a) and defined in Table 13.2, using the axes as defined above. Grain roundness is a distinct concept from shape: it is the degree to which edges and corners of rock fragments have been removed by weathering and abrasion. The forms shown in Figure 13.4(a) can exist in angular (unrounded) form as shown, or they can be transformed into rounded condition. Rounding has conventionally been assessed visually using charts such as shown in Figure 13.4(b). Roundness is an important property of sediment, it provides a quick and useful indication of the distance and duration of transit from the source rock. However, there is no simple relation between grain roundness and distance of transit from the source rock because of differences in resistance among lithologies, because rocks can round as they weather in place in floodplain and terrace deposits, and because rounded clasts may be recycled.

13.4 SURFACE VS. SUBSURFACE LAYERS IN GRAVEL-BED RIVERS

The surface layer of a gravel bed is commonly coarser than the underlying, subsurface layers (Figure 13.5). The size distribution of the subsurface gravel is commonly similar to that of the transported bedload (Parker and Klingeman 1982) or somewhat coarser (Leopold 1992, Lisle 1995). The framework grains of the surface are generally not larger than those of the underlying sediment, but the surface layer is typically deficient in the finer fractions of the distribution. In part, this can be explained by selective transport of finer grains exposed on the surface at flows too low to mobilize the entire bed. The paucity of interstitial fine sediment in the surface layer implies that while framework size can be estimated by sampling the surface

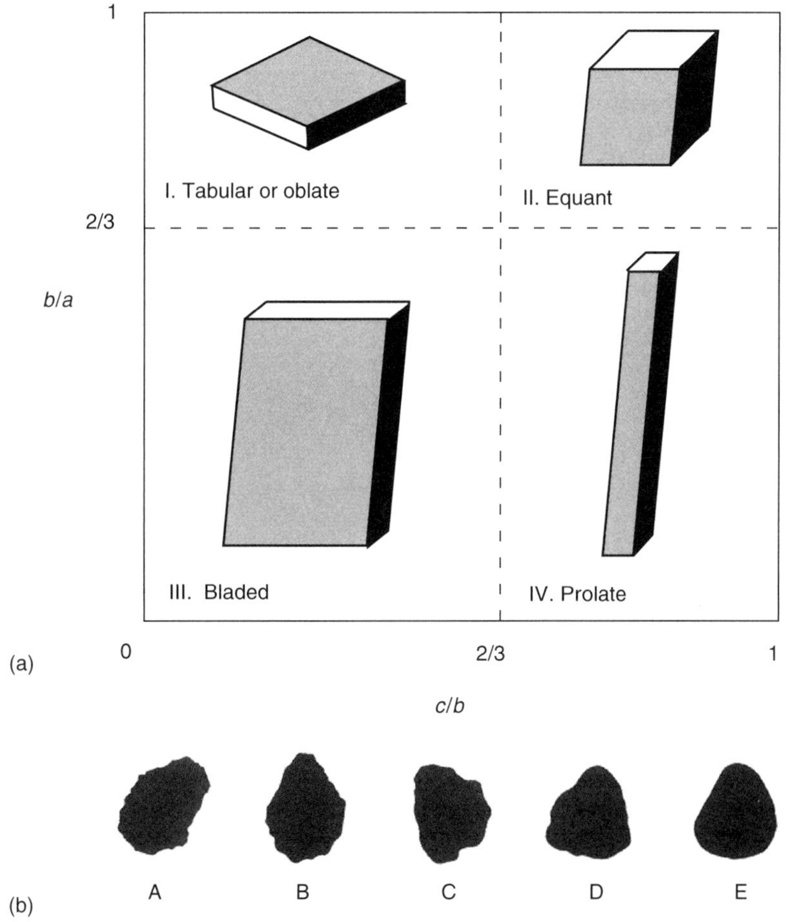

Figure 13.4 Grain shape and roundness. (a) Zingg shape classes, plotted by ratios of principal axes. (b) Roundness classes: A, angular; B, sub-angular; C, sub-rounded; D, rounded; E, well-rounded (modified from Pettijohn 1975)

Table 13.2 Zingg shape classes (adapted from Pettijohn 1975) (reproduced by permission of HarperCollins)

Class number	Ratio of axes		
	b/*a*	*c*/*b*	Shape
I	$>\frac{2}{3}$	$<\frac{2}{3}$	Oblate (discoidal)
II	$>\frac{2}{3}$	$>\frac{2}{3}$	Equiaxial, or equant (spherical)
III	$<\frac{2}{3}$	$<\frac{2}{3}$	Triaxial (bladed)
IV	$<\frac{2}{3}$	$>\frac{2}{3}$	Prolate (rod-shaped)

layer, matrix assessment requires subsurface sampling.

However, some coarse surface layers are active features in that they persist (or reestablish) despite frequent mobilization of the bed and are common features of gravel beds with active sediment transport. These surface layers were termed *pavements* by Parker and Klingeman (1982), as distinct from inactive coarse surface layers that result from the progressive winnowing of finer fractions from the surface layer, as might be encountered below a dam, which they termed *armor layers*. Gomez (1984) proposed similar distinctions, but argued that the terms should be used in an opposite fashion: armor for active surface layer,

Figure 13.5 Surface and subsurface texture in gravel bar in the Garcia River, California (photo by Kondolf, April 1992)

pavement for inactive surface layer, by analogy to the fact that armor (like medieval armor) can be removed, while pavement (like road surface) cannot. Our partial, non-random sampling of recent discourse suggests that a number of terms are being used, including 'active armor'. Many avoid the issue by identifying a 'coarse surface' layer, an approach with the virtue of avoiding inferences of degree of mobility when there are no direct observations of such.

Differences between surface- and subsurface grain sizes can help to evaluate variations in bed mobility associated with sediment supply (Dietrich *et al.* 1989, Buffington and Montgomery 1999b, Lisle *et al.* 2000). Increases in the bed-material load result in an increase in transport intensity at a given flow magnitude. This is accommodated by a decrease in surface particle size approaching that of the load or subsurface, but the adjustment is mediated by the magnitude of boundary shear stress exerted on the bed. The foregoing references detail the methods used to measure indices (e.g., Shields stress q^*) that are based on this adjustment. The index, q^*, is a ratio of bedload transport rate predicted from the particle size distribution of the load (commonly represented by the subsurface material) to that predicted from the size distribution of the bed surface, given a reference boundary shear stress (Dietrich *et al.* 1989). As armoring decreases, values of q^* generally increase towards 1, with the value of 1 signifying the absence of armoring. Shields stress is more commonly used to scale bed mobility (Yalin 1977). It is the ratio of the forces of traction and gravity acting on a representative size fraction (usually D_{50}) in a river bed and, more precisely $\tau^* = \tau / RD_{50}$, where τ^* is boundary shear stress and R is submerged specific gravity of sediment. For both indices, boundary shear stress is conventionally evaluated at bankfull stage. Computing Shields stress requires values of surface particle size and boundary shear stress; computing q^* requires values of surface particle size, bedload or subsurface particle size, and boundary shear stress.

In fish habitat studies, the distinction between surface and subsurface populations has not always been acknowledged. If habitat for fry or aquatic insects is the concern, then the surface population should be sampled. If intragravel condition for incubating salmonid embryos is the concern, then the subsurface population should be sampled (Kondolf 2000). If the size of framework gravel selected for spawning by a species is the concern, then either surface or subsurface sampling will yield reasonable results, unless the bed is so armored that the two populations are extremely different. In completed salmon redds (nests containing incubating salmon eggs), the surface and subsurface populations have already been mixed, although coarser gravels may have been concentrated at the base of the redd as lag deposits in the egg pocket.

13.5 SAMPLING AND ANALYZING SAND AND FINER-GRAINED SEDIMENT

Sampling sand and finer-grained sediments is relatively straightforward, so long as there is access to sample sites. If grain size is needed for input to transport equations, sand can be sampled from the exposed bed or bars with a shovel or trowel, from underwater sites (and under high-flow conditions) with various bed-material samplers including drag bucket, grab bucket, and vertical pipe type samplers (see Vanoni 1975: pp. 334–337, for a description). For these sediments, a liter is generally an adequate sample size. The number and location of samples will depend on the spatial homogeneity of the bed material.

The stratigraphy of channel deposits (the vertical and horizontal arrangements of sedimentary layers) and their grain size, lithology, etc., can be studied to yield information on depositional history and channel hydraulics, as described in Chapter 2.

Samples of sand (with particles up to fine gravel in size) are typically sieved in the laboratory to obtain particle size distributions. The samples are dried and passed through a series of progressively finer mesh sizes, and the amounts retained on each sieve are weighed to develop a particle size distribution for the sediment. Commonly, large field samples of sand and finer sediment are split into smaller subsamples for size analysis, as 100 g is an adequate sample size for sediments whose largest particle is smaller than 5 mm (e.g., see Bunte and Abt 2001b: p. 292). To measure particle size of sediments finer than sand requires using sedimentation apparatus, such as settling tubes, elutriation (washing lighter particles off, leaving heavier particles behind), and centrifuge separation (Figure 13.6) (Pettijohn 1975, Vanoni 1975).

13.6 METHODS OF SAMPLING OR DESCRIBING THE SURFACE OF GRAVEL BEDS

Methods for sampling the surface of gravel beds can be divided into those that sample the surface or subsurface. Surface methods include facies mapping, visual estimates, pebble counts, and photographic methods. Subsurface methods include shovel samples (from exposed bars), core samples collected with cylindrical samplers driven into the bed, core samples obtained by freezing interstitial water around a probe in the bed, potentially large samples obtained by backhoe, and those obtained by dredge.

Facies Mapping

The term *facies* refers to sedimentary deposits distinct in grain size and/or sedimentary structure, representing distinct local depositional environments (Pettijohn 1975). Others use a less sedimentologic term, *patch*, to refer to a mapable area of a stream bed that can be delineated on the basis of surface particle size (Seal and Paola 1995). These terms are basically synonymous and apply to areas whose long dimension is on the scale of a large fraction of channel width, or larger. We use *facies* in this chapter, but do not endorse either term. A facies may consist of a mixture of poorly sorted grains of many sizes, but it should be consistently so over the entire patch. Usually, more than one facies can be visually distinguished in a reach (on exposed bars or in clear water) and mapped. This approach can be used not only to distinguish sand from gravel, but also to distinguish different gravel–sand mixtures that may be present in the channel. Field trials described in the literature indicate that observers are capable of visually distinguishing distinct facies (Shirazi and Seim 1981), and in fact that visual identification tended to over-estimate differences in grain size among different facies of gravel–sand mixtures (Kondolf and Li 1992).

Facies maps can be extremely useful tools as descriptors of current conditions (from which relative proportions of different sized units can be measured), as baseline data against which to measure future change, or as a basis for comparing sediment conditions among channels. Their primary advantage is that they capture reach-wide variations in surface size and are not subject to the greater variability commonly encountered at smaller channel scales (e.g., pebble counts at cross sections), which can be affected by changes in channel morphology.

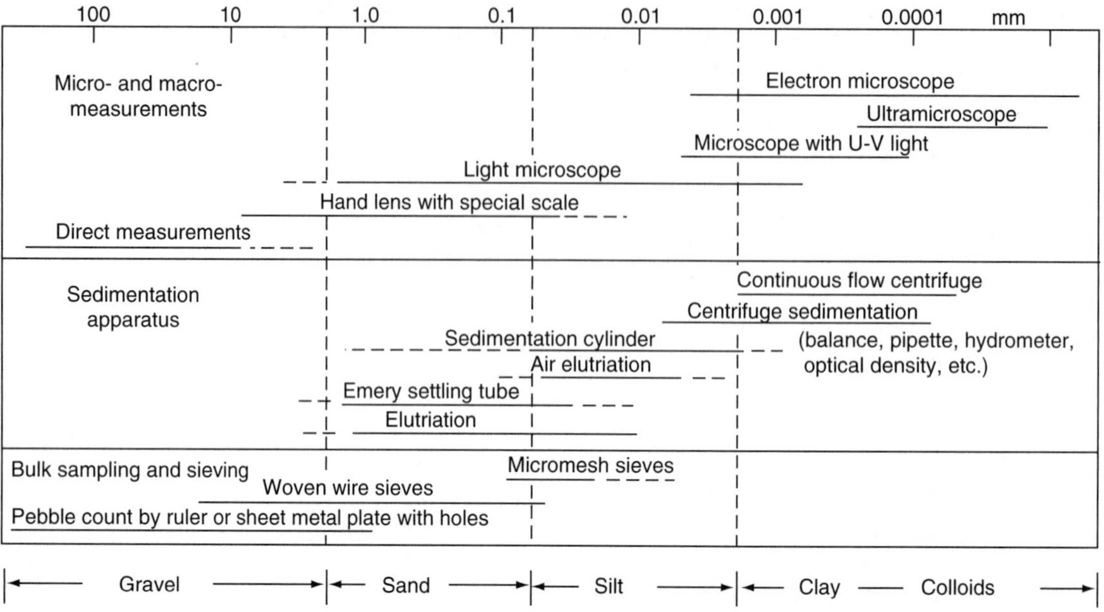

Figure 13.6 Methods of size analysis by grain size. Unlike traditional size distribution analysis in which a sample is taken back to the laboratory for analysis, the pebble count (Wolman 1954) involves direct measurement of particle size with a ruler or passing through holes in a sheet metal plate (a form of sieving) (modified from Pettijohn 1975)

Moreover, facies maps coupled with channel morphology are effective for interpreting channel processes. For example, Wolman and Schick (1967) used facies maps to document the influence of construction-derived sediment on Oregon Branch north of Baltimore, Maryland (Figure 13.7). Beaverdam Run, a tributary that was unaffected by the recent construction, consisted of cobble riffles and pools, with small deposits of fine humic sediments in convex portions of bends. By contrast, Oregon Branch below the construction area contained silt and sand deposits throughout, some as thick as 60 cm (Wolman and Schick 1967).

Buffington and Montgomery (1999a) developed a system for classifying textural patches using a number of possible ternary diagrams. To apply this approach, one first identifies the three dominant size categories present (e.g., silt, sand, gravel, cobble, boulder), ignoring any component less than 5% of the deposit, and then classifies the deposit based on the relative abundance of the three sediment types, as determined by visual estimation. For example, a deposit with more cobble than sand and more sand than gravel would be a 'gravelly sandy cobble'. As a second step, the particle size distributions of several facies patches are determined by pebble count (discussed below) or other sampling method.

The Pebble Count, or Grid Sampling

The pebble count (Wolman 1954) is a sampling of approximately 100 grains (stones) on the river bed (or gravel bar), on a grid or line. The pebble count can be conducted on an exposed gravel bar, by wading in shallow water, or, in greater water depths, by diving (Klingeman and Emmett 1982). The stone measured at each sample point is selected randomly by closing (or averting) the observer's eyes as the finger falls to the bed (to avoid bias to larger particles). As an alternative to a grid, sampling points can be selected by picking grains encountered in front of the observer's boots at regular intervals as one proceeds across the bed. So long as the two methods yield truly random samples, they yield equivalent results in terms of stones selected. However, Marcus *et al.* (1995) noted that when sampling underwater some operators favored larger particles because their fingers were displaced from a vertical descent by the current. Bunte and Abt (2001a) proposed use of a

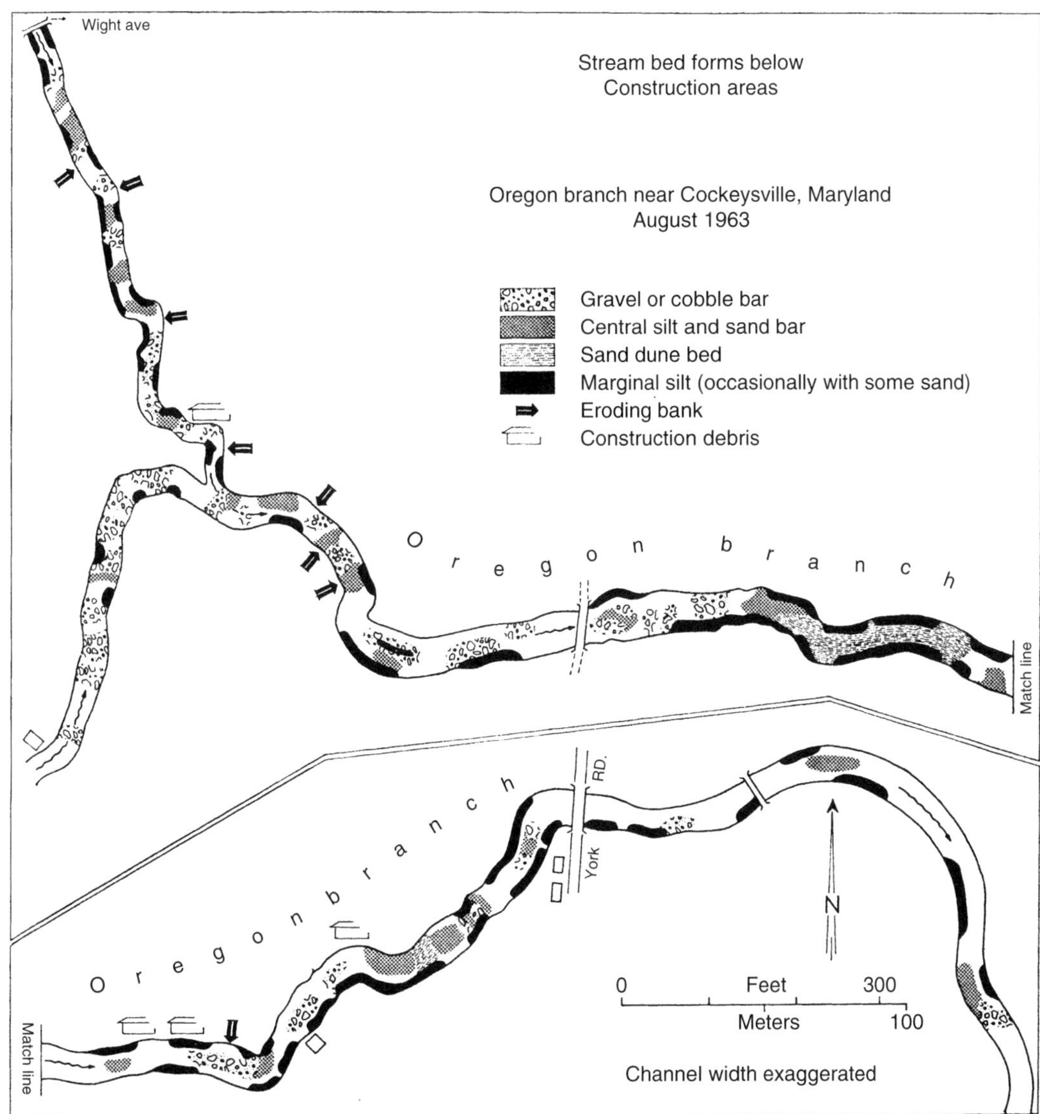

Figure 13.7 Facies map for Oregon branch, near Cockeysville, Maryland, showing extent and pattern of silt and sand deposits downstream of rapidly eroding construction area, in contrast to the limited fine-sediment deposits in the channel of Beaverdam Run, a tributary unaffected by recent construction activity (reproduced from Wolman and Schick 1967)

60 cm × 60 cm frame with an adjustable wire grid to locate the sample points more objectively.

The intermediate axes of the stones are measured either with a pocket ruler and recorded within predetermined size classes, or passed through a template in which squares have been cut in the sizes of the grain size classes, analogous to sieve openings (Hey and Thorne 1983). For well-rounded stones, the two methods are virtually equivalent. However, flattened, elongated clasts can pass through square template openings with a diameter smaller than their actual *b*-axis length, because they can orient diagonally through the openings, resulting in smaller measured sizes for the same stones (Church *et al.* 1987).

Template measurements are more comparable to sieve measurements and less prone to observer error in identifying the *b*-axis to measure. Sieve size may have more utility than the *b*-axis in relating to physical processes; empirical bedload transport formulae, for example, are scaled in part by sieve size, as distinct from the *b*-axis. The *c*-axis, which is usually aligned roughly normal to the bed surface, is more closely related to hydraulic interactions than the *b*-axis, which is aligned roughly parallel to the bed. Thus, we recommend using templates where possible. In any event, when reporting methods for pebble counts, the method used should be clearly stated, and the shape of particles can be described or quantified using standard sedimentological particle shape factors (Pettijohn 1975).

As the stones are sampled, their sieve (or *b*-axis) sizes are recorded in grain size classes (in mm) that increase by powers of $2^{0.5}$, also termed 'half-phi' classes after the phi scale of Krumbein (1934) described earlier. The grains are recorded on the row identified by the lower end of the size range, by analogy to sediments collected on sieves. A 2-mm sieve collects all grains smaller than the next largest sieve size (4 mm for sieves following the phi scale) but larger than 2 mm. Similarly, in a pebble count, a stone with intermediate axis of 52 mm would fall in the 45 mm class because it is smaller than 64 mm but larger than 45 mm. It is not possible to accurately measure sand and very small gravel by hand, so the smallest fraction is lumped together as fines below a certain size. When the finger encounters sediment finer than 4 mm, it is recorded simply as '<4 mm' (Wolman 1954). Other cutoffs can be used, such as 8 mm (Kondolf 1997). The results are recorded in the field book as tick marks, yielding a histogram in the field book (Figure 13.8). A 52-mm stone would be recorded by a tick on the row labeled '45 mm'; a 44-mm stone would be recorded by a tick on the row labeled '32 mm'. The total in each class divided by the total sample number (easily calculated on a spreadsheet, hand calculator, slide rule, or by arithmetic) yields cumulative percentages for each size class. These are the cumulative percentages *finer* than the next largest class size (i.e., the class size immediately above it on the table), analogous to the total retained on a sieve of that size and all finer sieves. In the example shown in Figure 13.8, adding the percentages from the bottom up yields 33.6% on the 45 mm row. This is the cumulative percentage finer than 64 mm. Grain sizes can be repeated in a column to the right of the cumulative percentages, with the sizes shifted downward by one row so the adjacent columns can be easily read as 'Cumulative percent finer than *size*'.

In some river beds, stones may be interlocked in the gravel matrix and difficult to remove. This is especially a problem with larger stones in armored cobble beds, or with boulders. In such cases, it may be impractical to remove every stone for a complete inspection and accurate measurement, so the observer can partially excavate the stone and estimate the size with a ruler. With the large sizes, the particle size classes are so widely separated in size (arithmetically) that often one can be fairly certain of the size class of most such embedded particles unless they fall near the boundary between classes. In the field notes, such an embedded particle can be recorded with an 'E' (instead of a tick mark) in the appropriate size class (Figure 13.8). The percentage of embedded stones in the sample can be calculated, providing a rough measure of the degree to which bed particles are interlocked, as well as a measure of this source of error in sampling (Kondolf 1997).

The pebble count is conducted over a patch of gravel comprising a single facies or population, 'a zone or area considered homogeneous' (Dunne and Leopold 1978: p. 666). If only one facies can be distinguished in a reach, the grain size distribution can be applied to the entire reach. For large homogenous areas, such as gravel bars on large rivers, which can be tens or hundreds of meters long, the pebble count can be conducted over part of the bed and its result applied to the entire previewed, homogeneous, feature, or the pebble count can be conducted such that its sampled particles are drawn from over the entire feature, with the distance between sampled pebbles commensurating with sample size. The first approach is based on the identification of the entire feature as 'homogeneous' and the corollary that a subsample should be representative of the entire population. The second approach defines the sampling universe (facies) and attempts to give each individual (pebble) an equal probability of being sampled. For one facies, the two approaches should be equivalent, but if there are systematic, undetected variations over the facies the second approach may more accurately represent the entire facies.

In the case where two or more distinct facies exist in the reach to be characterized, the different facies can be mapped and measured, and separate pebble counts conducted on each. If a composite grain size distribution for the entire reach is desired, the proportions of the bed occupied by each distinct bed-material facies can be measured, pebble counts conducted on each,

Rush Creek at IFIM site 8/28/87

XS 36 GMK, SSC

AVK

Stratified populations A,B,C (sand)

Population 'A'					
Size class (mm)	Number of rocks	Number of rocks	%	Cum %	Finer than (mm)
256	II	2	1.6	100	360
180	IIIE	5	4	98.4	256
128	IIII IIEE IIII E	16	12.8	94.4	180
90	IIII IIII EIII IEII IEEE IEII	30	24	81.6	128
64	IIII IIII EIII IIII IEEI EIII	30	24	57.6	90
45	IIII IIII IIII IIII	19	15.2	33.6	64
32	IIII I	6	4.8	18.4	45
22.6	IIII II	7	5.6	13.6	32
16	IIII	4	3.2	8.0	22.6
11.3	II	2	1.6	4.8	16
8	I	1	0.8	3.2	11.3
< 8	III	3	2.4	2.4	8
		$\Sigma = 125$			
		$\Sigma > 8$ mm = 122			

Total embedded = 14, all > 64 mm
= 11% of all rocks

Figure 13.8 Field notes from a pebble count on Rush Creek, California, showing field-generated histogram. Note sample points on embedded rocks designated by E's instead of regular tick marks. As per convention, initials of persons measuring stones are indicated by a stick figure pictograph, the note taker by a book pictograph

and a weighted average grain size distribution computed (Figure 13.9).

To address the pebble count's inability to sample particles smaller than about 4 mm, Fripp and Diplas (1993) and Petrie and Diplas (2000) proposed a hybrid technique in which the pebble count sample is truncated around 10 mm and the fine tail is defined by a completely different technique, areal sampling by adhesion to clay, and the two samples are merged.

Visual Estimates

Visual estimates, often grandly termed 'ocular assessments', involve estimating, by eye alone, the sizes of substrate particles, or estimating percentages in different broad size classes. Casual descriptions of bed-material size such as '2-inch gravel' are, in effect, visual estimates.

Visual estimations of grain size have been used more formally and systematically by fisheries biologists, and are the basis for many of the published descriptions of gravel sizes used for salmonid spawning, typically reported as a range of sizes preferred by the studied species, such as '1-to-4-inch gravel' (e.g., Greeley 1932, Hazzard 1932, Cope 1957, Hunter 1973). Visual estimates are the basis of the substrate code used in the Instream Flow Incremental Methodology (IFIM) developed by the US Fish and Wildlife Service Instream Flow Group (IFG) (Bovee 1982). The first, generalized version of this code used a scale from 1 to 8 in ascending order of coarseness: plant detritus, clay, silt, sand, gravel,

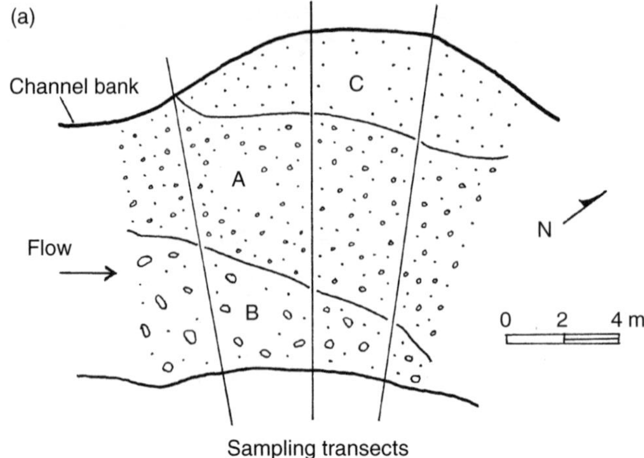

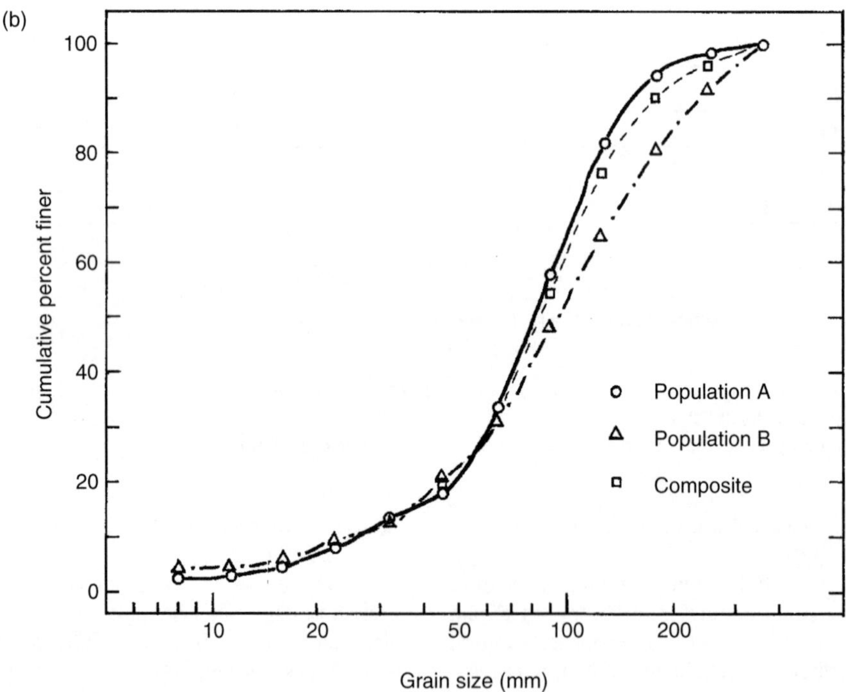

Figure 13.9 Bed material at a sample site on Rush Creek, California. (a) Facies map of distinct gravel–sand mixtures and sand deposits identified on the bed. The middle cross section shown was a sampling transect used in an aquatic habitat study (Kondolf and Li 1992). The adjacent two cross sections were established to provide enough sampling points for grid-based pebble counts. (b) Grain size distributions for facies A and B, and composite distribution for the two based on their respective bed areas in the study reach (reproduced from Kondolf 1997)

cobble, boulder, and bedrock, with intermediate sized substrates reported using decimals, such as 5.2 to designate gravel with 20% cobble. The code was later revised to incorporate more finely divided substrate classes and embeddedness (Bovee 1982). The Alaska Fish and Game Department modified the code to account for upwelling in studies on the Susitna River system (Vincent-Lang *et al.* 1984), and Wilson *et al.* (1981) measured gravels in the field with a scale but still estimated the relative proportions of different grades visually. The IFIM method is widely used, so visual estimates of this kind are among the most widely applied methods of bed-material sampling.

Despite the widespread use of visual estimates, we are unaware of any systematic studies demonstrating that these subjective estimates of percentages of various size classes in the bed are reproducible among different investigators. Moreover, even if these estimates are accurate, the results are usually reported in the form of a range of sizes (e.g., 1-to-3-inch gravel), 'dominant' and 'subdominant' size class, or as percentages of classes such as '80% cobble, 10% sand, and 10% silt'. Thus, these estimates are not readily compared with sediment sizes reported in the engineering and geomorphic literature, in which statistics are drawn from standard size distributions.

Mapping the parts of the bed occupied by distinct facies and conducting adequately sized pebble counts on each facies is an approach that can be used to provide reproducible grain size information for aquatic habitat studies, in lieu of visual estimates at many points. For a bed with numerous facies present, the pebble count approach will be more time-consuming than visual estimate at each point, but for a bed with only one facies, the pebble count approach is faster than visual estimates at many points and in both cases offers the advantage of yielding data consistent with those reported in the engineering and geomorphic literature (Kondolf and Li 1992). The Buffington and Montgomery (1999a) approach to mapping textural facies described above could be useful in these applications.

Photographic Grid Methods

Photographic methods for determining grain size distributions have been described in several papers in the geomorphic literature. Kellerhals and Bray (1971) photographed a 0.25-m^2 area of gravel bed under a grid, measured particles under grid points on the photograph and in the field, then obtained bulk samples. Sizes measured from photographs were somewhat smaller than those obtained by field measurements or sieving, an effect attributed to partially obscuring or foreshortening of the *b*-axis in the photographs.

Adams (1979) measured particles under grid points on photographs, then spray painted the bed surface and collected painted particles for sieving. He found mean sizes measured from photographs to be slightly larger (by about 9% in phi units) than sieve results. However, sieving of all surface particles is an area-by-weight method, which is not equivalent to the grid-by-number method employed for the photographic method (Kellerhals and Bray 1971). Ibbeken and Schleyer (1986) described procedures for 'photo-sieving' whereby the surface of the deposit is photographed, and the image is enlarged and digitized, yielding "almost [exact] sieve equivalents even for populations with very poorly sorted shaped distributions".

Church *et al.* (1987) found a mean negative bias of 22% (phi units) in comparisons of median sizes determined by grid on photographs and field measurements (grid-by-number) from samples obtained from a range of deposits on a single gravel bar. They attributed the bias to:

1. partial hiding, a variable factor that must be empirically determined for each study site, and
2. angle of imbrication (the downstream tilt commonly observed in individual grains in natural stream bed gravels).

Imbrication angle has been shown to vary with clast size, shape, and sphericity (Church *et al.* 1987).

In fisheries literature, Burns (1978) found that for Chinook salmon gravels with 18–20% fine sediment (by analysis of bulk samples), photographic estimates ranged from about 7% to 33%. However, his bulk gravel samples ranged sixfold in weight, implying a similar range in depth of sampling; so the bulk samples probably also reflected the different surface–subsurface mixtures. Chapman *et al.* (1984) photo-estimated the percentage of gravel larger than 132 mm and smaller than 76 mm in chinook salmon spawning grounds on the Columbia River at deep-water sites where other approaches might be impractical, and at sites where bulk samples were obtained, but reported no comparison of the two methods.

Recent efforts to use remote sensing and image processing analysis to estimate grain size in exposed gravel beds using a filtering program to identify individual grains and measure their *b*-axes are promising (Lane 2002). The challenge now is to calibrate the techniques under conditions of different grain size, sorting, lithology, particle shape, particle imbrication, and to develop protocols for grain sizing underwater facies and particle lower than 8 mm. Butler *et al.* (2001) developed a protocol based on recognizing particle boundaries from grey thresholding and watershed segmentation algorithm. They emphasized that orthorectified imagery is not necessary but grey thresholding calibration is required with reference to grain size measurement from redrawn particles on the photos. Moreover, such procedures also require reference field measures in order to correct imagery percentiles, which tend to be underestimated because of particle characteristics (see discussion of photo-sieving above).

Recent work on the gravel bars of the Ain River indicate that grain size percentiles calculated on photos from particle redrawing are underestimated compared to field-measured percentiles (paint and pict method) because of partial burial of particles in the structure (Figure 13.10a). Automatic image extraction tended to underestimate percentile values when compared with percentiles based on redrawing particles, whether thresholds were set visually or based on histogram breaks. The percentile error between photo and field observations is greater than that measured between automated and manual grain measurement on photos, with some imaged–extracted percentiles underestimated, others overestimated, depending on sedimentary structure. Thus, automatic measurement of percentiles from photos can be a good techniques for calculating field-measured percentiles (Figure 13.10a, right diagram), bearing in mind the three main sources of errors in particle boundary recognition:

(i) coarse particle sharing because of surface roughness of particles,
(ii) fusing of neighboring particules, and
(iii) incorrect subdividing of large into small particles in sandy areas or on the shaded borders of large particles (Figure 13.10b).

The errors are not random but depend on sedimentary facies, with the first type error mostly occurring in coarse, poorly sorted sediments, the second in fine, poorly sorted sediments, and the third in well-sorted, coarse-grained sediments (Piégay, personal communication, 2002; Rollet *et al.* 2002). Errors can thus be estimated by standard regression as a function of the sorting and percent finer than 8 mm in the deposit.

The principal advantages of photographic size analysis are the preservation of the image of the bed surface and saving effort in the field. These advantages were exploited by Klein (1987), who used digitized images to monitor changes in surface bed material at 24 sites in the Redwood Creek Basin; over 13 000 grid sampling points were used in the study, a task made considerably easier by digitization. As discussed above, the principal disadvantage of the method is the potential bias in measurement of axes from the photographic image.

13.7 SUBSURFACE SAMPLING METHODS

Bulk Core Sampling

Bulk core sampling involves directly removing a sample from the bed, usually within a predetermined area and down to a predetermined depth. Gravel exposed on a bar can be easily sampled by shovel, or better yet for adequate sample size, a backhoe. In flowing water, bulk samples are commonly obtained by driving a cylindrical core sampler into the bed and removing (by hand) the material within, geomorphologists have used bottomless 50-cm oil drums in various forms to obtain sufficiently large samples, such as the 140–240 kg samples collected by Wilcock *et al.* (1996b). The 'cookie-cutter' sampler is a 50-cm drum sampler with an underwater sample box with mesh screen to collect fine material washed downstream (Klingeman and Emmett 1982), and the 'barrel' sampler is a 46-cm drum sampler fitted with a 152-cm long hood of filter mesh to collect fine sediment (Milhous *et al.* 1995).

When removing the gravel from drum samplers, it is possible to remove the surface layer first and analyze it separately. Given the difference between surface and subsurface layers, we recommend that these layers be sampled and, if warranted by the study goals, analyzed separately. Likewise, the subsurface may be stratified into distinct units deposited by separate events. Depending on the study goals, one may want to avoid mixing sediments from such different layers.

Studies of salmon spawning and other fisheries habitat resources usually require that samples be obtained under water. In these studies, the most popular bulk core sampler has been the FRI or McNeil sampler, constructed from a 50-cm drum with a 15-to-30-cm diameter pipe welded on the bottom (Figure 13.11). The smaller pipe is worked into the bed, the gravel removed by hand, and the

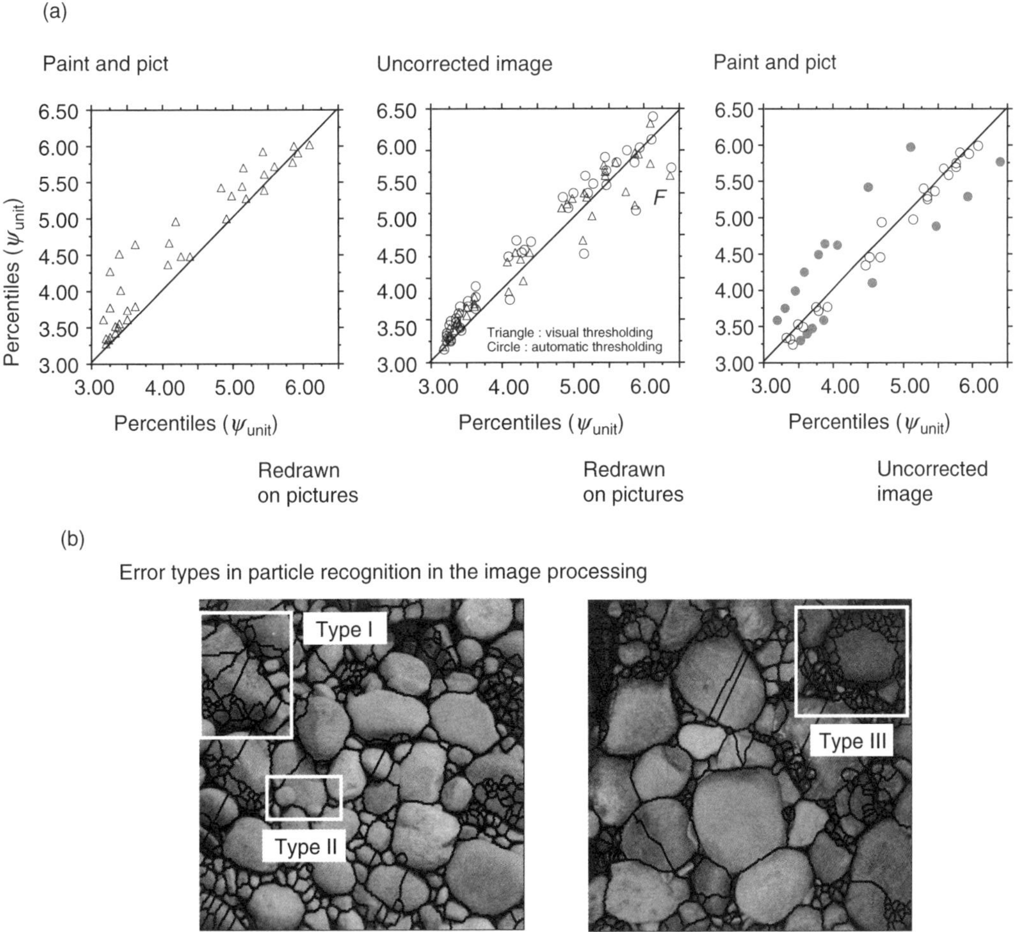

Figure 13.10 (a) Comparison of the grain size percentiles ($\psi_{10}, \psi_{16}, \psi_{25}, \psi_{50}, \psi_{75}, \psi_{84}, \psi_{90}$) performed by paint and pict field measures, manual redrawing of grain boundaries and uncorrected image analysis extraction (visual and automatic thresholding). Six samples shown, the line is the $X = Y$ line of perfect agreement. (b) Types of error due to automatic particle boundary recognition using watershed segmentation algorithm (*Micromorph* software) (*Source*: Piégay unpublished data)

muddy water within the sampler retained to permit suspended fine sediments to be sampled (McNeil and Ahnell 1964). The small pipe reduces the sample size, potentially to less than the minimum required to adequately sample the grain sizes present. Moreover, we suspect that there are some serious edge effects, especially in gravels that include particles coarser than 50 mm, as the pipe edge is likely to hit a big rock and cannot continue downward unless the rock is moved out of the way and either included in the sample or discarded. This problem exists with any cylindrical core sampler in gravel, but is more serious as the grain size approaches the pipe diameter. Other variants of cylindrical core samplers have included 50-cm drums with the top and bottom removed, and usually shortened to permit the operator's arms to reach the bottom of the sampler (e.g., Chambers *et al.* 1954, 1955, Orcutt *et al.* 1968), a 60-cm length of 46-cm diameter well casing with a serrated lower edge and handles attached to the top (Horton and Rogers 1969), a 53-cm length of 35-cm diameter pipe with a serrated lower edge (Van Woert and Smith 1962), a 25-cm diameter Hess-type core sampler (Shirazi *et al.* 1981), and a 75-cm length of

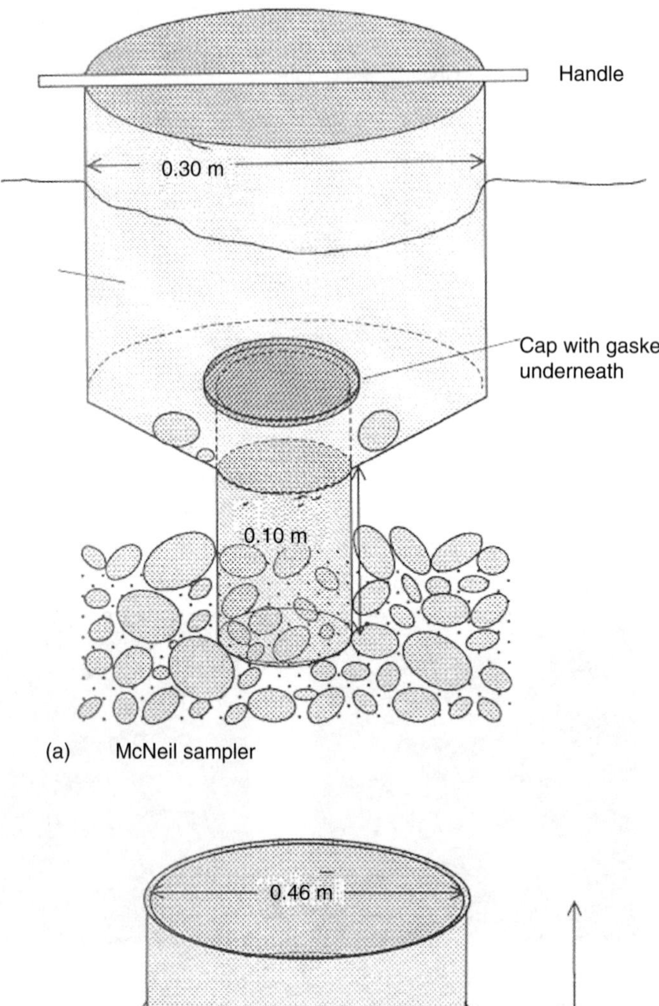

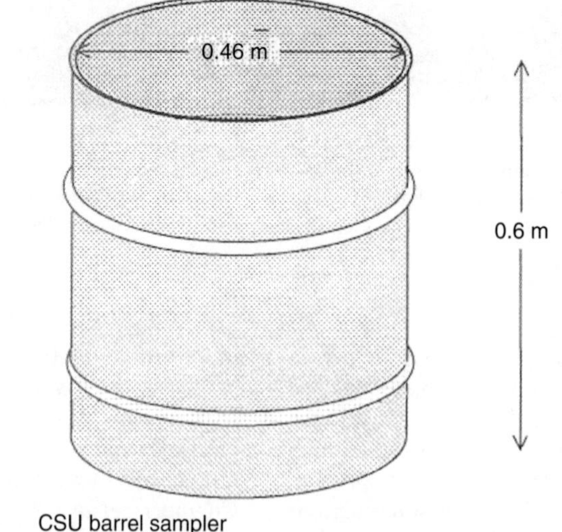

Figure 13.11 Diagram of the McNeil (FRI) sampler with sample in collecting basin. The sampler is driven into streambed, and sample is removed by hand from the core and placed in the collecting basin or directly into a bucket. The cap on the core is to permit suspended fine sediment in the basin to be retained for settling. An alternative to the cap (e.g., Kondolf *et al.* 1993, Vining *et al.* 1985) is to collect depth-integrated samples of the suspended sediment within the basin, and to analyze them for total suspended solids (reproduced from Platts *et al.* 1983)

15-cm diameter galvanized stove pipe (Peterson 1978). Yet another variant has been pointed shovel samplers fitted with hoods to retain fine sediment (Curtin 1978).

Freeze-core Sampling

Freeze-core sampling involves driving steel probes into the bed, discharging a cooling agent (such as liquid CO_2 or nitrogen) into the probes to freeze the interstitial water adjacent to the probe, and withdrawing the probes (with gravel samples frozen in to them) from the bed with a tripod-mounted winch. The first versions of the method used a single probe, later versions used three probes (Everest *et al.* 1980). The method was developed largely to obtain gravel samples that preserved vertical stratification of the sediments, especially with respect to the vertical infiltration of fine sediments into salmon redds. However, laboratory experiments have shown that driving the probes into the bed can disrupt the existing stratification (Beschta and Jackson 1979).

Freeze-core samples tend to have a 'ragged edge', with larger particles protruding from the frozen mass (Figure 13.12) implying that all fractions of the distribution are not sampled proportionately. Most importantly, however, freeze-core samples are typically less than 10 kg, too small to accurately represent gravels that include particles of 64 mm and greater (Church *et al.* 1987).

Figure 13.12 Photograph of frozen core extracted from sockeye salmon redd in Quartz Creek, Kenai Peninsula, Alaska (photo by Kondolf August 1986)

Comparing Bulk Core and Freeze-core Sampling

The ragged edge of freeze-core samples would imply that these samples would have fewer fines than bulk core samples of the same gravels. However, comparisons of the two methods by various authors have yielded mixed results, with some studies showing freeze-core samples to be finer, some coarser. In a systematic comparison of shovel, bulk core (McNeil and Ahnell 1964) and freeze-core sampling, Young *et al.* (1991) found that the bulk core samples most frequently approximate the true substrate composition.

Bulk core sampling is simple (although labor intensive), can yield large samples, and does not suffer from the ragged edge of freeze-core sampling. Thus, for most purposes, the bulk core sampling approach is more appropriate. Rood and Church (1994) developed a *hybrid* bulk-cylindrical-freeze-core apparatus whose samples were unbiased with respect to grain size distributions.

13.8 SAMPLE SIZE REQUIREMENTS

Adequate Sample Sizes for Bulk Gravel Samples

In general, big samples reduce bias and uncertainty in the various tools and methods to sample bed material, aside from reducing the variance if the material were sampled without bias. The question of an adequate size for volumetric samples of coarse sediments is not new. Wentworth (1926, p. 10) recommended that samples should be "large enough to include several fragments which fall into the largest grade present in the deposit". He recognized that "... it is rarely practicable for the geologist to collect samples as large as those demanded by the strict requirements of accuracy" and suggested 'practical' sample sizes along with ideal sample sizes for various grain sizes (Table 13.3).

Table 13.3 Ideal and 'practical' sample sizes of Wentworth (1926)

Maximum grain size (mm)	Ideal minimum sample size (kg)	Suggested practical sample size (kg)
64–128	256	32
32–64	32	16
16–32	4	8
8–16	0.5	4

Mosley and Tinsdale (1985) reviewed sample size criteria of the American Society for Testing and Materials (ASTM 1978), the British Standards Institution (BSI 1975), and the International Standards Organization (ISO 1977), and found the minimum sample sizes were inconsistent. For example, for a 50-mm gravel, the American Society of Testing Materials standard was 100 kg, while the BSI standard was 35 kg.

Shirazi *et al.* (1981) recommended that the diameter of a bulk core sample should be 2–3 times the size of the largest particle and that the fraction falling in the largest size interval should not exceed 5–10% of the total sample weight. Mosley and Tinsdale (1985) collected 28 bulk gravel samples from one location in the Ashley River, New Zealand, and concluded that accurate determination of mean grain size of this gravel (median size about 16 mm) required a sample of about 100 kg, but "samples in which the weight of the largest stone is less than 5% of the total weight have unbiased estimates of" the median size. However, the size needed for accurate representation will be greater as sorting decreases and to represent the coarse tail of the distribution (rather than simply the median).

Church *et al.* (1987) reviewed the sample size problem, and concluded that the largest clast should constitute no more than 0.1% of the sample by bulk weight. However, if the largest clast is 100 mm, this would dictate a sample 1300 kg in size. Accordingly, much as Wentworth (1926) presented ideal and practical sizes, Church *et al.* (1987) have, in practice, used the 0.1% criterion for sizes up to 32 mm, thereafter using a 1% criterion up to 128 mm, resulting in samples weighing 150–350 kg.

Sample size has been inadequate in many studies. For example, in spawning gravels of chinook salmon, particles 90 mm or larger are commonly encountered. To accurately represent the coarsest fraction, Wentworth's rule would imply that bulk samples should be about 30 kg, while Church *et al.*'s (1987) recommendation would call for samples weighing over 200 kg. Many spawning gravel samples reported in the literature (especially freeze-core samples) were smaller than 30 kg, thus probably do not accurately represent the coarser grades. However, the egg pockets in chinook salmon redds and the entire redd of smaller species may consist of considerably less than 30 kg of gravel, so larger samples would, by necessity, include particles from outside the egg pocket or redd. This situation raises the question of what is being sampled. In the case of a small egg pocket, the entire population may be obtainable, so that sample size criteria, which were designed to obtain representative samples from an unobtainable population, become irrelevant. Platts and Penton (1980) utilized multiple freeze-core probes and heavy equipment to sample an entire, large redd, but this approach is hardly practical for most studies. One approach to this problem is to lump small samples from many redds together into a large, composite sample. This procedure would mask variability in gravel size among redds, but much of the apparent variability may be due to problems in representing the larger size fractions in small samples, so composite size distributions may be more accurate measures of the size distribution. Many studies of spawning gravel have reported only averaged, composite size distributions. If obtained from relatively homogeneous stream channel conditions, these composites probably reflect the gravel population well, but composites from a variety of channel types will not reflect the actual size distribution at any site.

Sample size can affect the size distribution obtained. In some cases, larger D_{50} values have been obtained from larger samples (Ferguson and Paola 1997). In gravels with a few large particles distributed throughout, the size distribution of a given sample may look very different depending upon whether it happened to include one of those big rocks. For example, a single 150 mm rock might constitute 20% of the entire sample. In such a sample, percentage values for the other grades would be decreased by 1/5 if the large particle were included, or increased by 1/4 if it were excluded. The influence of occasional large particles on the values of other size grades in spawning gravels has been widely recognized, and many authors in the fisheries literature have dealt with the problem by excluding large particles from the analysis. For example, Chambers *et al.* (1954, 1955) excluded rocks larger than 152 mm, McNeil and Ahnell (1964) excluded rocks larger than 102 mm, and Adams and Beschta (1980) excluded rocks larger than 51 mm. Adams and Beschta noted that by excluding large rocks, the variance in percentages of fine sediment within a single riffle was reduced. Tappel and Bjornn (1983) found that size distribution curves were straighter on log-probability paper if rocks larger than 25 mm were excluded.

Church *et al.* (1987) recommend that grain size distributions should be computed and compared only for the ranges that have been representatively sampled. For pebble counts, this implies a lower truncation point (e.g., 4 or 8 mm); for bulk samples, this implies an upper truncation point that is a function of sample size and the standard selected.

The implications of excluding large rocks depend on what is to be done with the data. If the study is designed to document variations in fine-sediment content over space or time, the approach can be justified as an alternative to collecting impractically large samples. However, size data drawn from such truncated curves may not accurately reflect framework sizes used by fish, and they certainly will not accurately reflect grain roughness. Similarly, computation of the percentage of fine material will also be affected by truncation. If the implications of truncation are not explicitly recognized, results from one study may be misapplied to another site. This is especially true when using biological indices such as 'percent finer than' a given size. Such indices make sense (for comparing among samples) only if the entire distribution is adequately sampled, or if the upper truncation value is constant.

The large bulk samples needed to satisfy sample size requirements mean that it becomes unwieldy to bring the adequately sized samples (hundreds of kilograms) back to the lab to sieve. Therefore, some field sieving is usually necessary. Typically the procedure is to extract, sun-dry, and weigh a big sample; sieve and weigh all the fractions coarser than a threshold size such as 8 mm; and split and weigh the (well-mixed) remaining sample until a subsample of a few kilograms remains, which is taken back to the lab to be dried and run through finer sieves. For large samples, more than one splitting and weighing step can be done in the field. Initially, all of the largest rocks are individually passed through the template, and the remainder of the sample is either split and sieved or all of it is sieved by passing the sample through rocker sieves with large screens (typically with a sieve size up to 64 mm) down to the size threshold. In the field, the sample can be dried in the sun (in warm, dry weather) or in buckets over a campfire or on a camp stove. Wet sieving is an alternative in wet weather or to process large volumes quickly, though we find it troublesome to handle the fine sediment and water mixture. Wilcock *et al.* (1996b) used wet sieving to process numerous Helley–Smith bedload samples on a raft and to process large (ca. 250 kg) bed-material samples, as discussed in the Trinity River case study in this chapter. For coarser fractions, dampness is not important because surface retention of water (by weight) is negligible, but for sediment finer than 8 mm, water content should not be ignored. The weight of the fine (e.g., <8 mm) material should corrected for water content if weighed in the field.

The large effort in obtaining statistically robust subsurface samples motivates schemes to reduce the size and number of bulk samples for measuring a composite particle size distribution for a reach of river. Assuming, as has been observed (Mosley and Tindale 1985, Lisle and Madej 1992), that there is a correlation between surface- and subsurface size distributions, the bed can be stratified according to mapped facies, subsurface material sampled in selected facies, and a weighted-average, or composite, size distribution computed (Lisle and Madej 1992). A composite size distribution is found by weighting the fraction in each size class by the fraction of bed area occupied by each facies and summing over all facies. An observed relation between surface and subsurface sizes at the facies scale suggests a simpler scheme. As one might expect, we have found that the average subsurface size distribution underlies the facies having the average surface size distribution (Lisle and Madej 1992). Accordingly, an average subsurface sample can be obtained in the facies whose surface grain size distribution is equivalent to the weighted average for the reach, although it would be prudent to amalgamate subsamples of subsurface material from a number of locations underlying the the facies with 'average' size distributions.

Sample Size and Reproducibility of Pebble Counts

The very large bulk sample sizes indicated for coarse gravel motivated development of the pebble count (Wolman 1954) as a way of obtaining a sufficiently large sample without having to collect samples that were impossibly large and heavy. Wolman (1954) found that a count of 100 stones produced consistent median grain sizes for multiple counts by one operator and among different operators, Brush (1961) found that 60 stones were sufficient, and Mosley and Tinsdale (1985) concluded from their experience that 70 stones were sufficient. However, to accurately represent the tails of the distribution requires larger sample sizes. Fripp and Diplas (1993) recommended 200–400 stones based on error estimations using the binomial distribution. Rice and Church (1996) used the bootstrapping method to randomly draw replicate samples from a sample of over 3000 stones, and found that beyond 400 stones sampling effort is not rewarded by increases in the precision of estimated parameters, so unless working with strict precision criteria, it makes practical sense to stop at this point. For coarse, clean gravels, coarse percentiles maybe sampled with higher precision at a given sample size than fine percentiles. Thus, estimates of D_{84} and D_{95} are often nearly as good as D_{50} estimates, while D_5 and D_{16} might remain poor, in contrast to the expectation of the same errors for equivalent percentages expected with the normal distribution (Rice and Church 1996). (This example is further developed in Chapter 20.) In river beds with very poorly sorted bed material, larger samples may be needed to adequately capture the greater variability (Wolman 1954). In gravel or cobble beds with interstitial fine sediment present, the minimum sample size (i.e., 100) should apply to grains larger than the 4 or 8 mm fine-sediment cutoff to insure an adequate sample of framework grains. Petrie and Diplas (2000) used the multinomial distribution to estimate sample size requirements for the full size distribution to given levels of accuracy.

Inter-operator variability was investigated by Hey and Thorne (1983), who concluded that random errors decrease with increasing sample size (>100 stones), such that difference among operators becomes statistically significant. Similarly, Marcus *et al.* (1995) documented statistically significant operator bias, which they attributed to differences in measurement of grain diameter and stone selection. Wohl *et al.* (1996) found statistically significant differences in results from multiple operators, presumably due to differences in stone selection or stone measurement.

Differences in stone selection can be attributed to a variety of factors, including the finger not being vertical and thus prone to over-sample large stones protruding from the bed (Hey and Thorne 1983), downstream displacement of the finger by water currents (Marcus *et al.* 1995), looking at the bed before allowing the finger to descend (a problem with fixed grids if the tapes are above the bed surface and the sampling point is selected visually), and the finger touching more than one stone simultaneously. Whether sampled by grid or pacing, the sampling point should serve only as the starting point for the finger's blind descent to the bed. If the finger touches more than one stone simultaneously (perhaps lodging between two stones), bias is introduced unless the operator consistently uses one point on the finger (such as the right corner of the fingernail) as the sample point. At the outset of the sampling program, decisions should be made about how to classify situations such as a thin layer of sand over a larger stone, etc. In cobble and boulder beds, templates cannot be used because particles are too large and frequently embedded, and the pacing technique is not truly random because the operator's instinct to preserve his shins will influence his pacing. We suspect that one common source of error is probably the failure to fully close or avert the eyes as the finger descends to the bed and the resultant attraction to larger (more visually discernable and more easily handled) particles.

Differences in measurement of particles may be due in part to differences in identification of the appropriate place to measure the *b*-axis: it should be measured perpendicular to the long (*a*) axis at the stone's widest point, as this is the dimension that would limit the size of sieve through which it could pass. Using a template should minimize these measurement differences.

It is unclear the extent to which some of the published studies have restricted pebble counts to 'homogeneous' populations. If more than one geomorphic feature is included in the sample, the potential for error is greater because of potential to sample different proportions of two different populations. The minimum sample sizes of 60–400 stones cited above should properly apply to sampling a single population. That is, to adequately sample three populations requires measuring an adequate number of stones (be it 60 or 400) in each.

If a gravel deposit comprises lithologies whose densities are different, larger particles of a lighter lithology may behave hydraulically more like small particles of a denser lithology, and may be seen sorted by flows with smaller, denser grains. In conducting the pebble count, the lithologies of the sampled stones should be noted along with particle size, and at least 100 stones of the dominant lithology should be collected. For example, the bed material of the lower Carmel River, California, consists mostly of granitic and metamorphic clasts with a specific gravity of about 2.7, but clasts of Monterey Formation (a Tertiary marine siltstone) with much lower specific gravity also occur. In pebble counts made to characterize this bed material, at least 100 non-Monterey clasts were sampled, thereby yielding an adequate sample of the dominant lithology (Kondolf and Matthews 1986).

13.9 COMPARABILITY OF PEBBLE COUNTS AND BULK SAMPLES

If there were no difference between the surface layer and subsurface size distributions, the surface could be considered a random slice through the deposit and the pebble count would yield a random sampling of grains. The pebble count is a random point sampling procedure, so its results are theoretically equivalent to bulk sampling and sieve analysis for sediments with constant density (Kellerhals and Bray 1971). However, the surface layer is typically deficient in fines relative to the subsurface population. Thus, by virtue of the real differences between surface and subsurface layers in gravel bed rivers, the pebble count is usually sampling a different deposit than the subsurface part of a bulk sample. Wolman (1954) noted that pebble counts tend to yield coarser grain size distributions than bulk samples of the same gravel deposit because the former are commonly deficient in fine sediments. This is illustrated by comparing pebble count and bulk samples for recently deposited gravel and sand on the Middle Yuba River, California (Figure 13.13). The pebble count and bulk sample distributions deviate at the fine tail, reflecting the deficiency in fine sediment at the surface. If particles <4 mm are excluded from the bulk sample analysis, the resulting curve tracks the pebble count more closely.

Leopold (1970) proposed that results of pebble count analyses be adjusted to compensate for a "bias... towards larger sizes which, because of their area, are more likely to be picked up". However, so long as grain volume is proportional to grain weight (true with constant density), there is no bias in a random point count. If larger grains are more likely to be encountered, it is because they occupy a greater part of the cross-sectional area of the slice (the surface) and thus a greater part of the volume of the three-dimensional deposit. Thus, there is no theoretical justification for decreasing the actual percentages observed for larger stones so long as the pebble count is conducted correctly (and the count is truly blind and thus random).

13.10 SAMPLING STRATEGY

A field scientist embarking on sampling a riverbed is commonly faced with a number of general sampling issues. These include:

1. mixing bank and bed materials,
2. sample size,
3. what the sample is supposed to represent, and
4. tradeoffs between consistency and relevancy.

These issues are presented to the sampler in every field problem and should be addressed separately, rather than simply accepting a pre-packaged sampling protocol.

Choosing a method described in a manual and used by predecessors provides some assurance that it has been proven and the data so gathered will be accepted

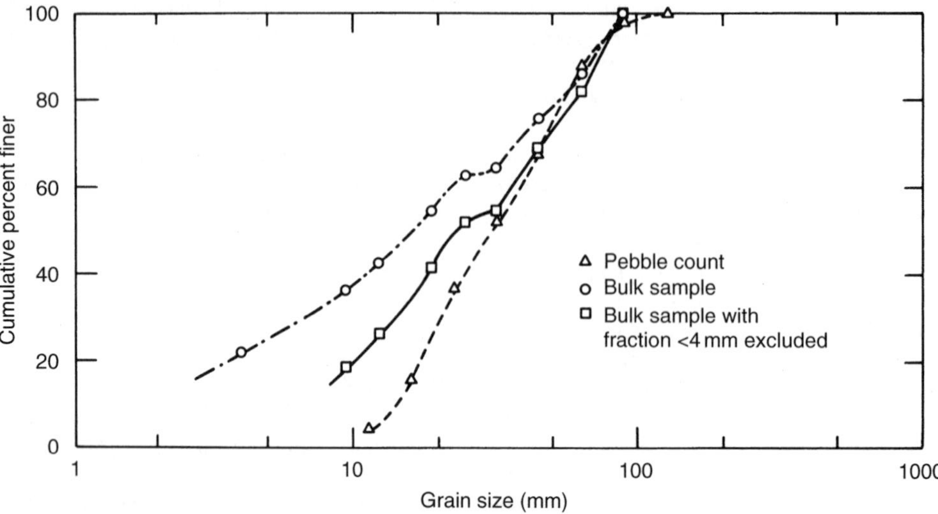

Figure 13.13 Grain size distributions of recently deposited sediment along the middle Yuba river, California, by pebble count and bulk sample (obtained with shovel from exposed bars). Also shown is size distribution for bulk sample with fraction <4 mm (*Source*: Kondolf 1988)

by others. However, the heterogeneity of gravel beds and the variety of problems requiring bed-material data have motivated a variety of sampling methods, some of which are still evolving. Sampling gravel beds is notoriously demanding for fluvial geomorphologists requiring, in some cases, the collection of tons of bed material from various locations in the channel and passing it through numerous sieves of low capacity. Sampling locations must be carefully chosen to target areas relevant to the problem, avoid bias, and obtain a large-enough sample to adequately reduce the variance of a highly variable population, while dealing with the practical problems of taking samples that inherently exist under flowing water. These constraints mean that, while 'quick and dirty' is likely to be wasted effort, there is no room for extra work that does not satisfy the purposes of the sampling program.

We believe that a bed-sampling program should pre-eminently meet the needs of the problem for which it is intended. River studies can be of regional or even national scale, but most often they entail a reach, basin, or biogeographic region. Of course, internal consistency is necessary for comparing spatial and temporal variations within the scope of a study. But if agencies press for agency-wide consistency for purposes beyond those of individual studies or programs, field scientists still have an obligation to understand and select the methods best suited to address the question posed, rather than simply defer to agency guidelines. If methods are chosen correctly to suit the needs of a particular problem (e.g., fine sediment in spawning gravels, hydraulic roughness, or channel mobility), over time these data can also prove useful for more comprehensive analyses addressing the same sort of problem, and useful degrees of consistency will naturally arise even as methods improve.

The challenge is to array methods in the field in a sampling scheme that accomplishes the goals of the study. Three qualities should guide the choice of sampling scheme: accuracy, precision, and consistency. The obvious reason for accuracy is to measure the part of the river that is meaningful to the problem. To typify spawning habitat, for example, we sample bed material where fish spawn, such as riffle crests. At the same time, we want to avoid bias, perhaps by sampling near actual redds instead of where we predict fish are likely to spawn. Most often, we sample at a number of sites in order to arrive at a representative average of the population to be sampled. These considerations lead to the first essential step: to define the sample population, that is, all of the areas of the riverbed that we want to represent.

The need for precision can also influence definition of sample population. At the outset, acceptable error should be determined quantitatively or qualitatively from the goals of the study. Sample error is a function of population variance and sample size (Benjamin and Cornell 1970). Population variance can only be limited by redefining a more homogeneous sample population. For example, if you want to measure downstream fining of sediment inputs or otherwise compare changes in particle size from one reach to another or one river to another, you can sample some consistent hydraulic or sedimentary environments, such as riffle crests or bar heads, and thereby remove some of the variation created by local channel form. Beyond that, sample variance can be reduced only by increases in sample size. Guidelines for determining sample sizes of bed material are provided by Church *et al.* (1987) and Ferguson and Paola (1997). In some cases, stratifying the bed according to bed-material size and sampling according to these designations can reduce sample error while limiting the number of samples.

Consistency means not only using the same criteria for sample location and measurement within a study, such as exemplified above for downstream fining, but also to match the accuracy, precision, and scale of bed-material samples to that of other related data. To compute Shields stress, for example, the entire bed surface should be sampled to determine bed-surface D_{50} if boundary shear stress is to be measured from reach averages of hydraulic radius and channel gradient. The error in any one of the three parameters should not greatly outweigh the error of the others.

In practice, these considerations often lead to tensions between objectivity, practicality, and professional judgment in choosing sampling schemes and locations. At the risk of bias, an experienced scientist can accurately stratify a bed according to the scale and characteristics of variations that bear on the problem, and thereby improve both accuracy and precision of measurements. For example, if you want an average grain roughness for a reach, then evenly spaced transects over the bed might be appropriate if there is no recognizable organization of bed material into riffles and pools. If some organization is apparent, then selecting transects in proportion to the relative area of recognizable strata can lower the variance.

River data, especially bed-material size data, are messy, and sending legions of novices into the field with a set of instructions is doomed to amplify the noise of uncertainty and variability. For this reason, it is essential that lead scientists devote time in the field to choosing the locations of bed-material samples, as well as the methods, and evaluate if the needs of the study are being met as the data are obtained.

To illustrate these considerations, we evaluate the following commonly used bed-material sampling schemes according to their application, adequacy of sample size, and bias.

Zig-zag Counts

In the zig-zag method (Bevenger and King 1995), the observer walks diagonally along the stream bed, traveling downstream in straight lines from left bank to right bank and back again, along which 100 pebbles are 'randomly selected' for measurement at a spacing of 3.5 ft (1.1 m). The stated aim of the method is to randomly sample "... numerous meander bends and all associated habitat features..as an integrated unit rather than as individual cross sections" (Figure 13.14), implying that a composite grain size for the entire reach was sought. The zig-zag method lumps data points from the bank and various features in the bed, which may have distinct grain size populations. The zig-zag method was proposed as a tool to evaluate fine-sediment content (and its changes over time) in relation to upstream land use (Bevenger and King 1995).

Lumping points from the bed and banks would be most appropriate in evaluating roughness. However, this measure of roughness could not be used directly to predict a stage–discharge relation in a natural channel because other sources of roughness such as sinuosity, obstructions, and bar-pool topography are not considered. Besides having limitations to support analyses of processes, it has several drawbacks as a monitoring tool. First, the sample size of 100 pebbles was the minimum recommended by Wolman (1954) to sample a homogeneous area of bed material, not a composite of areas such as bed and banks. Thus, the sample size is inadequate to yield accurate size data and to develop complete size distribution curves (Kondolf 1997). Trial runs showed that zig-zag counts were not reproducible by different observers (Bevenger and King 1995). Moreover, there are three sources of variation that cannot be resolved if the zig-zag method is used to assess fine-sediment variations over time:

1. The actual particle size of bed and bank material could vary between surveys;
2. The relative areas of beds and banks could vary due to bank erosion or accretion;
3. The relative sampled areas of bed and banks could vary between measurements.

It is unlikely that these areas will be sampled according to their areal extent by accident of

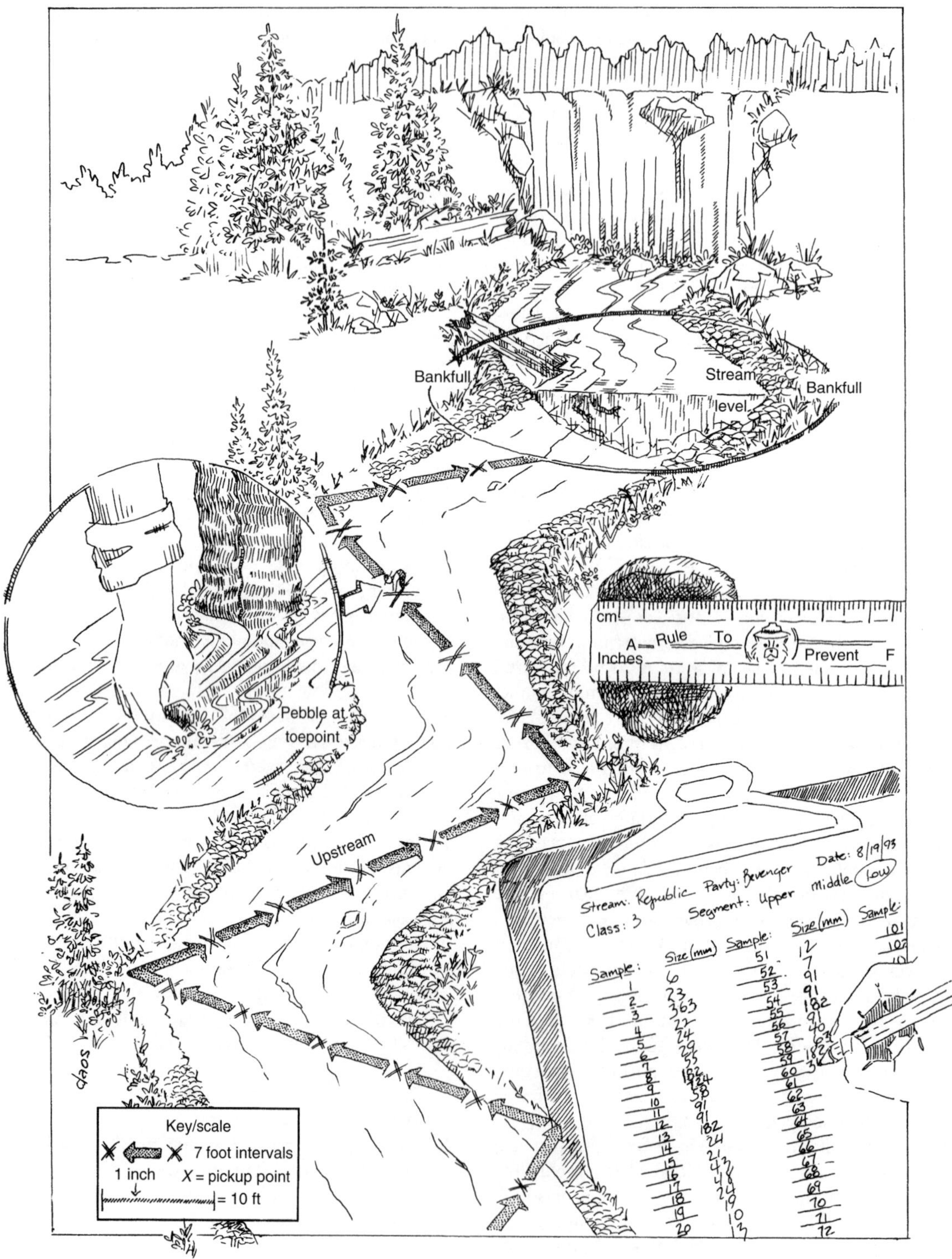

Figure 13.14 Diagram of sampling by the zig-zag method (reproduced from Bevenger and King 1995)

randomly selecting zigs and zags. The inability to decipher the added noise due to multiple sources of variation would severely hamper this method as a monitoring tool.

Despite these drawbacks, the zig-zag method has evidently had strong appeal (among non-geomorphologists) and has been adopted by some agencies. However, we do not recommend its use, as it does not share the advantages of a true pebble count (i.e., it does not yield an adequate sample size or reproducible grain size distributions), because it does not necessarily sample various facies equally or proportionally, and fundamentally because the geomorphic significance of a composite sample mixing bank material and various channel features is unclear.

Bank-to-bank Transects

Rosgen (1996) offers another variant on the reach-wide pebble count. A reach 20 channel-widths in length is divided into 10 cross sections extending from bank to bank, at each of which 10 points are randomly sampled for a total sample size of 100. Cross sections are distributed among riffles and pools in proportion to the abundance of riffles and pools in the reach, so if riffles make up 20% of the reach length, two of the 10 cross sections are located in riffles. The samples begin and end with the 'bankfull' bank top along each cross section. The size distribution resulting from this counting procedure is then used as one of the inputs to a stream classification system (Chapter 7).

The Rosgen (1996) method overcomes some but not all of the limitations of the zig-zag method. An improvement is that transects are selected to sample channel units (pools and riffles) according to their relative area in the channel. However, this presumes that distinguishing pools and riffles provides adequate stratification to delineate distinct facies. A more accurate stratification could be done by directly mapping surface facies. The Rosgen method suffers some of the same drawbacks as the zig-zag method in its inadequately small sample size and mixing bed and bank materials in a count. The resulting size distribution would have only limited application other than as an input to the stream classification.

13.11 SPECIALIZED TECHNIQUES AND APPLICATIONS OF BED SEDIMENT SAMPLING RELATED TO AQUATIC HABITAT

The principles and methods of sediment sampling and analysis derive from the field of sedimentology, geomorphology, and engineering, but concern over impacts of land use and water development on habitat for salmon and trout has created a 'boom' in sampling of bed material as part of biological studies. Here we review some specialized techniques applied to fish habitat.

Measurement of Fine-sediment Accumulation in Pools: V^*

Fine bed material in excess of that which can be stored in the matrix of gravel bed channels can be winnowed from the bed surface and accumulate during low flow in zones of low boundary shear stress, such as pools. There it can form thick patches and significantly fill the residual pool volume. Residual pool volume is the volume of a pool (disregarding fine bed material) below the elevation of the downstream riffle crest (Bathurst 1981). The fraction of residual pool volume filled with fine bed material (V^*) was developed as a measure of the in-channel supply of excess fine bed material, which can be sensitive to land use in the catchment and affect a variety of aquatic species (Lisle and Hilton 1992, 1999, Hilton and Lisle 1993). V^* is a dimensionless parameter that is independent of pool size. Fine sediment in pools signifies a reduction in pool habitat and, more importantly, an abundance of fine material that can affect benthic and intergravel habitats.

The particle size of fine sediment in pools varies among channels but usually ranges from fine sand to fine gravel. Although silt commonly constitutes a large proportion of the sediment load, a small proportion is stored in pools, usually in minor fractions with sand. Silt is carried in suspension and has a short residence time in the channel, whereas sand is transported intermittently as bedload and suspended load and therefore spends more time stored in pools between peak flows. More silt may be stored in low-gradient channels, but here bed material is usually so dominated by sand that there is not enough gravel to form an armor layer underlying pools.

V^* is best measured during low flow when the water surface of the pool is nearly horizontal (an assumption in the calculations), the channel is easily wadeable or navigable, and the bed is visible. In each pool, water depths are sounded and the thickness of fine material is probed with a graduated steel rod at roughly 50 locations total along 4–8 transects across the pool, depending on the complexity of pool topography and the distribution of fine patches

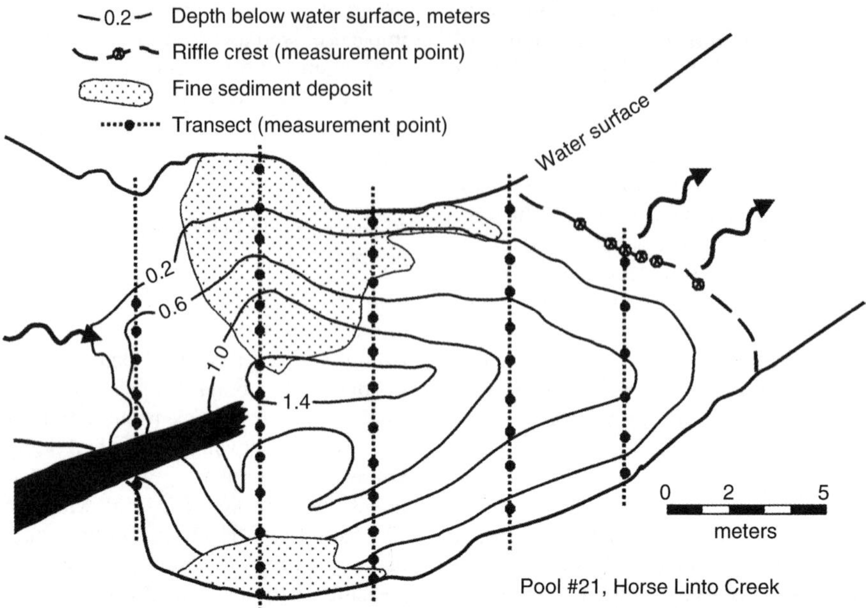

Figure 13.15 Measuring fine sediment in pools to calculate V^*. Water depth is measured along transects in the pool and subtracted by water depths at the riffle crest to compute residual depths. Fine-sediment thicknesses are probed along the same transects and, in some cases, augmented by more points over the deposits. Residual pool volume and fine-sediment volume are then computed (Hilton and Lisle 1993; www.rsl.psw.fs.fed.us/)

(Figure 13.15) (Hilton and Lisle 1993). Volumes of water and fine material are computed within the boundaries of the residual pool by summing the volumes contained between adjacent transects. V^* can be calculated on-site with a portable computer using a program available at a website (www.rsl.psw.fs.fed.us/). An experienced team of three can measure and compute a value of V^* in a pool in a wadeable channel in approximately one hour.

Like many parameters for natural channels, V^* can be highly variable and a single value is of little use. A number of pools (8–20 depending on variation in V^* in the reach) must be measured to obtain a reach-mean value of V^*. This can be a simple average, but more often is computed by weighting the value for each pool by its volume. Weighted mean V^* is simply the sum of fine-sediment volumes divided by the sum of residual pool volumes with fines removed. Attention must be paid to reduce variability in an unbiased way and to assure an adequate sample for the purposes of the investigation. Variability can be reduced by discarding pools containing significant volumes of large woody debris (which may increase V^*) and setting a lower limit of area or depth of pools to include in the sample. V^* tends to be more variable in small pools. A 20% statistical error can be obtained usually with a sample of about 8–20 pools, the sample being greater for lower V^* and greater channel complexity (Hilton and Lisle 1993). We believe that this error is low enough to detect significant changes in V^* for most applications, but whatever the acceptable error, it should be determined beforehand. Then an adequate sample can be assured by calculating mean V^* and its standard error as pools are measured in the field.

The most effective and statistically powerful application of V^* is for monitoring in a reach of channel, where factors other than sediment supply (e.g., flow regime, lithology) remain essentially constant. Fine bed material stored on the bed surface has very short residence times, and changes in supply from the basin register quickly as changes in storage in pools (Lisle and Hilton 1999). High values of V^* signify large chronic or recent inputs of fine sediment to a channel.

Because a number of factors affect V^*, using V^* to interpret channel condition requires supporting information about basin condition:

- Lithology of sediment sources strongly affects V^*. In fines-rich lithologies, like the weathered granitics and highly fractured sandstones and shales of the Franciscan Formation in northwestern California, values of $V^* < 0.1$ are interpreted as low (a reference for undisturbed landscapes), values of $0.1 < V^* < 0.2$ as moderate, and values of $V^* > 0.2$ as high (Lisle and Hilton 1999). Values of V^* in fines-poor lithologies are usually less than 0.05 and are not sensitive to sediment supply because the coarse component provides the interstices to store the fine component. In such streams, large sediment inputs may reduce residual pool volume with gravel rather than fine sediment. Reference values of V^* for other lithologies are poorly known, and many basins are underlain by a variety of lithologies that contribute to the sediment load.
- Flow regime may affect V^*. Snowmelt-dominated regimes, which usually have low variability of flow and long duration of flows (e.g., the snowmelt recessional curve) that selectively transport fine bed material, can be expected to flush fine sediment from the bed surface and have low values of V^*. Rainfall-dominated regimes, which have high variability and sharp recessional curves, have the highest reported values of V^*.
- Poor sediment sorting promotes sorting of fine sediment from the rest of the bed material and its deposition in pools. When sorting increases ($\sigma_{sg} < 4$) and dominant gravel sizes become finer than medium pebbles, the distinction between framework and matrix particles may break down and inhibit sorting of bed material into fine and coarse patches. The method is applicable to armored channels where the interface between a layer of fine sediment overlying a gravel bed can be detected with confidence. Such conditions are not common in lowland channels with beds of predominantly sand.

Other factors have little or uncertain effect. For example, large woody debris and other form roughness can increase deposition of fine sediment on a gravel bed (Buffington and Montgomery 1999b), but the effect on V^* appears weak or variable (Lisle and Hilton 1999). Hilton and Lisle (1993) recommended avoiding pools with large relative volumes of LWD in order to remove any possible complicating effects of LWD on V^*. There is no conclusive evidence that variations in channel slope or alternation between bar-pool and step-pool topographies affect V^*.

In summary, V^* is most applicable in basins whose lithology produces abundant fine sediment. Such lithologies include coarse-crystalline igneous rocks, poorly indurated sandstone, and schist. They do not include limestone, many aerenaceous sedimentary rocks, and basalt and other aphanitic igneous and metamorphic rocks. Channels may be single-thread, pool-riffle or step-pool. Channel size is limited only by the practicality of probing pools during low flow. At a coarse scale, visually inspecting the amount of fines stored in a pool can give a quick read on the severity of the problem of fine sediment in the context of an ecological or water-supply issue. A reconnaissance of pools can be enough to decide whether to adopt or discard V^* or other sediment measurements in a monitoring program. Our experience is that in a channel with $V^* \leq 0.1$, fine bed material in pools is characteristically confined to small and discontinuous deposits in eddies; outside of pools, a fine mode may not be evident among surface interstices (e.g., the Beaverdam Run tributary to Oregon branch shown in Figure 13.7). In such cases, fine-sediment supply would probably not be a critical issue, nor would V^* be an appropriate monitoring parameter unless large inputs of fine material were anticipated. A channel with $V^* \geq -0.2$ characteristically has large patches of fines occupying much of the area of pools; fine patches are evident elsewhere in the channel and surface interstices may be noticeably filled (e.g., Oregon Branch shown in Figure 13.7). However, ecologically important supplies of fine sediment may not be clearly evident on riffle armors, which can be effectively winnowed even in sediment-rich channels (Lisle and Madej 1992).

Assessing Salmonid Spawning Gravel Quality

The size of available streambed gravels can limit the success of spawning by salmonids (Groot and Margolis 1991). The bed material may be too coarse for spawning fish to move, a problem particularly common where dams eliminate supply of smaller, mobile gravels (e.g., Parfitt and Buer 1980). Excessive levels of interstitial fine sediment may clog spawning gravels. These effects have been documented downstream of land uses that increase sediment yields, such as timber harvest and road construction (Cederholm and Salo 1979, Everest *et al.* 1987, Meehan 1991).

Because of these problems, there is frequently a need to assess the quality of spawning gravels to determine whether gravel size limits spawning success.

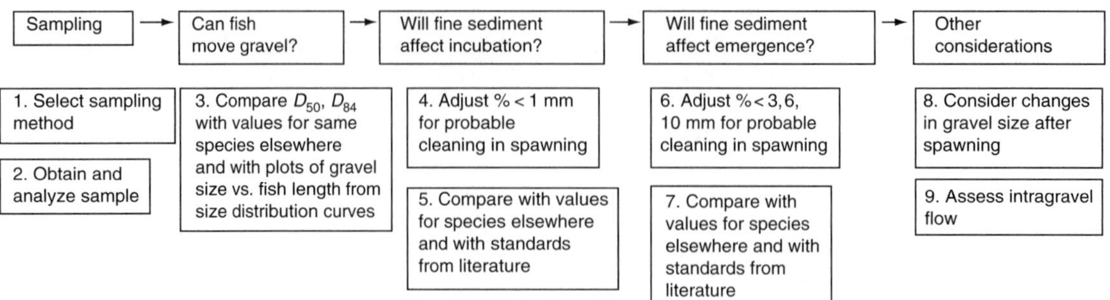

Figure 13.16 Flow chart illustrating nine discrete steps in evaluating salmonid spawning gravel quality (reproduced from Kondolf 2000)

Any such assessment involves comparison of gravel size on-site with information on gravel size suitability from laboratory studies or field observations elsewhere. Much of the recent literature on salmonid spawning gravels has been devoted to the search for a single statistic drawn or computed from the streambed particle size distribution to serve as an index of gravel quality (e.g., Lotspeich and Everest 1981, Shirazi and Seim 1981, 1982, Beschta 1982). These single-variable statistics are easier to report than full size distributions and provide convenient independent variables against which to compare incubation and emergence success in field and laboratory studies. However, a natural gravel mixture cannot be fully described by any single statistic, because gravel requirements of salmonids differ with life stage, and thus the appropriate descriptor will vary with the functions of gravel at each life stage (Kondolf 2000).

To assess whether gravels are small enough to be moved by a given salmonid to construct a redd, the size of the framework gravels (the larger gravels that make up the structure of the deposit) is of interest, and the D_{50} or D_{84} of the study gravel should be compared with the spawning gravel sizes observed for the species elsewhere. To assess whether the interstitial fine-sediment content is so high as to interfere with incubation or emergence, the percentage of fine sediment of the potential spawning gravel should be adjusted for probable cleansing effects during redd construction, and then compared with rough standards drawn from laboratory and field studies of incubation and emergence success. An assessment should also consider that the fine-sediment content of gravel can increase during incubation by infiltration, the gravels may become armored over time, or downwelling and upwelling currents may be inadequate. These considerations are incorporated in a nine-step, life-stage-specific assessment approach (Figure 13.16), whose steps are described below.

Sample the gravel and develop a size distribution (Steps 1 and 2). The sampling method depends upon the purpose of the assessment. If the concerns are limited to whether the fish can move the gravels, pebble counts may be adequate, although such values (obtained from the surface layer) may be larger than those from bulk samples, because the latter would be influenced by interstitial fine sediment in the subsurface. More commonly, however, the fine-sediment content is also of concern, in which case subsurface samples must be obtained. Because of the drawbacks of freeze-core sampling discussed earlier, bulk core samples (of adequate size) are preferable. Pebble counts directly yield size distributions, but bulk subsurface samples must be passed through sieves and weighed to obtain size distributions (Vanoni 1975). In either case, the size distribution should be plotted as a cumulative frequency curve; to compare multiple distributions, box-and-whisker plots can be plotted from percentile values drawn from the cumulative distributions.

Determine whether gravel is movable by spawning fish (Step 3). Whether the framework gravels are too large for the fish to move can be determined by comparing the D_{50} or D_{84} with those reported for the species elsewhere and with the maximum movable size predicted by Figure 13.17, which suggests that spawning fish can move gravels with a median diameter up to about 10% of their body length. In some channels, gravels may be compacted or cemented, rendering otherwise suitable sizes unsuitable. No widely accepted or easily applied method has been developed to quantify this phenomenon, so it must be evaluated qualitatively.

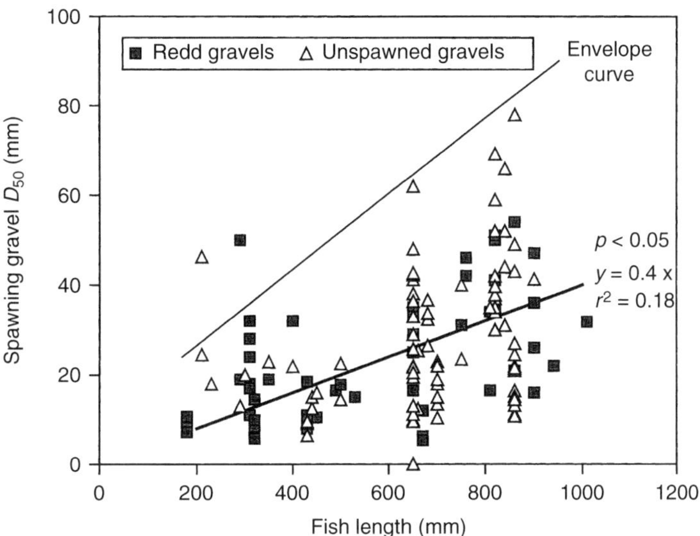

Figure 13.17 Median grain size of salmonid spawning gravel plotted against length of spawning fish, indicating that a spawning fish can use gravels with a median diameter up to about 10% of her body length. Based on data from published studies and field work on spawning gravels of chinook and sockeye salmon, rainbow and brown trout. Solid squares denote samples from redds, open triangles are 'unspawned gravels', which are potential spawning gravels sampled from the undisturbed bed near redds (reproduced from Kondolf 2000; see Kondolf *et al.* 1993 for sources of data)

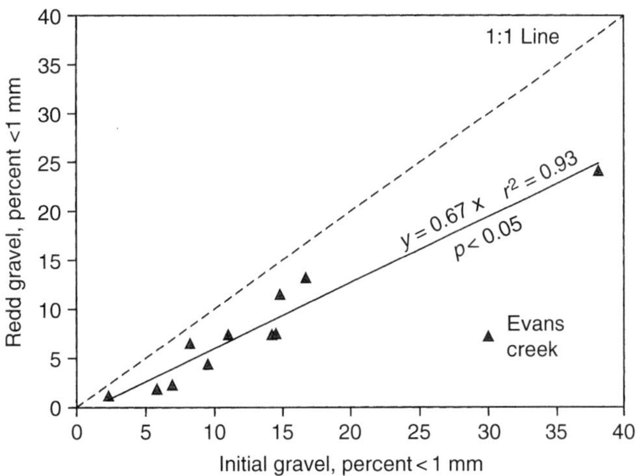

Figure 13.18 Cleansing of gravels during spawning, as reflected in percentages of sediment finer than 1 mm in redds and potential (comparable but unspawned) gravels. The data points for Evans Creek are excluded from the regression (reproduced from Kondolf 2000; see Kondolf *et al.* 1993 for sources of data)

Determine whether fine-sediment content is excessive for incubation (Steps 4 and 5). The question is whether the amount of sediment finer than 1 mm is so great that gravel permeability, and thus intragravel flow, is negatively affected. The percentage finer than 1 mm should be drawn from the grain size distribution curves and adjusted downward (using Figure 13.18) to reflect the probable cleaning effect of redd construction before evaluating fine-sediment content. The resulting values can be compared with values reported from redds elsewhere and with standards drawn from laboratory and field studies of incubation and emergence in Table 13.4 (showing values for 50% survival) and from conclusions drawn from field observations by McNeil and Ahnell (1964) and Cederholm and Salo (1979) that less than 12–14% of gravels should be finer than 1 mm (or, as published in these papers, 0.83 mm, reflecting the size sieves available to McNeil and Ahnell in their original work) for successful incubation. (It is prudent to bear in mind that these relations between percent fines and incubation success may have been based on insufficient sample sizes or inconsistent truncation.)

Determine whether fine-sediment content is excessive for emergence (Steps 6 and 7). To assess whether the fine sediment will block the upward migration of fry, the percentage finer than 3, 6, or 10 mm can be compared with values reported from redds elsewhere and with standards drawn from laboratory and field studies of incubation and emergence. However, while the fine-sediment (<1 mm) threshold for incubation effects can be estimated at 12–14%, the upper limits of the (larger) fine sediments affecting emergence (percents less than 3–10 mm) are more difficult to select,

Table 13.4 Standards for fine-sediment content in spawning gravels (defined as fine-sediment percentages corresponding to 50% emergence of salmonid fry), drawn from published field and laboratory studies (reproduced from Kondolf 2000) (reproduced by permission of the American Fisheries Society)

Reference	Species[1]	Maximum percentage finer than indicated[2] Grain size (mm) for 50% emergence				
		0.83	2	3.35	6.35	9.5
Hausle and Coble (1976)	Brook trout		10			
Weaver and White (1985)	Bull trout					40
Weaver and White (1985)	Bull trout					16
Bjornn (1969)	Chinook				15, 26	
Tappel and Bjornn (1983)	Chinook				40	
McCuddin (1977)	Chinook				30, 35	
Koski (1975, 1981)	Chum			27		
Cederholm and Salo (1979)	Coho	7.5, 17				
Koski (1966)	Coho	21		30		
Phillips *et al.* (1975)	Coho			36		
Tagart (1984)	Coho	11				
Irving and Bjornn (1984)	Cutthroat				20	
Irving and Bjornn (1984)	Kokane				33	
Irving and Bjornn (1984)	Rainbow				30	
NCASI (1984)	Rainbow	12			40	
Bjornn (1969)	Steelhead				25	
Tappel and Bjornn (1983)	Steelhead				39	
McCuddin (1977)	Steelhead				27	
Phillips *et al.* (1975)	Steelhead			25		
Average		13.7	10.0	29.5	30.3	28.0
Standard deviation		4.7	0.0	4.2	7.4	12.0

[1]*Scientific names*: brook trout, *Salvelinus fontinalis*; bull trout, *S. confluentus*; chinook salmon, *Oncorhynchus tshawytscha*; chum salmon, *O. keta*; coho salmon, *O. kisutch*; cutthroat trout, *O. clarki*; kokanee, *O. nerka*; rainbow trout (non-anadromous) and steelhead (anadromous), *O. mykiss*.

[2]Results of different experimental runs separated by commas.

showing considerable variability (Table 13.4). As with the percentage of sediment less than 1 mm, the percentages less than 3, 6, or 10 mm should be adjusted downward to reflect the probable cleaning effect of redd construction, but the effects of redd building on these sizes are more variable than upon the percentage finer than 1 mm (Kondolf *et al.* 1993).

Consider changes in gravel size after sampling (Step 8). Potential changes in sediment yield and local sediment transport capacity should be evaluated at the watershed scale to identify potential sources of fine sediment during the incubation period and to evaluate the potential for bed scour or coarsening. Field studies to monitor changes in fine-sediment percentages over the course of the incubation season (Adams and Beschta 1980, Lisle and Eads 1991) may be appropriate. Because the future applicability of gravel size data collected may be compromised by long-term changes in bed-material size, monitoring of bed-material sizes in future years may also be appropriate.

Evaluate intragravel flow conditions (Step 9). Intragravel flow depends both on the gravel permeability and the hydraulic gradient. The former is affected by fine-sediment content and thus is partly addressed in Steps 4 and 5. The hydraulic gradient is more complex to evaluate, as it depends on flow level, channel bed geometry, and possibly on large scale groundwater circulation patterns. Standpipe measurements, dye studies, or examination of the channel bed geometry can all be used to shed light on intragravel flow conditions (e.g., Terhune 1958, Barnard and McBain 1994).

Measuring Infiltration of Fine Sediment into Spawning Gravels

Whatever the concentration of fine sediment, the available evidence suggests that fish can typically flush enough from spawning gravel during redd construction to provide initially adequate intergravel flow of oxygenated water to incubating embryos (Kondolf *et al.* 1993). Fine sediment transported by subsequent flows, however, can infiltrate and fill intergravel pores (e.g., Carling and McCahon 1987), and higher flows that mobilize the gravel can deposit new layers of bed material containing abundant fines (Lisle 1989). Therefore, the critical measure of spawning habitat is not the initial size composition of spawning gravel, but the gravel composition during the incubation period. We describe three methods for implanting and recovering gravels in order to measure changes in gravel composition caused by sediment-transporting events. Excavations are usually done in spawning gravels but not in active redds to avoid destroying incubating ova.

Solid-walled containers (buckets or cans) filled with clean gravel and buried flush with the bed surface can be used to measure infiltration of fine sediment (Lisle and Eads 1991). Surface particles are removed and replaced after burial of the containers in order to replicate conditions in which sediment naturally enters the bed. The containers are retrieved after the allotted time and their contents sieved in order to measure the volume of infiltrated sediment. Loosely fitting collars installed around the containers can facilitate rapid removal and replacement. This method is relatively easy to perform. It is most appropriate where scour and fill of the bed is minimal and most infiltrating sediment (fine sand or larger) is coarse enough to be predominantly influenced by gravity rather than intergravel flow once it penetrates the bed surface. In a similar manner, porous containers (such as plastic mesh boxes) filled with clean gravel can be buried below the bed surface. These have the advantage of capturing fine sediment carried by intergravel flow. Their disadvantage is that some of the sample can be lost through holes in the containers during retrieval, and, like all containers buried just below the surface, can be lost or rendered ineffective by scour and fill of the bed.

The final method solves the problems of intergravel transport and scour and fill, but installation is more demanding (Lisle and Eads 1991; personal communication from George Sterling, University of Alberta). A collapsed bag sewn onto a steel rim is buried under an unbounded column of clean gravel and later pulled vertically out of the bed, enclosing the overlying infiltrated gravel (Figure 13.19). The bed is first excavated from inside an open cylinder down to the desired depth. The collapsed bag is placed open-end-up in the pit, and cables attached to the rim are extended to the surface. Clean gravel is poured back into the pit, the cylinder removed, and surface layer replaced. To retrieve the sample, the cables are drawn upwards with a chain hoist mounted overhead. In the process, the bag rises rim-first through the gravel, capturing the sample. Scour chains can be installed alongside to measure the contribution of scour and fill to changes in gravel composition, and freeze tubes can be installed within the sample to measure the stratigraphy of fine deposits. As in the other methods, the sample is sieved to measure the influx of fine sediment. The advantage of this method is that it can measure the introduction of fine sediment by scour and fill.

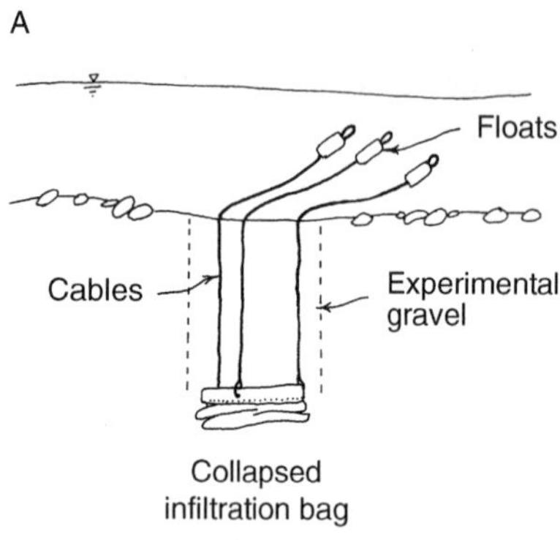

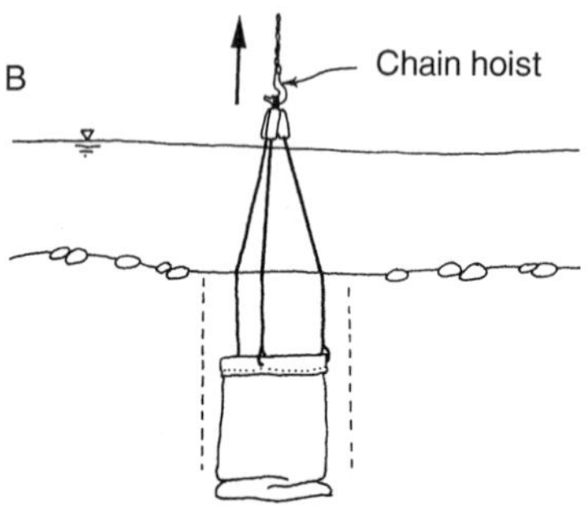

Figure 13.19 The collapsed-bag technique for obtaining an unbounded sampled of bed material infiltrated by fine sediment (*Source*: Lisle and Eads 1991)

13.12 CASE STUDY: DOWNSTREAM FINING ON THE PIAVE RIVER, ITALY

Case Study Description

Downstream fining is widely observed in rivers, with implications for fluvial processes and ecology. Studies of downstream fining have addressed the fundamental processes responsible for fining (generally regarded as a combination of abrasion and sorting) and general downstream patterns in grain size and sorting, with particular attention to factors such as tributary inputs that affect grain size patterns and can produce steps in the downstream pattern (e.g., Rice and Church 1998).

The Piave River drains 3900 km^2 of mostly limestone and dolomite, flowing about 222 km from its headwaters in the dolomites through the Venetian PreAlps and Venetian Plain to the Adriatic Sea. Nineteen tributaries join the river, and the downstream continuity of bedload sediment transport is interrupted by four dams and in-channel gravel mining. Since the turn of the 20th century, the effects of bank protection and channel incision (largely from mining and dams) have been to reduce channel width 65% and braiding index by half (3.0–1.5). To document the downstream pattern of bed-material size and to examine the influence of tributaries and dams, Surian (2002) sampled bed material along a 115-km study reach.

Case Study Methods and Results

Surian (2002) conducted pebble counts (by grid) at 35 sites, spaced at regular intervals of approximately 3–3.5 km. To obtain samples representative of the coarsest active bed material, he sampled exposed lateral, mid-channel, and point bars at low flow. He measured 120–130 particles per site to accurately characterize D_{50}, though not necessarily the tails of the distribution.

D_{50} ranged from 13 to 83 mm, with a weak downstream reduction. An exponential model for the decay in D_{50} had a low coefficient of determination ($r^2 = 0.02$). The fining coefficient for D_{50} was 0.027 km^{-1}, considerably lower than most values reported in the literature (Surian 2002: p. 146). Surian identified 15 steps, i.e., places where the grain size increases in the downstream direction. To distinguish real increases from sampling error, he estimated confidence intervals using the bootstrap technique. At a 95% confidence interval, six of the 15 steps were considered significant, and two others were also accepted as real given the low overlap of confidence intervals. Of these eight steps (shown in Figure 13.20), two were associated with dams, two with tributaries, three with other lateral bedload sources, and one step was not associated with these factors. Drawing upon published rates of abrasion, Surian concluded that sorting was probably the more important mechanism for downstream fining in the Piave River.

13.13 CASE STUDY: DETERMINING CHANGES IN FINE-SEDIMENT CONTENT DURING FLUSHING FLOWS, TRINITY RIVER, CALIFORNIA

Case Study Site Description

Since construction of the Trinity and Lewiston Dams in 1961, about 80% of runoff from the upper Trinity River has been exported to the Sacramento River, reducing high flows such that fine sediment delivered from tributaries accumulated in the bed of the Trinity without being flushed out, degrading spawning gravels and other habitats for anadromous fish. As part of a legally mandated effort to restore fish popu-

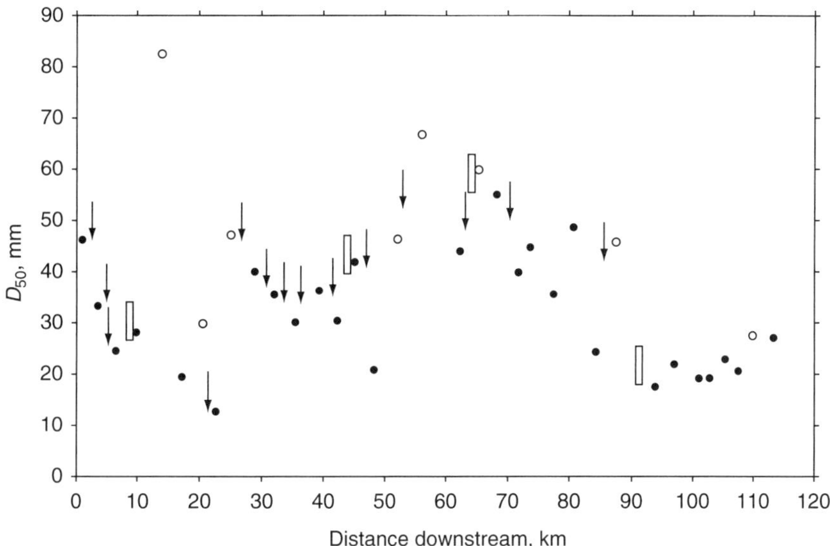

Figure 13.20 Downstream variation in D_{50} along the Piave River, showing locations of discontinuities or steps (where grain size increased downstream) by circles, and factors potentially producing steps: tributary confluences (arrows), and barrages (rectangles) (reproduced from Surian 2002, with permission of Elsevier Science)

lations in the Trinity River, the US Bureau of Reclamation made a series of controlled, experimental, high-flow releases (flushing flows) from Trinity and Lewiston Dams in each of three years, 1991, 1992, and 1993: $76\,m^3\,s^{-1}$, $164\,m^3\,s^{-1}$, and $80\,m^3\,s^{-1}$ (USFWS 1999). Flows of $80\,m^3\,s^{-1}$ and $164\,m^3\,s^{-1}$ were both well below the $Q_{1.5}$ (the flow with a return period of 1.5 years as an annual maximum) on the pre-dam flood frequency curve, but correspond to approximately the Q_3 and Q_7 (annual peak flows with return periods of 3 and 7 years, respectively) on the post-dam curve (USFWS 1999). Wilcock *et al.* (1995, 1996a,b) documented the effect of the flushing flows on spawning gravel quality and channel form at two study reaches heavily used by spawning salmon, Poker Bar and Steelbridge (Figure 13.21).

Case Study Methods and Results

During the flushing flows, Wilcock *et al.* (1996 a,b) measured vertical velocity profiles and sampled bedload in transport. Before and after the flushing flows they placed tracer gravels and bedload traps to document bed mobility, surveyed cross sections, visually estimated fine sediment stored in the bed surface over a 3-km study reach, and measured changes in fine sediment in the bed at the study sites through repeated visual observation, pebble counts, and bulk core sampling. They made measurements during the 1993 flow only at the Poker Bar site, as this flow was essentially the same as the 1991 flow.

Visual estimates of surficial fine sediment in 3-km reach. Wilcock *et al.* floated the reach from the confluence of Grass Valley Creek to the Steelbridge study site, visually estimating percentage of fine sediment on the bed, before the 1992 release and after the 1993 release. They chose this approach as the only practicable way to quantify (albeit roughly) the volume of fine sediment stored in the bed elsewhere in the study reach, information needed for overall sediment routing calculations. They computed sediment storage values for each of six subreaches, bounded by large pools (some of which have been dredged to reduce the river's sand load). Prior to the 1992 release, percentage fine-sediment ranged from 13.6 to 43.5 in the subreaches, while after the 1993 release fine sediment ranged from 13.4% to 27.6%, with all subreaches showing reductions. The overall change in surficial fine-sediment content can be read from the cumulative frequency plot of occurrence (in areal extent) of various classes of fine-sediment percentage. Note that this is not a standard cumulative size distribution plot, in which coarse sediment mixtures would plot to the right of finer.

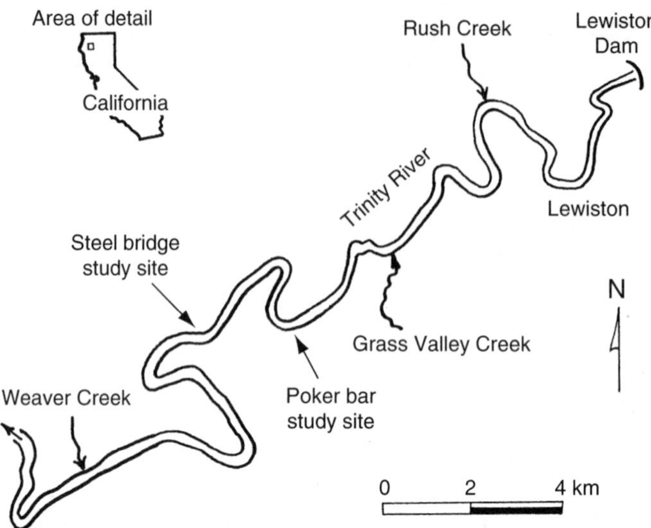

Figure 13.21 Location map of the Trinity River downstream of Trinity and Lewiston Dams, showing study reaches of Wilcock *et al.* (1995, 1996a,b) (reproduced from Wilcock *et al.* 1996a)

Visual estimates and pebble counts at detailed study sites. In 1991, Wilcock *et al.* visually estimated variations in bed roughness for hydraulic modeling by classifying the sediment as sand, gravel, cobble, or boulder. In 1992, they expanded the visual estimates with the goal of detecting changes in bed texture (especially fine-sediment content), and (at regular transect points) estimated the percentage of the bed covered by sediment <8 mm (termed the 'percent embedded' after terminology in the fisheries literature), as well as the D_{50} and D_{90} (to nearest phi class), and used the same operator for all observations. The estimated error was ±10% for percentage embedded and ±1 phi unit for D_{50} and D_{90}. To characterize surficial sediment, Wilcock *et al.* conducted pebble counts along the cross sections, measuring 100 stones per cross section in 1991 and 200 stones per cross section in 1992 and 1993. The bed material was relatively consistent across the channel in these study reaches, justifying the use of a single pebble count for a section-wide characterization. The 1991 release did not produce significant change in the substrate at either Poker Bar or Steelbridge. The 1992 release resulted in decreased surficial fine sediment at nearly all cross sections at both sites, as measured by visual estimates of embeddedness and by pebble count, although the two methods produced different results at the three downstream-most sections (Figure 13.22).

Bulk sampling at detailed study sites. To better quantify changes in fine-sediment content at the study sites, in 1991 and 1992, Wilcock *et al.* collected three types of bulk samples from the cross sections used annually by spawning salmon: pre-release, post-release at pre-release locations, and new post-release samples next to the original samples, the latter to control for the effect of pre-release sampling on the post-release sediment composition. They also inserted tracer gravels in the pre-release sample sites (see Chapter 14 for review of tracer gravel tools). After surveying the bed elevation at the sample point, they inserted a metal cylinder into the bed as deep as possible and removed all sediment down to the bottom of the sampler. In 1991, they used a 30-cm diameter sampler (standard FRI/McNeil sampler) and sampled down to 30 cm in depth, which yielded samples of 13–30 kg (mean 22 kg). They dried these samples in buckets on a camp stove prior to sieving the coarse fractions (>8 mm) in their entirety, weighing the finer fraction and retaining a split for sieving. In 1992, they enlarged their sample size by using a 59-cm-diameter cylinder (a bottomless 55-gal drum) and sampling as deep as 40 cm, which yielded sample sizes 112–281 kg (mean 182 kg). To process the large 1992 samples, Wilcock *et al.* wet-sieved on site, counting coarse particles (>8 mm) and measuring the volume of sediment less than 8 mm, converting to mass using average weights per particle or volume by size class measured in 1991.

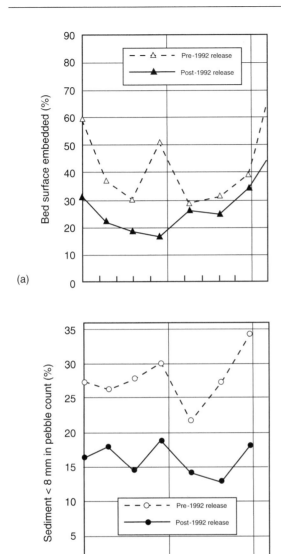

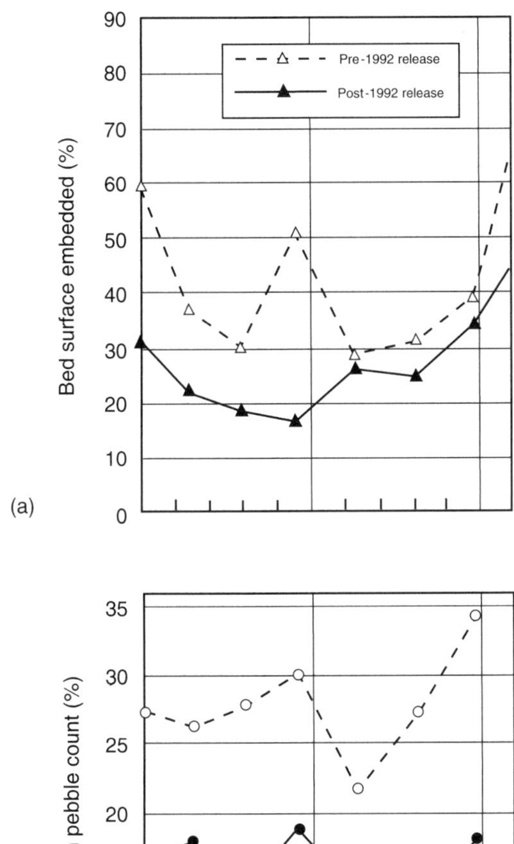

Figure 13.22 Fine sediment in the bed vs. downstream distance at the Poker Bar study site on the Trinity River in 1992, before and after experimental flushing flows. The open symbols and dashed lines represent pre-release conditions and the closed symbols and solid lines represent post-release conditions. (a) Visually estimated embeddedness, or percent of surface covered by fine sediment. (b) Percent finer than 8 mm as measured in pebble counts (adapted from Wilcock *et al.* 1995, used by permission)

Bulk sample results were less consistent than the surface sampling results. At Poker Bar Cross Section 2, the pre-release % < 8 mm ranged from 23% to 34%, the post-release from 26% to 35%. At Steelbridge Cross Section C, the pre-release % < 8 mm ranged from 16 to 26 mm, the post-release from 11% to 29%. Results from various methods of sediment sampling are summarized for Poker Bar Cross Section 2 in Figure 13.23, displaying the range of results possible from these diverse methods.

The Trinity River case study illustrates the use of on-site sieving to process adequately large bulk core samples of bed material, the use of on-site wet sieving to process multiple bedload samples, and estimation of surficial fine-sediment content over a long river reach to put detailed site measurements in a larger reach context. Each technique served a purpose, and taken together provided information needed to develop recommendations for sediment management in the river. No single technique would have been suitable for the different questions involved.

13.14 CASE STUDY: APPLICATION OF V^* TO FRENCH AND BEAR CREEKS, CALIFORNIA

V^* is most applicable to monitoring annual variations in a reach of channel and for evaluating the transport and spread of well-defined inputs of fine sediment. Two examples from northwestern California illustrate these uses:

French Creek

French Creek drains 60.4 km^2 in the Klamath Mountains, flowing into the Scott River near Etna, California. Much of its basin is underlain by deeply weathered granitic soils that erode to sand sized sediments. Erosion of logged areas before 1990 contributed large volumes of fine sediment to the channel, filling pools and armor interstices in riffles with coarse sand, and negatively affecting habitat for a native population of anadromous salmonids. From 1991 to 1994, landowners collaborated with the Klamath National Forest on an erosion-control program to reduce sediment supply. Because the problem was sand entering a gravel-bed channel, V^* was selected to test the effectiveness of the program to reduce the in-channel supply of sand. The monitoring program eventually focused on a mainstem reach including 10 pools not far downstream of the sediment sources.

Soon after the erosion-control program was implemented in 1992, fines volume decreased by more than

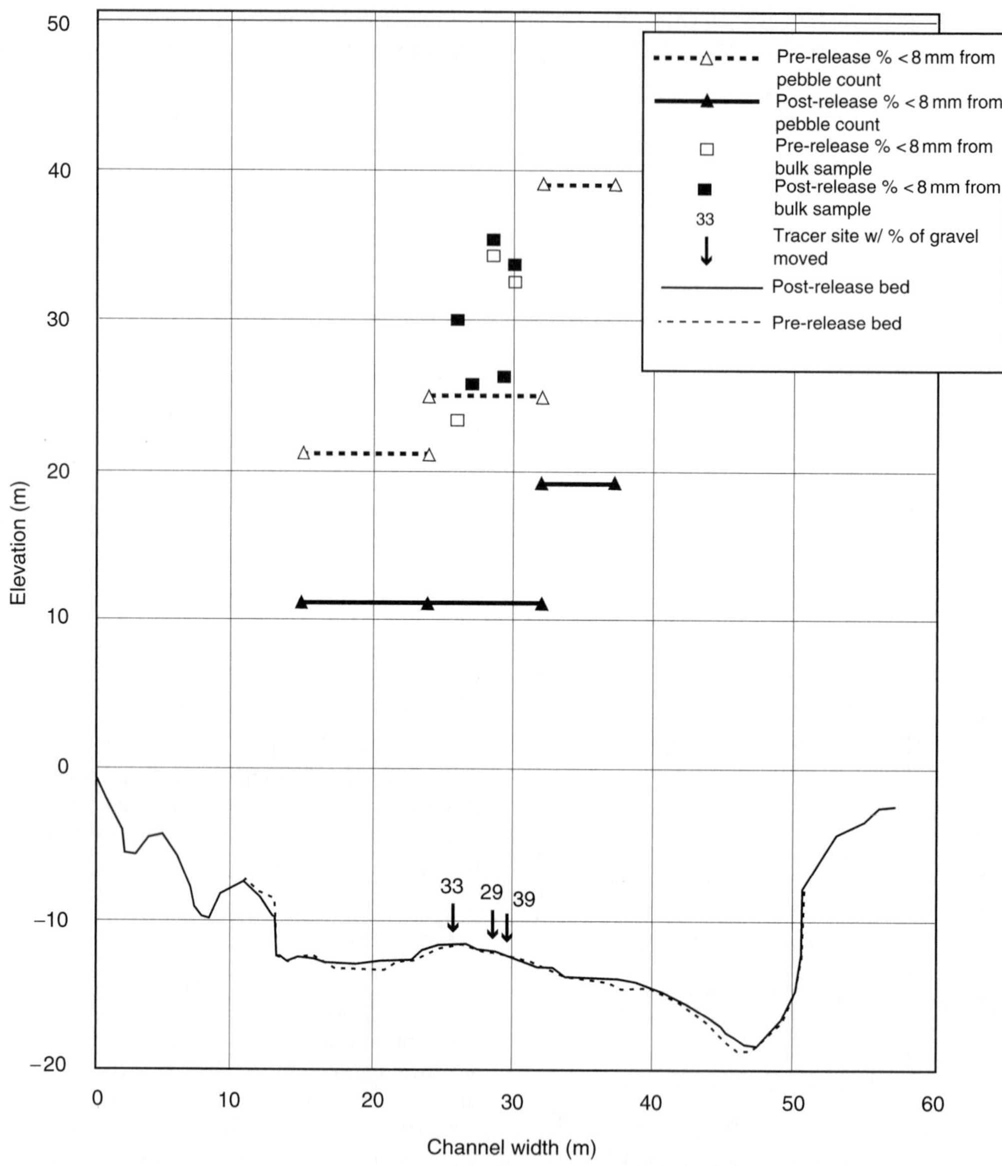

Figure 13.23 Bed elevation and sediment conditions at Cross Section 2 of the Poker Bar study site on the Trinity River, before and after experimental flushing flow releases in 1992. Dashed lines and open symbols denote pre-release conditions; solid lines and filled symbols denote post-release conditions. Pre- and post-release cross sections are shown in m relative to an arbitrary datum and have been vertically exaggerated five times. Horizontal lines represent pebble counts, with the extent of the horizontal line indicating the cross-sectional extent of the facies sampled, and the vertical position reflecting the percent less than 8 mm. Percent less than 8 mm are also shown for all bulk samples: four pre-release samples and six post-release samples. The location of the tracer sites are shown with the percent tracer (by mass) which moved during the release. (adapted from Wilcock *et al.* 1995, used by permission)

one-half as scoured-pool volume (residual volume minus fine-sediment volume) remained essentially unchanged (Figure 13.24, after Lisle and Hilton 1999). Values of V^* decreased to approximately one-third the initial value. A large rain-generated flood in January, 1997 (recurrence interval = 14.5 years in the Scott River) caused fines volume and V^* to nearly double and scoured-pool volume to decrease, but in subse-

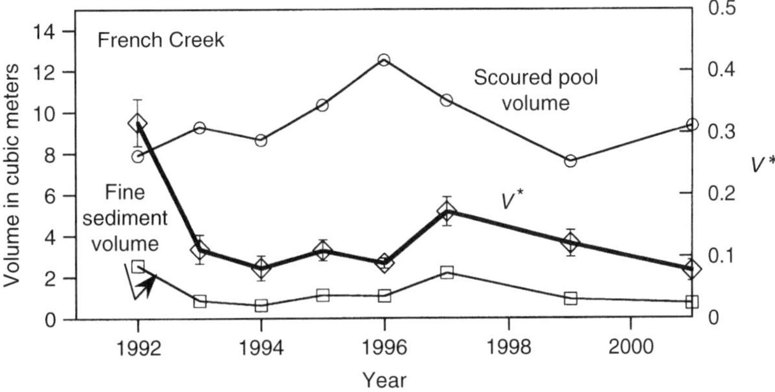

Figure 13.24 Annual variations in $V*$, fine-sediment volume, and scoured-pool volume in French Creek, 1991–2001 (adapted from Lisle and Hilton 1999)

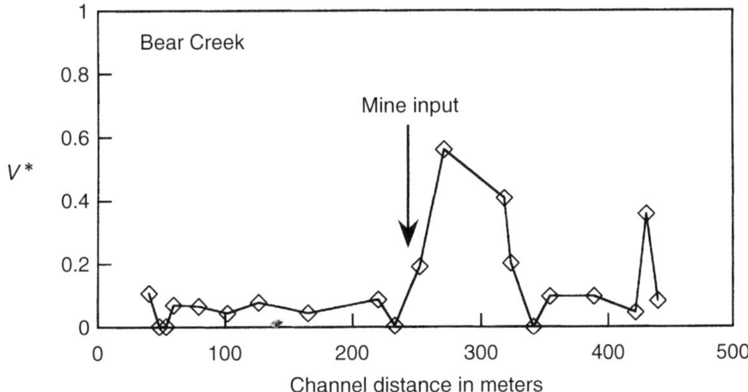

Figure 13.25 Variations in $V*$ with longitudinal distance along Bear Creek, showing influence of previously unknown, illegal mining operation on fine-sediment loading (*Source*: Lisle and Hilton 1992)

quent years $V*$ has decreased again and pool volume has begun to recover. The current background value of $V*$ in French Creek (≤ 0.1) is equal to the reference value for channels draining weathered granite (Lisle and Hilton 1999). Therefore, the effectiveness of the erosion-control program was clearly demonstrated: in-channel supplies of sand, which was the dominant sediment input, were reduced to background levels immediately after erosion was controlled, in spite of a major flood that could have greatly increased sediment inputs.

Bear Creek

Bear Creek drains a 20-km^2 basin also in the Klamath Mountains and enters the South Fork Trinity River near Hyampom, California. Like French Creek, it is underlain predominantly by weathered granite, but unlike French Creek, its basin is nearly undisturbed. In 1991, $V*$ was measured in a reach of channel including 19 pools to help establish reference values for $V*$. In the upstream portion of the reach, $V*$ was indeed low (<0.1), but at the halfway point, $V*$ increased dramatically to >0.5, and then decreased downstream (Figure 13.25) (Lisle and Hilton 1992). Upon further inspection, Lisle and Hilton (1992) discovered the input of fine sediment from a small illegal mining operation upslope from the point of the large inflection of $V*$. Although it was not the purpose of the study, the $V*$ values permitted detection of a sediment source and evaluation of its effect on the channel.

13.15 CASE STUDY: ASSESSING SPAWNING GRAVEL QUALITY IN THE COLORADO RIVER AND TRIBUTARIES BELOW GLEN CANYON DAM

Case Study Site Description

Since closure of Glen Canyon Dam on the Colorado River in 1963, a popular rainbow trout (*Oncorhynchus mykiss*) sport fishery has developed in the Colorado River downstream in Grand Canyon National Park. The fishery is especially productive in the tailwater reach between the dam and the Paria River confluence (Figure 13.26), where consistently cold water releases have produced a nearly complete change in species composition from native warm water fishes to introduced rainbow trout. The trout spawn both in the mainstem and tributaries. Spawning habitat is limited to some large mainstem gravel bars and to gravel deposits on tributary reaches downstream of migration barriers, many of which are 'pocket gravels' within boulder-dominated channels (Figure 13.27). The quality of these spawning gravels had not been assessed prior to the study described here.

Case Study Methods

As part of a larger research effort to examine the fish resources (native and exotic) of the Colorado River system in the Grand Canyon, and the effects of fluctuating flows upon them (Maddux *et al.* 1987), Kondolf *et al.* (1989) sampled trout redds and potential spawning gravels in the Colorado River and tributaries below Glen Canyon Dam in 1985 (Figure 13.26). Mainstem gravels were sampled by shovel from three gravel bars exposed by low river levels, removed to the laboratory, dried, and sieved. Tributary samples were obtained within a few hundred meters of the Colorado River confluence, using a 25-cm PVC-bucket with the bottom removed. The latter sampling method was dictated by logistics of the remote tributary sites, the small size of the trout redds, and the often limited extent of gravel deposits in which they occurred. Samples were obtained from both redd and potential

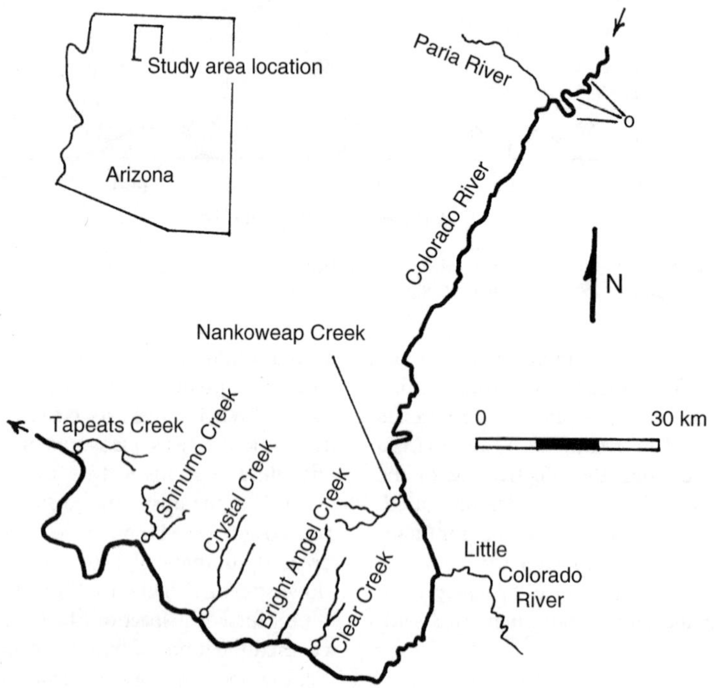

Figure 13.26 Location map of Colorado River and tributaries, Grand Canyon National Park, Arizona. Spawning gravel sample locations (on Nankoweap, Clear, Crystal, Shinumo, and Tapeats Creeks, and on the mainstem) are shown in open circles. Other important tributaries also shown for reference (reproduced from Kondolf 2000)

Figure 13.27 Pocket gravel on lower Nankoweap Creek, view from right bank (photograph by Kondolf, December 1985; reproduced from Kondolf 2000)

spawning gravels, dried by sun or in buckets over a campfire, and sieved and weighed on-site, except for subsamples of the fine fraction retained for laboratory sieving.

Case Study Results

Cumulative size distributions for all samples are reported in Kondolf *et al.* (1989), curves for mainstem and Nankoweap Creek gravels are shown in Figure 13.2, and box-and-whisker plots of all samples in Figure 13.3. These distributions illustrate the wider range of gravel sizes in the tributaries, reflecting greater variability in hydraulic conditions. These distributions also illustrate the larger percentage of fine sediments in the mainstem gravels, as reflected in the fine tails of the mainstem distributions. Because most samples were taken at or adjacent to redds, the ability of the fish to move the gravels was generally not at issue. The D_{50} values were similar to those reported in the literature for rainbow trout elsewhere (Figure 13.3, Table 13.5) and fall within the range of D_{50} values expected for these spawning females, which average 40-to-45-cm long (13.20).

In assessing the fine-sediment content, the potential spawning gravels had less than 7% finer than 1 mm, and the redds even less, values well below the standard from laboratory studies and other values reported for rainbow trout even before adjusting for the probable cleaning effect of redd digging (Table 13.5). In the mainstem, the redd gravels had less than 7% finer than 1 mm, but the one sample of potential spawning gravel had 15% finer than 1 mm. This would exceed the standard of 12%, but with the expected effects of spawning taken into account (Figure 13.18), the percentage of fine sediment in the redd gravels would be less than 10%. Thus, the quality of these gravels was quite good.

Table 13.5 Grain size statistics for rainbow trout spawning gravels reported in the literature and in the Colorado River and tributaries, Grand Canyon National Park (from Kondolf *et al.* 1989); n is the number of samples, D_{50} is the median size, D_g is the geometric mean diameter, sg is the geometric sorting coefficient, sk is the geometric skewness coefficient (see text for definitions), Nd indicates no data

Location	Reference	Fish size	n	D_{50} (cm)	D_g	sg (mm)	sk (mm)	Percent less than	
								0.85 mm	3.4 mm
Lardeau River, British Columbia	Hartman and Galbraith (1970)	75	6	23.5	14.7	3.6	−0.37	6.3	15.2
N. Fork Boise River, Idaho	Platts *et al.* (1979)	30	45	20	12.4	6.5	−0.25	7	24
Missouri River, Montana	Spoon (1985)	44	27	12.5	8.3	4.6	−0.27	11.1	Nd
Beaver Creek, Montana	Spoon (1985)	44	19	15	9.3	4.9	−0.3	9.8	Nd
Colorado River, Arizona	Kondolf *et al.* (1989)								
tributary redds		40	9	33.4	24.5	3.04	−0.27	1.5	7.7
tributary unspawned		40	8	21.9	14.9	4.21	−0.25	6.5	17.8
mainstem redds (22–23)		45	2	10.5	5.6	5.4	−0.4	6.8	15.2
mainstem unspawned (24)		45	1	16	5.2	10.5	−0.47	15	25

Case Study Discussion

Although their quality was good, the extent of potential tributary spawning gravels was limited because of the small size of the channels and the often patchy distribution of gravels. Some of these gravels may be inaccessible at low river stage because of migration barriers. Mainstem gravels were limited at the time of sampling (1985), and the extent of suitable gravel bars probably continues to decrease and the grain size to coarsen as smaller, mobile gravels are transported from the reach by high flows without replacement from upstream. In repeated visual observations over 2 years, Maddux *et al.* (1987) noted large variations in the fine-sediment content of tributary gravels, presumably reflecting changes wrought by flash floods. Thus, repeated gravel sampling might be warranted here. Intragravel flow conditions were not evaluated.

13.16 CONCLUSION: SELECTING APPROPRIATE SAMPLING METHODS

Let the punishment fit the crime! This chapter's overriding conceptual idea is that the purposes of the study must be clearly articulated and the methods chosen should logically follow from the questions posed. When selecting a sampling method, we should ask ourselves why are we collecting these data? Precisely how do we plan to use them? And what confidence limits are required to answer the questions we pose of the data?

As suggested in Table 13.6, different sampling methods lend themselves to different questions. For example, the pebble count (Wolman 1954) was developed to quantify grain size for surface roughness, to avoid the need for inconveniently large bulk samples. It is a simple and useful technique for measuring the size distribution of the surface layer (e.g., for estimating grain roughness, predicting bed mobilization thresholds, assessing framework size of spawning gravels, or tracking changes in surficial fine-sediment content in specific geomorphic channel units), but it does not address subsurface size distributions (including interstitial fine-sediment content), nor does it address the distribution of surficial sediments beyond the unit sampled (e.g., the bar, riffle, or other geomorphic feature on which the pebble count is performed).

To calculate bedload transport, subsurface bulk samples are needed if we assume that the subsurface grain size distribution approximates that of the bedload. But bulk sampling is labor intensive and cannot be done over the entire bed. Moreover, since there is uncertainty whether the subsurface material in a particular channel does in fact represent bedload, or if it is too coarse, it might be safer to sample the surface layer and use Parker's (1990) surface-based equation. To assess interstitial fine-sediment content of gravels to assess the quality of spawning gravels (or for other ecological studies) requires subsurface bulk samples. Again there is the problem that they are too labor intensive to do everywhere, so they might be done in specific units such as important spawning gravels and assumed to apply to other spawning beds. However, in other cases there is a need to extrapolate these results over a broader reach of river. Rather than assume that the unsampled 99.9% of the channel bed is represented by the 0.1% sampled, a very different approach is to map the facies (as in Figure 13.7), one providing less detailed information (no size distribution curves) about any one point, but useful information over a wide area. In the Oregon Branch example in Wolman and Schick (1967), the question concerned the effects of increased sediment yield from part of the catchment, and the facies map showed where in the channel the sediment deposited and the proportion of the bed surface covered by fine sediment. It can be particularly powerful to combine facies mapping with pebble counts or bulk sampling of specific facies units, with the former providing an overall context and the latter providing specific size distribution data.

ACKNOWLEDGMENTS

This chapter was improved significantly by review comments from Hervé Piégay and Stephen Rice, who suggested many excellent, substantive revisions. Peter Wilcock shared graphic materials from the Trinity River study. Erin Lutrick assisted greatly in figure preparation.

Table 13.6 Advantages and disadvantages of various tools for sampling bed material

Objective	Method	Advantages/disadvantages	Reference
• Estimate grain roughness	Pebble count	Produces reproducible estimate of surface grain size	Wolman (1954)
• Calculate bed mobility threshold	Pebble count or bulk sample	As the grain size of the surface layer controls the gravel bed mobilization threshold, the surficial size distribution is relevant, and easily obtained via pebble count	
• Calculate bedload transport assuming subsurface sediment reflects bedload	Bulk sampling	If subsurface sediment reflects bedload in transit, subsurface sample needed. Captures more mobile fine-grained tail where present	Parker and Klingeman (1982)
• Calculate bedload transport without assuming subsurface reflects bedload	Pebble count	Where the relation between subsurface grain size distribution and size of bedload in transit is unknown, can use surface-based equation. This may underestimate transport by missing finer fraction where present	Parker (1990)
• Assess suitability of framework size of salmonid spawning gravels	Pebble count or bulk sample	Pebble count is faster and provides reasonable estimate	Kondolf (2000)
• Assess interstitial fine-sediment content of spawning gravels	Bulk sample	Fine-sediment content not visible from surface inspection alone, as gravels that seem 'clean' on the surface may contain high interstitial fine-sediment subsurface	Kondolf (2000)
• Assess vertical distribution of fine sediments in gravel	Freeze-core sample	Provides vertical stratification, but laboratory observations show stratification may be disrupted by insertion of probe	Everest *et al.* (1980), Adams and Beschta (1980)

REFERENCES

Adams, J. 1979. Gravel size analysis from photographs. *Journal of the Hydraulics Division, American Society of Civil Engineers* 105: 1247–1285.

Adams, J.N. and Beschta, R.L. 1980. Gravel bed composition in Oregon coastal streams. *Canadian Journal of Fisheries and Aquatic Sciences* 37: 1514–1521.

ASTM (American Society of Testing and Materials). 1978. Standard methods of sampling aggregates, ANSI/ASTM D75-71.

Bagnold, R.A. 1941. *The Physics of Blown Sand and Desert Dunes*, London: Methuen, 265 p.

Barnard, K. and McBain, S. 1994. Standpipe to determine permeability, dissolved oxygen, and vertical particle size distribution in salmonid spawning gravels. *Currents* (US Fish and Wildlife Service Fish Habitat Relationships Technical Bulletin) 15: 1–12.

Bathurst, J.C. 1981. Discussion of bar resistance of gravel-bed streams. *Journal of the Hydraulics Division, ASCE* 107 (HY10): 1276–1278.

Benjamin, J.R. and Cornell, C.A. 1970. *Probability, Statistics, and Decision for Civil Engineers*, New York: McGraw-Hill, 682 p.

Beschta, R.L. 1982. Comment on 'stream system evaluation with emphasis on spawning habitat for salmonids' by Mostafa A. Shirazi and Wayne K. Seim. *Water Resources Research* 18: 1292–1295.

Beschta, R.L. and Jackson, W.L. 1979. The intrusion of fine sediments into a stable gravel bed. *Journal of the Fisheries Research Board of Canada* 36: 204–210.

Bevenger, G.S. and King, R.M. 1995. A pebble count procedure for assessing watershed cumulative effects, USDA Forest Service Research Paper RM-RP-319, Rocky Mountain Forest and Range Experiment Station, Fort Collins, Colorado.

Bjornn, T.C. 1969. Embryo survival and emergence studies, Job Completion Report, Job No. 5, Project F-49-R-7, Idaho Fish and Game Department.

Bovee, K.D. 1982. A guide to stream habitat analysis using the instream flow incremental methodology, US Fish and Wildlife Service FWS/OBS-82/86, Instream Flow Information Paper 12.

Brush, L.M. 1961. Drainage basins, channels and flow characteristics of selected streams in Central Pennsylvania, US Geological Survey Professional Paper 282-F.

BSI (British Standards Institution). 1975. British standard methods for sampling and testing of mineral aggregates, sands and fillers. Part I. Sampling, size, shape and classification, BS 812 Part I: 1975.

Buffington, J.M., Montgomery, D.R. 1999a. A procedure for classifying textural facies in gravel-bed rivers. *Water Resources Research* 35(11): 1903–1914.

Buffington, J.M., Montgomery, D.R. 1999b. Effects of sediment supply on surface textures of gravel-bed rivers, *Water Resources Research 35*(11): 3523–3530.

Bunte, K. and Abt, S.R. 2001a. Sampling frame for improving pebble count accuracy in coarse gravel-bed streams. *Journal of the American Water Resources Association* 37 (4): 1001–1014.

Bunte, K. and Abt, S.R. 2001b. Sampling surface and subsurface particle-size distributions in wadable gravel- and cobble-bed streams for analyses in sediment transport, hydraulic, and streambed monitoring, USDA Forest Service, Rocky Mountain Research Station, General Technical Report RMRSS-GTR-74.

Burns, D.C. 1978. Spawning bed sedimentation studies in northern California streams. *California Fish and Game* 56: 253–270.

Butler, J.B., Lane, S.N. and Chandler, 2001. Automated extraction of grain-size data for gravel surfaces using digital image processing. *Journal of Hydraulic Research*, 39: 519–529.

Carling, P.A. and McCahon, C.P. 1987. Natural siltations of brown trout (*Salmo trutta*) spawning gravels during low-flow conditions. In: Craig, J.F. and Kemper, J.B., eds., *Regulated Streams: Advances in Ecology*, New York: Plenum Press, pp. 229–244.

Carling, P.A. and Reader, N.A. 1982. Structure, composition and bulk properties of upland stream gravels. *Earth Surface Processes and Landforms* 7: 349–365.

Cederholm, C.J. and Salo, E.O. 1979. The effects of logging road landslide siltation on the salmon and trout spawning gravels of Stequaleho Creek and the Clearwater River Basin, Jefferson County, Washington, 1972–1978, Fisheries Research Institute, University of Washington, Seattle, Report No. FRI-UW-7915.

Chambers, J.S., Allen, G.H. and Pressey, R.T. 1955. Research relating to study of spawning grounds in natural areas, Annual report to US Army Corps of Engineers, Contract DA-35026-Eng-20572 (available from Washington State Department of Fisheries, Olympia).

Chambers, J.S., Pressey, R.T., Donaldson, J.R. and McKinley, W.R. 1954. Research relating to study of spawning grounds in natural areas, Annual report to US Army Corps of Engineers, Contract DA-35026-Eng-20572 (available from State of Washington, Department of Fisheries, Olympia).

Chapman, D.W., Weitkamp, D.E., Welsh, T.L. and Schadt, T.H. 1984. Effects of minimum flow regimes on fall Chinook spawning at Vernita Bar, 1978–82, Don Chapman Consultants and Parametrix, Report to Grant County Public Util. Dist., Ephrata WA.

Church, M.A., McLean, D.G. and Wolcott, J.F. 1987. River bed gravels: sampling and analysis. In: Thorne, C.R., Bathurst, J.C., Hey, R.D., eds., *Sediment Transport in Gravel Bed Rivers*, New York: John Wiley and Sons, pp. 43–79.

Cope, O.B. 1957. The choice of spawning sites by cutthroat trout. *Proceedings of Utah Academy of Science, Arts, and Letters* 34: 73–79.

Curtin, G.C. 1978. A tubular-scoop sampler for stream sediments. *Journal of Geochemical Exploration* 10: 193–194.

Dietrich, W.E., Kirchner, J.W., Ikeda, H. and Iseya, F. 1989. Sediment supply and development of coarse surface layer in gravel bedded rivers. *Nature* 340: 215–217.

Dunne, T. and Leopold, L.B. 1978. *Water in Environmental Planning*, New York: W.H. Freeman and Company, 666 p.

Everest, F.L., McLemore, C.E. and Ward, J.F. 1980. *An Improved Tri-tube Cryogenic Gravel Sampler*, US Forest Service: Pacific Northwest Forest and Range Exp. Station, Portland, Oregon, 8 p. (PNW-350).

Everest, F.L., Beschta, R.L., Scrivener, J.C., Koski, K.V., Sedell, J.R. and Cederholm, C.J. 1987. Fine sediment and salmonid production – a paradox. In: Salo, E.O. and Cundy, T.W., eds., *Streamside Management: Forestry and Fishery Interactions*, Seattle: College of Forest Resources, University of Washington, pp. 98–142.

Ferguson, R.I. and Paola, C. 1997. Bias and precision of percentiles of bulk grain size distributions. *Earth Surface Processes and Landforms* 22: 1061–1077.

Folk, R.L. 1968. *Petrology of Sedimentary Rocks*, Austin, Texas: Hemphill's, 170 pp.

Fripp, J.B. and Diplas, P. 1993. Surface sampling in gravel streams: American Civil Engineers. *Journal of Hydraulic Engineering* 119: 473–490.

Gomez, B. 1984. Typology of segregated (armor/paved) surfaces: some comments. *Earth Surface Processes and Landforms* 9: 19–24.

Greeley, J.R. 1932. The spawning habits of brook, brown, and rainbow trout and the problem of egg predators. *Transactions of the American Fisheries Society* 62: 239–248.

Griffiths, J.C. 1967. *Scientific Method in the Analysis of Sediments*, New York: McGraw-Hill, 508 p.

Groot, C. and Margolis, L. 1991. *Pacific Salmon Life Histories*, Vancouver: University of British Columbia Press, 564 p.

Hartman, G.F. and Galbraith, D.M. 1970. *The Reproductive Environment of the Gerrard Stock Rainbow Trout*, Fisheries Management Publication No 15, Victoria, BC: Department of Recreation and Conservation, Fisheries Research Section, 51 p.

Hausle, D.A. and Coble, D.W. 1976. Influence of sand in redds on survival and emergence of brook trout (*Salvelinus fontinalis*). *Transactions of the American Fisheries Society* 105: 57–63.

Hazzard, A.S. 1932. Some phases of the life history of the eastern brook trout *Salvelinus fontinalis* (Mitchill). *Transactions of the American Fisheries Society* 62: 344–350.

Hey, R.D. and Thorne, C.R. 1983. Accuracy of surface samples from gravel bed material. *Journal of Hydrological Engineering* 109: 842–851.

Hilton, S. and Lisle, T.E. 1993. Measuring the fraction of pool volume filled with fine sediment, Albany, California: USDA Forest Service, Pacific Southwest Research Station, PSW Research Note PSW-RN-414, 11 p.

Horton, J.L. and Rogers, D.W. 1969. The optimum streamflow requirements for king salmon spawning in the Van Duzen River, Humboldt County, California, Administrative Report 69-2, California Department of Fish and Game, Water Projects Branch, Sacramento.

Hunter, J.W. 1973. A discussion of game fish in the state of Washington as related to water requirements, Unpublished Report available from Fishery Management Div., Wash. State Dept. Game, 66 p.

Ibbeken, H. 1983. Jointed source rock and fluvial gravels controlled by Rosin's law: a grain size study in Calabria, South Italy. *Journal of Sedimentary Petrology* 53: 1213–1231.

Ibbeken, H., and Schleyer, R. 1986. Photo-sieving: a method for grain-size analysis of coarse-grained, unconsolidated bedding surfaces. *Earth Surfaces Processes and Landforms* 11: 59–77.

Inman, D.L. 1952. Measures for describing the size distribution of sediments. *Journal of Sedimentary Petrology* 22: 125–145.

Irving, J.S. and Bjornn, T.C. 1984. Effects of substrate size composition on survival of kokanee salmon and cutthroat trout and rainbow trout embryos, Compl. Rept. For Cooperative Agreement No. 12-11-204-11, Supplement No. 87, Boise, Idaho, USDA Forestry Service Int. For. Range Exp. Sta.

ISO (International Standards Organization). 1977. Liquid flow measurements in open channels-bed material sampling, ISO 4364-1977 (E).

Kellerhals, R. and Bray, D.I. 1971. Sampling procedures for coarse fluvial sediments. *Journal of Hydraulics Division of the American Society of Civil Engineers* 97: 1165–1179.

Klein, R.D. 1987. Stream channel adjustments following logging road removal in Redwood National Park, MS Thesis, Humboldt State University, Arcata, California.

Klingeman, P.C. and Emmett, W.W. 1982. Gravel bed-load transport processes. In: Hey, R.D., *et al.*, eds., *Gravel Bed Rivers*, Chichester, UK: Wiley and Sons, pp. 141–179.

Kondolf, G.M. 1988. Salmonid spawning gravels: a geomorphic perspective on their distribution, size modification by spawning fish, and application of criteria for gravel quality. Ph.D. Thesis, The Johns Hopkins University, Baltimore.

Kondolf, G.M. 1997. Application of the pebble count: reflections on purpose, method, and variants. *Journal of the American Water Resources Association* (formerly *Water Resources Bulletin*) 33(1): 79–87.

Kondolf, G.M. 2000. Assessing salmonid spawning gravels. *Transactions of the American Fisheries Society* 129: 262–281.

Kondolf, G.M. and Adhikari, A. 2000. Weibull vs. lognormal distributions for fluvial gravels. *Journal of Sedimentary Research* 70(3): 456–460.

Kondolf, G.M., Cook, S.S., Maddux, H.R. and Persons, W.R. 1989. Spawning gravels of rainbow trout in the Grand Canyon, Arizona. *Journal of the Arizona-Nevada Academy of Science* 23: 19–28.

Kondolf, G.M. and Li, S. 1992. The pebble count technique for quantifying surface bed material in instream flow studies. *Rivers* 3: 80–87.

Kondolf, G.M. and Matthews, W.V.G. 1986. Transport of tracer gravels on a coastal California River. *Journal of Hydrology* 85: 265–280.

Kondolf, G.M. and Wolman, M.G. 1993. The sizes of salmonid spawning gravels. Water Resources Research 29: 2275–2285.

Kondolf, G.M., Cook, S.S., Maddux, H.R. and Persons, W. R. 1989. Spawning gravels of rainbow trout in Glen and Grand Canyons, Arizona. *Journal of the Arizona-Nevada Academy of Science* 23: 19–28.

Kondolf, G.M., Sale, M.J. and Wolman, M.G. 1993. Modification of gravel size by spawning salmonids. *Water Resources Research* 29: 2265–2274.

Koski, K.U. 1966. The survival of coho salmon (*Oncorhynchus kisutch*) from egg deposition to emergence in three Oregon coastal streams, MS Thesis, Oregon State University, Corvallis.

Koski, K.V. 1975. The survival and fitness of the two stocks of chum salmon (*Oncorhynchus keta*) from egg deposition to emergence in a controlled-stream environment at Big Beef Creek, Ph.D. Thesis, University of Washington.

Koski, K.V. 1981. The survival and quality of two stocks of chum salmon (*Oncorhynchus keta*) form egg deposition to emergence. *Rapp. P.-v. Reun. Cons.int. Explor. Mer.* 178: 330–333.

Krumbein, W.C. 1934. Size Frequency Distribution of Sediments. *Journal of Sedimentary Petrology* 4: 65–77.

Krumbein, W.C. 1941. Measurement and Geologic Significance of Shape and Roundness of Sedimentary Particles. *Journal of Sedimentary Petrology* 11: 64–72.

Krumbein, W.C. and Pettijohn, F.J. 1938. *Manual of Sedimentary Petrography*, New York: Appleton-Century, Inc., pp. 31–33.

Lane, S.N. 2002. The measurement of gravel-bed river morphology. In: Mosley, M.P., ed., *Gravel Bed Rivers V*, Wellington: New Zealand Hydrological Society, pp. 291–311.

Leopold, L.B. 1970. An improved method for size distribution of stream-bed gravel. *Water Resources Research* 6: 1357–1366.

Leopold, L.B. 1992. Sediment size that determines channel morphology. In: Billi, P., Hey, R.D., Thorne, C.R. and Tacconi, P., eds., *Dynamics of Gravel-bed Rivers*, Chichester, UK: John Wiley and Sons, pp. 297–311.

Lisle, T.E. 1989. Sediment transport and resulting deposition in spawning gravels, north coastal California. *Water Resources Research* 25(6): 1303–1319.

Lisle, T.E. 1995. Particle size variations between bed load and bed material in natural gravel bed channels. *Water Resources Research* 31(4): 1107–1118.

Lisle, T.E. and Eads, R.E. 1991. Methods to measure sedimentation of spawning gravels, US Forest Service Research Note PSW-411, US Forest Service Research Pacific Southwest Research Station, Berkeley, California.

Lisle, T.E. and Hilton, S. 1992, The volume of fine sediment in pools: an index of the supply of mobile sediment in stream channels. *Water Resources Bulletin* 28(2): 371–383.

Lisle, T.E. and Madej, M.A. 1992. Spatial variation in armouring in a channel with high sediment supply. In: Billi, R.D.H.P., Thorne, C.R., Tacconi, P., eds., *Dynamics of Gravel-bed Rivers*, Chichester, UK: John Wiley and Sons, pp. 277–291.

Lisle, T.E. and Hilton, S. 1999. Fine bed material on pools of natural gravel bed channels. *Water Resources Research* 35 (4): 1291–1304.

Lisle, T.E., Nelson, J.M., Pitlick, J., Madej, M.A. and Barkett, B.L. 2000. Variability of bed mobility in natural gravel-bed channels and adjustments to sediment load at the local and reach scales. *Water Resources Research* 36 (12): 3743–3756.

Lotspeich, F.B. and Everest, F.H. 1981. A new method for reporting and interpreting textural composition of spawning gravel, US Forest Service Research Note PNW-369.

Maddux, H.R., Kulby, D.M., deVos, J.C., Jr., Persons, W. R., Staedicke, R. and Wright, R.L. 1987. Effects of varied flow regimes on aquatic resources of Glen and Grand Canyons. Final Report prepared for US Department of Interior, Bureau of Reclamation under contract No. 4-AG-40-01810, 291 p.

Marcus, W.A., Ladd, S.C., Stoughton, J.A. and Stock, J.W. 1995. Pebble counts and the role of user-dependent bias in documenting sediment size distributions. *Water Resources Research* 31(10): 2625–2631.

McCuddin, M.E. 1977. Survival of salmon and trout embryos and fry in gravel-sand mixtures, MS Thesis, University of Idaho, Moscow, ID (as cited by Chapman and McLeod 1987).

McNeil, W.J. and Ahnell, W.H. 1964. Success of pink salmon spawning relative to size of spawning bed materials, US Fish and Wildlife Service, Special Scientific Report – Fisheries 469.

Meehan, W.R. 1991. *Influences of Forest and Rangeland Management on Salmonid Fishes and Their Habitats*, American Fisheries Society Special Publication 19.

Milhous, R.T., Hogan, S.A., Abt, S.R. and Watson, C.C. 1995. Sampling river-bed material: the barrel sampler. *Rivers* 5: 239–249.

Mosley, M.P. and Tinsdale, D.S. 1985. Sediment variability and bed material sampling in gravel bed rivers. *Earth Surface Procedures and Landforms* 10: 465–482.

NCASI (National Council of the Paper Industry for Air and Stream Improvement). 1984. A laboratory study of the effects of sediments of different size characteristics on survival of rainbow trout (*Salmo gairdneri*) embryos to fry emergence. *National Council of the Paper Industry for Air and Stream Improvement Technical Bulletin* 429.

Orcutt, D.R., Pulliam, B.R. and Arp, A. 1968. Characteristics of steelhead trout redds in Idaho streams. *Transactions of the American Fisheries Society* 97: 42–45.

Otto, G.H. 1939. A modified logarithmic probability graph for interpretation of mechanical analyses of sediments. *Journal Sedimentary Petrology* 9: 62–76.

Parfitt, D. and Buer, K. 1980. *Upper Sacramento River Spawning Gravel Study*, Red Bluff: California Department of Water Resources, Northern Division.

Parker, G. 1990. Surface-based bedload transport relation for gravel rivers, *Journal of Hydraulic Research* 28(4): 417–436.

Parker, G. and Andrews, E.D. 1985. Sorting of bed load sediment by flow in meander bends, *Water Resources Research* 21(9): 1361–1373.

Parker, G. and Klingeman, P.C. 1982. On why gravel bed streams are paved. *Water Resources Research* 18: 1409–1423.

Peterson, R.H. 1978. Physical characteristics of Atlantic salmon spawning gravels in some New Brunswick streams, Fisheries and Oceans Canada, Biological Station, Fisheries and Marine Service Technical Report 785, St. Andrews, New Brunswick.

Petrie, J. and Diplas, P. 2000. Statistical approach to sediment sampling accuracy. *Water Resources Research* 36(2): 597–606.

Pettijohn, F.J. 1975. *Sedimentary Rocks*, Third Edition, New York: Harper & Row, 628 p.

Phillips, R.W., Lantz, R.L., Claire, E.W. and Moring, J.R. 1975. Some effects of gravel mixtures on emergence of coho salmon and steelhead trout fry. *Transactions of the American Fisheries Society* 104: 461–466.

Platts, W.S., Megahan, W.F. and Minshall, G.W. 1983. Methods for evaluating stream, riparian, and biotic conditions, USDA Forest Service General Technical Report INT-138, 70 p.

Platts, W.S. and Penton, V.E. 1980. A new freezing technique for sampling salmonid redds, Paper INT-248, USDA Forest Service Intermountain Research Station, 22 p.

Platts, W.S., Shirazi, M.A. and Lewis, D.H. 1979. Sediment particle sizes used by salmon for spawning with methods for evaluation, US Environmental Protection Agency Report No. EPA600/3-79-043, 32 p.

Rice, S. and Church, M. 1996. Sampling surficial fluvial gravels: the precision of size distribution percentile estimates, *Journal of Sedimentary Research* 66(3): 654–665.

Rice, S. and Church, M. 1998. Grain size along two gravel-bed rivers: statistical variation, spatial pattern and sedimentary links. *Earth Surface Processes and Landforms* 23: 345–363.

Rollet, A.J., Kauffman, B., Piégay, H. and Lacaze, B. 2002. Automated grain size percentiles assessment using image analysis: example of the bars of the Ain River France. In: *Symposium of the British Hydrological Society: River Bed Patches: Hydraulics, Ecology and Geomorphology, Department of Geography, Loughborough University, 8th May 2002* (unpublished poster).

Rood, K. and Church, M. 1994. Modified freeze-core technique for sampling the permanently-wetted streambed. *North American Journal of Fisheries Management* 14: 852–861.

Rosgen, D.L. 1996. *Applied River Morphology*, Pagosa Springs, Colorado: (privately published by) Wildland Hydrology.

Seal, R. and Paola, C. 1995. Observations of downstream fining on the North Fork Toutle River near Mount St. Helens, Washington. *Water Resources Research* 31: 1409–1419.

Shirazi, M.A. and Seim, W.K. 1981. Stream system evaluation with emphasis on spawning habitat for salmonids. *Water Resources Research* 17: 592–594.

Shirazi, M.A. and Seim, W.K. 1982. Reply. *Water Resources Research* 18: 1296–1298.

Shirazi, M.A., Seim, W.K. and Lewis, D.H. 1981. Characterization of spawning gravel and stream system evaluation. In: *Salmon-spawning Gravel: A Renewal Resource in the Pacific Northwest?* Report 39, Pullman: Washington State University, State of Washington Water Research Center, pp. 227–278.

Spoon, R.L. 1985. Reproduction biology of brown and rainbow trout below Hauser Dam, Missouri River, with reference to proposed hydroelectric peaking, MS Thesis, Montana State University, Bozeman, 144 p.

Surian, N. 2002. Downstream variation in grain size along an Alpine River: analysis of controls and processes. *Geomorphology* 43: 137–149.

Tagart, J.V. 1984. Coho salmon survival from egg deposition to emergence. In: Walton, J.M. and Houston, D.B., eds., *Proceedings of the Olympic Wild Fish Conference, Port Angeles, Washington, March 1983*, pp. 173–181.

Tappel, P.D. and Bjornn, T.C. 1983. A new method of relating size of spawning gravel to salmonid embryo survival. *North American Journal of Fisheries Management* 3: 123–135.

Terhune, L.B.D. 1958. The Mark VI groundwater standpipe for measuring seepage though salmon spawning gravel. *Journal of the Fisheries Research Board of Canada* 15(5): 1027–1063.

Trask, P.D. 1932. *Origin and Environment of Source Sediments of Petroleum*, Houston: Gulf Publications & Co. (as cited by Inman 1952).

Tukey, J.W. 1977. *Exploratory Data Analysis*, Reading, Massachusetts: Addison-Wesley.

USFWS. 1999. Trinity River flow evaluation, Final report to the Secretary of the Interior by the US Fish and Wildlife Service, Arcata, and Hoopa Valley Tribe, Hoopa, California, 307 p. + appendices.

Van Woert, W.F. and Smith, E.J., Jr. 1962. Upper Sacramento River tributary study: relationship between stream flow and usable spawning gravel, Cottonwood and Cow Creek, Tehama and Shasta Counties, Calif. Dept. Fish and Game, Redding (unpublished office report).

Vanoni, V.A. 1975. *Sedimentation Engineering*, New York: American Society of Civil Engineers.

Vincent-Lang, D., Hoffman, A., Bingham, A.E., Estes, C., Hilliard, D., Stewart, C., Trihey, E.W. and Crumley, S. 1984. An evaluation of chum and sockeye salmon spawning habitat in sloughs and side channels of the Middle Susitna River. In: *Aquatic Habitat and Instream Flow investigations (May–October 1983)*, Report to Alaska Authority, Anchorage: Alaska Dept. Fish and Game (Chapter 7).

Vining, L.J., Blakely, J.S. and Freeman, B.M. 1985. *An Evaluation of the Incubation Life-phase of Chum Salmon in the Middle Susitna River, Alaska, Rep. 5*, Anchorage: Alaska Department of Fish and Game.

Weaver, T.M. and White, R.G. 1985. Coal Creek fisheries monitoring study no. III, Report to the USDA Forestry Service, Flathead National Forest, Bozeman, MT, Coop. Fish Res. Unit, Montana State University.

Wentworth, C.K. 1926. Methods of mechanical analysis of sediments. *University of Iowa Studies in Natural History* II: 3–52.

Wilcock, P.R., Kondolf, G.M., Matthews, W.V.G. and Barta, A.F. 1996a. Specification of sediment maintenance flows for a large gravel-bed river. *Water Resources Research* 32: 2911–2921.

Wilcock, P.R., Barta, A.F., Shea, C.C., Kondolf, G.M., Matthews, W.V. and Pitlick, J. 1996b. Observations of flow and sediment entrainment on a large gravel-bed river. *Water Resources Research* 32(9): 2897–2909.

Wilcock, P.R., Kondolf, G.M., Barta, A.F., Matthews, W.V. G. and Shea, C.C. 1995. Spawning gravel flushing during trial reservoir releases on the Trinity River: field observations and recommendations for sediment maintenance flushing flows, Report CEDR-05-95, Center for Environmental Design Research, University of California at Berkeley.

Williams, H., Turner, F.J. and Gilbert, C.M. 1982. *Petrography: An Introduction to the Study of Rocks in Thin sections*, 2nd edition, San Francisco: W.H. Freeman.

Wilson, W.J., Trihey, E.W., Baldrige, J.E., Evans, C.D., Thiele, J.G. and Trudgen, D.E. 1981. *An Assessment of Environmental Effects of Construction and Operation of the Proposed Terror Lake Hydroelectric Facility, Kodiak Island, Alaska*, Report to Kodiak Elec. Assn, Inc., Arctic Environment Information and Data Center, University of Alaska, Anchorage.

Wohl, E.E., Anthony, D.J., Madsen, S.W. and Thompson, D.M. 1996. A comparison of surface sampling methods for coarse fluvial sediments. *Water Resources Research* 32 (10): 3219–3226.

Wolman, M.G. 1954. A method of sampling coarse river-bed material. *Transactions American Geophysical Union* 35: 951–956.

Wolman, M.G. and Schick, A.P. 1967. Effects of construction on fluvial sediment, urban and suburban areas of Maryland. *Water Resources Research* 3: 451–464.

Yalin, M.S. 1977. *Mechanics of Sediment Transport*, Oxford: Permagon Press, 298 p.

Young, M.K., Hubert, W.A. and Wesche, T.A. 1991. Biases associated with four stream substrate samplers. *Canadian Journal Fisheries and Aquatic Sciences* 48: 1882–1886.

Zingg, T. 1935. Beiträge zur Schotteranalyse. *Min. Petrog. Mitt. Schweiz.* 15: 39–140 (as cited by Pettijohn 1975).

14

Use of Tracers in Fluvial Geomorphology

MARWAN A. HASSAN[1] AND PETER ERGENZINGER[2]
Department of Geography, University of British Columbia, Vancouver, Canada
Geographisches Institut, FU Berlin, B.E.R.G., Berlin

14.1 INTRODUCTION

In this chapter, we present a review of sediment tracing techniques in gravel and sand bed rivers. The primary objective is to assess critically the use of tracers to obtain quantitative information of bedload sediment transport, with general references to washload and suspended load. Since most of the tracing techniques are site-specific and to avoid tedious repetition, each method is briefly described. For more technical information the reader should refer to the original material. Although we are aware that the topic has been reviewed in many countries, we mainly provide a summary of material either originally in, or translated into, the English literature. After reviewing each tracing method, we present an evaluation of the techniques available for studying sediment transport in a fluvial environment.

Although considerable attention has been paid to specific influences that are known to determine river sediment transport rates and patterns, there still is a large inconsistency between data collected in the field and the results of theoretical and empirical models. This inconsistency is due to the large number of interrelated variables that affect sediment transport, and also due to the difficulties involved in field measurements. The numerous different field techniques have evolved to meet the complexities associated with understanding the processes of fluvial transport. A broad division can be made between the use, on the one hand, of traps and samplers in which moving sediment is collected during a flood event and, on the other hand, tracing the movement of individual grains between and during floods.

Characteristics of sediment movement would be much easier to obtain if it were possible to trace the movement of either the entire bed material or a given mass of individual grains. However, given inherent difficulties in field sampling, this is not possible. So tracers are particularly valuable to fluvial geomorphology as they provide a means of overcoming technical and sampling problems. Tracers are defined as marked particles that are introduced into streams in order to obtain general information on the movement of sediment in rivers. Such labeling must permit traced sediment to be detected, as well as operate within the fluvial environment the same way as the natural material. Tracers provide an opportunity to study the general characteristics of sediment movement under varying flow conditions without the need for a detailed kinematic study of the sedimentary regime (Crickmore *et al.* 1990). They reveal the long term action of the fluvial system under varying flow conditions, sediment supply, and channel morphology. Therefore, naturally or artificially labeled sediment can be used within channel reaches for tracing the paths and timing the rate of sediment movement.

The overall objective of using tracers is to obtain information on the fluvial transport of sediment. Tracers can be used to provide information on the rate and direction of sediment transportation, particle entrainment, periods of rest and movement of particles, step length of individual particles, residence time, flow competence, virtual rate of sediment movement, relations between distance of movement and flow strength, effect of physical characteristics on distance of travel, downstream fining, depth of the active layer, impact of particle sedimentological environment on distance of movement, sediment sources and depositional areas, volume of mobile sediment, and wearing rate. Also, tracers can yield useful information on the dispersion of contaminants within the fluvial system.

There are reports of bedload tracing programs in the field and laboratory since the late 1930s. To our knowledge, Einstein (1937) was the first to use tracers in a flume, while Takayama (1965) and Leopold *et al.* (1966) were pioneers in using painted tracers in the field. Since then, tracing natural clasts has become a common technique in many studies of river sediment transport. The most common method is to paint a clast so that it stands out from the rest of the bed material. However, the data are limited to material moving on the bed surface and hence the recovery rates of painted stones are extremely variable (see Hassan *et al.* 1984, Hassan and Church 1992), and usually low (about 20%). Most studies (e.g., Leopold *et al.* 1966, Church 1972, Schick and Sharon 1974, Ashworth 1987) have emphasized the lack of simple relations between distance of travel and particle size. Since the late 1950s, new tracer techniques have been developed for both field and laboratory studies. The most productive investigations include those using radioactive tracers (e.g., Ramette and Heuzel 1962, Crickmore and Lean 1962a,b, Hubbell and Sayre 1964, Sayre and Hubbell 1965, Crickmore 1967, Stelczer 1968, Michalik and Bartnik 1994), fluorescent coatings (e.g., Crickmore 1967, Nordin and Rathbun 1970, Rathbun *et al.* 1971, Rathbun and Kennedy 1978), ferric coatings (Nir 1964), metal collars (Butler 1977), natural magnetism (e.g., Ergenzinger and Custer 1983, Spieker and Ergenzinger 1990), enhanced magnetism (e.g., Arkell *et al.* 1983, Sear 1992), inserted magnets (e.g., Hassan *et al.* 1984, Hassan and Church 1992, Schmidt and Ergenzinger 1992, Ferguson *et al.* 1996, Wathen *et al.* 1997, Ferguson and Wathen 1998), and radio transmitters (Ergenzinger *et al.* 1989, Chacho *et al.* 1989, Emmett *et al.* 1990, Busskamp 1993, 1994).

14.2 GENERAL OVERVIEW OF TRACER TECHNIQUES

There are two tracer techniques. One technique uses passive tracers, which must be seen or sensed by an observer or a detector. The other technique uses active tracers, which emit waves or rays detected by a spectrometer or receiver. Each approach has advantages and disadvantages (Table 14.1), including different recovery rates (Table 14.2). Passive tracers include exotic and painted particles, fluorescent paint, radioactive elements, ferruginous, and magnetic. The radio transmitter is the only available active tracing technique.

Table 14.1 Summary of advantages, disadvantages, and key research questions of various tracer methods

Method	Advantages	Disadvantages	Key questions
Exotic (material and minerals)	Cheap, easy to apply, visual inspection of the surface	Limited to the surface, low recovery rate	Downstream fining, wearing rate, sources and destinations, spatial dispersion
Painting	Cheap, easy to apply, visual inspection of the surface	Low recovery rate, limited to the surface, paint abrades after a few events	Travel distance, flow competence, sources and destination, virtual velocity, spatial dispersion
Fluorescent	Non-toxic, simple to inject, cheap, long life, large quantity can be traced, useful for measurements on slowly evolving systems	Difficult to detect, limited to quantitative information, repeated studies are difficult in the case of long life dye, wears out quickly, adheres to untagged particles, affected by temperature, light, and salinity	Travel distance, flow competence, sources and destination, virtual velocity, spatial dispersion
Radioactive	High recovery rate, very powerful, detects both surface and buried particles	Toxic, hazard to environment and public, difficult to handle and inject, expensive, needs special laboratory and field equipment, limited to areas with no radioactive background, licensing constraints	Travel distance, flow competence, entrainment, virtual velocity, burial depth, 3-D dispersion, volume of mobile sediment, sources and destinations

(*Continues*)

Table 14.1 (*Continued*)

Method	Advantages	Disadvantages	Key questions
Iron oxide coating	Easy to apply, cheap	Limited to particles >11 mm, affected by noise from scrap metal, recovery may substantially disturb the bed, labor intensive, difficult to apply in deep water, coating abrades after few events	Travel distance, virtual velocity, depth of burial, 3-D dispersion, flow competence, sources and destinations, volume of mobile sediment
Metal strips or plugs	Easy to apply, cheap	Affected by background noise, collar separation from pebbles, limited to large particles, recovery may substantially disturb the bed, labor intensive, difficult to apply in deep water	As in iron oxide coating
Iron core	Long time expectancy, easy, cheap, can locate both buried and surface particles	Limited to large particles, low recovery rate, affected by iron mineral and scrap metal in the background, labor intensive, difficult to apply in deep water, might alter particle density, recovery may substantially disturb the bed	Travel distance, virtual velocity, depth of burial, 3-D dispersion, downstream fining, wearing rate, flow competence, sources and destinations, volume of mobile sediment
Inserted magnets	High recovery rate, can locate both buried and surface particles, cheap, easy to apply, long life expectancy	Labor intensive, limited to sizes >11 mm, affected by background noise, difficult to apply in deep water, recovery may substantially disturb the bed	As in iron core
Natural magnetic	No cost for tracers, unlimited number of tracers, easy to apply, long life expectancy	Difficult to trace individual particles, detection system is fixed, needs field equipment and well-trained technician, affected by background noise, recovery may substantially disturb the bed	As in iron core
Artificial magnetic	High recovery rate, easy to apply, long life expectancy	Moderately expensive, affected by background noise, recovery may substantially disturb the bed, physical particle characteristics are different from natural material	As in iron core
Magnetic enhancement	All sizes can be tagged, reasonable recovery rate, locate both buried and exposed particles	Cost of baking, requires furnace, high iron content in material, recovery may substantially disturb the bed, cracking under thermal stress, labor intensive, moderately expensive	As in iron core
Radio transmitters	Very powerful, high recovery rate, can locate both buried and surface particles, continuous information on particle position during flood	Very expensive, limited to large material, limited number of tracers	Step length, rest duration, entrainment, travel distance, burial depth, particle trajectory, flow competence, sources and destinations, volume of mobile sediment, virtual velocity

Table 14.2 Recovery rates of several bedload tracing experiments. Based partially on Hassan *et al.* (1984), Hassan and Church (1992), and Sear (1996)

Method	Reference	Tracers size range (mm)	Site	Recovery rate (%)
Exotic	Mosley (1978)	8–300	Tamaki River	5
	Kondolf and Matthews (1986)		Carmel River	
Painting	Einstein (1937)	17–24	Flume	High
	Takayama (1965)	22–128	Hayakawa	10–23
	Takayama (1965)	22–128	Fukugawa	21–27
	Takayama (1965)	22–128	Okawa	32–40
	Leopold *et al.* (1966)	75–150	Arroyo de les Frijoles	0–88
	Keller (1970)	Pebbles–cobbles	Dry Creek	41–65
	Church (1972)	Pebbles–cobbles	Ekalugad Rivers	High
	Slaymaker (1972)	Cobbles	Nant Calefwr	85–100
	Slaymaker (1972)	Cobbles	Nant Y Grader 9A	85–100
	Slaymaker (1972)	Cobbles	Nant Y Grader 9B	85–100
	Schick and Sharon (1974)	32–512	Nahal Yael	2.5–57
	Laronne and Carson (1976)	4–256	Seales Brook	5
	Thorne and Lewin (1979)	Pebbles–cobbles	Severn	40–79
	Leopold and Emmett (1981)	47–91	White Clay Creek	High
	Hassan *et al.* (1984)	45–180	Nahal Hebron	31–34
	Ashworth (1987)[a]	24–238	Allt Dubhaig	30–96
	Ashworth (1987)[a]	24–171	Feshie River	40–84
	Ashworth (1987)[a]	35–200	Lyngsdalselva	26–89
	Petit (1987)	Pebbles	La Rulles	High
	Tacconi *et al.* (1990)[b]	16–128	Virginio Creek	5–9
	Carling (1992)[b]	15–130	Carl Beck	98
	Carling (1992)[a]	15–130	Great Eggleshope	98
	Sear (1992, 1996)	Pebbles–cobbles	North Tyne	35–100
	Thorne (1996)[c]	Pebbles–cobbles	Swale	86
Radioactive	Ramette and Heuzel (1962)	25–75	Rhone	100
	Stelczer (1968, 1981)	8–34	Danube	100
	Michalik and Bartnik (1986)	2–25	Wisloka River	100
	Michalik and Bartnik (1986)	2–25	Dunajac River	100
Ferric coating	Nir (1964)	52–240	Nahal Zin	4.4
Metal strips	Butler (1977)	34–116	Horse Creek	35
Iron core	Hassan *et al.* (1984)	45–180	Nahal Hebron	31–34
	Schmidt and Ergenzinger (1992)	50–170	Lainbach	17–92
Magnetic: inserted and artificial	Froehlich (1982)	Pebbles–cobbles	Homerka	High
	Ergenzinger and Conrady (1982)	Cobbles	Buonamico	100
	Hassan *et al.* (1984)	45–180	Nahal Hebron	90–93

(*Continues*)

Table 14.2 (*Continued*)

Method	Reference	Tracers size range (mm)	Site	Recovery rate (%)
	Reid *et al.* (1984)	29	Turky Brook	100
	Hassan (1990)	45–180	Nahal Og	55–56
	Hassan and Church (1992)	16–512	Harris Creek	75
	Hassan and Church (1992)	16–180	Carnation Creek	80
	Lekach (1992)	45–180	Nahal Yael	100
	Schmidt and Ergenzinger (1992)[d]	60–137	Lainbach	25–100
	Hassan *et al.* (1995, 1999)	18–90	Metsemotlhaba	22–28
	Haschenburger (1996)	16–200	Carnation Creek	>80[e]
	Wathen *et al.* (1997)[f]	23–362	Allt Dubhaig	50–100
Enhanced	Arkell *et al.* (1983)	5.6–22.4	Plynlimon	63
Magnetism	Sear (1992, 1996)	<22	North Tyne	5
Natural magnetism	Ergenzinger and Custer (1983)	>5	Squaw Creek	
Radio	Ergenzinger *et al.* (1989)[g]	85–130	Lainbach	100
	Chacho *et al.* (1989)[h]	60–100	Toklat	100

Comments
[a] See also Ashworth and Ferguson (1989).
[b] Reported in Hassan and Church (1992).
[c] Reported in Sear (1996).
[d] See also Gintz *et al.* (1996).
[e] For complete surveys.
[f] See also Ferguson *et al.* (1996) and Ferguson and Wathen (1998).
[g] See also Schmidt and Ergenzinger (1992) and Busskamp (1993, 1994).
[h] See also Emmett *et al.* (1990).

Both exotic and artificial tracers have been used to further our understanding of sediment transport in rivers. Given favorable circumstances, the mineralogical characteristics of exotic particles could serve or be tagged as natural tracers. However, their use is limited and there is a need to tag non-exotic sediment. In the latter case, artificial tracers are collected from a given experimental site and then processed in order to distinguish them from the remainder of the bed material. Each can be marked with an identification code specific to the individual clast before being reseeded back into the river. Comparing positions of clasts before, during, and after a flow event can shed valuable light on the dispersion of individual clasts within the active bed layer.

There are aspects common to all tracer studies (Figure 14.1). Each experiment entails defining objectives, selecting methods and materials that will effectively allow realization of the objectives, determining sample size and recovery rate, injecting back into the river, detecting, data collecting, data analyzing, and interpreting. The successful use of tracers depends largely on the selection of methods and work procedures. Both study objectives and channel characteristics determine the type of tracers to be used, the duration of experiments, and data analyses. If the objective is to study downstream sediment sorting, then the experiment should be carried on for a long time; therefore a long life tracer should be selected, such as magnetic tracers in gravel bed rivers. In some cases, where it is important to repeat an experiment in the same channel, a short half-life of fluorescent sand or radioactive tracers can be used. If the focus of the study is the behavior of sediment within the active bed surface layer, then one should use magnetic or radioactive methods that are not limited to surface detection. In selecting a tracing method, channel size and morphology should be considered. Painted and magnetic tracers are suitable in small ephemeral channels or in shallow water where conditions promote a thorough search of the channel. In contrast, deep waters limit the possibility of recovering magnetically tagged

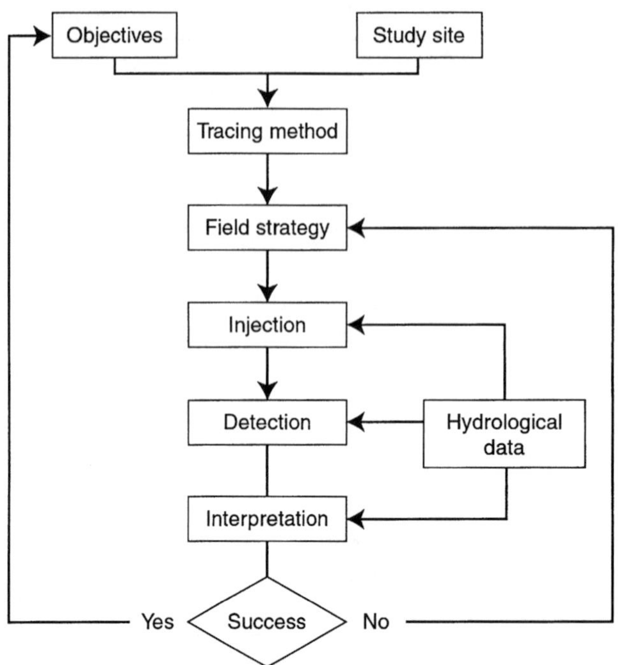

Figure 14.1 Flow chart of tracer experimental design

particles, so radioactive tracers or radio transmitters should be used.

Selecting a tagging technique also should ensure that the tracers are easily distinguishable but still accurately represent the natural sediment and its movement. This task is not easily achieved due to the fact that the tracing technique may change some of the particle's physical characteristics, such as its density, and hence its movement. In selecting a tagging technique one should consider the following:

1. suitability, including potential background noise,
2. cost,
3. recovery rate,
4. detectability at low concentration levels,
5. decay rate,
6. ease of handling in both the laboratory and the field, and
7. safety for both the handler and environment.

Recovery Rate and Population Size

Very little is known about the population size needed to represent the movement of sediment in a stream channel. Researchers have rarely addressed this issue and hence the labeled population size has typically ranged from a few particles up to several hundred. Appropriate sample size depends on study objectives and site characteristics. For example, reasonable estimates of mean travel distance may be obtained from a small sample of the order of a few tens of particles. However, in order to characterize the travel distance of individual particles, a sample of the order of 1000 is needed (see Hassan and Church 1992). The same logic applies to channel size and length of study reach. For small streams, the travel distance of individual particles may be obtained using a sample of the order of a few tens of tracers, whereas, much larger samples are needed in large rivers (Hassan and Church 1992). However, the question of adequate sample size is still largely an open question in fluvial geomorphology and needs to be further studied.

All of the above is correct provided that the recovery rate of the tracers is very high. However, the recovery rate depends on the tracing technique, sample texture, site characteristics, and flow conditions. For techniques that are limited to the bed surface (e.g., paint, exotic tracers), the recovery rate is usually low and depends on the texture of tracers; it is low for small particles and high for coarse fractions (e.g., Leopold *et al.* 1966, Laronne and Carson 1976). In comparison, radioactive and magnetic techniques enable the detection of buried tracers and are therefore likely to provide high recovery rates. In cases where the bed material contains natural magnetic particles, the recovery rate of magnetic particles is likely to be low. For a given fluvial system, the recovery rate will also depend on the magnitude of flow events. Small flow events are likely to disperse tagged stone near the bed surface while larger flow events are likely to result in deeper burial. Deeply buried particles are more difficult to recover than near surface particles because they may be beyond the sensitivity of the detection system.

Injection

Once the tagging technique is selected and the tracers prepared, the manner in which the sediment is introduced into the fluvial system should be determined. This will vary from one experiment to another and depends mainly on the objectives and the physical conditions of the study site. In all cases, the tagged sediment should be introduced in a way that reduces the effects of artificial seeding and corresponds with natural sediment transport. Since this task is difficult to achieve, researchers treat the initial movement following seeding with caution (e.g., Hassan and Church 1992, Schmidt and Ergenzinger 1992). The volumetric method is one way to seed particles into a river. A trench across the width of the channel is excavated to the estimated depth of the active layer, the sediment is tagged, and then returned to the trench (Wilcock *et al.* 1996a,b). However, a large proportion of the tagged sediment is likely to remain in place except in large flows (e.g., Hassan 1988). In the case of sand, the material is usually seeded at a point or across the channel width (e.g., Sayre and Hubbell 1965, Rathbun and Kennedy 1978). Several methods have been used for seeding pebble and cobble tracers:

1. seeding on the bed surface along lines across the channel (e.g., Leopold *et al.* 1966, Hassan *et al.* 1984, Hassan *et al.* 1999);
2. replacing a local stone found in the bed (e.g., Leopold and Emmett 1981, Ashworth and Ferguson 1989, Schmidt and Ergenzinger 1992);
3. selecting random particles, tagging and returning them to the same places (Hassan *et al.* 1991, Hassan 1993);
4. dumping tracers into a river (e.g., Mosley 1978);
5. volumetric placement in a trench (Hassan 1988, Wilcock *et al.* 1996a,b);
6. marking in situ (Ritter 1967).

In all these techniques, the positioning of the tagged particles on the bed surface is artificial and therefore likely to influence sediment movement during at least the first transporting flow event. If carried out successfully, however, method '6' and possibly '3' seem to avoid this problem.

Detection

The detection strategy and method depend largely on the study objectives and the tagging technique. In selecting the detection method one must ensure that both buried and surface particles can be detected. This will provide information on sediment dispersion within the active bed layer. Detection methods include visual identification, magnetic or metal detection, radioactive detection, neutron activation, and radio wave detection. Some detection methods permit individual identification of tracer particles such as paint, magnets, and radio, while some do not (i.e., radioactive material). This affects the type of recovered information, and hence data analysis and interpretations. The following detection strategies have been used:

1. *Individual location of particles after a flow event*: The bed is searched between flow events, providing information on net particle movement during an entire flow event. Particles are then located visually or detected by metal or magnetic detectors.
2. *Fixed automatic detection of tagged particles*: This system provides automatic detection of the passage of the tracers during a flow event. Two magnetic detection systems of this kind are available, as described in Section 14.8. A detecting device can also be mounted on a boat or a low bridge.
3. *Sample and search*: Samples are taken from different parts of the bed. Particles are located visually as in the case of fluorescent dye or by a detector for radioactive material.
4. *Periodic dynamic detection of the tagged particles*: The dispersion area of the tagged particles is surveyed during a transporting flow event. Such information can be used to prepare dispersal maps of the tagged particles based on time intervals during a flow event. To do so, the detection device should be mounted on a boat such that the entire dispersal area can be searched. Tracers are

detected by scintillation detector or radio transmitters.

5. *Automatic dynamic detection of tagged particles*: This method allows for the continuous positioning of the tagged particles and provides very detailed information on their movement during a flow event. Radio transmitters are used in locating these tracers.

Interpretations

Data analyses and interpretations depend largely on the study objectives, experimental design, and tracing method, which all control the quality of the data. Qualitative and quantitative methods of data analysis have been used. There is some subjectivity in the qualitative analysis, which may contain considerable errors and lead to misinterpretations of data. In contrast, the objective of a quantitative analysis is to obtain information on the movement of sediment, such as direction of dominant movement, spatial dispersion, the relative mobility of sediment between events or different sites, and estimates on the depth of the active bed layer.

Crickmore *et al.* (1990) described three quantitative methods for data interpretation and sediment discharge calculations: time integrated, steady dilution, and spatial integration. Both, the injection method and the study objectives determine what method to use for a particular case. In the time-integrated method, a single dose of tracers is introduced into the flow and then the concentration at a given downstream cross-section is monitored continuously. The distance between the injection point and the monitoring station should be long enough to allow sufficient mixing of sediment over the study site. In cases of slow moving sediment such as gravel, the distance should be short but decreasing with particle size. In the steady dilution method, the tracers are injected continuously and the concentration is measured at a downstream station. Since the concentration is measured after a steady concentration has been reached, a single measurement is required for the sediment discharge calculations (Crickmore *et al.* 1990). The main concern with this method is determining when the steady concentration is reached. This depends on flow conditions, size of mobile sediment, and intensity of sediment transport. Spatial integration is the most widely used method in sediment, transport studies. Sediment discharge is obtained using the velocity of the tracers and the depth of the active layer. The velocity of the tracers is determined from the positional shift of the centroid of the spatial distribution as mapped at time intervals (Crickmore *et al.* 1990). Usually, sediment cores taken over the study area determine the mean depth of the active layer. The method is based on the assumption that the sediment transport is close to uniform over the study reach, an assumption that is difficult to satisfy in nature (De Vries 1973, Crickmore *et al.* 1990). Theoretically, this method is applicable to both suspended and bedload modes of sediment transport. This method has been used to calculate sediment transport in sand bed rivers (Sayre and Hubbell 1965—see Section 14.11) and gravel bed rivers (Hassan *et al.* 1992, Haschenburger and Church 1998).

Sources of error in all stages of the experiment should be examined and resolved. In the case of artificial placement of tracers, data collected after the first flow event should be handled with caution. This problem could be overcome by in situ marking, and to certain extent by random selection of particles from the bed surface, returning them to the same positions.

Hydraulic Data

Most tracer studies include the collection of hydraulic and geomorphic data from the study site. Such information is necessary for data interpretation and extrapolation. The hydraulic data are collected in order to establish relations between sediment transport characteristics and flow parameters. Appropriate hydraulic and geomorphic data depend largely on study objectives and site characteristics. Hydraulic data that have been used includes: flow hydrograph, peak discharge, water depth, mean velocity, water surface slope, shear stress, and stream power. In some fluvial systems, for example desert streams, it is difficult to obtain detailed hydrological data and researchers use general hydraulic variables such as peak discharge. Tracer data such as travel distance, burial depth, and virtual velocity have been related to hydraulic variables which include peak discharge, shear stress, and stream power (e.g., Sayre and Hubbell 1965, Leopold *et al.* 1966, Reid *et al.* 1984, Hassan *et al.* 1992, Hassan and Church 1994, Haschenburger and Church 1998).

Measured geomorphic characteristics include: channel size, bedforms, bed texture, surface structures, bank stability, and sediment supply from adjacent slopes and upstream tributaries. Both channel bedforms (e.g., bars, pools, and riffles) and surface structures (e.g., pebble clusters, imbrications) impact particle entrainment probability and hence the distance of travel. Both the initial position before entrainment and the final position after movement should be described. The outcome of the study depends, also, on the amount of sediment supply from slopes and channel banks. Sediment supply to the channel will likely influence relations between flow parameters and movement characteristics.

14.3 EXOTIC PARTICLES

The simplest form of tracing is probably the injection of exotic bed material into a river. Mosley (1978) introduced a truckload (about $3\,m^3$) of limestone aggregate into the Tamaku River in New Zealand. The limestone aggregates had a golden color, easily distinguishable from the greywacke transported by the river. The aggregates ranged in size from sand up to boulders, similar to the size range found in the river. After a mobilizing event, the channel bed was searched and the movement distance of particles larger than 8 mm was recorded. The recovery rate was low, approximately 5% of the volume.

Kondolf and Matthews (1986) used a similar method to trace the movement of white dolomite fragments introduced into the Carmel River, California, as rip-rap to combat bridge erosion. Since the rip-rap was introduced into the flow at a defined point, variation in the downstream distribution of tracers provided some useful information. As the introduction of tracer material into the flow was dependent on rip-rap erosion, the values of distances moved were artificially lowered by the newly introduced particles. The main problem faced when dealing with the introduction of exotic rock is whether the characteristics of the new material—density, shape, roundness, and size distribution—are comparable with those of the natural material found in the channel (Table 14.1). In the case of Kondolf and Matthews' (1986) study, the introduced material was more angular than that of the natural material, while the opposite was true in Mosley (1978).

Natural labeling of sediment may be provided by the presence of mining waste, which could be divided into mineral and clastic tracers (e.g., Lewin and Macklin 1987, Knighton 1989, Macklin and Lewin 1989, Macklin *et al.* 1992, Hattingh and Rust 1993, Langedal 1997). The minerals, mostly sand or finer material, can be used for a long term study of spatial dispersion and residence time in floodplains, channels, and terraces; whereas, clast sizes can be used in the channels and bars. In using such material, we need to make sure that the physical characteristics of the mining material is comparable with that of the natural material found in the channel. Chapter 3 covers this topic.

14.4 PAINTED PARTICLES

In this method, particles taken from the channel are painted to stand out against the rest of the bed material. Each particle is identified by number. The entire tracer population can be painted one color, or colors may differ according to class size or point of insertion, or other conventions. After each flow the entire length of the study area is searched for painted particles. Those found on the bed surface are recorded, their morphological and sedimentological environments are described, and they are then replaced on the bed surface to await the next flood.

Leopold *et al.* (1966) found a correlation between the strength of flow events and the recovery rates of painted tracers. The relation between clast size and recovery was further established by Laronne and Carson (1976), who found that the recovery rate ranged between 100% for large particles and as low as 0.5% for the smallest.

14.5 FLUORESCENT PAINT

In fluorescent-dye tagging an amount of sediment is labeled with a dye which, upon stimulation by light of a suitable wavelength, emits light of a wavelength characteristic of the dye. Techniques for coating grains have not been standardized and several methods are found in the literature. Ingle (1966) gives a detailed description of coating techniques. The fluorescent method is suitable for sand tracing, but can also be used for larger material (Yano *et al.* 1969). Sediment has to be washed, thoroughly air-dried, and then sieved into size fractions, each of which can be painted a different color to yield insights into differential mobility. Fluorescent dyes suitable for sediment tracing experiments include rhodamine (red), auramine (yellow), eosine (green-yellow), primulin (dark blue), fluorescein (green), and anthracine (blue-violet) (Shteinman *et al.* 1997).

To create the fluorescent tags, a given amount of sediment is placed in a motor-driven cement mixer (e.g., Kennedy and Kouba 1970, Rathbun *et al.* 1971). One of either acetone, agar, ethanol or chloroform solutions is then added with a small amount of resin to the mixture. The selection of the solution type depends on the dye type and the desired life expectancy of the tracer. For example, agar lasts for a few days while chloroform can last for a few weeks (Shteinman *et al.* 1997). As the grains tumble, they are coated with a thin layer of solution. The dye solution should be added until fluorescence of the material reaches an adequate level. The mixture should then be spread out on a polyethylene sheet to dry and the solution to evaporate. This work should be conducted outside in a strong breeze. The resulting coated aggregates are placed between a rubber roller, or other similar device, to separate them to single grains. In order to avoid the tendency of the coated particles to cluster, one can add detergent to the material before injection into the river.

After each flow, core samples are taken along the study reach. The core samples permit one to assess the vertical and downstream dispersion of the traced material. The collected samples should be air-dried, examined under ultraviolet light to determine the presence of the traced material, and tracers counted by their number per unit weight of the bulk material. To represent the spatial dispersion of the tagged particles, a large number of samples is recommended. Visual analysis of the samples can be laborious and hence expensive. Some instruments have been developed for automatic counting of the tagged particles but their efficiency is not clear (for more information see Nelson and Coakley 1974).

14.6 RADIOACTIVE TRACERS

Radioactive tracing was first used in the late 1950s (e.g., Hours and Jeffry 1959, Lean and Crickmore 1960, Hubbell and Sayre 1964, Sayre and Hubbell 1965). The presence of radioactively labeled sediment is determined by detecting the radiation given off by the tag. A wide range of radioisotopes allows a choice of tracers with half-lives that match study objectives. For example, radioactive tagging with a short half-life can be used for repetitive studies on the same fluvial system.

Processes for attaching radioisotope tags to sediment particles are reviewed in detail by Caillot (1970), Ariman *et al.* (1960), and Petersen (1960). The radioactive tracers can be grouped into three categories according to the method of labeling. The first involves the use of manufactured radioactive grains of glass of the same density and shape as the bed material. The second consists of applying the isotope to the outside or within the natural grains. In the outside application technique, the isotope is incorporated into a glue, forming a thin layer on the particle surface. This approach has been widely used in studies of bedload transport in gravel bed rivers. The third method involves the introduction of a natural radioactive mineral to the fluvial system.

The most commonly used isotopes are ^{46}Sc (half-life of 83.9 days), ^{110}Ag (half-life of 253 days), ^{51}Cr (half-life of 27.8 days), and ^{140}La (half-life of 40.2 h). Isotopes such as ^{60}Co are considered to be dangerous because of their long half-life (Nelson and Coakley 1974). For safety reasons it is recommended that the radioactive material be introduced into an isolated section of the river. A crystal or plastic scintillator optically coupled to a photomultiplier tube can be used to detect the radiation emitted by the tagged material.

To study the movement of coarse particles the isotope is introduced into the particles themselves. A small hole is drilled into the particle, then the isotope is inserted, and the hole is sealed with cement. Since the movement of coarse bedload is very sporadic and slow, long-lived isotopes may be appropriate. This method has been used in several studies concerned with the movement of coarse material and has proved to be successful (Ramette and Heuzel 1962, Stelczer 1968, 1981, Michalik and Bartnik 1986, 1994). However, the procedure is so time consuming that the total number of tagged particles is limited to a few tens per site.

According to Crickmore *et al.* (1990), the radioactive technique is the most versatile tracing method that can be used in silt, sand, and gravels. Due to its toxic nature, however, this method has not proved as popular as first expected (Crickmore *et al.* 1990). Radioactive tracers allow in situ, continuous detection of exposed and buried particles and make possible immediate data processing and hence evaluation of the tracking strategy (Crickmore *et al.* 1990). They also offer the opportunity for detailed mapping of the tracers over large areas and a range of time intervals. The main disadvantages are their toxicity and the hazard they pose to the environment, difficulties in handling, and their expense.

14.7 FERRUGINOUS TRACERS

Iron Oxide Coating

This was used in one of the first attempts made to locate buried clasts. Nir (1964) painted synthetic concrete cobbles with an iron oxide coating. As shown in Table 14.2, recovery rates were low primarily due to the low sensitivity of the detection equipment. However, with modern detectors, this problem can be overcome.

Metal Strips

Butler (1977) tagged particles for relocation with a metal detector by wrapping strips of aluminum around them. However, attachment was a problem, and several of the aluminum collars were found to have broken away from their pebbles, contributing to the large loss rates reported in Table 14.2. In addition, the presence of iron minerals in the background rock seriously interfered with the signals received by the detector.

Iron Core Tracers

The preparation and use of iron tracers is the same as described below for magnetic tracers (Section 14.8). The method is effective in locating shallow buried particles down to a few centimeters using a metal detector (Schmidt and Ergenzinger 1992). In order to increase the recovery rate, iron pieces of the order of 3 cm are recommended. However, this limits the size of the particles that can be used and might alter the particle density (Bunte and Ergenzinger 1989). With modern detectors, however, small pieces of iron can be detected.

14.8 MAGNETIC TRACERS

Inserted Magnets

Magnets are inserted into pre-drilled holes situated at the center of gravity of each particle. The magnet used in most studies is ceramic and can be manufactured in different sizes and shapes. The magnetic field is unaffected by changes of the environment, weather or matrix. In order to increase the recovery rate the size of the inserted magnet should be as large as possible; however, it is limited by the particle size and density. A large magnet is apt to alter the particle density and change its behavior relative to natural ones. After insertion of the magnets, the cavity in each particle is filled with transparent epoxy. To identify particles, numbers can be inserted just inside the cavity and then covered again by the transparent epoxy. The drilling and the magnet insertion should not affect the particle strength or significantly alter its density (Hassan *et al.* 1984). Magnetic locators can be used to find the tagged stones. This method permits the location of buried particles up to 1 m underneath the bed surface, as well as on the surface (Hassan *et al.* 1995, Gintz *et al.* 1996). It is, however, a very tedious and time consuming job to dig and record the tagged particles, and it disturbs the channel bed extensively.

Natural Magnetic Tracers

Natural sediment with a high magnetic content can be used as a tracer. Ergenzinger and co-workers used natural magnetic cobbles and pebbles in Squaw Creek to estimate bedload transport rates during floods. The subsection *Automatic Detection of Magnetic Tracers* gives a description of this study.

Artificial Magnetic Tracers

If no naturally magnetic material is present where studies are to be made, it is necessary to use artificially magnetized tracers. Reid *et al.* (1984) manufactured synthetic magnetic clasts made from resin and crushed barites with a ferrite rod core. These would make ideal classical tracers if it were not for the expense and the uniformity of shape and size when produced in bulk from molds. The motion of the particles during a flood was registered by their passage over a sensor that distorts the magnetic field causing a change in the inductance of the coils installed across the river-bed. A metal detector was used to locate the tracers after a flow event.

To examine the impact of particle size and shape on travel distance and burial depth, Schmidt and Ergenzinger (1992), Gintz (1994), and Gintz *et al.* (1996) used colored concrete. They manufactured artificial cobbles of about the same weight and density as natural sediment, but in different shapes. The concrete was formed in moulds with a magnetic core in the center. Due to difficulties in drilling and inserting magnets in small particles, Hassan *et al.* (1995) and Hassan and Church (1992) manufactured small stones made of resin, magnets, and small pieces of lead shot, to adjust overall density. Using the artificial stones they were able to tag and release clasts as small as 8 mm.

Artificial Magnetic Enhancement

It has been noticed that, after forest fires, the magnetic content of soil particles is enhanced to a level that can be detected (Rummery *et al.* 1979). This phenomenon has been used to trace the sources of suspended sediment in drainage basins (Oldfield *et al.* 1979, Walling *et al.* 1979). This magnetic tracing method is based on the enhancement of natural magnetism by high level heating of naturally iron-rich fluvial pebbles and their reintroduction to the stream bed for subsequent tracing. Oldfield *et al.* (1981) heated clasts to temperatures ranging between 200 and 1150 °C and found that 900 °C yielded optimum results in terms of magnetic enhancement. Major changes in the bulk density of the material were observed for temperatures greater than 1000 °C. The best results are obtained by rapid heating with the sample inserted into a preheated oven close to the peak temperature of 900 °C. Heating time ranges between 20 min for small particles and up to 2 h for large particles. Rapid cooling of the samples, either in air or water, gives rise to higher levels of magnetism (Oldfield *et al.* 1981). Through the heat treatment the particle mineralogy is altered and the magnetism is enhanced up to 300 times its original power, a level that can be detected and distinguished from the bed material. The method has been used to trace bedload in small forest ditches in the Welsh uplands (Arkell *et al.* 1983) with a recovery rate of 63%. However, magnetic enhancement yielded very low recovery rates in the North Tyne River (Table 14.2, Sear 1992, 1996), which can be attributed to the larger size of that system. This demonstrates the need to consider channel scale in selecting a suitable tracing method.

Automatic Detection of Magnetic Tracers

Automatic detection systems can track the movement of natural, artificial, or inserted magnetic tracers. The underlying principle is that when a magnet passes over an iron-cored coil of wire, a measurable electronic pulse is generated.

In the Buonamico River, Calabria, Ergenzinger and Conrady (1982) inserted magnets into holes drilled in pebbles, and a magnetic detector was used to monitor their passage. A similar system was used to detect the passage of naturally magnetic cobbles and pebbles past a fixed point during flow events in Squaw Creek, Montana (Ergenzinger and Custer 1983). The Squaw Creek system (Figure 14.2) consisted of four wire coils, 1.4 m apart, connected in series and protected from water with several layers of silicon. The wires were wrapped around a l-m long, 2-cm diameter iron bar. Each coil consisted of 9000 windings of 0.2 mm copper wire. Hassan (1988) achieved similar detection results by using a pipe 10 cm in diameter with 3000 windings of 0.2 mm copper wire. The Squaw Creek system was bolted in a slotted concrete block 1.25 m long, 0.2 m wide, and 0.15 m high. The slot was covered with aluminum sheet metal to protect the detector from the impact of the passing stones. The wires of the detectors were placed in copper tubing that ran under a log that had been installed across the channel width and were connected to an amplifier, filter, and flat bed recorder in a shelter on the riverbank. In Squaw Creek, two detectors were installed below a log that forced the pebbles and cobbles to overpass the sensors. As a result of the smooth surface of the aluminum sheet metal and the positioning of the detector immediately below the log, no sediment accumulated on the system.

This type of system provides in situ continuous measurements of bedload movement during flow events, and an unlimited number of stones can be recorded. However, it is an expensive method and requires a considerable knowledge of electronics. Furthermore, as the signals are a function of particle velocity, magnetic content, and distance from the sensor, it is difficult to calibrate the signals and convert them to the number of stones. Using a sophisticated system of demodulation and electronic data processes, Bunte *et al.* (1987) and Spieker and Ergenzinger (1990) were able to detect stones as small as 3 cm in diameter.

Reid *et al.* (1984) used a commercially built system that worked in the same fashion as a metal detector. The system consisted of two elongated unscreened coils, each 2.3 m long. The sensors were fully balanced over the entire width of the channel. The passage of the tracers over the sensor distorted the magnetic field and produced a change in the inductance of the coils. The detected signal was amplified and demodulated to produce a change in voltage that was recorded on a chart. To avoid double registration of tracers and the influence of particles settling on or very close to the system, a self-balancing system that tuned out the influence of such particles after a predetermined time interval was built into the circuit. The system operated automatically and was activated by circuit closure in the mercury tilt-switch that was attached to the water stage recorder. Two sensors were installed, 11 m apart, on a straight reach in Turkey Brook, England.

The main advantage of both systems is the automatic detection of the traced particles. In addition, the system of Reid *et al.* (1984) allows for individual detection of the tracers after a flow event. However, the system is expensive and is fixed in one position.

14.9 ACTIVE TRACERS: RADIO TRANSMITTERS

This method permits the active detection of tracers during flow events. Within each tracer set, the stones

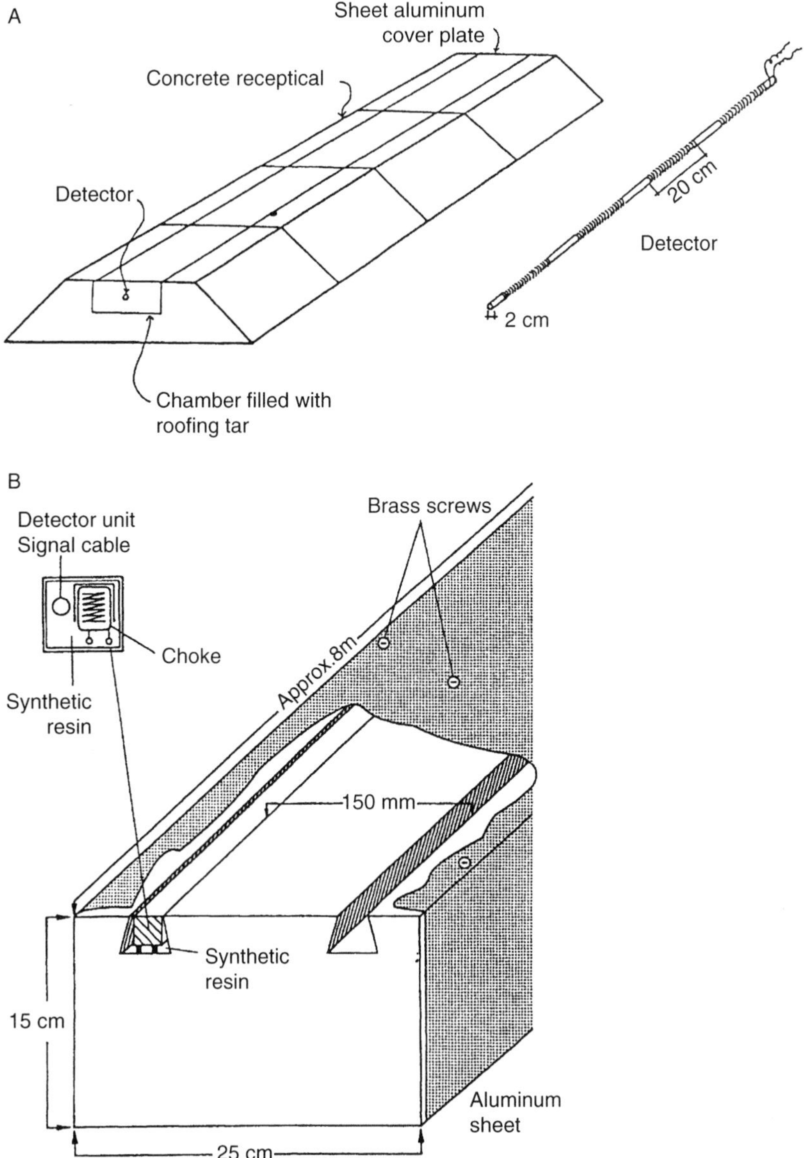

Figure 14.2 (A) Schematic diagram of the automatic magnetic detecting system (from Ergenzinger and Custer 1983) (copyright by the American Geophysical Union). (B) Cross-section of the detector log (from Spieker and Ergenzinger 1990) (reproduced by permission of IAHS Press, Wallingford, UK)

emit slightly different frequencies so that each can be separated from the other. Two systems of this kind are available: the Ergenzinger *et al.* (1989) system (described in Section 14.14) and the system of Chacho *et al.* (1989, 1994) (see also Emmett *et al.* 1990) used in the Toklat River, Alaska. The Toklat radio tracking system consisted of a radio transmitter, which included an antenna and battery, a radio receiver, and directional antenna. The transmitters were 18×72 mm in size and lasted about 10 months. For the Toklat River, nine tracers were equipped with radios transmitting at different frequencies. Also, two transmitters were equipped with a motion sensor that emitted a signal to indicate whether the particle was in rest or in motion.

The Lainbach system (e.g., Ergenzinger *et al.* 1989) consisted of a transmitter, an antenna, receiver, and a data logger to store data. The transmitter emitted 2 m long wave signals, operated at a frequency of 150 MHz, and could be received on the riverbank (Figures 14.3A and B). A waterproof capsule containing a transmitter, lithium battery, antenna, and mercury switch was inserted inside a hole drilled into the center of a pebble. The size of the capsule was 65 mm long and 20 mm in diameter, and the life expectancy of the battery was about 3 months. A plug over the hole allowed battery replacement. The function of the antenna and the mercury switch was to change the emitted signals as the particle rotated.

The tagged particle also can be located during and after a flow event using antennas. There are three types of antenna: stationary, mobile, and search antennas. The stationary antennas, 2 m long, are located on the riverbank (Figure 14.3). These antennas, 5 m apart, are used to follow the passage of the tagged particles. The mobile antenna, mounted on a tripod, is carried along the study reach and used to maintain continuous contact with the particle. This antenna allows one to locate a particle to within 1 m from a distance as far as 100 m from the tracer. Finally, after flow, a special search antenna can be used to locate the exact particle position.

14.10 SUSPENDED LOAD AND WASHLOAD TRACING

In comparison with the bed material load, tracing and detecting suspended and wash loads is a more complex problem. Five characteristics of the fine fractions make their tracing difficult (Coakley and Long 1990, Crickmore *et al.* 1990):

1. Since fine sediment is usually carried in suspension, its movement is rapid and dispersed over large areas. Such movement requires the tagging of large amounts of sediment, taking large numbers of samples, or using a dynamic detection system. The challenge is to keep enough tracers at one time in the water body to be detected. For example, Krone (1957) found that the minimum detectable amount of radiation is three radioactive particles per one square centimeter of water. For this reason

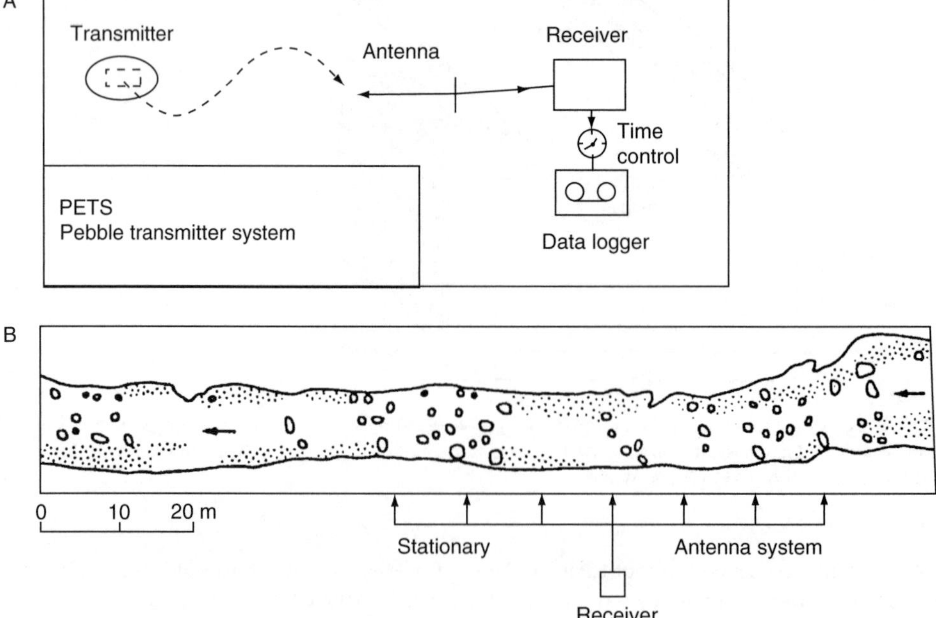

Figure 14.3 (A) Sketch of the pebble radio system (from Ergenzinger *et al.* 1989) (reproduced by permission of Gebr. Borntraeger Verlagsbuchhandlung). (B) The antenna system along the Lainbach study site (from Ergenzinger *et al.* 1989) (reproduced by permission of Gebr. Borntraeger Verlagsbuchhandlung)

it is not surprising that few tracer experiments with suspended material have been conducted. Of these a majority were conducted in estuaries (e.g., Schulz 1960) or used to estimate diffusion constants (e.g., Griesseier 1962).
2. The cohesive nature of the clay and silt causes flocculation, and labeling may lead to changes in the chemical properties and hence behavior of the traced sediment. However, Crickmore *et al.* (1990) suggest that the strong tendency of the fine material to flocculate ensures that the silt and clay are entrained and deposited in bulk. Therefore, despite potential changes in chemical properties, the tracer particles are expected to behave in a similar way to that of the natural material.
3. Coating or attaching tags onto the fine sediment may change the physical characteristics of the particle (e.g., shape, density) and impact its movement. Not preserving the physical properties of the fine sediment can change the settling velocity of the fine material by several orders of magnitude (e.g., Caillot 1983). Therefore, it is important that the tagging not significantly increases the size or change the shape of the natural fine material.
4. It is difficult to produce artificial tracers that have similar physical and chemical properties as natural fine sediment.
5. In the case of radioactive tracers, the wide dispersal of sediment might create health problem.

In terms of long and medium duration experiments, there is no acceptable technique for tracing fine sediment (Crickmore *et al.* 1990). However, fine sediment plays a major role in the dispersal of adsorbed toxic chemical contaminants in water ecosystems (e.g., Coakley and Long 1990, Crickmore *et al.* 1990). Hence, an improved knowledge of their transport and deposition is of importance. Due to their mode of movement, fines are likely to settle over banks, in floodplains, and deltas. Tracking such material requires a dynamic sampling program. To our knowledge, the only dynamic tracing technique for clay and silt is the radioactive one, which has not been used widely because of health issues (Nelson and Coakley 1974).

As with coarse sediments, both artificial and natural tracers are available for tracking fine material. Procedures used to trace coarse sediment described in Section 14.2 are applicable here (Figure 14.1). Artificial tracers are tagged or manufactured, materials such as glass or commercial powder, designed to simulate the movement of sediment in a specific experiment. In selecting an artificial tracer, the geochemical and physical properties, and hydrodynamic conditions should be taken into consideration. Radioactive, neutron activatable elements, and fluorescent dye are the most commonly used artificial tracers in fine sediment studies (Nelson and Coakely 1974, Coakely and Long 1990). Both the fluorescent and radioactive methods have been described in Sections 14.5 and 14.6, respectively. However, fluorescent dyes are difficult to attach to the silt and clay sizes and therefore are rarely used (Louisse *et al.* 1986, Coakley and Long 1990).

In the neutron activatable method, a chemical element is fixed to the sediment and its concentration is measured after activation in a reactor. When selecting a chemical element, care should be taken to ensure that it is not naturally present in the study site (Nelson and Coakley 1974, Crickmore *et al.* 1990). The sediment is labeled by an element with a high thermal neutron, which after irradiation is detectable above the natural radioactive background. Labeling elements is an expensive process and the method requires a reactor for neutron activation (Crickmore *et al.* 1990). Studies have used, among other elements, cobalt, tantalum, iridium, rhenium, and gold (de Groot *et al.* 1973, Crickmore *et al.* 1990). The major advantage of this method over radioactive tagging is its relative safety (Coakley and Long 1990). More details of the technique can be found in de Groote *et al.* (1970).

The fingerprinting technique has been used to define sediment sources and transfer within a drainage basin (e.g., Oldfield *et al.* 1979, Peart and Walling 1986, 1988, Foster *et al.* 1998). This method is based on:

1. the selection of physical and/or chemical properties which are distinguishable from the parent material, and
2. the comparison of measurements of the same properties of the suspended sediment with the equivalent values obtained for the source material (Walling and Woodward 1992).

The ideal tracer properties should be independent of soil type and lithology (Walling and Woodward 1992).

Fallout radionuclides have been used to study a wide range of topics within the drainage basin, including: suspended sediment sources (e.g., Peart 1993, 1995, Collins *et al.* 1997, Bonniwell *et al.* 1999), sedimentation rates and sediment yield (e.g., Walling *et al.*

1999, Hasholt *et al.* 2000), gully erosion (e.g., Olley *et al.* 1993), floodplain sedimentation (e.g., Walling and He 1997, Owens *et al.* 1999a,b), sediment budget (e.g., Owens *et al.* 1997), residence time of fine sediment (e.g., Bonniwell *et al.* 1999), redistribution of soil (e.g., Loughran *et al.* 1990, Ritchie and McHenry 1990, Walling and Quine 1990), and lake and reservoir sedimentation (e.g., Hasholt *et al.* 2000). On reaching the earth's surface, by wet and dry fallout, they are adsorbed onto the soil and become an effective tracer of the movement of fine sediment (Ritchie *et al.* 1970, Campbell *et al.* 1982, Walling and Bradley 1989, Loughran *et al.* 1990, Ritchie and McHenry 1990). Three fallout radionuclides have been used: ^{137}Cs, unsupported ^{210}Pb, and ^{7}Be (see Chapter 9 of this volume for a more detailed treatment of radionuclide methods).

Magnetic properties of the soil have been used to trace sediment sources and fine sediment transport in alluvial streams. The upper part of the soil profile has a higher concentration of magnetic content than the lower part or the parent material (Oldfield *et al.* 1979). The enhanced magnetic properties are due to pedogenic processes and burning (Mullins 1977). Magnetic measurements provide a means of distinguishing suspended sediment sources within a drainage basin. The enhanced magnetism (see the subsection *Artificial Magnetic Enhancement*) helps to differentiate between the surface and subsurface soil and hence the suspended sediment sources (Oldfield *et al.* 1979, Oldfield and Clark 1990).

Finally, sediment mineralogy and chemistry have been used to identify sediment sources within a drainage basin (e.g., Lund *et al.* 1972, Klages and Hsieh 1975, Wall and Wilding 1976). Wall and Wilding (1976) asserted that the mineralogy of sediments derived from agricultural areas should resemble the soil surface; whereas, the mineralogy of channel bank material should be similar to the subsurface soil and parent material. They found that mica, vermiculite–chlorite, expandable clay minerals, and calcium carbonate are important indicators for sediment sources. However, a study by Nabhan *et al.* (1969) indicated similar clay mineralogy for suspended sediment at three sites along the Nile River.

14.11 CASE STUDY: ARROYO DE LOS FRIJOLES, NEW MEXICO, USA (Leopold *et al.* 1966)

The labeled tracing in Arroyo de los Frijoles was started in 1958 and ended in 1962. The main objective of the program was to obtain a quantitative estimate on the mode and the rate of sediment transport in gravel-bearing sand bed streams in New Mexico. A group of rocks consisting of 24 particles were located along lines across the channel width. The initial spacing between the particles was one of three values, 0.61, 0.30, and 0.15 m. Depending on the particle size and flow magnitude, the recovery rate ranged from 0% to 88% (Table 14.2). In addition to the tracers, the study made use of scour chains, recurrent cross-section surveys, and geomorphic observations.

The effect of particle spacing and particle size on percentage of particles moving for a given flow condition is shown in Figures 14.4A and B. The results show that a larger flow is needed to move particles that are close to one another than when spaced far apart. As the spacing between particles increases, the particle–particle interaction decreases and the interference becomes negligible for spacing larger than about 8 particle diameters. The results of the effect of spacing on particle mobility led Langbein and Leopold (1968) to introduce the concept that a gravel bar is a kinematic wave caused by particle interaction. Figure 14.4C shows the relation between distance of movement and particle weight. Leopold *et al.* (1966) stated that the distance of movement is weakly related to its size and that, especially in large events, the retrieved tracers seemed to have moved downstream *en masse* rather than following any other predicted longitudinal distribution. Many studies have now reported the lack of relation between distance of travel and particle size (e.g., Church 1972, Schick and Sharon 1974, Hassan and Church 1992). Leopold *et al.* (1966) proposed a relation between mean travel distance of coarse particles and unit discharge.

The Arroyo de los Frijoles was a pioneering and influential study because:

1. the large number of tracers were used,
2. the long period of observation,
3. the wide range of flow conditions and channel morphology, and
4. the use of several interdependent methods relevant to the tracer dispersal.

14.12 CASE STUDY: NORTH LOUP RIVER, NEBRASKA, USA (Hubbell and Sayre 1964, Sayre and Hubbell 1965)

This study examined the transport and dispersion of sand in the North Loup River, Nebraska, using particles tagged with the iridium 192 isotope. The site was

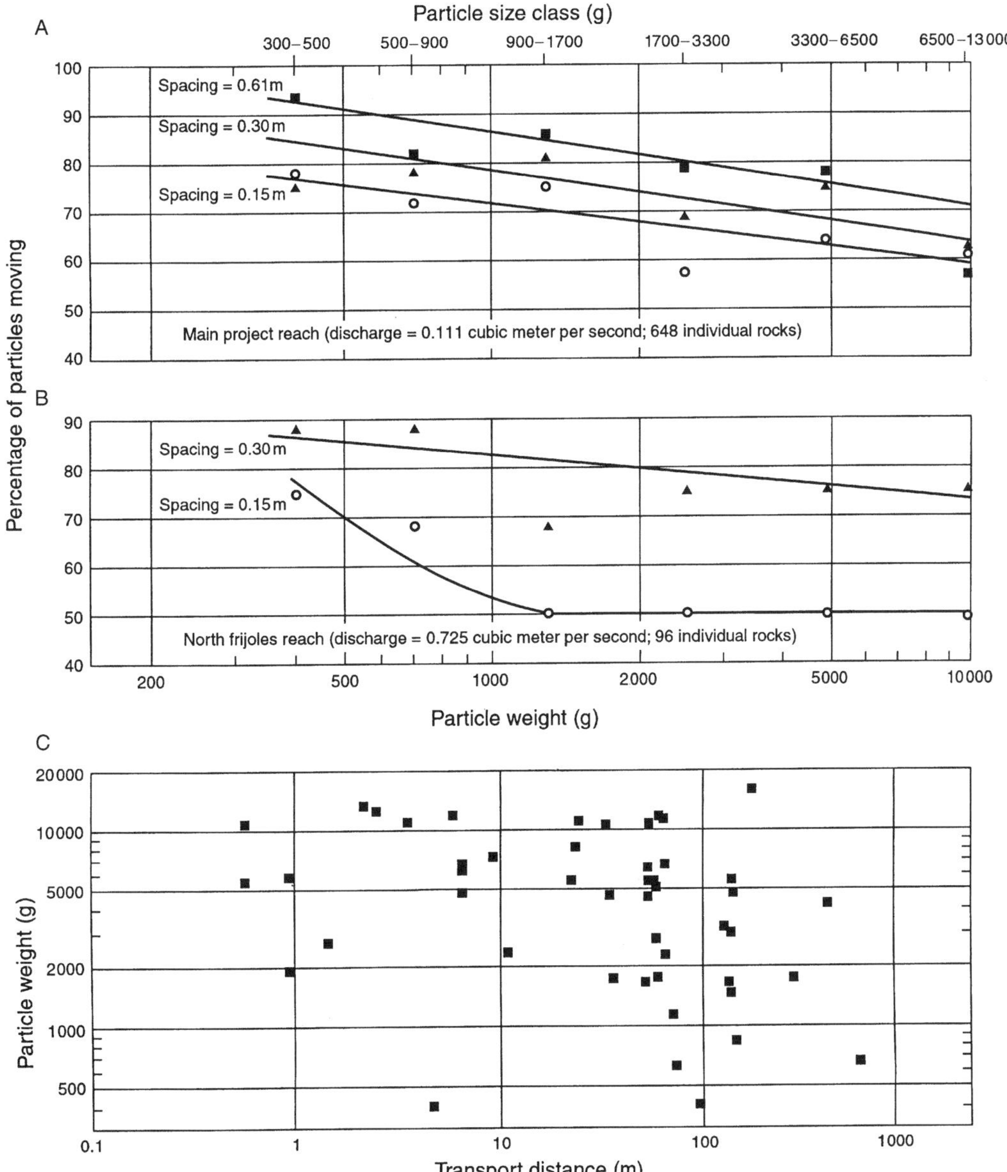

Figure 14.4 (A), (B) The effect of initial spacing on particle moving for a given discharge and particle size (from Leopold *et al.* 1966). (C) Travel distance of tagged particles as a function of particle weight (from Leopold *et al.* 1966)

selected because the flow is relatively constant for several weeks annually, the bed is covered by large dunes, the test reach is straight and narrow, and very few people live adjacent to the reach. Iridium 192 was selected for the following reasons:

1. the emitted gamma rays can be detected underwater,
2. the 74-day half-life was about the intended length of the experiment, and
3. the tracer is commercially available.

The movement of the tagged particles was detected using a scintillation detector that can be used over a wide range of counting rates. The transport and dispersion of the tagged particles was measured periodically by monitoring the radioactivity along the channel, while their vertical distribution was observed by taking bottom core samples. Detailed calibrations were made for each experiment in order to convert the observed radioactivity into a concentration of tracers, expressed in weight per unit volume of sediment (for more details see Sayre and Hubbell 1965). A total of 40 lb (18.144 kg) of tracer particles were introduced to the study reach by injecting 2 lb (0.907 kg) lots on the bed surface at 2 ft (about 60 cm) intervals across the entire width of the channel. The tracers were injected into the bed using a movable funnel tube mounted to a boat (for details see Sayre and Hubbell 1965).

Figure 14.5A presents two examples of the longitudinal distribution of tracer particles at various dispersion times. These distributions were obtained by converting the field count to concentrations using the calibration curves. Typical vertical distributions that show the variation in particle concentration with depth are presented in Figure 14.5B. To characterize the dispersion process, a concentration distribution function was derived from which the bedload discharge was calculated.

Sayre and Hubbell (1965) suggested that bedload discharge (Qs) could be calculated as

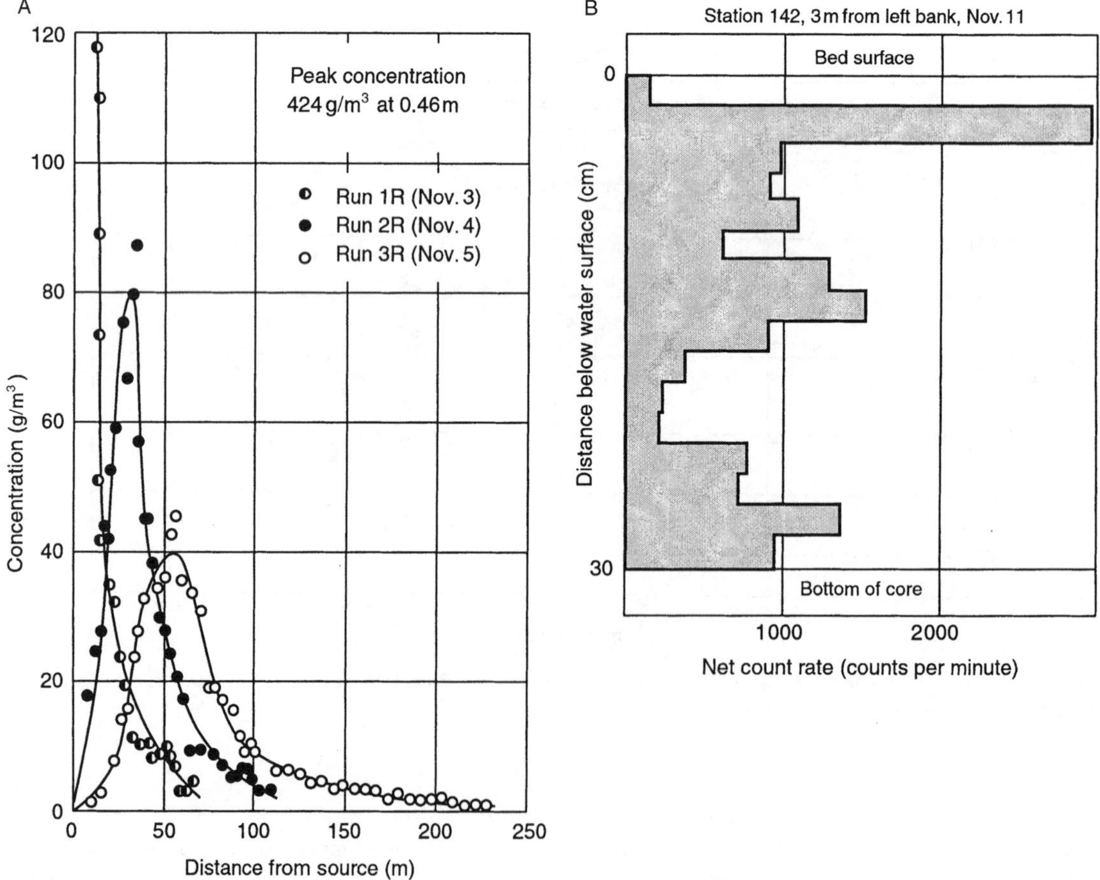

Figure 14.5 (A) Longitudinal distribution of tagged particles concentration along the right side of the channel, November 3–8, 1960 (from Sayre and Hubbell 1965). (B) Vertical distribution of tagged particles at a selected vertical, November 10–11, 1960 (from Sayre and Hubbell 1965)

$$Qs = \gamma_s(1 - \lambda)Bd\frac{x}{t}$$

where x is the mean distance from the source, t the time interval, d the average depth over which the tracer particles are distributed beneath the bed surface, B the width of the channel, γ_s the bulk specific weight of the bed material, and λ is porosity. In fact, the x/t ratio is the virtual rate of particle travel—defined as the mean velocity of the bed particles—which was first introduced by Einstein (1937). Sayre and Hubbell's bedload formula is the first in which the mean travel distance and the area of the active layer (*Bd*) are involved (Stelczer 1981). Recently, this method has been applied in tracer studies in gravel bed rivers (e.g., Hassan *et al.* 1992, Haschenburger and Church 1998).

14.13 CASE STUDY: NAHAL HEBRON, NEGEV, ISRAEL (Schick *et al.* 1987, Hassan *et al.* 1991, Hassan and Church 1992, 1994)

This study examined the three-dimensional (3-D) dispersion of coarse material and its relation to the mechanism of the scour layer in Nahal Hebron, an ephemeral stream in the Negev Desert. Two tracing techniques were used: an automatic magnetic system similar to that of the Buonamico River, Calabria (Ergenzinger and Conrady 1982) and inserted magnets and iron cores. In addition, scour chains and cross-section surveys were used. The first magnetic group was picked up randomly from the bed surface, tagged and returned to the same position, the second magnetic group and the iron groups were located along lines crossing the channel, and the third magnetic group was seeded in a trench excavated across the channel. After one transporting flow event some were found buried while others remained on or within the surface layer. Of the first group of 282 tagged particles, 66% were found on the bed surface and 34% were buried (Figure 14.6A). Figure 14.6A demonstrates, flood by flood, a clear pattern of sediment exchange between exposed and buried particles. The vertical exchange of the particles is subject to local influences such as channel morphology (e.g., pools, riffles, and bars), slope, and shielding by other particles. Hassan (1990) related the depth of burial of the tagged particles to the depth of fill and scour in the study site.

The distance of movement varied substantially between particles and between flow events. The study, like many others (e.g., Leopold *et al.* 1966, Church 1972, Schick and Sharon 1974, among others), confirmed the lack of a simple relation between travel distance and particle size. The distribution of distance of movement for complete flow events was examined. Figures 14.6B and C provide examples of travel distance distribution after a flow event in Nahal Hebron (Hassan *et al.* 1991). Based on data collected from Nahal Hebron and other rivers, bed surface structure appeared more important in controlling the travel distance than the particle size (Church and Hassan 1992, Hassan and Church 1992). The distribution of burial depth of tagged particles in Nahal Hebron fits the exponential function well (Figures 14.6D and E) and serves as the basis for a model describing the vertical mixing of sediment (Hassan and Church 1994).

Nahal Hebron was an important study for the following reasons. Firstly, several tracing techniques were tested in the river. Secondly, the use of the magnetic tracers provided information on the burial depth and vertical mixing of coarse particles. Thirdly, three different injection techniques were used. Finally, relevant to the tracer dispersal, several independent methods were used.

14.14 CASE STUDY: LAINBACH, GERMANY (Ergenzinger *et al.* 1989, Schmidt and Ergenzinger 1992, Busskamp 1994, Gintz *et al.* 1996)

Extensive work on bedload transport using ferruginous, magnetic, and radio transmitter tracers has been conducted in the Lainbach experimental site (Schmidt and Ergenzinger 1992, Busskamp 1993, Busskamp 1994, Gintz *et al.* 1996). The magnetic and ferruginous results have been published in Schmidt and Ergenzinger (1992) and Gintz *et al.* (1996), and hence we concentrate here on the radio transmitter data.

Radio transmitters were inserted into seven natural cobbles and with the stationary antennas located along the study reach, continuous tracking of the tagged stones was possible. Movement of the radio tracers was recorded during six flow events during the summers of 1988 and 1989. The first movement of the tracers was disregarded because of the artificial positioning of the particles before the flow event. The system provided information on the initial condition of sediment movement, step length, particle velocity, and rest period length. Figure 14.7A shows, as described by Einstein (1937), that the particles' displacement included phases of movement (single step) and phases of rest (rest period). The observed step duration and length were very short and varied with bed

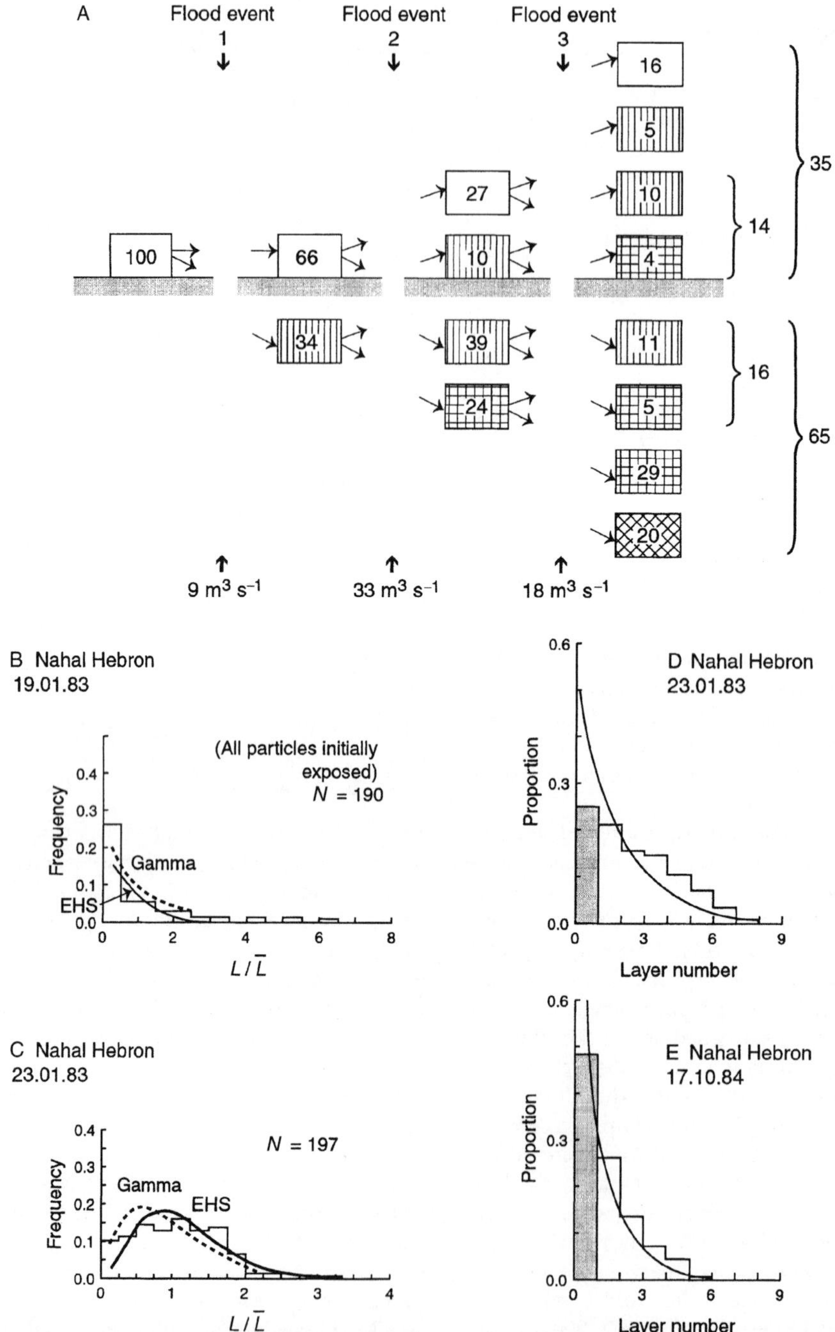

Figure 14.6 (A) Vertical exchange of tagged particles within the active layer as a result of flow events in Nahal Hebron (from Schick *et al.* 1987) (reproduced by permission of the Geological Society). Boxes denote sediment exchange between the surface and subsurface, and do not represent burial depth. (B), (C) Travel distance distribution of all tagged particles in Nahal Hebron (from Hassan *et al.* 1991) (copyright by the American Geophysical Union). EHS is the Einstein–Hubbell–Sayre distribution, $L/\bar{L}$ represents the scale distance of movement, and layer number is the scaled burial depth. (D), (E) Burial depth distribution of tagged particles in Nahal Hebron (from Hassan and Church 1994) (copyright by the American Geophysical Union)

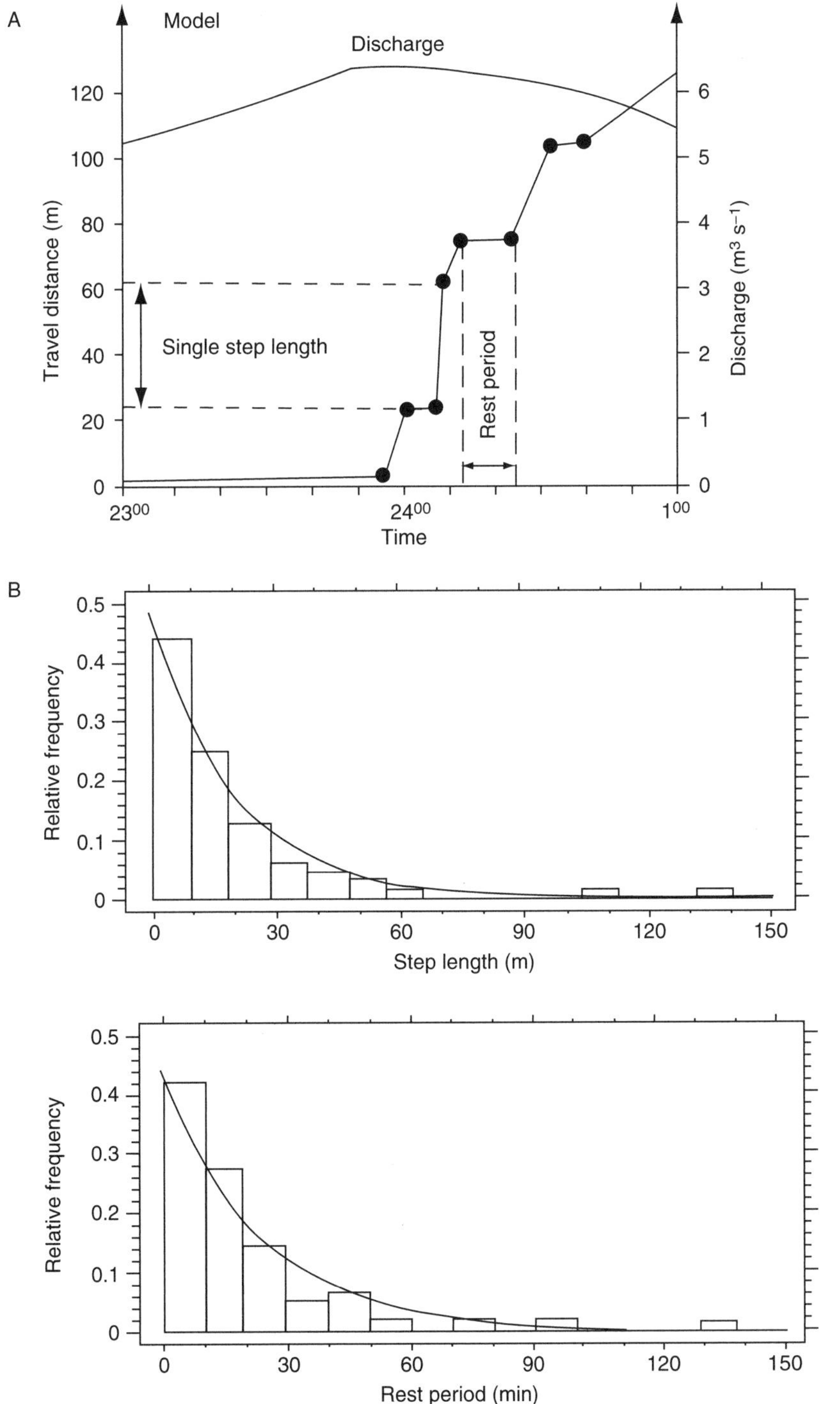

Figure 14.7 (A) Step length and rest period measurements in the Lainbach (from Busskamp 1994) (reproduced by permission of Springer-Verlag GmBH & Co. KG). (B) Distributions of step lengths and rest periods in the Lainbach (from Schmidt and Ergenzinger 1992) (reproduced by permission of John Wiley and Sons, Ltd.)

characteristics and flow condition. They also found that the distributions of both the step lengths and the rest periods (Figure 14.7B) followed a gamma function (Ergenzinger and Schmidt 1990, Schmidt and Ergenzinger 1990, 1992, Busskamp 1993, 1994). Schmidt and Ergenzinger (1992) reported that the exponential function yielded better results than the gamma function; however, the exponential is a special case of the gamma function.

The Lainbach experimental site is considered to be important for the following reasons:

1. several tracing techniques were developed and tested in the river;
2. the use of the radio tracers provided information on the rest period and step length of coarse particles;
3. a large number of tracers were used in the study;
4. the long duration of observation;
5. the studies covered a wide range of flow conditions and channel morphology.

14.15 CONCLUDING REMARKS

In this chapter we have reviewed several techniques that have been used in tracing bed material load in a fluvial environment. Selecting a suitable technique depends on its performance in the field, determined mainly by the recovery rate and depth of detection relative to the thickness of the active layer. Generally speaking, methods that are reasonably successful for large material are unsuited for fine material and vice versa.

Visual detection of exotic and painted particles is probably the simplest, cheapest, and the most easily applied method of tracing sediment. However, the recovery rate is relatively low and stands around 50% in the case of the first event. Given the fact that the method is limited to the bed surface, its representation of sediment transport through the scouring layer is poor. On the whole, it may be concluded that these methods should be used only to get rough and qualitative estimates of sediment transport. For more exact knowledge on sediment transport, it is advisable to use other methods.

The radioactive technique is one of the most reliable and powerful tracing techniques. Using this technique, the tagged particles can be detected even in a large volume of non-active natural sediment. This method has been used in tracing both sand and gravel size material. Its toxicity and danger to the environment is the main disadvantage. Presently, for public safety reasons, it is difficult to obtain permission to use the radioactive tracers. Therefore, the fluorescent method for tracing sand material is recommended.

Magnetic tracing techniques were developed as an alternative to the paint and exotic methods for tagging coarse sediment. Apart from the radioactive method, Table 14.2 shows that the magnetic techniques outperformed the others. Given the limitations of the natural magnetic and artificial magnetic enhancement methods, inserted magnets and artificial magnetic particles are the most widely used. The method is very good for a fluvial system with a few discrete events during the rain season. If the movement is rapid and the time interval between events is short, then the possibility of obtaining data event by event is very limited.

Radio tracers are a very attractive alternative to both paint and magnetic tracing techniques. Unfortunately, high cost limits the number of tracers and hence the obtained information obtained is limited, although very valuable. To represent the movement of natural material in a fluvial system, this method should be used in connection with other tracing techniques such as the magnetic approach.

ACKNOWLEDGMENTS

This paper benefited from valuable comments and suggestions made by Simon Berkowicz, Mordecai Briemberg, G. Mathias Kondolf, Hervé Piégay, Ian Reid and Hamish Weatherly. We are also particularly grateful for the insightful comments of Michael Church.

REFERENCES

Ariman, J.J., Svasek, J.N., and Verkerk, B. 1960. The use of radioactive isotopes for the study of littoral drift. *Philips Technical Review* 21: 157–166.

Arkell, P.L., Leeks, G., Newson, M. and Oldfield, F. 1983. Trapping and tracing: some recent observations on supply and transport of coarse sediment from Wales. In: Collinson, J.D. and Lewin, J., eds., *Modern and Ancient Fluvial Systems*, IAS Special Publication 6, pp. 107–119.

Ashworth, P.J. 1987. Bedload transport and channel changes in gravel bed rivers, Ph.D. Thesis, University of Sterling, 352 p.

Ashworth, P.J. and Ferguson, R.I. 1989. Size selective entrainment of bedload in gravel bed streams. *Water Resources Research* 25: 627–634.

Bonniwell, E.C., Matisoff, G., Whiting, P.J. 1999. Determining the times and distances of particle transit in a mountain stream using fallout radionuclides. *Geomorphology* 27: 75–92.

Bunte, K. and Ergenzinger, P. 1989. New tracer techniques for particles in gravel bed rivers. *Bulletin de la Societe Geographique de Liege* 25: 85–90.

Bunte, K., Custer, S., Ergenzinger, P. and Spieker, R. 1987. Messung des grobgeschiebetranportes mit der magnettracertechnik. *Deutsche gewasserkundliche Mitt.* 31: 60–67.

Butler, R.P. 1977. Movement of cobbles in a gravel bed stream during a flood season. *Geological Society of America, Bulletin* 88: 1072–1074.

Busskamp, R. 1993. Erosion, Einzellaufwege und Ruhephasen: Analysen und Modellierungen der Stochastischen Parameter des Grobgeschiebetransportes, Ph.D. Thesis, Free University of Berlin, 137 p.

Busskamp, R. 1994. The influence of channel steps on coarse bedload transport in mountain torrents: case study using the radio tracer technique 'PETSY'. In: Ergenzinger, P. and Schmidt, K.-H., eds., *Dynamics and Geomorphology of Mountain Rivers*, Lecture Notes in Earth Sciences 52, Berlin: Springer-Verlag, pp. 129–140.

Caillot, A. 1970. Les méthodes de marquage des sédiments par des indicateurs radioactifs. *La Houille Blanche* 7: 661–674.

Caillot, A. 1983. *Guidebook on Nuclear Techniques in Hydrology*, IAEA Technical Report Series No. 91.

Campbell, B.L., Loughran, R.J. and Elliot, G.L. 1982 Caesium-137 as an indicator of geomorphological processes in a drainage basin. *Journal Australian Geog. Studies* 20: 49–64.

Chacho, Jr., Edward, F., Burrows, R.L. and Emmett, W.W. 1989. Detection of coarse sediment movement using radio transmitters. In: *Proceedings of the XXIII Congress on Hydraulics and the Environment, IAHR, Ottawa, Canada*, pp. B-367–B-373.

Chacho, E.F. Jr., Emmett, W.W. and Burrows, R.L. 1994. Monitoring gravel movement using radiotransmitters. In: *Hydraulic Engineering* 94, New York: ASCE Publication, pp. 785–789.

Church, M. 1972. Baffin Island sandurs: a study of arctic fluvial processes. *Geological Survey of Canada, Bulletin* 216: 208 p.

Church, M. and Hassan, M.A. 1992. Size and distance of travel of unconstrained clasts on a streambed. *Water Resources Research* 28: 233–241.

Coakley, J.P. and Long, B.F.N. 1990. *Tracing the Movement of Fine Grained Sediment in Aquatic Systems: A Literature Review*, Scientific Series No. 174, Environment Canada, Inland Waters Directorate, 21 p.

Collins, A.L., Walling, D.E. and Leeks, G.J.L. 1997. Source type ascription for fluvial suspended sediment based on a quantitative composite fingerprinting technique. *Catena* 29: 1–27.

Crickmore, M.J. and Lean, G.H. 1962a. The measurement of sand transport by means of radioactive tracers. *Proceedings of the Royal Society of London, Series A* 266: 402–421.

Crickmore, M.J. and Lean, G.H. 1962b. The measurement of sand transport by the time-integration method with radioactive tracers. *Proceedings of the Royal Society of London, Series A* 270: 27–47.

Crickmore, M.J. 1967. Measurement of sand transport in rivers with special reference to tracer methods. *Sedimentology* 8: 175–228.

Crickmore, M.J., Tazioli, G.S., Appleby, P.G. and Oldfield, F. 1990. *The Use of Nuclear Techniques in Sediment Transport and Sedimentation Problems*, Technical Documents in Hydrology, IHP-III Project 5.2, Paris: UNESCO, 170 p.

de Groot, A.J., Allersman, E., de Bruin, M. and Houtman, J.P.W. 1970. Cobalt and tantalum tracers measured by activation analysis in sediment transport studies. In: *Isotope Hydrology*, Vienna: IAEA, pp. 885–898.

de Groot, A.J., Allersman, E., de Bruin, M. and Houtman, J.P.W. 1973. Tracer techniques in sediment transport, IAEA Technical Report Series, No. 145, pp. 151–166.

De Vries, M. 1973. Applicability of fluorescent tracers. In: *Tracer Techniques in Sediment Transport*, IAEA Technical Report Series No. 145.

Einstein, H.A. 1937. Bedload transport as a probability problem, Ph.D. Thesis. In: Shen, H.W., ed., *Sedimentation*, 1972, Colorado State University, App. C.

Emmett, W.W., Burrows, R.L., Edward, F. and Chacho, E F. Jr. 1990. Coarse particle transport in a gravel-bed river. In: *Third International Workshop on Gravel-bed Rivers, Firenze, Italy, 24–28 September*, 1990.

Ergenzinger, P. and Conrady, J. 1982. A new tracer techinque for measuring bedload in natural channels. *Catena* 9: 77–80.

Ergenzinger, P. and Custer, S.G. 1983. Determination of bedload transport using naturally magnetic tracers: first experiences at Squaw Creek, Gallatin County, Montana. *Water Resources Research* 19: 187–193.

Ergenzinger, P. and Schmidt, K.-H. 1990. Stochastic elements of bedload transport in a step-pool mountain river. In: *Hydrology in Mountainous Regions II*, IAHS Publication No. 194, Wallingford: IAHS, pp. 39–46.

Ergenzinger, P., Schmidt, K.-H. and Busskamp, R. 1989. The pebble transmitter system (PETS): first results of a technique for studying coarse material erosion, transport and deposition. *Z. Geomorph. N.F.* 33: 503–508.

Ferguson, R.I. and Wathen, S.J. 1998. Tracer-pebble movement along a concave river profile: virtual velocity in relation to grain size and shear stress. *Water Resources Research* 34: 2031–2038.

Ferguson, R.I., Hoey, T.B., Wathen, S.J. and Werritty, A. 1996. Field evidence for rapid downstream fining of river gravels through selective transport. *Geology* 24: 179–182.

Foster, I.D.L., Lees, J.A., Owens, P.N. and Walling, D.E. 1998. Mineral magnetic characterization of sediment sources from an analysis of lake and floodplain sediments in the catchments of the Old Mill Reservoir and Slapton Ley, south Devon, UK. *Earth Surface Processes and Landforms* 23: 685–703.

Froehlich, W. 1982. *The Mechanism of Fluvial Transport and Waste Supply into the Stream Channel in a Mountainous Flysch Catchment*, Polska Akademia Nauk 143, Warsaw.

Gintz, D. 1994. Transportdistanzen und raumliche Verteilung von Grobgeschieben in Abhangigkeit von Geschiebeeigenschaften und Gerinnemorphologie, Ph.D. Thesis, Free University of Berlin, 106 p.

Gintz, D., Hassan, M.A. and Schmidt, K.-H. 1996. Frequency and magnitude of bedload transport in a mountain river. *Earth Surface Processes and Landforms* 21: 433–445.

Griesseier, H. 1962. Luminophoren und radioisotope im diensteder kustenforschung forschn. *Fortschr.* 36: 326–330.

Haschenburger, J.K. 1996. Scour and fill in a gravel-bed channel: observations and stochastic models, Ph.D. Thesis, University of British Columbia, Vancouver.

Haschenburger, J.K. and Church, M. 1998. Bed material transport estimated from the virtual velocity of sediment. *Earth Surface Processes and Landforms* 23: 791–808.

Hasholt, B., Walling, D.E. and Owens, P.N. 2000. Sedimentation in arctic proglacial lakes: Mittivakkat Glacier, south-east Greenland. *Hydrological Processes* 14: 679–699.

Hassan, M.A. 1988. The movement of bedload particles in a gravel bed stream and its relationship to the transport mechanism of the scour layer, Ph.D. Thesis, Hebrew University of Jerusalem, 203 p. (in Hebrew).

Hassan, M.A. 1990. Scour, fill, and burial depth of coarse material in gravel bed streams. *Earth Surface processes and Landforms* 15: 341–356.

Hassan, M.A. 1993. Structural controls of the mobility of coarse material in gravel-bed channels. *Israelian Journal of Earth Sciences* 41: 105–122.

Hassan, M.A. and Church, M. 1992. The movement of individual grains on the streambed. In: Billi, P., Hey, R.E., Thorne, C.R. and Tacconi, P., eds., *Dynamics of Gravel Bed Rivers*, New York: Wiley, pp. 159–175.

Hassan, M.A. and Church, M. 1994. Vertical mixing of coarse particles in gravel bed rivers: a kinematic model. *Water Resources Research* 30: 1173–1185.

Hassan, M.A., Schick, A.P. and Laronne, J.B. 1984. The recovery of flood-dispersed coarse sediment particle, a three dimensional magnetic tracing method. In: Schick, A.P., ed., *Channel Processes – Water, Sediment and Catchment Controls*, Catena Supplement 5, pp. 153–162.

Hassan, M.A., Church, M. and Schick, A.P. 1991. Distance of movement of coarse particles in gravel bed streams. *Water Resources Research* 27: 503–511.

Hassan, M.A., Church, M. and Ashworth, P.J. 1992. Virtual rate and mean distance of travel of individual clasts in gravel-bed channels. *Earth Surface Processes and Landforms* 17: 617–628.

Hassan, M.A., Schick, A.P. and Shaw, P.A. 1995. Movement of pebbles on a sandbed river, Botswana. In: *Application of Tracers in Arid Zone Hydrology*, IAHS, Publication No. 232, Wallingford: IAHS, pp. 437–442.

Hassan, M.A., Schick, A.P. and Shaw, P.A. 1999. The transport of gravel in an ephemeral sandbed river. *Earth Surface Processes and Landforms* 24: 623–640.

Hattingh, J. and Rust, I.C. 1993. Flood transport and deposition of tracer heavy minerals in a gravel-bed meander bend channel. *Journal of Sedimentary Petrology* 63: 828–834.

Hours, R. and Jeffry, P. 1959. Application des isotopes radioactifs à l'étude des mouvements de sédiments et des galets dans les cours d' eau et en mer. *La Houille Blanche* 14: 318–347.

Hubbell, D.W. and Sayre, W.W. 1964. Sand transport studies with radioactive tracers. *ASCE, Journal of Hydraulics Division* 90: 39–68.

Ingle, J.C., Jr. 1966. *The Movement of Beach Sand.* Amsterdam: Elsevier, 221 p.

Keller, E.A. 1970. Bedload movement experiments, Dry Creek, California. *Journal of Sedimentary Petrology* 40: 1339–1344.

Kennedy, V.C. and Kouba, D.L. 1970. *Fluorescent Sand as a Tracer of Fluvial Sediment*, US Geological Survey Professional Paper 562-E, pp. E1–E13.

Klages, M.G. and Hsieh, Y.P. 1975. Suspended solids carried by the Gallatin River of southwestern Montana. II. using mineralogy for inferring sources. *Journal of Environmental Quality* 4: 68–73.

Knighton, A.D. 1989. River adjustment to change in sediment load: the effects of tin mining on the Ringarooma River, Tasmania, 1875–1984. *Earth Surface Processes and Landforms* 14: 333–359.

Kondolf, M.G. and Matthews, W.V.G. 1986. Transport of tracer gravels on a coastal California river. *Journal of Hydrology* 85: 265–280.

Krone, R.B. 1957. *Silt Transport Studies Utilizing Radioisotopes. Hydrological Engineering and Sanitary Engineering*, Berkeley: University of California, 112 p.

Langbein, W.B. and Leopold, L.B. 1968. *River Channel Bars and Dunes: Theory of Kinematic Waves*, US Geological Survey Professional Paper, 422L.

Langedal, M. 1997. The influence of a large anthropogenic sediment source on the fluvial geomorphology of Knabeana-Kvina Rivers, Norway. *Geomorphology* 19: 117–132.

Laronne, J.B. and Carson, M.A. 1976. Interrelationship between morphology and bed material transport for a small gravel-bed channel. *Sedimentology* 23: 67–85.

Lean, G.H. and Crickmore, M.J. 1960. *The Laboratory Measurement of Sand Transport Using Radioactive Tracers*, Wallingford, England: Department of Science and Industrial Research, Hydraulic Research Station, 26 p.

Lekach, J. 1992. Bedload movement in a small mountain watershed in an extremely arid environment, Ph.D. Thesis, Hebrew University, Jerusalem (in Hebrew).

Leopold, L.B. and Emmett, W.W. 1981. Some observation on movement of cobbles on a stream bed. In: *Erosion and Sediment Transport Measurements*, IAHS, Late papers – Poster Session, Florence, pp. 49–59.

Leopold, L.B., Emmett, W.W. and Myrick, R.M. 1966. *Channel and Hillslope Processes in a Semi-arid Area, New Mexico*. US Geologicalo Survey Professional Paper 352G, pp. 193–253.

Lewin, J. and Macklin, M.G. 1987. Metal mining and floodplain sedimentation. In: V. Gardiner, ed., *International Geomorphology 1986*, Part 1, Chichester, England: Wiley, pp. 1009–1027.

Loughran, R.J., Campbell, B.L. and Elliott, G.L. 1990. The calculation of net soil loss using caesium-137. In: Boardman, J., Foster, I.D.L. and Dearing, J.A., eds., *Soil Erosion on Agricultural Land*, New York: Wiley, pp. 119–126.

Louisse, C.J., Akkerman, R.J. and Suylen, J.M. 1986. A fluorescent tracer for cohesive sediment. In: *Proceedings of the International Conference on Measuring Techniques of Hydraulics Phenomena in Offshore, Coastal and Inland Waters, ACE/IAHR, London, England*, pp. 367–390.

Lund, L.J., Kohnke, H. and Paulet, M. 1972. An interpretation of reservoir sedimentation. II. Clay mineralogy. *Journal of Environmental Quality* 1: 303–337.

Macklin, M.G. and Lewin, J. 1989. Sediment transfer and transformation of an alluvial valley floor: the river South Tyne, Northumbria, UK. *Earth Surface Processes and Landforms* 14: 233–246.

Macklin, M.G., Rumsby, B.T. and Newson, M.D. 1992. Historical floods and vertical accretion of fine-grained alluvium in the Lower Tyne Valley, north England. In: Billi, P., Hey, R.E., Thorne, C.R. and Tacconi, P., eds., *Dynamics of Gravel Bed Rivers*, New York: Wiley, pp. 573–589.

Michalik, A. and Bartnik, W. 1986. Beginning of bedload motion in rivers. In: *Third International Symposium on River Sedimentation, The University of Mississippi, March 31–April 4, 1986*, Volume 3, pp. 177–186.

Michalik, A. and Bartnik, W. 1994. An attempt at determination of incipient bedload motion in mountain streams. In: Ergenzinger, P. and Schmidt, K.-H., eds., *Dynamics and Geomorphology of Mountain Rivers*, Lecture Notes in Earth Sciences 52, Berlin: Springer-Verlag, pp. 289–300.

Mosley, M.P. 1978. Bed material transport in the Tamaki River near Dannevirke, North Island, New Zealand. *New Zealand Journal of Science* 21: 619–626.

Mullins, C.E. 1977. Magnetic susceptibility processes of the soil and its significance in soil science: a review. *Journal of Soil Science* 28: 223–246.

Nabhan, H.M., Sys, C. and Stoops, G. 1969. Mineralogical study of the suspended matter in the Nile water. *Pedologie* 19: 30–38.

Nelson, D.E. and Coakley, J.P. 1974. *Techniques for Tracing Sediment Movement*, Environment Canada, Scientific Series No. 32, 40 p.

Nordin, C.F., Jr. and Rathbun, R.E. 1970. Field studies of sediment movement using fluorescent tracers. In: *World Meteorological Organization Symposium on Hydrometry, Koblenz*, 12 p.

Nir, D. 1964. Les processus érosifs dans le Nahal Zine (Neguev septentrional) pendant les saisons pluvieuses. *Annales de géographie* 73: 8–20.

Oldfield, F. and Clark, R.L. 1990. Lake sediment based studies of soil erosion. In: Boardman, J., Foster, I.D.L. and Dearing, J.A., eds., Soil Erosion on Agricultural Land, New York: Wiley, pp. 201–228.

Oldfield, F., Rummery, T.A., Thompson, R. and Walling, D.E. 1979. Identification of suspended sediment sources by means of magnetic measurements: some preliminary results. *Water Resources Research* 15: 211–217.

Oldfield, F., Thompson, F.R. and Dickson, D.P.E. 1981. Artificial enhancement of stream bedload: a hydrological application of superparamagnetism. *Physics of the Earth and Planetary Interiors* 26: 107–124.

Olley, J.M., Murray, A.S., Mackenzie, D.H. and Edwards, K. 1993. Identifying sediment sources in a gullied catchment using natural and anthropogenic radioactivity. *Water Resources Research* 29: 1037–1043.

Owens, P.N., Walling, D.E., He, Q., Shanahan, J. and Foster, I.D.L. 1997. The use of caesuim-137 measurements to establish a sediment budget for the Start catchment, Devon, UK. *Hydrological Sciences Journal* 42: 405–423.

Owens, P.N., Walling, D.E. and Leeks, G.J.L. 1999a. Use of floodplain sediment cores to investigate recent historical changes in over bank sedimentation rates and sediment sources in the catchment of the Ouse, Yorkshire, UK. *Catena* 36: 21–47.

Owens, P.N., Walling, D.E. and Leeks, G.J.L. 1999b. Deposition and storage of fine-grained sediment within the main channel system of the river Tweed, Scotland. *Earth Surface Processes and Landforms* 24: 1061–1076.

Peart, M.R. 1993. Using sediment properties as natural tracers for sediment source: two case studies from Hong Kong. In: Peters, N.E., Hoehn, E., Leibundgut, Ch., Tase, N. and Walling, D.E., eds., *Tracers in Hydrology*, Wallingford: IAHS, pp. 313–318.

Peart, M.R. 1995. Fingerprinting sediment sources: an example from Hong Kong. In: Foster, I.D.L., Gurnell, A.M. and Webb, B.W., eds., *Sediment and Water Quality in River Catchments*, New York: Wiley, pp. 179–186.

Peart, M.R. and Walling, D.E. 1986. Fingerprinting sediment sources: the example of a drainage basin in Devon, UK. In: Hadley, R.F., ed., *Drainage Basin Sediment Delivery*, Wallingford: IAHS, pp. 41–55.

Peart, M.R. and Walling, D.E. 1988. Techniques for establishing suspended sediment sources in two drainage basins in Devon, UK: a comparative assessment. In: Bordas, M.P. and Walling, D.E., eds., *Sediment Budgets*, Publication No. 174, Wallingford: IAHS, pp. 269 279.

Petit, F. 1987. The relationship between shear stress and the shaping of the bed of a pebble-loaded river, La Rulles, Ardenne. *Catena* 14: 453–468.

Petersen, B.R. 1960. Some radioactive surface labeling methods. *Ingeniφren* 4: 99–102.

Ramette, M.M. and Heuzel 1962. Le Rhône à Lyon: étude de l'entrainement des galets à l'aide de traceurs radioactifs. *La Houille Blanche* 6: 389–399.

Rathbun, R.E. and Kennedy, V.C. 1978. *Transport and Dispersion of Fluorescent Tracer Particles for the Dune-bed*

Condition, Atrisco Feeder Canal near Bernalillo, New Mexico. US Geological Survey Professional Paper 1037, 95 p.

Rathbun, R.E., Kennedy, V.C. and Culbertson, J.K. 1971. *Transport and Dispersion of Fluorescent Tracer Particles for the Flat-bed Condition, Rio Grande Conveyance Channel, near Bernardo, New Mexico*, US Geological Survey Professional Paper 562-I, 56 p.

Reid, I., Brayshaw, A.C. and Frostick, L.E. 1984. An electromagnetic device for automatic detection of bedload motion and its field applications. *Sedimentology* 31: 269–276.

Ritchie, J.C., McHenry, J.R. and Hawks, P.H. 1970. The use of fallout caesium-137 as a tracer of sediment movement and deposition. In: *Proceedings of the Mississippi Water Resources Conference*, pp. 149–162.

Ritchie, J.C. and McHenry, J.R. 1990. Application of radioactive fallout caesium-137 for measuring soil erosion and sediment accumulation rates and patterns. *Journal of Environmental Quality* 19: 215–233.

Ritter, J.R. 1967. *Bed Material Movement, Middle Fork Eel River, California*, US Geological Survey Professional Paper 575, pp. C219–C221.

Rummery, T.A., Oldfield, F., Thompson, R. and Newson, M. 1979. Magnetic tracing of stream bedload. *Geophysical and Astronomical Society* 57: 278–279.

Sayre, W.W. and Hubbell, D.W. 1965. *Transport and Dispersion of Labeled bed Material: North Loup River, Nebraska*, US Geological Survey Professional Paper 433-C, 48 p.

Schick, A.P. and Sharon, D. 1974. *Geomorphology and Climatology of an Arid Watershed*, Jerusalem: Department of Geography, Hebrew University, 161 p.

Schick, A.P., Lekach, J. and Hassan, M.A. 1987. Vertical exchange of coarse bedload in desert streams. In: Frostick, L.E. and Reid, I., eds., *Desert Sediments: Ancient and Modern*, Geological Society Special Publication No. 35, London, pp. 7–16.

Schmidt, K.-H. and Ergenzinger, P. 1990. Radiotracer und magnettracer – die leistungen neur meßsysteme fur die fluviale dynamik. *Die Geowissenschaften* 8: 96–102.

Schmidt, K.-H. and Ergenzinger, P. 1992. Bedload entrainment, travel lengths, step lengths, rest periods – studies with passive (iron, magnetic) and active (radio) tracer techniques. *Earth Surface Processes and Landforms* 17: 147–165.

Schulz, H. 1960. Verwendung radioaktiver leitstoffe zur untersuchung der sandund schlickwanderung in astuarien und kustengewassern. Dt. Gewasserk Mitt., Sonderhelft.

Sear, D.A. 1992. Sediment transport processes in pool-riffle sequences in a river experiencing hydropower regulation. In: Billi, P., Hey, R.E., Thorne, C.R. and Tacconi, P., eds., *Dynamics of Gravel Bed Rivers*, New York: Wiley, pp. 629–650.

Sear, D.A. 1996. Sediment transport processes in pool-riffle sequences. *Earth Surface Processes and Landforms* 21: 241–262.

Shteinman, B., Berman, T., Inbar, M. and Gaft, M. 1997. A modified fluorescent tracer approach for studies of sediment dynamics. *Israelian Journal of Earth Sciences* 46: 107–112.

Slaymaker, H.O. 1972. Patterns of present sub-aerial erosion and landforms in mid-Wales. *Inst. Br. Geogrs. Trans.* 55: 47–68.

Spieker, R. and Ergenzinger, P. 1990. New developments in measuring bedload by magnetic tracer technique. In: Walling, D.E., Yair, A. and Berkowicz, S., eds., *Erosion, Transport, and Deposition Processes*, IAHS Publication No. 189, Wallingford: IAHS, pp. 169–178.

Stelczer, K. 1968. Investigation of bedload movement. In: *Current Problems in River Training and Sediment Movement, Symposium, Budapest*, 9 p.

Stelczer, K. 1981. *Bedload Transport: Theory and Practice*, Littleton, Colorado: Water Resources Publication, 295 p.

Takayama, S. 1965. Bedload movement in torrential mountain streams. *Tokyo Geographical Paper* 9: 169–188 (in Japanese).

Thorne, C.R. and Lewin, J. 1979. Bank processes, bed material movement and planform development in a meandering river. In: Rhodes, D.D. and Williams, G.P., eds., *Adjustment of the Fluvial System*, Iowa: Kendall/Hunt Publishing, pp. 117–137.

Wall, G.J. and Wilding, L.P. 1976. Mineralogy and related parameters of fluvial suspended sediment in northwestern Ohio. *Journal of Environmental Quality* 5: 168–173.

Walling, D.E. and Bradley, S.B. 1989. The use of caesium-137 measurements to investigate sediment delivery from cultivated areas in Devon, UK. In: Bordas, M.P., Walling, D.E. eds., *Sediment Budgets*, IAHS Publication No. 174, Wallingford: IAHS, pp. 325–335.

Walling, D.E. and He, Q. 1997. Use of fallout ^{137}Cs in investigations of overbank sediment deposition on river floodplains. *Catena* 29: 263–282.

Walling, D.E. and Quine, T.A. 1990. Use of caesuim-137 to investigate patterns and rates of soil erosion on arable fields. In: Boardman, J., Foster, I.D.L. and Dearing, J.A., eds., *Soil Erosion on Agricultural Land*, New York: Wiley, pp. 33–53.

Walling, D.E. and Woodward, J.C. 1992. Use of radiometric fingerprints to derive information on suspended sediment sources. In: Bogen, J. and Walling, D.E., eds., *Erosion and Sediment Monitoring Programmes in River Basins*, IAHS Publication No. 210, Wallingford: IAHS, pp. 153–164.

Walling, D.E., He, Q. and Blake, W. 1999. Use of ^{7}Be and ^{137}Cs measurements to document short and medium term rates of water induced soil erosion on agricultural lands. *Water Resources Research* 35: 3865–3874.

Walling, D.E., Peart, M., Oldfield, F. and Thompson, R. 1979. Identifying suspended sediment sources by magnetic measurements on filter paper residues. *Nature* 281: 110–113.

Wathen, S.J., Hoey, T.B. and Werritty, A. 1997. Quantitative determination of the activity of within-reach sediment storage in a small gravel-bed river using transit time and response time. *Geomorphology* 20: 113–134.

Wilcock, P.R., Barta, A.F., Shea, C.C., Kondolf, G.M., Matthews, W.V.G. and Pitlick, J.C. 1996a. Observations of flow and sediment entrainment on a large gravel-bed river. *Water Resources Research* 32: 2897–2909.

Wilcock, P.R., Kondolf, G.M., Matthews, W.V.G. and Barta, A.F. 1996b. Specification of sediment maintenance flows for a large gravel-bed river. *Water Resources Research* 32: 2911–2921.

Yano, K., Tsuchiya, Y. and Michiue, M. 1969. Tracer studies on the movement of sand and gravel. In: *Proceedings of the XII Congress*, Volume 2, IAHR, Kyoto, Japan.

15

Sediment Transport

D. MURRAY HICKS[1] AND BASIL GOMEZ[2]
[1] *NIWA, New Zealand*
[2] *Indiana State University, USA*

15.1 INTRODUCTION

Life would be much simpler for river engineers and fluvial geomorphologists if all channels had rigid boundaries and the water remained free of suspended material. Loosen the boundaries and throw sediment into a river, however, and one is immediately faced with a host of problems, issues, and questions to which there are often no exact answers. How far will this bank erode? How will scour affect that bridge pier? How much time will pass before a reservoir fills with sediment? What impact will an influx of turbid water have on river biota? To address these and many other questions knowledge of the physical properties of the entrained sediment and measurements and/or calculations of sediment transport are required.

The tools available to acquire this information are many and varied; they must accommodate the question under consideration, different modes of sediment transport, the physical limitations of working in rivers during floods, when most sediment transport occurs (Nelson and Benedict 1950), and the fact that sediment transport in rivers invariably tends to show substantial spatial and temporal variability (Ashmore and Day 1988, Meade *et al.* 1990, Church *et al.* 1999). Consequently, in many situations, there may be no perfect tool available and several approaches must be applied to increase confidence in a result (Wren *et al.* 2000), or at least to set upper and lower bounds to it.

The mode of sediment transport, that is, whether the sediment is moving in a rolling or saltating mode or in suspension (Abbott and Francis 1977), is a primary discriminator. Suspended load, which is fine-grained and dispersed throughout the flow field, demands a different measurement approach than does the coarser bedload, which is confined to a narrow zone immediately above the bed. The supply of sediment is also important. For example, in many rivers the concentration of suspended material tends to be limited by the supply of fine sediment to the channel rather than by the capacity of the flow to support it in suspension (e.g., Hicks *et al.* 2000). While the bedload is more commonly constrained by a river's transport capacity, transport rates may also be limited by sediment availability (e.g., Milhous 1973, Hayward 1980, Jackson and Beschta 1982, Gomez 1991, Lenzi *et al.* 1999).

The modes of transport, the processes that disperse sediment within a river, and the factors that affect sediment supply all contribute to substantial spatial and temporal variation in the sediment load. Since measuring the sediment load everywhere continuously is impractical (if not impossible), this variation must be sampled. Thus, an appropriate sampling strategy is a key component of any approach that attempts to measure sediment transport by direct means.

We begin by reviewing the fundamental concepts of transport mode, sediment supply, capacity and competence. Next, we focus on the tools available for determining the suspended load, the bedload, and assessing the total sediment load. We then consider the case of sedimentation in reservoirs, which retain much or all of the inflowing sediment and, thus, afford a unique opportunity to measure time-averaged sediment transport. Finally, we discuss sediment monitoring programme design.

15.2 BASIC CONCEPTS

The capacity of a river determines the maximum concentration of sediment (i.e., mass of sediment per unit volume of water or per unit area of bed) that can be moved downstream. This is limited by the ability of the flowing water to disperse the sediment, either through turbulence or by traction (Bagnold 1966). Flow competence relates to the maximum size of sediment that can be moved by a given flow condition (Nevin 1946). Often, the supply of sediment to a river channel is less than its sediment transport capacity and the river is termed 'supply limited'. A variety of intrinsic (e.g., the cohesive strength of the bed and bank material) or extrinsic (e.g., the efficacy of sheet or other erosion processes) factors may combine to limit sediment supply (Nanson 1974, Walling 1974). Competence can also limit sediment supply, for example, where the bed has developed an armor layer (Gomez 1983).

Traditionally, the sediment load of a river has been subdivided by source or by mode of transport (Einstein *et al.* 1940, Figure 15.1A). By source, the total load is split between bed material load and washload. The bed material load is derived from the river bed and is typically sand- or gravel-sized; its concentration is directly related to a river's transport capacity. The washload consists of sediment that has been flushed into the river from upland sources and is sufficiently fine-grained that the river is always competent to entrain it in suspension. Consequently, only trace quantities of washload material are found in the bed material, even if the washload dominates the total load. Typically, the washload comprises clay, silt, and up to fine sand grades, although in steep, headwater streams it may also be considered to include coarse sand and even pebbles trapped in the interstices between boulders (in fact, any sediment that would be suspended if exposed to the flow). Generally, washload concentration is dependent on the relative rates of supply of water and sediment to the channel. Being finer-grained, it is rarely capacity limited; indeed, when the concentration of mud-rich washload becomes sufficiently large (several hundred thousand parts per million), the fluid properties change from those of a water flow to those of a hyperconcentrated or debris flow (Costa 1988).

By mode of transport, the sediment load is divided into suspended load and bedload. The suspended load is dispersed in the flow by turbulence and is carried for considerable distances without touching the bed. It is usually fine sand, silt, and clay; in terms of source, it is largely derived from the washload and the finer fractions of the bed material. The bedload is typically coarser sediment moving in almost continuous contact with the bed, rolling, sliding, or saltating under the tractive force exerted by the water flow. In practice, particularly where sand comprises a large proportion of the total load, the boundary between bedload and suspended load blurs. Downstream through a drainage basin, the bed material generally becomes finer through the action of sorting and abrasion; in consequence, the suspended load increasingly dominates over the bedload.

To illustrate these concepts of load classification, Figure 15.1B compares size-gradings of the bed material, bedload, and suspended load sampled from the Shotover River, which drains a 1000-km^2 basin in the South Island, New Zealand. Note that the bed material is bimodal, containing a dominant gravel mode that matches the gravel bedload, a significant fine-medium sand mode that matches the coarser suspended load mode, and negligible quantities of sediment in the clay to very fine sand range (i.e., finer than 0.125 mm). The latter fraction, however, constitutes approximately 60% of the suspended load and may be regarded as the washload.

The differences between the supply and mode of transport of the suspended and bedloads are reflected in the different approaches employed to determine them. Because the suspended load often is related more to the sediment supply than transport capacity, it must generally be measured directly. In contrast, the bedload, which is typically controlled by the transport capacity, may (in theory, if not in practice) be more readily estimated using a theoretical or empirical approach.

15.3 SUSPENDED LOAD SAMPLING AND MONITORING

Overview

The tools used to determine the suspended load vary with the problem to be addressed. Is the primary concern the sediment discharge or concentration? Is the size-grading of the suspended load important? Is information required on kinematic aspects of the suspended load, such as the time required for a stream to clear on a flood recession? A key control on the approach used is the time-base of the problem: does the problem require continuous time-series data, event-based results, or simply long-term statistics, such as the mean annual sediment yield? This is important because the effort required to conduct a single 'instantaneous' measurement of the suspended load, properly sampled across the channel, is impractical to sustain on an ongoing basis. Thus, trade-offs,

A

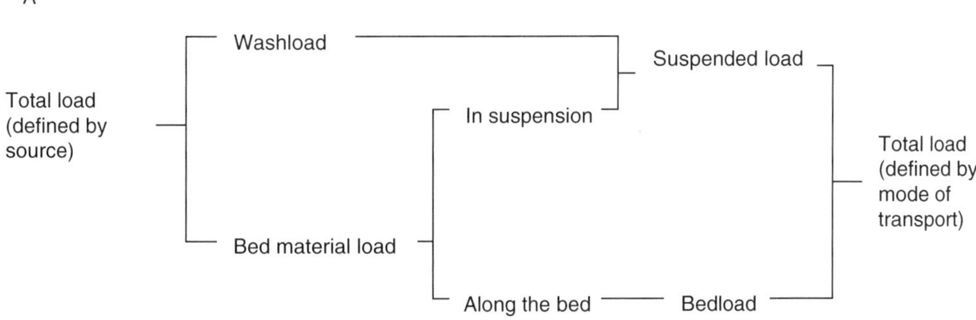

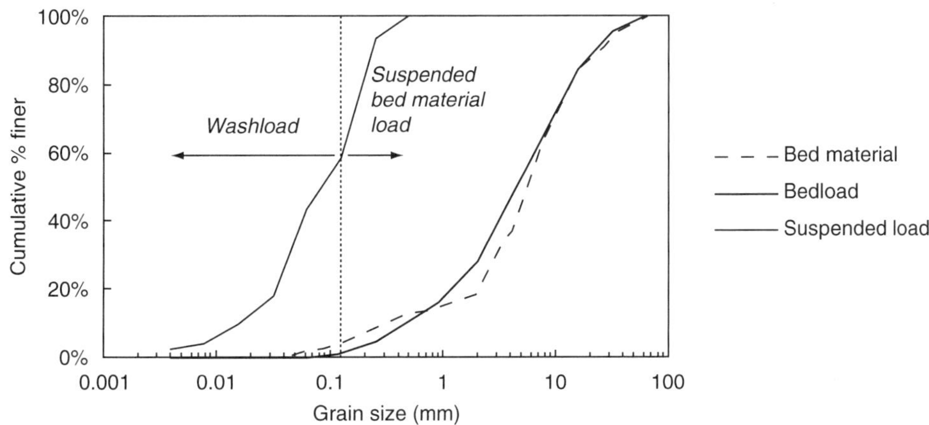

Figure 15.1 (A) Breakdown of stream sediment load in terms of sediment source and mode of transport. (B) Size-grading of suspended load (average from eight samples), bedload (average of nine gaugings) and bed material (average of nine samples) for Shotover River, New Zealand

or simplifications, have to be made when temporal detail is required (Wren *et al.* 2000).

In this section, we first review the requirements of a single suspended sediment 'gauging' which adequately samples the cross-section spatial variability in load. We then look at strategies for collecting and analysing data on a continuous basis. Next, we consider the sediment rating and related approaches, where the interest is only in aspects of the long-term average load. We then consider sediment yields on an event basis and address methods for determining suspended load particle size. Finally, we discuss synoptic sampling, where a spatial overview may be required on suspended sediment concentrations, perhaps during an extreme flood.

Suspended Sediment Gaugings

A suspended load gauging requires the spatially distributed measurement of both sediment concentration and water velocity. Strictly, the suspended sediment discharge, or flux (q_s), past a single vertical in a river cross-section is determined from

$$q_s \quad \int c_s(z) v_s(z)\, dz \tag{1}$$

where c_s and v_s are the concentration and downstream velocity of the suspended sediment. Both c_s and v_s vary with depth (z direction). The variation of c_s with depth depends on the intensity of turbulence and the fall velocity of the sediment. For a given

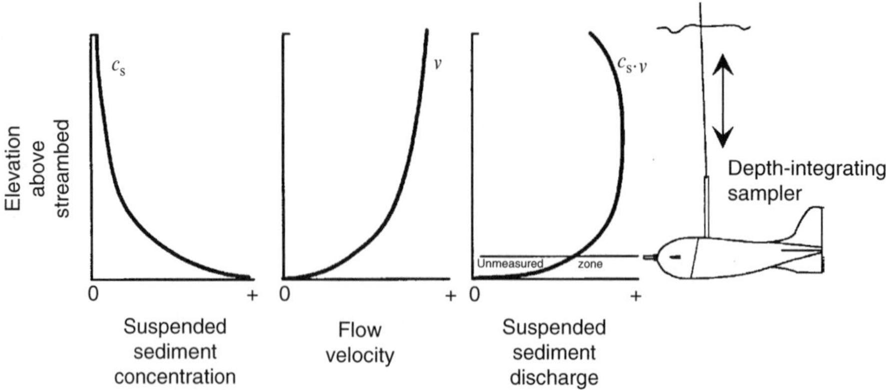

Figure 15.2 Vertical profiles of suspended sediment concentration (c_s), streamwise velocity (v) and sediment discharge ($c_s\ v$ product). A depth-integrating sampler mechanically integrates the $c_s\ v$ profile if it is traversed from streambed to water surface at a uniform rate, but it misses a narrow zone next to the bed

level of turbulence, finer sediment (silt and clay), with a lower fall velocity, tends to be mixed more uniformly over the flow depth, while coarser sediment (sand fractions) tends to be concentrated near the bed. In practice, it is usually assumed that the sediment moves at the same velocity as the water, i.e., $v_s \quad v$ (where v is the streamwise fluid velocity), even though there may be a significant slip velocity in the case of sand grains (Aziz 1996). With this approximation,

$$q_s \quad \int c_s(z)v(z)\ \mathrm{d}z \qquad (2)$$

Practically, the integral in Equation (2) can be determined in two ways. The first is to collect point-samples of water and to make point velocity measurements at intervals over the flow depth, plot profiles of concentration and velocity (Figure 15.2), then integrate the product $c_s(z)v(z)$. The point-samples must be collected with a properly designed 'point-sampler', such as the US P-61 (Figure 15.3B). These cable-suspended samplers, comprising a brass bomb and an internal glass sample bottle, have a solenoid-operated valve to control the water-sample capture and are designed so that they sample isokinetically (i.e., they accept a sample at the ambient water velocity). If this was not the case, the flow lines entering the sampler nozzle would be distorted and the sampled concentration would be biased from the true concentration (particularly for sand grade sediment). Typically, 6–8 points are required to define a point-sampled profile.

The second, and simpler, way is to use a 'depth-integrating' sampler, such as the US DH-48. This is similar to a point-sampler except that the inlet nozzle is kept open while the sampler is traversed at a uniform rate from the water surface to the bed and back again. As it traverses the flow depth, it samples isokinetically and so performs a mechanical integration of the $c_s(z)v(z)$ product. The mass of sediment collected in the sample bottle, when multiplied by the ratio (vertical area)/(sampler nozzle area sampling time), provides a direct estimate of q_s at the sampling vertical (in units gs ^{1}m ^{1}). However, it is more usual to determine q_s from the product of the sediment concentration in the sample bottle at the end of the traverse, termed the 'discharge-weighted mean concentration', c_q, and the unit water discharge as determined with a current meter. Note that c_q is not the same as the mean concentration in the sampling vertical, which is defined as

$$\int c_s(z)\ \mathrm{d}z \Big/ \int \mathrm{d}z$$

With the depth-integrating approach, the speed with which the sampler traverses the water column must be fast enough that the sample bottle is not filled before the traverse is complete. In some circumstances this may be too fast to adequately sample the turbulence-driven fluctuations in near-bed sand concentration (Hicks and Duncan 1997) or to avoid contorting the streamlines entering the sampler nozzle (Edwards and Glysson 1999). In such conditions, the point-sampling

Figure 15.3 Sampling and measuring devices. (A) The Helley–Smith (1971) pressure-difference bedload sampler. (B) The US P-61 suspended sediment sampler. (C) The vortex bedload trap in Torlesse Stream (Hayward and Sutherland 1974) (photograph supplied by TRH Davies) (reproduced by permission of Tim Davies). (D) The Birkbeck-type (Reid *et al.* 1980) pit trap installed in Nahel Eshtemoa (Reid *et al.* 1998); the slotted metal covers were installed to extend the lifetime of the trap during storm events. The US P-61 suspended sediment sampler, which incorporates a solenoid valve that controls nozzle operation, collects a time-integrated sample from a specific point in the channel. Like the Helley–Smith bedload sampler, it is deployed from a cable. Depth-integrated samples may be collected using US D-74 or US DH-48 samplers (not illustrated). The US DH-48 sampler is mounted on a wading rod, and the Helley–Smith sampler may be deployed in similar fashion. Vortex and Birkbeck-type bedload traps are permanent installations that may be used to continuously monitor the mass of accumulating sediment. Information about many sediment sampling and measuring devices may be obtained from the FISP Home Page http://fisp.wes.army.mil

approach, or a combination approach involving depth-integrating over limited depth spans using a point-sampler, is more accurate.

Point-samples may also be collected by pumping up to a surface container, and multiple lines can be used to simultaneously sample an array of points in the vertical (e.g., Van Rijn and Gaweesh 1992). With pumping, however, care is required to maintain isokinetic flow through the intake nozzle and to have velocities up the riser line substantially greater than the sediment fall velocity, otherwise a false concentration of suspended sand will be sampled. This is less an issue with silt and clay-sized sediment.

Variations in suspended sediment load across channel may be substantial, at least for the sand fractions of the total suspended load, which are less well mixed than the silt and clay fractions. This is dealt with by sampling at multiple verticals, preferably spaced either at equal intervals of channel width or so that sub-sections contain equal portions of the total water discharge. The total suspended sediment discharge for a section is typically found by multiplying the discharge-weighted mean concentration at each sampling vertical by the water discharge in the sub-section that each vertical represents (as obtained during an accompanying flow gauging).

At a greater level of detail, the sampling strategy for suspended sediment discharge gaugings requires choices to be made of sampler types, nozzle sizes, sample-bottle sizes, traverse methods and rates, and the number and location of sampling verticals. Details about the standard samplers and methods developed in North America by the Federal Interagency Sedimentation Project (FISP) can be found in technical manuals such Edwards and Glysson (1999) as well as on the world wide web at http://fisp.wes.army.mil. Hydrological field teams from New Zealand's National Institute of Water and Atmospheric Research use notebook computers loaded with data from recent water discharge gaugings to identify the optimum sampling strategy using FISP methods (Hicks and Fenwick 1994).

Most suspended sediment samplers sample only to within 75–100 mm of the bed (Figure 15.2). With the depth-integrating sampler approach, the suspended sediment concentration in the unsampled zone is implicitly assumed equal to the mean concentration in the sampled zone. This is reasonable if the sediment is well mixed through the vertical, as silt and clay invariably are, but may underestimate the sand load since sand tends to be concentrated near the bed. Procedures for adjusting the mean concentration and size distribution of the suspended load to incorporate the unsampled zone are given in Colby and Hembree (1955) and Stevens and Yang (1989). This requires information on the concentration and size distribution of sediment in the measured zone, the flow velocity, and the bed material size distribution.

Continuous Monitoring

A single suspended sediment gauging, as described above, may require tens of samples and several hours to complete. Needless to say, it is usually not a measurement that can be repeated on an ongoing basis, and it becomes impractical when the flow rate or sediment concentration change rapidly. If the need is for continuous (or near-continuous) data on suspended sediment load (such as to define concentration-exceedence probabilities or sediment yields during discrete runoff events), then the usual approach is to collect 'index' samples from a single location. Such samples may be depth-integrated from a fixed vertical or collected from a single point. Calibration measurements are required to establish a relation between the point concentration and the cross-section mean concentration.

Index samples may be collected by hand, however, in remote locations or in 'flashy' small basins they are more often collected by an automatic pumping sampler. Auto-samplers with on-board processors, or when coupled to a programmable data-logger, may be programmed to sample under various strategies, including fixed time, fixed stage-change, or fixed flow-volume (flow-proportional) bases. The main disadvantage of auto-samplers is their relatively small number of sample bottles—typically 24–28 for the more portable of samplers. Other disadvantages include mechanical break-down and limited pumping head. Typically, the pumping head capability is about 5–6 m, which constrains their application to smaller streams (although some have been modified with booster pumps, e.g., Gray and Fisk 1992). Also, auto-samplers do not sample isokinetically and so the concentrations of sand may be biased. To a degree, this bias can be removed empirically by the point to cross-section mean calibration process.

If an extended and detailed time-series record of the suspended load is required, then sensors provide an economical option. Optical sensors are particularly attractive if the primary interest in the sediment information concerns a water clarity issue, and they have been widely used to date (e.g., Walling 1977, Gippel 1989, 1995, Lawler and Brown 1992, Lewis 1996). Since these do not sense sediment concentration directly, they require that a further calibration relation be established between the optical signal and the local suspended sediment concentration. The optical signal depends both on sediment concentration and particle characteristics, notably particle size and shape. Light scattering from clay particles (4 μ), which are platy in shape, is 20 times more effective than from the same mass concentration of coarse silt (Foster *et al.* 1992, Figure 15.4A), thus optical sensors are more sensitive to washload than to suspended bed material load. To a lesser degree, the optical signal is also sensitive to particle composition (e.g., organic particles give a different signature compared with mineral particles) and to colour-producing dissolved organic substances (Gippel 1995). As discussed later, suspended sediment particle size (and other properties for that matter) can vary temporally through events and seasonally, leading to scatter in the relation between the optical signal and sediment concentration. However, at least at sites where washload dominates the suspended load, this scatter is small compared with the range in concen-

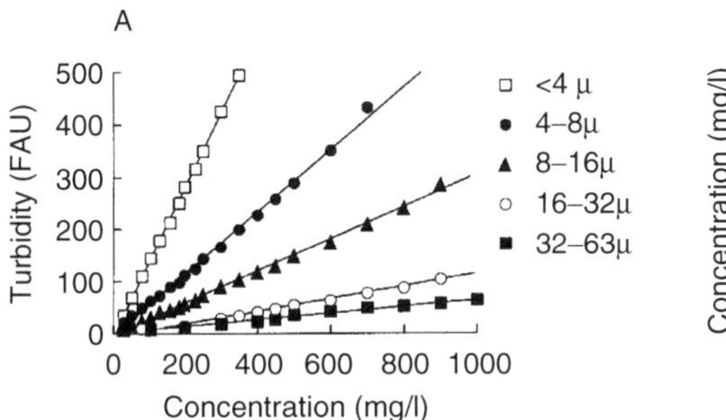

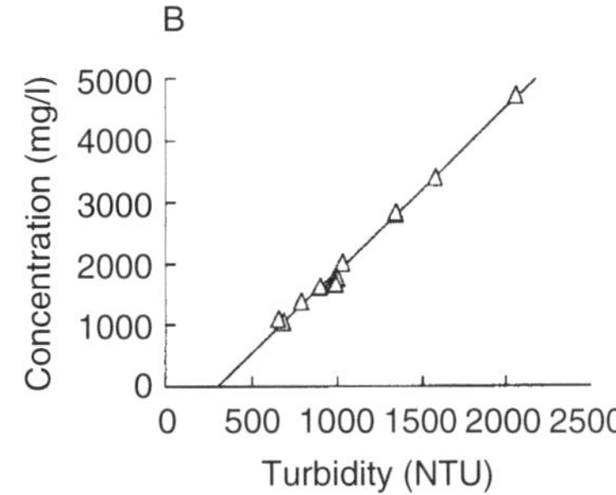

Figure 15.4 (A) Relations between turbidity (as measured by a Partech S100 dual-path sensor) and stream suspended sediment concentration for five size fractions (after Foster *et al.* 1992) (reproduced by permission of IAHS Press, Wallingford, UK). (B) Relation between turbidity (as measured by an OBS-3 back-scattering sensor) and suspended sediment concentration for Waipaoa River, New Zealand. The Waipaoa's suspended load is dominantly silt and clay

tration and a good calibration function is usually achieved (Figure 15.4B).

The optical sensors may be either transmissivity (attenuance) or back-scattering (nephelometric) types. Traditionally, the back-scattering types have been better suited to monitoring sediment loads since with these the signal-to-noise ratio increases as sediment concentration increases (the reverse occurs with transmissivity sensors, which monitor light transmission over a fixed path). Even so, until recently these have been limited in range to 2000 Nephelometric Turbidity Units (NTU), which typically corresponds to about 5000 mg/l. In the last several years, a new generation of 'smart', self-ranging sensors, of both transmissivity and back-scattering types, have become available. For the first time, these provide adequate ranges (some as high as 200 000 mg/l) and sensitivities to monitor the full range of concentration found in most streams and rivers.

A nuisance often encountered with optical sensors is lens bio-fouling. This can be controlled with varying success by hand-cleaning, mechanical wipers, algae-repelling polymer coats, and jets that squirt chemicals or simply water onto the lens. Within limits, dual-path transmissivity sensors do not 'see' fouling because they sense the relative transmissivity over two different path lengths. Another disadvantage with optical sensors is that with only a small proportion of clay in suspension, they will not register coarser sediment—thus they are not suitable where the sand load is the primary interest. In such cases, an option is to deploy an optical sensor in parallel with an acoustic back-scatter sensor. Green *et al.* (1999) show how the two together can be used to monitor sand and silt in mixtures, since the acoustic back-scattering (given a sound source of the appropriate frequency) is greater from sand sizes (Thorne *et al.* 1993). Indeed, acoustic sensors appear to offer many of the advantages of optical sensors while being less vulnerable to bio-fouling.

Suspended Sediment Ratings

If the main interest lies in determining the long-term average suspended sediment yield, then either the 'sediment rating' or 'direct estimation' (Cohn 1995) approaches offer considerable economies of sampling effort and obviate the need for continuous records of sediment concentration. The secret to their successful implementation, though, lies in a well-designed and implemented sampling strategy.

A sediment rating aims to represent the suspended sediment concentration as a continuous function of water discharge. There are two main approaches. The first recognises that there is no unique relation between suspended sediment concentration, C, and water discharge, Q, and so aims to model the conditional mean concentration (as a function of water discharge) over the time period of interest. The conditional mean relation is estimated by sampling a series of concurrent measurements of water discharge and discharge-weighted sediment concentration. This relation is then combined with the water discharge record $Q(t)$ for the same period in order determine the sediment yield. In terms of accuracy, it matters little if the full flow time-series is used or that it is compressed into a flow-duration table, providing that in the flow duration table the flow range is divided into small intervals or at least the high flow range is well detailed (Miller 1951, see also Walling 1977). The greatest sources of error arise from the method used to model the relation and from the sampling strategy. The second approach involves attempting to explicitly model suspended sediment concentration with an empirically derived multivariate relation, relating sediment concentration not only to water discharge but to other controls or processes affecting the sediment supply, such as season, long-term trend, hysteresis of sediment delivery during storms, and so on. With this approach, time-series information is required on all of the independent variables in order to generate a long-term average sediment yield. The first approach is more common, and we focus on it here.

Modelling the C–Q relation. The traditional approach to deriving a rating model has been to plot concurrent measurements of C against Q on log–log graphs. There are several good reasons for this:

(i) log–log plots accommodate the large ranges of discharge and sediment concentration in rivers,
(ii) the data scatter tends to be homoscedastic (i.e., independent of discharge), and
(iii) the underlying relation typically shows a simple power form $C \quad aQ^b$ (a and b are empirical coefficients), which is linear on a log–log plot.

At first, such rating equations tended to be eye-fitted, but with the arrival of personal computers and statistical-analysis packages, linear regression of the log-transformed data became widely used. Ferguson (1986, 1987) pointed out that by using log data, the linear regression procedure modelled the geometric conditional mean, rather than the desired arithmetic conditional mean, thus he proposed correcting the coefficient a by the factor $\exp(s^2)$ (where s is the standard error of the estimate in natural log units), which is based on the assumption that the data scatter about the modelled line is log-normally distributed. Cohn *et al.* (1989) showed that if the residuals distribution was not log-normal, then Ferguson's bias-correction factor could be substantially in error. They provided an alternative method of correcting for log–log bias based on a maximum likelihood estimator. Duan's (1983) empirical 'smearing' estimator is also used for the same purpose. Crawford (1991) showed that both of the latter two correction factors improved the accuracy of the log-linear least-squares approach.

Even with appropriate bias correction, however, independent assessments of sediment yield have shown that sediment-rating assessed yields have still been in error by factors as large as 10 (Walling and Webb 1988). Sometimes, such apparently poor results can arise because the simple power law model, while appearing to fit the overall dataset reasonably well, gives a poor fit to the high discharge end of the relation (which may only be a short tail of sparse data on the right-hand end of the log–log plot, but transports the bulk of the long-term load). For example, there are cases where the C–Q relation is curved in log–log space, reflecting a tendency for concentration to increase less rapidly at higher discharges. In such cases, other curve fitting techniques such as non-linear regression (Singh *et al.* 1986) or locally weighted scatterplot smoothing (LOWESS) (Cleveland 1979, Hicks *et al.* 2000) perform better. Essentially, LOWESS constructs a 'running' linear regression fit to the data, using a limited window (or band) of discharge and weighting each data point in inverse relation to its distance from the window centre.

Figure 15.5 plots sediment-rating data for the Shotover River, South Island, New Zealand, where a continuous concentration record was generated from a turbidity record by relating turbidity to concentration. Figure 15.5A shows the full C–Q bivariate distribution over a 6-month period, the conditional-mean concentration trend (over 50 flow bands), and a simple regression model of the log-transformed data. Note (i) how the regression model, weighted to the more numerous data at lower flows, misses the high flow tail of the plot, and (ii) the erratic form of the conditional-mean trend at higher flows, as the number of sample points per flow band decreases. Compared to the true yield (733 000 t, estimated from the full time-series of concentration and discharge records), the yield estimated by the simple regression model was 693 000 t. This increased to 716 000 t when the log bias correction factor of Ferguson (1987) was applied. Figure 15.5B shows a stratified random sample of 100 points (50 at flows above 100 m^3/s and 50 at lower flows), designed to simulate a series of gaugings, plus ratings fitted to these points using simple log–log regression and LOWESS. The yields estimated by these models were 748 000 t (after log bias correction) and 743 000 t, respectively.

Sampling strategies. An often major source of error in rating relations are the data themselves. One assumption with the rating method is that there is no bias in the data collection—for example, that, for a given discharge band, there is no preference to collect samples when the concentration is less than the long-term mean concentration for those discharges. Particularly in smaller streams, there is a

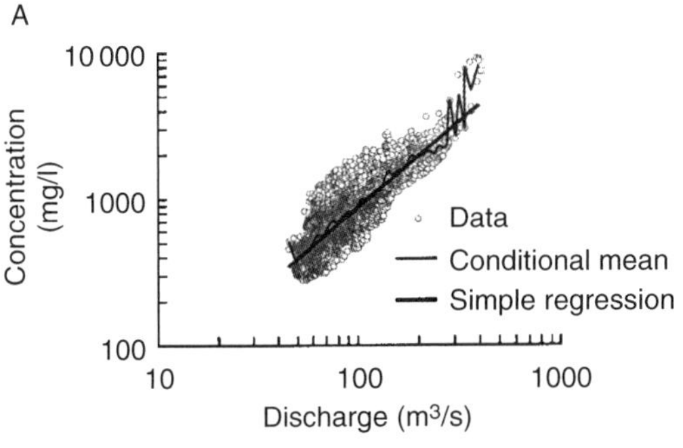

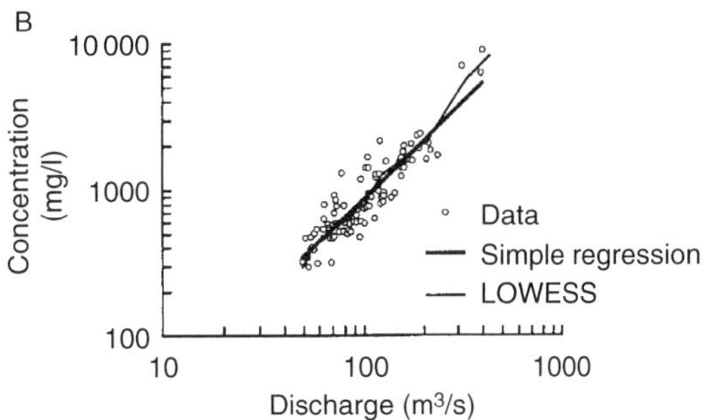

Figure 15.5 (A) Relation between suspended sediment concentration and water discharge for 6 months of hourly data from Shotover River, New Zealand, with rating relations modelled by conditional-mean concentration and simple regression. (B) Relation for a stratified random sample of 100 points, with ratings fitted to these points using simple regression and LOWESS

tendency for concentrations to be lower on flood recessions, due to an initial 'flush' of readily available sediment from the channel (e.g., Christian and Thompson 1978, Walling and Webb 1988). If the rising and falling stages of these streams are not sampled in proportion to their relative frequencies-of-occurrence, for example, because the remoteness of a site prevents field parties from arriving until after the flood peak, then the sampling, and the resultant sediment rating, will be biased to the lower concentrations. If this bias can be identified, then it can possibly be removed by developing a multivariate rating, or more simply by splitting the data and deriving separate ratings (e.g., for rising and falling stage, or for separate seasons). Thus, it is clear that sampling should cover adequately and representatively a wide range of discharges, rising and falling stages, and different seasons. Note that in some cases, there may be a trend or non-stationary signal in the sediment supply, such as may follow from catchment landuse change or as a catchment recovers from an extreme event such as a storm or earthquake that induces catastrophic erosion (e.g., Kelsey 1980). In these cases there may be no stable, long-term sediment rating.

In practice, many existing suspended sediment datasets have been compiled with no plan or strategy that relates directly to sediment sampling considerations, and unwitting bias may have been introduced. For example, in New Zealand, for many years suspended sediment was gauged in conjunction with flow gaugings, which were scheduled for the purpose of checking stage–discharge ratings and had no particular need for rising/falling stage or seasonal representation. Moreover, there may be only, say, 10–20 measurements with which to construct a sediment rating. In such cases, the yield estimate will have a substantially greater uncertainty due to the small number of samples. However, determining the level of uncertainty is not a simple matter, since the uncertainty due to sample numbers is difficult to separate from bias introduced by the sampling strategy and method of fitting the rating curve.

If at all possible, it is better to avoid sampling bias by designing an objective sampling strategy (Olive and Rieger 1988). Unbiased strategies include regular time-interval, random time-interval, flow-weighted probability, or load-weighted probability. Simulations have shown (e.g., Thomas 1988, Walling and Webb 1988) that regular/random time-based sampling can provide poor results, since the chances of intercepting flood flows are slim. Flow- and load-weighted sampling work better, particularly where simple regression is used to derive the rating model, because they tend to result in reasonably uniform densities of data points over the flow range (e.g., at high discharges, the greater frequency of sampling tends to balance the lower frequency of occurrence). Load-weighted sampling ensures data-points in the 'most effective' discharge range, i.e., that which transports the bulk of the long-term load. Thomas's (1985) selection-at-list-time (SALT) sampling method (see following section) can also be used to schedule sediment-rating data. With this, a 'first estimate' rating is used to estimate the sediment load from the current flow, and the probability of collecting a sample in a given time interval is then assigned in proportion to this estimated load. The 'estimator' rating can be tuned as data are collected. Automatic samplers, controlled by programmable data-loggers coupled to a stage recorder, permit such flexible sampling strategies to be realised. In many situations, this benefit outweighs the disadvantage of having to conduct manual sampling to establish a relation between the point concentration at the sampler intake and the cross-section mean. Thus, auto-samplers should be seriously considered where the sediment-rating method is to be applied.

Direct Estimation Approach

Unlike the sediment-rating approach, which seeks to define a continuous relation between concentration and discharge and then to use this to estimate a long-term average load, a more direct approach is to sum a series of selectively sampled load estimates (Cohn 1995). A simple example is the flow-interval (Verhoff *et al.* 1980) or load-interval (Walling and Webb 1981) approach. This involves dividing the discharge range over the period of interest into equal intervals (or strata) and determining the mean of all sampled loads that fall in each interval. The sums of these discrete conditional means, weighted by the proportions of time that the discharge is in the interval, provide an estimate of long-term average yield. Advantages of this approach are that no data transformations are required and the variance of the yield may be estimated.

As with the sediment-rating approach, however, the key to its successful implementation is a sampling strategy that is unbiased yet delivers sufficient samples during runoff events to ensure a reasonable accuracy of the yield estimate. Thomas (1985) and Thomas and Lewis (1993a, 1995) developed several random sampling strategies that vary the sampling rate to reflect changing flow or load conditions. These rely on use of a programmable data-logger to adjust the sampling by an auto-sampler in real-time. The SALT strategy (Thomas 1985) varies the probability of sampling according to the value of an auxiliary variable, which in this case is the value of the load generated from a 'first estimate' sediment rating. The time-stratified strategy (Thomas and Lewis 1993a) involves collecting random samples within time intervals set by the current flow conditions. Flow stratified sampling (Thomas and Lewis 1995) is similar, except that the sampling 'strata' are selected in terms of stage or discharge intervals. Clearly, this strategy will optimise sampling for the flow-interval method. Thomas and Lewis (1995) found that the relative success of these different strategies varied with the size and complexity of runoff events.

Event Suspended Sediment Yields

In certain situations, there is greater interest in suspended sediment yields during individual runoff events than in the long-term average yield. One example is the filling of sediment retention dams, which are frequently placed in basins undergoing urban development or timber harvesting in order to limit sediment exports. In such situations, knowledge of the magnitude–frequency distribution of the event sediment yields is important for designing sediment trap capacities and for setting limits on sediment releases to downstream waterways. Event-yield magnitude–frequency relations are also useful discriminators of landuse effects on sediment yields.

Determining a magnitude–frequency relation for event sediment yields at a site ideally requires continuously monitoring stream sediment loads (or reservoir deposition—see a later section) for a period of years. With a time-series of stream loads, discrete sediment-yield events can be totalled on the basis of discrete quickflow events. The analysis then proceeds in like fashion to undertaking a peaks-over-threshold (partial duration) analysis for event peak flows (e.g., Haan 1977): the event yields above a threshold size are ranked, assigned a return period (T) using a 'plotting formula', e.g., T m/n (where m is the event rank and n the number of years of record), then modelled with an appropriate distribution.

Where no long record of continuous sediment load data is available, an alternative approach is to establish a relation between event sediment yield and some correlated index of the event magnitude that is more easily monitored. Event peak flow typically correlates well with event sediment yield and can be used for this purpose (e.g., Neff 1967, Hicks 1994—Figure 15.6A). Given a long flow record, this relation can be used to simulate a many-year, unbroken series of event sediment yields from which a magnitude–frequency distribution can be extracted. Alternatively, the relation can be used to estimate the event yields associated with peak flows of given return period.

As with 'instantaneous' sediment ratings, care is required in modelling the event-yield vs. peak-flow relation and in extrapolating it outside the range of the data. Unrealistic extrapolation can induce large errors in the yields estimated for extreme events. Parker and Troutman (1989) outline an approach that incorporates the uncertainty in the event-yield vs. peak-flow relation. They used a log–Pearson type III distribution to model the probability density function of the annual peak flows (Y) and a quadratic regression relation between the logarithms of annual flood peaks and associated sediment yields (Y_s). They assumed a normal distribution of the errors in the Y vs. Y_s regression relation in order to estimate the conditional probability density of the event sediment yield given a peak value. Finally, they combined the functions for the probability of Y and the conditional probability of Y_s given Y to derive a function which, when integrated numerically, predicted the sediment yield for a given return period. While Parker and Troutman (1989) dealt with annual maxima events, their approach could potentially also be applied to a peaks-over-threshold series.

An event sediment-rating relation can be compiled over one or two years, allowing that a good range of events is sampled and that the relation remains 'stationary' over the period of flow record used for the return period analysis (i.e., the sediment sources and erosion processes in the basin do not change appreciably, such as might occur during a landuse conversion). Automatic pumping samplers or turbidity sensors are well suited for such short-term event-sampling deployments. When coupled with a data logger and a stage recorder, auto-samplers can be set to sample through flood events when the stage exceeds a threshold value. If programmed to sample on a flow-proportional basis, the event sediment yield can be computed directly from the product of average sampled concentration and average flow over the event (in this case, too, samples through an event can be composited before analysis, allowing greater economy of sample bottles). Otherwise, if the sampling interval is suitably short, the event sediment yield can be computed directly by integrating the product of sediment concentration and water discharge, or else the yield can be estimated using a stratified or variable-probability type sampling strategy (Thomas and Lewis 1995).

As already discussed, use of both auto-samplers or turbidity sensors requires a phase of manual sampling to calibrate point measurements to cross-section mean values. Lewis (1996) compares strategies for collecting auto-samples for calibrating turbidity vs. concentration relations during runoff events.

The effect of landuse on event sediment yield magnitude–frequency relations was demonstrated by Hicks (1994) using storm yield data from four small basins around Auckland, New Zealand. The overall position of a data series on the magnitude–frequency plot (Figure 15.6B), reflecting the sediment yield per unit area during an event of given return period, indicates the overall availability of sediment in the basin. This was higher by an order-of-magnitude in the basin undergoing urban development, due to the ready sediment supply from earthworks and road construction. The steepness of the data series reflects the continuity of the sediment supply during larger events. The plot was steepest for the urbanising basin, where sediment was abundant, flattest for the mature urban basin, where sediment became exhausted during large events, and had intermediate slopes for pasture and market-gardening basins. The mature urban basin yielded more sediment than the pasture basin during sub-monthly events, but the reverse was true with less frequent events. Such a plot is more informative than a simple comparison of average-annual yields.

Event-based sediment yields may also be obtained from analysis of reservoir strata, as discussed in Section 15.6.

Suspended Sediment Particle Size

Particle-size information of suspended sediment is important for two main reasons. First, it is a primary control on entrainment and deposition, and so affects the degree of mixing within a river cross-section, downstream sorting, and settling in reservoirs, lakes and backwaters. Second, it influences the capacity of the sediment to adsorb and transport contaminants such as heavy-metals. Finer sediment fractions (i.e., clays) tend to constitute platy silicate minerals that have larger specific surface areas and exhibit greater cation-exchange capacity (Ongley *et al.* 1982, Horowitz 1985, Walling and Moorehead 1989, Ongley *et al.* 1990).

Typically, the type of particle-size information required, and hence the method of analysis, varies with the application. Where the size data are required for hydraulic analyses, such as to predict settling in a reservoir or dispersion within a stream, a fall-speed-based measure is required. However, if the actual physical dimension of the sediment grains is important, such as when considering sediment effects on fish, machinery, turbidity and acoustic properties, and contaminant adsorption, then a direct measurement is required. Sometimes, both fall-speed and physical size data are pertinent, as, for example, where contaminant-carrying sediment must be removed from a waterway by settling. Since suspended grains, particularly those in the clay grade, often tend to cluster in

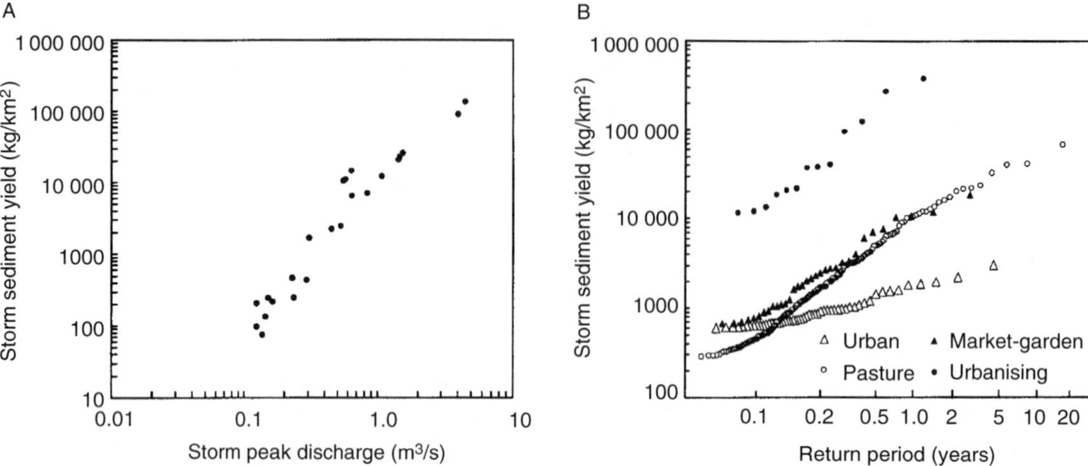

Figure 15.6 (A) Relation between suspended sediment yield and peak water discharge during storm runoff, Alexandra Stream, Auckland, New Zealand. (B) Magnitude–frequency relations for storm sediment yields at four small basins under different landuses, Auckland, New Zealand (after Hicks 1994). The return period is scaled in terms of the extreme value distribution

flocs (or aggregates), a key question is whether the flocs should or should not be dispersed before analysis. If flocs are not to be dispersed, then a serious consideration is whether in situ size analysis is required, since the stability of a distribution of flocculated sediment between the river and laboratory bench is questionable, particularly if samples are left to settle. Several texts (e.g., Guy 1969, Vanoni 1975, Allen 1990) detail size analysis techniques; a brief overview follows.

Several fall-speed-based size-analysis methods are used for stream suspended sediment samples. Most are based on monitoring changes in sediment concentration or accumulation with time at a point in a settling chamber. Depending on the method, the sample is either introduced to the top of the chamber or is thoroughly mixed in the chamber before settling starts. The resultant data yield settling rates, which are converted into frequency distributions by mass of equivalent diameters of spherical quartz grains. Traditionally, the most common method has been the pipette method, wherein samples are extracted with a pipette (Krumbein and Pettijohn 1938). A variant is the bottom-withdrawal tube, wherein the accumulated sediment in a neck at the base of the chamber is withdrawn. Its advantage is that a smaller mass of sediment is required (0.5 g), but it is limited in its ability to resolve sand sizes (Guy 1969). Modern instruments automate these manual methods. For example, the SEDIGRAPH (Coakley and Syvitski 1991) uses X-rays to monitor sediment concentration in a settling chamber, while the rapid sediment analyser (RSA) (De Lange *et al.* 1997) records the weight of sediment accumulating on a pan at the chamber base. Both the RSA and the visual accumulation (VA) tube (Guy 1969) are designed for sand-size fractions. Since fall-speed analyses usually require a sediment mass of one to several grams (Vanoni 1975), the volume of sample that needs to be taken from the river to provide this mass will depend on the suspended sediment concentration. Porterfield (1972) provides nomographs for estimating sampling requirements for particle-size analysis. Hydrometers are rarely used for suspended sediment analysis, owing to the relatively large masses of sediment required.

Several techniques directly measure grain dimensions. The traditional technique is sieving. Dry sieving is suitable for fine sand and coarser sediment (125 μ). Its utility rapidly collapses for finer sediment due to problems with sieve pore clogging and the effect of air currents within the vibrating sieve stack, which hinder the settling of fine sediment (M. Church, pers. comm.). Such problems are diminished by wet sieving, which is useful in the coarse silt range. Air-jet and sonic sieving are adjuncts to standard mechanical dry sieving analysis (Malhotra 1967). Laser-diffraction spectroscopy is based on the principle that particles of a given size diffract light through a given angle, which increases with decreasing particle size (Agrawal *et al.* 1991). Laser backscatter devices (e.g., Philips and Walling 1995a, Galai 1997) record the time required for a laser beam to traverse an arc across individual particles. Particle size may also potentially be determined using phase Doppler anemometry (Cioffi and Gallerano 1991, Bennett and Best 1995). Analysis of digital imagery offers exciting advances in size analysis, since this allows analysis of particle shape as well as size, can identify flocculation, and may potentially be used to identify constituent minerals from their shape signature. We stress, however, that except for quartz spheres, the different assumptions and approaches used with these techniques do not yield exactly the same result as a settling analysis. Thus, comparisons between the different techniques should be examined critically (McCave and Syvitski 1991).

A key advantage of most of the modern instrument-based methods is that they require only very small masses of sediment—an important consideration when dealing with dilute sediment concentrations. Also, some are portable (e.g., they can be set up on the stream bank), and some can even be deployed in situ (e.g., Bale and Morris 1991, Phillips and Walling 1995a, Gentien *et al.* 1995).

A key decision to be made before undertaking suspended sediment particle-size analysis is whether to break-up (i.e., disperse) particle flocs. The difference between un-dispersed ('native' or 'effective') size distributions and dispersed (or 'ultimate') distributions is often substantial, with factor-of-ten reductions in the median diameter being common (e.g., Walling and Moorehead 1989, Figure 15.7). In many applications, e.g., where the aim is to obtain fall-speed information, the sediment should be analysed as it occurs in the field. Traditionally, when only laboratory analysis was possible, samples were commonly duplicated or spilt, with one being kept for 'native' analysis and the other dispersed by chemical means (usually with a solution of sodium hexametaphosphate, or 'CALGON'). Nowadays, ultra-sonic vibration also provides an effective dispersing mechanism. The problem with the native sample analysis, though, is that the original floc distribution may be altered between the river and laboratory bench, particularly if samples are left to settle in their bottles for an extended period before analysis (Phillips and Walling 1995b). Thus, where particle flocs are important, analysing in situ, or as near in situ as possible, is desirable.

While less is known about spatial and temporal variation in suspended sediment size-grading than about bulk loading, existing information indicates that the variation can be substantial (Walling and Moorehead 1989). Spatial variability includes the effects of climate, basin lithology, sediment delivery processes from the basin slopes, in-channel sediment sources, and in channel processes such as sorting, abrasion, and turbulence. Temporal variability in particle size is influenced by turbulence, water discharge (to a degree, the size of suspended sediment tends to increase as competence increases), and by factors that cause temporal variation in sediment supply, such as delivery dynamics within a basin during a flood and seasonal effects on erosion processes and sediment sources. Phillips and Walling (1995a) note that the

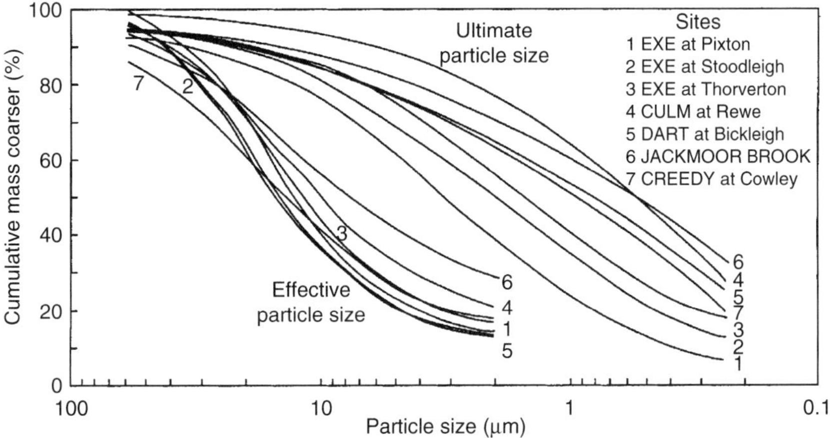

Figure 15.7 Comparison of native (effective) and dispersed (ultimate) size distributions of suspended sediment sampled at seven sites in the Exe River Basin, England (after Walling and Moorehead 1989) (reproduced by permission of Kluwer Academic Publishers and D.E. Walling)

particle-size characteristics of a single 'instantaneous' sample (such as collected over a few seconds by a point-sampler) cannot be considered representative of the load over a flood. Thus, as with the bulk sediment load, the variability in particle size demands an appropriate sampling strategy, and the first task is to identify the main factors causing the variability. A great advantage of in situ particle-size sensors is that they can integrate temporal variation.

Synoptic Sampling

Synoptic sampling of suspended sediment can provide an inexpensive indication of relative sediment sources within a drainage basin. Single-stage samplers may be used to collect synoptic datasets of near-surface sediment concentrations. These are passive sampling devices that are plumbed so that they fill, then seal, when the stage exceeds a set level (Edwards and Glysson 1999). Typically, they are deployed in vertical arrays so that a series of samples can be collected on either the rising or falling stage of an event (e.g., Gray and Fisk 1992). Their simplicity and cheapness means that they are relatively easy to deploy in numbers through an entire drainage basin.

A synoptic perspective on suspended sediment concentrations in near-surface waters can also be gained from remotely sensed imagery (Gomez *et al.* 1995—see Chapter 6). Reflectance of the visible and near-infrared wavelengths is sensitive to scattering of radiation by fine-grained sediment particles. This permits near-surface suspended sediment concentrations to be determined from Thematic Mapper (TM) data after appropriate calibration using reflectance data for a range of native sediment–water mixtures (Witte *et al.* 1981, Ritchie and Cooper 1988, Mertes *et al.* 1993). We caution that such synoptic mapping of near-surface sediment concentration may not necessarily reflect the true distribution of sediment yield across a basin over an event, due to the effects of different travel times and dispersion of sediment along tributaries, phase differences in rainfall and erosion processes across the basin, and the absence of flow-weighting.

15.4 BEDLOAD SAMPLING, MEASUREMENT AND PREDICTION

Overview

Since the inception of interest in bedload sampling in the 19th century (Humphreys and Abbot 1861, Davies 1900), the intent of most systematic bedload sampling programmes has been to characterise the bedload transport regime of the river in question by documenting the amount and size of sediment moved by different flows, and to determine how these parameters change downstream (e.g., Swiss Federal Authority 1939, McLean and Church 1999, McLean *et al.* 1999). Since the continuity of bedload transport is typically not maintained along a river, such information is crucial for identifying reaches in which the transport capacity will potentially be either larger or smaller than the supply, and hence the conditions under which scour or fill of the river bed and other adjustments to channel geometry may occur. Restoration or enhancement of a river corridor in a manner that actively promotes the formation of viable aquatic habitats, the specification of channel maintenance or flushing flows, and the effective in-stream management of sediment resources also require quantitative knowledge of bedload transport (American Society of Civil Engineers Task Committee 1992, Andrews and Nankervis 1995, Kondolf and Wilcock 1996, Wilcock *et al.* 1996).

As with suspended load, the information needs and approach adopted to quantify bedload will vary with the situation. For example, it will depend on whether information is required on bedload transport rates past a single section or whether a broader-scale view is required, such as the average pattern of aggradation or degradation within a reach. At this juncture it is also important to note that systematic bedload sampling is a time-consuming and expensive undertaking that presently is neither facilitated by commonly accepted equipment nor governed by a consistent set of protocols. Moreover, a practicable solution to the complex problem of predicting bedload transport also remains tantalisingly out of reach. For these reasons, we provide an overview of the available equipment and procedures rather than explicitly advocating the use of particular instruments and methodologies. We focus on describing two commonly utilised approaches for gaining knowledge of the bedload transfer through a reach of a river under a range of flow conditions:

(i) field sampling or measurement, and
(ii) application of a formula.

In this context, 'sampling' involves collecting discrete samples of bedload across a channel section for limited intervals of time; 'measurements' involve recording the continuous or time-integrated bedload over the whole cross-section or reach by way of a trap, repeat surveys of river bed morphology, or by monitoring tracer grains.

Other promising methods for monitoring bedload, including Doppler sonar and other acoustic techniques, continue to be developed and refined (e.g. Lowe 1988, Thorne *et al.* 1989), but only methods and techniques for investigating bedload that have wide currency are discussed here. For this reason too, model studies, the application of which is constrained to specific cases and the utility of which is exemplified by Young and Davies' (1990) Froude-scaled model of the braided North Branch of the Asburton River and Shvidchenko and Kopaliani's (1998) undistorted scale model of the Laba River, are also not addressed.

Bedload Sampling

Sampler types. Many of the earliest sampling devices were of the basket type. In Europe, they were patterned on Mülhofer's (1933) design. These samplers retain sediment primarily by screening it from the flow, but also because there is a reduction in flow velocity within the sampler. The minimum size of particle retained is determined by the size of the screen openings, which are not intended to retain sand. Basket samplers typically have a large capacity and were designed specifically to accommodate very coarse gravel particles.

Pressure-difference samplers were developed originally for use in sand-bed channels (Schaank 1937), but later designs were also developed for gravel-bed rivers (Helley and Smith 1971). They are designed so that the entrance velocity and the ambient flow velocity are equal. This is accomplished by constructing the sampler with walls that diverge towards the rear, creating a pressure drop at the sampler exit. Sediment is retained in a mesh bag mounted behind the sampler. The Helley–Smith sampler's simple design (Figure 15.3A) has proved immensely attractive, and it is probably the most popular sampling device in use today. A modified version of the Helley–Smith sampler, the US BLH-84, which has a nozzle expansion ratio of 1.4 (compared to the original expansion ratio of 3.2), has tentatively been recommended for use by the Technical Committee of the Federal Inter-Agency Sedimentation Project. A version of the modified sampler, with a 305 mm wide 152 mm high entrance, has been developed for use in rivers with coarse gravel beds (Childers *et al.* 1989). The Delft Nile sampler is a pressure difference sampler that has been specially designed for sand-bed rivers (Van Rijn and Gaweesh 1992, Gaweesh and Van Rijn 1994).

Sampling and hydraulic efficiency. The presence of a sampling device on the river bed necessarily alters the pattern of the flow and sediment transport in its vicinity. Thus, bedload samplers must be calibrated to determine their efficiency under different hydraulic and sediment transport conditions. Hydraulic efficiency is a measure of the degree to which the flow is accelerated or retarded by a sampler, defined as the ratio of the mean velocity of the flow through the sampler entrance to the mean velocity of the flow through the area occupied by the sampler entrance in the absence of the sampler (Hubbell 1964). Sampling efficiency indicates the extent to which a sampler either over- or under-samples the material in transport, defined as the ratio of the mass of the bedload collected during a specified time period to the true mass of bedload that would have passed through the entrance width in the same time period in the absence of the sampler (Hubbell 1964).

Studies conducted by Einstein (1937) and Engel and Lau (1981) indicate that depending on factors such as the sediment size, prevailing transport rate and flow conditions, sampling time and degree to which the sampler is filled, the sampling efficiency of basket samplers typically varies between 30% and 60%. The sampling efficiency of pressure-difference samplers appears equally variable (Gibbs and Neill 1973, Engel 1983). Laboratory tests suggest that for a relatively narrow, well-defined range of flow and sediment transport conditions, the sampling efficiency of the original Helley–Smith sampler may be approximately 100%, but that it varies appreciably outside this range (Hubbell *et al.* 1981). The sampler has a commonly accepted, field-determined sampling efficiency of 100% (Emmett 1980). Geometrically scaled versions of the original Helley–Smith sampler, which has a 76-mm square nozzle, appear to preserve the hydraulic and sampling efficiencies of the original design (Druffel *et al.* 1976, Hubbell *et al.* 1985). The Delft Nile sampler is reported to have a sampling efficiency close to 100% for particles larger than 0.4 mm (Gaweesh and Van Rijn 1994).

In practice, determining the hydraulic efficiency of a bedload sampler has proved to be a relatively simple task, whereas determining the sampling efficiency is considerably more complex. Two factors confound the issue. First, the true bedload transport rate may only be defined indirectly (Hubbell *et al.* 1981). For example, transport rates typically vary with time and from point to point along and across a channel (e.g., Gomez 1983). Thus, strictly speaking, it is not possible to directly compare the transport rate determined at different points, irrespective of how closely the two points are located. It is also important to appreciate that a 'bedload' sampler does not sample bedload alone, rather a sampler retains all the material moving in a zone immediately above the bed (the thickness of the zone is equivalent to the height of the sampler entrance), regardless of the mode of sediment dispersion (Einstein 1948). Second, to obtain a single, constant measure of sampler efficiency directly, via a comparison of the mean at-a-point bedload transport rate, b, with measurements of the true mean bedload transport rate, t, the relation between b and t must be linear (De Vries 1973). Typically, this has proved not to be the case (Einstein 1948, Hubbell 1987, Thomas and Lewis 1993b). Indeed, the sampler calibration process remains incomplete because none of the tests performed to date on any sampling device has provided definitive results.

A century of experience has demonstrated that bedload samplers rarely provide an unbiased estimate of the prevailing transport rate but, in the face of a plethora of disputable facts about sampler performance, two important details about the operation of Helley–Smith-type samplers have emerged. First, sampler performance declines appreciably if the sample bag is overfilled (40% full) and if the mesh is clogged by fine sediment or organic debris (Druffel *et al.* 1976, Johnson *et al.* 1977, Beschta 1981). Second, deviations from original design specifications give rise to discernible differences in sampler performance (Pitlick 1988, Ryan and Porth 1999). It is also likely that errors due to the imperfect design or calibration of a sampler are small as compared with the errors that arise from the adoption of poor sampling techniques (Gomez *et al.* 1991).

Bedload sampling strategy and practice. Since bedload transport rates vary across channel and with time, appropriate temporal and spatial sampling strategies are required to minimise error in the estimates of the mean bedload transport rate. Temporal strategies involve three elements: the sampling time (the length of time the sampler remains on the bed); the sampling interval (the length of time that elapses between consecutive samples); and the sampling period (the sum of the sampling times and sampling intervals). Also, we note that a distinction is often made between 'bedload transport rate' (local, mean values) and 'bedload discharge' (the total or unit-width mean rate).

Any number of random samples may, in theory, provide an estimate of the prevailing mean bedload transport rate, though the magnitude of the errors involved may be expected to decrease as sample size increases (Csoma 1973, DeVries 1973, Carey and Hubbell 1986). However, it is almost impossible to obtain a truly independent random sequence of bedload samples under field conditions, and sequential samples likely will be serially correlated. Each observation in an auto-correlated time-series repeats part of the information contained in previous observations; thus, more sequential samples than independent random samples are required to provide the same information about the true mean. Nesper's (1937) experience prompted him to comment that, at 'low' transport rates, between 10 and 15 sequential samples were probably required to provide an acceptable indication of the mean transport rate, while about 30 samples were required at 'high' transport rates. Gomez *et al.* (1990) evaluated errors associated with at-a-point sampling where bedforms (dunes) were present. Their analysis suggests that 21 sequential samples are required to obtain an estimate of the mean at-a-point bedload transport rate that falls within 50% of the true mean rate, at the 99% confidence level. This assumes that the sampling period is long enough to allow at least one primary bedform to migrate past the sampling point. Effects due to non-stationarity may be minimised by ensuring that the sampling interval does not coincide with the period of the bedforms that are present.

Note that the preceding discussion properly refers to the case of fully established motion, where the entire bed locally is taking part in the transport process. This is typically the case in sand and mixed sand and gravel-bed rivers, but may not be true in gravel-bed rivers where conditions are near the threshold of motion and transport normally occurs at low rates (e.g., Andrews 1994). In this case, the transport is highly variable even in the short term, so serial correlation is apt to be low between successive samples, and many samples are still required to characterise the variability. The sampling time will dictate the actual number of samples; if this time is protracted, it may absorb much of the short-term variability.

Reliable estimates of the streamwide bedload discharge obtained using sampling devices are dependent upon good at-a-point knowledge across the full width of the channel. The statistics of the sediment transport regime provide information on the number of times the sediment transport across the channel must be sampled in order to obtain a reliable value for the time-averaged bedload transport rate that conforms to reasonable limits (Kuhnle 1998). Gomez and Troutman (1997) showed that sampling errors decrease as the number of samples collected increases and the number of traverses of the channel over which the samples are collected increases. Assuming sampling is conducted at a pace which allows a number of bedforms to pass through the sampling cross-section, bedload sampling schemes typically should involve four or five traverses of a river, and the collection of 20–40 samples at a rate of five or six samples per hour. The objective is to reduce both random and systematic errors, and hence minimise the total error involved in the sampling process, by ensuring that spatial and temporal variability in the transport process is addressed.

Regardless of the manner in which the computational exercise is performed, the message is clear: the collection of reliable bedload data is a time-consuming process since the sampling period will, of necessity, be lengthy. A corollary of this is that the sampling time may also be longer than is practicable with conventional sampling devices, which do not have the capacity to accommodate large amounts of sediment. For this reason, a sampling time that is of the order of 30 s typically is used. Since flow unsteadiness affects the rate of bedload transport, it is also assumed that the flow remains steady for the duration of sampling. In practice, since many rivers respond rapidly to precipitation inputs, this may prove an untenable requirement in all but snowmelt dominated runoff regimes.

Care should be taken when using the sediment from a bedload sampler to characterise the size distribution of the bedload. This is because even the composite of a series of samples may be smaller than the minimum weight required to avoid bias and to achieve good precision in the calculated grain size percentiles (Ferguson and Paola 1997). A final practical comment is that great care should be exercised when raising and lowering bedload samplers, particularly when they are deployed with a cable. An 'over catch' will result if the sampler is allowed to 'shovel' into the bed or the face of a dune when it contacts the bed. Also, the bedload sample may be flushed or spilt if the sampler is allowed to rotate downstream or tip downwards during retrieval.

Bedload Traps

Unlike the data obtained from bedload samplers, data obtained from bedload traps are usually regarded as exact. A trap is a cavity sunk into the streambed, with its upstream lip flush with the surface (Figures 15.3C and D). The bedload falls into the trap and is retained in the well of the trap. Assuming overfilling is not a problem, trap efficiencies of the order of 100% are to be expected if the opening is wide enough to prevent overpassing of saltating particles (Poreh *et al.* 1970, Habersack *et al.* 1998). Traps also have a distinct advantage over samplers in as much as, if the trap spans the entire width of the river, it is not only possible to catch all the bedload that passes through the measuring section in a given period of time but also to continuously measure the rate at which sediment accumulates.

The simplest traps consist of lined pits or slots in the streambed in which the bedload collects over one or more events (Church *et al.* 1991). The bedload yield for the period in question is either determined volumetrically by surveying the deposit (as is normally undertaken in reservoirs) or by manually excavating and weighing the sediment (Hansen 1973, Newson 1980). More sophisticated traps incorporate pressure sensors that continuously weigh the mass of sediment in situ (Reid *et al.* 1980) or use a pump or conveyor belt to transfer it to a weighing station on the streambank (Dobson and Johnson 1940, Einstein 1944, Leopold and Emmett 1997). Other traps are designed so as to generate a vortex that ejects the sediment as it accumulates (Parshall 1952, Robinson 1962, Milhous 1973, Hayward and Sutherland 1974, Tacconi and Billi 1987), or separate the bedload from the fine sediment and water (Lenzi *et al.* 1999). So long as they are not overfilled, traps invariably provide reliable data, but the limiting factor in their deployment is that they are both difficult and expensive to install.

Bedload Tracer (see Chapter 14)

As discussed in detail in Chapter 14, tracer particles may provide an alternative or useful adjunct to the use of sampling devices or traps. Particles can be marked by painting (Laronne and Carson 1976), inserting magnets (Ergenzinger and Conrady 1982, Hassan *et al.* 1984), or using the inherent magnetic properties and enhanced natural magnetic remanence of sediment (Arkell *et al.* 1983, Ergenzinger and Custer 1983). An electromagnetic sensor installed in the river bed can monitor the inception, intensity, and duration of transport of electromagnetic tracers (Custer *et al.* 1987, De Jong and Ergenzinger 1998). In the past, radionuclides have also been used as a tracer (Sayre and Hubbell 1965). If tracers can be recovered downstream, they provide information on initial motion, depth of burial, distance travelled and the proportion and size of particles moved, typically on a flood-by-flood basis (Hassan and Church 1992, 1994, Church and Hassan 1992). Detailed information about the statistics of particle movements (and improved recovery rates) is obtained by inserting radio transmitters (Ergenzinger and Schmidt 1995, Emmett *et al.* 1996). Calculation of the transport rate using a tracer requires an explicit statement of the relation between transport rate, entrainment and displacement length (Wilcock 1997b), or the virtual velocity of sediment (Stelczer 1981, Haschenburger and Church 1998).

Morphological Methods

The emphasis of the above techniques is on providing short-term data. However, at the event or intra-event scale, the most pronounced feature of the bedload transport process is its spatial and temporal variability (Gomez 1991). This variability reflects both variations in the flow conditions (i.e., capacity) and variations in the supply or availability of bed material.

Factors influencing the bed material supply include event magnitude, the translation and dispersion of sediment waves (including bedforms), the presence of an armour layer, and the occurrence of patches (Gomez *et al.* 1989, Parker 1990, Seal *et al.* 1993, Lisle 1995, Garcia *et al.* 1999, Lenzi *et al.* 1999, Lisle *et al.* 2000). From the perspective of characterising the bedload transport regime in a particular reach, this spatial and temporal variability of bedload transport may be sufficiently complicated that it requires many measurements to reduce the variance in the observed data to an acceptable level. For this reason, the direct collection of quantitative data may always be an impractical method of estimating bed material transport rates in large rivers and over engineering timescales of 10–100 years.

'Morphological' methods offer an alternative approach for determining bedload discharge. In essence, these are based on the continuity relation for bedload transport, which requires that the rate of change in the mean level of a segment of river bed is proportional to the difference between the transport in and out of the segment. Knowledge of the bed level change and the transport across one segment boundary permits computation of the transport rate across the other boundary. Natural situations arise where this relation can be exploited. Attempts have been made to determine transport rates from bedform statistics by comparing sequential bed profiles (Simons *et al.* 1965, Willis and Kennedy 1977). It remains, though, that the geometry and movement of bedforms is highly variable and no unique and quantitative method has been developed for describing dunes, which are also imperfect sediment traps (Moll *et al.* 1987, Gabel 1993, Mohrig and Smith 1996, Nikora and Hicks 1997). At a larger scale, however, evaluating bed material transport by the analysis of the morphological changes that occur along a river reach can provide a viable alternative to direct sampling or measurement (Popov 1962a,b).

Neill (1971, 1987) developed an approach for defining the relation between morphological change and bed material transport in a systematically migrating meander bend. To derive a transport estimate, knowledge of the volume of sediment mobilised per unit length of channel and the average distance of travel (approximated as one-half the meander wavelength) is required. Church *et al.* (1986) described a more generalised approach in which knowledge of the changes in the volume of sediment stored in a reach and an estimate of the transport at one section permit the transport to be estimated throughout the reach. Carson and Griffiths (1989) used a similar approach to estimate gravel transport during flood in the large, braided Waimakariri River, New Zealand. McLean and Church (1999) compared two approaches which were used to estimate the annual gravel load of the lower Fraser River. The assumptions, procedures and limitations involved in the latter approach have also been discussed by Martin and Church (1995), who used it to estimate bed material transport in an 8-km long reach of the Vedder River.

Channel change is relatively easy to measure, and the magnitude of exchanges of sediment between the channel and floodplain and other storage sites can be discerned accurately from photographs and field surveys (Ferguson and Ashworth 1992, Lane *et al.* 1995). Inasmuch as it yields information of comparable quality to that of direct measurements, and requires less field effort, the morphological approach is relatively robust (Martin and Church 1995). Given the rapid advances that are being made in the use of GIS software in combination with differential GPS, digital photogrammetry, airborne laser altimetry, and other instruments that make it possible to obtain high resolution, spatially resolved data rapidly, morphological methods are also likely to be more widely used in the future (e.g., Lane 1998).

Bedload Formulae

Formulae predict bedload transport capacity under given flow conditions. Their ability to do this is predicated on the assumption that it is possible to equate the rate at which bedload is transported to a specific set of hydraulic and sedimentological variables. Indeed, the underlying physics appear quite straightforward (Du Boys 1879, Bagnold 1966). Ignoring the problems caused by variations in the supply or availability of sediment, it has long been recognised that, even at a constant discharge, bedload transport rates fluctuate (Figure 15.8A). It is also apparent that if many observations are made over a time period that is long enough to delimit the entire range of transport rates, a reliable estimate of the mean rate can be obtained (Einstein 1937). Consequently, field (and laboratory) data that are integrated over lengthy time periods and across the whole width of the channel often yield coherent relations (Figure 15.8B). The availability of such data seemed to confirm the existence of a bedload function and to demonstrate that bedload transport indeed occurred in accordance with established principles (Müller 1937). This, coupled with the realisation that sampler calibration was by no means a straightforward task (Nesper 1937, Einstein 1937), helped foster the view that the prediction of bedload discharge was a viable proposition. In practice, however, the complexities of the interrelations between the conditions governing bedload transport confound the issue (Gilbert 1914).

Most formulae either describe a relation that has been either defined empirically on the basis of laboratory or field data or derived from basic mechanical or physical principles. From the outset, two issues have contributed to the profusion of bedload transport formulae. First, there is no consensus about the fundamental hydraulic and sedimentological quantities involved. Second, dissatisfaction with the performance of a particular formula (which was often inspired by its poor performance against data that were not included in the initial analysis) encouraged attempts to develop new relations. It remains a source of some discomfort that there appear to be more bedload formulae than there are reliable data sets by which to test them (Gomez and Church 1989).

There have been a number of major reviews that use field data to compare bedload transport formulae (Table 15.1). None provided definitive results. In consequence, no one, or even a small group of formulae, has been universally accepted or recognised as being especially appropriate for practical application and it is important to remember this point when applying models such as HEC6 or FLUVIAL12. In the face of such overwhelming indecision, an interested party has little option but to make an intuitive selection on the basis of the similarity of the conditions for which a particular formula was derived and those in the river in question. The indexes developed by Williams and Julien (1989) and Bechteler and Maurer (1991) may assist with this process. However, since there is no reason to suppose that any formula will necessarily provide complete correlation, caution dictates that the results of several formulae be compared.

In applying any one-dimensional equilibrium transport formulae, local hydraulic parameters should be utilised as the use of mean values represents a channel-wide integration before the transport calculation. The effect may be important since most formulae are non-linear. However, there are obviously problems involved in maintaining strict observance to the form of any formula, not least, for example, because local bed shear stress cannot be directly measured and will vary widely across a gravel-bed river (Dietrich and Whiting 1989, Wilcock *et al.* 1994). In coarse-grained channels, the point at which motion is initiated may depend more on the relative size than the absolute size of the bed material. That is not to say that the fundamental effect of particle weight is eliminated, but rather that because of effects due to sheltering, protrusion, grading and shape, it becomes less dominant. Starting with Einstein (1950), several models have sought to account for these factors (Parker *et al.* 1982, Parker 1990, Andrews and Smith 1992). Accounting for surface structure in gravels is also a significant obstacle to the assured application of any bedload formula in the field (e.g., Jackson and Beschta 1982, Lisle and Madej 1992, Seal *et al.* 1993, Hassan and Church 2000). Structure may reflect the flow history and its creation or disruption is known to influence bedload transport rates (Gomez 1991). Moreover, because the summary effects of structure are not readily measured, assumptions about fundamental parameters, such as the Shields' Number, are not easily made.

The selection of an appropriate formula for use in sand-bed rivers, where relative size effects are not an issue and the constraints on the availability of sediment are relaxed, may be more straightforward. A variety of field data suggest that the unit discharge of sand varies approximately with the fifth power of mean velocity and inversely with particle diameter (Posada and Nordin 1993). Several theoretical relations, such as that developed by Engelund and Hansen (1967), conform with such a trend, although all incorporate one or more empirically derived coefficients.

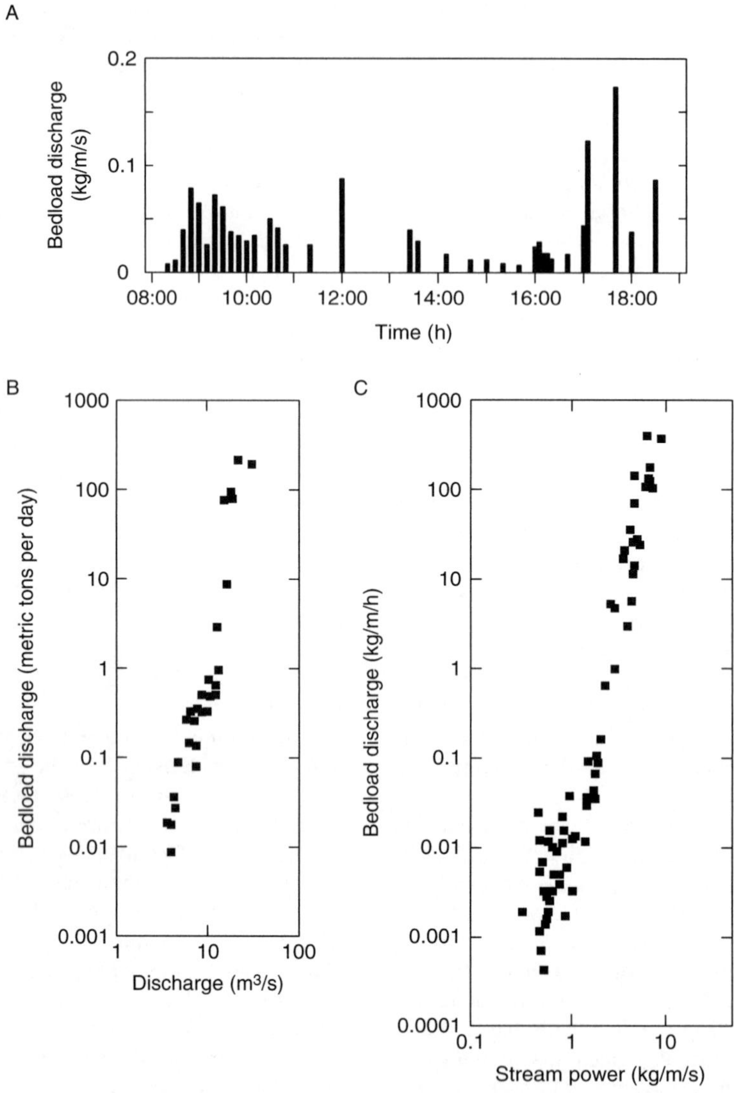

Figure 15.8 (A) Temporal variations in bedload transport rates observed at virtually constant discharge (0.62–0.63 m^3/s) in Torlesse Stream, New Zealand, 30th August 1973 (data from Hayward 1980). (B) Observed relation between bedload transport rate and water discharge in the Enoree River, South Carolina, 1939–1940 (data from Dobson and Johnson 1940). (C) Observed relation between bedload transport rate and stream power in Oak Creek, Oregon, 1971 (data from Milhous 1973). Note that all of these data were obtained using traps that spanned the entire channel width

Bedload Rating Curves

Though it requires verification with field data, the appeal of a formula is that it produces a rating that may, in principle, be used in conjunction with discharge data to compute bedload yield over a specified time period. Though there is often considerable scatter and the data rarely extend across the entire range of flow conditions, field data may also be used to define a rating curve. Simple functions are typically used (e.g., Wilcock *et al.* 1996, Moog and Whiting 1998), with the transport rate (q_b) commonly por-

Table 15.1 Summary of major reviews of bedload sediment transport formulae

Authors	Formulae examined	Methodology	Recommended, representative or preferred formulae
Vanoni *et al.* (1961)	Du Boys-Straub, Einstein, Einstein-Brown, *Laursen, Meyer-Peter, Meyer-Peter & Müller, Schoklitsch 1934, Shields	Comparison of sediment-rating curves	None specified
Shulits and Hill (1968)	Du Boys-Straub, Casey, Einstein, Elzerman-Frijlink, Haywood, Kalinske, *Larsen, Meyer-Peter, Meyer-Peter & Müller, Rottner, Schoklitsch 1934, Schoklitsch 1943, Shields, USWES	Determination of limits of agreement between calculated bedload transport rates	Du Boys-Straub, Meyer-Peter & Müller, Schoklitsch 1934
ASCE Task Committee (1971)	Blench, Du Boys-Straub, *Colby, Einstein, Einstein-Brown, *Engelund-Hansen, *Inglis-Lacey, *Larsen, Meyer-Peter, Meyer-Peter & Müller, Schoklitsch 1934, Shields	Comparison of sediment-rating curves	*Colby, *Engelund–Hansen, *Toffaleti
White *et al.* (1973)	*Ackers-White, Bagnold 1956, *Bagnold 1966, Bishop, *Simons-Richardson, *Blench, *Einstein, Einstein-Brown, *Engelund-Hansen, *Graf, *Inglis, Kalinske, *Laursen, Meyer-Peter & Müller, Rottner, Shields, *Toffaleti, Yalin	Comparison of discrepancy ratios	*Ackers-White, *Engelund–Hansen, Rottner
Mahmood (1980)	*Ackers-White, *Colby, *modified Colby, Einstein, *modified Einstein, *Engelund-Hansen, *Laursen, Meyer-Peter & Müller, Mahmood, Shen-Hwang, *Toffaleti, *Yang	Comparison with the modified Einstein procedure	Shen-Hwang, *Toffaleti
Gomez and Church (1989)	Ackers-White, Ackers-White-Day, Ackers-White-Sutherland, Du Boys-Straub, Bagnold 1980, Einstein, Meyer-Peter, Meyer-Peter & Müller, Parker, Schoklitsch 1934, Schoklitsch 1943, Yalin 1963	Comparison of mean and local bias	Ackers-White-Day, Bagnold 1980, Einstein, Parker

*Total load formulae.

trayed as a power function ($q_b \quad aQ^b$) of discharge, Q (where the parameters a and b are estimated by linear regression to the log-transformed variables). However, there is no prescribed method for undertaking such an analysis. Irrespective of whether a formula, field data, or a combination are used to construct a bedload rating curve, it is common practice to compute transport rates for the sand and gravel fractions separately (e.g., Wilcock 1997a, 1998). This is because all particle sizes present on the bed are rarely in motion at once and the bedload size distribution only infrequently approaches that of the bed material (Gomez 1995, Lisle 1995).

15.5 TOTAL LOAD

Determining the total sediment load requires matching complementary methods of determining bedload and suspended load. It is important that this matching is done over time and spatial scales that are consistent with the component methods; also, double-accounting of size-fractions that overlap the suspended and bedloads should be avoided. Primarily, the approach for determining total sediment load depends on whether the streambed material is sand or gravel.

For sand-bed streams, there are three total-load approaches: formula, sampling and a combination. So-called 'total load' formulae (e.g., Engelund and Hansen 1967, Toffaleti 1968) actually only determine the bed material load, and they are appropriate where there is no washload and an unrestricted supply of sand from the bed. These require input data on flow hydraulics and bed material size characteristics. Where there is washload and/or restrictions on the sand supply, one approach is to sample the suspended load and to compute the suspended and bedloads in the unmeasured zone (Figure 15.2), as in the 'modified

Einstein' and related approaches (Colby and Hembree 1955, Stevens and Yang 1989). Alternatively, both the bedload and suspended load can be sampled. Ideally, the sampling of both modes should be synchronous, using a device such as the Delft Nile sampler (Van Rijn and Gaweesh 1992, Gaweesh and Van Rijn 1993). This combines a bedload sampler specially designed for sand beds with a vertical array of pumping point-samplers. Combined (although not synchronous) bedload and suspended load sampling over sand-beds has also been conducted using the Helley–Smith type bedload samplers and depth-integrating suspended sediment samplers (e.g., Andrews 1981). With this approach, it is necessary to correct for any double-accounting of sand fractions in the depth range intercepted by the bedload sampler. In a naturally contracted section or turbulence flume, sand bedload is forced into suspension and can be treated as suspended load. Colby and Hembree (1955) and Hubbell and Matejka (1959) employed this procedure to determine the total sediment load of rivers in the Nebraska Sand Hills.

The combined sampling approach is also an option for gravel-bed streams. Again, any double-accounting of size fractions intercepted by both the suspended and bedload samplers in the near-bed zone needs to be addressed, and it is important that appropriate sampling strategies are employed so that the two load components can be combined over matching spatial and temporal scales. Alternatively, the bedload component can be determined by formula. The morphological method for 'bedload' actually determines the time-averaged bed material load; this needs to be combined with the washload component of the suspended load over the same time frame.

15.6 ESTIMATING SEDIMENT YIELDS FROM RESERVOIR SEDIMENTATION

Reservoirs present special needs for sedimentation information (such as their rate of infilling and the rate of depletion of bed material load to the channel and coastline downstream), but they also afford unique and robust opportunities for measuring the total sediment load of the inflowing river(s), both on an event basis and on a long-term average basis. Because they are backwaters, reservoirs trap part of the washload of inflowing streams as well as the bed material load (Brune 1953, Maneux *et al.* 2001), but the degree of entrapment of each size fraction depends upon the hydraulic conditions through the reservoir, which vary with time. For this reason, it is often easier to determine a reservoir sediment budget, and the total inflowing sediment load, by combining reservoir sedimentation volumes with the suspended load in the outflow.

Techniques for surveying sedimentation volumes in reservoirs are well detailed in several texts (e.g., Vanoni 1975, Morris and Fan 1998). Traditionally, there have been two basic approaches for defining reservoir bed topography: with contours or cross-sections. Accurately defining contours can require an extremely detailed survey, while the cross-section (or range line) approach generally requires less effort and has the advantage that sections (marked by monuments) can be precisely relocated. With either approach, various surveying methods can be used, depending on the size of the reservoir and the depth of water. Typically, water depth is sounded by boat either with a weighted line or an echo-sounder, while horizontal position may be measured from a tagline, a Total Station system, or differential GPS (DGPS). The most modern approach is to use a DGPS and sounder integrated with hydrographic software. Indeed, this approach has proven to be sufficiently accurate, rapid, and reliable that it has become cost-effective to collect enough data to develop a digital elevation model (DEM) of the reservoir bed (e.g., Schall and Fisher 1996, Sullivan 1996). Whether using contours, cross-sections, or rectangular DEMs, the reservoir volume (and volume changes due to sedimentation) can be calculated using an interpolation method such as Simpson's rule or the prismoidal formula (Vanoni 1975). A DEM combined with readily available topographic/GIS software permits easy computation of volume changes and also mapping of sedimentation depths. The interval between surveys will depend largely on the rate of sedimentation, but typically this may be 5–10 years.

To reconcile sedimentation volumes with other sediment budget information (which is usually measured in units of mass flux), it is necessary to determine the bulk density (or specific weight) of the reservoir deposits. Measuring this is relatively easy if a reservoir is periodically dry, which facilitates the extraction of cores or in situ measurements. A simple method to use with either cores or in situ excavations is the 'sand cone' approach (e.g., Vanoni 1975), wherein the mass of sediment removed from a hole is weighed and the volume of the hole is determined by refilling it with sand. The density of submerged deposits may be measured in situ using probes that are pushed into the bed sediment, such as the Gamma Probe (e.g., McHenry 1971), measured in the laboratory from core samples, or estimated based on analysis of the grain size in sediment cores and an empirical relation between bulk density and grain size of freshly deposited sediment (e.g., Vanoni 1975). The bulk density of reservoir sediment increases with time due to consolidation. Various semi-empirical approaches have been developed to estimate this increase in density, or alternatively to estimate the settling of a sedimentary layer due to consolidation (e.g., Lane and Koelzer 1953, Miller 1953, Gill 1988).

Reservoir outflows, because their suspended loads are typically fine-grained and well mixed, are usually well suited to continuous monitoring with turbidity sensors. Sediment ratings for reservoir outflows often show wide data scatter owing to phase lags between the water discharge and sediment concentration peaks and to artificial manipulation of the outflow discharge.

Stratigraphic and sedimentological analysis of reservoir deposits can provide detailed information on event sediment yields and long-term average yields. Laronne and Wilhelm (2001) observed that the deposits from individual flood events could be recognised as couplets. For example, when describing reservoir deposits in Israel's Arava Rift Valley, they noted that the lower part of the couplet was an ungraded traction deposit (often sand), formed early in the flood when the reservoir inflows were high and the reservoir was often at a relatively low level. The upper part was a graded silt and clay, formed by settling from suspension.

The stratigraphy of a reservoir can be sampled either from cores or from excavations if the reservoir dries out. Individual couplets within sequences can be correlated spatially, and their depths can be mapped and then integrated to determine the total volumes of discrete deposits. With adequate dating control on the stratigraphic record (which is usually straightforward since the reservoir construction date is usually well documented), the series of event sediment yields may then be transformed into a probability density function or a magnitude–frequency relation similar to that shown in Figure 15.6B. With a record spanning many decades, long-term changes in the mean annual sediment yield and probability density function of event yields may be related to factors such as catchment landuse change or climate change (Laronne 1990, Seydell 1998). Stratigraphic mapping can also be used to hindcast bed levels and stage–volume curves during earlier phases of the reservoir's life, which can, in turn, be used to hindcast volumes of flood-water inflows should a record of these not be available from an upstream hydrological station (Laronne and Wilhelm 2001).

15.7 KEY POINTS FOR DESIGNING A SEDIMENT MEASUREMENT PROGRAMME—A SUMMARY

The basic steps in designing a sediment measurement programme are set out in Table 15.2. First it is necessary to define the purpose(s) of the programme, since this largely determines the basic measurement approach. With this is decided, the measurement 'tools' can be selected with the aid of Tables 15.3 and 15.4. Some prior knowledge of the relative importance of suspended load and bedload and of the size-grade of the bedload (whether sand or gravel) will help to focus the measurement effort and choice of tools.

The suspended load needs to be sampled, since typically it is limited by the supply of fine sediment to the channel rather than by physical transport capacity. A key consideration is whether the problem at hand requires near-continuous data, event-based information, or simply long-term statistics such as the mean annual yield (Table 15.3). Continuous data require index or point-samples collected manually or by auto-sampler, or else records from optical or acoustic sensors. Such point measurements need to be related to the cross-section mean sediment concentration. If the sampling purpose is only to determine the average annual sediment yield, then the sediment rating or direct estimation methods are more economical alternatives to continuous monitoring. Suspended sediment ratings either attempt to explicitly model sediment concentration as a function of all significant controlling factors, or, more commonly, are used to model the conditional mean concentration over the period of interest as a function of water discharge. Care is required in fitting statistical models to sediment-rating relations (the rating model should suit the form of the bivariate concentration—water discharge distribution), in correcting for bias induced by data transformations, and with sampling strategies for compiling rating datasets. Direct load estimation methods employ unbiased, selective sampling strategies, implemented in real-time by stream-bank data-loggers and auto-samplers, to directly estimate the average sediment load over a sampling period.

Suspended sediment yields for discrete runoff events can be related to indices such as event peak flow. Event-yield magnitude–frequency relations are useful for discriminating landuse effects on sediment supply. Synoptic sampling of near-surface waters, either with manual samples or remote sensing, provides useful relative indicators of basin-wide sediment sources.

Table 15.2 Key steps and considerations when designing a sediment measurement programme

Step		Examples/considerations
1	Decide main purpose of measurements	Statistics of instantaneous sediment load Annual-average total sediment load Erosion/deposition in a river reach or reservoir Scientific study of fluvial processes in a river reach Other/a combination of the above
2	Identify nature of sediment load of primary interest	Suspended load, bedload, or total load Expected suspended/bedload ratio Composition of bedload—e.g., sand or gravel
3	Decide basic temporal sampling approach appropriate to purpose determined in step 1	Continuous sampling Event-based measurements Statistical sampling to determine only annual average loads
4	Choose measuring approach to suit outcome of steps 1–3 and accuracy requirements	In situ sensors for suspended load and bedload Manual samplers for bedload and suspended load Automatic suspended sediment samplers Bedload traps Surveys of erosion/deposition in river reaches or reservoirs Bedload tracers Bedload or total load formulae Remote sensing of suspended load Merging bedload and suspended load measurements For bedload, use a combination of methods to reduce uncertainty
5	Select basin-scale spatial sampling strategy to suit purpose from step 1	Network of measurement stations in river basin Inflows/outflows of reach of interest Synoptic sampling
6	Design at-a-section spatial sampling strategy to suit measurement approach from step 4	Sampling verticals Point vs. cross-section mean calibration relations
7	Design temporal sampling strategy to suit approaches decided in steps 3 and 4	Duration of discrete measurements (e.g., bedload samples) Time base for auto-sampling (e.g., fixed time, flow-proportional, etc.) Time interval between measurements/surveys Duration of measurement programme
8	Determine requirements for analysis of particle size	In situ or laboratory measurement Effective or ultimate size distribution of suspended load Adequate mass sampled for analysis technique
9	Identify supplementary data needs (e.g., for computing sediment discharge using rating relations or formulae; for converting sediment volumes to masses)	Sediment mineral and/or bulk density Water discharge records Flow hydraulic data, e.g., channel geometry, slope, roughness

Suspended sediment particle size influences entrainment, mixing, deposition, downstream sorting, and the capacity of sediment to adsorb and transport contaminants. Methods for determining particle size are based either on analysis of fall-speed or on direct measurement of physical dimensions. The method used depends on the problem. A decision is required whether to determine the effective particle-size distribution or the ultimate distribution, after particle flocs have been dispersed. In situ sensors avoid the problem of having the effective distribution change between stream and laboratory.

It is relatively easy to start a programme of suspended sediment monitoring; knowing when to stop it

Table 15.3 Tools, typical applications, and constraints on information about suspended sediment (SS)

Information requirement	Application	Tools	Constraints
'Instantaneous' SS concentration or load	Determine cross-section mean SS load or discharge-weighted mean concentration under given, steady flow conditions	Point sampling to produce concentration and velocity profiles or depth-integrated sampling with matching water discharge gauging, all at multiple verticals	Time consuming
Continuous SS concentration or load	Continuous records of SS load, concentration, or turbidity for determining statistics such as ranges, exceedance probabilities, mean, annual variability, etc.	Single-point index sampling with manual sampler, auto-sampler, optical or acoustic sensor	Requires relations calibrating point values to cross-section mean SS concentration
Long-term average SS concentration or load	Long-term average SS yield—e.g., for reservoir sediment inflows	*C* vs. *Q* sediment rating combined with either flow duration table or flow time-series; multivariate rating and appropriate time-series data; direct load estimation using stratified or variable probability sampling strategies with data-loggers and auto-samplers	Accuracy limited by sampling strategy, number of samples, rating model fitting; Requires auto-sampler, data-logger, calibration relations
Event-based SS concentration or load	Event sediment yields or peak concentrations, e.g., for predicting inflows to small reservoirs and water clarity in estuaries	Storm sediment yield ratings, reservoir stratigraphy; event-yield magnitude–frequency analysis	Need point to cross-section mean calibration relations when relying on continuous sampling; need bulk density measurements of reservoir sediments
Synoptic sampling	Mapping relative sediment sources across basins	Multi-spectral analysis of satellite/aerial imagery; manual sampling; single-stage samplers	Requires calibration of SS concentration to image signature; near-surface data only; synoptic map may not represent event average due to phase differences in sediment supply and transport from tributaries
Particle size by settling analysis	Entrainment, mixing, deposition issues	Pipette; bottom-withdrawal tube; hydrometer; VA tube; RSA; sedigraph	Manual methods time consuming; requires minimum mass of sediment; sand and finer fractions analysed separately
Particle size by physical size analysis	Machinery damage, sediment filtering, contaminant adsorption and transport	Wet sieves; laser-diffraction devices; laser back-scatter devices; microscopic image analysis	Non-standard measurements among devices; basic distributions often by grain count not by sediment mass; cannot be directly compared to settling analysis results

(*Continues*)

Table 15.3 (*Continued*)

Information requirement	Application	Tools	Constraints
Effective/ultimate particle size	Adsorbed contaminant transport and management; sediment settling and water clarity issues	Chemical and dispersing agents and ultrasonic devices; in situ laser sensors	Undispersed sample properties may alter between sampling and laboratory

Table 15.4 Tools, typical applications, and constraints on information about bedload

Information requirement	Application	Tools	Constraints
'At-a-point' bedload transport rate	Characterisation of temporal variability in bedload transport rates and estimation of mean rate under given, steady flow conditions; determination of conditions for incipient motion of a given size fraction	Basket and pressure difference samplers; compartmentalised, continuously recording pit traps	Sampling accuracy limited by number of samples obtained; pit traps expensive to construct
'Stream-wide' bedload transport rate	Estimation of mean bedload transport rate under given, steady flow conditions; Determination of bedload discharge; Characterisation of patterns of scour and fill; Construction of i_b v Q rating	Samplers and traps; bedform surveys; formulae	Time consuming and labor intensive; formulae require calibration/verification with field data
Bedload yield	Estimation of bedload yield on an event, seasonal, or multi-year basis	Traps; surveys of sedimentation basins or reservoirs; morphological methods; i_b v Q rating; tracers and scour chains	Morphological methods require information on bedload across reach boundary; rating curves require calibration with field data. Limited recovery of tracers

is less straightforward. Studies of long records of continuous sampling have shown that annual sediment yields, at least in small catchments, typically have an approximately log-normal distribution and high variability (Renard and Lane 1975, Van Sickle 1981). Day (1988) analysed long records (up to 30 years) from Canadian rivers and found that mean characteristics of the suspended sediment yield stabilised (with a stable standard error on the mean) after approximately 10 years. Thus, assuming no longer-term trend or non-stationary signal exists, such as induced by landuse change or catastrophic climatic and tectonic events, a decadal time-span for monitoring is suggested.

Bedload may be determined by field sampling and measurement or by formula (Table 15.4), although the approach should be selected on a site-specific basis. The hydraulic and sampling efficiencies of bedload samplers vary with their design, and the sampling efficiency is difficult to establish definitively. However, in practice such deficiencies are less important than having the correct sampling strategy to overcome the considerable spatial and temporal variation in bedload transport observed in rivers, particularly those with gravel beds. Thus, some prior knowledge of the transport conditions is required to set a sampling strategy that keeps random and systematic errors to acceptable levels and minimises the total error involved in the sampling process. It remains that, strictly speaking, at-a-point and cross-channel sampling programmes provide only estimates of the mean bedload transport rate and mean bedload discharge. Moreover, it may not always be possible to collect quantitative bedload measurements because of scale considerations. For example, there are practicable limits to the size of a river in which samplers can be deployed. Bedload traps are more exact devices, but are limited in size and are expensive, thus they are generally limited to research applications in narrow channels.

Morphological methods provide a reasonably robust estimate of the time- and space-averaged bedload, even on large rivers, provided the field conditions are appropriate for the method and some independent means of confirming the transport estimate is available. While there are many bedload formulae, none has been universally accepted. They all apply more or less to a limited range of conditions, and none reproduce the short-term fluctuations in bedload transport rates seen in nature.

Thus, excepting traps, none of the existing methodologies for estimating bedload is inherently reliable. Indeed, Hubbell's (1964, p. 2) observation that "no single apparatus or procedure, whether theoretical or empirical, has been universally accepted as completely satisfactory for the determination of bedload discharge" remains current. Thus, caution dictates that a combination of techniques be used to estimate bedload discharge and their results compared. Carson and Griffiths (1987) and McLean and Church (1999) provide an indication as to how this might be done. End-users should also be aware that to effectively address many environmental and management issues, a more comprehensive (and inevitably longer-term) perspective on sediment transfers within a basin is typically required than is provided by a site-specific characterisation of a river's bedload transport regime.

Determining the total sediment load requires matching methods for determining suspended load and bedload. These should have consistent time bases, and some correction may be required to avoid double-accounting size fractions that appear both in the bedload and suspended load. Simple 'total load' formulae are appropriate only for sand-bed channels lacking washload.

Reservoirs offer unique and robust opportunities for measuring the sediment load of their inflowing rivers, both on a long-term average basis, from periodic surveys of bed levels, and on an event basis from sedimentological/stratigraphical analysis of their bed sediment. Older reservoirs can provide detailed records of catchment sediment yield that may capture the impacts of changes in landuse or climate.

15.8 CASE EXAMPLE: SEDIMENT BUDGET FOR UPPER CLUTHA RIVER, NEW ZEALAND

The Clutha River drains 20 500 km^2 of mainly schist terrain in South Island, New Zealand, and has a mean flow near the coast of 565 m^3/s. The upper river is used for hydro-electricity generation, with dams built at Roxburgh in 1957 and upstream at Clyde in 1992 (Figure 15.9A). Most of the runoff into the upper Clutha is sourced from the wetter, northwest corner of the basin and passes through three large natural lakes. Sediment is derived from tributaries downstream of these lakes, particularly from the Shotover River, which has the largest (1088 km^2) and steepest catchment and receives the highest annual rainfall (2000 mm). Sediment loads in the upper Clutha system have been monitored since the 1960s, the main purposes being to quantify inputs to existing and planned hydro-reservoirs and, at least in the early years, to clarify sediment source areas to establish the practicality of reducing the sediment supply using soil conservation measures. The following summarises this monitoring programme and the results obtained for the period up to 1992, when the Clyde Dam was commissioned.

Suspended sediment has been gauged (i.e., using depth-integrating samplers at multiple verticals) at flow recording sites on all of the major tributaries (Figure 15.9A), with the aim of determining mean annual yields via the sediment-rating approach. The most recent yield estimates, using LOWESS to fit ratings to the log-transformed datasets, show a total suspended load from tributaries upstream of Lake Roxburgh of 1.97 10^6 t/year, with the Shotover River supplying 67% of this. For some gaugings at each site, duplicate samples were bulked and analysed for particle size (usually with a bottom-withdrawal tube), with the results averaged to estimate a representative size-grading. The relations between suspended sediment concentration and water discharge are comparatively poor on the Kawarau and Clutha Rivers, since both receive much of their flows as clear water from the natural lakes, thus over the period 1977–1980, daily index samples were collected from sites that covered the inflows and outflows to Lake Roxburgh and the future Lake Dunstan. Depth-integrated multi-vertical gaugings were used to develop relations between index sample concentration and cross-section mean concentration and also to measure the particle size of the inflowing and outflowing suspended loads. The results confirmed that the main source of sediment was from the Shotover River, via the Kawarau River, and showed that the average trap efficiency of suspended sediment entering Lake Roxburgh was 80% (Jowett and Hicks 1981).

Bedload was sampled in the Shotover River over a range of flows using a 150-mm oriface-width Helley–Smith sampler operated from a motorised cableway. Each bedload measurement involved repeat traverses of 20 verticals, and all samples of the sandy-gravel bedload were analysed for particle size (Figure 15.1B). The gauged bedload discharges were used to verify (albeit with considerable data scatter) a bedload rating (Figure 15.9B) derived using the approach of Wilcock (1997a, 1998), which involves slightly different transport functions for the sand and gravel fractions and relates the threshold of motion of sand and gravel to their relative proportions. The bedload yield of the Shotover so estimated was 0.26 10^6 t/year, which is equivalent to 20% of the suspended load. Because of the cost of bedload sampling and because the Shotover was the dominant source of suspended load, the bedloads of the other tributaries were not sampled but were assumed equal to 20% of their suspended loads, based on the Shotover result.

Deposition in Lake Roxburgh has been monitored by cross-section survey at approximately 5-yearly intervals since 1961, when the reservoir storage volume was 101 10^6 m^3 (Webby *et al.* 1996). The earliest surveys used tagline and sounding line, while the latest surveys use differential GPS and echo-sounder. Cores collected along the length of the lake were used by Thompson (1976) to determine the particle size of the trapped sediment and to estimate an overall bulk sediment density of 1.27 t/m^3 via empirical relations given in Vanoni (1975). Using this bulk density, the average mass entrapment rate over the period 1961–1989 was 1.80 10^6 t/year. When combined with the trap efficiency information provided by the index sampling programme (Jowett and Hicks 1981), this indicates a robust measure of the mean annual sediment inflows to Lake Roxburgh of 2.15 10^6 t/year, which compares very favourably with the total sediment inflow of 2.36 10^6 t/year estimated independently from the tributary data. Moreover, after adjusting bedload inputs from the tributaries for abrasion (after Adams 1979)—which transforms part of the bedload to fine suspended load—a close match is achieved between the sediment inflows by size fraction (based on the tributary data) and the rate of entrapment by size fraction, at least for the sand and gravel fractions that are efficiently trapped in the reservoir (Figure 15.9C). Such agreement by independent methods lends confidence to the overall sediment budget determination, particularly to the bedload results obtained for the Shotover River where the agreement between sampled bedload discharges and formula predictions was not ideal. It also highlights the use of reservoirs as large-scale, long-term sampling devices.

ACKNOWLEDGEMENTS

We thank Mike Church, Jonathan Laronne and thevolume editors for their constructive review comments. The authors' collaborative efforts were supported by the National Science Foundation (SBR-9807195) and the New Zealand Foundation for Research, Science and Technology (C09612).

REFERENCES

Abbott, J.E. and Francis, J.D.R. 1977. Saltation and suspension trajectories of solid grains in a water stream. *Philosophical Transactions of the Royal Society of London* 284A: 225–254.

Adams, J. 1979. Wear of unsound pebbles in river headwaters. *Science* 203: 171–172.

Agrawal, Y.C., McCave, I.N. and Riley, J.B. 1991. Laser diffraction size analysis. In: Syvitski, J.P.M., ed., *Principles, Methods, and Application of Particle Size Analysis*, Cambridge: Cambridge University Press, pp. 119–128.

Allen, T. 1990. *Particle Size Measurement*, Fourth Edition, London: Chapman & Hall.

American Society of Civil Engineers Task Committee. 1992. Sediment and aquatic habitat in river systems. *Journal of Hydraulic Engineering* 118: 669–687.

Andrews, E.D. 1981. Measurement and computation of bed-material discharge in a shallow sand-bed stream, Muddy Creek, Wyoming. *Water Resources Research* 17: 131–141.

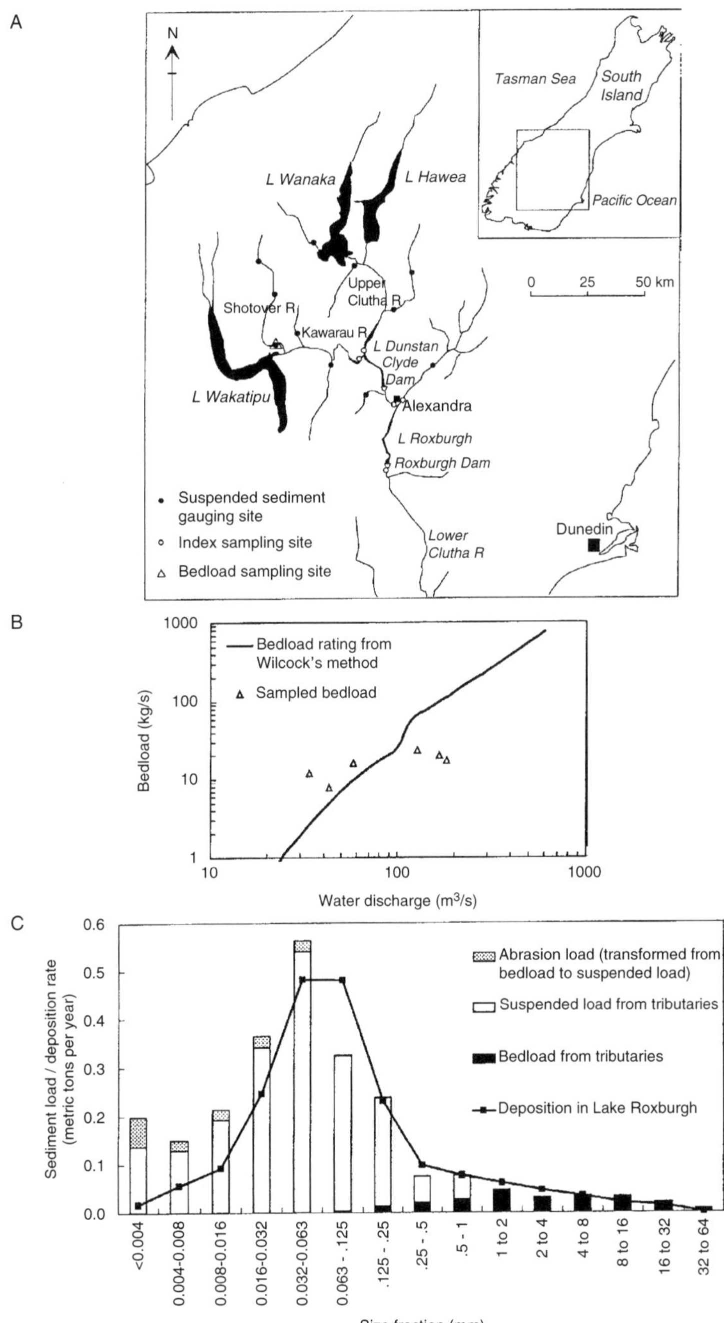

Figure 15.9 (A) Clutha River Basin, South Island, New Zealand, showing hydro-dams, suspended sediment and bedload gauging sites on tributaries and index sampling sites on mainstem channels. (B) Bedload rating for Shotover River, based on Wilcock's (1997a) method, compared with bedload discharges measured with the Helley–Smith sampler. (C) Annual average sediment inflows to Lake Roxburgh by size fraction, based on sediment sampling in tributaries and adjusted for abrasion (i.e., transformation of coarse bedload into suspended load), compared with average deposition rate in Lake Roxburgh by size fraction, based on reservoir surveys. See text for explanation

Andrews, E.D. 1994. Marginal bed load transport in a gravel bed stream, Sagehen Creek, California. *Water Resources Research* 30: 2241–2250.

Andrews, E.D. and Smith, J.D. 1992. A theoretical model for calculating marginal bedload transport rates of gravel. In: Billi, P., Hey, R.D., Thorne, C.R. and Tacconi, P., eds., *Dynamics of Gravel-bed Rivers*, Chichester: John Wiley and Sons Ltd., pp. 41–48.

Andrews, E.D. and Nankervis, J.M. 1995. Effective discharge and the design of channel maintenance flows for gravel-bed rivers. *American Geophysical Union, Geophysical Monograph* 89: 151–164.

Arkell, B., Leeks, G.J.L., Newson, M.D. and Oldfield, F. 1983. Trapping and tracing: some recent observations of supply and transport of coarse sediment from upland Wales. *International Association of Sedimentologists Special Publication* 6: 107–119.

ASCE Task Committee. 1971. Sediment transportation mechanics: sediment discharge formulas. *Journal of the Hydraulics Division, American Society of Civil Engineers* 97(HY4): 523–567.

Ashmore, P.E. and Day, T.J. 1988. Spatial and temporal patterns of suspended-sediment yield in the Saskatchewan River Basin. *Canadian Journal of Earth Sciences* 25: 1450–1463.

Aziz, N.M. 1996. Error estimate in Einstein's suspended sediment load method. *Journal of Hydraulic Engineering* 122: 282–285.

Bagnold, R.A. 1966. *An Approach to the Sediment Transport Problem from General Physics*, US Geological Survey Professional Paper 422-I, 37 p.

Bale, A.J. and Morris, A.W. 1991. In situ size measurements of suspended particles in estuarine and coastal waters using laser diffraction. In: Syvitski, J.P.M., ed., *Principles, Methods, and Application of Particle Size Analysis*, Cambridge: Cambridge University Press, pp. 197–208.

Bechteler, W. and Maurer, M. 1991. Reliability theory applied to sediment transport formulae. In: Soulsby, R. and Bettes, R., eds., *Sand Transport in Rivers*, Rotterdam: A.A. Balkema, pp. 223–228.

Bennett, S.J. and Best, J.L. 1995. Particle size and velocity discrimination in a sediment-laden turbulent flow using phase Doppler anemometry. *Journal of Fluids Engineering* 117: 505–511.

Beschta, R.L. 1981. Increased bag size improves Helley–Smith bedload sampler for use in streams with high sand and organic matter transport. *International Association of Hydrological Sciences Publication* 133: 17–25.

Brune, R.C. 1953. Trap efficiency of reservoirs. *Transactions of the American Geophysical Union* 34(3): 407–418.

Carey, W.P. and Hubbell, D.W. 1986. Probability distributions for bedload transport. In: *Proceedings of the 4th Federal Inter-Agency Sedimentation Conference* 2, pp. 131–140.

Carson, M.A., and Griffiths, G.A. 1987. Bedload transport in gravel channels. *Journal of Hydrology (New Zealand)* 26: 1–151.

Carson, M.A. and Griffiths, G.A. 1989. Gravel transport in the braided Waimakariri River: mechanisms, measurements and predictions. *Journal of Hydrology* 109: 201–220.

Childers, D., Huanjing, G. and Gangyan, Z. 1989. Field comparison of bedload samplers from the United States of America and the Peoples Republic of China. In: *Proceedings of the 4th International Symposium on River Sedimentation* 2, pp. 1309–1316.

Christian, R. and Thompson, S.M. 1978. Loop rating and grading of suspended sediment in the Mararoa. *Journal of Hydrology (New Zealand)* 17: 50–53.

Church, M.A. and Hassan, M.A. 1992. Size and distance of travel of unconstrained clasts on a streambed. *Water Resources Research* 28: 299–303.

Church, M.A., Ham, E., Hassan, M. and Slaymaker, O. 1999. Fluvial clastic sediment yield in Canada: scaled analysis. *Canadian Journal of Earth Sciences* 36: 1267–1280.

Church, M.A., Wolcott, J.F. and Fletcher, W.K. 1991. A test of equal mobility in fluvial sediment transport: behavior of the sand fraction. *Water Resources Research* 27: 2941–2951.

Church, M.A., Miles, M. and Rood, K. 1986. Sediment transfer along Mackenzie River: a feasibility study, Sediment Survey Section Report IWD-WNR(NWT)-WRB-SS-871, Environment Canada, Inland Waters Directorate, Western and Northern Region.

Cioffi, F. and Gallerano, F. 1991. Velocity and concentration profiles of solid particles in a channel with movable and erodible bed. *Journal of Hydraulic Research* 29: 387–401.

Cleveland, W.S. 1979. Robust locally weighted regression and smoothing scatterplots. *Journal of the American Statistical Association* 74: 829–836.

Coakley, J.P. and Syvitski, J.P.M. 1991. SediGraph Technique. In: Syvitski, J.P.M., ed., *Principles, Methods, and Application of Particle Size Analysis*, Cambridge: Cambridge University Press, pp. 129–142.

Cohn, T.A. 1995. Recent advances in statistical methods for the estimation of sediment and nutrient transport in rivers. Rev. Geophys. 33(Suppl.) (American Geophysical Union).

Cohn, T.A., DeLong, L.L., Gilroy, E.J., Hirsch, R.M. and Wells, D.K. 1989. Estimating constituent loads. *Water Resources Research* 25: 937–942.

Colby, B.R. and Hembree, B.H. 1955. *Computations of Total Sediment Discharge, Niobrara River Near Cody, Nebraska*, US Geological Survey Water Supply Paper 1357, 187 p.

Costa, J.E. 1988. Rheologic, geomorphic, and sedimentologic differentiation of water floods, hyperconcentrated flows, and debris flow. In: Baker, V.R., Kochel, R.C. and Patton, P.C., eds., *Flood Geomorphology*, Chichester: John Wiley and Sons Ltd., pp. 113–122.

Crawford, C.G. 1991. Estimation of suspended-sediment rating curves and mean suspended-sediment loads. *Journal of Hydrology* 129: 331–348.

Csoma, J. 1973. Reliability of bed-load sampling. *Proceedings of the International Association for Hydraulics Research* 2: 97–107.

Custer, S.G., Bugosh, N., Ergenzinger, P.E. and Anderson, B.C. 1987. Electromagnetic detection of pebble transport in streams: a method for measurement of sediment-transport waves. *Society of Economic Paleontologists and Mineralogists Special Publication* 39: 21–26.

Davies, A.P. 1900. *Hydrography of Nicaragua*, US Geological Survey 20th Annual Report 1898–99, pp. 563–637.

Day, T.J. 1988. Evaluation of long term suspended sediment records for selected Canadian rivers. In: *Sediment Budgets, Proceedings of the Porto Alegre Symposium, December 1988*, International Association of Hydrological Sciences Publication No. 174, pp. 101–107.

De Jong, C. and Ergenzinger, P.J. 1998. On the nature of spatial and temporal variations of bedload transport in a gravel-bed stream. *Zeitschrift für Geomorphologie Supplementband* 112: 29–54.

De Lange, W.P., Healy, T.R. and Darlan, Y. 1997. Reproducibility of sieve and settling tube textural determinations for sand-sized beach sediment. *Journal of Coastal Research* 13: 73–80.

De Vries, M. 1973. On measuring discharge and sediment transport. *Delft Hydraulics Laboratory Publication* 106.

Dietrich, W.E. and Whiting, P.J. 1989. Boundary shear stress and sediment transport in river meanders of sand and gravel. In: Ikeda, S. and Parker, G., eds., *River Meandering*, Washington, DC: American Geophysical Union, pp. 1–50.

Dobson, G.C. and Johnson, J.W. 1940. Studying sediment loads in natural streams. *Civil Engineering* 10: 93–96.

Druffel, L., Emmett, W.W., Schneider, V.R. and Skinner, J.V. 1976. *Laboratory Hydraulic Calibration of the Helley–Smith Bedload Sediment Sampler*, US Geological Survey Open-File Report 76-752.

Du Boys, P.F.D. 1879. Le Rhin et les rivières à lit affouillable. Annales des Ponts et Chaussées, 5e Serie, Tome 18: 141–195.

Duan, N. 1983. Smearing estimate: a non-parametric retransformation method. *Journal of the American Statistical Association* 78: 605–610.

Edwards, T.K. and Glysson, G.D. 1999. *Field Methods for Measurement of Fluvial Sediment*, US Geological Survey Techniques of Water-resources Investigations, 89 p. (Book 3, Chapter C2).

Einstein, H.A. 1937. Die Eichung des im Rhein verwendeten Geschiebefängers [Calibrating the bedload trap used in the Rhine]. *Schweizerische Bauzeitung Heft* 110.

Einstein, H.A. 1944. Bedload transportation in Mountain Creek, Soil Conservation Service Technical Paper, SCS-TP-55, US Department of Agriculture, 54 pp.

Einstein, H.A. 1948. Determination of rates of bedload movement. In: *Proceedings of the 1st Federal Inter-Agency Sedimentation Conference*, pp. 75–93.

Einstein, H.A. 1950. The bedload function for sediment transportation in open channels. *US Department of Agriculture, Soil Conservation Service, Technical Bulletin* 1026: 78 p.

Einstein, H.A., Anderson, A.G. and Johnson, J.W. 1940. A distinction between bed load and suspended load. *Transactions of the American Geophysical Union* 21: 628–632.

Emmett, W.W. 1980. *A Field Calibration of the Sediment-trapping Characteristics of the Helley–Smith bedload Sampler*, US Geological Survey Professional Paper 1139.

Emmett, W.W., Burrows, R.L. and Chacho, E.F. 1996. Coarse-particle transport in a gravel-bed river. *International Journal of Sediment Research* 11: 8–21.

Engel, P. 1983. Sampling efficiency of the VUV bedload sampler, National Water Research Institute of Canada Report H82-377.

Engel, P. and Lau, L. 1981. The efficiency of basket-type bedload samplers. *International Association of Hydrological Sciences Publication* 133: 27–34.

Engelund, F. and Hansen, E. 1967. *A Monograph on Sediment Transport in Alluvial Streams*, Copenhagen: Teknisk Vorlag, 62 p.

Ergenzinger, P.J. and Conrady, J. 1982. A new tracer technique for measuring bedload in natural channels. *Catena* 9: 77–80.

Ergenzinger, P.J. and Schmidt, K.H. 1995. Single particle bedload transport: first results from new approaches. *Zeitschrift für Geomorphologie Supplementband* 100: 193–203.

Ergenzinger, P.J. and Custer, S.G. 1983. Determination of bedload transport using natural magnetic tracers: first experience at Squaw Creek, Gallatin County, Montana. *Water Resources Research* 19: 187–193.

Ferguson, R.I. 1986. River loads underestimated by rating curves. *Water Resources Research* 22: 74–76.

Ferguson, R.I. 1987. Accuracy and precision of methods for estimating river loads. *Earth Surface Processes and Landforms* 12: 95–104.

Ferguson, R.I. and Paola, C. 1997. Bias and precision of percentiles of bulk grain size distributions. *Earth Surface Processes and Landforms* 22: 1061–1077.

Ferguson, R.I. and Ashworth, P.J. 1992. Spatial patterns of bedload transport and channel change in braided and near-braided rivers. In: Billi, P., Hey, R.D., Thorne, C.R. and Tacconi, P., eds., *Dynamics of Gravel-bed Rivers*, Chichester: John Wiley and Sons Ltd., pp. 477–496.

Foster, I.D.L., Millington, R. and Grew, R.G. 1992. The impact of particle size controls on stream turbidity measurements; some implications for suspended sediment yield estimation. In: Bogen, J., Walling, D.E. and Day, T.J., eds., *Erosion and Sediment Transport Monitoring Programmes in River Basins*, International Association of Hydrological Sciences Publication No. 210, pp. 51–62.

Gabel, S.L. 1993. Geometry and kinematics of dunes during steady and unsteady flows in the Calamus River, Nebraska, USA. *Sedimentology* 40: 237–269.

Galai. 1997. *WCIS-100 Particle Size Analyzer Operation Manual*, Migdal Haemek, Israel: Galai Production Ltd.

Garcia, C., Laronne, J.B. and Sala, M. 1999. Variable source areas of bedload in a gravel-bed stream. *Journal of Sedimentary Research* 69: 27–31.

Gaweesh, M.T.K. and Van Rijn, L.C. 1994. Bed-load sampling in sand-bed rivers. *Journal of Hydraulic Engineering* 120: 1364–1384.

Gentien, P., Lunven, M., Lehaitre, M. and Duvent, J.L. 1995. In situ depth profiling of particle sizes. *Deep-sea Research* 42: 1297–1312.

Gibbs, C.J. and Neill, C.R. 1973. Laboratory testing of model VUV bedload samplers, Research Council of Alberta Report REH/73/2.

Gilbert, G.K. 1914. *The Transportation of Debris by Running Water*, US Geological Survey Professional Paper 86.

Gill, M.A. 1988. Planning the useful life of a reservoir. *Water Power and Dam Construction* (May): 46–47.

Gippel, C.J. 1989. The use of turbidimeters in suspended sediment research. *Hydrobiologia* 176/177: 465–480.

Gippel, C.J. 1995. Potential of turbidity monitoring for measuring the transport of suspended solids in streams. *Hydrological Processes* 9: 83–97.

Gomez, B. 1983. Temporal variations in bedload transport rates: the effect of progressive bed armouring. *Earth Surface Processes and Landforms* 8: 41–54.

Gomez, B. 1991. Bedload transport. *Earth Science Reviews* 31: 89–132.

Gomez, B. 1995. Bedload transport and changing grain size distributions. In: Gurnell, A.M. and Petts, G., eds., *Changing River Channels*, Chichester: John Wiley and Sons Ltd., pp. 177–199.

Gomez, B. and Troutman, B.M. 1997. Evaluation of process errors in bed load sampling using a dune model. *Water Resources Research* 33: 2387–2398.

Gomez, B. and Church, M. 1989. An assessment of bedload sediment transport formulae for gravel-bed rivers. *Water Resources Research* 25: 1161–1186.

Gomez, B., Hubbell, D.W. and Stevens, H.H. 1990. At-a-point bedload sampling in the presence of dunes. *Water Resources Research* 26: 2717–2731.

Gomez, B., Mertes, L.A.K., Phillips, J.D., Magilligan, F.J. and James, L.A. 1995. Sediment characteristics of an extreme flood: 1993 upper Mississippi River valley. *Geology* 23: 963–966.

Gomez, B., Naff, R.L. and Hubbell, D.W. 1989. Temporal variations in bedload transport rates associated with the migration of bedforms. *Earth Surface Processes and Landforms* 14: 135–156.

Gomez, B., Emmett, W.W. and Hubbell, D.W. 1991. Comments on bedload sampling in small rivers. In: *Proceedings of the 5th Federal Inter-Agency Sedimentation Conference* 1, pp. 2.65–2.72.

Gray, J.R. and Fisk, G.G. 1992. Monitoring radionuclide and suspended-sediment transport in the Little Colorado River Basin, Arizona and New Mexico, USA. *International Association of Hydrological Sciences Publication* No. 210, pp. 505–516.

Green, M.O., Dolphin, T.J., Swales, A. and Vincent, C.E. 1999. Transport of mixed-size sediments in a tidal channel. In: Kraus, N.C. and McDougal, W.G., eds., *Coastal Sediments '99*, Volume 1, Reston, Virginia: A.S.C.E., pp. 644–658.

Guy, H.P. 1969. *Laboratory Theory and Methods for Sediment Analysis*, US Geological Survey Techniques of Water Resources Investigations, 58 p. (Book 5, Chapter C1).

Haan, C.T. 1977. *Statistical Methods in Hydrology*, Ames, Iowa: The Iowa State University Press, 378 p.

Habersack, H.M., Nachtnebel, H.-P. and Laronne, J.B. 1998. Hydraulic efficiency of a slot sampler: flow velocity measurements in the Drau River, Austria. In: Klingeman, P.C., Beschta, R.L., Komar, P.D. and Bradley, J.B., eds., *Gravel-bed Rivers in the Environment*, LLC, Highland Ranch, Colorado: Water Resources Publications, pp. 749–754.

Hansen, E.A. 1973. In-channel sedimentation basins, a possible tool in trout habitat management. *Progressive Fish Culturist* 35: 136–142.

Haschenburger, J.K. and Church, M.A. 1998. Bed material transport estimated from the virtual velocity of sediment. *Earth Surface Processes and Landforms* 23: 791–808.

Hassan, M.A. and Church, M.A. 1992. The movement of individual grains on the streambed. In: Billi, P., Hey, R.D., Thorne, C.R. and Tacconi, P., eds., *Dynamics of Gravel-bed Rivers*, Chichester: John Wiley and Sons Ltd., pp. 159–175.

Hassan, M.A. and Church, M.A. 1994. Vertical mixing of coarse particles in gravel bed rivers: a kinematic model. *Water Resources Research* 30: 1173–1186.

Hassan, M. and Church, M.A. 2000. Experiments on surface structure and partial sediment transport on a gravel bed. *Water Resources Research* 36: 1885–1895.

Hassan, M.A., Schick, A.P. and Laronne, J.B. 1984. The recovery of flood-dispersed coarse sediment particles, a three-dimensional magnetic tracing method. *Catena* 5 (Suppl.): 153–162.

Hayward, J.A. 1980. *Hydrology and Stream Sediment from Torlesse Stream Catchment*, Special Publication 17, Tussock Grasslands and Mountain Lands Institute, Lincoln College, 236 p.

Hayward, J.A. and Sutherland, A.J. 1974. The Torlesse Stream vortex-tube sediment trap. *Journal of Hydrology (New Zealand)* 13: 41–53.

Helley, E.J. and Smith, W. 1971. Development and calibration of a pressure-difference bedload sampler, US Geological Survey Open-File Report.

Hicks, D.M. 1994. Landuse effects on magnitude-frequency characteristics of storm sediment yields: some New Zealand examples. *International Association of Hydrological Sciences Publication* No. 224: 395–402.

Hicks, D.M. and Fenwick, J.K. 1994. *Suspended Sediment Manual – Field, Laboratory and Office Procedures for Collecting and Processing Suspended Sediment Data*, NIWA Science and Technology Series No. 6, Christchurch, New Zealand: National Institute of Water and Atmospheric Research, 84 p.

Hicks, D.M. and Duncan, M.J. 1997. The efficiency of depth-integrating samplers in sampling the suspended sand load in gravel bed rivers. *Journal of Hydrology* 201: 138–160.

Hicks, D.M., Gomez, B. and Trustrum, N.A. 2000. Erosion thresholds and suspended sediment yields, Waipaoa River Basin, New Zealand. *Water Resources Research* 36: 1129–1142.

Horowitz, A.J. 1985. *A Primer on Trace Metal Sediment Chemistry*, US Geological Survey Water Supply Paper 2277.

Hubbell, D.W. 1964. *Apparatus and Techniques for Measuring Bedload*, US Geological Survey Water-Supply Paper 1748.

Hubbell, D.W. 1987. Bedload sampling and analysis. In: Thorne, C.R., Bathurst, J.C. and Hey, R.D., eds., *Sediment Transport in Gravel-bed Rivers*, Chichester: John Wiley and Sons Ltd., pp. 89–120.

Hubbell, D.W. and Matejka, D.Q. 1959. *Investigations of Sediment Transportation, Middle Loup River at Dunning, Nebraska, With Application of Data from Turbulence Flume*, US Geological Survey Water-Supply Paper 1476.

Hubbell, D.W., Stevens, H.H., Skinner, J.V. and Beverage, J.P. 1985. New approach to calibrating bedload samplers. *Journal of Hydraulic Engineering* 111: 677–694.

Hubbell, D.W., Stevens, H.H., Skinner, J.V. and Beverage, J.P. 1981. Recent refinements in calibrating bedload samplers. In: *Water Forum '81*, New York: American Society of Civil Engineers, pp. 128–140.

Humphreys, A.A. and Abbot, H.L. 1861. *Report Upon the Physics and Hydraulics of the Mississippi River; Upon the Protection of the Alluvial Region Against Overflow; and Upon the Deepening of the Mouths Based Upon Surveys and Investigations*, Professional Paper of the Corps of Topographical Engineers, US Army, 4, 456 p.

Jackson, W.L. and Beschta, R.L. 1982. A model of two-phase bedload transport in an Oregon Coast Range stream. *Earth Surface Processes and Landforms* 7: 517–527.

Johnson, C.W., Engleman, R.L., Smith, J.P. and Hanson, C. L. 1977. Helley–Smith bedload samplers. *American Society of Civil Engineers, Journal of the Hydraulics Division* 103: 1217–1221.

Jowett, I.G. and Hicks, D.M. 1981. Surface, suspended and bedload sediment – Clutha River system. *Journal of Hydrology (New Zealand)* 20: 121–130.

Kelsey, H.M. 1980. A sediment budget and an analysis of geomorphic processes in the Van Duzen River Basin, northern coastal California 1941–1975. *Geological Society of America Bulletin Part II* 91: 1119–1216.

Kondolf, G.M. and Wilcock, P.R. 1996. The flushing flow problem: defining and evaluating objectives. *Water Resources Research* 32: 2589–2599.

Krumbein, W.C. and Pettijohn, F.J. 1938. *Manual of Sedimentary Petrography*, New York: D. Appleton-Century Co.

Kuhnle, R. 1998. Statistics of sediment transport in Goodwin Creek. *Journal of Hydraulic Research* 124: 1109–1114.

Lane, S.N. 1998. The use of digital terrain modelling in the understanding of dynamic river channel systems. In: Lane, S.N., Richards, K.S. and Chandler, J.H., eds., *Landform Monitoring, Modelling and Analysis*, Chichester: John Wiley and Sons Ltd., pp. 311–342.

Lane, S.N., Richards, K.S. and Chandler, J.H. 1995. Morphological estimation of the time-integrated bed load transport rate. *Water Resources Research* 31: 761–772.

Lane, E.W. and Koelzer, V.A. 1953. Density of sediments deposited in reservoirs, Report No. 9, St Paul US Engineering District.

Laronne, J.B. and Carson, M.A. 1976. Interrelations between bed morphology and bed material transport for a small gravel-bed channel. *Sedimentology* 23: 67–85.

Laronne, J.B. 1990. Probability distribution of event sediment yields in the Northern Negev, Israel. In: Boardman, J., Foster, I. and Dearing, J., eds., *Soil Erosion in Agricultural Land*, London: Wiley, pp. 481–492.

Laronne, J.B. and Wilhelm, R. 2001. Shifting stage–volume curves: predicting event sedimentation rate based on reservoir stratigraphy. In: Anthony, D., Ethridge, F., Harvey, M., Laronne, J.B. and Mosley, M.P., eds., *Applying Geomorphology to Environmental Management*, Highlands Ranch, Colorado: Water Resources Publications, pp. 33–54.

Lawler, D.M. and Brown, R.M. 1992. A simple and inexpensive turbidity meter for the estimation of suspended sediment concentrations. *Hydrological Processes* 6: 159–168.

Lenzi, M.A., D'Agostino, V. and Billi, P. 1999. Bedload transport in the instrumented catchment of the Rio Cordon. Part I. Analysis of bedload records, conditions and threshold of bedload entrainment. *Catena* 36: 171–190.

Leopold, L.B. and Emmett, W.W. 1997. *Bedload and River Hydraulics Inferences From the East Fork River, Wyoming*, US Geological Survey Professional Paper 1583, 52 p.

Lewis, J. 1996. Turbidity-controlled suspended sediment sampling for runoff-event load estimation. *Water Resources Research* 32: 2299–2310.

Lisle, T.E. 1995. Particle size variations between bed load and bed material in natural gravel bed channels. *Water Resources Research* 31: 1107–1118.

Lisle, T.E. and Madej, M.A. 1992. Spatial variation in armouring in a channel with high sediment supply. In: Billi, P., Hey, R.D., Thorne, C.R. and Tacconi, P., eds., *Dynamics of Gravel Bed Rivers*, Chichester: John Wiley and Sons Ltd., pp. 277–296.

Lisle, T.E., Nelson, J.E., Pitlick, J., Madej, M.A. and Barkett, B.L. 2000. Variability of bed mobility in natural gravel-bed channels and adjustments to sediment load at local and reach scales. *Water Resources Research* 36: 3743–3755.

Lowe, R.L. 1988. Measuring sediment dynamics. In: Seymor, R.J., ed., *Nearshore Sediment Transport*, New York: Plenum Press, pp. 91–93.

Mahmood, K. 1980. Verification of sediment transport functions – Missouri River, US Army Corps of Engineers, Missouri River Division Sediment Series Report 19.

Malhotra, V.M. 1967. The Alpine airjet sieve: a new method of determining the fineness of cement. *Indian Concrete Journal* 41: 305–309.

Maneux, E., Probst, J.L., Veyssey, E. and Etcheber, H. 2001. Assessment of dam trapping efficiency from water residence time: application to fluvial sediment transport in the Adour, Dordogne, and Garonne River Basins (France). *Water Resources Research* 37: 801–811.

Martin, Y. and Church, M.A. 1995. Bed-material transport estimated from channel surveys: Vedder River, British Columbia. *Earth Surface Processes and Landforms* 20: 347–361.

McCave, I.N. and Syvitski, J.P.M. 1991. Principles and methods of geological particle size analysis. In: Syvitski, J.P.M., ed., *Principles, Methods and Application of Particle Size Analysis*, Cambridge: Cambridge University Press, pp. 1–21.

McHenry, J.R. 1971. Discussion of sediment measurement techniques. B. Reservoir deposits. *Journal of the Hydraulics Division, American Society of Civil Engineers* 97: 1253–1257.

McLean, D.G., Church, M.A. and Tassone, B. 1999. Sediment transport along lower Fraser River. 1. Measurement and hydraulic computations. *Water Resources Research* 35: 2533–2548.

McLean, D.G. and Church, M.A. 1999. Sediment transport along lower Fraser River. 2. Estimates based on the long term gravel budget. *Water Resources Research* 35: 2549–2559.

Meade, R.H., Yuzyk, T.R. and Day, T.J. 1990. Movement and storage of sediment in rivers of the United States and Canada. In: Wolman, M.G. and Riggs, H.C., eds., *Surface Water Hydrology*, Boulder: Geological Society of America, pp. 255–280.

Mertes, L.A.K., Smith, M.O. and Adams, J.B. 1993. Estimating suspended sediment concentrations in surface waters of the Amazon River wetlands from Landsat images. *Remote Sensing of the Environment* 43: 281–301.

Milhous, R.T. 1973. Sediment transport in a gravel-bottomed stream, Unpublished Ph.D. Thesis, Oregon State University, Corvallis, Oregon, 232 p.

Miller, C.R. 1951. Analysis of flow-duration sediment rating curve method of computing sediment yield, US Bureau of Reclamation Technical Report, Denver, Colorado, 55 p.

Miller, C.R. 1953. *Determination of Unit Weight of Sediment for Use in Sediment Volume Computations*, Memorandum, Denver, Colorado: Bureau of Reclamation, United States Department of the Interior, February 1953.

Mohrig, D. and Smith, J.D. 1996. Predicting the migration rate of subaqueous dunes. *Water Resources Research* 32: 3207–3217.

Moll, J.R., Schilperoort, T. and Leeuw, A.J. 1987. Stochastic analysis of bedform dimensions. *Journal of Hydraulic Research* 25: 465–479.

Moog, D.B. and Whiting, P.J. 1998. Annual hysteresis in bed load rating curves. *Water Resources Research* 34: 2393–2399.

Morris, L. and Fan, J. 1998. *Reservoir Sedimentation Handbook: Design and Management of Dams, Reservoirs, and Watersheds for Sustainable Use*, New York: McGraw-Hill, 704 p.

Mülhofer, L. 1933. Untersuchungen über der Schwebstoff und Geschiebeführung des Inn nächst Kirchbichl, Tirol [Investigations into suspended load and bedload of the River Inn, near Kirchbichl, Tirol], *Die Wasserwirtschaft Heft*: 1–6.

Müller, R. 1937. Ueberprufung des Geschiebegesetzes und der Berechnungsmethode der Versuchsanstalt für Wasserbau an der E.T.H. mit Hilfe der direkten Geschiebemessungen am Rhein [Verification of the bedload law and calculation method of the Hydraulic Experiment Station E.T.H. by direct bedload measurements on the Rhine], *Schweizerische Bauzeitung Heft*: 110.

Nanson, G.C. 1974. Bedload and suspended-load transport in a small, steep mountain stream. *American Journal of Science* 274: 471–486.

Neff, E.L. 1967. Discharge frequency compared to long-term sediment yields. *International Association of Hydrological Sciences Publication* No. 75: 236–242.

Neill, C.R. 1971. River bed transport related to meander migration rates. *Journal of the Waterways, Harbors and Coastal Engineering Division, American Society of Civil Engineers* 97: 783–786.

Neill, C.R. 1987. Sediment balance considerations linking long-term transport and channel processes. In: Thorne, C.R., Bathurst, J.C. and Hey, R.D., eds., *Sediment Transport in Gravel-bed Rivers*, Chichester: John Wiley and Sons Ltd., pp. 225–240.

Nelson, M.E. and Benedict, P.C. 1950. Measurement and analysis of suspended-sediment loads in streams. *Transactions of the American Society of Civil Engineers* 116: 891–918.

Nesper, F. 1937. Ergebnisse der Messungen über die Geschiebe – und Schlammführung des Rheins an der Brugger Rheinbrücke [Results of bedload and silt movement observations on the Rhine at the Brugg Bridge]. *Schweizerische Bauzeitung Heft* 110.

Nevin, C. 1946. Competency of moving water to transport debris. *Geological Society of America Bulletin* 57: 651–674.

Newson, M. 1980. The geomorphological effectiveness of floods: a contribution stimulated by two recent events in mid-Wales. *Earth Surface Processes and Landforms* 5: 1–16.

Nikora, V.I. and Hicks, D.M. 1997. Scaling relations for sandwave development in unidirectional flow. *Journal of Hydraulic Engineering* 123: 1152–1156.

Olive, L.J. and Rieger, W.A. 1988. An examination of the role of sampling strategies in the study of suspended sediment transport. *International Association of Hydrological Sciences Publication* No. 174: 259–267.

Ongley, E.D., Yuzyk, T.R. and Krishnappan, B.G. 1990. Vertical and lateral distribution of fine-grained particulates in prairie and cordilleran rivers: sampling implications for water quality programs. *Water Research* 24: 303–312.

Ongley, E.D., Bynoe, M.C. and Percival, J.B. 1982. Physical and geochemical characteristics of suspended solids, Wilton Creek, Ontario. *Hydrobiologia* 91: 41–57.

Parker, G. 1990. Surface-based bedload transport relation for gravel rivers. *Journal of Hydraulic Research* 28: 417–436.

Parker, G., Klingerman, P.C. and McLean, D.G. 1982. Bedload and size distribution in paved gravel-bed streams. *Journal of the Hydraulics Division, American Society of Civil Engineers* 108: 544–571.

Parker, R.S. and Troutman, B.M. 1989. Frequency distribution for suspended sediment loads. *Water Resources Research* 25: 1567–1574.

Parshall, R.L. 1952. Model and prototype studies of sand traps. *Transactions of the American Society of Civil Engineers* 117: 204–212.

Phillips, J.M. and Walling, D.E. 1995a. Measurement in situ of the effective particle-size characteristics of fluvial suspended sediment by means of a field-portable laser backscatter probe: some preliminary results. *Marine and Freshwater Research* 46: 349–357.

Phillips, J.M. and Walling, D.E. 1995b. An assessment of the effects of sample collection, storage and resuspension on the representativeness of measurements of the effective particle size distribution of fluvial suspended sediment. *Water Research* 29: 2498–2508.

Pitlick, J. 1988. Variability of bed load measurements. *Water Resources Research* 24: 173–177.

Popov, I.V. 1962a. A sediment balance of river reaches and its use for the characteristics of channel processes. *Soviet Hydrology* 3: 249–266.

Popov, I.V. 1962b. Application of morphological analysis to the evaluation of the general channel deformation of the River Ob. *Soviet Hydrology* 3: 267–324.

Poreh, M., Sagiv, A. and Seginer, I. 1970. Sampling efficiency of slots. American Society of Civil Engineers, *Journal of the Hydraulics Division* 96: 2065–2078.

Porterfield, G. 1972. *Computation of Fluvial Sediment Discharge*, US Geological Survey Techniques of Water Resources Investigations, 66 p. (Book 3, Chapter C3).

Posada G,L. and Nordin, C.F. 1993. *Total Sediment Loads of Tropical Rivers*, Hydraulic Engineering 93, New York: American Society of Civil Engineers, 1, pp. 258–262.

Reid, I., Layman, J.T. and Frostick, L.E. 1980. The continuous measurement of bedload discharge. *Journal of Hydraulics Research* 18: 243–249.

Reid, I., Laronne, J.B. and Powell, M.D. 1998. Flash-flood bedload dynamics of desert gravel-bed streams. *Hydrological Processes* 12: 543–557.

Renard, K.G. and Lane, L.J. 1975. Sediment yield as related to a stochastic model of ephemeral runoff, Report ARS-S-40, Agriculture Research Service, Washington, DC.

Ritchie, J.C. and Cooper, C.M. 1988. Comparison of measured suspended sediment concentrations with suspended sediment concentrations estimated from Landsat MSS data. *International Journal of Remote Sensing* 9: 379–387.

Robinson, A.R. 1962. Vortex tube sand trap. *Transactions of the American Society of Civil Engineers* 127: 391–425.

Ryan, S.E. and Porth, L.S. 1999. A field comparison of three pressure-difference bedload samplers. *Geomorphology* 30: 307–322.

Sayre, W.W. and Hubbell, D.W. 1965. *Transport and Dispersion of Labelled Bed Material, North Loup River, Nebraska*, US Geological Survey Professional Paper 433C.

Schaank, E.M.H. 1937. Discussion of: J. Smetana, Appareil pour le jaugeage du débit solide entrainé sur le fond du cours d'eau [An instrument for the measurement of bedload in rivers]. In: *Proceedings of the 1st Meeting*, International Association for Hydraulic Structures Research, Appendix 4, pp. 93–120.

Schall, J.D. and Fisher, G.A. 1996. Hydrographic surveying using global positioning techniques. In: Albertson, M.L., Molinas, A. and Hotchkiss, R., eds., *Proceedings of the International Conference on Reservoir Sedimentation, Colorado State University, Fort Collins, Colorado*, pp. 247–253.

Seal, R., Paola, C. and Parker, G. 1993. The effect of local patchiness of gravel grain size distributions on bed load transport in braided rivers. In: Wang, S.S.Y., ed., *Advances in Hydro-science Engineering*, University of Mississippi, pp. 1331–1338.

Seydell, I. 1998. Evaluating the effect of historical land use on sediment yield, Ruhama Basin, Israel, Unpublished Vertieferarbeit thesis, Department of Civil Engineering, Technical University, Darmstadt, 57 p.

Shulits, S. and Hill, R.D. 1968. Bedload formulas, US Department of Agriculture, Agricultural Research Service Report ARS-SCW-1.

Shvidchenko, A.B. and Kopaliani, Z.D. 1998. Hydraulic modelling of bed load transport in gravel-bed Laba River. *Journal of Hydraulic Engineering* 124: 778–785.

Simons, D.B., Richardson, E.V. and Nordin, C.R. 1965. *Bedload Equation for Ripples and Dunes*, US Geological Survey Professional Paper 462-H, 9 p.

Singh, K.P., Durgunoglu, A. and Ramamurthy, G.S. 1986. Some solutions to underestimation of stream sediment yields. *Hydrological Science and Technology: Short Papers* 2(4): 1–7.

Stelczer, K. 1981. *Bed-load Transport*. Littleton: Water Resources Publications, 295 p.

Stevens, H.H., Jr. and Yang, C.T. 1989. *Summary and Use of Selected Fluvial Sediment Discharge Formulas*, US Geological Survey Water Resources Investigations Report 89-4026, 123 p.

Sullivan, S.A. 1996. DGPS and GIS improve lake sedimentation survey procedures. In: Albertson, M.L., Molinas, A. and Hotchkiss, R., eds., *Proceedings of the International Conference on Reservoir Sedimentation*, Fort Collins, Colorado: Colorado State University, pp. 255–262.

Swiss Federal Authority for Water Utilization. 1939. Untersuchungen in der Natur über Bettbildung, Geschiebe- und Schwebestoffhrung [Field research on bed formations, bedload and suspended load movement], Mitteilung Nr. 33 des Amtes für Wasserwirtschaft, 114 p.

Tacconi, P. and Billi, P. 1987. Bed load transport measurements by the vortex-tube trap on Virginio Creek, Italy. In: Thorne, C.R., Bathurst, J.C. and Hey, R.D., eds., *Sediment Transport in Gravel-bed Rivers*, Chichester: John Wiley and Sons Ltd., pp. 583–616.

Thomas, R.B. 1985. Estimating total suspended sediment yield with probability sampling. *Water Resources Research* 21: 1381–1388.

Thomas, R.B. 1988. Monitoring baseline suspended sediment in forested basins: the effects of sampling on suspended sediment rating curves. *Hydrological Sciences Journal* 33: 499–514.

Thomas, R.B. and Lewis, J. 1993a. A comparison of selection at list time and time-stratified sampling for estimating suspended sediment loads. *Water Resources Research* 29: 1247–1256.

Thomas, R.B. and Lewis, J. 1993b. A new model for bed load sampler calibration to replace the probability-matching method. *Water Resources Research* 29: 583–597.

Thomas, R.B. and Lewis, J. 1995. An evaluation of flow-stratified sampling for estimating suspended sediment loads. *Journal of Hydrology* 170: 27–45.

Thompson, S.M. 1976. *Clutha Power Development, Siltation of Hydro-electric Lakes*, Wellington: Ministry of Works and Development, 39 p.

Thorne, P.D., Williams, J.J. and Heathershaw, A.D. 1989. In situ measurements of marine gravel threshold transport. *Sedimentology* 36: 61–74.

Thorne, P.D., Hardcastle, P.J. and Soulsby, R.L. 1993. Analysis of acoustic measurements of suspended sediments. *Journal of Geophysical Research* 98: 899–910.

Toffaleti, F.B. 1968. A procedure for computation of the total river sand discharge and detailed distribution, bed to surface, Technical Report No. 5, Committee on Channel Stabilization, United States Army Corps of Engineers, Vicksburg, Mississippi.

Van Rijn, L.C. and Gaweesh, M.T.K. 1992. A new total sediment load sampler. *Journal of Hydraulic Engineering* 118: 1686–1691.

Van Sickle, J. 1981. Long-term distributions of annual sediment yields from small watersheds. *Water Resources Research* 17: 659–663.

Vanoni, V.A., ed. 1975. *Sedimentation Engineering*, Manuals and Reports on Engineering Practice No. 54, New York: American Society of Civil Engineers, 745 p.

Vanoni, V.A., Brooks, N.H. and Kennedy, J.F. 1961. Lecture notes on sediment transportation and channel stability, W.M. Keck Laboratory of Hydraulics and Water Resources Report KH-R-1.

Verhoff, F.H., Yaksich, S.M. and Melfi, D.A. 1980. River nutrient and chemical transport estimation. *Journal of the Environmental Engineering Division, ASCE* 106(EE3): 591–608.

Walling, D.E., 1974. Suspended sediment and solute yields from a small catchment prior to urbanisation. In: Gregory, K.J. and Walling, D.E., eds., *Fluvial Processes in Instrumented Watersheds*, Institute of British Geographers Special Publication 6, pp. 169–192.

Walling, D.E. 1977. Limitations of the rating curve technique for estimating suspended sediment loads, with particular reference to British rivers. *International Association of Hydrological Sciences Publication* No. 122: pp. 34–48.

Walling, D.E. and Webb, B.W. 1981. The reliability of suspended sediment load data. *International Association of Hydrological Sciences Publication* No. 133: pp. 177–194.

Walling, D.E. and Webb, B.W. 1988. The reliability of rating curve estimates of suspended sediment yield: some further comments. *International Association of Hydrological Sciences Publication* No. 174: pp. 337–350.

Walling, D.E. and Moorehead, P.W. 1989. The particle size characteristics of fluvial suspended sediment: an overview. *Hydrobiologia* 176/177: 125–149.

Webby, M.G., Walsh, J.M. and Goring, D.G. 1996. Sediment redistribution in Lake Roxburgh, New Zealand. In: Albertson, M.L., Molinas, A. and Hotchkiss, R., eds., *Proceedings of the International Conference on Reservoir Sedimentation*, Fort Collins, Colorado: Colorado State University, pp. 1867–1881.

White, W.R., Milli, H. and Crabbe, A.D. 1973. Sediment transport; an appraisal of available methods, Report 119, UK Hydraulics Research Station, Wallingford, England.

Wilcock, P.R. 1997a. *A Method for Predicting Sediment Transport in Gravel-bed Rivers*, Baltimore: Department of Geography and Environmental Engineering, The Johns Hopkins University, 59 p.

Wilcock, P.R. 1997b. Entrainment, displacement and transport of tracer gravels. *Earth Surface Processes and Landforms* 22: 1125–1138.

Wilcock, P.R. 1998. Two-fraction model of initial sediment motion in gravel-bed rivers. *Science* 280: 410–412.

Wilcock, P.R., Barta, A.F. and Shea, C.C.C. 1994. Estimating local boundary shear stress in large gravel-bed rivers. In: *Hydraulic Engineering 94*, New York: American Society of Civil Engineers, pp. 834–838.

Wilcock, P.R., Kondolf, G.M., Matthews, W.V.G. and Barta, A.F. 1996. Specification of sediment maintenance flows for a large gravel-bed river. *Water Resources Research* 32: 2911–1921.

Williams, D.T. and Julien, P.Y. 1989. Applicability index for sand transport equation. *Journal of Hydraulic Engineering* 115: 1578–1581.

Willis, J.C. and Kennedy, J.F. 1977. Sediment discharge of alluvial streams calculated from bed-form statistics, Iowa Institute of Hydraulic Research Report 202, 132 p.

Witte, W.G., Whitlock, C.H., Usry, J.W., Morris, W.D. and Gurganus, E. 1981. Laboratory measurements of physical, chemical and optical characteristics of Lake Chicot sediment waters, National Aeronautical and Space Administration Technical Paper 1941, 27 p.

Wren, D.G., Barkdoll, B.D., Kuhnle, R.A. and Derrow, R.W. 2000. Field techniques for suspended-sediment measurement. *Journal of Hydraulic Engineering* 126: 97–104.

Young, W.J. and Davies, T.R.H. 1990. Prediction of bedload transport rates in braided rivers: a hydraulic model study. *Journal of Hydrology* (*New Zealand*) 29: 75–92.

16

Sediment Budgets as an Organizing Framework in Fluvial Geomorphology

LESLIE M. REID[1] AND THOMAS DUNNE[2]
[1]*USDA Forest Service Pacific Southwest Research Station, Arcata, CA, USA*
[2]*Donald Bren School of Environmental Science and Management and Department of Geological Sciences, University of California, CA, USA*

16.1 INTRODUCTION

A river's character is strongly influenced by the amount and timing of the water and sediment provided to it, and a change in sediment or water supply usually provokes a change in the river. When a river's character changes, the activities the river had supported are often disrupted. Applied fluvial geomorphology is largely devoted to understanding and designing strategies for coexisting with changing fluvial systems. To do so, it is usually necessary to understand the river's sediment regime. Theoretical fluvial geomorphology focuses on understanding the mechanics of fluvial processes and the evolution of fluvial landforms. Whenever a fluvial feature changes form, there has been a local imbalance in the movement of sediment to and from the site. In this case, too, an understanding of the sediment regime is usually central to addressing the questions posed.

Both theoretical and applied fluvial geomorphologists address questions of how changes in catchment conditions affect channels, how long the effects will last, and what the sequence of responses will be. Answers to these questions require an understanding of how a river collects, transports, and deposits sediment, and sediment budgets are tools for building that understanding.

Sediment budgets provide frameworks for organizing and interpreting information about sediment regimes, they identify the information needed to address particular questions, and they assist in comparing conditions across catchments and displaying likely outcomes of management options. Magnitudes of a budget's components can be compared in order to establish management priorities or design efficient sediment control or monitoring strategies.

Previous chapters describe a rich array of tools for answering questions concerning rivers, but with an array of choices comes the challenge of identifying the tools appropriate for particular applications. Preliminary sediment budgets can provide information to guide selection of the analytical tools needed for more detailed analyses, and the framework provided by simple, qualitative sediment budgets can aid interpretation of results of the more detailed studies. For other applications, more sophisticated sediment budgets are themselves the strategy for addressing the problems posed. This chapter discusses the nature of sediment budgets, provides examples of how they have been used, and describes an approach for designing and constructing useful budgets.

The Sediment Budget Defined

A sediment budget describes the input, transport, storage, and export of sediment from a geomorphic system. That system can be anything from an individual hillslope segment to the Amazon Basin, from a channel reach to a mountain range. Sediment budgets can be designed to describe the magnitude of a process or response rate, its location, and its timing, or to explore the influences contributing to a morphologic change. They can be used to compare the likely outcomes of different land-management options or climatic changes, or to evaluate the significance and implications of climatic, tectonic, or land-use changes that have already occurred. Sediment budgets provide a framework for organizing both qualitative information about process interactions and quantitative information about process rates. Budgets can take many forms, describe many scales, and incorporate varied levels of precision. The most commonly used sediment budgets take the form of flowcharts that describe relationships between sediment sources and transport processes; these have been the starting point for many geomorphological studies. At the other extreme, long-term monitoring projects have provided precise measurements of particular budget components (e.g., Caine and Swanson 1989).

Whether qualitative or quantitative, all sediment budgets are conceptually underlain by a basic continuity equation for sediment:

$$\text{sediment input to a landscape element} = \text{sediment output} + \text{change in sediment storage}$$

where all terms are expressed as quantities per unit time. The basic equation can be refined in many ways: changes in grain size can be accounted for by constructing equations for different size classes, for example, and specific processes can be identified.

Given the various forms sediment budgets may take and the variety of problems to which they can be applied (Table 16.1), it is clearly not useful to think of sediment budgeting as a singular tool to be applied using a uniform protocol. Instead, sediment budgeting represents a general approach to geomorphic problem solving, and the methods most useful for each budget depend on the intended application for that budget.

History and Applications

Geomorphologists have long used the concept that imbalances between sediment supply and transport capacity cause aggradation and degradation, and by the late 1800s sediment production and transport rates were being measured. Hill (1896), for example, integrated the results of landslide surveys in New Zealand to demonstrate that landslides could influence landscape evolution. Two decades later, Gilbert (1917) incorporated a sediment budget into his analysis of the effects of sediment from hydraulic mining in the Sierra Nevada on navigation in San Francisco Bay.

In 1960s the idea of systematically quantifying the balance between sediment inputs, transport rates, and storage changes began to spread widely through geomorphology. Many methods for quantifying components of sediment regimes had been developed by then, and long-term monitoring and air-photo records were becoming available. At first, quantification of sediment budgets simply involved systematic accounting of process measurements. For example, Jäckli (1957) and Rapp (1960) accumulated measurements of rock material transfers in two European mountain ranges. Leopold *et al.* (1966) measured sediment inputs and some aspects of particle movement and storage in an ephemeral stream in New Mexico.

Sediment budgeting soon proved useful for analyzing the landscape-scale effects of land use. Both Haggett (1961) in Brazil and Trimble (1977) in the Southern Appalachian Mountains of the US found large disparities between landscape-averaged estimates of soil erosion following colonization and the subsequent amounts of fluvial sediment transport in neighboring lowlands. Both authors interpreted the disparities to indicate that huge volumes of sediment must be stored on footslopes and valley floors, and that the storage elements would continue to contribute fluvial sediment long after hillslopes restabilize. These studies emphasized the connections between sediment fluxes through various landscape elements and highlighted the importance of studying changes in sediment storage. Trimble (1983) then retrospectively quantified changes in hillslope-derived sediment input, valley-floor sediment storage, and sediment yield of a catchment in Wisconsin as land was first converted to agricultural use, soil conservation measures were then implemented, and forest eventually reclaimed parts of the catchment.

Field-based monitoring studies began to be coupled with modeling to interpret, extend, and generalize results, and increasingly sophisticated tools were developed to extend the spatial and temporal scales of process measurements. As methodological and conceptual difficulties were surmounted, the organizing power of the sediment budget concept became more evident, and the approach is now widely applied to quantify landform evolution under both natural and modified conditions (Table 16.1).

Sediment budgets play a key role in basic and applied geomorphological studies over a wide range of scales and levels of complexity. For example, Flemings and Jordan (1989) used a model of mountain building, isostasy, and crustal flexure to analyze the partitioning of sediment between an evolving orogen, the adjacent sedimentary basin, and export downstream. Burbank (1992) documented the large-scale sediment budget of deposition in the Ganges foreland basin to identify controls on uplift of the Himalayas.

Table 16.1 Examples of sediment budgets used to address issues in fluvial geomorphology. Explanation of symbols provided at the end of the table

Reference	Problem addressed	Why	Precision	Regime	Time	Method
		PNF	QNS	DESY	AELS	MFACSH
Spatial focus: catchment response						
Gellis *et al.* (2001)	Prioritize subbasins for rehabilitation by erosion potential	.N.	..S	.E..	...S	.FA...
Hovius *et al.* (1997)	Proportion of sediment yield from landslides	.N.	.N.	.E..	A...	.FA...
Page *et al.* (1994)	Sediment contribution to lake from cyclone	.N.	.N.	DESY	.E..	.FA.S.
Pearce and Watson (1986)	Effects of earthquakes on sediment input and transport	P..	.N.	.ESY	.E..	.FA...
Phillips (1986)	Evaluate effectiveness of soil conservation strategies	..F	.N.	DESY	..L.	.F
Phillips (1991)	Influence of upper-basin sediment on lower basin	PN.	.N.	DESY	..L.	.F
Reneau and Dietrich (1991)	Relation between hillslope erosion and sediment yield	P..	.N.	.E.Y	A...	.F..SH
Roberts and Church (1986)	Effects of logging on sediment input and channel form	.N.	.N.	.ES.	..L.	.FA...
Smith and Swanson (1987)	Sediment routing after tephra deposition by an eruption	.N.	.N.	.ES.	.E.S	MF....
Springer *et al.* (2001)	Effect of storm	.N.	.N.	.ESY	.E..	.F....
Wasson *et al.* (1998)	Effects of land use on sediment yield	P..	.N.	.ESY	...S	.FA.SH
Spatial focus: channel system response						
Abernethy and Rutherfurd (1998)	Use distribution of bank erosion types to plan restoration	.NF	Q.S	DE..	...S	.F.C..
Benda (1990)	Effect of debris flows on downstream channel form	.N.	.N.	DESY	...S	.FA...
Brizga and Finlayson (1994)	Is upstream land use causing downstream aggradation?	.N.	Q..	DES.	..L.	.FA..H
James (1999)	Extent of channel recovery from mining debris inputs	.NF	..S	DES.	..L.	MFA..H
Knighton (1991)	Effect of mining on downstream channels	P.F	.N.	DESY	..L.	.FA...
Le Pera and Sorriso-Valvo (2000)	Changes in sediment character along a channel	.N.	.N.	D...	...S	.F....
Liébault and Piégay (2001)	Effect of land-use change on form of downstream channel	.N.	.N.	.ES.	..L.	.FACS.
Madej and Ozaki (1996)	Extent of channel recovery from the 1964 flood	P..	.N.	DESY	.E..	MFA...
Marron (1992)	Downstream distribution of introduced mining debris	P..	.N.	DES.	..L.	.FA.S.
Marutani *et al.* (1999)	Significance of temporary sediment storage	P..	.N.	.ESY	...S	.FA...

(*Continues*)

Table 16.1 (*Continued*)

Reference	Problem addressed	Why	Precision	Regime	Time	Method
		PNF	QNS	DESY	AELS	MFACSH
Trimble (1983)	Effects of land use on downstream conditions	P..	.N.	DESY	..L.	.F.C..
Trimble (1993)	Develop strategy for catchment rehabilitation	..F	..S	DESY	..L.	.F.C...
Spatial focus: response of a particular reach						
Brooks and Brierley (1997)	What caused the channel form to change?	P..	Q..	.ES.	..L.	.F..S.
Collins and Dunne (1989)	Effect of gravel mining on channel form	P..	.N.	.ES.	..L.	.FAC.H
Davis *et al.* (2000)	Basis for designing appropriate gravel harvest rate	.N.	..S	DES.	...S	.F.C.H
Dunne *et al.* (1998)	Exchanges of sediment between channel and floodplain	.N.	.N.	DESY	A...	MFAC..
Gilbert (1917)	Will hydraulic mining debris affect shipping channels?	..F	.N.	.ESY	..L.	.F.C..
Kesel *et al.* (1992)	Describe original sediment regime in lower Mississippi River	P..	.N.	.ESY	A...	.....H
McLean *et al.* (1999)	Variation of particle transport modes through a reach	.N.	.N.	D.SY	A...	M.....
Nakamura *et al.* (1997)	Effect of channelization on a wetland	P..	.N.	D.S.	..L.	MFA...
Parker (1988)	Suspended sediment budget for channel reach	.N.	.N.	DESY	...S	M.....
Pitlick and Van Steeter (1998)	River management strategies to improve fish habitat	..F	.N.	DES.	..L.	MF.C..
ten Brinke *et al.* (1998)	Extent of sand deposition during floods	.N.	.N.	D.S.	.E..	.FA...
Van Steeter and Pitlick (1998)	Effect of dams on channel form	PN.	.N.	DES.	..L.	M.A..H
Walling *et al.* (1998)	Extent of deposition of suspended sediment load	.N.	.N.	..S.	...S	MF..S.
Wathen and Hoey (1998)	Behavior of sediment wave and effects on channel form	.N.	.N.	DES.	.E..	MFA...
Wiele *et al.* (1996)	Distribution of sands introduced by tributary	.N.	.N.	DES.	.E..	MFAC..
Wilcock *et al.* (1996)	What dam release regime will most benefit bed material?	..F	.N.	.ES7.	A...	...C..
Wohl and Cenderelli (2000)	Effect of sediment release on downstream conditions	.N.	.N.	DESY	.E..	MF....
Spatial focus: specific land use, landform, etc.						
Harvey (1992)	Factors controlling a gully's form and evolution	PNF	.N.	DESY	A...	MFA...
Megahan *et al.* (1986)	Sediment input during forest road construction	.N.	.N.	.E.Y	..L.	M.....

(*Continues*)

Table 16.1 (*Continued*)

Reference	Problem addressed	Why	Precision	Regime	Time	Method
		PNF	QNS	DESY	AELS	MFACSH
Page and Trustrum (1997)	Effect of land-use changes on lake sedimentation	P..	.N.	...Y	A.L.	....S.

Explanation of symbols used:

Why: purpose

P	Past	Explain how a condition developed
N	Present	Describe interactions within present system
F	Future	Forecast future forms, conditions, or outputs

Precision

Q	Qualitative	Depends primarily on qualitative results
N	Quantitative	Depends primarily on quantitative results
S	Semi-quantitative	Incidental quantification, rankings, etc.

Component of sediment regime

D	Distribution	Conclusions involve spatial distribution
E	Erosion	Erosion specifically evaluated
S	Storage	Sediment storage specifically evaluated
Y	Sediment yield	Yield specifically evaluated

Time considered

A	Average	Generalized, or long-term average
E	Particular event	Effect of a particular triggering event
L	Land-use activity	Timescale selected to evaluate land use
S	Specific period	Conclusions referenced to particular period

Method

M	Monitoring	Monitoring was carried out for the study
F	Field work	Field measurements or observations
A	Air photos	Air photos for measurement or interpretation
C	Modeling	(Includes use of published equations)
S	Stratigraphy	Analysis of datable deposits
H	Historical records	(Includes archived monitoring records)

Church and various colleagues (Church and Ryder 1972, Church and Slaymaker 1989, Church *et al.* 1999) illustrated the importance of lagged and indirect responses in erosion and sedimentation during and after glaciation. Questions about the response of rivers to perturbations such as land use (Trimble 1974), dam construction and gravel mining (Kondolf and Swanson 1993), volcanic eruptions (Lehre *et al.* 1983), and sea-level rise (Meade 1982, Allison *et al.* 1998) have also been explored by systematically accounting for input and output of sediment. Other studies have examined the exchange of sediment between channels and their floodplains (Marron 1992, Dunne *et al.* 1998) and between catchment hillslopes and floodplains (Marutani *et al.* 1999, Gomez *et al.* 1999), and have evaluated the imbalance between rates of sediment generation from bedrock and its export from drainage basins over various time periods (Clapp *et al.* 2000, 2001; Kirchner *et al.* 2001). Such studies provide information needed to motivate, plan, and interpret more detailed studies of sediment production and transport processes and their driving agents.

Because sediment influences many ecosystem and watershed processes, sediment budgeting can also be used to explore non-geomorphological issues. Graf (1994), Malmon (2002), and Malmon *et al.* (2002), for example, studied the migration of radionuclides through channels and floodplains of Los Alamos Canyon, New Mexico, paying particular attention to the disparate trajectories of coarse sediment that contains little contaminant and the more reactive fine sediment.

Sediment budgeting has long contributed to management of sediment-related problems. Studies have described the effects of logging on sediment regimes through long-term monitoring (Swanson *et al.* 1982) and field-based surrogate measures of process rates (Roberts and Church 1986). Other studies have quantified the effects of specific activities such as road construction (Megahan *et al.* 1986) and road use (Reid and Dunne 1984). Such studies aid erosion control efforts by identifying the most important influences on and sources of sediment production. Sediment budgets for river channels have been quantified to establish appropriate extraction rates for gravel (Collins and Dunne 1989, Davis *et al.* 2000). Sediment budgeting has also been used to design strategies for catchment-scale sediment control (Phillips 1986, Trimble 1993, Gellis *et al.* 2001) and riparian restoration (Abernethy and Rutherfurd 1998), and to plan reservoir releases to maintain habitat for particular species (Wilcock *et al.* 1996, Pitlick and Van Steeter 1998).

Sediment budgets are increasingly used to aid regulatory oversight of land-use activities. Under the US Clean Water Act, for example, "Total Maximum Daily Load" allocations and sediment control plans must be developed for non-point-source sediment in catchments found to be impaired by such sediment. To do so, the amount and sources of anthropogenic sediment must be determined. Sediment budgeting can also aid preparation of environmental impact assessments, which usually must predict the effects of planned projects on sediment production and on cumulative environmental impacts. Off-site cumulative impacts often result from changes in erosion, transport, or deposition of sediment.

16.2 BACKGROUND: THE SEDIMENT SYSTEM

A sediment system can be examined from many points of view, and each of these could be represented by a sediment budget. Which point of view is most useful depends on the intended application. To understand the variety of approaches possible, the components of a catchment's sediment production and transport system must first be reviewed.

Entire books have been written about specific aspects of the sediment system, and other chapters in this book discuss sediment transport (Chapters 13 and 15) and channel change (Chapters 10, 11, and 19). Here we briefly summarize concepts that are particularly relevant to sediment budgeting. Reid and Dunne (1996) provide more detailed descriptions of methods.

Hillslope Processes and Sediment Delivery to Streams

Sediment in a catchment originates from bedrock, atmospheric deposition, and biological activity. Bedrock becomes sediment through physical and chemical weathering, during which some of the original material is removed by dissolution. Heimsath *et al.* (1997) have defined the "soil production rate" as the rate per unit area at which soil material is converted from bedrock; Small *et al.* (1999) refer to this quantity as the "regolith production rate".

As weathering progresses, a particle may remain in place as saprolite or be dislodged (eroded) and transported downslope as colluvium. Erosion rates are generally described as a net loss of sediment per unit area, while transport rates represent the discharge of sediment per unit width of hillslope or through a channel cross section. Net erosion occurs only where transport into an area is less than transport out. Rates are reported in terms of either volume or mass, but values reported by mass can be compared more readily.

Ordinarily, a sequence of disparate hillslope processes (Table 16.2) moves a sediment particle downslope until it is finally delivered to a channel. The rate of sediment production to stream channels has been defined as the rate of colluvial sediment transport across a line corresponding to the streambank (Reid and Dunne 1996)—note that the words "production" and "delivery" can refer to transfer between any landscape elements, so the context for the usage must be considered if confusion is to be avoided.

Table 16.2 Examples of methods used to evaluate erosion, colluvial sediment transport, and primary sediment production to channels. Additional methods, including direct monitoring, are usually available. The major controlling variables are listed in parentheses after the process name to aid stratification and facilitate use of data from analogous sites. Expected accuracies are estimated for typical conditions; an expected accuracy of *H* indicates that the estimated value is expected to be between 0.6 and 1.6 times the actual value; *M*, 0.4–2.5 times; and *L*, less than 0.4 to more than 2.5 times. Accuracies can be increased through more detailed work or long-term monitoring or decreased through use of reconnaissance methods. References are selected to provide further information about a process, demonstrate methods for evaluating it, or illustrate its incorporation into a sediment budget

Examples of analysis methods by process	Sample references
Dissolution (controlling variables: topography, climate, bedrock, soil depth, vegetation)	
Monitor discharge and concentration of non-organically derived solutes in outflow and precipitation. May be able to develop relation between concentrations and specific conductivity, which is readily monitored. Define relation between concentration and discharge—this may vary seasonally and by solute. Apply this relation to the annual hydrograph to calculate a year's total non-organic solute yield, and subtract solute input from precipitation (expected accuracy: *H*)	Chapters 12 and 20; Janda (1971), Lewis and Grant (1979), Saunders and Young (1983), Clayton and Megahan (1986), Caine and Thurman (1990), Small *et al.* (1999), Hodson *et al.* (2000)
Soil creep (gradient, climate, soil type, soil depth, vegetation)	
Difficult to monitor accurately; few measurements exist. Method 1: apply values of creep rate measured at similar sites, and multiply estimated creep discharge per unit width by the length of colluvial streambank (*L*). Method 2: sediment production from creep transport is by way of other processes such as bank erosion and streambank landslides, so assess these processes instead (*M*)	Saunders and Young (1983), Auzet and Ambroise (1996), Reid and Dunne (1996), Clarke *et al.* (1999)
Burrowing (gradient, species, soil type, vegetation)	
Production is by transport of burrow tailings across streambanks; measure deposit volumes and consider the seasonal distribution of burrowing (*H*). Delivery by overland flow may be important. Must know burrow patterns to assess on-slope transport; these vary by species (*L*)	Chapter 20; Thorn (1978), Meentemeyer *et al.* (1998), Gabet (2000)
Tree-throw (gradient, storm size, vegetation type and age)	
Identify fall modes (i.e., breakage or tree-throw) for each vegetation type, and use age of associated vegetation to identify fall-age diagnostics (e.g., time to shedding of twigs or loss of bark). Field sample to estimate delivery ratios and number of contributing rootwads by age per unit channel length (*H*). For transport rate, sample frequency per unit area, rootwad volumes (less root volume), and displacement of mounds from scars (*H*). Consider history of wind storms	Chapters 10 and 20; Santantonio *et al.* (1977), Schaetzl *et al.* (1989), Norman *et al.* (1995), Small (1997), Reid and Hilton (1998)
Earthflows (gradient, seasonal rainfall, bedrock, vegetation)	
Map visible flows on air photos; less visible flows require fieldwork. Delivery is by bank erosion, gully erosion, and streambank landslides. Method 1: use methods described below to assess rates of delivery processes (*M*). Method 2: estimate surface velocity near toe by measuring displacement of survey markers or features visible on sequential air photos, assume a characteristic velocity profile (or measure it using inclinometer tubes), and apply the resulting unit discharge to the measured flow cross section at the streambank (*H*). Also assess gully erosion if present (see below)	Chapters 6, 10, and 20; Van Asch and Van Genuchten (1990), Zhang *et al.* (1991a,b, 1993), Nolan and Janda (1995), Swanston *et al.* (1995)

(*Continues*)

Table 16.2 (*Continued*)

Examples of analysis methods by process	Sample references
Deep-seated slides (gradient, seasonal rainfall, bedrock, vegetation)	
Map and date using sequential aerial photos; field sample to measure sediment delivery (compare scar and deposit volumes), evaluate slide not visible on photos, and date small scars from vegetation ages. Assess as for earthflows if movement is chronic or intermittent; if removal of deposits is intermittent, evaluate temporary storage. May be able to define relationships between scar area and volume and between topographic setting and delivery ratio. Calculate production as frequency × volume × delivery ratio for each land stratum. Consider abnormally wet or dry seasons when interpreting average rates (*H*)	Chapters 6, 10, and 20; Chandler and Brunsden (1995), Ibsen and Brunsden (1996), Mantovani *et al.* (1996), Wiles *et al.* (1996), Corominas and Moya (1999), Fantucci and Sorriso-Valvo (1999)
Shallow slides (gradient, landform, storm rainfall, bedrock, vegetation, earthquakes)	
Map and date using sequential aerial photos; field sample to measure sediment delivery (compare scar and deposit volumes), evaluate slides not visible on photos, and date small scars using vegetation ages. May be able to define relationships between scar area and volume and between topographic setting and delivery ratio. Calculate production as frequency × volume × delivery ratio for each land stratum. Consider storm history when interpreting average rates (*H*)	Chapters 6, 9, 10, and 20; Roberts amd Church 1986, Page *et al.* (1994), Mantovani *et al.* (1996), Reid and Dunne (1996), Hovius *et al.* (1997), Reid (1998), Corominas and Moya (1999)
Rockfalls and rock slides (gradient, weather, bedrock exposure, bedrock type, earthquakes)	
Map and date large falls using sequential aerial photos; older falls might be dated using vegetation ages or lichenometry. More diffuse falls might be assessed by measuring volumes accumulated on a datable surface such as a snowpack, or from repeated ground-based stereophotography of rock faces. Consider proximity to stream to estimate delivery; blocks may be too large for transport (*H*)	Chapters 6, 10, and 20; Bull *et al.* (1994), Wieczorek and Jäger (1996), André (1997), Matsuoka and Sakai (1999)
Streambank slides (gradient, flow distribution, channel size and form, bedrock, vegetation)	
Assess most inner-gorge and valley-wall failures as for shallow slides (*H*). If undercut valley walls are long-term sources, assess wall retreat using air photos or vegetation (*H*). Small failures in alluvium and colluvium can often be analyzed with other bank erosion processes. Otherwise, identify characteristic failure sizes and temporal and spatial patterns of failure along channels of different kinds; consider recent storm history when estimating average rates (*H*)	Chapters 7, 10, 11, and 20; Thorne and Tovey (1981), Kelsey *et al.* (1995), Couper and Maddock (2001)
Debris flow erosion (hillslope gradient, storm rainfall, channel gradient, bedrock, vegetation)	
Under steady state, only erosion of colluvium and bedrock is primary; otherwise, also evaluate remobilization of channel deposits. Map and date visible scars using air photos; measure widths in the field if obscured by trees. Identify other flows in the field from debris deposits and date using vegetation. Determine characteristic erosion depths from scarp heights and from depths of soil and channel deposits at analogous sites. Consider storm history when interpreting average rates (*H*)	Chapters 6, 7, 10, 11, and 20; Benda (1990), van Steijn (1996), Yoshida *et al.* (1997), Cenderelli and Kite (1998), Springer *et al.* (2001)

(*Continues*)

Table 16.2 (*Continued*)

Examples of analysis methods by process	Sample references
Primary streambank erosion (gradient, peak-flow size, channel size, soil type, vegetation)	
Under steady state, only colluvial and bedrock erosion is primary; otherwise, also evaluate erosion of alluvium because it alters the distribution of stored sediment. Stratify by channel type. Estimate bank retreat rates in large channels with visible banks using sequential air photos and field measurements of bank height (*H*). Otherwise, field sample to estimate proportion of banks eroding (i.e., without vegetation). Rates are often difficult to assess without monitoring, but might be estimated from datable vegetation, scarp depths, or deposit volumes (*M*). Estimates have also been made by applying an estimated creep discharge to the susceptible bank area (*L*)	Chapters 4, 6, 7, 10, 11, and 20; Roberts and Church (1986), Kesel *et al.* (1992), Trimble (1994), Reid and Dunne (1996), Zonge *et al.* (1996), Barker *et al.* (1997), Stott (1997), Dunne *et al.* (1998), Cohen and Brierley (2000), Prosser *et al.* (2000), Couper and Maddock (2001)
Primary channel erosion (channel gradient, peakflow size, channel size, bedrock)	
Under steady state, only colluvial and bedrock erosion is primary. Long-term, steady-state channel lowering is of the same order as the mean hillslope lowering rate, but channel area is small relative to hillslope area so production from this source is relatively small. Where steady state cannot be assumed, incision of alluvium must also be evaluated (Table 16.3). Most reliably assessed by dating terrace surfaces to calculate incision rate since the surfaces were formed (*H*)	Chapters 4, 7, 9–11, and 20; Clayton (1997), Seidl *et al.* (1997), Trimble (1997), Ward and Carter (1999), Reneau (2000)
Tunnel erosion (gradient, landform, soil type)	
Examine channel heads to ascertain presence of tunnels (soil pipes); identify upslope extent by trenching or observing collapse features. Sediment delivery is most reliably assessed by monitoring effluent from multiple tunnels (*M*). Otherwise, measure deposit volumes on hillslopes (ascertain that their presence does not indicate an atypical feature) and apply to the full distribution of tunnels (*L*), or apply published measurements from analogous features (*L*)	Chapters 12, 15, and 20; Jones (1987), Ziemer (1992), Garland and Humphrey (1992), Page *et al.* (1994), García-Ruiz *et al.* (1997), Terajima *et al.* (1997)
Gullying (gradient, catchment area, peakflow size, channel size, soil type, vegetation, compaction)	
Map and date gullies in open terrain using sequential air photos; use field measurements to identify relations between length or area and volume, and use these to estimate volume changes through time (*H*). Otherwise, field sample for distribution, frequency, size, and age. Date using associated vegetation, eye-witness accounts, or age of causal features. Estimate sediment production through time from retreat rate and volume–length relationships (*H*)	Chapters 2–4, 6, 9, 10, and 20; Swanson *et al.* (1989), Harvey (1992), Archibold *et al.* (1996), DeRose *et al.* (1998), Nachtergaele and Poesen (1999), Vandekerckhove *et al.* (2000)
Rilling (gradient, slope length, storm rainfall, soil type, vegetation, compaction)	
Assess distribution considering controlling variables and season. Monitor or field sample rill dimensions before and after wet season or storms of different sizes, or through year. Estimate delivery ratio from size distribution and volume of deposits (compare to soil texture), or from sediment concentration measurements. If using an erosion equation, test by comparing predictions and measurements (*H*)	Chapters 12, 15, 17, and 20; Collins and Dunne (1986), Smith and Swanson (1987), Harvey (1992), Slattery *et al.* (1994), Flanagan and Nearing (1995), Morgan *et al.* (1998)

(*Continues*)

Table 16.2 (*Continued*)

Examples of analysis methods by process	Sample references
Sheet and rainsplash erosion (gradient, slope length, storm rainfall, soil type, vegetation, compaction)	
Define distribution by controlling variables and season. Field sample root exposure on datable plants or monitor erosion pins (*H*). These measurements combine effects of sheet, rainsplash, dry ravel, and wind erosion, so use the spatial and temporal distribution of each to interpret results. If using an erosion equation, test by comparing predictions and measurements (even short-term data will indicate whether results are approximately correct). Estimate delivery ratio from size distribution and volume of deposits (compare to soil texture), or from sediment concentration measurements; erosion equation results should be checked against field evidence (*H*). Surface erosion can also be quantified by sampling runoff from small, definable catchments (*H*)	Chapters 12, 15, 17, and 20; Wischmeier and Smith (1978), Carrara and Carroll (1979), Trimble (1983), Reid and Dunne (1984), Collins and Dunne (1986), Smith and Swanson (1987), Flanagan and Nearing (1995), Quine *et al.* (1997), Renard *et al.* (1997), Morgan *et al.* (1998), Brazier *et al.* (2000), Yanda (2000)
Wind erosion (storm size, soil moisture, soil type, vegetation)	
Define distribution by controlling variables and season. Measure root exposure on datable plants or monitor erosion pins (*H*). These measurements combine effects of sheet, rainsplash, dry ravel, and wind erosion, so consider their spatial and temporal distributions to interpret results. If using an erosion equation, compare predictions and measurements. Estimate delivery ratio from intersection of travel paths with streams, or calculate input from measured aerial deposition (*M*)	Chapters 17 and 20; Skidmore and Woodruff (1968), Potter *et al.* (1990), Leys and McTainsh (1996), Phillips *et al.* (1999), Borówka and Rotnicki (2001), Goossens *et al.* (2001)
Dry ravel (gradient, soil moisture, soil type, temperature, vegetation)	
Define distribution by controlling variables and season. Measure root exposure on datable plants or monitor erosion pins or accumulation in troughs. Such measurements combine effects of sheet, rainsplash, dry ravel, and wind erosion, so consider their spatial and temporal distributions to interpret results. Compare grain sizes of deposits and sources to estimate delivery (*H*)	Chapter 20; Megahan *et al.* (1983)
Construction, tillage, engineering, etc. (gradient, type of project, soil type, bedrock)	
Only direct mechanical displacement of sediment is considered here; secondary processes are considered above. Aerial photos, maps, plans, interviews with equipment operators, and field observations can indicate distribution and timing of effects and location of displaced material with respect to streams (*H*). Estimate tillage transport from topographic discontinuities at field edges, and delivery by applying the resulting rate to adjacent streams or assessing bank erosion there (*H*)	Chapter 20; Riley (1990), Vandaele *et al.* (1996), Quine *et al.* (1997), Phillips *et al.* (1999)
Deposition on hillslopes and swales (gradient, landform, soil type, vegetation, sediment input)	
For non-discrete processes, field sample accumulation depths around datable plants or structures, or measure seasonal accumulations atop leaf litter (*H*). Stratigraphic dating methods can be used for long-term accumulations (*H*). For discrete processes, measure volumes of deposits and date using sequential aerial photographs or ages of associated plants (*H*)	Chapters 2, 3, 6, 9, 10, and 20; Megahan *et al.* (1986), Page *et al.* (1994), Vandaele *et al.* (1996), Phillips *et al.* (1999), Beuselinck *et al.* (2000)

In an accounting of primary sediment input, any particle can be delivered to the stream system only once. For example, soil creep often moves sediment to the base of a slope, where bank erosion carves away the encroaching sediment. In this case, sediment delivery is evaluated as either the rate of streambank erosion or the soil creep discharge at the channel margin, but these rates cannot be summed because both processes involve the same particles. Similarly, sediment production cannot be calculated by summing rates of landsliding and soil creep where creep transports colluvium to bedrock hollows that are episodically evacuated by landslides (Reneau and Dietrich 1991, Dunne 2000). A portion of the sediment derived from colluvium may be deposited downstream and later reenter the channel through erosion of alluvium. Such re-entry (described in Table 16.3) represents remobilization from temporary storage rather than primary sediment delivery. In contrast, sediment introduced by channel incision into bedrock represents primary sediment delivery (Table 16.2).

Colluvial transport processes are of two kinds. "Chronic" processes include transport by rainsplash, soil creep, sheetwash, wind, and other mechanisms that recur frequently at the same sites. "Discrete" processes, in contrast, are events that can be counted, such as landslides and tree-throws. Table 16.2 describes methods for measuring a selection of process rates and attributes and lists examples of studies that have evaluated or are relevant to assessment of each process. Other methods are also available, including direct, long-term monitoring for most processes listed.

Chronic processes are usually evaluated by determining average rates and applying those to the areas affected. For example, the rate of sheetwash erosion on rangelands might be estimated from measurements of root exposure around datable vegetation, by monitoring surface lowering around erosion pins, or by modeling using the USLE (Wischmeier and Smith 1978, Renard *et al.* 1997) or WEPP (Flanagan and Nearing 1995). The estimated rate would then be assumed to apply throughout the area of similar land-use activity, topography, and soil type. Only part of the eroded sediment is delivered to channels, however, and this amount varies with conditions in and around the eroding sites. The delivery ratio can be estimated for different site types using methods such as monitoring sediment transport in overland flow during a few storms, applying erosion models, or comparing the grain size of deposits to that of the eroding material. Sediment production rates are then calculated by multiplying erosion rates by sediment delivery ratios for each site type and applying these values to the distribution of site types present.

Rates of discrete processes, such as landslides, usually are evaluated by applying the measured spatial and temporal frequency of events to the area susceptible. Shallow landslide scars, for example, ordinarily are counted on sequential air photos to determine the number of slides per unit area per unit time. Fieldwork is usually necessary to define a relationship between scar area and scar volume and to determine the proportion of the landslide debris characteristically delivered to streams, and is also useful for estimating the frequency of landslides too small to detect on photos. Because shallow landslides are generally triggered by infrequent, large storms, the dependence of areal landslide density on the magnitude of triggering events may need to be defined to determine whether the sampling period is long enough to estimate average rates validly (e.g., Figure 16.1). For many applications, only the relative rates between different land uses or landforms need be known, and results from a single major storm often can provide this information.

Analysis of other process rates generally follows similar patterns (Table 16.2). The success of each rate analysis depends on:

1. having a well-defined objective that identifies the information required;
2. using a sampling design that permits valid characterization of the process;
3. recognizing the area and time period over which the estimate applies.

Wherever possible, rates should be estimated using multiple methods and should be checked for consistency; this is particularly important if rates are to be modeled mathematically in areas or under conditions for which the model has not been adequately tested.

Sediment Transport in Channels

Changes in hillslope sediment transport rates arouse particular concern when sediment reaches a channel, and so can affect off-site conditions. Sediment entering a channel either is transported downstream or alters channel morphology where it enters. Such alterations modify the channel bed and the sediment transport rate through the reach, and may even change the dominant transport mode. For example, long-term sediment accumulation in first-order steepland channels can convert them into unchanneled swales (zero-order basins), and eventually deepen the fill enough to

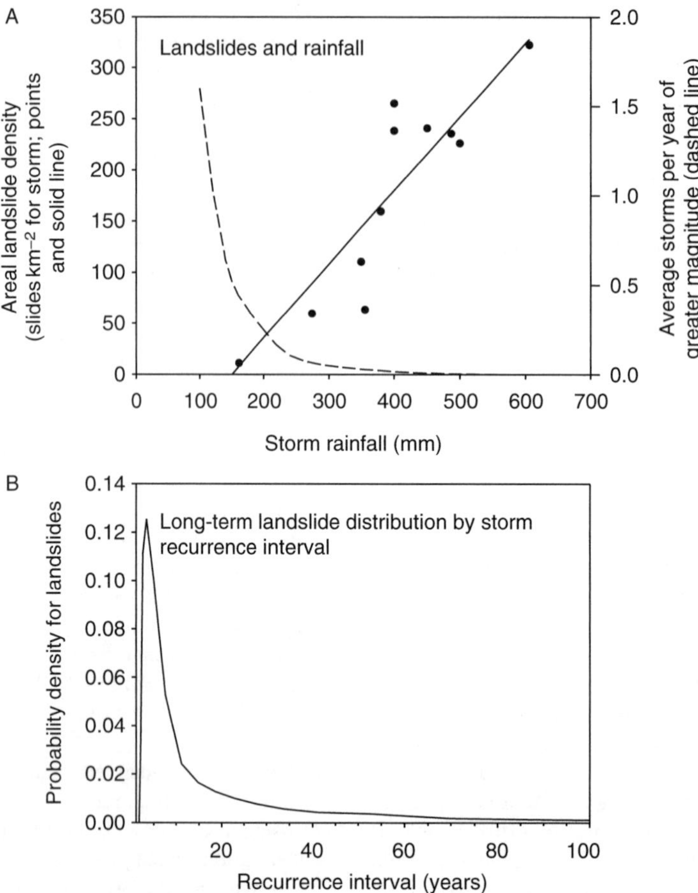

Figure 16.1 Assessing the long-term distribution of landslides as a function of storm recurrence interval for colluvial landslides on the Te Arai land system under the Waipaoa Station rainfall regime, North Island, New Zealand. (A) Areal landslide density and storm frequency as a function of storm rainfall. (B) Long-term distribution of landslide sediment inputs as a function of storm recurrence interval

generate shallow landslides (Dietrich *et al.* 1982); sediment transport shifts from fluvial, to soil creep, to landsliding. In other cases, channel flows gradually remove newly introduced sediment, leaving only the largest clasts to weather in place. How rapidly the sediment is removed depends on characteristics of both the sediment input and the channel.

Streams transport sediment in three ways. The largest grains are rolled or jostled along the bed as "bedload", while the smallest particles are continually suspended in the flow (washload). Intermediate grains are entrained repeatedly by eddies and move predominantly as suspended load. These intermediate sizes return to the bed when flow slows and are referred to as the "bed material suspended load"; most sediment in the streambed represents size fractions moved as bedload (especially in gravel-bed channels) or bed material suspended load (in sand-bed channels). The transport mode for a particular grain varies with flow and with channel characteristics (e.g., McLean *et al.* 1999).

Travel times for different components of the sediment load vary widely. Washload can exit a 1500-km^2 catchment during the same storm that eroded the sediment from a headwater hillslope, while bedload may require many decades to move the same distance. Matisoff *et al.* (2002b) have demonstrated that concentrations of the isotopes ^{7}Be, ^{137}Cs, and ^{210}Pb on suspended sediment can be used to measure the speeds and distances of fine sediment transport in single events.

The most accurate estimates of sediment transport rates in channels are provided by well-designed and maintained monitoring stations with records long enough to produce representative results. However, most sediment budgets must be constructed too rapidly to allow determination of transport rates by monitoring if records do not already exist (Table 16.3). Instead, rates ordinarily are estimated using

Table 16.3 Examples of methods used to evaluate sediment transport and storage in channels, erosion of alluvial sediment, and sediment yield. Additional methods, including direct monitoring, are usually available. The major controlling variables are listed in parentheses after the process name to aid stratification and facilitate use of data from analogous sites. Expected accuracies are estimated for typical conditions; an expected accuracy of *H* indicates that the estimated value is expected to be between 0.6 and 1.6 times the actual value; *M*, 0.4–2.5 times; and *L*, less than 0.4 to more than 2.5 times. Accuracies can be increased through more detailed work or long-term monitoring or decreased through use of reconnaissance methods. References are selected to provide further information about a process, demonstrate methods for evaluating it, or illustrate its incorporation into a sediment budget

Examples of analysis methods by process	Sample references
Bedload (controlling variables: channel gradient and form, flow distribution, grain size, sediment input, bedrock)	
Where the coarse load is trapped in a lake or low-gradient reach, estimate transport by measuring changes in depositional landforms through time using bathymetric measurements or air photos (*H*). Otherwise, bedload transport equations are usually applied. Equations must be carefully selected to be appropriate for the conditions being assessed (*M*). Bedload sampling data are available for a few stations, but records are usually sparse and short.	Chapters 7, 11–15, 17, 18, and 20; Collins and Dunne (1989), Wilcock *et al.* (1996), Reid and Dunne (1996), McLean and Church (1999), Ham and Church (2000)
Suspended load (channel gradient and form, flow distribution, grain size, sediment input, bedrock)	
Measure suspended sediment concentrations over a range of flows to define a sediment rating curve and apply the resulting curve to annual hydrographs (*H*). Sediment transport equations for suspendible bed material load are useful if input-dependent washload is not large (*M*).	Chapters 7, 12, 15, 17, 18, and 20; Walling and Webb (1982), Reid and Dunne (1984, 1996), Parker (1988), McLean *et al.* (1999), Asselman (2000), Singer and Dunne (2001)
Sediment attrition (transport rate, grain size, rock type)	
Tumbling-mill experiments can indicate grain size changes per unit travel distance (*H*). If different lithologies are present, use changes in relative abundance to estimate relative breakdown rates (*H*). Compare sediment stored for different lengths to assess importance of dissolution (*M*)	Chapters 13 and 20; Collins and Dunne (1989), Le Pera and Sorriso-Valvo (2000)
Bed aggradation (channel gradient and form, flow distribution, grain size, sediment load)	
Land surveys or surveys for bridge planning can be repeated; local residents can describe recent changes; and engulfed artifacts, woody debris, or plants can indicate the extent and timing of aggradation, as can changes in flood severity. Long-term gauging data can document changes in bed elevation. Recently aggraded bed material often is finer grained and can sometimes be probed to determine the depth of a buried gravel armor layer. Comparison of geometry in aggraded and non-aggraded channels can indicate the magnitude and distribution of changes. Establish timing from personal accounts, vegetation ages, and comparison of sequential air photos. Estimate bar aggradation rates by multiplying the areas of bars deposited by average bar heights. (*H* to *M*)	Chapters 4, 7, 10, 11, 13 and 20; Smith and Swanson (1987), Collins and Dunne (1989), Benda (1990), Knighton (1991), Madej and Ozaki (1996), Wiele *et al.* (1996), Brooks and Brierley (1997), Heritage *et al.* (1998), Wathen and Hoey (1998), Lisle and Hilton (1999), James (1999), Sloan *et al.* (2001)

(Continues)

Table 16.3 (*Continued*)

Examples of analysis methods by process	Sample references
Bank and floodplain aggradation (channel gradient and form, flow history, grain size, sediment load, vegetation)	
Measure deposit depths around datable plants or structures, or date deposits using methods described in other chapters (*H*). Data from sediment traps or erosion pins can indicate relation between deposition and flood size, as can post-flood observations of deposition atop litter layers; data from large floods are needed to estimate long-term rates (*H*). Many methods for assessing bed aggradation can be applied to banks. Several-decade-long cores can be dated with ^{137}Cs concentration profiles; profiles of ^{210}Pb attached to clay particles can provide longer (~100 years) records; and ^{14}C dating produces records dating back thousands of years. Standard stratigraphic methods can be used to correlate cores at different locations.	Chapters 2, 3, 6, 7, 9–12, 14, 15, and 20; Marron (1992), Nakamura *et al.* (1995), Asselman and Middelkoop (1998), ten Brinke *et al.* (1998), Walling *et al.* (1998), Walling and He (1998), Goodbred and Kuehl (1998), Gomez *et al.* (1999), Rumsby (2000), Lecce and Pavlowsky (2001), Steiger *et al.* (2001)
Channel erosion of alluvial sediments (channel gradient and form, flow distribution, grain size, sediment load, bedrock)	
Compare channel geometry to that of unaffected channels (*H*). Land surveys or cross sections surveyed for bridge planning can be resurveyed if available. Calculate river-bed elevation trends at gauging stations from low-flow stage records and flow-depth measurements. Residents can describe recent changes, and undercut vegetation or exposed bridge piers may provide data. Timing is usually established from personal accounts or comparison of sequential air photos. Evaluate erosion rates from shifting of large channels by multiplying the areas of bank eroded by the average bank height. (*H* to *M*).	Chapters 4, 7, 10, 11, and 20; Dunne (1977), Collins and Dunne (1989), Booth (1990), Erskine *et al.* (1992), James (1997, 1999), Heritage *et al.* (1998), Knighton (1999), Wohl and Cenderelli (2000), Gonzalez (2001), Liébault and Piégay (2001)
Sediment yield (catchment size, flow distribution, sediment input, bedrock, vegetation, topography)	
Where catchments drain into lakes or ponds, yield can be estimated from rates of lake sedimentation if the trap efficiency is known and the bathymetry has been monitored or can be reconstructed (*H*). Measurements or calculations of sediment transport at the mouth of a catchment provide an estimate of yield (*H* to *M*). Nearby catchments with similar characteristics are expected to have similar sediment yields (*M*).	Chapters 2, 9, 11, 14, and 20; Brune (1953), Megahan *et al.* (1986), Evans (1997), Hill *et al.* (1997), Page and Trustrum (1997), Wilby *et al.* (1997), Lloyd *et al.* (1998), Evans and Church (2000), Verstraeten and Poesen (2000), Cisternas *et al.* (2001)

transport equations, as described in Chapter 15, or by measuring the volume of sediment deposited in an effective sediment trap over a known time (e.g., Ham and Church 2000, Davis *et al.* 2000).

Transport equations can provide useful estimates of non-washload components if equations are selected that are calibrated over the range of conditions appropriate for the application. Reid and Dunne (1996) tabulate published comparisons between predicted and observed results and identify the equations that appear to be most reliable for various bed materials and channel sizes. Results are usually more accurate for sand-bedded than for gravel-bedded channels, but even the most reliable equations generally are accurate only within a factor of 2. Because washload is influenced more by sediment availability than by flow properties, transport equations are not useful if this component of the load is important. Instead, short-term monitoring results can be used to produce sediment rating curves, which can then be combined with calculated or measured hydrographs to estimate total suspended sediment loads. As with any monitoring-based method, errors are introduced if the monitoring period is unrepresentative or if estimates are made by extrapolation beyond the conditions measured. However, if applied carefully, the method is the best available for predictions of washload, and probably of all suspended load. Although most frequently used for larger catchments, the approach can also be applied to small, ephemeral drainages such as road-surface catchments (e.g., Reid and Dunne 1984).

During transport, sediment grains are subject to fracture, abrasion, weathering, and dissolution, contributing to widely observed downstream decreases in grain size and, often, shifts in particle composition (e.g., Le Pera and Sorriso-Valvo 2000). Downstream fining is also influenced by size-dependent transport, so attrition rates are not directly calculable from downstream size trends. Minimum attrition rates for particular lithologies can sometimes be estimated by evaluating downstream trends in the proportions of clasts of different lithologies. Where source materials differ greatly in durability, such compositional surveys are needed to identify the major sources for downstream bed materials; relative input rates do not necessarily reflect relative influence on downstream bed composition (Dietrich and Dunne 1978). Breakdown rates have also been estimated by measuring changes in clast size distribution as a function of "travel distance" in rock tumblers that have been modified to provide realistic rates of particle interaction (Kuenen 1956, Collins and Dunne 1989).

Different components of the sediment load influence downstream environments in different ways during transport. High suspended loads complicate water purification procedures and strongly influence aquatic ecosystems by limiting photosynthesis, decreasing the ability of visually feeding organisms to find their prey, abrading gill tissues, and disrupting behavior. The severity of such impacts depends on both the magnitude and duration of exposure (Newcombe and Jensen 1996). A variety of aquatic organisms inhabit the bed surface and interstices within the bed material of a stream, and this community is disrupted when bed material moves. Excessive scour can destroy the eggs of salmonids and other species that nest within gravel stream beds; this may occur where increased sediment loads contribute to a fining of the bed material, or where in-stream mining modifies bar morphology, thereby increasing the susceptibility to scouring as high flows rearrange the destabilized bars (Harvey and Lisle 1999).

Channel and Floodplain Sediment Storage

Periods of significant sediment transport in small to medium channels are interspersed with much longer periods of low transport, during which most of the transportable sediment is held in temporary storage in the channel bed, bars, and floodplains. Durations of temporary storage vary by depositional feature and by location in a catchment. Small amounts of even washload size material can be trapped within the bed material during transport or can infiltrate the bed as flows recede, and fine sediment that is decanted overbank can settle quickly onto the floodplain. Storage within floodplains is favored in rivers with high concentrations of particularly fine-grained sediment (Gomez *et al.* 1999) and frequent, sustained overbank flows (Dunne *et al.* 1998). Sediment deposited on floodplains generally remains until eroded by channel migration, so the residence time in storage is greater where channel migration is slow (Malmon *et al.* 2002).

Silt and sand are also deposited on stream banks. Residence times can be very long if banks are well vegetated and sediment is remobilized only by bank erosion. Elsewhere, as flow recedes, slumping and rilling can redistribute some of the newly deposited sediment back into the channel. Silt and sand can accumulate to aggrade stream beds if sediment loads are particularly high or transport capacity is perennially or seasonally low (e.g., Knighton 1999). Where aggradation is triggered by an altered balance between input and transport capacity, pools commonly fill first (e.g., Lisle and Hilton 1999, Wohl and Cenderelli 2000).

Gravel is most frequently deposited within the bank-full channel. These deposits can become incorporated into the floodplain as a channel migrates. However, unless the channel is aggrading, most coarse sediment temporarily resides in bars until the next bed-mobilizing flow occurs and moves the clasts farther downstream (Hassan *et al.* 1991). Widespread increases in inputs of coarse sediment to a channel network can produce temporary waveforms of gravel that can be either mobile or stationary (Jacobson and Gran 1999, Lisle *et al.* 2001).

Ponds, lakes, and reservoirs interrupt sediment transport. Coarser sediments are often trapped in these environments, while some finer sediment may remain in suspension long enough to emerge with the outflow. The trap efficiency for sediment depends on the ratio between the inflow rate and the volume of the impoundment (Brune 1953). In small ponds, trap efficiency can vary significantly over short periods (Verstraeten and Poesen 2000). Alluvial fans and major breaks in slope along a channel profile can also serve as long-term sediment sinks (e.g., McLean *et al.* 1999). At such sites, where transport capacity decreases and channels are not confined, bars accumulate rapidly, with associated rapid channel shifting or braiding (e.g., Dunne 1988).

No landscape is unchanging, but many change slowly enough that "steady-state" rates of sediment production and deposition can be assumed over moderate timescales. Over the long-term, the evolution of landforms would need to be considered (e.g., progressive erosion may decrease the hillslope gradient to the point that the average erosion rate decreases), while over a shorter period, weather patterns would need consideration (e.g., 10-year-old flood deposits may provide a temporary sediment source). On average, though, if no areas of chronic aggradation or incision exist downstream, sediment contributed to a stream system under steady-state conditions roughly balances the sediment exported from the catchment.

In catchments with rapidly evolving landforms or changing conditions, or over short periods for which seasonal and inter-annual variations are important, this simplified view must be expanded to account for changes in sediment storage (Trimble 1977). A large storm or anthropogenic alteration of vegetation cover, channel morphology, or flow regime can trigger major changes in sediment input, transport, and storage. A change in any of these components provokes compensating changes in the others, which themselves provoke additional changes. Long periods may be required for re-equilibration of the system, and different portions may respond out of phase with one another (e.g., Womack and Schumm 1977, Trimble 1983, Madej and Ozaki 1996). Under these conditions, both input to and output from channel storage need to be evaluated. Evaluation methods for erosion from storage are similar to those for erosion of hillslope materials (Table 16.3); results produce estimates of alluvial sediment input due to incision or changes in channel form. Rates of aggradation on streambeds, banks, and floodplains can be assessed using stratigraphic and dating methods described in previous chapters and in Table 16.3.

Sediment input and transport rates also vary over long periods as landforms evolve and climatic and tectonic conditions change. Many landscapes include landforms deposited by processes that are no longer active, and these forms now interact with the current process regime. Streams may be eroding into glacial outwash terraces, for example, or Pleistocene landslide deposits may locally constrict valleys. In such cases, the average erosion rate for this sediment is not balanced by an equivalent rate of deposition (e.g., Church and Slaymaker 1989), and the remobilized sediment effectively constitutes a new sediment input to the modern channel.

Where there has been a change in sediment transport rates or sediment input, channel form often changes through infilling, incision, or altered rates and modes of channel migration. If channel form changes, flood frequency also changes, and valley-bottom structures and transportation networks may be affected. Altered channel form also strongly affects aquatic ecosystems by modifying the distribution and quality of habitat. Animals that spawn on streambeds are often particularly sensitive to accumulations of fine particles in or on the bed (Phillips 1986, Everest *et al.* 1987). Accelerated accumulation of bed material within gravel rivers can lead to pool filling that impairs recreational uses and fish rearing habitat. Estuaries commonly are sites of sediment accumulation, which can interfere with harbor use and impair coastal fisheries (e.g., Nichol *et al.* 2000).

Accelerated removal or impoundment of bed material can lead to the degradation and armoring of channel beds (Collins and Dunne 1990, Kondolf and Swanson 1993, Bravard *et al.* 1999, Liébault and Piégay 2001). Such degradation can undermine bridge piers and other in-channel structures.

The Catchment: Integrating the Sediment System

Different parts of a catchment participate in the sediment regime in different ways. Low-order channels are often the major conduits for sediment input both because they are most closely connected with hillslopes and because they account for most of the drainage density. Downstream, channels are often inset into their own deposits. These terraces and floodplains can prevent hillslope sediment from reaching the channel directly, and channels at these locations may simply rework sediment initially contributed from hillslopes upstream. Opportunities for deposition generally increase downstream as alluvial valleys widen and their gradients decrease.

The "sediment yield" is the rate of sediment output from a catchment. Because sediment yields vary with catchment size, comparisons between catchments are usually made in terms of sediment yield per unit area of catchment. Sediment yields per unit area frequently decrease downstream, both because average hillslope gradients decrease with increasing drainage area and because long-term aggradation is more likely downstream. The "sediment delivery ratio" for a catchment is the proportion of sediment eroded from hillslopes that is exported from the catchment. Early work within uniform physiographic regions of generally low relief (e.g., Maner 1958, Roehl 1962) indicated that the sediment delivery ratio is characteristically less than one and decreases with increasing drainage area. This pattern is consistent with a downstream increase in sediment diverted into long-term storage. Although the original relationships have not been widely tested, they have been used elsewhere to estimate catchment sediment yield from evaluations of hillslope sediment supply. This is probably not an accurate prediction for most basins, except in the crudest sense. More research is needed to verify and elucidate such generalized predictions of sediment delivery ratios. Milliman and Syvitski (1992) have demonstrated that such a negative relationship is detectable at the scale of whole continents, although its predictive power is low.

Long-term sediment yields, especially for small to moderate-sized catchments, are most readily evaluated using measurements of sedimentation in reservoirs or ponds (Table 16.3). Repeated bathymetric surveys allow calculation of infill volumes per unit time, and total yield can then be estimated if the sediment-trapping efficiency is known. In-stream monitoring of sediment discharge also can provide estimates of sediment yield.

Although logistical and sampling difficulties have precluded much comparison of catchment sediment yields to define their reproducibility and transferability, yields are generally expected to be similar for similar-sized catchments within an area of relatively uniform physiography, geology, climate, land use, and vegetation cover, unless large, discrete sediment sources are present. For example, Dunne and Ongweny (1976) used average values for the forested, cultivated, and grazed parts of a catchment, developed from a few gauged subcatchments, to identify major sources of sediment threatening the useful life of a reservoir. In this case, new sediment sampling surveys confirmed the calculations, which were based on decade-old measurements. Such an approach requires careful consideration of potential sources of variation between catchments.

Sediment yields are often estimated by summing sediment input rates from individual processes. Such calculations are expected to be reliable under steady-state conditions if long-term aggradation is accounted for. Where conditions have recently changed, however, differences in mobility for different grain sizes cause lags between a change in hillslope process rate, the response of the suspended sediment yield, and the response of the bedload yield.

Even under steady-state conditions, rates of sediment input and transport vary considerably through time at several scales. Climates with pronounced wet seasons generally have strong seasonal sediment inputs, and channels dominated by snowmelt are often highly predictable in the timing of high sediment loads. Seasonal land-use activities also influence the timing of sediment outputs (e.g., Collins *et al.* 1998). Year-to-year variation is imparted by variations in storm intensity and by occurrence of other events that can influence erosion rates, such as wildfires, earthquakes, droughts, and episodic land-use activities (e.g., Rice 1982).

The timing of sediment transport varies through a catchment. Many headwater channels cannot move clasts coarser than pebbles during ordinary floods, and gravel and cobbles often are effectively trapped by woody debris in forest streams. At these sites, bed material might be mobilized only when debris jams fail (Mosley 1981) or during particularly large floods or debris flows (Benda and Dunne 1987). Farther downstream, transport capacities are usually high enough to mobilize bed material during ordinary bank-full events. Still farther downstream, breakdown of clasts and sequestering of the larger clasts lead to fining of the sediment load, until the largest rivers often transport primarily sand. Sand can be transported even at relatively low flows, so some transport occurs nearly continuously.

16.3 DESIGNING A SEDIMENT BUDGET

Construction of a useful sediment budget requires assessment of those parts of the sediment regime that are relevant to the particular application. No two applications have exactly the same goals or setting, so there is no single codifiable method for constructing sediment budgets. Sediment budgets vary widely in scope, approach, and methods (Tables 16.1 and 16.4), and much of the art of budget construction lies in deciding which form of budget will be most effective for addressing the particular question posed. This flexibility can be a problem in settings where the adequacy of a result is judged by whether it was obtained using standard procedures or where procedural manuals are expected to compensate for uneven levels of expertise. In such settings, it is critical for sediment budget analysts to present a strong conceptual model of the sediment budget and to provide clear, well-documented explanations of the basis for the methods to be used.

Although by no means comprehensive, Table 16.4 lists a variety of options that can be selected for each of 11 attributes of a sediment budget. Selection of appropriate options requires that several basic questions be addressed during design of the budgeting strategy (Table 16.5).

Identifying the Study Objectives

The success of a sediment budget depends strongly on the investigator's skill in defining the focal question and identifying the information needed to answer that question. To do so, the overall purpose for the inquiry must be understood. Are results of the sediment budget to be used to explain the shape of a landform? To describe existing conditions? To identify the cause of an existing condition? To design a plan to modify an existing condition? To predict the outcome of future actions? To provide a basis for regulatory oversight? To better understand interactions between specific processes? A single inquiry may have multiple purposes, but careful definition of the primary purpose allows the overall strategy to be optimized for that goal.

If the budget is to be constructed as part of a broader project, it is important that the goals of and motivations for the project be understood. Such understanding often can improve the utility of the resulting budget, but it can also protect those constructing the budget if the project is controversial. If a particular outcome would be preferred by project managers or interest groups, the geomorphologists' professional integrity and reputation rest on their ability to maintain technical independence. Documen-

Table 16.4 Examples of options for sediment budget design. A particular sediment budget would be characterized by one or more options for each numbered attribute

1. *Purpose of budget*:

Explain landform
Explain change or impact
Describe effect of activity
Describe effect of event
Prioritize, plan remediation
Describe system
Compare systems
Predict system response

2. *Focal issue*:

Land-use activity
Land-use effects
Background
Particular event
Particular impact

3. *Form of results*:

Absolute amounts
Relative amounts
Description of interactions
Locations
Timing of response

4. *Spatial organization*:

Distributed by sites
Generalized by strata
Conceptual
Lumped
Hypothetical

5. *Temporal context*:

Reconstruct past
Describe present
Predict future

6. *Duration considered*:

Event-specific
Specified duration
Long-term average
Land-use activity
Synthetic average

7. *Precision*:

Qualitative
Order-of-magnitude
Precise

8. *Part of sediment regime*:

Weathering
Hillslope transport
Hillslope storage
Erosion
Delivery to channels
Channel storage
Channel transport
Sediment attrition
Sediment yield
Morphology

9. *Landscape element*:

Hillslopes
Catchment
Specific landform
Land-use activity site
Channel reach
Channel system
Particular process
Particular land-use sites
Administrative unit

10. *Material*:

All
Non-dissolved
Suspended sediment
Bedload
Sand
Gravel
Organic material

11. *Method*:

Modeling
Existing evidence
Inference
Analogy
Historical records
Air photos
Remote sensing
Stratigraphic analysis
Monitoring

Table 16.5 Questions useful for guiding design of sediment budgets

Technical questions	
1	What is the overall goal of the study or project of which the sediment budget is to be a part?
2	What kinds of decisions or conclusions are expected to follow from the study's results?
3	What information is needed to make those decisions or conclusions?
4	Are approaches other than sediment budgeting capable of providing that information?
5	What is the minimum level of precision needed to make the decisions or conclusions?
6	What is the minimum portion of the sediment regime that must be understood to make the decisions or conclusions?
7	To what area must the results apply?
8	To what period must the understanding apply?
Logistical questions	
9	How much time is available for the study?
10	How much funding is available for the study?

tation and procedures should be maintained at the highest standard in such cases, and rigorous, independent technical review should be sought and heeded. Such input is useful both during the initial steps of designing the investigation and after results are produced.

Once the primary goal is identified, it is useful to specify how sediment budget results would contribute to meeting that goal. This can be done by first identifying the kinds of conclusions or decisions to be made once results are available and then evaluating how different kinds of results might influence those decisions or conclusions. For example, if the ultimate goal of a project is to design effective measures to reduce turbidity in a trout stream, it will be necessary to decide which sediment sources to control and how to control them. For this application, sediment budget results might be used simply to identify and prioritize controllable sources of fine sediment. However, the budget could also be constructed to aid design of sediment control measures by evaluating the importance of various influences on those process rates and describing the distribution of processes throughout the catchment. The strategy used for budget construction would differ according to which of these applications is selected.

Necessary and Sufficient Precision

An evaluation of how the sediment budget results are to be used also leads to definition of the minimum level of precision required. If the result is intended to guide sediment control efforts, for example, a relative ranking of sediment sources based on order-of-magnitude estimates of input might be sufficient to achieve the purpose. In contrast, a study designed to describe the relation between sediment production and sediment yield as a function of catchment size would require more precise estimates. The necessary precision can be estimated by identifying the range in potential values for results that would not alter the decisions and conclusions to follow from the budget. If the range is wide, precision can be low.

Although many investigations would benefit from increased precision, for many other studies the attainable precision is higher than that actually needed to answer the relevant questions. Pursuit of unnecessary levels of precision in such cases represents an inappropriate allocation of effort and saps resources from other aspects of analysis where effort might be more usefully applied.

Components to be Analyzed

It is sometimes asserted that sediment budget construction is too complicated to be practical for most applications. This appearance of complexity may have arisen in part because some sediment budgets have been constructed for ill-defined purposes and so embrace greater complexity than the applications actually require. Most applications require exploration of only a portion of the overall sediment regime. Targeting of sediment sources for control, for example, requires assessment only of sediment input rates to channels, while evaluation of gravel-mining influences focuses instead on changes in sediment storage in and downstream of the affected reaches. If the intent of the budget is to determine the relative importance of a particular kind of source, it may be sufficient to evaluate the input rate from that source relative to the total sediment yield (e.g., Hovius *et al.* 1997).

Identification of the portion of the sediment regime requiring study is easiest once a conceptual model has been developed for the sediment system in the area. Such models have generally been in the form of flow charts (e.g., Dietrich and Dunne 1978, Reid and Dunne 1996) and tables (e.g., Kesel *et al.* 1992, Clapp *et al.* 2001), but underlying each of these is the continuity equation for sediment transport, which simply states that, for some time interval, input equals output plus the change in storage. Constructing the relevant continuity equation for a particular application is useful because it requires identification of relationships that need to be evaluated, discloses the implications of disregarding particular components of the budget, and identifies the information needed to balance the budget.

Different formulations of the equation are useful for different applications. Benda and Dunne (1997), for example, model the stochastic nature of sediment supply to reaches of channel throughout a network from landslides, debris flows, and soil creep. To do so, they use a version of the equation modified to apply to individual reaches of third or higher order at specific times:

$$Q_i(k,t) + I(k,t) - Q_o(k,t) = \Delta V(k,t)/\Delta t \quad (1)$$

The terms representing input (in m^3 $year^{-1}$) into the channel segment k during year t are $Q_i(k,t)$, the fluvial transport (suspended and bed load) from upstream, and $I(k,t)$, the sum of sediment supplied to the channel segment during the year by the processes illustrated in Figure 16.2. $Q_o(k,t)$ is the corresponding export (m^3 $year^{-1}$) from segment k during year t, and the final term represents the change in the volume of sediment (V, m^3) stored in segment k during year t (represented by Δt, year).

For applications involving other kinds of information, the equation can be modified to specify the information required. For example, Dunne *et al.* (1998) examined interactions between the Amazon River and its floodplain (Figure 16.3), so the equation used to organize the study separated the term describing storage into four parts, including deposition on bars within and adjacent to the channel (D_{bar}), diffuse overbank deposition (D_{ovrbk}), deposition in floodplain channels attached to the main channel (D_{fpc}), and deposition on the bed and banks ($A_c\rho_b\Delta z/\Delta t$, where A_c and Δz are, respectively, the area and

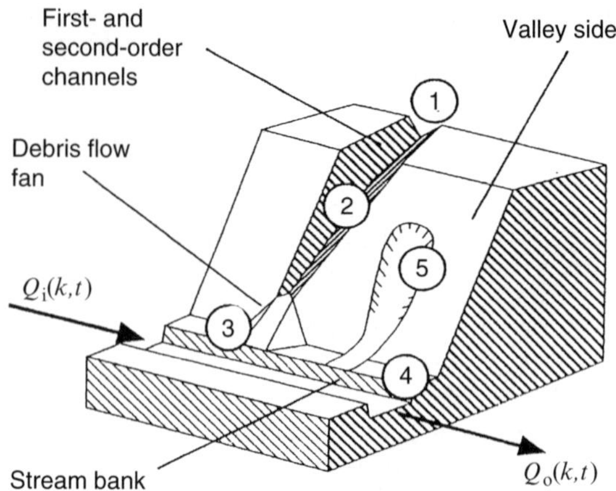

Figure 16.2 Conceptual model of the sediment budget of third- and higher-order channel segments in the Oregon Coast Range. Sediment input processes include: (1) shallow landsliding and debris flows in first- and second-order channels, (2) fluvial erosion and transport in first- and second-order channels, (3) bank erosion of debris flow fans and terraces, (4) soil creep along toeslopes of hillsides, and (5) landslides from streamside hollows. $Q_i(k,t)$ and $Q_o(k,t)$ represent the annual fluxes of sediment load into and out of the kth segment in year t. From Benda and Dunne (1997) (reproduced by permission of the American Geophysical Union)

average elevation change of the channel bed and banks in the reach, ρ_b is the bulk density of the bed material, and Δt is the time interval of the computation):

$$Q_u + \sum_i Q_{trib_i} + E_{bk} = Q_d + D_{bar} + D_{ovrbk} + D_{fpc} + A_c\rho_b\frac{\Delta z}{\Delta t} + \varepsilon \quad (2)$$

Q_u, Q_d, and Q_{trib} are, respectively, the annual fluxes of suspended and bedload sediment at the upstream and downstream ends of each channel reach and from the i tributaries entering the reach, E_{bk} is bank erosion, and ε is the error; each term has units of Mt $year^{-1}$. This equation, too, was formulated to apply to particular reaches, but in this case the results define the average annual balance of sediment transport of each grain size for each reach.

Once the underlying equation is defined, flowcharts are useful for organizing specific information about processes. Preliminary information about major erosion and transport processes is usually available for a study area or for similar settings. This information can be used to identify potential sediment inputs,

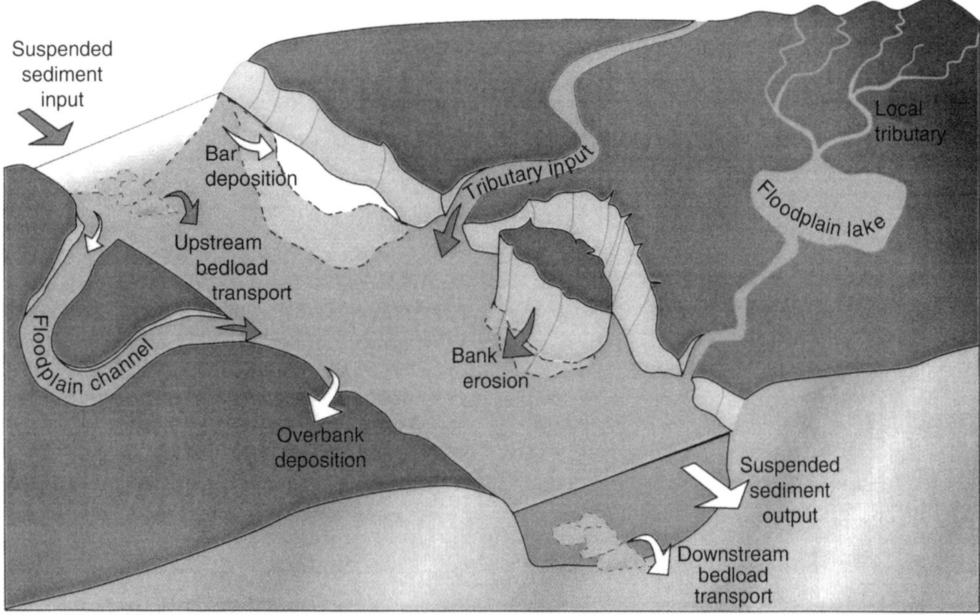

Figure 16.3 Components of the budget of channel-floodplain sediment exchanges for ~200-km-long reaches of the Amazon River, Brazil. Modified after Dunne *et al.* (1998) (reproduced by permission of the Geological Society of America)

outputs, and storage changes in the area and to diagram interactions between transport processes and storage elements, with primary focus on aspects of the sediment regime on which the study is to concentrate (e.g., Figure 16.4). If location is important, the flowchart can be organized to indicate spatial relationships (e.g., Figure 16.5). The conceptual model described by the flowchart then provides the framework for further analysis, always keeping in mind that the model is a hypothesis to be tested by field observations and analysis.

The preliminary flowchart can be used to identify which parts of the sediment regime need to be evaluated to meet the project's goals. If the analysis is intended to define the influence of a land-use activity on the sediment yield, the amount and kind of sediment produced by that activity would be evaluated relative to background and ambient sediment yields. In contrast, if localized channel aggradation is the focal concern, the amount of sediment contributed to the reach might be quantified and constraints on transport identified. If control of excess sedimentation is desired, a study might define the relative importance of sediment sources throughout the catchment. Sediment budgets can be constructed to describe the same sediment system from any of these perspectives, or from others relevant to other problems.

Spatial Scale of Analysis

The most useful spatial analysis strategy for a particular study depends on the kind of area to which results are to be applied. The relevant areas might be a particular catchment, a specific kind of landform found in a region, a particular channel reach, a hypothetical "typical" hillslope, or a politically defined district. Each of these areas requires a different strategy for spatial analysis.

If the budget is intended to explain conditions at a particular site, details of that site are critical to the problem. The location of tributary inputs in a channel reach, for example, may strongly influence the behavior of the reach. Similarly, for budgets developed to explore landform evolution, information concerning process rates must be distributed over that landform. Geographic information systems (GIS) are useful for constructing these spatially distributed sediment

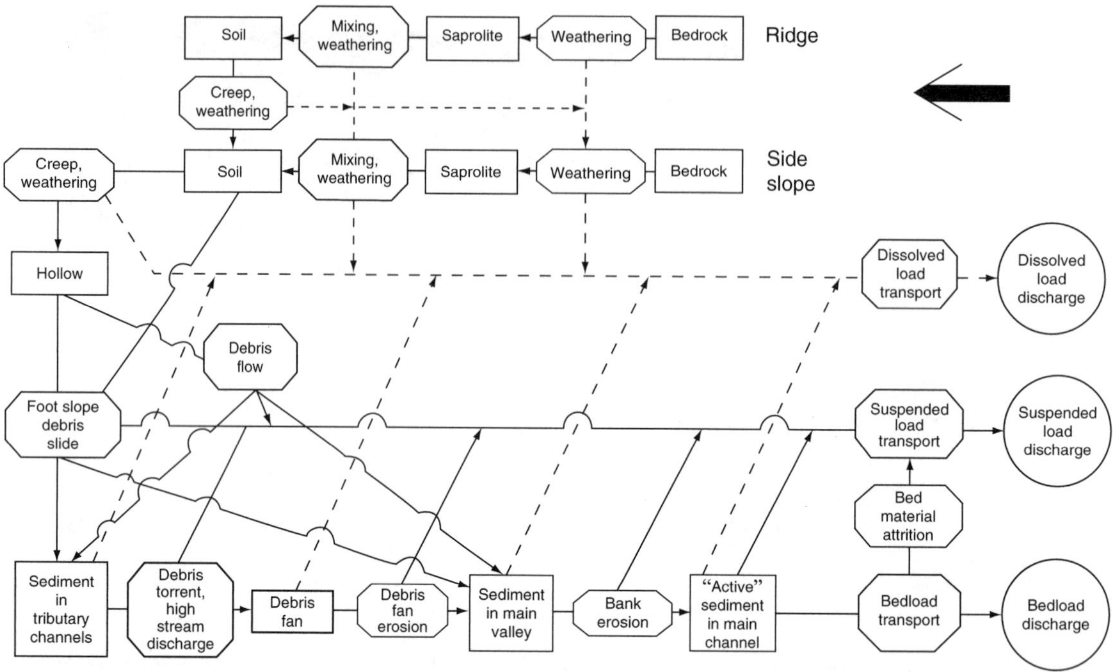

Figure 16.4 Conceptual model of the sediment budget of a small mountainous watershed in the Oregon Coast Range. Rectangles represent storage elements, octagons indicate transfer processes, and circles represent outputs. Solid lines indicate the transfer of sediment and dotted lines represent the migration of solutes. From Dietrich and Dunne (1978) (reproduced by permission of Borntraeger Publishers, http://www.schweizerart.de)

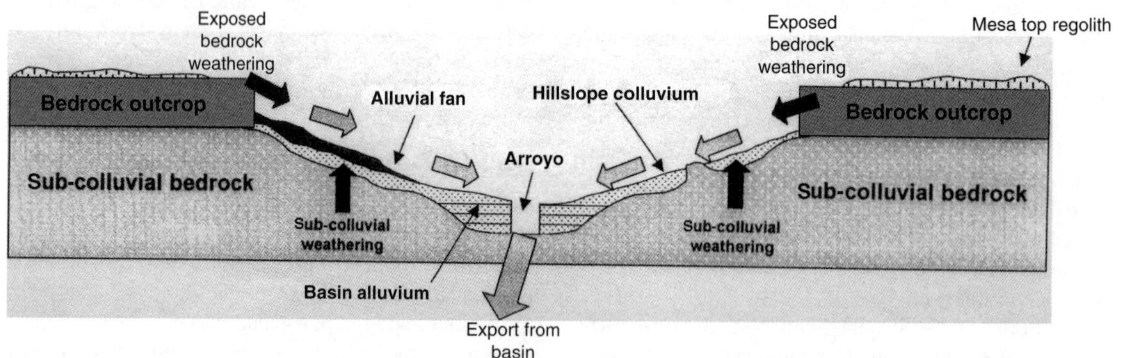

Figure 16.5 Schematic cross section of a small catchment in New Mexico showing the general flow paths of sediment from sources (weathering of bedrock outcrops and subcolluvial bedrock) to various temporary sediment "reservoirs" or "stores" (hillslope colluvium, alluvial fan, and valley-floor alluvium), and out of the basin. Black arrows represent mobilization by bedrock weathering. Gray arrows indicate sediment transport processes. The cross section is approximately 1 km across, and the relief is approximately 30 m. From Clapp *et al.* (2001) (reproduced by permission of Academic Press)

budgets. For many applications, this level of spatial specificity is unnecessary, and results can be presented as averages for particular land types, land uses, subcatchments, or entire catchments. Budgets can even be constructed for hypothetical settings, as is often useful when likely outcomes of different planning options are to be compared.

A preliminary evaluation of the likely spatial distribution of processes usually simplifies budget construction by allowing the study area to be "stratified" into areas that are likely to behave uniformly with respect to a particular process. This approach is useful for any scale of problem; even a sediment budget constructed for an individual gully relied on sampling of process rates on different site types within the gully (Reid 1989). Variables useful for stratification are generally those that control the rate or distribution of processes (Tables 16.2 and 16.3), and often include geologic substrate and vegetation type for hillslope processes and channel order and geologic substrate for channels. In practice, 3–20 stratification units for each process are usually sufficient to facilitate analysis without over-simplifying the problem, and a single stratification scheme is often applicable to multiple processes. Various methods have been used for stratification, ranging from visual delineations using aerial photographs to automated methods using GIS-based information or SPOT imagery (e.g., Giles 1998, Fernández *et al.* 1999). For most applications, statistically based sampling of conditions within stratification units is the most efficient approach to evaluating process rates and distribution; rarely are complete inventories necessary.

Stratification allows generalization of results across wider areas and estimates of values for particular parts of an area, and is useful for construction of budgets for hypothetical conditions. In each case, results are calculated according to the distribution of strata—which implicitly represents the distribution of controlling variables—in the area of interest.

Temporal Scale of Analysis

Appropriate temporal scales can be selected for sediment budgets by considering the intended applications and the timescales over which conditions change in the study area. A budget designed to examine the effects of land use on landsliding might evaluate landslide distribution after a single major storm on lands undergoing different uses. In contrast, a budget intended to estimate a long-term average would assess rates over a period long enough to either evaluate or average out year-to-year variations.

Budgets that evaluate changes in average sediment input relative to background conditions must consider two timescales, the first to assess long-term average natural rates, the second to provide an analogous estimate of current rates. Where current conditions are changing rapidly, as is the case where land-use patterns are shifting, definition of a "long-term average" for current conditions requires assessment of the hypothetical response of the current land-use pattern to the long-term average distribution of triggering events. If landslide rates are defined as a function of storm size, for example, landslide incidence can be evaluated for the distribution of storms expected over a 100-year period to calculate the average sediment input expected if current conditions were to be maintained for 100 years (Figure 16.1).

Comparison of current to background conditions requires estimates of process rates active under conditions no longer present. Where nearby catchments remain in a relatively pristine condition, sediment budgets for pristine and disturbed conditions can be compared directly. Comparisons are most useful for catchments of similar size, and the less-disturbed catchments tend to be small. Consequently, processes characteristic of downstream reaches often cannot be directly compared using this strategy.

More commonly, pristine examples are unavailable, and catchments or hillslopes undergoing different intensities of land use are examined instead. Analysis of trends along the gradient of land-use intensities can then allow inferences about predisturbance conditions. A gradient of hillslope conditions usually exists even where land use is uniformly distributed between catchments, so rates of hillslope processes usually can be compared even where comparison of downstream processes is not possible.

Past conditions and process rates in downstream channels usually must be evaluated using evidence left by those processes or from records and accounts of earlier conditions. Stratigraphic analysis of floodplain deposits can provide considerable information about disturbance-related changes in sediment regime and channel response (e.g., Lecce and Pavlowsky 2001). Where changes in land use are recent, interviews with long-term residents can provide valuable information about past conditions.

Budgets can also be designed to forecast future conditions. Because many erosion and sediment transport processes are strongly influenced by large, infrequent events, exact predictions usually are not possible. Instead, such budgets generally describe the likely outcome, given the expected distribution of events based on a probability analysis.

The nature of on-going changes also influences the temporal scale appropriate for a sediment budget. If a sediment system is recovering after a major event or adjusting to a land-use change, the budget generally would need to incorporate a broad enough temporal scale to evaluate the nature and progress of the system's response.

Selection of Analysis Methods

Examples of analysis methods are listed in Tables 16.2 and 16.3. The methods appropriate for a particular problem depend on the nature and context of the problem, but the choice is also influenced by logistical constraints. If answers are required quickly, analysis must depend largely on existing information, sequential aerial photographs, and field observations of evidence of past process rates. The depth of root exposure on datable vegetation might be measured to assess past surface erosion rates, for example, or isotope concentrations might be measured in sediment deposits to evaluate changes in deposition rate (Walling and He 1997, Panin *et al.* 2001, Matisoff *et al.* 2002a). If more time is available for the study, it may be useful to monitor process rates. Few studies employ only one method; different methods are used to evaluate different components of the budget or to provide multiple estimates for a single component.

Most sediment budget analyses use aerial photographs. Multiple photo sets spanning a 50-year period now exist for most locations, and some kinds of landscape changes can be documented by comparison of sequential sets. Landsliding rates can be estimated for periods spanned by aerial photographs by calculating the frequency of new scars, for example, and rates of migration can sometimes be measured for large channels. Some changes in channel form can be inferred from observed changes in riparian cover (Grant 1988). Aerial photographs are also useful for aiding landscape stratification and for planning fieldwork. Satellite imagery, now extending back for more than 30 years, can disclose alterations of land cover, changes in the position and form of large rivers, and broad patterns of variation across the landscape. Old maps and survey records, which can date from hundreds of years ago in Japan and Europe, can also indicate changes in land use and in channel location and character. Planning departments for cities, counties, and land-management agencies may have GIS coverage for some attributes in the area.

Useful information can also be provided by water quality reports, bridge surveys, flood zoning reports, reservoir surveys, and stream gauging records. Local experts can also be identified and interviewed; these people often know of relevant information that might be otherwise overlooked and usually can identify individuals who have observed recent changes. Affected landowners, for example, usually can provide detailed descriptions of recent channel changes.

Information is also useful from similar settings in other areas. In some cases, measured process rates can be transferred directly to other areas of similar character, but even where rates are not transferable, information on the nature of and interactions between processes may be transferred. When analogy is used to estimate process rates or interpret process interactions, similarities and differences between the study area and the measurement site need to be carefully evaluated.

Published equations and models can be used to evaluate some process rates, but use of any model or equation requires that the underlying assumptions be valid for the area in question and that appropriate data be available. No model or equation can be assumed valid for a particular application, so a model or equation should not be applied without understanding its assumptions and limitations and the conditions for which it was constructed. Results should be tested against those of other analytical methods.

Fieldwork is essential for refining the conceptual framework originally established for the budget and for checking aerial photographic interpretations. Evaluation of most chronic sediment sources requires fieldwork, and fieldwork often reveals unexpected measurement opportunities. Fieldwork also allows interviews with local observers and experts at the sites of interest; general recollections can become very specific in the presence of identifiable landmarks. Fieldwork is most useful if it is approached both with a prioritized list of tasks to be accomplished and with the willingness to abandon that list if more effective opportunities are found for answering the important questions. If possible, fieldwork should be scheduled for periods when important processes are likely to be active. Dry-season fieldwork, for example, is rarely useful for evaluating the distribution and significance of overland flow and sheet erosion.

Monitoring is sometimes useful during budget construction. Long-term average process rates can be defined through monitoring either if the monitoring duration is long enough to account for temporal variations in rate (e.g., Trimble 1999) or if monitoring produces a relationship between process rate and driving variables that allows the long-term rate to be calculated from a known distribution of driving variables (Reid and Dunne 1984, Clayton and Megahan 1986, Reid 1998). Short-term monitoring also can be useful for testing event-based modeling outputs. Comparison of modeled and monitored results for the range of sampled events indicates the level of confidence that can be placed on modeled results for unsampled events. Short-term monitoring can also reveal differences in process rates between particular site-types or treatments. For any of these applications, enough sites should be monitored to provide adequate confidence that results are characteristic of the relevant site-type during the monitoring period. Statistical analysis of preliminary results can identify the necessary sample size.

Integrating the Results

Sediment budgets commonly incorporate disparate kinds of information, and each information source usually represents a different temporal or spatial scale and a different level of data quality. The overall budget must reconcile these differences to produce an internally consistent, interpretable result.

Particular care must be taken to avoid mismatching timescales within a budget. Sediment budget results cannot be compared or components of a single budget combined if they represent different time periods. For example, sediment budgets commonly incorporate monitoring data, modeling results, and retrospective rate estimates. If a budget is to be checked by comparing results to 2 years of sediment yield measurements, each kind of information would need to be evaluated in such a way that results apply to that 2-year period. Such a comparison would require that travel times for sediment also be evaluated.

Differences in spatial analysis scales are usually accounted for by stratification. A single budget, for example, might include an air-photo inventory of road-related landslides throughout a catchment and modeled sheet erosion rates from road surfaces on two soil types. The sediment input for the entire catchment would be calculated in a different way for each source. Overall rates for both sources would vary through time as the road system developed, so inputs would be calculated per unit length of road. The average annual landslide delivery would be calculated as the total landslide delivery divided by the road-kilometer-years present during the period for which aerial photos are available. Sheet erosion would be calculated by applying the modeled rates for each soil type to the road-kilometer-years present for that soil type during the period of air-photo coverage. Results could then be combined to estimate either the total input from these sources over the period of air-photo coverage or the combined average rate per unit length of road per year.

In general, inventory data can be used directly after suitable spatial and temporal averaging, while information characterizing particular land strata or site types is applied according to the distribution of those site types. Data that are randomly sampled without regard to site type characterize the area as a whole and cannot be used to characterize portions of the sample area unless the random sampling disclosed relationships between rates and controlling variables. This pattern is also true for sampling through time. A process rate evaluated as a long-term average cannot be assumed to apply to a particular interval within the analysis period.

Auditing the Sediment Budget

An answer is not useful if it is not possible to determine whether it is likely to be true. Most sediment budgets represent a complex mix of calculations, mapping, measurements, and qualitative inferences, so standard methods of error analysis are rarely applicable. Instead, results usually are tested by comparing estimated to measured sediment yields, assessing the reliability of each of the methods used, or carrying out sensitivity analyses. The effectiveness of each approach depends on the kind of error present.

The most serious errors generally result from overlooking important processes, and this can be avoided only through careful fieldwork. It is useful to begin with a complete list of potential processes, identify the evidence needed to demonstrate their presence, and determine whether such evidence is present. Technical review sometimes can identify missing components if reviewers are familiar with similar areas. Occasionally, comparison of the summed components of the budget with a known output reveals an imbalance, but uncertainties in components are often large enough that a shortfall cannot be correctly diagnosed. Assessment of the reliability of component methods and sensitivity analysis are ineffective for addressing this problem.

Important errors also occur when decisions are founded on budget results that are mistakenly assumed to be precise and accurate. In many cases, a sensitivity analysis would have revealed the uncertainty in the budget, and decisions could have been tempered to reflect that uncertainty or further work done to decrease it.

Large errors in individual budget components have occurred when modeling results were relied on without field testing or when short-term rates were assumed to represent long-term averages. Each of these methods is inherently unreliable and can be identified through technical reviews. Where such methods are considered necessary, it is important that the associated uncertainty be evaluated and reported.

Problems have also arisen when a difficult-to-evaluate component was estimated by subtracting the other components from a measured total. This approach implicitly assumes that all components have been identified and that the cumulative error in the sum is small enough that the difference between the sum and the total is meaningful (Kondolf and Matthews 1991). If such an approach is used, a sensitivity analysis should be carried out to identify the potential error in the result, and the presence or absence of all potential budget components should be carefully verified.

Another problem is beginning to appear with increasing frequency as budgeting is more widely applied: profound errors can be introduced when those preparing the budget have insufficient understanding of geomorphological processes. In a recent example, half the estimated bank erosion along major tributaries in a severely impacted catchment was arbitrarily assumed to be "natural", and recently accelerated aggradation on banks and floodplains was simply ignored because rates varied between sites and because a hypothesized future increase in bank erosion was assumed to compensate for the increased aggradation. Problems arising from lack of expertise can be addressed through a careful technical review, although the problems would ideally be circumvented earlier by ensuring that those constructing the budget are qualified to do such work.

Sediment budgeting is increasingly used in support of land-management planning, and in this context there usually is considerable financial interest in the results. Technical review from independent experts is particularly important in this case. The credibility of the budget can then be evaluated on the basis of both the content of the reviews and how the reviews are received: if those preparing the budget fail to address issues raised by review, the budget is clearly inadequate.

Although no single approach to testing budget results can ensure that the result is accurate, each is useful. Where estimated and measured sediment yields agree, it is likely that the major components are not severely over- or under-estimated, although compensating errors can occur. Methods of "fingerprinting" deposited or transported sediments to identify their provenance (Collins *et al.* 1998, Hill *et al.* 1998) can be used to test portions of the overall budget. For any such test to be valid, clearly the estimates of sediment yield must be completely independent of analysis of budget components.

Even technically valid sediment budgets can be unsuccessful if the question addressed by the budget is not relevant to the underlying problem. Central to formulation of a useful question is development of a strong understanding of what the problem to be addressed actually is. For problems associated with land-use activities, identified technical problems are often mere symptoms of underlying social, political, or economic problems (Rossi 1998), and technical solutions that do not take into account the underlying causes will not be relevant or workable.

Assessing Uncertainty

The reliability of specific methods used in budget construction usually can be assessed from the performance of a method at other sites or from other knowledge of process rates. In some cases, reliability can be expressed as a confidence interval, while in others, only a maximum likely error can be estimated. If multiple methods are used to estimate the same budget component, discrepancies between methods indicate the maximum potential accuracy for the suite of methods used.

In some cases, formal error propagation analysis is possible for parts of a budget. The sediment budget for the Amazon River (Figure 16.3; Dunne *et al.* 1998), for example, was structured in such a way as to allow such analysis. Equation (2) was first simplified to

$$Q_u + \sum_i Q_{trib_i} - Q_d = \frac{\Delta V}{\Delta t} + \varepsilon \qquad (3)$$

where Q_u, Q_d, and Q_{trib} are, respectively, the annual fluxes of suspended and bedload sediment at the upstream and downstream ends of each channel reach and from the i tributaries entering the reach, and $\Delta V/\Delta t$ represents the rate of change of the total volume of sediment in storage in the reach. Sediment rating curves and flow duration curves available for each station on the main channel and each tributary were then used to analyze error propagation for the fluvial transport terms, allowing the uncertainties in estimating $\Delta V/\Delta t$ to be evaluated. The standard errors of the $\Delta V/\Delta t$ terms for individual reaches were significantly different from zero for the sand fraction in many reaches where the geomorphic and hydrologic setting suggested that net erosion or deposition would occur. This was not generally the case for the larger silt–clay fraction, although subtle trends in the net storage of this fraction did correlate with the same geomorphic and hydrologic patterns. Also, the standard error of the storage estimate for the entire 2000-km-long floodplain reach (200 Mt year^{-1}) was significantly different from zero for both size fractions, and it agreed approximately with the storage of 500 Mt year^{-1} estimated by quantifying each term in Equation (2). However, the paucity of information available for specific storage fluxes prevented estimation of uncertainties for terms describing individual exchanges between the channel and the floodplain.

The various kinds of information used to construct a sediment budget ordinarily incorporate different kinds and levels of uncertainty, so a standard calculation of uncertainty usually is not possible for the overall result. Instead, the sensitivity of the result to likely levels of uncertainty in the budget components can be assessed by recalculating the result for their estimated ranges of uncertainty. For some components the uncertainty will be represented by a 95% confidence interval; for others it may reflect a maximum likely error or complete removal of the component from consideration. Such calculations can indicate which components of the budget require the most careful analysis.

It is often useful to distinguish between a result and the conclusions based on that result. For example, a result might be a tabulation of sediment input by process, while the conclusion drawn from the result might be that road-surface erosion is the most useful target for sediment control. It is then possible to identify how much of a change in the result would be necessary to cause modification of the conclusion. The range in the result over which the conclusion is not changed defines the operationally significant tolerance interval around the result. If the calculations leading to the result are routine enough to be amenable to Monte Carlo simulation, a method similar to that of generalized likelihood uncertainty estimation (Beven and Freer 2001) can be used to estimate the probability that the actual result supports the conclusion drawn (Reid and Page 2002, Figure 16.6).

16.4 EXAMPLES

Because of the wide variety of methods and strategies available for constructing sediment budgets, it is useful to examine several examples to illustrate how particular options were selected to address specific questions (Table 16.6). The first example represents a reconnaissance-level, order-of-magnitude budget, while the second incorporates more detailed analysis.

Evaluating Sediment Production from a Hurricane in Hawaii

Hurricane Iniki hit the 1325-km^2 island of Kauai, northernmost of the major Hawaiian Islands, in September 1992. Flooding has been a problem on the island in the past, and officials feared that changes caused by the hurricane would increase future flood danger. They needed to know how and where flood hazard was affected and how it could be reduced. One aspect of concern was decreased channel capacity due to aggradation from Iniki-related erosion. A sediment budget study was undertaken to determine whether significant aggradation was likely to identify the endangered areas and to plan hazard reduction measures (Reid and Smith 1992).

The impacts to be evaluated were those of the hurricane, so the budget had to be event-based (Table 16.7). Results did not need to be precise; comparison of the order-of-magnitude of sediment inputs to "normal" values would be sufficient to determine whether a problem is likely. Results needed to be spatially distributed because sites at risk had to be identified. However, vulnerable communities and structures are concentrated at the mouths of catchments, so spatial resolution by the 18 major catchments was adequate, and calculations within each catchment could be spatially generalized.

The relevant standard of comparison for this application ordinarily would be the volume of sediment that the rivers can remove without undergoing significant morphological change. If Iniki contributed more than this volume, channels might flood because of aggradation, while a lower sediment input would not cause aggradation. Aggradation was not a major problem before Iniki, and rainstorms with recurrence intervals of 100 years or less have triggered significant landsliding and produced sediment loads of at least the same order as an average year's sediment yield. We thus assumed that downstream aggradation would not occur if the hurricane contributed less than an average year's sediment load to a river. The standard of comparison then became the average annual sediment input in years without hurricanes. This standard is not well defined; no measurements exist of sediment yields on Kauai. Various estimates suggest that sediment yields range between 300 and 3000 t km^{-2} year^{-1}, and the distribution of old landslide scars suggests that years with intense storms have produced considerably higher yields.

Landslides were mapped by comparing 1:12 000 color infrared aerial photographs taken before and after the storm. Most landslides displaced only the

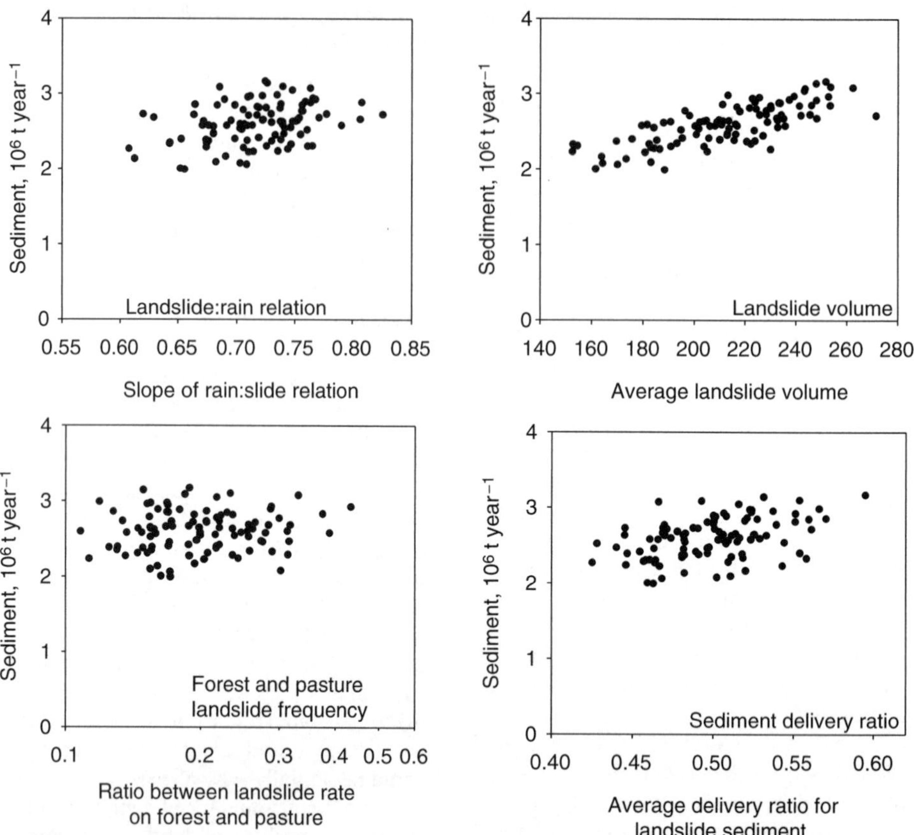

Figure 16.6 Distribution of estimates of sediment yield from shallow landslides in the Waipaoa catchment, North Island, New Zealand, using Monte Carlo simulation to incorporate uncertainty in 32 variables. Errors are selected randomly according to their assumed distributions. Results are displayed as functions of four variables from the Te Arai land system. Scatter in the y-dimension represents the combined influence of errors in the 32 variables; the randomly generated values for each of four variables among the 32 are plotted along the x-axes. The analysis indicates that results are most sensitive to errors in landslide volume, and allows estimation of the confidence interval for the calculated result. Modified after Reid and Page (2002)

soil profile, and their average depth was estimated from average soil depths. The hurricane was relatively dry, so excess sheetwash erosion could have occurred only in areas bared by the hurricane, which were restricted to landslide scars. Trees blown down into streams carried sediment with their roots, and these were mapped from a helicopter in streams wider than 5 m. The average frequency of blown-down trees was then extrapolated to smaller drainages. The volume of sediment carried by each rootwad was estimated from field observations. Sheetwash erosion rates, depths of soil removed by landslides, and rootwad volumes were represented by likely maximum values so that results would represent the maximum potential input.

The orders-of-magnitude of the estimated storm inputs were then compared to expected average annual inputs (Table 16.7). Sediment inputs from Iniki were found to be potentially significant only in the two watersheds that contained large debris flows, but even there the storm-related sediment input was of the same order as an average year's input and so not likely to cause a problem. On this basis, it was recommended that channel cross sections be monitored periodically at potentially vulnerable locations, but that major mitigation efforts for sediment were not necessary. In all, approximately 10 h of helicopter time, one day of fieldwork, and a week of office work were used to construct the budget.

Table 16.6 Examples of strategies selected for two sediment budgeting applications. See Table 16.4 for other options

Sediment budget attribute	Hurricane Iniki budget (Reid and Smith 1992)	Clearwater road budget (Reid *et al.* 1981)
1 Purpose of budget	Prioritize, plan remediation	Prioritize, plan remediation
2 Focal issue	Particular event and impact	Particular land-use activity and impact
3 Form of results	Relative amounts	Absolute amounts
4 Spatial organization	Distributed by catchments	Hypothetical
5 Temporal context	Describe present	Describe present
6 Duration considered	Event-specific	Synthetic average
7 Precision	Order-of-magnitude	Precise
8 Part of regime	Erosion	Delivery to channels
9 Landscape element	Catchment	Land-use activity sites
10 Material	Non-dissolved	Non-dissolved
11 Method	Modeling, Air photos, Existing evidence	Monitoring, Air photos, Existing evidence

Table 16.7 An order-of-magnitude sediment budget for sediment contributed by Hurricane Iniki to catchments and hydrologic zones on the island of Kauai, Hawaii (adapted from Reid and Smith 1992). Expected annual sediment inputs are on the order of 1000 t km^{-2} year^{-1}. Symbols: — indicates <1 t km^{-2}; +, 1–10 t km^{-2}; ++, 10–100 t km^{-2}; + + +, 100–1000 t km^{-2}

Watershed or zone	Increased sediment input from hurricane			
	Sheet erosion	Landslides	Uprooting	**Total**
1 Wainiha	++	+++	+	+++
2 Lumahai	++	+++	+	+++
3 Waioli	—	—	+	+
4 Hanalei	—	—	+	+
5 Kalihiwai	—	—	+	+
6 Kilauea	—	—	+	+
7 Anahola	—	+	+	++
8 Kapaa	—	+	+	+
9 Wailua	—	+	+	++
10 Hanamaulu	—	—	+	+
11 Huleia	—	—	+	+
12 Waikomo	—	—	+	+
13 Lawai	—	—	+	+
14 Wahiawa	—	—	+	+
15 Hanapepe	++	++	+	++
16 Canyon zone	—	—	+	+
17 Waimea	+	++	+	++
18 Na Pali zone	++	++	+	++

Prioritizing Erosion Control on Roads in the Olympic Mountains, Washington, USA

The 375-km^2 Clearwater watershed, located in the Olympic Mountains of Washington State, is intensively managed for timber production by the Washington Department of Natural Resources. The Clearwater River is an important producer of salmon, and earlier work suggested that the presence of roads is associated with impairment of spawning habitat by intrusion of fine-grained sediment (Cederholm *et al.* 1981). Department staff needed to know the most important sources of road-related sediment to select effective erosion-control measures, so a sediment budget was constructed to evaluate the relative importance of various road-related sediment sources in the area (Reid *et al.* 1981). Work on the budget required less than one person-year distributed over a 2-year period.

The only portion of the sediment regime that required analysis was sediment production to streams. The budget focused on fine-grained sediment, which had been identified as the major problem, and sources not related to roads could be excluded. Results could be in the form of long-term average inputs and so did not need to be related to particular sites or time periods. The budget could therefore be spatially generalized. Because relative values were the major concern, only a moderate level of precision was needed.

Road-related landslide, sidecast erosion, gully, and debris flow rates were evaluated for two watersheds using road construction records and three sequential sets of aerial photographs. Delivery ratios for these sources were assessed by measuring the volumes of sediment deposits at a selection of field sites, and secondary erosion on landslide scars was estimated from root exposure and erosion-pin measurements. Road-surface erosion was evaluated by sampling effluent from 10 culverts during 17 storms of various sizes and defining relations between sediment concentration and discharge for different intensities of road use. Similar measurements on a paved road segment allowed isolation of roadcut and ditch erosion rates, and these were also estimated from erosion-pin measurements and by measuring root exposure on roadcut vegetation. Only the erosion pins required a lengthy monitoring period, and they turned out to be redundant because root exposure measurements provided analogous data over a much longer effective sampling period.

Culvert hydrographs were reconstructed for unsampled storms using unit hydrographs. The sediment rating curves then allowed estimation of the sediment yield for each storm over an 11-year period, and an average annual yield was calculated from the storm yields (Reid and Dunne 1984). Delivery ratios for road-surface, ditch, and roadcut sediment were estimated by determining the proportion of culverts that contribute flow directly to streams in the area. Sediment production rates from landslides, sheetwash, and roadcut erosion were then calculated as a rate per kilometer for an average distribution of road types in a hypothetical 20-km^2 watershed, taking into account the distribution of road types and use intensities present in the area. Expected yields for any specific watershed could be calculated using the road distribution actually present. Because relations were quantified between road-surface sediment yields, storm intensities, and traffic levels, average yields can be estimated for particular years, and future yields can be estimated from projected use levels.

Results clearly indicate the sediment sources in most need of control: road-surface erosion and landsliding each produce 10 times as much sediment as other sources (Table 16.8), while roadcut erosion is

Table 16.8 Road-related production of sediment finer than 2 mm in a hypothetical 20 km^2 watershed in the Clearwater Basin (after Reid *et al.* 1981). Road density is 2.5 km km^{-2} and includes an average distribution of road-use intensities: 6% heavily used, 5% moderately used, 39% lightly used, and 50% abandoned. Heavily used roads fall into the "temporary non-use" category at night and on weekends

Source	Fine sediment production ($t\,km^{-2}\,year^{-1}$)
Landslides	40
Debris flows[a]	6.6
Gullies	0.4
Sidecast erosion	2.8
Secondary surface erosion on slide scars	12
Rills on landslide scars	3.2
Roadcut erosion[b]	(4.0)[b]
Road surface and roadcut	
Heavy use	36
Temporary non-use	5.2
Moderate use	5.2
Light use	3.7
Abandoned	0.6
Total	116

[a] Only valley-wall erosion is listed here; the triggering landslide is included in the landslide category.
[b] Roadcut erosion is included in the values for "road surface and roadcut" but is listed separately here to allow comparison.

relatively unimportant. Results also demonstrate the importance of road use in generating sediment and suggest that curtailment of use during wet weather could significantly decrease sediment input.

16.5 CONCLUSIONS

Recent studies demonstrate the utility of the sediment budgeting approach in addressing a wide range of theoretical and applied problems in fluvial geomorphology (Table 16.1). The most efficiently constructed and effective budgets have been those designed to incorporate the kind and precision of information necessary and sufficient to address the questions posed. Only with a carefully defined focus can the appropriate options for budget construction be selected from the wide variety available, and only with carefully defined objectives can the reliability of the resulting budget be assessed.

REFERENCES

Abernethy, B. and Rutherfurd, I.D. 1998. Where along a river's length will vegetation most effectively stabilise stream banks? *Geomorphology* 23(1): 55–76.

Allison, M.A., Kuehl, S.A., Martin, T.C. and Hassan, A. 1998. Importance of floodplain sedimentation for river sediment budgets and terrigenous input to the ocean: insights from the Brahmaputra-Jamuna River. *Geology* 26: 175–178.

André, M.-F. 1997. Holocene rockwall retreat in Svalbard: a triple-rate evolution. *Earth Surface Processes and Landforms* 22(5): 423–440.

Archibold, O.W., de Boer, D.H. and Delanoy, L. 1996. A device for measuring gully headwall morphology. *Earth Surface Processes and Landforms* 21(11): 1001–1005.

Asselman, N.E.M. 2000. Fitting and interpretation of sediment rating curves. *Journal of Hydrology* 234(3/4): 228–248.

Asselman, N.E.M. and Middelkoop, H. 1998. Temporal variability of contemporary floodplain sedimentation in the Rhine-Meuse delta, The Netherlands. *Earth Surface Processes and Landforms* 23(7): 595–609.

Auzet, A.-V. and Ambroise, B. 1996. Soil creep dynamics, soil moisture and temperature conditions on a forested slope in the granitic Vosges Mountains, France. *Earth Surface Processes and Landforms* 21(6): 531–542.

Barker, R., Dixon, L. and Hooke, J. 1997. Use of terrestrial photogrammetry for monitoring and measuring bank erosion. *Earth Surface Processes and Landforms* 22(13): 1217–1227.

Benda, L.E. 1990. The influence of debris flows on channels and valley floors in the Oregon Coast Range, USA. *Earth Surface Processes and Landforms* 15(5): 457–466.

Benda, L. and Dunne, T. 1987. Sediment routing by debris flows. In: Beschta, R.L., Blinn, T., Grant, G.E., Ice, G.G. and Swanson, F.J., eds., *Erosion and Sedimentation in the Pacific Rim*, International Association of Hydrological Sciences Publication 165, pp. 213–223.

Benda, L. and Dunne, T. 1997. Stochastic forcing of sediment storage and routing in channel networks. *Water Resources Research* 33: 2865–2880.

Beuselinck, L., Steegen, A., Govers, G., Nachtergaele, J., Takken, I. and Poesen, J. 2000. Characteristics of sediment deposits formed by intense rainfall events in small catchments in the Belgian Loam Belt. *Geomorphology* 32(1/2): 69–82.

Beven, K. and Freer, J. 2001. Equifinality, data assimilation, and uncertainty estimation in mechanistic modelling of complex environmental systems using the GLUE methodology. *Journal of Hydrology* 249(1–4): 11–29.

Booth, D.B. 1990. Stream-channel incision following drainage-basin urbanization. *Water Resources Bulletin* 26: 407–417.

Borówka, M., II. and Rotnicki, K. 2001. Budget of the eolian sand transport on the sandy barrier beach (a case study of the Łeba Barrier, southern Baltic coast, Poland). *Zeitschrift für Geomorphologie N.F.* 45(1): 55–79.

Bravard, J.-P., Kondolf, G.M. and Piégay, H. 1999. Environmental and societal effects of channel incision and remedial strategies. In: Darby, S.E. and Simon, A., eds., *Incised River Channels*, New York: John Wiley and Sons, pp. 303–341.

Brazier, R.E., Beven, K.J., Freer, J. and Rowan, J.S. 2000. Equifinality and uncertainty in physically based soil erosion models: application of the GLUE methodology to WEPP—the Water Erosion Prediction Project—for sites in the UK and USA. *Earth Surface Processes and Landforms* 25(8): 825–846.

Brizga, S.O. and Finlayson, B.L. 1994. Interactions between upland catchment and lowland rivers: an applied Australian case study. *Geomorphology* 9(3): 189–202.

Brooks, A.P. and Brierley, G.J. 1997. Geomorphic responses of lower Bega River to catchment disturbance, 1851–1926. *Geomorphology* 18(3/4): 291–304.

Brune, G.M. 1953. Trap efficiency of reservoirs. *Transactions of the American Geophysical Union* 34(3): 407–418.

Bull, W.B., King, J., Kong, F., Moutoux, T. and Phillips, W.M. 1994. Lichen dating of coseismic landslide hazards in alpine mountains. *Geomorphology* 10(1–4): 253–264.

Burbank, D.W. 1992. Causes of recent Himalayan uplift deduced from deposited patterns in the Ganges Basin. *Nature* 357(6380): 680–683.

Caine, N. and Swanson, F.J. 1989. Geomorphic coupling of hillslope and channel systems in two small mountain basins. *Zeitschrift für Geomorphologie* 33(2): 189–203.

Caine, N. and Thurman, E.M. 1990. Temporal and spatial variations in the solute content of an alpine stream, Colorado Front Range. *Geomorphology* 4(1): 55–72.

Carrara, P.E. and Carroll, T.R. 1979. The determination of erosion rates from exposed tree roots in the Piceance Basin, Colorado. *Earth Surface Processes* 4: 307–318.

Cederholm, C.J., Reid, L.M. and Salo, E.O. 1981. Cumulative effects of logging road sediment on salmonid populations in the Clearwater River, Jefferson County, Washington. In: *Proceedings of the Conference on Salmon Spawning Gravel: A Renewable Resource in the Pacific Northwest?* State of Washington Water Research Center Report 39, Washington: Pullman, pp. 38–74.

Cenderelli, D.A. and Kite, J.S. 1998. Geomorphic effects of large debris flows on channel morphology at North Fork Mountain, Eastern West Virginia. *Earth Surface Processes and Landforms* 23(1): 1–20.

Chandler, J.H. and Brunsden, D. 1995. Steady state behaviour of the Black Ven Mudslide: the application of archival analytical photogrammetry to studies of landform change. *Earth Surface Processes and Landforms* 20(3): 255–275.

Church, M., Ham, D., Hassan, M.A. and Slaymaker, O. 1999. Fluvial sediment yield in Canada: scale analysis. *Canadian Journal of Earth Sciences* 36: 1267–1280.

Church, M. and Ryder, J.M. 1972. Paraglacial sedimentation: a consideration of fluvial processes conditioned by glaciation. *Geological Society of America Bulletin* 83: 3059–3072.

Church, M. and Slaymaker, O. 1989. Disequilibrium of Holocene sediment yield in glaciated British Columbia. *Nature* 337(6206): 452–453.

Cisternas, M., Araneda, A., Martínez, P. and Pérez, S. 2001. Effects of historical land use on sediment yield from a lacustrine watershed in central Chile. *Earth Surface Processes and Landforms* 26(1): 63–76.

Clapp, E.M., Bierman, P.R., Nichols, K.K., Pavich, M. and Caffee, M. 2001. Rates of sediment supply to arroyos from upland erosion determined using in situ produced cosmogenic ^{10}Be and ^{26}Al. *Quaternary Research* 55: 235–245.

Clapp, E.M., Bierman, P.R., Schick, A.P., Lekach, J., Enzel, Y. and Cafee, M. 2000. Sediment yield exceeds sediment production in arid region drainage basins. *Geology* 28: 995–998.

Clarke, M.F., Williams, M.A.J. and Stokes, T. 1999. Soil creep: problems raised by a 23 year study in Australia. *Earth Surface Processes and Landforms* 24(2): 151–176.

Clayton, J.L. and Megahan, W.F. 1986. Erosional and chemical denudation rates in the southwestern Idaho Batholith. *Earth Surface Processes and Landforms* 11(4): 389–400.

Clayton, K.M. 1997. The rate of denudation of some British lowland landscapes. *Earth Surface Processes and Landforms* 22(8): 721–732.

Cohen, T.J. and Brierley, G.J. 2000. Channel instability in a forested catchment: a case study from Jones Creek, East Gippsland, Australia. *Geomorphology* 32(1/2): 109–128.

Collins, A.L., Walling, D.E. and Leeks, G.J.L. 1998. Use of composite fingerprints to determine the provenance of the contemporary suspended sediment load transported by rivers. *Earth Surface Processes and Landforms* 23(1): 31–52.

Collins, B.D. and Dunne, T. 1986. Erosion of tephra from the 1980 eruption of Mount St. Helens. *Geological Society of America Bulletin* 97: 896–905.

Collins, B.D. and Dunne, T. 1989. Gravel transport, gravel harvesting, and channel-bed degradation in rivers draining the southern Olympic Mountains, Washington, USA. *Environmental Geology and Water Science* 13(3): 213–224.

Collins, B.D. and Dunne, T. 1990. *Fluvial Geomorphology and River-gravel Mining*, Special Publication 98, Division of Mines and Geology, California Department of Conservation, 29 p.

Corominas, J. and Moya, J. 1999. Reconstructing recent landslide activity in relation to rainfall in the Llobregat River Basin, Eastern Pyrenees, Spain. *Geomorphology* 30 (1/2): 79–93.

Couper, P.R. and Maddock, I.P. 2001. Subaerial river bank erosion processes and their interaction with other bank erosion mechanisms on the River Arrow, Warwickshire, UK. *Earth Surface Processes and Landforms* 26(6): 631–646.

Davis, J., Bird, J., Finlayson, B. and Scott, R. 2000. The management of gravel extraction in alluvial rivers: a case study from the Avon River, southeastern Australia. *Physical Geography* 21(2): 133–154.

DeRose, R.C., Gomez, B., Marden, M. and Trustrum, N.A. 1998. Gully erosion in Mangatu Forest, New Zealand, estimated from digital elevation models. *Earth Surface Processes and Landforms* 23(11): 1045–1053.

Dietrich, W.E. and Dunne, T. 1978. Sediment budget for a small catchment in mountainous terrain. *Zeitschrift für Geomorphologie Supplement Band* 29: 191–206.

Dietrich, W.E., Dunne, T., Humphrey, N.F. and Reid, L.M. 1982. Construction of sediment budgets for drainage basins. In: Swanson, F.J., Janda, R.J., Dunne, T. and Swanston, D.N., eds., *Sediment Budgets and Routing in Forested Catchments*, Pacific Northwest Forest and Range Experiment Station General Technical Report PNW-141, US Department of Agriculture, Forest Service, pp. 5–23.

Dunne, T. 1977. Evaluation of erosion conditions and trends. In: Kunkle, S.H., ed., *Guidelines for Watershed Management*, U.N. Food and Agriculture Organization Conservation Guide 1, Rome, Italy, pp. 53–83.

Dunne, T. 1988. Geomorphological contributions to flood-control planning. In: Baker, V.R., Kochel, R.C. and Patton, P.C., eds., *Flood Geomorphology*, New York: Wiley and Sons, pp. 421–438.

Dunne, T. 2000. Critical data requirements for prediction of erosion and sedimentation in mountain drainage basins. *Journal of the American Water Works Association* 34: 795–808.

Dunne, T., Mertes, L.A.K., Meade, R.H., Richey, J.E. and Forsberg, B.R. 1998. Exchanges of sediment between the flood plain and the channel of the Amazon River in Brazil. *Geological Society of America Bulletin* 110: 450–467.

Dunne, T. and Ongweny, G.S.O. 1976. A new estimate of sediment yields in the upper Tana catchment. *The Kenya Geographer* 2: 109–126.

Erskine, W., McFadden, C. and Bishop, P. 1992. Alluvial cutoffs as indicators of former channel conditions. *Earth Surface Processes and Landforms* 17(1): 23–37.

Evans, M. 1997. Temporal and spatial representativeness of alpine sediment yields: Cascade Mountains, British Columbia. *Earth Surface Processes and Landforms* 22(3): 287–295.

Evans, M. and Church, M. 2000. A method for error analysis of sediment yields derived from estimates of lacustrine sediment accumulation. *Earth Surface Processes and Landforms* 25(11): 1257–1274.

Everest, F.H., Beschta, R.L., Scrivener, J.C., Koski, K.V., Sedell, J.R. and Cederholm, C.J. 1987. Fine sediment and salmonid production: a paradox. In: Salo, E.O. and Cundy, T.W., eds., *Streamside Management: Forestry and Fishery Interactions*, University of Washington Institute of Forest Resources Contribution No. 57, Seattle, Washington, pp. 98–142.

Fantucci, R. and Sorriso-Valvo, M. 1999. Dendrogeomorphological analysis of a slope near Lago, Calabria (Italy). *Geomorphology* 30(1/2): 165–174.

Fernández, C.I., del Castillo, T., Hamdouni, R.E. and Montero, J.C. 1999. Verification of landslide susceptibility mapping: a case study. *Earth Surface Processes and Landforms* 24(6): 537–544.

Flanagan, D.C. and Nearing, M.A. 1995. USDA WEPP: Technical documentation, USDA-ARS National Soil Erosion Laboratory Report No. 10, West Lafayette, Illinois.

Flemings, P.B. and Jordan, T.E. 1989. A synthetic stratigraphic model of foreland basin development. *Journal of Geophysical Research* 94: 3851–3866.

Gabet, E.J. 2000. Gopher bioturbation: field evidence for non-linear hillslope diffusion. *Earth Surface Processes and Landforms* 25(13): 1419–1428.

García-Ruiz, J.M., Lasanta, T. and Alberto, F. 1997. Soil erosion by piping in irrigated fields. *Geomorphology* 20(3/4): 269–278.

Garland, G. and Humphrey, B. 1992. Field measurement of discharge and sediment yield from a soil pipe in the Natal Drakensberg, South Africa. *Zeitschrift für Geomorphologie N.F.* 36: 15–23.

Gellis, A.C., Cheama, A. and Lalio, S.M. 2001. Developing a geomorphic approach for ranking watersheds for rehabilitation, Zuni Indian Reservation, New Mexico. *Geomorphology* 37(1/2): 105–134.

Gilbert, G.K. 1917. *Hydraulic-mining Debris in the Sierra Nevada*, USDI Geological Survey Professional Paper 105, 154 p.

Giles, P.T. 1998. Geomorphological signatures: classification of aggregated slope unit objects from digital elevation and remote sensing data. *Earth Surface Processes and Landforms* 23(7): 581–594.

Gomez, B., Eden, D.N., Hicks, D.M., Trustrum, N.A., Peacock, D.H. and Wilmshurst, J. 1999. Contribution of floodplain sequestration to the sediment budget of the Waipaoa River, New Zealand. In: Marriott, S.B. and Alexander, J., eds., *Floodplains: Interdisciplinary Approaches*, Geological Society of London Special Publication 165, pp. 69–88.

Gonzalez, M.A. 2001. Recent formation of arroyos in the Little Missouri Badlands of southwestern North Dakota. *Geomorphology* 38(1/2): 63–84.

Goodbred, S.L. and Kuehl, S.A. 1998. Floodplain processes in the Bengal Basin and the storage of Ganges-Brahmaputra River sediment: an accretionary study using ^{137}Cs and ^{210}Pb geochemistry. *Sedimentary Geology* 121: 239–258.

Goossens, D., Gross, J. and Spaan, W. 2001. Aeolian dust dynamics in agricultural land areas in Lower Saxony, Germany. *Earth Surface Processes and Landforms* 26(7): 701–720.

Graf, W.L. 1994. *Plutonium in the Rio Grande: Environmental Change and Contamination in the Nuclear Age*, New York: Oxford University Press, 329 p.

Grant, G. 1988. The RAPID technique: a new method for evaluating downstream effects of forest practices on riparian zones, USDA Forest Service Pacific Northwest Research Station General Technical Report PNW-220, 36 p.

Haggett, P.E. 1961. Land-use and sediment yield in an old plantation tract of the Serra do Mar, Brazil. *Geographical Journal* 127(1): 50–62.

Ham, D.G. and Church, M. 2000. Bed-material transport estimated from channel morphodynamics: Chilliwack River, British Columbia. *Earth Surface Processes and Landforms* 25(10): 1123–1142.

Harvey, A.M. 1992. Process interactions, temporal scales and the development of hillslope gully systems: Howgill Fells, northwest England. *Geomorphology* 5(3–5): 323–344.

Harvey, B.C. and Lisle, T.E. 1999. Scour of chinook salmon redds on suction dredge tailings. *North American Journal of Fisheries Management* 19(2): 613–617.

Hassan, M.A., Church, M. and Schick, A.P. 1991. Distance of movement of coarse particles in gravel bed streams. *Water Resources Research* 27(4): 503–512.

Heimsath, A.M., Dietrich, W.E., Nishiizumi, K., Finkel, R.C. 1997. The soil production function and landscape equilibrium. *Nature* 388: 358–361.

Heritage, G.L., Fuller, I.C., Charlton, M.E., Brewer, P.A. and Passmore, D.P. 1998. CDW photogrammetry of low relief fluvial features: accuracy and implications for reach-scale sediment budgeting. *Earth Surface Processes and Landforms* 23(13): 1219–1233.

Hill, B.R., DeCarlo, E.H., Fuller, C.C. and Wong, M.F. 1998. Using sediment "fingerprints" to assess sediment-budget errors, North Halawa Valley, Oahu, Hawaii, 1991–1992. *Earth Surface Processes and Landforms* 23(6): 493–508.

Hill, B.R., Fuller, C.C. and DeCarlo, E.H. 1997. Hillslope soil erosion estimated from aerosol concentrations, North Halawa Valley, Oahu, Hawaii. *Geomorphology* 20(1/2): 67–79.

Hill, H. 1896. Denudation as a factor of geologic time. *Transactions and Proceedings of the New Zealand Institute* 28(11th of New Series): 666–679.

Hodson, A., Tranter, M. and Vatne, G. 2000. Contemporary rates of chemical denudation and atmospheric CO_2 sequestration in glacier basins: an arctic perspective. *Earth Surface Processes and Landforms* 25(13): 1447–1472.

Hovius, N., Stark, C.P. and Allen, P.A. 1997. Sediment flux from a mountain belt derived by landslide mapping. *Geology* 25(3): 231–234.

Ibsen, M.-L. and Brunsden, D. 1996. The nature, use and problems of historical archives for the temporal occurrence of landslides, with specific reference to the south coast of Britain, Ventnor, Isle of Wight. *Geomorphology* 15(3/4): 241–258.

Jacobson, R.B. and Gran, K.B. 1999. Gravel sediment routing from widespread, low-intensity landscape disturbance, Current River Basin, Missouri. *Earth Surface Processes and Landforms* 24(10): 897–917.

James, L.A. 1997. Channel incision on the lower American River, California, from streamflow gage records. *Water Resources Research* 33(3): 485–490.

James, A. 1999. Time and the persistence of alluvium: river engineering, fluvial geomorphology, and mining sediment in California. *Geomorphology* 31(1–4): 265–290.

Janda, R.J. 1971. An evaluation of procedures used in computing chemical denudation rates. *Geological Society of America Bulletin* 82: 67–80.

Jäckli, H. 1957. Gegenwartsgeologie des bundnerischen Rheingebieters – ein Beitrag zur exogenen Dynamik Alpiner Gebirgslandschaften. Beiträge zur Geologie der Schweiz, Geotechnische Serie, Leiferung 36.

Jones, J.A.A. 1987. The effects of soil piping on contributing areas and erosion patterns. *Earth Surface Processes and Landforms* 12(3): 229–248.

Kelsey, H.M., Coghlan, M., Pitlick, J. and Best, D. 1995. *Geomorphic Analysis of Streamside Landslides in the Redwood Creek Basin, Northwestern California*, USDI Geological Survey Professional Paper 1454-J, 12 p.

Kesel, R.H., Yodis, E.G. and McCraw, D.J. 1992. An approximation of the sediment budget of the lower Mississippi River prior to major human modification. *Earth Surface Processes and Landforms* 17(7): 711–722.

Kirchner, J.W., Finkel, R.C., Riebe, C.S., Granger, D.E., Clayton, J.L., King, J.G. and Megahan, W.F. 2001. Mountain erosion over 10 yr, 10 k.y., and 10 m.y. time scales. *Geology* 29(7): 591–594.

Knighton, A.D. 1991. Channel bed adjustment along mine-affected rivers of northeast Tasmania. *Geomorphology* 4(3/4): 205–219.

Knighton, A.D. 1999. The gravel-sand transition in a disturbed catchment. *Geomorphology* 27(3/4): 325–341.

Kondolf, G.M. and Matthews, W.G.V. 1991. Unmeasured residuals in sediment budgets: a cautionary note. *Water Resources Research* 27: 2483–2486.

Kondolf, G.M. and Swanson, M.L. 1993. Channel adjustments to reservoir construction and gravel extraction along Stony Creek, California. *Environmental Geology* 21: 256–269.

Kuenen, P.H. 1956. Experimental abrasion of pebbles. 2. Rolling by current. *Journal of Geology* 64: 336–368.

Le Pera, E. and Sorriso-Valvo, M. 2000. Weathering, erosion and sediment composition in a high-gradient river, Calabria, Italy. *Earth Surface Processes and Landforms* 25(3): 277–292.

Lecce, S.A. and Pavlowsky, R.T. 2001. Use of mining-contaminated sediment tracers to investigate the timing and rates of historical flood plain sedimentation. *Geomorphology* 38(1/2): 85–108.

Lehre, A.K., Collins, B.D. and Dunne, T. 1983. Post-eruption sediment budget for the North Fork Toutle River drainage. June 1980–June 1981. *Zeitschrift für Geomorphologie Supplement Band* 46: 143–163.

Leopold, L.B., Emmett, W.W. and Myrick, R.M. 1966. *Channel and Hillslope Processes in a Semiarid Area, New Mexico*, USDI Geological Survey Professional Paper 352-G, pp. 193–253.

Lewis, W.M., Jr. and Grant, M.C. 1979. Relationships between stream discharge and yield of dissolved substances from a Colorado mountain watershed. *Soil Science* 128(6): 353–363.

Leys, J.F. and McTainsh, G.H. 1996. Sediment fluxes and particle grain-size characteristics of wind-eroded sediments in southeastern Australia. *Earth Surface Processes and Landforms* 21(7): 661–671.

Liébault, F. and Piégay, H. 2001. Assessment of channel changes due to long-term bedload supply decrease, Roubion River, France. *Geomorphology* 36(3/4): 167–186.

Lisle, T.E., Cui, Y., Parker, G., Pizzuto, J.E. and Dodd, A. M. 2001. The dominance of dispersion in the evolution of bed material waves in gravel-bed rivers. *Earth Surface Processes and Landforms* 26: 1409–1420.

Lisle, T.E. and Hilton, S. 1999. Fine bed material in pools of natural gravel bed channels. *Water Resources Research* 35 (4): 1291–1304.

Lloyd, S.D., Bishop, P. and Reinfelds, I. 1998. Shoreline erosion: a cautionary note on using small farm dams to determine catchment erosion rates. *Earth Surface Processes and Landforms* 23(10): 905–912.

Madej, M.A. and Ozaki, V. 1996. Channel response to sediment wave propagation and movement, Redwood Creek, California, USA. *Earth Surface Processes and Landforms* 21: 911–927.

Malmon, D.V. 2002. Sediment trajectories through a semi arid valley, Ph.D. Dissertation, University of California, Santa Barbara, California.

Malmon, D.V., Dunne, T. and Reneau, S.L. 2002. Predicting the fate of sediment and pollutants in river floodplains. *Environmental Science and Technology* 36(9): 2026–2032.

Maner, S.B. 1958. Factors influencing sediment delivery rates in the Red Hills physiographic area. *EOS American Geophysical Union Transactions* 39: 669–675.

Mantovani, F., Soeters, R. and Van Westen, C.J. 1996. Remote sensing techniques for landslide studies and hazard zonation in Europe. *Geomorphology* 15(3/4): 213–225.

Marron, D.C. 1992. Floodplain storage of mine tailings in the Belle Fourche River system: a sediment budget approach. *Earth Surface Processes and Landforms* 17(7): 675–685.

Marutani, T., Kasai, M., Reid, L.M. and Trustrum, N.A. 1999. Influence of storm-related sediment storage on the sediment delivery from tributary catchments in the Upper Waipaoa River, New Zealand. *Earth Surface Processes and Landforms* 24(10): 881–896.

Matisoff, G., Bonniwell, E.C. and Whiting, P.J. 2002a. Soil erosion and sediment sources in an Ohio watershed using beryllium-7, cesium-137, and lead-210. *Journal of Environmental Quality* 31(1): 54–61.

Matisoff, G., Bonniwell, E.C. and Whiting, P.J. 2002b. Radionuclides as indicators of sediment transport in agricultural watersheds that drain to Lake Erie. *Journal of Environmental Quality* 31(1): 62–72.

Matsuoka, N. and Sakai, H. 1999. Rockfall activity from an alpine cliff during thawing periods. *Geomorphology* 28(3/4): 309–328.

McLean, D.G. and Church, M. 1999. Sediment transport along lower Fraser River. 2. Estimates based on the long-term gravel budget. *Water Resources Research* 35 (8): 2549–2559.

McLean, D.G., Church, M. and Tassone, B. 1999. Sediment transport along lower Fraser River. 1. Measurements and hydraulic computations. *Water Resources Research* 35(8): 2533–2548.

Meade, R.H. 1982. Sources, sinks, and storage of river sediment in the Atlantic drainage of the United States. *Journal of Geology* 90: 235–252.

Meentemeyer, R.K., Vogler, J.B. and Butler, D.R. 1998. The geomorphic influences of burrowing beavers on streambanks, Bolin Creek, North Carolina. *Zeitschrift für Geomorphologie N.F.* 42(4): 453–468.

Megahan, W.F., Seyedbagheri, K.A. and Dodson, P.C. 1983. Long-term erosion on granitic roadcuts based on exposed tree roots. *Earth Surface Processes* 8(1): 19–28.

Megahan, W.F., Seyedbagheri, K.A., Mosko, T.L. and Ketcheson, G.L. 1986. Construction phase sediment budget for forest roads on granitic slopes in Idaho. In: Hadley, R., ed., *Drainage Basin Sediment Delivery*, International Association of Hydrological Sciences Publication 159, pp. 31–39.

Milliman, J.D. and Syvitski, J.P.M. 1992. Geomorphic/tectonic control of sediment discharge to the ocean: the importance of small mountainous rivers. *Journal of Geology* 100: 525–544.

Morgan, R.P.C., Quinton, J.N., Smith, R.E., Govers, G., Poesen, J.W.A., Auerswald, K., Chisci, G., Torri, D. and Styczen, M.E. 1998. The European Soil Erosion Model (EUROSEM): a dynamic approach for predicting sediment transport from fields and small catchments. *Earth Surface Processes and Landforms* 23(6): 527–544.

Mosley, M.P. 1981. The influence of organic debris on channel morphology and bedload transport in New Zealand forest stream. *Earth Surface Processes and Landforms* 6(6): 571–580.

Nachtergaele, J. and Poesen, J. 1999. Assessment of soil losses by ephemeral gully erosion using high-altitude (stereo) aerial photographs. *Earth Surface Processes and Landforms* 24(8): 693–706.

Nakamura, F., Maita, H. and Araya, T. 1995. Sediment routing analyses based on chronological changes in hillslope and riverbed morphologies. *Earth Surface Processes and Landforms* 20: 333–346.

Nakamura, F., Sudo, T., Kameyama, S. and Jitsu, M. 1997. Influences of channelization on discharge of suspended sediment and wetland vegetation in Kushiro Marsh, northern Japan. *Geomorphology* 18(3/4): 279–289.

Newcombe, C.P. and Jensen, J.O.T. 1996. Channel suspended sediment and fisheries: a synthesis for quantitative assessment of risk and impact. *North American Journal of Fisheries Management* 16(4): 693–727.

Nichol, S.L., Augustinus, P.C., Gregory, M.R., Creese, R. and Horrocks, M.H. 2000. Geomorphic and sedimentary evidence of human impact on the New Zealand coastal landscape. *Physical Geography* 21(2): 109–132.

Nolan, K.M. and Janda, R.J. 1995. *Movement and Sediment Yield of Two Earthflows, Northwestern California*, USDI Geological Survey Professional Paper 1454 F, 12 p.

Norman, S.A., Schaetzl, R.J. and Small, T.W. 1995. Effects of slope angle on mass movement by tree uprooting. *Geomorphology* 14(1): 19–27.

Page, M.J., Trustrum, N.A. and Dymond, J.R. 1994. Sediment budget to assess the geomorphic effect of a cyclonic storm, New Zealand. *Geomorphology* 9(3): 169–188.

Page, M.J. and Trustrum, N.A. 1997. A late Holocene lake sediment record of the erosion response to land use change in a steepland catchment, New Zealand. *Zeitschrift für Geomorphologie N.F.* 41(3): 369–392.

Panin A.V., Walling, D.E. and Golosov, V.N.. 2001. The role of soil erosion and fluvial processes in the post-fallout redistribution of Chernobyl-derived caesium-137: a case study of the Lapki catchment, Central Russia. *Geomorphology* 40(3/4): 185–204.

Parker, R.S. 1988. Uncertainties in defining the suspended sediment budget for large drainage basins. In: Bordas, M.P. and Walling, D.E., eds., *Sediment Budgets*, International Association of Hydrological Sciences Publication 174, pp. 523–532.

Pearce, A.J. and Watson, A.J. 1986. Effects of earthquake-induced landslides on sediment budget and transport over a 50-year period. *Geology* 14: 52–55.

Phillips, J.D. 1986. The utility of the sediment budget concept in sediment pollution control. *Professional Geographer* 38: 246–252.

Phillips, J.D. 1991. Fluvial sediment budgets in the North Carolina Piedmont. *Geomorphology* 4: 231–241.

Phillips, J.D., Slattery, M.C. and Gares, P.A. 1999. Truncation and accretion of soil profiles on coastal plain croplands: implications for sediment redistribution. *Geomorphology* 28(1/2): 119–140.

Pitlick, J. and Van Steeter, M.M. 1998. Geomorphology and endangered fish habitats of the upper Colorado River. 2.

Linking sediment transport to habitat maintenance. *Water Resources Research* 34(2): 303–316.

Potter, K.N., Zobeck, T.M. and Hagan, L.J. 1990. A micro-relief index to estimate soil erodibility by wind. *Transactions of the American Society of Agricultural Engineers* 33 (1): 151–155.

Prosser, I.P., Hughs, A.O. and Rutherfurd, I.D. 2000. Bank erosion of an incised upland channel by subaerial processes: Tasmania, Australia. *Earth Surface Processes and Landforms* 25(10): 1085–1102.

Quine, T.A., Govers, G., Walling, D.E., Zhang, X., Desmet, P.J.J., Zhang, Y. and Vandaele, K. 1997. Erosion processes and landform evolution on agricultural land—new perspectives from Caesium-137 measurements and topographic-based erosion modelling. *Earth Surface Processes and Landforms* 22(9): 799–816.

Rapp, A. 1960. Recent developments of mountain slopes in Kärkevagge and surroundings, northern Scandinavia. *Geografiska Annaler* 42(2/3): 65–200.

Reid, L.M. 1989. Channel incision by surface runoff in grassland catchments, Ph.D. Dissertation, University of Washington, Seattle, Washington.

Reid, L.M. 1998. Calculation of average landslide frequency using climatic records. *Water Resources Research* 34(4): 869–877.

Reid, L.M. and Dunne, T. 1984. Sediment production from forest road surfaces. *Water Resources Research* 20(11): 1753–1761.

Reid, L.M. and Dunne, T. 1996. *Rapid Evaluation of Sediment Budgets*, Reiskirchen, Germany: Catena Verlag GMBH, 164 p.

Reid, L.M., Dunne, T. and Cederholm, C.J. 1981. Application of sediment budget studies to the evaluation of logging road impact. *Journal of Hydrology (New Zealand)* 20 (1): 49–62.

Reid, L.M. and Hilton, S. 1998. Buffering the buffer. In: Ziemer, R.R., ed., *Proceedings of the Conference on Coastal Watersheds: the Caspar Creek Story*, USDA Forest Service Pacific Southwest Research Station General Technical Report PSW-GTR-168, Albany, California, pp. 71–80.

Reid, L.M. and Page, M.J. 2002. Magnitude and frequency of landsliding in a large New Zealand catchment. *Geomorphology* 49 (1/2): 71–88.

Reid, L.M. and Smith, C.W. 1992. The effects of Hurricane Iniki on flood hazard on Kauai, Report to the Hawaii Department of Land and Natural Resources, Honolulu, Hawaii, 26 p.

Renard, K.G., Foster, G.R., Weesies, G.A., McCool, D.K., Yoder, D.C., coordinators. 1997. *Predicting Soil Erosion by Water: A Guide to Conservation Planning with the Revised Universal Soil Loss Equation (RUSLE)*, Agriculture Handbook 703, US Department of Agriculture, 404 p.

Reneau, S.L. and Dietrich, W.E. 1991. Erosion rates in the southern Oregon Coast Range: evidence for an equilibrium between hillslope erosion and sediment yield. *Earth Surface Processes and Landforms* 16(4): 307–322.

Reneau, S.L. 2000. Stream incision and terrace development in Frijoles Canyon, Bandeleier National Monument, New Mexico, and the influence of lithology and climate. *Geomorphology* 32(1/2): 171–194.

Rice, R.M., 1982. Sedimentation in chaparral: how do you handle unusual events? In: Swanson, F.J., Janda, R.J., Dunne, T. and Swanston, D.N., eds., *Sediment Budgets and Routing in Forested Catchments*, US Department of Agriculture, Forest Service, Pacific Northwest Forest and Range Experiment Station General Technical Report PNW-141, pp. 39–49.

Riley, S.J. 1990. Monitoring the erosion of an expressway during its construction: problems and lessons. *Hydrological Sciences Journal* 35: 365–381.

Roberts, R.G. and Church, M. 1986. The sediment budget in severely disturbed watersheds, Queen Charlotte Ranges, British Columbia. *Canadian Journal of Forest Research* 16: 1092–1106.

Roehl, J.W. 1962. Sediment source areas, delivery ratios, and influencing morphological factors. *International Association of Scientific Hydrology Publication* 59: 202–213.

Rossi, G. 1998. Erosions differ. *Zeitschrift für Geomorphologie N.F.* 42(3): 297–305.

Rumsby, B. 2000. Vertical accretion rates in fluvial systems: a comparison of volumetric and depth-based estimates. *Earth Surface Processes and Landforms* 25(6): 617–632.

Santantonio, D., Hermann, R.K. and Overton, W.S. 1977. Root biomass studies in forest ecosystems. *Pedobiologia* 17 (1): 1–13.

Saunders, I. and Young, A. 1983. Rates of surface progresses on slopes, slope retreat and denudation. *Earth Surface Processes and Landforms* 8: 473–501.

Schaetzl, R.J., Johnson, D.J., Burns, S.F. and Small, T.W. 1989. Tree uprooting: review of terminology, process, and environmental implications. *Canadian Journal of Forest Research* 19(1): 1–11.

Seidl, M.A., Finkel, R.C., Caffee, M.W., Hudson, G.B. and Dietrich, W.E. 1997. Cosmogenic isotope analyses applied to river longitudinal profile evolution: problems and interpretations. *Earth Surface Processes and Landforms* 22(3): 195–209.

Singer, M.B. and Dunne, T. 2001. Identifying eroding and depositional reaches of valley by analysis of suspended sediment transport in the Sacramento River, California. *Water Resources Research* 37(12): 3371–3381.

Skidmore, E.L. and Woodruff, N.P. 1968. *Wind Erosion Forces in the United States and Their Use in Predicting Soil Loss*, USDA Agricultural Research Service Agricultural Handbook 346.

Slattery, M.C., Burt, T.P. and Boardman, J. 1994. Rill erosion along the thalweg of a hillslope hollow: a case study from the Cotswold Hills, Central England. *Earth Surface Processes and Landforms* 19(4): 377–385.

Sloan, J., Miller, J.R. and Lancaster, N. 2001. Response and recovery of the Eel River, California, and its tributaries to floods in 1955, 1964, and 1997. *Geomorphology* 36(3/4): 129–154.

Small, T.W. 1997. The Goodlett-Denny mound: a glimpse at 45 years of Pennsylvania treethrow mound evolution with implications for mass wasting. *Geomorphology* 18(3/4): 305–314.

Small, E.E., Anderson, R.S. and Hancock, G.S. 1999. Estimates of the rate of regolith production using ^{10}Be and ^{26}Al from an alpine hillslope. *Geomorphology* 27(1/2): 131–150.

Smith, R.D. and Swanson, F.J. 1987. Sediment routing in a small drainage basin in the blast zone at Mount St. Helens, Washington, USA. *Geomorphology* 1: 1–13.

Springer, G.S., Dowdy, H.S. and Eaton, L.S. 2001. Sediment budgets for two mountainous basins affected by a catastrophic storm: Blue Ridge Mountains, Virginia. *Geomorphology* 37(1/2): 135–148.

Steiger, J., Gurnell, A.M., Ergenzinger, P. and Snelder, D. 2001. Sedimentation in the riparian zone of an incising river. *Earth Surface Processes and Landforms* 26(1): 91–108.

Stott, T. 1997. A comparison of stream bank erosion processes on forested and moorland streams in the Balquhidder catchments, central Scotland. *Earth Surface Processes and Landforms* 22(4): 383–399.

Swanson, F.J., Fredriksen, R.L. and McCorison, F.M. 1982. Material transfer in a western Oregon forested watershed. In: Edmonds, R.L., ed., *Analysis of Coniferous Forest Ecosystems in the Western United States*, Stroudsburg, Pennsylvania: Hutchinson Press, pp. 233–266.

Swanson, M.L., Kondolf, G.M. and Boison, P.J. 1989. An example of rapid gully initiation and extension by subsurface erosion: coastal San Mateo County, California. *Geomorphology* 2(4): 393–404.

Swanston, D.N., Ziemer, R.R. and Janda, R.J. 1995. *Rate and Mechanics of Progressive Hillslope Failure in the Redwood Creek Basin, Northwestern California*, USDI Geological Survey Professional Paper 1454 E, 16 p.

ten Brinke, W.B.M., Schoor, M.M., Sorber, A.M. and Berendsen, H.J.A. 1998. Overbank sand deposition in relation to transport volumes during large-magnitude floods in the Dutch sand-bed Rhine River system. *Earth Surface Processes and Landforms* 23(9): 809–824.

Terajima, T., Sakamoto, T., Nakai, Y. and Kitamura, K. 1997. Suspended sediment discharge in subsurface flow from the head hollow of a small forested watershed, northern Japan. *Earth Surface Processes and Landforms* 22(11): 987–1000.

Thorn, C.E. 1978. A preliminary assessment of the geomorphic role of pocket gophers in the alpine zone of the Colorado Front Range. *Geografiska Annaler* 60A(3/4): 181–187.

Thorne, C.R. and Tovey, N.K. 1981. Stability of composite river banks. *Earth Surface Processes* 6(5): 469–484.

Trimble, S.W. 1974. *Man-induced Soil Erosion on the Southern Piedmont, 1700–1970*, Ankeny, Iowa: Soil Conservation Society of America, 180 p.

Trimble, S.W. 1977. The fallacy of stream equilibrium in contemporary denudation studies. *American Journal of Science* 277: 876–887.

Trimble, S.W. 1983. A sediment budget for Coon Creek basin in the driftless area, Wisconsin, 1853–1977. *American Journal of Science* 283: 454–474.

Trimble, S.W. 1993. The distributed sediment budget model and watershed management in the Paleozoic plateau of the upper Midwestern United States. *Physical Geography* 14 (3): 285–303.

Trimble, S.W. 1994. Erosional effects of cattle on streambanks in Tennessee. *Earth Surface Processes and Landforms* 19(5): 451–464.

Trimble, S.W. 1997. Contribution of stream channel erosion to sediment yield from an urbanizing watershed. *Science* 278: 1442–1444.

Trimble, S.W. 1999. Decreased rates of alluvial storage in the Coon Creek Basin, Wisconsin, 1975–93. *Science* 285: 1244–1246.

Van Asch, T.J.W. and Van Genuchten, P.M.B. 1990. A comparison between theoretical and measured creep profiles of landslides. *Geomorphology* 3(1): 45–55.

Van Steeter, M.M. and Pitlick, J. 1998. Geomorphology and endangered fish habitats of the upper Colorado River. 1. Historic changes in streamflow, sediment load, and channel morphology. *Water Resources Research* 34(2): 287–302.

van Steijn, H. 1996. Debris-flow magnitude–frequency relationships for mountainous regions of Central and Northwest Europe. *Geomorphology* 15(3/4): 259–273.

Vandaele, K., Vanommeslaeghe, J., Muylaert, R. and Govers, G. 1996. Monitoring soil redistribution patterns using sequential aerial photographs. *Earth Surface Processes and Landforms* 21(4): 353–364.

Vandekerckhove, L., Poesen, J., Oostwoud Wijdenes, D., Gyssels, G., Beuselinck, L. and de Luna, E. 2000. Characteristics and controlling factors of bank gullies in two semi-arid mediterranean environments. *Geomorphology* 33(1/2): 37–58.

Verstraeten, G. and Poesen, J. 2000. Estimating trap efficiency of small reservoirs and ponds: methods and implications. *Progress in Physical Geography* 24(2): 219–251.

Walling, D.E. and He, Q. 1997. Use of fallout Cs-137 in investigations of overbank sediment deposition on river floodplains. *Catena* 29(3/4): 263–282.

Walling, D.E. and He, Q. 1998. The spatial variability of overbank sedimentation on river floodplains. *Geomorphology* 24(2/3): 209–224.

Walling, D.E., Owens, P.N. and Leeks, G.J.L. 1998. The role of channel and floodplain storage in the suspended sediment budget of the River Ouse, Yorkshire, UK. *Geomorphology* 22(3/4): 225–242.

Walling, D.E. and Webb, B.W. 1982. Sediment availability and the prediction of storm-period sediment yields. *International Association of Hydrological Sciences Publication* 137: 327–337.

Ward, P.A., III. and Carter, B.J. 1999. Rates of stream incision in the middle part of the Arkansas River Basin based on late Tertiary to mid-Pleistocene volcanic ash. *Geomorphology* 27(3/4): 205–228.

Wasson, R.J., Mazari, R.K., Starr, B. and Clifton, G. 1998. The recent history of erosion and sedimentation on the Southern Tablelands of southeastern Australia: sediment flux dominated by channel incision. *Geomorphology* 24(4): 291–308.

Wathen, S.J. and Hoey, T.B. 1998. Morphological controls on the downstream passage of a sediment wave in a gravel-bed stream. *Earth Surface Processes and Landforms* 23(8): 715–730.

Wieczorek, G.F. and Jäger, S. 1996. Triggering mechanisms and depositional rates of postglacial slope-movement processes in the Yosemite Valley, California. *Geomorphology* 15(1): 17–31.

Wiele, S.M., Graf, J.B. and Smith, J.D. 1996. Sand deposition in the Colorado River in the Grand Canyon from flooding of the Little Colorado River. *Water Resources Research* 32(12): 3579–3596.

Wilby, R.L., Dalgleish, H.Y. and Foster, I.D.L. 1997. The impact of weather patterns on historic and contemporary catchment sediment yields. *Earth Surface Processes and Landforms* 22(4): 353–364.

Wilcock, P.R., Kondolf, G.M., Matthews, W.V.G. and Barta, A.F. 1996. Specification of sediment maintenance flows for a large gravel-bed river. *Water Resources Research* 32(9): 2911–2921.

Wiles, G.C., Calkin, P.E. and Jacoby, G.C. 1996. Tree-ring analysis and Quaternary geology: principles and recent applications. *Geomorphology* 16(3): 259–272.

Wischmeier, W.H. and Smith, D.D. 1978. *Predicting Rainfall Erosion Losses—A Guide to Conservation Planning*, USDA Agriculture Handbook No. 537, 58 p.

Wohl, E.E. and Cenderelli, D.A. 2000. Sediment deposition and transport patterns following a reservoir sediment release. *Water Resources Research* 36(1): 319–333.

Womack, W.R. and Schumm, S.A. 1977. Terraces of Douglas Creek, northwestern Colorado: an example of episodic erosion. *Geology* 5: 72–76.

Yanda, P.Z. 2000. Use of soil horizons for assessing soil degradation and reconstructing chronology of degradation processes: the case of Mwisanga Catchment, Kondoa, central Tanzania. *Geomorphology* 34(3/4): 209–226.

Yoshida, K., Kikuchi, S., Nakamura, F. and Noda, M. 1997. Dendrochronological analysis of debris flow disturbance on Rishiri Island. *Geomorphology* 20(1/2): 135–145.

Zhang, X., Phillips, C. and Pearce, A. 1991a. Surface movement in an earthflow complex, Raukumara Peninsula, New Zealand. *Geomorphology* 4: 261–272.

Zhang, X., Phillips, C. and Marden, M. 1991b. Internal deformation of a fast-moving earthflow, Raukumara Peninsula, New Zealand. *Geomorphology* 4: 145–154.

Zhang, X., Phillips, C. and Marden, M. 1993. A comparison of earthflow movement mechanisms on forested and grassed slopes, Raukumara Peninsula, North Island, New Zealand. *Geomorphology* 6: 175–187.

Ziemer, R.R. 1992. Effect of logging on subsurface pipeflow and erosion: coastal northern California, USA. In: Walling, D.E., Davies, T.R. and Hasholt, B., eds., *Erosion, Debris Flows and Environment in Mountain Regions*, International Association of Hydrological Sciences Publication 209, pp. 187–197.

Zonge, K.L., Swanson, S. and Myers, T. 1996. Drought year changes in streambank profiles on incised streams in the Sierra Nevada Mountains. *Geomorphology* 15(1): 47–56.

Part VI

Discriminating, Simulating and Modeling Processes and Trends

17

Models in Fluvial Geomorphology

STEPHEN E. DARBY[1] AND MARCO J. VAN DE WIEL[2]
[1]*Department of Geography, University of Southampton, Southampton, UK*
[2]*Institute of Geography and Earth Sciences, University of Wales, Aberystwyth, Wales, UK*

17.1 INTRODUCTION

Models are conceptions of physical reality that result in qualitative or quantitative predictions. This simple definition masks the fact that the complexity of natural systems, together with gaps in our knowledge, means that the development of a model involves an approach wherein physical reality is in some way restructured to a form that fits the available resources and permits prediction (ASCE Task Committee 1998). It is these models of restructured reality that are actually used and, as a general premise, there is often a discrepancy between physical reality (the 'problem') and the model. It is, therefore, incumbent on potential users to judge if in fact the model can provide predictions that are accurate enough to solve their particular problem. The primary aim of this chapter is to provide the information and guidance necessary to help users in making this judgement. The specific contents and scope of this chapter in relation to this aim are described at the end of this introduction. However, it is initially helpful to recognize the different motives of the various individuals who have formulated the fluvial geomorphological models that have been published and used by practitioners.

In fact, fluvial geomorphological modelling tools have been developed to address problems within two broad themes: pure scientific research and river engineering. Fluvial geomorphologists have traditionally focussed on the study of river landforms and the processes that create them (Kirkby 1996). However, progress in understanding these areas has been hampered because rates of landform change are typically much less than the scales at which changes in natural systems are readily observable. Hence, geomorphologists often use models simply as study tools to increase their understanding of the landscape. This is because models provide the basis for aggregating from the scales of observation to the scales of interest, as well as a pivotal link between the study of process and the study of landforms (Kirkby 1996). For example, many different boundary conditions exist in natural alluvial channels (Richards 1996, Lane 1998). The availability of empirical data for analysis of process–form interactions in rivers is, therefore, limited and models are increasingly being used as a supplement (e.g. Howard 1994, Darby and Thorne 1996a, 1997, Tucker and Slingerland 1997, Bradbrook *et al.* 1998, Hodskinson and Ferguson 1998, Lane 1998, Nicholas and Walling 1998, Nicholas *et al.* 1999).

Application of fluvial geomorphological knowledge to a diverse range of practical river management problems has also been increasing. For example, erosion can lead to loss of agricultural land (Lohnes 1991) and damage to irrigation intakes (Pokrefke *et al.* 1998), flood defences and river crossing infrastructure (Simon and Downs 1995, Johnson *et al.* 1999). On the other hand, deposition can result in a loss of flood carrying capacity that promotes increased risk of flooding, or necessitates expensive dredging in navigable waterways (Darby and Thorne 1994), as well as causing damage to salmonid spawning habitats (Reiser 1998). These problems are particularly serious when accelerated erosion and sedimentation occurs in response to disruption of the natural channel morphology, or perturbation of the fluxes of water and sediment supplied from the catchment upstream. Such perturbations are often initiated by human activities such as channelization (Brookes 1988), river regulation (Petts 1980, 1984), gravel mining (Kondolf and Swanson 1993, Kondolf 1997, Bravard *et al.* 1999) and land use changes (Leeks 1992, Kuhnle *et al.* 1996), but they can also be caused by changes in climate, vegetative cover, or tectonic changes (Galay 1983, Schumm 1999). Widespread human exploitation of river resources through activities such as river regulation and channelization for flood control, navigation and hydropower have resulted in severe damage to river ecosystems across the face of the planet (Gore 1996). Awareness of these negative impacts has now increased and there is now a strong trend of river restoration to rehabilitate aquatic ecosystems (e.g. Brookes and Shields 1996). For this reason, there is a strong demand for robust and reliable modelling tools to predict how river systems are affected by human or naturally induced environmental changes, and how system components interact with human activities and management constraints.

Models are, therefore, used to provide an insight into the functioning of the natural environment, as well as forecasting tools to underpin sustainable river management. While the priority given to the pure and applied objectives of the discipline will differ according to the needs of diverse user groups, in the best practice there is often no conflict between them (Kirkby 1996). As Thorne (1995) puts it, there is an 'unbreakable thread' that runs between pure research and practical applications. However, while modelling tools are undoubtedly useful to a wide and diverse range of possible users, this diversity can sometimes create problems. In particular, users who are predominantly interested in practical applications may have been trained in ecology, hydrology, planning or management, but often do not have the experience to use geomorphological modelling tools with confidence. There is clear potential for models to be applied inappropriately, or without detailed knowledge of the strengths and weaknesses of diverse modelling approaches.

To meet this need we herein provide a systematic review of different types of modelling tools, providing guidance on their strengths, weaknesses and scope. Specifically, we review four categories of models, namely, conceptual models, statistical or empirical models, analytical models and numerical simulation models. Although they are undoubtedly valuable modelling tools, we have excluded physical or hardware models from our review since we do not have any experience in this field. In any case, detailed reviews of physical models have already been provided by Ashmore (1982), Schumm *et al.* (1987), Shen (1991), Ashworth *et al.* (1994), Peakall *et al.* (1996) and Warburton and Davies (1998), among others. We next present a conceptual framework with criteria for selecting the types of models appropriate for diverse user requirements, together with information required to develop an enhanced understanding of the strengths and weaknesses of specific models within each category (Table 17.1). This may help potential users in selecting between different broad modelling approaches, but reference must still be made to the guidelines detailed in Section 17.6 for the specific details of the modelling process. Finally, we use specific detailed case studies to show how the proposed conceptual framework can be used to 'steer' the direction of both pure and applied fluvial geomorphological modelling research applications. It should be noted that while some attention is inevitably given to the individual submodels of flow, sediment transport and bank migration processes that control the formation of river morphology, the main discussion is restricted to broader models of channel geomorphology. Detailed reviews of flow and sediment transport modelling as topics in their own right are provided elsewhere in this volume (see Chapters 8, 18 and 19).

17.2 CONCEPTUAL MODELS

Conceptual models are an important category of tools that provide qualitative descriptions and predictions of landform and landscape evolution. A wide range of conceptual models have been developed, with applications covering the full spectrum of geographical scales from a specific river reach, up to entire landscapes. The best known example of a fluvial geomorphological conceptual model is W.M. Davis' theory of landscape development, the geographical cycle (Figure 17.1). Davis' (1899) theory is based on an organic analogy drawn from the Darwinian theory of evolution. The theory describes how landscapes evolve as a manifestation of structure, process and time (Sack 1992). Given certain assumptions regarding structure and process, Davis believed that landscapes evolve through successive phases of youth, maturity and old age in a deterministic sequence (Sack 1992). A key feature of this conceptual model is that landscapes are assumed to have certain diagnostic features that enable the stage of landscape evolution to be identified. Since each stage represents part of a fixed evolutionary sequence, identification of a landscape's present stage provides the opportunity to obtain information both on its past (postdiction) and future (prediction) configuration (Sack 1992).

The simplicity with which this type of model can be applied has meant that similar models of landform evolution have been developed to predict the

Table 17.1 Summary of the characteristics, advantages and limitations of different fluvial geomorphological modelling strategies

Model category	Typical applications	Advantages	Limitations	Model scale
Conceptual	• Reconnaissance studies	• Rapid assessment method—good for large areas and scoping studies	• Requires basic training	Conceptual models are available across a wide range of scales (bar to catchment)
	• Qualitative forecasting	• Relatively simple—requires few resources and minimal background data	• Qualitative results only	
	• Qualitative postdiction			
Empirical/statistical	• Channel design	• Simple—these models are easy to understand and use	• Site specific technique—care is required to avoid misapplication	Individual cross-sections representative of short river reaches
	• Quantitative forecasting	• Input data are usually readily available	• No information on rates of change	
	• Quantitative postdiction		• Requires estimate of formative discharge	
	• Palaeohydrology		• Dimensionally inconsistent	
Analytical	• Channel design	• Improved physical basis means these models are often valid across a range of environments	• No information on rates of change	Individual cross-sections representative of short river reaches
	• Quantitative forecasting	• Input data requirements are usually manageable	• Requires estimate of formative discharge	
			• Models can be quite complex	
Numerical simulation	• Channel design	• When calibrated, valid in a wide range of environments	• Models are very complex and require specialist training	In theory any, but heavily constrained by data requirements
	• Quantitative forecasting	• Provides detailed predictions of transient adjustments	• Input data requirements are very large	

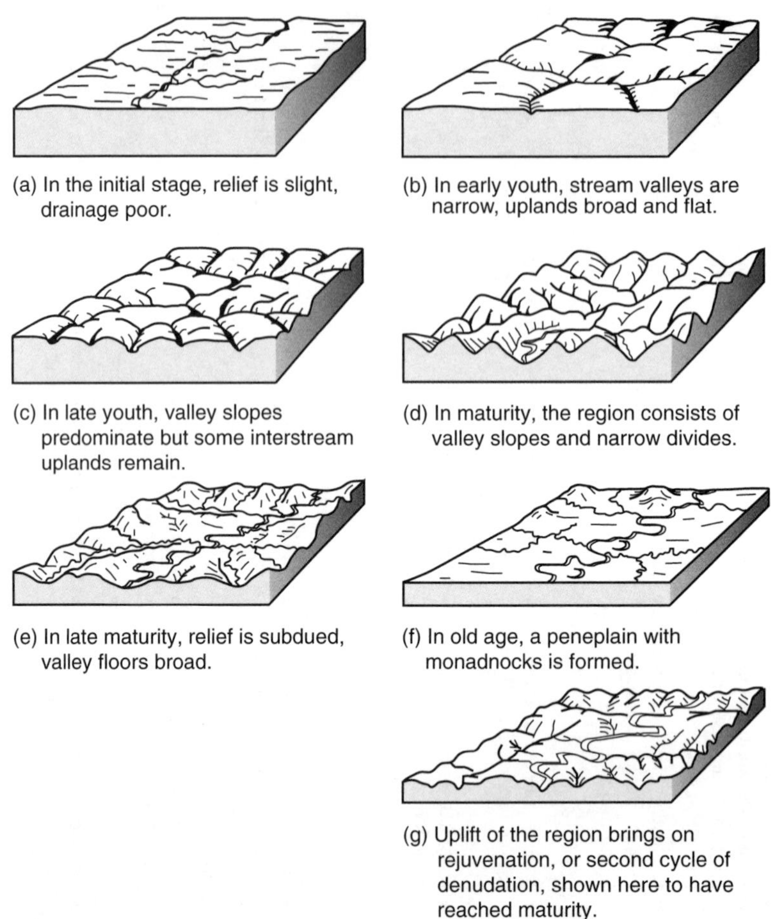

(a) In the initial stage, relief is slight, drainage poor.
(b) In early youth, stream valleys are narrow, uplands broad and flat.
(c) In late youth, valley slopes predominate but some interstream uplands remain.
(d) In maturity, the region consists of valley slopes and narrow divides.
(e) In late maturity, relief is subdued, valley floors broad.
(f) In old age, a peneplain with monadnocks is formed.
(g) Uplift of the region brings on rejuvenation, or second cycle of denudation, shown here to have reached maturity.

Figure 17.1 Illustration of W.M. Davis' (1899) 'geographical cycle' in which landscape evolution is portrayed as a sequence of erosional transformations (progressing through stages of 'youth', 'maturity' and 'old age' of an initially elevated landmass)

evolution of a wide range of fluvial landforms. Examples include the evolution of drainage networks (Glock 1931), river planform (Keller 1972, Thompson 1986, Slingerland and Smith 1998), the evolution of cross-sectional shape and longitudinal profile of incised channels (Schumm *et al.* 1984, Simon 1989, see Chapter 5) and arroyos (Elliott *et al.* 1999), and the morphology of step-pool systems in mountain streams (Chin 1999), to name but a few. A characteristic feature of conceptual fluvial geomorphological models is that they rely on the technique of space-for-time substitution in their development (Paine 1985, Schumm 1991). Inherent in these models is the fact that the distribution of characteristic landforms across space represents the passage of time exclusively. When using space-for-time substitution it is, therefore, important to compare features produced by the same processes that are operating under the same physical conditions. For example, the evolution of an incised channel in alluvium can be determined by surveying cross-sections at several locations where the channel is in alluvium, but one cannot combine data or compare channels in weak alluvium with channels in resistant alluvium or bedrock and expect to find meaningful results (Schumm 1991). Hence, considerable care is required in applying conceptual models based on space-for-time substitution, to ensure the modelling application replicates the conditions under which the model was developed.

In more recent years, conceptual models have also been linked with tools for quantitative analysis to develop composite modelling tools that are more powerful and robust. Particularly exciting in this respect is the increasing use of geographical information systems (GIS) to store spatial data and spatial characteristics of fluvial systems. When linked to simple logical rules derived from conceptual models, it is possible to develop quantitative models that allow for complex analyses and manipulation of data over a broad range of spatial scales. Examples of this type of composite modelling approach include: linking GIS with incised channel evolution models (Simon and Downs 1995); modelling large woody debris distributions in drainage basins (Wallerstein *et al.* 1997); prediction of chinook salmon spawning habitat (Geist and Dauble 1998); modelling channel-reach morphology in mountain drainage basins (Montgomery and Buffington 1997); catchment-scale sediment transport modelling (Viney and Sivapalan 1999), and development of conceptual models to support stream restoration (Shields *et al.* 1998) and wetland rehabilitation (O'Neill *et al.* 1997).

The main advantage of conceptual models relative to other modelling approaches (Table 17.1) lies in their relative simplicity and ease of application. However, these factors are simultaneously the main limitation. Hence, conceptual models provide qualitative insight into the nature of the problem but cannot be used for quantitative forecasting. It must also be remembered that conceptual models are based on the ordering of a set of empirical observations. Care must, therefore, be taken to avoid misapplication of conceptual models to fluvial systems dissimilar in character to those for which the model was originally derived. In general, conceptual models are most often used as a first step in the modelling process, enabling the model user to develop a broad understanding of the system before she or he attempts to apply more complex approaches geared towards quantitative forecasting (see Section 17.6). Conceptual models are also often used to identify problem river reaches for more detailed investigation as a part of reconnaissance studies (e.g. Simon and Downs 1995).

17.3 STATISTICAL AND EMPIRICAL MODELS

Statistical models have played a major role in the analysis of fluvial systems over the last 40 years (Rhoads 1992, also see Chapter 20 in this volume). These models have been developed using functional relationships between dependent morphological variables and the independent variables of sediment load and discharge. These functional relationships are applied at the scale of individual cross-sections, or cross-sections representative of short reaches of river. The widespread adoption of statistical models since the Second World War is associated with a quantitative revolution that has now superseded the historical–descriptive paradigm embodied in Davis' 'cycle of erosion' and in the other qualitative conceptual models reviewed in Section 17.2. However, as analytical (Section 17.4) and numerical simulation (Section 17.5) models have become increasingly fashionable, in recent years statistical and empirical models have been strongly criticized (Strahler 1980, Thornes and Ferguson 1981) for their lack of a solid theoretical base. As with any regression study, it must be recognized that the fitted relationships of these models may not extrapolate outside the range of conditions studied, and that an occasional predictor may only be significant through sampling fluke (Ferguson 1986). Despite this, statistical models are often misused. Typically, models are erroneously applied to rivers with characteristics that are not similar to those used to derive the equations. Alternatively, these models have sometimes been used to estimate palaeo-discharge values from observed channel dimensions, even though the channel dimensions are *dependent* variables in these models [see Equations (17.1)–(17.3)].

Nonetheless, the hydraulic geometry of streams represents the first major application of the quantitative paradigm in fluvial geomorphology (Rhoads 1992) and these models continue to be in widespread use today. The power functions that constitute the basic equations of hydraulic geometry (Leopold and Maddock 1953) are well known:

$$W = aQ^b \qquad (17.1)$$

$$D = cQ^f \qquad (17.2)$$

$$V = kQ^m \qquad (17.3)$$

where W is the flow width (m), D the flow depth (m), V the flow velocity (m s^{-1}) and Q is the flow discharge (m^3 s^{-1}). The exponents b, f and m describe relative rates of change in W, D, or V either as discharge changes through time at a particular location (at-a-station hydraulic geometry) or as discharge of a constant frequency changes in the downstream direction (downstream hydraulic geometry). The coefficients a, c and k are scale factors that define the magnitudes of W, D and V when $Q = 1$ (Rhoads 1992).

Many authors have compiled data to develop empirical downstream hydraulic geometry relationships of the type shown above. These studies have indicated that, on the whole, channel width and depth are related in a regular way to discharge as it varies over a billionfold range from laboratory channels to the largest rivers in the world (Ferguson 1986). This suggests that transient changes in channel dimensions at a point are minor compared to systematic, downstream, trends and that discharge is the dominant control on equilibrium morphology. Still, much scatter is observed in these relationships, and other factors are intuitively relevant (Ferguson 1986). The best use for empirical models of downstream hydraulic geometry may well be in the identification of factors relevant in the process of morphological adjustment. However, multivariate approaches were used only later in the history of hydraulic geometry studies (see below).

Early studies, following Leopold and Maddock (1953), instead tended to be bivariate in approach. Examples of these studies are numerous, and they have been used in a variety of analyses of channel form (Williams 1978). A particular feature of these studies [see Richards (1977) for a full list] is that while they show considerable variation in the values of the fitted coefficients and exponents, they all exhibit systematic downstream variation in hydraulic geometry variables in response to increasing discharge in the downstream direction. Thus, these studies demonstrate the essential uniformity of river morphology, and also represent a considerable database of river channel behaviour (Rhoads 1992).

Attempts to reduce the degree of scatter in early hydraulic geometry models led to the adoption of the multivariate approach (see Hey 1978). Wolman and Brush (1961), Caddie (1969) and Maddock (1969) all carried out experiments which involved varying the type and concentration of sediment load as well as discharge. Subsequent studies have continued to highlight the fundamental importance of the amount and type of sediment transported by the river (e.g. Schumm 1960, Wilcock 1971, Hey and Thorne 1986). Similar research has also identified the effects of vegetation on channel morphology (Zimmerman *et al.* 1967, Charlton *et al.* 1978, Andrews 1984, Hey and Thorne 1986). In recent years, research has also focussed on extending the range of environments beyond the sand-bed rivers that were the focus of attention in early studies. In particular, Bray (1982) in Canada and Hey and Thorne (1986) in Great Britain have made significant contributions to the gravel-bed river data set. Most recently, the wide availability of data collected over many years, coupled with advances in computer processing power, has allowed the opportunity to develop multivariate hydraulic geometry equations based on very large data sets (e.g. Julien and Wargadalam 1995).

Despite their widespread availability and use, statistical models have a number of technical and conceptual limitations (Thornes 1977). Primary amongst the conceptual difficulties is that channel changes are, by definition, transient in nature, but the various hydraulic geometry models listed above define only steady-state behaviour. A second area of difficulty lies in the undoubted existence of thresholds and discontinuities in the regression relationships involved (Thornes 1977). However, in using standard power functions [Equations (17.1)–(17.3)] it is generally assumed that these relationships are smooth and continuous. With respect to technical problems, Benson (1965) has shown that the use of a common variable on both sides of the regression equation (discharge is defined as the product of width, depth and velocity) can lead to spurious correlation. Traditional regression models are dimensionally inconsistent and are limited in that they only describe simple input–output relations between the states of the independent and dependent variables (Rhoads 1992). More recent hydraulic geometry investigations have attempted to address some of these difficulties through the rigorous application of more sophisticated statistical and analytical techniques (Miller 1984, Rhoads 1992), and these advances have established improved conceptual and technical foundations for these models.

In summary, statistical models, especially simple ones, tend to have less stringent data requirements, albeit with a weaker theoretical base, than their analytical and numerical counterparts (Rhoads 1992). Statistical models generally do not incorporate physical reasoning and are often dimensionally imbalanced (Hey and Heritage 1988). Critically, applications of these models are limited to the domain of the data used to estimate the model and are scale dependent (Rhoads 1992). Despite their limitations, statistical models have made a substantial contribution to our understanding of fluvial systems. Bivariate and multivariate regression models of channel and flow geometry have generated insight into relationships amongst various components of fluvial systems, and such models also serve as empirical tests of theoretical models (Rhoads 1992, see below). The simplicity and ease of application of these models have led to their widespread use in practical applications, most commonly as a preliminary step in the design of geomorphologically stable restoration reaches (e.g. Brookes and Sear 1996, Shields 1996).

17.4 ANALYTICAL MODELS

The limitations associated with the various empirical and statistical models reviewed above have led river scientists and engineers to seek models that are based more on the physical processes involved in the establishment of channel morphology. River engineers, in particular, have developed models that have a more powerful predictive element than had been the case previously. For the purposes of this review, it is possible to classify these analytical models into categories based on the type of approach adopted in their formulation. It should be noted that, like the empirical models reviewed in Section 17.3, most of these analytical models are used to predict equilibrium morphology at the scale of the river cross-section, though Pizzuto (1992) has developed an analytical modelling approach linked to the scale of the watershed.

Extremal Hypothesis Approaches

In seeking to develop analytical models with an increased physical basis, there is almost universal agreement that sediment transport and alluvial friction are significant and should be included in any rational approach to deriving regime-type morphological relationships (Bettess *et al.* 1988). However, even when neglecting planform and bedform changes, natural rivers have at least three degrees of freedom (width, depth and slope) of adjustment (Hey 1978). Hence, sediment transport and alluvial friction relationships are, by themselves, insufficient to enable a solution for width, depth and slope, even assuming that the sediment transport and flow resistance submodels presently available describe these processes adequately (White *et al.* 1982, Gomez and Church 1989). Extremal hypotheses have, therefore, been proposed to provide the extra relationship necessary to close the system and enable the channel morphology to be determined. The term 'extremal hypothesis' refers to the fact that the models are based on the assumption that the equilibrium channel morphology corresponds to the morphology that maximizes or minimizes the value of a specific parameter. Examples of extremal hypotheses include the minimization of energy dissipation rate (Yang *et al.* 1981), minimization of stream power (Chang 1980, 1988) or unit stream power (Yang and Song 1979), and the maximization of friction factor (Davies and Sutherland 1983) or sediment transport rate (White *et al.* 1982). Griffiths (1984) showed that these various extremal hypotheses are closely related, and under certain conditions essentially equivalent. However, he viewed them as providing more of an illusion of progress than actual progress itself.

The main criticism of analytical modelling tools based on the use of extremal hypotheses is that they are similar to empirical and statistical modelling approaches in that they simply present a method of calculating steady-state channel dimensions while not suggesting a mechanism by which this is achieved (Bettess *et al.* 1988). Hence, these hypotheses offer an essentially metaphysical method of predicting steady-state channel dimensions, which offers no explanatory power (Ferguson 1986). The theoretical justifications for extremal hypotheses are also still not entirely clear (ASCE Task Committee 1998). Yang (1971) originally proposed analogies between river elevation and temperature and between potential and thermal energy to deduce his 'law of least time rate of energy expenditure' from the thermodynamic principle of minimum rate of energy production. However, this principle is valid only in the range of linear thermodynamic processes (Davy and Davies 1979), whereas energy transformations in rivers are often highly non-linear. Davies and Sutherland (1983) also attack the theoretical basis of extremal hypotheses on the grounds that an assumed analogy between laminar and turbulent flow used to derive the hypotheses is fundamentally unjustified.

A promising analytical modelling approach based on the use of an extremal hypothesis has been developed by Millar and Quick (1993, 1998). Their models were developed to determine the influence of bank stability on the stable width and depth of gravel-bed rivers with non-cohesive (Millar and Quick 1993) and cohesive (Millar and Quick 1998) bank materials. These models were based on the central assumption that equilibrium channel morphology corresponds to a condition that is equivalent to the maximum bed load transporting capacity (White *et al.* 1982). However, unlike previous approaches (see references in preceding paragraph), which did not explicitly account for the effect of bank stability, Millar and Quick included a mechanistic bank failure analysis directly into the modelling approach. This, therefore, provided a much stronger theoretical basis than previous approaches. Millar and Quick (1998) were able to use a calibrated version of their model to assess the effect of bank vegetation on bank stability, which is important in designing environment-friendly stable channels.

Verification exercises undertaken by the respective authors have indicated that model predictions based on extremal hypotheses provide global, if not exacting, agreement with a wide range of observations (ASCE Task Committee 1998). In an independent assessment of the predictive capabilities of extremal hypotheses, Wang Shiqiang *et al.* (1986) compared model predictions with empirical data from 203 sand-bed rivers and canals and 59 gravel-bed rivers. They found that the various extremal hypotheses achieved a considerable degree of predictive success. Mean discrepancy ratios (Me) for six different extremal hypotheses ranged from 0.84 to 1.33 and from 0.74 to 1.38 for sand-bed and gravel-bed rivers, respectively. Of the tested hypotheses, the principles of minimum stream power (Me = 1.07) or maximum sediment concentration (Me = 1.05) gave the best agreement with field data. However, with the theoretical justification of such hypotheses unclear, application of extremal hypotheses certainly requires a clear understanding of the physical constraints presented by geological or other boundary conditions (ASCE Task Committee 1998). Recent experimental work by Simon (1992) and Abrahams *et al.* (1994) suggests that extremal hypotheses have not yet been properly tested under a full range of imposed conditions and constraints. Judgement on the apparent predictive success of these methods must, therefore, at least be partially reserved.

Tractive Force Methods

Tractive force modelling approaches have a strong theoretical basis, since they employ the basic laws of mechanics to obtain expressions that specify the geometry of stable channel cross-sections. The basis of the approach, which was initiated in the late 1940s by the US Bureau of Reclamation (Glover and Florey 1951, Lane 1955), is to consider the magnitude of the critical tractive stress for sediment entrainment. A 'threshold' channel form is then computed that, for a pre-specified channel gradient (Carson and Griffiths 1987), can convey the flow discharge without attaining the critical stress. The various tractive force methods are all, therefore, based on various methods of solving the fluid momentum balance to obtain the local boundary shear stress, coupled with an entrainment criterion for the sediment particles that make up the channel perimeter (ASCE Task Committee 1998).

Central to the tractive force method is the need to solve the force balance along all parts of the channel perimeter, including the sloping banks of the channel. Lane (1955) was able to obtain the approximate result that the critical shear stress for material resting on the bed and banks is

$$\frac{\tau_{cs}}{\tau_c} = \sqrt{\left(1 - \frac{\sin^2\alpha}{\sin^2\phi}\right)} \tag{17.4}$$

where τ_{cs} and τ_c are the critical fluid shear stresses for entrainment on a sloping bank and on a plane bed, respectively (Pa), α the side slope angle of the bank (degrees) and ϕ is the friction angle of the perimeter sediments (degrees). Lane (1955) used a highly simplified version (assuming steady, uniform flow) of the fluid momentum balance to calculate the mean fluid shear stress using

$$\tau = \gamma_w y_m S \tag{17.5}$$

where τ is the boundary shear stress (Pa), γ_w the unit weight of the fluid (N m^{-3}), y_m the depth of flow above a point on the bed (m) and S is the energy slope, assumed equal to the slope of the bed for steady, uniform flow. Based on the above assumptions, Lane (1955) was able to show that

$$\frac{y_m}{D_m} = \cos\left(\frac{z\tan\alpha}{D_m}\right) \tag{17.6}$$

where D_m is the flow depth at the foot of the bank (m) and z is the lateral co-ordinate (m). This equation is valid for steady, uniform, flow and for non-cohesive sediment particles that do not vary within the channel. The equation shows that the shape of the bank is required to have the profile of a cosine curve in order to maintain the predicted fluid shear stress below the predicted critical shear stress. This profile enables entrainment to begin simultaneously at all points around the channel perimeter when the fluid shear stress reaches the critical value. To develop a regime-type functional relationship linking the stable channel dimensions to flow discharge and sediment properties, a flow resistance equation is introduced to solve the problem (Ferguson 1986). For example, Henderson (1966) used the Manning–Strickler equation to predict that

$$W = 1.1 d_s^{-0.15} Q^{0.46} \tag{17.7}$$

where d_s is the representative particle size of the boundary material (mm).

In principle, the tractive force method is an attractive approach for modelling the equilibrium channel cross-section dimensions. This is because the method relies on a rational treatment of channel boundary erosion in terms of explicit consideration of the applied fluid forces and the resisting forces influenced by the physical properties of the boundary material. In practice, however, there are limitations with the method. Calculation of the critical shear stress for entrainment of the boundary material is notoriously problematic and a likely source of error (e.g. Gomez and Church 1989, Buffington and Montgomery 1997). Ferguson (1986) notes that the precise values of the coefficients and exponents in the functional relationship [Equation (17.7)] actually vary depending on the assumed value of the boundary material friction angle and the selected flow resistance equation. The friction angle strongly influences critical shear stress, but is difficult to estimate accurately in practice (e.g. Buffington *et al.* 1992), while the limitations of flow resistance equations, all of which are empirically calibrated, are well known (e.g. Thorne and Zevenbergen 1985, Robert 1990, Bathurst 1997). In terms of practical limitations, the boundary material in natural streams is heterogeneous. So it is not clear what the 'correct' value of the representative particle size (d_s) actually is.

Unfortunately, the tractive force method as developed by Glover and Florey (1951) and Lane (1955) provides predictions that are inconsistent with observations. In theory, a threshold channel does not allow for bed load transport (Parker 1978, Diplas 1990). This result is contrary to observations from natural streams and flume experiments, which attest to the coexistence of a mobile bed and stable banks (ASCE Task Committee 1998). In fact, this inconsistency is a remnant of the simplified form of the fluid momentum balance used by Glover and Florey (1951) and Lane (1955). By introducing the lateral turbulent diffusion of downstream momentum into the momentum balance, Parker (1978) was able to reconcile the existence of sediment movement within a stable channel. However, several terms are still neglected in calculating the applied fluid force, including the effects of convective momentum exchanges through secondary flows and the effects of fluid lift.

More recently, a number of advances have been made in attempts to improve the physical basis of the tractive force approach. Lateral distributions of fluid shear stress have been more accurately predicted using sophisticated solutions of the fluid momentum balance which include most of the important momentum exchange terms (e.g. Parker 1978, Pizzuto 1990). Advances have also been made in predicting the critical threshold for entrainment of non-cohesive sediments, particularly along sloping banks (Kovacs and Parker 1994) and for widely graded sediments (Wiberg and Smith 1987). As a result, various authors have built on Parker's (1978) model to include the effects of sediment heterogeneity (Ikeda *et al.* 1988), bank vegetation (Ikeda and Izumi 1990) and suspended load (Ikeda and Izumi 1991). Verification exercises undertaken by the respective authors have, as for extremal hypotheses, indicated that model predictions provide global, if not exacting, agreement with observations (ASCE Task Committee 1998).

The tractive force model, in the form proposed by Parker, has recently been refined by Diplas and Vigilar (1992). The main differences from the previous work are that the governing equations are solved numerically and that the bank geometry was not assumed, but instead became part of the solution (ASCE Task Committee 1998). As a result, the channel shape turned out to be different from a cosine curve, having a larger top width and centre depth (Diplas and Vigilar 1992, Vigilar and Diplas 1997, 1998). To make the findings more amenable to practical applications, Vigilar and Diplas (1998) have developed stable channel design equations and plots based on their modelling results.

In summary, tractive force theory provides a useful method of predicting the equilibrium dimensions of river cross-sections composed of non-cohesive materials. The restrictive nature of the theory indicates that careful application of tractive force models is required in practice. The true value of the tractive force approach appears to rest in its emphasis on prediction of channel form through an understanding of the mechanics of flow and sediment transport in non-cohesive channels. Furthermore, all of the quantitative models for predicting stable channel dimensions reviewed in Sections 17.3 and 17.4 are faced with a major difficulty in terms of practical applications. The functional form of these equations, which relates stable channel dimensions to a single value of the flow discharge, means that it is necessary to specify a 'dominant' discharge value that is representative of the naturally fluctuating discharge regime (Ferguson 1986, Carson and Griffiths 1987). In practice, specification of a dominant discharge is a difficult and complex task, and some authors do not believe that it is possible at all (Pickup and Warner 1976). For further information about the identification of the dominant discharge, readers are referred to articles by Andrews and Nankervis (1995) and Thorne *et al.* (1996).

17.5 NUMERICAL MODELS

Numerical models differ from their conceptual, empirical and analytical counterparts in that they are multidimensional, capable of dealing with both spatial and temporal dimensions. In a numerical model, physical space is represented by a grid or mesh consisting of a finite number of points. Spatial physical properties or characteristics (e.g. landform elevation, water depth, roughness, flow velocity, etc.) are represented on this grid by a set of discrete values. Representation of the physical processes relevant to a particular problem is achieved in two steps. First, the relevant processes are identified and described in mathematical form (i.e. a set of governing process equations is formulated). Second, a numerical algorithm is developed in order to solve or approximate the governing equations over the discretized grid. The time dimension is also discretized into timesteps, and temporal change or evolution is represented by changes in the values on the grid (Table 17.2).

The use of numerical simulation models in fluvial geomorphology has rapidly increased during the last 20 years. Early models were predominantly restricted to one spatial dimension, but advances in computational hardware now provide the capability of dealing with two or three dimensions. It is these advances that are unlocking the real potential of numerical modelling for fluvial geomorphology. Numerous models have been developed for various applications. These can be categorized into four groups of increasing complexity (see Table 17.3). The models mentioned in Table 17.3 cover a whole range of spatial and temporal scales, varying from detailed predictions of flow field in just a few metres of flow reach to the evolution of several kilometres of floodplain over several tens or hundreds of years.

Table 17.2 Representation of the real world in a numerical simulation model

Real world	Model representation
Space dimensions	Grid (discretization)
Time dimension	Timesteps (discretization)
Physical properties	Discrete values on grid
Physical processes	Governing equations
Evolution	Numerical algorithm solves equations and changes values on grid

Concepts of Numerical Modelling

Translating the relevant physical processes into a set of governing equations that can be solved by a numerical algorithm is the key element in numerical modelling. Mathematical descriptions of physical processes can either be derived theoretically (e.g. the equations of fluid motion) or empirically (e.g. most sediment transport equations), but it is the latter approach that is most common in geomorphology (Kirkby 1996). Subsequent solution of the governing equations by means of a numerical algorithm can be achieved in a number of different ways. Exactly which processes are realized, and how this is done, very much depends on the problem under investigation. However, it is important to realize that the predictive capability of a model is largely influenced by the adequacy of the descriptions of the physical processes and by the techniques used for solving the equations.

Representing fluid motion is one of the inherent challenges of all numerical models in fluvial geomorphology. The basic laws governing the motion of fluids are the 'conservation of mass' and 'conservation of momentum'. Mathematically these laws are expressed by a set of non-linear differential equations (Tritton 1988):

$$\frac{\partial \rho}{\partial t} + \nabla \cdot (\rho u) = 0 \tag{17.8}$$

$$\rho \frac{\partial u}{\partial t} = -\nabla p + \mu \nabla^2 u + F \tag{17.9}$$

where u is the velocity vector (m s^{-1}), t the time coordinate (s), ρ the fluid density (kg m^{-3}), p the pressure (Pa), μ the viscosity coefficient (m^2 s^{-1}), and F is the external forces (Pa). In their general form, applying to all fluid motion, these equations are called the Navier–Stokes equations. Theoretically, these equations can be solved exactly, in all three spatial dimensions, and accounting for all turbulence effects, if only the resolution of the grid and the timestep of the calculations are fine enough. However, even for the simplest problems the required resolution would lead to a very large number of grid points and unacceptable computing times (Hervouet and Van Haren 1996, Lane 1998).

Table 17.3 Examples of applications of numerical modelling in fluvial geomorphology

Reference	Modelled processes					Application domain	Illustrated on[a]	Physical scale (m × m)	Grid dimensions	Number of gridcells
	Flow	Solute transport	Sediment transport	Bed-level change	Planform change					
Kalkwijk and De Vriend (1980)	+					Flow in river channels	Experimental data	118 × 6	2	
Vreugdenhill and Wijbenga (1982)	+					Flow in river channels	River Maas, Netherlands	4500 × 600	2	3000
De Vriend and Geldof (1983)	+					Flow in river bends	River Dommel, Netherlands	243 × 6	2	740
Smith and MacLean (1984)	+					Flow in river bends	Experimental data	6.6 × 1	2	
Hodskinson and Ferguson (1998)	+					Flow in river bends	River Dean, UK	20 × 6	3	18 000
Lien *et al.* (1999)	+					Flow in river bends	Experimental data	29 × 6	2	3605
Gee *et al.* (1990)	+					Flow on floodplains	River Fulda, Germany	24 000 × 1000	2	860
Bates *et al.* (1992)	+					Flow on floodplains	River Culm, UK	11 000 × 450	2	1090
Thomas and Williams (1995)	+					Flow on floodplains	Experimental data		3	2 354 688
Bates *et al.* (1997)	+					Flow on floodplains	Missouri River, Nebraska	60 000 × 2000	2	10 000
Bates *et al.* (1998)	+					Flow on floodplains	Hypothetical river	12 000 × 700	2	7310
Olsen and Stokseth (1995)	+					Flow in rivers with large roughness	River Sokna, Norway	80 × 30	3	
Fischer-Antze *et al.* (2001)	+					Flow in vegetated channels	Experimental data	25.5 × 1	3	63 000
Fennema and Chaudhry (1989, 1990)	+					Dambreak	Hypothetical river	200 × 200	2	1681

(*Continues*)

Table 17.3 (*Continued*)

Reference	Modelled processes					Application domain	Illustrated on[a]	Physical scale (m × m)	Grid dimensions	Number of gridcells
	Flow	Solute transport	Sediment transport	Bed-level change	Planform change					
Shettar and Murthy (1997)	+					Flow at channel division	Experimental data	2.4 × 0.3	2	2420
Bradbrook *et al.* (1998)	+					Flow at channel confluence	Experimental data	1 × 0.3	3	77 000
Lane and Richards (1998)	+					Flow in braided channels	Proglacial river, Switzerland	50 × 30	2	150 000
Nicholas and Sambrook Smith (1999)	+					Flow in braided channels	Ferpecle River, Switzerland	75 × 25	3	300 000
Lin and Shiono (1995)	+	+				Solute transport in compound channel	Experimental data		3	
Steward *et al.* (1998)	+	+				Pollutant transport on floodplain	Rivers Culm and Severn, UK	40 000 × 2000	2	12 000
Moulin and Ben Slama (1998)	+	+				Transport of heavy metals	Hypothetical river	400 × 100	2	6037
Lopez and Garcia (1998)	+		+			Sediment transport through vegetation	Experimental data	2.1 × 0.13	2	
Olesen (1987)	+		+	+		Bed load transport	Experimental data	100 × 2.3	2	320
Wu *et al.* (2000)	+		+	+		Sediment transport	Experimental data	80 × 2.44	3	39 930
Nicholas and Walling (1998)	+		+	+		Suspended load on floodplain	River Culm, UK	600 × 700	2	16 800
Hardy *et al.* (2000)	+		+	+		Suspended load on floodplain	River Culm, UK	11 000 × 450	2	9825
Hoey and Ferguson (1994)	+		+	+		Downstream fining in gravel rivers	Alt Dubhaig, Scotland	2800 × 10	1	29
Olsen and Kjellesvig (1998)	+		+	+		Bed scour around cylindrical obstructions	Experimental data	23 × 8.5	3	78 400

(*Continues*)

Table 17.3 (*Continued*)

Reference	Modelled processes					Application domain	Illustrated on[a]	Physical scale (m × m)	Grid dimensions	Number of gridcells
	Flow	Solute transport	Sediment transport	Bed-level change	Planform change					
Vigilar and Diplas (1997, 1998)	+		+	+		Design of stable mobile bed channels	Experimental data	15 × 0.5	2	
Howard (1992, 1996)					+	Long-term meander planform evolution	Hypothetical river		1	
Sun *et al.* (1996)					+	Long-term meander planform evolution	Hypothetical river	20 000 × 5000	1	
Alabyan (1996)	+				+	Bank erosion in meandering channels	Seyakha River, Russia		1	16
Murray and Paola (1994)	+		+	+	+	Evolution of braided channel	Hypothetical river		2	11 000
Pizzuto (1990)	+			+	+	Width adjustment in straight channels	Experimental data		2	3000
Darby *et al.* (1996)	+		+	+	+	Width adjustment in straight channels	Forked Deer River, Tennessee	13 500 × 40	1	
Simon and Darby (1997a)	+		+	+	+	Width adjustment in straight channels	Hypothetical river	4000 × 35	1	30
Mosselman (1992)	+		+	+	+	Width adjustment in meandering channels	Ohre River, Czechoslovakia	400 × 54	2	1725
Nagata *et al.* (2000)	+		+	+	+	Width adjustment in meandering channels	Experimental data	4.5 × 0.4	2	

Notes: These examples serve to illustrate the versatile use of numerical models in fluvial geomorphology. The list is not exhaustive, nor is it intended to be.
[a]The term 'experimental data' refers to data derived from flume experiments and other physical laboratory setups.

Nearly all fluvial geomorphological numerical models rely on solving the Navier–Stokes flow equations in one form or another. The flow equations are usually solved using either finite difference techniques or finite element techniques (Sewell 1988), although spectral methods can be used as well (e.g. Olsen and Stokseth 1995, Pinelli *et al.* 1997). Solution techniques differ mainly in their ease of implementation, computational efficiency and conservation properties, although they can also cause minor variations in model output (Bates *et al.* 1996). In general, finite difference techniques are easier to implement, whereas finite element techniques use less computing time, achieve higher order accuracy and allow more freedom in assembling the computational grid, which is convenient when dealing with highly irregular geometries (Rice 1983, Sewell 1988).

Spatially, geomorphological processes can be viewed in a complex hierarchical context: every geomorphic system consists of a series of ever smaller, lower-level systems, but is at the same time part of a sequence of ever larger, higher-level systems. Depending on the scale of the system and the objective of the investigation, certain levels will be dominant whereas others play a secondary role and can be ignored (De Boer 1992). Numerical simulation of geomorphological processes implicitly involves three levels of the spatial hierarchy. The largest of these is the area under investigation, which is represented by a 2- or 3-dimensional grid. Generally, none of the geomorphological processes on this level are explicitly incorporated in the model, as prediction and study of these processes are usually the purpose of the model. The second level is represented by the individual grid element or cell. The cell forms the core of the numerical model as this is the level on which the processes are explicitly modelled. The third and smallest level of processes is commonly referred to as the subgrid-scale level. Subgrid-scale processes are modelled implicitly by aggregating their effects on the grid element level. Usually, this requires assumptions of the spatial and temporal occurrence

Table 17.4 Inherent limitations of fluvial geomorphological models (Haff 1996)

Limitation	Examples
Model imperfection	Model imperfection refers to the fact that incremental ‘improvement’ in models at laboratory scale do not necessarily add to our ability to make predictions at larger scales. For example, sediment transport is difficult to predict because the uniqueness of each natural sediment bed makes model implementation increasingly difficult as the model becomes more ‘realistic’ (e.g. Gomez and Church 1989)
Omission of significant processes	The larger the scale of the fluvial system, the greater is the chance that more than one important process will be omitted. For example, Tetzlaff and Harbaugh (1989) use a fluvial sedimentation model to simulate alluvial fan evolution, but in some locations debris flows may dominate fan construction (e.g. Whipple and Dunne 1992)
Unknown initial conditions	Initial conditions refer to the distribution of grain sizes, bank material characteristics, bed topography, etc. These conditions are often known only approximately, or in some cases not at all. As far as predictive power is concerned, local site-specific data collection can be at least as important as model choice or model refinement (e.g. ASCE Task Committee 1998)
Sensitivity to initial conditions	Fluvial systems are non-linear, so there can exist a sensitivity to initial conditions that effectively prohibits detailed prediction of system evolution (e.g. Howard 1994, Howard *et al.* 1994, Murray and Paola 1994)
Unresolved heterogeneity	In large-scale fluvial systems it may be impossible to define a meaningful averaging volume for each computational cell. Heterogeneity appears in factors such as vegetative cover, soil type, etc.
External forcing	In fluvial systems, external forcing may be due to increases of discharge resulting from storms or dam releases. Predictive capabilities are limited if unpredictable external forcing can occur. In many cases forcing can only be incorporated statistically if the distribution of events is known (e.g. Patton and Baker 1977)

of subgrid-scale processes (e.g. turbulence) to be made (see Table 17.4). Subgrid-scale processes are treated as such because their explicit modelling would be too demanding on computational resources, or because they are improperly understood.

Numerical grids can differ in many ways: dimension (1, 2 or 3), shape of the elements (triangular, quadragonal, hexagonal), co-ordinate system (Cartesian, cylindrical, curvilinear). These grid attributes are generally determined by the numerical techniques used for solving the mathematical equations and, given a certain model, cannot be influenced by the user. However, the user is usually faced with constructing a grid which captures the physical world. The resolution of the grid affects the internal working of the model and its output (Olsen and Kjellesvig 1998, Hardy *et al.* 1999). It is often thought that the accuracy of model prediction increases with increasing grid resolution. This hypothesis is powered by the idea that a finer grid results in improved representation of the physical world and improved stability of the numerical algorithm (Hardy *et al.* 1999). According to this reasoning, very high predictive accuracy can be achieved if only the necessary computational resources are invested. However, recent research suggests that this hypothesis only holds true up to a certain level and that there is a limit to grid refining, beyond which further increases in spatial resolution will not result in a significant improvement of predictive accuracy (Farajalla and Vieux 1995, Bates *et al.* 1996, Hardy *et al.* 1999). In terms of the hierarchical levels it can be said that this limit is reached when the grid resolution captures all the essential characteristics and variability of the explicitly modelled processes on the grid element level. At that stage further improvement can only be made by explicitly modelling the subgrid-scale processes, that is by 'upgrading' them to the grid element level. However, this is often computationally unachievable, as has been shown for the explicit modelling of turbulence (Hervouet and Van Haren 1996, Lane 1998).

In numerical models the temporal dimension is discretized in timesteps. The temporal scale of processes to be simulated exercises considerable control over the structure of the model and the spatial grid. Simulations over long time periods, i.e. hundreds of years, usually require large timesteps to be computationally efficient, which reflects back into the choice of grid resolution and the notion of what processes can be explicitly modelled on the grid element level and what should be aggregated on the subgrid-scale. When dealing with long timescales, physical processes are often simplified to the point of heuristic summaries of empirical observations (Howard 1996). For some processes the calculations can be decoupled from the numerical computation of the flow field, that is the representative grid values are recalculated at a slower rate, thus saving computational resources. Usually, this concerns processes which also operate at a slower pace in the physical world, like channel width adjustment (Mosselman 1992).

All models require input data. In the case of fluvial models these usually include topography, discharge and bed roughness. Additional input requirements are dependent on the application and can involve elements like bank and floodplain roughness, vegetation, infiltration rate, and so on. It seems trivial to note that the accuracy of the input data influences the output of the model. Nonetheless, this is a point of importance as some data may be difficult or expensive to obtain accurately. Moreover, nearly all field data which are used as input are obtained from a relatively sparse collection of point measurements. Spreading these values over the spatial grid requires assumptions about their spatial (and sometimes temporal) distribution and usually involves some sort of interpolation routine, which introduces yet another source of uncertainty (see Table 17.4).

Frequently, a model uses parameters which can only be guessed at within a certain range (friction being the most typical example). While running the model using a known data set, these parameters are adjusted until an acceptable agreement between model prediction and observed data is found. This process is known as calibration. The parameter values used to obtain the optimal result are then recommended to be used in other applications of the model. This calibration process is not undisputed (Beven and Binley 1992, Bates *et al.* 1998). When several parameters are adjusted during calibration, the uniqueness of an optimal setting is not guaranteed. There might be other combinations of calibration parameters which result in equally acceptable predictions, a condition known as 'model equifinality' (Beven 1996). Furthermore, the obtained calibrated parameters may mask systematic errors in model predictions. Alternative calibration schemes, which partially address these problems, have been proposed (Beven and Binley 1992, Bates *et al.* 1998, Hankin and Beven 1998, Campbell and Cox 1999). It should be noted, however, that the influence of the calibration parameters on the model results can be outweighed by other sources of uncertainty, such as grid construction (Hardy *et al.* 1999) or the limited accuracy of input data.

Once calibrated, a model is validated by running it against another known data set and checking the predicted results versus observed data. If this comparison is satisfactory, the model is said to perform well; if not, the model will be checked for errors and recalibrated. Both calibration and validation require complete data sets, in which some entity, for which predictions can be made, is known. The existence of, or access to, such data sets for natural fluvial systems is not always guaranteed (Bates *et al.* 1997). Very often, therefore, models are calibrated and validated using laboratory data, which may undermine their applicability in natural systems (ASCE Task Committee 1998).

Recent numerical models generally provide a graphical user-interface (GUI) which enhances the ease of use. Numerical models provide a vast amount of data as output, which can be summarized in tables, analyzed statistically or visualized graphically. This often very detailed output can be overwhelming and can easily inspire unjustified faith in its accuracy. For this reason it is important that the user, when interpreting model results, is aware of the underlying assumptions and inherent limitations of a model.

Disadvantages and Benefits over Other Techniques

Numerical modelling is inherently dependent on an adequate representation of the physical world. Accurate translation of physical processes into a realistic set of governing equations and an accurate solution algorithm can be problematic, while obtaining the vast amount of necessary field data can be difficult and expensive. Numerical models can be very demanding on computer resources depending on the construction of the model, duration and complexity of the simulated processes and the required level of accuracy. In many cases, these disadvantages are outweighed by the benefits. Numerical models provide both qualitative and quantitative output, which can readily be visualized in different forms. Moreover, the ability to perform repeated simulations, thus allowing sensitivity tests and what-if scenarios, can be a valuable aid in environmental planning and land or resource management. Possibly, the greatest danger in using numerical models lies with the faith they inspire in their output, neatly organized in tables and visualized in fancy graphical representations. It is tempting to forget that one is using a model: an abstraction of the physical world that embraces many assumptions and uncertainties. The user should be aware of these assumptions and realize the limitations of the model predictions.

17.6 MODELLING APPLICATIONS IN FLUVIAL GEOMORPHOLOGY

Modelling tools are used in a wide range of applications in fluvial geomorphology. Examples of practical modelling applications include prediction of the impacts of flushing flows on aquatic habitat (Milhous 1998a,b), prediction of erosion and sedimentation impacts on land loss (Lohnes 1991) and planform adjustment (Mosselman 1995), as well as the prediction of scour in the vicinity of bridges and other river crossing infrastructure (Melville and Sutherland 1988, Johnson *et al.* 1999). The wide range of modelling tools means that selection of an appropriate model, geared to the demands of a specific application, is a difficult task. However, there are certain generic indicators of model quality, as well as inherent limitations, that can be used as reference points in selecting from this diverse range of modelling tools.

Kirkby (1996) argues that a good model is characterized by an explicit physical basis, simplicity, generality, richness and the potential for scaling up or down. Table 17.5 summarizes these criteria, providing a framework that may be helpful when attempting to assess whether or not a particular model reaches acceptable minimum standards. However, Table 17.5 should not be used to determine whether an individual model is 'better' or 'worse' than another model developed by a competing author. Such comparative assessment exercises are very difficult to undertake, especially for complex numerical simulation models and, unless considerable care is taken, the results of such exercises can be rather arbitrary and misleading. For example, the ASCE Task Committee (1998) tested discrete numerical models of bank erosion and channel widening developed by a range of authors (Pizzuto 1990, Wiele 1992, Li and Wang 1993, Kovacs and Parker 1994) using a common data set obtained from a laboratory study (Ikeda 1981). The Task Committee deliberately avoided assessing the relative performance of the various models because the models deviated in the numerical values of empirical coefficients used in the various process equations used by each of the aforementioned authors.

Table 17.5 indicates that model quality is the product of a series of criteria and, though important, is not simply dependent upon the formulation of complete, appropriate and realistic process laws. In fact, models consist of process laws or predictive rules, as well as a set of input data to characterize the river reach or system for a specific application. Model quality is, therefore, highly constrained by the quality of input data, both in terms of accuracy of data acquisition and the spatial scale at which the data are, or can be, acquired. A good example of the latter is provided by recent developments in the use of computational fluid dynamics (CFD) modelling to investigate fluid flow processes in river channels (e.g. Lane 1998, Lane and Richards 1998, Bates *et al.* 1998, see Section 17.5). In the development and application of CFD models, there has been a trend to increase the spatial resolution (the number of cells representing the spatial area of interest) in the expectation of improved insights into temporal and spatial processes (Hardy *et al.* 1999). Unfortunately, the spatial resolution at which a CFD model is applied affects the solution of the equations and thus the simulation results. Furthermore, data are rarely available at a sufficient level of detail to provide a data value for each cell in a numerical model, so that many values must be interpolated, leading to errors (Kirkby 1990).

Issues of data quality and quantity are implicit within several of the criteria listed in Table 17.5. For models with a low physical basis, parameter values are usually calibrated by model optimization, whereas models with a stronger physical basis may utilize parameters that are universal across a wide range of theory (Kirkby 1996). Models with a weak physical basis, therefore, require large data sets in order to calibrate parameter values and achieve a level of accuracy acceptable for forecasting purposes. Complex models generally have high demands for input data, but large numbers of parameters tend to provide opportunities for achieving a good fit between simulated and observed data. This includes cases where it is qualitatively plain that the 'right' answer is being produced by the 'wrong' set of processes (Kirkby 1996). Selecting a model with an optimal 'quality' level often involves achieving a balance between providing a strong physical basis and reducing model complexity, in order to optimize input data quality. In trading off the increased physical basis of a model with its increasing complexity, a useful notion is the concept of model 'richness'. Model richness is analogous to the net information gain of the model (Kirkby 1996). For example, some complex and highly distributed landscape evolution models (e.g. Tucker and Slingerland 1997) have very large input data requirements, but when used to forecast the sediment yield at the outlet of the watershed their net information gain is strongly negative.

For many practical applications, particularly those where the level of expertize of the user is low relative to the complexity of the model, practical criteria additional to those listed in Table 17.5 may also be significant factors in assessing the overall 'quality' of a model. In particular, model portability, accessibility, cost and usability are, in practice, often just as important as the criteria listed in Table 17.5. These issues are almost exclusively concerned with quantitative models, which are usually implemented and used in the form of computer software packages. It is incumbent upon the developers of such packages to ensure that the modelling software is readily available (e.g. via the Internet), well documented, and portable across a range of different hardware platforms. This is important in facilitating scientific dialogue and exchange and in promoting the dissemination of modelling tools for practical applications. Furthermore, model developers and users need to consider whether or not data entry and interpretation of output data are facilitated by a friendly graphical interface.

In determining minimum standards of model quality, it is important to recognize that there are inherent limitations to fluvial modelling. Several authors have questioned the nature of prediction in the geosciences in general (Tetzlaff 1989, Oreskes *et al.* 1994), and in fluvial geomorphology (Baker 1988, 1994, Haff 1996) and hydrology (Anderson and Woessner 1992, Konikow and Bredehoeft 1992, Rojstaczer 1994) in particular. Schumm (1985, 1991) has pointed out that there are several distinctive features of geomorphological systems that make prediction inherently difficult (Haff 1996). Table 17.5 summarizes these sources of uncertainty and error as they apply to quantitative models in fluvial geomorphology.

Beven (1996) has also discussed problems of equifinality and uncertainty in geomorphological modelling. The problem of equifinality in this context refers to modelling scenarios wherein agreement between modelled and observed data can be obtained by a wide variety of parameter sets. As a result, this leads to uncertainty in inference and prediction (Beven 1996). It will be apparent that this problem is closely related to the issue of model complexity described above, wherein equifinality is associated primarily with complex models that have complex input data requirements. To overcome this difficulty, carefully designed validation and verification exercises are required to demonstrate that agreement between observed and modelled data is assessed both for (non-distributed) bulk parameters, and for spatially distributed parameters (see Section 17.5).

While the above discussion perhaps provides some guidance on acceptable minimum standards for modelling tools, potential users are still faced with the problem of selecting a modelling tool appropriate

Table 17.5 Generic indicators of model quality (from Kirkby 1996)

Fundamental questions	Criteria	Comments
What should a model provide?	Understanding	Models help to promote deeper understanding of the natural environment, and underpin dialogue between the development of theory and critical experiments
	Forecasting potential	Forecasting is important in practical applications, and as a means of testing the validity and range of understanding
What makes a good model?	Physical basis	When models have a strong physical basis, this provides consistency with other theories. This supports their validity, and provides more users within the scientific community
	Simplicity	Models should be as simple as possible, so that they can be understood and communicated. It is difficult to construct a model in which more than three dominant processes interact at a time
	Generality and richness	Good models should be transferable to other geographical areas. Richness refers to the net information gain of the model
	Potential for scaling up or down	There is usually only scope for model application over a range of scales if the model has an explicit physical basis

for a specific application. The wide range of models associated with the treatment of practical fluvial geomorphological problems makes it essential that practitioners adopt a broad and rational approach to such problems (ASCE Task Committee 1998). Such an approach is needed to analyze the majority of problems that arise with the assurance that important factors are not overlooked, appropriate techniques are applied, and hence effective solutions are developed. A generic framework for selecting and using modelling tools in a range of fluvial geomorphological applications is, therefore, now proposed.

17.7 GENERIC FRAMEWORK FOR FLUVIAL GEOMORPHOLOGICAL MODELLING APPLICATIONS

The generic framework presented here is based on a procedure recommended by the ASCE Task Committee (1998). While this Task Committee were concerned with the specific context of modelling river width adjustment, it is still a valid approach that is based on amassing and utilizing a range of methods and techniques appropriate to a specific problem. This framework recognizes that while each case is unique, there are a number of generic elements that are relevant for the majority of specific situations (ASCE Task Committee 1998). Figure 17.2 indicates that the framework consists of a series of methodological steps. The first of these steps, problem identification, is perhaps the most crucial in that it involves the formulation of a clear set of objectives for the modelling application. A fundamental part of problem identification involves determining the resources (time, money, skill level of personnel, etc.) available for the project. The problem should be formulated in terms of whether the modelling objectives are geared towards understanding existing behaviour, or whether predictions of future system behaviour are required. For practical applications, key questions relate to the definition of who or what are affected by the 'problem', and what level of analysis and response is appropriate.

Having completed problem identification, the second step (Figure 17.2) may, if appropriate, comprise a reconnaissance visit to the site and river reaches upstream or downstream (ASCE Task Committee 1998). This is required in order to identify channel characteristics, bank conditions, bank materials, extent of existing or expected bank erosion problems, nature of the flow and bed materials, presence and nature of any vegetation, and presence and condition of any engineering structures. Appropriate

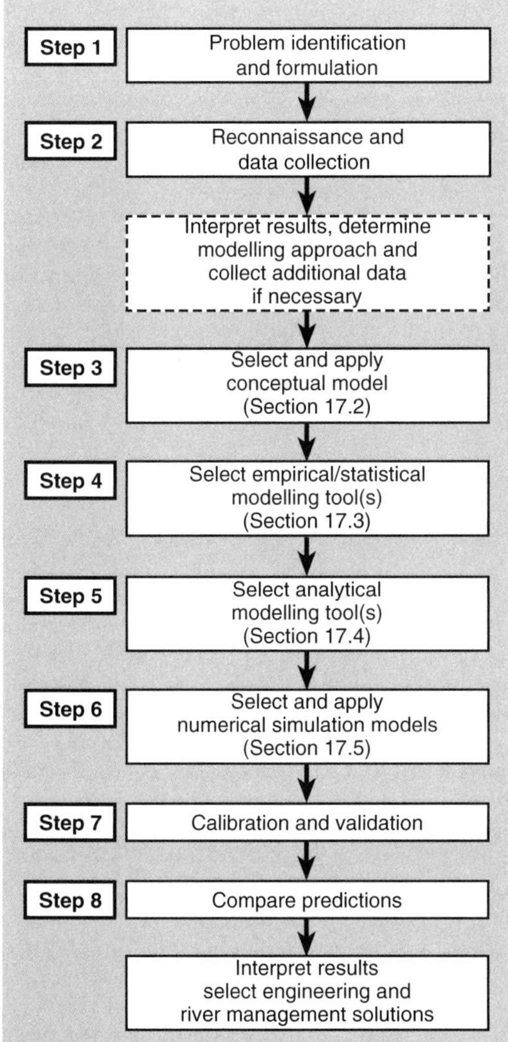
Step 1
Problem identification and formulation
Step 2
Reconnaissance and data collection
Interpret results, determine modelling approach and collect additional data if necessary
Step 3
Select and apply conceptual model (Section 17.2)
Step 4
Select empirical/statistical modelling tool(s) (Section 17.3)
Step 5
Select analytical modelling tool(s) (Section 17.4)
Step 6
Select and apply numerical simulation models (Section 17.5)
Step 7
Calibration and validation
Step 8
Compare predictions
Interpret results select engineering and river management solutions

Figure 17.2 Proposed generic framework for applying modelling tools in practical applications

stream reconnaissance techniques are described by Kellerhals *et al.* (1976), Downs and Brookes (1994) and Thorne *et al.* (1996), among others. Reconnaissance should be viewed as the necessary means by which the modeller familiarizes himself or herself with the characteristics of the problem, as they are manifest in the field. The reconnaissance can also help to elucidate whether the problem is simple or complex, and which aspects of the problem might necessitate selection of specific modelling tools in subsequent steps of the framework. Reconnaissance is, therefore, a key step in developing a strategic view of the problem and planning an appropriate modelling strategy.

After reconnaissance, the user addresses the problem through the application of progressively more complex modelling tools (Figure 17.2). This allows the modeller to absorb manageable increments of knowledge at each step of the process. This step-by-step approach is flexible in its application, depending on the complexity of a particular problem. For very simple problems, users may decide that simple models (Steps 3 and 4) are sufficient to solve the problem without the need for unnecessary investment in complex modelling approaches (Steps 5 and 6). However, for complex problems every step would be used, as understanding developed from application of complex simulation models is often enhanced by attempts to break down the system to its simplest possible components.

Step 3, therefore, involves the selection and application of an appropriate conceptual model (Section 17.2). Conceptual models aid in the identification of dominant processes and trends and help in forming a structure for subsequent, more detailed, modelling. Following on from this, Step 4 consists of the selection and application of relatively simple empirical or statistical models (Section 17.3). Step 4 differs from previous steps in that quantitative rather than qualitative predictions are obtained. If the complexity and severity of the problem merits further analysis, Steps 5 and 6 involve the selection and application of analytical (Section 17.4) and numerical simulation (Section 17.5) models, respectively.

Figure 17.2 shows that application of the quantitative modelling tools in Steps 4–6 normally requires validation (Step 7) of the models concerned against any available field data (see Section 17.6). Assuming that validation confirms that model predictions are reliable, users can then compare the predictions (Step 8) obtained from the various approaches and interpret the results (Step 9). If the user is undertaking a practical modelling application, appropriate engineering or management solutions can then be selected.

17.8 CASE STUDIES

The operation of the generic framework (Figure 17.2) can be illustrated using two examples of typical 'problems'. These case studies illustrate the operation and flexible nature of the framework itself, as well as highlighting specific modelling tools actually used by the various authors concerned.

Example Problem 1: Managing Incised River Channels

Incised river channels are ubiquitous features of disturbed landscapes (Simon and Darby 1999). Rejuvenation of fluvial networks by channel incision often leads to further network development and increased drainage density as gullies migrate into previously non-incised surfaces (Howard 1999). Incised river systems tend to have very high sediment yields, with erosion from the channel banks being a particularly important source of sediment (Simon and Darby 1999). Incision reduces the diversity of physical aquatic habitat and lowering of local ground water tables as a result of incision can have far-reaching effects on floodplain and wetland flora and fauna (Bravard *et al.* 1999). Incised channel systems pose a real challenge to engineers, ecologists, managers, and planners because they are extremely dynamic. Since incised channels convey flows that are even more erosive than the same flow within a non-incised alignment, river-crossing and other in-stream structures are susceptible to failure. Similarly, rapid bank erosion associated with the mass failure of overly heightened banks poses significant land management problems along incised river systems.

For all these reasons modelling tools are commonly used to predict the behaviour of incised channels in support of a range of practical land management, ecological and river engineering and management applications. In terms of identifying a specific 'problem' for the purposes of illustrating the generic framework (Step 1 of Figure 17.2), a key concern relates to the need to assess the magnitude, distribution, and potential for channel erosion along incised river systems. This information is required to estimate the potential for riparian land loss, and to obtain insight into ways of mitigating excessive erosion (Watson and Biedenharn 1999).

A number of stream reconnaissance schemes (Step 2 of Figure 17.2) have been developed for the express purpose of obtaining a strategic overview of conditions along unstable river systems (Simon and Downs 1995, Thorne *et al.* 1996, Johnson *et al.* 1999). Many of these schemes involve the application of conceptual modelling tools (Step 3 of Figure 17.2). In north Mississippi, Schumm *et al.* (1984) developed an incised channel evolution model to predict sequences of channel adjustment (see Chapter 5). Similar models have been proposed by Simon (1989) and Harvey and Watson (1986), amongst others. Identification of the various stages of evolution, mapped over a broad spatial area, enables the determination of stable and unstable reaches, and the limits between different zones of instability. This is useful in directing planning for the construction of erosion control structures. For example, grade-control structures are effective in limiting stream erosion in reaches upstream of a nickpoint (Stage 1 reaches). However, if they are installed downstream of nickpoints, in late-stage (Stages 4, 5 or 6) reaches, they serve only to trap sediment transmitted from eroding reaches upstream. This can help to promote channel recovery upstream (Watson and Biedenharn 1999), but it also initiates bed degradation and bank erosion in sediment starved reaches downstream (Simon and Darby 1997b). Thus, mapping the stage of evolution at catchment scale can help to identify those locations where structures would be most effective, providing a reach-scale focus for the application of more complex modelling tools.

However, empirical and analytical models of the type discussed in Sections 17.3 and 17.4 are not appropriate for the analysis of incised channels. This is because these modelling tools are designed to predict steady-state conditions, whereas incised rivers are very dynamic. In this example, Step 6, which involves the application of numerical simulation models, therefore becomes the next step in the procedure (see Figure 17.2). In recent years, there has been considerable interest in the development of these models, and tools appropriate for two key scales of analysis are now available.

A range of modelling tools appropriate for reach-scale applications have been developed. In the past, many of these numerical simulation models of erosion and sedimentation (e.g. Thomas 1982) assumed that the banks were non-erodible and width was, therefore, fixed with respect to time. However, this assumption is untenable in the context of incised channels, because the very high banks along these rivers are usually unstable with respect to mass failure under gravity. To address this limitation a range of width adjustment models have been developed recently (ASCE Task Committee 1998). These models can be divided into two categories, those that employ some form of extremal hypothesis (e.g. Chang 1988, Yang *et al.* 1988), and those that use mechanistic stability analyses to account explicitly for bank failure and widening (e.g. Pizzuto 1990, Li and Wang 1993, Kovacs and Parker 1994, Darby and Thorne 1996b, Mosselman 1998, Nicholas *et al.* 1999).

The ASCE Task Committee (1998) has comprehensively reviewed all of these models, and the individual details for each are not repeated here. However, it is

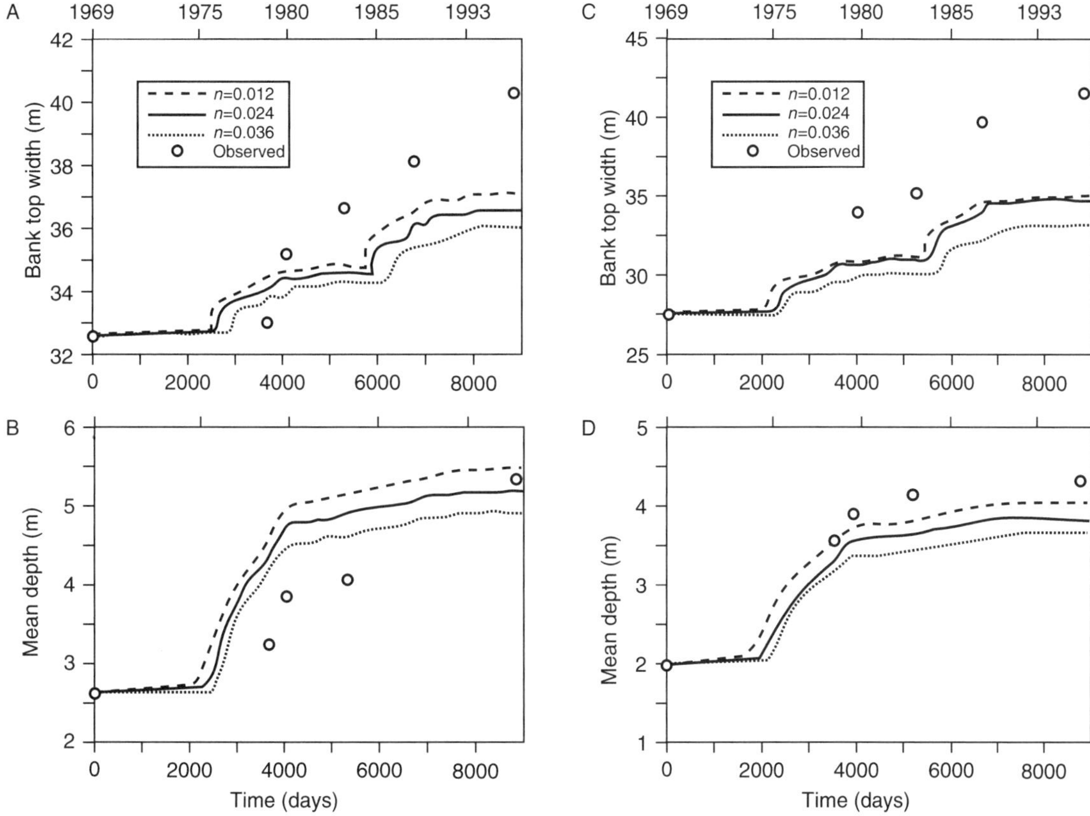

Figure 17.3 Results of field testing of the Darby-Thorne channel widening model at two sites (Chestnut Bluff and Crossroads) on the South Fork Forked Deer River, Tennessee. (A) Widening at Chestnut Bluff; (B) deepening at Chestnut Bluff; (C) widening at Crossroads; (D) deepening at Crossroads [Each graph shows observed (open circles) and predicted (lines) parameter values, with a range of predicted values corresponding to a range of Manning n values (from Darby 1998) (reproduced by permission of John Wiley and Sons, Ltd.)]

useful to showcase the model of Darby and Thorne (1996b) to highlight some of the problems and features of the mechanistic width adjustment modelling approaches. This model was developed to predict changes in channel width caused by cohesive bank erosion along straight incised channels. Predictions are obtained by first applying a simple 1-dimensional step-backwater flow model to estimate flow depths and energy slopes along the modelled reach. This requires an estimate of hydraulic roughness at each model cross-section, which is obtained by relating bed-material grain size to Manning's n via the Strickler (1923) equation. Output from the flow model is next used to predict bed-material transport rates at each model cross-section using an empirically calibrated sediment transport formula (Engelund and Hansen 1967). The model then solves the mass sediment continuity equation to predict changes in bed level, averaged across each cross-section. These changes in bed elevation are used to change the bank height. A simple bank stability analysis is then used to determine whether or not the predicted bank height is stable with respect to mass failure. With the known magnitudes of bed-level and channel width changes resulting from these processes, the model updates the bed topography and width at each cross-section and repeats the iteration cycle until the end of the simulation is reached.

Darby *et al.* (1996) tested this model using data over a 24-year time period obtained from a reach of a channelled stream in West Tennessee (Figure 17.3). It is noteworthy that despite the highly deterministic basis of this somewhat complex model, and although in qualitative terms the agreement between predicted and observed patterns of channel widening is reasonable, there are considerable discrepancies between predicted and observed data. This was attributed in part to the use of a 1-dimensional flow and sediment transport model, which does not provide sufficient spatial resolution to assess changes in bed topography close to the banks. Furthermore, the bank stability analysis used in the model is somewhat idealized (see Darby *et al.* 1996). While this might suggest that more complex flow, sediment transport and bank erosion models might yield better results this is not necessarily the case. In this example, the availability of data to calibrate more sophisticated models is restricted and the application of more complex models would, therefore, involve considerable uncertainty. Hence, present knowledge is such that only tentative predictions of width adjustment can be made using this type of modelling approach. Despite their more restricted theoretical basis, those approaches that are based on extremal hypotheses have, therefore, been used in more engineering applications than the more mechanistic width adjustment models, which are at present used essentially as research tools.

A number of modelling tools are also available for applications at the scale of the entire drainage basin (e.g. Kirkby 1971, Hirano 1975, Ahnert 1976, Willgoose *et al.* 1991, Tucker and Slingerland 1997, Howard 1999). These tools are all essentially advection–dispersion models. On the one hand, mass wasting and soil erosion from the land surface are treated as dispersive terms in the governing equations. On the other hand in-channel erosion, transport and deposition are treated as advective terms. Whether or not the landscape incises is a function of the relative efficacy of sediment supplied from dispersive processes on the slopes versus the rate at which sediment is removed by fluvial transport. These tools are, at present, used primarily for research applications, but their spatially explicit nature means that they have significant potential to be used as tools in the management of incised channel systems. For example, they can be used to indicate the effects of changes in land cover on the development of the river network and long-term water and sediment yield.

Thus, while there is a wide range of numerical modelling tools appropriate for this specific application, many of them are not yet well developed for engineering applications. Instead, the current state of the art places emphasis on composite modelling technologies that combine GIS with conceptual channel evolution and stream reconnaissance techniques. One of the best examples of this type of approach is the interdisciplinary evaluation methodology proposed by Simon and Downs (1995). Their methodology is similar to the framework proposed herein (Figure 17.2) as it employs rapid stream reconnaissance to assess the stability of specific channel sites (see Chapter 11 for a detailed description of this example). By conducting reconnaissance over a large number of sites throughout the drainage basin, spatial stability trends can be identified through GIS-based database management tools. The methodology is based on the identification of channel characteristics that are diagnostic of channel stability or instability. Hence, during a visit to an individual field site, the user would record the values associated with each of the diagnostic variables and then sum these values to calculate a channel-instability index to represent the site as a whole. Since the diagnostic variables employed in the methodology are readily observable, users who have had a relatively small amount of training can complete the site evaluation rapidly. In turn, this allows many sites to be visited across a wide region or throughout an individual drainage basin. Simon and Downs (1995) suggest that ranking the stability of sites across the drainage basin then provides a rational basis for selecting critical sites at which the expense of applying more complex numerical simulation tools is justified. Alternatively, displaying the results in the form of a GIS can help identify spatial trends of instability that can aid interpretation of cause and hence selection of appropriate mitigation strategies.

Example Problem 2: Secondary Flows and River Meander Dynamics

While all rivers are worthy of investigation, meandering rivers have been the subject of extensive research due to their common occurrence, aesthetic appeal, and the intriguing question as to the mechanism of formation of the sinuous planform. Furthermore, meandering rivers migrate across their floodplains through deposition on the point bar at the inner bank matching erosion at the outer bank. As a result, meander migration can adversely impact floodplain dwellers and users through erosion of agricultural land and by causing damage to riparian infrastructure. In terms of problem identification (Step 1 of Figure 17.2), there is a clear priority in furthering our understanding of the process-form linkages between the fluid flows driving morphological change, and the resulting meander migration rates. Particularly important in this regard is the relationship between meander morphology, secondary flow patterns and meander migration rate. This example problem is, therefore, based on the pure scientific application of modelling tools. In this context, stream reconnaissance (Step 2 of Figure 17.2) is not a necessary component of the proposed generic framework, and we may pass directly on to application of relevant conceptual models.

The nature of secondary flow patterns in idealized meander bends is well represented in conceptual models (Step 3 of Figure 17.2) that are based on combinations of theory and empirical observations. Secondary flow at river bends is of the strong, skew-induced type (Prandtl 1952), that is radial forces act on the body of the water in addition to the tangential and frictional forces (Markham and Thorne 1992). Thomson (1876) was among the first to explain the cause of these secondary currents. He proposed that centrifugal forces act on the water column as it flows through the bend, the strength of the centrifugal force varying with velocity vertically through the water column. As a result, relatively fast flowing water near the surface moves towards the outer bank region, causing a build-up of water adjacent to the bank. Conversely, the water surface at the inner bank is drawn down and the two effects together cause a tilting of the free surface and an inwards-acting pressure gradient force. Near the surface the centrifugal force dominates the pressure gradient force and the water is driven outwards, but near the bed the pressure force is dominant and the fluid flows inward, causing the formation of a secondary flow cell that occupies the entire cross-section (Markham and Thorne 1992).

Subsequent empirical investigations have modified this conceptual model of secondary flow in meander bends. First, observations in laboratory flumes (Einstein and Harder 1954) and in natural channels (Markham and Thorne 1992) have identified the existence of an additional secondary flow cell located near the outer bank. Bathurst *et al.* (1977) attributed the formation of this cell to an effect produced by the interaction of the main cell with the outer bank. Second, observations of secondary flow over the point bar near the inner bank have indicated that there is outward flow throughout the entire water column (Hickin 1978, Dietrich and Smith 1983, Markham and Thorne 1992). This is caused by water shoaling over the point bar causing a decrease in downstream water surface slope along the inner bank and a consequent reduction of the cross-stream pressure gradient force (Dietrich and Smith 1983). An overview of the revised conceptual model of secondary flow in meander bends is shown in Figure 17.4. The geomorphological significance of this model is that the secondary flows redistributes the primary flow, such that the shear stress maximum is located near the outer bank just downstream of the bend apex. This explains how the secondary flow field helps to redistribute primary flow momentum in a way that promotes meander migration. Hooke (1997) has reviewed conceptual models of planform development and meander migration.

Empirical models (Step 4 of Figure 17.2) have been used to explain meander migration in terms of large-scale flow separation processes in bends (Markham and Thorne 1992). Flow separation occurs where boundaries turn away from the main flow causing the streamlines to diverge. The exact point at which the flow separates is closely linked to the shape and roughness of the boundary (Markham and Thorne 1992). In particular, a strong relationship between the onset of flow separation in pipes and the ratio of radius of curvature to width (R/W) has been found (Bagnold 1960). As R/W declined to about 2, inner bank separation was found to occur, altering the apparent geometry in such a way that the flow resistance

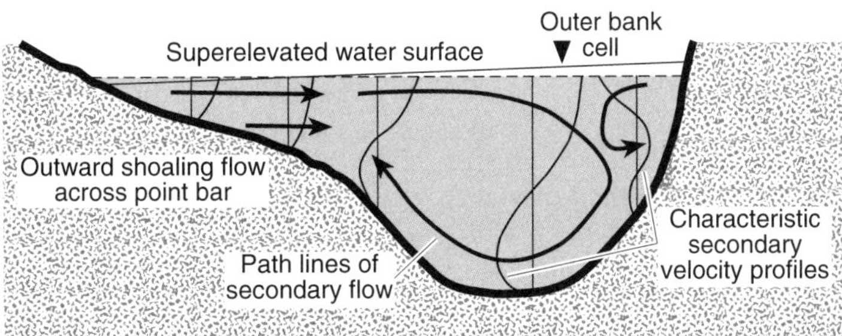

Figure 17.4 Secondary flow at a river bend apex showing the main skew-induced cell restricted to the thalweg, the outer bank cell and shoaling-induced outward flow over the upper point bar (from Markham and Thorne 1992) (reproduced by the permission of John Wiley and Sons, Ltd.)

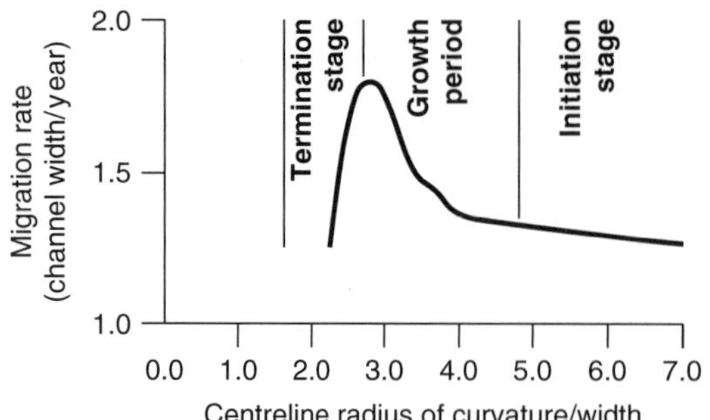

Figure 17.5 Meander migration rate and evolutionary stage as a function of radius of curvature to width ratio (from Markham and Thorne 1992, based on Hickin 1977) (reproduced by permission of John Wiley and Sons, Ltd.)

decreases to a minimum value. In rivers, separation occurs at the inside bank downstream of the apex of the point bar and on the outside of the channel close to, but upstream of, the bend apex (Carey 1969, Leeder and Bridges 1975). Hickin and Nanson (1975) and Hickin (1977, 1978) have used these observations to develop an empirical model of meander migration that relates migration rates to the value of R/W (Figure 17.5). Their data from the Squamish and Beatton Rivers in British Columbia, together with similar studies (e.g. Hooke and Harvey 1983, Biedenharn *et al.* 1989) from other environments, show that meander migration rates have a maximum value associated with R/W values of between around 2 and 3. This is consistent with the observed correspondence of minimum friction factor related to flow separation. This empirical model, therefore, appears to have a reasonably solid theoretical basis, and has potential to be a useful tool when calibrated with for a particular stream system.

Analytical models (Step 5 of Figure 17.2) of river bend morphology have focussed on the interaction between flow, bed topography and sediment load (e.g. Allen 1970, Engelund 1974, Bridge 1977, 1992, Parker and Andrews 1985). These models are concerned with the cross-sectional form of meander bends and in particular the transverse slope and resulting depth of the scour pool adjacent to the outer bank. These models are all fundamentally based on specifying a force balance on the moving bed load grains, coupled to a description of the downstream and transverse velocity field. The models differ in the ways the respective authors make simplifying assumptions regarding the terms in the equations of depth-averaged fluid motion, in the vertical profile of downstream flow velocity, in the description of secondary flow, and in the force balance on bed load grains (Bridge 1992). For example, initial efforts at modelling equilibrium bed topography in bends were restricted to steady, uniform bend flow (e.g. Rozovskii 1961, Engelund 1974, Falcon and Kennedy 1983), while more recent efforts have been directed to the case of non-uniform flow (e.g. Engelund 1974, Struiksma *et al.* 1985, Odgaard 1989, Johannesson and Parker 1989). In addition, efforts have also been made to include the effects of sorting within the analysis (e.g. Allen 1970, Parker and Andrews 1985, Bridge 1992). Probably the most comprehensive of these approaches is the model proposed by Bridge (1992). Importantly this is the only one that has been subjected to rigorous testing across a range of different river environments (1992). These results indicate that the Bridge (1992) model predicts bed topography in natural river bends quite well. Nonetheless, the assumptions embodied within all these models restrict their application to certain specific types of river channels, as listed in Table 17.6, and this must be recognized when using these models for engineering applications.

Two specific limitations with these analytical models warrant brief further discussion. First, Dietrich and Smith (1984) have argued that there are several deficiencies in the governing flow and force balance equations used therein. In particular, they argue that these models neglect convective flow accelerations that are of primary importance in controlling the spatial distribution of boundary shear stress within the bend. The second limitation is that no account is taken of bank erodibility. Thorne and Osman (1988) have discussed the implications of this constraint. For easily erodible banks, the depth of the scour pool adjacent to the outer bank is limited because the bank became prone to mass failure promoting the development of a wider, shallower, cross-section. This effect is neglected in models that take no account of bank erosion, so that unrealistically deep scour pools are predicted as a result. This effect is evident in comparisons of predicted versus observed cross-sections (Figure 17.6) taken from Muddy Creek, Wyoming (Bridge 1992). This figure clearly shows that the depth of the scour pool adjacent to erodible outer banks can be significantly over-predicted.

Numerical simulations (Step 6 of Figure 17.2) of flow and meander migration in bends are becoming increasingly popular as more detailed explanations of the physical processes involved in the formation and evolution of meander bends are sought. A number of models of meander flow and sediment transport have

Table 17.6 Summary of restrictive assumptions embodied in the Bridge (1992) meander bed topography model

Assumption	Notes
Radius of curvature ≫ flow depth	A reasonable assumption for many bends
Channel width ≪ distance between crossovers	A reasonable assumption for many bends
Channel width ≪ centreline radius of curvature	May not apply in very tight bends in wide channels
Channel width ≫ centreline flow depth	A reasonable assumption at most formative discharges
Width and depth should not vary appreciably in the downstream direction	This is particularly restrictive in many natural bends with irregular planforms or where localized bank erosion has occurred
Primary flow velocity ≫ transverse flow velocity	This assumption may not hold close to steep, outer banks where secondary flow velocity magnitudes can be quite high. This implies the model should not be applied within the near-bank zone

been proposed (Blondeaux and Seminara 1985, Nelson and Smith 1989, Johannesson and Parker 1989). There has been particular interest in the use of high-resolution models (e.g. Lane 1998) to improve understanding of secondary flow patterns in bends. For example, Hodskinson and Ferguson (1998) investigated the morphological controls on flow separation in meander bends. They used a 3-dimensional, fixed-lid CFD model to simulate a series of idealized channel bends. The simulations demonstrated that the existence and extent of concave bank flow separation can be significantly influenced by changes in bend planform, point bar topography and upstream planform. These models are useful over relatively short time and spatial scales, but they require a great deal of input data and technical expertize. At present, this has restricted their use in more general morphological models used to account for changes in channel planform resulting from bank erosion and accretion processes. However, as hardware and software develops, it is likely that this type of flow model will, in the future, be used more frequently for morphological modelling problems.

In the meantime, a number of numerical meander migration models have been proposed. The simplest of these models (e.g. Ikeda *et al.* 1981, Parker *et al.* 1982, Howard 1992, Sun *et al.* 1996) calculate meander migration using functional relationships of the form

$$\xi = \varepsilon(U + U_b) \tag{17.10}$$

where ξ is the outer bank erosion rate, U is the mean flow velocity, U_b is the near-bank flow velocity and ε is a proportionality coefficient whose value is determined by calibration. Models based on Equation (17.10) are attractive in that they are relatively simple and they do provide useful qualitative results when simulating the long-term (>100 years) migration of meandering rivers. However, there are a number of key deficiencies that make these models inappropriate for use in engineering applications over shorter timescales. First, they utilize highly idealized versions of the governing flow equations to obtain the required estimate of near-bank velocity. Specifically, because the governing equations are depth-averaged, they cannot describe the secondary flow field accurately. To compensate for this effect, they artificially force the secondary flow field to adapt to the local bed topography, but this provides an approximation that is too crude for many applications (Johannesson and Parker 1989). Second, although the proportionality coefficient is considered to be a 'catch-all' parameter representing the relative erodibility of the bank materials, in practice its value must be determined by calibration. This requires detailed historical data on meander planform changes which are not often readily available. A final deficiency is that these models normally assume that the erodibility coefficient is constant both through space and time. This is most unlikely in the context of natural meandering streams, which create erodible point bars and resistant fine-grained deposits (in ox-bow lakes) as they migrate across the floodplain. The model by Sun *et al.* (1996) is probably the most useful of this category of models precisely because these authors allow the local erodibility of the floodplain to vary in relation to the formation of these discrete deposits.

Recent modelling efforts (Mosselman 1998, Nagata *et al.* 2000) have attempted to address these

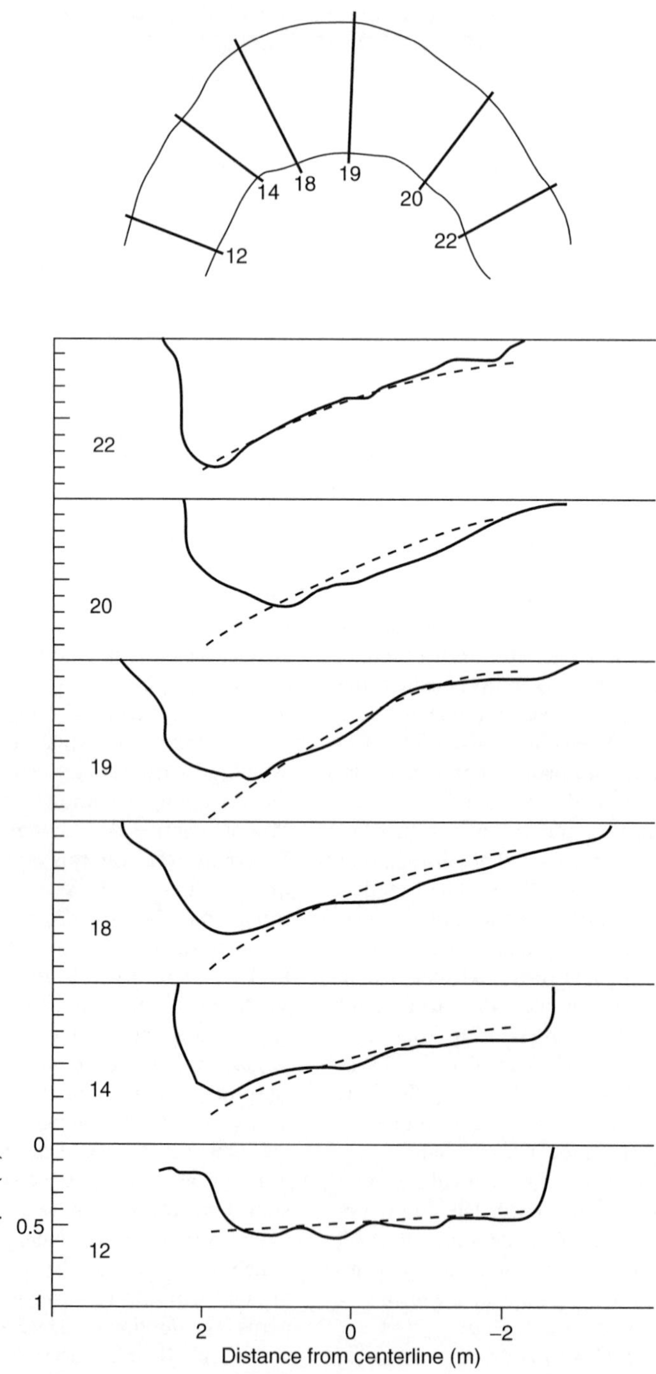

Figure 17.6 Comparison of observed (solid curves) and predicted (dashed curve) equilibrium bed topography for Muddy Creek, Wyoming (from Bridge 1992) (copyright by the American Geophysical Union). Note that the depth of the pool at the erodible outer bank is significantly over-predicted at cross-sections 18–20 near the bend apex

deficiencies by including much more realistic descriptions of the flow, sediment transport and bank erosion processes to develop modelling tools suitable for engineering applications. However, these numerical models are highly complex and require significant training, as well as large amounts of input data. These models utilize a three-step solution procedure. In the first step, the flow is computed while keeping the bed and bank configuration fixed. Sediment transport and bank erosion rates resulting from the predicted flow field are then computed. Second, bed level changes are computed from the sediment transport flux gradients and the input of bank erosion products. Finally, bankline changes are computed from the bank erosion rates (Mosselman 1998). The predictive ability of these models is, therefore, related in part to the predictive abilities of each of the submodels used in these three modules. In fact, limitations with these models are, therefore, to be expected because they utilize empirically calibrated sediment transport equations. Likewise, each model utilizes a bank erosion submodel tailored for specific physical environments. Both bank erosion submodels simulate the undermining and collapse of the bank material under gravity in response to erosion of the bed at the toe of the bank (see Figure 17.7), but while Nagata *et al.* (2000) simulated non-cohesive bank materials, Mosselman's (1998) model is tailored to cohesive banks.

A key feature of both these meander migration models is that neither have yet been rigorously tested using field data, though preliminary simulations using flume data indicate that the predictive ability of the

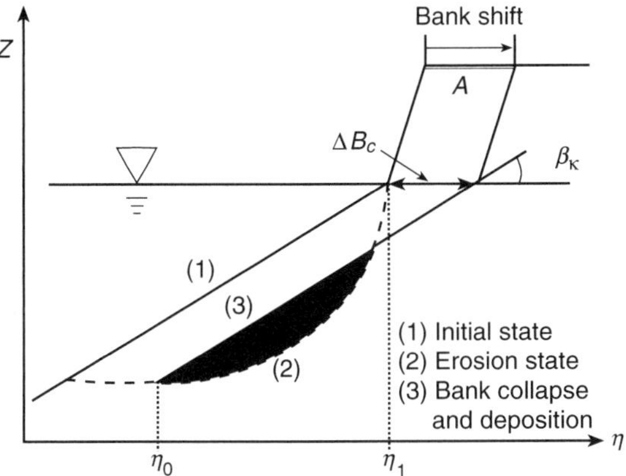

Figure 17.7 Illustration of the bank slumping model used in the meander migration model proposed by Nagata *et al.* (2000) (reproduced by permission of the American Society of Civil Engineers)

Nagata *et al.* (2000) model is quite good (Figure 17.8). These models must, therefore, be regarded as being in an early stage of development and are at present research, rather than engineering, modelling tools. Nonetheless, these approaches represent the state of the art in simulating meander migration and offer considerable potential. This is because they offer generic frameworks into which improvements in individual submodels (e.g. use of 3-dimensional flow models, improved bank erosion models, etc.) can be readily introduced. In this respect, the key advance in the meander migration model developed by Mosselman (1998) is that it utilizes a channel boundary fitted co-ordinate system. This allows the grid to become non-orthogonal during planform adjustment and is, therefore, well suited for simulating natural meandering rivers in which localized bank failure can lead to non-uniform channel width variation (and hence significant grid distortion) in the streamwise direction.

17.9 CONCLUSION

This chapter has shown that modelling tools are used for a wide variety of both pure and practical applications in fluvial geomorphology. It is, therefore, unsurprising that a great many different types of models have been developed (Sections 17.2–17.5). Excluding physical models from consideration, these types of models were herein classified into four main categories (conceptual or theoretical, statistical or empirical, analytical and numerical simulation models). Within each of these categories a selection of models was reviewed, highlighting the strengths, weaknesses, capabilities and limitations of each discrete approach.

That there is considerable diversity in the range of modelling tools, both within and between different categories, indicates that users are faced with significant practical difficulties when selecting a modelling tool or tools to address a specific application. Fortunately, certain generic indicators of model quality (Table 17.5) can be identified, and these can be used to help decide if a particular model meets acceptable quality standards for a specific application. Similarly, certain inherent limitations of models can be identified (Table 17.4). A particular constraint on the overall quality of a specific model appears to be the amount of input data required to obtain a prediction, relative to the amount of output data obtained. Furthermore, the scale at which input data is or can be acquired, relative to the scale at which the model is applied, is another key limiting factor. These constraints mean that there is often (though not always) a trade off between choosing

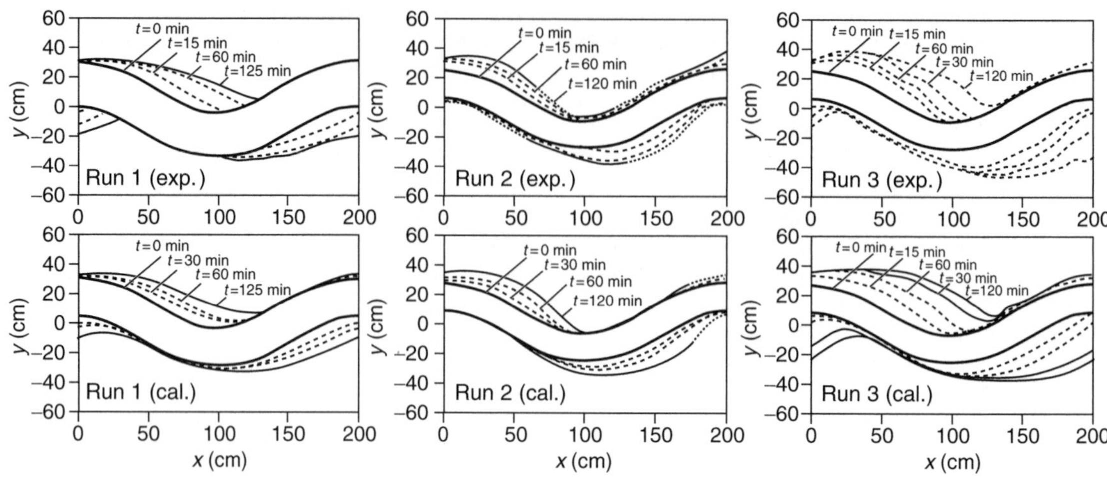

Figure 17.8 Comparison of numerical simulation results (lower plots) against observations for three laboratory flume experiments (from Nagata *et al.* 2000) (reproduced by permission of the American Society of Civil Engineers)

models with a strong physical basis and complex data requirements, and choosing models which have less demanding data requirements and a commensurately lower theoretical basis.

It is important to recognize that ranking the quality of individual models is difficult, time consuming, and often ultimately flawed. Instead of providing recommendations regarding the selection of different types of fluvial geomorphological models, a flexible generic framework for model applications has been proposed instead. In this framework, clear problem formulation provides the basis for a rational modelling approach. For modelling applications involving relatively simple problems, conceptual or statistical models alone may be sufficient to generate reliable predictions and develop the required level of understanding of the problem. For more complex problems, application of a spectrum of modelling strategies ranging from simple to complex is required. The application of multiple modelling tools provides the user with overlapping sets of predictions that leads to an enhanced level of understanding, as well as increased confidence in the predictions themselves.

ACKNOWLEDGEMENTS

Marco J. Van de Wiel was supported by a University of Southampton High Performance Computing Bursary.

REFERENCES

Abrahams, A.D., Li, G. and Atkinson, J.F. 1994. Step-pool streams: adjustment to maximum flow resistance. *Water Resources Research* 31: 2593–2602.

Ahnert, F. 1976. Brief description of a comprehensive three-dimensional process-response model of landform development. *Zeitschrift für Geomorphologie* 25: 29–49.

Alabyan, A. 1996. A computer model of bank erosion based on secondary flow simulation. In: Ashworth, P.J., Bennett, S.J., Best, J.L. and McLelland, S.J. eds., *Coherent Flow Structures*, Chichester, England: John Wiley and Sons, pp. 567–580.

Allen, J.R.L. 1970. Studies in fluviatile sedimentation: a comparison of fining-upwards cyclotherms, with special reference to coarse-member composition and interpretation. *Journal of Sedimentary Petrology* 40: 298–323.

American Society of Civil Engineers Task Committee. 1998. River width adjustment. II. Modelling. *Journal of Hydraulic Engineering* 124: 903–917.

Anderson, M.P. and Woessner, W.W. 1992. The role of the postaudit in model validation. *Advances in Water Resources* 15: 167–173.

Andrews, E.D. 1984. Bed-material entrainment and hydraulic geometry of gravel-bed rivers in Colorado. *Bulletin of the Geological Society of America* 95: 371–378.

Andrews, E.D. and Nankervis, J.M. 1995. Effective discharge and the design of channel maintenance flows for gravel-bed rivers. In: Costa, J.E., Miller, A.J., Potter, K.W. and Wilcock, P.R., eds., *Natural and Anthropogenic Influences in Fluvial Geomorphology*, Washington, DC: American Geophysical Union, pp. 151–164.

Ashmore, P.E. 1982. Laboratory modelling of gravel braided stream morphology. *Earth Surface Processes and Landforms* 7: 201–225.

Ashworth, P.J., Best, J.L., Leddy, J.O. and Geehan, G.W. 1994. The physical modelling of braided rivers and deposition of fine-grained sediment. In: Kirkby, M.J., ed., *Process Models and Theoretical Geomorphology*, Chichester, England: John Wiley and Sons, pp. 115–139.

Bagnold, R.A. 1960. *Some Aspects of the Shape of River Meanders*, US Geological Survey Professional Paper 282E, Washington, DC.

Baker, V.R. 1988. Geological fluvial geomorphology. *Bulletin of the Geological Society of America* 100: 1157–1167.

Baker, V.R. 1994. Geomorphological understanding of floods. *Geomorphology* 10: 139–156.

Bates, P.D., Anderson, M.G., Baird, L., Walling, D.E. and Simm, D. 1992. Modelling floodplain flow with a two dimensional finite element scheme. *Earth Surface Processes and Landforms* 17: 575–588.

Bates, P.D., Anderson, M.G., Price, D.A., Hardy, R.J. and Smith C.N. 1996. Analysis and development of hydraulic models for floodplain flows. In: Anderson, M.G., Walling, D.E. and Bates, P.D., eds., *Floodplain Processes*, Chichester, England: John Wiley and Sons, pp. 215–254.

Bates, P.D., Horritt, M. and Hervouet, J.M. 1998. Investigating two-dimensional, finite element predictions of floodplain inundation using fractal generated topography. *Hydrological Processes* 12: 1257–1277.

Bates, P.D., Horritt, M.S., Smith, C.N. and Mason, D. 1997. Integrating remote sensing observations of flood hydrology and hydraulic modelling. *Hydrological Processes* 11: 1777–1795.

Bathurst, J.C. 1997. Environmental river flow hydraulics. In: Thorne, C.R., Hey, R.D. and Newson, M.D., eds., *Applied Fluvial Geomorphology for River Engineering and Management*, Chichester, England: John Wiley and Sons, pp. 69–93.

Bathurst, J.C., Thorne, C.R. and Hey, R.D. 1977. Direct measurements of secondary currents in river bends. *Nature* 269: 504–506.

Benson, M.A. 1965. Spurious correlation in hydraulics and hydrology. *Journal of the Hydraulics Division of the American Society of Civil Engineers* 91: 35–43.

Bettess, R., White, W.R. and Reeve, C.E. 1988. On the width of regime channels. In: White, W.R., ed., *River Regime*, Chichester, England: John Wiley and Sons, pp. 149–161.

Beven, K. 1996. Equifinality and uncertainty in geomorphological modelling. In: Rhoads, B.L. and Thorn, C.E., eds., *The Scientific Nature of Geomorphology*, Chichester, England: John Wiley and Sons, pp. 289–313.

Beven, K. and Binley, A. 1992. The future of distributed models: model calibration and uncertainty prediction. *Hydrological Processes* 6: 279–298.

Biedenharn, D.S., Combs, P.G., Hill, G.J., Pinkard, C.F. and Pinkstone, C.B. 1989. Relationship between channel migration and radius of curvature on the Red River. In: Wang, S.S.Y., ed., *Sediment Transport Modelling*, New York: American Society of Civil Engineers, pp. 536–541.

Blondeaux, P. and Seminara, G. 1985. A unified bar-bend theory of river meanders. *Journal of Fluid Mechanics* 112: 363–377.

Bradbrook, K.F., Biron, P.M., Lane, S.N., Richards, K.S. and Roy, A.G. 1998. Investigation of controls on secondary circulation in a simple confluence geometry using a three-dimensional numerical model. *Hydrological Processes* 12: 1371–1396.

Bravard, J.P., Kondolf, G.M. and Piégay, H. 1999. Environmental and societal effects of channel incision and remedial strategies. In: Darby, S.E. and Simon, A., eds., *Incised River Channels*, Chichester, England: John Wiley and Sons, pp. 303–341.

Bray, D.I. 1982. Regime equations for gravel-bed rivers. In: Hey, R.D., Bathurst, J.C. and Thorne, C.R., eds., *Gravel-bed Rivers*, Chichester, England: John Wiley and Sons, pp. 517–544.

Bridge, J.S. 1977. Flow, bed topography, grain size and sedimentary structure in open-channel bends: a three-dimensional model. *Earth Surface Processes* 2: 401–416.

Bridge, J.S. 1992. A revised model for water flow, sediment transport, bed topography and grain size sorting in natural river bends. *Water Resources Research* 28: 999–1013.

Brookes, A. 1988. *Channelized Rivers*, Chichester, England: John Wiley and Sons.

Brookes, A. and Shields, F.D., Jr., eds., 1996. *River Channel Restoration*. Chichester, England: John Wiley and Sons.

Brookes, A. and Sear, D.A. 1996. Geomorphological principles for restoring channels. In: Brookes, A. and Shields, F.D., Jr., eds., *River Channel Restoration*, Chichester, England: John Wiley and Sons, pp. 75–101.

Buffington, J.M. and Montgomery, D.R. 1997. A systematic analysis of eight decades of incipient motion studies, with special reference to gravel-bedded rivers. *Water Resources Research* 33: 1993–2029.

Buffington, J.M., Dietrich, W.E. and Kirchner, J.W. 1992. Friction angle measurements on a naturally formed gravel stream bed: implications for critical boundary shear stress. *Water Resources Research* 28: 411–425.

Caddie, G.H. 1969. *An Analysis of Some Data from Natural Alluvial Channels*, United States Geological Survey Professional Paper 650C, Washington, DC, pp. 188–194.

Campbell, E.P. and Cox, D.R. 1999. A Bayesian approach to parameter estimation and pooling in nonlinear flood event models. *Water Resources Research* 35: 211–220.

Carey, W.C. 1969. Formation of flood plain lands. *Journal of the Hydraulics Division of the American Society of Civil Engineers* 95: 981–994.

Carson, M.A. and Griffiths, G.A. 1987. Bedload transport in gravel channels. *Journal of Hydrology (New Zealand)* 26: 1–151.

Chang, H.H. 1980. Geometry of gravel streams. *Journal of the Hydraulics Division of the American Society of Civil Engineers* 105: 1443–1456.

Chang, H.H. 1988. *Fluvial Processes in River Engineering*, New York: Wiley Interscience.

Charlton, F.G., Brown, P.M. and Benson, R.W. 1978. The hydraulic geometry of some gravel rivers in Britain, Report IT180, Hydraulics Research Station, Wallingford, England.

Chin, A. 1999. The morphologic structure of step-pools in mountain streams. *Geomorphology* 27: 191–204.

Darby, S.E. 1998. Modelling width adjustment in straight alluvial channels. *Hydrological Processes* 12: 1299–1321.

Darby, S.E. and Thorne, C.R. 1994. Fluvial maintenance operations in managed alluvial rivers. *Aquatic Conservation: Marine and Freshwater Ecosystems* 4: 130–148.

Darby, S.E. and Thorne, C.R. 1996a. Modelling the sensitivity of channel adjustments in destabilized sand-bed rivers. *Earth Surface Processes and Landforms* 21: 1109–1125.

Darby, S.E. and Thorne, C.R. 1996b. Numerical simulation of widening and bed deformation of straight sand-bed rivers. I. Model development. *Journal of Hydraulic Engineering* 122:184–193.

Darby, S.E., Thorne, C.R. and Simon, A. 1996. Numerical simulation of widening and bed deformation of straight sand-bed rivers. II. Model evaluation. *Journal of Hydraulic Engineering* 122: 194–202.

Davies, T.R.H. and Sutherland, A.J. 1983. Extremal hypotheses for river behaviour. *Water Resources Research* 19: 141–148.

Davis, W.M. 1899. The geographical cycle. *Geographical Journal* 14: 481–504.

Davy, B.W. and Davies, T.R.H. 1979. Entropy concepts in fluvial geomorphology: a reevaluation. *Water Resources Research* 15: 103–106.

Dietrich, W.E. and Smith, J.D. 1983. Influence of the point bar on flow through curved channels. *Water Resources Research* 19: 1173–1192.

De Boer, D.H. 1992. Hierarchies and spatial scale in process geomorphology: a review. *Geomorphology* 4: 303–318.

De Vriend, H.J. and Geldof, H.J. 1983. Main flow velocity in short river bends. *Journal of Hydraulic Engineering* 109: 991–1011.

Dietrich, W.E. and Smith, J.D. 1984. Bed load transport in a river meander. *Water Resources Research* 20: 1355–1380.

Diplas, P. 1990. Characteristics of self-formed straight channels. *Journal of Hydraulic Engineering* 116: 707–728.

Diplas, P. and Vigilar, G.G. 1992. Hydraulic geometry of threshold channels. *Journal of Hydraulic Engineering* 118: 597–614.

Downs, P.W. and Brookes, A. 1994. Developing a standard geomorphological approach for the appraisal of river projects. In: Kirby, C. and White, W.R., eds., *Integrated River Basin Development*, Chichester, England: John Wiley and Sons, pp. 299–310.

Einstein, H.A. and Harder, J.A. 1954. Velocity distribution and the boundary layer at channel bends. *Transactions of the American Geophysical Union* 35: 114–120.

Elliott, J.G., Gellis, A.C. and Aby, S.B. 1999. Evolution of arroyos: Incised channels of the Southwestern United States. In: Darby, S.E. and Simon, A., eds., *Incised River Channels*, Chichester, England: John Wiley and Sons, pp. 153–185.

Engelund, F. 1974. Flow and bed topography in channel bends. *Journal of the Hydraulics Division of the American Society of Civil Engineers* 100: 1631–1648.

Engelund, F. and Hansen, E. 1967. *A Monograph on Sediment Transport in Alluvial Streams*, Copenhagen: Teknisk Forlag.

Falcon, M.A. and Kennedy, J.F. 1983. Flow in alluvial river curves. *Journal of Fluid Mechanics* 133: 1–16.

Farajalla, N.S. and Vieux, B.E. 1995. Capturing the essential spatial variability in distributed hydrological modelling: Infiltration parameters. *Hydrological Processes* 9: 55–68.

Fennema, R.J. and Chaudhry, M.H. 1989. Implicit methods for two-dimensional unsteady free-surface flows. *Journal of Hydraulic Research* 27: 321–332.

Fennema, R.J. and Chaudhry, M.H. 1990. Explicit methods for 2-D transient free-surface flows. *Journal of Hydraulic Engineering* 116: 1013–1034.

Ferguson, R.I. 1986. Hydraulics and hydraulic geometry. *Progress in Physical Geography* 10: 1–31.

Fischer-Antze, T., Stoesser, T., Bates, P. and Olsen, N.R.B. 2001. 3D numerical modelling of open-channel flow with submerged vegetation. *Journal of Hydraulic Research* 39: 303–310.

Galay, V.J. 1983. Causes of river bed degradation. *Water Resources Research* 19: 1057–1090.

Gee, D.M., Anderson, M.G. and Baird, L. 1990. Large scale floodplain modelling. *Earth Surface Processes and Landforms* 15: 513–523.

Geist, D.R. and Dauble, D.D. 1998. Redd site selection and spawning habitat use by fall chinook salmon: the importance of geomorphic features in large rivers. *Environmental Management* 22: 655–669.

Glock, W.S. 1931. The development of drainage systems: a synoptic view. *Geographical Review* 21: 475–482.

Glover, R.E. and Florey, Q.L. 1951. *Stable Channel Profiles*, Denver, Colorado: US Bureau of Reclamation.

Gomez, B. and Church, M. 1989. An assessment of bed load sediment transport formulae for gravel bed rivers. *Water Resources Research* 25: 1161–1186.

Gore, J.A. 1996. Foreword. In: Brookes, A. and Shields, F.D., eds., *River Channel Restoration*, Chichester, England: John Wiley and Sons, pp. xiii–xv.

Griffiths, G.A. 1984. Extremal hypotheses for river regime: an illusion of progress. *Water Resources Research* 20: 113–118.

Haff, P.K. 1996. Limitations on predictive modelling in geomorphology. In: Rhoads, B.L. and Thorn, C.E., eds., *The Scientific Nature of Geomorphology*, Chichester, England: John Wiley and Sons, pp. 337–358.

Hankin, B.C. and Beven, K.J. 1998. Modeling dispersion in complex open channel flows: Fuzzy calibration. *Stochastic Hydrology and Hydraulics* 12: 397–412.

Hardy, R.J., Bates, P.D. and Anderson, M.G. 1999. The importance of spatial resolution in hydraulic models for floodplain environments. *Journal of Hydrology* 216: 124–136.

Hardy, R.J., Bates, P.D. and Anderson, M.G. 2000. Modelling suspended sediment deposition on a fluvial floodplain using a two-dimensional finite element model. *Journal of Hydrology* 229: 202–218.

Harvey, M.D. and Watson, C.C. 1986. Fluvial processes and morphological thresholds in incised channel restoration. *Water Resources Bulletin* 3: 359–368.

Henderson, F.M. 1966, *Open Channel Flow*, New York: Macmillan.

Hervouet, J.-M. and Van Haren, L. 1996. Recent advances in numerical methods for fluid flows. In: Anderson, M.G., Walling, D.E. and Bates, P.D., eds., *Floodplain Processes*, Chichester, England: John Wiley and Sons, pp. 183–214.

Hey, R.D. 1978. Determinate hydraulic geometry of river channels. *Journal of the Hydraulics Division, American Society of Civil Engineers* 104: 869–885.

Hey, R.D. and Thorne, C.R. 1986. Stable channels with mobile gravel beds. *Journal of Hydraulic Engineering* 112: 671–689.

Hey, R.D. and Heritage, G.L. 1988. Dimensional and dimensionless regime equations for gravel-bed rivers. In: White, W.R., ed., *River Regime*, Chichester, England: John Wiley and Sons, pp. 1–8.

Hickin, E.J. 1977. Hydraulic factors controlling channel migration. In: Davidson-Arnott, R. and Nickling, W., eds., *Research in Fluvial Systems*, Norwich, England: Geobooks, pp. 59–72.

Hickin, E.J. 1978. Mean flow structure in meanders of the Squamish River, British Columbia. *Canadian Journal of Earth Sciences* 15: 1833–1849.

Hickin, E.J. and Nanson, G.C. 1975. The character of channel migration on the Beatton River, Northeast British Columbia, Canada. *Bulletin of the Geological Society of America* 86: 487–494.

Hirano, M. 1975. Simulation of developmental process of interfluvial slopes with reference to graded form. *Journal of Geology* 83: 111–123.

Hodskinson, A. and Ferguson, R.I. 1998. Numerical modelling of separated flow in river bends: model testing and experimental investigation of geometric controls on the extent of flow separation at the concave bank. *Hydrological Processes* 12: 1323–1338.

Hoey, T.B. and Ferguson, R. 1994. Numerical simulation of downstream fining by selective transport in gravel bed rivers: Model development and illustration. *Water Resources Research* 30: 2251–2260.

Hooke, J.M. 1997. Styles of channel change. In: Thorne, C. R., Hey, R.D. and Newson, M.D., eds., *Applied Fluvial Geomorphology for River Engineering and Management*, Chichester, England: John Wiley and Sons, pp. 237–268.

Hooke, J.M. and Harvey, A.M. 1983. Meander changes in relation to bend morphology and secondary flows. In: Collinson, J. and Lewin, J., eds., *Modern and Ancient Fluvial Systems*, Oxford: Blackwell, pp. 121–132.

Howard, A.D. 1992. Modelling channel migration and floodplain development in meandering streams. In: Carling, P.A. and Petts, G.E., eds., *Lowland Floodplain Rivers*, Chichester, England: John Wiley and Sons, pp. 1–42.

Howard, A.D. 1994. A detachment-limited model of drainage basin evolution. *Water Resources Research* 30: 2261–2285.

Howard, A.D. 1996. Modelling channel evolution and floodplain morphology. In: Anderson, M.G., Walling, D.E. and Bates, P.D. eds., *Floodplain Processes*, Chichester, England: John Wiley and Sons, pp. 15–62.

Howard, A.D. 1999. Simulation of gully erosion and bistable landforms. In: Darby, S.E. and Simon, A. eds., *Incised River Channels*, Chichester, England: John Wiley and Sons, pp. 277–299.

Howard, A.D., Dietrich, W.E. and Seidl, M.A. 1994. Modeling fluvial erosion on regional to continental scales. *Journal of Geophysical Research* 99: 13 971–13 986.

Ikeda, S. 1981. Self-formed straight channels in sandy beds. *Journal of the Hydraulics Division of the American Society of Civil Engineers* 107: 389–406.

Ikeda, S. and Izumi, N. 1990. Width and depth of self-formed straight gravel rivers with bank vegetation. *Water Resources Research* 26: 2353–2364.

Ikeda, S. and Izumi, N. 1991. Stable channel cross section of straight sand rivers. *Water Resources Research* 27: 2429–2438.

Ikeda, S., Parker, G. and Sawai, K. 1981. Bend theory of river meanders. 1. Linear development. *Journal of Fluid Mechanics* 112: 363–377.

Ikeda, S., Parker, G. and Kimura, Y. 1988. Stable width and depth of straight gravel rivers with heterogeneous bed materials. *Water Resources Research* 24: 713–722.

Johannesson, H. and Parker, G. 1989. Secondary flow in mildly sinuous channels. *Journal of Hydraulic Engineering* 115: 289–308.

Johnson, P.A., Gleason, G.L. and Hey, R.D. 1999. Rapid assessment of channel stability in vicinity of road crossing. *Journal of Hydraulic Engineering* 125: 645–651.

Julien, P.Y. and Wargadalam, J. 1995. Alluvial channel geometry: theory and applications. *Journal of Hydraulic Engineering* 121: 312–325.

Kalkwijk, J.P.T. and De Vriend, H.J. 1980. Computation of the flow in shallow river bends. *Journal of Hydraulic Research* 18: 327–342.

Keller, E.A. 1972. Development of alluvial stream channels: A 5 stage model. *Bulletin of the Geological Society of America* 83: 1531–1536.

Kellerhals, R. Church, M. and Bray, D.I. 1976. Classification and analysis of river processes. *Journal of the Hydraulics Division of the American Society of Civil Engineers* 102: 813–829.

Kirkby, M.J. 1971. Hillslope process-response models based on the continuity equation. *Special Publication, Institute of British Geographers* 3: 15–30.

Kirkby, M.J. 1990. The landscape viewed through models. *Zeitschrift für Geomorphologie* 79: 63–81.

Kirkby, M.J. 1996. A role for theoretical models in geomorphology? In: Rhoads, B.L. and Thorn, C.E., eds., *The Scientific Nature of Geomorphology*, Chichester, England: John Wiley and Sons, pp. 257–272.

Kondolf, G.M. 1997. Hungry water: Effects of dams and gravel mining on river channels. *Environmental Management* 21: 533–551.

Kondolf, G.M. and Swanson, M.L. 1993. Channel adjustments to reservoir construction and in stream gravel mining, Stony Creek, California. *Environmental Geology and Water Science* 21: 256–269.

Konikow, L.F. and Bredehoeft, J.D. 1992. Ground-water models cannot be validated. *Advances in Water Resources* 15: 75–83.

Kovacs, A. and Parker, G. 1994. A new vectorial bedload formulation and its application to the time evolution of straight river channels. *Journal of Fluid Mechanics* 267: 153–183.

Kuhnle, R.A., Bingner, R.L., Foster, G.R. and Grissinger, E.H. 1996. Effect of land use changes on sediment transport in Goodwin Creek. *Water Resources Research* 32: 3189–3196.

Lane, E.W. 1955. Design of stable channels. *Transactions of the American Society of Civil Engineers* 120: 1234–1260.

Lane, S.N. 1998. Hydraulic modelling in hydrology and geomorphology: a review of high resolution approaches. *Hydrological Processes* 12: 1131–1150.

Lane, S.N. and Richards, K.S. 1998. High resolution, two-dimensional spatial modelling of flow processes in a multi-thread channel. *Hydrological Processes* 12: 1279–1298.

Leeder, M.R. and Bridges, P.H. 1975. Flow separation in meander bends. *Nature* 253: 338–339.

Leeks, G.J.L. 1992. Impact of plantation forestry on sediment transport processes. In: Billi, P., Hey, R.D., Thorne, C.R. and Tacconi, P., eds., *Dynamics of Gravel-bed Rivers*, Chichester, England: John Wiley and Sons, pp. 651–670.

Leopold, L.B. and Maddock, T. 1953. *The Hydraulic Geometry of Stream Channels and Some Physiographic Implications*, US Geological Survey Professional Paper 252, Washington, DC.

Li, L. and Wang, S.S.Y. 1993. Numerical modelling of alluvial stream bank erosion. In: Wang, S.S.Y., ed., *Advances in Hydro-Science and Engineering*, Oxford, Mississippi: University of Mississippi, pp. 2085–2090.

Lien, H.C., Hsieh, T.Y., Yang, J.C. and Yeh, K.C. 1999. Bend-flow simulation using 2D depth-averaged model. *Journal of Hydraulic Engineering, ASCE* 125: 1097–1108.

Lin, B. and Shiono, K. 1995. Numerical modelling of solute transport in compound channel flows. *Journal of Hydraulic Research* 33: 773–788.

Lohnes, R. 1991. A method for estimating land loss associated with stream channel degradation. *Engineering Geology* 31: 115–130.

López, F. and García, M. 1998. Open-channel flow through simulated vegetation: suspended sediment transport modeling. *Water Resources Research* 34: 2341–2352.

Maddock, T. 1969. *The Behaviour of Straight Open Channels with Movable Beds*, US Geological Survey Professional Paper 622A, Washington, DC, pp. 1–70.

Markham, A.J. and Thorne, C.R. 1992. Geomorphology of gravel-bed river bends. In: Billi, P., Hey, R.D., Thorne, C.R. and Tacconi, P., eds., *Dynamics of Gravel-bed Rivers*, Chichester: John Wiley and Sons, pp. 433–456.

Melville, B.W. and Sutherland, A.J. 1988. Design method for local scour at bridge piers. *Journal of Hydraulic Engineering* 114: 1210–1226.

Milhous, R.T. 1998a. Numerical modelling of flushing flows in gravel-bed rivers. In: Klingeman, P.C., Beschta, R.L., Komar, P.D. and Bradley, J.B., eds., *Gravel-bed Rivers in the Environment*, Highlands Ranch, Colorado: Water Resources Publications, pp. 579–608.

Milhous, R.T. 1998b. Modelling of instream flow needs: the link between sediment and aquatic habitat. *Regulated Rivers: Research and Management* 14: 79–94.

Millar, R.G. and Quick, M.C. 1993. Effect of bank stability on geometry of gravel rivers. *Journal of Hydraulic Engineering* 119: 1343–1363.

Millar, R.G. and Quick, M.C. 1998. Stable width and depth of gravel-bed rivers with cohesive banks. *Journal of Hydraulic Engineering* 124: 1005–1013.

Miller, T.K. 1984. A system model of stream channel shape and size. *Bulletin of the Geological Society of America* 95: 237–241.

Montgomery, D.R. and Buffington, J.M. 1997. Channel-reach morphology in mountain drainage basins. *Bulletin of the Geological Society of America* 109: 596–611.

Mosselman, E. 1992. Mathematical Modelling of Morphological Processes in Rivers with Erodible Cohesive Banks, Ph.D. Thesis, Technische Universiteit Delft, Delft, Netherlands.

Mosselman, E. 1995. A review of mathematical models of river planform changes. *Earth Surface Processes and Landforms* 20: 661–670.

Mosselman, E. 1998. Morphological modelling of rivers with erodible banks. *Hydrological Processes* 12: 1357–1370.

Moulin, C. and Slama, E.B. 1998. The two-dimensional transport module SUBIEF. Applications to sediment transport and water quality processes. *Hydrological Processes* 12: 1183–1195.

Murray, A.B. and Paola, C. 1994. A cellular model of braided rivers. *Nature* 371: 54–56.

Nagata, N., Hosoda, T. and Muramoto, Y. 2000. Numerical analysis of river channel processes with bank erosion. *Journal of Hydraulic Engineering* 126: 243–252.

Nelson, J.M. and Smith, J.D. 1989. Flow in meandering channels with natural topography. In: Ikeda, S. and Parker, G., eds., *River Meandering*, Washington, DC: American Geophysical Union, pp. 321–377.

Nicholas, A.P. and Sambrook Smith, G.H. 1999. Numerical simulation of three-dimensional flow hydraulics in a braided channel. *Hydrological Processes* 13: 913–929.

Nicholas, A.P. and Walling, D.E. 1998. Numerical modelling of floodplain hydraulics and suspended sediment transport and deposition. *Hydrological Processes* 12: 1339–1355.

Nicholas, A.P., Woodward, J.C., Christopoulos, G. and Macklin, M.G. 1999. Modelling and monitoring river response to environmental change: the impact of dam construction and alluvial gravel extraction on bank erosion rates in the Lower Alfios Basin, Greece. In: Grown, A.G. and Quine, T.A., eds., *Fluvial Processes and Environmental Change*, Chichester, England: John Wiley and Sons, pp. 117–137.

Odgaard, A.J. 1989. River meander model. I. Development. *Journal of Hydraulic Engineering* 115: 1433–1450.

O'Neill, M.P., Schmidt, J.C., Dobrowolski, J.P., Hawkins, C.P. and Neale, C.M.U. 1997. Identifying sites for riparian wetland restoration: application of a model to the Upper Arkansas River Basin. *Restoration Ecology* 5: 85–102.

Olesen, K.W. 1987. Bed topography in shallow river bends, Delft Communications on Hydraulic and Geotechnical Engineering Report No. 87-1, T.U. Delft, Delft, Netherlands.

Olsen, N.R.B. and Stokseth, S. 1995. Three-simensional numerical modelling of water flow in a river with large bed roughness. *Journal of Hydraulic Research* 33: 571–581.

Olsen, N.R.B. and Kjellesvig, H.M. 1998. Three-dimensional numerical flow modeling for estimation of maximum local scour depth. *Journal of Hydraulic Research* 36: 579–590.

Oreskes, N., Shrader-Frechette, K. and Belitz, K. 1994. Verification, validation and confirmation of numerical models in the earth sciences. *Science* 263: 641–646.

Paine, A.D.M. 1985. 'Ergodic' reasoning in geomorphology. *Progress in Physical Geography* 9: 1–15.

Patton, P.C. and Baker, V.R. 1977. Geomorphic response of central Texas stream channels to catastrophic rainfall and runoff. In: Doehring, D.O., ed., *Geomorphology in Arid and Semi-arid Regions*, Publications in Geomorphology, Binghamton, New York: State University of New York, pp. 189–217.

Parker, G. 1978. Self-formed straight rivers with equilibrium banks and mobile bed. Part 2. The gravel river. *Journal of Fluid Mechanics* 89: 127–146.

Parker, G. and Andrews, E.D. 1985. Sorting of bed load sediment by flow in meander bends. *Water Resources Research* 21: 1361–1373.

Parker, G., Sawai, K. and Ikeda, S. 1982. Bend theory of river meanders. Part 2. Non-linear deformation of finite amplitude bends. *Journal of Fluid Mechanics* 115: 303–314.

Peakall, J., Ashworth, P.J. and Best, J. 1996. Physical modelling in fluvial geomorphology: principles, applications and unresolved issues. In: Rhoads, B.L. and Thorn, C.E., eds., *The Scientific Nature of Geomorphology*, Chichester, England: John Wiley and Sons, pp. 221–253.

Petts, G.E. 1980. Morphological changes of river channels consequent upon headwater impoundment. *Journal of the Institute of Water Engineers and Scientists* 34: 374–382.

Petts, G.E. 1984. Sedimentation within a regulated river. *Earth Surface Processes and Landforms* 9: 125–134.

Pickup, G. and Warner, R.F. 1976. Effects of hydrologic regime on magnitude and frequency of dominant discharge. *Journal of Hydrology* 29: 51–76.

Pinelli, A., Vacca, A. and Quarteroni, A. 1997. A spectral multidomain method for the numerical simulation of turbulent flows. *Journal of Computational Physics* 136: 546–558.

Pizzuto, J.E. 1990. Numerical simulation of gravel river widening. *Water Resources Research* 26: 1971–1980.

Pizzuto, J.E. 1992. The morphology of graded gravel rivers: a network perspective. *Geomorphology* 5: 457–474.

Pokrefke, T.J., Abraham, D.A., Hoffman, P.H., Thomas, W.A., Darby, S.E. and Thorne, C.R. 1998. Cumulative erosion impacts analysis for the Missouri River master water control manual review and update study, Technical Report CHL-98-7, US Army Corps of Engineers Waterways Experiment Station, Vicksburg, Mississippi.

Prandtl, L. 1952. *Essentials of Fluid Dynamics*, London: Blackie.

Reiser, D.W. 1998. Sediment in gravel bed rivers: ecological and biological considerations. In: Klingeman, P.C., Beschta, R.L., Komar, P.D. and Bradley, J.B., eds., *Gravel-bed Rivers in the Environment*, Highlands Ranch, Colorado: Water Resources Publications, pp. 199–228.

Rhoads, B.L. 1992. Statistical models of fluvial systems. *Geomorphology* 5: 433–455.

Rice, J.R. 1983. *Numerical Methods, Software and Analysis: ISML Reference Edition*. New York: McGraw-Hill.

Richards, K. 1977. Channel and flow geometry: a geomorphological perspective. *Progress in Physical Geography* 1: 65–102.

Richards, K. 1996. Samples and cases: generalisation and explanation in geomorphology. In: Rhoads, B.L. and Thorn, C.E., eds., *The Scientific Nature of Geomorphology*, Chichester, England: John Wiley and Sons, pp. 171–190.

Robert, A. 1990. Boundary roughness in coarse-grained channels. *Progress in Physical Geography* 14: 42–70.

Rojstaczer, S.A. 1994. The limitations of ground water models. *Journal of Geological Education* 42: 362–368.

Rozovskii, I.L. 1961. *Flow of Water in Bends of Open Channels*, Israel Program for Scientific Translations, Jerusalem.

Sack, D. 1992. New wine in old bottles: the historiography of a paradigm change. *Geomorphology* 5: 251–263.

Schumm, S.A. 1960. *The Shape of Alluvial Channels in Relation to Sediment Type*, US Geological Survey Professional Paper 352B, Washington, DC.

Schumm, S.A. 1985. Explanation and extrapolation in geomorphology: seven reasons for geologic uncertainty. *Transactions of the Japanese Geomorphological Union* 6: 1–18.

Schumm, S.A. 1991. *To Interpret the Earth: Ten Ways to be Wrong*, Cambridge, England: Cambridge University Press.

Schumm, S.A. 1999. Causes and controls of channel incision. In: Darby, S.E. and Simon, A., eds., *Incised River Channels*, Chichester, England: John Wiley and Sons, pp. 19–33.

Schumm, S.A., Harvey, M.D. and Watson, C.C. 1984. *Incised River Channels: Morphology, Dynamics and Control*, Littleton, Colorado: Water Resources Publications.

Schumm, S.A., Mosley, M.P. and Weaver, W.E. 1987. *Experimental Fluvial Geomorphology*, New York: John Wiley and Sons.

Sewell, G. 1988. *The Numerical Solution of Ordinary and Partial Differential Equations*, Boston, Massachussetts: Academic Press.

Shen, H.W. 1991. *Movable Bed Physical Models*, Boston: Kluwer Academic.

Shettar, A.S. and Murthy, K.K. 1997. A numerical study of division of flow in open channels. *Journal of Hydraulic Research* 34: 651–675.

Shields, F.D. 1996. Hydrologic and hydraulic stability. In: Brookes, A. and Shields, F.D., Jr., eds., *River Channel Restoration*, Chichester, England: John Wiley and Sons, pp. 75–101.

Shields, F.D., Knight, S.S. and Cooper, C.M. 1998. Rehabilitation of aquatic habitats in warmwater streams damaged by channel incision in Mississippi. *Hydrobiologia* 382: 63–86.

Simon, A. 1989. A model of channel response in disturbed alluvial channels. *Earth Surface Processes and Landforms* 14: 11–26.

Simon, A. 1992. Energy, time, and channel evolution in catastrophically disturbed fluvial systems. *Geomorphology* 5: 345–372.

Simon, A. and Darby, S.E. 1997a. Process-form interactions in unstable sand-bed river channels: a numerical modelling approach. *Geomorphology* 21: 85–106.

Simon, A. and Darby, S.E. 1997b. Disturbance, channel evolution and erosion rates: Hotophia Creek, Mississippi. In: Wang, S.S.Y., Langendoen, E.J. and Shields, F.D., Jr., eds., *Management of Landscapes Disturbed by Channel Incision*, Oxford, Mississippi: University of Mississippi, pp. 476–481.

Simon, A. and Darby, S.E. 1999. The nature and significance of incised river channels. In: Darby, S.E. and Simon, A., eds., *Incised River Channels*, Chichester, England: John Wiley and Sons, pp. 3–18.

Simon, A. and Downs, P.W. 1995. An interdisciplinary approach to evaluation of potential instability in alluvial channels. *Geomorphology* 12: 215–232.

Slingerland, R. and Smith, N.D. 1998. Necessary conditions for a meandering river avulsion. *Geology* 26: 435–438.

Smith, J.D. and MacLean, S.R. 1984. A model for flow in meandering streams. *Water Resources Research* 20: 1301–1315.

Steward, M.D., Bates, P.D., Price, D.A. and Burt, T.P. 1998. Modelling the spatial variability in floodplain soil contamination during flood events to improve chemical mass balance estimates. *Hydrological Processes* 12: 1233–1255.

Strahler, A.N. 1980. Systems theory in physical geography. *Physical Geography* 1: 1–27.

Strickler, A. 1923. Beitrage zur frage der geschwindigheitsformel und der rauhigkeitszahlen fur strome, kanale und geschlossene leitungen. *Mitteilungen des Eidgenossicher Amtes fur Wasserwirtscaft*, Bern, Switzerland.

Struiksma, N., Olsen, K.W., Flokstra, C. and De Vriend, H. J. 1985. Bed deformation in curved alluvial channels. *Journal of Hydraulic Research* 23: 57–78.

Sun, T., Meakin, P., Jossang, T. and Schwarz, K. 1996. A simulation model for meandering rivers. *Water Resources Research* 32: 2937–2954.

Tetzlaff, D.M. 1989. Limits to the predictive stability of dynamic models that simulate clastic sedimentation. In: Cross, T.A., ed., *Quantitative Dynamic Stratigraphy*, Englewood Cliffs, New Jersey: Prentice-Hall, pp. 55–65.

Tetzlaff, D.M. and Harbaugh, J.W. 1989. *Simulating Clastic Sedimentation*, New York: Van Nostrand Reinhold.

Thomas, W.A. 1982. Mathematical modelling of sediment movement. In: Hey, R.D., Bathurst, J.C. and Thorne, C.R., eds., *Gravel-bed Rivers*, Chichester, England: John Wiley and Sons, pp. 487–508.

Thomas, T.G. and Williams, J.J.R. 1995. Large eddy simulation of a symmetrical trapezoidal channel at a Reynolds number of 430000. *Journal of Hydraulic Research* 33: 825–842.

Thompson, A. 1986. Secondary flows and the pool-riffle unit: a case study of the processes of meander development. *Earth Surface Processes and Landforms* 11: 631–641.

Thomson, J. 1876. On the origins and winding of rivers in alluvial plains. *Proceedings of the Royal Society of London* 25: 5–8.

Thorne, C.R. 1995. Editorial – geomorphology at work. *Earth Surface Processes and Landforms* 20: 583–584.

Thorne, C.R. and Zevenbergen, L.W. 1985. Estimating mean velocity in mountain rivers. *Journal of Hydraulic Engineering* 111: 612–624.

Thorne, C.R. and Osman, A.M. 1988. Riverbank stability analysis. II. Applications. *Journal of Hydraulic Engineering* 114: 151–172.

Thorne, C.R., Allen, R.G. and Simon, A. 1996. Geomorphological river channel reconnaissance for river analysis, engineering and management. *Transactions of the Institute of British Geographers* 21: 469–483.

Thornes, J.B. 1977. Hydraulic geometry and channel change. In: Gregory, K.J., ed., *River Channel Changes*, Chichester, England: John Wiley and Sons, pp. 91–100.

Thornes, J.B. and Ferguson, R.I. 1981. Geomorphology. In: Wrigley, N. and Bennett, R.J., eds., *Quantitative Geography: A British View*, London: Routledge and Kegan Paul, pp. 284–293.

Tritton, D.J. 1988. *Physical Fluid Dynamics*, Second edition, Oxford, England: Oxford Science Publications.

Tucker, G.E. and Slingerland, R. 1997. Drainage basin responses to climate change. *Water Resources Research* 33: 2031–2047.

Vigilar, G.G. and Diplas, P. 1997. Stable channels with mobile bed: formulation and numerical solution. *Journal of Hydraulic Engineering* 123: 189–199.

Vigilar, G.G. and Diplas, P. 1998. Stable channels with mobile bed: model verification and graphical solution. *Journal of Hydraulic Engineering* 124: 1097–1108.

Viney, N.R. and Sivapalan, M. 1999. A conceptual model of sediment transport: application to the Avon River Basin in Western Australia. *Hydrological Processes* 13: 727–743.

Vreugdenhill, C.B. and Wijbenga, J.H.A. 1982. Computation of flow patterns in rivers. *Journal of the Hydraulics Division, ASCE* 108: 1296–1310.

Wallerstein, N., Thorne, C.R. and Doyle, M.W. 1997. Spatial distribution and impact of large woody debris in northern Mississippi. In: Wang, S.S.Y., Langendoen, E.J. and Shields, F.D., eds., *Management of Landscapes Disturbed by Channel Incision*, Oxford, Mississippi: University of Mississippi, pp. 145–150.

Wang Shiqiang, White, W.R. and Bettess, R. 1986. A rational approach to river regime. In: Wang, S.S.Y., Shen, H.W. and Ding, L.Z., eds., *Proceedings of the 3rd International Symposium on River Sedimentation*, Oxford, Mississippi: University of Mississippi, pp. 167–176.

Warburton, J. and Davies, T.R.H. 1998. Use of hydraulic models in management of braided gravel-bed rivers. In: Klingeman, P.C., Beschta, R.L., Komar, P.D. and Bradley, J.B., eds., *Gravel-bed Rivers in the Environment*, Highlands Ranch, Colorado: Water Resources Publications, pp. 513–542.

Watson, C.C. and Biedenharn, D.S. 1999. Design and effectiveness of grade control structures in incised river channels of North Mississippi, USA. In: Darby, S.E. and Simon, A., eds., *Incised River Channels*, Chichester, England: John Wiley and Sons, pp. 395–422.

Wu, W., Rodi, W. and Wenka, T. 2000. 3D numerical modelling of flow and sediment transport in open channels. *Journal of Hydraulic Engineering, ASCE* 126: 4–14.

Whipple, K.X. and Dunne, T. 1992. The influence of debris-flow rheology on fan morphology, Owens Valley, California. *Bulletin of the Geological Society of America* 104: 887–900.

White, W.R., Bettess, R. and Paris, E. 1982. Analytical approach to river regime. *Journal of the Hydraulics Division of the American Society of Civil Engineers* 108: 1179–1193.

Wiberg, P.L. and Smith, J.D. 1987. Calculations of the critical shear stress for motion of uniform and heterogeneous sediments. *Water Resources Research* 23: 1471–1480.

Wiele, S.M. 1992. A computational investigation of bank erosion and midchannel bar formation in gravel-bed rivers, Ph.D. Thesis, Minneapolis, Minnesota: University of Minnesota.

Wilcock, D.N. 1971. Investigation into the relations between bedload transport and channel shape. *Bulletin of the Geological Society of America* 82: 2159–2176.

Willgoose, G., Bras, R.L. and Rodriguez-Iturbe, I. 1991. A coupled channel network growth and hillslope evolution model. 1. Theory. *Water Resources Research* 27: 1671–1684.

Williams, G.P. 1978. *Hydraulic Geometry of River Cross-sections – Theory of Minimum Variance*, US Geological Survey Professional Paper 1029, Washington, DC.

Wolman, M.G. and Brush, L.M. 1961. *Factors Controlling the Size and Shape of Stream Channels in Coarse Noncohesive Sands*, United States Geological Survey Professional Paper 282G, Washington, DC, pp. 183–210.

Yang, C.T. 1971. Potential energy and stream morphology. *Water Resources Research* 7: 312–322.

Yang, C.T. and Song, C.C.S. 1979. Theory of minimum rate of energy dissipation. *Journal of the Hydraulics Division of the American Society of Civil Engineers* 105: 769–784.

Yang, C.T., Song, C.C.S. and Woldenberg, M.J. 1981. Hydraulic geometry and minimum rate of energy dissipation. *Water Resources Research* 17: 1014–1018.

Yang, C.T., Molinas, A. and Song, C.S. 1988. GSTARS – Generalized stream tube model for alluvial river simulation. In: Fan, S., ed., *Twelve Selected Computer Stream Sedimentation Models Developed in the United States*, Washington, DC: Federal Energy Regulatory Commission.

Zimmerman, R.C., Goodlett, J.C. and Comer, G.H. 1967. The influence of vegetation on channel form of small streams. *International Association of Scientific Hydrology* 75: 255–275.

18

Flow and Sediment-transport Modeling

JONATHAN M. NELSON[1], JAMES P. BENNETT[1] AND STEPHEN M. WIELE[2]
[1] *US Geological Survey, Lakewood, CO, USA*
[2] *US Geological Survey, Tucson, AZ, USA*

18.1 INTRODUCTION

Overview

Predicting the response of natural or man-made channels to imposed supplies of water and sediment is one of the most difficult practical problems commonly addressed by geomorphologists. This problem typically arises in three different situations. In the first situation, geomorphologists are attempting to understand why a channel, or class of channels, has a certain general form; in a sense, this is the central goal of fluvial geomorphology. In the second situation, geomorphologists are trying to understand and explain how and why a specific channel will evolve or has evolved in response to altered or unusual sediment and water supplies to that channel. For example, this would include explaining the short-term response of a channel to an unusually large flood, or predicting the response of a channel to long-term changes in flow or sediment supply due to dams or diversions. Finally, geomorphologists may be called upon to design or assess the design of proposed man-made channels that must carry a certain range of flows and sediment loads in a stable or at least quasi-stable manner. In each of these three situations, the problem is really the same: geomorphologists must understand and predict the interaction of the flow field in the channel, the sediment movement in the channel, and the geometry of the channel bed and banks. In general, the flow field, the movement of sediment making up the bed, and the morphology of the bed are intricately linked; the flow moves the sediment, the bed is altered by the erosion and deposition of sediment, and the shape of the bed is critically important for predicting the flow. This complex linkage is precisely what makes understanding channel form and process such a difficult and interesting challenge.

Until about the mid-1960s, channel form and response were evaluated primarily through qualitative understanding of process coupled with detailed empirical observation. These approaches gave rise to several powerful tools that are still in use, including regime theory and hydraulic geometry relationships. These tools provided geomorphologists with predictive methodologies for channel form and response. However, as the understanding of processes in channels has increased, so too has the detail of the questions being asked with regard to channel morphology and response to disturbance. Over the last 30 years or so, the need for more precise predictive tools has led researchers in both geomorphology and engineering to formulate quantitative models of the coupled flow/sediment/bed system. These approaches are based on the capability to predict the flow field accurately, so their evolution in accuracy and detail over the last three decades has largely been determined by developments in computational methods for flow prediction. The techniques, which both complement and extend more classical techniques in fluvial geomorphology, offer powerful tools to geomorphologists trying to understand or predict stable channel forms and channel adjustments to altered flow and sediment supply.

In this chapter, a brief overview of techniques for predicting flow, sediment transport, and bed evolution is presented, emphasizing the physical processes that are captured by various approaches. The goal of the chapter is not to provide recipes for constructing such models, although several components will be discussed in detail, and the industrious reader should find enough detail here and in the references to construct such a model. Rather, this material should be used as a guide in understanding these approaches and in selecting appropriate models for specific problems. Where user-friendly models with appropriate interfaces are available, the reader is directed to sources for the models. This is a rapidly developing field in geomorphology and engineering, so discussion of specific models and algorithms is avoided for the most part, with the knowledge that most of these approaches are evolving over time and statements made herein about specific models may soon be incorrect. On the other hand, the difference in processes captured by various approaches is emphasized, so the reader may be able to judge which models or algorithms should be used for applications, both now and in the future, as more and more flexible models are available.

The Coupled Model Concept

The key to the development of computational techniques for flow, sediment transport, and bed evolution is the observation that, in almost all situations of practical interest, the timescales associated with the flow are much shorter than the timescales of bed and bank evolution. Another way to state this is that, even when the bed is evolving, the process is so slow that the flow can be computed as if the bed and banks were not changing in time. This allows partial decoupling of the flow computation from the sediment motion and bed evolution. In other words, it is possible to compute the flow field first, without simultaneously solving for the sediment-transport field and the bed morphology. Thus, one can compute the flow based on the input discharge (which may vary in time), use the flow solution to compute the sediment-transport patterns, and evaluate those sediment-transport patterns to deduce local rates of erosion and deposition on the channel bed and banks. Given these rates and a specified time step, one can predict the topographic evolution some short time into the future. Provided this time step is small enough, it is possible to then recompute the flow and continue to iterate on the flow, the sediment-transport field and the bed evolution, predicting the changes in each as a function of time. If the bed is not perfectly stable, but evolves in time, the flow patterns will change as time progresses even in the absence of discharge variations; they will change in response to the change in the channel morphology. Note that this intuitive methodology is the same across a range of actual modeling techniques from the simplest one-dimensional model to complex three-dimensional turbulence-resolving models; each model exploits the separation in timescales between the flow and the bed evolution to allow iterative, rather than simultaneous, solution of the governing relations. Thus, although the general problem requires the simultaneous solution of the flow field, the sediment-transport field, and the channel geometry, almost all practical problems can be solved with the much simpler iterative procedure.

Although the details of the methodology and specific applications are yet to be discussed here, the potential utility of the coupled model concept in geomorphology should be clear. The method allows one to examine the stability of a channel over time, using hypothetical or real initial geometry, which is key to understanding both stable channel forms and the adjustment of channels to anthropogenic or natural changes in flow and sediment supply. The accuracy with which one can carry out these predictions depends critically on the accuracy of the various components of the coupled flow/sediment/bed modeling, and on knowing what physics must be incorporated in the models to address certain classes of problems. With this in mind, the remainder of this chapter deals with the particulars of such models and hopefully will help readers to delineate the applicability and potential accuracy of various treatments.

18.2 FLOW CONSERVATION LAWS

Conservation of Mass and Momentum

The conservation equations governing fluid and sediment motion are the fundamental building blocks of all coupled flow/sediment-transport/bed evolution models, but various models use versions of the full equations that are reduced by neglecting certain terms, or more commonly, by integrating over one or more dimensions to develop averaged equations. The most important things to note in going through this exercise are the approximations that are required in order to develop certain methods; these will be explicitly noted in the text as will the physical meaning of the approximations. The first approximation to be used here is that, throughout, the flow will be assumed to be incompressible. This is a good assumption as long as the flow velocities are much less than the speed of sound, a condition that is well satisfied in channel flows. Using this assumption, conservation of mass and momentum for the flow are given by the following:

$$\nabla \cdot \vec{u} = 0 \qquad (18.2.1)$$

$$\frac{\partial \vec{u}}{\partial t} + \vec{u} \cdot \nabla \vec{u} = -\frac{1}{\rho} \nabla P + \vec{g} + \nu \nabla^2 \vec{u} \qquad (18.2.2)$$

where $\vec{u}$ is the vector velocity, ρ the fluid density, P the pressure, and ν is the fluid kinematic viscosity. These equations describe fluid motion in general; the only assumption made in deriving them is that the fluid is incompressible. In general, solving these equations in this full form in natural flows is difficult and impractical. Usually, the equations that are actually used to compute flow solutions are reduced forms of the above equations developed by temporal or spatial averaging, or through scaling the equations to discover which terms are most important and retaining only those terms in the numerical solution.

The primary reason that these equations are difficult to solve for most natural flows is turbulence. With the exception of flows characterized by appropriate combinations of low velocity, small scale, and/or high fluid viscosity (characterized by the Reynolds number, see Tennekes and Lumley 1972, pp. 1–26), flows are unstable to perturbations and are characterized by three-dimensional variability across a wide range of time and length scales. For example, even if one creates a simple channel flow with a smooth bottom, rectilinear channel shape, and steady discharge, the velocity at any point in the flow will vary in time for typical length and timescales due to turbulent eddies. In addition to adding substantially to the complexity of the flow, these variations give rise to important momentum fluxes, changing even the time-averaged character of the flow significantly. To avoid the necessity of computing the variations in flow associated with turbulence, by far the majority of computational models used for natural flows use the so-called Reynolds equations. These equations are developed by splitting the vector velocity into a time-mean part (or an ensemble-averaged part) and a time-varying part (or the variation about the ensemble average). For a detailed description of this procedure and the reasoning behind it, the reader is referred to Tennekes and Lumley (1972, pp. 28–33) or any other beginning text on turbulence. In a Cartesian coordinate system with z positive upwards, the Reynolds momentum equations for the x, y, and z directions are given by:

$$\frac{\partial \bar{u}}{\partial t} + \bar{u}\frac{\partial \bar{u}}{\partial x} + \bar{v}\frac{\partial \bar{u}}{\partial y} + \bar{w}\frac{\partial \bar{u}}{\partial z} = -\frac{1}{\rho}\frac{\partial P}{\partial x} + \nu\nabla^2\bar{u} - \frac{\partial \overline{u'^2}}{\partial x} - \frac{\partial \overline{u'v'}}{\partial y} - \frac{\partial \overline{u'w'}}{\partial z} \qquad (18.2.3)$$

$$\frac{\partial \bar{v}}{\partial t} + \bar{u}\frac{\partial \bar{v}}{\partial x} + \bar{v}\frac{\partial \bar{v}}{\partial y} + \bar{w}\frac{\partial \bar{v}}{\partial z} = -\frac{1}{\rho}\frac{\partial P}{\partial y} + \nu\nabla^2\bar{v} - \frac{\partial \overline{u'v'}}{\partial x} - \frac{\partial \overline{v'^2}}{\partial y} - \frac{\partial \overline{v'w'}}{\partial z} \qquad (18.2.4)$$

$$\frac{\partial \bar{w}}{\partial t} + \bar{u}\frac{\partial \bar{w}}{\partial x} + \bar{v}\frac{\partial \bar{w}}{\partial y} + \bar{w}\frac{\partial \bar{w}}{\partial z} = -\frac{1}{\rho}\frac{\partial P}{\partial z} - g + \nu\nabla^2\bar{w} - \frac{\partial \overline{u'w'}}{\partial x} - \frac{\partial \overline{v'w'}}{\partial y} - \frac{\partial \overline{w'^2}}{\partial z} \qquad (18.2.5)$$

where u, v, and w are the velocity components in the x, y, and z directions, and where overbars represent time (or ensemble) averages and primes represent deviations from that average (e.g., $u = \bar{u} + u'$). Strictly speaking, time averaging would cause the first term in each momentum equation to be identically zero, but in practice, the time required to compute the average of a turbulent quantity is often less than the timescale associated with externally imposed unsteadiness. For example, in a channel flow with slowly varying discharge, it may be possible to construct a time average over the turbulence using an averaging time, much smaller than the time over which discharge variations occur. For ensemble averages, where one averages over many realizations of the same flow, the inclusion of the unsteady term in the equations is not problematic. For example, if one makes measurements of velocity in a turbulent wave boundary layer, it is possible to average over many waves to determine the ensemble averaged behavior of the flow; the departure from that average over a specific wave or time series of waves yields the turbulent variability. The last three terms on the right-hand side of the above equations arise as a result of the momentum fluxes due to turbulent fluctuations. These terms are very important for transferring momentum within the flow, especially near boundaries or anywhere strong shears occur in the flow. For further discussion of Reynolds stresses and their generation in turbulent shear flows, the reader is referred to Tritton (1977, pp. 244–248).

Applying the same averaging procedure to the conservation of mass equation yields

$$\frac{\partial \bar{u}}{\partial x} + \frac{\partial \bar{v}}{\partial y} + \frac{\partial \bar{w}}{\partial z} = 0 \qquad (18.2.6)$$

The original four equations expressing conservation of mass and momentum had four unknowns: the three components of velocity and the pressure. The number of unknowns matched the number of equations, so this was a well-posed problem. However, the four

Reynolds averaged mass and momentum equations yield more than four unknowns because of the appearance of the momentum fluxes associated with the turbulent fluctuations. This is the so-called closure problem of turbulence.

Reynolds Stresses and Turbulence Closures

The quantities involving time or ensemble averages of products of time-varying quantities shown in Equations (18.2.3)–(18.2.5) are referred to as Reynolds stresses. Although they are called stresses, it is important to remember that these terms arise due to advective transport of momentum. However, because they appear in the Reynolds averaged momentum equations in a manner analogous to viscous stresses, they are referred to as stresses and are often parameterized in terms of the mean flow using concepts developed for viscous stresses. Rewriting Equations (18.2.3)–(18.2.5) in terms of the components of the Reynolds stress tensor yields the following:

$$\frac{\partial \bar{u}}{\partial t} + \bar{u}\frac{\partial \bar{u}}{\partial x} + \bar{v}\frac{\partial \bar{u}}{\partial y} + \bar{w}\frac{\partial \bar{u}}{\partial z} = -\frac{1}{\rho}\frac{\partial P}{\partial x} + \nu\nabla^2\bar{u} + \frac{\partial \tau_{xx}}{\partial x} + \frac{\partial \tau_{yx}}{\partial y} + \frac{\partial \tau_{zx}}{\partial z} \qquad (18.2.7)$$

$$\frac{\partial \bar{v}}{\partial t} + \bar{u}\frac{\partial \bar{v}}{\partial x} + \bar{v}\frac{\partial \bar{v}}{\partial y} + \bar{w}\frac{\partial \bar{v}}{\partial z} = -\frac{1}{\rho}\frac{\partial P}{\partial y} + \nu\nabla^2\bar{v} + \frac{\partial \tau_{xy}}{\partial x} + \frac{\partial \tau_{yy}}{\partial y} + \frac{\partial \tau_{zy}}{\partial z} \qquad (18.2.8)$$

$$\frac{\partial \bar{w}}{\partial t} + \bar{u}\frac{\partial \bar{w}}{\partial x} + \bar{v}\frac{\partial \bar{w}}{\partial y} + \bar{w}\frac{\partial \bar{w}}{\partial z} = -\frac{1}{\rho}\frac{\partial P}{\partial z} - g + \nu\nabla^2\bar{w} + \frac{\partial \tau_{xz}}{\partial x} + \frac{\partial \tau_{yz}}{\partial y} + \frac{\partial \tau_{zz}}{\partial z} \qquad (18.2.9)$$

where the Reynolds stresses are defined as follows:

$$\begin{aligned} \tau_{xx} &= -\rho\overline{u'^2} \\ \tau_{zz} &= -\rho\overline{w'^2} \\ \tau_{zz} &= -\rho\overline{w'^2} \\ \tau_{xz} &= \tau_{zx} = -\rho\overline{u'w'} \\ \tau_{xy} &= \tau_{yx} = -\rho\overline{u'v'} \\ \tau_{yz} &= \tau_{zy} = -\rho\overline{v'w'} \end{aligned} \qquad (18.2.10)$$

Generally, the Reynolds stresses are much greater than viscous stresses in natural channel flows, and the viscous stresses are neglected in the momentum equations. Thus, the terms in the above equations involving ν, the kinematic viscosity, are negligibly small and are omitted from the equations.

In order to solve the above equations, one must either rewrite the Reynolds stresses in terms of the mean flow quantities or provide some other manner by which these terms may be evaluated using additional relations. The most common method in simulating natural flows is to relate the Reynolds stresses to the mean flow quantities by analogy with the relation between viscous stress and the rate of strain tensor. This leads to the concept of eddy viscosity, which assumes a proportionality between the Reynolds stresses and the components of the rate of strain. While there is good justification for this kind of approach in situations where the flow is dominated by one length and velocity scale, as in a simple boundary layer, the concept is generally only a crude approximation for real, complex flows in nature. Nevertheless, many approaches are based on this concept, and there are a number of ways of estimating the spatial structure and values for eddy viscosity using simple dimensional arguments or more complex reasoning. For example, some models use the eddy viscosity concept, but evaluate the local eddy viscosity using advection–diffusion equations for the turbulent kinetic energy and the length scale of the turbulence; this allows treatment of situations where the local flow parameters are not accurate predictors of local turbulence structure. There are also a variety of closure approaches that are not predicated on the existence of an eddy viscosity. For example, it is possible to manipulate the momentum equations to develop expressions for each of the Reynolds stresses. However, these introduce more unknowns that must in turn be parameterized or estimated. A more complete discussion of turbulence closure techniques is beyond the scope of this chapter, but the reader is referred to the review by Rodi (1993) for an excellent discussion.

If the existence of a scalar, isotropic eddy viscosity, K, is assumed, the Reynolds stress terms in Equations (18.2.7)–(18.2.9) may be replaced by the following relations:

$$\begin{aligned}
\tau_{xx} &\cong 2\rho K \frac{\partial \bar{u}}{\partial x} \\
\tau_{zz} &\cong 2\rho K \frac{\partial \bar{w}}{\partial z} \\
\tau_{yy} &\cong 2\rho K \frac{\partial \bar{v}}{\partial y} \\
\tau_{xz} = \tau_{zx} &\cong \rho K \left[\frac{\partial \bar{u}}{\partial z} + \frac{\partial \bar{w}}{\partial x}\right] \\
\tau_{xy} = \tau_{yx} &\cong \rho K \left[\frac{\partial \bar{u}}{\partial y} + \frac{\partial \bar{v}}{\partial x}\right] \\
\tau_{yz} = \tau_{zy} &\cong \rho K \left[\frac{\partial \bar{v}}{\partial z} + \frac{\partial \bar{w}}{\partial y}\right]
\end{aligned} \tag{18.2.11}$$

Substituting the above relations, Equations (18.2.6)–(18.2.9) once again become a closed set of equations, with unknowns consisting of the Reynolds averaged velocities and pressure. However, in order to solve these equations, an eddy viscosity still needs to be determined. As noted above, there are many ways to do this, but one of the most common is based on extending the well-posed relations for simple steady, uniform boundary layers to more complex flows in channels. This extension is based on the observation that flows in unstratified channels are dominantly boundary-layer-like in character. In simple boundary layers, the local turbulence is well described by the local boundary shear stress and distance from the boundary. Indeed, this result stems directly from simple dimensional analysis (e.g., Tennekes and Lumley 1972). This result is complicated only slightly when one considers the effect of finite depth. The shear velocity is defined in terms of the local boundary shear stress and the fluid density as follows:

$$u_* = \left[\frac{(\tau_{zx})_B}{\rho}\right]^{1/2} \tag{18.2.12}$$

where B denotes evaluation at the bed. Dimensional analysis yields the result that the eddy viscosity, K, can be written in the following form:

$$K = k u_* h \kappa(\xi) \tag{18.2.13}$$

where k is an empirical constant of proportionality called von Karman's constant ($\approx$0.408, see Long *et al.* 1993) and $\kappa(\xi)$ is a shape function giving the vertical distribution of K between the bed and the water surface, using $\xi = z/h$, where h is the local flow depth and z is distance from the boundary. For the choice of a parabolic distribution of eddy viscosity, as given by

$$\kappa(\xi) = \xi(1 - \xi) \tag{18.2.14}$$

the velocity profile in the boundary layer will be logarithmic, as follows:

$$\bar{u} = \frac{u_*}{k} \ln \frac{z}{z_0} \tag{18.2.15}$$

where z_0, the so-called roughness length, is a constant of integration that depends on the boundary shear stress, the fluid viscosity, and/or the size of the roughness elements on the bed (see Middleton and Southard 1984, or any text on wall-bounded shear flows for a discussion of roughness lengths). In practice, experimental evidence suggests that Equation (18.2.14) is not the best choice, although it may be quite accurate close to the boundary. While there are several other possibilities suggested in the literature, there is not much evidence to suggest that more complicated structure functions are verifiably better than simply using Equation (18.2.14) from the bed up to one-fifth of the flow depth and using a constant value above that level, i.e.,

$$\begin{aligned}
\kappa(\xi) &= \xi(1 - \xi), \quad \xi < 0.2 \\
\kappa(\xi) &= 0.16, \quad \xi \geq 0.2
\end{aligned} \tag{18.2.16}$$

This choice for κ yields a logarithmic velocity profile near the bed and a parabolic one well away from the bed, and was first described by Rattray and Mitsuda (1974).

In applying models that use the simple eddy viscosity closure described above, it is absolutely critical to note that this form of the eddy viscosity is strictly correct only in a steady, uniform boundary layer. While natural rivers and streams are predominantly boundary-layer-like in nature and are commonly steady over time steps used in most models, they can be decidedly nonuniform, introducing free shear layers and wakes for which these eddy viscosity closures are inappropriate. One immediate shortcoming of the model above is that it predicts zero flux of momentum due to turbulence in regions where the boundary shear stress is zero. In a simple shear layer bounding a separation zone in a river, this suggests that, as the boundary shear stress must change sign somewhere in the region of between upstream and downstream flow, there must be a surface across which no momentum is transferred by turbulence. This is wrong; if these effects are important, a different closure must be employed. Nevertheless, these simple closures perform adequately in a wide variety of natural flows. The most important point here is that, when using a closure of a certain type, one must keep in mind the potential errors in that closure and which physical processes are likely to be well treated and which processes are likely to be poorly treated.

Hydrostatic Assumption

Up to this point, each of the three components of velocity has been treated equally and the terms in the momentum equations for u, v, and w have been treated in the same manner. However, in many flows of interest, both vertical velocities and vertical accelerations are small, and the vertical equation of motion [(18.2.5) or (18.2.9)] can be accurately approximated by retaining only the pressure gradient and gravitational terms:

$$-\frac{1}{\rho}\frac{\partial P}{\partial z} - g = 0 \qquad (18.2.17)$$

This assumption is referred to as the hydrostatic assumption, as it results in the pressure being distributed hydrostatically in the vertical, meaning that the pressure is equivalent to the overlying weight of fluid per unit area at any point. This simplification is a good one provided vertical accelerations are small, meaning that bed slopes are relatively small along the direction of the flow. For flows with strong vertical acceleration produced by abrupt bed variations (as may be caused by bedrock or man-made structures), this assumption will be locally inaccurate.

In situations where Equation (18.2.17) is a suitable approximation for Equation (18.2.9), the pressure gradients in the horizontal equations of motion can be written in terms of the water surface elevation, E, by integrating Equation (18.2.17) in z and differentiating the result in each of the horizontal directions to obtain:

$$-\frac{1}{\rho}\frac{\partial P}{\partial x} = -g\frac{\partial E}{\partial x} \qquad (18.2.18)$$

$$-\frac{1}{\rho}\frac{\partial P}{\partial y} = -g\frac{\partial E}{\partial y} \qquad (18.2.19)$$

These relations simplify solution of the equations, because they reduce determining the pressure at each (x,y,z) location in the flow to determining only the water surface elevation at each horizontal (x,y) location.

Coordinate Systems

All the equations above have been cast in a simple Cartesian coordinate system. In practice, flow solutions are computed in a wide variety of coordinate systems, including Cartesian, orthogonal curvilinear, and general coordinate systems for finite difference solutions and a variety of structured and unstructured grids for finite element solutions. The primary advantage of general or unstructured grids is that they allow the coordinate system to be fitted precisely to the flow domain. The disadvantage is that they increase computational complexity considerably, and in cases where the bed and banks of the channel are evolving in time, the coordinate system must be recomputed at every time step, which is time consuming. In addition, most finite element solutions conserve mass only in a global sense; they typically are poor at enforcing mass conservation locally (Oliveira *et al.* 2000). This problem can be mitigated by careful construction of the flow grid, but it is difficult to avoid entirely, especially in channels with strong spatial accelerations produced by topography or channel curvature. Oliveira *et al.* (2000) found errors in local mass conservation of up to 85% after only 3 days of simulation applying standard finite element methods to the Tagus estuary. In channel flows, errors of this magnitude result in solutions that are not good representations of the real flow, and certainly could not be used to accurately compute the movement of sediment or other constituents within the flow.

Developing a variety of commonly used coordinate systems is not within the scope of this chapter, but it is worth mentioning one specific orthogonal curvilinear system that has been widely used in modeling river flows. This coordinate system is essentially a generalization of a cylindrical coordinate system where the curvature of the coordinate system is allowed to vary in the streamwise direction. This so-called "channel-fitted" coordinate system has been used widely over the last 50 years or so, although most early applications involved only an incomplete set of equations. The system was formally derived and the full equations were published by Smith and McLean (1984). If the radius of curvature of the channel centerline is defined as R and s, n and z are defined as the streamwise, cross-stream, and vertical coordinates,

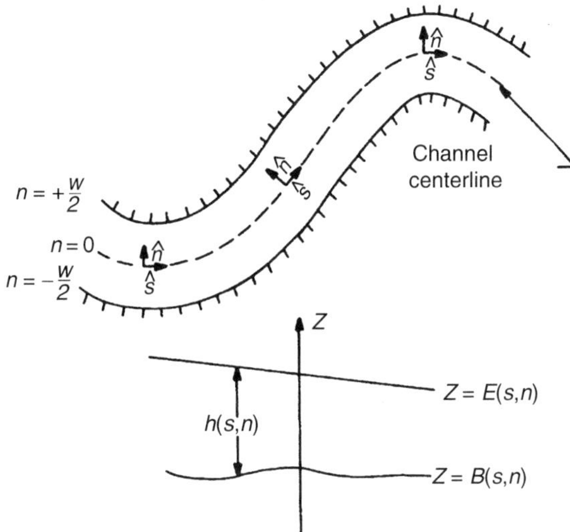

Figure 18.1 Schematic of the channel-fitted coordinate system

respectively, as shown in Figure 18.1, the hydrostatic assumption is employed, and the viscous stresses are assumed to be negligibly small, and $N = n/R$, then the continuity and momentum equations in this coordinate system are given by the following:

$$\frac{1}{1-N}\frac{\partial u}{\partial s} - \frac{v}{(1-N)R} + \frac{\partial v}{\partial n} + \frac{\partial w}{\partial z} = 0 \qquad (18.2.20)$$

$$\frac{\partial u}{\partial t} + \frac{u}{(1-N)}\frac{\partial u}{\partial s} + v\frac{\partial u}{\partial n} + w\frac{\partial u}{\partial z} - \frac{uv}{(1-N)R} = \frac{-g}{1-N}\frac{\partial E}{\partial s} + \frac{1}{\rho}\left[\frac{1}{1-N}\frac{\partial \tau_{ss}}{\partial s} + \frac{\partial \tau_{ns}}{\partial n} + \frac{\partial \tau_{zs}}{\partial z} - \frac{2\tau_{ns}}{(1-N)R}\right] \qquad (18.2.21)$$

$$\frac{\partial v}{\partial t} + \frac{u}{(1-N)}\frac{\partial v}{\partial s} + v\frac{\partial v}{\partial n} + w\frac{\partial v}{\partial z} + \frac{u^2}{(1-N)R} = \frac{-g}{1-N}\frac{\partial E}{\partial n} + \frac{1}{\rho}\left[\frac{1}{1-N}\frac{\partial \tau_{ns}}{\partial s} + \frac{\partial \tau_{nn}}{\partial n} + \frac{\partial \tau_{zn}}{\partial z} + \frac{\tau_{ss} - \tau_{nn}}{(1-N)R}\right] \qquad (18.2.22)$$

$$-\frac{1}{\rho}\frac{\partial P}{\partial z} - g = 0 \qquad (18.2.23)$$

If the existence of a scalar, isotropic eddy viscosity is assumed, we can rewrite Equation (18.2.11) in the channel-fitted coordinate system, resulting in the following expressions for the six independent components of the deviatoric Reynolds stress tensor:

$$\begin{aligned}
\tau_{ss} &= 2\rho K\left[\frac{1}{1-N}\frac{\partial u}{\partial s} - \frac{v}{(1-N)R}\right] \\
\tau_{ns} &= \rho K\left[\frac{1}{1-N}\frac{\partial v}{\partial s} + \frac{u}{(1-N)R} + \frac{\partial u}{\partial n}\right] \\
\tau_{zs} &= \rho K\left[\frac{1}{1-N}\frac{\partial w}{\partial s} - \frac{\partial u}{\partial z}\right] \\
\tau_{nn} &= 2\rho K\left[\frac{\partial v}{\partial n}\right] \\
\tau_{zn} &= \rho K\left[\frac{\partial w}{\partial n} + \frac{\partial v}{\partial z}\right] \\
\tau_{zz} &= 2\rho K\left[\frac{\partial w}{\partial z}\right]
\end{aligned} \qquad (18.2.24)$$

where the overbars denoting Reynolds averaging of the equations have been omitted for simplicity. If the radius of curvature of the channel centerline goes to infinity, meaning that the channel is straight, Equations (18.2.20)–(18.2.24) revert back to the standard momentum equations with x and y oriented streamwise and cross-stream, respectively. However, if the channel is curved, the u and v velocity components in the s–n–z coordinate system still correspond to streamwise and cross-stream velocities, as the s direction is always streamwise. Clearly, this would not be true if a Cartesian system were used; the orientation of the x and y components of velocity with respect to the channel would change with position. Thus, the channel-fitted coordinate system is in some sense the natural one, as it divides local velocity vectors into streamwise and cross-stream components. This system is also the one typically used in analyzing field measurements in channels, because those measurements are frequently taken perpendicular to and parallel to sections that are themselves perpendicular to the channel centerline.

The first, and perhaps most confusing, step in applying the channel-fitted coordinate system is determining the channel centerline and the radius of curvature of that centerline. This is not a purely mathematical process; it requires some consideration of what one is trying to capture in the channel-fitted coordinate system. As long as the numerics are correct and the full equations are used, the flow solution should be essentially independent of the coordinate system. Thus, one could use a Cartesian coordinate system for a curved channel, or even a curved coordinate system for a straight channel. However, if one chooses a coordinate system that follows the path of the channel, at least approximately, two advantages arise: first, the number of grid points required is minimized and second, the convective accelerations associated with the curvature of the channel appear primarily in centripetal acceleration terms, rather than in differential terms in the governing equations. This latter consideration is the key to choosing the channel centerline for the coordinate system. Basically, one wants to find a centerline that captures the average curvature of the flow streamlines, which are approximately the same as the large-scale curvature of the banks. Because the flow "averages" the effects of the local banks over a length scale comparable with width, one can digitize a centerline for the coordinate system (which need not correspond exactly to the channel centerline) in two ways. Either one may digitize the centerline with points that are closer together than the channel width and then filter the resulting curve over distances of about a channel width, or one may simply choose a number of points, each about a channel width apart. In either case, the radius of curvature is easily found by noting that, if θ is the angle between the down valley direction and the local tangent to the centerline, the radius of curvature is given by

$$R = \left[\frac{\partial \theta}{\partial s}\right]^{-1} \tag{18.2.25}$$

When generating a channel-fitted coordinate system, the centerline defining the coordinate system should be drawn to approximate the average streamline curvature in the reach of interest as well as possible. It is not appropriate to take a precise channel centerline defined by a detailed (i.e., with spatial resolution much smaller than a channel width) survey of the banks, as the resulting detailed centerline may have local curvature values that are very poor approximations to the average streamline curvature.

Spatial Averaging

In many cases, solution of the full momentum equations is not warranted either by the nature of the questions to be addressed in a given study, or as a result of the kind and amount of data available. For example, applying a three-dimensional model to several hundred channel widths of a given river for a study of floodplain inundation when cross-sections of bathymetry are available only every 10 channel widths is not reasonable, because getting good results with a three-dimensional model would require more topographic data. Generally, more complete models that yield more precise results require much more input information in order to be applied relative to simpler models. In many cases, accurate results for a given purpose can be found using a simple model with relatively sparse topographic data. The two most common ways of developing simpler models are scaling analyses and spatial averaging. Scaling analysis refers to the concept of using the time and length scales of the flow to determine the most important terms in the governing equations and to develop simpler equations by retaining only these terms. This is a powerful tool for certain flows, but it generally results in a model that is specifically applicable to only a certain flow or class of flows. Spatial averaging is a method whereby one or more dimensions are removed from the model equations by integrating or averaging over those dimensions. For example, development of a one-dimensional flow model requires averaging the momentum equations over a channel cross-section, so that instead of solving for the velocity at every point in the channel, the model solves only for the cross-sectionally averaged velocity at each model cross-section. Note that while model simplicity is gained by spatial averaging, detail is lost.

The most common applications of spatial averaging result in one-dimensional models, two-dimensional models that treat the channel flow in planform (vertically averaged models), and two-dimensional models that treat the flow in the streamwise-vertical plane (cross-stream averaged models). While treating each of these in any detail is beyond the scope of this introduction to modeling flow and sediment transport, a single example illustrates some of the issues that arise in developing spatially averaged equations. Using $\langle\ \rangle$ to represent vertical averaging, the vertical average of the u velocity component is defined as follows:

$$\langle u \rangle = \frac{1}{h}\int_{B}^{E} u\,\mathrm{d}z \tag{18.2.26}$$

Applying this same operator to Equations (18.2.20)–(18.2.22), the following vertically averaged continuity and horizontal momentum equations arise in the channel-fitted coordinate system (again, note that the standard Cartesian relations are easily found from the following by letting R go to infinity):

$$\frac{1}{1-N}\frac{\partial}{\partial s}(\langle u \rangle > h) - \frac{\langle v \rangle h}{(1-N)R} + \frac{\partial}{\partial n}(\langle v \rangle h) = 0 \tag{18.2.27}$$

$$\frac{1}{1-N}\frac{\partial}{\partial s}(\langle u^2\rangle h)+\frac{\partial}{\partial n}(\langle uv\rangle h)-\frac{2\langle uv\rangle h}{(1-N)R}$$
$$=-\frac{gh}{1-N}\frac{\partial E}{\partial s}$$
$$+\frac{1}{\rho}\left[\frac{1}{1-N}\frac{\partial}{\partial s}(\langle\tau_{ss}\rangle h)+\frac{\partial}{\partial n}(\langle\tau_{ns}\rangle h)-\frac{2\langle\tau_{ns}\rangle h}{(1-N)R}\right]$$
$$+\frac{1}{\rho}\left[\frac{1}{1-N}(\tau_{ss})_B\frac{\partial B}{\partial s}+(\tau_{ns})_B\frac{\partial B}{\partial n}-(\tau_{zs})_B\right] \tag{18.2.28}$$

$$\frac{1}{1-N}\frac{\partial}{\partial s}(\langle uv\rangle h)+\frac{\partial}{\partial n}(\langle v^2\rangle h)+\frac{(\langle u^2\rangle-\langle v^2\rangle)h}{(1-N)R}$$
$$=-\frac{gh}{1-N}\frac{\partial E}{\partial n}$$
$$+\frac{1}{\rho}\left[\frac{1}{1-N}\frac{\partial}{\partial s}(\langle\tau_{ns}\rangle h)+\frac{\partial}{\partial n}(\langle\tau_{nn}\rangle h)-\frac{\langle\tau_{ss}-\tau_{nn}\rangle h}{(1-N)R}\right]$$
$$+\frac{1}{\rho}\left[\frac{1}{1-N}(\tau_{ns})_B\frac{\partial B}{\partial s}+(\tau_{nn})_B\frac{\partial B}{\partial n}-(\tau_{zn})_B\right] \tag{18.2.29}$$

These equations, which have been used in variety of models for flow and bed evolution (Smith and McLean 1984, Nelson and Smith 1989a,b, Shimizu *et al.* 1991), introduce a new kind of closure problem that is analogous to the turbulence closure problem introduced by Reynolds averaging. Terms that arise due to vertical correlations such as $\langle uv\rangle$, $\langle u^2\rangle$, and $\langle v^2\rangle$ cannot be expressed in terms of simple vertically averaged variables like $\langle u\rangle$ and $\langle v\rangle$ except where the velocities have no vertical structure whatsoever, so that $\langle uv\rangle=\langle u\rangle\langle v\rangle$ and $\langle u^2\rangle=\langle u\rangle^2$, and so forth. However, this is not generally true. For example, for a logarithmic velocity profile, the difference between $\langle u^2\rangle$ and $\langle u\rangle^2$ depends on the ratio of the roughness length to the flow depth and is typically on the order of 5% or 10%. In almost all vertically averaged models, the correlations are neglected, and one assumes that the equalities that hold for the case of no vertical structure are accurate in cases with vertical structure. However, some important effects can be excluded when this assumption is used. For example, in long meander bends with weak topography, the term $\langle uv\rangle$ has been shown to be at least partially responsible for the movement of the high-velocity region of the flow from the inner bank at the upstream part of the bend to the outer bank at the downstream part of the bend (Shimizu *et al.* 1991). This is because helical cross-stream flow moves high velocity fluid outward near the surface of the flow and low velocity fluid inward near the bed, resulting in a net momentum flux toward the outer bank. This effect is overwhelmed by topographic steering of the flow in shorter bends with point bars, but it is potentially an important effect in some natural flows. Even though this effect is dependent on vertical structure, it can be treated to some degree in vertically averaged models using dispersion coefficients. Similarly, when spatial averaging is carried out, spatial correlations between variables that appear as a result of the averaging process can generally be treated at least to some approximate degree.

Dispersion Coefficients

A general definition of a dispersion or correlation coefficient between two variables is given by the following:

$$\alpha_{ab}=\frac{\langle ab\rangle}{\langle a\rangle\langle b\rangle} \tag{18.2.30}$$

where $\langle\ \rangle$ may represent vertical averaging or some other spatial average (e.g., cross-sectional). Using this definition, Equations (18.2.28) and (18.2.29) may be rewritten in terms of only $\langle u\rangle$ and $\langle v\rangle$ along with the dispersion coefficients α_{uu}, α_{vv}, and α_{uv}. The values of these coefficients may be set theoretically or empirically. In either case, the coefficients allow at least approximate treatment of momentum fluxes that would otherwise be neglected. Another way to treat the correlation terms in averaged equations is to separate each variable into an averaged part and a deviation from that average, in parallel to the development of the Reynolds momentum equations. For example, if we use primes to denote departures from the vertical average, such as $u(z)=\langle u\rangle+u'(z)$, we can rewrite Equations (18.2.28) and (18.2.29) as follows:

$$\frac{1}{1-N}\frac{\partial}{\partial s}\left(\langle u\rangle^2h\right)+\frac{\partial}{\partial n}(\langle u\rangle\langle v\rangle h)-\frac{2\langle u\rangle\langle v\rangle h}{(1-N)R}+F'$$
$$=-\frac{gh}{1-N}\frac{\partial E}{\partial s}$$
$$+\frac{1}{\rho}\left[\frac{1}{1-N}\frac{\partial}{\partial s}(\langle\tau_{ss}\rangle h)+\frac{\partial}{\partial n}(\langle\tau_{ns}\rangle h)-\frac{2\langle\tau_{ns}\rangle h}{(1-N)R}\right]$$
$$+\frac{1}{\rho}\left[\frac{1}{1-N}(\tau_{ss})_B\frac{\partial B}{\partial s}+(\tau_{ns})_B\frac{\partial B}{\partial n}-(\tau_{zs})_B\right] \tag{18.2.31}$$

$$\frac{1}{1-N}\frac{\partial}{\partial s}(\langle u\rangle\langle v\rangle h)+\frac{\partial}{\partial n}(\langle v^2\rangle h)$$
$$+\frac{\left(\langle u\rangle^2-\langle v\rangle^2\right)h}{(1-N)R}+G'=-\frac{gh}{1-N}\frac{\partial E}{\partial n}$$
$$+\frac{1}{\rho}\left[\frac{1}{1-N}\frac{\partial}{\partial s}(\langle\tau_{ns}\rangle h)+\frac{\partial}{\partial n}(\langle\tau_{nn}\rangle h)-\frac{\langle\tau_{ss}-\tau_{nn}\rangle h}{(1-N)R}\right]$$
$$+\frac{1}{\rho}\left[\frac{1}{1-N}(\tau_{ns})_B\frac{\partial B}{\partial s}+(\tau_{nn})_B\frac{\partial B}{\partial n}-(\tau_{zn})_B\right] \tag{18.2.32}$$

where the new terms are defined by

$$F'=\frac{1}{1-N}\frac{\partial}{\partial s}\left(\langle u'^2\rangle h\right)+\frac{\partial}{\partial n}(\langle u'v'\rangle h)-\frac{2(\langle u'v'\rangle)h}{(1-N)R} \tag{18.2.33}$$

and

$$G'=\frac{1}{1-N}\frac{\partial}{\partial s}(\langle u'v'\rangle h)+\frac{\partial}{\partial n}\left(\langle v'^2\rangle h\right)+\frac{\left(\langle u'^2\rangle-\langle v'^2\rangle\right)h}{(1-N)R} \tag{18.2.34}$$

In cases where simple structure functions can be supplied for u and v based on measurements or theoretical arguments, these "extra" terms arising from correlations can be evaluated approximately. If these terms are set to zero, it is important to have an understanding of what kinds of processes are being neglected in the formulation. Situations where spatial correlations are important can often be treated without solving the full equations.

Bed Stress Closure

Whenever the equations of motion are averaged in the direction perpendicular to a boundary, closures for stress terms at that boundary must be supplied. In the vertically averaged equations used as an example, the boundary shear stress terms that arise in the horizontal momentum equations must be expressed in terms of $\langle u\rangle$ and $\langle v\rangle$. There are many ways to do this, including using Manning's or Chezy's closure, as discussed below, but the most common in multidimensional models is to use a drag coefficient closure:

$$\tau_B=\rho C_d\left(u^2+v^2\right) \tag{18.2.35}$$

Splitting this into component parts yields:

$$(\tau_{zs})_B=\rho C_d\sqrt{\langle u\rangle^2+\langle v\rangle^2}\langle u\rangle \tag{18.2.36}$$

and

$$(\tau_{zn})_B=\rho C_d\sqrt{\langle u\rangle^2+\langle v\rangle^2}\langle v\rangle \tag{18.2.37}$$

There are many other choices of bottom stress closure, but most can be directly related to this one. For example, if the flow is assumed to have a vertical structure

$$u=u_*f(z,z_0) \tag{18.2.38}$$

then the drag coefficient can be shown to be a function only of flow depth and z_0:

$$C_d=\left[\frac{1}{h}\int_{z_0}^{h}f(z,\ z_0)\,\mathrm{d}z\right]^{-2} \tag{18.2.39}$$

Closures for lateral shear stresses at banks can be handled in a similar manner. Using this closure or others that are similar, the vertically averaged horizontal momentum equations and the continuity equation can be written entirely in terms of the vertically averaged u and v velocity components and the water surface elevation (if the flow is assumed hydrostatic). This is a well-posed system of equations and unknowns, so a solution is straightforward. Although these assumptions are often not explicitly stated, any model developed from spatial averaging of the full equations requires specification of dispersion coefficients and closures for stresses at boundaries.

18.3 SEDIMENT-TRANSPORT RELATIONS

In order to determine the rates of transport of sediment traveling as bedload or in suspension, information from the flow model is typically used as input to a variety of empirical, semi-empirical, or theoretical relations for predicting sediment flux. Computations of local fluxes can be used with the equation for conservation of sediment mass to predict local erosion and deposition. However, relatively small errors in local fluxes can make a significant difference in the local rates of erosion and deposition, and errors in methods for computing sediment fluxes are often large. Choosing a method that can be calibrated with measured data, or that was developed in situations with similar grain sizes and flow characteristics, is the best way to build confidence in predictions. As in most complex problems in physical science, progress is almost always made in the interplay between careful field measurement and modeling efforts.

Bedload Transport

Bedload transport refers to grain motion near the bed consisting of rolling and hopping grains; these grains typically are moving with horizontal velocities less than the speed of the flow through most of their trajectory. Although there have been a few notable attempts to develop purely theoretical relations for bedload sediment entrainment and motion, even these models rely heavily on empirical data, and most predictions of bedload flux are made using empirical equations. As a result, it is especially important to understand how a given relation was developed and calibrated when choosing a method for computational prediction. The review paper by Gomez (1991) provides a good overview of methods for predicting and measuring bedload transport and also points out some of the physical characteristics that make developing a general model difficult. Formulas for predicting bedload flux as a function of properties of the flow (velocity, boundary shear stress, stream power, viscosity, fluid density, etc.) are usually dependent on grain size and density, and may also depend on sorting or other properties of the bed itself. One of the simplest bedload equations is that developed by Meyer-Peter and Müller (1948) and it will serve as an example of these equations for the purposes of this chapter. Defining nondimensional transport and boundary shear stress as follows:

$$(q_b)_* = \frac{q_b}{\left[\left(\frac{\rho_s - \rho}{\rho}\right) g D^3\right]^{1/2}} \qquad (18.3.1)$$

where q_b is the volumetric bedload flux per unit width, D is the grain size, g is the gravitational constant, ρ_s is the sediment density, ρ is the fluid density and

$$\tau_* = \frac{\tau_b}{[(\rho_s - \rho) g D]} \qquad (18.3.2)$$

the Meyer-Peter and Müller (1948) equation can be written as

$$(q_b)_* = 8[\tau_* - 0.047]^{3/2} \qquad (18.3.3)$$

Thus, the Meyer-Peter and Müller (1948) bedload equation yields the bedload flux as a function of only the boundary shear stress, grain size, and the particle and fluid densities. Many users apply the so-called modified Meyer-Peter–Müller equation, given by

$$(q_b)_* = 8[\tau_* - (\tau_*)_c]^{3/2} \qquad (13.3.4)$$

where $(\tau_*)_c$ is the nondimensional form of the Shields critical shear stress. The Shields critical shear stress is defined to be that value of shear stress for which significant sediment motion begins to occur for a given grain size. The reader is referred to Middleton and Southard (1984) for an in-depth review of this quantity and methods for determining the value of critical shear stress. Many other bedload equations also used this concept. Although critical shear stress was originally developed for the case of well-sorted beds that could be considered uniform in size, the concept has been generalized and extended to the case of mixed-grain-size beds by several researchers (e.g., Wiberg and Smith 1987). Using a critical shear stress developed for beds of mixed sizes is the commonest way to deal with poorly sorted sediment beds in sediment-transport models, but it is important to note that this treatment does not correctly parameterize many of the details of mixed-grain transport. This is especially true if small-scale spatial sorting occurs, or if the texture or structure of the bed evolves during flow events in other ways. Recent progress on more complete parameterization of mixed-grain transport appears likely to lead to better models. For further discussion on this and related topics, the reader is referred to Wilcock (1997, 2001).

Suspended Load Transport

Suspended load is carried by the flow both near and well above the bed, depending on the grain size and the turbulence levels in the flow, as characterized by the Rouse number [see the discussion accompanying Equation (18.5.8) and Middleton and Southard 1984]. Sediment particles moving in suspension travel at approximately the horizontal speed of the flow. In many rivers and streams, suspended load, which is typically finer and faster moving than bedload, is a greater contributor to the overall sediment load of the channel than bedload. However, the bedload is often still very important for understanding the geomorphology of the channel, because permanent bed and bank features are often dominantly made up of the grain sizes carried as bedload. Furthermore, the quantity of suspended load may not be as tightly coupled to the hydraulics (flow characteristics) of the channel compared with bedload, because the amount of suspended material in transport may be governed primarily by the amount of fine material supplied to the channel. Thus, hysteresis in suspended load relations is much more common than in bedload relations. This characteristic can make suspended load more difficult to estimate, especially for the finest sizes in suspension.

In some cases, it is possible to calculate the flux of suspended load using an empirical total load equation, such as that proposed by Engelund and Hansen (1967). However, in most cases, models use some form of the advection–diffusion equation to treat suspended load transport. As in the case of momentum, turbulence produces advective transport of suspended sediment. Following Reynolds averaging and assuming a gradient-transport closure with a scalar isotropic eddy viscosity, the advection–diffusion equation for suspended sediment in vector form is given by

$$\frac{\partial c_s}{\partial t} + (\vec{u} - \vec{w}_s) \cdot \nabla c_s = \nabla \cdot K \nabla c_s \qquad (18.3.5)$$

where c_s is the concentration of suspended material, $\vec{w}_s$ the settling velocity (positive downward) and K is the eddy diffusivity. For steady, uniform flow and an eddy diffusivity of the form given in Equation (18.2.14), this equation can be solved directly to yield the Rouse profile as discussed below or see Middleton and Southard (1984, p. 219), provided that an appropriate lower boundary condition for the sediment concentration is supplied. For more complex flows, the flow solution can be inserted along with the appropriate diffusivity and the equation can be solved numerically for the distribution of suspended sediment. For steady, uniform flows, the boundary condition at the bed is generally taken as a simple reference concentration [as a function of boundary shear stress, for example, see Garcia and Parker (1991)]. For more complex flows, the lower boundary condition is set by using a boundary condition on upward flux from the bed as a function of boundary shear stress. The form of the reference flux condition for nonuniform flows is derived directly from generalizing the reference concentration for uniform flows into an upward flux boundary condition. Thus, for example, in a situation where the boundary shear stress goes to zero at a point in a nonuniform flow, the upward flux off the bed is assumed to be zero, and the actual concentration at the bed is set by the settling of grains already in suspension in the flow.

In situations with both high concentrations and high concentration gradients, corrections to the eddy diffusivity must be made due to the stratifying effect of the suspended sediment. The reader is referred to McLean (1992) for an in-depth discussion of stratification corrections.

Erosion Equation

Once the flux of bedload and suspended load are computed, determination of the local erosion or deposition on the bed is straightforward. Applying conservation of sediment mass, the rate of erosion or deposition on the bed is given by the so-called erosion equation:

$$\frac{\partial B}{\partial t} = -\frac{1}{c_b}\left[\nabla \cdot \vec{Q}_s + \frac{\partial}{\partial t}\int_B^E c_s \, \mathrm{d}z\right] \qquad (18.3.6)$$

where $\vec{Q}_s$ is the local vector sediment flux and c_b is the concentration of sediment in the bed (typically about 0.65, i.e., unity minus the porosity).

Gravitational Corrections to Sediment Fluxes

When sediment moves as bedload over a laterally sloping bed, the sediment will not move in the direction of the near-bed flow and bottom stress, but will be deflected somewhat downslope due to the action of gravity. The degree of deflection is roughly related to the ratio of drag forces on the particle to gravitational forces on the particle, with low values of that ratio corresponding to greater downslope deflection of the particle path. Because gravitational forces are proportional to particle volume, whereas drag forces are proportional to particle area, larger particles typically experience greater deflections than smaller ones. This explains, for example, why coarse grains are preferentially sorted down the sloping faces of point bars relative to finer particles. There are several published gravitational correction models, and all are fairly similar. Nelson (1990) showed that the bedload gravitational corrections developed by Engelund (1974), Kikkawa *et al.* (1976), Hasegawa (1984), and Parker (1984) could all be written in the following form:

$$Q_n = Q_s\left[\frac{\tau_s}{\tau_n} + \Gamma f\left(\frac{\tau_c}{\tau_b}\right)\frac{\partial B}{\partial n}\right] \qquad (18.3.7)$$

where τ_s and τ_n refer to the streamwise and cross-stream components of the boundary shear stress, Q_s and Q_n are the streamwise and cross-stream components of bedload sediment flux, Γ is a coefficient, and f is a simple function of the ratio of critical to boundary shear stress. For details of the values of Γ and f, the reader is referred to Nelson (1990) or the original publications listed above. These corrections are all developed assuming no correction needs to be made along the direction of the boundary shear stress, but this assumption is questionable and awaits more careful experimental examination. Nelson (1990) proposed a method of gravitational correction based on the creation of a gravitational pseudo-stress which is added in a vector sense to the boundary shear stress. This formulation also reduces to Equation (18.3.7) for the case of small angles and cross-stream corrections only, but also treats corrections in an approximate manner for bed slopes oriented arbitrarily with respect to the boundary shear stress.

Gravitational corrections are extremely important in bed evolution models as they play a critical role in determining the lateral slopes of bars. Unless transport and erosion and deposition are completely dominated by suspended load, a correction for the influence of gravity is a necessity for accurate prediction of bar morphology.

18.4 NUMERICAL METHODS

A full discussion of the various numerical methods used in computing flow, sediment transport, and bed evolution would be difficult to cover in a book, much less a chapter or a chapter section. Because the intent of this book is to provide an overview of tools in geomorphology, not tools in computational fluid mechanics, the subject of numerical techniques will be given short shrift herein, although certain common algorithms will be referred to briefly in subsequent sections. Nevertheless, this is an important part of constructing coupled models for predicting channel behavior, and particular care must be taken in choosing algorithms. There are two primary issues, somewhat related, that require special attention in choosing algorithms: stability and numerical dispersion.

Stability, or more precisely the lack of it, is easy to observe in model results. Poorly designed algorithms for computing flow and/or bed evolution lead to unrealistic results that rapidly become more unrealistic as one iterates toward a steady solution or steps the model forward in time for unsteady solutions. Stability considerations for the flow computations alone are generally outlined by the author of the flow computation method. Stability considerations for coupled flow/sediment/bed models are altogether more subtle and depend on a number of considerations. First, the time step of bed evolution must be chosen such that bed evolution is slow relative to the timescales associated with the flow field, as this is really the basic premise of the semi-coupled modeling approach. If large changes in the bed and/or bank geometry occur within a single flow time step, the solution is almost certain to be unstable. Second, the numerical techniques must be chosen such that artificial phase lags between flow and sediment parameters are not introduced. This may seem complicated but actually relies on basic common sense. Consider the following example: if a one-dimensional model is used on a low-Froude number flow through a simple channel constriction, the cross-sectionally averaged velocity (which is all one computes in a true one-dimensional channel model) will be maximum at the constriction. If that velocity is used to compute bedload sediment transport, it will also be a maximum at the constriction, assuming typical relations between velocity, bed stress, and sediment flux. Because the flux is maximum at the constriction, the spatial gradient in sediment flux is zero at that point. Because the spatial gradient of the flux is directly related to erosion and deposition [Equation (18.3.6)], the constriction will neither expand nor contract further. However, noting that the flux must be less than the value at the constriction both upstream and downstream of it, a paradox arises. If the spatial gradient in the flux is computed at the constriction throat using the value at the throat and the one immediately upstream, erosion is predicted to occur at the constriction. If the value at the constriction and the value immediately downstream are used, deposition is predicted to occur at the constriction. Both results are wrong and will lead to runaway expansion or contraction of the constriction. This can be dealt with in a number of simple ways, but the example shows how phase lags introduced between the flow and sediment-transport parameters can lead to instabilities in the bed that are not real. Numerical methods must be chosen to avoid artificial instability of the flow field as well as the coupled flow/bed/sediment system.

Excessive numerical dispersion is typically not as obvious to the user as a stability problem. One of the important physical elements of modeling flow and sediment transport is the treatment of the movement of mass and momentum due to true diffusion or to advective processes that can be treated as diffusion-like (notably the transfer of momentum and mass by turbulence). Although a detailed mathematical discussion of this topic is outside the scope of this chapter, one of the basic problems of treating continuous systems with discretized equations is that commonly some artificial transfer of mass and/or momentum can occur as a result of the discretization process. This is referred to as numerical dispersion or numerical viscosity. The magnitudes of these effects are strongly dependent on the numerical scheme chosen and the actual numerical grid. Ideally, one would like numerical dispersion to be vanishingly small relative to the real processes of dispersion that one is trying to treat in the numerical solution, thereby ensuring that the model results are consistent with real world observations. Unfortunately, numerical dispersion has an added benefit for models that tend to be unstable in that it effectively increases the stability of the model solutions. Accordingly, it is not unusual to see model results where the values of diffusivities are an order of magnitude or more larger than real world values. In fact, it is not uncommon to see diffusivities (especially lateral diffusivities in two- or three-dimensional flow models) assigned unrealistically high values strictly to provide model stability. These models produce artificially smooth distributions of velocity and stress, and generally cannot provide accurate predictions of sediment flux or bed morphology. The hallmarks of this kind of approach for two- or three-dimensional models are separation eddies that are very short relative to real world values, rapid spreading of shear layers in the streamwise direction, and near-bank shears that are low relative to observations. Typically, models with very large values of numerical dispersion show insensitivity to the parameters of the model governing momentum exchange (e.g., drag coefficient, Manning's *n*, turbulent diffusivity). Models that use unrealistically high values of diffusivity often are unable to produce stable solutions when using realistic values of diffusivity.

Although the problems of stability issues and numerical dispersion are especially important in coupled models for flow, sediment transport, and bed evolution, there are many other considerations to be made in developing numerical techniques for such approaches. Fortunately, there are many excellent texts on this subject; for specific examples of different numerical solution techniques, the reader is referred to the excellent text by Patankar (1980). Furthermore, for well-written algorithmic elements that are useful in a variety of different approaches (e.g., tridiagonal solvers, matrix inverters, alternating direction implicit solvers, mesh generators, etc.), the reader is encouraged to explore numerical recipes (Press *et al.* 1986), and the algorithms in the libraries of standard applications (e.g., IMSL, MatLab, etc.).

18.5 ONE-DIMENSIONAL MODELS

As has already been pointed out, in many cases, solution of the full momentum equations is not warranted by either the nature of the questions to be addressed in a given study, or by the kind and amount of data available. As noted above, applying a three-dimensional model to several hundred channel widths of a river for a study of floodplain inundation when cross-sections of bathymetry are available only every 10 channel widths is not reasonable. In this situation, a one-dimensional model is probably more appropriate. Development of a one-dimensional flow model requires averaging the momentum equations over a channel cross-section, so that instead of solving for the velocity at every point in the channel, the model solves only for the cross-sectionally averaged velocity, flow rate, or discharge at each model cross-section. Recall that while model simplicity is gained by spatial averaging, detail is lost. Nevertheless, one-dimensional models are suitable for a wide range of important problems, and they are simple to develop and use.

One-dimensional Processes

One-dimensional models capture a relatively small fraction of the processes that are active in rivers and streams, but the key to their overall success and utility is that they can make predictions over long length and timescales. Because these models predict only cross-sectionally averaged quantities, they cannot predict vertical or cross-stream flow structure. They handle the response of the flow to expansions and contractions in the channel quite well, correctly predicting the streamwise free-surface response to these features. One of the most common uses of one-dimensional models is for predicting water surface levels for various hydrographs, and these techniques are still the most commonly used for predicting inundation levels during flood events. Because they treat flow expansion and contraction well, one-dimensional mobile-bed models are appropriate for determining cross-sectionally averaged scour or fill.

There are a variety of one-dimensional models that incorporate two-dimensional processes through empirical relations. Generally, these models are applicable for the situations for which they are calibrated, and they can be useful when carefully applied, but extending them outside of their immediate range of applicability is prone to error. Typically, models that attempt to treat two-dimensional processes (such as bar formation) introduce several additional coefficients or parameters and are often more complex than a simple two-dimensional approach. One example of this is the so-called stream tube method, where the flow in a channel is reduced to one-dimensional flow in a suite of stream tubes that span the channel. Coefficients or parameters accounting for momentum exchange between the tubes must be incorporated, and it is questionable whether these models are any simpler than a correct two-dimensional application, which the present authors would recommend.

Unsteady One-dimensional Flow Models

For application as a one-dimensional model, the continuity equation (18.2.1) and x direction momentum equation (18.2.3) are integrated over the flow depth and across the channel width. The water surface elevation, z, can be represented using depth, h, or cross-section area, A, and the downstream flow rate by the cross-section-average velocity, U, or downstream discharge, Q. A convenient choice (Cunge *et al.* 1980) is z and Q and the continuity equation becomes

$$\frac{\partial z}{\partial t} + \frac{1}{W}\frac{\partial Q}{\partial x} = 0 \qquad (18.5.1)$$

where W is the top-width of the wetted cross-section. In this notation, the x direction momentum equation becomes

$$\frac{\partial Q}{\partial t} + \frac{\partial}{\partial x}\left(\frac{Q^2}{A}\right) + gA\frac{\partial y}{\partial x} + gAS_f = 0 \qquad (18.5.2)$$

where S_f, called the friction slope, represents all of the energy loss terms of Equation (18.2.3). Because, in this form, it no longer truly represents conservation of momentum, Cunge *et al.* (1980) point out that Equation (18.5.2) is more properly called the "dynamic equation".

In unsteady one-dimensional flow, S_f is commonly assumed to be the same as for a steady-uniform flow with comparable depth and velocity so that common uniform flow equations such as Manning's or Chezy's can be used in its determination. From Manning's equation

$$S_f = \frac{Q|Q|W^{1.33}}{A^{3.33}}\left(\frac{n}{\chi}\right)^2 \qquad (18.5.3)$$

where n is Manning's roughness coefficient, $\chi = 1.0$ in SI units and $\chi = 1.49$ in English units and the channel is assumed to be wide enough that W approximates the wetted perimeter. Similarly, from Chezy's equation

$$S_f = \frac{Q|Q|W}{C^2A^3} \qquad (18.5.4)$$

where C is Chezy's coefficient. Further, from consideration of a fully developed turbulent flow over a plane bed covered by sand grains of some representative dimension, k_s, it can be shown (Bennett 1995) that

$$\frac{C}{\sqrt{g}} = \frac{1}{k}\ln\left[\frac{11h}{k_s}\right] \qquad (18.5.5)$$

where k_s is called Nikuradse's grain roughness. In fact, if the fixed roughness elements consist of uniform sand or gravel grains of diameter D, $k_s \approx 2D$ (Chang 1988, p. 50). Equation (18.5.5) provides a convenient means for relating the physical dimensions of the channel boundary roughness elements to Chezy or Manning coefficients (van Rijn 1984b).

In certain cases it is appropriate to ignore one or more of the terms of Equations (18.5.1) and (18.5.2) and to combine them into a single differential equation that is more amenable to solution than the original pair. The solutions to these simplified equations form a class called flood-routing models. The characteristics of the solutions and the conditions under which it is appropriate to make the simplifying assumptions are discussed in standard open-channel hydraulic textbooks, for example, Chaudhry (1993). There is copious hydraulic engineering literature dealing with the solution of the general formulation of the coupled equations (18.5.1) and (18.5.2). For prismatic channels (constant cross-section and uniform slope), they can be converted into two ordinary differential equations that must be solved for conditions at the next time step along the paths over which small disturbances would propagate upstream and downstream from a fixed point in space. This solution technique is called the method of characteristics. Another more general class, called explicit solutions, is formulated to enable resolution of flow properties at the next time level at a particular location only from conditions at its nearest neighbors at the present, or known time level. This contrasts with a final class, called implicit solutions, which require the simultaneous solution of $2N$ (generally nonlinear) equations for a channel reach consisting of N individual cross-sections ($N-1$ spatial subdivisions). At least for subcritical flow, the latter techniques are the most general, the most robust, and the most forgiving in terms of the size of the time-stepping increment. Although they are subject to a constraint on the size of the time step, called the Courant condition, the explicit techniques are probably preferable when supercritical flow or mixed subcritical and supercritical flow is expected. Chaudhry (1993) presents an overview of computer programming techniques for these solution methods and discusses their consistency and stability.

There are a number of public domain one-dimensional unsteady flow models available from government agencies. In general, these are appropriate for simulating unsteady, vertically homogeneous flow in networks of interconnected one-dimensional channels such as rivers with tributaries, tidally influenced barge canals, and delta distribution systems. For example, the US Army Corps of Engineers provides a program called UNET for DOS microcomputers (see http://www.hec.usace.army.mil/) that also provides the user with the ability to apply several external and internal boundary conditions, including flow and stage hydrographs, rating curves, gated and uncontrolled spillways, pumps, and bridges and culverts. The US Geological Survey (USGS) has three models (see http://water.usgs.gov/software/) for implementation on DOS and UNIX systems that can be applied in similar situations. These include BRANCH (Schaffranek 1987), FEQ (Franz and Melching 1997), and FOURPT (DeLong *et al.* 1997). BRANCH is generally used by USGS for computing discharge at backwater-affected stream gaging stations. The National Weather Service supports a model called FLDWAV (see http://hsp.nws.noaa.gov/oh/tt/soft/hsoft.shtml), also for dendritic systems, that focuses on routing extreme events such as natural and dam-break floods. Generally, these models are for treating flow but, as will be discussed in more detail below, the extension to treating flow, sediment transport, and bed evolution is straightforward.

Steady One-dimensional Flow Models

In the situation where the upstream discharge, Q is invariant in time, or when a hydrograph can be treated as stepwise steady state, Equation (18.5.1) reduces to $Q=\text{constant}$ and Equation (18.5.2) becomes an ordinary differential equation. The hydraulic engineering literature concerning solution of this differential equation is copious and is well summarized by Chaudhry (1993). In general, one-dimensional sediment-transport models contain algorithms that compute steady-state flow in nonprismatic channels using one or more of the techniques discussed there. It is desirable, but not necessary, for these algorithms to deal automatically with transitions between sub- and supercritical flow and to compute flow in arbitrary networks of channels. A general stand-alone steady-state one-dimensional flow analysis package called HEC-RAS with a convenient graphical user interface (GUI) is available (US Army Corps of Engineers 1997, see also //www.hec.usace.army.mil/). This package can handle mixed subcritical and supercritical flow and channel networks and the effects of flood plain obstructions such as bridges, culverts, and weirs.

Kinematic Wave Models

As noted above, there are a variety of assumptions that can be made to reduce Equations (18.5.1) and (18.5.2) to a single differential equation. One special case that has been widely employed will be discussed here. The retention of all the terms in the one-dimensional flow equations is meaningful only if the longitudinal variations in channel shape can be represented in sufficient detail. In many situations of interest, detailed data are unavailable, and model cross-sections are often so far apart that the local role of convective accelerations in the momentum balance is entirely obscured. In such cases, retention of the convective acceleration terms is probably harmless, but this should not be confused with a meaningful increase in modeling sophistication or accuracy. If local detail in channel topography is unavailable, local detail in model results will also be missing, and there is really no point in using a complete model. With this in mind, an approximate solution for unsteady flow that neglects local convective accelerations can be easily constructed using a few widely spaced cross-sections or even a single averaged cross-section. Such a model is unlikely to give accurate local predictions of stage or velocity, as local variations in the channel are not incorporated, but the approach can predict flood wave propagation fairly accurately if sufficient data are available to calibrate channel roughness. Essentially, this method neglects local convective accelerations, but absorbs the energy losses associated with them into the reach-averaged channel roughness. With this simplification, the one-dimensional continuity and momentum equations can be combined into a single diffusion equation to route unsteady discharge (Lighthill and Whitham 1955):

$$\frac{\partial h}{\partial t} + S_*^{0.5}\frac{\partial Q_k}{\partial A}\frac{\partial h}{\partial x} + \frac{Q_k}{2bS_*^{0.5}}\frac{\partial^2 h}{\partial x^2} = 0 \qquad (18.5.6)$$

where h is local depth, t the time, Q_k the discharge at a given stage under steady flow, A the cross-sectional area, x the downstream dimension, b channel width at stage h, and

$$S_* = 1 - \frac{\frac{\partial h}{\partial x}}{S} \qquad (18.5.7)$$

S_* is often set to unity, but can be significant for routing of unsteady discharge over long distances and if the wave amplitude is large. With a reach-averaged model, local stage for a given discharge is determined from local stage-discharge relations.

By representing the channel shape of the Colorado River in Grand Canyon with a simple reach-averaged cross-section, Wiele and Smith (1996) were able to compute an accurate prediction for flood wave propagation through the Canyon; an example of such a computation is shown in Figure 18.2. The figure shows a single snapshot in time of the water discharge as a function of streamwise location. For the case shown, Glen Canyon Dam was operated in a manner to produce a smoothly fluctuating flow with a period of 1 day. As the discharge wave propagates downstream, the wave front steepens and decreases in amplitude, in good agreement with observations.

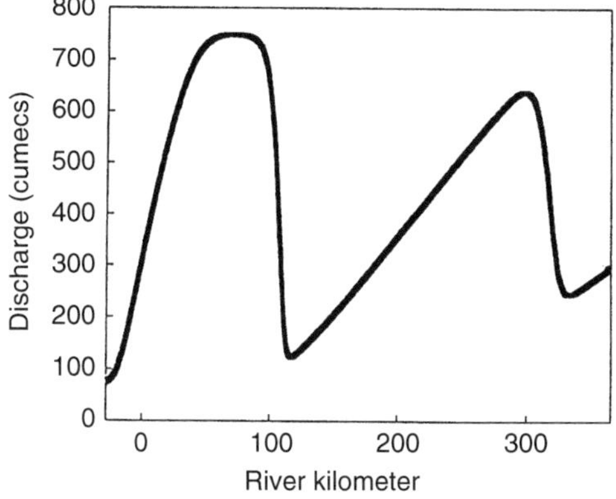

Figure 18.2 Kinematic wave simulation of discharge transients in the Colorado River in Grand Canyon. River kilometer 0 is at Lee's Ferry, and the left-hand edge of the plot corresponds to the location of Glen Canyon Dam

One-dimensional Sediment-transport Models

Combination of one-dimensional flow model with sediment bedload and suspension relations and the erosion equation produces a sediment-transport model. Such a model should contain components for computing flow velocity and depth, bedload and suspended sediment transport, and for bed elevation and size composition accounting. A general one-dimensional sediment-transport model designed for computing bed evolution should deal primarily with the sediment sizes found in the channel bed. In some situations, this restricts consideration to particles of sand size ($0.0625\,\text{mm} < D < 2.0\,\text{mm}$) and larger, but generally a wider distribution of sizes is required in natural channels. The flow model should at least have the capability to handle hydrographs by sequential steady-state flow simulation using one of the methods mentioned above. In such a model, and in its multidimensional counterparts discussed below, bedload and suspended transport are typically treated separately and by individual size subdivisions. Generally, bedload subdivisions should be no larger than increments of one ϕ ($\phi = -\log_2 D$, where D is in mm) and suspended transport increments should be no larger than half ϕ. Most one-dimensional mobile-bed models use the cross-sectionally integrated version of the erosion equation (18.3.6) to track channel bottom elevation changes as well as the composition of the bed sediment surface. The latter algorithm should also provide a mechanism to simulate the process of armoring (coarsening of the surface layer by winnowing away of the finer fractions that are still found in the subsurface layer). In addition, it is desirable for a one-dimensional sediment-transport model to provide the capability to route wash load (particles with sizes not necessarily present in the bed). The relevance of the model is enhanced if it can use flow hydraulic and bed sediment parameters and predict bedform type and geometry, which, in turn, enables the determination of the accompanying alluvial resistance to flow. Ideally, the flow modeling component should be able to deal with at least simple channel networks (for example, tributaries, islands, and distributaries) and in some instances to treat truly unsteady flow. Finally, for long-term geomorphic modeling it is desirable for a one-dimensional sediment-transport model to be capable of considering erodibility of bank materials and to adjust channel width, by simulating both erosion and deposition processes.

As discussed above, bedload transport is typically treated using an empirical or semi-empirical bedload equations. For the purposes here, that may be assumed to be the Meyer-Peter and Müller (1948) equation (18.3.4), although specific choices of the equation employed should be made with the size of material and the flow conditions in mind.

A general treatment of suspended sediment in a one-dimensional model requires cross-sectional averaging of Equation (18.3.5) to yield a one-dimensional advection–diffusion equation. However, this is generally difficult, because the interaction of velocity shear and steep gradients in the suspended sediment profile makes choices of dispersion coefficients in the average equations both problematic and extremely important. In many cases, modelers opt to treat the suspended sediment as being in local equilibrium, and to determine erosion and deposition rates from the equilibrium transport. There are many cases where this procedure is obviously wrong, especially if the settling times of a large proportion of the sediment in suspension is long relative to the timescales associated with flow nonuniformity. For example, if a parcel of sediment-laden water is advected from a location of high boundary shear stress to a place of relatively low boundary shear stress, and then back to a region of high stress, the equilibrium transport hypothesis would dictate that much of the suspended sediment would be deposited between the first and second locations. However, if the parcel of water is moved through these locations in a time much shorter than the average time required for a particle to settle, most of the material would not settle out. At the intermediate location, the concentration may remain high despite the low value of boundary shear stress. However, if sizes that settle quickly relative to advection times are treated, the equilibrium transport hypothesis is workable.

Assuming steady, uniform flow and equilibrium transport conditions in the downstream direction, the advection–diffusion equation (18.3.5) for suspended sediment can be solved analytically in the vertical direction to yield

$$c_s(z) = C_a\left(\frac{a}{h-a}\frac{h-z}{z}\right)^{w_s/ku_*} \tag{18.5.8}$$

where c_s is the concentration at elevation z above the bed and w_s is the fall velocity of the sediment. In Equation (18.5.8), a is the height above the bed at which the reference concentration C_a is specified; following McLean (1992), the elevation adopted here is the saltation layer thickness (the thickness of the layer in which the bedload or contact load moves). Equation (18.5.8) is known as the Rouse equation and the ratio v_s/ku_* is the Rouse number. Garcia and Parker (1991) compared seven relations from the literature for predicting C_a. They found the functional form of Smith and McLean (1977)

$$C_a = \frac{c_b\gamma_0 T_*}{1+\gamma_0 T_*} \tag{18.5.9}$$

and one from van Rijn (1984a) to perform about equally well and to be superior to the other five relations. In Equation (18.5.9), as noted above, c_b is the volume concentration of sediment in the bed, $\gamma_0 = 0.004$ is a dimensionless parameter, and $T_* = (\tau_*/\tau_{*c}) - 1$, the transport strength.

To illustrate the sensitivity of these relationships to particle size, consider an example of an equilibrium 1 m deep flow over a sand bed with $k_s = 2$ mm and slope of 0.0005. Assuming that the skin friction boundary shear stress is given by the depth slope product, and that the downstream velocity is logarithmically distributed [Equation (18.2.15)], one expects a mean velocity of 1.51 m/s for this flow. Assuming that the particles in the bedload layer (approximately 5 diameters thick) travel at the average water velocity in the layer, for 1-mm particles, this average is 0.68 m/s and, from Equation (18.3.3), the bedload transport rate is 1.78×10^{-5} m^2/s. Defining the average suspended load particle velocity as the suspended transport rate (the depth integral of the product uc_s) divided by the depth average concentration, Equations (18.5.8) and (18.5.9) yield 0.83 m/s for the particle velocity and 2.58×10^{-5} m^2/s for the suspended transport rate. For this flow and the 1-mm particles, the bedload and suspended particles have about the same discharge rates and transport velocities, the latter being approximately half the water velocity. For 0.08 mm particles, the comparable bedload layer velocity is 0.51 m/s and the bedload transport rate is 2.01×10^{-5} m^2/s, neither being significantly different from the 1 mm values. However, due to the near uniform vertical distribution of these smaller particles in suspension, the average particle velocity for the suspended load is 1.46 m/s, nearly equal to the average flow velocity, and the suspended transport rate is 0.0331 m^2/s, three orders of magnitude greater than for the larger particles. Considering the variation in the suspended load transport across only this portion of the sand-size range, it is easy to understand why it is important, in general, to treat the two transport processes separately.

Because wide or bimodal bed sediment size distributions are frequently encountered in practice, because simulation of sorting phenomena is often crucial, and because for given hydraulic conditions the mobility and capability for suspension of individual particles of different sizes change rapidly and drastically across the particle size range of sand in particular, it is also necessary to compute sediment-transport by individual size classes. In programming a sediment-transport model, it is impractical to deal with bed material, bedload, and suspended transport size distributions differently, so that the half-phi minimum subdivision mentioned above for suspended transport effectively establishes the particle size discretization protocol for the model. This means that for a well-sorted natural sand (narrow size distribution) the model would deal primarily with three particle sizes whereas for a poorly sorted one it would handle five or six. Following McLean (1992) the bedload transport capacity for size class i is determined by multiplying the sediment flux computed from a bedload equation by the fraction of that size in the bed, f_i, and replacing D in Equation (18.3.1) with D_i, the median particle diameter for the size class. For predicting both bedload and suspended transport, the $(\tau_*)_{cr}$ [Equation (18.3.3)] for the median diameter, D_{50}, of the mixture is employed. In calculating C_a, McLean (1992) modifies Equation (18.5.9) similarly but uses an f_i that biases the concentration slightly towards the finer sizes for smaller values of T_*. No matter how the size fractionation of suspension is treated, if there is suspended transport, the size distribution of the total load is finer than the bed material distribution.

When dealing with sand-bed channels, McLean (1992) points out that the effects of bedforms also need to be considered. This is because bedforms affect the overall resistance characteristics of the flow and they determine the fraction of the total channel bottom shear stress that is available to transport sediment. When bedforms are present, the total bottom shear stress is commonly divided into two components. The first, called form drag, is due to the pressure distribution over the length of the bedform and is assumed not to contribute to sediment transport. The second, called grain shear stress, is due directly to the movement of the fluid over the sediment surface and is assumed to determine the bedload transport rate. If the mean velocity of the flow over the bedforms is known, the method of Engelund and Hansen (1967) is commonly used for separating the two components. Alternatively, if bedform geometry and size are known, the drag coefficient closure method of Nelson *et al.* (1993) is more intuitively appealing. Finally, Zyserman and Fredsoe (1994) conclude that when bedforms are present the suspended load can best be determined when C_a is evaluated using grain shear stress and the Rouse number is computed using total shear stress.

The existence of the different types of bedforms and the geometry of a particularly common type, called dunes, is often determined as a function of median size of the bed sediment and the value of the transport strength as determined from the grain shear stress, see for example, van Rijn (1984b). Bennett (1995) presents an algorithm that incorporates these interrelated processes and predicts alluvial channel bedform geometry. Because the bedform geometry is dependent on grain shear stress that in turn is a function of the geometry, this is necessarily an iterative procedure. In this computation, the effect of the bedforms on the shape of the longitudinally averaged velocity profile is incorporated and, hence, the resulting overall resistance to flow can also be predicted.

The final essential element of a one-dimensional sediment-transport model is bed elevation accounting provided through implementation of the bed sediment conservation of mass equation (18.3.6). By tracking cross-section geometry evolution and bottom sediment size composition changes, this provides necessary feedback to controlling the sediment-transport processes. Hirano (1971) originated the "active layer" concept for dealing with bed sediment size class accounting. Bennett and Nordin (1977) conceptualize this process as occurring in three layers. The lowermost, which has a size composition as specified at the beginning of the simulation period, and extends downward indefinitely or to some limiting bedrock elevation, is called the original material layer. The second, which may not be present, is called inactive deposition and consists of material derived from the upper layer as a result of net deposition at the cross-section or material that has been re-worked in the upper layer but for some reason is currently excluded from it. The mix of sizes in the upper, or active layer, determines the f_i's used in the fractional-size transport simulation calculations outlined above and the thickness of the layer is related to the physics of the surface phenomenon, that is, to the bedform height or sediment exchange thickness for upper-regime flow. During a simulation time step, sediment can be added to or removed from the upper layer only. Before the next time step, the amounts of sediment in all the layers, the resulting bed elevation, and the size composition of the upper two layers must be recalculated. This conceptualization also allows realistic simulation of armoring. If some size of particle present in significant quantities in the upper layer cannot be transported by the existing flow, then the thickness of the upper layer can be recomputed such that it contains just enough of that size of particle (and any larger ones, if present) to comprise eventually a layer one particle diameter thick. During subsequent time steps, no more sediment is incorporated into the upper layer unless the flow becomes able to move particles of that size. More recently, Parker *et al.* (2000) have conceptualized a probabilistic approach that would track vertical bed sediment size composition changes without defining any layers. However, using this concept in digital modeling practice, the bed would still have to be discretized vertically, producing the same general effect.

One-dimensional models can readily accommodate longitudinal changes in width (sometimes called active width) of the movable portion of the channel bottom. It is also relatively simple to incorporate consideration of vertical changes (downward narrowing) in the horizontal distance between fixed banks at individual cross-sections into such models. However, because one-dimensional models cannot consider lateral variations in transport capacity due to velocity and depth variation near banks or to curvature in alignment, it is difficult to incorporate into them physically realistic mechanisms for simulation of active width changes.

Eleven one-dimensional sediment-transport models are evaluated in the proceedings edited by Fan (1993); recent versions of three of them are briefly discussed here. FLUVIAL-12 (Chang 1998) has the option of solving the dynamic flow equations (18.5.1) and (18.5.2) or using the sequential steady-state methodology discussed above; it accepts tributary discharges and flow resistance is computed from user-supplied Manning coefficients or Brownlie's (1983) formulas. GSTARS (Yang *et al.* 1998) uses the sequential steady-state approach to flow computation, can handle subcritical, supercritical, or mixed hydraulic regimes, and computes flow resistance from user supplied Manning, Chezy, or Darcy–Weisbach coefficients. HEC-6 (US Army Corps of Engineers 1993) also uses the sequential steady-state approach to flow computation for networks of subcritical channels and uses Manning coefficients or a method based on relative roughness (h/D) by Limerinos (1970) to compute flow resistance. All these models provide bed elevation accounting and armoring algorithms, compute sediment transport by size class, consider one or more size classes smaller than 0.0625 mm, and offer a choice of one of several semi-empirical total load equations from the literature for computation of sand and gravel transport. GSTARS provides semi-two-dimensional simulation by conducting sediment-transport computations separately in a user-selectable number of stream tubes that each carries equal fractions of the channel discharge. Although no transfer of sediment between stream tubes occurs during a time step, some cross-channel dispersion must happen as a result of the adjustment in bed material layer composition accompanying the lateral movement of stream tube boundaries in the cross-sections induced by the variations in stage accompanying discharge fluctuations. Both FLUVIAL-12 and GSTARS have the capability to adjust channel width based on considerations of minimization of stream power expenditure. In FLUVIAL-12, scour and fill thickness varies laterally as a function of the local tractive force and due to channel curvature-induced secondary circulation.

Example One-dimensional Model Application

A detailed example of the application of a simple one-dimensional model is provided by model results for the channel of the Snake River downstream of Boise, Idaho. This reach, which is part of Deer Flats National Wildlife Refuge, is characterized by a number of permanent islands, most of which are set aside for use as migrant waterfowl habitat. Within this reach, managers expressed concern that river discharge modified by irrigation-mandated storage and release patterns could reduce the transport capacity of the river to the point that some of the back-channels between the islands and the shore could silt up and permit access to the islands by predators. Part of this problem was addressed through the application of a flow and sediment-transport model for networks of interconnected one-dimensional channels to a segment of the Snake River containing Blind Island. The goal was to predict approximate changes in back-channel bed elevation in response to a known supply rate of medium sand. The example was constructed using the 20 cross-sections shown in Figure 18.3. Initial bed composition was $D_{50} = 90$ mm and $\sigma_g = 1.09$. Upstream water discharge was 700 m^3/s and the sediment supply rate was 0.026 m^3/s (100 g/m^3) of sand with $D_{50} = 0.5$ mm and $\sigma_g = 1.5$. The simulation network consists of an approach channel, channels to the left

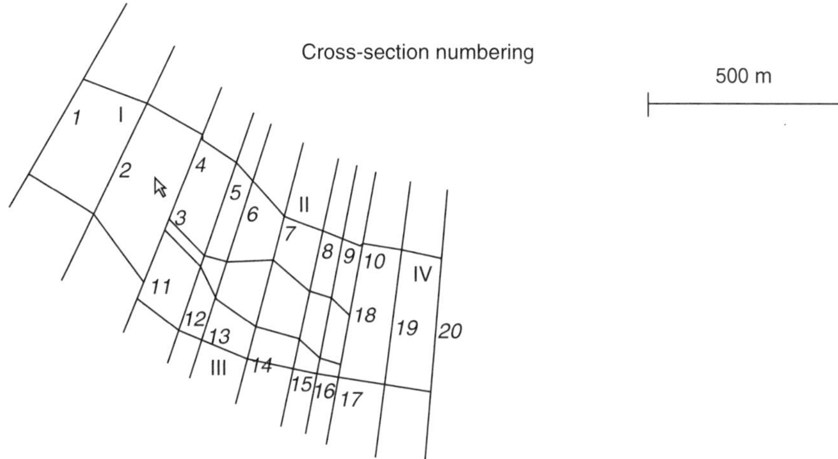

Figure 18.3 Blind Island cross-section locations and numbering; channel reaches are denoted by Roman numerals. Lateral demarcations indicate active width of channel segments

and right around Blind Island, and an exit channel; the four channels are designated by Roman numerals in the figure. Figures 18.4 and 18.5 show the results of a 10-day simulation. Bed elevations did not change enough to alter the flow distribution around the island and initial and final discharge distributions were approximately the same with about 540 m^3/s (approximately 77% of the total) flowing through the left channel segment (II) and about 160 m^3/s through the right segment (III). There was 1.05 m of fall through the approximately 1 km simulation reach. Figures 18.4(a)–(d) show initial and final longitudinal water surface, bed elevation, and bed sediment size profiles for segments II and III. Figure 18.5(a) and (b) are time series plots of transport amounts and D_{50} through channel segments II and III. For the input size distribution used in the simulation, all transport is by bedload. Transport through segment II is more than an order of magnitude greater than through segment III and the plots demonstrate that equilibrium (transport rate out equal rate in) was achieved during the second day of the simulation period. Approximately 4500 m^3 of the incoming sediment supply (20% of the 10-day input) was deposited throughout the simulation reach. The simulated final bed D_{50} was on the order of 80 mm with $\sigma_g = 1.15$ for all cross-sections. Because equilibrium was established early during the simulation period with negligible change in either back-channel bed elevation or size composition, it is unlikely that the imposed combination of flow and sediment supply rates would cause the back-channel to be closed by siltation.

18.6 TWO-DIMENSIONAL MODELS

In many cases, one-dimensional models may efficiently represent large-scale flow and sediment-transport processes. However, if specific questions about at-a-point flow, sediment transport, and erosion and deposition must be answered, a two- or three-dimensional model is required. For example, if the questions to be addressed are related to the position and amplitude of bars within the channel reach of interest, generally a two-dimensional model is necessary, as a one-dimensional model cannot predict the local flow and transport structure that gives rise to bar evolution. Generally, if the flow field of interest includes steering of the flow around islands or bars, or if there is significant cross-stream variability in the flow, at least a two-dimensional model should be applied to predict the details of local sediment transport or changes in bed morphology. Notably, in some cases where one-dimensional models yield cross-sectionally averaged velocities that are incapable of entraining sediment, two-dimensional computations will show a high-velocity thalweg in the channel in which sediment is in motion.

Two-dimensional Processes

In going from a one-dimensional to a two-dimensional model, three critical improvements are gained. First, instead of predicting only the cross-sectionally averaged component of downstream velocity and bed stress, the model predicts the value of vertically averaged downstream velocity and bed stress at many

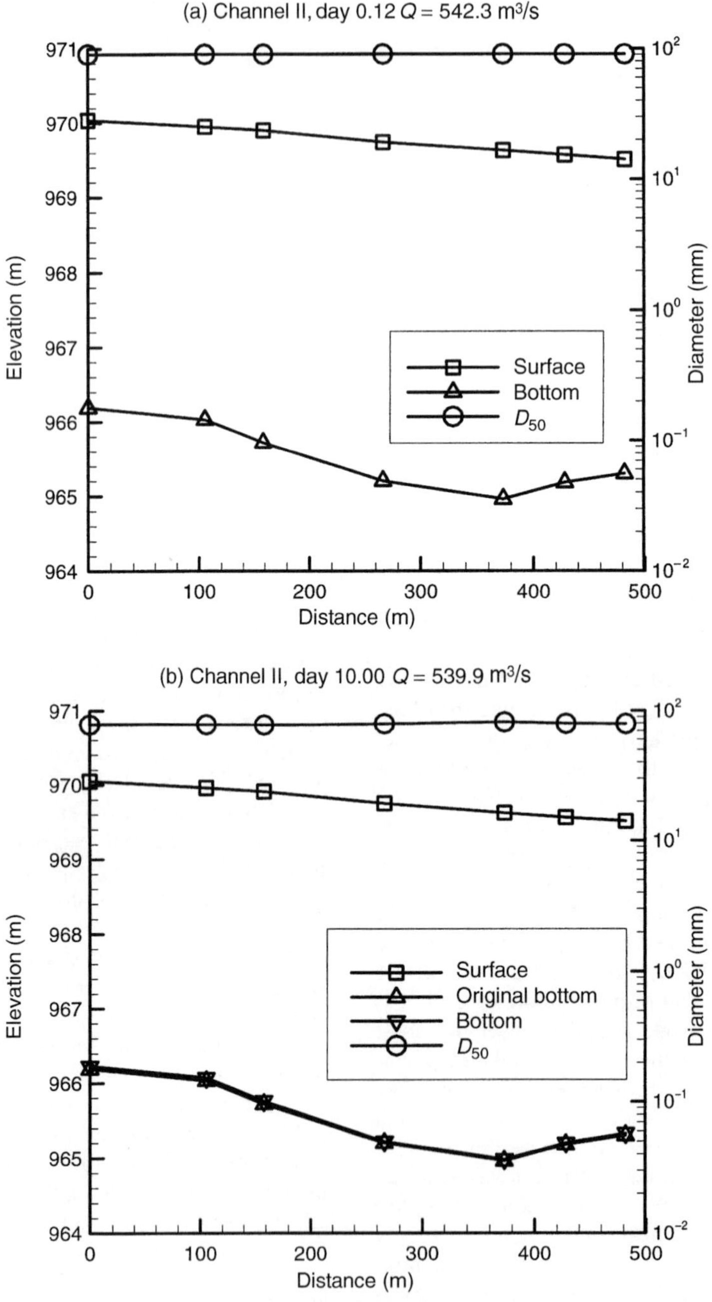

Figure 18.4 Longitudinal profiles of bed- and water-surface elevation at the beginning and end of the 10-day simulation period for segments II [cross-sections 4–10, parts (a) and (b)] and III [cross-sections 11–17, parts (c) and (d)] of the Snake River, Blind Island reach. Each profile shows the bed-surface D_{50} at the time noted in the caption above it and the initial and ending bed elevation. In portions of parts (b) and (d), the symbols for the bed profiles appear as stars because two sets of triangles are superimposed where there was little or no bed elevation change during the simulation period

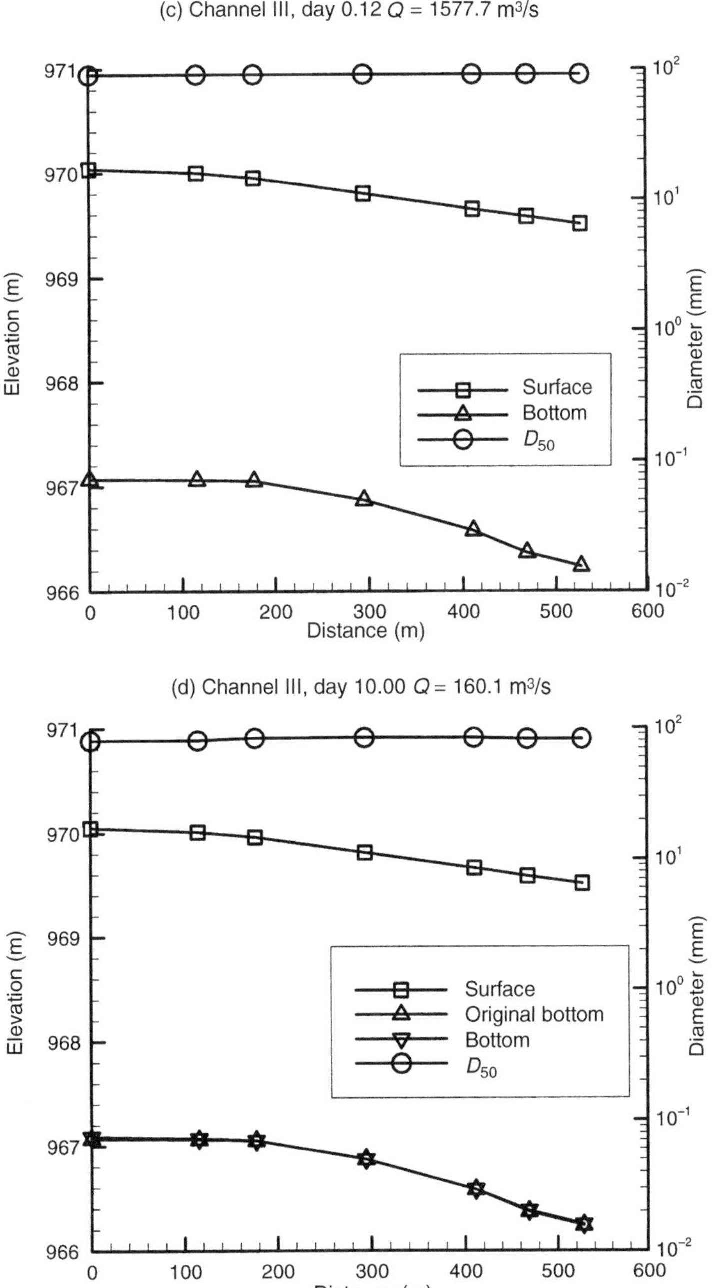

Figure 18.4 (c) and (d)

points across the channel. This means that the model can explicitly treat situations with large cross-stream velocity gradients and flow separation, which is particularly important when computing sediment transport. Second, the model also predicts the cross-stream components of vertically averaged velocity and bed stress at each point in the computational grid. As already noted above, this means that a two-dimensional

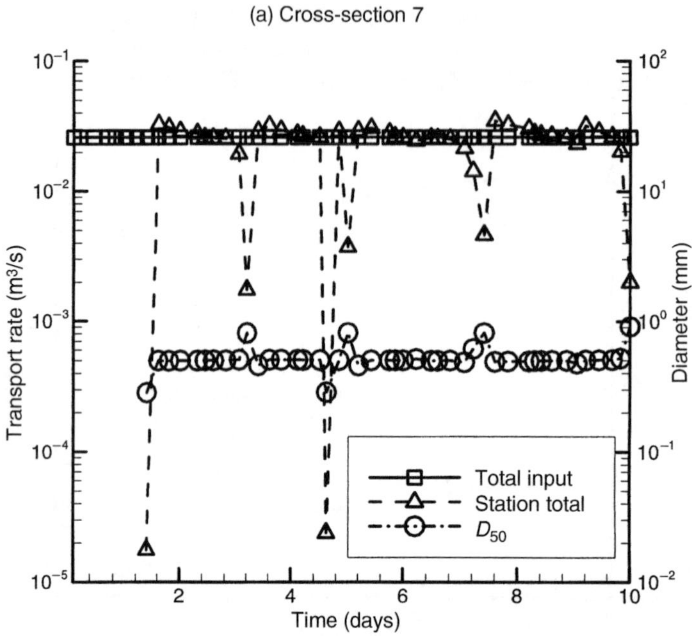

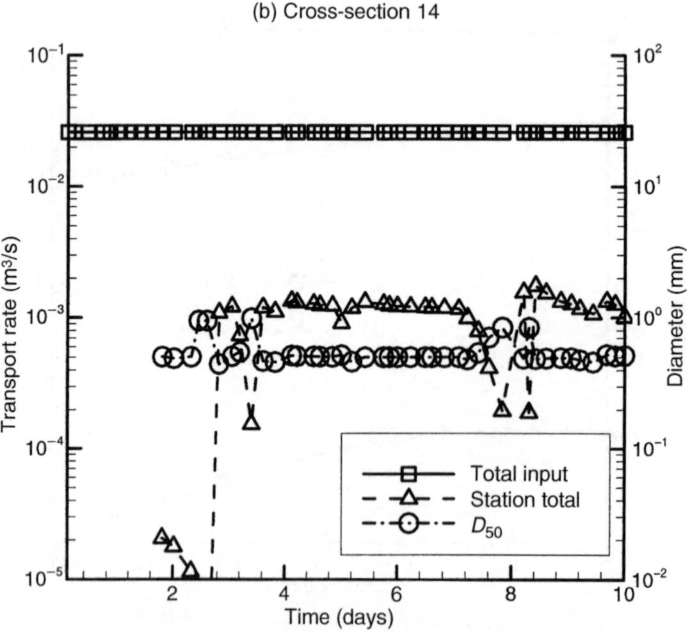

Figure 18.5 Time series of sediment transport (Station: Total) and D_{50} of transported sediment for segments II (a, cross-section 7) and III (b, cross-section 14) of the Snake River, Blind Island reach, during the 10-day simulation period. All transport in both segments is by bedload. The input sediment transport at section 1 is also shown on both parts of the figure

model can handle steering of the flow around bars and islands. This capability is critically important for prediction of the evolution and stability of bars in rivers, as the basic instability leading to these is often associated with the interaction of topographic steering of the flow and the sediment transport (Nelson and Smith 1989b). Finally, two-dimensional models allow prediction of cross-stream structure in the water surface elevation, while one-dimensional approaches do not. In many cases, superelevation of the water surface due to channel curvature or bathymetric variability results in cross-stream gradients in water surface elevation that are much larger than downstream components, so if accurate local water edge elevation is required, at least a two-dimensional model must be applied.

These three basic enhancements have several corollaries. Because it yields spatially localized quantities, a two-dimensional approach can be used to predict near-bank velocities, stresses, and sediment evacuation rates, which may be critical for predicting bank erosion. In addition, because these approaches give detailed information in a planform sense, they are useful for evaluating fields other than simply sediment-transport and bed evolution. For example, two-dimensional approaches are currently becoming the standard for habitat modeling. Habitat evaluation for many riparian species typically requires physical variables including vertically averaged velocity, depth, substrate, and so forth. A one-dimensional model could only evaluate habitat on a cross-sectional basis, which is not sufficient in most cases, as streamwise variations are typically unimportant relative to lateral ones.

It is important to point out that the enhanced predictive capabilities of a two-dimensional approach do not come without a price. Typically, input data (primarily topography) required for two-dimensional modeling application are substantially more detailed than those required for a one-dimensional model. Similarly, two-dimensional models are more computationally intensive and require a great deal more field data for verification or testing. In situations where one-dimensional models are sufficient for answering the research question, or where data of sufficient detail to warrant a two-dimensional application are not available, there is no point in applying the two-dimensional approach.

Two-dimensional Flow Models

There are a wide variety of steady and unsteady two-dimensional flow models available. The available models are more or less evenly split between finite difference and finite element solutions, with the advantages/disadvantages already discussed above. Readily available government models include FESWMS (available at www.usgs.gov/software) and HEC2D (available at www.hec.usace.army.mil) among others. There are also a variety of commercial two-dimensional modeling packages available, including SMS (available at http://www.scisoft-gms.com/index.html) and certain versions of MIKE (available at http://www.dhi.dk/index.htm).

Many two-dimensional mobile-bed models can handle hydrographs by varying the discharge in time without including the unsteady term in the momentum equations (the flow is assumed "quasi-steady"). This is a reasonable assumption only if the unsteady term in the momentum equation can be shown to be small relative to the other terms in the equation. Although there are quite a few two-dimensional flow models available for use, there are still only a few coupled flow, sediment-transport, and bed evolution models.

Two-dimensional Sediment-transport Models

The sediment-transport components of two-dimensional models are developed essentially in parallel with those discussed above in the section on one-dimensional models, with the notable exception that sediment fluxes are computed in both streamwise and cross-stream directions, and a gravitational correction is typically incorporated. As in the case of one-dimensional models, developing a vertically averaged solution for suspended sediment flux is difficult except for the simplest case of vanishingly small Rouse numbers. There are essentially three common techniques. First, some two-dimensional models that treat suspended sediment advection–diffusion do so by assigning vertical structure functions for velocity and solving the three-dimensional advection–diffusion equation for sediment concentration. Second, some models assign vertical structure functions for both velocity and sediment concentration, and then solve a vertically averaged version of Equation (18.3.5) using the vertical structure functions to assign dispersion coefficients. Finally, some approaches simply assume the suspended sediment and the velocity are uniformly distributed in the vertical, and solve the vertically averaged version of Equation (18.3.5) assuming that the dispersion coefficients are zero. The last approach gives approximately correct results only for low values of the Rouse number; in most cases either of the other two is a better choice.

Bar Evolution

Some of the very first applications of two-dimensional models for flow, sediment transport, and bed evolution were carried out in order to predict the formation of certain simple bar types. Shimizu and Itakura (1985, 1989) showed that a two-dimensional flow model could be used to predict the formation of alternating bars in simple channels. Figure 18.6 shows the equilibrium results of both computational and experimental studies of bed evolution, where the experimental results are those of Hasegawa (1984). In the experimental case, the bed was initially flat, and in the computational case, the bed was flat with the exception of a single small perturbation. In both cases, transport was exclusively as bedload. As shown in Figure 18.6, the two-dimensional model was remarkably accurate in predicting the wavelength and amplitude of bed adjustment of alternate bars. This agreement is found in spite of some obvious differences between the computed and measured flow. Even though the general flow pattern is similar, the measured flow tends to show less smooth distributions of velocity.

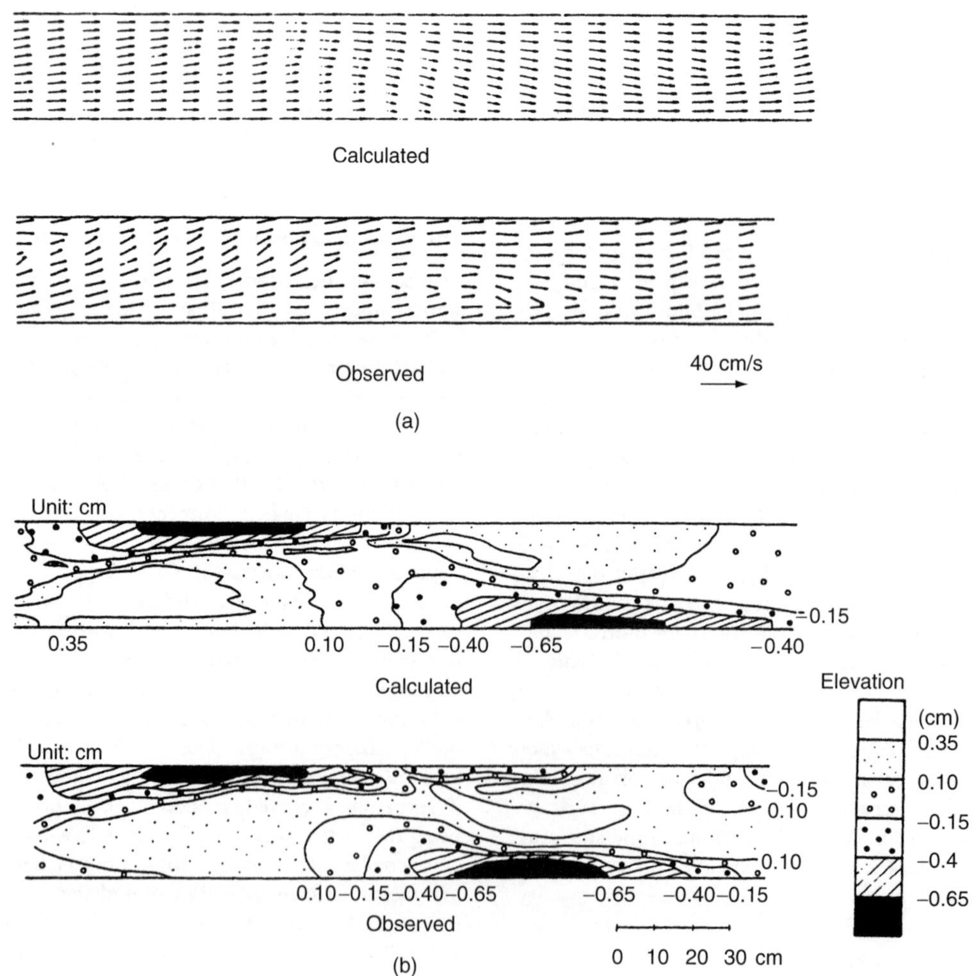

Figure 18.6 Plan view of computed and measured vertically averaged velocity (a) and bed topography (b) for run ST-1 of Hasegawa (1984). Reproduced from Shimizu and Itakura (1989)

In the same study, Shimizu and Itakura (1985, 1989) also showed that the growth and stability of point bars could be treated with a two-dimensional flow model provided that an empirical correction for the presence of secondary flows due to channel curvature was made in the computation for cross-stream sediment flux. This same methodology was employed by Struiksma (1985), who used a depth-averaged model to investigate bed evolution in the Waal River. Thus, even some of the first applications of two-dimensional flow and bed evolution models recognized that at least some three-dimensional processes had to be included in an approximate fashion to get reasonable results.

Following the initial development of two-dimensional mobile-bed models in the mid-1980s, a variety of such approaches were developed and applied by various investigators, and commercial packages for two-dimensional flow and bed evolution became available (e.g., certain versions of MIKE, developed by the Danish Hydraulic Institute, and SEDIBO, developed at Delft Hydraulics). Two-dimensional approaches have also been extended to treat the evolution of channel planform (e.g., Duan *et al.* 1997, Nagata *et al.* 2000). Over the past 15 years or so, two-dimensional approaches have made the transition from relatively rare use primarily as research tools to much more common use as practical tools for river management and engineering.

Example of Two-dimensional Model Application

One application of a multi-dimensional model to environmentally related sand supply issues is to the Colorado River in the Grand Canyon (Wiele and Smith 1996, Wiele *et al.* 1996, 1999). Reduction in sand supply by the closure of Glen Canyon Dam has affected the riparian environment downstream and generated an interest in the capability to predict the response of sand bars to sand supply and dam releases. Of particular interest are recirculation zones in the lee of channel constrictions that have the capacity to store large volumes of sand. These recirculation zones cannot be directly represented by a one-dimensional calculation, thus requiring the addition of the cross-stream dimension. Although there are certain regions of the flow that generate strong three-dimensional effects, the gross circulation patterns in these areas can be well represented by a two-dimensional model. Although the addition of the vertical dimension could potentially improve the model predictions, the computational overhead can be prohibitive in a time-evolution model and model results suggest these effects are not significant, provided one is primarily interested in the reach-scale depositional patterns rather than the detailed morphology of local features. Thus, this situation is an ideal example of a situation where a two-dimensional model is clearly preferred over a one-dimensional approach, and where the questions to be answered are sufficiently large-scale that a three-dimensional model was not warranted.

A 6-day flood on an upstream tributary, the Little Colorado River, increased the mainstem discharge and greatly increased the local sand supply, resulting in massive deposition in the mainstem channel. A modeled reach from Wiele *et al.* (1996) is shown in Figure 18.7. In this reach, the flow field, sediment-transport

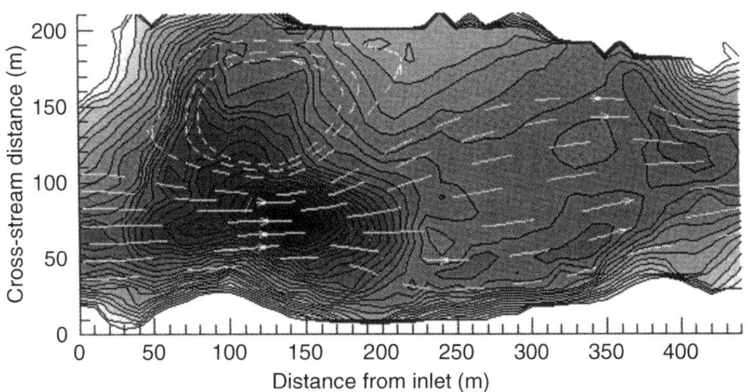

Figure 18.7 Shaded contours of bed topography of a short reach of the Colorado River in Grand Canyon overlain with streamlines of vertically averaged flow from a two-dimensional model

field and bed evolution were predicted using a two-dimensional steady flow model coupled to a bedload equation and a solution of the advection–diffusion equation for suspended sediment. The advection–diffusion solution used a logarithmic structure function for the flow velocity, as described above. The input sediment load from the Little Colorado River was computed directly from suspended sediment rating curves and gage records. The coupled model predicted rapid deposition in the reach immediately downstream of the confluence in response to overloading of the mainstem flow with suspended sediment. The bed evolution predicted by the model shows good agreement with five cross sections measured in the reach before and after the tributary flood (see Wiele *et al.* 1996). The model results predicted a filling of the depositional sites within 3 days of the start of the 6-day flooding event. Subsequently, a high-discharge dam release in 1996 designed to test the efficiency with which bars could be deposited by manipulation of dam releases built bars in a different pattern than occurred during the 1993 flood on the Little Colorado River. In this case, lower sand concentrations and higher water discharge resulted in the sand deposition being focused at the reattachment point near the reach inlet (Wiele *et al.* 1999) rather than being deposited throughout the channel. Thus, the spatial pattern and magnitude of deposition is quite sensitive to the mainstem concentration, as one might expect.

Application of the two-dimensional model provides a physically based means of inferring processes and also provides an estimate of deposition rates and the effect of variable mainstem sand concentrations on sand deposits. The model has been used to compare deposition rates and locations during known events, such as those described above, with deposition rates and locations that would have occurred with historically high and low sand concentrations. Thus, modeling helped to demonstrate that the morphological response of the channel to high flow was critically dependent on the volume and sizes of sediment stored in the channel and provided a way to make a linkage between channel response and sediment availability.

18.7 THREE-DIMENSIONAL MODELS

Three-dimensional coupled models for flow, sediment transport and bed evolution are still relatively rare. This is due to the difficulty of constructing full three-dimensional flow solutions in complex domains and also because computational limitations arising from the fact that bed evolution models generally require hundreds of thousands of iterative calculations. Thus, full three-dimensional coupled models are still considered to be prohibitively slow except for very specialized small-scale calculations, such as scour near structures. However, there is a class of erodible bed models that treat a good deal of the three-dimensional processes while avoiding the numerical overhead of a full three-dimensional solution, which will be discussed below. These approaches provide a method for treating some three-dimensional processes without incurring the penalties that a fully three-dimensional model would. As computational resources continue to increase, the applicability and utility of truly three-dimensional approaches will greatly expand, with a commensurate expansion in the understanding of certain processes such as nonhydrostatic effects that are inescapably three-dimensional and cannot be captured adequately by simpler models.

Three-dimensional Processes

In addition to the obvious improvement of predicting velocity components and stresses throughout the flow, three-dimensional models introduce three distinctly important physical processes that are not captured in one- or two-dimensional models. First, and perhaps most importantly, three-dimensional approaches allow the prediction of secondary flows. Secondary flows are defined as flows with no net discharge, acting perpendicular to the streamlines of the vertically averaged flow. The most common example is the helical flow found in meander bends, but there are others. Secondary flows are commonly driven by channel (or streamline) curvature or by gradients in normal Reynolds-stress components. Secondary flows produce a difference in vector direction of the flow over the flow depth. In the case of a meander bend, helical flow is produced that is directed toward the center of curvature near the bed and away from the center of curvature near the surface of the flow. This pattern results in a tendency for sediment deposition near the inner bank of channel bends (point bars) as discussed in fluvial geomorphology texts. Secondary flows are also responsible for many similar effects that are not quite so obvious, and they play an important role in the evolution and stability of river bars through an interplay with topographic steering and gravitational effects on sediment transport. Except for the simple case of vanishingly low Rouse number suspension (in which secondary flows produce no net advection of sediment), secondary flows play a critically important role in determining erodible bed behavior.

The second enhancement produced by a three-dimensional model is the precise treatment of momentum fluxes that vary in the vertical. These are sometimes referred to as "redistribution of momentum" effects. The helical flow discussed briefly above provides an ideal example. Because rivers tend to behave at least somewhat like simple two-dimensional boundary layer flows, it is typical for velocity to increase away from the bed. If a cross-stream helical flow is present, it will tend to advect low streamwise momentum in one direction and high streamwise momentum in another direction, resulting in a net lateral flux of streamwise momentum that cannot be predicted from the vertically averaged velocity field. This topic is directly related to the specification of dispersion coefficients, which are one way to attempt to capture the redistribution of momentum effects in a lower-dimensional model. Fully three-dimensional models automatically treat these effects correctly. Redistribution of momentum effects are important for generating velocity maxima below the free surface, as shown, for example, by Shimizu *et al.* (1991), and can significantly alter the stress patterns on the channel bed and banks.

The final enhancement found in a three-dimensional model is the treatment of nonhydrostatic effects. These effects are notoriously difficult to treat in any approximate manner in a lower-dimensional model, and solution of the full three-dimensional equations is required. Fortunately, bed slopes in the direction of flow tend to be relatively gentle, and the hydrostatic approximation is good in many situations. However, if flow is to be accurately predicted in regions of steep downstream bed slopes (e.g., over dunes, bedrock obstructions, etc.), then nonhydrostatic effects cannot be neglected. Generally, bedforms are treated parametrically in one- and two-dimensional coupled models, but if they are explicitly treated in a three-dimensional approach, accurate prediction of the flow requires a nonhydrostatic pressure distribution. The other notable situation where nonhydrostatic effects play a significant role is in flows over and around man-made structures, such as bridge piers. Coupled models for local scour near piers must incorporate nonhydrostatic effects.

Three-dimensional Flow Models

As the dimension of the model increases, the necessity to understand the approximations or assumptions that go into the model decreases, and the need to spend more time on the numerical approaches increases, both in development and computation. This is especially true if the three-dimensional model consists of a solution of Equations (18.2.1) and (18.2.2), or a so-called direct numerical simulation. In that case, there is no need for a turbulence closure, although the grid spacing must be small enough to treat viscous dissipation of the turbulence. Needless to say, these models are not used to compute sediment transport and bed evolution. However, there are models that compute fully unsteady three-dimensional flow fields without the prescription of a standard turbulence closure. These models generally use a closure that is used to treat only fluctuations (and dissipation) that occur at a scale smaller than the model grid scale. These models are called large-eddy simulations, as they do compute the larger scales of the turbulence field directly, but treat the smaller ones parametrically. These models are especially promising for complex problems in flow and sediment transport, and coupled models for flow and bed morphology are being developed and tested by various researchers (e.g., Shimizu and Schmeeckle, pers. comm.).

The three-dimensional flow models that have been applied to sediment transport and bed evolution as of this writing (and to these authors' knowledge) assume steady flow. As discussed above, these can be applied to hydrographs in situations where quasi-steadiness can be assumed. These steady models can be roughly divided into two types. First, there are a few full solutions to the steady momentum equations for certain simple geometries. For example, Olsen and Melaaen (1993) carried out coupled three-dimensional flow-sediment-bed evolution calculations for scour around a circular cylinder. Shimizu *et al.* (1991) used a three-dimensional model, but assumed that the flow was hydrostatic and did not compute bed deformation from the full three-dimensional model, although they did discuss some implications in that regard. Second, there are several three-dimensional models that are based on a so-called 2.5-dimensional technique. In this method, the three-dimensional model solution is made up of a two-dimensional (vertically averaged) solution along with a separate computation for secondary flows and, in some cases, redistribution of momentum effects. An approach of this type was discussed in detail by Nelson and Smith (1989a,b) in the context of bed evolution calculations. Their method incorporated secondary flows generated by both channel and streamline curvature (i.e., the method predicted secondary flows even in straight channels if topographic nonuniformity was present). Shimizu *et al.* (1991) used the same methods, but iterated on the vertically averaged solution to capture the redistribution of momentum effects and were able to show that a 2.5-dimensional approach was sufficient to capture these effects parametrically. Currently, the 2.5-dimensional approach is the most common method for three-dimensional mobile-bed calculations because it captures some important three-dimensional features without requiring a full three-dimensional solution. As computer resources continue to improve, these models are likely to be replaced with true three-dimensional solvers.

Three-dimensional Sediment-transport Models

Three-dimensional sediment-transport models are developed and applied almost identically to two-dimensional ones. Bedload is computed in the same manner, as are gravitational corrections to bedload fluxes. As full three-dimensional velocity fields are available, the advection–diffusion equation can be readily solved for the suspended sediment field, provided that the model incorporates some kind of eddy diffusivity closure.

Bar Evolution

As in the case of two-dimensional approaches, the first application of coupled three-dimensional models addressed the formation of simple bar forms. Nelson and Smith (1989b) used a 2.5-dimensional approach to predict the evolution and stability of point bars in curved channels and alternating bars in straight channels. Their model was cast in the channel-fitted coordinate system, but some terms in the equations were dropped due to scaling arguments. In Figures (18.8) and (18.9), the evolution of both the bottom stress field and the topography is shown for a channel with a sine-generated planform shape and an initially flat bed. The solution at the bottom of the figure is the initial condition, and the upper panel is the equilibrium condition, with the intervening figures evenly spaced in time. For the initial flat bed, the vector boundary shear stress has a clear component toward the inner bank, which is produced by the secondary flow. This produces a deposit on the inner bank of the bend and scour on the outer bank, as shown in Figure 18.9. As time progresses, the growth of the point bar tends to steer the flow along the inner bank outward and also produces a lateral slope which deflects sediment flux downslope. For the case shown, all transport was by bedload, so this correction has a significant impact on the pattern of sediment flux. Thus, development of the point bar is initially driven by curvature, but is stabilized by a combination of topographic steering and gravitational effects on sediment flux. The ultimate position of the point bar is determined by the balance of cross-stream convergences of sediment, which are in phase with the channel curvature, and downstream convergences, which are out of phase with the channel curvature. In Figure 18.10, calculated and observed bathymetric contours are shown for a weakly sinuous channel constructed by Whiting and Dietrich (1987). Although there are clear differences between the modeled and measured equilibrium topography, primarily in regions of steep slope, overall the comparison is favorable. For this weakly sinuous case, the point bar is almost entirely downstream of the bend apex. The success of the relatively simple 2.5-dimensional coupled model for a variety of simple bar types indicated the potential of the technique. Since this early work, a variety of more general models based on the same concept have been applied in a number of practical situations.

Example of Three-dimensional Model Application

The final example in this chapter uses a 2.5-dimensional model to predict the details of deposition in lateral separation zones in rivers. Lateral separation zones occur in rivers and streams where bank curvature causes separation of the downstream flow from the bank, producing a region bordered by relatively slow upstream flow near the bank and by strong lateral shear along its riverward margin. These regions, also referred to as lateral separation eddies, are efficient traps of sediment and organic material, and play important roles for riparian habitat and, in some cases, for recreational use (Schmidt and Graf 1990).

As already discussed in the section on two-dimensional flow modeling, a two-dimensional flow can predict the presence of lateral separation zones, and can also predict deposition within them for cases when the mainstem sediment concentration is relatively high. However, because there is a separation streamline between the mainstem and the eddy region, a vertically averaged two-dimensional model can predict transport across the eddy boundary (i.e. across

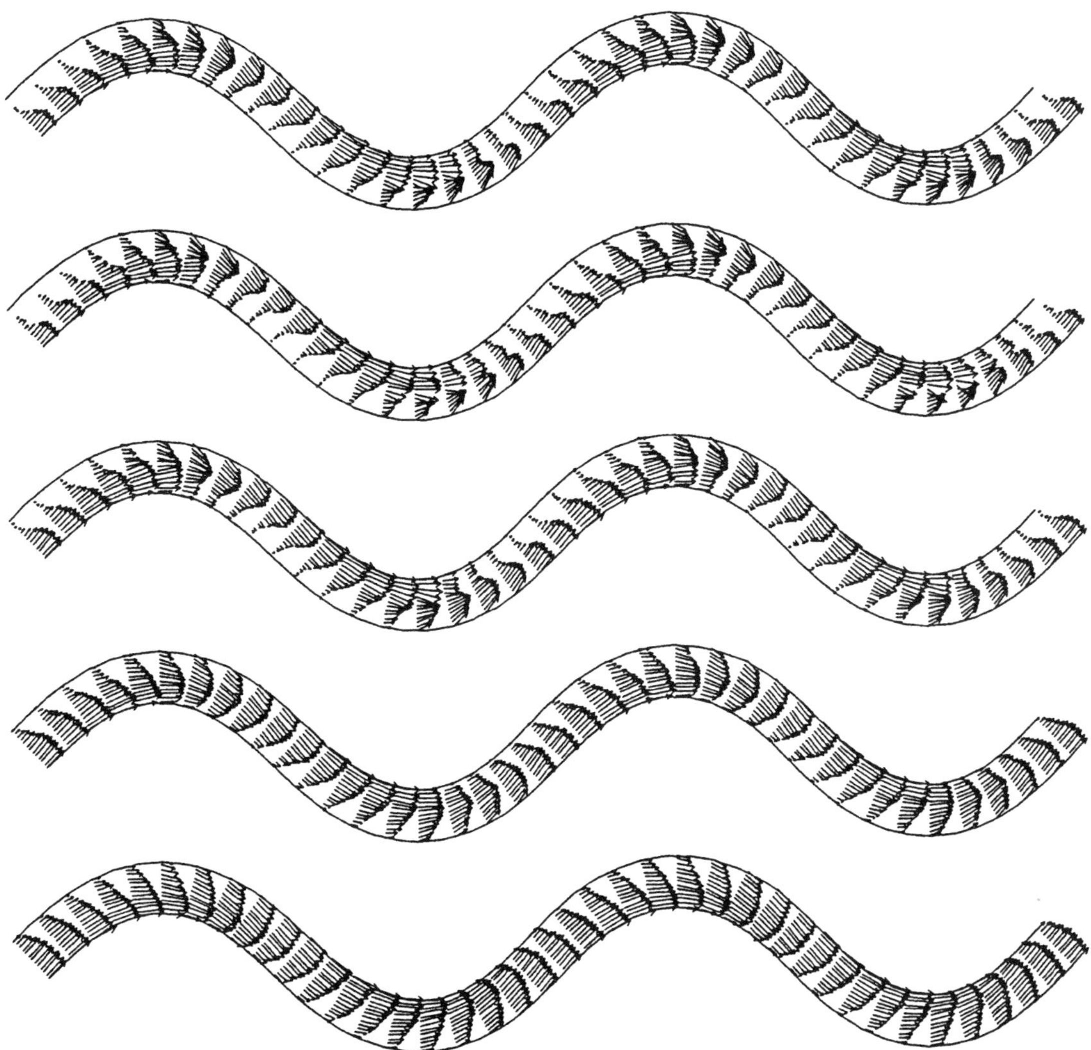

Figure 18.8 Vector boundary shear stress patterns during the development of a point bar in simple sine-generated channel. The width-to-depth ratio is 12.5 and the down valley meander wavelength is 10 widths. The contour interval is one-fourth of the initial depth. Reproduced from Nelson and Smith (1989b)

the reattachment streamline) only by diffusion. This is true because, by definition, there is no component of flow across the streamline joining the separation point and the reattachment point. However, for the case of bedload or suspended load with significant vertical structure, this is incorrect. Laboratory observations show that there are strong three-dimensional effects producing advection directly into the eddy. This effect is principally a product of secondary flows generated at the riverward margin of the eddy, which tends to produce flow into the eddy near the bed and out of the eddy near the surface. Thus, for lateral separation deposits formed by bedload or suspended load distributed nonuniformly in the vertical, this secondary flow creates a strong capability for capturing sediment.

To construct an appropriate model that includes the effect of secondary flows, these authors combined

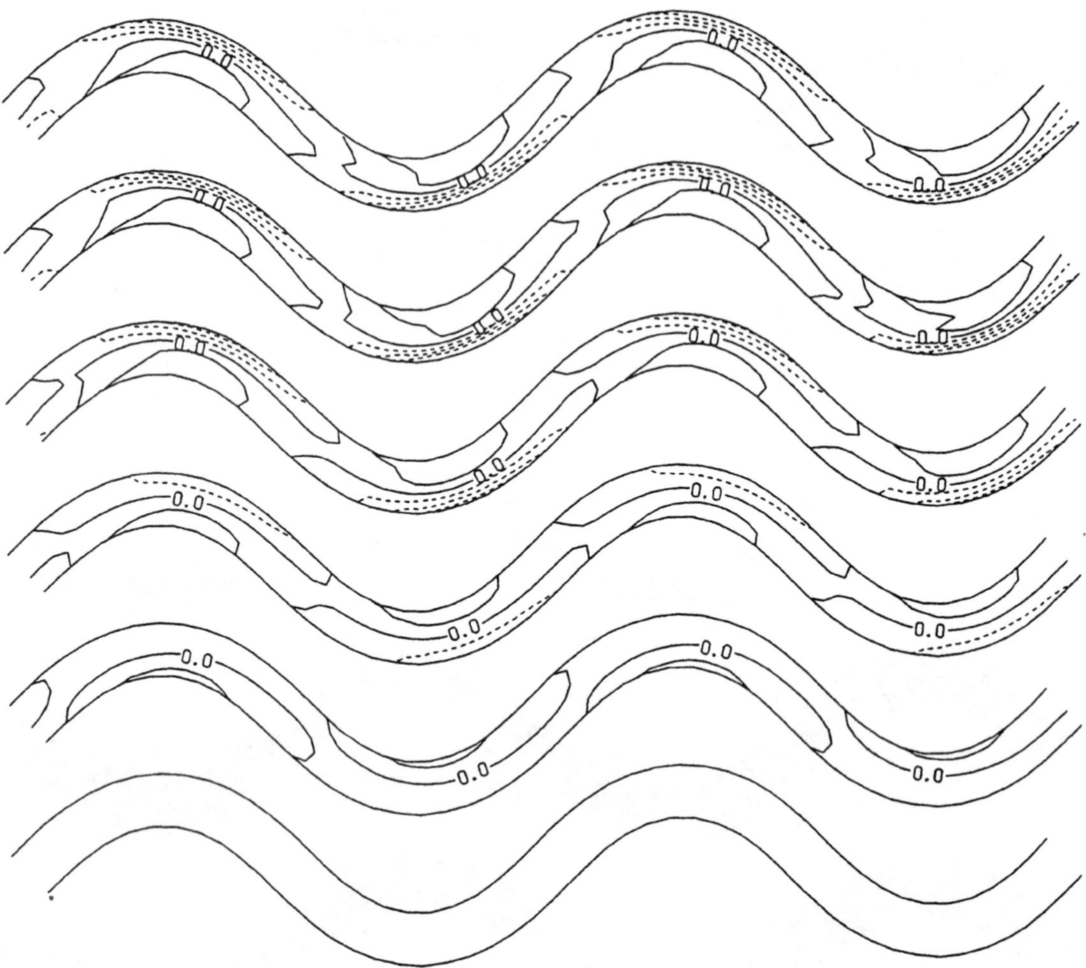

Figure 18.9 Bed elevation contours for the conditions given in the caption of Figure 18.8. Reproduced from Nelson and Smith (1989b)

a solution of the full vertically averaged equations with the method for computing vertical structure and secondary flows developed by Nelson and Smith (1989b). The model equations were solved using the SIMPLE method proposed by Patankar (1980). Although this model does not treat hydrostatic effects, or the redistribution of downstream momentum, it appears to be the simplest approach that can predict the streamline-curvature-driven secondary flows in eddies. The turbulence closure uses a simple vertical distribution of eddy viscosity with a correction for the high lateral shear present along the riverward margin of the eddy.

For this example (from Nelson 1996), laboratory data provide a clear demonstration of the method and, due to the ease of comprehensive data collection in the laboratory, also provides an opportunity to test the accuracy of model predictions. In Figure 18.11, laser-Doppler measurements of streamwise flow along the high-shear region of a flat-bedded lateral separation zone are shown with the predictions of the flow model. The computational model yields accurate predictions of the mean flow in the eddy. In Figure 18.12, the temporal evolution of the initially flat laboratory bed is shown over a period of 2 h, at which time the bed was near equilibrium. For the case shown, all

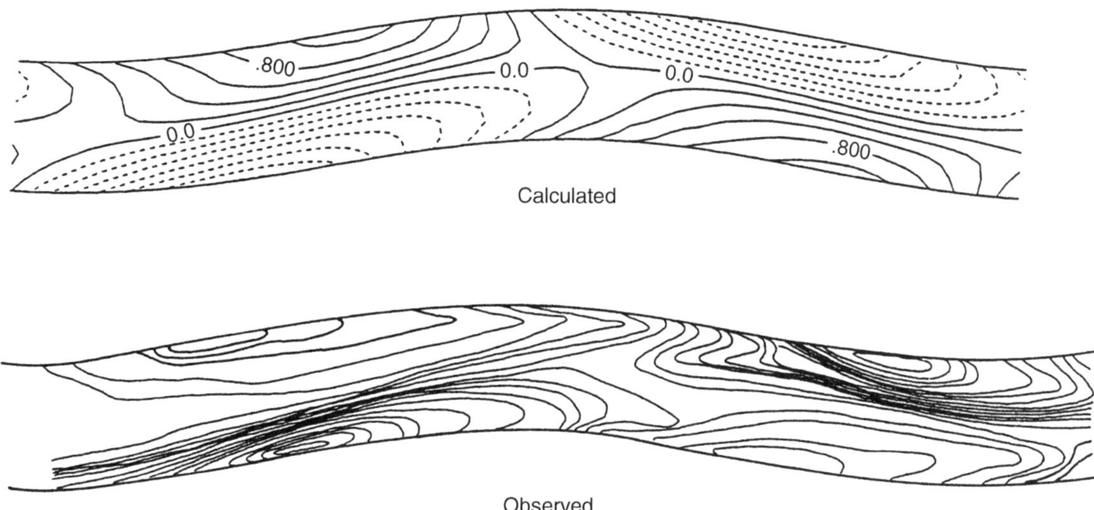

Figure 18.10 Comparison between measured and predicted equilibrium topography for Whiting's run S-25 (Whiting and Dietrich 1987). Contours are drawn at 2 mm intervals in both cases

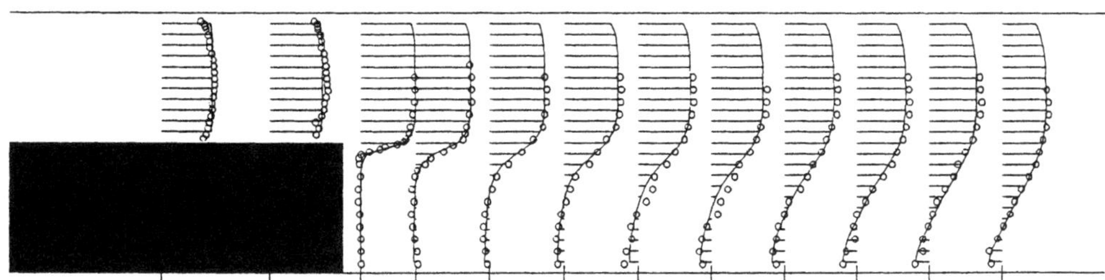

Figure 18.11 Plan view of predicted (lines) and measured (symbols) values of downstream velocity for the experimental lateral separation eddy. Flow is from left to right, and the channel width increases by a factor of 2 downstream of the obstruction producing lateral flow separation. Maximum velocity is 63 cm/s

transport occurred as bedload. The initial response of the sediment to the flow is to deposit a bar immediately inside the region of high shear bounding the eddy zone. As time progresses, this deposit grows and moves into the eddy and a deposit forms in the main channel in response to the decelerating flow. These predictions are in good agreement with the observed changes in bed elevation, both in terms of magnitude and timing.

In Figure 18.13, computed bed elevation is shown after 2 h of evolution for the cross-section located at the point of maximum elevation of the separation zone deposit for both the case where the 2.5-dimensional model is applied (i.e., including the effects of secondary flows) and for the case where secondary flows are neglected (a two-dimensional approach). The corresponding measured bed topography is also shown as individual points. Neglecting secondary flows produces approximately the correct form of the deposit but seriously underpredicts the growth rate. The need for a three-dimensional model for making accurate predictions of the morphological evolution of the channel is clear. This example shows just how carefully the choice of modeling approaches must be made. The two-dimensional approach discussed above can predict the presence of lateral separation zones and should even make accurate predictions of their deposits for low Rouse numbers, as secondary flows will produce little, if any, augmentation of sediment trapping in the eddy for that case. However, that approach yields poor

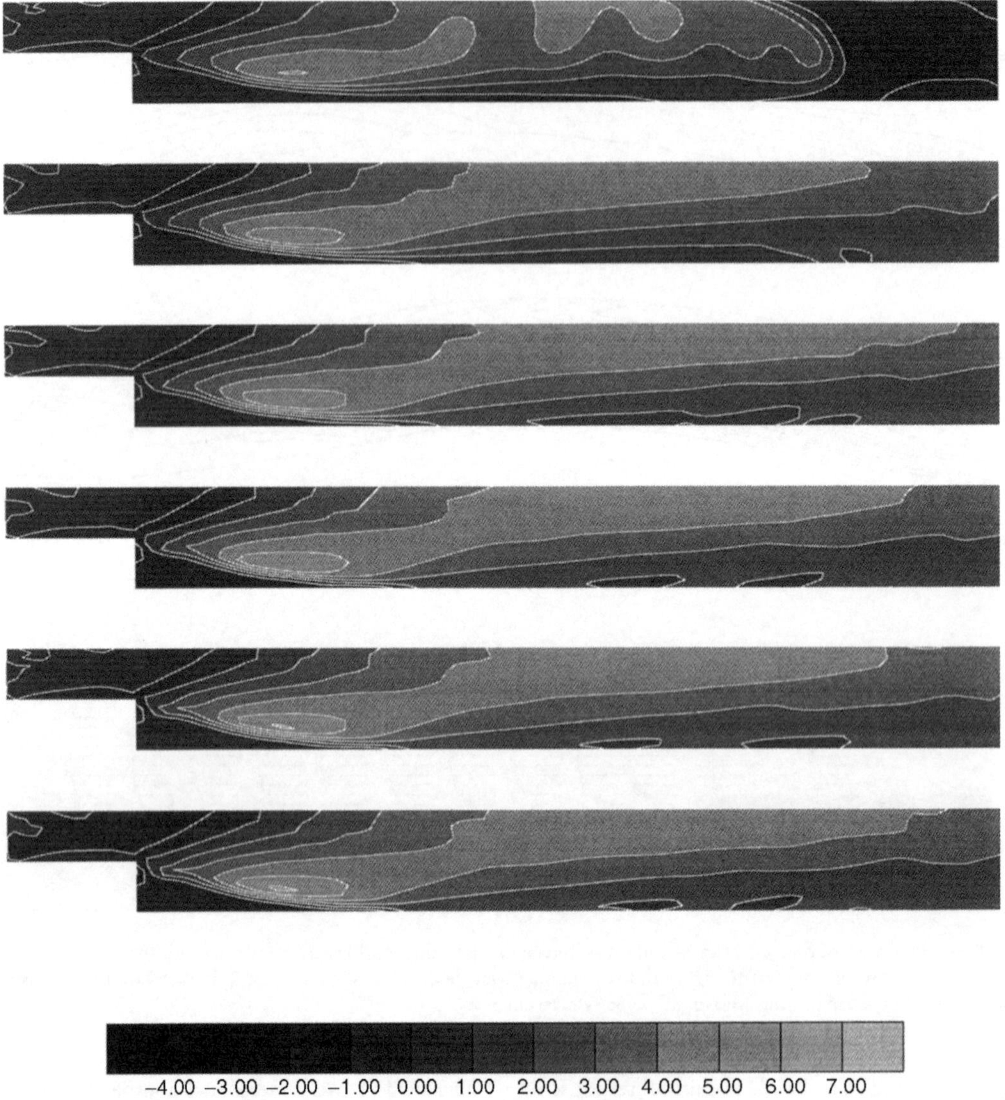

Figure 18.12 Temporal evolution of an initially flat bed in a simple eddy. The bed is shown at 20 min increments starting from the top and contour intervals are 1 cm

results for higher Rouse numbers or dominantly bedload transport situations because of the importance of secondary flows. Thus, choice of a two- or three-dimensional model depends critically on the transport characteristics.

The 2.5-dimensional model used in this example has also been applied by the first author in a variety of natural channels, including the Colorado River (see Figure 18.14), the Green River in Utah, the Snake River in Oregon and Idaho, the Klamath River in Oregon, the Platte River in Nebraska, and others. In each case, the model has been used to address specific practical issues in river mechanics. Thus, although 2.5-dimensional models are still not commonly available, they are suitable for use in real channels and can deal with a wide variety of flow, sediment trans-

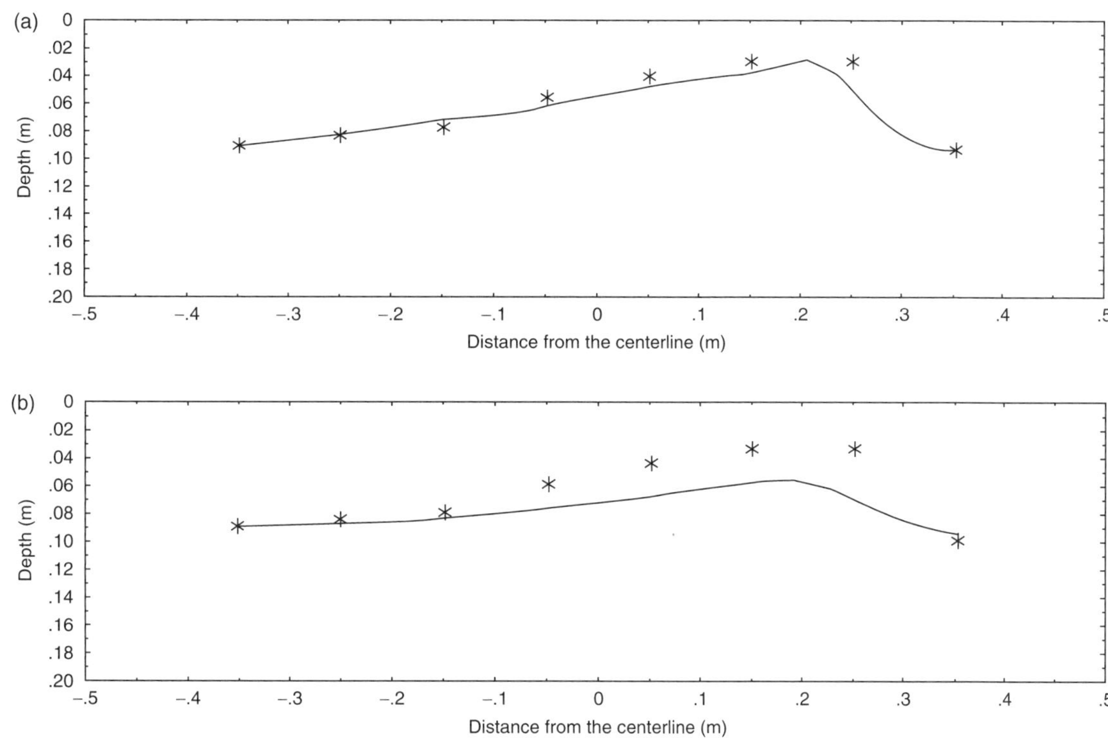

Figure 18.13 Cross-sections across the maximum bar height after 2 h of temporal evolution for the full model (a) and a two-dimensional model (b). Lines reflect the topography predicted from a computational model and stars show the measured values from laboratory experiment

port, and bed evolution issues. Nevertheless, fully three-dimensional models, including large-eddy simulations, are likely to supplant this technique as computational resources continue to increase.

18.8 CONCLUSIONS AND FUTURE DIRECTIONS

Computational modeling of flow, sediment transport, and bed evolution has made dramatic progress over the past 20 years. Over this period, most advances have been dictated by improvements in the ability to predict complex flows. In other words, the flow computation has regulated progress. However, at this point, as truly complete flow models are beginning to become available, it appears likely that research will return to the details of the sediment-transport process and how one should model it, especially in complex flows. Currently, most techniques for predicting sediment motion are extensions of methods that are strictly valid only for steady uniform flows. For example, parameterizing the forces on sediment particles that lead to bedload motion in terms of boundary shear stress assumes that all the local variability near the bed can be captured in the boundary shear stress. Generally, this is not true in complex flows and bedload motion can occur even where the mean boundary shear stress is zero.

Continued improvements in the ability to predict erodible bed behavior will require developing more precise predictors for sediment transport. Currently, even some of the most common bed features in nature are difficult to treat in coupled models. For example, the evolution, stability, and adjustment to changing flow conditions of bedforms have yet to be predicted from a coupled flow and sediment transport model. This and similar problems require both state-of-the-art computational flow models and more accurate relations between near-bed flow and sediment motion than are currently available. However, it

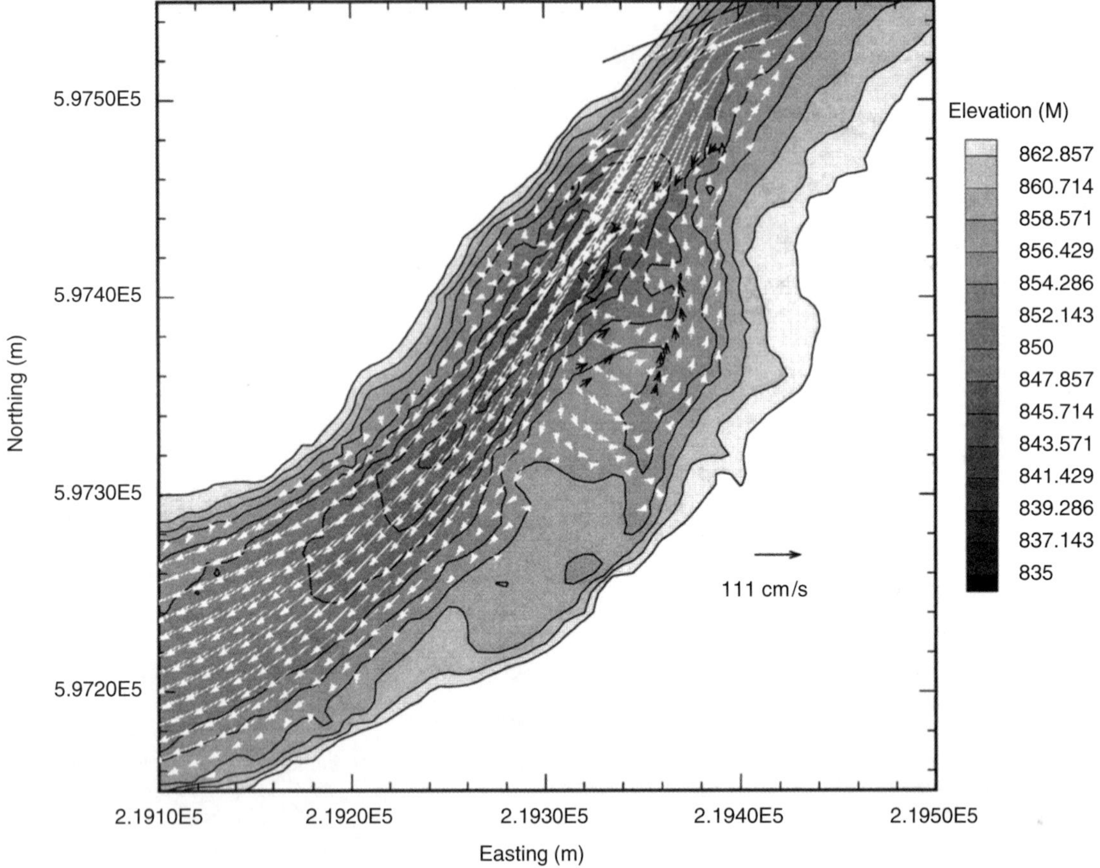

Figure 18.14 Map of bed contours, computed flow (white vectors), and measured flow (black vectors) in Eminence Break eddy in the Colorado River in Grand Canyon

appears that increasing computational power in conjunction with better experimental techniques for visualizing and measuring complex flows are likely to produce exciting advancements in this field in the near future.

REFERENCES

Bennett, J.P. 1995. Algorithm for resistance to flow and transport in sand-bed channels. *Journal of Hydrologic Engineering, ASCE* 121(8): 578–590.

Bennett, J.P. and Nordin, C.F., Jr. 1977. Simulation of sediment transport and armoring. *Hydrological Science Bulletin* XXII 4: 555–570.

Brownlie, W.R. 1983. Flow depth in sand-bed channels. *Journal of Hydrologic Engineering, ASCE* 109(7): 959–990.

Chang, H.H. 1988. *Fluvial Processes in River Engineering*, New York: John Wiley and Sons, 432 p.

Chang, H.H. 1998. *FLUVIAL-12 Mathematical Model For Erodible Channels, Users Manual*, San Diego, CA, 54 p.

Chaudhry, M.H. 1993. *Open-Channel Flow*, Englewood Cliffs, NJ: Prentice-Hall, Inc., 483 p.

Cunge, J.A., Holly, F.M., Jr. and Verwey, A. 1980. *Practical Aspects of Computational River Hydraulics*, London: Pitman Publishing Limited, 420 p.

DeLong, L.L., Thompson, D.B. and Lee, J.K. 1997. The computer program FourPt (Version 95.01) – a model for simulating one-dimensional, unsteady, open-channel flow, US Geological Survey Water-Resources Investigations Report 97-4016, 69 p.

Duan, G., Jia, Y. and Wang, S. 1997. Meandering process simulation with a two-dimensional numerical model. In: *Proceedings Conference on Management of Landscapes*

Disturbed by Channel Incision, Mississippi: University of Mississippi, pp. 389–394.

Engelund, F. 1974. Flow and bed topography in channel bends. *Journal of the Hydraulics Division, ASCE* 100 (HY11): 1631–1648.

Engelund, F. and Hansen, E. 1967. *A Monograph On Sediment Transport In Alluvial Streams*, Copenhagen: Technical University of Denmark, 63 p.

Fan, S. 1993. *Proceedings, The Second Bilateral Workshop On Understanding Sedimentation Processes And Model Evaluation*, US Federal Energy Regulatory Commission, 2 volumes, 747 p.

Franz, D.D. and Melching, C.S. 1997. Full Equations (FEQ) model for the solution of the full, dynamic equations of motion for one-dimensional unsteady flow in open channels and through control structures, US Geological Survey Water-Resources Investigations Report 96-4240. 258 p.

Garcia, M. and Parker, G. 1991. Entrainment of bed sediment into suspension. *Journal of Hydrologic Engineering, ASCE* 117(4): 414–435.

Gomez, B. 1991. Bedload transport. *Earth-Science Reviews* 31: 89–132.

Hasegawa, K. 1984. Hydraulic research on planimetric forms, bed topographies, and flow in alluvial channels, Ph.D. Dissertation, Hokkaido University, Sapporo, Japan.

Hirano, M. 1971. River bed degradation with armoring. *Proceedings of JSCE* 195: 55–65 (in Japanese).

Kikkawa, H., Ikeda, S. and Kitagawa, A. 1976. Flow and bed topography in curved open channels. *Journal of Hydraulics Division, ASCE* 102(HY9): 1327–1342.

Lighthill, M.J. and Whitham, G.B. 1955. On kinematic waves. I. Flood movement in long rivers. *Proceedings of the Royal Society of London* 229: 1178 p.

Limerinos, J.T. 1970. Determination of the Manning coefficient for measured bed roughness in natural channels. *USGS Water Supply Paper* 1898-B: 20 p.

Long, C.E., Wiberg, P.L. and Nowell, A.R.M. 1993. Evaluation of von Karman's constant from integral flow parameters. *Journal of Hydrologic Engineering, ASCE* 119(10): 1182–1190.

McLean, S.R. 1992. On the calculation of suspended load for noncohesive sediments. *Journal of Geophysical Research* 97(C4): 5759–5770.

Meyer-Peter, E. and Müller, R. 1948. Formulas for bed-load transport. In: *Proceedings of the 2nd Congress IAHR, Stockholm, Sweden*, pp. 39–64.

Middleton, G.V. and Southard, J.B. 1984. *Mechanics of Sediment Movement*, Tulsa, SEPM, 401 p.

Nagata, N., Hosoda, T. and Muramoto, Y. 2000. Numerical analysis of river channel processes with bank erosion. *Journal of Hydrologic Engineering, ASCE* 126(4): 243–252.

Nelson, J.M. 1990. The initial instability and finite-amplitude stability of alternate bars in straight channels. *Earth-Science Reviews* 29: 97–115.

Nelson, J.M. 1996. Predictive techniques for river channel evolution and maintenance. *Water, Air and Soil Pollution* 90: 321–333.

Nelson, J.M. and Smith, J.D. 1989a. Flow in meandering channels with natural topography. In: Ikeda, S. and Parker, G., eds., *River Meandering*, AGU Water Resources Monograph 12, Washington, DC, pp. 69–102.

Nelson, J.M. and Smith, J.D. 1989b. Evolution and stability of erodible channel beds. In: Ikeda, S. and Parker, G., eds., *River Meandering*, AGU Water Resources Monograph 12, Washington, DC, pp. 321–377.

Nelson, J.M., McLean, S.R. and Wolfe, S.R. 1993. Mean flow and turbulence fields over two-dimensional bedforms. *Water Resources Research* 29(12): 3935–3954.

Oliveira, A., Fortunato, A.B. and Baptista, A.M. 2000. Mass balance in Eulerian–Lagrangian transport simulations in estuaries. *Journal of Hydrologic Engineering, ASCE* 126 (8): 605–614.

Olsen, N.R.B. and Melaaen, M.C. 1993. Three-dimensional calculation of scour around cylinders. *Journal of Hydrologic Engineering, ASCE* 119(9): 1048–1054.

Parker, G. 1984. Discussion of: Lateral bedload transport on side slopes (by S. Ikeda, Nov., 1982). *Journal of Hydraulics Division, ASCE* 110(2): 197–203.

Parker, G., Paola, C. and Leclair, S. 2000. Probabilistic Exner sediment continuity equation for mixtures with no active layer. *Journal of Hydrologic Engineering, ASCE* 126 (11): 818–826.

Patankar, S.V. 1980. *Numerical Heat Transfer and Fluid Flow*, Washington, DC: Hemisphere, 197 p.

Press, W.H., Flannery, B.P., Teukolsky, S.A. and Vetterling, W.T. 1986. *Numerical Recipes: The Art of Scientific Computing*, New York: Cambridge Press.

Rattray, M., Jr. and Mitsuda, E. 1974. Theoretical analysis of conditions in a salt wedge. *Estuarine and Coastal Marine Science* 2: 375–394.

Rodi, W. 1993. *Turbulence Models and Their Application in Hydraulics – A State of the Art Review*, Third edition, Delft: International Association for Hydraulic Research, 104 p.

Schaffranek, R.W. 1987. *Flow Model for Open-channel Reach or Network*, US Geological Survey Professional Paper 1384, 12 p.

Schmidt, J. and Graf, J.B. 1990. US Geological Survey Professional Paper 1493.

Shimizu, Y. and Itakura, T. 1985. Practical computation of two-dimensional flow and bed deformation in alluvial streams, Civil Engineering Research Institute Report, Hokkaido Development Bureau, Sapporo.

Shimizu, Y. and Itakura, T. 1989. Calculation of bed variation in alluvial channels. *Journal of Hydrologic Engineering, ASCE* 115(3): 367–384.

Shimizu, Y., Yamaguchi, H. and Itakura, T. 1991. Three-dimensional computation of flow and bed deformation. *Journal of Hydrologic Engineering, ASCE* 116(9): 1090–1108.

Smith, J.D. and McLean, S.R. 1977. Spatially averaged flow over a wavy surface. *Journal of Geophysical Research* 84 (12): 1735–1746.

Smith, J.D. and McLean, S.R. 1984. A model for flow in meandering streams. *Water Resources Research* 20(9): 1301–1315.

Struiksma, N. 1985. Prediction of 2-d bed topography in rivers. *Journal of Hydrologic Engineering, ASCE* 111(8): 1169–1182.

Tennekes, H. and Lumley, J.L. 1972. *A First Course in Turbulence*, Cambridge, MA: MIT Press, 300 p.

Tritton, D.J. 1977. *Physical Fluid Dynamics*, New York: Van Nostrand Reinhold, 362 p.

van Rijn L. 1984a. Sediment transport. Part II. Suspended load transport. *Journal of Hydrologic Engineering, ASCE* 110(11): 1613–1641.

van Rijn L. 1984b. Sediment transport. Part III. Bed forms and alluvial roughness. *Journal of Hydrologic Engineering, ASCE* 110(12): 1733–1754.

US Army Corps of Engineers. 1993. HEC-6, *Scour and Deposition in Rivers and Reservoirs, User's Manual*, Davis, CA: Hydrologic Engineering Center, 286 p.

US Army Corps of Engineers. 1997. *HEC-RAS River Analysis System User's Manual, Version 2.0*, Davis, CA: Hydrologic Engineering Center, 220 p.

Whiting, P.J. and Dietrich, W.E. 1987. Experiments on bar morphology and dynamics in straight and low-sinuousity channels. *Eos Trans.* AGU 68(44): 1293.

Wiberg, P.L. and Smith, J.D. 1987. Calculations of critical shear stress for motion of uniform and heterogeneous sediments. *Water Resources Research* 23(8): 1471–1480.

Wiele, S.M., Graf, J.B. and Smith, J.D. 1996. Sand deposition in the Colorado River in the Grand Canyon from flooding of the Little Colorado River. *Water Resources Research* 32(12): 3579–3596.

Wiele, S.M. and Smith, J.D. 1996. A reach-averaged model of diurnal discharge wave propagation down the Colorado River through the Grand Canyon. *Water Resources Research* 32(5): 1375–1386.

Wiele, S.M., Andrews, E.D. and Griffin, E.R. 1999. The effect of sand concentration on depositional rate, magnitude, and location in the Colorado River below the Little Colorado River. In: Webb, R.H., Schmidt, J.C., Marzolf, G.R. and Valdez, R., eds., *The Controlled Flood in Grand Canyon*, Monograph 110, American Geophysical Union, pp. 131–145.

Wilcock, P.R. 1997. The components of fractional transport rate. *Water Resources Research* 33: 247–258.

Wilcock, P.R. 2001. The flow, the bed and the transport interaction. In: Mosley, M.P., ed., *Gravel-bed Rivers V*, Christchurch: New Zealand Hydrological Society, pp. 183–220.

Yang, C.T., Trevino, M.T. and Simoes, F.J.M. 1998. *User's Manual for GSTARS 2.0*, Denver, CO: US Bureau of Reclamation, 244 p.

Zyserman, J.A. and Fredsoe, J. 1994. Data analysis of bed concentration of suspended sediment. *Journal of Hydrologic Engineering, ASCE* 120(9): 1021–1042.

19

Numerical Modeling of Alluvial Landforms

JAMES E. PIZZUTO
Department of Geology, University of Delaware, Newark, DE, USA

19.1 INTRODUCTION

Overview

The goal of this chapter is to introduce some current methods for modeling fluvial processes. Models are emphasized that encompass a broad range of river channel behavior. As a result, well-established models limited to predicting changes in streambed elevation (for example) are discussed only briefly, while more holistic models are described in greater detail.

The introduction is designed to outline some general principles of modeling fluvial processes, so readers can appreciate how models can be used and the information required to complete a meaningful modeling study. General features of numerical models are first introduced and then specific issues related to modeling rivers are discussed. The information models can provide is summarized, and the process of modeling is introduced: what steps are needed to produce a useful model, and what kind of information is required? The mathematical basis of modeling river channels is discussed briefly, but generally without equations.

What is a Numerical Model?

Numerical models are hypotheses about how nature works, expressed in quantitative form. They may initially be expressed as equations, as a series of rules, or as a computational algorithm. It is assumed here that hypotheses are ultimately implemented as a computer program.

Numerical models are created to solve many different types of problems. Some of these are purely scientific, while others are practical. For example, a numerical model could be written to better understand why rivers meander or braid. Such a model might only include those processes relevant to meandering and braiding (others being held constant or ignored), and it might lack precision. Such a model would have little practical value because it could not accurately be applied to any real river channel, but its results could provide better understanding of an important scientific question. A more practical numerical model might be written to predict the migration patterns of a restored meandering river, which could be a very important engineering problem. These different models would be evaluated using different criteria: the first model would be evaluated for its ability to reproduce some general features of meandering or braided channels. The model could be successful even if it could not accurately reproduce the behavior of any particular field example. The meander migration model, on the other hand, should be precise to be useful. The entire purpose of modeling, in this case, is to accurately predict the future positions of a specific river channel.

In this chapter, both the scientific accuracy and practical utility of some current models of fluvial processes are discussed so the reader can better understand how different models could be used.

Mathematical Basis of River Channel Models

Models of fluvial processes are based on at least four types of equations:

1. Hydraulic equations to predict water surface elevations, flow velocities, and forces exerted by the flow on the channel perimeter. These equations include turbulence parameters, coefficients that represent frictional resistance caused by bars, bedforms, channel curvature, and other parameters whose values are often not precisely known.

2. Equations to predict the occurrence of morphologic elements within the channel, such as bedforms, bars, patches, bed material clusters, and armoring.
3. Sediment transport equations to predict the ability of the flow to erode, deposit, and transport sediment.
4. Equations to predict topographic changes of the bed, banks, and floodplain.

Other equations could also be part of specific models. For example, a model describing the growth of riparian vegetation might be important in some applications.

Without adjusting model predictions to fit specific circumstances (a process known as calibration), these equations are not particularly accurate. Sediment transport rates in complex natural rivers with a wide variety of grain sizes can rarely be achieved within a factor of 2 (Chang 1988), although sediment transport rates in relatively straight rivers with well-sorted sediment can be predicted with greater accuracy (Andrews 1981). Predicting friction caused by bedforms, vegetation, and channel curvature is equally imprecise. Predicting the extent of bank erosion without calibration is even more difficult. The reader should appreciate that although many fluvial processes are understood in considerable detail, and quantitative models may be accurate (meaning that they represent the relevant processes correctly), they are frequently imprecise. Precise predictions will always require calibration, and may in some cases not be attainable.

The equations themselves are not sufficient to define a modeling problem that can be solved: initial and boundary conditions are always necessary. Initial conditions refer to the variables at the beginning of the period of time to be computed: the future configuration of a river channel cannot be determined unless the present configuration is known. For models of river morphology, initial conditions could include specifying the shape of the channel and its floodplain, the spatial distribution of characteristic bed and bank sediment types, the distribution of riparian vegetation, the initial depth of flow, and the initial velocity field. Boundary conditions refer to important variables at the edges of the computational domain. These might include specifying water surface elevations, velocities, water and sediment discharges, and rates of erosion and deposition at upstream or downstream boundaries. It is essential to understand that the model itself cannot, by definition, provide boundary conditions because the boundary conditions are external to the model. Clearly, a well-posed modeling problem will require considerable field data if a particular river reach is to be modeled.

The Process of Modeling

Figure 19.1 (after Anderson and Woessner 1991) presents steps in a protocol for modeling studies. The steps are primarily designed for a practical application where future conditions are to be predicted. Important steps include:

1. Establish the purpose of the modeling study. Without a clearly defined purpose, the study cannot succeed.
2. Use field observations to develop a conceptual model of the problem to be solved. This is particularly important for modeling fluvial processes, as rivers are influenced by many different processes, and all of them cannot be modeled simultaneously. Field data are needed to determine which processes are key, and which can be neglected.
3. The conceptual model is then rendered into mathematical form that can be solved numerically in a computer program.
4. Model design refers to the implementation of the computer program for a specific case. It involves developing a computational grid, selecting boundary conditions, relevant parameters, and other data that may be characteristic of a particular site or problem. At this stage, detailed field data are always required for a site-specific modeling study.
5. Calibration involves adjusting the model design to reproduce a set of field observations. This is a standard practice in modeling and, though problematical, is nearly always required to demonstrate that the model is accurate enough to solve the problem that has been posed.
6. Verification is the process of applying the model to an independent set of field observations not used for calibration. This step provides an additional demonstration of the model's accuracy.
7. Prediction involves specifying future conditions. This usually satisfies the basic purpose of the modeling study.
8. A postaudit represents an evaluation of the model some time after the initial modeling study. The goal of a postaudit is to compare the predictions made in step 7 with new field data. These data allow the assumptions of the model to be evaluated. If the model's performance is not adequate, then the entire modeling process may need to be repeated, with appropriate corrections made at each step.

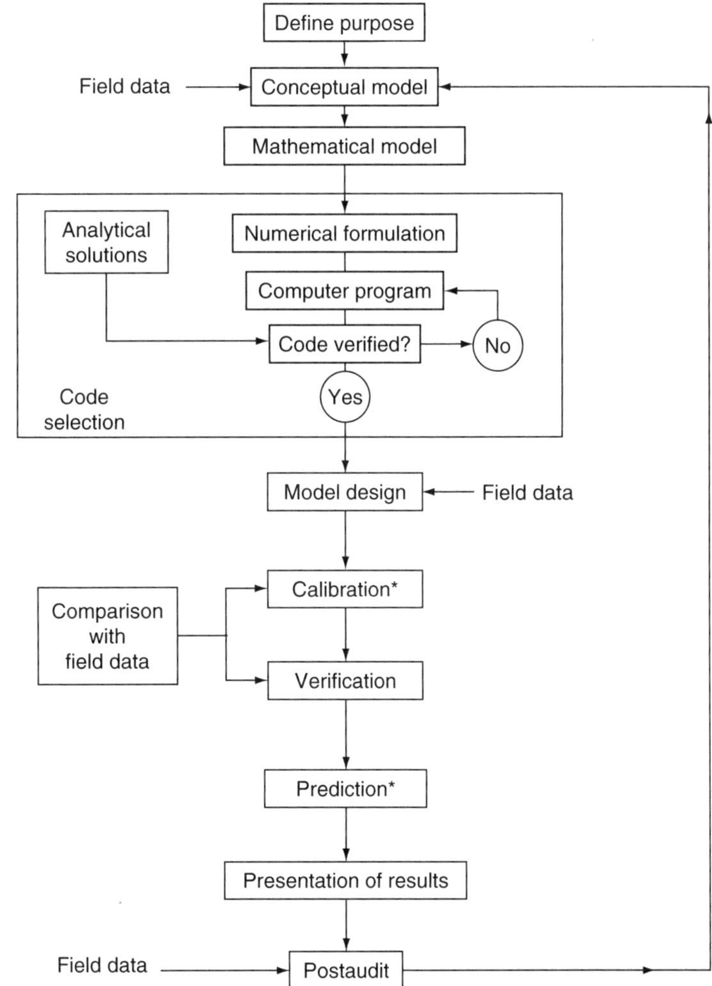

Figure 19.1 Steps in a protocol for model application (reproduced from Anderson and Woessner 1991)

Anderson and Woessner (1991) particularly emphasize the utility of postaudits in modeling studies. They describe four postaudits of groundwater modeling studies made at least 10 years after the initial predictions were made. At that time, these were the only postaudit studies of groundwater modeling available in the literature. All four postaudits revealed serious inadequacies in model predictions. Unlike models of fluvial processes, the physical basis for groundwater modeling is well established, so Anderson and Woessner's (1991) conclusions raise important questions about the viability of long-term model predictions of river behavior.

Why Rivers Are Difficult to Model?

The protocol described in Figure 19.1 was proposed for groundwater modeling studies by Anderson and Woessner (1991). Several of the steps are problematical when applied to rivers. It is important to highlight and discuss these so to put modeling studies of rivers in an appropriate context.

Field observations provide important guidance in model development according to the protocol of Figure 19.1. However, rivers evolve over characteristically long timescales, and suitable field observations may be difficult to obtain during the duration of a particular modeling study. For example, if significant bedload transport or bank erosion occurs only a few times per year, then several years of data will be necessary for model calibration. If the duration of the modeling study is too short for such data to be obtained, then historical data (obtained from the geologic record, aerial photographs, historical maps, and past hydrologic records, often combined in a sediment budget) may provide the only basis for calibration. Unfortunately, historical data are rarely detailed enough for model calibration as it is usually performed.

The Models Discussed in This Chapter

The rest of the chapter is devoted to examples of numerical modeling covering a range of spatial and temporal scales (Table 19.1). Spatial scales range from the reach scale to an entire valley to the watershed scale, where models of entire networks are discussed. Temporal scales range from models of steady state channel morphology to models that predict network morphology over millions of years of geologic time.

Models discussed in this chapter only apply to alluvial rivers: bedrock channels are not discussed. This reflects the current state of the art, because models of alluvial channels are much more fully developed than models of bedrock rivers.

19.2 MODELING CHANGES IN STREAMBED ELEVATION

Changes in streambed elevation are caused by erosion or deposition on the streambed, which in turn are caused by imbalances in the local sediment budget. For example, if more sediment is supplied to an area of the bed than is removed, then deposition will occur. If more sediment is removed than is supplied, then erosion will occur.

Models that predict changes in streambed elevation are relatively well established. As discussed in Chapters 15 and 18, these models couple hydrodynamic computations with predictions of sediment transport (both bedload and suspended load) within the channel. The extent of erosion and deposition on the bed is then determined by model predictions of the local sediment budget (or, more precisely, by solving the sediment continuity equation as discussed in Chapter 18). A useful review of these kinds of models is presented by Bradley *et al.* (1998). Excellent examples of model predictions of changes in streambed elevation in Grand Canyon are presented by Wiele *et al.* (1996, 1999).

Models of changes in streambed elevation do not have wide applicability because they neglect bank erosion, interactions with the floodplain, and other processes that are of nearly universal importance in alluvial rivers. As a result, these models can only be used in situations where channel and valley morphology are unlikely to change significantly during the period to be modeled. Semi-alluvial channels confined by bedrock such as the Colorado River in Grand Canyon, or channels whose banks are fixed over short timescales by trees, engineering structures, or inerodible bank sediments present promising situations for the application of these models.

Table 19.1 Models discussed in this chapter

Subject	Spatial scale	Temporal scale
Changes in stream bed elevation	Reach[a] to many kilometers	Event[b] to millenia
Quasi-equilibrium cross-sectional morphology	Cross-section only	Not applicable
Bank erosion	Reach to many kilometers	Event to decades
Meander evolution	Many kilometers	Decades to millenia
Braided channel mechanics	Many kilometers	Decades to millenia
Short-term floodplain sedimentation	A few kilometers	Event
Long-term evolution of floodplains with meandering rivers	Many kilometers	Decades to millenia
Bed material waves and the basin scale sediment routing problem	Entire watershed	Decades to millenia
Evolution of fluvial landscapes	Entire watershed	Millenia

[a]A reach is a section of river approximately 10–100 widths in length.
[b]Refers to an individual high flow event.

19.3 MODELING QUASI-EQUILIBRIUM CROSS-SECTIONAL CHANNEL GEOMETRY

Overview

Geomorphologists (Davis 1899, Mackin 1948) and engineers (Blench 1969) have long hypothesized that rivers and canals attempt to create a form that is in equilibrium with prevailing discharge, sediment supply, and other constraints. Leopold and Maddock (1953) noted that, for most natural rivers, width, depth, slope, and velocity, when plotted against discharge of a constant recurrence interval in a watershed, define power functions with nearly universal exponents. Leopold and Maddock (1953) referred to these functions as hydraulic geometry equations, and they have been widely interpreted to represent evidence for the existence of a quasi-equilibrium state (Wolman 1955), though the hydraulic geometry equations appear to be satisfied even when fluvial systems are changing rapidly (Knox 1976).

The models described in this section can be considered theories to explain the empirical observations represented by the hydraulic geometry equations. These theories are based on conservation equations, such as conservation of fluid or sediment mass, and additional hypotheses regarding the behavior of the channel. The independent variables typically include sediment characteristics, presence or absence of vegetation, bankfull discharge, and discharge of sediment at bankfull discharge. The theories seek to predict, given these independent variables, the reach-averaged width, depth, and slope of a stable, quasi-equilibrium channel.

Vigilar and Diplas' (1997) Model of Stable Gravel Channel Cross-sections

Vigilar and Diplas (1997, 1998) present a method for determining the width, depth, and slope of gravel channels. Banks are assumed to be non-cohesive, unvegetated, and composed of unstratified sediment identical to that of the bed material and bedload. The channel is assumed to be straight, with active bedload transport, negligible suspended load transport, and banks that neither erode nor deposit. More specifically, the sediment on the banks is hypothesized to be at the threshold of sediment motion. Vigilar and Diplas (1997) model is the most recent of many attempts at solving this problem, including efforts by Parker (1978a), Pizzuto (1990), and Kovacs and Parker (1994).

Based on previous theoretical (Parker 1978a,b, Pizzuto 1990, Kovacs and Parker 1994) and laboratory studies (Stebbings 1963, Ikeda 1981, Diplas 1990), Vigilar and Diplas (1997) suggest that a straight gravel channel will have a curving bank region and a flat bed region. In their most general approach, Vigilar and Diplas (1997) assume that the following are specified: steady discharges of water (Q) and sediment (Q_s), critical Shields stress for sediment entrainment on a flat bed (τ_{*c}), the 50th and 90th grain size percentile sizes (D_{50} and D_{90}), the ratio of lift to drag force on sediment particles (β), and the submerged coefficient of friction of the bed material (μ). To determine the force exerted by the flow on the cross-section, Vigilar and Diplas (1997) solve a cross-channel momentum equation that includes cross-channel turbulent diffusion of momentum but neglects transport of momentum by secondary currents. An iterative procedure is used to determine the shape of the banks, and to select a wide enough bed region and a slope that is steep enough to transport the specified water and sediment discharges.

Vigilar and Diplas (1997) test their model in several ways (Vigilar and Diplas 1998). The shape of the bank profile is compared with laboratory data, and predicted values of width and depth are compared with laboratory and field data (Figures 19.2). Because data are not available that include bedload transport and channel morphology, Vigilar and Diplas (1997) do not test the ability of their model to predict the channel slope.

Other Approaches

Many other attempts at predicting the quasi-equilibrium width, depth, and slope of rivers may be found in the literature. Ikeda and Izumi (1990) present a theory that accounts for bank vegetation, while Parker (1978a,b), Ikeda and Izumi (1991), and Pizzuto (1984b) account for suspended sediment. Lane (1955) and Diplas and Vigilar (1992) present theories for threshold channels, i.e., channels whose bed material is at the threshold of sediment motion, and which therefore transport no bedload. Pizzuto (1984b) and Millar and Quick (1998) account for the presence of cohesive bank sediments.

Methods that optimize the channel geometry for a variety of different conditions represent a different approach to the physical model of Vigilar and Diplas (1997). Chang (1980, 1988) suggests that quasi-equilibrium channels optimize their morphology to minimize stream power, while Millar and Quick

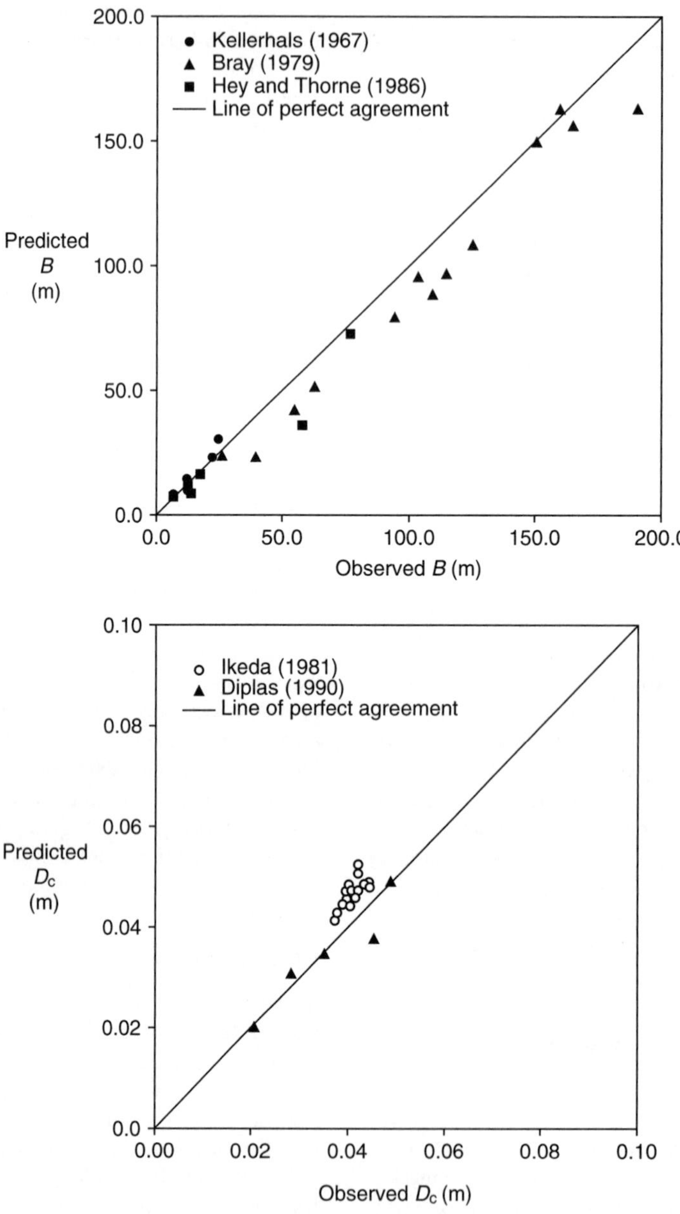

Figure 19.2 Width (B) and depth (D_c) predicted using Vigilar and Diplas' (1997) model compared with field data (after Vigilar and Diplas 1998)

(1993, 1998) propose that streams adjust their morphology to maximize bedload transport. Cao and Knight (1998) present a method that accounts for secondary currents while optimizing channel geometry for maximum bedload transport.

Evaluation and Discussion

Available models of equilibrium cross-sectional geometry have identified a variety of important processes that control river channel morphology, but they have also important weaknesses. They invariably are based on a constant bankfull discharge and constant supply of sediment, which are considerable idealizations. They assume that channels are straight and symmetrical, which is rarely achieved in nature. They also rarely account for the full degrees of freedom available to a river, including plan form changes, development of bars, ripples and dunes, mobile armor, etc. Furthermore, often important input variables are not available; this is particularly true for bedload transport. If bankfull bedload transport measurements are not available, many methods can only predict two variables, which are usually taken to be width and depth. Under these circumstances, the slope is usually treated as a known variable. This simplification may be adequate for short timescales of a few years, but, given time, the slope of a river may be adjusted in addition to its width and depth, so slope should be treated as a variable as well.

19.4 MODELING BANK EROSION

Overview

Bank erosion is an important practical problem. Retreating river banks can endanger bridges and other engineering structures, destroy property, and modify the sediment budget of a stream. The ability to predict and control bank erosion is therefore of considerable interest. In this section, strategies for modeling different processes of bank erosion are briefly reviewed. Interested readers should consult recent comprehensive review papers, such as ASCE (1998a,b) for more detailed discussion of these topics.

Strategies for Modeling Bank Erosion

Models of bank erosion can be either based on the physics of specific erosional processes, or else a parametric approach can be adopted which relates bank erosion rates to controlling parameters using empirical coefficients. Process-based methods differ depending on whether the bank sediments are non-cohesive or cohesive. Parametric methods simply relate bank erosion rates to the near-bank velocity or shear stress.

Riverbanks composed of sand and gravel without vegetation are considered non-cohesive. Sediment on non-cohesive riverbanks is only held in place by frictional forces acting at grain-to-grain contacts. Erosion of non-cohesive sediment on riverbanks is widely believed to occur by entrainment of single particles, rather than by mass movement of large numbers of particles. The ASCE (1998a,b) reviewed four numerical models that predict erosion rates of non-cohesive riverbanks. Kovacs and Parker's (1994) model is probably the most elegant, employing a generalized vectorial bedload transport model to compute bedload transport on steeply sloping river banks, and an integral formulation of the conservation equation of sediment mass to determine the velocity of the front of erosion at the water's edge. All four numerical models reviewed by the ASCE are tested with data from Ikeda's (1981) flume experiments, the only available data describing the evolution of eroding non-cohesive banks. This lack of data highlights an obvious problem with these models: vegetation, moisture, and small amounts of mud nearly always produce some cohesion in riverbank soils, and as a result, models of non-cohesive bank erosion can rarely be applied to natural rivers.

Cohesive banks erode by a bewildering variety of different processes, including particle-by-particle erosion and mass failure processes such as rotational slip, tensile failures, and beam failures (Thorne and Tovey 1981, Pizzuto 1984a). Cohesive bank erosion is often facilitated by freeze–thaw cycles that loosen the soil (Wolman 1959, Lawler 1986), or by desiccation cracks formed by periods of wetting and drying (Thorne and Tovey 1981). Mechanistic models have been developed to predict the occurrence of mass failure on cohesive banks (these are reviewed in detail by the ASCE 1998a,b). These are generally 2-dimensional models that evaluate bank stability in terms of the strength of the soil and the geometry of the bank. Fluvial processes are only invoked to remove failed bank materials supplied to the toe of the slope. However, only a few of these mechanistic mass failure models have been incorporated into models designed to predict rates of bank erosion.

Darby and Thorne (1996) developed a numerical model (discussed in detail in Chapter 17) for predicting bank erosion and bed deformation in straight sand-bed channels with cohesive banks. In an application of the model to the Forked Deer River from 1969 to 1993, Darby *et al.* (1996) note that "although the model was able to qualitatively predict trends of widening and deepening, quantitative predictions were not reliable".

Parametric methods predict rates of bank erosion by correlating rates of observed bank retreat (determined from maps, aerial photographs, or ground surveys) with the "near-bank velocity" (Odgaard 1987, Pizzuto and Meckelnburg 1989) or shear stress (Hasegawa 1989). Figure 19.3 provides an example

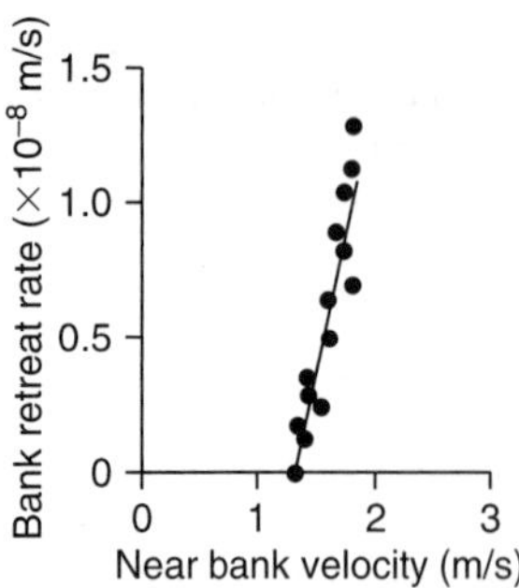

Figure 19.3 Rate of meander bed retreat as a function of the near-bank velocity (reproduced from Pizzuto and Meckelnburg 1989)

from the Brandywine Creek in southeastern Pennsylvania (Pizzuto and Meckelnburg 1989), where annual average rates of bank retreat are correlated with values of the near-bank velocity estimated using the hydrodynamic model of Ikeda *et al.* (1981) for a channel forming discharge. The empirical coefficients obtained in these studies are often strongly influenced by the nature of the bank vegetation. For example, Pizzuto and Meckelnburg (1989) and Odgaard (1987) found that large trees reduced the rate of migration of eroding cutbanks at meander bends.

Evaluation of Methods for Modeling Bank Erosion

Although many important processes of bank erosion have been identified, detailed process-based models of bank erosion suffer from too many limitations to be widely used to predict bank erosion rates. Most models of cohesive bank failure, for example, are only 2-dimensional: they fail to predict the downstream dimension of failed bank soils. Once the bank has failed, the fate of the slumped or toppled material is also difficult to specify. Failed blocks may disaggregate nearly instantly, or in some cases they may persist for years. It is also difficult to determine which failure process will dominate erosion at a particular site. Given these difficulties, empirical parametric methods provide the most practical means of modeling rates of bank erosion at present. Field data are necessary for calibrating these models to a particular site. However, once the relevant field data have been obtained, these methods can be coupled with sophisticated hydrodynamic models to determine rates of bank retreat, at least for relatively short distances and over relatively small temporal scales.

19.5 MODELING THE EVOLUTION OF MEANDERS

Models for simulating the evolution of meanders have been relatively well developed during the last 20 years. Nearly all existing models are based on the linearized hydrodynamic models of Ikeda *et al.* (1981) and Johanneson and Parker (1989). The hydrodynamic models predict the magnitude of the near-bank velocity in meander bends based on a reduced and linearized version of the St. Venant equations written in curvilinear coordinates. Once the velocity near the bank is computed, the extent of lateral migration of the channel is assumed to be proportional to the near-bank velocity. Channels are assumed to have constant widths, and the discharge and reach-averaged depth are also assumed to be constant. The sediment supply is not explicitly considered, but is implicitly treated as a constant as well. Deposition on the inner bank is not treated explicitly; because the width is assumed to be constant. Thus, when bank erosion occurs, the channel is simply shifted laterally.

Howard and Knutson (1984) published one of the first extensive modeling studies of meander evolution (Figure 19.4). They concluded that realistic patterns of meander migration can only be predicted when local meander migration rates are related not only to local curvature, but also to curvature of the channel upstream. Stolum (1996) argued that models of meander evolution illustrate processes of self-organization. Models of meander migration and evolution have also been coupled with floodplain sedimentation models to investigate floodplain development on meandering rivers (these are discussed in a subsequent section).

Despite the convincing patterns produced by these models (Figure 19.4), they have not been extensively tested. As discussed above, Pizzuto and Meckelnburg (1989) and Odgaard (1987) have verified that bank erosion is proportional to near-bank velocity, at least over short spatial and temporal scales. Furbish (1991, 1988) demonstrated that upstream bends influence patterns of meander migration downstream, as predicted by the linearized hydrodynamic model. Howard and Hemberger (1991) noted that meanders produced by a model based on the flow equations of Ikeda *et al.* (1981) differ systematically from natural meanders. Stolum (1998) demonstrated that meanders created using Johanneson and Parker's (1989) hydrodynamic model appear to be realistic (Figure 19.5), and he also demonstrated that the fractal scaling properties of simulated meanders were similar to the fractal scaling properties of natural meanders.

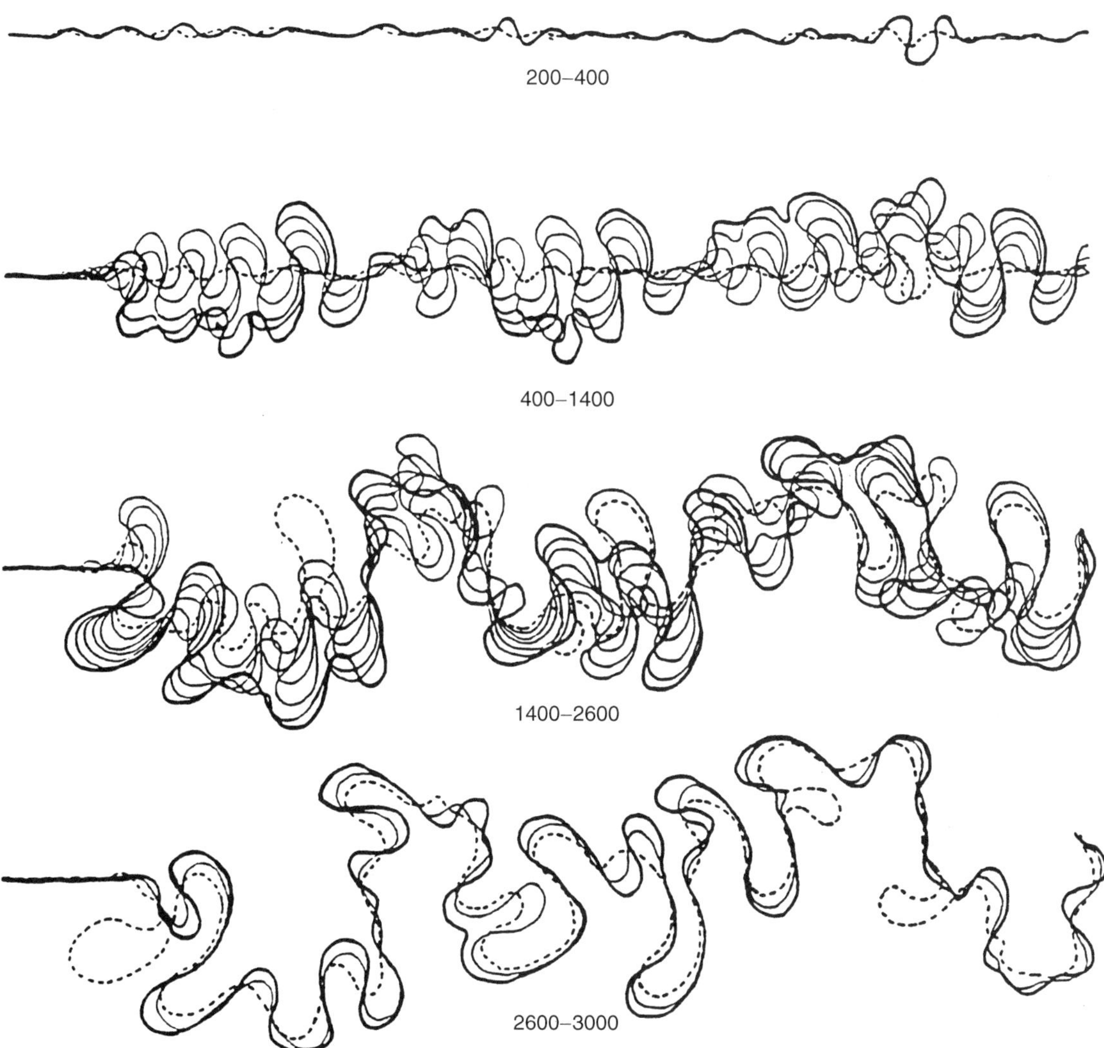

Figure 19.4 Successive centerlines of simulated streams, displayed in increments of 200 iterations. Downstream is to the right. The curves are identified by inclusive iteration numbers. The first centerline of each sequence is identified by a dashed line, while the last centerline is identified by a bold line (reproduced from Howard and Knutson 1984). The scale of the figure is not given by Howard and Knutson (1984)

19.6 MODELING BRAIDED CHANNELS

Braided channels have a complex morphology that changes quickly, and as a result, modeling of braided channels has proven difficult. Murray and Paola (1994), however, have produced a successful cellular model that will spontaneously produce a braided channel from an initial channel without a braided planform.

Murray and Paola's (1994) model is based on a series of rules for routing sediment and water between the cells that comprise the model domain. Water is routed into downstream cells according to the local bed slope. Sediment is routed into downstream cells according to sediment transport equations such as:

$$Q_s = KQ^m \quad (19.1)$$

$$Q_s = K(QS)^m \quad (19.2)$$

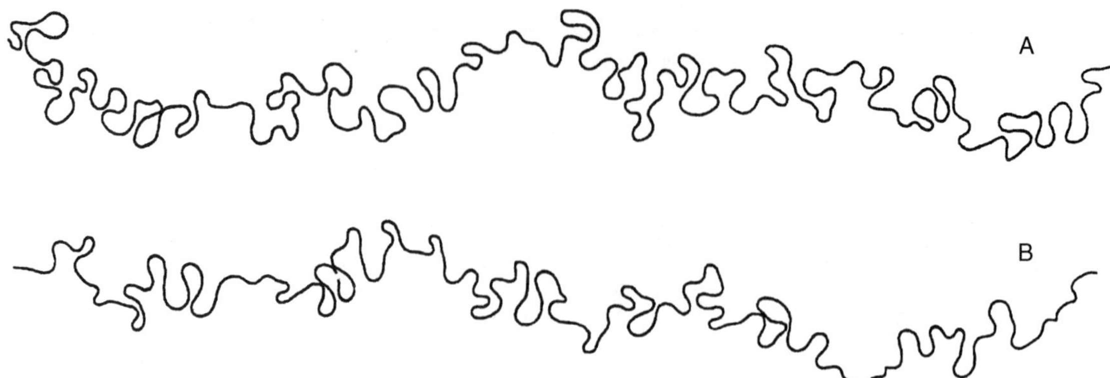

Figure 19.5 Qualitative comparison of simulated and actual free meandering river planforms. (A) Reach of Purus river, Brazil, approximately 200 km in length. (B) Simulated river reach with a length equivalent to 570 widths (reproduced from Stolum 1998)

$$Q_s = K[Q(S + C)]^m \tag{19.3}$$

where Q_s is the sediment transport rate, Q is the local discharge, and K, m, and C are constants. Any of the equations for the sediment transport rate will produce braiding, as long as the exponent m in the transport law is greater than 1. Murray and Paola (1994) conclude from their simulations that braiding develops when bedload transport leads to excess scour in flow convergences and deposition in divergences. This requires that the flow be sufficiently unconstrained laterally, that it can change its width freely, and, if discharge or stream power is used to parameterize sediment flux, that the exponent in the sediment-flux law be >1. The model results suggest that braiding is a simple, robust result of bedload transport under these conditions. The ubiquity of braiding in model runs suggests that braiding may be *the* fundamental instability of laterally unconstrained free-surface flow over cohesionless beds.

Murray and Paola's (1994) model is a good example of a scientific, rather than a practical, tool. Although the model has provided interesting insights into the origins and nature of braiding, the model cannot be used to predict the evolution of any particular river due to the simplified nature of the hydrodynamic and sediment transport equations used in the model.

19.7 MODELING SHORT-TERM FLOODPLAIN SEDIMENTATION

Initial efforts to compute the extent of sedimentation on floodplains relied on simple representations of floodplain geometry and depositional processes (James 1985, Pizzuto 1987, Marriott 1992). Recently, attempts have been made to incorporate more realistic floodplain morphology and a more complete representation of overbank hydraulics and sediment transport processes. Siggers *et al.* (1999) compared floodplain sedimentation rates over timescales of decades to a century with hydraulic parameters computed for a 12-year flood using a 2-dimensional hydrodynamic model. They obtained a weak negative correlation between velocity and the percentage of silt and clay of sediments deposited on the floodplain. Their results also suggested that "whilst hydraulics are the driving force behind all sediment transport, a consideration of these forces alone only permits a partial insight into the processes operating". Siggers *et al.* (1999) also noted that "... suspended-sediment concentrations for successive flood events are determined by factors upstream, and are not simply related to the bulk flow characteristics".

Nicholas and Walling (1997) used a simplified 2-dimensional hydraulic model that ignored convective acceleration and time-dependent terms in the governing equations. Their hydrodynamic model provided input for a 2-dimensional, vertically averaged sediment transport model that included terms for diffusion, convection, and erosion/deposition. The model was calibrated using observations of sediment accumulation during a flood in 1992. After calibration, Nicholas and Walling (1997) concluded that the model predictions "compare favorably with those of previous models of overbank deposition" (e.g., James 1985, Pizzuto 1987). Middlekoop and van der Perk

(1998) used the 2-dimensional hydrodynamic model WAQUA and a suspended-sediment transport model called SEDIFLUX. SEDIFLUX represents convection of suspended sediment and deposition, but ignores diffusion, erosion, and bedload transport. Middlekoop and van der Perk (1998) used SEDIFLUX to predict patterns of deposition caused by a flood in December 1993, on the Waal River in the Netherlands

The studies reviewed above illustrate some of the strengths and weaknesses of current models of short-term floodplain sedimentation. Practical hydraulic models are only 2-dimensional, and they neglect important 3-dimensional effects that occur during floods (Sellin and Ervine 1993, Willetts and Hardwick 1993, Ervine and Jasem 1995). Models of sedimentation are even more simplified. For example, all of the models reviewed above ignore bedload transport and erosional processes. Nonetheless, useful modeling results can be obtained for short-term processes (1) if deposition of suspended sediment is the dominant process, and (2) if sufficient resources are available to obtain all the field data needed for model calibration and for determining appropriate boundary conditions.

19.8 MODELING THE LONG-TERM EVOLUTION OF FLOODPLAINS WITH MEANDERING RIVER

The apparent success of models that predict the evolution of meandering rivers (as described above) has encouraged several authors to extend these models to simulate the evolution of floodplains. Three of these efforts are briefly reviewed below.

Sun *et al.* (1996) developed a model that generates floodplain deposits of differing lithology. Deposits consist of an initial, uneroded floodplain, point bar deposits created when the channel migrates laterally, and channel fill deposits representing deposition in oxbow lakes. Sun *et al.*'s (1996) model allows these different units to have different erodibilities, allowing the influence of different lithologic units on channel morphology to be evaluated.

Gross and Small (1998) developed a model that simulates the formation of four sedimentary facies and associated landforms in evolving floodplains with meandering rivers. The four facies are channel fill, levee, crevasse splay, and floodplain fill deposits. Channel fill deposits are formed by lateral migration of meandering channels, which is modeled using the theory of Johanneson and Parker (1989). Levee and floodplain fill deposits are produced by diffusion of sediment from the main channel to the floodplain. Crevasse splay deposits are created by a probabilistic algorithm related to the depth of overbank flow. Floods are selected from a log-Pearson type III distribution. The model does not account for consolidation of floodplain deposits, and avulsion processes were not also included in Gross and Small's (1998) results. Gross and Small (1998) used the model to simulate the creation of facies of the Oligocene and Miocene Frio Formation of southeast Texas. The model simulated 2300 years of sedimentation. The model was evaluated by comparing the frequency distributions of facies geometries and transition probabilities produced by the model with those observed in the Frio Formation. Gross and Small (1998) note that "the development and application of the geologic process model confirm its feasibility and potential application to subsurface characterization . . . While the overall agreement is encouraging, it is recognized that many aspects of the model formulation require further field investigation, and particular aspects of the geologic process simulation model are not as yet verifiable".

Howard (1992, 1996) coupled the Johanneson and Parker (1989) model for flow and bed topography in meandering channels with simple models of meander migration and floodplain deposition. Howard's models treat the rate of bank migration as being proportional to the near-bank velocity. Floodplain deposition is divided into two components, an overbank deposition component and a point bar component. The overbank deposition rate, Φ, is computed by

$$\Phi = (E_{\max} - E_{\text{act}})(\upsilon + \mu\, e^{-D/\lambda}) \qquad (19.4)$$

where $E_{\max}$ is a maximum floodplain height, E_{act} the local floodplain height, υ the position-independent deposition rate of fine sediment, μ the deposition rate of coarser sediment by overbank diffusion, λ a characteristic diffusion length scale, and D is the distance to the nearest channel. Howard (1992) notes that "this model is assumed to provide a crude representation of both deposition very close to the channel (banks and levees) as well as more distant overbank sedimentation". The point bar component is accounted for by making the initial floodplain elevation prior to overbank deposition equal to the near-bank channel-bed elevation as determined from Johanneson and Parker's (1989) model of bed topography in meander bends. Howard's model produces interesting floodplain topography that at least superficially resembles natural floodplain topography (Figure 19.6).

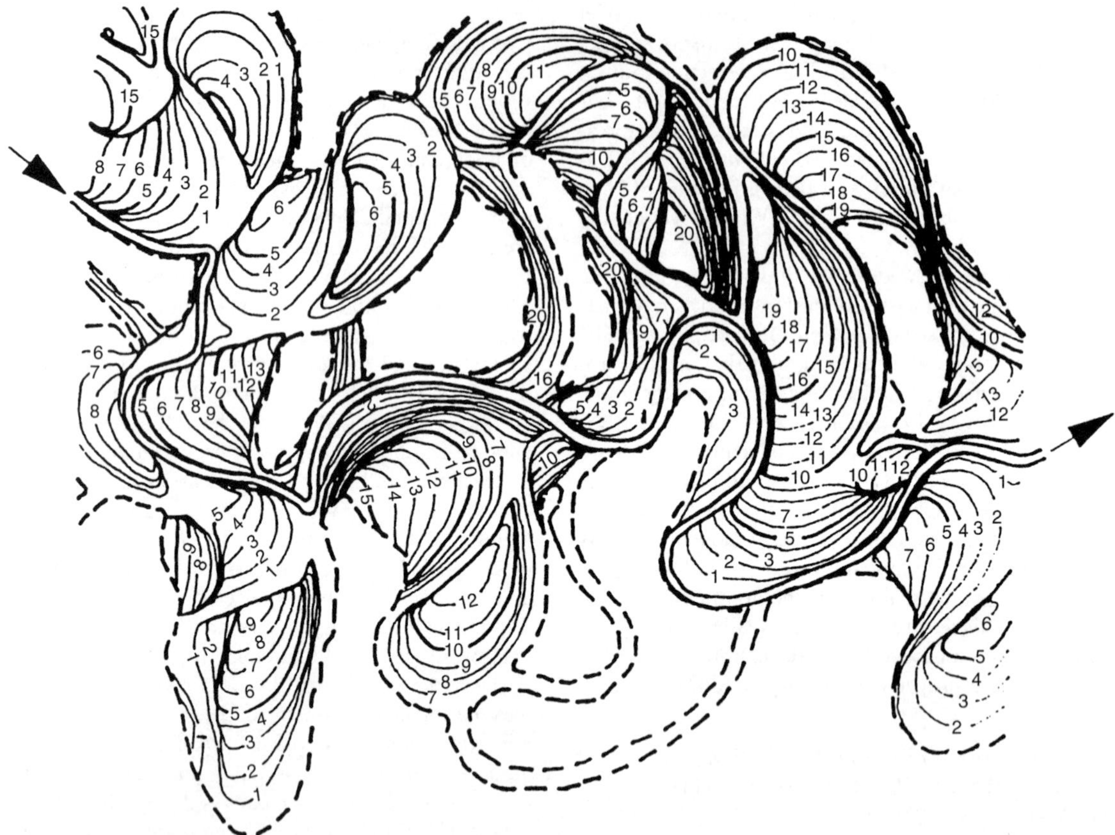

Figure 19.6 Meandering river floodplain computed by Howard (1992). Contours indicate the ages of floodplain deposits in units of hundreds of iterations (reproduced from Howard 1992)

All of the models discussed above show considerable promise, but have yet to be compared in detail with well-documented field examples. Gross and Small's (1998) model was only tested using subsurface data obtained from cores. Thus, they should be considered research tools at present, rather than as models that can be used with confidence to predict floodplain evolution at a particular site.

19.9 BED MATERIAL WAVES AND THE BASIN-SCALE SEDIMENT ROUTING PROBLEM

Lisle *et al.* (2001) define bed material waves as temporary zones of sediment accumulation created by large sediment inputs. Bed material waves may be initiated by landslides related to land use changes (Kelsey 1980) or climatic events (Benda and Dunne 1997a), dam failures (Pitlick 1993), forest fire, disposal of mining waste (Pickup *et al.* 1983) and other causes. Routing the sediment stored in bed material waves through a watershed remains an important problem in watershed management, as many anthropogenic impacts on watersheds create excess sediment yield, many of which at least initially create bed material waves.

There have been many studies of bed material wave phenomena. Relevant references are reviewed by Nicholas *et al.* (1995) and Lisle *et al.* (1997). Benda and Dunne (1997a,b) developed a model for routing sediment supplied by landsliding and debris flows through watersheds. They propose interesting relationships between climatic fluctuations, the frequency of landsliding and debris flow initiation, and sediment yield. Their model, however, is not tested with laboratory or field data.

Cui *et al.* (2001) developed a 1-dimensional numerical model that has been applied to an extensive series of laboratory studies (Lisle *et al.* 2001) to bed material waves in Redwood Creek, California (Madej and Ozaki 1996, Sutherland *et al.* 2002), and to watershed scale sediment routing (Cui *et al.* 2001). The model computes the evolution of the longitudinal profile, as well as the surface and subsurface grain size distribution of the bed material. Bed material abrasion is also included. The model only applies to bed material waves in gravel-bed rivers.

Figure 19.7 illustrates the evolution of a bed material wave observed in a flume study (Lisle *et al.* 2001) and computed using Cui *et al.*'s (2001) numerical

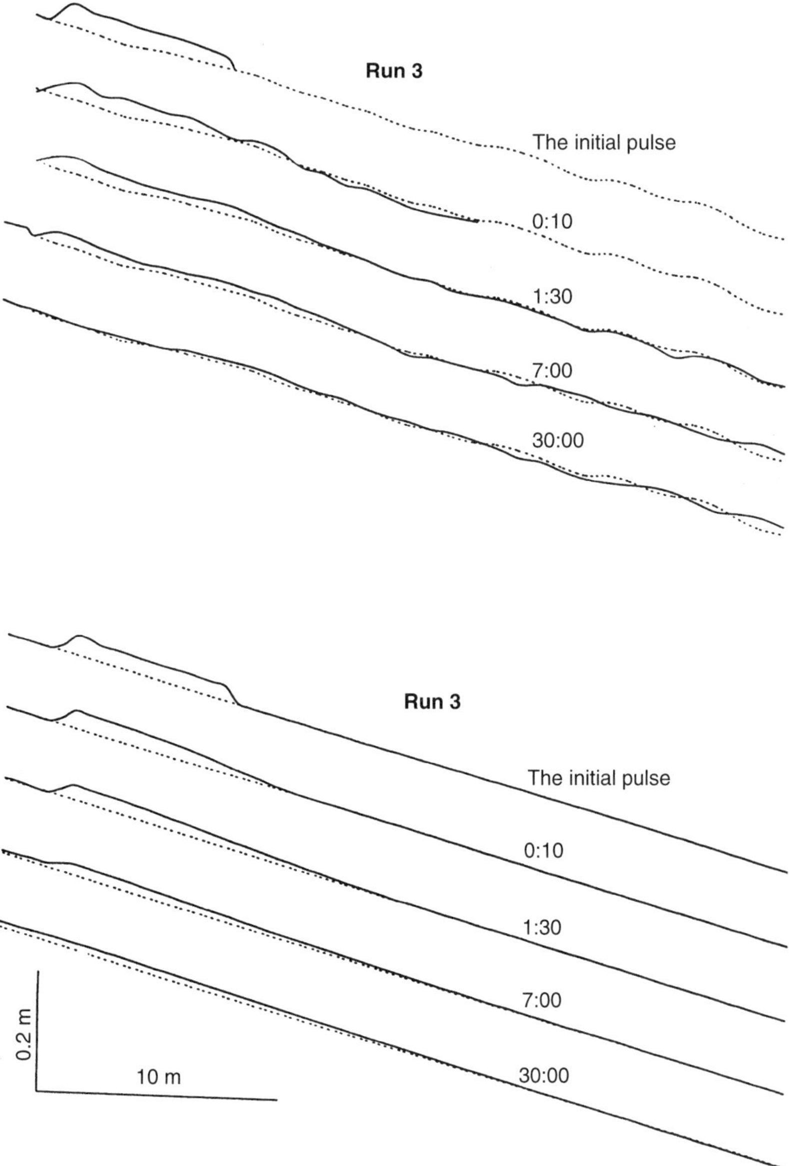

Figure 19.7 Longitudinal profiles illustrating the evolution of a sediment wave in a laboratory flume. The dashed line represents the equilibrium slope before the sediment was introduced. Upper sequence shows the observed profiles, while the lower sequence shows model computations (from Cui, pers. comm.)

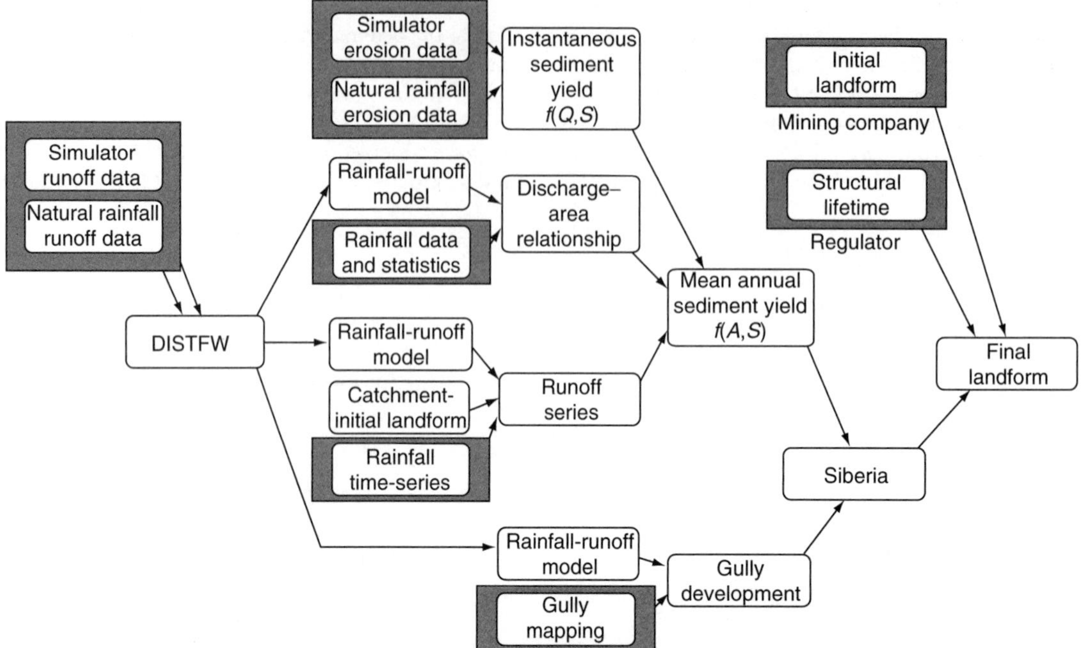

Figure 19.8 Schematic of the calibration process and use of SIBERIA to simulate the evolution of the above-ground storage of mining waste at the Ranger Uranium Mine (reproduced from Willgoose and Riley 1998)

model (Cui, pers. comm.). The flume results were obtained by first creating an equilibrium channel. Next, the water was turned off, and the initial pulse was placed in the channel by hand. The sediment in the initial bed material wave was substantially coarser than the original sediment feed, and the height of the wave was approximately equal to the original equilibrium water depth. After the initial pulse was emplaced, the water was turned on again and the evolution of the bed material wave was observed. The model successfully reproduces the rapid dispersion of the bed material wave into a sheet that becomes progressively longer and thinner (Figure 19.7).

Cui *et al.*'s (2001) model appears to provide robust predictions of the evolution of 1-dimensional bed material waves. Because it is only 1-dimensional, however, it does not apply when:

1. sediment is stored on floodplains or elevated bars,
2. the channel width is influenced by the presence of the additional sediment, or
3. the channel is meandering or braided.

Table 19.2 Limitations of numerical models reviewed in this chapter

Limitations	Possible solutions
Models are difficult to use	Develop user-friendly interface; only experts should use them
Conceptual models of system behavior poorly developed	Detailed field work, further model development
Parameterization of sediment transport processes is imprecise	Monitoring studies to calibrate transport models
Long timescales make model calibration imprecise or impossible	Calibrate model (at least partially) using stratigraphic data
Suitable boundary/initial conditions are unavailable	Detailed field work

19.10 MODELING THE EVOLUTION OF FLUVIAL LANDSCAPES

Several models have been developed in recent years to compute the evolution of entire fluvial landscapes. Chase (1992) developed a simple algorithm using cellular automata to represent landscape evolution. Diffusional, erosional, and depositional processes were all modeled using simple rules applied to a grid. The statistical properties of landscapes generated by the model were compared with that of real landscapes. Howard (1994) developed a detachment-limited model of drainage basin evolution that includes terms representing weathering, rainsplash, mass movement, and non-alluvial and alluvial transport. Howard (1997) used his model to explain the evolution of badlands near Caineville, Utah. Willgoose *et al.* (1991a,b) developed a model of drainage basin evolution that specifically predicts the locations of river channels and determines the location of the channel head (the upstream limit of channelized flow) as a function of time.

Landscape evolution models to date have rarely been applied to practical problems. Willgoose and Riley (1998), however, have recently used the model of Willgoose *et al.* (1991a,b), reformulated as SIBERIA, to assess the long-term stability of the Ranger Uranium mine in Northern Territory, Australia. The rehabilitation of the mine will involve shaping waste rock and ore dumps consisting of more than 100 million tones, and mill tailings. The mill tailings must be structurally stable for a minimum period of 1000 years. Willgoose and Riley (1998) used SIBERIA to assess the stability of a proposed engineering design that would store the mining waste above the ground.

The application of the model involved extensive calibration and field testing (Figure 19.8). Natural and simulated rainfall events were used to determine 10 calibration parameters for models of instantaneous runoff and sediment transport processes. The calibrated instantaneous models were then used to estimate the parameters for time-averaged hydrologic and sediment transport processes used by SIBERIA. This model can only represent time-averaged, rather than instantaneous processes because of limitations in computing power. For example, SIBERIA only predicts the mean annual sediment yield, rather than the yield produced by individual storms. Parameters that govern gully development were obtained from studies of a natural area with "similar geologic material" as the mine waste rock. Rainfall records from a nearby weather station were used to create a rainfall series to drive the predicted erosion over 1000 years.

The results of the simulations (Figure 19.9) indicated several potential problems with the original mine reclamation design. Willgoose and Riley (1998) predicted that after 1000 years:

1. steep slopes will suffer severe degradation on the order of 5–7 m,
2. a number of valleys will dissect the central region of the cap rock,
3. eroded material will create deposits about 5 m deep on the margin of the waste rock dumps, and
4. little erosion (less than 500 mm) will occur away from the gullies.

Willgoose and Riley (1998) note that the design morphology of the rehabilitated mine differs consider-

Table 19.3 Ability of numerical models reviewed in this chapter to solve site-specific problems

Subject	Relative utility	Comments
Modeling changes in streambed elevation	High	Calibration required, channel form must be constant
Quasi-equilibrium cross-sectional morphology	Poor	Conceptual basis poorly established
Bank erosion	Moderate	Detailed field work required for calibration, identification of relevant controlling processes
Meander evolution	High	Relatively untested
Braided channels	Poor	Existing models schematic
Short-term floodplain sedimentation	Moderate	Models still in development stage
Long-term evolution of floodplains with meandering rivers	Moderate	Models schematic, relatively untested, but promising
Bed material waves	Poor-high	Useful if only bed evolution is involved, detailed calibration required
Evolution of fluvial landscapes	Moderate	Difficult to calibrate, relatively untested

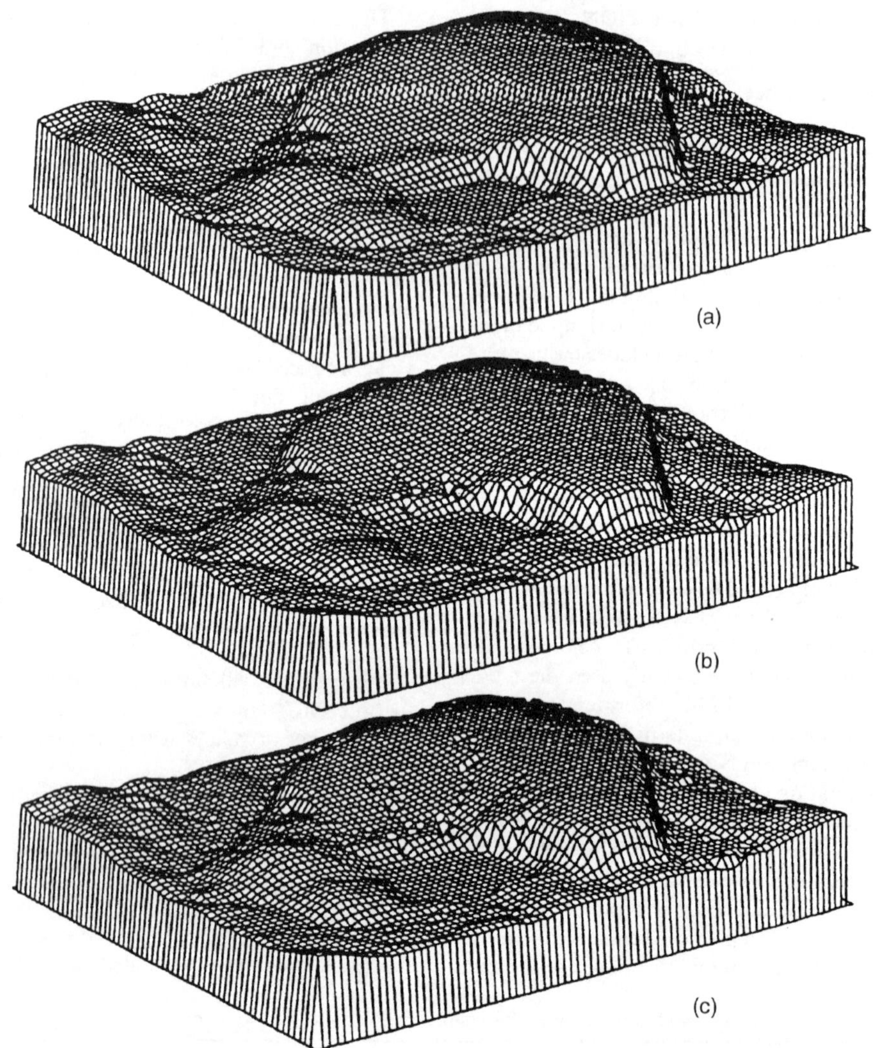

Figure 19.9 Simulated evolution of above-ground storage at the Ranger Uranium mine of 0 (a), 500 (b) and 1000 (c) years (from Willgoose and Riley 1998). The area shown is about 2500 m^2 and has a maximum relief of approximately 20 m

ably from natural topography, and that improvements could be achieved by adopting a design that mimics topographic characteristics of natural catchments with a quasi-equilibrium form.

19.11 SUMMARY

The models reviewed above have several limitations in common (Table 19.2). Few of the models have been routinely used in practical applications [although the 1-dimensional bed material wave model of Cui *et al.* (2001) is currently being used in several projects]. Few of the models are commercially available, all are very difficult to use, and few have been thoroughly tested in a wide variety of field settings. Most of these models require data for calibration, parameterization, or initial and boundary conditions that are rarely available. Finally, it is important to recognize that the conceptual basis for many of the models discussed above is still being actively debated. Scientists still do not agree on the essential controlling mechanisms for many of the processes simulated by these models.

These observations have important implications for how such models should be used (Table 19.3). First, detailed field studies should always accompany any modeling study that is designed to predict the behavior of any particular river. These field studies should not simply seek to provide parameters needed for modeling, but they should be comprehensive enough so the modeling team can understand what processes control the river's behavior and so that relevant historical influences on river channel form can be identified. Broadly trained fluvial geomorphologists are perhaps best suited for these types of investigations. Second, modeling should only be performed by those intimately familiar with the mathematical and physical basis for the models being used. General experience with numerical methods will not be sufficient to use these models properly, as all of the models are based on limited empirical data and simplifying assumptions that are often not readily appreciated or understood.

Clearly, numerical modeling of river geomorphology is in its infancy as a discipline applied to practical problems. However, some of the models reviewed here can provide useful information to managers if used carefully (Table 19.3). For example, changes in streambed elevation, patterns of meander migration, and the evolution of 1-dimensional sediment waves may currently be evaluated with some confidence if enough field data are available for calibration. Processes of bank erosion, short- and long-term floodplain sedimentation, and simulation of fluvial landscapes may be simulated with less confidence; specific applications should be based on extensive fieldwork for model calibration and to identify the relevant geomorphic processes at a particular site. The cross-sectional morphology of quasi-equilibrium channels and the dynamics of braided channels cannot be simulated with confidence at present. Further research is needed to establish useful conceptual models and to quantify the relevant processes of erosion, deposition, and transport.

REFERENCES

ASCE (American Society of Civil Engineers Task Committee on Bank Erosion and River Width Adjustment). 1998a. I. Processes and mechanisms. *Journal of Hydraulic Engineering* 124: 881–902.

ASCE (American Society of Civil Engineers Task Committee on Bank Erosion and River Width Adjustment). 1998b. II. Modeling. *Journal of Hydraulic Engineering* 124: 903–917.

Anderson, M.P. and Woessner, W.W. 1991. *Applied Groundwater Modeling: Simulation of Flow and Advective Transport*, London: Academic Press, 381 p.

Andrews, E.D. 1981. Measurement and computation of bed-material discharge in a shallow sand-bed stream, Muddy Creek, Wyoming. *Water Resources Research* 17: 131–141.

Benda, L. and Dunne, T. 1997a. Stochastic forcing of sediment supply to channel networks from landsliding and debris flow. *Water Resources Research* 33: 2849–2863.

Benda, L. and Dunne, T. 1997b. Stochastic forcing of sediment routing and storage in channel networks. *Water Resources Research* 33: 2865–2880.

Blench, T. 1969. *Mobile-bed Fluviology: A Regime Treatment of Canals and Rivers*, Edmonton: University of Alberta Press, 168 p.

Bradley, J.B., Williams, P.T. and Walton, R. 1998. Applicability and limitations of sediment transport modeling in gravel-bed rivers. In: Klingeman, P.C., Beschta, R.L., Komar, P.D. and Bradley, J.B., eds., *Gravel-bed Rivers in the Environment*, Highlands Ranch, CO: Water Resources Publications, pp. 543–578 (838 p.).

Cao, S. and Knight, D.W. 1998. Design for hydraulic geometry of alluvial channels. *Journal of Hydraulic Engineering* 24: 484.

Chang, H.H. 1980. Geometry of gravel streams. *Journal of Hydraulic Engineering* 106: 1443–1456.

Chang, H.H. 1988. *Fluvial Processes in River Engineering*, New York: John Wiley, 432 p.

Chase, C.G. 1992. Fluvial landsculpting and the fractal dimension of topography. *Geomorphology* 5: 39–57.

Cui, Y., Dietrich, W.E. and Parker, G. 2001. Routing bedload sediment through river networks draining steep uplands. In: Eos Trans, AGU, 82(47), Fall Meet. Suppl., Abstract H51F-09.

Darby, S.E. and Thorne, C.R. 1996. Numerical simulation of widening and bed deformation of straight sand-bed rivers. I. Model development. *Journal of Hydraulic Engineering* 122: 184–193.

Darby, S.E., Thorne, C.R. and Simon, A. 1996, Numerical simulation of widening and bed deformation of straight sand-bed rivers. II. Model evaluation. *Journal of Hydraulic Engineering* 122: 194–202.

Davis, W.M. 1899. The geographical cycle. *Geographical Journal* 14: 481–504.

Diplas, P. 1990. Characteristics of self-formed straight channels. *Journal of Hydraulic Engineering* 118: 707–728.

Diplas, P. and Vigilar, G. 1992. Hydraulic geometry of threshold channels. *Journal of Hydraulic Engineering* 118: 597–614.

Ervine, D.A. and Jasem, H.K. 1995. Observations on flows in skewed compound channels. *Proceedings of the Institution of Civil Engineers Water, Maritime and Energy* 112: 249–259.

Furbish, D.J. 1988. River-bend curvature and migration: How are they related? *Geology* 16: 752–755.

Furbish, D.J. 1991. Spatial autoregressive structure in meander evolution. *Geological Society of America Bulletin* 103: 1576–1588.

Gross, L.J. and Small, M.J. 1998. River and floodplain process simulation for subsurface characterization. *Water Resources Research* 34: 2365–2376.

Hasegawa, K. 1989. Universal bank erosion coefficient for meandering rivers. *Journal of Hydraulic Engineering* 115: 744–765.

Howard, A.D. 1992. Modelling channel migration and floodplain sedimentation in meandering streams. In: Carling, P.A. and Petts, G.E., eds., *Lowland Floodplain Rivers, Geomorphological Perspectives*, New York: John Wiley and Sons, pp. 1–41.

Howard, A.D. 1994. A detachment-limited model of drainage basin evolution. *Water Resources Research* 30: 2261–2285.

Howard, A.D. 1996. Modelling channel evolution and floodplain morphology. In: Anderson, M.G., Walling, D.E. and Bates, P.D., eds., *Floodplain Processes*, London: John Wiley and Sons.

Howard, A.D. 1997. Badland morphology and evolution: interpretation using a simulation model. *Earth Surface Processes and Landforms* 22: 211–227.

Howard, A.D. and Hemberger, A.T. 1991. Multivariate characterization of meandering. *Geomorphology* 4: 161–186.

Howard, A.D. and Knutson, T.R. 1984. Sufficient conditions for river meandering: a simulation approach. *Water Resources Research* 20: 1659–1667.

Ikeda, S. 1981. Self-formed channels in sandy beds. *Journal of Hydraulic Engineering* 107: 389–406.

Ikeda, S. and Izumi, N. 1990. Width and depth of self-formed straight gravel rivers with bank vegetation. *Water Resources Research* 26: 2353–2364.

Ikeda, S. and Izumi, N. 1991. Stable channel cross-sections of straight sand rivers. *Water Resources Research* 27: 2429–2438.

Ikeda, S. Parker, G. and Sawai, K. 1981. Bend theory of river meanders. Part 1. Linear development. *Journal of Fluid Mechanics* 112: 363–377.

James, C.S. 1985. Sediment transfer to overbank sections. *Journal of Hydraulic Research* 23: 435–452.

Johanneson, H. and Parker, G. 1989. Linear theory of river meanders. In: Ikeda, S. and Parker, G., eds., *River Meandering*, Water Resources Monograph 12, Washington, DC: American Geophysical Union, pp. 181–214.

Kelsey, H.M. 1980. A sediment budget and an analysis of geomorphic process in the Van Duzen River Basin, north coastal California. *Geological Society of America Bulletin* 91: 190–195.

Knox, J.C. 1976. Concept of the graded stream. In: Melhorn, W. and Flemal, R., eds., *Theories of Landform Development*, London: Allen and Unwin, pp. 168–198.

Kovacs, A. and Parker, G. 1994. A new vectorial bedload formulation and its application to the time evolution of straight river channels. *Journal of Fluid Mechanics* 267: 153–183.

Lane, E.W. 1955. Design of stable channels. *Transactions of the American Society of Civil Engineers* 120: 1234–1260.

Lawler, D.M. 1986. River bank erosion and the influence of frost: a statistical examination. *Trans. Inst. Br. Geogr. N.S.* 11: 227–242.

Leopold, L.B. and Maddock, T. 1953. *The Hydraulic Geometry of Stream Channels and Some Physiographic Implications*, US Geological Survey Professional Paper 252, 57 p.

Lisle, T.E., Cui, Y., Parker, G. and Pizzuto, J.E. 2001. The dominance of dispersion in the evolution of bed material waves in gravel bed rivers. *Earth Surface Processes and Landforms* 26: 1409–1420

Lisle, T.E., Pizzuto, J.E., Ikeda, H., Iseya, F. and Kodama, Y. 1997. Evolution of a sediment wave in an experimental channel. *Water Resources Research* 33: 1971–1981.

Mackin, J.H. 1948. Concept of the graded river. *Geological Society of America Bulletin* 59: 463–512.

Madej, M.A. and Ozaki, V. 1996. Channel response to sediment wave propagation and movement, Redwood Creek, California, USA. *Earth Surface Processes and Landforms* 21: 911–927.

Marriott, S. 1992. Textural analysis and modelling of a flood deposit: river Severn, U.K. *Earth Surface Processes and Landforms* 17: 687–697.

Middlekoop, H. and van der Perk, M. 1998. Modelling spatial patterns of overbank sedimentation on embanked floodplains. *Geografiska Annaler* 80A: 95–109.

Millar, R.G. and Quick, M.C. 1993. Effect of bank stability on geometry of gravel rivers. *Journal of Hydraulic Engineering* 119: 1343–1363.

Millar, R.G. and Quick, M.C. 1998. Stable width and depth of gravel-bed rivers with cohesive banks. *Journal of Hydraulic Engineering* 124: 1005–1013.

Murray, A.B. and Paola, C. 1994. A cellular model of braided rivers. *Nature* 371: 54–57.

Nicholas, A.P., Ashworth, P.J., Kirkby, M.J., Macklin, M.G. and Murray, T. 1995. Sediment slugs: large-scale fluctuations in fluvial sediment transport rates and storage volumes. *Progress in Physical Geography* 19: 500–519.

Nicholas, A.P. and Walling, D.E. 1997. Modelling flood hydraulics and overbank deposition on river floodplains. *Earth Surface Processes and Landforms* 22: 59–77.

Odgaard, A.J. 1987. Streambank erosion along two rivers in Iowa. *Water Resources Research* 23: 1225–1236.

Parker, G. 1978a. Self-formed straight rivers with equilibrium banks and mobile bed. Pt. 1. The sand-silt river. *Journal of Fluid Mechanics* 89: 109–125.

Parker, G. 1978b. Self-formed straight rivers with equilibrium banks and mobile bed. Pt. 2. The gravel river. *Journal of Fluid Mechanics* 89: 127–146.

Pickup, G., Higgins, R.J. and Grant, I. 1983. Modelling sediment transport as a moving wave – the transfer and deposition of mining waste. *Journal of Hydrology* 60: 281–301.

Pitlick, J. 1993. Response and recovery of a subalpine stream following a catastrophic flood. *Geological Society of America Bulletin* 105: 657–670.

Pizzuto, J.E. 1984a. Bank erodibility of sand-bed streams. *Earth Surface Processes and Landforms* 9: 113–124.

Pizzuto, J.E. 1984b. Equilibrium bank geometry and the width of shallow sand-bed streams. *Earth Surface Processes and Landforms* 9: 199–207.

Pizzuto, J.E. 1987. Sediment diffusion during overbank flows. *Sedimentology* 34: 301–317.

Pizzuto, J.E. 1990. Numerical simulation of gravel river widening. *Water Resources Research* 26: 1971–1980.

Pizzuto, J.E. and Meckelnburg, T.S. 1989. Evaluation of a linear bank erosion equation. *Water Resources Research* 25: 1005–1013.

Sellin, R.H.J. and Ervine, D.A. 1993. Behaviour of meandering two-stage channels. *Proceedings of the Institution of Civil Engineers Water, Maritime, and Energy* 101: 99–111.

Siggers, G.B., Bates, P.D., Anderson, M.G., Walling, D.E. and He, Q. 1999. A preliminary investigation of the integration of modelled floodplain hydraulic with estimates of overbank floodplain sedimentation derived from Pb-210 and Cs-137 measurements. *Earth Surface Processes and Landforms* 24: 211–231.

Stebbings, J. 1963. The shape of self-formed model alluvial channels. *Proceedings of the Institution of Civil Engineers* 25: 485–510.

Stolum, H. 1996. River meandering as a self-organization process. *Science* 271: 1710–1713.

Stolum, H. 1998. Planform geometry and dynamics of meandering rivers. *Geological Society of America Bulletin* 110: 1485–1498.

Sun, T., Meakin, P., Jossang, T., and Schwarz, K. 1996. A simulation model for meandering rivers. *Water Resources Research* 32: 2937–2954.

Sutherland, D.G., Ball, M.H., Hilton, S.J. and Lisle, T.E. 2002. Evolution of a landslide-induced sediment wave in the Navarro River, California. *Geological Society of America Bulletin* 114: 1036–1048.

Thorne, C.R. and Tovey, N.K. 1981. Stability of composite riverbanks. *Earth Surface Processes and Landforms* 6: 469–484.

Vigilar, G.G. and Diplas, P. 1997. Stable channels with mobile bed: formulation and numerical solution. *Journal of Hydraulic Engineering* 123: 189–199.

Vigilar, G.G. and Diplas, P. 1998. Stable channels with mobile bed: model verification and graphical solution. *Journal of Hydraulic Engineering* 124: 1097–1108.

Wiele, S.M., Graf, J.B. and Smith, J.D. 1996. Sand deposition in the Colorado River in the Grand Canyon from flooding of the Little Colorado River. *Water Resources Research* 32: 3579–3596.

Wiele, S.M., Andrews, E.D. and Griffin, E.R. 1999. The effect of sand concentration on deposition rate, magnitude, and location in the Colorado River below the Little Colorado River. In: Webb, R.H., Schmidt, J.C., Marzolf, G. R. and Valdez, R.A., eds., *The Controlled Flood in Grand Canyon*, Geophysical Monograph 110, Washington, DC: American Geophysical Union, pp. 131–145 (367 p.).

Willetts, B.B. and Hardwick, R.I. 1993. Stage dependency for overbank flow in meandering channels. *Proceedings of the Institution of Civil Engineers, Water, Maritime, and Energy* 101: 45–54.

Willgoose, G. and Riley, S. 1998. The long-term stability of engineered landforms of the Ranger Uranium mine, Northern Territory, Australia: application of a catchment evolution model. *Earth Surface Processes and Landforms* 23: 237–259.

Willgoose, G., Bras, R.L. and Rodriguez-Iturbe, I. 1991a. A coupled channel network growth and hillslope evolution model. 2. Nondimensionalization and applications. *Water Resources Research* 27: 1685–1696.

Willgoose, G., Bras, R.L. and Rodriguez-Iturbe, I. 1991b. Coupled channel network growth and hillslope evolution model. 1. Theory. *Water Resources Research* 27: 1671–1684.

Wolman, M.G. 1955. *The Natural Channel of Brandywine Creek*, US Geological Survey Professional Paper 271, 56 p.

Wolman, M.G. 1959. Factors influencing erosion of cohesive river banks. *American Journal of Science* 257: 204–216.

20

Statistics and Fluvial Geomorphology

PIERRE CLÉMENT[1] AND HERVÉ PIÉGAY[2]
[1]*Laboratoire de Géographie Physique de l'Environnement de l'Université Lyon 2, Bronc, France*
[2]*UMR 5600 du CNRS, Lyon, France*

20.1 INTRODUCTION

Why a chapter about statistics in a book on fluvial geomorphology? An examination of the literature in this field over the last decade suggests that while other sciences have embraced a variety of statistical tools, geomorphic studies have used mostly basic regression analysis and other "classical inferential statistics" (Table 20.1). Fluvial geomorphology has lagged behind its sister disciplines both because of its historical-descriptive tradition in the first half of the 20th century, and because the strong influence of physics has resulted in a mechanistic approach. Thus, the discipline has not relied on concepts of variability as much as social sciences or biology, where variation within and between groups is so high that it must be assessed and understood before making progress in the field. Geomorphologists have applied the laws of physics and mechanics to explain river processes and have tended to view statistics as a secondary tool to address variability, such as spatial and temporal complexity (e.g., event frequency) and highly variable attributes in physical laws (e.g., velocity, grain size). The classic text by Leopold *et al.* (1964) illustrates the use of statistics in this field in the 1950s and 1960s (Table 20.2). The leading concepts such as drainage organization and magnitude and frequency of flow and sediment transport were based on the pioneering works of researchers, many of whom were working in engineering and earth sciences (Horton 1945, Strahler 1952, 1954, Wolman and Miller 1960). Statistical analyses were applied to develop relations among climate, flow characteristics, and channel form, and to evaluate regional controls and scale effects. Regression techniques were also used to define rating curves relating a quantity that is difficult to measure, like suspended sediment concentration, to a variable, like the discharge or stage that is relatively easy to record.

This "mix of physical arguments and pure empiricism" (Rhoads 1992) confronts geomorphologists with the problem of relating theoretical or experimental results that are usually expressed in dimensionless forms, with empirical ones, and frequently introduces scaling problems. If laboratory or field experiments help to understand the physical laws controlling channel forms and processes, they are often uniscalar and atemporal, one would say reductionist, and unable to consider holistically all the complexity of geomorphological phenomena controlled by climatic, geologic, and topographic contexts existing at the earth's surface and also human impacts that occurred variously in space and time.

However, the systems approach had emerged long ago (Strahler 1954) but was often taken up only theoretically or partially documented (Chorley and Kennedy 1971, Schumm 1977). The study of systematic relationships within physical units, such as watersheds and channels, appears as a complementary approach of physically based advances in considering the forms and processes in their temporal and spatial diversity.

The aim of this chapter is then to provide a partial review of the statistical tools available, their use in fluvial geomorphology and their limits, and to give some examples that illustrate how some of these tools can be used to answer geomorphological questions. We consider fluvial geomorphology in its widest sense and incorporate consideration of the floodplain and watershed systems.

Table 20.1 Use frequency of statistical tools in manuscripts published in four international journals in geomorphology between 1987 and 1997

	n	Classical statistics (%)	Multivariate statistics (%)	Stochastic models (%)	Other statistics[a] (%)
Catena	61	5.00	0.00	0.00	0.00
ESPL*	223	9.87	0.90	0.90	0.90
Geomorphology	129	9.30	3.10	0.00	0.00
Zeitschrift**	58	17.24	0.00	3.45	3.45

[a]Fractal analysis, geostatistics, and chronostatistics.
*Earth Surface Processes and Landforms.
**Zeitschrift für Geomorphologie.

Table 20.2 Use of statistical analysis in graphical illustrations (excluding photographs) in textbooks

	Fluvial geometry (%)	Fluvial processes (%)	Total (%)	Bivariate form (%)
Leopold *et al.* (1964)	23	13	42	83
Gregory and Walling (1973)	10	17	42	63
Bravard and Petit (1997)	19	25	57	66

20.2 A BRIEF OVERVIEW OF STATISTICS

Statistics can be defined as a set of mathematical techniques used to collect, characterize, summarize, and classify numerical data, test differences between groups, and provide predictions. It is commonly used to interpret phenomena for which an exhaustive study of all the acting factors and populations is not possible due to their great number or complexity. Statistical analyses allow to interpret data or samples of data from large populations, and to examine their variability. Fluvial geomorphologists are dealing with complex objects at various scales, such as in channel features, channel beds and reaches, valleys, watersheds, regions, and even continents. These units are multidimensional, at least quadridimensional since they vary according to space and time. They can be characterized by attributes whose values can be continuous (magnitude or rank, ratios, intervals) or nominal (qualitative).

The first need is often to evaluate the basic variability of the different attributes, simplify these attributes by measuring their size and homogeneity according to their distributions (Table 20.3). This poses the problem of the definition of such objects: What sets of attributes can discriminate subgroups in the most certain way? Where can we draw their spatial and temporal limits that are often rather fuzzy? Then we are interested in characterizing their structures and their functioning by defining relationships between their attributes (Table 20.3). These relations are supposed to be causal as far as they can be explained by physical laws—although correlations are not always causal. Discrepancies from deterministic models have to be brought to the fore and eventually accounted for by further processing (Table 20.3). Geomorphic approaches can also be supported by stochastic modeling and by particular statistics for highlighting spatial and temporal structures of the data.

It is not our objective to give a detailed summary of the statistical possibilities as many statistical textbooks (Wilson and Kirkby 1980, Williams 1984, Wrigley 1985, Saporta 1990, Lebart *et al.* 1995) and software guidelines or websites offer a wide range of statistical options, potentially useful for environmental scientists. The main issue is to show how statistical tools can help assist in solving a geomorphological problem, in formulating and validating hypotheses, and in highlighting advantages and disadvantages of various approaches. Tables 20.4–20.6 give an overview of studies in which statistical tools have been used in different ways for geomorphological purposes.

Descriptive Statistics

The aim of descriptive statistics is to give a synthetic view of the data by summarizing them using a few values (Tables 20.3 and 20.4). The conclusions are then supported by data (not samples), which are not inferred to a wider population. But this examination is often a preliminary step required as many statistical

Table 20.3 Summary of some statistical tools that can be used in fluvial geomorphology according to research objectives and data characteristics

	Objectives	V_1	V_2	V_n	$V_1 \times V_n$	$V_n \times V_n$
1	Description (data exploration and simplification)	Mean, median, standard deviation, percentile		PCA (Q), CA (q), MCA (q, Q), Clusters (*k*-means, hierarchical classification)		Co-Inertia analysis (q, Q)
2	Differences between variables		*t*-test (Q–Q), ANOVA (q/Q), χ^2 (q–q), Correlation (Q–Q), Non-parametric tests (Spearman correlation, Kruskal–Wallis test)	ANOVA (q/Q)	Discriminant function analysis (Q/q)	
3	Deterministic models		Simple regression (Q/Q)		Multiple regression (Q/Q), Regression on factorial components (Q/Q)	CCA (Q/Q)
4	Stochastic models	Probability laws (normal, Gumbel, Poisson)	Simple logistic regression, Markov chain		Multiple logistic regression (qQ or qq or q/qQ)	
5	Measurement of spatial and temporal structures		Autocorrelation, periodicity, and threshold within a chronological or spatial continuum, Spectral analysis, Fractal analysis			

q, qualitative data; Q, quantitative data; V, variable, statistics can be univariate (V_1), bivariate (V_2) or multivariate (V_n); PCA, Principal Component Analysis; CA, Correspondance Analysis; MCA, Multiple Correspondance Analysis; CCA, canonical correspondance analysis (CA–CA coupling).

Table 20.4 Overview of statistical tools used in fluvial geomorphology (except simple and multiple linear regressions)

Type of statistical tools	Type of data	References
t-test	Channel width and depth at two dates	Rhoads and Miller (1991)
	D_{50} measured by three operators	Wohl *et al.* (1996)
	Grain size measured at different sites	Dawson (1988)
ANOVA	D_{50} and D_{84} measured by three operators	Wohl *et al.* (1996)
χ^2 test	Grain size distributions (classes)	Wohl *et al.* (1996)
Polynomial regression (order 2)	Y = sample density, X = depth of sampling	Reneau and Dietrich (1991)
PCA	Magnetic properties of sediments	Yu and Oldfield (1993)
	Mineralogy characteristics of sediments	Llorens *et al.* (1997)
	Geochemical concentrations in sediment samples	Passmore and Macklin (1994)
Discriminant function analysis	X_i = set of channel indicators, Y = human disturbance	Woodsmith and Buffington (1996)
	X_i = morphological descriptors, Y = reaches	Gurnell (1997)
	X_i = morphometric variables of meander pattern, Y = meander types/models	Howard and Hemberger (1991)
	X_i = hydraulic geometry descriptors, Y = reaches according to bank stability	Ridenour and Giardino (1995)
Canonical correspondence analysis	X_i = potential causal factors, Y = drainage basin characteristics	Ebisemiju (1988)
Co-Inertia analysis	X_i = field variables, X_j = temporal variables	Piégay *et al.* (2000)
Distribution modeling	Prediction of individual step distribution	Schmidt and Ergenzinger (1992)
	Prediction of velocity distribution within a channel reach	Lamouroux *et al.* (1995)
Logistic regression	Y = channel adjustment; X = geology/channel gradient/land use/management	Downs (1995)
	Y = channel pattern, X = overbank sediment thickness	Piégay *et al.* (2002)
	Y = probability that a tributary is a significant sediment source, X_1 = relative basin area of the tributary, X_2 = index of the tributary's sediment delivery potential based on stream power in the tributary	Rice (1998)
	Y = channel instability; X = mobility index (slope, discharge, D_{50})	Bledsoe and Watson (2001)
Markov chain	Sediment budget	Kelsey *et al.* (1987) Malmon *et al.* (2002)
	Longitudinal succession of in-channel features	Grant *et al.* (1990)
	Fluctuating velocity profile	Kirkbride and Ferguson (1995)
Fractal analysis and geostatistics	Spectral analysis of channel width and stream gradient	Nakamura and Swanson (1993)
	Autocorrelation coefficients for channel width and depth	Robison and Beschta (1990)
	Semivariogram of bed microprofile statistics	Clifford *et al.* (1992), Madej (1999)
	Fractal analysis for drainage network organization	Ibamez *et al.* (1994), Gao and Xia (1996)

procedures assume that data follow the normal law of distribution—the so-called bell-shaped curve—which is more likely obtained by using large samples. Data distributions usually can be compared to known ones through graph visualization or various dispersion parameters.

With recent increases in computational capabilities, multidimensional statistics can now summarize large data sets and their structure. Factorial methods compute the main axis of multidimensional scatter plots and produce simplified graphics of the descriptive elements identified. Depending on the data characteristics (continuous vs. categorical variables), various techniques are used including Principal Component Analysis (PCA), Correspondence Analysis (CA), and Multiple Correspondence Analysis (MCA). Among factorial techniques, PCA is the most common. It examines a set of continuous attribute variables (e.g., channel width, depth, grain size, sinuosity) measured at different sites (e.g., cross-sections, reaches) and identifies the key associations between them by reducing a large number of correlated variables to a smaller, more manageable set of factors. The principal components can be defined as new variables, independent from the others, that summarize the correlations of the measured variables. Correspondence analysis is a weighted PCA of a contingency table, whereas MCA examines the relations between categorical variables of different characters that are reduced to dichotomous variables (absence vs. presence).

The factors identified in regression analysis can be used to develop inferential statistics, or add supplementary data to factorial plots to visualize associations. A further exploratory approach is to combine two sets of variables—*a* and *b*—characterizing a similar set of stations—*S*—to identify their common structure. Such multivariate procedure, called inter-battery analysis (Co-Inertia analysis), searches the Co-Inertia axis that maximize the covariance of projection coordinates of the data sets *a*–*S* and *b*–*S* for which each structure was previously studied with factorial analysis. It can be used, e.g., to identify for given river reaches a co-structure between two groups of variables, one describing channel and the other floodplain (see Tucker 1958 or Chessel and Mercier 1993 for details).

Cluster analysis (e.g., *k*-means method, hierarchical classifications) is used with similar data sets but to answer different questions. Such tools use distance computation algorithms to distinguish groups of variables or stations. They are not based on the distribution theories of classical statistics and they can then be used when the variables are not independent. Factorial and cluster analysis are often complementary. Factorial analysis can be used to simplify a data set and its n meaningful factors can be ordered by cluster analysis.

Classical Inferential Statistics

Inferential statistics (Tables 20.3 and 20.4) extend characteristics of samples to the total population and are based on tests and probabilistic models that validate or invalidate an a priori hypothesis (the null hypothesis H_0).

First, these methods test independence between two sets of variables (V_1 vs. V_2; V_1 vs. V_n). Parametric or non-parametric tests can be selected according to the type and the size of the sets, but also according to the form of the distribution of the variable (normal, log-normal...). A correlation test is a convenient exploratory tool. Visualization of results is obtained through ordered coefficient matrices where variables are resorted on both axes so that the highest positive coefficients are the closest to the central diagonal; density shading is attributed to coefficient classes. Partial correlations, in matrix form or not, take into account the influence of collinear variables on the relationship between a pair of so-called dependent and independent variables. Student's t-test, analysis of variance (ANOVA-test), and χ^2 test are also used to evaluate dependence between data sets (e.g., sample vs. population whose distribution is known or assumed; sample vs. sample; spatial unit vs. spatial unit...). For example, an ANOVA highlights the variability of a continuous variable (e.g., channel depth) according to the modality of a categorical variable (e.g., several meaningful groups of sampling reaches). It distinguishes two parts: the variability of the measure within each group (e.g., a set of reaches) and the variability of the measure between the groups (e.g., between different meaningful sets of reaches). The greater this second variability is relative to the first, the more likely the independence between the groups.

In the case of nominal and ordinal variables, comparison of group distributions can be based on non-parametric tests on a pairwise basis (Table 20.3). Although they have no underlying distribution assumptions and do not require large samples, these tools are rarely encountered in the geomorphological literature, probably because of their lower concluding flexibility and power. Testing differences between groups can also be performed in an exploratory perspective using multivariate analysis. Discriminant analysis classifies observations into two or more groups known beforehand on the basis of quantitative variables. The aim is to identify discriminant factorial axes that maximize the inertia (e.g., the variance) between the groups. Discriminant function can be established and used for discriminating purposes when using additional data. Within-class distributions must be approximately normal; the discriminant function that separates the two groups may be linear or quadratic (parametric method). If not, non-parametric techniques are used.

Deterministic and Stochastic Modeling

Deterministic models—whose behavior is predicted by mathematical functions—can give an estimated value of one or more variables from one or more known variables. The appropriate tool depends on the type of variables, the number of individuals, and the number of independent/dependant variables. Simple and multiple linear regressions are widely used, but canonical correspondence analysis is less known.

In the canonical correspondence analysis, the link between two sets of continuous variables is tested. This technique has similarities with Co-Inertia described previously, but is less flexible and not as easy to interpret because inferential assumptions are required. If variables can be divided into two sets, canonical correlation provides a suitable simplifying tool as the model successively finds pairs of linear combinations from each set (canonical variables) such that the correlation between the canonical variables is maximized. Each of the canonical variables is uncorrelated with all the similar variables in the other pairs. The analysis can go on so as to find other sets of canonical variables uncorrelated with the first pair, the limit of combination being the number of variables in the smaller set.

In stochastic (probabilistic) models, the output does not correspond to an estimated value of a given variable but an occurrence of a given variable, varying from 0 to 1. They are useful tools to generate models in which predicted variables are categorical, to predict distributions (e.g., normal, exponential, gamma distributions, mixed distributions...), and recurrence intervals (e.g., peak flows). Logistic regression is commonly used in medical sciences to distinguish two groups of individuals (e.g., healthy and unhealthy persons, treated and untreated persons). This tool models the link between a categorical variable, usually dichotomous, and categorical or continuous explanatory variables. It can be usefully applied in fluvial geomorphology to predict the probability of occurrence of a specific spatial entity according to its human-induced or natural characteristics (Table 20.4). Transition probabilities over time or space (Markov chain) can also be used to assess cascading phenomenon within the river system (Table 20.4). This approach is convenient when processes are partly understood and where interdependence of variables makes the definition of functional links difficult. Randomness is assumed in sequential states or events (equal probability) and frequency distributions must be known. For example, in sedimentology, depositional sequences can be simulated in space and time. Transitions between deposition units are not totally independent from the previous ones so that a memory effect may be apparent. The probabilities of these transitions must be specified in adequate matrices whose number depends on the length of such a memory.

Temporal Series and Geostatistics

Many types of tools exist to describe temporal and spatial patterns, such as fractal analysis, spectral analysis, autoregressive moving average models (ARMA), autocorrelation measurement but also segmentation and threshold tests (Tables 20.3 and 20.4). The aim is to evaluate a tendency, a cyclicity, homogeneous segments, or some thresholds in the series, and to assess and model the complexity of spatial or temporal information.

A fractal can be defined as a spatial object comprising elements that exhibit a similar pattern over all scales. It is possible to define a fractal dimension (D_f) that corresponds to the rate at which the element complexity changes with the scale. Repetitions of pattern can be assessed using spectral analysis, the finite Fourier transforms, to estimate the power spectrum of a signal (Hardisty 1993) plotted in a periodogram. Time series and spatial continua can be composed by a sum of basic sinusoids, each having an amplitude and a frequency. Autocorrelation analysis is another way to evaluate a periodicity and trends in spatial and temporal data. Spatial autocorrelation can be defined as a similarity between values as a function of their spatial position—e.g., geometrical characteristics along a stream profile. Positive spatial autocorrelation occurs when the values measured on neighboring plots are more similar than the others. Tests such as the I of Moran and the c of Geary were developed to measure the autocorrelation structure of geomorphological data (see review of Aubry and Piégay 2001). The c of Geary has no upper limit when the lower limit is 0. Zero means that the autocorrelation is high and 1 means it is low. Between 0 and 1, the autocorrelation is positive (the values of the neighboring entities are more similar than those of the other entities) whilst when c varies between 1 and $+\infty$, it infers negative autocorrelation (the values of the neighboring entities are more different than those of the other entities).

Other approaches can be used to focus on trend breaks in spatial and temporal continuum and also identify distinct homogeneous segments (Brunel 2000). A break in a temporal series can be defined as a change in a probability law of the series at a given time t. Different tests, such as the test of Lee and Heghinian (1977), the test of Pettit (1979) or the U statistics of Buisband (1984) can identify such breaks. For example, the test of Pettit is a non-parametric test based on the Mann–Whitney test, with the null hypothesis being the absence of a break in the series X_i of size N. The statistics $U_{t,N}$ considers that for each time period t with a value between 1 and N, the two series of time X_i and X_j, for $i = 1$ at time t and for $j = t+1$ at N, are issued of the same population. The segmentation test (Hubert 1989) is another way to describe non-stationary series by detecting several breaks and then homogeneous segments between the breaks. Two steps are usually distinguished, the first one corresponding to the segmentation. The operator can define the number of segments required and their minimal size, or an optimization algorithm can be used to identify the best segmentation amongst all the possible ones. The second one corresponds to the statistical validity of the identified segments. A segmentation can be validated for example if the mean of neighboring segments is significantly different.

Significance Level

Inferential tests are used to reject null hypothesis concerning independence of groups of data and to test for significance. The "p-value" can be defined as the probability of getting a test statistic value of a given magnitude if the null hypothesis of no difference is true. Conventionally, 0.05 and 0.01 probability have been used as thresholds to reach conclusions. In regression models, it is possible to test whether the constants α (slope of the line of regression) and β (Y-intercept) but also the coefficient of determination r^2 are significant. Testing if r^2 coefficient is significantly greater than 0 using the F-test procedure serves to estimate the predictive power of the regression model.

In the cases where inferential tests do not exist or cannot be applied, other procedures can be used whose recent development has been possible because of increases in computer capacity (e.g., permutational and randomization tests). Such tests are more flexible in term of assumptions than classical tests and also provide p-values. Because the distribution of the statistics under H_0 is built with some data and not some samples, the tests concern only the data and not a population from which these data could be inferred. The goal of the permutation test is to generate all the possible values of a given statistic (all the permutations of the values amongst the individuals) in order to calculate the p-value associated with the observed value of the statistics. When the censing of all the permutations is not possible ($n > 10$), such tests can be approximated by a randomization test which is based on a limited number of permutations rather than doing all of the possible ones (Manly 1991).

20.3 QUALITY OF DATA AND SAMPLES

One of the main issues in using data and then applying statistics is data quality, i.e., precision and accuracy of the measurements, their validity and reliability, and the representativeness of the sampling. "What is the error of detection when measuring Cs-137 or Pb-210 activity?" is an important question when testing for differences between sites or samples. "What is the root mean square (RMS) error when georectifying satellite images or air photos?" must be addressed when overlapping images and reaching conclusions about channel changes. The question "Is there any bias in the measure due to techniques/protocols used or operators?" must be addressed to determine whether relationships among different data sets are geomorphically meaningful. We can also use statistics to calibrate, establish corrections to allow two different measures of the same object or process to be compared. A related problem is the heterogeneity in precision among data sets and the lack of knowledge of error resulting from data derived from diverse sources, such as field surveys, different scale maps, remotely sensed imagery, and statistical data from various public services.

ANOVA, χ^2 tests or t-tests are the tools most commonly used to assess whether the measurements are biased by operators or by methods. For example, Thévenet *et al.* (1998) measured geometrical volumes of accumulations composed of both woody debris and air as the following product: width $\times$ height $\times$ length. Before determining a linear model linking large

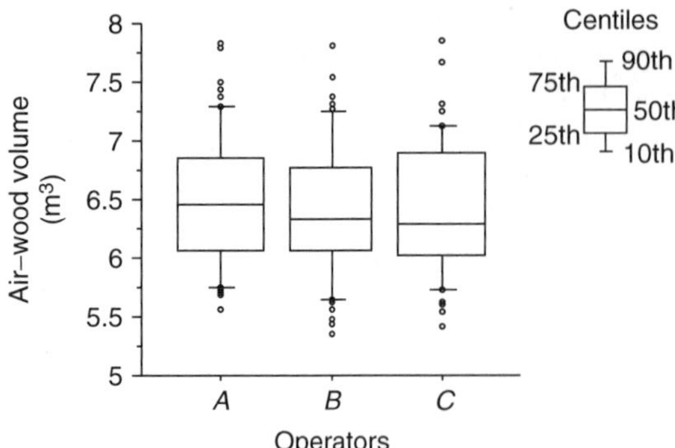

Figure 20.1 Box plots of the "air–wood" volume of LWD accumulations calculated from the data collected by three operators on gravel bars of the Drôme River (France)

woody debris (LWD) mass and air–wood volume, it was necessary to confirm that there was no estimator bias. Measures were done by three operators, and an ANOVA test performed to test the independence between them. The null hypothesis was "no significant difference between the three operators" at $\alpha = 5\%$. The null hypothesis was accepted, meaning that there was less than a 5% chance that the measure is different from one operator to another, validating the procedure of field data measurement (degrees of freedom = 2; $F = 0.661$; $p = 0.52$) (Figure 20.1).

Field sampling of sediment in gravel-bed rivers has been widely debated, especially since introduction of the pebble count technique by Wolman (1954). Issues include the best method of sampling (bulk vs. sieving) and operator bias. Wohl *et al.* (1996) characterized the variability among replicates of a sampling method, among four methods and among operators. Three types of tests were used: *t*-tests to evaluate differences in the D_{50} for each operator (e.g., veteran, experienced, and novice); ANOVA to determine whether any of the methods yield statistically different distribution parameter estimates (D_{50} and D_{84}); χ^2 test to evaluate differences in phi class distributions. Three of the four methods produced values of D_{50} and D_{84} that were not statistically different. However, grain size distributions by different operators yielded samples that were statistically different (D_{50}, D_{84}, distribution of size classes, variance).

Because data generally cannot be collected exhaustively both in space and time, adequate sampling techniques are required in both dimensions, which implies variability and discontinuity. The choice of tool from the broad array available should be informed by the aim of the study, and the availability, type, and degree of confidence in the data. The sampling strategy is of greater importance for many reasons, such as randomness, representativeness, method standardization, feasibility, and cost. Scaled stratification within geographical information systems (GIS) must also be considered.

Most of the time, it is not possible to consider an entire population or an entire area, often infinite (e.g., grain size of gravels on a bar, velocity on channel cross-sections, floodplain elevation on a given site). Sampling strategies are needed to efficiently and cost-effectively extract a representative sample that accurately reflects characteristics of the population or the area. When the entire population or area is known, probability sampling can be used, including random, systematic, stratified or clustered procedures. However, in fluvial geomorphology, it is often difficult to know the distribution of a population and thereby predetermine the sample size necessary to obtain an estimate of given precision.

One topic that has inspired considerable literature about sampling methods is the measurement of size of coarse-grained sediments. Because we do not know the mean and the standard deviation of the grain size of the gravel population, also because its distribution does not follow the normal law, we cannot determine with accuracy the best sampling size. Resampling procedures, such as bootstrap simulation techniques (Lebart *et al.* 1995), can be helpful in determining the best sample; this method produces confidence intervals for the parameter of interest without requiring any distribution assumptions. Such techniques were used by Rice and Church (1996a) to determine percentile standard errors in Wolman counts so as to evaluate the sample size needed to maximize the precision of grain size estimated within a gravel-bed river. At each of two sites studied, the *b*-axis of around 3500 particles were measured and the procedure was applied for 20 runs from $n = 50$ to $n = 1000$ and each time repeated 200 times to estimate a standard error for D_5, D_{16}, D_{25}, D_{50}, D_{75}, D_{84}, and D_{95}. They also calculated the theoretical normal percentile standard errors and compared it with the bootstrap percentile standard errors (Figure 20.2). They obtained two main results:

(i) While D_{50} standard errors were consistently low, fine-tail percentile errors were underestimated by the normal model and coarse-tail percentile errors are overestimated, demonstrating that for a given precision, it would be necessary to collect respect-

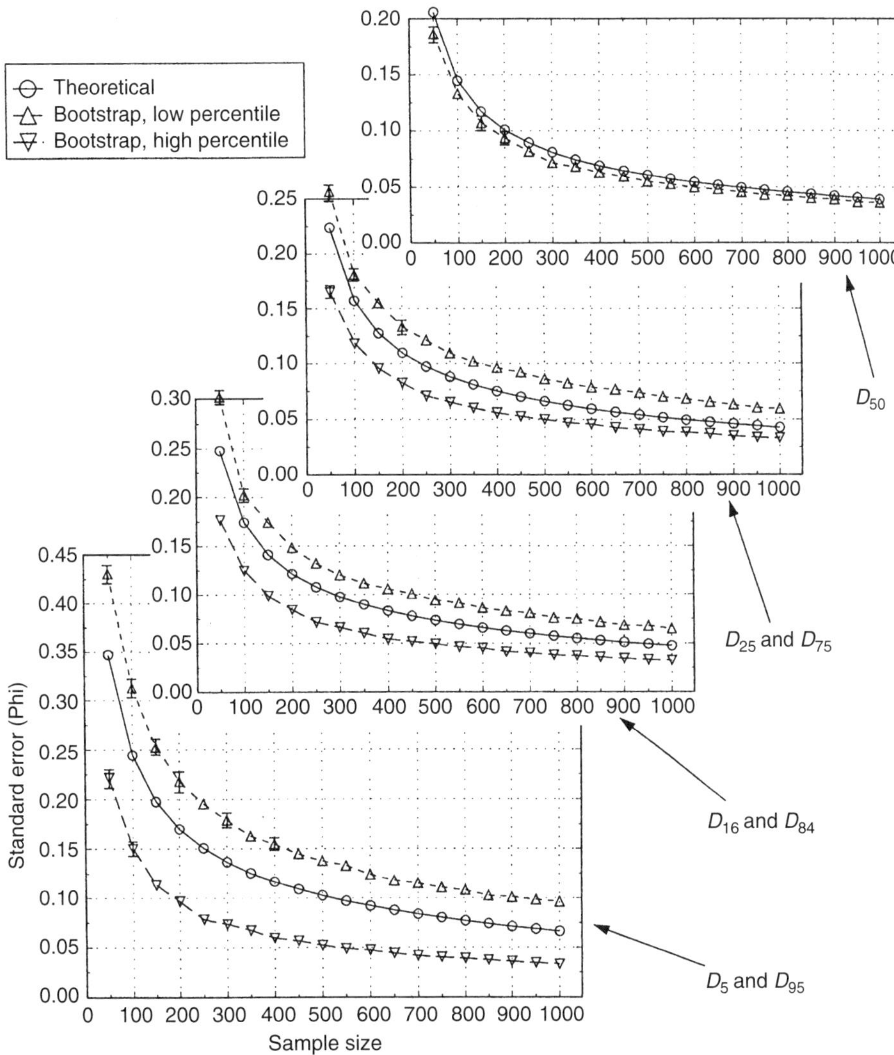

Figure 20.2 Comparison of bootstrap and normal percentile standard errors (Mamquam River site, British Columbia). Error bars indicate the 95% confidence intervals about the mean bootstrap results based on 10 replications. Error bars are shown only where they exceed symbol dimensions (reproduced from Rice and Church 1996a)

ively more and less particles than that expected by a normal distribution to characterize the distribution tails;

(ii) for sample sizes exceeding 300–400 particles, the marginal gains in precision were small relative to the additional sampling effort.

Spatial autocorrelation functions are also useful to calibrate a sampling design when it concerns geographic area. An assumption of many statistical techniques is the spatial independence of the data values collected: the deviation of the *Y* values must be independent of the values of the other variables. In this context, it is important to determine the lag of the spatial dependency, in other words, the distance above which the heterogeneity of the values are independent of their spatial position. If we know the lag, we can define a systematic sampling for which the grid width overpasses the lag and we can then use the classical statistic tests on the sample inferential to the entire population. We did this procedure before developing a logistic regression model to assess occurrence of gullies according to various geographical parameters (e.g., slope, altitude, land use). A sample of pixels was extracted from GIS covers in order to assess the spatial autocorrelation of the variables controlling the gully occurrence. The question was to determine a lag above which the altitude or slope values were effectively independent from the others. The c test of Geary was performed for these two variables for 70 lag classes with a step of 50 m each (Figure 20.3). In order to assess the lag of the positive spatial autocorrelation, some p-values were calculated from randomization tests (1000 random permutations of the two values z_i and z_j for each class) and are plotted as a joined function of the autocorrelation function (Figure 20.3a). For a given threshold (e.g., $\alpha = 0.05$), the values were then distinguished as being significantly autocorrelated or not and this confidence interval was then plotted on the graph of the c of Geary function (Figure 20.3b). Once we observed a sharp change between a sequence of low p-values and a sequence of high p-values, we considered that the lag was reached. In this case, the lag was reached over 2450–2500 m for the altitude and 2550–2600 m for the slope. In order to respect the assumption of independence in the modeling process, systematic sampling should be conducted within a grid where each sample should be separated from the others by 2600 m.

20.4 DISCRIMINATION OF FORMS AND PROCESSES

One of the main tasks in fluvial geomorphology is to distinguish spatial entities (bars, channel reaches, parts of the floodplain, sedimentary facies) according to their specific characteristics. Many options are available, some referring to a wider population while others focus on the data without making any distribution assumption.

Standard Parametric and Non-parametric Difference Tests

The topological characteristics of networks are one field of application for non-parametric tests. Distribu-

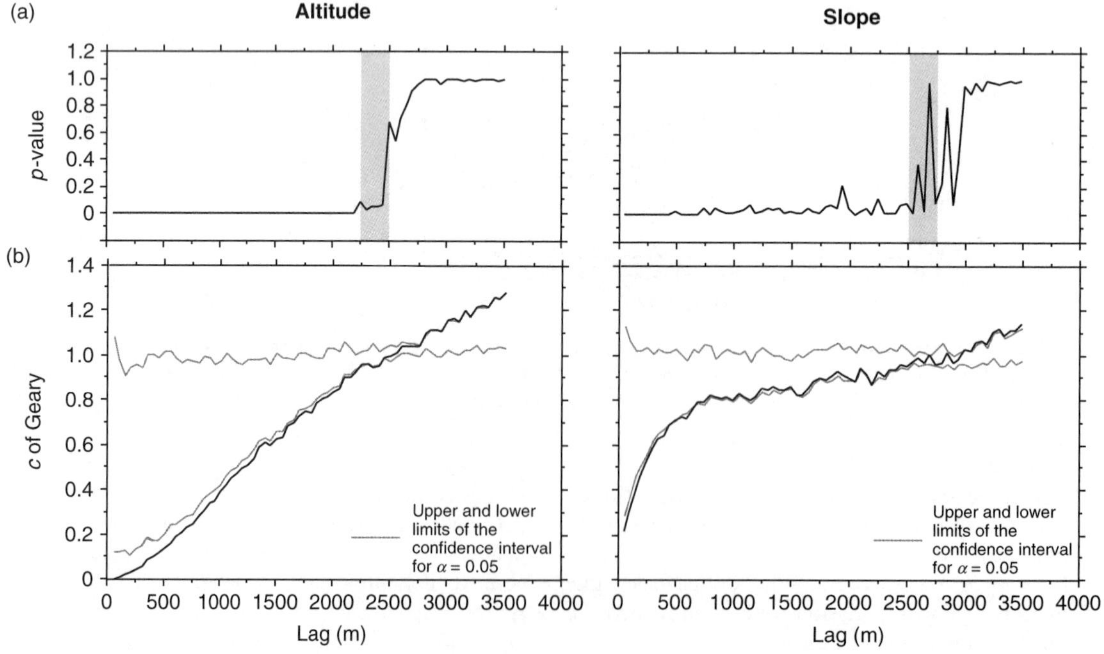

Figure 20.3 Determination of the threshold above which the values are not autocorrelated: example of elevations and slopes of the Bine and Soubrion catchments (France) (samples of 50 × 50 m pixels extracted on a digital elevation model). (a) p-values of randomization test (1000 random permutations of the neighboring values at each 50 m step). When the p-value averages 0, the positive autocorrelation is high. (b) c of Geary and its confidence interval under the null hypothesis at $\alpha = 0.05$

tions of source and tributary source link lengths were examined with a Kolmogorov–Smirnov test in tidal creeks of northern Australia. Knighton *et al.* (1992) compared the size and extent of channels in different years, thereby documenting that network evolution through time followed an exponential growth. We used χ^2 tests to assess the variability in grain size distribution between sites on two streams of the Massif Central (France). We compared the morphology and grain size of three reaches to evaluate the potential effects of woody debris storage on sediment deposition. Site D_1 had no woody debris, whilst D_{2a} averaged 20 kg m^{-1} of river length and D_{2b}, located immediately downstream of D_{2a}, averaged 38 kg m^{-1}. We randomly sampled the bed and determined the dominant grain size in each reach. The hypothesis was that grain size distribution was different between reaches as a function of woody debris amount. The χ^2 test confirmed that the three grain size distributions were different, with D_1 the most heterogeneous and D_{2a} and D_{2b} having higher frequency of one or two classes (Figure 20.4). D_{2a} had a high frequency of sandy plots associated with side channel jams, while D_{2b}, within which woody debris formed channel dams, had bars composed of 8–32 mm gravel.

To elevate the role of local disturbances on bed sediment distribution in low order streams, Rice and Church (1996b) tested the hypothesis that the systematic downstream reduction of grain size—the negative exponential model—is precluded by colluvial inputs and log jams whose distribution is random both in space and age. First an ANOVA showed significant differences between the surficial D_{50} of sites within each study reach and the same method established the textural differences between a reach decoupled from lateral slope inputs and one that was not.

Multivariate Analysis

These analyses can simplify preliminary data such as to watershed morphometric characteristics in Guyana savannas (Ebisemiju 1989) or geochemical signatures of surficial deposits in northern England (Passmore and Macklin 1994). In the latter example, because elemental compositions, notably heavy mineral concentrations (Pb, Zn) depend on geological conditions and historical mining operations, this approach discriminated deposits according to their provenance from geologically distinct sub-catchments.

Multivariate analyses can also be useful discriminating tools when the geomorphological question is posed at a large geographical scale, and the correlations between the variables or the controlling factors are not well known. Figure 20.5 illustrates how various multivariate techniques can be used in a complementary way.

Discriminant analyses were used to throw light on European valleys where water mills have made river

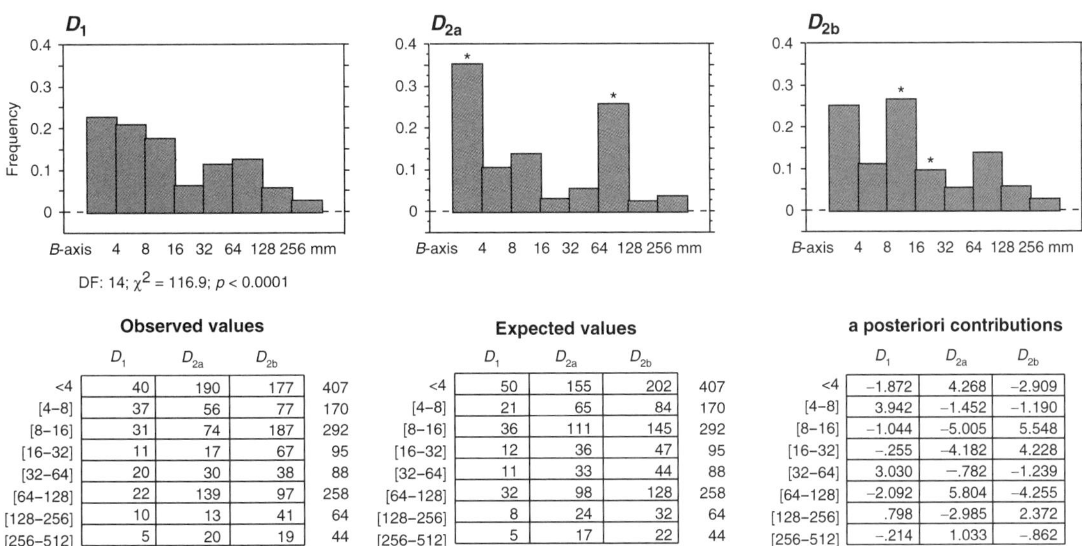

Observed values

	D_1	D_{2a}	D_{2b}	
<4	40	190	177	407
[4–8]	37	56	77	170
[8–16]	31	74	187	292
[16–32]	11	17	67	95
[32–64]	20	30	38	88
[64–128]	22	139	97	258
[128–256]	10	13	41	64
[256–512]	5	20	19	44
	176	539	703	1418

Expected values

	D_1	D_{2a}	D_{2b}	
<4	50	155	202	407
[4–8]	21	65	84	170
[8–16]	36	111	145	292
[16–32]	12	36	47	95
[32–64]	11	33	44	88
[64–128]	32	98	128	258
[128–256]	8	24	32	64
[256–512]	5	17	22	44
	176	539	703	1418

a posteriori contributions

	D_1	D_{2a}	D_{2b}
<4	−1.872	4.268	−2.909
[4–8]	3.942	−1.452	−1.190
[8–16]	−1.044	−5.005	5.548
[16–32]	−.255	−4.182	4.228
[32–64]	3.030	−.782	−1.239
[64–128]	−2.092	5.804	−4.255
[128–256]	.798	−2.985	2.372
[256–512]	−.214	1.033	−.862

Figure 20.4 Grain size distributions of three reaches of the Doulon stream (Central Massif, France) and results of the χ^2 test: observed/expected values, a posteriori contributions

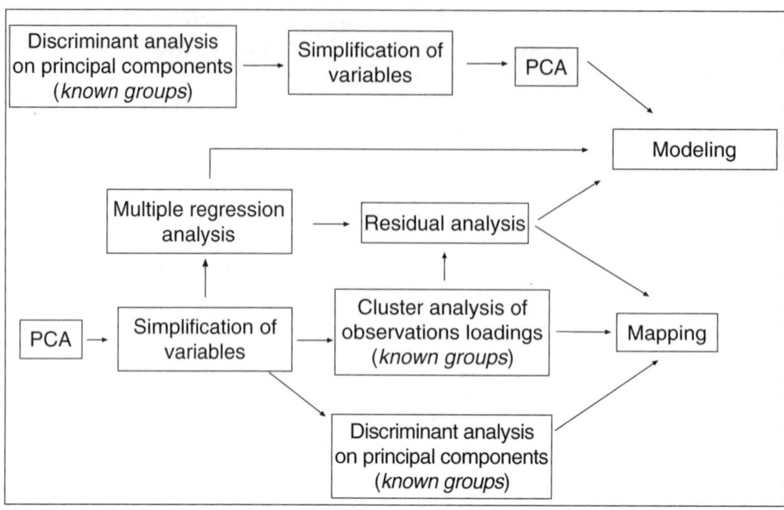

Figure 20.5 Examples of combined procedures: multivariate analysis, simplification, and discrimination tools used in multiple statistical step procedures

long profiles uneven. Some questions may be considered: Were there preferential conditions for mill construction? Where were they more difficult to keep running? What channel modifications did they bring about? A medium scale approach was employed on the Solnan River (155 km^2, Bresse, France) where 17 sections were selected according to elevation contours from the 1/25 000 map in the Jura piedmont. They were described by 12 variables dealing with sinuosity, slope, valley width, and stream power indexes, and by three categories: reaches with sluices still operated for flood control, reaches with destroyed mill sites, and reaches without control structures. Present time distribution is more simply described than the former one for which discriminant variables are often closely related and deal largely with channel characteristics influenced by human control such as gradient and sinuosity.

A normalized PCA was performed using the variables previously discriminated. As shown by the first factorial map of observation loadings (Figure 20.6a), mills were located in different sets of conditions: either steep sinuous reaches where diversion canals were more efficient (Reaches 6, 10, 11), or downstream reaches with abundant flow (Reaches 12, 15, 17). Disappearance of mills seems to be related to the difficulty of maintaining sites in reaches with erodible sands, high sinuosity, and major tributary confluences. Reaches without mills typically exhibit extreme conditions such as a wide and rectilinear valley and high stream power. As shown by Figure 20.6b, the presence of mills was often associated with lower channel sinuosity and a higher bed gradient, potentially decreasing stability, and that mill disappearance was followed by a quick return to sinuous patterns.

Multivariate techniques can also be used to distinguish between two sets of variables, describing a set of spatial entities, those that are strongly correlated, suggesting causal links and clusters, and providing input to the modeling process. Inter-battery analysis can be used in this way to analyze two sets of data without classical assumptions.

In the Eygues River Basin (1150 km^2), a tributary of the Rhône River, the regional variability of channel changes was studied in 20 sub-catchments (10–100 km^2) using inter-battery analysis (Liébault *et al.* 2002). Channel morphology and watershed data were matched in order to assess the covariation of the variables describing the two scales: the basin and the reach. The aim was to determine the basin variables that could predict channel morphology. The position of each sub-watershed on factorial maps was determined both by channel morphology and watershed characteristics. Three groups were discriminated and atypical catchments were identified. For example, one of the three groups is characterized by high channel width, fine grain size, and high drainage density.

The same approach was conducted along the Ain River, France (3640 km^2) on terrestrial plugs defined as floodplain areas separating the main channel from the permanent aquatic zone of former

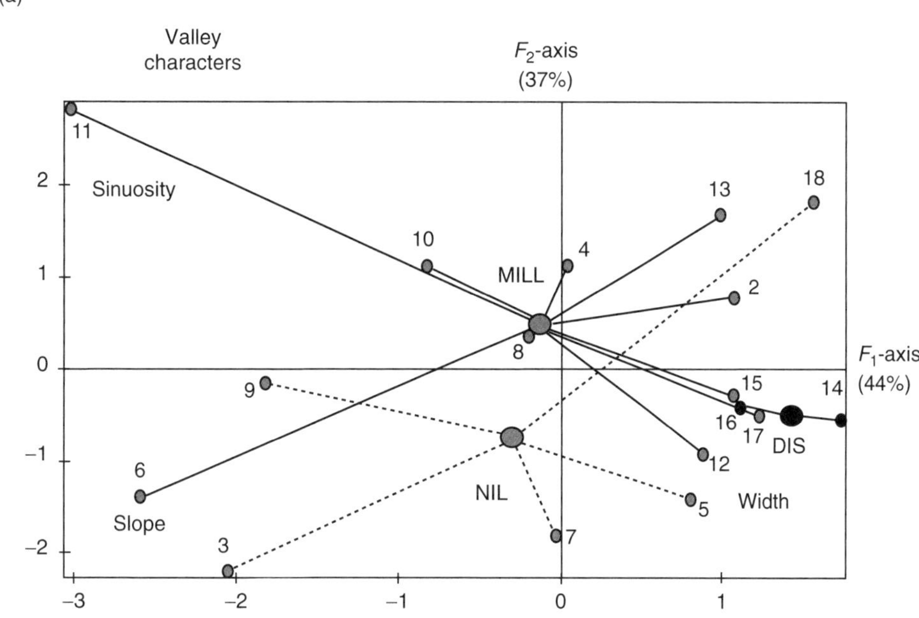

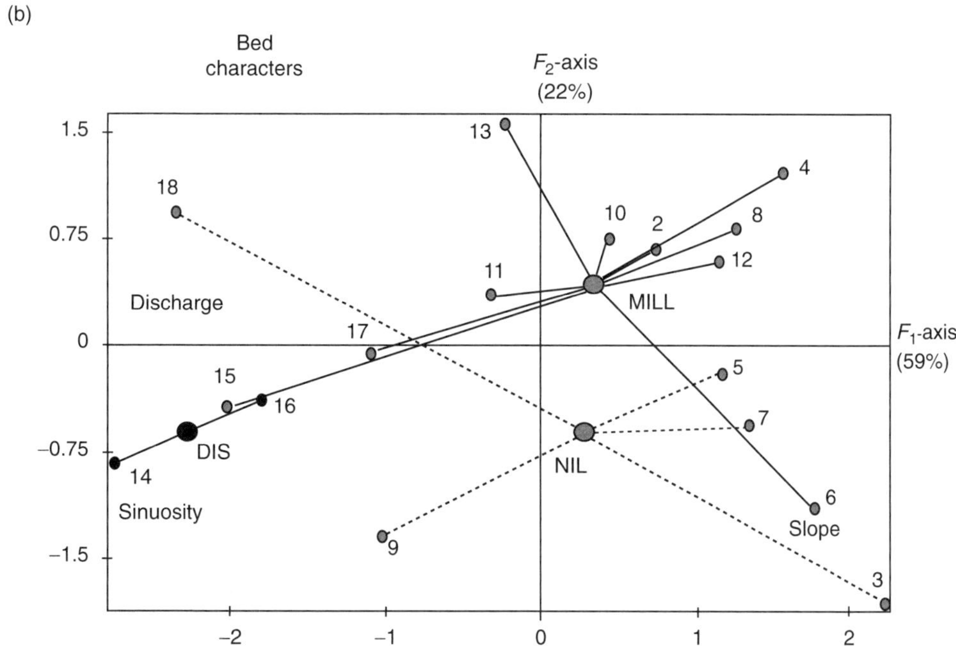

Figure 20.6 (a) Morphological conditions of mill location (Solnan River, Jura piedmont, France). Reaches are numbered sequentially downwards. (b) Influence of mill sites on bed morphology of the Solnan River. NIL: natural reaches. MILL: reaches including mills. DIS: reaches formerly with mills

channels (Piégay *et al.* 2002). Two data sets were built:

1. field data describing the floodplain biogeomorphology;
2. large scale structural data based on aerial photography analysis and historical documents describing the environmental changes (e.g., main channel aggradation, degradation, shifting) and dating fluvial forms (e.g., cut-offs).

An inter-battery analysis was performed to identify the co-structure of structural variables and field variables. Figure 20.7a must be interpreted with Figure 20.7b, on which are plotted the modalities of the structural and field variables on the first Co-Inertia map. The sampling plots on the plugs correspond to the position of the end of the arrows, and the associated channel characteristics are plotted at their beginning. Two main groups of plugs can then be distinguished according to their original geomorphic

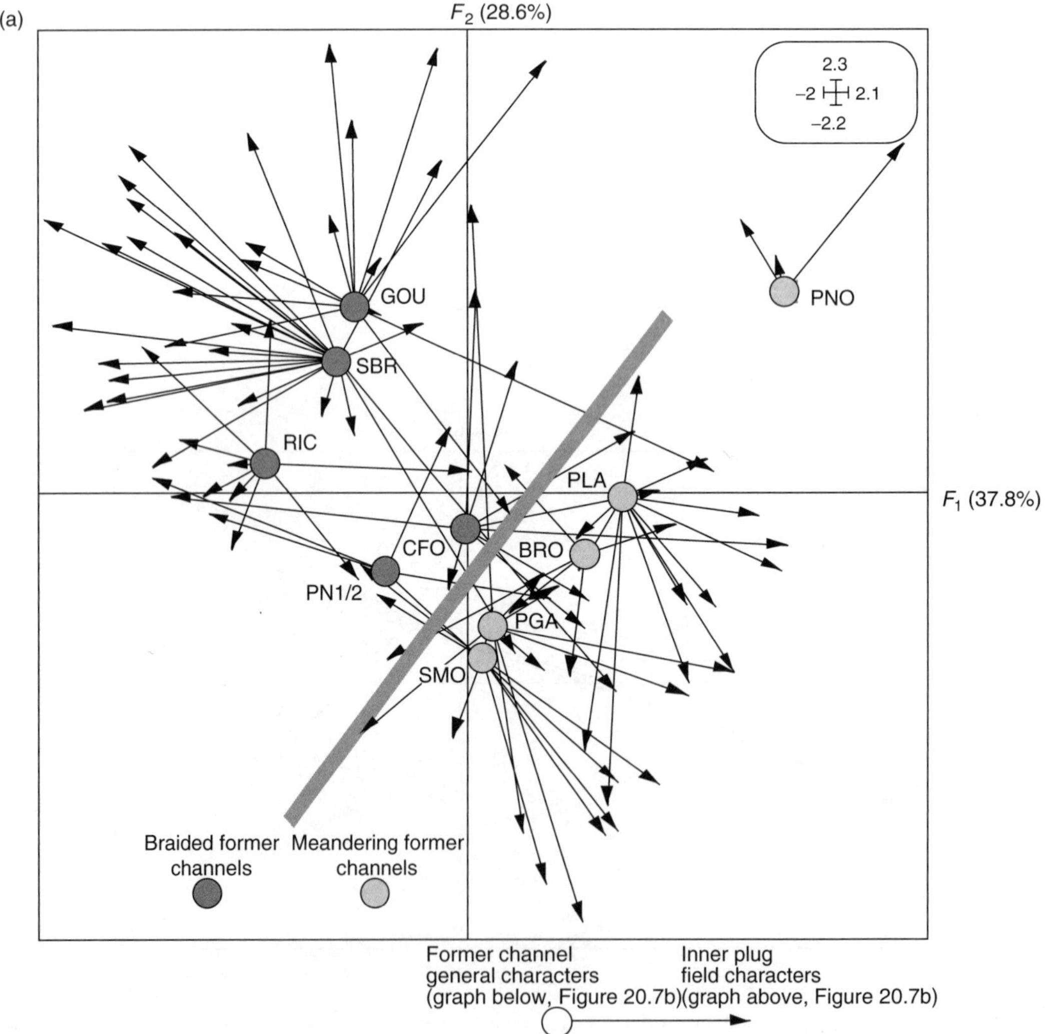

Figure 20.7 Co-Inertia analysis between field data and structural data describing the plugs of the former channels of the Ain River: (a) match of the two scatters of the first factorial map, and (b) projection of the modalities of the structural and field variables on the first factorial map (from Piégay *et al.* 2002) (reproduced by permission of John Wiley and Sons, Ltd.)

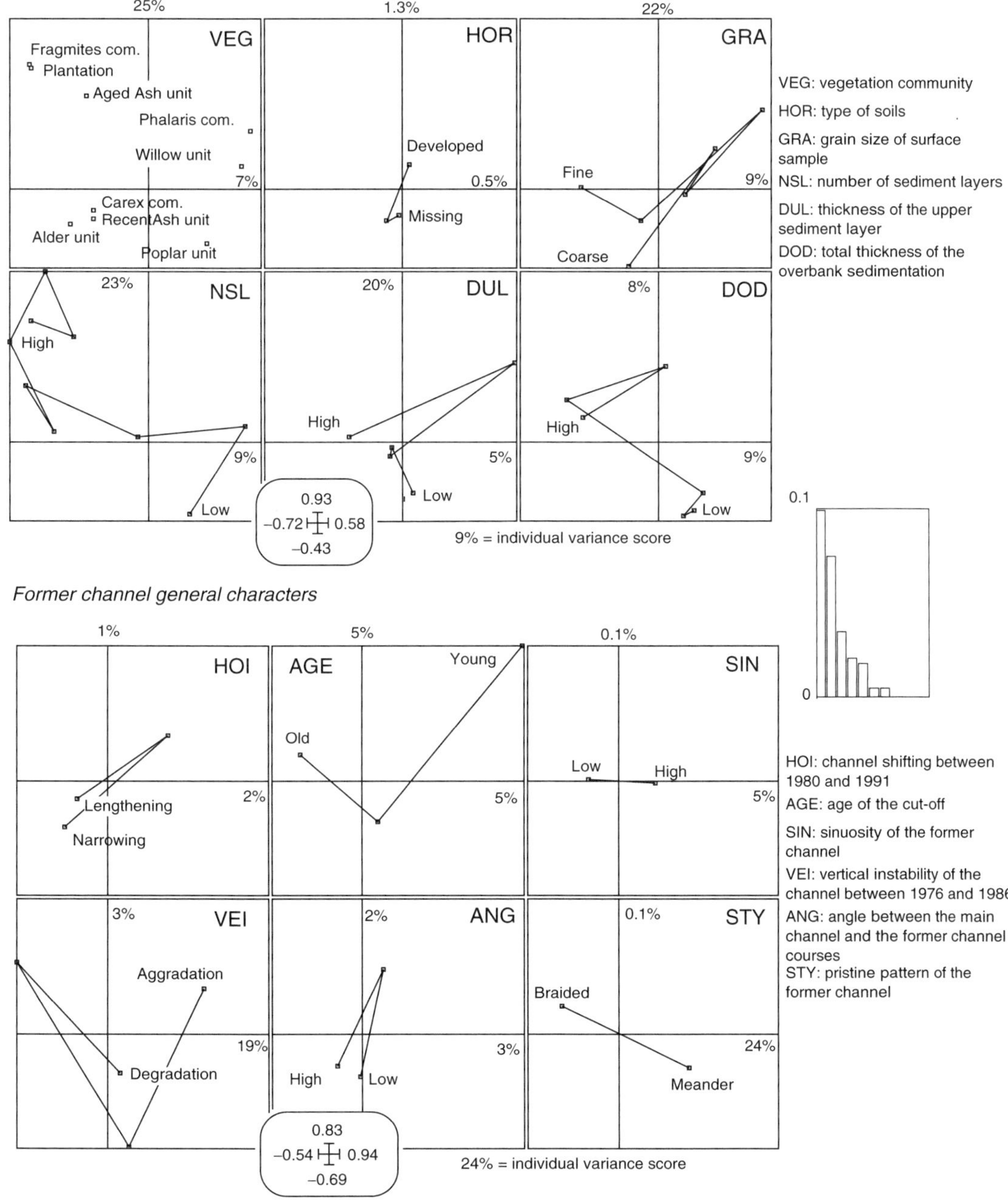
(b)
Inner plug field characters
25%
VEG
Fragmites com.
Plantation
Aged Ash unit
Phalaris com.
Willow unit
7%
Carex com.
Recent Ash unit
Alder unit
Poplar unit
1.3%
HOR
Developed
0.5%
Missing
22%
GRA
Fine
9%
Coarse
VEG: vegetation community
HOR: type of soils
GRA: grain size of surface sample
NSL: number of sediment layers
DUL: thickness of the upper sediment layer
DOD: total thickness of the overbank sedimentation
23%
NSL
High
9%
Low
20%
DUL
High
5%
Low
8%
DOD
High
9%
Low
0.93
−0.72
0.58
−0.43
9% = individual variance score
0.1
0
Former channel general characters
1%
HOI
Lengthening
Narrowing
2%
5%
AGE
Young
Old
5%
0.1%
SIN
Low
High
5%
HOI: channel shifting between 1980 and 1991
AGE: age of the cut-off
SIN: sinuosity of the former channel
VEI: vertical instability of the channel between 1976 and 1986
ANG: angle between the main channel and the former channel courses
STY: pristine pattern of the former channel
3%
VEI
Aggradation
19%
Degradation
2%
ANG
3%
High
Low
0.1%
STY
Braided
24%
Meander
0.83
−0.54
0.94
−0.69
24% = individual variance score

Figure 20.7 (b)

pattern (STY): the plugs of the former meandering channels and the plugs of the former braided channels. Other variables, such as the angle between the main channel and the former channel (ANG) or channel shifting (HOI) and degradation (VEI), do not exhibit clear patterns on the first Co-Inertia map. The difference between the two groups defined above is also related to age (i.e., gradient from old to medium-aged). Among the braided former channels, the oldest exhibit fine grain size, thick overbank deposits, and numerous sediment facies. Their vegetation includes both hardwood and semiaquatic communities. The youngest (PN1/2, CFO) have characteristics similar to those of meanders: fine to medium grain size, and moderately thick overbank deposits.

20.5 SIMPLE AND MULTIPLE LINEAR REGRESSIONS

Bivariate Analysis

Channel forms and processes. Bivariate analysis has been the most commonly used statistical tool to analyze channel form, especially since the influential work of Leopold and Maddock (1953). Even over the last three decades of the 20th century [as illustrated by the textbooks of Gregory and Walling (1973) and Bravard and Petit (1997)], increasing frequency of use has not been matched by increasing variety of statistical tools (Table 20.2). Multivariate methods have been either in the form of multiple regression equations or ratios in bivariate regressions, with most relations following a log–log or semi-log form (see Tables 20.5 and 20.6). Other tools have been rarely used. Such is the case of polynomial regression, probably because of the limited confidence in the corresponding equations, although graphical interpretation may be useful.

Channel geometry has been often described by simple regressions, such as the classical set of relationships of Leopold and Maddock (1953) (Table 20.5). Residual partition has sometimes been performed on graphs according to environmental conditions: for instance, bank material or vegetation in the power function relating width–depth and discharge (Schumm 1960, Ferguson 1986a). Such models are extended to hydraulic and sedimentological variables, often in a predictive way, meeting the requirements of the widely used least squares regression analysis (Mark and Church 1977). The classical set of relations between discharge and width or depth or velocity (Leopold and Maddock 1953) was reconsidered exhaustively by Rhoads (1992) in the light of different bivariate models and sub-models. Validity conditions and criticism of the various estimation procedures were stated in relation to measurement constraints. Rhoads also compared multivariate models of channel geometry adjustments, and introduced variations of discharge and bed material properties as regressor variables.

Bivariate regressions between sediment discharge and some flow characteristic can be used to define thresholds of bed material motion by determining the abcissae intercept graphically. Hydraulic variables may be discharge, power or bed shear stress (Shields 1936, Reid and Frostick 1986, Gomez and Church 1989, Kunhle 1992, Wilcock 1993). Motion thresholds according to particle diameter classes could be assessed in the same manner. However, this process is rarely successful because particles do not move progressively according to their size, but also according to other criteria such as size mixture, shape, and bed structure or position within the longitudinal profile. This interfering or masking influence is also responsible for the scatter in relations between particle size and travel length; other approaches such as the use of stochastic concepts are needed to describe bedload transport (Schmidt and Ergenzinger 1992).

Partitioning has been used to characterize pool–riffle sequences. The power equations linking width with discharge have different coefficients for riffles vs. pools, while the exponents are very similar (Richards 1976). A long debated problem is the velocity or shear stress reversal over these forms between low and high discharges. Bidimensional relations can be defined in a simple way such as between discharge and shear stress for each section (Petit 1987) or stream power and bedload transport (Sear 1996): where the regression lines cross defines the threshold at which transport becomes more efficient in pools. Sear also illustrated this efficiency by the relation between mean distance traveled by particles >20 mm and excess stream power, and used the Shield entrainment function to demonstrate higher entrainment thresholds over riffles (Figure 20.8). Dispersion of data as shown by residuals from the regression line was explained by the effect of different textural and structural features of the bed sediments, which changed as discharge increased and flow type changed. Some of the sedimentological differences—grain size, bed strength, structure, and cluster components—were demonstrated through Mann–Whitney tests for population differences.

Simple regressions have also been widely used to predict process such as sediment transport. A well-known predictive form is the power function relating

Table 20.5 Variables used in defining channel geometry through bivariate or multiple regressions (arithmetic form in italics; otherwise log-transformed)

Response variable	Regressor (+multiple)
Valley length, gradient, specific power	Channel gradient
Valley slope	*Drainage area*
Valley width	Meander wavelength
Bed or valley elevation	Distance
Bed or valley elevation	Distance, *distance*
Channel width, depth	Discharge (mean, max. or dominant). Drainage area + frequency of discharge occurrence
Bankfull width	Drainage area, bankfull discharge
Bankfull depth	Drainage area
Change in width/change in depth	Channel gradient
Maximum or mean depth	Discharge + median bed particle size
Critical bank height	Bank angle
Cross-sectional area of channel	Drainage area + frequency of discharge occurrence
Width/depth	Percentage of silt–clay (weighted mean), channel gradient/Froude number
Channel gradient	Channel width, discharge (mean, max., bankfull or dominant)
Channel gradient	Drainage area + transported load median. Median bed particle size + bankfull discharge. Discharge + bedload transport + median bed particle size
Channel slope	*Mean annual discharge*
Knickpoint migration rate	Basin area
Plunge pool depth	Waterfall height + mean discharge
Meander length	Drainage area. Width
Meander wavelength	*Bankfull discharge*
Meander wavelength	Bankfull channel width, discharge (mean, max., dominant or bankfull)
Meander amplitude	*Meander wavelength*
Meander amplitude	Channel width, discharge (mean, max., dominant or bankfull)
Curvature radius	*Meander wavelength*
Sinuosity	Width/depth. Percentage of silt–clay (weighted mean). Valley slope/channel slope
Critical bank height, Stream bank angle	Discharge
Wandering intensity	Peak discharge/bankfull discharge
Alluvium area	Drainage area, valley width
Weight of particules held on sieve	Particule diameter
Roughness	Bed particle diameter (D_{50})
Particule diameter	Distance downstream
Flatness/roundness index of particules	Distance
Number of pools per reach	Channel width/watershed area
Bankfull depth/width	Number of LWD pieces per reach
Depth/bankfull depth	discharge/bankfull discharge; discharge + grain size

References: Leopold and Maddock (1953), Hack (1957), Leopold and Wolman (1957), Schumm (1960, 1977), Dury (1964), Stall and Yang (1970), Gregory and Walling (1973), Parker (1976), Dunne and Leopold (1978), Richards (1982), Hey and Thorne (1983), Young (1985), Ferguson (1986a,b), Newbury and Gaboury (1993), Yodis and Kesel (1993), Simon (1995), Wharton (1995), Woodsmith and Buffington (1996), Bravard and Petit (1997), and Lecce (1997).

Table 20.6 Variables used in defining channel processes through bivariate (linear or polynomial) or multiple regressions (arithmetic form in italics; otherwise log-transformed)

Response variable	Regressor (+ multiple)
Form parameter of velocity distribution	Relative roughness + Froude number + width variability
Velocity	Mean annual discharge. Drainage area + frequency of discharge occurrence
Pulse period	Dimensionless stream power index
Mean stream power	*Distance downstream*
Power index	Bed particle diameter (D_{50})
Moved particle diameter	Shear stress
Shields parameter	Reynolds number
$1/\sqrt{Froude}$	*Depth*/D_{84}
Specific load	Drainage area. *Mean relief/length*
Sediment delivery ratio	Drainage area
Overbank sedimentation rate	Lateral distance to the channel
Transport rate	*Shear stress*
Mean distance of movement, virtual rate of movement	Excess power
Mean travel distance	Excess power + mean grain size
Sand discharge	Velocity
Average annual sediment discharge	*Percentage of forest cover, crop land*
Sand discharge, suspended sediment discharge	Discharge
Sediment load/flood	*Effective runoff volume, normalized peak discharge*
Suspended sediment annual discharge	Runoff + discharge peakedness + area + stream slope + *percentage of different sol types* + *percentage of different land use*
Sediment delivery rate/unit area/day	Discharge + eroding bank length + *percentage of different sol types* + *stage type*
Suspended sediment concentration	Discharge, 1st peak discharge. *Time relation to hydrograph peak* + instant discharge + *antecedent flow level* + *index of flood intensity*
Bedload transport rate. Discharge/bankfull discharge	Discharge, bed shear stress
Point bar volume change	Annual bank caving
Alluvium area	Cross-sectional stream power. Cross-sectional stream power + valley width

References: Leopold *et al.* (1964), Graf (1971), McPherson (1971), Gregory and Walling (1973), Dawson (1988), Petit (1989), Tropeano (1990), Hassan *et al.* (1992), Kesel *et al.* (1992), Kunhle (1992), Park (1992), Smith *et al.* (1993), Lamouroux *et al.* (1995), Gintz *et al.* (1996), Bravard and Petit (1997), Lecce (1997), Rickenmann (1997), Walling and He (1998).

discharge and sediment transport. Observed results can be compared to curves obtained through various published formulae in order to find out the sediment discharge equations, which are most appropriate to local conditions. Alternatively, the observed results can be plotted against predicted ones and departures from the line of perfect agreement can be examined (Nakato 1990). Individual and mean predicting ability of the formula can be estimated through deviations by discrepancy ratios between the values predicted by the

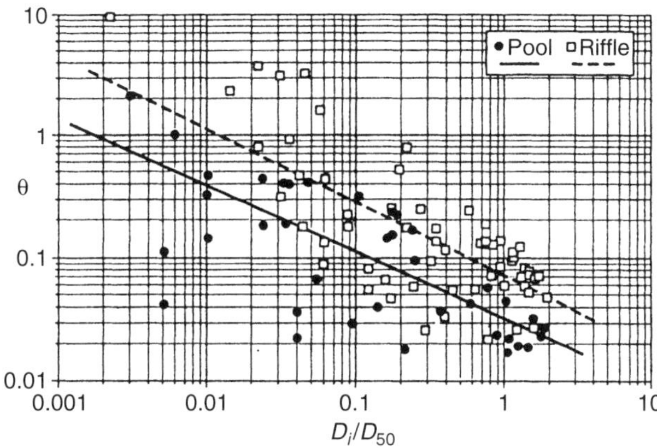

Figure 20.8 Bivariate plot linking the Shields parameter θ with the D_i/D_{50} ratio. Two models were fitted, one for the pools and one for the riffles observed in the North Tyne (from Sear 1996) (reproduced by permission of John Wiley and Sons, Ltd.)

equation and the observed ones (White *et al.* 1973). Agreement between sets of observations and calculated values for various bedload formulae may be assessed more rigorously through ANOVA procedures (Gomez and Church 1989). Satisfactory formulae are those for which the *F* scores indicate no significant differences between calculated and observed mean and trend parameters.

The shape of the graphical relation between predicted values and residuals is also instructive if different trends appear on both sides of some threshold. For instance, high positive residuals may indicate some radical change in flow dynamics such as the destruction of the armored bed layer above a discharge or power value close to or at bankfull stage (Batalla and Sala 1995, Batalla 1997).

Partitioning according to hydrograph stages is widely used when dealing with sediment concentrations (Kunhle 1992), and can be extended to flood sequences (Park 1992). Large, and even anomalous, residuals may exhibit some time trend or hysteresis effect. This is common in sediment rating curves and reflects temporal variations in sediment availability during flood events or on a seasonal basis.

Spatial comparisons. Regional distinctions can be obtained through comparison of coefficients and exponents in regression equations. This approach is common in studies relating watershed area and sediment yields (Poulos *et al.* 1996) where predictive equations are derived from various data bases. The weakest point of such an approach is the accuracy of different methods used to define sediment budgets. Similarly, piedmont sediment accumulations have been linked to their upstream drainage areas, for instance, in alluvial fans from Japan and the southwestern USA (Oguchi and Ohmori 1994).

Residual distances can be used to classify observations according to group characteristics within some geographical unit. Alluvial fan types were distinguished in this way in southeast Spain (Silva *et al.* 1992). A bidimensional plot of high residuals from fan gradient–drainage area and fan area–drainage area was interpreted according to regional knowledge of geology, tectonics, and geomorphic evolution of the Guadalentin depression.

Large deviations in general channel geometry–discharge relations (*Q*/width, *Q*/depth) have been used as indices of local sensitivity to bed modification and in identifying areas where channel design or river restoration are required (Wharton 1995).

Multiple Regression Analysis

If samples are large enough, multivariate regressions can be employed, sometimes following a stepwise procedure. Standardized regression coefficients β are recommended in order to identify the dominant variables. Multicolinear variables must be excluded. Dummy variables can be introduced in order to improve multiple correlations, but subjectivity is involved in the transformation of quantitative variables to discrete numerical ones.

As an example, valley widths appear to be more influential than mean stream power in the statistical explanation of alluvium accumulation in the period following European settlement in the Wisconsin

Driftless Area (Lecce 1997). However, more insights are possible through a combination of other qualitative and quantitative relationships linked to the existence and functioning of meander belts in medium-sized tributaries.

Relationships established between the peak flow and the suspended matter concentration can be improved by adding supplementary variables such as the season, the falling or rising flood stage to produce more powerful models. However, determination coefficients in multivariate regressions may not increase significantly because of scatter in the data, and the introduction of inappropriate variables. Some influences are difficult or impossible to measure in the field, e.g., bed structure (Hassan *et al.* 1992), velocity pulses (Hoey 1992), or roughness of migrating bed forms during floods.

Basins can be considered on a comparative basis as black or gray boxes from which sediment output is delivered by erosion processes. Rickenmann (1997) used multiple predictive equations to relate total bedload transport to water volumes and peak discharges over a threshold of $0.5\,m^3\,s^{-1}$ in Pre-Alpine Swiss watersheds.

Bivariate and multivariate analyses (Tables 20.5 and 20.6) have some advantages such as pedagogical efficiency—the scatter plot gives an instantaneous view of the results, a simplicity of the techniques and generation of new questions, and a capacity to give simultaneously some clue to a wide range of questions, including causal analysis, threshold assessment, group discrimination, and also prediction. Several populations can be distinguished by discriminant lines. The slope–discharge plot within which river patterns are distinguished is the best example (Leopold and Wolman 1957). Simple projection of a categorical variable on to a plot is often useful for explaining the residuals of a given relationships. A model can be established to predict one variable from n others. Many possibilities exist to add several independent variables from the standard multiple regression ($Y = a_iX_i + b$) to the transformation of the independent variables. Walling and He (1998) established an exponential model predicting the floodplain sedimentation rate from the lateral distance to the channel according to the flood depth and the mean sediment concentration (Figure 20.9).

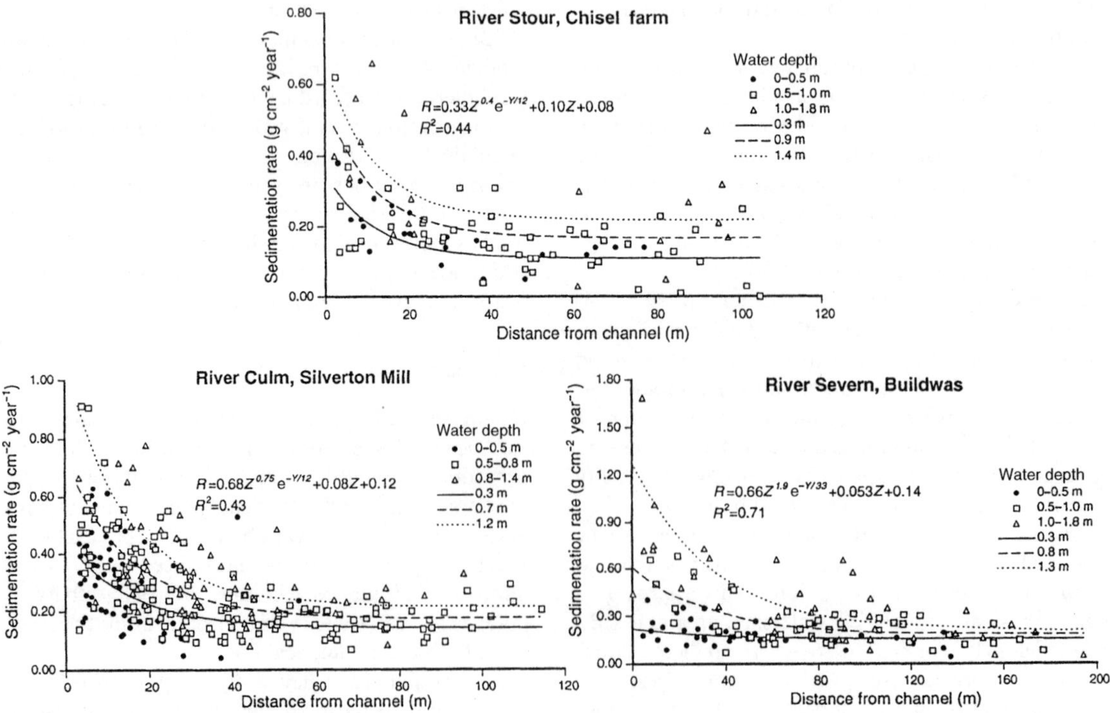

Figure 20.9 Prediction of floodplain sedimentation rate from the lateral distance to the channel according to the flood depth and the mean sediment concentration (from Walling and He 1998) (reproduced by permission of Elsevier Science)

20.6 MULTIVARIATE ANALYSIS AND MODELING

Multivariate analyses are especially powerful when they are included in a multiple-step statistical analysis that allows not only discrimination of sampling sites but also the definition of deterministic models for prediction (see Figure 20.5). Once the large data set is summarized by factorial analysis, regressions on the components can be done and the linear models can be used for predictions, as was done on data from French southern Pre- Alpine Mountain streams. One objective was to predict channel narrowing observed over the last decades on these streams from other characteristics, such as geometry and grain size, and to identify factors controlling such processes at the catchment scale (see Chapter 17). The model was defined using data from the Eygues Basin but was used to predict channel narrowing of the tributaries of the Drôme and the Roubion, two neighboring rivers. First, a normalized PCA was used to synthesize the channel parameters (Figure 20.10a). The first map constitutes a good summary. The first component distinguished wide channels with fine bed materials (e.g., Sauve, Bentrix; Rieu Sec) from narrow, coarse-grained channels while the second component distinguished steep channels with poorly sorted beds from deep channels. The first two components were used as regressors after transformation $[(X+5)^{-0.5}]$ to predict the channel narrowing observed between 1945 and 1995 on air photos. A scatter plot "observed–predicted channel narrowing" was then produced (Figure 20.10b). The equation showed that narrowing mainly occurred in reaches characterized by embeddedness, coarse grain sizes, and steep gradients. The geomorphic interpretation is that narrowing is observed in high-energy tributaries, with narrow valleys that are located closer to the basin sediment sources. These channels first experienced a decrease in bedload supply with land-use changes in the basin in the early 20th century. The sediment moving into the channel was then rapidly exported, the channel was slightly degraded and the coarse bars were colonized by vegetation. Because they have a limited capacity to store gravel, they are mostly conveyor channels rather than depositional reaches. These channels are now narrowed, slightly degraded, paved, and embedded. Unlike the sequence of channel incision and widening observed in the loess region of the Mississippi inner delta, channel incision in the French Pre-Alps is associated with channel narrowing.

Grain size and channel form parameters were also measured on the Drôme and Roubion tributaries and they were added to the normalized PCA as supplementary stations. This approach did not change the previous calculation but produced coordinates of the first and second components. We then predicted their narrowing using the model performed on the Eygues tributaries and compared these values with the observed values. The model fits well, suggesting that it may be broadly applicable across a large geographical region with roughly similar hydrology, geomorphology, and land-use history.

In an example from Eastern Nigeria, morphometric properties of sub-catchments were related to their relief, soils, and vegetation cover (Ebisemiju 1988). Canonical correspondence analysis was used to distinguish three patterns of association: texture of dissection vs. soil and vegetation characteristics, network size vs. stage of basin relief evolution, bifurcation ratio vs. basin relief. Simplification and independence of identified patterns are certainly advantages but, as in other methods, matching available data and statistical requirements, such as normality and multicolinearity, may be problematic. As in other multivariate analysis, interpretation can be difficult due to the intricate relations between variables, which are sometimes ambiguous, especially if they are indices or ratios.

20.7 STOCHASTIC MODELING

Distribution Modeling

When deterministic models behave poorly or are not well suited to describe some distributions, stochastic models may be suitable alternatives. The distribution of step length of tagged particle movement has been described by exponential or gamma functions, with the former model accounting for the duration of rest periods (Schmidt and Ergenzinger 1992). For local water velocities, which vary in space and time within reaches, Lamouroux *et al.* (1995) assumed the relative velocity distribution to be a mixture of Gaussian (centered) and exponential (decentered) distributions. The probability density of the relative velocity was modeled through a maximum likelihood method, with a shape parameter s measuring this mixing. Agreement between predicted and observed frequencies was assessed by calculation of unexplained variance for the velocity classes. The best predictors of the shape parameter were determined by using stepwise forward linear regression on averaged variables describing flow conditions (Lamouroux *et al.* 1995; Figure 20.11).

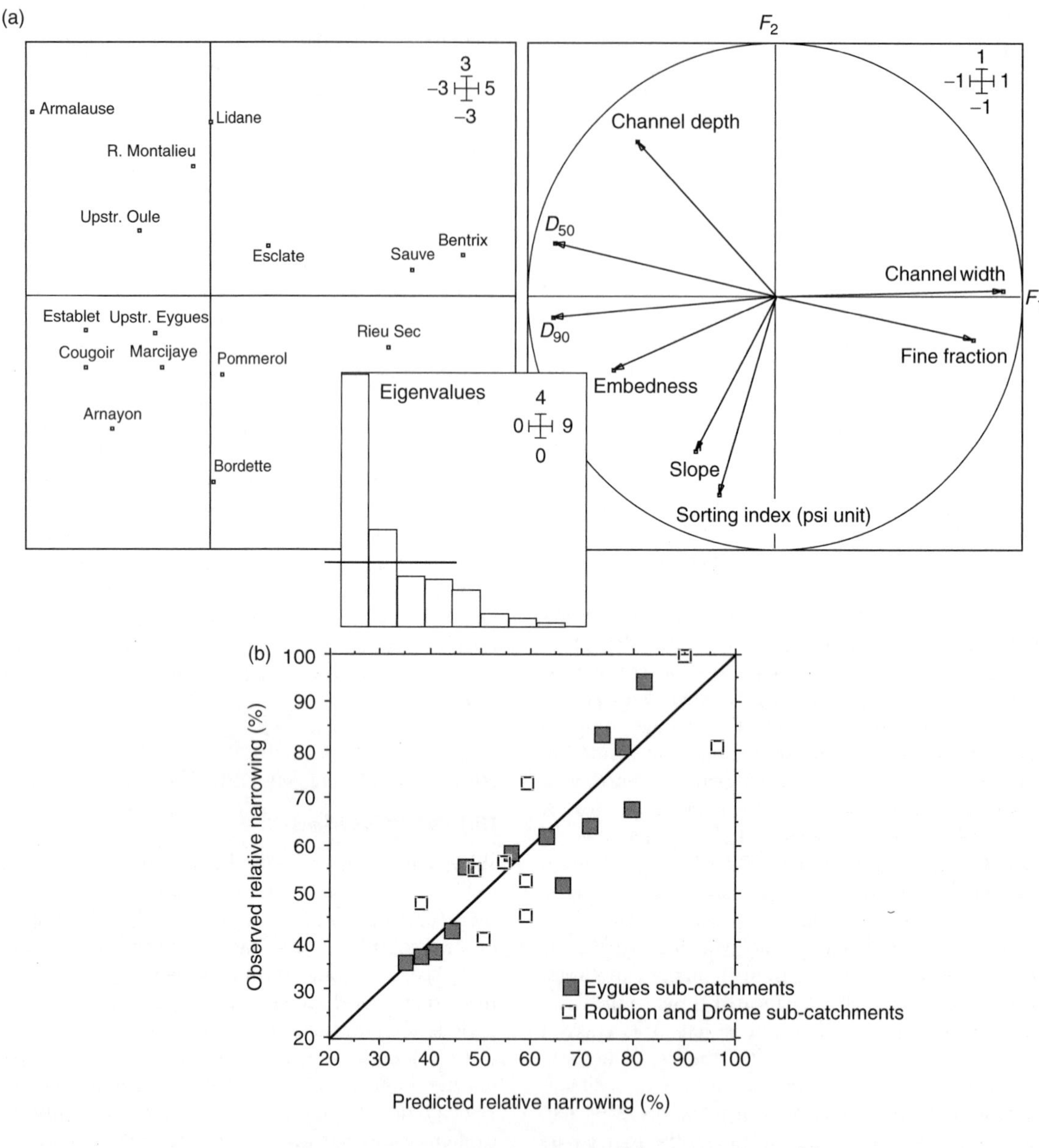

$$\text{Narrowing} = -82.95 + 128.02\,(F_1+5)^{-0.5} + 176.55\,(F_2+5)^{-0.5}$$
$$r^2 = 0.82;\ p = 0.0002$$

Figure 20.10 Prediction of channel narrowing from its grain size and geometry characteristics: (a) projection of the studied streams and variables describing the channel morphology on the first factorial of a normalized PCA and graph of the eigenvalues; (b) observations vs. predictions of channel narrowing between 1945 and 1995 from the first two components of the PCA. The model is performed on the Eygues sub-catchments (gray squares) and validated on the Roubion and Drôme sub-catchments (white squares)

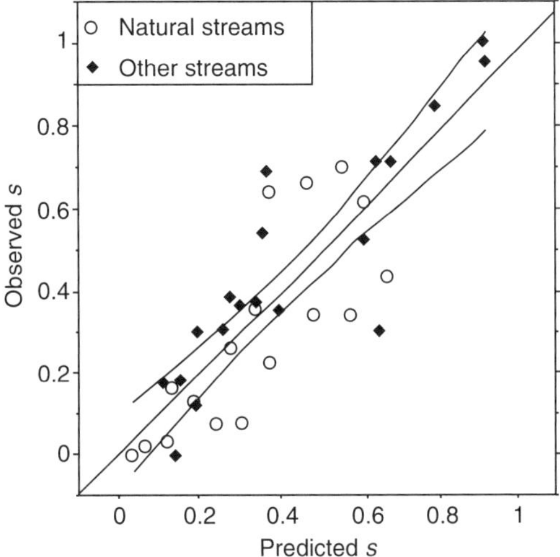

Figure 20.11 Observations vs. predictions of s from Froude number (*Fr*) and roughness/depth ratio (*D/H*), showing the 95% confidence interval of the linear regression of observations on predictions and the line of perfect agreement. Natural streams are those where neither morphology nor discharge were altered by human activities (from Lamouroux *et al.* 1995) (reproduced by permission of American Geophysical Union)

Markov Chain

A Markov chain approach can be used to assess transition probabilities in fluvial sediment transport, determined by stochastic flood events, and storage in reaches. Probabilities of particle transitions from one "storage reservoir" to others were derived from computed residence times in a stream of northern California by Kelsey *et al.* (1987). Applications included estimating flushing times of sediment out of a reach, and changes in sediment masses stored in the four sediment storage reservoirs. This approach yields better results with a longer hydrological data record (in this case 35 years), and with rapid rates of morphological changes in the channel, as in this example.

The Markov chain approach has also been used to highlight the longitudinal distribution of channel units (e.g., pools, riffles, rapids, cascades, steps) along stream reaches indicating that channel units occur in non-random two-unit sequences (Grant *et al.* 1990). Two-unit sequences have been described by a matrix of transition probability where each cell is the probability with which a given morphological unit is followed downstream by another given morphological unit (e.g., a pool followed by a riffle). Preferred sequences, defined as the positive difference between the observed sequences and those expected from random sequence, demonstrated that steps, cascades, and rapids are frequently followed by pools in Lookout Creek (Oregon). The sequences are slightly different in the steeper French Pete Creek (Oregon), where the cascade–pool and rapid–pool sequences are infrequent, but riffle-cascade sequences are common, reflecting the higher gradient and supply of large boulders from debris flows (Figure 20.12).

The question of the nature of flow turbulence has been tackled through different statistical approaches: either turbulence is regarded as a spatially independent and temporally random phenomenon, thus without memory, or as a structurally coherent one. In the first case, descriptive parameters such as standard deviations of velocity or shear stress can be used, which is not appropriate if interactions between layers, internal structures and periodic behavior are considered. If so, transitions are inherited from previous states. Temporal fluctuations in streamwise and vertical velocities at different depths over a gravel bed can be analyzed in terms of a Markov process after coding positive or negative differences with average horizontal velocity or directions of vertical movement. Statistical properties of the Markov chain are then tested against the null hypothesis of equiprobability linked to absence of spatial structure. This hypothesis is invalidated by the demonstration of some more frequent states and transitions (Kirkbride and Ferguson 1995).

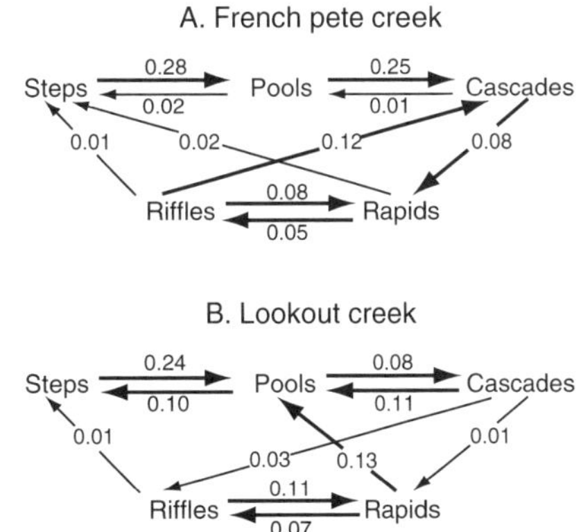

Figure 20.12 (Sequence relation) diagrams for French Pete Creek (A) and Lookout Creek (B) (Western Cascades, Oregon). Numbers shown are differences between observed and random transitional probabilities. Bolder arrows indicate transitional probabilities >0.05 (Grant *et al.* 1990; reproduced by permission of Geological Society of America)

Logistic Regression Modeling

Only a few examples have been published of the use of these techniques in fluvial geomorphology, but they consider different types of questions. Rice (1998) studied a set of torrential basins and assessed their potential bedload supply delivery to the main stem. A dichotomous variable, considered as the dependant variable, was then created with modalities 0 (insignificant delivery of sediment) and 1 (significant delivery of sediment). A logistic model was then developed to predict the probability that a torrent contributes significantly to the bedload supply of the main stem. The independent variables were relative basin area and the product of absolute basin area and slope. Another example focused on the sedimentation of former channels of the Ain River (France). Piégay *et al.* (2000) showed that the depth of sediment deposited in former channels does not depend on the age of the forms but on their geometry. While former meandered channels are the youngest, they exhibit higher sedimentation than former braided channels. A logistic regression model was applied to predict the probability that a former channel has originated from a meander channel or a braided channel according to its overbank sediment thickness.

Downs (1994, 1995) used logistic regression to predict the probability of channel adjustment (e.g., widening, shifting, deposition) from continuous (channel slope) and categorical variables (channel environment, channel regulation) describing some 285 reaches within the Thames Basin (England). He identified predictive models of channel response to natural and human controls. For example, the probability of in-channel deposition increases as slope decreases because the ability of the channel to transport sediment is reduced, but this probability is less in sand/gravel and urbanized basins than in basins with other land uses and substrate types. Silt deposition is the primary sedimentation process in the studied area.

Bledsoe and Watson (2001) used logistic regression to predict thresholds of channel pattern and instability, using the logistic curve as a visual tool to highlight the sensitivity of channel to shifting as a function of specific stream power relative to grain size (Figure 20.13).

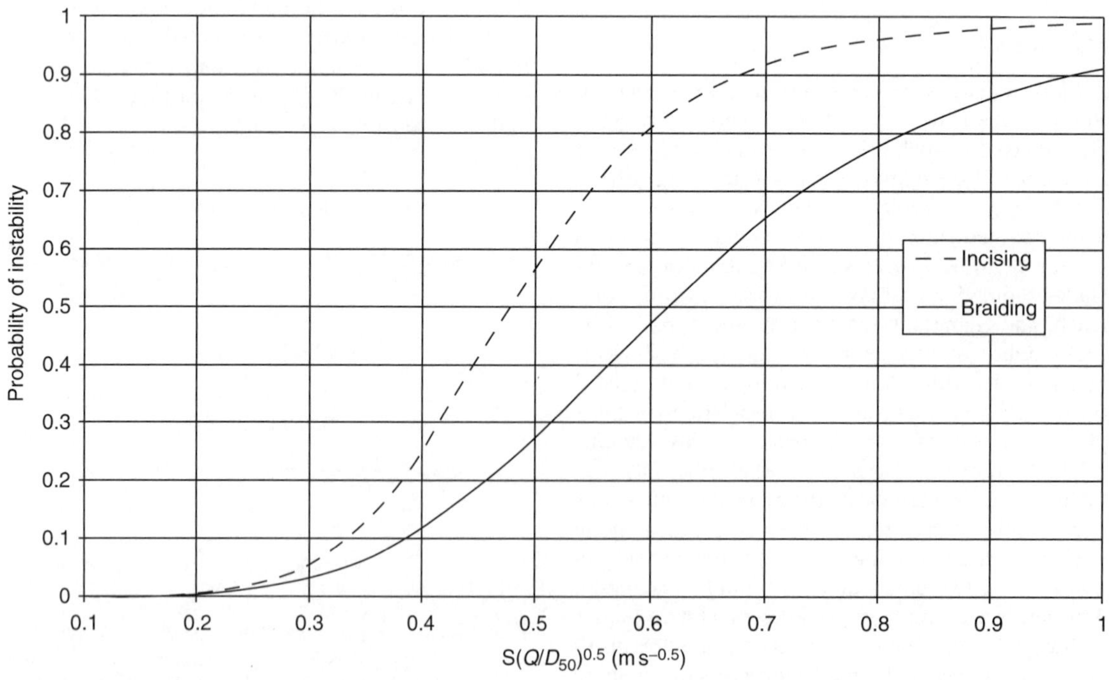

Figure 20.13 Logistic regression model predicting the probability of channel instability from three basic variables: the channel slope, S, the annual flood, Q, and the median grain size, D_{50}. Two particular channel features are distinguished: the incised reaches and the braided reaches (from Bledsoe and Watson 2001) (reproduced by permission of Elsevier Science)

20.8 TEMPORAL SERIES AND GEOSTATISTICS IN FLUVIAL GEOMORPHOLOGY

Analysis of temporal series has not been widely used in fluvial geomorphology, mostly because long-term series are usually not available on which such statistics can be applied to assess thresholds or determine periods. Most of the chronological approaches have been developed on hydrological records, notably to assess the flood recurrence interval of the peak flows based on Gumbel or Pearson probabilities (see Gordon *et al.* 1992 for an introduction). Studies of non-stationarity of discharge series can be used to detect changes occurring during the last decades and century due to human or climate modifications within catchments. The gaging station of Luc-en-Diois (Drôme, France) has recorded daily discharge (Q_d) since 1907. The series of annual peak flows was studied for possible breaks and homogeneous time sequences using the Buisband, Lee and Heghinian, and Pettit tests, and using the Hubert segmentation algorithm. The segmentation was performed for 2, 3, 4, and 5 segments, each of which was required to include at least 5 events. For the annual peak flow series, the test of Pettit identified a possible change in the 1930s without clear break according to other statistics. When using a variable describing the form of morphogenic peak flow events ($Q_{di} > Q_{d1.5}$), such as the residuals of the linear relationship between the flood water volume and the flood duration, most of the stationarity tests used (mainly Hubert, Pettit, Lee and Heghinian) validated a break in the trend in the 1930s (Figure 20.14). Using different variables to describe peak flows and using different threshold tests pointed to a statistically valid hydrological change in the basin in the middle of the 20th century. The change was less pronounced in magnitude than in

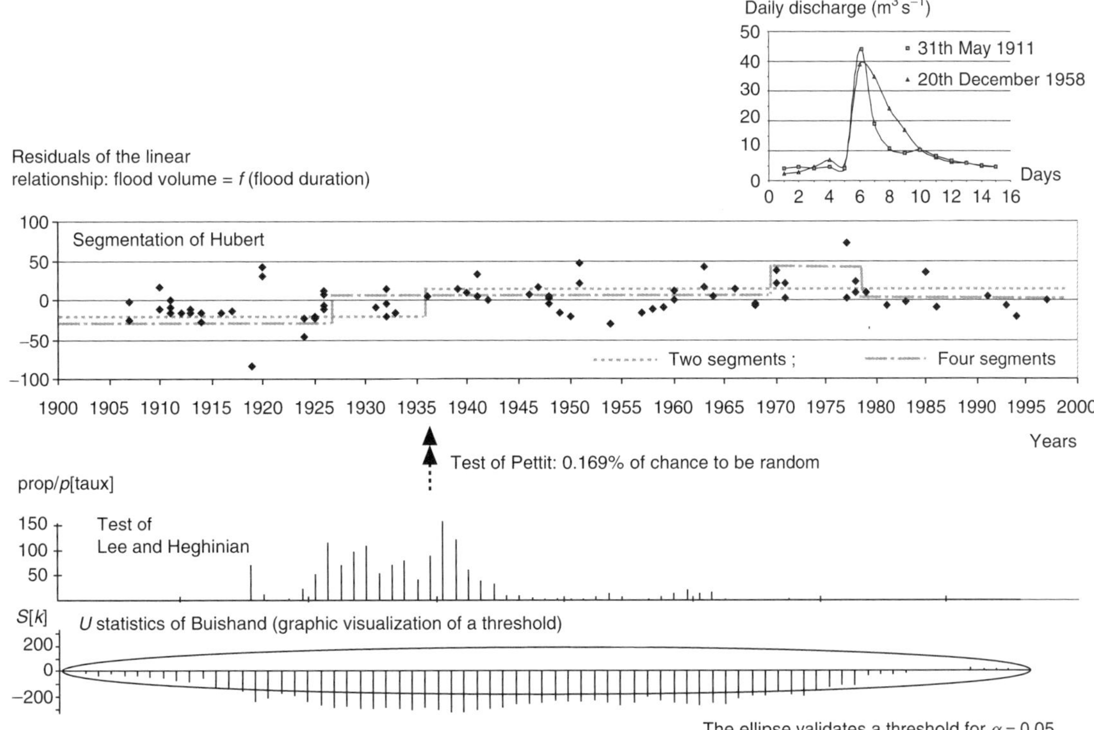

Figure 20.14 Threshold tests and Hubert segmentation performed on the series of the daily discharge of the Luc-en-Diois gaging station (1907–1997). The extracted variable is the residuals of the relationship between the flood volume and the flood duration. Each selected flood event is higher than $Q_{1.5}$ in recurrency and the flood duration is calculated using a threshold discharge of $10\,m^3\,s^{-1}$ as base level

the shape of the flood hydrographs. Floods were flashier, with sharper peaks before the 1930s than after, consistent with the hydrologic response expected from catchment afforestation and cessation of grazing, which occurred two to three decades before.

When studying spatial structure, various tests can be performed, including threshold tests. We applied the Hubert segmentation on the first factorial factors of a normed PCA summarizing variables describing longitudinal pattern of the channel shifting of the Willamette River (Oregon) (Figure 20.15). Maps of the channel reach (from Eugene to the confluence with the Columbia River) at four points in time (1850, 1895, 1932, and 1995) were cut into 1-km long sections. The channel maps were overlayed using a GIS system and we extracted channel change variables (e.g., channel narrowing, eroded floodplain area, constructed floodplain area) during the three periods (1850–1895, 1895–1932, and 1932–1995). The segmentation analysis on the first factorial axis defined homogeneous reaches in terms of channel shifting (e.g., the spatial structure) whatever the temporal trend (Figure 20.15). When using the segmentation in four segments (statistically validated on the Wald threshold $\alpha = 0.05$), from km 17 to km 80 (Saint Paul), the river was characterized by a very stable channel, whereas from km 81 to km 100 (ca Salem), as well as from km 150 to km 223 (downstream Eugene), the river channel is highly mobile whatever the period concerned. Between km 100 and km 150, the pattern is a little more contrasted longitudinally but much less mobile than in its two neighborhood segments. The second factorial axis provides the spatio-temporal changes: downstream from Salem, no real change occurred in channel shifting, whereas the channel underwent narrowing from 1895 to 1932 between km 110 (Salem) to km 160 (Albany), and of Monroe (km 200). More recently (1932–1995), the channel narrowed in the reach from km 110 to km 160.

The fractal analysis, which is another way to tackle with spatial structure by introducing a scaling perspective, has mainly focused on drainage network (Gao and Xia 1996). Amongst the 13 papers published in *Water Resources Research* between 1987 and 1997 dealing with fractal analysis, 11 concerned drainage networks.

In another application of these techniques, Nestler and Sutton (2000) used fractals to characterize cross-sectional distributions of area and energy as a function of scale to evaluate effects of river regulation on aquatic habitat. For a cross-section of the Missouri River, the authors plotted an energy–area graph showing the modification of historical habitats by regulation works (Figure 20.16). Under intermediate flows ($906\,m^3\,s^{-1}$), the existing conditions no longer contain large-scale habitat components (oval pictograms) that were present in past.

Spatial autocorrelation functions have not been widely used to describe spatial structures of fluvial forms, but the few examples have mainly addressed the regularity of fluvial facies (pool, riffle) along the long profiles. Madej (1999) used Moran's I to detect the presence and scale of significant spatial autocorrelation of bed elevations and also to evaluate the distance above which a bed elevation value is statistically independent from its neighboring ones. An innovative approach is illustrated by Clifford *et al.*'s (1992) use of both autocorrelation functions and periodograms to evaluate the geographical scale of roughness elements (e.g., grains and bedforms). Then they integrated their results into hydraulic roughness formulae to predict mean velocity.

Aubry and Piégay (2001) described examples of using spatial autocorrelation functions to describe longitudinal complexity of channel geometry (e.g., trend or repetition of characteristics) and spatial structure of basins (elevation, geological features). Nevertheless, the limits of these techniques argues for simultaneous use of several functions to establish robust conclusions. We used three sets of simulated data (absence of spatial structure, periodic spatial structure, linear gradient) to compare different autocorrelation functions (Geary's c, non-ergodic covariance, and correlation). Because each considers the local variance differently, the three functions provide different patterns. In the case of linear gradient, the c of Geary grows as a parabolic branch underlining the existing trend, while the non-ergodic covariance is bounded, and the non-ergodic correlation is 0 whatever the distance. As a consequence, the lag distance strongly depends on the statistics used; from 28 km (c of Geary) to 10–15 km (non-ergodic covariance) for longitudinal variation of channel incision. In the case of two-dimensional grid data, omnidirectional analysis can provide an autocorrelation lag which is lower than those provided by one of the directional analysis when the variable is characterized by a geographical orientation. Thus, simultaneous use of the different statistics and use of various directions (rather than a simple omnidirectional analysis) are advisable.

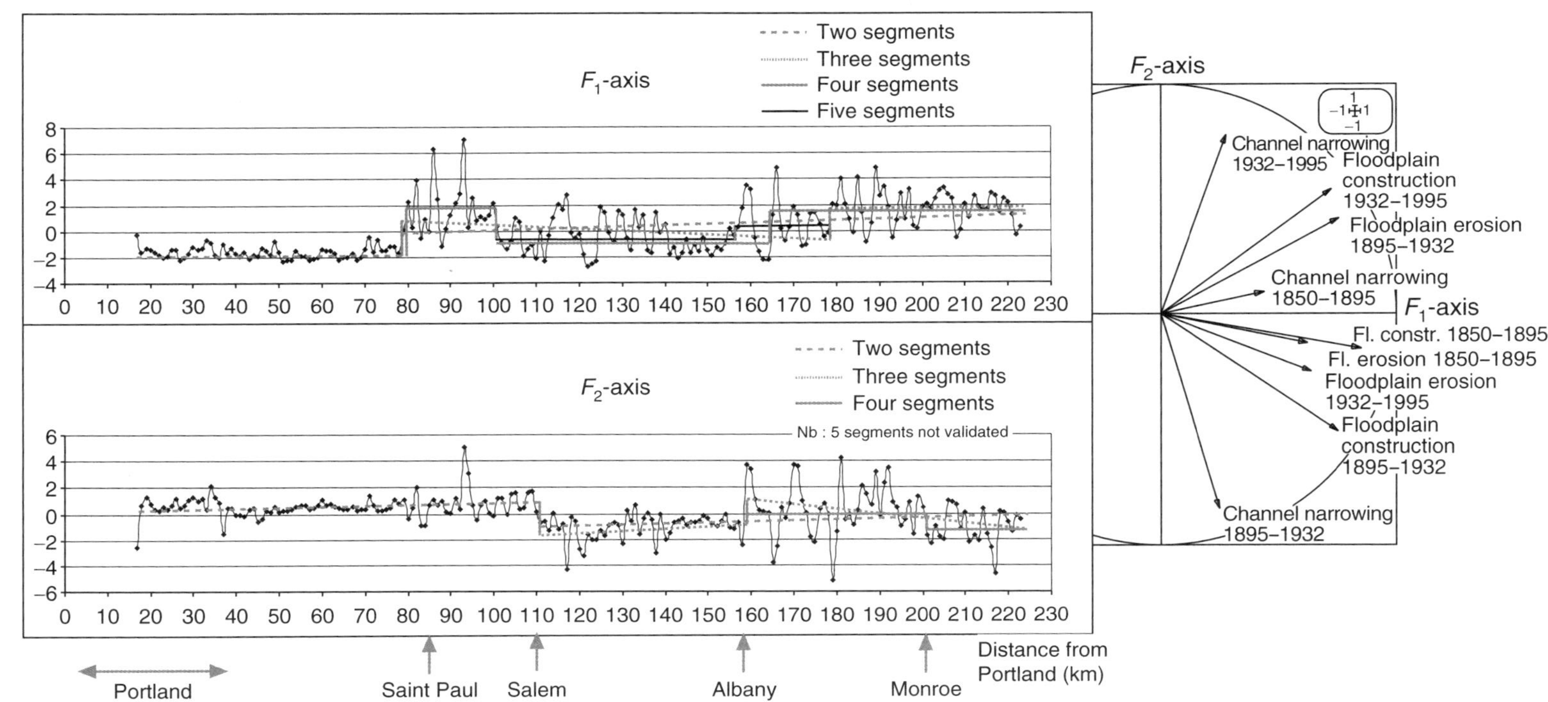

Figure 20.15 Hubert segmentation performed on the longitudinal continuum of the Willamette River between its confluence with the Columbia to Eugene (223 km). Each elementary channel segment is 1 km long and described from four different maps. A previous data table of 207 slices × 9 variables describing channel planform changes have been summarized by a normed PCA to two main factors which are then segmented using the segmentation test of Hubert (1989).

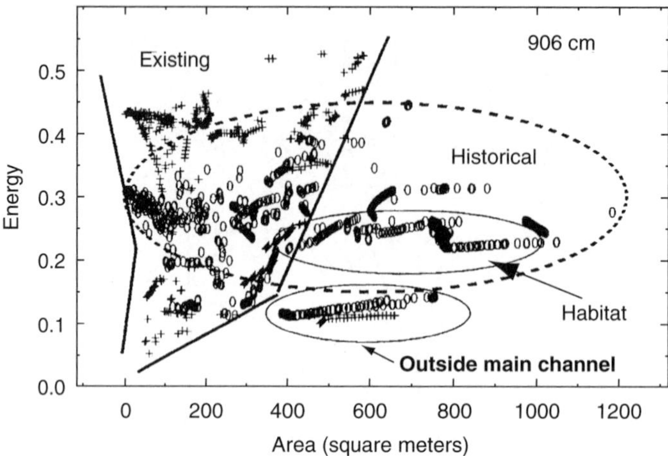

Figure 20.16 Energy—area envelops for a Missouri River cross-section (km 1254.3) for intermediate flow ($903\,m^3\,s^{-1}$), highlighting the loss of large-scale low energy habitat components (oval pictograms) between 2000 and the pre-regulation period (from Nestler and Sutton, 2000) (reproduced by permission of John Wiley and Sons, Ltd.)

20.9 RELEVANCE AND LIMITATIONS OF STATISTICAL TOOLS

Tables 20.7 and 20.8 summarize the general advantages and constraints of statistical tools in fluvial geomorphology. With increasing amounts of information (considerably increased by automatic recording and environmental data banks), classified objects must be compared both in space and time (e.g., sediment yields in different basins and at different flows). Differences and ordering must be statistically validated before any interpretation.

Fluvial geomorphology deals with deterministic and random events. The stochastic nature of the physical processes involved, for instance, in sediment entrainment and transport, is illustrated by Einstein's (1942) bedload formula. Randomness may come from convergence (different combinations of the so-called independent variables may give similar results) or from interference between subsystems.

Inherent Errors

Inherent errors are linked to observations and measurements in as much as field data collection is hampered by spatial variability, temporal variability, and resource availability. Variations in sediment supply rates may reflect the passage of bed forms and groups of particles. Sampling should last long enough to take short-term fluctuations into account, and when using point samplers, sampling should be dense enough to characterize variations across the streams (Gomez and Church 1989). Inevitably measurement errors are not always precisely or equally known for all the included variables. Moreover, they can be correlated with random errors, such as in regressions that associate discharge with width, depth, and velocity (Rhoads 1992).

Assumptions and Accuracy

If sampling is representative, the real population distribution can be assessed through distribution laws. For instance, distances of movement of tagged bed particles during floods can be compared with the two-parameter gamma distribution and the Einstein–Hubbell–Sayre model, and agreement or disagreement between models can be verified through χ^2 tests (Gintz *et al.* 1996). Linearization of data through some transformation is often required according to theoretical considerations or to previous observations. Geometric progressions in data sets, such as length, area, and slope angle, entail log, square root or dimensionless normalizations. Consequently, statistical and graphical residual analysis is recommended in order to verify some requirements such as linearity and homoscedasticity. They may also indicate the existence of thresholds.

When dealing with fluvial processes, time distribution laws are not always easy to define, especially in time-trend analysis because of inadequate or short records. Extreme events are absent or are not sufficiently represented. Stratigraphical lacustrine records

Table 20.7 Relevance of statistical tools

Pedagogy	Statistics and graphic extensions can highlight phenomenon, at first sight unclear, because of simplification and ordering capability. It is very useful to do comparisons between geomorphic forms and events
Objectivity	Statistics permit reduction of subjectivity and enlarge discussion of results regarding other controlling factors
Flexibility	The number of statistical tools is so large that it is always possible to find one which might help you to interpret your information if it is numerical. Moreover, it is possible to create specific statistics for a given question. With the increase in computer capacity, randomization tests considerably enlarge the statistical relevance
Prediction	Statistics could produce models that evaluate sensitivity of fluvial systems to disturbance
Multiple-scale capability	Statistics can consider elements/objects at different temporal and spatial scales and can also examine interaction of scales. Moreover, statistics enable large-scale, holistic approaches
Interdisciplinary	Statistics are widely used in environmental sciences (geology, geography, ecology, hydrology) and are consequently a mean for interdisciplinary collaboration, which is presently a key concept for success in applied research

Table 20.8 Constraints and limitations of statistical tools

In the formulation of scientific hypothesis	The use of statistics may impose a specific framework for sampling, data collection, interpretation, and validation of hypotheses. Is the scientific question confirmed by statistical tools or do the tools guide the scientific question?
Assumption and accuracy of tools	Standard statistical tools are based on preliminary hypotheses (e.g., linearity, homoscedasticity, independence, normality). Some tests are robust, others are more sensitive concerning the maintenance of these assumptions. There are some constraints for using statistical tools when there are gaps into the data, or when we get a mixture of categorical and continuous data
Results	The quality of the results is not dependent on whether or not one uses statistics but on the quality and originality of the geomorphological question one poses and solves
Prediction	Models are empirical, and not necessarily related with the physical laws which control the geomorphological processes. As a consequence, they are often limited in context and use. Numerical models which are based on physical laws, may need many assumptions and simplifications to be accurate, which also limits their use to a particular functional context
Psychology	The use of statistics may be suspect. Where is the verification? It is often said that statistics can demonstrate all and can hide the poorness of the results and the data

may act as substitutes for discharge or rainfall records in recurrence interval definition. Such events can be considered as partial duration series. A limiting condition is that exceedance probabilities may not be defined for all the accumulated layers if they do occur more often than once a year. This restricts their use to dry areas (Laronne 1990). Long records may include events that are caused by climatic exceptions not following the "common" law (for instance, cyclones out of their ordinary tracks). They can be inhomogeneous in the case of climatic changes, especially in transition zones that are more sensitive to any climatic shift, or human modification of vegetation cover.

Statistical distributions of variables within populations are close to mathematical models that can be determined through probability theory, thus providing useful simplifications. Linear regression is a common example in which assumptions are made about the scatter distribution of the probable value of the dependent variable corresponding to a single value of the independent one. Distinction between dependent and independent variables may be hypothetical and is often based upon qualitative field knowledge. The interaction of fluvial variables makes the latter distinction somehow artificial so that other names are preferred such as response variable and regressor. Some authors have included normalized or dimensionless data to produce better fits or better agreements with theoretical statements (Church and Mark 1980). Most of the relations follow a log–log or semi-log form (Tables 20.5 and 20.6). Consequently, correction factors should be applied as when using detransformed log functions (Ferguson 1986b). However, statistical bias may be obtained through transformations specially when data are not homogeneous in their distribution. The use of dimensionless variables may cause spurious correlations if some common scaling appears on both sides of the equation (Rhoads 1992).

Relationships between variables verify, or sometimes describe, physical laws in process–response systems, often hydraulic relations; functional or structural analysis should be used instead of least squares regression (Mark and Church 1977). In morphological system analysis, such relationships are highly empirical. When they derive from empirical observations and measurements, they are valid for defined geographical areas or the value range of the regressor variables (Hey and Thorne 1983). Some of them do not show any clear causality because of multicolinearity: one or several other factors may determine the introduced variables. Such is the case for the correlation between channel width and meander wavelength that are both dependent on flow dynamic characteristics (Chorley and Kennedy 1971).

Residual Significance

Although coefficients can be statistically valid because of large samples, relations are often blurred by data scattering that poses the problem of residual interpretation and of further processing. Sometimes only larger residuals are commented upon and eventually withdrawn from regressions if found to be exceptions (Lecce 1997). Data sets can be partitioned according to various criteria, such as distance from sources, planform—e.g., braiding and meandering, lithology, bank deposits, or vegetation (Hey and Thorne 1983, Ferguson 1986a). These categories are considered to be discriminant variables although corresponding statistics are not employed. Selection of bounds is consequently subjective while degrees of freedom are reduced. In a more general way, residuals can be classified through cluster analysis.

The significance of noise, if not due to errors in data collection (in such cases, they can help in improving the defective methods), may be due to hidden relationships. This is often linked to the existence of gray boxes. Introduction of new or more adapted variables in the model is then performed if possible. However, the more complex the statistical tools, the more they require a large number of data for validation.

Regional Heterogeneity and Scale Effects

Reconstitution and prediction relations are ultimately sought in applied geomorphology. The utilization of multivariate analysis in preliminary studies can indicate which are the most significant and predictive variables before extending the area of research. However, the choice may be refuted because of regional heterogeneity. Some methods may be better suited than others: for instance, additive error power functions (Rhoads 1992). Extrapolation must be validated, therefore, according to probability laws.

An equilibrium state within a hydrosystem may exist between its elements and forms, as well as the interaction of its components (Ebisemiju 1988). Mass equilibrium is more rapidly attained in small watersheds (Church and Mark 1980), which implies that catchment comparisons should consider size effects. Strong correlations could be interpreted as evidence of such an equilibrium. However, complexity arises from unequal responses to process changes in the various subsystems so that deviations could be clues either of inherited features or of different, sometimes unsuspected, behavior. Multiple regressions should be used only in defining response equations in which the net influence of regressors on interacting variables is described (Hey 1978).

20.10 CONCLUSIONS

Statistics is a universal tool whose language is understood by scientists whatever their discipline. By allowing the scientist to interpret environmental data with high temporal and spatial variability, statistics complements physical and experimental-based methods, facilitates interpretation of fluvial forms and processes in their diversity (regional, longitudinal, size, time). Variability is not measurement imprecision or experimental error, but is a fundamental attribute of environmental data, which merits specific approaches for assessment and understanding. Statistical tools can support causal research, predictions, and help provide an experimental framework where hypotheses can be formulated, tested, and validated, allowing to produce laws and then theories.

Application of statistical tools in fluvial geomorphology has advantages of reducing subjectivity, eliminating assumptions, facilitating comparison between different spatial and temporal data sets of large sizes, refining data collection and eventually disclosure of exceptions or new relations, prediction power, and improvement of system analysis. In a systems approach, statistical tools have become more applicable through the increasing diffusion of computational capacity and increasingly practical statistical software. Given the mixed nature of data sets increasingly used by fluvial geomorphologists to understand process and form, the more sophisticated statistical tools can offer benefits to research in the field. To develop more realistic descriptions of fluvial morphological systems, process–response systems, time and space trends, size effects, will require collection of sufficient data and more thought about their relevance. Appropriate statistical analyses can contribute to interpretation of these data.

However, the use of statistical tools to date had not been entirely satisfactory. As only one type of tool among many, statistics cannot solve every geomorphological question. In fluvial geomorphology, statistical applications to date have been dominated by linear regression, criticized as a "black-box" empirical approach. Another criticism of statistics concerns the triteness of many interpretations, and the seeming possibility of concluding everything and its contrary. When conducting these studies, errors can be made at the stage of experimental design and choice of the statistics used, usually traceable to misunderstanding of the purpose and limitations of the statistical analyses. This trap can generally be avoided by collaborating with statisticians. Other errors occur when interpreting the results, notably in causal interpretation when the researcher may ask the data to say what they cannot say.

ACKNOWLEDGMENTS

J. Buffington, P. Downs, G. M. Kondolf, N. Lamouroux, S. Rice, and E. Wohl read the manuscript and gave many useful comments. L. Ashkenas, S. Dufour, S. Gregory, N. Landon, F. Liébault and A. Segaud contributed to the data collections used to exemplify. P. Aubry and E. Leblois contributed to clarify the questions of spatial autocorrelation, also segmentation and threshold tests.

REFERENCES

Aubry, P. and Piégay, H. 2001. Pratique de l'analyse de l'autocorrélation spatiale en géomorphologie fluviale: définitions opératoires et test. *Géographie Physique et Quaternaire* 55(2): 115–133.

Batalla, R.J. and Sala, M. 1995. Effective discharge for bedload transport in a subhumid Mediterranean sandy gravel-bed river (Arbucies, NE Spain). In: Hickin, E.J., eds., *River Geomorphology*, Chichester: John Wiley and Sons, pp. 93–103.

Batalla, R.J. 1997. Evaluating bed-material transport equations using field measurements in a sandy gravel-bed stream, Arbucies River, NE Spain. *Earth Surface Processes and Landforms* 22: 121–130.

Bledsoe, B.P. and Watson, C.C. 2001. Logistic analysis of channel pattern thresholds: meandering, braiding, and incising. *Geomorphology* 38: 281–300.

Bravard, J.P. and Petit, F. 1997. *Les cours d'eau. Dynamique du système fluvial.* Paris: Armand Colin, 222 p.

Brunel, S. 2000. *Utilisation de l'hydrologie quantifiée pour l'étude des changements fluviaux et de l'éventuelle discrimination de leurs causes climatiques et anthropiques dans le cas de 10 cours d'eau français*, Unpublished D.E.A. Interface Nature Société, University of Lyon 3 and Cemagref, 157 p.

Buisband, T.A. 1984. Tests for detecting shift in the mean of hydrological time series. *Journal of Hydrology* 73: 51–69.

Chessel, D. and Mercier, P. 1993. Couplage de triplets statistiques et liaisons espèces-environnement. In: Lebreton, J.D. and Asselain, B., eds., *Biométrie et Environnement* Masson, Paris, pp. 15–43.

Chorley, R.J. and Kennedy, B.A. 1971. *Physical Geography. A Systems Approach*, London: Prentice-Hall, 370 p.

Clifford, N.J., Robert, A. and Richards, K.S. 1992. Estimation of flow resistance in gravel-bedded rivers: a physical explanation of the multiplier of roughness length. *Earth Surface Processes and Landforms* 17: 111–126.

Church, M. and Mark, D.M. 1980. On size and scale in geomorphology. *Progress in Physical Geography* 4: 342–390.

Dawson, M. 1988. Sediment size variation in a braided reach of the Sunwapta River, Alberta, Canada. *Earth Surface Processes and Landforms* 13: 599–618.

Downs, P.W. 1994. Characterisation of river channel adjustments in the Thames basin, south-east England. *Regulated Rivers: Research and Management* 9: 151–175.

Downs, P.W. 1995. Estimating the probability of river channel adjustment. *Earth Surface Processes and Landforms* 20: 687–705.

Dunne, T. and Leopold, L.B. 1978. *Water in Environmental Planning*, San Francisco: W.H. Freeman, 818 p.

Dury, G.H. 1964. *Principles of Underfit Streams*, US Geological Survey Paper 425A, 67 p.

Ebisemiju, F.S. 1988. Canonical correlation analysis in geomorphology with particular reference to drainage basin characteristics. *Geomorphology* 1: 331–342.

Ebisemiju, F.S. 1989. A morphometric approach to gully analysis. *Zeitschrift für Geomorphologie* 33: 307–322.

Einstein, H.A. 1942. Formulas for the transportation of bed-load. *Transactions of American Society of Civil Engineering* 107: 561–577.

Ferguson, R.I. 1986a. Hydraulics and hydraulic geometry. *Progress in Physical Geography*: 1–29.

Ferguson, R.I. 1986b. River loads underestimated by rating curves. *Water Resources Research* 22: 74–76.

Gao, J. and Xia, Z. 1996. Fractals in physical geography. *Progress in Physical Geography* 20: 178–191.

Gintz, D., Hassan, M.A. and Schmidt, K. 1996. Frequency and magnitude of bedload transport in a mountain river. *Earth Surface Processes and Landforms* 21: 433–445.

Gomez, B. and Church, M. 1989. An assessment of bedload sediment transport formulae for gravel bed rivers. *Water Resources Research* 25: 1161–1186.

Gordon, N.D., MacMahon, T.A. and Finlayson, B.L. 1992. *Stream Hydrology, an Introduction for Ecologists*, Chichester: John Wiley and Sons, 526 p.

Graf, W.H. 1971. *Hydraulics of Sediment Transport*, New York: McGraw-Hill, 544 p.

Gregory, K.J. and Walling, D.E. 1973. *Drainage Basin Form and Process. A Geomorphological Approach*, London: Edward Arnold, 458 p.

Grant, G.E., Swanson, F.J. and Wolman, M.G. 1990. Pattern and origin of stepped-bed morphology in high-gradient streams, Western Cascades, Oregon. *Geological Society of America Bulletin* 102: 340–352.

Gurnell, A.M. 1997. Adjustments in river channel geometry associated with hydraulic discontinuities across the fluvial-tidal transition of a regulated river. *Earth Surface Processes and Landforms* 22: 967–985.

Hack, J.T. 1957. *Studies of Longitudinal Stream Profiles in Virginia and Maryland*, Geological survey professional paper 294-B, pp. 45–91.

Hardisty, J. 1993. Time series analysis using spectral techniques: oscillatory currents. *Earth Surface Processes and Landforms* 18: 855–862.

Hassan, M.A., Church, M. and Ashworth, P.J. 1992. Virtual rate and mean distance of travel of individual clasts in gravel-bed channels. *Earth Surface Processes and Landforms* 17: 617–627.

Hey, R.D. 1978. Determinate hydraulic geometry of river channels. *Journal of Hydraulics Division* 104(HY6): 869–885.

Hey, R.D. and Thorne, C.R. 1983. Hydraulic geometry of gravel-bed rivers. In: *Proceedings, 2nd International Symposium on River Sedimentology, Nanjing*, pp. 713–723.

Hoey, T. 1992. Temporal variations in bedload transport rates and sediment storage in gravel-bed rivers. *Progress in Physical Geography* 16: 319–338.

Horton, R.E. 1945. Erosional development of streams and their drainage basin; hydrophysical approach to quantitative morphology. *Geological Society of America Bulletin* 56: 275–370.

Howard, A.D. and Hemberger, A.T. 1991. Multivariate characterization of meandering. *Geomorphology* 4: 161–186.

Hubert, P. 1989. Segmentation des séries hydrométéorologiques: application à des séries de précipitations et de débits de l'Afrique de l'ouest. *Journal of Hydrology* 110: 349–367.

Ibanez, J.J., Perez-Gonsalez, A., Jimenez-Balesta, R., Saldana, A. and Gallardo-Diaz, J. 1994. Evolution of fluvial dissection landscapes in mediterranean environments. Quantitative estimates and geomorphological, pedological and phytocenotic repercussions. *Zeitschrift für Geomorphology* NF 38(1): 105–120.

Kelsey, H.M., Lamberson, R. and Madej, M.A. 1987. Stochastic model for the long-term transport of stored sediment in a river channel. *Water Resources Research* 23: 1738–1750.

Kesel, R.H., Yodis, E.G. and McCraw, D.J. 1992. An approximation of the sediment budget of the Lower Mississippi River prior to major human modification. *Earth Surface Processes and Landforms* 17: 711–722.

Kirkbride, A.D. and Ferguson, R. 1995. Turbulent flow structure in a gravel-bed river: Markov chain analysis of the fluctuating velocity profile. *Earth Surface Processes and Landforms* 20: 721–733.

Knighton, A.D., Woodroffe, C.D. and Mills, K. 1992. The evolution of tidal creek networks, Mary River, Northern Australia. *Earth Surface Processes and Landforms* 17: 167–190.

Kunhle, R.A. 1992. Bed load transport during rising and falling stages on two small streams. *Earth Surface Processes and Landforms* 17: 191–197.

Lamouroux, N., Souchon, Y. and Herouin, E. 1995. Predicting velocity frequency distributions in stream reaches. *Water Resources Research* 31: 2367–2375.

Laronne, J.B. 1990. Probability distribution of event sediment yields in the Northern Negev, Israël. In: Boardman, J., Foster, D.L. and Dearing, J.A., eds., *Soil Erosion on Agricultural Land*, Chichester: John Wiley and Sons, pp. 481–492.

Lebart, L., Morineau, A. and Piron, M. 1995. *Statistique Exploratoire Multidimensionnelle*, Paris: Dunod, 439 p.

Lee, A.F.S. and Heghinian, S.M. 1977. A shift of the mean level in a sequence of independent normal random variables: a Bayesian approach. *Technometrics* 19(4): 503–506.

Lecce, S.A. 1997. Spatial patterns of historical overbank sedimentation and floodplain evolution, Blue River, Wisconsin. *Geomorphology* 18: 265–277.

Leopold, L.B. and Maddock, T. 1953. *The Hydraulic Geometry of Streams Channels and Some Physiographic Implications*, US Geological Survey Professional Paper 252, pp. 1–57.

Leopold, L.B. and Wolman, M.G. 1957. *River Channel Patterns; Braided, Meandering and Straight*, US Geological Survey Professional Paper 282-b, pp. 39–85.

Leopold, L.B., Wolman, M.G. and Miller, J.P. 1964. *Fluvial Processes in Geomorphology*, San Francisco: W.H. Freeman, 522 p.

Liébault, F., Clément, P., Piégay, H., Rogers, C.F., Kondolf, G.M. and Landon, N. 2002. Contemporary channel changes in the Eygues Basin, southern French Prealps:

causes of regional variability according to watershed characteristics. *Geomorphology* 45: 53–66.

Llorens, P., Queralt, I., Plama, F. and Gallart, F. 1997. Studying solute and particulate sediment transfer in a small Mediterranean mountaimous catchment subject to land abandoment. *Earth Surface Processes and Landforms* 22: 1027–1035.

Madej, M.A. 1999. Temporal and spatial variability in thalweg profiles of a gravel-bed river. *Earth Surface Processes and Landforms* 24: 1153–1169.

Malmon, D.V., Dunne, T. and Reneau, S.L. 2002. Predicting the fate of sediment and pollutants in river floodplains. *Environmental Science and Technology* 36: 2026–2032.

Manly, B.F.J. 1991. *Randomization and Monte Carlo Methods in Biology*, London: Chapman & Hall, 281 p.

Mark, D.M. and Church, M. 1977. On the misuse of regression in Earth Science. *Mathematical Geology* 9: 63–75.

McPherson, H.J. 1971. Downstream changes in sediment character in a high energy mountain stream channel. *Artic and Alpine Research* 3: 65–79.

Nakamura, F. and Swanson, F.J. 1993. Effects of coarse woody debris on morphology and sediment storage of a mountain stream system in western Oregon. *Earth Surface Processes and Landforms* 18: 43–61.

Nakato, T. 1990. Tests of selected sediment-transport formulas. *Journal of Hydraulic Engineering* 116: 362–379.

Nestler, J. and Sutton, V.K. 2000. Describing scales of features in river channels using fractal geometry concepts. *Regulated Rivers: Research and Management* 16: 1–22.

Newbury, R. and Gaboury, M. 1993. Exploration and rehabilitation of hydraulic habitats in streams using principles of fluvial behaviour. *Freshwater Biology* 29: 195–210.

Oguchi, T. and Ohmori, H. 1994. Analysis of relationships, among alluvial fan area, source basin area, basin slope, sediment yield. *Zeitschrift für Geomorphologie* 38: 405–420.

Park, J. 1992. Suspended Sediment Transport in a Mountainous Catchment. *Sci. Rept., Inst. Geosci., University of Tsukuba, A* 13: 138–190.

Parker, G. 1976. On the cause and characteristic scales of meandering and braiding in rivers. *Journal of Fluid Mechanics* 76: 457–480.

Passmore, D.G. and Macklin, M.G. 1994. Provenance of fine-grained alluvium and late Holocene land-use change in the Tyne Basin, northern England. *Geomorphology* 9: 127–142.

Petit, F. 1987. The relationship between shear stress and the shaping of the bed of a pebble-loaded river, La Rulles, Ardennes. *Catena* 14: 453–468.

Petit, F. 1989. Evaluation des critères de mise en mouvement et de transport de la charge de fond en milieu naturel. *Bulletin de la Société Géographique de Liège* 25: 91–111.

Pettit, A.N. 1979. A non-parametric approach to the change-point problem. *Applied Statistics* 28(2): 126–135.

Piégay, H., Bornette, G., Citterio, A., Hérouin, E., Moulin, B. and Statiotis, C. 2000. Channel instability as control factor of silting dynamics and vegetation pattern within perifluvial aquatic zones. *Hydrological Processes* 14(16/17): 3011–3029.

Piégay, H., Bornette, G. and Grante, P. 2002. Assessment of silting-up dynamics of 11 cut-off channel plugs on a free-meandering river (the Ain River), France. In: Allison, R.J., ed., *Applied Geomorphology, Theory and Practice*, Chichester: John Wiley and Sons, pp. 227–247.

Poulos, S.E., Collins, M. and Evans, G. 1996. Water-sediment fluxes of Greek rivers, southeastern Alpine Europe: annual yields, seasonal variability, delta formation and human impact. *Zeitschrift für Geomorphologie* 40: 243–261.

Reid, I. and Frostick, L.E. 1986. Dynamics of bedload transport in Turkey Brook, a coarse-grained alluvial channel. *Earth Surface Processes and Landforms* 11: 143–155.

Reneau, S.L. and Dietrich, W.E. 1991. Erosion rates in the southern Oregon coast range: evidence for an equilibrium between hillslope erosion and sediment yield. *Earth Surface Processes and Landforms* 16: 307–322.

Rhoads, B.L. and Miller, M.V. 1991. Impact of flow variability on the morphology of a low-energy meandering river. *Earth Surface Processes and Landforms* 16: 357–368.

Rhoads, B.L. 1992. Statistical models of fluvial systems. *Geomorphology* 5: 433–455.

Rice, S. and Church, M. 1996a. Sampling surficial gravels: the precision of size distribution percentile estimates. *Journal of Sedimentary Research* 66(3): 654–665.

Rice, S. and Church, M. 1996b. Bed material texture in low order streams on the Queen Charlotte Islands, British Columbia. *Earth Surface Processes and Landforms* 21: 1–18.

Rice, S. 1998. Which tributaries disrupt downstream fining along gravel-bed rivers? *Geomorphology* 22: 39–56.

Richards, K. 1976. The morphology of riffle–pool sequence. *Earth Surface Processes and Landforms* 1: 71–88.

Richards, K. 1982. *Rivers: Form and Process in Alluvial Channels*, London: Methuen, pp. 358.

Rickenmann, D. 1997. Sediment transport in Swiss torrents. *Earth Surface Processes and Landforms* 22: 937–951.

Ridenour, G.S. and Giardino, J. 1995. Discriminant function analysis of computational data: an example from hydraulic geometry. *Physical Geography* 15: 481–492.

Robison, E.G. and Beschta, R.L. 1990. Coarse woody debris and channel morphology interactions for undisturbed streams in southeast Alaska. *Earth Surface Processes and Landforms* 15: 149–156.

Saporta, G. 1990. *Probabilités. Analyse des données et statistique*. Paris: Technip, 493 p.

Schmidt, K.H. and Ergenzinger, P. 1992. Bedload entrainment, travel lengths, step lengths, rest periods – studied with passive (iron, magnetic) and active (radio) tracer techniques. *Earth Surface Processes and Landforms* 17: 147–165.

Schumm, S.A. 1960. *The Shape of Alluvial Channels in Relation to Sediment Type*, US Geological Survey Professional Paper 352–B, pp. 17–30.

Schumm, S.A. 1977. *The Fluvial System*, New York: Wiley, 338 p.

Sear, D.A., 1996. Sediment transport processes in pool – riffle sequences. *Earth Surface Processes and Landforms* 21: 241–262.

Shields, N.D. 1936. Anwendung der Aehnlichkeitsmechanik und der Turbulenzforschung auf die Geschiebelerwegung. In: *Mitteilung der Preussischen Versuchanstalt fur Wasserbau und Schiffbau, Heft 26, Berlin.*

Silva, P.G., Harvey, A.M., Zazo, C. and Goy, J.L. 1992. Geomorphology, depositional style and morphometric relationships of Quaternary alluvial fans in the Guadalantin Depression (Murcia, Southeast Spain). *Zeitschrift für Geomorphologie* 36: 325–341.

Simon, A. 1995. Adjustment and recovery of unstable alluvial channels: identification and approaches for engineering management. *Earth Surface Processes and Landforms* 20: 611–628.

Smith, R.D., Sidle, R.C. and Porter, P.E. 1993. Effects on bedload transport of experimental removal of woody debris from a forest gravel-bed stream. *Earth Surface Processes and Landforms* 18: 455–468.

Stall, J.B. and Yang, C.T. 1970. Hydraulic geometry of 12 selected stream systems of the United States, University of Illinois Water Resources Center, Research Report 32, 73 p.

Strahler, A.N. 1952. Dynamic basis of geomorphology. *Geological Society of America Bulletin* 63: 923–938.

Strahler, A.N. 1954. Statistical analysis in geomorphic research. *Journal of Geology* 62: 1–25.

Thévenet, A., Citterio, A. and Piégay, H. 1998. A new methodology for the assessment of large woody debris accumulations on highly modified rivers (example of two French piedmont rivers). *Regulated Rivers: Research and Management* 14: 467–483.

Tropeano, D. 1991. High flow events and sediment transport in small streams in the tertiary basin area in Piedmont (Northwest Italy). *Earth Surface Processes and Landforms* 16(4): 323–340.

Tucker, L.R. 1958. An inter-battery method of factor analysis. *Psychometrika* 23: 111–136.

Walling, D.E. and He, Q. 1998. The spatial variability of overbank sedimentation on river floodplains. *Geomorphology* 24: 209–223.

Wharton, G. 1995. The channel-geometry method: guidelines and applications. *Earth Surface Processes and Landforms* 20: 649–660.

White, W.R., Milli, H. and Crabbe, A.D. 1973. Sediment transport: an appraisal of available methods, Report 119, UK Hydraulics Res. Stat., Wallingford.

Wilcock, P.R. 1993. Critical shear stress of natural sediments. *Journal of Hydraulic Engineering* 119: 491–505.

Williams, R.B.G. 1984. *Introduction to Statistics for Geographers and Earth Scientists*, New York: Macmillan (2 volumes).

Wilson, A.G. and Kirkby, M.J. 1980. *Mathematics of Geographers and Planners*, Oxford: Clarendon Press, 408 p.

Wohl, E.E., Antony, D.J., Masden, S.W. and Thompson, D.M. 1996. A comparison of surface sampling methods for coarse fluvial sediments. *Water Resources Research* 32: 3219–3226.

Wolman, M.G. 1954. A Method of Sampling Coarse River Bed Material. *Transactions, Am. Geophy. Union* 35: 951–956.

Wolman, M.G. and Miller, R.J.P. 1960. Magnitude and frequency of forces in geomorphic processes. *Journal of Geology* 68: 54–74.

Woodsmith, R.D. and Buffington, J. 1996. Multivariate geomorphic analysis of forest streams: implications for assessment of land-use impacts on channel condition. *Earth Surface Processes and Landforms* 21: 377–394.

Wrigley, N. 1985. *Categorical Data Analysis for Geographers and Environmental Scientists*, London: Longman.

Yodis, E.G. and Kesel, R.H. 1993. The effects and implications of base-level changes to Mississippi River tributaries. *Zeitschrift für Geomorphologie* 37: 385–402.

Young, R.W. 1985. Waterfalls: form and process. *Zeitschrift für Geomorphologie* 55(Suppl. Bd): 81–95.

Yu, L. and Oldfield, F. 1993. Quantitative sediment source ascription using magnetic measurements in a reservoir-catchment system near Nijar, S.E. Spain. *Earth Surface Processes and Landforms* 18: 441–454.

Part VII

Conclusion: Applying the Tools

21

Integrating Geomorphological Tools in Ecological and Management Studies

G. MATHIAS KONDOLF[1], HERVÉ PIÉGAY[2] AND DAVID SEAR[3]

[1]*Department of Landscape Architecture and Environmental Planning and Department of Geography, University of California, Berkeley, CA, USA*

[2]*CNRS – UMR 5600, Lyon, France*

[3]*Department of Geography, University of Southampton, Highfield, UK*

21.1 INTRODUCTION

Fluvial geomorphology can be useful to other disciplines, such as ecology (e.g., to provide a framework within which to analyze habitats) and engineers, as well as practitioners, such as planners and river managers (e.g., to understand risks and effects of flooding, or to regulate instream gravel extraction), and those who implement ecological restoration programs (e.g., through insights into the functioning of former ecosystems and constraints posed by human alterations). Geomorphological questions posed by other scientists and practitioners are often complex and merit being subdivided into a set of more specific questions. The physical, chemical, and biological interactions in river systems operate at multiple temporal and spatial scales implying that to understand relations or to solve problems typically requires application of multiple tools. Some of these tools are proper to geomorphology, while others were developed in allied fields (such as biology or engineering sciences) and are applied to geomorphological problems. These tools range widely in the temporal and spatial scales of application, from a few minutes or hours (the duration of the bedload movement during a flood event) to several centuries (the time needed for a fluvial system to adjust its geometry to a climate change), and from centimeters (benthic invertebrate habitat) to thousands of square kilometers (large river catchments).

Through the range of tools presented in this book, we have sought to provide a reference not only for the practicing geomorphologist and graduate student, but also to provide the manager and scientist working with geomorphologists with an idea of the range of approaches potentially available to address fluvial geomorphic problems. The purpose of this chapter is to provide a framework within which the tools can be used, and to present examples of application of geomorphic tools to problems in river management and restoration.

21.2 MOTIVATIONS FOR APPLYING FLUVIAL GEOMORPHOLOGY TO MANAGEMENT

"It should be possible to persuade decision-makers that incorporating historical or empirical geomorphic information into river management strategies is at least as valuable as basing decisions on precise, yet fallible mechanistic models" (Rhoads 1994). This statement captures the sense of potential for applied fluvial geomorphology that rose in concert with growth in environmental awareness and political will to recognize and account for the environment in land and water management. Since the late 1980s, applied fluvial geomorphology has risen up the operational and policy agendas of river management authorities, most recently propelled by the demands for "morphological" assessment in support of river restoration (Sear *et al.* 1995, Brookes and Shields 1996).

With increasing emphasis on environmental river management and interest in sustainable approaches to use of water (and other natural) resources, managers must base their decisions on insights from a variety of disciplines. Because fluvial geomorphology provides the overall framework within which habitats develop, ecological processes operate, floods propagate, and waters may (or may not) undergo purification en route to the river and downstream, geomorphological analyses are central to understanding many issues in river management, including maintenance and restoration of aquatic and riparian habitat, flood risk, and water quality. Specifically, fluvial geomorphologists are increasingly called upon to answer questions at different temporal and spatial scales than other disciplines have typically employed. Graf (1996) describes this recent resurgence of geomorphological application as the "return to its roots of a close association with environmental resource management and public policy", arguing that geomorphology is now mature enough after a period characterized by a focus on basic research, to begin applying this collective wisdom to issues of social concern.

The upsurge in the application of geomorphology has also been driven by the recognition of the costs, financial as well as environmental, of ignoring natural system processes and structure in river channel management (Gilvear 1999). Legislative and economic drivers aimed at reversing a trend of ecological degradation have begun to transform the way many agencies approach intervention in river systems. But as Newson (1988) has made clear, translation of science into policy frequently has long lead times, and uptake of policy at the operational level is probably much longer again. Furthermore, the trigger for any particular phase of uptake may be an externally imposed policy shift, which invites a subsequent scientific input, rather than a science advance that demands policy modification. The recent policy emphasis on sustainable river channel management (Raven *et al.* 2002) exemplifies a shift of stance driven by political pressure rather than scientific logic. Nevertheless, statutory requirements to take regard for "physiographic features" or "hydromorphology" (note the emphasis on the static descriptive nature of "geomorphology" in legislation, which lags 30–40 years behind the shift away from this position in the discipline) and the ecological integrity of river systems have focused attention on their natural form and function. Most recently, the rise of physical habitat restoration has provoked new research initiatives among engineers focusing on the hydraulic functions of river channel features, whilst ecologists are increasingly recognizing the value of geomorphology in describing and accounting for the habitat structure of aquatic systems (Jeffers 1998, Newson *et al.* 1998a, Newson and Newson 2000).

21.3 CHALLENGES FOR GEOMORPHOLOGISTS IN EMBRACING APPLIED QUESTIONS

The potential contribution of geomorphology to river channel management has yet to be fully realized, for four main reasons:

1. The awareness of the subject among the public and other environmental and engineering sciences is low. Whilst most people have heard of geologists and engineers, few (at least in Europe) have heard of geomorphologists! Public perception is misinformed, since geomorphology is often seen as an academic and descriptive discipline rather than a management-oriented predictive science. Graf (1996) argues that the acceptance of geomorphological research by society assists adoption among management authorities.
2. It has been difficult for geomorphology to establish its position within existing management structures. Where should it fit as an operational discipline? This is a management issue as much as a technical problem. In the UK, a management context could be suggested for fluvial geomorphology in association with either engineering or conservation—arguments for both are strong. The challenge is one of chronology rather than function, and does not imply that geomorphology is more or less ambiguous than other operational components. New elements imposed on existing management structures always struggle for assimilation.
3. The cost–benefit models (or other economically based option appraisal devices) used for project justification tend to undervalue or ignore the longer-term benefits provided by geomorphological contributions.
4. There is a relative lack of specialists prepared to apply the science in some countries. This is partly a reflection of inappropriate training curricula in the disciplines concerned, and partly an indication that a lack of perceived professional need militates against recruitment in the area concerned. Geomorphological programs in most universities are relatively small components of geological or geographical departments, and the number of students produced has historically not been large. Moreover, until quite recently, most academic fluvial

geomorphologists have focused on theoretical or historical questions rather than applied problems, and have historically made little effort to make their work accessible to other fields or applicable to practical problems.

21.4 MEETING THE DEMAND: GEOMORPHOLOGICAL TRAINING AND APPLICATION

As river managers and other scientific disciplines recognized a need for geomorphological input over the past two decades, the established field of geomorphology was not prepared to meet the demand. Instead, much of this demand was met by non-geomorphologists with little academic training (at least in geomorphology), and frequently using what might be termed "shortcuts". For example, non-geomorphologists have based channel reconstructions on relations between channel width and meander wavelength, and on predictions of "stable" channel configuration derived from a classification scheme, instead of undertaking a historical–geomorphological study of the river under consideration. Although these applications are often termed "geomorphically based", they lack an understanding of basin-scale influences or even channel-level process interactions that actually determine the success of the intervention (Sear 1994). Moreover, they typically involve applications of only the (limited) tools with which the non-geomorphologist has been exposed.

A "cookbook" approach to restoration, involving application of the channel classification system of Rosgen (1994), has proved enormously popular among managers and other non-geomorphologists in the US. In part, this has been because of the availability of one-week training courses where managers and staff learned the system, becoming overnight experts and apparently obviating the need for detailed geomorphic studies, and apparently satisfying the demand for integration of geomorphology into river management (Kondolf 1995). Most river restoration projects designed in this way have never been objectively evaluated, but of those that have undergone post-project appraisal, the track record has included a high proportion of failures (Smith 1997, Kondolf *et al.* 2001, also see Chapter 7, this volume).

More recently, academically trained geomorphologists have responded to the demand from managers by evolving classification and design methods using the basic research within the discipline (Fryirs and Brierley 1998, Newson *et al.* 1998a,b), and by conducting post-project appraisals of restoration as a basis for improving future designs (Downs and Kondolf 2002). The challenge remains, however, to educate a broad section of society as to the existence of the field and its real potential contributions to the management of rivers (Brookes 1995, Kondolf and Larson 1995), and to communicate to practitioners alternative approaches to the cookbook methods so popular now. Wilcock (1997) observed that while the scientific community has criticized restoration projects whose failures could be attributed to lack of substantial geomorphic study:

> Practitioners may be inclined to dismiss criticism from those who do not leave their handiwork on the landscape ... This view, however, misses the point that the job of science is not to address particular cases, but to find the general principles that apply to all cases. Also, it is likely that much of the criticism is basically correct (given the advantage of hindsight). The problem is not faulty criticism, nor that scientists do not have the right to criticize. The problem is that the critique is ineffective, if effectiveness is measured in terms of injecting better ideas and reliability into practice ... To be heard, the scientific community must come up with a message that is not only correct, but also simple, direct, and coherent. (Wilcock 1997: 454)

When we assess the performance of restoration projects designed and implemented by professionals without a solid background in fluvial geomorphology, we see that commonly the designers have not recognized basic but important controls on channel form, such as legacy effects of mining or flood control efforts, changed sediment supply from the catchment, or even the implications of the position of the reach within the larger drainage network (e.g., depositional reaches at the transition from piedmont uplands to coastal plain along the Atlantic Seaboard of the US). Reading the written justifications for such projects, it is clear that one of the shortcomings of the lack of substantive training is that one neither tends to ask the right questions, nor to use the full range of tools available. With limited training, one is likely to approach every problem in essentially the same way, and one is unlikely to step back from the manager's immediate concern (be it with bank erosion or degradation of fish habitat) to redefine the problem in terms of longer-term and catchment-scale processes that may be the underlying cause of the perceived reach-level problem. The advantages of taking a larger-scale geomorphic approach are illustrated in case studies presented in this chapter.

21.5 INTERACTIONS BETWEEN GEOMORPHOLOGISTS, STAKEHOLDERS AND RIVER MANAGERS

Interactions between geomorphologists (and for that matter, scientists in general) and managers can be tricky, because the goals and methods of operation of the two groups are frequently at odds. To get past the traditional behavioral and institutional barriers may require that each makes efforts to understand the perspective of the other. Managers and geomorphologists often approach problems in very different ways, as reflected in their different attitudes to issues such as uncertainty, the timescale of problem definition (managers typically being concerned with shorter timescales than the geomorphologists), the timescale within which an answer is desired, the spatial scale of problem definition (managers typically taking a site-specific view as opposed to the catchment-scale view needed to understand many geomorphic processes), and willingness to invest in substantive studies by qualified technical personnel.

Interactions between scientists and managers have been further complicated by the emergence in recent years of participation in decision-making by stakeholders and the public in general. Overall, this has probably been a good thing, as decision-makers are less likely to make unilateral decisions that run counter to local interests, but it poses an interesting challenge to geomorphologists, who must now communicate not only with elected officials and appointed administrators, but also with members of the public, whose scientific backgrounds and ability to understand sometimes complex interactions may vary widely. For the geomorphologists, this puts a premium on effective communication (public education). It has also created situations in which members of the public and interest groups have *voted* on what are essentially scientific questions. This perhaps is not very different from juries deciding questions of fact in legal cases that involve expert witness testimony, but it has resulted in some strange (at least to a trained scientist) conclusions being drawn.

As described by Wilcock *et al.* (2003), increasing use of models as a basis for decisions in natural resource management has led to more frequent interactions between managers and modelers in fluvial geomorphology, and these interactions have highlighted differences in objectives between managers and modelers. "The policy or legal context [may demand] a precision in model predictions that the available knowledge cannot support", such as the requirement of water law in states of the western US that in-stream water users claim only the minimum flow needed for a purpose, such as maintenance of channel form. Conflict may also arise between managers and modelers because management questions may require predictions that are more temporally and spatially explicit than possible or practical (Wilcock *et al.* 2003). However, models can serve useful functions besides prediction: "to educate managers about the ecosystem, to identify gaps in the current knowledge, to allocate scarce research dollars for future work, and to define plausible management scenarios that merit further evaluation" (Wilcock *et al.* 2003). For example, in framing the range of possible dam operation alternatives along the Colorado River below Glen Canyon Dam and their potential effect on a multitude of resources (ranging from extent of sand beaches to hydropower generation to native fish), models cannot produce specific predictions, but can increase understanding of how parts of the system work together (Schmidt *et al.* 1998, Wilcock *et al.* 2003).

Most rivers have been affected by human intervention to one degree or another, so their current conditions result from the interplay of the river and social systems (Figure 21.1). Within the river system, flow regime (Q) and sediment load (Q_s) from the basin are the independent variables that largely determine alluvial channel form, as reflected in the adjustment of dependent variables of width, depth, grain size, and pattern. This simple system can be made more complex by adding the biological and chemical elements and their relationships with the geomorphic elements. Human activities (a function of the social system) can affect both the independent variables (e.g., through urbanization and flashier runoff) and the channel form directly (e.g., by channelization, or in-channel sand and gravel mining), with resulting effects on water chemistry and aquatic and riverine ecology. Because rivers are dynamic systems, such actions typically beget reactions, such as channel incision, which in turn can affect human infrastructure or other uses (e.g., through undermining bridges and pipeline crossings) (Bravard *et al.* 1999). In response to such negative feedback from the river system, the social system tends to respond with countermeasures such as structures to control erosion of bed and banks, which in turn may produce further erosion elsewhere in the channel.

Although we speak here of the social system as a single entity, in reality, the human actors or "stakeholders" range widely in interests, motivation, and

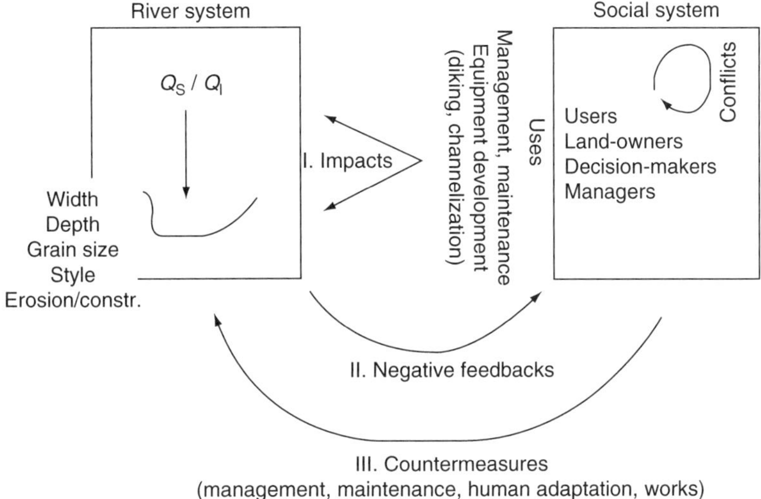

Figure 21.1 Interactions between the geomorphological river system and the social system: impacts, negative feedbacks, and countermeasures

power. Landowners, recreational users, resource managers, and elected decision-makers can act and react at different spatial and temporal scales, sometimes in complete contradiction to one another. Some conflicts occur on many rivers, such as those between canoeists and fishers, between hydroelectric companies and fish and wildlife agencies, between managers of upstream reaches and managers of downstream reaches, and between regional and local planners. Each has specific objectives and stakes, which may conflict with others.

In this environment, fluvial geomorphologists must pursue the science and encourage participatory planning and management to diagnose problems and propose solutions so they can be understood by the broadest community of actors. Applied fluvial geomorphic questions can generally be classed as relating to (I) impacts of human development on the river system, (II) the response of the river system to these human influences, or (III) the countermeasures taken by human actors to deal with the river response to development. The success of the solutions proposed will depend in large measure on how the geomorphologist interacts with the social system. At level II, it is essential for them to interact with other disciplines such as ecology, economy, and history to show the cascading consequences of geomorphological adjustments or functioning in terms of biodiversity or financial fluxes as well as job provided or lost. At level III, scenarios must be generated to project not only the river's geomorphologic response, but also resulting natural hazards, resource availability, user satisfaction, and sustainable development at the basin scale. Otherwise, the solutions proposed may be effective only at a short timescale.

Applied fluvial geomorphology is now called on to evaluate the river system's function, sensitivity to change, and its ecological potential. These concerns have arisen because the river is increasingly viewed as dynamic and supporting a variety of resources, and has to be managed sustainably to continue providing those resources. A sampling of such questions and concerns is presented in Table 21.1, and the case studies in this chapter. One class of questions asks: "How does the river work?" This question is typically posed by users who want to know if a projected action may trigger unwanted responses. Another class of questions asks: "Where is the river going?" This is typically posed to understand the consequences of ongoing river adjustments to past interventions. A final class of questions, "How can we improve the state of the river?", encompasses questions related to sustainable river management and restoration.

21.6 COMPONENTS OF GEOMORPHOLOGICAL STUDIES IN RIVER MANAGEMENT

An assessment of current geomorphological practice suggests that fluvial geomorphology provides management information in four key areas:

Table 21.1 Examples of geomorphological questions posed to help end-users such as aquatic ecologists or natural resources managers to answer their questions

Geomorphological questions	Reasons why geomorphological questions posed	End-users	Examples
How does the river work? Assessment of on-going processes and forms			
What is (or what will be if...?) the sediment transport in a given reach?	Rate of reservoir filling	Water resource managers	Polish Carpathians (Lajczak 1996)
	Flooding frequency increase	Risk managers	Waiho fan, New Zealand (Davies and McSaveney 2001)
	Gravel resource availability	Gravel miners, regulators	Rhône River, France (Petit *et al.* 1996)
What is the sensitivity of the river system to any modifications of runoff and sediment load?	Assess possible consequences of development	Natural resource managers, developers	Streams in Washington state (Moscrip and Montgomery 1997)
How do former river channels (e.g., oxbow lakes) vary in terms of sedimentation rate and geometry in a given reach?	Vegetation diversity and successional rates, lifespan of given states	Landscape ecologists, environmentalists	Ain River corridor, France (Piégay *et al.* 2000)
Where the river is going? Assessment of human impacts at various spatial and temporal scales			
What is the impact of a dam on the sediment transport and channel form downstream?	Changes in fish habitat	Aquatic ecologists, fisheries agencies	North Tyne: hydropower regulation impacts on spawning riffles and channel geometry (Sear 1995)
	Changes in channel geometry (narrowing, incision, aggradation)	Manager of natural hazard (flooding)	Hanjiang River, China (Xu 1997)
	Changes in vegetation mosaic in the riparian zone	Landscape/aquatic ecologists and environmentalists	Large dammed rivers in USA (Collier *et al.* 1996)
Have past human actions (e.g., engineering works/mining) induced channel changes downstream?	Increase in channel instability	Land managers	Action of river maintenance activities in the UK rivers (Sear *et al.* 1995)
	Geometry adjustment	River managers	Californian rivers (Kondolf 1997); rivers of England and Wales (Brookes 1987)
	Effects on biological communities	Ecologists, environmental planners	Redwood Creek Basin, Northwestern California (Ricks 1995); Pennsylvania streams (Wohl and Carline 1996)

What is the magnitude of current and potential channel incision following channel straightening or mining?	Sensitivity of bridges to undermining	Civil engineers	Streams in southern US (Simon and Downs 1995)
	Drop in groundwater	Aquatic ecologists, agriculture and water resource managers	Coal Creek, Colorado (Scott *et al.* 1999)
What is the effect of an in-channel mining site on the bedload transport?	Beach degradation downstream	Land managers, engineers	Fiume Seccu and Figarella in Corsica (Gaillot and Piégay 1999)
	Fish habitat degradation	Aquatic ecologists	Wooler water—massive channel incision (Sear and Archer 1998)
What are the effects of catchment afforestation/deforestation?	Channel geometry and associated flooding risks and bank erosion	Managers of natural hazards (flooding)	Conifer afforestation in upland humid temperate climates (Newson and Leeks 1987)
How can we improve the state of the river? Sustainable management and restoration			
What are the effects of restoration practices?	Monitoring	Planners and managers	Lowland UK rivers (Sear *et al.* 1998)
What are the channel types at the regional/national scale?	Monitoring	Planners	UK/France channel classifications (Newson *et al.* 1998a,b, Raven *et al.* 2002; Chapter 7, this volume)
What are the best maintenance practices to balance flood hazard and natural quality?	Mitigate regulation/maintenance practices	Managers, environmentalists	Sustainable river maintenance procedure for the UK rivers (Sear *et al.* 1995); French guidelines for riparian forest maintenance (Boyer *et al.* 1998). Danish streams (Iversen *et al.* 1993)
What are the geomorphological designs to promote on a given site?	Improve the landscape and ecological integrity	Managers, environmental planners	Mimmshall brook (Sear *et al.* 1994); Mississippi streams (Shields *et al.* 1995)
What is the lifespan of a given restored habitat (gravel bar)?	Aquatic habitat monitoring	Managers, environmental planners	Kissimmee River (Toth *et al.* 1995)

- *Assessment* — Establishing cause and effect in river management problems
- *Decision support* — Strategic decisions on when, and when not to intervene
  Operational guidance as to where and what intervention to adopt
- *Design* — Mainly used on enhancement and restoration projects
  Advice on type and dimensions of channel morphology, appropriate flushing flow regimes, sedimentology, and nature of adjustments
- *Post-project appraisal* — For a range of river management practices

In the UK, a decade of investment in geomorphological research and development has culminated in a suite of "standard" methods for incorporating geomorphological information into existing river management practice that provides a useful template for deploying the range of tools discussed within this volume (Environment Agency 1998). At their core lies the basic notion that geomorphology has contributions to make across the broad sector of river management, including strategic and operational management, the latter involves actions that modify watercourses.

An axiom of this approach is that it is essential to understand the cause of the management problem. The methods are designed to nest in a quasi-hierarchical fashion, collapsing from the catchment (strategic) over-view of physical habitat resource, down to the project level design and assessment. This framework involves deployment of a range of geomorphological tools to provide increasing levels of certainty in the interpretation of system functioning, in support of specific management goals. The approach is based on the view of the river network as a continuum, whereby reaches are classified according to the information recovered from the catchment under study. This prevents the imposition of rigid classifications, and recognizes the inherent value in the uniqueness of a river, whilst seeking to encourage standard approaches to the analysis of channel processes and the resulting forms and habitats. While these generic methodologies may be applied to a range of river management projects, there will of course be more specific studies required by individual projects. In this case, the tools described in this book should be considered within the context of the specific project brief. For example, the setting of flushing flows for watercourses may require measurement of sediment transport in relation to specific discharges.

Table 21.2 provides examples of the scale and nature of the information produced by these different methods of data collection and assessment. Within each scale of survey, different tools are required for the collection of the relevant data. For example, with the evolution of remote sensing technology and data post-processing, much of the catchment and network scale topography and morphology may be generated without full ground survey; although calibration and ground-truthing will still be necessary.

The following section takes each scale of geomorphological analysis in Table 21.2, and elaborates through case studies, the application of different tools to solve specific management problems.

21.7 GEOMORPHOLOGICAL ASSESSMENT AT THE CATCHMENT SCALE

Catchment Baseline Survey

At the largest scale, catchment baseline surveys aim to provide catchment inventories of the geomorphological sensitivity of the river network, and conservation status of each reach. The information from catchment surveys is used strategically to target investment in rehabilitation or conservation designations based on physical habitat diversity. Figure 21.2 illustrates an output from a catchment scale survey of the river Britt, a low gradient groundwater dominated stream in southern England. An important component in both catchment surveys and fluvial audits (discussed below) is visualization of the data sets. Application of GIS to spatial geomorphological data is a powerful tool in itself both for analysis and presentation, the latter an important consideration when communicating results to non-specialist audiences. The use of a GIS also permits integration of other data sets that may be relevant to a particular study, for example, the presence and structure of macrophyte, invertebrate, and fish communities.

The method of field data collection and desk-top study may be standardized in terms of the type of information collected through the use of data entry forms which can be mounted on digital hand-held platforms for direct data entry. The advantage of standardized approaches is that they can be β-tested and are replicable (and therefore accountable) with clearly defined outputs (Table 21.2). Their main dis-

Table 21.2 A framework for incorporating geomorphological tools within river management (after Newson and Sear 1993)

Stage	Planning/project		Project	Project	Project
	Geomorphological assessment		*Geomorphological dynamic assessment*	*Geomorphological channel design*	*Geomorphological post-project appraisal*
	Catchment baseline study	Fluvial audit			
Aims	Overview of the river channel morphology and classification of geomorphological conservation value	Overview of the river basin sediment system typically aimed at addressing specific sediment related management problems and identifying sediment source, transfer and storage reaches within in the river network	To provide quantitative guidance on stream power, sediment transport, and bank stability processes through a specific reach with the aim of understanding the relationships between reach dynamics and channel morphology	To design channels within the context of the basin sediment system and local processes	To assess the degree of compliance between design expectations and outcomes in terms of geomorphological processes, dimensions and morphology
Scale	Catchment (size <1–3000+ km^2)	Catchment (size <1–3000+ km^2) to channel segment	Project and adjacent reach	Project reach	Project reach
Methods	Data collation, including reconnaissance fieldwork at key points throughout catchment. Integration of data within GIS	Detailed studies of sediment sources, sinks, transport processes, floods, and land use impacts on sediment system. Historical and contemporary data sets derived from desk-based study. Integration of data within GIS	Field survey of channel form and flows; hydrological and hydraulic data, bank materials, bed sediments (GA/FA if not available)	Quantitative description of channel dimensions and location of features, substrates, revetments, etc. (GDA/FA/GA if not available)	Review of project aims/expectations. Compliance audit of channel against design. Re-survey of project data sets. Field survey approach
Core information	*Characterization* of river lengths on basis of morphology and sensitivity to management intervention	Identifies *range of options* and "potentially destabilizing phenomena" (PDP) for sediment-related river management problems	Sediment transport *rates* and morphological *stability/trends.* "Regime" approach where appropriate	The "appropriate" *features* and their *dimensions* within a functionally designed channel	*Extent of changes or conformity* to original project design and recommendations for mitigation options

(*Continues*)

Table 21.2 (*Continued*)

Stage	Planning/project		Project	Project	Project
Procedure	*Geomorphological assessment*		*Geomorphological dynamic assessment*	*Geomorphological channel design*	*Geomorphological post project appraisal*
	Catchment baseline study	Fluvial audit			
Outputs	15–30 page report; GIS including photographs detailing conservation value and sensitivity of reaches to management actions	GIS; time chart of potentially destabilizing phenomena; report including recommendations for further geomorphological input where necessary	Quantitative guidance as to intervention (or not) and predicted impacts on reach and beyond. Identification of causes of specific problems where possible	Plans, drawings, tables and report suitable as input to quantity surveying and engineering costings. Justification for design. Explicit consideration of channel dynamics and sediment transport	Plans, tables, report. Assessment of project performance in terms of geomorphological processes and morphology/physical habitat. Recommendations for input into adaptive management.
Destination	Feasibility studies for rehab/restoration.	Investment/management staff, river managers or policy forums, Project steering groups	River managers and project steering groups	River managers and project steering groups	River managers and project steering groups

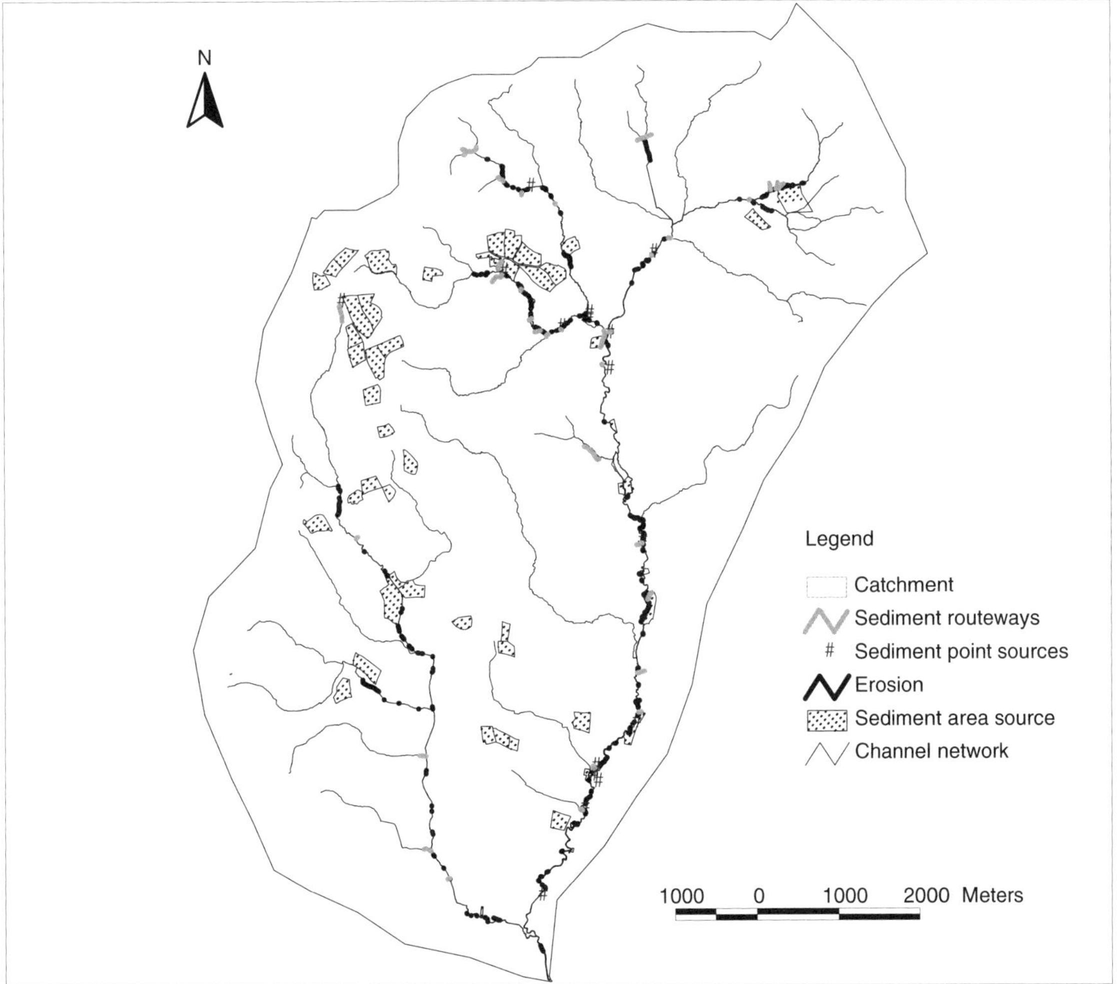

Figure 21.2 Example of a GIS output from a catchment-scale geomorphological assessment; the river Britt, a groundwater dominated river in southern UK

advantages are their inflexibility in the face of unique conditions.

The Fluvial Audit

An audit represents the second level of geomorphological analysis of a river catchment and is aimed at providing an interpretation of the functioning of the river system in terms of a sediment budget, and establishing a link between this functioning and the morphology and physical habitats of the system. Like the catchment survey, the fluvial audit is a field-based, reconnaissance survey, undertaken in a structured framework to provide consistency and ease of data entry and analysis within the GIS environment. The field survey is continuous, and includes inventories of features, coupled with assessments of materials and processes operating within the river corridor (Sear *et al.* 1995). Estimates of sediment supply and storage are calculated from measures of sediment deposits within the channel and floodplain (see Chapters 2, 9–11, 13, and 16), whilst supply from bank erosion is informed from measures of bank morphology and historical rates of erosion determined from historical surveys, maps, and remotely sensed data (see Chapters 4 and 6). In this way, historical information is integrated with contemporary survey to establish a process-based classification based on channel activity (both vertical and lateral). Classification of the data within the GIS (see Chapters 7 and 8) facilitates identification of zones of sediment storage, supply, and transfer. In addition, information on the types of erosion and deposition processes are displayed that may be used to guide erosion control measures (Sear *et al.* 1995). This catchment scale information should ultimately seek to provide a calibrated sediment budget (though often without fluxes in European rivers) that can inform river managers of the sources of sediments that may be causing sedimentation problems or the most probable impact on the sediment system (see Chapter 5) of undertaking a given channel modification. Ideally, another layer in the information includes field-determination of physical habitats and associated biological communities (Newson and Newson 2000). This information not only provides guidance on the relationships between the geomorphology of the river and the floodplain and channel ecology, but also establishes a framework for integrating different disciplines.

Case Study: Strategic Assessment of a Proposed Plan for Rehabilitating Salmonid Spawning Habitat Using Catchment Survey Methods

Deer Creek drains 540 km^2 (average discharge 9 $m^3 s^{-1}$), originating on the western slopes of Mount Lassen, the southernmost volcano of the Cascades range, and flows southwestward through rugged bedrock canyons to the Sacramento Valley floor, where it traverses its alluvial fan for about 17 km before joining the Sacramento River at Vina, California. Historically, flood flows would overflow the current channel of Deer Creek and flow across the floodplain, on which land use is primarily livestock grazing and orchards. The multiple distributaries appearing on the topographic map hint at the history of channel aggradation and avulsion by which such alluvial fans are built (Figure 21.3).

Adult spring-run chinook salmon traverse the lower, alluvial reach en route to remote canyons cut into volcanic rocks that offer them suitable spawning and rearing habitat. Fall-run chinook use the alluvial reach for spawning, and spring-run, winter-run, and other salmonids, mostly from other drainages, use the alluvial reach for rearing.

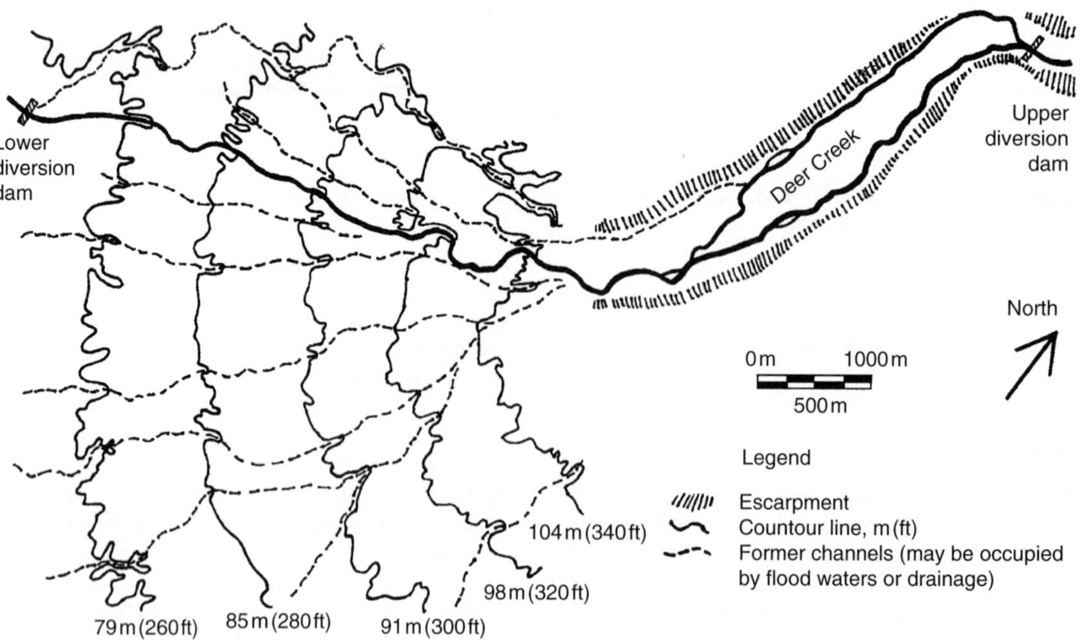

Figure 21.3 Distributaries of Deer Creek (based on 1:24 000 US Geological Survey topographic maps)

Habitat conditions in the alluvial reach of Deer Creek are not ideal. There is relatively little shade because vegetation is lacking along most banks, the channel form is very simple, and the gravel sizes (D_{50} decreases from 190 mm near the foot of the mountains to 55 mm about 3 km upstream of the Sacramento River confluence) are larger than ideal for chinook salmon spawning (Kondolf and Wolman 1993). The channel lacks complex features that provide good habitat, such as tight bends and secondary circulations, undercut banks, scour pools, woody debris jams, and a well-developed pool-riffle sequence. To partially address these problems, planning documents for salmon restoration programs in the early 1990s proposed planting riparian vegetation along the low-flow channel banks and constructing spawning riffles by regrading the bed and adding smaller-sized gravels to the creek.

In 1949 the US Army Corps of Engineers straightened and cleared the channel, and constructed levees along much of the lower 11 km of Deer Creek. The manager of the Stanford-Vina ranch, immediately downstream of the proposed levees, protested against the project, accurately predicting that the proposed flood control project would deliver to his ranch "... more water and at greater velocity than under existing conditions" (Robson 1948). Given its then entirely agricultural land use, the need for federally funded flood control along Deer Creek is not obvious. Despite the apparent lack of benefit, and despite the protest of the downstream landowner, the project was completed and responsibility for maintaining the levees was given to the county, with responsibility for the channel bed to the state. In the early 1980s, the California Department of Water Resources cleared the channel of accumulated gravel and vegetation to maintain channel capacity. During a large flood in January 1997, Deer Creek broke through its left bank levee about 8 km upstream of the Sacramento River confluence and flowed across the floodplain, some water flowing directly into the Sacramento, some flowing back into Deer Creek (Figure 21.4). The channel had broken through its levee in the same place in the early 1980s. After the recent flooding, and the history of levee failure, there was strong interest among local residents and land owners to find a more sustainable solution to flooding along Deer Creek.

A geomorphic study conducted as part of the Deer Creek Watershed Management Plan (DCWC 1998) reviewed historical maps and aerial photographs, measured bed material size along the channel, and evaluated current conditions in a longer-term and larger-scale context. Examination of aerial photographs from 1939, prior to the flood control project, showed highly complex channel forms, with frequent pool-riffle alternations, meander bends, and gravel bars, with trees overhanging the channel, undercut banks, and log jams (Figure 21.5). Aerial photographs taken after the flood control project showed a marked simplification of the channel and thus loss of habitat (Figure 21.5). The geomorphic study concluded that habitat in Deer Creek was limited mostly by the effects of the flood control project, not only because of the direct impacts of clearing and subsequent maintenance, but also because the levees constrain high flows within the low-flow channel, so that a given flow is deeper (and thus exerts greater shear stress on the channel bed) than was the case when floods overflowed onto the floodplain. As a result, very high shear stresses in the channel wash out spawning gravels, complex channel forms, and riparian vegetation. The report also concluded that the proposed riparian plantings and spawning riffle construction along and in the channel would probably wash out in the high shear stresses of the constrained levee system.

In this example, a geomorphic analysis suggested that the habitat problems previously identified along Deer Creek—lack of shading, simplified channel form, and reported lack of suitable spawning gravels—were all symptoms of an underlying cause: the effects of a flood control project built five decades before. Rather than spending restoration funds on treating the symptoms (planting trees and adding gravel to the channel) the report recommended addressing the problem and redesigning the flood infrastructure to allow Deer Creek to flood its (still mostly agricultural) floodplain in a controlled way, protecting vulnerable structures by ring levees. This approach would have the advantages of providing better flood protection to the structures on the floodplain, while reducing shear stresses in the low-flow channel, allowing bars to build and riparian vegetation to establish within the currently "maintained" (cleared) floodway, thereby improving habitat in the channel of Deer Creek. A feasibility study for the Deer Creek floodplain project has recently been identified for funding by the California Bay Delta Ecosystem Restoration program, and a three-dimensional modeling study is currently underway as part of a Ph.D. thesis. Fortunately, the Deer Creek floodplain remains mostly agricultural so there is a better opportunity here to reconnect the floodplain and channel than along most streams in California.

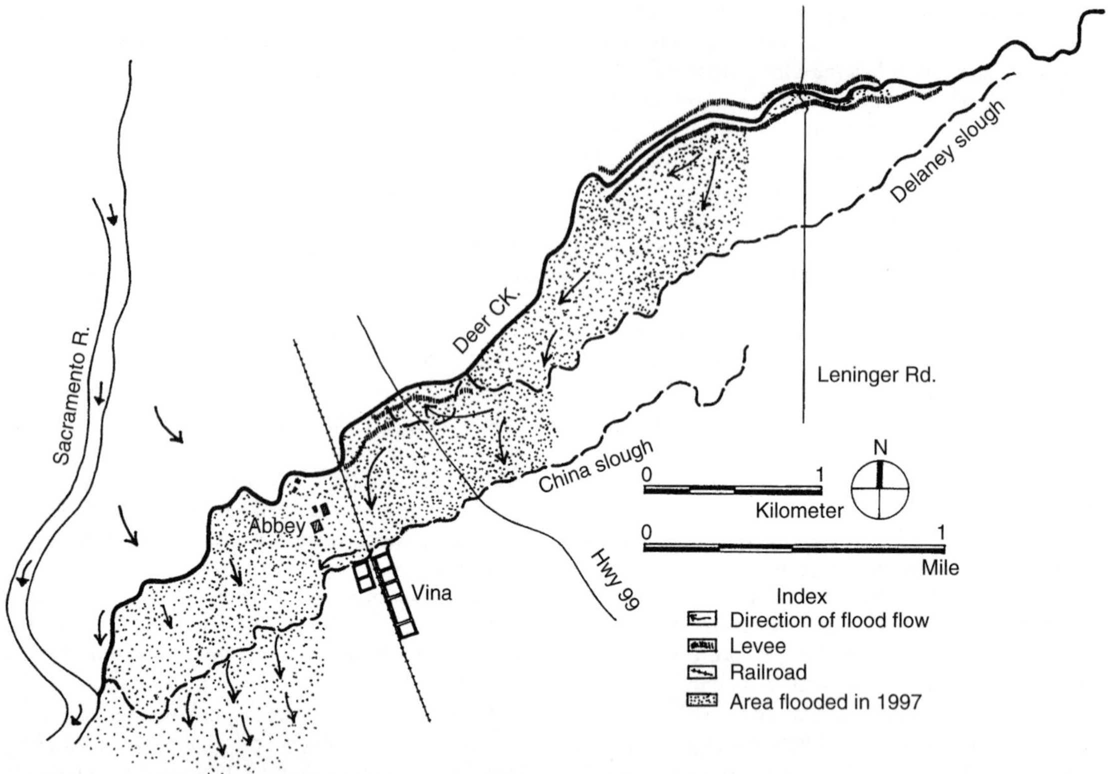

Figure 21.4 Extent of flooding in 1997 along Deer Creek and its floodplain. Some of the overflow returned to the main channel at Hwy 99, the rest followed China slough to flow directly into the Sacramento River. Based on contemporary observations, high water marks documented shortly after the flood, and post-flood interviews with residents

21.8 GEOMORPHOLOGICAL DYNAMIC ASSESSMENT AT THE REACH SCALE

The remaining levels of geomorphological assessment relate to reach-scale, problem-focused management, often associated with specific schemes (e.g., design of river restoration projects, bank erosion control measures, etc.). The nature of the questions posed is more specific, and entail deployment of tools necessary for quantifying system functionality (See Chapters 6, 11–20). Thus, a Geomorphological Dynamics Assessment may quantify bank stability and sediment transfer within a design or "problem" reach, while establishing it within the broader catchment context by applying a fluvial audit catchment survey. Numerous examples exist in the geomorphological literature of what could be termed "Geomorphological Dynamics Assessment". Specific examples include Sear *et al.* (1994), Kondolf and Wilcock (1996), and Thorne *et al.* (1996). The range of tools deployed within differing studies is dependent on project budget and availability. In large-scale restoration programs, where sediment load and hydrodynamics are crucial factors to quantify, then the sums of money may justify expensive pre-project monitoring programs and model calibration. In many cases, however, the budgets are more modest, and the tools used must be carefully selected in order to provide the most robust answers. For example, the sedimentation of a small rural land drainage channel is unlikely to attract the level of investment in geomorphology that is required in order to understand the sediment dynamics of a major hydropower or restoration program. A crucial point to consider here is the validity of the information obtained,

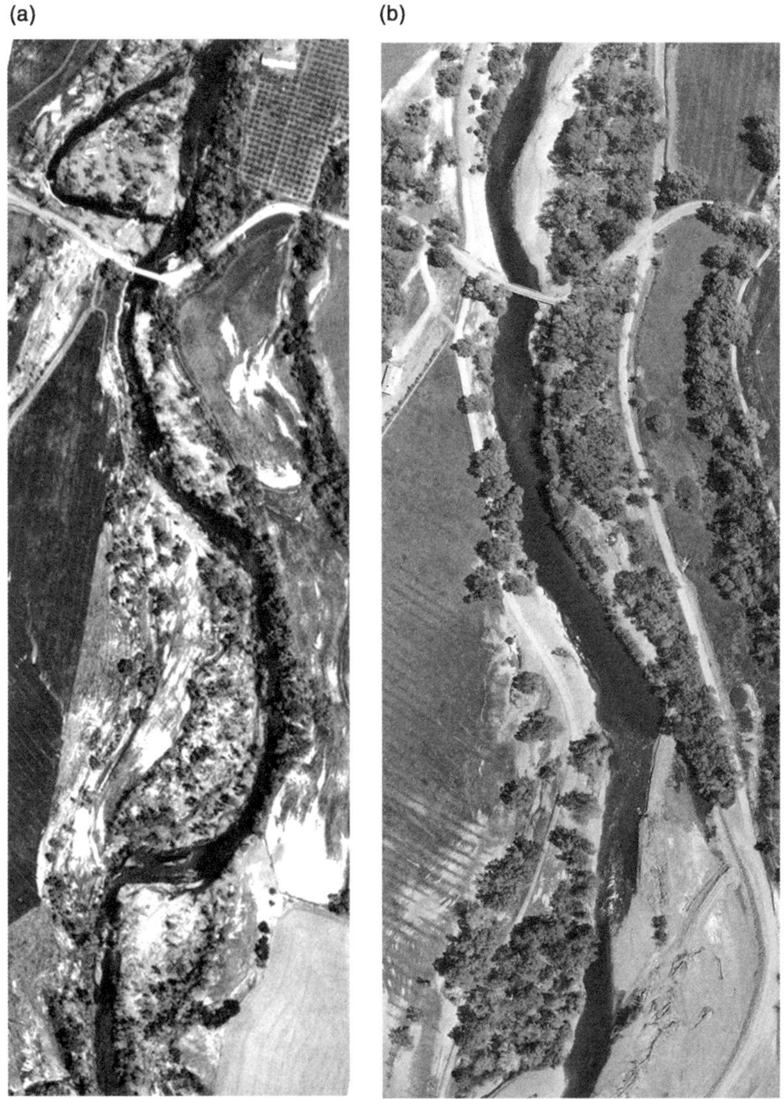

Figure 21.5 Details from aerial photographs of Deer Creek near Leininger bridge in 1939 and 1997. Flow is from top to bottom, length of river shown is about 1100 m. (a) The 1939 channel was more complex, sinuous, and with more frequent bed variations, gravel bars, and undercut banks with adjacent riparian vegetation. (b) The 1997 channel is less sinuous, wider, and less complex, reflecting the effects of the 1949 flood control project, and subsequent maintenance. Also visible in the 1997 photo detail is the levee breach along the left (south) bank

particularly when legal challenge is possible. However, geomorphologists must avoid "advocacy science" (Graf, pers. comm. 2002), and if more investment is necessary to answer a problem, then the science case must be made to the stakeholders.

21.9 GEOMORPHOLOGICAL CHANNEL DESIGN

Aims

This approach uses geomorphological principles to develop an appropriate channel design. In practice, the tools deployed will depend on the nature of the design problem and the type of river system under study. For example, the restoration of a channel for physical habitat enhancement in an urban setting may be constrained in terms of what is possible in comparison to a similar scheme undertaken in relatively undeveloped landscapes. Similarly, low-energy cohesive channels may require more detailed design consideration compared with higher energy alluvial streams that are in effect able to design themselves (Sear *et al.* 1998). Approaches to geomorphological design may be based on the derivation of local hydraulic geometry relationships or from analogue reaches within the same or adjacent basins. In many situations, however, development of the catchment and modification of the hydrology and channel form may be so extensive that such approaches are not possible. In these situations, modeling of the channel form may be attempted provided that effective calibration is made (See Chapters 17–19). Recent consideration of the process of geomorphological channel design has highlighted the role of both field survey and modeling in quantifying and reducing levels of uncertainty, and communicating these to the other disciplines associated with the process.

Case Study: Setting Riparian Channel Widths. An example of Geomorphological Channel Design

An interdisciplinary problem. Riparian zones have been recognized as critical environments for biodiversity in river systems (Odum 1978, Naiman *et al.* 1993). Riparian zones influence aquatic ecosystems because living tree roots and branches, as well as dead wood, create channel (and thus habitat) complexity, because leaves and invertebrates falling into the channel from the canopy provide food, and because shading by trees influences water temperature (Meehan *et al.* 1977, Vannote *et al.* 1980, Gregory *et al.* 1991). Riparian vegetation diversity in turn depends on channel migration and rejuvenation of floodplain habitats (Pautou and Wuillot 1989, Marston *et al.* 1995, Bornette *et al.* 1998). In addition, well-vegetated riparian zones can also serve to filter sediments, nutrients, and contaminants from runoff, thereby improving water quality in rivers. Accordingly, riparian buffer strips or streamside management zones are commonly prescribed, often for multiple objectives, and riparian zones have been the object of conservation and restoration efforts (Nilsson 1992, Petersen *et al.* 1992, Goodwin *et al.* 1997, Landers 1997, Hunter *et al.* 1999).

In addition to the inherent ecological values of a vegetated riparian zone, setting human structures back from the active channel can minimize conflicts between dynamic river behavior and human development. Engineering measures to protect human structures, such as dikes, bank protection, and channel straightening or deepening, affect channel geometry and bedload transport, often with negative consequences for habitat (Brookes 1988, Petts 1989). The "streamway" concept is to leave a wide belt within which the river channel can freely move and flood; for a meandering river this zone can correspond to the meander belt (Palmer 1976, Piégay *et al.* 1994, 1996, Brookes 1996). The regional agency in charge of water management in the Rhône Basin, France (SDAGE RMC 1997) defined the streamway as a floodplain band within which channel migration is useful for the ecosystems and bedload supply.

One of the questions posed by managers and decision-makers to scientists in order to maintain an optimal river corridor which can support, over a long period of time, many uses and high biodiversity is: "What is the riparian width to be preserved or to be restored?". It is an interdisciplinary question, which can be shared in many sub-disciplinary questions (Budd *et al.* 1987). What is the width of the riparian zone within which most of the riparian vegetation species (Spackman and Hughes 1995) or most of the avian communities (Keller *et al.* 1993) are observed? What is the wooded corridor width to be preserved to mitigate the effects of logging on stream habitat, invertebrate community composition and fish abundance (Davies and Nelson 1994)? What is the width of the riparian zone required for plants to take up most of the nutrients (e.g., nitrates) coming from neighboring agricultural areas (Pinay and Décamps 1988, Petersen *et al.* 1992)? What width of wooded corridor is needed to trap sediments and woody debris from floodwaters to protect floodplain farmland against damage from sediment and woody debris? This is an important question in the lower Rhône Valley, where tributaries traverse alluvial plains with vineyards, which are sensitive to scouring and burial under sediments during floods, but not to inundation (Piégay and Bravard 1997).

What streamway width should be preserved for ecological or long-term socio-economical purposes (Malavoi *et al.* 1998, Piégay and Saulnier 2000)? The last question needs not only geomorphological and ecological analysis, but also economic and policy analysis to evaluate relative costs of bank erosion versus bank protection construction and maintenance, the price of the land, and the annual value of its agricultural production (Combe 1991, Piégay *et al.* 1997).

Applying geomorphic tools to assess channel migration. The magnitude and frequency of the channel shifting may be of interest at a local scale to a land owner, at a larger scale to decision-makers and managers designing a streamway for ecological conservation or to minimize future conflicts between human settlement and bank erosion processes. From a geomorphological point of view, bank erosion is a natural process, which contributes to the overall physical functioning of the river. If we counteract this process, it may involve cascading changes in channel geometry, and may affect other human uses. In actively shifting channels (wandering, meandering, and braided channels), the geomorphological approach is first applied at the scale of a homogeneous reach (from 10^0 to 10^2 km) where bank erosion occurs. For example, if the river is characterized by a meander train, it is first important to work at the scale of the whole train rather than at the single curve scale.

Two main approaches include historical studies of channel mobility, increasingly using GIS technology (Downward *et al.* 1994, Marston *et al.* 1995). Historical maps and air photos are scanned, georeferenced, and rectified. Using several temporal series, it is possible to overlay the different channel courses, documenting temporal and spatial variation of the channel migration rate, channel cut-off frequency and character, and the areas of newly eroded and constructed floodplain, and assessing the sensitivity to erosion of individual reaches (Figure 21.6).

Analysis of a series of air photos of the Ain River showed that from 1945 to 2000, the surface area of the unvegetated, active channel reduced from 630 to 450 ha, and riparian forest established in the formerly open channel (Piégay and Saulnier 2000). Erosion of the floodplain surface averaged 7 ha year^{-1} over a 40-km reach between 1980 and 1996; 8.3 ha year^{-1} between 1996 and 2000. Fortunately, only 6% of the eroded areas were occupied by agriculture, most of the remaining areas consisted of the riparian forest established on the former active channel. In this context of in-channel forest establishment and low human pressure on the riparian zone, the streamway concept (preventing development within the river corridor) has potential to succeed. The streamway zone width is based on historical analysis of channel change, with different patches distinguished according to their probability of being eroded in the next three decades. Such mapping has been done on few tens of rivers in France. The Ain River is one of the most advanced example (Figure 21.6). Channel shifting has been mapped over a 40-km reach over the last five decades, as a basis for determining the streamway. River managers have identified land ownership within the streaming band and are now determining guidelines for managing actions such as forest harvest, bridge construction, or extension of existing mining sites.

Because this approach is based on expert information, historical channel mobility as a basis to predict future mobility and future advances on numerical modeling to simulate channel evolution should improve the process. At this stage of the research process, such tools are available but not robust enough to be used in an applied sense. Numerical modeling has progressed such that it is now possible to simulate channel meandering or braiding (see Chapter 19, this volume). Models of meander migration are based on a general relationship relating lateral bank erosion rate to the near-bank velocity, channel depth, and bank erodability. Previously limited to neck cut-offs, recent models incorporate not only chute cut-offs from probability function but also spatial variability of bank erodability (exposure of resistant floodplain or terrace sediments, bedrock outcrops). Moreover, Howard (1996) also simulated the meander belt width which can be defined as the maximum floodplain width occupied by the meandering stream during the simulation time. He proposed different sceneries including chute cut-offs and also resistant valley walls or oxbow plugs (Figure 21.7). These models can help set streamway width, as well as other objectives, such as assessing the residence time of contaminants in the floodplain sediments.

Recent work combining planform models and bank erosion models (Rinaldi and Casagli 1999) may improve model predictions further. Tools to assess bank erosion at a short timescale, such as traditional erosion pins, terrestrial photogrammetry, and Photo-Electronic Erosion Pin (PEEP) system, can be easily applied to a wide range of environments at low cost, and provide accurate results even for bank retreat of few millimeters (Lawler 1993). However, they yield information at the site scale, and must be somehow scaled up to provide meaningful insights at a larger scale (Figure 21.8).

Applying geomorphic tools to assess potential incision. Use of buffer strips to improve water quality is increasingly popular among managers, but the efficiency of such measures depends strongly on the potential of the reach to affect biogeochemistry of shallow groundwater, which is partly controlled by channel geometry. For example, nitrates in shallow groundwater can be transformed by soil (denitrification

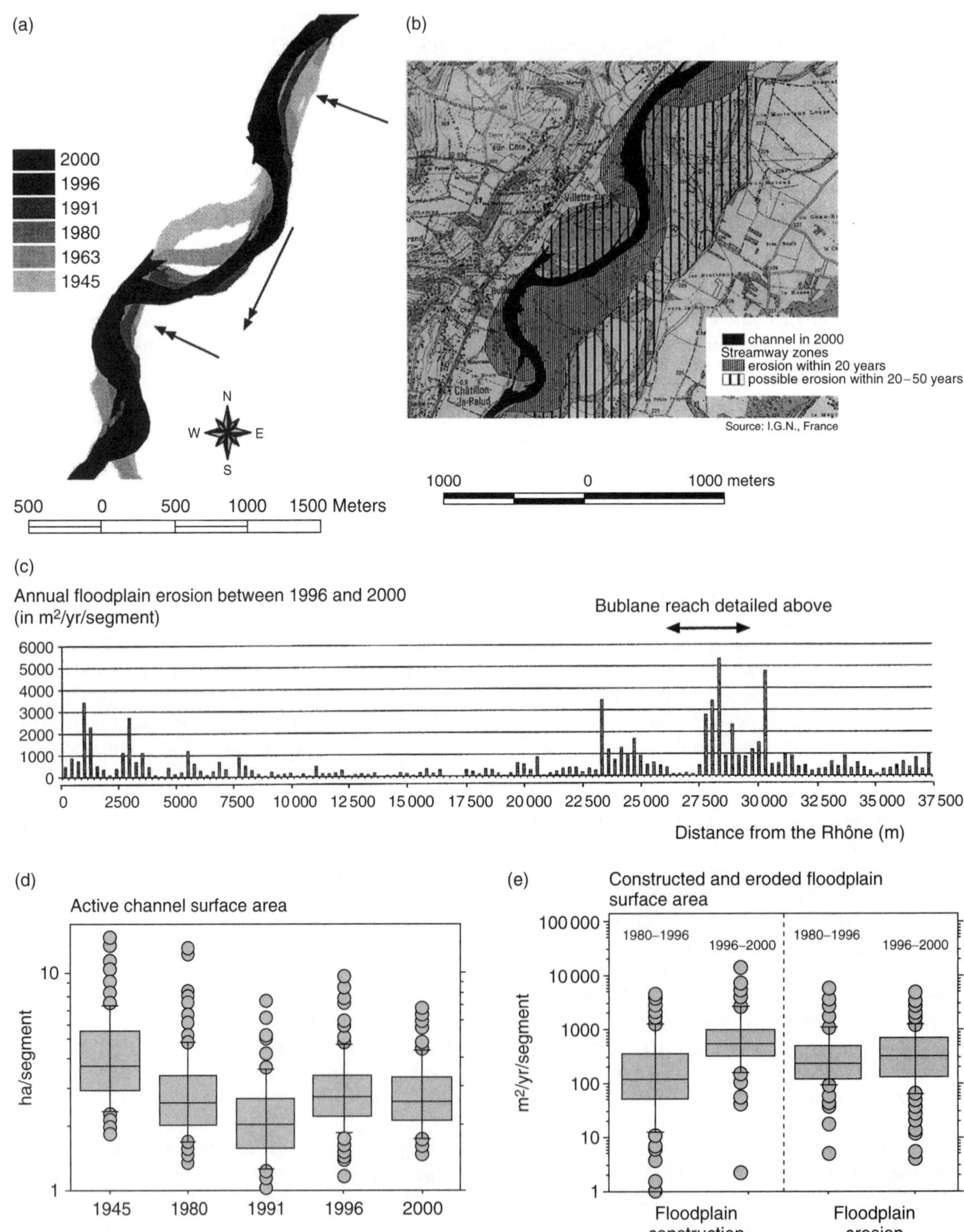

Figure 21.6 Retrospective analysis of the Ain channel mobility. View of the different channels (a), streamway differentiated according to sensitivity of zones to erosion (b), diagnosis graph showing longitudinal trends in terms of recent eroded floodplain surfaces (c), as well as temporal trends in terms of channel surface area (d), and eroded floodplain surfaces versus created floodplain surfaces (e)

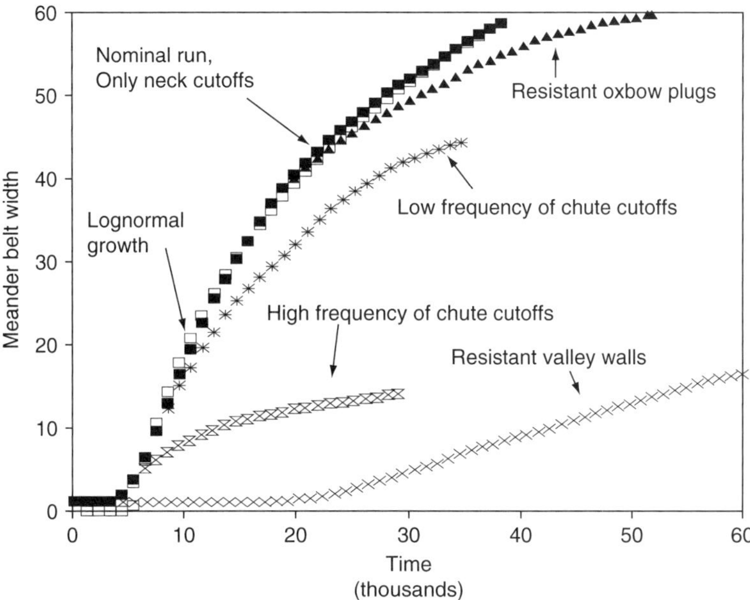

Figure 21.7 Cumulative meander belt width versus simulation time (arbitrary units). Filled boxes for simulated meandering with uniform bank erodability and no lateral constraints. Open boxes are a logarithmic growth curve fit to simulate results. Other curves show effects on growth rate of chute cut-offs, resistant clay plugs, and resistant valley walls (from Howard 1996) (reproduced by permission of John Wiley and Sons, Ltd.)

process based on bacteria activity) and root systems (nitrates are absorbed by plants). However, if the channel incises, and the alluvial water table drops, the denitrification process is reduced or inactive. Accordingly, when planning riparian restoration projects with such water quality objectives, it is essential to assess the sensitivity of the channel to incision and its current status.

The current status, future trends, and potential for incision are important attributes to assess in establishing riparian buffer zones and for other purposes. First, we can examine evidence for recent and ongoing incision, such as historical cross-sections, bridge drawings, and field evidence (Simon 1992, Piégay and Peiry 1997, Landon 1999). Historical documents demonstrated incision over the downstream 160 km of the Arno River, with a maximum of 10 m in the lower Valadarno during the contemporary period (Figure 21.9a) (Rinaldi *et al.* 1997). Cross-sections for 1890–1920 showed incision from basin afforestation and reduced sediment delivery from the catchment, while cross-sections from 1945 to 2000 showed incision caused by channel regulation and mining (Figure 21.9b).

We can also evaluate whether incision is continuing. As described in Chapter 11 for channels in loess, if at least two topographical surveys are available, it is possible to predict (by extrapolation) the elevation of the channel bed at time t_n from an exponential function. The more detailed the temporal series and the more extensive the survey, the better this approach can predict the duration of adjustment and the stages attained in different reaches. A channel survey upstream and downstream from the study reach can indicate the existence or absence of headcuts or destabilized banks from regressive and progressive erosion. A detailed historical analysis of human factors that could produce incision can provide information on probable future trends.

The inherent sensitivity of the channel to incision can be evaluated from the stratigraphy of the valley sediments and the elevation of the bedrock. In several rivers in the French Alps, a few meters of alluvial gravel overlie fine-grained lacustrine silts, sand, and peat. Once the channel incises through the gravel, the underlying fine sediments erode rapidly (Peiry *et al.* 1994). By contrast, incision through the gravel bed on the Roubion River from gravel mining and reduced sediment supply from the catchment has exhumed boulders deposited by large floods during the last glacial period, effectively armoring the bed. Similarly, exhumed moraines or bedrock outcrops can limit incision. For the bed material now exposed, grain size

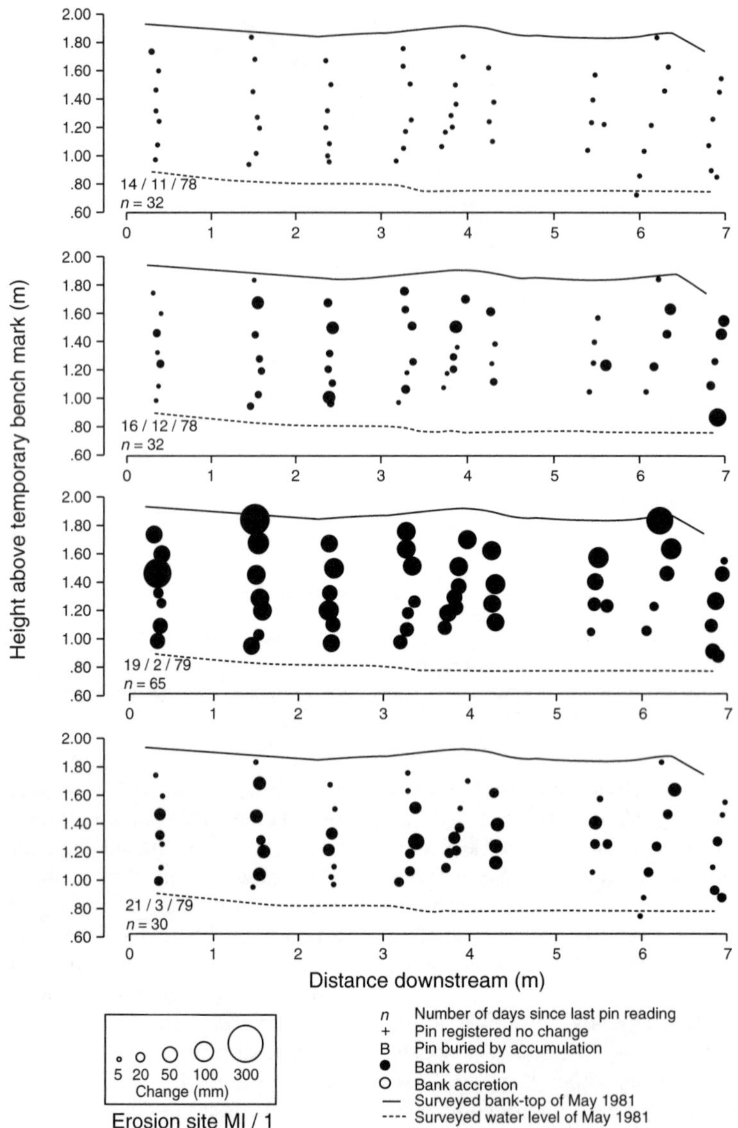

Figure 21.8 Example of information derived from erosion pins: variation on the vertical face of a bank of erosion rates for the river Ilston (South Wales). Erosion value is given for each pin and mapped as a proportional circle. The bank is viewed as a vertical face (from Lawler 1993) (reproduced by permission of John Wiley and Sons, Ltd.)

can be measured, the critical shear stress required to move it estimated, and compared with the expected range of discharges, thereby assessing the sensitivity of reaches to winnowing and armoring. These approaches can be useful on reaches with reduced sediment supply, such as downstream of dams (Komura and Simons 1967).

21.10 GEOMORPHOLOGICAL POST-PROJECT APPRAISAL: EXAMPLE OF THE RIVER WYLYE RESTORATION PROJECT

This final phase in the approach again deploys a range of tools to determine the success or otherwise of a river management program, and aims to feedback

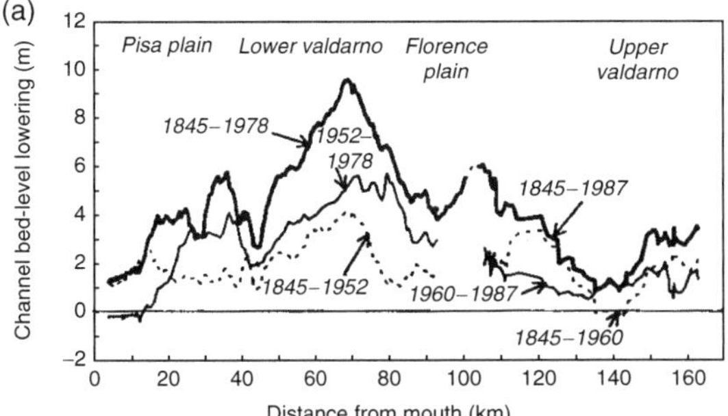

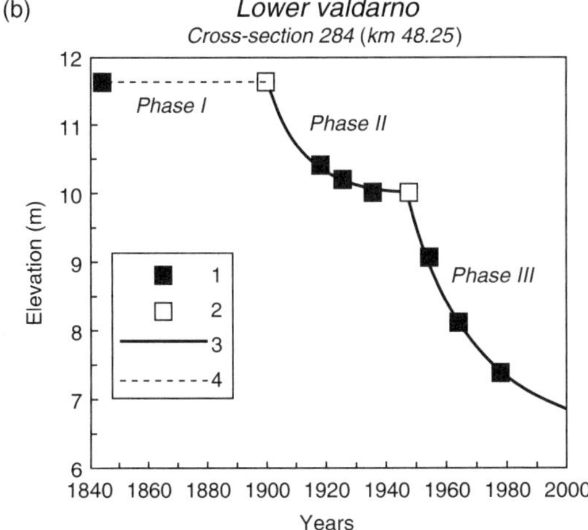

Figure 21.9 Retrospective analysis of channel degradation: the Arno River between 1845 and 1987: (a) summary of lowering over the downstream 160 km course of the river; (b) summery of lowering through time at cross-section 284 (km 48.25) (from Rinaldi *et al.* 1997) (reproduced by permission of University of Mississippi)

into the adaptive management process (Downs and Kondolf 2002). Geomorphological post-project appraisal is, however, often overlooked and underfunded by river managers who see it as an expensive "luxury" rather than as a valuable tool in itself. Those published studies demonstrate that at worse, the information may lead to a need to intervene in a scheme, but even in these cases, the information derived has value in terms of lessons learned that can be input into other projects.

Restoration of groundwater dominated rivers in the UK is largely undertaken in order to provide enhanced physical habitat diversity for different life-stages of salmonids, or as part of a wider program of rehabilitation of past channelization. Typical rehabilitation options include narrowing of the channel to encourage flushing of silt from the gravels, and the re-introduction of riffles. Typically, these schemes are undertaken without any reference to geomorphological processes, but are intuitively designed by people with some knowledge of fisheries. As part of a wider study of the sediment dynamics and physical habitat of the river Wylye (see Figure 21.2), a geomorphological post-project appraisal was undertaken in order to establish the geomorphological impact and performance of a range of rehabilitation schemes.

Setting the performance criteria for such schemes depends on establishing their original aims. In the case of the restoration schemes on the Wylye, the main aims were to flush silt, create riffles for salmonids, and to restore a physically diverse habitat. Assessing such criteria can prove problematic since few schemes have quantified targets (e.g., establish silt levels at <10% by weight of bulk sampled gravels), as was the case in this instance. Instead, an alternative approach was adopted that sought to establish the performance of the restored channels as measured by three criteria:

(a) channel geomorphology and erosion/deposition processes,
(b) substrate heterogeneity, and
(c) hydraulic habitat.

The performance of each rehabilitated reach was measured against an adjacent control reach that had not been rehabilitated. In addition, a reference condition site was measured in order to provide a suitable analog for assessing overall success of each scheme (Figure 21.10).

Channel geomorphology and processes were recorded through geomorphological mapping of each site, locating the features (pools, riffles, etc.) and processes (sediment storage and erosion). Channel form, channel geometry, and water surface elevation were recorded at cross-sections spaced at every bankfull channel width using standard Total Station surveying. Water surface slope was determined through the reach for the time of survey. Physical habitat was quantified for 20 randomly selected sections through each reach. At each section, five measures were made at points located in the channel center, edges, and mid-way between these points. These measures included; average velocity (0.6 depth), flow depth, and substrate. Flow velocity was measured using an electromagnetic current meter that is not mechanically affected by submerged aquatic macrophytes. Substrate was estimated

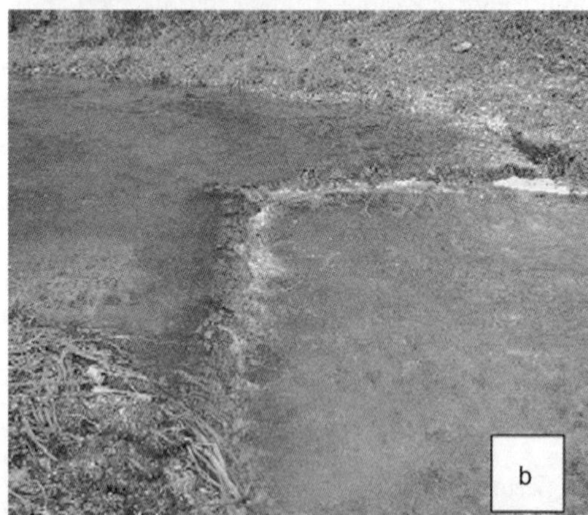

Figure 21.10 River Wylye geomorphological post-project appraisal. (a) A reach subjected to dredging in the 1950s for land drainage. (b) A rehabilitated reach, using soft engineering to manipulate channel form. (c) The semi-natural reference condition site

visually as flow depths precluded manual sampling. A total of 100 spot readings were therefore made for each reach.

The geomorphological maps were used to generate indices of geomorphological and physical biotope (flow type), diversity, and patchiness (sensu, Newson and Newson 2000). Diversity scores are estimated as the product of the number of different features and the total number of features within a reach. The values are normalized by reach length and multiplied up to a standard 100 m length to give scores in terms of 100 m channel sections. Patchiness is simply the number of different features recorded; again normalized by reach length and multiplied up to 100 m lengths. The hydraulic and substrate data were used to generate summary statistics and distributions for comparison.

The process level analysis was based on an assessment of the sediment transport and stream power characteristics of each reach. Three criteria were used:

1. ability to mobilize median surface bed material which was seen as a test of the overall stability of the river bed;
2. sediment continuity through the reach, which was used as a test of the sustainability of the reach in terms of sediment transfer (reaches in equilibrium should convey as much sediment on entry as exit) and a way of assessing the impacts of rehabilitation;
3. presence of significant fluvial or geotechnical bank erosion in the reach.

Estimates of stream power, critical entrainment threshold for the median (D_{50}) particle motion and sediment transport rates ($\mathrm{kg\,m^{-1}\,s^{-1}}$) were all established for each cross-section, in each reach for bankfull conditions using standard one–dimensional hydraulic and sediment transport modeling (See Chapter 18).

From the information collected at each site, the following specific conclusions can be drawn regarding the geomorphological process regimes. First, at all of the sites except one semi-natural reach, bed substrates are immobile at bankfull and lower discharges. This supports the findings of the wider "fluvial audit" that had highlighted the absence of bed morphology derived from scour and deposition of coarse sediments relative to other stream types. Secondly, the impact of rehabilitation has been to increase sediment transport capacity and maximum mobile particle size, but not sufficiently to generate a self-sustaining coarse sediment morphology (bars, pools, riffles). Rather, sediment conveyance is limited at most sites to at most fine gravels (<4 mm) and sands (<2 mm). Rehabilitation, has maintained sediment continuity with as much transport capacity into as out of the reach. Thirdly, at bankfull discharges, all the channels are competent to mobilize fine sediments (<2 mm). The observed accumulation of fines within each reach is therefore related to local zones of lower transport capacity such as channel margins and backwaters, or where flows are locally over-deep; for example, the pools between the rehabilitated riffles at one site. The relative roughness of vegetated channel margins and areas of flow recirculation downstream of meander bends are susceptible sites for fine sediment accumulation. This is corroborated by field observations of vegetated and unvegetated fine sediment berms at most sites.

In terms of channel geometry, the impacts of rehabilitation are again site specific, but in general they reduce bankfull depth (one of the design aims) and result in higher and more varied width:depth ratios. In this respect, they are probably moving towards the typical cross-section of natural chalk streams (Sear *et al.* 1999).

The physical features that contribute to channel form and habitat are an important measure of rehabilitation effectiveness, especially when they are the result of physical processes. Most rehabilitation schemes are based on the assumption that high physical habitat diversity or a specific suite of physical habitats will create improvements in biodiversity or specific target species. In practice, few studies have explicitly made this link. The analysis of physical habitat diversity undertaken for the river Wylye, revealed the following points; first the presence of coarse woody debris and riparian trees significantly increases the total number and type of geomorphological features present in a given length of channel. Thus, while rehabilitation of sites on the Wylye has increased both the frequency and type of geomorphological features found in a reach, they have not optimized these when considered in relation to a semi-natural analog stream (Figure 21.11). Secondly, the control reaches on the Wylye, which may be considered as typical of reaches that have been subjected to dredging, are shown to have an impoverished geomorphology relative to semi-natural reaches in the same river (Figure 21.11). However, rehabilitation has not achieved the same balance of features as those found in semi-natural chalk streams. Overall, there are too few pools and berms, and too many runs. Woody debris, though present, is currently

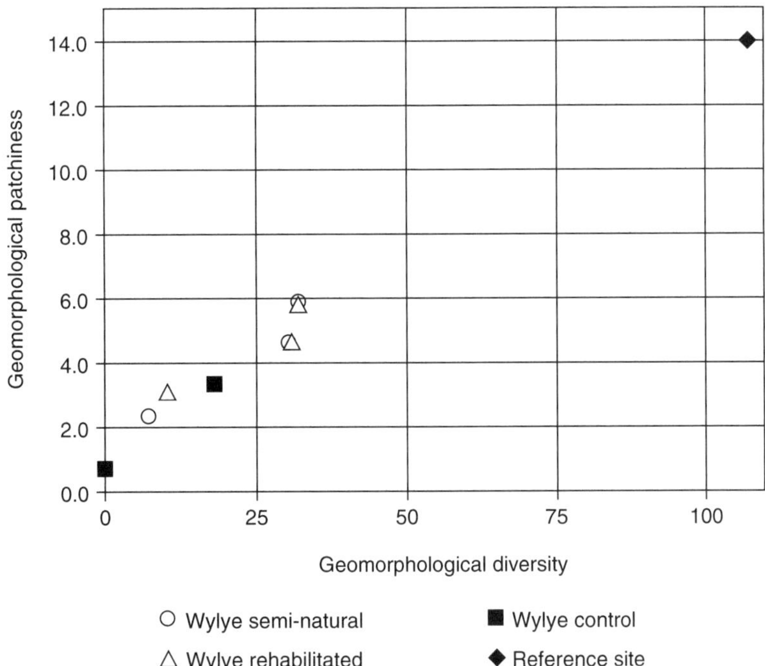

Figure 21.11 Geomorphological diversity and patchiness for control, rehabilitated, and reference sites illustrating the effect of coarse woody debris on physical habitat diversity

limited to bankside features or "island" type features that are not common in semi-natural chalk streams.

Overall, the rehabilitation has not significantly increased bed mobility or bank erosion. It has increased fine sediment transport capacity, but at the same time can increase the opportunities for accumulation due to the creation of a more varied hydraulic habitat. In terms of channel geometry cross-section form remains simple and relatively uniform. Only where riffle creation has been undertaken does the long profile show significant changes over control conditions. It is here where hydraulic conditions vary most and where fine sediment loads have increased over control conditions. Rehabilitation has decreased hydraulic variability whilst increasing depth variability at two sites, but it is very clear that each site has reacted differently. At no sites did physical habitat diversity approach that found in a semi-natural stream. Future rehabilitation programs should seek to create more varied physical habitat through the use of large woody debris. However, without treatment of the catchment-scale problems of fine sediment delivery, such rehabilitation projects will be subject to sedimentation. A more strategic approach that includes the catchment-scale issues is to be advocated.

21.11 CONCLUSIONS

The framework presented within this chapter is only one of many evolving within different regions around the world (Rosgen 1994, Brierley *et al.* 2002, Raven *et al.* 2002). Some, such as the river styles approach developed by Brierley *et al.* (2002), share a similar hierarchical structure in an attempt to integrate catchment-scale and reach-scale levels of investigation. Others, however, are tailored to provide specific outputs for a specific purpose (See Chapter 7). What is common to all is that the application of geomorphological tools must be undertaken within a clear conceptual framework designed to identify the geomorphic principles relevant to the management requirements. Furthermore, it is also vital in most applications, to interface with other relevant disciplines. GIS and the transfer of technology between disciplines are useful vehicles towards achieving these goals.

Social demands are complex, with multiple stakeholders and conflicts amongst them. With river management agencies increasingly considering longer-term perspective and larger spatial scales, the opportunity for geomorphologists to participate in the assessment of specific issues and to propose solutions is increasing (Piégay *et al.* 2002).

Land-owners, flood managers, ecosystem/nature conservancy managers, land managers and planners, civil engineers, as well as ecologists and other scientists can benefit from understanding geomorphological controls upon habitat dynamics and complexity, environmental chemistry, the complexity of fluvial forms and processes operating in the river system, basin-scale water and sediment transfer, and biogeochemical cycles.

Geomorphology programs in universities are now training more students who can operate at a practical level, and who typically work for management agencies or private companies conducting geomorphological studies and engineering designs. Moreover, interdisciplinary teams of scientists are increasingly common, and the traditional boundaries between disciplines are eroding as new fields such as ecogeomorphology, hydrogeomorphology, and ecohydrology develop.

There is a clear need to better explain to end-users the geomorphic basis of management-oriented classifications, and the tools geomorphologists use for different applications.

Fluvial geomorphic tools can support design planning, river bank protection, sediment supply, evaluation of impacts of proposed actions, sensitivity of systems, channel maintenance, ecological restoration and conservation.

At this stage of the evolution of the discipline and its increasing application to solving problems, there are strong needs to articulate the benefits of the geomorphological approach, to identify indicators and metrics to monitor and assess the efficiency of measures, to learn from experience in river interventions, to develop more collaborations within the geomorphological communities in order to make profit of the different experiences, and to use more of the (complementary) tools available. The development of models is a key challenge, as there is a need to simplify them, then to adapt them to the local cases, and to test their performance and calibrate them by using retrospective information. For each river of application, the development of conceptual models should be encouraged providing a schedule within which some hypothesized links can be tested. Finally, even if the geomorphological analyses and predictions are correct, that does not guarantee a successful project. Other factors such as cost efficiency, water resources demands, and social consequences must be considered. Interdisciplinary teams and scenario elaboration (prospective approaches) can help improve the chances of success of future projects.

Whilst the increasing use of geomorphology is encouraging, a further problem lies in ensuring that the information is translated into policy and improved practices.

Information derived using tools such as those described in this book is only valuable if the people commissioning the work understand its value and utility. Perhaps after all, among the most powerful tools available to the geomorphologist is the ability to educate non-specialists!

ACKNOWLEDGMENTS

Manuscript preparation was partly supported by the Beatrix Farrand Fund of the Department of Landscape Architecture and Environmental Planning, University of California, Berkeley. The authors would like to acknowledge funding from EU LIFE in Rivers Project, Environment Agency and English Nature. The input from Dr. Sally German on the river Wylye project is gratefully acknowledged.

REFERENCES

Bornette, G., Amoros, C., Piégay, H., Tachet, J. and Hein, T. 1998. Ecological complexity of wetlands within a river landscape. *Biological Conservation* 85: 35–45.

Boyer, M., Piégay, H., Ruffinoni, C., Citterio, A., Bourgery, C. and Caillebote, P. 1998. Guide Technique SDAGE no. 1 – La gestion des boisements de rivière: dynamique et fonctions de la ripisylve, Unpublished report, Agence de l'Eau Rhône Méditerranée Corse, 49 p.

Bravard, J.P., Kondolf, G.M. and Piégay, H. 1999. Environmental and societal effects of river incision and remedial strategies. In: Simon, A. and Darby, S., eds., *Incised River Channels*, Chichester, UK: John Wiley and Sons, pp. 303–341 (442 p.; Chapter 12).

Brierley, G.J., Fryirs, K., Outhet, D. and Massey, C. 2002. Application of the River Styles framework as a basis for river management in New South Wales, Australia. *Applied Geography* 22: 91–122.

Brookes, A. 1987. River channel adjustments downstream from channelization works in England and Wales. *Earth Surface Processes and Landforms* 12: 337–351.

Brookes, A. 1988. *Channelized Rivers: Perspectives for Environmental Management*, Chichester, UK: John Wiley and Sons, 326 p.

Brookes, A. 1995. Challenges and objectives for geomorphology in UK river management. *Earth Surface Processes and Landforms* 20: 593–610.

Brookes, A. 1996. Floodplain restoration and rehabilitation. In: Anderson, M.G., Walling, D.E. and Bates, P.D., eds., *Floodplain Processes*, Chichester, UK: John Wiley and Sons, pp. 553–576 (658 p.).

Brookes, A. and Shields, F.D., Jr., eds., 1996. *River Restoration: Guiding Principles for Sustainable Projects*, Chichester, UK: John Wiley and Sons, 433 p.

Budd, W.W., Cohen, P.L., Saunders, P.R. and Steiner, F.R. 1987. Stream corridor management in the Pacific Northwest. 1. Determination of stream-corridor widths. *Environmental Management* 11: 587–597.

Collier, M., Webb, R.H. and Schmidt, J.C. 1996. *Dams and Rivers. Primer on the Downstream Effects of Dams*, US Geological Survey, Circular 1126, Tucson, Arizona, 94 p.

Combe, P.M. 1991. Etude préalable à la mise en place d'une gestion intégrée de la basse vallée de l'Ain. Volume 4. Enjeux économiques, Unpublished report, GRAIE, Conseil Général de l'Ain, Agence de l'Eau RMC, 98 p.

Davies, P.E. and Nelson, M. 1994. Relationships between riparian buffer widths and the effects of logging on stream habitat, invertebrate community composition and fish abundance. *Australian Journal of Marine and Freshwater Research* 45: 1289–1305.

Davies, T.R. and McSaveney, M.J. 2001. Anthropogenic fanhead aggradation, Waiho River, Westland, New Zealand. In: Mosley, M.P., ed., *Gravel Bed Rivers V*, Wellington: New Zealand Hydrological Society, pp. 531–553 (642 p.).

DCWC (Deer Creek Watershed Environmental Conservancy). 1998. *Deer Creek Watershed Management Plan.* Vina, California: DCWC (June 1998).

Downs, P.W. and Kondolf, G.M. 2002. Post-project appraisal in adaptive management of river channel restoration. *Environmental Management* 29(4): 477–496.

Downward, S.R., Gurnell, A.M. and Brookes, A. 1994. A methodology for quantifying river channel planform change using GIS: variability in stream erosion and sediment transport. In: *Proceedings of the Canberra Symposium*, pp. 449–456.

Environment Agency. 1998. *River Geomorphology: A practical guide*, Guidance Note 18, Environment Agency, Tothill St., London, UK: National Centre for Risk Analysis and Options Appraisal, Steel House, 56 p.

Fryirs, K. and Brierley, G.J. 1998. The character and age structure of valley fills in upper Wolumla Creek, South Coast, New South Wales, and Australia. *Earth Surface Processes and Landforms* 23: 271–287.

Gaillot, S. and Piégay, H. 1999. Impact of gravel-mining on stream channel and coastal sediment supply, example of the Calvi Bay in Corsica (France). *Journal of Coastal Research* 15(3): 774–788.

Gilvear, D.J. 1999. Fluvial geomorphology and river engineering: future roles utilizing a fluvial hydrosystems framework. *Geomorphology* 31: 229–245.

Goodwin, C.N., Hawkins, C.P. and Kershner, J.L. 1997. Riparian restoration in the Western United States: Overview and perspective. *Restoration Ecology* 5(4s): 4–14.

Graf, W.L. 1996. Geomorphology and policy for restoration of impounded American Rivers: what is 'natural'? In: Rhoads, B.L. and Thorn, C.E., eds., *The Scientific Nature of Geomorphology*, Chichester, UK: John Wiley and Sons, pp. 443–473.

Gregory, S.V., Swanson, F.J., McKee, W.A. and Cummins, D.W. 1991. An ecosystem perspective of riparian zones: focus on links between land and water. *Bioscience* 41(8): 540–551.

Howard, A.D. 1996. Modelling channel evolution and floodplain morphology. In: Anderson, M.G., Walling, D.E. and Bates, P.D., eds., *Floodplain Processes*, Chichester, UK: John Wiley and Sons, pp. 15–62 (658 p.).

Hunter, J.C., Willett, K.B., Mc Koy, M.C., Quinn, J.F. and Keller, K.E. 1999. Prospects for preservation and restoration of riparian forests in the Sacramento Valley, California, USA. *Environmental Management* 24: 65–75.

Jeffers, J.N.R. 1998. Characterisation of river habitats and prediction of habitat features using ordination techniques. *Aquatic Conservation: Marine and Freshwater Ecosystems* 8: 529–540.

Iversen, T.M., Kronvang, B., Madsen, B.L., Markmann, P. and Nielsen, M.B. 1993. Re-establishment of Danish streams: restoration and maintenance measures. *Aquatic Conservation: Marine and Freshwater Ecosystems* 3: 73–92.

Keller, C.M.E., Robbins, C.S. and Hatfield, J.S. 1993. Avian communities in riparian forests of different widths in Maryland and Delaware. *Wetlands* 13: 137–144.

Komura, S. and Simons, D.B. 1967. River bed degradation below dams. *Journal of the Hydraulics Division, ASCE* 93: 1–14.

Kondolf, G.M. 1995. Geomorphological stream channel classification in aquatic habitat restoration: uses and limitations. *Aquatic Conservation: Marine and Freshwater Ecosystems* 5: 127–141.

Kondolf, G.M. 1997. Hungry water: effects of dams and gravel mining on river channel. *Environmental Management* 21: 533–551.

Kondolf, G.M. and Larson, M. 1995. Historical channel analysis and its application to riparian and aquatic habitat restoration. *Aquatic Conservation: Marine and Freshwater Ecosystems* 5: 109–126.

Kondolf, G.M. and Wilcock, P.R. 1996. The flushing flow problem: defining and evaluating objectives. *Water Resources Research* 32(8): 2589–2599.

Kondolf, G.M. and Wolman, M.G. 1993. The sizes of salmonid spawning gravels. *Water Resources Research* 29: 2275–2285.

Kondolf, G.M., Smeltzer, M.W. and Railsback, S.F. 2001. Design and performance of a channel reconstruction project in a coastal California gravel-bed stream. *Environmental Management* 28: 761–776.

Lajczak, A. 1996. Modelling the long-term course of non-flushed reservoir sedimentation and estimating the life of dams. *Earth Surface Processes and Landforms* 21: 1091–1108.

Landers, D.H. 1997. Riparian restoration: current status and the reach to the future. *Restoration Ecology* 5: 113–121.

Landon, N. 1999. *L'évolution contemporaine du profil en long des affluents du Rhône moyen. Constat régional et analyse d'un hydrosystème complexe, la Drôme*. PhD Thesis, University of Paris IV–Sorbonne, 560 p.

Lawler, D.M. 1993. The measurement of river bank erosion and lateral channel change: a review. *Earth Surface Processes and Landforms* 18: 777–821.

Malavoi, J.R., Bravard, J.P., Piégay, H., Hérouin, E. and Ramez, P. 1998. Détermination de l'espace de liberté des cours d'eau, Guide technique no. 2, SDAGE RMC, unpublished report, Agenve de l'eau R.M.C., 39 p.

Marston, R.A., Girel, J., Pautou, G., Piégay, H., Bravard, J.P. and Arneson, C. 1995. Channel metamorphosis, floodplain disturbance and vegetation development: Ain River, France. *Geomorphology* 13: 121–131.

Meehan, W.R., Swanson, F.J. and Sedell, J.R. 1977. Influence of riparian vegetation on aquatic ecosystems with particular references to salmonid fishes and their food supply. In: *Importance, Preservation and Management of Riparian Habitat*, Gen. Tech. Rep. RM-43, USDA Forest Service, pp. 137–145.

Moscrip, A.L. and Montgomery, D.R. 1997. Urbanization, flood frequency and salmon abundance in Puget Lowland streams. *Journal of the American Water Resources Association* 33: 1289–1297.

Naiman, R.J., Décamps, H. and Pollock, M. 1993. The role of riparian corridors in maintaining regional biodiversity. *Ecological Applications* 3(2): 209–212.

Newson, M.D. 1988. Upland land use and land management – policy and research aspects of the effects on water. In: Hooke, J.M., ed., *Geomorphology in Environmental Planning*, Chichester, UK: John Wiley and Sons, pp. 19–32.

Newson, M.D. and Leeks, G.J.L. 1987. Transport processes at the catchment scale – a regional study of increasing sediment yields and its effects in Mid-Wales, UK. In: Thorne, C.R., Bathurst, J.C. and Hey, R.D., eds., Sediment Transport in Gravel-bed Rivers, Chichester, UK: John Wiley and Sons, pp. 187–223.

Newson, M.D. and Newson, C.L. 2000. Geomorphology, ecology and river channel habitat: mesoscale approaches to basin-scale challenges. *Progress in Physical Geography* 24: 195–217.

Newson, M.D. and Sear, D.A. 1993. River conservation, river dynamics, river maintenance: contradictions? In: White, S., Green, J. and Macklin, M.G., eds., *Conserving our Landscape*, Joint Nature Conservancy, pp. 139–146.

Newson, M.D., Clark, M.J., Sear, D.A. and Brookes, A.B. 1998a. The geomorphological basis for classifying rivers. *Aquatic Conservation: Marine and Freshwater Ecosystems* 8: 415–430.

Newson, M.D., Harper, D.M., Padmore, C.L., Kemp, J.L. and Vogel, B. 1998b. A cost-effective approach for linking habitats, flow types and species requirements. *Aquatic Conservation: Marine and Freshwater Ecosystems* 8: 431–446.

Nilsson, C. 1992. Conservation management of riparian communities. In: Hansson, L., ed., *Ecological Principles of Nature Conservation*, London: Elsevier, pp. 352–372 (applications in temperate and boreal forests).

Odum, E.P. 1978. Ecological importance of riparian zone. In: *National Symposium on Strategies for Protection and*

Management of Floodplain Wetlands and other Riparian Ecosystems, pp. 2–4.

Palmer, L. 1976. River management criteria for Oregon and Washington. In: Coates, D.R., ed., *Geomorphology and Engineering*, Stroudsburg, Pennsylvania: Dowden, Hutchinson and Ross, pp. 329–346.

Pautou, G. and Wuillot, J. 1989. La diversité spatiale des forêts alluviales dans les îles du Haut-Rhône français. *Bulletin d'Ecologie* 20(3): 211–230.

Petersen, R.C., Petersen, L.B. and Lacoursière, J. 1992. A building block model for stream restauration. In: Calow, P., Petts, G.E. and Boon, P.J., eds., *River Conservation and Management*, Chichester, UK: John Wiley and Sons, pp. 293–309.

Petit, F., Poinsard, D. and Bravard, J.P., 1996. Channel incision, gravel mining and bedload transport in the Rhône River upstream of Lyon, France ("canal of Miribel"). *Catena* 26: 209–226.

Peiry, J.L., Salvador, P.G. and Nouguier, F. 1994. L'incision des rivières des Alpes du Nord: état de la question. *Revue de Géographie de Lyon* 69: 47–56.

Petts, G.E. 1989. Historical analysis of fluvial hydrosystems. In: Petts, G.E., Möller, H., Roux, A.L., eds., *Historical Change of Large Alluvial Rivers, Western Europe*, Chichester, UK: John Wiley and Sons, pp. 1–19.

Piégay, H. and Bravard, J.P. 1997. Response of a Mediterranean riparian forest to a 1 in 400 year flood, Ouvèze River, Drôme-Vaucluse, France. *Earth Surface Processes and Landforms* 22: 31–43.

Piégay, H. and Peiry, J.L. 1997. Long profile evolution of a mountain stream in relation to gravel load management: example of the middle Giffre river (French Alps). *Environmental Management* 21(6): 909–920.

Piégay, H. and Saulnier, D. 2000. The streamway, a management concept applied to the French gravel bed rivers. In: Nolan, T.J. and Thorne, C.R., eds., CD-Rom Gravel Bed Rivers 2000, Conference, Christchurch, New Zealand, 27 August–3 September, http://geog.canterbury.ac.nz/services/carto/intro.htm (poster published).

Piégay, H., Barge, O. and Landon, N. 1996. Streamway concept applied to River mobility/human use conflict management. In: *First International Conference on New/Emerging Concepts for Rivers. Proceedings Rivertech 96*, International Water Resources Association, pp. 681–688.

Piégay, H., Bornette, G., Citterio, A., Hérouin, E., Moulin, B. and Statiotis, C. 2000. Channel instability as control factor of silting dynamics and vegetation pattern within perifluvial aquatic zones. *Hydrological Processes* 14(16/17): 3011–3029.

Piégay, H., Bravard, J.P. and Dupont, P. 1994. The French water law: a new approach for alluvial hydrosystem management (French Alpine and Perialpine stream examples). In: Marston, R.A. and Hasfurther, V.R., eds., *Annual Summer Symposium of the American Water Resources Association, Effects of Human-induced Changes on Hydrologic Systems*, Jackson Hole, Wyoming, USA: American Water Resources Association, pp. 371–383.

Piégay, H., Cuaz, M., Javelle, E. and Mandier, P. 1997. A new approach to bank erosion management: the case of the Galaure River, France. *Regulated Rivers: Research and Management* 13: 433–448.

Piégay, H., Dupont, P. and Faby, J.A. 2002. Questions of water resources management: feedback of the French implemented plans SAGE and SDAGE (1992–1999). *Water Policy* 4(3): 239–262.

Pinay, G. and Décamps, H. 1988. The role of riparian woods in regulating nitrogen fluxes between the alluvial aquifer and surface water. A conceptual model. *Regulated Rivers: Research and Management* 2: 507–516.

Raven, P.J., Holmes, N.T.H., Charrier, P., Dawson, F.H., Naura, M. and Boon, P.J. 2002. Towards a harmonised approach for hydromorphological assessment of rivers in Europe: a qualitative comparison of three survey methods. *Aquatic Conservation: Marine and Freshwater Ecosystems* 12(4): 405–424.

Rhoads, B.L. 1994. Fluvial Geomorphology. *Progress in Physical Geography* 18(1): 103–123.

Ricks, C.L. 1995. Effects of channelization on sediment distribution and aquatic habitat at the mouth of Redwood Creek Basin, Northwestern California. In: Nolan, K.M., Kelsey, H.M. and Marron, D.C., eds., *Geomorphic Processes and Aquatic Habitat in the Redwood Creek Basin, Northwestern California, Washington*, US Geological Survey Professional Paper, pp. Q1–Q17.

Rinaldi, M., Simon, A. and Billi, P. 1997. Disturbance and adjustment of the Arno River, Central Italy. II. Quantitative analysis of the last 150 years. In: Wang, S.S.Y., Langendoen, E.J. and Shield, F.D., Jr., eds., *Management of Landscapes Disturbed by Channel Incision*, Oxford, Mississippi: The University of Mississippi, pp. 601–606 (1134 p.).

Rinaldi, M. and Casagli, N. 1999. Stability of streambanks formed in partially saturated soils and effects of negative pore water pressures: the Sieve River (Italy). *Geomorphology* 26: 253–277.

Robson, F.T. 1948. Letter to Colonial Joseph S. Eorlinski, US Army Corps of Engineers, Sacramento District Engineer, 20 May 1948.

Rosgen, D.L. 1994. A classification of natural rivers. *Catena* 22: 169–199.

Schmidt, J.C., Webb, R.H., Valdez, R.A., Marzolf, G.R. and Stevens, L.E. 1998. Science and values in river restoration in the Grand Canyon. *Bioscience* 48: 735–747.

Scott, M.L., Shafroth, P.B. and Auble, G.T. 1999. Responses of riparian cottonwoods to alluvial water table declines. *Environmental Management* 23: 347–358.

SDAGE RMC. 1997. *Schéma Directeur d'Aménagement et de Gestion des Eaux du bassin Rhône Méditerranée Corse (Master Plan for Water Management and Development of the Rhône Méditerranée Corse Basin)*, Comité de Bassin Rhône Méditerranée Corse, 3 volumes, 1 atlas, 15 guidebooks.

Sear, D.A. 1994. River restoration and geomorphology. *Aquatic Conservation: Marine and Freshwater Ecosystems* 4: 169–177.

Sear, D.A., Darby, S.E. and Thorne, C.R. 1994. Geomorphological approach to stream stabilisation and restoration: case study of the Mimmshall Brook, Hertfordshire, U.K. *Regulated Rivers Research and Management* 9: 205–223.

Sear, D.A. 1995. The effects of 10 years river regulation for hydropower on the morphology and sedimentology of a gravel-bed river. *Regulated Rivers: Research and Management* 10: 247–264.

Sear, D.A., Newson, M.D. and Brookes, A. 1995. Sediment related river maintenance: the role of fluvial geomorphology. *Earth Surface Processes and Landforms* 20: 629–647.

Sear, D.A., Briggs, A. and Brookes, A. 1998. A preliminary analysis of the morphological adjustment within and downstream of a lowland river subject to river restoration. *Aquatic Conservation: Marine and Freshwater Ecosystems* 8(1): 167–184.

Sear, D.A., Armitage, P.D. and Dawson, F.D.H. 1999. Groundwater dominated rivers. *Hydrological Processes* 11(14): 255–276.

Sear, D.A. and Archer, D. 1998. The geomorphological impacts of gravel mining: case study of the Wooler Water, Northumberland U.K. In: Klingeman, P., Komar, P.D. and Hey, R.D., eds., *Gravel-bed Rivers in the Environment*, Boulder, Colorado: Water Resources Press, pp. 325–344 (332 p.).

Shields, F.D., Knight, S.S. and Cooper, C.M. 1995. Incised stream physical habitat restoration with stone weirs. *Regulated Rivers: Research and Management* 10: 181–198.

Simon, A. 1992. Energy, time, and channel evolution in catastrophically disturbed fluvial systems. *Geomorphology* 5: 345–372.

Simon, A. and Downs, P.W. 1995. An interdisciplinary approach to evaluation of potential instability in alluvial channels. *Geomorphology* 12: 215–232.

Smith, S. 1997. Changes in the hydraulic and morphological characterstics of a relocated stream channel. Unpublished Masters of Science Thesis, University of Maryland, Annapolis.

Spackman, S.C. and Hughes, J.W. 1995. Assessment of minimum stream corridor width for biological conservation: species richness and distribution along mid-order streams in Vermont, USA. *Biological Conservation* 71: 325–332.

Thorne, C.R., Allen, R.G. and Simon, A. 1996. Geomorphological stream reconnaissance for river analysis, engineering and management. *Transactions of the Institute of British Geographers* 21: 455–468.

Toth, L.A., Albrey, A.D., Brady, M.A. and Muszick, D.A. 1995. Conceptual evaluation of factors potentially affecting restoration of habitat structure within the channelized Kissimmee River ecosystem. *Restoration Ecology* 3: 160–180.

Vannote, R.L., Minshall, G.W., Cummins, K.W., Sedell, J.R. and Cushing, C.E. 1980. The river continuum concept. *Canadian Journal of Fisheries and Aquatic Science* 37: 130–137.

Wilcock, P.R. 1997. Friction between science and practice: the case of river restoration. *Eos, Transactions, American Geophysical Union* 78(41): 454.

Wilcock, P.R., Schmidt, J.C., Wolman, M.G., Dietrich, W.E., Dominick, D., Doyle, M.W., Grant, G.E., Iverson, R.M., Montgomery, D.R., Pierson, T.C., Schilling, S.P. and Wilson, R.C. 2003. When models meet managers: examples from geomorphology. In: Wilcock, P.R. and Iverson, R.M., eds., *Prediction in Geomorphology*, Geophysical Monograph 135, Am. Geophys. Union, pp. 27–40 (DOI: 10.1029/135GM03).

Wohl, N.E. and Carline, R.F. 1996. Relations among riparian grazing, sediment loads, macroinvertebrates, and fishes in three central Pennsylvania streams. *Canadian Journal Fisheries and Aquatic Sciences* 53: 260–266.

Xu, J. 1997. Evolution of mid-channel bars in a braided river and complex response to reservoir construction: an example from the middle Hanjiang River, China. *Earth Surface Processes and Landforms* 22: 953–965.

Index

Page numbers in *italic* refer to illustrations; those in **bold** type refer to tables.

Tools in Fluvial Geomorphology. Edited by G. Mathias Kondolf and Hervé Piégay
 ISBN: 0-471-49142-X

Index compiled by Connie Tyler